Developmental Plasticity and Evolution

DEVELOPMENTAL PLASTICITY AND EVOLUTION

Mary Jane West-Eberhard

OXFORD

UNIVERSITY PRESS

2003

OXFORD
UNIVERSITY PRESS

Oxford University Press, Inc., publishes works that further
Oxford University's objective of excellence
in research, scholarship, and education.

Oxford   New York
Auckland   Cape Town   Dar es Salaam   Hong Kong   Karachi
Kuala Lumpur   Madrid   Melbourne   Mexico City   Nairobi
New Delhi   Shanghai   Taipei   Toronto

With offices in
Argentina   Austria   Brazil   Chile   Czech Republic   France   Greece
Guatemala   Hungary   Italy   Japan   Poland   Portugal   Singapore
South Korea   Switzerland   Thailand   Turkey   Ukraine   Vietnam

Published by Oxford University Press, Inc
198 Madison Avenue, New York, New York 10016

www.oup.com

Oxford is a registered trademark of Oxford University Press

Library of Congress Cataloging-in-Publication Data
West-Eberhard, Mary Jane.
    Developmental plasticity and evolution / Mary Jane West-Eberhard.
       p. cm.
    Includes bibliographical references (p.).
    ISBN-13 978-0-19-512234-3; 978-0-19-512235-0 (pbk.)
    ISBN 0-19-512234-8; 0-19-512235-6 (pbk.)
       1. Adaptation (Biology) 2. Phenotype. 3. Evolution (Biology)
       4. Developmental biology.   I. Title
QH546 W45 2002
578.4–dc21   2001055164

9 8 7 6
Printed in the United States of America
on acid free paper

Part I (critique and concepts), to my Professor, Dick Alexander, inspired teacher of evolutionary ideas.

Part II (origins and transitions), to Jessica, Anna, and Andrés.

Part III (alternative phenotypes) to Chloe C. and (in memoriam) Earl C. West, adaptively flexible parents and grandparents.

Part IV (major themes), to Bill.

Preface

This is a book about development and evolution primarily for biologists interested in evolutionary theory. Jared Diamond, in his preface to *Guns, Germs, and Steel* (1998), notes that authors are regularly asked to summarize a long book in one sentence. For this book, the sentence is: "The universal environmental responsiveness of organisms, alongside genes, influences individual development and organic evolution, and this realization compels us to reexamine the major themes of evolutionary biology in a new light."

This book began more modestly, as a book on alternative phenotypes. It was to contain a giant collection of complex polymorphisms and polyphenisms, showing once and for all how common and important they are in evolution. It soon became clear that the principles involved were really much broader—that they could be extended to apply to development and evolution in general. So I was faced with a decision: the extended project would be much larger and more difficult but at the same time more exciting. It would demand an enormous amount of time, and expertise I did not possess, and yet held the promise of resolving some of the most important problems of biology, such as how to relate nature and nurture, genes and environment, within a fundamentally genetic theory of evolution.

I decided that the time was ripe for a synthesis of development and evolution and that someone schooled in evolutionary biology needed to attempt it. A single unified treatment demands a single-authored book, for multiauthored volumes always divide in order to conquer a complicated subject. It was easy to decide in favor of the larger job. I have the good fortune of working at a research institute where major undertakings are possible even if they mean a period of invisible work, where the library services are excellent, and where there is a diversity of colleagues active in relevant fields. There seemed to be no excuse not to take this on. Yet if I had realized then, in the late 1980s, how long this would take, and how soon this topic would experience an explosion of interest and activity that would severely tax the comprehension of a single writer, I wonder if I would have had the courage to begin.

In the intervening years, the need for a book on developmental *evolutionary* biology seems to have increased rather than diminished. The initiative for connecting development and evolution has come largely from developmental biology, with the rise of evolutionary developmental biology. Important as it is, "evo-devo" naturally reflects the concerns of modern developmental biology with regulatory genes, body plans, and morphology, as seen in a few model organisms. Attention to variation and selection within populations, speciation, developmental plasticity, and the origin of behavioral, physiological, and life history traits is missing. I suspect that many evolutionary biologists fail to see the relevance of evolutionary developmental biology to their research. One purpose of this book is to fill that gap, by addressing the traditional questions of Darwinian evolutionary biology from a developmental point of view.

The developmental point of view taken here may be unrecognizable in developmental biology, descended as it is from what used to be called embryology, via its marriage to molecular biology, to form what is now a laboratory science where development refers to gene expression and tissue differentiation, primarily during early development, and primarily in multicellular animals. If a developmental approach is to permeate evolutionary biology, development has to be defined more broadly, to include the ontogeny of all aspects of the phenotype, at all levels of organization, and in all organisms. Under the broad definition of development taken in this book, a developmental evolutionary biologist is one who wonders about how organisms got to be the way they are, in both an evolutionary and a proximate, ontogenetic sense. Although some evolutionary biologists claim that you can study adaptive evolution in terms of selection alone, without attention to proximate mechanisms, I find that curiosity regarding individual development is almost inevitable during research on the origins of traits, and it is surprising that development has been missing as a natural element of evolutionary biology for so long.

Anyone who tries to generalize beyond a single specialized area has to deal with the justified skepticism of experts whose territories are overlapped. I first encountered this in 1973, when I dared to present a lecture on the evolution of social behavior that went beyond the single genus of social wasps that was the subject of my doctoral thesis. At the mention of another subfamily of wasps, one that I had not actually observed myself, I was vehemently criticized by one specialist in that group as lacking the authority for such audacious speculation. I can only imagine the horror of such a reader at the pretensions of a book like this one. My defense now is the same as it was then: I can read. If there were no attempt, however flawed, to generalize and synthesize, beginning with the findings of specialized empirical research, there would be little point in doing the research to begin with. Nevertheless, incursions into the minefields of alien subjects are at high risk of error. I apologize in advance for having certainly missed major important references in many areas. Still, a reader may wonder what my particular qualifications are and how I came to be broadly interested in development and evolution.

My primary research is on the evolution of insect societies. In the social insects (wasps, ants, bees and termites), social role can be environmentally manipulated. In the adult females of social wasps, behavioral differentiation is related to ovarian development, which reflects hormonal condition, and nutrition, and so there are mechanistic links between behavior, morphology, social environment, and reproductive success. Behavioral development is thus tied, hormonally, to evolution—to the causes of differential reproductive success, or natural selection. It is a natural setting for an interest in development in relation to evolution.

The condition-dependent differences between workers and queens led me to think about the evolution of alternative phenotypes in general, and to think about the relevance of environmentally mediated gene-expression differences to the divergence, without genetic isolation, between two morphs. It is only a small step from thinking about that to thinking about mosaic evolution in general. How, as I was once asked by my colleague Jeremy Jackson, are alternative phenotypes different from a hand and a foot? How, indeed, do the different parts of organisms evolve to be semi-independently specialized relative to other parts? Thinking in this way led to the discovery of several contradictions (listed in chapter 1) between traditional evolutionary explanations and what must have occurred to produce adaptively flexible phenotypes. Although the contributions of gene expression patterns to patterns of evolutionary change seemed obvious from a study of individual development, no definite connection between the two had been incorporated into the coherent theory of evolution that I had learned and that has continued to dominate the literature on genetics and evolutionary biology.

As I began to develop a list of general principles of development and evolution, the next step was to realize that while the principles were simple, backing them up with facts would be a challenge. It was reasonable to assume, for example, that hormonally regulated alternatives such as worker and queen would represent different sets of expressed genes, but I remember another colleague, Rob Colwell, asking me how I knew this was really true. So I resolved early to attempt to actually document each conceptual claim.

The result is not a review, but an extensively documented set of ideas. The concepts outlined here are simple, and many of them have been discussed before. I believe that the contribution of this book is to bring together a set of insights; to present them as a coherent theory of development and evolution; to show them to be compatible with the genetic theory of evolution that is so well supported by previous work; and to support the argument with reference to empirical research.

The text is designed so that individual chapters can stand more or less alone. Each is outlined in the Chapter Contents at the beginning of the book, and a brief abstract is given at the beginning of each chapter. In addition, key terms are italicized in the text where they are defined, and the defined terms are in bold in the index, which can therefore be used as a glossary. In the bibliography, books published or reprinted in more than one edition are listed with their date of first edition followed by, in brackets, the date of the edition cited here. Facsimile volumes are cited by original date, with date of facsimile publication given only in the bibliography. When I did not personally consult a primary reference, I give the primary source and state that it was "cited by" or "from" a secondary source. Both primary and secondary sources are listed in the bibliography.

MJW-E
August, 2001

Acknowledgments

For a scientist used to writing articles and isolated reviews, doing a long book is like running a marathon for a sprinter. A friend (A. S. Rand) has described the experience as getting farther and farther from the beginning, but not closer and closer to the end. Vladimir Nabokov's son, Dimitri, who edited a collection of his father's letters, described deadlines that passed, as did mine, "like telegraph poles glimpsed from a speeding train." I often felt as if I were trying to run through deep mud, sustained only by the illusion of being almost finished.

In some ways this has been a team project of the Smithsonian Tropical Research Institute (STRI), my home institution, to which I am extraordinarily indebted for long-term support and encouragement. I have been aided by a series of excellent librarians, most especially by Vielka Chang-Lau and the STRI library staff, especially Angel Aguirre, who promptly and patiently provided materials and answers to many urgent inquiries. Fellow staff-scientists William Eberhard, N. G. Smith, and William Wcislo brought many key references to my attention and provided copies. Annette Aiello tracked down some obscure nineteenth century references thanks to her mastery of literature on butterflies. I especially acknowledge the encouragement and patience of STRI Director Ira Rubinoff and Deputy Director Tony Coates, whose always positive, while demanding, support of this long effort could serve as a model of effective scientist management.

Donna Conlon prepared the figures, to which she brought an unusual combination of artistic ability, biological sophistication, and forbearing patience where several versions were needed to reach a final form. We were aided by expert work with computer graphics by Lina Gonzalez at STRI. Bruce Goldman, David Norris, and David Wake provided information for figures (23.6, 5.19, and 9.5, respectively) that is not available in any single book or article. Those figures reflect their critical mastery of broad fields. The following people generously produced figures tailored for use in this book from their own published and unpublished research: Susan Alberts, Chris Brönmark, Bill Eberhard, Douglas Emlen, Larry Frank, Wulfila Gronenberg, William H. Hamilton III, Katherine Holekamp, Hubert Herz, Jennifer Jarvis, Jim Lloyd, Fred Nijhout, Dan Otte, David Pfennig, Hal Reed, Louise Roth, Richard Rowe, Paul Sherman, David Stern, Karen Warkentin, and Greg Wray. Authors who provided originals of published figures are gratefully acknowledged in the figure captions.

The help and encouragement of colleagues and friends have been indispensible. Fred Nijhout read a complete early draft at a stage when few could have endured it and supplied many perceptive suggestions regarding content and style. A rewritten version was then critiqued in its entirety by four readers: John Bonner, to whom I am especially grateful for rapid feedback and advice on readability, a well as suggestions on sections relating to conceptual and historical issues in developmental biology (Kate Clark facilitated electronic text transfers); William Eberhard, who provided extremely detailed critical comments on all aspects; William Wcislo, who from the outset understood what I was trying to do and provided, over a period of years, many crucial references and ideas; and George Williams, whose approach to adaptive evolution I thought would be more different from mine than it turned out (happily) to be, and who went out of his way to travel to places where we could discuss the issues that arose.

In addition, individual chapters and chapter excerpts were critiqued by specialists in particular areas:

Part I, A. Eberhard Friedlander, E.G. Leigh, F. Joyce, S. Paulsen, K. Harms.
Chapter 4, M. Kidwell, D. Lisch.
Chapter 8 (on Darwin's theory of development and evolution), R. F. Johnston, F. Joyce, E. Mayr.
Chapter 12 (reversion), I. D. Gauld.

Chapter 13 (heterochrony), D. Stern.

Chapter 15 (cross-sex transfer), J. Doebley, S. Emerson, R. Levin, D. Piperno, S. Shuster, K. Summers, E. F. Smith, K. Wells.

Chapter 16 (correlated shifts), V. L. Roth.

Chapter 17 (combinatorial molecular evolution), E. Davidson, J. Gerhart, M. Kirschner.

Chapter 18 (learning), P. Marler, B. Capaldi.

Chapter 19 (recurrence), R. Leschen, T. Price, R. Johnston, the late B. Alexander, J. M. Carpenter, G. Melo, and C. D. Michener.

Part III (on alternative phenotypes), M. Bell, W. Bradshaw, J. Brockmann, B. Goldman, D. Pfennig, S. Paulson, S. Shuster, P. Wimberger.

Chapter 24 (gradualism), V. L. Roth.

Chapter 25 (homology), J. M. Carpenter, H. Greene, R. Leschen, V. L. Roth.

Chapter 26 (environmental modifications), F. Joyce, D. Smillie, V. L. Roth, J. Podos, R. Strathmann, K. Winter.

Chapter 27 (speciation), R. D. Alexander, R. Anstey, M. Bell, P. R. Grant, W. Bradshaw, H. Lessios, J. Feder, J. Endler, R. Payne, K. Ross, D. Shoemaker, K. and M. Tauber, W. Tschinkel, P. Wimberger, T. K. Wood. A very early version of the section on socially parasitic ants was critiqued by B. Bolton, A. Buschinger, A. Bourke, G. Elmes, D. Futuyma, K. MacKay, S. Rissing, P. Ward, and E.O. Wilson.

Chapter 28 (adaptive radiation), M. Bell, P. Grant, M. Gross, I. Fleming, K. Y. Kaneshiro, H. Carson, T. Shelly, T. B. Smith.

Chapter 29 (macroevolution), R. Anstey, A. Cheetham, D. Raup, V. L. Roth, R. R. Strathmann.

Chapter 30 (punctuation), M. Bell, R. Anstey, A. Cheetham, P. L. Cook, J. B. C. Jackson, D. Raup, V. L. Roth, J. Winston, and B. MacFadden.

I also benefitted from comments by seminar participants at the University of Kansas and Duke University on some early drafts of chapters, from suggestions by Jessica Eberhard on titles and content, and from stimulating meetings at The Santa Fe Institute. Several other colleagues provided extraordinarily helpful explanations and references in response to particular inquiries: Hugh Iltis, John Doebley, and Marshall D. Sundberg on the evolution of maize; George Barlow, Michael Bell, Bill Pearcy, and Ian Fleming on the biology of fishes; Jarmila Kukalova Peck on insect development in relation to fossils; Richard Burian on the history and philosophy of evolutionary studies of development; Bruce Lyon on terminology of alternative tactics; Elizabeth F. Smith on cross-sex transfer; and Lynn H. Caporale on genome evolution.

Will Provine provided enlightened discussions of concepts and also gave me copies of books of historical importance to add to my personal library, including classical works by Baldwin, Weismann, and others. I treasure the books and the conversations we had about them.

Particular books and articles proved especially important in thinking about the topics covered here, and I referred to them repeatedly. These are: papers by Shapiro (1976) and Turner (1977) on alternative phenotypes, by Clarke (1966) on morph ratio clines, and by Leigh van Valen on many topics in evolutionary biology; Mayr's 1963 book on speciation, which continues to provide a standard reference and summary of ideas; Futuyma's textbooks (1986a, 1998) on evolutionary biology, which accurately and perceptively summarize the genetic theory of evolution in most common current use; and books by Gould (1977), Raff and Kaufman (1983), Matsuda (1987), Stone and Schwartz (1990a), Mayr and Provine (1980), Deuchar (1975), John and Miklos (1988), Davidson (1986), and Büning (1994), to which I turned repeatedly to clarify questions about the content and history of molecular and developmental biology in relation to evolution. I read Gerhart and Kirschner (1997) in manuscript and enjoyed sharing with those authors the cross-cultural experience of attempting similar syntheses starting from very different scientific backgrounds. I benefitted greatly from their insights.

Conversations with several people especially helped to reinforce and sustain my enthusiasm for this project: during early stages of writing, the late Caryl P. Haskins, Ellen Ketterson, Jorge Lobo, C.D. Michener, Roy Pearson, Stan Rachootin, Bill Sheehan, Glenn Tupper, Ilse Walker, Bill Wcislo, Paul Wilson, and Alexei Yablokov; and, more recently, John Bonner, Sarah Hrdy, Ellen Larsen, Fred Nijhout, and Stan Rand. Kirk Jensen has been a wise, patient, and supportive editor. I thank him and the staff at Oxford University Press, especially Lisa Stallings.

Anna Eberhard Friedlander provided advice regarding manuscript preparation, and transferred a film provided by W. J. Hamilton III into electronic format, video, and prints. She, Lynne Hartshorn, and The Smithsonian librarians tracked down some difficult-to-find references.

Many people have assisted in preparation of the bibliography, including especially G. Hills, D. Mills, the late S. Holder, J. Coates, M. Denton, and L. Arneson. My deepest debt is to Lynne Hartshorn, who twice traveled from the United States to Costa Rica, taking time off from her own busy life, to come to my aid. Her expert help was critical in the final exhausting twenty-day sprint to the finish.

And of course there is my family. I have not found the words to sufficiently acknowledge the help of my husband Bill Eberhard, whose informed and mercilessly rigorous criticisms of ideas and manuscripts I have come to regard as words of endearment; and our offspring, Jessica, Anna, and Andrés, for whom this book has been just one unusually long episode in a lifetime of maternal distraction (not to say neglect), which I like to think is partly responsible for their creative independence and cheerful patience with their elders.

Outline of Contents

I FRAMEWORK FOR A SYNTHESIS

   1. Gaps and Inconsistencies in Modern Evolutionary Thought   3

   2. Material for a Synthesis   21

   3. Plasticity   34

   4. Modularity   56

   5. Development   89

   6. Adaptive Evolution   139

   7. Principles of Development and Evolution   159

   8. Darwin's Theory of Development and Evolution   188

II THE ORIGINS OF NOVELTY

   9. The Nature and Analysis of Phenotype Transitions   197

   10. Duplication   209

   11. Deletion   218

   12. Reversion   232

   13. Heterochrony   241

   14. Heterotopy   255

   15. Cross-sexual Transfer   260

   16. Quantitative Shifts and Correlated Change   296

   17. Combinatorial Evolution at the Molecular Level   317

   18. Phenotypic Recombination Due to Learning   337

   19. Recurrence   353

III ALTERNATIVE PHENOTYPES

   20. Alternative Phenotypes as a Phase of Evolution   377

   21. Divergence without Speciation   394

   22. Maintenance without Equilibrium   417

   23. Assessment   440

IV DEVELOPMENTAL PLASTICITY AND THE MAJOR THEMES OF EVOLUTIONARY BIOLOGY

   24. Gradualism   471

   25. Homology   485

   26. Environmental Modifications   498

27. Speciation   526
28. Adaptive Radiation   564
29. Macroevolution   598
30. Punctuation   617
31. One Final Word: Sex   630
    Literature Cited   639
    Author Index   745
    Taxonomic Index   759
    Subject Index   767

Chapter Contents

PART I   FRAMEWORK FOR A SYNTHESIS

CHAPTER 1   GAPS AND INCONSISTENCIES IN MODERN EVOLUTIONARY THOUGHT   3

    Introduction   3

    Six Points of Confusion and Controversy   6
        1. The Unimodal Adaptation Concept and the Multimodal Products
           of Development and Plasticity   6
        2. The Cohesiveness Problem   8
        3. Proximate and Ultimate Causation   10
        4. The Problem of Continuous versus Discrete Variation and Change   11
        5. Problematic Metaphors   13
        6. The Genotype–Phenotype Problem   16

    Toward a Solution   18

CHAPTER 2   MATERIAL FOR A SYNTHESIS   21

    Introduction   21

    Previous Insights on Development and Evolution   21

    A Unified Theory of Phenotypic Development and Evolution   28

    Definitions of Key Terms   30

CHAPTER 3   PLASTICITY   34

    Introduction   34

    The Meaning of Plasticity   34

    Mechanisms of Plasticity   37

    Phenotypic Accommodation   51

    The Evolutionary Importance of Mechanisms   54

CHAPTER 4   MODULARITY   56

    Introduction   56

    Modularity as Plasticity   58

    Hierarchy and Integration as Aspects of Modularity   60

    Application of the Modularity Concept at Different Levels of Organization   61

    Metamorphosis and Life Cycle Modularity   66

    The Genetic Architecture of Modular Traits   67

    Modular Traits as Subunits of Gene Expression   70

    Limitations of the Modularity Concept   81

    Complementarity   83

    Landmarks in the Evolution of Modularity   84

Hypermodularity and Somatic Sequestration   86
General Evolutionary Consequences of Increased Modularity   86

CHAPTER 5   DEVELOPMENT   89
Introduction   89
Continuity of the Phenotype   90
The Dual Nature of All Regulation   98
Gene–Environment Equivalence and Interchangeability   116
The Organization of Development by Switches   129
Complementarity, Continued   135
Consequences for Selection and Evolution   138

CHAPTER 6   ADAPTIVE EVOLUTION   139
Introduction   139
Prerequisites for Evolution by Natural Selection   142
The Origins of Novelty   143
Genetic Accommodation   147
Genes as Followers in Evolution   157
A Developmental Definition of Adaptive Evolution   158

CHAPTER 7   PRINCIPLES OF DEVELOPMENT AND EVOLUTION   159
Introduction   159
Evolutionary Consequences of Plasticity   160
Evolutionary Consequences of Modularity   163
Consequences of Hierarchical Organization   174
Consequences of Regulatory Complexity   175
Does Plasticity Accelerate or Retard Evolution?   178
Does Plasticity per se Evolve?   178
Does Behavior Take the Lead in Evolution?   180
Evolvability   182
Developmental Plasticity as a Solution to the Cohesiveness Problem   183

CHAPTER 8   DARWIN'S THEORY OF DEVELOPMENT AND EVOLUTION   188

PART II   THE ORIGINS OF NOVELTY

CHAPTER 9   THE NATURE AND ANALYSIS OF PHENOTYPIC TRANSITIONS   197
Introduction   197
Missing Chapters   198
Developmental Recombination as a Complex Response to a Simple Input   200
Important Distinctions   200
Problems in Interpretation   202
In Praise of Anomalies   205

CHAPTER 10   DUPLICATION   209
Introduction   209
Duplication and the Rule of Independent Selection   210
Gene Duplication   211
Duplication in the Origin of Novel Morphologies   212

Duplication in the Origin of Novel Behaviors   212

Concerted Evolution and Diversification in Multigene and Multiphenotype Families   214

CHAPTER 11   DELETION   218

Introduction   218

Melanophore Deletion in the Midas Cichlid   218

Deletion in the Evolution of the Arthropod Body Plan   219

Life-Stage Deletions   219

Deletion in the Evolution of Behavior   222

Deletion of the Male Phenotype in Unisexual Flowers and Fishes   223

Deletion of Intermediates   224

What Happens to the Genes?   230

CHAPTER 12   REVERSION   232

Introduction   232

Examples   232

The Developmental and Genetic Basis of Atavisms and Reversions   237

Pleiotropy and Silent Genes   238

One-Step and Gradual Reversions   239

CHAPTER 13   HETEROCHRONY   241

Introduction   241

Behavioral Heterochrony   244

Socially Induced Heterochrony in the Evolution of Termites   248

Life-History Heterochrony in Vertebrates   248

Heterochrony in Plants   250

Gradual versus One-Step Heterochrony   252

CHAPTER 14   HETEROTOPY   255

CHAPTER 15   CROSS-SEXUAL TRANSFER   260

Introduction   260

The Organization and Reorganization of Sex Expression   260

Darwin's Theory of Cross-sexual Transfer   262

Kinds of Evidence for Cross-sexual Transfer   263

Alternative Explanations for Sexual Monomorphism   263

Cross-sexual Transfer in Plants   265

Cross-sexual Transfer in Animals   270

Female Hormones and Neurotransmitter Substances in Male Semen and Accessory Glands   277

Alternative Reproductive Tactics   277

Cross-sexual Transfer of Parental Care   282

Cross-sexual Transfer of Switch Mechanisms   291

The Social Environment as an Inducer of Cross-sexual Transfer   293

CHAPTER 16   QUANTITATIVE SHIFTS AND CORRELATED CHANGE   296

Introduction   296

Correlated Extremes   297

The Two-Legged Goat Effect in Domestic and Natural Populations   298

Trade-offs   302

Quantum Shifts and Environmental Extremes   315

CHAPTER 17   COMBINATORIAL EVOLUTION AT THE MOLECULAR LEVEL   317

Introduction   317

Combinatorial Evolution in Regulatory Molecules   318

Combinatorial Evolution in the Genome   320

Phenotypic Recombination by RNA Splicing   323

Genetic Accommodation at the Molecular Level   324

Combinatorial Evolution and DNA Sequence Conservation   325

Molecular Terminology and the Definitions of Evolution and the Gene   327

Speculations   329

CHAPTER 18   PHENOTYPIC RECOMBINATION DUE TO LEARNING   337

Introduction   337

Learning in Relation to Selection and Evolution   338

Learning as a Developmental Source of Evolved Correlations   338

The Genetic Accommodation of Learned Traits   339

Mimicry of Natural Selection by Learning   339

Evolved Components of Learning   341

Learned Components of Evolved Traits   342

Learning and Individual Differences in the Evolution of Specialization   344

Social Competition and Learning   349

The Importance of Forgetting   350

CHAPTER 19   RECURRENCE   353

Introduction   353

Historical Discussions of Recurrence   354

Problems in the Interpretation of Recurrent Similarity   357

Patterns of Recurrence   363

Environmentally Correlated Recurrence   368

Consequences of Recurrence for Systematics and Phylogenetics   369

The Evolutionary Significance of Recurrence   373

PART III   ALTERNATIVE PHENOTYPES

CHAPTER 20   ALTERNATIVE PHENOTYPES AS A PHASE OF EVOLUTION   377

Introduction   377

Terms and Distinctions   377

Alternative Phenotypes as Models for Relating Development and Evolution   379

Phenomena Easily Confused with Alternative Phenotypes   380

Historical Misconceptions about Alternative Phenotypes   383

How Alternatives Facilitate Evolution   392

CHAPTER 21   DIVERGENCE WITHOUT SPECIATION   394

Introduction   394

Specieslike Aspects of Alternatives   395

Evidence of Postorigin Divergence   400

Why Alternatives May Foster Divergence More Effectively Than Speciation   401

Phenotype Fixation and Developmental Character Release   404

Genetic Assimilation Revisited   415

Conclusions   416

CHAPTER 22   MAINTENANCE WITHOUT EQUILIBRIUM   417

Introduction   417

Matching Models to Modes of Regulation   418

Maintenance of Conditional Alternatives   429

Alternative Phenotypes and Maintenance of Genetic Polymorphism   434

Conclusions   439

CHAPTER 23   ASSESSMENT   440

Introduction   440

Terminology   442

Selected Examples   443

Learning and Assessment   462

How Complex Mechanisms of Assessment Originate and Evolve   464

The Evolution of Assessment Involving Choice   466

PART IV   DEVELOPMENTAL PLASTICITY AND THE MAJOR THEMES
OF EVOLUTIONARY BIOLOGY

CHAPTER 24   GRADUALISM   471

Introduction   471

Modern Permutations of the Gradualism Controversy   473

What the Gradualism Controversy Is Not   474

Fisher's Solution, or Why the Neo-Darwinian Resolution of the Gradualism Controversy
Was Unsatisfactory   476

Nine Modern Beliefs about Gradualism Reexamined in the Light of
Developmental Plasticity   476

Is Darwinian Gradualism Falsified by a Developmental Evolutionary Biology?   481

Conclusions   482

CHAPTER 25   HOMOLOGY   485

Introduction   485

Cladistic and Broad-Sense Homology   486

The Criteria versus the Definition of Homology   488

Iterative or "Paralogous" Homology   490

"Mixed" Homology   490

Levels of Analysis and the Perception of Homology   494

Multiple Developmental Pathways and the Homology Concept   494

Conclusions   497

CHAPTER 26   ENVIRONMENTAL MODIFICATIONS   498

Introduction   498

The Entrenchment of Environmental Elements in Development   500

The Environmental Induction of Novelty   503

The Superior Evolutionary Prospects of Environmentally Induced Traits   503

Recurrent Extreme Environments and Phenotypic Innovation   505

Evidence for Environmental Initiation of Reorganizational Novelty   508

Environmental Influence and the Paleontological Time Scales of Evolutionary Change   518

Conclusions   524

CHAPTER 27   SPECIATION   526

Introduction   526

Developmental Plasticity and Speciation: Theory   528

Developmental Plasticity and Speciation: Kinds of Evidence   530

Examples   531

Speciation by Fixation of Parallel Alternative Phenotypes in the Two Sexes   538

Plasticity and Abrupt Sympatric Speciation   551

Other Proposed Examples of Sympatric Speciation   554

Alternative Phenotypes and Speciation in Clines   560

Learning, Sexual Selection, and Speciation   562

Conclusions   562

CHAPTER 28   ADAPTIVE RADIATION   564

Introduction   564

Binary Radiations   566

Multidirectional Radiations   573

Synergism of Plasticity and Other Factors in Adaptive Radiation   591

Grounds for Generalization   593

Predictions   596

CHAPTER 29   MACROEVOLUTION   598

Introduction   598

Intraspecific Macroevolution Compared with Previous Macroevolution Concepts   599

How Developmental Plasticity Facilitates Intraspecific Macroevolution   602

Evidence   603

Sexually Selected Flexibility and Macroevolutionary Trends   608

Systema Naturae, or Why All Phyla Are Old   609

Why Molecular Biology Cannot Solve the Macroevolution Problem   615

CHAPTER 30   PUNCTUATION   617

Introduction   617

Plasticity and Punctuation   619

Two Fossil Examples   620

Morphological Stasis Is Not Evolutionary Stasis   627

Conclusions   629

CHAPTER 31   ONE FINAL WORD: SEX   630

LITERATURE CITED   639

AUTHOR INDEX   745

TAXONOMIC INDEX   759

SUBJECT INDEX   767

PART I

FRAMEWORK FOR A SYNTHESIS

1

Gaps and Inconsistencies in Modern Evolutionary Thought

How does developmental plasticity fit within a genetic theory of evolution? This question remains largely unanswered, yet evolving organisms are universally responsive to the environment as well as to genes. The result is a theory of evolution fraught with contradictions. There is no one-to-one relation between phenotypes and genes, yet a gene mutation is often visualized as the originator of a new phenotypic trait. Development is seen as a homeostatic, canalizing, constraining force, yet it is the originator of all adaptive change. Proximate mechanisms are excluded from ultimate (evolutionary) explanations, yet a proximate mechanism (development) produces the variation that is screened by selection. Evolution is often portrayed as a gradual shift in the dimensions of quantitative traits, yet biologists wish to explain the origin and spread of the discrete and genetically complex entities we call "traits." The conceptual gap that should be filled by development has been filled instead with metaphors, such as genetic programming, blueprints for organisms, and gene-environment interaction. Here I put the flexible phenotype first, as the product of development and the object of selection, and examine the consequences for the genetic theory of evolution. Theoretical incorporation of developmental plasticity calls for changes in thinking about virtually every major question of evolutionary biology, and resolves some of its most persistent controversies.

Introduction

One of the oldest unresolved controversies in evolutionary biology—and a source of many bitter arguments and failed revolutions—concerns the relation between nature and nurture in the evolution of adaptive design. In modern evolutionary biology there is still a gap between the conclusions of a genetical theory for the origin and spread of new traits, and the observed nature of the traits being explained, the manifest phenotypes, always products of genes and environment.

This gap is especially clear in discussions of adaptively flexible morphology and behavior. How are complex adaptively flexible traits constructed during evolution? Remarkable adaptability is shown by humble plants and animals. I see a young female *Polistes* wasp approach the dominant queen of her colony and adopt a subordinate position that condemns her to permanent sterility and serfdom.

Yet she has done the right thing in Darwinian terms, for she genetically profits by helping to rear her despotic sister's young. How should we explain such behavior? Should we visualize the spread of genes for flexibility and altruism, and the eventual construction of a genetic capacity for environmental assessment and social judgment? Or is the evolutionary construction of such complex abilities something other than an accumulation of modifier alleles, selected gradually and independently, one by one?

It is not surprising that students of human behavior have been among the first to complain about the failure of evolutionary biology to deal effectively with complex adaptive plasticity. Anthropologists, for example, have good reason to question the explanations of a strongly gene-centered sociobiology. Human behavior is *essentially* circumstantial. We know intuitively that our phenotypes are molded by our environments—by

3

mothers, fathers, schoolteachers, economics, and accidents of history. But in this respect human nature is like every other phenotype of every other animal or plant. A phenotype is a product of both genotype and environment. If this is true, then how can students of social evolution so often predict cultural patterns and insect behavior from models based on genes alone? To state the problem in more general terms: if recurrent phenotypes are as much a product of recurrent circumstances as they are of replicated genes, how can we accept a theory of organic evolution that deals primarily with genes? How does the systematic incorporation of environmental influence evolve?

The failure of sociobiology to give completely satisfactory answers to these questions is only a local symptom of a systemic ill: the same questions could be raised about the adaptive flexibility of all living things. Nor is gene-centered thinking a problem peculiar to evolutionary biology. The recognition of prions as a previously unknown type of infectious pathogen—one that can be either genetically (mutationally) or environmentally acquired, via infection with a protein-manipulating (template) protein—was greeted with decades of skepticism within medicine and molecular biology (Prusiner, 1998). Among the obstacles that prion researchers had to overcome was a strong reluctance to accept the idea that the structural plasticity of proteins allows them to be altered through intracellular interactions with other proteins, to yield structural (folding) properties formerly thought to be mandated exclusively by genes (Prusiner, 1998).

Within evolutionary biology, as in molecular biology, an emphasis on genetics is understandable, at least in the light of history. The great strength of the twentieth-century synthetic theory of evolution has been to use genetics to advance the study of evolution. By far the greatest progress toward a formal coherent theory of evolution has been achieved in genetics. Even among biologists who work primarily with whole organisms, observe only phenotypes, and can only imagine their genetic underpinnings, genetic models take precedence when it comes to conceptualizing evolutionary transitions.

In defense of this historical emphasis on genetics, Mayr (1982a, p. 832) argues that early progress in genetics required the banishment of development from discussions of heredity. Certainly this helped early geneticists to avoid Lamarkian errors. It also enabled them to put aside the developmental anomalies or "macromutations" that were seen by some, such as William Bateson (1894), as evidence against Darwinian gradualism. But, as so often happens, a

temporary expediency became a reflexive belief. In experimental genetics, environmental sensitivity came to be considered noise rather than a subject for research, or an important factor in evolution (see Sang, 1961; Kirkpatrick, 1996, p. 134). The influential theoretical geneticist Sir Ronald Fisher has been quoted as dismissing complex intraspecific polyphenisms, or conditionally expressed alternative morphological traits, as unfortunate defects in the adaptive operation of genes, saying, "It is not surprising that such elaborate machinery should sometimes go wrong" (Wigglesworth, 1961, p. 107). Phenotypic plasticity is still just noise for many biologists: "For the functional anatomist interested in the evolution of the digestive system, phenotypic plasticity can be a major nuisance because it may obscure important adapted patterns" (McWilliams et al., 1997).

The habit of identifying environmental effects primarily in order to dismiss them predates genetics in the practice of taxonomy. It has long been accepted that if variation is merely environmental, plastic, or polymorphic, it is an unreliable species character. Occasionally, the *capacity* to produce a particular phenotype, deduced from a study of related species, is considered a trait, as in the concept of "underlying synapomorphy" (shared developmental potential due to common ancestry; Saether, 1979, 1990). Needless to say, this idea is fraught with controversy, even though there is much experimental and comparative evidence to support it (see chapter 19). The problem is even worse in paleontology than in phylogenetic analysis of living specimens. In fossil material, the morphs of a polymorphism are more difficult to distinguish from species or higher taxa. So in the three major branches of science that were the foundations of the modern synthesis—genetics, systematics, and paleontology—phenotypic plasticity has been an intractable nuisance, rather than a subject of productive interest.

Given this history, it is perhaps not surprising that the areas of biology that deal most directly with the mechanisms of plasticity have been among those most estranged from modern evolutionary theory and genetics (see Hamburger, 1980; Bennet and Huey, 1990; Burian, 1986, 1997). This includes embryology, molecular and cellular developmental biology, comparative psychology (animal learning), and physiology, including endocrinology and neurobiology. Their estrangement was reinforced within developmental biology by early resistance to Mendelian Theory among embryologists (Burian, 1986), and within evolutionary biology by a strong distinction, in discussions of adaptation, between proximate and ultimate causation (Mayr, 1961), a problem further discussed below. In sum, both developmental and evolutionary biology embraced elaborate sets of attitudes that seemed to conspire against a major role for the study of plasticity and development in relation to evolution. As a result, we have a solid understanding of the genetic consequences of selection for adaptive evolution, and a poor understanding of how the environment influences evolution through its effects on development and of the developmental causes of variation.

Meanwhile, some remarkable discoveries regarding phenotypic flexibility have accumulated, even in areas of evolutionary biology where the main focus of attention has been on genes. Modern kin-selection research, for example, was born under the title "The *genetical* evolution of social behaviour" (Hamilton, 1964a,b; emphasis mine) and marched forward under the banner of the selfish gene (Williams, 1966; Dawkins, 1976). While debate raged over genetic determinism and human sociobiology, evolutionary biologists went about testing the predictions of kin-selection theory. The result was not a strengthened belief in the genetic determinism of social behavior, as critics feared. Rather, it was an unprecedented growth of evidence for adaptive phenotypic flexibility: ingenious research documented an astounding capacity for learned kinship discrimination in animals, not only in primates and social insects but also in such unlikely creatures as tadpoles and isopods (Fletcher and Michener, 1987, review the first two decades of this work).

Nonhuman animals proved capable of fine-tuned assessment of other parameters as well. They make strategic adjustments of morphology, behavior, and life-history tactics in accord with the relative reproductive and fighting capacity of competitors and with the changing costs and payoffs for investing in offspring of different sexes or ages under different parental circumstances. Animal decisions were proven over and over again to conform to the quantitative predictions of theory regarding situation-dependent behavior. The result was not massive documentation of genetic determinism in behavior, but rather the opposite—massive documentation of a heretofore widely underestimated capacity for adaptive condition-sensitive behavior and development. Yet the emphasis on genetic reasoning was such that these remarkable data on phenotypic flexibility were discussed almost entirely in terms of fitness effects and genes. There was no explicit recognition of the fact that Hamilton's formulation was actually being used as a model

of development—a description of the advantageous switch point between alternative developmental pathways (e.g., see Crozier, 1992, cf. West-Eberhard, 1992b). The developmental implications of the findings were largely overshadowed by attention to the genetic details.

In the latter part of the twentieth century there was an upsurge of interest in development and phenotypic plasticity, beginning with the publication of the landmark text by Raff and Kaufman on *Embryos, Genes, and Evolution* (1983) and two interdisciplinary conferences that encouraged integration of ideas about development and evolution (Maynard Smith et al., 1985; Bonner, 1982). These events drew attention to the need for a new synthesis.

Subsequent publications have resurrected heterochrony as a unifying theme (McKinney, 1988; McKinney and McNamara, 1991; see also Gould, 1977) or featured phenotypic plasticity (Matsuda, 1987; Schlicting and Pigliucci, 1998). Many authors have drawn attention to the general importance of development and plasticity in certain very broad areas of evolutionary thought, including in discussions of homology (Hall, 1994; Wagner, 1989b, 1995); morphogenesis (Thomson, 1988; Atchley and Hall, 1991); macroevolution (Valentine and Campbell, 1975); speciation (e.g., Meyer, 1989b; West-Eberhard, 1986, 1989); and the evolution of individuality (Buss, 1987). A recent textbook of vertebrate paleontology (Carroll, 1997) integrates modern developmental genetics into a coherent view of evolution using vertebrates as examples. Some of the most ambitious efforts to show the general relationship of development to evolution have come from biologists originally trained not in evolution but in molecular, cellular, and developmental biology (e.g., Raff and Kaufman, 1983; Hall, 1992; John and Miklos, 1988; Nijhout, 1991b; Kauffman, 1993; Gerhart and Kirschner, 1997). These publications echo some thoroughly ignored prescient voices of the past (e.g., Conrad, 1983; Whyte, 1965) whose insights regarding development and evolution were discussed in mathematical or philosophical terms outside the usual ken of biology. Wake (1996b) gives an overview of attempts at synthesis and their limitations in crossing the cultural border between developmental biology and evolutionary biology.

The result so far has been a piecemeal synthesis and a heightened awareness that the modern neo-Darwinian approach may be incomplete, without a broad treatment from the point of view of evolutionary biology as a unified science. In the absence of a major attempt at synthesis, this seeming renaissance of developmental insight could end as a passing fad. Interest in development and evolution has come and gone repeatedly in the history of biology (see Gottlieb, 1992) without leaving a trace in the mainstream of evolutionary thought.

Six Points of Confusion and Controversy

What has delayed a sweeping synthesis of development and evolution, now that the technical and conceptual difficulties that confronted early genetics have been overcome? Before listing some positive steps from the past, and then attempting a more general solution, it is worth considering the complexity of the formidable task at hand. Darwinians after Darwin have embraced a whole set of interrelated ideas, which now forms a coherent package of mutually reinforcing concepts. Their internal consistency imparts a kind of selective blindness to facts that do not fit. In general, ideas that are obviously compatible with the fundamentally genetic theory of evolution, such as the theory of speciation by geographic, gene-pool isolation and the roles of mutation and genetic drift in the initiation of novelty, have been emphasized. Other ideas that are less obviously compatible with a genetic explanation, such as the suggestion that developmental anomalies could facilitate evolutionary change or that learning could influence the evolution of morphological diversity, have been relatively neglected. Uncertainty as to how to accommodate such factors within a genetic framework has sidetracked the connections that might have been made between development and evolutionary theory.

There is no intrinsic barrier to including development in the modern theory of evolution. The problems that need to be solved are not so much deficiencies of theory as they are inconsistencies between formal theory and habits of thought. The inconsistencies are revealed by what practicing biologists actually say and do. Here I list six of the most obvious inconsistencies and show how they form an occult barrier to a working synthesis of development and evolution.

1. The Unimodal Adaptation Concept and the Multimodal Products of Development and Plasticity

Darwin's theory of inheritance included development alongside transmission, what we would now

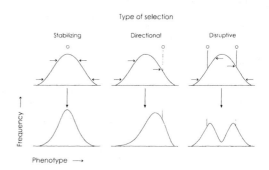

Fig. 1.1. The three types of selection. In a common textbook description, evolution is seen as a shift in the frequency distribution of a quantitative trait under selection that either augments, shifts, or reduces the ancestral population mode (similar diagrams in Mather, 1973; Grant, 1977; Curtis, 1983; Willson, 1984; Endler, 1986; Futuyma, 1986a, 1998; Ridley, 1993).

call trait expression alongside trait genetics (see chapter 8). At the same time, he greatly emphasized the importance of gradual change, and this became a key issue for evolutionary biology. With the populational thinking of the modern synthesis, evolutionary change came to mean a gradual shift in the mode of a continuous distribution (figure 1.1). The focus of population and applied genetics on quantitative, continuous variation reinforces this shifting-mean view, as further discussed in the following section. The focus in paleontology on "trends," or directional quantitative shifts in morphology, also emphasizes a shifting mean of a continuous distribution (see chapter 29, on macroevolution). This shifting-mode depiction of evolution persists into the present, in part because it is a convenient way to show gradual change. It is not an erroneous view of evolution, and I argue further below that most evolution by natural selection probably involves such quantitative shifts. Nonetheless, by itself, quantitative change is only part of the story of evolution, for it does not address the question of the origin of discrete (qualitatively different) novelties. Simpson's (1944) "quantum evolution," punctuated equilibrium, and the associated idea of speciational macroevolution (chapter 29) also describe quantitative change in continuously variable traits, or trends. But they do not address the question of the origin of discrete phenotypic traits, except as threshold consequences of quantitative trends.

Populational thinking, then, has been translated into quantitative variation and shifting mean values of metric traits. It meant the demise of the typology that saw individuals as manifestations of an idealized species-typical form, replacing this with the image of a distribution around a mean. But it established a new typology, that of the unimodally adapted population. This overlooks two important properties of the phenotype and of adaptive evolution, both of them related to development: (1) the ability of organisms to facultatively switch among specializations, producing polymorphisms, polyphenisms, and different behavioral and morphological modes; and (2) the common occurrence of distinctive life-stage variants during ontogeny, with larvae, juveniles, and adults commonly having very different adaptive morphologies and behaviors. As phrased by Buss (1987), the synthetic theory is "a 'theory of adults'—one which has failed to address the diversity of ontogeny" (p. 65). This emphasis has been so strong that de Beer (1958, ex Bonner, 1988, p. 179) called larval divergence "clandestine evolution." It is also a theory of adults that insufficiently addresses the polymodal diversity of the adults themselves.

The unimodal adaptation concept, combined with other strong and compatible currents in early twentieth-century biology such as Cannon's (1932) idea of physiological homeostasis came to be used as a unifying principle for all of biology (see Emerson, 1954), and Waddington's (1942) idea of canalization, plus the idea of stabilizing selection (e.g., Schmalhausen, 1949 [1986], Dobzhansky, 1970), put evolutionary theory on a track that has made it difficult to reinstate development as an innovative factor in evolution, even after the early issues of Lamarkian inheritance and speciation by macromutation were resolved. The idea of a single optimum or primary adaptive state affects even how one thinks about discrete, competing alternative genetic alleles. It suggests that competition among discrete entities must eventually lead to the persistence of only one, since they are unlikely to have equal values under selection. Thus, polymorphism came to be seen as unstable, and (except under special circumstances such as heterozygote advantage or frequency dependence) selection is expected to produce a single adaptive norm.

This intuitively reasonable conclusion is reinforced by models demonstrating that adaptive polymorphisms require "rather severe conditions" of frequency dependence for their evolution (Nei, 1975, pp. 76–77). Disruptive selection is effective in producing the high variance represented by a bimodal distribution "only if coupled with assortative mating among animals of like phenotype," or inbreeding (Futuyma, 1979, p. 371). And elaborate polymorphic forms within a species are unlikely because "the process of divergence is constantly opposed by interbreeding between the divergent polymorphic types" (Grant, 1971, p. 120). Some have reasoned that these genetically determined polymorphisms are simple to evolve when compared to the great difficulty of the evolution of complex environmentally cued polyphenisms:

"The origin of a fixed adaptation is simple. The population merely needs to have or to acquire some genetic variation in the right general direction. The origin of a facultative response is a problem of much greater magnitude since it implies evolution not only of two distinct complex forms but also the sensing and control mechanisms allowing them to be adaptively expressed. (Williams, 1966, pp. 81–82).

This kind of skepticism has led in the past to disparagement of conditional alternatives as phenomena worthy of attention: "A high degree of predictability of environmental change is a necessary condition for the evolution of conditional strategies and . . . the very liberal use of conditional strategies in adaptive explanations is unwarranted" (Wade, 1984, p. 707).

The unimodal adaptation concept shows the power of an internally consistent set of ideas to reinforce a single line of thought, even when this means overlooking a large body of contradictory facts. The idea of unimodal adaptation is complemented and reinforced, for example, by seeing a species as a cohesive coadapted gene complex (Mayr, 1988). The well-adapted species is protected from genetically disruptive hybridization (which is known to produce maladaptive phenotypes) by reproductive isolating mechanisms (Mayr, 1963). Competing divergent adaptations, then, would require a genetic revolution associated with speciation. "Every species lives on an adaptive peak and the problem of speciation is how to reach new not previously occupied, adaptive peaks" (Mayr, 1976, p. 147; for a history of the genetic-revolution concept and its variants from 1954 to 1988, see Provine, 1989, in Giddings et al., 1989).

Without the genetic revolution associated with speciation, populations are seen as characterized by stasis (Eldredge and Gould, 1972)—they do not acquire persistent new characteristics (or, presumably, multimodal novelties alongside their single adaptive peak). In the view of many evolutionary biologists (e.g., see Giddings et al., 1989), the drastic reorganization required for speciation-related divergence most likely occurs in a small isolated population in a new and different environment, where genetic change due to drift and/or selection can avoid the diluting influx of genes from other populations having contrasting adaptations. Behind this is the concept of unimodal adaptation, the vision of a population variable to be sure, but centered, barring speciation, around a single stable or slowly oscillating mode (see Cheetham, 1986a).

By this view, all biological diversification depends upon speciation. Yet this is inconsistent with the patterns of intraspecific diversity that every biologist can see. In virtually every species, juveniles have one set of morphological and behavioral adaptations, and adults another. Sexual dimorphism in structure, behavior, and physiology is legion. The striking differences between queens and workers of social insects are familiar alternatives producible by a single genotype. Every biology student memorizes the life cycle polymorphisms of plants and host-alternating parasites—intraspecific phenotypic divergence of an extreme kind. And alternative tactics of courting, fighting, foraging, and territorial defense are known to every behavioral ecologist. As discussed in subsequent chapters, this kind of diversification is possible because *developmental* divergence permits mosaic evolution in different directions within the same population.

2. The Cohesiveness Problem: Development as a Conservative Force versus Development as the Source of All Change

Related to the unimodal adaptation concept is the idea that development is a conservative or homeostatic factor in evolution, one that prevents rather than promotes change. A belief in the stabilizing role of development is consistent with an equilibrium approach to evolutionary theory that begins with the Hardy-Weinberg equilibrium and treats the causes of change, such as mutation, selection, and drift, as departures from equilibrium. The impression given by such an approach, whether intended or not, is that the natural tendency of populations is to resist change unless perturbed by some definite force or chance event. By this view, "[e]volution involves substantial adaptive impedimenta," and the remarkable numbers of well-designed organisms we observe may be due to "serendipity that allows adaptation to have the upper hand so often in our world" (Kirpatrick, 1996, pp. 143–144).

The Hardy-Weinberg equilibrium has been the starting point for evolutionary genetics in the training of biologists for more than sixty years (e.g., see Dobzhansky, 1937 [1982]; Simpson and Beck,

1965; Dobzhansky et al., 1977; Futuyma, 1986a; Ridley, 1993), one whose "importance cannot be overemphasized" (Futuyma, 1998, p. 236). Given this background, it is not surprising that it took a frequency-dependent equilibrium theory of "evolutionary *stable* strategies" to convince evolutionists that more than one complex behavior could coexist within a population. Even though the equilibrium assumptions behind these models turned out to be unrealistic, they effectively promoted the ideas. The result, as in the case of kin-selection theory, was the discovery of a large array of condition-sensitive alternative behaviors, explainable by equilibrium models only if the equal-payoff equilibrium point functions as a conditional or developmental switch point (see chapter 22).

In keeping with an equilibrium, stability approach to evolution, the stabilizing role of development has been described in terms of "developmental homeostasis" (Dobzhansky and Wallace, 1953) and "canalization" (Waddington, 1940), defined as the evolved ability to maintain a single phenotypic norm in the face of deviation-inducing genetic and environmental effects during ontogeny. The traditional emphasis on stability, homeostasis, equilibrium, canalization, and resistence to change has its modern counterpart in the concept of "developmental constraints" (Gould and Lewontin, 1979; Alberch, 1983a; Cheverud, 1984b; Edelman, 1988; Gould, 1989a; Kauffman, 1983a; Maynard Smith et al., 1985; Wake, 1982), a term that seems to emphasize the conservative rather than the variable side of development. Some authors defend this concept as broad enough to include diversifying effects of development (e.g., Wake, 1982; Stearns, 1986, p. 37; Gould, 1989a). But the conservative emphasis of the term is inevitable: the word "constraint" evokes a mental image of limitation and stability more readily than it does one of variation and change. P.J. Taylor (1987) notes how thinking about evolution is transformed if one considers development a facilitator of change rather than a constraint. *Robustness*, or ability to be maintained in the face of potentially disruptive factors, is the most recent addition to the lexicon of stability but, unlike earlier concepts, is seen as achievable by plasticity (e.g., Gerhart and Kirschner, 1997).

Thoday (1955, p. 318) noted the "paradox of terminology" in evolutionary biology, where a language of stasis and stability has long dominated the science of change. Even Schmalhausen, a champion of "the value of instability," or the capacity for variation crucial to adaptive modification, subtitled his best-known book *The Theory of Stabilizing Selection* (1949 [1986]). The stability obsession reached its ironic heights in the title of a paper by Haldane (1954) called "The Statics of Evolution," an essay on factors that retard evolution.

Some authors argue that change is difficult because of the cohesive coadapted gene pool (e.g., Mayr, 1988, pp. 423–438; Carson, 1985; Maynard Smith, 1983b; Eldredge and Gould, 1972). A. T. Ohta (1989) and Provine (1989) provide brief histories of this influential idea, which is traceable to Lerner's book *Genetic Homeostasis* (1954). Mayr (1988) summarized his view of the "cohesion of the genotype" as follows:

> Free variability is found only in a limited portion of the genotype. Most genes are tied together into balanced complexes that resist change. The fitness of genes tied up in these complexes is determined far more by the fitness of the complex as a whole than by any functional qualities of individual genes. (p. 424)

Mayr's (1988) argument recognizes coadapted phenotype subunits but emphasizes their cohesiveness and possible links, not their semi-independence. Thus, a species has a set of adaptations, each one intricately accommodated by the others into a complexly integrated and delicately balanced whole under selection at the level of the individual. Evolution threatens individual integration because it requires intricate readjustments of so many interconnected parts (see also Rollo, 1994). Kauffman (1993, p. 72) calls the problem of cohesiveness in relation to change the "complexity catastrophe"— the constraint of epistatic interaction on adaptive improvement. As in the debate that apposes saltation and gradualism, the complexity catastrophe reflects the tension in biology between the complementary properties of integration and atomization, in this case with regard to the possibility of change.

The cohesiveness problem has a long history in both embryology and evolutionary biology. Early authors (e.g., Darwin, 1859 [1872], p. 205; Lillie, 1927) recognized the need for special explanation of multiple simultaneous directions of evolution of different complex traits within a single individual, traits produced by a single genome that is unchanging during ontogeny. Some recent authors, while emphasizing the importance of developmental constraints, have recognized that constraints do not make it wholly impossible to alter one trait adaptively without producing excessive maladaptation in others (Maynard Smith et al., 1985, p. 266), though the authors do not specify how this is achieved.

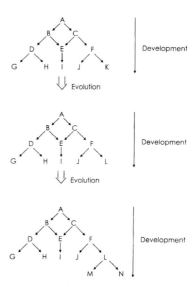

Fig. 1.2. The expectation of terminal addition as a consequence of the hierarchical nature of development. "Each arrow represents a process and each letter a resulting structure. An alteration early in the hierarchy would be unlikely to be compatible with life. It is easier to imagine an alteration or addition at the end being compatible with survival" (Maynard Smith, 1983b, p. 41). After Maynard Smith (1983b).

The cohesiveness idea has profoundly affected views of development and evolution. The genetic theory of evolution has not sufficiently addressed this problem because it sees the origins of novelty as primarily genetic or mutational, leaving development as a black box. The cohesiveness idea leads to the conviction that terminal additions or modifications late in ontogeny are most feasible (e.g. Maynard Smith, 1983b, p. 41; Arthur, 1984). A belief in terminal addition reflects a vision of the individual phenotype and its development as a cohesive interconnected gene net (Figure 1.2; Maynard Smith, 1983b; see also Stebbins, 1977). The earlier the change in such a net, the greater the disturbance propagated through the whole.

The idea that developmental cohesiveness restricts evolution persists even though it has long been clear that innovation does not occur exclusively by terminal addition. The idea of terminal addition, and its corollary, von Baer's biogenetic law (ontogeny recapitulates phylogeny), is applied to both the ontogeny of individuals and that of particular traits (Wenzel, 1993). Many exceptions were cited as counterexamples more than seventy years ago by developmental biologists, in the debate over recapitulation (see Garstang, 1929 [1985]; Gould, 1977; for a concise history, see Gould, 1979). Profound change during early ontogeny occurs in many animals, including mollusks, beetles, sea urchins, and fish (Thomson, 1988; G. C. Williams, 1992; Raff et al., 1990). In one of the rare tests of the assumption of terminal addition, Mabee (1993) surveyed a large number of characters in centrarchid fishes and found that only 52% of them evolved via terminal addition.

When recapitulation of adult structures occurs during the ontogeny of an individual, it demonstrates the capacity for internal rearrangements, not terminal addition to a cohesive unchanging phenotype, because it implies that features originating in the adult have come to be expressed in earlier stages (Gould, 1977, 1979). But the final blow to the ideas of cohesiveness, terminal addition, and conservation of early ontogeny is dealt by molecular developmental biology. Molecules crucial to early embryology can be reorganized during evolution to function in different ways (e.g., see Davidson, 1990, p. 384). In short, the cohesive terminal addition view of development has been replaced in developmental biology by a view that emphasizes the mosaic or modular nature of phenotype organization and its relation to the dissociability and reorganization of traits. This view has not fully permeated evolutionary biology. Some cladists invoke the assumption of terminal addition to establish polarity, or direction of evolution (discussion in Wheeler, 1990; Wenzel, 1993), and disruption of cohesiveness by speciation is still considered by many to be a requirement of evolutionary change (see chapters 27, 29, 30).

Part I offers a way out of this peculiar trap where the science of change finds change difficult to explain. A simple model of phenotypic structure shows how quantitative genetic variation and the continuous variation around a mode, so often documented when biologists measure observable traits, are related to the evolution of multimodal or discrete phenotypes governed by switches. Ontogeny is a condition-sensitive, bifurcating process that allows and even promotes polymodal adaptation. Not only can species differences originate prior to speciation and contribute to the origin of reproductive isolation or speciation itself (chapter 27), but patterns of intraspecific diversification influence the form of radiations (chapter 28) and the origins of macroevolutionary change (chapter 29). These findings represent a radical departure from the cohesiveness concept and the related idea that phenotypic diversification depends on speciation to break the cohesiveness of coadapted gene complexes.

The dynamic role of development in the production of selectable variation is profoundly significant beyond evolutionary theory. All areas of biology concerned with the phenotype, including those dedicated entirely to mechanism, medicine, or such psychological phenomena as motivation and learning, depend on an adequate concept of phenotypic structure and variable development. The realization that genes affect phenotypes only if they are expressed, for example, means that development must assume prominence in any intelligent discussion of genetic issues. The naked ignorance of supposing that a genome map could represent a blueprint for an organism is exposed by realizing that all gene expression depends on preexisting phenotypic structure and specific *conditions* as surely as upon specific genes. The phenotype is cohesive, but it is also eminently changeable. The chapters that follow discuss the creative role of developmental variation primarily in relation to issues in evolutionary biology. But the implications beyond evolutionary biology are clear.

3. Proximate and Ultimate Causation

The answer to "why" an organism is, or behaves in, a certain way can be answered either in terms of mechanisms (proximate causes) or in terms of selection and evolution (termed "ultimate causes" by Mayr, 1961). This distinction is designed to prevent confusion between levels of explanation in biology. But it was an easy step from this important point to the idea that the mechanisms of development have nothing directly to do with evolution, or that they are the focus of a different research approach, one not primarily concerned with evolution and justifiably left aside by those primarily interested in selection and adaptation (e.g., see Tinbergen, 1951; Alcock and Sherman, 1994). Any such position runs the risk that in evolutionary biology the mechanisms of developmental plasticity will be shelved among the proximate causes of design, isolated from the ultimate, evolutionary causes. The problem is exacerbated because the plasticity of organisms—their ability to respond to environmental circumstances, to assess complex situations, and to learn—has always seemed a more elusive property than their morphologies or their genes.

Much discussion at cross purposes regarding proximate and ultimate causation (e.g., Sherman, 1988, 1989; Crozier, 1992; West-Eberhard, 1992b) has transpired without either side stating clearly what needs to be said: the proximate–ultimate distinction has given rise to a new confusion, namely, a belief that proximate causes of phenotypic variation have nothing to do with ultimate, evolutionary explanation. For evolutionary biology, proximate mechanisms represent more than just different levels of analysis or research styles. They are *the* causes of the variation upon which selection acts. Even in ethology, where environmental influence on the development of behavior can be directly observed, environmental effects were far from being a universal subject of interest. A leading ethologist (Tinbergen, 1951, ex Fagen, 1981) stated categorically that "[t]he study of ontogeny of behavior cannot be expected to contribute much to the study of evolution" (p. 15). There is a severe barrier to synthesis when a prominent and learned evolutionist (Wallace, 1986) queries, "Can embryologists contribute to an understanding of evolutionary mechanisms?" and answers with a resounding and provocative No: "Embryologists . . . need not explain how somatic development might affect the evolution of these developmental programs. Except for achieving success in reproduction, they do not" (p. 150). Wallace's challenge to developmental evolutionary biology is to demand an explanation of how purely phenotypic events, somatic not genetic changes, can affect evolution. Answering that challenge is one of the main themes of this book.

Among the consequences of neglect of mechanisms in modern evolutionary biology are the problems that arise when the black box of mechanism is filled with imaginary devices. By imagining that there is a one-to-one relationship between genes and phenotype, for example, biologists lose sight of the role of the environment and polygenic influence in development and evolution. Imagining that there are two types of genes—structural genes and regulatory genes—one fails to realize that the same gene can play either a structural or a regulatory role. And imagining that complex alternative phenotypes are controlled by chromosomally linked gene sets, or supergenes, one fails to realize how easily genes on different chromosomes can have their expression coordinated by hormones and other means, and how easily the components of a genetic mosaic can be independently altered. All of these imagined mechanisms are discussed in subsequent chapters and shown to be fundamentally wrong. Confused concepts are virtually inevitable if data on real mechanisms are ignored, because one test of the adequacy of a theory is mechanistic feasibility. The confusion sired by imaginary mechanisms demonstrates the importance of proximate causation for understanding evolution.

But more important than errors due to neglect of development have been the decades of missed opportunity to understand how universal developmental properties of phenotypes, such as plasticity and modularity, affect virtually all of the phenomena of interest to evolutionary biologists.

4. The Problem of Continuous versus Discrete Variation and Change

Darwin placed such great emphasis on gradual change and the importance of small variations that he was willing to stake the validity of his theory on this one point (Darwin, 1859 [1872], p. 135; see chapter 24 on gradualism). So it is not surprising that critics of Darwinism have long focused on the evidence against gradualism. In 1883, at the height of controversy over Darwin's theory, the Royal Society of London established a committee eventually known as "The Evolution Committee," where bitter controversy over continuous and discontinuous variation divided the members (see Provine, 1971). Among those active in the debate was William Bateson, who in *Materials for the Study of Variation* (1894) collected hundreds of examples of developmental anomalies proving the existence of discontinuous variation and, therefore, the feasibility of saltatory phenotypic change. In the fascinating period that followed, the discovery of particulate Mendelian inheritance was hailed by Bateson and others as support for saltation, being evidence of the potential importance of discontinuous variation in evolution, while a prominent school of biometricians defeuded the importance of continuous, quantitative variation (see Provine, 1971). As I discuss in chapter 24 on gradualism, this was more than an argument over kinds of variation: if large variants could instantly produce a new form or species, then Darwin's selection would be seen as relatively unimportant, compared to developmental innovation, in determining the form of organisms and the course of evolutionary change. Thus, the gradualism debate quickly and unfortunately became a face-off between selectionists and developmentalists.

This was only the beginning of a perennial debate in evolutionary biology about the relative importance of continuous versus discrete variation, selection versus development as the cause of adaptive form. While gradualism has prevailed in descriptions of evolutionary change, many valid points raised by Bateson (1894), Goldschmidt (1940), and others, founded on study of discontinuous developmental phenomena, have been put aside without serious consideration. The gradualism controversy always reappears, like a recurrent nightmare, to haunt Darwinians. It is replayed once more in a recent controversy between modern neo-Darwinians (e.g., see Charlesworth et al., 1982; Simpson's 1984 introduction to his 1944 book) and punctuationists (e.g., see Gould, 1981). Although that discussion really concerns variable *rates* of evolution, not whether or not it occurs by large or small steps (see chapter 30), the gradualism issue resurfaced because large-step regulatory changes were mentioned as potentially more rapid than small-step gradual change mediated by selection on mutations of small effect. The predominant neo-Darwinian view, as concisely summarized by Charlesworth et al. (1982), is that "selection is regarded as the main guiding force of phenotypic evolution" (p. 474).

The controversy over particulate inheritance and gradualism was in large part resolved by quantitative genetics, which emphasizes the additive qualities of the small particulate contributions corresponding to single genes, and shows how polygenic particulate variation can produce continuous distributions within populations, and gradual change. This resolution of discrete, particulate inheritance with gradualism, though central to the modern neo-Darwinian synthesis (see Charlesworth et al., 1982), did not do away with the existence within species of discrete phenotypic variation not directly corresponding to the discrete particles called genes and not obviously or necessarily explained in terms of cumulative small genetic effects. These two kinds of variation continue to foster two views of evolution and a focus on two kinds of data, both of them real.

Confusion regarding the relationship between the continuous and the discrete has led to an odd set of unconscious contradictions in modern biology. Most biologists deal with phenotypic units larger than single molecules and not assignable to a single gene. The traits and characters described for organisms—what Simpson (1944) called "the 'unit characters' of morphologists and taxonomists" (p. 51)—are discrete. They are generally qualitatively distinctive, or at least somewhat discontinuous, relative to others. Research on adaptation often examines the functional significance of these qualitatively different *discrete* phenotypic traits. Meanwhile, quantitative genetics deals entirely with *continuous* variation. And while this empirical study of *continuous* variation documents the genetical change called evolution, *theoretical* genetics, in another corner of evolutionary biology, constructs genetic models where the units are, again, *discrete* alternative alleles. I suspect that few

evolutionary biologists have paused to wonder how the evolution of the complex characters we call traits is related to the continuously variable kind of change studied in quantitative genetics, and precisely how that continuous variation is related, in turn, to the particulate inheritance of competing alleles. There are hidden gaps in our thinking about the continuous and the discrete.

Sometimes a partial resolution of the continuous–discrete problem has been mistaken for a complete one, and this in effect has shelved the real problem and left it unsolved. For example, speciation has been declared by some biologists to be the basis of all phenotypic discontinuities (see chapter 30 on punctuation). This implies that discontinuities do not originate within species, and leaves continuous, quantitative variation the only kind of variation that could be important for evolution within populations. This line of thought minimizes the importance of divergence *within* populations in the form of polyphenisms, polymorphisms, and the complex developmental anomalies that some see as potentially adaptive novelties, and it leaves the significance of these things unexplained. It reinforces the unimodal adaptation concept already discussed, which sees as a difficulty or exception the persistence of polymorphisms and other departures from unimodal equilibria. Meanwhile, developmental biology concerns itself with the discrete products of bifurcating developmental pathways but is little concerned with variation or its adaptive significance. It is no wonder, given this profound duality in approach, that the quantitative genetics of continuous variation and the developmental biology of the discrete have long been estranged (Buss, 1987; but see Atchley and Hall, 1991; Cheverud, 1982; Cheverud and Moore, 1994; Nijhout and Paulson, 1997).

The unresolved problem of the continuous and the discrete is hidden like a Freudian complex beneath many of the long-standing controversies of evolutionary biology. I believe that this problem reflects a deeper one, the unperceived tension between compartmentalization and connectedness—subdivision and integration—in biological organization itself. A solution is proposed in the form of a theory of phenotypic development and organization, given in part I. Continuous variation characterizes the dimensions of measurable, discrete traits. Yet all traits have both modular qualities and connectedness with others—modularity and connectedness or continuity are *complementary*, but differently manifested and analyzed, properties of all traits. Both discreteness and connectedness are products of the switch points that structure phenotypic development. Modularity is reflected in developmental reorganization via the shuffling or recombination of discrete phenotypic parts. Connectedness is reflected in the accommodation of change and the emergence of integration at different levels of organization. Switches, by integrating environmental and genomic influence, allow for quantitative genetic fine-tuning of plasticity and gene expression. The result is a theory of the phenotype based on the complementarity of continuous and discrete variation (see chapter 4).

The *principle of complementarity* regards connectedness and discreteness in phenotypic organization as developmentally related: both continuous and discrete variation traces to regulation by switches, which both coordinate expression within traits and divide expression between traits, both creating and breaking genetic correlations, and determining both the discrete variation (modularity) and continuous variation (in dimensions) of traits. The thresholds of switches are continuously variable and polygenic. So the discrete, switch-controlled variants of development can be directly related to the continuous variation that underlies quantitative genetic change and is the foundation of neo-Darwinian evolutionary biology.

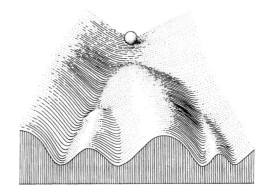

Fig. 1.3. Waddington's epigenetic landscape. The surface contains all potential values of a single quantitative trait producible by a genotype under all possible environmental conditions considered at time 0 (before any development has occurred). The path followed by the ball, as it rolls toward the viewer, corresponds to the developmental history of changes in value of the trait, as it develops along channels with alternative potential pathways, some of which (e.g., the one at the lower left) can only be reached over a threshold. After Waddington (1974), as reprinted in Waddington (1975).

5. Problematic Metaphors

The conceptual vacuum caused by lack of an adequate theory of development and evolution has been filled by metaphors. In many cases, these metaphors vividly represent important qualities of development. This explains the popularity of metaphors as theory surrogates. But metaphors are inevitably deficient as general guides to thinking about development and evolution (Nijhout, 1990; Mahner and Kary, 1997; Oyama, 2000). Their partial accuracies and their ultimate weaknesses indicate the task at hand.

Waddington's *epigenetic landscape* (figure 1.3) is a famous metaphor designed to connect development with genetics and evolution. It shows the bifurcating pathways that characterize development and uses a three-dimensional diagram to represent the developmental trajectory of a phenotype (a rolling ball) over time. Waddington did not always label the axes of his diagrams, so it is easy to suppose that they show the dynamic and flexible developmental canalization of complex phenotypic traits, including switches at bifurcation points. In fact, the model is incomplete, and potentially misleading, because developmental potentialities change as development proceeds. When environ-

mental inputs are included, more channels are possible and the shape of the landscape (e.g., the relative depth of alternative channels) is constantly changing due to changes in both genomic and environmental inputs as decision points are passed. Waddington's diagram is static. It shows only potentials defined genetically at birth. All that the environment can do, in Waddington's scheme, is deflect development into a new genetically specified path (figure 1.4).

Oddly, Waddington depicted mutations as exerting their effects from outside the genetically defined landscape, like an environmental factor. This is because Waddington saw mutations and environmental changes as disturbances whose effects are resisted or canalized by the sloping contours of the epigenetic surface to produce normal, adaptive development rather than disruptive change. In genetic assimilation, an originally anomalous environmental effect eventually leads to a new pathway of genetic canalization, such that environmental influence is seen as canceled, not incorporated, as a determinant of normal development. The form of the landscape is entirely specified by genes (figure 1.5).

The epigenetic landscape is limited and even inconsistent as a model of development in its failure

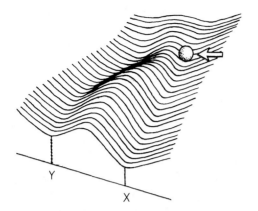

Fig. 1.4. The environment and the epigenetic landscape. In Waddington's metaphorical model of development, an environmental factor (arrow) can affect a developmental trajectory (path followed by the ball) at a decision point (between paths X and Y) but, unlike a gene, is not a determinant of the shape of the landscape. After Waddington (1959), as reprinted in Waddington (1975).

tion without interaction. The computer analogy allows input from the environment. But, as in Waddington's epigenetic landscape, the rules and outcomes of interactions are completely defined by the genes. Development begins with genetic instructions, and evolved reprogramming begins with genetic change.

Mercilessly, Stent (1985) calls the genetic program idea a result of "the noxious impact of molecular biology on embryology," even while recognizing that his own view of development is supported by data from the "vanguard techniques" of molecular research. The genetic program metaphor encourages the idea that each decision point in development or behavior, and each environmental input, acts under a genetic directive or a set of genetic rules, like "in condition x do A, in condition Y do B." But there are no such rules in the genome,

to deal with the dynamic effects of environmental inputs on the epigenetic pathways. Although the landscape has a flexible-looking surface, in fact it is a rigid and static representation of development. Change in the contours requires evolutionary change. To take environmental influence into account, the diagram would need to show how environmental vectors influence canalization, as do genetic ones, by impinging upon a flexible and changing phenotypic structure. To accurately show the effect of mutations, they would have to be represented as new lines among the guyropes of genetic influence.

The *genetic program* metaphor is attributed to Brenner, who used the word "program" in his first paper on the genetics of nematodes in the 1970's (Lewin, 1984). To his credit, he emphasized the "implications of this language even when used metaphorically" (Lewin, 1984, p. 1357). The genetic program has become the predominant metaphor for the organization of development and evolution, which describes development as programmed by the genes and reprogrammable by genetic change during evolution. An entire set of instructions for development is contained in the genes. This metaphor, a comparison with computer programming, is less closed to environmental influence than is the blueprint metaphor (e.g., Eigen, 1992), an image of rigid genetic control, instruc-

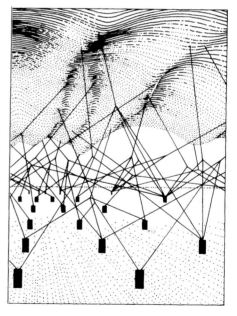

Fig. 1.5. Genetic determination of Waddington's epigenetic landscape: "The pegs in the ground represent genes; the strings leading from them, the chemical tendencies which the genes produce. The modelling of the epigenetic landscape, which slopes down from above one's head towards the distance, is controlled by the pull of these numerous guyropes which are ultimately anchored to the genes" (Gilbert, 1991, p. 149) after Waddington (1956).

only a set of templates for molecules that will become part of the phenotype. The behavior of those molecules depends as much on the nature of the environmentally supplied materials that compose them as on the genetic template that contributes to their organization. Adaptive flexibility, or behavior-as-if-by-rules, is a property of phenotypes, not of genotypes, and phenotypic organization incorporates environmental directives as well as genetic ones (see chapter 5). The genetic program metaphor does not suggest the possibility that environmental elements are partly or entirely responsible for the development (or nondevelopment) of a phenotypic trait. It is a metaphor that leads to thinking in terms of genetic directives.

Beyond crude imaginings regarding decision rules, the programming analogy has little heuristic value, as it does not give specific guidance as to how to visualize control of development. What does it mean, for example, when an author states that development is "programmed" or "reprogrammed" at some point during ontogeny? This usually means simply that some major phenotypic reorganization is observed or some new switch point introduced. The role of a genetic program, as opposed to the role of a new environmental input, in that reorganization is entirely imagined or assumed, justified at best by data showing some genetic difference between the before and after states of the phenotype. The developmental role of the genes specified may not be known. The possibility that environmental elements are entirely responsible for the observed change or its initiation given a preexisting responsive structure is not considered.

It does not take much reflection to realize that the genome does not contain a complete set of instructions for development, or even "an outline of the information needed to create a human being " (Baltimore, 2001) or any other organism. Humans and many other vertebrates only acquire the complex phenotypic ability to communicate using particular signals, for example, if they have information from other individuals (Marler, 1998). Critical cues for hatching, germination, metamorphosis, or sexual maturation may come from outside the genes in the form of a change in environmental day length, temperature, or humidity (e.g., Tauber et al., 1986). Even sex determination can depend on environmental information (Bull, 1983). The genetic program metaphor, and other similar ones such as the genetic blueprint and recipe metaphors, misses the point that in development the input:output rules identified with a program, a blueprint, or a recipe are properties of the organismic hardware,

of the phenotype, and as such they are products of both genetic inputs and environmental building blocks and cues that may be as specific and dependable as gene products themselves.

The complete-instruction metaphors are particularly problematic because they reinforce the misconception held by many that the genome is a complete set of instructions for making an organism. This image has little resemblance to the decentralized way that responsive structure is made during ontogeny, with the form and function of tissues and organs depending importantly on their circumstances and uses as they are being formed (see chapter 3).

Genotype–environment interaction has also become a metaphor for development. In quantitative genetics "genotype-by-environment interaction" describes the result of a quantitative analysis of the causes of phenotypic variation in a population, some of it due to genetic variance and some to environmental or nongenetic factors. It has been paraphrased as "genetic variation for plasticity" (Pigliucci et al., 1995). The word "interaction" refers to the contribution to the phenotypic variance of the differing effects of different genotypes in different environments. It is easy to presume from the words alone that because quantitative genetics takes into account both genotype and environment and the interactions between them, it is an adequate, even a quantitative theory of development. But genotype–environment interaction is misleading as a description of development because genes do not interact directly with the external environment during development. All interaction is indirect, via effects of both factors on a preexisting phenotype (see chapter 5).

There are many other metaphors. For Dawkins (1982), the genome is a recipe; the phenotype, a cake. A less lifelike image for the dynamic activity of genes as they participate in development can hardly be imagined. Even so, genome-as-recipe is better than the replicator-vehicle dichotomy, which sees the phenotype as bearer of genes whose condition-dependent somatic activities are obscured. Mayr (1968, p. 379) compares development to an orchestra. The genome is the score, its instructions reinforced by the conductor turning musicians (gene loci) off and on. Mayr combined the musical and the cybernetic to write, "The activity of an orchestra . . . is just as much controlled by the score as the development of an organism is controlled by its genetic program" and concluded that given this mixture of metaphors "we already have, in principle, 'a general theory of development'" (p. 380). In view of the weaknesses of these metaphorical stopgaps, the already mentioned conclusion of Fontana et al. (1994) that biology still lacks a theory of organization, seems more accurate. The need for a conceptual framework for the study of organization lies at the heart of unsolved problems in both ontogeny and phylogeny (Fontana et al., 1994).

Part of Darwin's success as a theoretician undoubtedly lay in his ability to treat metaphors with caution. Darwin (1859) remarked that "analogy may be a deceitful guide" and predicted progress when terms "cease to be metaphorical and will have a plain significance" (p. 484). Darwin could be accused of relying upon an anthropomorphic metaphor with the term natural "selection" if it were not for the fact that artificial selection of domesticated plants and animals is so closely comparable to the process hypothesized to occur in nature that even today human artificial selection experiments serve to illuminate the process of natural selection. Perhaps because of his own meticulous empiricism, Darwin's largely forgotten theory of biological organization is more accurate than modern metaphorical ones in relating development to inheritance and evolution (see chapter 8).

If metaphors mislead, then how are we to visualize the phenotype in relation to development, environment, selection, and genes? Beginning with the concepts discussed in the first section of chapter 2, we can begin to design a theory of the phenotype based on established principles—to construct a vision of phenotypic development based on properties of the phenotype itself. Metaphors are not only potentially misleading, but they are also dull substitutes for reality, because they reduce exquisitely dynamic phenomena to the lifeless images of a computer, a blueprint, or a cake. Why not describe real mechanisms and then use development as a metaphor for itself?

6. The Genotype–Phenotype Problem

Perhaps the most remarkable inconsistency, and the most urgent to resolve, is the practice of phrasing evolutionary explanations almost entirely in terms of genes, while the phenomena we endeavor to explain are phenotypes, always products of both environment and genes. This has led to a number of contradictions that collectively might be called the genotype–phenotype problem.

One symptom of this problem is a conflict of opinions about the definition of selection. Many authors (e.g., Lewontin, 1970, 1974; Futuyma, 1986a) define selection in terms of phenotypes, with differential reproduction or fitness effects associated with phenotypic differences, whereas others (e.g., Dawkins, 1982; G. C. Williams, 1966, 1992; Endler, 1986) argue that selection acts on genes or that genotypic differences are required for selection to occur. This has led to confusion regarding cause and effect in evolution. Here I adopt a phenotypic definition of selection. The idea that the gene, or the genotype, rather than the phenotype, is the object of selection is more than a trivial or arbitrary matter of terms. It can lead to contradictions between theory and reality so absurd that they might well lead an outsider to question the rationality of biologists. For example, the misleading belief that the genotype is the focus of selection invites the idea that an individual genotype has a measurable fitness that is somehow intrinsic to itself (the naked genes) rather than being *entirely* a secondary consequence of the fitness or of reproductive success of the collective phenotypes it produces across life stages and in different environments. O'Donald (1982), for example, writes

I assume that in a particular generation a particular individual is stuck with the fitness he has got: he cannot change it. This is entailed by the assumption that the fitness parameters are constant for particular genotypes. Thus I shall not consider models in which an individual might change his fitness in the course of his life by seeking some other environment or adapting some other strategy of behaviour [plasticity]. Such models would always be special cases [!]. They would not be models of evolution unless genetically some individuals could alter their fitness and others could not. (p. 67; bracketed additions mine)

Here, the wish to strip the genotype bare of its flexible, environmentally responsive phenotype in a discussion of selection is particularly striking because O'Donald accurately states the conditions for the evolution of adaptive flexibility while nonetheless dismissing it as a special case unworthy of general consideration. Adaptive flexibility evolves, of course, precisely because "genetically some individuals" *do* in fact "alter their fitnesses" more effectively than do others, just as O'Donald would require. So the fitness effects of plasticity can be included in the calculation of individual fitness under a phenotypic concept of selection. Differential representation of genes in the next generation depends entirely on their correlation with phenotypic success (individual fitness, including contributions of developmental plasticity).

Setting aside the plasticity and development of the phenotype so as to visualize fitness as a property of a gene or a genotype, as in the passage just cited, would seem to reduce evolution to its genetic essence. But it has the effect of excising the phenotype along with development and environmental effects from the causal chain of evolutionary change. When condition-sensitive development is seen as the cause of selectable variation, several otherwise puzzling phenomena are explained. It is because of selection on *phenotypes* that environmental elements such as carbon and iron become incorporated into normal ontogeny and phenotypic structure. And it is for this reason that major adaptive phenotypic change can be environmentally induced (see chapters 6 and 26).

A consequence of the genotype–phenotype problem is further incongruity between what is observed and what is explainable. Widespread use of adaptive phenotypic flexibility to test genetic models of optimization and adaptation exists alongside theoretical skepticism that adaptive flexibility could be widespread (e.g., see Wade, 1984). Thus, a powerful demonstration of the adaptively flexible nature of phenotypes, used to test theory, persists alongside theoretical denial that such flexibility could exist. As a result, there is almost complete ignorance regarding how such flexible organization actually originates and evolves. Adaptive flexibility is taken for granted, not only in the study of social behavior as described at the beginning of this chapter, but also in research on foraging decisions, life-history strategies, and sex allocation, where conditional decisions are routinely incorporated into models and used to demonstrate the applicability of their predictions (see, e.g., on kin selection: Hamilton, 1964a,b; on energy allocation: Stephens and Krebs, 1986; Perrin and Sibly, 1993; on sex allocation: Charnov, 1982, 1993; on life-history tactics: Stearns, 1992, Charnov, 1993; Roff 1992). All of the models of adaptive decisions in these areas contribute to understanding of adaptive flexibility by predicting patterns of behavior and ontogeny in variable environments. But they require an appropriate regulatory system, one capable of environmental assessment and adaptive response.

How, precisely, does such regulation evolve? The requisite condition sensitivity is simply assumed, perhaps visualized as a property of a genetic program or just unconsciously lumped among the effects of a single mutant allele (Crozier, 1992), as if the same allele were responsible for both a novel behavior and its finely tuned conditional expression. Dawkins (1976), for example, thought in terms of imaginary genes that have one effect in

adults and another in the young, depending upon circumstances. With assumptions of this kind, it is possible to pass lightly over the question of how adaptive, condition-sensitive regulation is organized and evolves. Explanation of phenotype evolution is fundamentally a problem of explaining the evolution of phenotype development. Yet evolutionary discussions routinely skip from phenotype to gene, unwittingly leaving development aside. Developmental and phenotypic plasticity is noise in a genetics lab, a tool in a field study, and a cryptic assumption in an allocation model. Only recently has it become the focus of research.

Due to strong convictions regarding the essentially genetic nature of evolution, alongside the impossibility of ignoring adaptive flexibility when working with organisms, we have learned to live with a set of concepts that function happily together only when condition sensitivity is ignored or taken for granted, whichever is more convenient. The definition of evolution is, and must be, explicit regarding the requirement for gene-frequency change. Nonetheless, a *phenotypic* concept of evolution is in common use, even though it is at odds with the genetic definition endorsed by all. Evolution, by the formal definition, must involve genetic change. But we are not consistent in this demand. We do not, for example, usually insist on gene-frequency information to validate the claim that evolutionary change has occurred when this is indicated by comparative study of phenotypes alone—in systematics, phylogenetics, ethology, ecology, and paleontology. Nor are we even-handed in the search for causes of change when we choose to investigate them. If breeding experiments reveal a genetic component to change, many authors call the phenotypic differences "genetic" without any attention to possible additional environmental causes. Yet if we accept the dual nature of the phenotype—the undeniable fact that the phenotype is a product of both genotype and environment, and the equally undeniable fact that *phenotypes* evolve, there is no escape from the conclusion that evolution of a commonly recognized sort can occur without genetic change. The problem here is obviously one of a definition that is out of step with what most biologists observe and strive to explain—the evolution of phenotypes, products of both genes and environments.

The genetic definition of evolution has a profound influence on research programs in evolutionary biology. While there is universal recognition that environmentally influenced development forms the phenotypes that evolve, "it has been the kind of formal, polite recognition accorded to a stranger at an otherwise intimate party" (Futuyma,

1986a, p. 440). We acknowledge development with a reflexive nod, then proceed as if it did not exist.

Another consequence of the genetic definition is the idea that environmentally mediated change is trivial for evolution. This is a dismissal of environmental factors that is more than just a matter of emphasis. It reflects a conviction, even among some prominent evolutionists, that environmental effects are noise in the essentially genetic evolutionary process. Gould (1984) has gone so far as to propose environmental susceptibility as an operational criterion of nonadaptation, oddly overlooking the facts that adaptations are often exquisitely responsive to environmental conditions, and that environmental susceptibility itself can evolve (for a more practicable approach to the problem of nonadaptation, see Emerson, 1984).

Lewontin addresses the genotype–phenotype problem in his book *The Genetic Basis of Evolutionary Change* (1974):

[I]t is the evolution of the phenotype that interests us. Population geneticists, in their enthusiasm to deal with the changes in genotype frequencies that underlie evolutionary changes, have often forgotten that what are ultimately to be explained are the myriad and subtle changes in size, shape, behavior, and the interactions with other species that constitute the real stuff of evolution. Thus Dobzhansky's dictum that "evolution is a change in the genetic composition of populations" . . . must not be understood as defining the evolutionary process, but only as describing its dynamical basis. A description and explanation of genetic change in populations is a description and explanation of evolutionary change only insofar as we can link those genetic changes to the manifest diversity of living organisms in space and time. To concentrate only on genetic change, without attempting to relate it to the kinds of physiological, morphogenetic, and behavioral evolution that are manifest in the fossil record and in the diversity of extant organisms and communities, is to forget entirely what it is we are trying to explain in the first place.

When contemplating the genetic basis of evolutionary change, Lewontin encounters an epistemological paradox: "nongenetic sources of phenotypic variation interfere with genetic measures," and

The substance of evolutionary change at the phenotypic level is precisely in those characters for which individual gene substitutions make

only slight differences as compared with variation produced by the genetic background and environment. What we can measure is by definition uninteresting and what we are interested in is by definition unmeasurable. (p. 23)

In short, we are a schizophrenic group, believing on the one hand that phenotypes are the real stuff of evolution—the observable qualities that we seek to explain—and yet ultimately treating selection and evolution as something else—a change in gene frequencies—whose study appears always to be obscured by the flexible nature of the phenotype. Lewontin concludes that this "paradox is not yet resolved, but recent advances in molecular biology and in population genetic theory . . . offer some hope of a solution" (p. 23). That is, if the variable and genetically complex phenotype gets in the way, the solution is more and better genetics!

I propose a different solution to Lewontin's paradox: if the phenotype is the real stuff of evolution, why not take a hard look at the nature of the phenotype itself and see what it may reveal about evolution?

Toward a Solution

Now is a good time for an evolutionary synthesis that connects knowledge of the genome to the facts of phenotypic plasticity and development. Recent progress in behavioral ecology and ethology has advanced understanding of phenotypic flexibility, demonstrating remarkably fine-tuned adaptively appropriate phenotype expression of complex traits and life-history patterns. New methods in paleontology go beyond conventional morphology, making it possible, for example, to deduce extinct patterns of diet and behavior from the chemical composition of bones and teeth. At the same time, the black box of genetic influence has been opened by laboratory studies that connect behavioral and morphological phenotypes, through endocrinology and neurobiology, to the molecular mechanisms of development and gene expression. For the first time in the history of evolutionary biology, it is possible to trace the manifest phenotype to its origins in the expressed genes—the genomic elements that actually change under selection. Meanwhile, comparative biology has been transformed by cladistic methods that have vastly improved our ability to trace evolutionary transitions between phenotypic states in order to track the sequence of phenotypic changes during evolution.

The piece that is missing from a synthesis of development and neo-Darwinism is an adequate theory of phenotype organization that incorporates the influence of the environment. Fontana et al. (1994) characterize neo-Darwinism as concerned with the dynamics of alleles within populations of individuals, as determined by mutation, selection, and drift. Environmentally influenced development is missing from this scheme. Fontana et al. argue that understanding of how mutation can give rise to phenotypic novelty requires an understanding of "how the organizations upon which the process of natural selection is based arise," given that selection, which is not a generative process, cannot occur until there are entities to select. In other words, evolutionary explanation requires a theory of phenotype organization that biology still lacks. "The need for a conceptual framework for the study of organization lies at the heart of unsolved problems in both ontogeny and phylogeny" (p. 212). I believe that there is a connection between the neglect of environmental influence in development and the lack of an adequate theory of biological organization.

Theories of organization do not often begin with the phenotype, much less with the role of the environment in development. Some discussions of biological organization and evolution that claim very broad generality deal almost exclusively with the ability of living things to self-replicate and to vary genetically. They focus heavily on the properties of DNA and the evolution of the genome (e.g., Eigen, 1992; Kauffman, 1993). Here I put the evolution of the genome in the background, to focus first on the phenotype. Can such a strong emphasis on the phenotype be justified?

A focus on the phenotype is justified if you are primarily interested in adaptive evolution by means of natural selection. Natural selection affects genes only if they are expressed—only if they differentially influence phenotypes. In other respects, the genome has a life of its own and marches to its own rhythm during evolution. The factors that initiate change in the genome itself may or may not produce phenotypic change. Many drastic experimental manipulations of the genome have no phenotypic effects at all. In fruitflies (*Drosophila melanogaster*), at least 50% of all experimentally induced gross chromosomal rearrangements have no phenotypic effects of any detected kind, and 60% of a well-studied group of genes, the 18S and 28S RNA genes, are believed to be nonfunctional (John and Miklos, 1988, pp. 297, 312). Many molecular biologists believe that a large proportion of the eukaryote genome is so-called "junk DNA" (references in John and Miklos, 1988). Some of this DNA may have cryptic functions. But unexpressed and selectively equivalent neutral elements of the genome and silent substitutions (those not affecting amino acids) are sheltered from selection, and they evolve at their own rates, independent of the phenotype (see Gillespie, 1991).

Among the processes that can cause genomic change without phenotypic effects are, in addition to drift, transposition (accompanied by insertion and sometimes duplication), production of untranscribed pseudogenes (e.g., by reintegration of reverse transcripts of functional genes), production of minisatellite regions due to replication slippage, duplication of small DNA segments, unequal sister chromatid exchange during germ-line mitosis or meiosis, "gene conversion" or change of one allele into the form of another via an inter- or intrachromosomal sequence transfer (sometimes called "molecular drive" because it could push the favored allelic form toward fixation without selection on the phenotype), and various patterns of sequence amplification (from John and Miklos, 1988). Given so many genome-altering phenomena that may go on with no exposure to selection, it is easy to understand why molecular biologists identify selection with *lack* of change (due to its allele-conserving effects), whereas evolutionary biologists identify selection with *change* (because they consider only expressed DNA and not the less stable portions of the genome where a multitude of processes cause and preserve uncorrected change).

The realization that the genome has an evolutionary life of its own is one of the cardinal discoveries of molecular biology (Kimura, 1985; John and Miklos, 1988). The idea of neutral genes was needlessly resisted by evolutionary biologists accustomed to think of genome evolution entirely as a reflection of selection on phenotypes. Discussions of molecular evolution are still sometimes confused by lumping under the heading of "molecular evolution" both the proteins, which are aspects of the phenotype, and genomic DNA, which may or may not be expressed and subject to effects of selection on the phenotype. Neutral or nearly neutral alleles may serve as a reservoir of genetic variation with the potential to influence development in a fluctuating environment (Kimura, 1985; Ohta, 1992a,b; Gillespie, 1991). Then and only then are these alleles important for development and adaptive evolution.

Here I consider the genome primarily in its role as a template for molecules that affect the phenotype, and whose replicates can be passed to subsequent generations. Because the genome affects the phenotype at nearly every turn, it obviously must play an important role in any phenotypic theory of organization. The problem in the past has been not too much attention to the genome, but too little attention to other factors of development, namely, the roles of the already organized phenotype and of inputs from the environment. The vocabulary of evolutionary biology has been enriched by such terms as reaction norms, plasticity, and developmental constraints. But a more fundamental reform is in store. It will have to begin with an adequate model of the phenotype in relation to development and genes, one that shows the causal relations among genome, environment, selection, and phenotypic evolution.

It is both an advantage and a challenge to attempt such a synthesis during what has become the Age of the Gene. While molecular studies of gene expression make it possible to actually connect phenotypes to genes by tracing the details of their developmental pathways, gene identification and mapping may reinforce the false impression that there are genes "for" complex traits. The one-gene, one-phenotype error, recognized decades ago within evolutionary biology (e.g., see Dobzhansky, 1970), has resurfaced twice, first as the selfish gene in behavioral ecology and sociobiology, where it became quite common to speak of genes "for" such traits as altruism or cheating (e.g., Dawkins, 1976), and second, as the mapped mutation in molecular biology, which gives the impression of genes "for" such traits as deafness or alcoholism. Whereas selfish genes "for" complex traits were only imagined, now they are localized on chromosomes and given biochemical names.

Genome mapping seems to achieve the ultimate goal in the specification of a genetic cause. Yet the phenotypic traits whose map birthdays are triumphantly announced in the popular press and in leading journals of science are virtually always condition sensitive and polygenic, complex phenotypes such as obesity, intelligence, alcoholism, baldness, or foraging behavior in bees. The one-gene, one-phenotype illusion comes from a methodology that depends on some pathological or artificially selected mutation, or a naturally occurring major gene effect, that reveals only one component of a complex developmental mechanism, usually one that is influenced by multiple genes and environmental factors as well. The identification and localization of one of the genetic factors, while a laudable accomplishment, is only the beginning of an explanation of why some individuals possess a trait

and others do not. This is a question of development, not just isolated individual genes.

This book looks broadly at the role of phenotypic flexibility, development, and plasticity in evolution. Especially, it emphasizes the role of the environment as an agent of development, not just selection, in the evolution of all forms of life. Part I outlines a conceptual framework, a general theory of phenotype organization to relate plasticity to genes, development, selection, and evolution. Part II traces the origins of novelties in the developmental reorganization of preexisting phenotypes. Part III deals with alternative phenotypes as a phase of evolution that follows the origin of novelty, showing how flexibility facilitates the persistence and gradual modification of novel traits. And part IV shows how a developmental perspective affects how we think about virtually every major theme in evolutionary biology.

I hope three major points will emerge from this discussion. First, environmental induction is a major initiator of adaptive evolutionary change. The origin and evolution of adaptive novelty do not await mutation; on the contrary, genes are followers, not leaders, in evolution. Second, evolutionary novelties result from the reorganization of preexisting phenotypes and the incorporation of environmental elements. Novel traits are not de novo constructions that depend on a series of genetic mutations. Third, phenotypic plasticity can facilitate evolution by the immediate accommodation and exaggeration of change. It should no longer be regarded as a source of noise in a system governed by genes, or as a "merely environmental" phenomenon without evolutionary importance.

A deep look at the evolutionary role of development reaches beyond the issue of nature and nurture to illuminate such themes as the patterns of adaptive radiation, the organization of societies, and the origin of intelligence. For it is undoubtedly the assessment and management of environmental and social contingencies that has led to the evolution of situation-appropriate regulation, with the eventual participation of the sophisticated device we call "mind." Indeed, seeing judgment and intelligence among other mechanisms of adaptive flexibility helps explain why learned aspects of human behavior so closely mimic evolved traits (see chapter 18). But this is not primarily a book about humans. It is a book about beetles, butterflies, buttercups, birds, and myriad other organisms whose patterned responses show the fundamental relationship between developmental plasticity and evolutionary change.

2

Material for a Synthesis

[T]he modern synthesis . . . was incomplete without a chapter dealing with the effects of selection on the gene-controlled variability of developmental processes.

Viktor Hamburger

Previous insights regarding development and evolution have been piecemeal solutions to specialized problems, but when taken together they provide a solid beginning for a comprehensive theory. A unified theory is presented here in brief outline, along with definitions of terms that are basic for a discussion of development and evolution.

Introduction

The inconsistencies discussed in chapter 1 point toward two fundamental problems in need of solution: how to relate the environmental influence inherent in phenotype development to the genetic emphasis of evolutionary theory—Lewontin's dilemma—and how to view the diverse phenomena of plasticity and development so as to illuminate evolutionary thinking in new ways—Wallace's challenge.

This chapter briefly describes some important, previously recognized connections among phenotypic flexibility, development, and evolution. It then defines key concepts for the chapters that follow.

Previous Insights on Development and Evolution

Important contributions toward a synthesis of development and evolution have accumulated over a period of many years. Some insights appear repeatedly in that cycle of inspiration and amnesia that characterizes important discoveries ahead of their times (for a concise review, see Hall, 1992, pp. 171–174). Some of these insights deal with the phenomenology of development and evolution—evidence that certain behavioral and developmental phenomena have influenced evolution in particular groups or in particular

ways. These ideas, long familiar to evolutionary biologists, are the starting points for any attempt at a modern synthesis. Each of them will reappear again in later chapters.

Independent Evolution of Juveniles and Adults

It does not require great sophistication in biology to realize that juveniles and adults have distinctive, divergent adaptations. Familiar extreme examples are the caterpillar and the butterfly, the tadpole and the frog. In such metamorphosing species, the juvenile has a dramatically different morphology, behavior, and ecology from that of the adult. Some hypermetamorphic insects show a striking series of differently specialized larval stages (figure 2.1), and it is probably true of most organisms that juveniles and adults have different, evolved characteristics appropriate to their different niches, if for no other reason than the different requirements for dispersal, respiration, feeding, and defense that confront individuals of differing size (Schmidt-Nielsen, 1984; see also McKinney and McNamara, 1991). As a corollary of this, different life stages evolve semi-independently. Thus, immature stages may evolve and diversify, undergoing their own adaptive radiations (figure 2.2).

Many authors have been impressed with the conservatism of certain aspects of early development. The extreme statement of this is von Baer's law, by which development begins with a general-

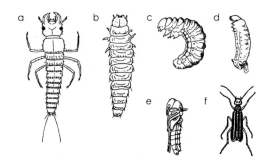

Fig. 2.1. Divergent life-stage specializations in a hypermetamorphic beetle (*Epicauta vittata*; Coleoptera: Meloidae). (a) First-instar triungulin larva, an active, mobile egg hunter and predator of grasshopper egg capsules. (b) Second-instar caraboid larva, a sedentary feeder; (c) Late-second-instar scarabaeoid larva; (d) immobile hibernating "coarctate" larva, sometimes called a pseudopupa; (e) pupa; (f) adult. Not to scale. After Riley (1878).

ized or undifferentiated form and proceeds toward a specialized one, with the result that the earliest embryonic stages of related organisms are identical and young embryos are undifferentiated generalized forms, not forms resembling adult ancestors (recapitulation does not occur). This so-called law is broken in many taxa, where there is striking divergent, semi-independent evolution of larvae (the "clandestine evolution" of de Beer, 1958). In many groups, briefly reviewed by G. C. Williams (1992), larvae not only have their own extreme specializations, (e.g., to parasitism, dispersal, or defense) but also are morphologically more distinctive and species specific in form than are the adults of the same species. Examples include mollusks (Coe, 1949), beetles (Peck, 1986), sea urchins (Sinervo and McEdward, 1988) and fishes (G. C. Williams, 1992). Some juvenile insects have instar-specific behavior as well (Rowe, 1992; see chapter 4).

Heterochrony

Heterochrony is an evolved shift in the timing of expression of a trait or set of traits. Gould (1977) gives a history and a review of the concept, with many examples and a clarification of terms. There is considerable evidence for heterochrony in both plants and animals (e.g., see Lord and Hill, 1987; Gould, 1977; Raff and Kaufman, 1983; McKinney, 1988; McKinney and McNamara, 1991; chapter 13). By some definitions, virtually every evolutionary change in the phenotype can be termed hete-

rochrony, but the original use of the term to describe major rearrangements in the order of life-stage-specific events at the whole-organism level is the usage I refer to here as having contributed to the understanding of how development can be reorganized in adaptive evolution (see Wake, 1996b; chapter 13). Although the term "heterochrony" is most commonly applied to morphology, the same phenomenon undoubtedly occurs in behavior, as pointed out in a discussion of avian social behavior by Lawton and Lawton (1986). Classic examples of heterochrony are the retention of ancestral primate fetal characteristics in adult humans (Bolk, cited by Raff and Kaufman, 1983, p. 174) and the attainment of sexual maturity and adult size in axolotls (*Ambystoma*) retaining larval gills, tail, and skin. The major proposals seeking to account for the mechanistic roots of evolution in ontogeny have centered on heterochrony (Raff and Kaufman, 1983, p. 173).

Developmental Origins of Adaptive Novelties

Beginning with the earliest critics of Darwinian gradualism (e.g., Bateson, 1894), developmental

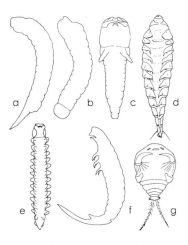

Fig. 2.2. Larval diversity in the Hymenoptera (Insecta). First-instar larvae are shown. (a) *Pygostolus falcatus* (Braconidae); (b) *Brachistes atricornis* (Braconidae); (c) *Opius fletcheri* (Braconidae); (d) *Perilampus hyalius* (Perilampidae); (e) *Macrocentrus figuensis* (Braconidae); (f) *Kleidotoma marshalli* (Eucoilidae); (g) *Synopeas rhanis* (Platygasteridae). After Gauld and Bolton (1988), by permission.

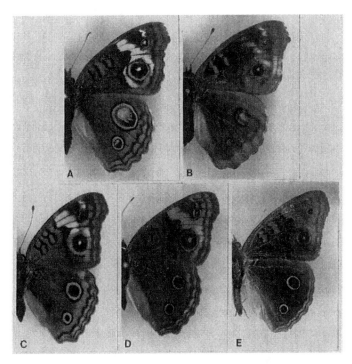

Fig. 2.3. Experimentally induced and naturally occurring phenocopies. (A) Normal phenotype of *Precis coenia*; (B) cold-shock-induced phenotype of *P. coenia*; related species (C) *P. evarete* and (D) *P. genoveva*; (E) the naturally occurring *nigrosuffusa* form of *P. coenia*. Form B is a phenocopy of form E, and also resembles the related species *genoveva* (D). Form E occurs as a geographically distinctive population in some areas and as an anomaly within populations of form A. Thus, the phenocopy (B) resembles forms occasionally found as anomalies in nature (due to either genetic or environmental variation; E); as geographic variants within species (E) and as species characteristics in closely related species (D). This shows that natural variation in frequency of occurrence of a phenotype could be due to variation in frequency of environmental induction, as well as to variation in underlying gene frequencies as usually assumed. From Nijhout (1991b), © H. F. Nijhout, by permission. Courtesy of H. F. Nijhout.

anomalies have been cited as a phenomenon of potential evolutionary significance. These assertions have long been controversial (see Gould's [1982b] introduction to a reprinting of Goldschmidt, 1940). There is now substantial, even massive, evidence that morphological novelties can originate as regulatory novelties (see references in Alberch, 1982, p. 25; West-Eberhard, 1989). A classic example is the origin of neotenous adults in salamanders, where a hormonally mediated switch blocks metamorphosis but not sexual maturation, producing an adult with larval morphology but fully developed gonads (review in Gould, 1977). Some phylogenetic transitions parallel developmental changes found within species, an observation that is consistent with the idea that evolutionary change is based on intraspecific developmental change. Examples include not only heterochronic shifts (neotenous

adults occur in some salamanders as a facultative alternative phenotype) but also omissions of life-stage characteristics or developmental sequences of varying length (e.g., Raff and Kaufman, 1983; Thomson, 1988); alterations in metamorphosis (Matsuda, 1979, 1987); and correlated shifts in multiple quantitative traits traceable to a regulatory change at some stage of development (Raff and Kaufman, 1983; Thomson, 1988).

Most of the evidence for regulatory origin of evolutionary novelties comes from comparative study, but it is reinforced by experimental manipulations of development, for example, in the production of *phenocopies*—environmentally induced phenotypes that resemble genetically determined or naturally occurring characteristics found in related species (e.g., Goldschmidt, 1935; Shapiro, 1976; Nijhout, 1991b; figure 2.3).

Developmental Gaps and Taxonomic Gaps

Students of development have long noted that mutations affecting early development have complex effects and can produce phenotypic discontinuities. These developmentally produced phenotypic gaps have often been compared to those separating taxa, especially higher taxa (above the species level), with the suggestion that phylogenetic gaps could have a sudden, developmental origin (e.g., Bateson, 1894; Goldschmidt, 1940; Thomson, 1988). This has been a prominent argument against Darwinian gradualism, where gaps between higher taxa are explained by progressive divergence (see Darwin, 1859 [1872], p. 87).

Behavioral Adaptability and Morphological Evolution

It is a common impression of naturalists (e.g., see Rau, 1933, pp. 269–307) and students of ethology and systematics (e.g., Baldwin, 1902; Evans, 1966; reviewed in Wcislo, 1989) that behavioral change often precedes and directs morphological change—that "behavior takes the lead in evolution." This is suggested when a behavior (movement) pattern widespread in a taxon is found in some species to be accompanied by specialized morphology that amplifies the effect of the behavior. For example, in *Drosophila*, 14 species of the *adiastola* species group have a courtship display in which the male stands before the female, curves his abdomen upward and forward like that of a stinging scorpion, and vibrates it near the female's head. In 13 of the 14 species, there is no obvious morphological modification of the male abdomen, but in one species (*D. clavisetae*, shown on the basis of chromosomal data to be "phylogenetically very advanced"; Carson, 1978, p. 105), the end of the male's abdomen bears a brush of long clavate hairs, which sweep over the female's head during display (Carson, 1978; West-Eberhard, 1983). Wcislo (1989) points out that behavior, including learned behavior, contributes to the "environment" in which morphological traits are selected. Of course, morphology can also precede and influence the direction of behavioral evolution (see chapter 28), so the two aspects of the phenotype can create a mutually reinforcing evolutionary trend in which it is false to dichotomize the two kinds of traits (Evans, 1966; Wcilso, 1989).

Mapping both kinds of traits onto a phylogeny can determine the polarity and sequence of changes (e.g., de Quieroz and Wimberger, 1993), although the behavioral lead may be underestimated in such studies by use of fixed rather than flexible behaviors as characters. The evolution of social specializations in wasps is known to have involved first behavioral then later, in some genera, morphological differentiation of queen and worker castes (phylogeny: Carpenter, 1991; caste information: Jeanne, 1980). Reasons for thinking that behavior may take the lead more often than morphology are discussed by Wcislo (1989) and West-Eberhard (1989). Mayr (1974a, p. 657) has expressed the belief that "behavioral shifts in the utilization of the animate and inanimate environment are by far the most important factors in macroevolution. They are involved in all major adaptive radiations and in the development of all major evolutionary novelties" (see also Roe and Simpson, 1958).

Genetic Assimilation and the Baldwin Effect

Genetic influence in the determination of traits can predominate over environmental influence when this is favored by selection in a population with genetic variation for degree of environmental sensitivity. When there is genetic variation for the expression of a trait, with constituitive (obligatory, in all individuals) expression at low frequency in a population where the trait is usually environmentally induced, obligatory expression may spread to fixation under persistent selection for the trait, as in Waddington's classic experiments on such phenotypes in *Drosophila*. Waddington termed the fixation of obligatory expression *genetic assimilation* (Waddington, 1953b). The Baldwin effect (Baldwin, 1896, 1902) is a broader concept (see chapter 6). It includes the possibility of selection for increased condition sensitivity rather than just increased genotypic influence on trait expression. The *Baldwin effect* is a process by which organic selection leads to evolutionary (genetic) change. Baldwin's *Organic selection*, the underlying mechanism of the Baldwin effect, is differential survival (selection) in which phenotypic accommodations to extreme conditions during individual ontogeny allow enhanced survival of appropriately responding individuals (Baldwin, 1902, p. 119). Generations of organic selection can lead to genetic (congenital or phylogenetic) change that makes the accommodation the norm in the population. Note that this does not imply that the advantageous response becomes genetically determined or genetically assimilated, only that the ability to produce the response becomes more common or fixed due to genetic change.

The most authoritative early discussion of the Baldwin effect is Baldwin (1902), which treats previous versions (Baldwin, 1896; Morgan, 1896a,b; Osborne, 1897a,b) and early criticisms. Some later authors discussed the Baldwin effect and similar ideas in a favorable light (e.g., see Cushing, 1941, 1944; Bateson, 1963; Evans, 1966; Gans, 1979; Wcislo, 1989). But most discussions (e.g., Simpson, 1953b; Mayr, 1963, Robinson and Dukas, 1999) have dismissed this idea as of minor or uncertain importance for evolution. I extensively argue in chapter 6 that there is good reason to resurrect a modern expanded version.

Canalization, Phylogenetic Inertia, and Constraints

Canalization (Waddington, 1942) is an evolved reduction in developmental flexibility that renders the development of an adaptive phenotype resistant to environmental and genetic (e.g., mutational) perturbations that would produce deviations from optimal form. Canalization is the developmental consequence of genetic assimilation. Thus, canalization can increase under selection, a classic illustration being the production of inherited (genetically determined) callosities in place of callosities requiring environmental induction. Canalization also refers to the increasing determination of cells during development, as they pass from totipotency in an early embryo to increasingly narrowed ("determined") form and function. Canalization, then, refers to the directedness of development. It does not necessarily imply rigidity: plasticity was discussed by Waddington as a device that can maintain directedness in the face of perturbations (called "developmental homeostasis" by Dobzhansky, 1970, p. 38). Canalization implies resistence to both environmental and genetic perturbations of development, so it implies resistence to evolutionary change, which depends upon genetic variation with a developmental effect on the phenotype. Canalization also refers to the bifurcate-pathway property of development:

[A]nimals are built up of sharply defined different tissues and not of masses of material which shade off gradually into one another . . . from the experimental point of view, it is usual to find that, while it may be possible to steer a mass of developing tissue into one of a number of pos-

sible paths, it is difficult to persuade it to differentiate into something intermediate between two of the normal possibilities. (Waddington, 1942 in Waddington, 1975, p. 18)

Mutations having a path-changing effect were called "switch genes" by Mather and De Winton (1941; cited by Waddington, 1942 in Waddington, 1975, p. 19).

Waddington visualized pathway canalization in his classic diagram of the "epigenetic landscape" (figure 1.3). This same image, of alternative discrete pathways as a general description of development, became generalized in studies of molecular regulatory genetics: "At various stages, depending in part on environmental signals, cells choose to use one or another set of genes, and thereby to proceed along one or another developmental pathway" (Ptashne, 1986, p. 1); "choice of alternate [*sic*: alternative] phenotypes is an almost universal part of ontogeny . . . switches determine which of the possible alternate [alternative] programs for morphological development will be selected from the repertoire of a single genome" (Raff and Kaufman, 1983, p. 283). This arrangement has evolutionary consequences: "Switch genes, through their involvement in binary developmental decisions, may lead to the wholesale reprogramming of development in both soma and germ line" (John and Miklos, 1988, p. 88; see the cited references for documentation of these conclusions using evidence from molecular and cell biology).

Constraints is a term commonly used to refer to the conservative and directive influence of development on evolution. "Developmental constraints" include any developmental bias in the production of variants—not only canalized pathways but also the broader limits placed on production of variation by the possibilities for shifts, cascades of effects, and accommodations inherent in condition-sensitive behavior and development (see Alberch, 1980; Maynard-Smith et al., 1985; Stearns, 1986). The concept of "developmental constraints" refers to the relative ease of transformation in evolution to a delimited set of neighboring or alternative forms (Kauffman, 1983a, p. 208). Alberch (1980), Wake (1982), and Maynard-Smith et al., (1985) give the empirical basis for this idea. Developmental constraints and "phylogenetic constraints" amount to the same thing (G. C. Williams, 1992; McKitrick, 1993; Watson et al., 1995).

A similar though less explicitly developmental theme is evident in discussions of the evolution of behavior. "Phylogenetic inertia," or inheritance of traits entrenched in ancestral populations, is cited by E. O. Wilson (1975, p. 32) as one of the "prime movers of social evolution," again explaining change in terms of conservative forces. Although inertia could be interpreted to accelerate directional change once it has started, as used it encompasses instead the idea of "preadaptation" and has been invoked to explain the restricted occurrence of certain phenomena (such as colonial life) to certain phylogenetic lineages. The idea of phylogenetic inertia implies a developmental constraint: due to their evolutionary histories individuals express only certain behaviors. Developmental variants that might allow evolution in a different direction do not arise. Constraints are sometimes contrasted with selection as factors responsible for the form of observed traits (e.g., Gould and Lewontin, 1979; see also Bateson, 1894).

Polymorphisms, Polyphenisms, and Divergence

Alternative phenotypes—contrasting complex phenotypes expressed as alternatives or options in the same life stage and sex by the same individual, or by different individuals of the same population, are common. Such alternative phenotypes occur in behavior, physiology, and morphology (for partial reviews, see Shapiro, 1976; Thornhill and Alcock, 1983; West-Eberhard, 1986, 1989; Moran, 1992b; Ryan et al., 1992; Wimberger, 1994; Halliday and Tejedo, 1995). In plants, alternative leaf forms (heterophylly) and seed and flower-color polymorphisms are comparable phenomena (see Cook and Johnson, 1968; Sorensen, 1978; Kay, 1978; Silvertown, 1984; Venable, 1985; Silvertown and Gordon, 1989). Alternatives are often cued by the physical environment (e.g., temperature, humidity, or photoperiod), developmental events or conditions (e.g., size at a particular stage), or social interactions. Sometimes these same-stage discontinuous variants parallel differences between species or higher taxa (West-Eberhard, 1986, 1989). This suggests that intraspecific diversification in ontogeny and behavior can be the basis for phylogenetic divergence.

Reaction Norms, Genotypes, and Selection

During the history of evolutionary biology since Darwin, most of the variation that has inspired interest in a synthesis of development and evolution in the writings of such authors as Bateson and

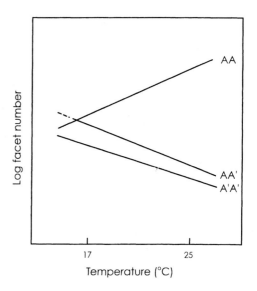

Fig. 2.4. Reaction norms. The concept of a reaction norm is a convenient summary of the response of a genotype to environmental variation. Three genotypes (AA, AA', and A'A') show different phenotypes (number of eye facets) in response to temperature change in *Drosophila*. Redrawn from Schmalhausen (1949 [1986]), after Hersh (1930).

Goldschmidt has been discontinuous, or qualitative variation. Perhaps one reason for the failure of these observations to have a profound effect on evolutionary thinking has been their awkwardness or unattractiveness for theoretical genetics. Population genetics, concerned with the spread of genes in populations, regards mutations of small effect as more likely to be important than mutations of large effect associated with the hopeful monsters of such developmental evolutionists as Goldschmidt (Charlesworth et al., 1982). Quantitative genetics, concerned with continuously variable traits, also focuses strongly on genes of small effect—the polygenes or quantitative trait genes of polygenically influenced, continuously variable traits (but see Falconer, 1981, on threshold characters).

The concept of a reaction norm bridges the gap between phenotypic plasticity and quantitative genetic studies of natural selection by connecting quantitative phenotypic plasticity and genotype (see Schlichting, 1986, 1989; Schlichting and Pigliucci, 1998). When an individual phenotype varies as a continuous function of some environmental variable, its reaction norm (Woltereck, 1909) is the full set of phenotypic responses to that variable, usually expressed as a curve on a graph of phenotype

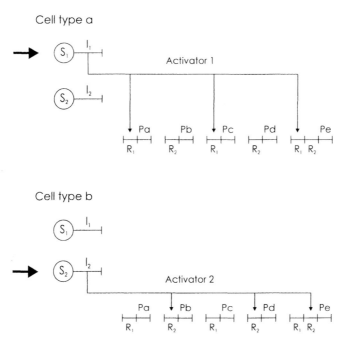

Fig. 2.5. Developmental linkage: the Britten-Davidson model of gene regulation. Pattern of gene expression differs among cell types, which respond differently to substances that impinge on the cells. A signal, environmental factor, or inducer (e.g., hormone) from outside the cell (heavy arrow) interacts with a sensor (S) to activate an integrator gene (I_1 or I_2), whose product (activator) then interacts with specific control or receptor elements (R) adjacent to structural or producer genes (P). Thus, chromosomally unlinked gene sets (e.g., Pa and Pc; Pb, Pd, and Pe) can be developmentally linked in their expression, and different combinations can be deployed in different cells (e.g., cell types a and b). After Britten and Davidson (1969), as summarized in Raff and Kaufman (1983).

value versus value of the environmental variable (figure 2.4). Studies of reaction norms demonstrate that (a) the varied curves of plastic responses correspond to genotypic differences (are heritable and subject to selection); and (b) heritability and effectiveness of selection on plastic traits depend on the range of environmental conditions (and, hence, phenotypes expressed) where selection occurs (see Gupta and Lewontin, 1982; Stearns, 1989). Because reaction norms vary among individuals, the environment is not only an agent of selection but also a determinant of the range of phenotypes exposed to selection (Alberch, 1980).

The Britten–Davidson Model of Gene Regulation

Britten and Davidson (1969; see also Britten and Davidson 1971; Davidson and Britten, 1979) proposed a model of eukaryote gene regulation based on evidence for cell-specific activation of gene expression (figure 2.5). Although this is a model restricted to processes of cell regulation at the level of genomic transcription, it has broad implications for how evolutionary biologists view genes in relation to the structure and evolution of phenotypes, as the authors clearly realized (see also Raff and Kaufman, 1983). Even though developmental molecular biology has progressed enormously since the invention of this model, the following points made by Britten and Davidson are of lasting importance:

1. Although differentiated cells have the same nuclear genomes their distinctiveness involves differential gene *expression*. This was supported by evidence that much of the genome in higher cell types is inactive; and different RNAs are synthesized in different cell types. So regulation of gene activity evidently underlies cell differentiation.
2. Differentiation and gene expression are mediated not (as in prokaryotes) by simple

"regulatory genes" (e.g., operons) but by multiple external signals (such as hormones, embryonic inductive signals, and other inputs) that can be polyfactorially, or polygenically influenced.

3. A given state of differentiation even at low (cellular) levels of organization involves large numbers of genes: cell phenotypic distinctiveness is a polygenic response.

4. The coexpressed battery of regulated "producer" or structural genes need not be contiguous or even on the same chromosome (cf. older concepts of genetic architecture, which saw functionally related genes as chromosomally linked "supergenes," e.g., Ford, 1964; chapter 4).

In general, these four properties of eukaryotic regulation were confirmed by subsequent work (see especially John and Miklos, 1988). Britten and Davidson (1969) pointed out that at higher grades of organization, evolution might indeed be considered principally in terms of changes in the regulatory systems and that "preexisting useful batteries of genes will tend to remain integrated in function. At the same time, there is the potentiality of formation of new integrative combinations" providing "[a] peculiar combination of conservatism and flexibility" (p. 356). "The model supplies an avenue for the appearance of novelty in evolution by combining into new systems the already functioning parts of preexisting systems" (p. 356).

In addition to these empirically based indications of a connection between development and evolution, there are reasons to consider the two topics related that could be called self-evident or axiomatic: because the phenotype is a product of development, it follows that any change in morphology or any other feature of the phenotype requires a commensurate change in development. Evolutionary change occurs through genetic modification of development. Development is inevitably linked to evolution by the fact that the phenotype variants that are the target of selection are products of developmental processes (in turn influenced by genotype). These axioms are acknowledged repeatedly in discussions of the evolutionary significance of development and behavior (e.g., see Alberch, 1980; Hamburger, 1980, p. 99; Raff and Kaufman, 1983, p. 337; Mayr, 1970b, p. 108; Wcislo, 1989; Thomson, 1988, pp. 5, 88).

Each of the phenomena discussed above has been treated as a general principle of development and evolution (see, e.g., on heterochrony: Gould,

1977, 1988; on regulatory novelties and evolutionary novelties: Bateson, 1894; on constraints and canalization: Waddington, 1942; Raff and Kaufman, 1983, p. 179; on the hierarchic branching of ontogeny in relation to hierarchic branching in phylogeny: Thomson, 1988; on behavior and evolution: Wcislo, 1989; on alternative phenotypes and phylogeny: West-Eberhard, 1989; on genetic assimilation: Waddington, 1975; on the Britten-Davidson model: Valentine and Campbell, 1975; Valentine and Erwin, 1987; Raff and Kaufman, 1983). Though general evolutionary significance has been claimed for these phenomena, they are usually discussed one at a time, as unconnected facts or concepts. The task of a comprehensive theory of development and evolution is to tie these phenomena together in a single conceptual framework. The obvious way to start is with a theory of the phenotype structure that can be related to development and genetics on one hand, and to selection and evolution on the other. There is presently no such theory of the phenotype.

A Unified Theory of Phenotypic Development and Evolution: Overview

The central theme of this book is the evolution of phenotypes—the expressed characteristics of living organisms. I define evolution as phenotypic change *involving* gene frequency change, not just gene-frequency change alone. I want to know what makes the phenotype change during evolution: how does change start and how do the genes get involved? The central argument I will make is that the secret to understanding evolution is to first understand phenotypes, including their development and their responsiveness to the environment. Of course, development and responsiveness involve genes, as does evolution. So I attempt to tie all of these things together—development, responsiveness, genetics, selection, and evolution—in a single scheme.

Part I proposes a general theory of phenotypes to replace the metaphors of the past. It shows how phenotypes are organized, including their relation to development and genes. Given this foundation, I then ask how evolution works. Why start with the phenotype and its development? Because that is where evolution starts. Development produces the variations that are screened by natural selection. Development also organizes gene expression, so it determines how the genes are exposed to selection and in what combinations. Adaptive evolution is a two-step process, first the production of phenotypic variation (by development), then selection. If the phenotypic variation has a genetic component, then selection produces evolutionary change. So evolution begins with genetically variable development.

The framework for relating phenotype, genotype, and development appears in chapters 4 and 5, which show how to relate gene action to phenotypic structure. The link is quite simple. Individual development always begins with an organized phenotype already in place, a principle I call continuity of the phenotype (chapter 5). Then, the parentally provided phenotype (e.g., that of an egg) changes during ontogeny due to inputs from genes and from the environment. The individual's genotype can never be said to control development. Development depends at every step on the preexistent structure of the *phenotype*, a structure that is complexly determined by a long history of both genomic and environmental influences.

The environmental sensitivity of development means that environmental variation can be a source of the selectable phenotypic variation that fuels adaptive evolution. This simple and unavoidable consequence of the nature of development requires a new view of how evolution works, described in chapter 6, on adaptive evolution. I show why genes are usually followers, not leaders, in evolutionary change. An adaptive novelty begins as a recurrent developmental change, whose recurrence may be due to either a mutation or a recurrent environmental induction (drift is also considered as a theoretical possibility). Then, most evolutionary genetic change, even following mutational origin, is at loci that influence the regulation of the trait (chapter 6). Recurrent expression allows selection to adjust the frequency and the form of the new trait and to eliminate negative sides effects, a process of quantitative genetic change that I call genetic accommodation. Chapter 6 discusses how this process of genetic accommodation differs from genetic assimilation and the Baldwin effect.

In the genetic accommodation of novelties, the degree of environmental influence may decline if a trait is advantageous in all circumstances. This can occur because genetic and environmental influence on switch points is interchangeable during development and evolution, a property of regulation I call interchangeability (chapter 5). Interchangeability of environmental and genotypic influence is responsible for the phenomenon of phenocopies. It is also responsible for the capacity for evolved increase or decrease in phenotypic plasticity and

canalization, and changes in frequency of conditional phenotypes in populations. Changes in frequency due to selection on switch mechanisms can lead to the fixation of an adaptive trait (chapters 21) and can affect speciation (chapter 27) and rates of morphological evolution (chapter 30). All of these evolutionary phenomena depend on the basic nature of developmental switches, which integrate environmental and genetic inputs, as discussed in chapter 4. Part I also shows the relation between continuous and discontinuous variation, both of which depend on their common dependence on regulation by switches, as mentioned in chapter 1 and more extensively discussed in chapter 4.

There is no hint of direct (Lamarckian) influence of environment on genome in this scheme—it is entirely consistent with conventional genetics and inheritence. By the view adopted here, evolutionary change depends upon the genetic component of phenotypic variation screened by selection, whether the phenotypic variants screened are genetically or environmentally induced. It is the genetic *variation* in a response (to mutation or environment) that produces a reponse to selection and crossgenerational, cumulative change in the gene pool. The role of environmental sensitivity in this process needs to be firmly incorporated as a fully recognized factor in evolution, not minimized or omitted out of fear of some Lamarckian overtone. Chapter 6 shows how to incorporate environmental influence along with genes in a view of development and evolution general for all phenotypes.

Other chapters show why the incorporation of the environment as a factor in evolution is necessary: chapter 7 summarizes how developmental plasticity facilitates evolutionary change. Adaptive phenotypic flexibility exaggerates novelties and may accommodate them without loss of function, a process I call phenotypic accommodation (chapter 3). Several other chapters (parts II, III, IV) show how major patterns of evolution, including of speciation, radiation, punctuation, macroevolution, and conditional polyphenisms, may be best understood, and in some cases *only* properly understood, when phenotypic plasticity and flexible development are taken into account. I present a flexible stem hypothesis that relates ancestral patterns of development and plasticity to patterns of interspecific divergence and adaptive radiation (chapter 28). Sympatric divergence in the form of conditional alternative phenotypes can influence speciation (chapter 27) and macroevolution (chapter 29). If flexibility and development are not considered, these phenomena may be misunderstood or not seen as expected results of events within species.

Modularity is another general property of phenotypic organization with broad implications. The developmental and genetic basis of modularity is described in chapter 4, and its evolutionary consequences are summarized in chapter 7. Modularity is the basis for the evolution of the internal diversity of the phenotype—its divergent life stages, organs, tissues, and behaviors and physiological responses. Why these can evolve semi-independently of each other is explained in chapter 7, and documented with many examples in chapter 21 (on divergence in alternative phenotypes) and chapter 29 (on macroevolution). Another consequence of modularity is combinatorial evolution, a major theme of this book. Part II shows how phenotypic evolution often occurs via *phenotypic recombination*, the ontogenetic and evolutionary reorganization of preexisting modular subunits. Because traits are regulated by switches (chapter 4), they can be turned off and on and recombined with other subunits during evolution. Part II shows the multitude of kinds of modular reorganziation that have evolved via phenotypic duplication, deletion, reversion, recurrence (parallelism), heterochrony, heterotopy, and sex transfer. Separate chapters treat phenotypic recombination at the molecular level (chapter 17) and evolved reorganization of behavior and morphology due to learning (chapter 18). These chapters show that combinatorial evolution is a major mode of evolution in all organisms and at all levels of organization. It occurs due to the fundamentally modular organization of the phenotype by switches, introduced in part I.

Part III is devoted entirely to alternative phenotypes, primarily polyphenisms. In the recent past, these were regarded as rarities unimportant for evolution. Why so much emphasis on them here? Alternative phenotypes illustrate the principles of development and evolution as seen in field studies of behavior and functional morphology. They show especially well how adaptive divergence can occur without speciation (chapter 21) and the nature of environmental assessment that allows condition-sensitive trait expression (chapter 23). I emphasize the importance of distinguishing between generalists and polyspecialists for predicting the results of selection, and the importance of alternative phenotypes for facilitating transitions between adaptive modes (chapter 20). Alternative phenotypes in fact prove to be extremely common. This is not surprising when one realizes that they are a necessary phase in the evolution of every novel discrete trait (chapter 20) and that they can be maintained in populations without equilibrium conditions formerly thought to render them difficult to evolve (chapter 22).

Part IV shows how all of this relates to some of the major themes and controversies of evolutionary biology since Darwin. These themes include gradualism versus saltation, speciation including sympatric speciation, macroevolution or large phenotypic change, and the evolution of complexity including the maintenance of sex. The chapters of part IV argue that a unified view of development, phenotypic plasticity, and evolution requires that we change how we think about all of these topics. These chapters are only brief discussions of complex topics with enormous literatures currently undergoing explosive growth. Yet even a cursory look shows that a developmental viewpoint simplifies and clarifies some of the key questions, and in some cases suggests new directions for research. For this reason, I have risked taking them on.

This is a fat book because it attempts to do all of this without resorting to mere assertions. At the risk of attempting too much, it seemed that the topic of evolution and development needed a broad look and an effort toward a comprehensive synthesis, in a single volume focused primarily on phenotypes and concerned with the traditional unresolved questions of evolutionary biology.

Definitions of Key Terms

One roadblock to a synthesis of development and evolution is inconsistency in the use of important words. This was evident in chapter 1. Within evolutionary biology, there is a surprising lack of agreement regarding such fundamental concepts as "selection" and "evolution." This has led to fruitless argument about the significance of development and plasticity in evolution. This section defines terms as I use them in this book, and in some cases briefly gives reasons for accepting particular meanings, at least provisionally. I do not review other definitions and do not attempt to discuss the very extensive debates over the meanings of these words (e.g., see Sober, 1984; Endler, 1986; Lloyd, 1988; Brandon, 1990, 1996). Subsequent chapters give specific definitions for many other terms and show how the concepts defined here fit together in a logically consistent framework for discussions of development, plasticity, and evolution. The ultimate justifications for the definitions listed here are in the findings of subsequent chapters. Especially, the phenotypic definition of selection, and genetic definition of evolution given below are fundamental for clarifying the relations between the immediate, developmental responses of organisms and their cross-generational, evolutionary change (Brandon, 1990, 1996; Lloyd, 1988).

Phenotype   The *phenotype* (Johanssen, 1911) includes all traits of an organism other than its genome. Lewontin (1992) and Mahner and Kary (1997) use the word "phenome" as a parallel with genome, to emphasize that the phenotype is the individual outside the genome. The enzyme products of genes are part of the phenotype, as are behaviors, metabolic pathways, morphologies, nervous tics, remembered phone numbers, and spots on the lung following a bout with the flu. That is, the phenotype can be adaptive or pathological, permanent or temporary, typical or atypical of a species. The word "phenotype" is sometimes used to denote a single phenotypic trait.

Phenotypic structure refers to the organization of the phenotype at any level of analysis, not just to an item of morphology.

Genotype   The *genotype* of an individual is the genetic makeup by which an individual or one of its traits can be characterized in genetic comparisons with other individuals or their phenotypic traits. In genetic studies, this term may be used to denote the inherited or genetic contribution to a phenotype, as inferred from breeding studies, or as a synonym of the word "individual" (meaning the genes of an individual). It sometimes refers to the expressed gene or set of genes that influence a particular phenotypic trait. The *genome*, by contrast, is the full complement of DNA present in a cell.

Selection   Darwinian selection is differential survival and reproduction (differential fitness) due to phenotypic differences among reproducing entities. *Inclusive fitness* (Hamilton, 1964a,b) gives a more general definition of fitness, including both differential reproduction and effects on the reproduction of relatives or, more broadly, co-carriers of the same genetic alleles.

Darwinian selection is the result of phenotypic differences. In this sense, selection "acts" on phenotypes. *Selection* as defined here does not require that phenotypic differences be due to genetic differences, but a *response to selection*, or *evolution*, does. Differential reproduction (selection) can be due to environmentally induced phenotypic differences and may have no evolutionary effect. Haldane (1958) distinguished between effective and ineffective selection, "which does not alter gene frequencies" (p. 16). Thus, selection may be strong (there may be marked variation in reproductive success) without genetic variation (West-Eberhard, 1979).

The phenotypic definition of selection recognizes that the phenotype is influenced by both genotype and environment and that selection depends on variation in phenotypes, not only on variation in genes. Nongenetic factors can cause selectable phenotypic variation and may thereby become integral parts of ontogeny. The phenotypic definition of selection circumvents the erroneous supposition that nongenetic contributions to the differential success of phenotypes are trivial for evolution.

Many discussions of selection concerns selection "on" particular traits. That is, a particular trait can be regarded as having a particular *fitness effect*, its positive or negative (or lack of) contribution to the lifetime reproductive success of the individual that bears the trait. While selection depends on phenotypes influenced by both environment and genes, selection can only "favor" (cause to increase in frequency across generations) certain traits through effects on the alleles that favor their production. It is in this indirect sense that selection acts on— affects the frequency of—genes. Falconer (1981, p. 24) recognizes this fundamental relationship between selection on phenotypes and the evolution of gene frequencies when he states:

> If the differences of fitness [defined by Falconer as contribution of offspring to the next generation] are in any way associated with the presence or absence of a particular gene in the individual's genotype, then *selection* [differential survival and reproduction by individuals] operates on [influences the frequency of] that gene. (p. 24; bracketed insertions mine)

A phenotypic definition of selection helps to clarify the relationship between phenotype and genotype in evolution. It is an essential adjunct to the theory of the phenotype proposed here, which attempts to relate genotypes, development, phenotypic variation, selection, and genetical evolution.

Evolution (Phenotypic Evolution)   The most general and essential property of *phenotypic evolution* is that it is cross-generational change in phenotypic frequencies or dimensions involving change in gene frequencies. Involvement of gene-frequency change is what gives evolution its cumulative, cross-generational aspect. Individuals and other entities (social groups, species) reproduce, but they do not make exact copies of themselves. Strictly speaking, the units that replicate themselves most precisely, and therefore have cross-generational effects that both reflect past differential reproduction and affect future reproduction, are genes. So genes are the most appropriate units of evolution. By this broad definition, evolution can include changes due to both natural selection and drift (stochastic changes not due to relative reproductive success).

Adaptive evolution is cross-generational change in phenotype frequencies involving gene-frequency change due to selection on heritable variation in phenotypes.

Units and Subunits of Selection   The *units of selection* are differential reproducers—any phenotypically variable differentially surviving or reproducing entities whose differences in cross-generational reproductive success depend on their own phenotypic characteristics, not on the reproduction of higher level units of which they are contributing parts. Thus, the genes or cells of a multicellular organism are not usually units of selection because they depend on a higher level reproducer (the individual) for cross-generational reproduction. The reproductive units may be individuals; groups where individual reproductive success depends on that of the group; or populations, species, or clades, which, on a long time scale, show differential multiplication and extinction due to their phenotypic differences.

Subunits of selection are switch-activated, coexpressed or coordinately used components of the phenotype, or phenotypic traits. They include the semidiscrete cells, tissues, organs, life stages, and alternative phenotypes (e.g., switch-controlled behaviors, morphologies, and physiological responses) or "traits" that compose a reproductive unit (e.g., an individual). A subunit of selection can have a fitness effect, a quantifiable lifetime contribution to the fitness of the unit of selection of which it is a part, and on whose reproductive success the subunit's survival and recurrence in the next generation depends.

Environment   *Environment* means the world outside a trait or individual of focal reference. The external environment, meaning the environment external to an individual, is sometimes broken down for special purposes into the physical environment, the biotic environment, the social environment, and so forth (e.g., Smillie, 1993; Wcislo, 1989; Williams, 1966; Brandon, 1990). The *internal* environment, meaning the environment within an individual, includes such factors as gene products, cells or growing tissues of different kinds, body

temperature, and so on. Williams (1966) and Buss (1987) call the internal environment the "somatic" environment, and the external environment the "ecological" (Williams) or the "extrasomatic" (Buss) environment. For a cell, important factors of the internal environment, inside the organism but outside the cell, might include hormones, viruses, a controlled temperature, and neighboring cells. For an individual, it might include parents and siblings as well as parasites, wind, and rain.

Some readers may object to the idea of social and internal environments because this enlarges the domain of "environment" to include even the products, or epistatic relationships, of genes. The same objection would apply with respect to the social environment provided by parents, or the biotic environment that results when a female lays her eggs in a particular plant or temperature condition. By being able to view the genes as contributors to the environments of other genes, one can better understand regulation and the relationship between external environmental factors (including genetically influenced ones such as parental care) and internal environmental factors (including genetically influenced ones like enzymes). The recognition of an internal environment puts the external environment and the genome on equal footing as architects of the internally orchestrated aspects of the phenotype, for all gene expression depends on conditions and materials of external-environmental origin within cells. This is more thoroughly discussed in chapter 4 on development.

Development   In this book, *development* is all phenotypic change during the lifetime of an individual or higher unit of organization. Since the phenotype, as defined here, includes every characteristic of an organism outside the genome, development encompasses ontogenetic change in all elements of structure other than of the genome, whether morphological, physiological, or behavioral. The term "development" will be used very broadly here to include both irreversible aspects of ontogeny, such as the production of a permanent organ or appendage, as well as reversible aspects, such as the expression of a behavior pattern or a physiological response.

This is a much broader concept of development than that of most developmental biologists. Although the word "development" is seldom explitly defined even in whole books on the subject, it is used in variously restrictive ways, sometimes to apply only to embryos, and, most often, to apply only to morphology. Some authors (e.g., Frank, 1996, p. 471) define development as the creation of a phenotype from the information encoded in the genome. But normal development often depends on specific information and materials from the environment, in addition to those from the genome. To solve the nature–nurture problem, one has to acknowledge the deterministic role of the environment, alongside the genes, in development and to consider the development of the entire phenotype.

Plasticity   *Plasticity* (responsiveness, flexibility) is the ability or an organism to react to an internal or external environmental input with a change in form, state, movement, or rate of activity. It may or may not be adaptive (a consequence of previous selection). Plasticity is sometimes defined as the ability of a phenotype associated with a single genotype to produce more than one continuously or discontinuously variable alternative form of mor-

phology, physiology, and/or behavior in different environmental circumstances (Stearns, 1989). It refers to all sorts of environmentally induced phenotypic variation (Stearns, 1989).

Plasticity includes responses that are reversible and irreversible, adaptive and nonadaptive, active and passive, and continuously and discontinuously variable. Many authors have proposed terms that distinguish among these variations on the theme of plasticity, as discussed in chapter 3. But I will adhere to the simple, inclusive definition given here. One can be trapped by specialized terms into having to make unnecesary distinctions at every turn, as important as they may be for certain points.

I use the word *lability* only to describe facility for *evolutionary* change in phenotype, as in "phenotypic plasticity correlates with evolutionary lability."

3

Plasticity

The first step toward a theory of development and evolution is to specify the properties of the phenotype that develops and evolves. Plasticity, or environmental responsiveness, is a universal property of living things. A brief look at a few of the mechanisms of plasticity shows why it is correct to regard plasticity itself as an evolving trait, for it is underlain by concrete and complex structures. Mechanisms of extreme plasticity involve the overproduction of variants followed by use of only a few (somatic selection), and such hyperflexibility characterizes plasticity at different levels of organization from the molecular to the social. The mechanisms of plasticity permit phenotypic accommodation, the integration and exaggeration of both developmental and evolutionary change without genetic change. Phenotypic accommodation is an important factor in the evolution of novel traits.

Introduction

A phenotype-centered view of evolution needs to start with a solid idea about the nature of the phenotype. This chapter and the next are devoted to two universal properties of phenotypes, plasticity, or responsiveness to environmental inputs; and modularity, or subdivision into semi-independent and dissociable parts (chapter 4). Of these two properties, plasticity is probably the more fundamental, for the ability to replicate, which distinguishes organic from inorganic nature, requires molecules which are interactive and precisely responsive—adaptively plastic. So plasticity must have been an early universal property of living things. The universality of modularity is a secondary, or "emergent" result of the universality of plasticity (see Wilczek, 2002, on emergent universality in physics). Any organism whose size, whether due to accretion or growth, is large enough to create internal environmental differences, such as those between the inner and the outer regions of a clump of material, has the potential for regional internal differentiation. As differentiation evolves to produce specialized parts and an internal division of labor, internal heterogeneity gives rise to conditional switches between developmental pathways. The result is a stucture characterized by somewhat discrete parts—modularity. Thus, given

plasticity as a universal property of living matter, modularity follows.

The present chapter describes some of the remarkable mechanisms of phenotypic plasticity. One reason to focus on mechanisms is to indicate the material basis for the evolution of plasticity, which is a product of concrete devices that are subject to genetic variation and selection. A cursory look at these mechanisms, however incomplete, by itself suggests the importance of plasticity in development and evolution, for the mechanisms of plasticity include some of the most ingenious and widely conserved creations of nature. Mechanisms of plasticity are further discussed in chapter 23, which describes how organisms assess environmental conditions when they adaptively switch between alternative developmental pathways.

The Meaning of Plasticity

Phenotypic plasticity has already been defined as the ability of an organism to react to an environmental input with a change in form, state, movement, or rate of activity. It is tempting to define plasticity as the responsiveness of the phenotype of a single genotype to environmental influence, but this is not strictly accurate, since the phenotype of an individual, including the ability to respond, exists as a parentally provided structure before its own genotype begins to act (see chapter 5). So an individual is not the product of a single genotype except in mutation-free clones. The words "responsiveness," "flexibility," "malleability," "deformability," and *developmental plasticity* are all synonyms of phenotypic plasticity as defined here.

Plasticity is a very old term for modifiability of morphology during development (Wilson, 1894) and for environmentally sensitive behavior (Baldwin, 1902). Wheeler (1910a) defined plasticity as "the power of an organism to adapt action to requirement without the guidance of a hereditary method of adjustment" or "action on the basis of individual, e.g., ontogenetic experience" (p. 531). Wheeler's definition implies that plasticity is necessarily active and adaptive. I include passive and nonadaptive responses as a reminder that organized action and adaptation can originate literally with a bump from the outside. What we perceive as an orderly formation may have begun as a change we would have classified as a deformation, and it could be favored by selection without modification of its mode of development.

Plasticity is, in effect, intra-individual variation (Evans, 1953; G. C. Williams, 1992). Definition of plasticity as intra-individual variation brings out the dual influence of environment and genes, for intra-individual variation during development is obviously the result of inputs from both sources, with the individual's genome a constant and its environments responsible for variation in its phenotype over time or topology.

Much narrower definitions of plasticity are used in some subdisciplines of biology and by particular authors (see West-Eberhard, 1989). At least one author (G. C. Williams, 1992) advocates elimination of the word "plasticity" entirely. When used, Williams and others (e.g., Sultan, 1987) would, like Wheeler, restrict the term to adaptive, active responsiveness. Williams (1966) calls nonadaptive responsiveness "susceptibility," and other authors (e.g., Smith-Gill, 1983) have proposed special terms to distinguish between adaptive and nonadaptive plasticity. Although many important consequences of plasticity, such as phenotypic accommodation of change (see below), imply adaptive plasticity, the line between nonadaptive and adaptive plasticity may occur without genetic change during evolution and may be difficult to distinguish in practice. Nonadaptively malnourished and therefore phenotypically distinctive larvae, for example, were likely the forerunners of workers in social insects, and they may have been immediately adaptive (having a positive effect on inclusive fitness) without their phenotypes or their degree of plasticity being modified under selection. Such morphological anomalies and misplaced behaviors, environmentally induced in new contexts, are regularly the sources of adaptive structures and social displays (see part II). So both adaptive and non-adaptive plasticity are important in evolution, and often it is difficult, given the onerousness of ascertaining adaptation (Williams, 1966), to confidently distinguish between them.

I see no reason to introduce special terms that will continually require distinctions that are impossible and, in many discussions, of little use: all novel adaptive phenotypes must originate before they can be molded by selection, and they need not be altered under selection to be adaptive (favored by selection). Many examples of environmental responsiveness have unknown fitness effects, and some, such as certain responses to manipulation by mates or parasites, and environmental contaminents (Hayes, 1997b) are known to be harmful, at least in some contexts in which they are observed. For these reasons, I prefer a fitness-neutral definition of plasticity (see also Schlichting and Pigliucci, 1998).

To draw a line between *active* and *passive* responses, with adaptive responses understood to be active, as in G. C. Williams's (1992) definition (above) might encourage the erroneous idea that all adaptive plasticity is an active rather than a passive response to the environment. Adaptive plasticity is clearly active when a plant responds by growing toward light, but it can also be passive, as when a baby's head is flattened as it passes through the birth canal.

Stearns (1989, p. 438) uses the word "plasticity" for an *irreversible* response to the environment, and "flexibility" for a *reversible* response (see also Clark, 1991). A widely used textbook of plant physiology (Salisbury and Ross, 1985) also restricts "plasticity" to mean irreversible environmentally mediated change, stating that the cell wall "stretches *plastically* (irreversibly, like bubble gum) rather than *elastically* (like a rubber balloon) as the cell grows" (p. 8). For most of the generalizations discussed here, this is not an important distinction, and when it is, this importance will be clear without a special term.

In still another definition, Bull (1987, p. 304) distinguishes between plasticity defined as *systematic* variation correlated with some environmental cue, and environmentally mediated random variation or developmental "noise." It is unclear whether "noise" means that the cue is unidentified (variation is random with respect to measured environmental variables), or that the cues are many and complex (and therefore incompletely unidentified), or both. As a hypothetical example of noise, Bull cites persistent variation in a genetically identical strain of insects reared under apparently identical conditions. This distinction would seem to depend too much on the perceptiveness or acumen of the observer. An observer with little interest in the evolutionary potential of environmental effects, and little feeling for the sensitivities of organisms, might, along with Fisher (see chapter 1; in Wigglesworth, 1961), consider virtually all environmental effects noise.

Two sets of models and vocabularies have grown up around what are perceived as two "kinds" of plasticity, one *continuously variable*, or *graded* plasticity, and the other *discontinuously variable*, or *discrete* plasticity. The term "reaction norm" is more often applied to graded plasticity, although it can also be applied to discrete, threshold responses (e.g., see Via et al., 1995). Smith-Gill (1983) calls continuous plasticity "phenotypic modulation," and the discrete plasticity represented by such phenomena as polyphenisms, "developmental conversion." Both continuous and discrete plasticity are products of condition-sensitive switches. A switch is characterized by a threshold potentially influenced by both genotype and environment (chapter 5). So a switch between alternative phenotypes or between successive events in a developmental pathway is a point where there can be plasticity in whether or not a trait is expressed or a process continued. Continuous variation in degree of plasticity of a quantitative trait is produced when the switches that govern the dimensions of the trait (e.g., those that mark the beginning and the end of limb growth) are environmentally sensitive. The basis for continuous and discrete variation is further discussed in chapter 5.

Smith-Gill (1983) regards discrete plasticity as programmed and adaptive, in contrast with continuous plasticity, which she sees as the result of "lack of mechanisms to completely block out or regulate external stimuli" (p. 54). While I agree with Smith-Gill that we should not assume that observed phenotypic plasticity is adaptive, both continuously variable and discrete plasticity are products of switches whose thresholds are subject to genetic variation and adaptive adjustment by natural selection (see chapter 6), and the adaptiveness of plasticity can be demonstrated using special techniques (e.g., see Schmitt et al., 1995, 1999; Pigliucci and Schmitt, 1999). All of the mechanisms described in the section on somatic selection below and all physiological mechanisms of homeostosis give rise to continuously variable accommodations to environmental change that are clearly adaptive. Allometric increases in size of adornments and weapons with environmentally influenced increase in body size are examples of graded plasticity that is known in some cases to be adaptive. For example, such allometries with slopes greater than 1 mean that investment in a continuously variable weapon such as the forceps of certain earwigs (Insecta: Dermaptera) is proportionately greater in the relatively large individuals that most often use these structures in fighting (Eberhard and Gutierrez, 1991; Eberhard et al., 2000).

Here I include adaptive and nonadaptive, active and passive, reversible and irreversible, and continuous and discontinuous responses in the single term "plasticity." All may play a role in normal development and evolution, and a given response may change category (between active and passive, nonadaptive and adaptive) from one developmental or evolutionary moment to the next. I do not use special terms for reversible and irreversible variation because so many generalizations apply equally to both. I prefer to use adjectives (adaptive, nonadaptive, reversible, irreversible) to emphasize these distinctions when necessary, in order to minimize special terminology and avoid the obligation to make distinctions when none is needed.

There may be potential for confusion regarding the meaning of plasticity due to the different language and perspectives of those who study plasticity as a property of individual organisms, and those who measure it by comparing populations (see Gordon, 1992). For an ethologist or a developmental biologist, plasticity is evident in the ontogeny and behavior of individuals in different circumstances. Environmental influence can be demonstrated, for example, by imposing different circumstances on different members of a single brood. In quantitative genetics, by contrast, genotype-by-environment interaction, along with environmental variation, is a component of the environmental variance of a quantitative trait. Here "genotype-by-environment interaction" does not mean, as it might to an ethologist, that a genotype interacts with the environment to produce a phenotypic response or that it exhibits plasticity. It is a measure that implies that genotypes *differ* in their plasticity, such that different genotypes have different responses to the same environmental conditions (different norms of reaction). Genotype-by-environmental interaction describes a statistical property of a population, not a developmental response of an individual.

Mechanisms of Plasticity

Hypervariability and Somatic Selection

In *somatic selection*, sometimes called epigenetic selection or developmental selection (Frank, 1996), large numbers of random, or possibly chaotic (see below), variants, modifications, movements, or positions are produced, and then some variants are selectively preserved or reinforced, while the remainder are unoccupied or eliminated (Steele, 1981; Sachs, 1988a). Individual variants act semi-independently, depending on local conditions, in accord with simple rules. The result is establishment of functional pattern without central coordination of elements. These mechanisms are highly plastic because the dispersed, independently responding elements can sense and adjust finely to a spatially heterogeneous field. The high degree of flexibility provided by overproduction of variants permits the active, dynamic circumvention of a genetically fixed stable or equilibrium state (Katz, 1987; Bonner, 1988, p. 54).

In somatic selection, or the induction of pattern in random variability, lack of stability maintains a system poised for directedness in any one of numerous directions. In contrast to the kind of flexibility represented by alternative phenotypes, flexibility through somatic selection may fail to produce recurrent responses, instead producing an array of finely tuned local adjustments. This would prevent selection from proceeding in any particular direction. As stated by Kirschner (1992), in discussing microtubules, this dynamic state allows small perturbations of the system to drive it to the most stable state, so systems far from equilibrium can operate with great precision. Such systems are more adaptable than those organized by a switch because they do not settle into a stable state that requires new inputs to be reversed. As a result, they readily accommodate both developmental and evolutionary change of many kinds. Gerhart and Kirschner (1997) call these mechanisms "exploratory systems," a term that draws attention to the trial-and-error behavior of elements that enter new regions without initial direction.

The idea that processes analogous to Darwinian selection occur during development has come up repeatedly in the writings of embryologists, immunologists, neurobiologists, and cell biologists. Roux (1891) visualized a struggle for nutrients among the parts of an organism, parallel to that among individuals in the Darwinian struggle for existence, and this inspired Weismann's (1904) idea of "intraselection," a subindividual process of accommodation and correlated change seen as promoting harmony and evolutionary change (see Buss, 1987; Dawkins, 1982, p. 169). The "competitive" quality of such processes as "material compensation" during growth and other kinds of developmental trade-offs, is discussed in chapter 16 (see also Buss, 1987). Hall (1984) discusses the "competitive interaction model" of relations among growing tissues, and some developmental biologists see regional differences as established by competiton among tissue domains, seen as showing hierarchies of suppression (e.g., see Berrill, 1971, p. 143).

Somatic selection is sometimes misconstrued as a hypothesis of Lamarckian inheritance capable of influencing heritable variation. Dawkins (1982) gives a concise critique of this idea. All of the examples of somatic selection described here are non-Darwinian selection: preserved variants are not inherited across generations. For this reason, the term "neural Darwinism" (Edelman, 1987) to refer to somatic selection in nervous system development is potentially misleading, even though it draws attention to a genuine similarity of neural development and Darwinian selection, both of which are characterized by differential survival within populations (of cells or individuals).

Some of the mechanisms of somatic selection are so poised to respond to local conditions that they are described as "self-organizing" (e.g., Kauffman, 1993; Bonabeau et al., 1997). The language of self-organization refers to "spontaneous" emergence of pattern since there is no central control of the pattern that results (see chapter 4). The words "spontaneous" and "self-organized" to describe the origins of order may give the impression of occurrence without external cause. But the emergence of pattern under somatic selection is as deterministic and environmentally influenced as any other aspect of development if one examines the behavior of individual responding elements. Thus, for example, while the foraging behavior of honeybees is not under central control by the queen or any property of the group that inhabits the hive, colonies selectively exploit the most profitable nectar sources due to the choice behavior of individual foragers (Seeley et al., 1991).

Self-organization is also sometimes described as giving rise to *emergent* functional properties greater than the sum of the parts. But this is a common characteristic of all coexpressed phenotypic subunits, however assembled or coordinated. An organ such as the hand has emergent properties beyond the sum of the bone, muscle, skin, and vascular tissues that compose it, but it is not an example of self-organization as the term is usually used. The term "self-organization" focuses on the emergence of *pattern* and complexity, whereas the term "somatic selection" describes the same processes but focuses at a lower level of organization, to emphasize the *mechanisms* of pattern formation.

Self-organization should not be confused with *self-assembly*, a process that has been compared to crystallization, where the structural properties of the components automatically produce a stereotyped pattern of organization without any interaction with the environment being necessary or influential, although concentration or temperature in a medium may trigger the process. The folding of proteins into their characteristic tertiary configurations is an example of self-assembly. In contrast to self-assembly, somatic selection mechanisms are exquisitely and adaptively sensitive to local conditions. Due to subdivision and dispersal of the responding elements, these mechanisms sense and adjust finely to a spatially heterogeneous field to produce variable but still functional structure.

In a discussion of self-organization in social insect colonies, Cole (1994; see also Ott et al., 1990 cited in Cole, 1994) argues that plasticity may be increased if the variation of elements is chaotic rather than random. "Chaos" is a kind of variability that differs from randomness in having, over the long term, deterministic pattern "constrained to an attractor" whose dimensions can be ascertained by measuring the observed variation. Pure random fluctuations, by contrast, are not so constrained but are space filling (Cole, 1991a, p. 254). Chaos implies deterministic constraints on variability, which presumably could evolve under selection on factors that affect the dimensions of the "attractor"—whatever these factors may prove to be in an organism. It is of interest to investigate the possibility of chaotic variation in biological systems showing variant overproduction and somatic selection, because it implies that the energy-costing production of variants may be to some degree designed under selection limiting the range or rhythmicity of variants produced. Cole suggests that systems with elements that are individually chaotic in their behavior can respond especially flexibly, because a chaotic process has embedded within it an infinite number of unstable periodic orbits. If a chaotic process is repeatedly perturbed, it can have a periodic output, and the range of possible patterned results is larger if individual behavior is chaotic than if it is initially random or periodic.

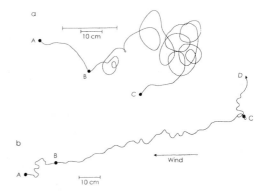

Fig. 3.1. Overproduction of movement variants (turns) in the search behavior of a male grain beetle (*Trogoderma variabile*): (a) in still air a two-second puff of female sex pheromone at time B causes searching movements during the following ten seconds (A-B); (b) in constant wind, after thirty seconds without pheromone (A-B), oriented movement occurs during 60 seconds (B-C) of exposure to pheromone, initiated at B and stopped at C. C-D is movement after pheromone is withdrawn. After Tobin and Bell (1986).

Examples of Somatic Selection and Similar Phenomena   The combination of variant overproduction and somatic selection underlies some of the most basic processes in eukaryotic development, including spindle formation in mitosis and the organization of nerve growth (Kirschner, 1992), and it characterizes fundamental behavioral processes, including orientation in certain protozoa (see examples in Fraenkel and Gunn, 1961), and trial-and-error learning (reviewed in Mackintosh, 1974). Some major kinds of somatic selection and similar phenomena, based on hypervariability and selective survival of functional configurations, are described in the following sections.

Animal Orientation   Selection of random or undirected movements has long been recognized as a mechanism of adaptive flexibility in studies of animal behavior, especially orientation and learning. The locomotion of paramecia, for example, is undirected in that the reorientation after having its path blocked by an obstacle is random rather than related to the path of incidence. This method of locomotion has been called "trial and error" and "selection of random movements" (see Fraenkel and Gunn, 1961, p. 44; figure 3.1). In many organisms oriented movement is accomplished by altering the length of randomly directed trajectories when they

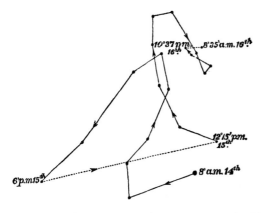

Fig. 3.2. Exploratory growth ("circumnutation") in a plant (*Lilium auratum*). The light-seeking movements of the stem of a plant are traced on a horizontal glass at different times during two days of growth in the dark. From Darwin (1880).

are associated with an increase or decrease of some stimulus such as light, a chemical, an object, or heat, that results in directional movement or klinokinesis (Fraenkel and Gunn, 1961); Evans, 1971, on polychaetes; Eberhard, 1990b).

Exploratory Growth in Plants   Trial and error locomotion has a vegetable counterpart, as demonstrated by Darwin in *The Power of Movement in Plants* (1880). Experiments showed that growing stems, roots, and leaves make erratic movements that Darwin termed "circumnutation" (figure 3.2). These movements continue until some event or stimulus, such as contact with a substrate, causes them to become channeled or directed. Although much is known about the intracellular events of plant morphogenesis, including the involvement of microtubules as sensing agents that respond to external forces (Wymer et al., 1996), the transduction of environmental stimuli into plant response is incompletely understood (Salisbury and Ross, 1985, p. 370; Cyr, 1994, p. 166).

One need only list the words used to describe growth in plants to indicate a diversity of plastic mechanisms. Plant growth involves searchlike up-and-down and side-to-side "nastic" movements evidently undirected by any external stimulus, and several kinds of tropism, or stimulus-oriented directional differential growth, the best studied being phototropism (light-oriented growth), gravitropism (gravity-oriented growth), and thigmotropism (response to contact with a solid object). In addition, there are tropisms apparently associated with

water and certain nutrients. Reaction wood—increased xylem production in response to mechanical stress in the branches and trunks of trees—reinforces the tree at points of excessive weight and strain. Its role in the origin of buttress roots is discussed in chapter 16.

In addition to these widespread phenomena, there are special flexible talents of plants that are associated with certain ways of life. For example, ground-germinating tropical vines (*Monstera gigantea*) grow directly toward potential host trees essential for support not by haphazard or random searching but by orientation toward the darkest sector of the horizon, thereby approaching the nearest tree (Strong and Ray, 1975). *Monstera* and other climbing aroids (Araceae) shift growth forms and degree of mobility in adaptively appropriate ways. They become sessile when they reach a supportive host and undergo a metamorphosis in the length and shape of internodes (Ray, 1992). The movement by growth of such plants through their environment, and these active responses to the conditions encountered are quite comparable to the behavior of animals (Ray, 1979). Botanists justifiably use such zoomorphic terms as foraging behavior (Slade and Hutchings, 1987, cited by Ray, 1992) and habitat selection (Bazzaz, 1991) to describe it.

Trial and Error Learning.   Trial and error learning, also called instrumental learning or operant conditioning, is performance of random, undirected behaviors with repetition of those rewarded. In many ways, trial and error learning is similar to trial and error locomotion (Changeux et al., 1984). It is taxonomically widespread in animals (Mackintosh, 1974). The evolutionary significance of trial and error learning is discussed in chapter 18, which treats the role of learning in phenotypic reorganization and selection, and chapter 28, on radiations.

Baldwin (1902, p. 115) was the first to my knowledge to realize that learning is a kind of somatic selection, or "functional selection from overproduced movements" (for a recent discussion, see Frank, 1996). As an example, he noted that a child learns to write from "excessively produced movements" and then gradually selects and fixes the slight successes made in the direction of correct writing.

In some birds, such as migratory populations of white-crowned sparrows (*Zonotrichia leucophrys oriantha*), males acquire several song dialects when young. Then, a process of selective attrition acts to narrow the learned pattern to one, as males hear increasingly more frequently the dialect of the re-

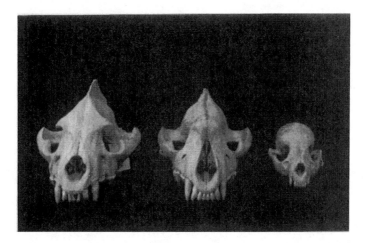

Fig. 3.3. Diet-dependent developmental change in the skull of wild spotted hyenas. Skulls collected following death of individuals (right to left) 3 months, 24 months, and 11 years old show marked change in attachment sites for feeding muscles (sagittal crest, cheekbones, and forehead). In animals raised in captivity, which feed relatively rarely on bone, these feeding structures are poorly developed; elderly captive hyenas have skulls that resemble those of cubs (Holekamp and Smale, 1998; Holekamp, personal communication). Courtesy of K. E. Holekamp.

gion where they settle to breed. In this way, they take advantage of an early sensitive period for song learning and are still able to match their song to that of territorial neighbors. Nonmigratory populations (*Z. leucophrys nuttalli*) have a different pattern of song learning, in accord with early certainty of adult-stage dialect for individuals who do not migrate from their natal region. Even when experimentally exposed to a variety of songs when young, they acquire imitations at a later age and have a shorter period of song plasticity (Nelson et al., 1995).

Developmental Plasticity of Muscle and Bone The responsiveness of phenotypic structure and development extends to materials often assumed to be rigidly formed. Bone is a good example (see Frazzetta, 1975). *Wolff's Law* (Wolff, 1892, ex Wimberger, 1991) refers to the malleability of bone under different regimes of use or stress during development. In some vertebrates, this flexibility produces morphological variants so marked and consistent that they are often interpreted as due to genetic variation (figure 3.3). Given our inability to perceive change in growing bones, it is easy to underestimate the extreme condition sensitivity of bone (Wimberger, 1991). A similar malleability characterizes the development of wood in trees, which responds to wind and position to produce species-specific variants so consistent in particular environments while absent in others that they could

easily be mistaken for genetically specified, constitutive traits (see chapter 16). Natural selection may mold the responsive phenotypic structure to make it more likely to produce certain responses.

The hypertrophy of muscle, bone, and callous tissues that are used or subjected to stress, and the atrophy of unused tissue are examples of somatic selection, where differential growth is involved rather than differential survival and/or multiplication of discrete entities. An important feature is the contrast with "patterned development," or development specified by prepatterns or programs (Sachs, 1988a).

The differential development of muscles under different exercise regimes is a familiar example of somatic selection by differential growth. Muscle activity patterns are also highly flexible and variable, and the loss of some patterns in favor of others is influenced by learning or practice (Galis, 1993a; Galis et al., 1994; figure 3.4). Wolff's law is less familiar. In some vertebrates this flexibility can produce marked morphological variants due to correlated changes in muscle and bone (Wimberger, 1991). As pointed out by Wimberger (1991, p. 1546) osteological variation has often been interpreted as the result of genetic variation rather than plasticity, perhaps due to the solid mineral aspect of bone.

The Immune System In human B cells that produce antibodies, special translocation events occur that

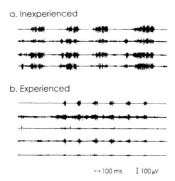

a. Inexperienced

b. Experienced

↦ 100 ms   I 100 μV

Fig. 3.4. The effect of experience on muscle action. Electromyograms show the activity of different muscles of the cichlid fish *Labrochromis ishmaeli* while crushing a crustacean prey: (a) a fish inexperienced with this prey type; (b) the same fish after 4 weeks of experience. From Galis et al. (1994), by permission of Academic Press Ltd., London.

bring together a set of three types of genes (of the VH, D, and JH gene classes) transmitted in germ-line DNA as separate genes but functioning together when combined in the B cells as one. Together, the set codes the V_H hypervariable domain of an immunoglobulin molecule. The germ line contains a large number of each of these three classes of genes that give rise to the V_H hypervariable domain. The particular combination is apparently random from a pool of 200–1500 possibilities, yielding literally millions of distinctive combinations, possibly in excess of 10^8. Somatic selection occurs when the presence of a foreign protein or antigen leads to the proliferation of clones of specific lymphocytes and, in turn, to production of large numbers of the appropriate antibody-producing B cells (concise summary by Committee on Research Opportunities in Biology, 1939).

Melanoblasts and Melanocytes Melanoblasts and melanocytes, which determine dark color patterns in a variety of animals, migrate and differentiate using cell movements and flexible responses resembling those of whole animals. Melanoblasts divide and proliferate at the base of the feather germ in birds and then migrate into the growing feather. There, as melanocytes, they synthesize melanin for a short time before they die (Price and Pavelka, 1996). Properties of both the epidermis and the melanocytes combine to determine whether or not the melanocyte survives long enough for melanin to be synthesized. Abundant experimental evidence indicates that melanocyte sensitivity to differentia-

tion and death is a labile trait, susceptible to many environmental and genetic influences (Price and Pavelka, 1996).

Neuron Growth and Nervous System Organization Neural growth and nervous system development are characterized by variant overproduction and somatic selection, or neuronal group selection (figure 3.5). The selective mechanisms are diverse and include neurite extension and retraction, synaptic reinforcement, cell migration, and cell death (Edelman, 1987, 1988). Growing nerve cells behave in some ways like single-celled organisms, and Edelman (1988) has described them as "gypsies," all of them migrating short or long distances during their careers. They send out and retract fingerlike processes as they move, seeming to use them to explore the surrounding space, until there is a sudden oriented response and a synaptic connection is formed. Then certain connections among cell groups are selected and others disappear.

Edelman recognizes several episodes of somatic selection during nervous system growth and function. A primary repertoire is built during early development. Then neuron groups that respond best

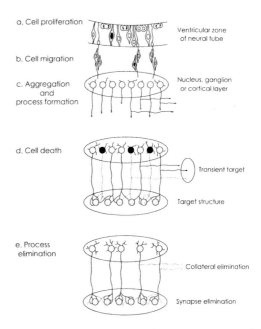

Fig. 3.5. Somatic selection during nerve growth. Following cell proliferation and migration, cell numbers are reduced by cell death, and overproduction of nerve processes and synapses is followed by selective elimination, leaving functional connections. After Cowan and O'Leary (1984).

to a given input are selected via increasing efficacy of utilized synaptic connections and suppression of others. The reorganization into groups involves a competitionlike accretion by different functional neuron groups of neurons to form definitive functional sets. By Edelman's view, this process of spatial reorganization based on experience never stops. Up to 70% of neurons in some regions of developing vertebrate nervous systems die before the structure of the region is complete (Edelman, 1988, p. 179). Variability and somatic selection, then, occur in several ways in the developing nervous system, by the differential persistence and retraction of exploratory nerve processes, by the differential "capture" of neurons by neighboring cell groups, and by differential cell death.

Microtubule Assembly in Mitotic Spindles   In a typical fibroblast cell, the microtubules of a cell have a relatively constant standing population of 500–1000, which is maintained by a "dynamic instability" in which they are continually being polymerized and depolymerized (Kirschner and Mitchison, 1986). In the formation of the mitotic spindles that move chromosomes during cell division, individual microtubules grow (polymerize) in random directions from the centrosome and then shrink back (depolymerize) to the cell center. Eventually, some of them are stabilized when they are "captured" by contacting the kinetochore of a chromosome, forming the familiar spindle fibers of mitotic metaphase. The kinetochore is described as acting like fly paper to capture the microtubules that have randomly collided with it, selectively stabilizing a functionally appropriate arrangement (Kirschner, 1992). Due to the adaptability of this process, specific pathways of morphogenesis do not need to be prescribed. The rapid turnover of microtubules in the mitotic cell generates many different configurations in a short time, most of which are unused.

This, like other mechanisms of somatic selection, appears on first sight to be an exceedingly inefficient process, yet it is eminently flexible and allows mitosis to start from many initial configurations, and to overcome experimentally imposed obstacles not normally encountered. As pointed out by Kirschner (1992), these properties could similarly accommodate evolutionary experiments in the configuration of dividing cells. Similar mechanisms of microtubule assembly exist in migrating cells and in developing nerves, allowing these cells to react and orient flexibly in response to extracellular signals (Mitchison and Kirschner, 1989, ex Kirschner, 1992). Gerhart and Kirschner (1997) state that vir-

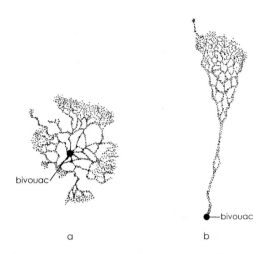

Fig. 3.6. Organization of predatory raids in an army ant (*Eciton burchelli*): (a) initial exploratory trails of hunting ants; (b) establishment of principal column following prey discovery. After Schneirla (1940).

tually every unusual and highly specialized intracellular structure in eukaryotes is based on some unique configuration of what were originally exploratory microfilaments, including microtubules, actin filaments, and intermediate filaments.

Individual Task Flexibility in Social Insect Colonies   In social insects, the multiple tasks of brood care, foraging, building, and defense are distributed in seemingly adaptive yet coordinated and condition-sensitive patterns. This occurs in very large colonies of ants and bees without central control or rigid individual programming (e.g., Grassé, 1959; Deneubourg et al., 1989; Gordon, 1996; Page and Mitchell, 1990, 1998; Bonabeau et al., 1997; Camazine et al., 1999). Characteristic group-level patterns emerge from the flexible behavior of interactive individuals engaging in initially randomly or chaotically variable patterns (Cole, 1991c). Variable movements are followed by conditional responses governed by a few simple rules. For example, species-specific patterns of army ant (*Eciton*) raiding columns emerge from a highly diverse initial array of pathways (figure 3.6) because (1) initially wandering foragers chemically mark both their outgoing and returning paths, (2) successful productive paths are marked more heavily and followed more frequently, and (3) individuals move toward a central trail if no guiding cue is before them (Deneubourg et al., 1989). In this way, differential use by individual searchers selects the successful pathways, abandons unsuccessful ones, and establishes a primary trunkway to the bivoac.

Insect colonies have a popular image of governance by blind, rigidly programmed instinct. In recent years this erroneous image has perhaps been reinforced in some peoples' minds through an identification of insect societies with genetic theories of sociality. But insect sociobiological theory from the beginning, and in contrast to many other fields within biology, had a prominent emphasis on conditional behavior where alternative tactics of reproduction, parent–offspring conflict, and reciprocity were shown to depend on cost–benefit evaluations rather than on genes alone (e.g., Hamilton, 1964a,b; West, 1967; Trivers, 1972, 1974; E. O. Wilson, 1971a, 1975).

Despite the claims of its detractors, sociobiology and behavioral ecology within evolutionary biology have taken the lead in studies of adaptive plasticity. It is now well known, for example, that the biased dispensation of aid to kin, basis of the so-called genetic evolution of altruism, in fact involves refined assessment of degree and probability of kinship and a corresponding adaptive adjustment of behavior in a very wide variety of organisms (chapter 23; Altmann et al., 1996). This was explicit in Hamilton's original (1964a,b) formulation, where relatedness, costs, and benefits were valued against each other in individual decisions regarding the performance of altruistic acts.

Additional examples of exploratory variant overproduction during development include exploratory tissue growth, and use and disuse in the formation of blood vessels and insect tracheae (reviewed in Gerhart and Kirschner, 1997). All of these mechanisms, involving an initial overproduction of variants or paths, followed by selection of those that work, share the quality of extreme flexibility in ability to adjust to unpredictably varied situations. Because the particular solution adopted is not genetically specified or transmitted to the next generation of individuals, each generation begins with its flexibility undiminished.

Hypotheses for the Evolution of Hypervariable Plasticity   All of these mechanisms of somatic selection have fostered evolutionary change or evolvability (see chapter 7; West-Eberhard, 1986, 1989, 1992a; Kirschner, 1992; Gerhart and Kirschner, 1997). But all of them can be explained as products of ordinary natural selection within populations, for all contribute to fitness-enhancing flexibility. Somatic selection clearly permits open responses (*sensu*

Mayr, 1974a), especially subject to conditional determination of multiple outcomes.

While this advantage in terms of flexibility would seem sufficient to explain somatic-selection mechanisms, there are several hypotheses to explain why certain ones are so widespread in organisms.

The Clade-Selection "Evolvability" Hypothesis Several authors (e.g., Cole, 1991a; Kirschner, 1992; Gerhart and Kirschner, 1997) have suggested that while the mechanisms of plasticity involving somatic selection may have originated under individual selection due to their immediate advantages, they may also have been favored by selection at higher levels such as the colony or clade, due to their contributions to evolutionary lability or "evolvability" itself. By this hypothesis, lineages with flexible mechanisms survived and multiplied differentially over the long course of evolutionary time due to their greater evolutionary lability.

The Cell-Lineage-Competition Hypothesis A recent theory regarding the origin of somatic selection is the cell-lineage-competition idea of Buss (1987), which holds that "the principal mechanisms of epigenesis invoke interactions between cell lineages such that one lineage furthers its relative rate of replication as a consequence of engaging in the interaction" (p. 96). Buss (p. 78) sees this pattern in higher Metazoans as a legacy from ancestral forms where, prior to sequestration of the germ line, genetically different cell lineages competed within cell groups, taken as proto-individuals, under true Darwinian selection affecting generational change in gene frequencies. This hypothesis is discussed in chapter 6, where I list several reasons for doubting the strength of this argument. Somatic selection in Metazoans is fundamentally non-Darwinian in nature, as noted by Buss and colleagues in a later publication (Fontana et al., 1994, p. 213).

The Nonadaptation Hypotheses Darwin (1880) hypothesized that the production of excess variants in some cases may not itself be an adaptation, or a product of selection at the individual level, but rather could be the outcome of noise or a side effect of development. Darwin considered that circumnutation "is so general, or rather so universal a phenomenon, that we cannot suppose it to have been gained for any special purpose. We must believe that it follows in some unknown way from the manner in which vegetable tissues grow" (p. 265). But Darwin did consider adaptive the *modifications* of circumnutation, such as its increased amplitude in the shoot tips of climbing plants, or the production of hooked tips upon contact with a support. Thus, in the terms of the present discussion, Darwin considered the variant overproduction represented by circumnutatory movements to be nonadaptive, but he considered that somatic selection among the patterns generated, whether "through innate causes or through the action of external conditions" (p. 265) could lead to special modifications of service to the plant and be favored by natural selection.

Some of these mechanisms may have originated and spread without having been selected as mechanisms of plasticity per se, as a side effect of selection for the particular materials themselves. Matter, whether organic or inorganic, has interactive properties—responsiveness, reactivity and interactivity, and the ability to be formed and deformed or broken by contact with other matter—properties that are themselves not necessarily evolved. When chemical and physical reactions and interactions happen within organisms, it is tempting to see them as genetically mandated or evolved, and when one bit of organic matter enlarges at the expense of another, we automatically think of Darwinian selection and genetically selfish manipulation. But this may be a misleading analogy. One reason for the persistent failure of biology to deal with the problem of nature and nurture, environment versus genes, may be that we have exaggerated the significance of the boundary between inorganic and organic reactions (see especially Kauffman, 1993).

Material interaction, movement, and compartmentalized activity likely characterized the intracellular and precellular molecular phenotype (see especially Fox, 1980, 1984). Cell-lineage interaction in Metazoan development may be not so much a legacy of ancient Darwinian battles between genetically distinctive cells (Buss, 1987) as a legacy, more ancient still, of the differentially interactive and responsive matter of which all life is composed. It is a legacy subject, of course, to gene-product influence.

Hormones

Hormones are signal molecules. They are produced in one part of an organism's body and diffuse or are carried, for example, by the blood in animals, to another part, where they elicit some developmental response or change in morphology, behavior, or physiological state. The effects of particular hormones are discussed in numerous chapters of this book. Here I describe some of the general prop-

erties of hormone systems that are important for evolution.

A *hormone system* refers to the signal–response arrangement involved in hormone action. It consists of some organ, such as the ovary, that produces a hormone, such as estrogen; a medium of transport, such as the blood; and one or more target tissues that respond to the presence of that particular hormone. This phenotypic complexity is important for evolution because it means that hormonal regulation of processes such as growth, sexual maturity, and behavior, is genetically complex, since it may involve organs and substances that are products of many genes. This means, in turn, that both hormone production and response sensitivity may be subject to polygenic variation and therefore subject to fine-tuned quantitative adjustments during evolution.

In addition, evolutionary adjustment of threshold responses can establish threshold effects that convert continuously variable or constant hormone titers (concentrations) into switches—discrete changes in the state of an organ or the rates of a process. Threshold adjustments can include limitation of sensitivity to a particular time or critical period or the amount of hormone present. Juvenile hormone control of differentiation in larval insects, for example, affects numerous binary developmental decisions, but there is no evidence that this involves varying concentrations of hormone. Instead, the response thresholds of different tissues vary (Nijhout, 1999b). Similar properties of environmental sensitivity and multiple effects characterize the hormone-like substances of plants, such as jasmonic acid (Creelman and Mullet, 1995).Thus, quantitative genetic variation can affect hormone systems in at least two different ways, by altering hormone output and by altering response sensitivity, thereby changing the threshold of a response. Threshold modulation enables the same hormone to adaptively influence many processes at once. The vast majority of developmental events in insect behavioral and morphological ontogeny are primarily controlled by just two hormones, ecdysteroids and juvenile hormones (Nijhout, 1999b, p. 218).

Alongside their sensitivity to quantitative genetic variation, hormone systems are often critically sensitive to environmental influence. Pituitary hormone production, for example, which controls gonad development, pelage change, and mating behavior in some mammals (see chapter 23), is sensitive to seasonal change in day length due to neuronal connections between the eyes, the brain (hypothalamus), and melatonin production by the pineal gland, which in turn influences the pituitary (Gorman et al., 2001). Since the links in the chain of control of a hormonally regulated trait are also subject to genetic variation, degree of environmental sensitivity (plasticity) can be adjusted under selection (Emlen, 1996, 2000).

Hormone systems are also coordinating mechanisms. A single organ, such as the pituitary just described, can affect the regulation of many different processes. Similarly, testosterone in birds is involved in regulation of sexual maturation, plumage coloration, parental behavior, and degree of aggressiveness (see chapter 15). And juvenile hormone controls a great variety of developmental events in insects (Nijhout, 1994a). Change in hormonal sensitivity of target organs can alter the temporal order of events, leading to heterochronic evolutionary change (see chapter 13). Hormonal changes can also alter the sex in which a particular trait is expressed, creating striking evolutionary novelties (see chapter 15).

Finally, hormones are of particular importance for development and evolution because their mode of action is to regulate gene expression. They can turn batteries of genes off and on as sets in what I call developmental linkage of traits (see chapter 4). By their influence on which genes are expressed together, hormones influence which genes are selected together (West-Eberhard, 1992a). In effect, hormone systems are a major developmental basis for genetic correlations among coexpressed traits.

For all of these reasons, hormones play a key role in the links among environmental effects, quantitative genetic variation, gene expression, selection, and evolution, and they are extensively discussed in the chapters that follow.

Homeostatic Mechanisms

The commonplace homeostatic physiology of plants and animals, which maintains the stability of the internal environment despite changing external environments, takes on a new look from an evolutionary point of view. The regulation of the internal environment in animals is achieved by mechanisms described by Nijhout (1994a, p. 25) as fluctuating more or less continuously as they adjust the animal's physiology to deal with current conditions. Usually, these workaday processes are associated with maintenance of the status quo. But it does not take much imagination to realize that they could also function to accommodate new conditions and structures as well as usual ones even though they are not especially evolved to do so. They would incidentally accommodate evolutionary change, acting immediately to buffer the phenotype from the effects of genetic or environmental stress-induced variations that initially may have both negative and positive effects on fitness. This phenotypic accommodation would give selection time for genetic accommodation, the evolved reduction of negative effects and enhancement of positive ones (see chapter 5).

In animals, the physiological processes that facilitate the phenotypic accommodation of change include mechanisms of temperature regulation, water balance, and immediate responses to predatory and social challenges, for example, the suite of responses evoked by a surge of adrenaline in an individual under attack. In plants, they include special metabolic enzyme systems allowing photosynthesis without desiccation in extreme conditions as in crassulacean acid metabolism (see chapter 17), stomatal opening and closing, mechanisms of photoinhibition, and thylakoid membrane organization within chloroplasts, which permits flexible use of different chloroplast photosynthetic pigments under different conditions (Salisbury and Ross, 1985; see below).

One need only consult a standard physiology text with an eye to accommodatory mechanisms to collect well-studied and beautiful examples of the remarkable mechanisms of flexibility. Here are some illustrations in plants (Salisbury and Ross, 1985):

Microfibril Plasticity in Plant Cell Walls   The primary wall of the plant cell is admirably adapted to growth. In response to growth-regulating chemicals, it softens in some way so that the microfibrils, cellulose molecules composing the skeleton of the wall, can slide past each other. In the terms used by Salisbury and Ross (1985, p. 6), the wall then stretches *plastically* (irreversibly, like bubble gum) rather than *elastically* (like a rubber balloon) as the cell grows. Some primary walls increase their area as much as 20 times during growth. Such plasticity could leave an indelible imprint on the morphology of the plant. This is one of the reasons why many botanists consider plants morphologically more plastic than animals (e.g., Bradshaw, 1965). Once the plant has stopped growing, this morphological plasticity of the cell wall is lost. But the rigid cell wall facilitates other kinds of physiological flexibility by providing support for the passages and membranes involved in fluid conduction, photosynthesis, and other processes (see below) that are the plant counterparts of animal behavior.

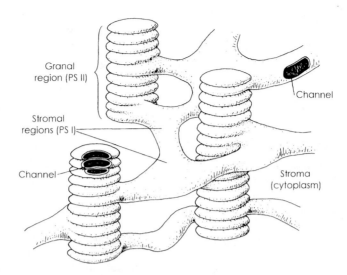

Fig. 3.7. Compartmentalized biochemistry in the chloroplast thylakoid membrane of a higher plant. Appressed granal regions contain photosystem II (PS II) complexes and their light-harvesting antennae (chlorophyll proteins); unappressed stromal regions contain photosystem I (PSI) complexes and ATP synthases. Granal–stromal proportions are facultatively varied in different light conditions. After Salisbury and Ross (1985).

Opening and Closing of Stomates   The movements of stomatal openings in the leaves of plants are a good example of adaptive responsiveness under local control and based on simple but ingenious biochemical mechanisms, without central plantwide coordination. Here, environmental variables—water and CO_2 availability—participate in regulation via feedback loops and become, in effect, elements of their own regulation.

Micromorphological Flexibility in Chloroplast Thylakoids and Other Internal Structures of Plants   In an article on "the grand design of photosynthesis," Anderson et al. (1995) discuss the exquisite acclimation of the photosynthetic apparatus to everchanging environmental stimuli. Light varies over a wide range of intensities and spectral qualities in full sun, early morning, and late evening and in cloudy and canopy shade. There are short sun flecks, long-lasting canopy gaps, and seasonal variations in light quality and quantity. Even within plants, there is a continuum of light intensity and quality experienced by different leaves, by different chloroplasts within leaves, and, within chloroplasts, by the stromal and the granal regions of the thylakoid membrane. Plants respond to these environmental variations with a multitude of coordinated signal cascades and networks not only to maintain efficient photosynthesis at low light levels, but also to avoid the adverse effects of excess

light. Their responses to extremes of low and high light are fully reversible, barring severe secondary effects. They include changes in photoreceptors that regulate gene expression at many levels during development, and metabolic signals derived from the physiological state of the plant (Anderson et al., 1995). But the most spectacular aspect of plant physiological flexibility from the viewpoint of an animal behaviorist is the extent to which the immediate responses of the plants involve *movements*. The first-defense avoidance mechanisms against high light, invoked to different extents in different species, include leaf movements, chloroplast movements, change in shape of chloroplasts, the lateral migration of phosphorylated compounds away from the photosynthetic apparatus to limit its overexcitation, and the physical folding of the thylakoid membrane (Anderson and Aro, 1994).

Among these mechanisms, the thylakoid apparatus (figure 3.7) is a particularly good illustration of the regulatory role of movements because it involves striking reversible changes in the conformation of a structure—the folding of a membrane network—within an organelle. In the photosynthetic systems of higher plants and green algae, granal regions of the thylakoid membrane contract relative to the stromal regions in sun-acclimated plants, and they accumulate photoinhibited photosystem II complexes with reduced light-harvesting antennae. The micromorphological movements modify biochemical interactions: the degree of folding of the membrane influences the degree of activation of processes associated with folded versus extended domains of the membrane network. Increased "stacking," or appression of the membrane grana, has been hypothesized to protect photosystem II from effects of high irradiance (Anderson and Aro, 1994).

As discussed in chapter 4, movements of the membrane affect gene expression as well (Anderson et al., 1995): only the unappressed domains are accessible to chloroplast ribosomes, which participate in expression (translation) of photosystem II reaction center proteins (Anderson and Aro, 1994). In sum, sustained high light causes plants to react with leaf movements, chloroplast movements, changes of chloroplast shape, and migration of light-harvesting chlorophylls away from the photosystem II complex, which limits its overexcitation (Anderson and Aro, 1994). Thus, while plants are rooted to one place and therefore do not show behavior in the sense of moving their bodies from place to place as do animals, they show considerable internal adaptive behavior. The thylakoid membrane illustrates modularity (compartmentalization of biochemical processes); connectedness, since the granal-stack and stromal domains are different regions of the same, continuous channeled membrane (figure 3.7); and adaptive plasticity in the form of environmentally influenced flexible morphology that affects biochemical interactions in ways favorable to the organism. Membrane folding also regulates gene expression.

Defensive Responses Among the most rapid flexible responses of plants is the ability to manufacture or release substances that deter the attacks of herbivores and pathogens (Adler and Harvell, 1990; Harvell, 1990; Ryan and Jagendorf, 1995). Conifers release large amounts of resin at sites attacked by fungus-bearing beetles, embalming their attackers at the site of the wound and simultaneously mobilizing defenses against infection by the pathogenic fungi. Some wounded plants release volatile substances that attract the parasitic enemies of their attackers, causing them to be attacked in turn. Others produce secondary metabolites that are toxic and distasteful to herbivores, and pathogenesis-related proteins that confer resistance to infections. Inducible plant defenses rival animal signaling systems in sophistication and complexity (Ryan and Jagendorf, 1995).

The customary view of physiology as homeostatic may overlook its role in the exaggeration, rather than the buffering, of phenotypic change. If provoked by recurrent or persistent environmental stressors, many homeostatic mechanisms adopt a new set point that in effect shifts the observed phenotypic norm to a new position, called "rheostasis" by Mrosovsky (1990). The physical appearance of plants grown in the shade, with their long, thin stems and pale leaves, is a familiar example of exaggerated form due to environmentally induced physiological responses. They are comparable in exaggeration and conditionality to the grotesquely exaggerated muscular morphology of some human body builders. Salisbury and Ross (1985) describe photomorphogenesis and the effects of light on germination, growth, and pigmentation in plants. The evolutionary role of such flexibility in the origin and accommodation of novelty is discussed in later chapters, especially chapter 16, which treats the origin of novelty by correlated shifts in continuously variable traits.

Developmental Plasticity and Coordination at the Molecular Level

Plasticity is one of the many phenomena for which it is artificial to draw a line between the genome and the phenotype. DNA is an active and adaptively responsive molecule that itself could be said to have a phenotype and a behavioral repertoire. This is illustrated by the so-called smart histospecific genes, segments of DNA that are hypothesized able to respond to multiple environmental and genomic inputs during embryogenesis (Davidson, 1990; figure 3.8). These genes have many attachment sites that enable them to be (1) responders to diverse inputs corresponding to the different binding sites indicated by the letters in figure 3.8, (2) activators, due to other sites that coordinate the expression of other genes, and (3) modulators, due to the action of other sites that inhibit or potentiate gene expression in particular circumstances. As a responder, the hypothetical smart gene of figure 3.8 is one of a set of genes that responds to input C (the cell cycle operator) and is also one of a different set that responds to T, the temporal regulator. The histospecific smart gene therefore has pleiotropic effects at the molecular level. The chicken β-globin gene, for example, has about 15 sites of DNA–protein interaction. Some of these sites bind temporal regulatory factors early in development, including erythrocyte-specific factors, amplitude modifiers, and negative regulators apparently used later in development. As an activator, the same gene can coordinate gene expression.

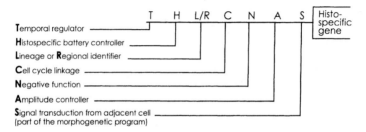

Fig. 3.8. Versatile condition-sensitive coordination and integration by a single gene. In the "smart" histospecific-gene model, each binding site (e.g., C, the cell cycle operator) is activated by a different input and then controls the expression of a particular set of genes (sometimes called a "gene battery") different from those controlled by other binding sites (T, H, L/R, N, A, S), resulting in different gene-expression combinations depending on different intracellular environments. After Davidson (1990).

Site H, the histospecifc battery controller, activates a gene battery required for its particular cell type. As a modulator of morphogenesis, site S, for signal transduction, could respond to a diffusible morphogen gradient, and N, for negative function, to a lateral inhibition factor that prevents two adjacent cells from expressing a gene that is used periodically in space. By their capacity to interact with a variety of molecules and perform multiple functions, such genes can potentially influence, or be activated by, many different factors or pathways, not all of them direct products of genes.

As pointed out long ago by Bonner (1974, p. 1), the earlier Britten-Davidson models, with "integrator genes" coordinating subunits of expression denoted "producer genes," gave a highly geneticized view of regulation. Bonner noted that there could be other, largely "cytoplasmic" means of coordination. The "smart" histospecific genes integrate signals from adjacent cells and cell-cycle states that are, in effect, part of their environments and whose presence or absence and state could conceivably reflect both genetic and nongenetic inputs affecting metabolism and growth.

Along with hormones, the "smart" genes offer one hypothetical mechanism, at the molecular level, for the interchangeability of genomic and environmental influences on development (see chapter 5). By virtue of their structure, they are versatile and integrative at the same time. Although Davidson (1990, p. 383) terms this a nonhierarchical model for the developmental control of gene expression, the versatility of the smart histospecific gene depends in part on its ability to assume both a controlling (activator) and a subordinate (responding) role in regulation. This regulatory scheme, like any other, is hierarchical in that there are elements that

are controllers and others that are modulated or controlled. The hierarchical role assumed by such a gene depends upon which binding sites are involved, so different levels of regulation can be occupied by the same molecule, depending on the context of its action. Similarly, a hormone system is both controller and controlled: it has a controlling and coordinating function through the hormones it produces, but at the same time its structure and function are influenced by various inputs from outside itself.

Alternative splicing—the formation of more than one mRNA from the same set of genomic exons via different pre-mRNA splicing patterns (reviewed by McKeown, 1992; figure 3.9)—is another example of phenotypic plasticity at the molecular level. Alternative splicing implies sensitivity of the splicing mechanism to differences in cell environments. Smith et al. (1990) hypothesize that "obvious candidates" as factors in the cell environments that might account for tissue- or stage-specific differences in splicing are snRNPs (small nuclear ribonuclear proteins), particles essential for pre-mRNA splicing and possibly varying in form among cell types. Another possibility is mutation in consensus sequences that influence splicing at boundaries of introns (e.g., those separating duplicated genes), which might then lead to duplicate copies being spliced in different ways either within or between cells (Sharp, 1994, p. 806; see also Smith et al., 1990, p. 173). The evolutionary consequences of alternative splicing are discussed in chapter 17.

Transposable elements, or transposons, are seen as bestowing both condition-sensitive and evolutionary flexibility on the genome (McClintock, 1984). Transposable elements are short segments

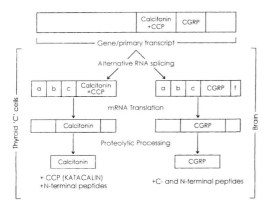

Fig. 3.9. Tissue-specific alternative splicing. Gene expression in eukaryotes often involves posttranscriptional splicing—the excision of primary transcript corresponding to introns, and the joining of exon products. In *alternative splicing*, primary exon transcripts are joined in different combinations under different conditions. In this example, alternative splicing from the same primary transcript (gene locus) produces calcitonin in thyroid C cells, and a different peptide (calcitonin-gene-related peptide, CGRP) in brain cells. After Thorndyke (1988).

of DNA that can insert themselves or be deleted at different positions in the genome and thereby alter gene expression. They are known to be activated in maize by various stresses to the genome, such as x-rays, ultraviolet rays, and growth in culture (Wessler, 1988, 1996). Noting their condition sensitivity, McClintock (1984) suggested that they may serve to generate diversity in life-threatening situations such as drought, extremes in temperature, or viral infection. Once activated, their impact is amplified because they can cause numerous nonautonomous elements in the genome to transpose, and this creates genetic diversity by adding nucleotides to coding or regulatory regions, altering proteins or tissue specificity of expression (Wessler, 1988). In plants, such somatic mutations can be inherited. There is evidence for inherited effects of "cut-and-paste" transposable elements in the human genome. A large fraction of the human genome is composed of interspersed repetitive sequences, most of which represent inactivated copies or "fossils" of transposable elements (Smit and Riggs, 1996). Although transposable elements can be horizontally transferred among species (Kidwell, 1993) and have many characteristics of selfish DNA, this interpretation does not exclude the possibility that they may

have beneficial developmental and evolutionary effects (Wessler, 1988).

Protein molecules participate in intracellular biochemical reactions as enzymes and catalysts, but this chemical-lab description belies their organism-like capacity for movement and interaction. These abilities are being revealed by new techniques for studying protein structure and behavior (e.g., see Wolynes, 1998). One of the best studied of the 100,000 known proteins is bacteriorhodopsin (BR), a molecule that functions as a light-driven protein pump, converting light into energy across a membrane in the halophilic archean *Halobacterium salinarum* (Réat et al., 1998; Frauenfelder and McMahon, 1998). The BR molecule consists of seven closely packed parallel α-helices that surround a central retinal region. Neutron spectroscopy and hydrogen–deuterium labeling that trace motions on a nano- to picosecond time scale have revealed the behavior of specific amino acids within the BR protein.

These studies suggest that structural flexibility is required for function and that amino acids clustered in the functionally central, retinal part of the protein are more rigid, with a smaller amplitude of motion, than are those of the periphery. Differential movement allows large conformational changes in peripheral regions of the molecule relative to the central region. Upon illumination, BR undergoes a catalytic cycle that causes protein translocation, followed by relaxation. Light is absorbed by the retinal part, and through conformational changes, a proton is transported from the cystolic to the extracellular side of the membrane. Myoglobin appears to function in a similar way: it has a relatively rigid central region containing the heme group with an iron atom at its center, and a relatively flexible moving periphery of seven α-helices. Motions are necessary for oxygen to enter the molecule for storage and transport, and it is the peripheral region that shows the greatest displacement or movement (see references in Frauenfelder and McMahon, 1998; Réat et al., 1998).

Another behaviorally agile molecule is the chaperonin GroEL, which mediates protein folding within bacterial cells (Ma and Karplus, 1998). This molecule has the shape of a cylinder with a large central cavity in which the polypeptide chain to be folded is sequestered and manipulated by changes of shape of the cavity. The cylinder is composed of rings containing seven subunits each, and the structure of the individual subunits facilitates the motions of bending and twisting required for the functioning of this device, which has been described as

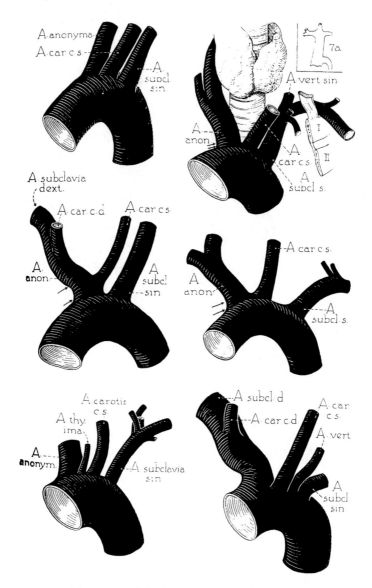

Fig. 3.10. Variation in arterial branches of the human aortic arch. Six common types of branching are shown. From McDonald and Anson (1940), *American Journal of Physical Anthropology*, © John Wiley and Sons, Inc. 1940. Reprinted by permission of Wiley-Liss, Inc., a subsidiary of John Wiley and Sons, Inc.

a "two-stroke motor" (Ma and Karplus, 1998, p. 8507). Thus, adaptive phenotypic plasticity and behavior occur even at the submolecular level, so behavior may take the lead in the evolution of molecular morphology as it sometimes does at higher levels of organization (see chapter 7).

Studies of protein plasticity, and its functional significance, are still in their infancy. But these examples allow one to imagine that changes in the amino acid composition of a protein could affect its flexibility and behavior. It is no wonder that genes, which specify and alter such responsive molecules, have such a profound influence on the phenotype and its evolution. And it is no wonder, given this potential for specialized movement and environmental manipulation, that specific environmental factors are as important to organisms as are their genes.

Phenotypic Accommodation

The Two-Legged Goat Effect in Development

Innovative phenotypes arise from preexisting phenotypes within developmentally variable populations. Fortunately for Darwinian arguments, variation is the norm in populations of organisms. But developmental variation in fitness-influencing traits has the potential to be disruptive. This has led to the idea of developmental homeostasis and evolved canalization, and the expectation that important traits will conform to a modal efficient design (see chapter 1). Biologists not actually engaged in the study of variation therefore may be surprised by the extreme variability observed even in vital organs such as the human heart and stomach. Variants that might be considered monstrous pathologies if seen in isolation occur as part of normal development in human populations (figures 3.10, 3.11). Yet development proceeds, and variant individuals survive without medical emergencies obviously related to these variations (Anson, 1951; Williams, 1956).

Phenotypic plasticity enables organisms to develop functional phenotypes despite variation and environmental change via *phenotypic accommodation*—adaptive mutual adjustment among variable parts during development without genetic change (term from West-Eberhard, 1998; see also Müller, 1989; Gerhart and Kirschner, 1997; Kirschner, 1992; Walker, 1996; West-Eberhard, 1989, 1992a). Phenotypic accommodation occurs regardless of the cause of variation, whether genetic or environmental, normal or pathological.

Abnormalities provide some of the best illustrations of phenotypic accommodation because they may be extreme, but the same principles are at work during normal development. Adaptive phenotypic accommodation enables individuals to maintain function despite unpredictable and sometimes large amounts of genetic variation in different aspects of morphology and physiology. So the much discussed stability of individual development, as well as the viability of higher level entities such as social groups and even ecosystems, may be due to adaptive phenotypic plasticity.

A dramatic example of phenotypic accommodation occurred in a famous developmental anomaly, a congenitally two-legged goat (Slijper, 1942a,b). A domestic goat born without forelegs adopted semi-upright posture and bipedal locomotion from the time of its birth. By the time it died due to an accident at the age of one year, it had

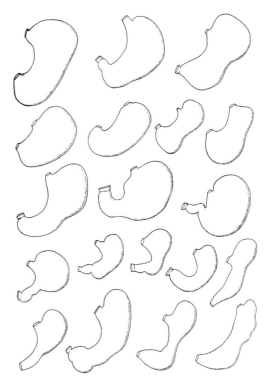

Fig. 3.11. Variation in form of the human stomach. From Anson (1951), by permission of W. B. Saunders Company.

developed several behavioral and morphological specializations similar to those of kangaroos and other bipedal mammals, including the ability to hop rapidly when disturbed, enlarged hind legs, a curved spine, and an unusually large neck. That such extreme accommodation can promote survival in nature was observed in a troop of baboons stricken by polio (W. J. Hamilton III, personal communication). One female, her front legs paralyzed, adopted semi-upright posture and bipedal locomotion similar to that of the two-legged goat (figure 3.12). A film loaned to me by Hamilton shows her walking semi-upright, then running with alternating steps, then racing forward with a series of strong, graceful running leaps like those of a broad jumper, always rising and landing with the right hind leg first, then pushing off strongly with the left.

Postmortem dissections of the two-legged goat revealed profound changes in the skeleton and muscle insertions compared to a normal goat of the same age, including changes in the hind leg bones, pelvic skeleton and musculature, and thoracic skeleton (figure 3.13). All of these changes are well-

Fig. 3.12. Phenotypic accommodation in behavior: compensatory upright posture and jumping locomotion of a bipedal chacma baboon (*Papio ursinus*). The female shown had undeveloped front legs used to stand and groom, but walked and jumped (as shown) using only the hind legs. Like the injuries of many other crippled individuals in her troop of 39 baboons, the withered condition of her front legs resembled polio injuries to humans. In 25 years of subsequent observations, no such injuries recurred. From film and observations by William H. Hamilton III, in the Moremi Game Reserve, Botswana, 1972–1973.

known consequences of increased muscle and bone loading and stress (Slijper, 1942a,b). Some of these novel features, such as the dorsoventrally flattened thorax and the long ischium (figure 3.13c), are very similar to those found in kangaroos, while the wide sternum (figure 3.13c) resembles that of an orangutang, a primate that walks upright and, like the bipedal goat, lacks a tail for support (Slijper, 1942a,b; Radinsky, 1987).

The point of the two-legged goat and baboon examples is not to argue that these handicapped individuals might have given rise to durable novelties, for it is unlikely in the extreme that such individuals would outperform normal individuals in nature. Rather, the point is to dramatize how a change in one aspect of the phenotype—in this case the front legs—can lead to correlated changes that show a degree of complexity and functional integration that we usually assume to require generations of natural selection and genetic change at many loci. Clearly, the phenotype can be elabo-

rately restructured, due largely to adaptive plasticity, without a proportional restructuring of the genome (in the case of the two-legged baboon, the change is thought to have been environmentally induced by disease; it likely involved no genomic change at all). The exaggeration of change by plasticity may help to explain otherwise puzzling facts, such as the small genetic distances between phenotypically very divergent species such as humans and chimpanzees (King and Wilson, 1975) and the general lack of correspondence between morphological evolution and the evolution of the genome (John and Miklos, 1988).

Phenotypic accommodation often begins with behavior. The morphological changes in skeletal and muscle structure of the two-legged goat are kinds of phenotypic change that can be produced through use (see section on development of muscle and bone, above). In an entomological version of the two-legged goat phenomenon, two female digger wasps with slightly defective wings, of a species

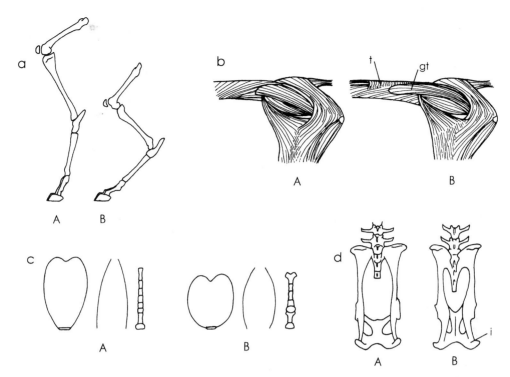

Fig. 3.13. Morphological innovation in a two-legged goat: development of (a) left hind leg, (b) pelvic musculature, (c) thoracic skeleton, and (d) pelvic skeleton of a normal (A) and a bipedal (B) specimen of the same age. Changes in the pelvic musculature (b) include a greatly thickened and elongated gluteal tongue (gt), whose anterior attachments are reinforced by the novel addition of numerous long, flat tendons (t). The form of the thoracic skeleton (c) is represented by a transverse section of the cranial part of the thorax (left), a horizontal section at the middle of the ribs (middle), and a ventral view of the sternum (right). The relatively long ischium (i) of the pelvis (d) and other changes (see text) resemble those of bipedal mammal species. After Slijper (1942a,b).

(*Sphex ichneumoneus*, Sphecidae) that usually preys upon katydids, switched to a smaller prey, tree crickets, which they could handle despite their handicaps (Brockmann, 1985a). This implies novel hunting tactics and habitat, and the defective wasps also provisioned a larger number of prey, compensating for the smaller prey size. This species shows impressive plasticity in many aspects of its provisioning and nesting behavior, including seasonal change in prey type evidently governed by change in prey availability (Brockmann, 1985a), and tool use (tamping the soil with a stone during nest closure), not an uncommon phenomenon in wasps (Brockmann, 1985b).

The *two-legged goat effect*—the exaggeration and accommodation of phenotypic novelties via adaptive plasticity—is a widespread phenomenon. Rachootin and Thomson (1981, p. 184) relate this capacity for accommodation to "a spectrum of self-organizing properties," which would include the kinds of mechanisms discussed in this chapter. Many examples of self-organization or hypervariability and somatic selection (see above) represent mechanisms of phenotypic accommodation, where fine-grained local responses by tiny elements of the phenotype achieve organized configurations that are exquisitely adjusted to internal conditions (Kirschner, 1992; see also below). Other authors (e.g., Galis, 1993a; Galis et al., 1994) refer to flexibility as a kind of "reserve capacity" for functional variation that accommodates novelties in both development and evolution.

Alberch (1982) gives many examples of phenotypic accommodation in vertebrate development. He cites a classical experimental demonstration by Twitty (1932): when the eye of a large salamander species is transplanted into the embryonic skull of a smaller species, the eye grows to its genetically determined large size, and the host develops a proportionally larger cartilaginous optic cup to perfectly accommodate the larger eye, as well as a concordant change in the tectal neuron population of the midbrain corresponding to an increase in the number of retinal ganglion cells associated with the grafted eye. Müller (1989) points out that many techniques in experimental embryology create abnormal phenotypes comparable to evolutionary novelties. Most of these are far more drastic abrupt changes than those found in nature, yet they are phenotypically accommodated.

Frazzetta (1975) discusses how "animals experiencing sudden changes . . . are buffered through compensations from related systems" (p. 150). His examples are particularly compelling because his favorite subject is bone, a tissue we think of as among the most rigid in nature. In fact, bone growth involves hyperflexibility via somatic selection, as discussed below, and some striking evolutionary innovations have originated due to correlated responses to stress involving bone (see chapter 16).

Unexpected confirmation of the power of phenotypic accommodation to compensate genetic defects has come from recent experiments in genetic engineering that use *genetic knockouts*, blockage of the expression of specific genes in order to study their effects on complex behaviors (e.g., Wilson and Tonegawa, 1997). Researchers have discovered that knockout effects are sometimes compensated by subsequent development, so knockouts have to be timed so that they do not occur too early in development, allowing the organism to respond in ways that obscure the effects of the gene (U. Mueller, personal communication). This demonstrates the power of phenotypic plasticity to accommodate genetic change, and a potential for extensive connectedness among gene effects (see Greenspan, 2001). The knockout technique would be potentially useful for the study of phenotypic accommodation as well as for the study of individual gene action.

The Two-Legged Goat Effect in Evolution

There is abundant evidence that the two-legged goat effect is important in evolution, in the phenotypic accommodation of novel traits and of potentially disruptive variants that occur during development (see especially Gerhart and Kirschner, 1997; Kirschner and Gerhart, 1998; chapter 16). Several authors have written of the phenotypic accommodation of evolutionary novelties, calling it

by various names: compensation (Darwin, 1859; Frazzettta, 1975), functional adaptation (Roux, 1895) epigenetic regulation (Alberch, 1982), and epigenetic accommodation (Rachootin and Thomson, 1981, p. 181).

Experimental studies of plasticity suggest that inducible changes in morphology similar in kind and exaggeration to those discussed here as anomalies, are not uncommon for established traits, and there are species where such plasticity has become incorporated into normal development. Various species of fish, insects, and birds show continuous plasticity in trophic morphology depending on diet. Cichlid fish (*Cichlasoma managuense*) fed relatively large prey developed a more pointed jaw, whereas those fed commercial flakes and nematodes had more rounded jaws (Meyer, 1987a,b), and the differences resemble those among related species having different diets. Grasshoppers fed soft versus hard leaves also develop differently shaped mandibles (Thompson, 1988, 1992), and caterpillars (*Pseudaletia unipuncta*, Lepidoptera) fed on tough grasses, compared with others reared on soft food, showed greater positive allometric growth of the head (Bernays, 1991, p. 84; see Bernays, 1986). The contrasting extremes produced under experimental phenotypic engineering (Ketterson and Nolan, 1992) are products of manipulating what is apparently continuous variation. If dietary extremes were simultaneously and constantly present, such plasticity could lead to a bimodality of structure without intermediates. It has been suggested that this kind of plasticity could have contributed to the evolution of trophic polymorphisms in the famous species flocks of African Lake cichlids (Meyer, 1987a; see also West-Eberhard, 1986, 1989; chapter 28, on adaptive radiations).

The Evolutionary Importance of Mechanisms

It has sometimes been claimed that for an adequate theory of plasticity and evolution, knowledge of mechanisms is unnecessary (e.g., Dunbar, 1982). Via (1993b), for example, points out that "quantitative genetic models are not meant to be mechanistic; they are blind to the precise genetic mechanisms that generated the patterns of genetic variance and covariance among characters" (p. 374).

This might be an acceptable approach if theory could remain "pure theory," in clear isolation from mechanisms and without inspiring imagined substitutes for them. But problems arise when the black box of mechanism is filled with imaginary devices, as when chromosomal linkage and balancing selection were assumed to underlie the coordinated expression and evolution of polyphenisms (see chapter 4). In the absence of knowledge of mechanism, imaginary mechanisms are virtually inevitable, because one criterion of theoretical plausibility is mechanistic feasibility.

It clearly matters, for example, how biologists imagine the roles of genes. By imagining that there is a one-to-one relationship between genes and phenotypes, biologists lost sight of the role of the environment in the development and evolution of phenotypes. By imagining that there are two types of genes, trait (structural) genes and regulatory genes, they may lose sight of the fact that the same gene can play either role and that the role assigned depends on the level of analysis. If we imagine that variation in quantitative traits is simple addition of numbers of trait genes or quantitative trait loci, the relationship between hormone titers and growth rates in the evolution of an allometric trait is obscured. Degree of development of a quantitative trait may depend not on numbers of genes or some aspect of their expression, but on amount of raw material available in the environment, or on the efficiency of an organism in sequestering it. The many genes that could influence foraging behavior—senses, responses, memory, ability to climb, and so on—could all be counted among the trait genes "for" the quantitative trait.

When mechanisms are left to the imagination, a confusing and contradictory collection of imagined mechanisms may result. Thus, for example, Via (1993b) relates reaction norms to development by reference to trait genes and regulatory genes, whereas Scheiner and Lyman (1991) and others (e.g., Schlicting and Pigliucci, 1993, 1998) refer to plasticity genes even while Via (1993a) doubts that plasticity per se can evolve. Are the regulatory genes of Via synonymous with her "modifier loci with environment-specific expression," which Scheiner (1993b, p. 371) sees as synonymous with his "plasticity genes"? Via et al. (1995, p. 215) note that those interested in "switch genes" have little interest in quantitative genetic models. Yet regulation by switches usually involves polygenic—quantitatively genetically variable—mechanisms (Falconer and Mackay, 1996; see chapter 4). As a result, the same models can be applied to regulation of both qualitative (threshold-determined) and quantitative (also threshold-determined) variation (Scheiner, 1993b; see chapter 4). An attempt to reach a consensus (Via et al., 1995) is a synthesis of confusion. For these reasons, this chapter has focused on concrete mechanisms of plasticity, and the next chapter describes how the mechanisms of plasticity give rise to both continuous and discrete variation in traits.

4

Modularity

Modular organization, like plasticity, is a universal property of phenotypes, the result of the universally branching nature of development. Switches divide the phenotype into somewhat discrete and semidissociable subunits of structure and gene expression or gene-product use at all levels of organization; even the genome has a modular structure that reflects patterns of expression and intragenomic activity (e.g., of exons, and transposable elements). Subunits of gene expression are also subunits of dissociation, selection, and evolution. Modular organization leads to coselection of coexpressed traits and the evolution of genetic correlations among them. The notion of the phenotype as a nested hierarchy of modular subunits implies both semi-independence and connectedness among subunits. Cohesiveness and discreteness are complementary properties of modular organization. Evolutionary increase in modularity can increase the flexibility and versatility of parts. Extreme modularity, as in the somatic sequestration of cells in arthropod imaginal discs, leads to extreme developmental independence of the parts and, consequently, to a capacity for extreme specialization of traits. The concept of units of selection needs to be revised to include subunits of selection (fitness effects) below the level of the individual but above the level of the gene, and to include different degrees of unit consolidation and independence under selection.

Introduction

Modularity, like the responsiveness that gives rise to it during development and evolution, is a universal property of living things and a fundamental determinant of how they evolve. *Modularity* refers to the properties of discreteness and dissociability among parts and integration within parts. There are many other words for the same thing, such as atomization (Wagner, 1995), individualization (Larson and Losos, 1996), autonomy (Nijhout, 1991b), dislocation (Schwanwitsch, 1924), decomposability (Wimsatt, 1981), discontinuity (Alberch, 1982), gene nets (Bonner, 1988), subunit organization (West-Eberhard, 1992a, 1996), compartments or compartmentation (Garcia-Bellido et al., 1979; Zuckerkandl, 1994; Maynard-Smith and Szathmáry, 1995; Kirschner and Gerhart, 1998), and compartmentalization (Gerhart and Kirschner, 1997).

One purpose of this chapter is to give consistent operational meaning to the concept of modularity in organisms. Seger and Stubblefield (1996, p. 118)

note that organisms show "natural planes of cleavage" among organ systems, biochemical pathways, life stages, and behaviors that allow independent selection of different ones. They ask, "What determines where these planes of cleavage are located" and suggest that a "theory of organic articulations" may give insight into the laws of correlation, without specifying what the laws of articulation may be. Wagner (1995, p. 282) recognizes the importance of modularity and proposes a "building block" concept of homology where structural units often correspond to units of function, but concludes (after Rosenberg, 1985) that "there exists no way to distinguish an adequate from an inadequate atomization of the organisms."

Here I propose that modularity has a specific developmental basis (see also West-Eberhard, 1989, 1992a, 1996; see also Larson and Losos, 1996). *Modular traits* are subunits of the phenotype that are determined by the switches or decision points that organize development, whether of morphology, physiology, or behavior. Development can be seen as a branching series of decision points (John

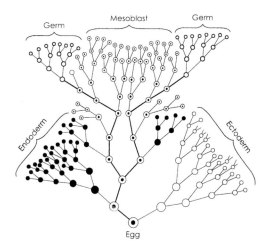

Germ   Mesoblast   Germ

Endoderm   Ectoderm

Egg

Fig. 4.1. The branching nature of development. Development has long been visualized as a branched sequence of decision points. In Weismann's diagram of the germ track of the nematode *Rhabditis* (*Caenorhabditis*) *nigrovenosa*, successive branches lead to major cell types (denoted by different types of circles and dots, and labeled after the 12th generation of cell divisions). The cells of the germ track (black nuclei) are connected by a heavy line. Each distinctive set of cells demarcated by a particular branch point constitutes a modular subunit of the developing phenotype. Subordinate branches determine lower level modular traits (hierarchical organization based on branching). After Weismann (1893) as reproduced in Gilbert (1991).

and Miklos, 1988; Raff, 1996; figure 4.1), including those caused by physical borders such as membranes or contact zones of growing or diffusing parts (e.g., see Meinhardt, 1982; see also chapter 5, on development). Each decision point demarcates the expression or use of a trait—a modular set—and subordinate branches demarcate lower level modular subunits, producing modular sets within modular sets (hierarchical organization of development; figure 4.1). Because these units of expression and use, which we call "traits," are subunits of natural selection, they become well-integrated functional subunits as well (chapter 7). Modular traits are more fundamentally defined by their developmental determination by switches than by their adaptedness or functional integration (cf. Wagner, 1996), for modular expression as a developmental innovation necessarily precedes functional integration and adaptation during evolution. This is evident from the fact that a variant must exist before it can be subject to selection. Of course,

selection may increase the degree of modularity—the degree of internal integration and coordinated expression of an adaptive trait. But modularity itself may or may not be adaptive, and it can occur without having been itself an object of selection. As discussed in chapter 7, coexpression leads to coevolution and cofunctionality, not the reverse.

Needham (1933) pioneered use of a modularity concept to describe the semi-independence and dissociability of such physiological and developmental processes as metabolism, enzymatic reactions, differentiation (including metamorphosis), and growth. He emphasized their disengagement, a word inspired by automobile parts ("they can be dissociated experimentally or thrown out of gear with one another"; p. 180). Dissociability is one of the important consequences of modular structure that derives from control of development by switches, broadly defined (see below).

The modular nature of phenotype structure is implicit in many kinds of biological analyses, such as taxonomic descriptions that deal with "traits" or with processes, such as seasonal molting or migration, that are turned on and off. In a recent discussion of homology, Wagner (1989a), refers to "parts" or "building blocks" of the body, traits that are found in more than one species, are morphologically well defined, and possess "a certain degree of developmental individuality" or "epigenetic autonomy" (see also Roth, 1991; Hall, 1994; Wenzel, 1992a; Raff, 1996).

While it is accurate to say that subdivisions of the phenotype have modular properties, it is potentially misleading to think of the phenotype as composed of modules, as if it were a collection of rigid independent parts. In some cases a modular subunit—a cell, a molecule, a leaf, a heart, a kidney, a tooth—can be dissected free, removed, transplanted, or replaced with a grafted counterpart from another individual. This is evidence that the perceived subdivisions of the phenotype are structurally, and often functionally, semi-independent. But it would invite confusion to describe the phenotype as composed of modules like the pieces of a jigsaw puzzle (as in Riedl, 1978, p. 112), for this would encourage neglect of the connectedness among parts. Organisms are only "partly compartmented" (Seger and Stubblefield, 1996). Limitations of the modularity concept, in view of the connectedness among phenotypic subdivisions, are discussed in a special section (below).

Here I am using a very broad concept of switches as organizers of modularity. Modular subdivisions of the phenotype may be sequential, as in life stages separated by metamorphosis. In others, they may

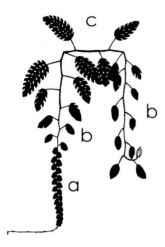

Fig. 4.2. Life-stage modularity and metamorphosis in a plant. Many plants switch between leaf forms during ontogeny (heteromorphy). The tropical vine *Monstera dubia* shows three leaf forms (a–c) during growth and may express an early-stage form (b) later in ontogeny. After Madison (1977).

be spatial, as in the position effects and inductions that are the switch points of morphogenesis. Metamorphosis occurs in invertebrates, vertebrates, and plants, such as those that have sharp ontogenetic changes in leaf morphology (Ray, 1990; figure 4.2). Cells are obviously modular, in being bounded by membranes and cell walls and in manifesting differences in gene expression (see below). In higher plants and green algae, a striking case of micromodularity occurs in the thylakoid membranes of chloroplasts, where a continuous membrane network is differentiated by assumption of regular domains of appressed (granal stacks) and nonappressed membranes. In morphogenesis a boundary between traits can be created by contact following diffusion or growth from two centers. Such a boundary represents a switch (a local stopping point) in the broad concept of a switch used here. In some cases, discrete modes equivalent to switch-controlled distributions are caused in natural populations by environmental discontinuities that influence a population of continuously plastic organisms such that only the extreme or quantitatively separate modal values are expressed (see chapter 11). This statistical modularity of continuous variation (called "continuous polymorphism" by Kennedy, 1961) is equivalent to switch-determined modularity in its ability to produce phenotypically discrete forms that are indepen-

dently subject to selection (Windig, 1993; see chapter 7).

Modular traits are recognizable by such qualities as (1) recurrence in time or in space of the same set of elements together, indicating that they are regulated (off, on, or rate change) as a set; (2) temporal or spatial discreteness relative to other structures or behaviors; (3) stereotypy of form; (4) coordinated expression as a unit by a known mechanism or regulatory factor that, when experimentally manipulated, affects the trait as a unit; (5) dissociability, or ability to be deleted or reexpressed as a unit; and (6) occurrence, with the same structure and location, in different individuals of the same species or higher taxon.

Recognition that the boundaries between modular traits are developmentally defined by switches and other developmental processes (below) strengthens the concept of modular organization by giving it a mechanistic basis that can be applied at all levels of organization. The massive evidence that switch-controlled traits are rearranged during evolution (discussed in part II) also supports this concept. It is clear that the phenotype has internal structure. An individual is not accurately described as a network, where every trait can be said to affect every other trait. Nor is it a single block of intergrading quantitative traits whose boundaries are just arbitrary points for convenient measurement. Developmental decision points are real, and they subdivide the phenotype into bounded, dissociable, genetically distinctive parts.

Modularity as Plasticity

Modularity contributes to phenotypic plasticity along with the mechanisms discussed in chapter 3. The modular organization of molecules, for example, permits somatic recombination in the production of variants by the immune system and during gene expression, when facultative recombination of exon products occurs via alternative splicing (see chapter 17). The functional versatility of particular molecules is due to modular submolecular structure, providing more than one discrete active site for interaction with other substances (Laoide et al., 1993; Davidson, 1990; see figure 3.8). Similarly, the boundaries around cells and intracellular organelles permit segregation of substances and functions and thereby permit diversity and divisions of labor among the responses of parts (Gerhart and Kirschner, 1997). And metamorphosis in plants and animals enhances developmental plasticity by

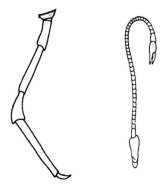

Fig. 4.3. Modularity as plasticity. Increased modularity and plasticity by multiplication of subunits (segments) in the legs of shrimp (Crustacea). Left, a simple walking leg; right, a multiarticulate leg. After Schmitt (1965).

allowing a separation of functions between stages devoted to growth and reproduction (Ray, 1990; Truman and Riddiford, 1999).

Morphological modularity increases morphological and behavioral flexibility as well. A multisegmented body or limb, for example, is more flexible than a unitary or less segmented one (figure 4.3). With the *goose-neck lamp principle* of design, increasingly subdivided modularity yields an approach to continuously variable flexibility based on discrete parts. Modular parts, furthermore, can locally respond to different contingencies and engage in different tasks. Frazzetta (1975) emphasized this, using as an example the innovative hinged jaw of a snake, where increased flexibility results from an increase from one subunit to two. Increased sophistication and diversity of animal behavior are associated with not only increased brain size but also an increasing number of brain subunits (Bossomaier and Snoad, 1995). The same principle applies to behavior, where modularity in behavioral sequences permits a great diversity of combinations to characterize the flexible repertoire of an individual (figure 4.4).

Extreme modular flexibility is found in the mechanisms sometimes called self-organizing (Kauffman, 1993; Gerhart and Kirschner, 1997; see chapter 3). In self-organization, the phenotype does not really organize itself. Rather, organization is highly flexible and locally responsive because a large number of modular subunits respond individually to local conditions according to simple, shared decision (switching) rules.

Ordinary phenotypic variation, as in crosses between unlike physiognomies in humans or between contrasting breeds of dogs, can result in outlandish combinations of traits. Yet the hybrid phenotypes hold together to accommodate a bizarre array of sizes and shapes of eyes, noses, jaws, teeth, ears, and heads, usually without serious functional failure. Modularity facilitates independent variation of parts, in effect marking boundaries around subspaces within which variation can occur without correspondingly large effects on neighboring parts. This facilitates semi-independent variation in such component structures as noses, jaws, and eyes within the coherent larger trait of the face (figure 4.5). Many homeostatic physiological devices that are continuously variable stabilizers of the phenotype can at the same time facilitate the phenotypic accommodation of evolutionary change.

Thomson (1987, pp. 205–206) notes that plasticity is a *general* function of compartmentalization, using as an example the contribution of compart-

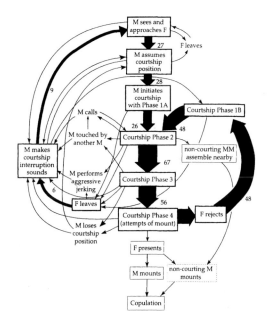

Fig. 4.4. Flexible courtship sequences in male of a grasshopper (*Syrbula admirabilis*). The modular organization of courtship as a number of somewhat independent phases permits condition-sensitive flexibility in the performance of courtship sequences. Numbers near arrows and thickness of arrows indicate frequency of pathway. From Otte (1972), by permission of Brill NV, Publishers. Courtesy of D. Otte.

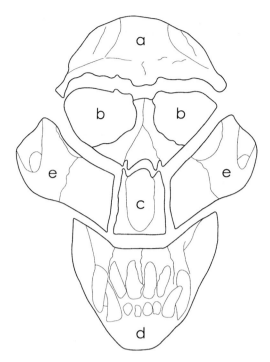

Fig. 4.5. Modularity, plasticity, and integration in the primate face. Skeletal subunits a–e are associated with the functional components of the face. Each lettered component is a developmentally and functionally related trait, and the facial components are highly integrated even while each is semi-independently variable. Integration of functionally and developmentally semi-independent components is likely facilitated by the plasticity of contiguous subunits of bone during growth. After Cheverud (1982).

1992) (figure 4.6). Not so familiar are its consequences for evolution. Hierarchical structure means that each trait is a decomposable mosaic of parts that are semi-independent in their regulation and function (Atchley and Hall, 1991). An example of internal modularity in a highly integrated trait, the a spider's web, is shown in figure 4.7. Chapter 5 provides further examples of independence and integration via hierarchical organization and discusses some common arrangements of switches and how they can be dissected into component switches corresponding to the smaller subunits of a mosaic trait. Subunits at any level can be developmentally deleted, multiplied, or shifted in the timing of their expression relative to others—at least, this is developmentally possible, though its effect on function is another question. Thus, hierarchical organization produces mosaic traits whose components have the potential to be reorganized somewhat independently of each other during development and evolution.

Failure to appreciate the hierarchical, mosaic nature of the phenotype has led to unnecessary controversy in some discussions of homology and evolutionary change, because homologous traits may change in some respects and not others, rather than behaving as a unit. As a result, homology is not an all-or-nothing phenomenon but rather a continuum of more or less similarity due to common ancestry (see chapter 25).

Integration, or coordinated development and function, is another facet of modularity. It may seem

mentalization to the interactive division of labor between the sexes. Similarly, the organization of metabolic cycles as a chain of semi-independent steps is a kind of physiological modularity that allows facultative interruption and feedback in what might otherwise be an unbreakable circular cascade.

Hierarchy and Integration as Aspects of Modularity

Hierarchical organization—an arrangement of parts within larger units composed of numerous smaller ones in nested sets—is an important aspect of modularity (Larsen, 1997). Hierarchical organization is a familiar property of organisms (Hall,

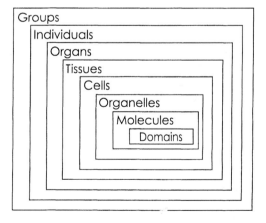

Fig. 4.6. Commonly recognized modular subunits at different levels of organization, showing their hierarchical relationships. The integration of a higher level implies the coordinated development, expression, or use of its lower level components.

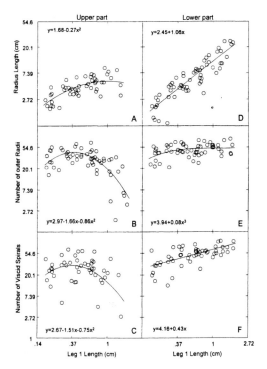

Fig. 4.7. Independent ontogenetic trajectories in independently expressed components of a highly integrated trait, the web of a spider, *Nephilengys cruentata* (Araneae: Tetragnathidae). As spiders grow (indicated by change in leg length), the dimensions of six measured web characteristics change, with the upper web region (A–C) showing quadratic trends (indicating decreasing growth rates), and the lower (D–F) parts of the web usually following linear trends. Subcomponents of the upper region have nonsimultaneous turning points, indicating further independence within regions (hierarchical organization of modular behaviors). From Japyassú and Ades (1998), with permission of Brill NV, publishers.

contradictory to attribute both dissociability and integration to the same cause. But switch mechanisms coordinate modular expression, and switches can be modulated at every level so that both integration and dissociability can be graduated up and down at particular points. This is discussed in a section on the anatomy of switch-controlled traits (below).

Wenzel (1993) points out that coordinated expression of separate units, as in the phenotypic elements controlled by a switch, results in a unique combination having emergent properties not evident from a reductionistic study of the parts in isolation. The stereotyped behavior of a rooster pecking at the ground, for example, may be part of a feeding sequence or part of a courtship display. The two sequences could be composed entirely of movements performed separately in other contexts. Yet each as a distinctive set has an emergent cohesive and functional quality of its own. Switches integrate the semi-independent subunits into a higher order coordinated trait.

One way to express the integration of modular traits is with reference to genetic and phenotypic correlations. Correlations among quantitative traits influenced by a common switch are statistical evidence of integration by switches. This approach was well development by Russian morphologists in the 1930s (see references in Ezhikov, 1934). One way to detect a switch-mediated phenotypic subunit and document that it is the product of a distinctive developmental pathway is to examine the independence of its correlation structure relative to other traits. Even when individual subunits show overlap in their quantitative dimensions, they may show lack of overlap in their correlation structure (Ezhikov, 1934), indicating that they represent developmentally independent traits.

The idea of switch-mediated integration is important for several concepts that relate development to evolution. One is the concept of phenotypic subunits as semi-independent subunits of selection (see below). Another is the relationship between modularity and connectedness in perceptions of how phenotypes are organized. There is always tension between the complementary effects of discreteness and continuity under selection, and this tension is reflected in theories of development and evolution, which have long centered on issues of gradualness and saltation and the relative importance of discrete versus continuous variation (see chapter 24). How a hierarchical view of the phenotype helps to resolve this tension is discussed at the end of this chapter. The emergent qualities of different levels of organization are one reason why biology needs to be studied at different levels, and why molecular biology, cell biology, or genetics alone cannot solve all of the important questions of evolutionary biology (see also Frankel, 1992).

Application of the Modularity Concept at Different Levels of Organization

Modularity is an aspect of phenotype organization at all levels, from the amino acid residues that compose a protein (figure 4.8), through the separation

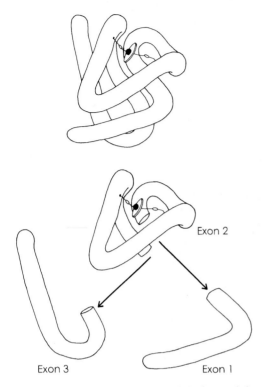

Exon 2

Exon 3   Exon 1

Fig. 4.8. Submolecular structure of the hemoglobin subunit. The three components are developmentally distinctive in that each is the product of an exon, and functionally distinctive in that the exon 2 product has the ability to bind to the heme group, whereas the other two exon products are required for reversible binding to oxygen. After Holland and Blake (1990).

of functions within and between cells (Anderson and Aro, 1994; Gerhart and Kirschner, 1997), the segmentation of body parts (García-Bellido et al., 1979), and other aspects of animal morphology (Riedl, 1978; Raff and Kaufman, 1983; Wagner and Altenberg, 1996), to the organization behavior and of societies with divisions of labor among individuals (Raff and Kaufman, 1983). Raff (1996) calls modularity "the basic pattern of order characteristic of organisms" (p. 325). Parkinson and Kofoid (1992) apply the concept, including the dissociability and evolutionary aspects, to the components of bacterial signaling proteins. Although the idea of protein subunits, or domains, as fundamental units of structure and function is not universally accepted for prokaryotes, domain structure is now considered the preeminent mode of organization of eukaryote proteins (Schultz et al., 1998; see chapter 17).

Although I focus here on the phenotype, defined (in chapter 2) as all features of organisms outside of the genome, the genome can be seen as having a phenotype of its own. The genomic phenotype is like the phenotype at all other levels of organization in having dynamic responsiveness (plasticity) and modularity (e.g., see especially Federoff, 1999; chapter 17). Comparisons of DNA structures within and between species in both prokaryotes and eukaryotes have revealed an underlying mosaic, or modular, structure to all genomes (Shapiro, 1992). This is evidenced in the behavior of chromosomes during crossing over, in which homologous chromosomal segments are recognized; and in gene transcription, also a segment-by-segment process where units of gene expression have borders in the form of promoter regions that determine where transcription begins and ends (for concise descriptions of these fundamental processes, see Futuyma, 1998). Promoters, in effect, function as switches, so the delineation of subunits of modular structure by switches applies to the genome level as well as to the phenotype at higher levels of organization. Noncoding regions of the genome also have modular structure. About 50% of the human genome, for example, is composed of repeated sequences, the majority of them products of the activity of transposable elements (Consortium, 2001; see chapter 17). One interpretation of the evolutionary significance of transposable elements is that they modularize the genome and thereby contribute to its plasticity and evolutionary lability (Federoff, 1999).

Some of the clearest examples of modularity are the alternative phenotypes of morphology, physiology, and behavior (see part III). Others are life-stage phenotypes in species where development is subdivided by metamorphosis into recognizably distinct phases, which may be characterized by behavioral as well as morphological distinctiveness (figure 4.9). The mechanisms of modularized development are extremely varied. Compartmentalized development from arthropod imaginal discs, for example, begins with a sequestration of undifferentiated islands of cells early in development (Nijhout, 1991b). Each imaginal disc is a cluster of epidermal cells in the terminal embryo that eventually gives rise during pupation to specific structures of the adult, for example, antennae and eyes, abdominal structures, single pairs of legs, and so forth. The master switch is the one that determines the clusters early in development. Similar cell clones constitute discrete regions of the insect wing (Nijhout, 1991b). In the development of wing coloration, the wing veins serve as diffusion bound-

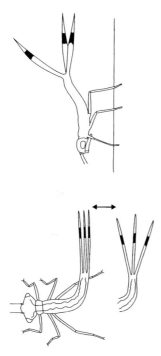

Fig. 4.9. Stage-specific behavioral displays in larval damselflies (*Xanthocnemis aealandica* (Zygoptera: Coenagrionidae). In the abdomen-bend and semaphore display (bottom), the larva alternately spreads and closes its caudal lamellae with a rhythmic period of about 2 seconds when facing another larva on a perch; the abdomen lift (top) is often followed by retreat of the performer. Territorial larvae perform these highly specialized displays only during the 6-8 larval instars (Rowe, 1985). Courtesy of R. J. Rowe.

Behavior is also organized in semidiscrete components of functionally and structurally specialized nature (S. D. Mitchell, 1990; figure 4.11). As in morphology, behavioral subunits are governed by switch, or decision points (e.g., see Miller, 1988). Gallistel (1980) synthetically reviews the voluminous evidence from classical neurobiology for a subunit and hierarchical view of behavior, which includes the organization of the brain, in humans (Edelman, 1988) as well as in ants (figure 4.12), suggesting that this may be a very widespread pattern in central nervous system functional morphology. Ethologists have perceived this at the behavioral level: "A cornerstone of ethological theory is the belief that behavior comes in discrete packets . . . segmented in time into actions that have a characteristic 'morphology' . . . they are the 'natural' units of behavior" (Barlow, 1977, p. 98; see also McBride 1971; Baerends et al., 1970; Marler, 1984; Brockmann, 1986; Otte, 1972; Wenzel, 1993; Greene, 1994). Some authors (e.g., Fentress, 1983, p. 27) consider behavioral units only "useful fictions at best" (see also critical discussions by Wenzel, 1993; Eberhard, 1990a), and this problem is especially acute when subunit traits (e.g., particular kinds of movements) are found in different combinations or contexts (Eberhard, 1990b). But all of these authors end up using a concept of more-or-less defined behavioral units.

As I argue here, it would be foolish to deny the modular properties of phenotypic organization just because there are connections and indistinct bor-

aries for compartmentalized color patterns (figure 4.10). In vertebrates, cell migration is a common feature of development, yet compartmentalization occurs when whole sets of motile cells respond to the same condition-dependent developmental signal to form a tissue or a discrete structure (Gerhart and Kirschner, 1997). There are genetic and phenotypic correlations among traits produced from different imaginal discs in *Drosophila*, because all show similarities in growth rates during the larval stages, but the correlations are greatest between traits derived from the same imaginal disc, relative to those between traits derived from different imaginal discs (Cowley and Atchley, 1990), a pattern that confirms the compartment hypothesis of Crick and Lawrence (1975) and Lawrence and Morata (1976) (both cited in Cowley and Atchley, 1990).

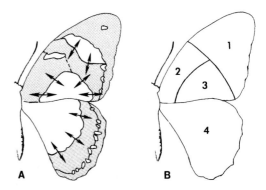

Fig. 4.10. Compartmentalized coloration and phenotypic recombination in butterfly wings. Diverse patterns, shown in figure 4.18, arise from two types of pattern formation: black elements (A) can change in size, and background color can change independently in four areas (B, 1–4). From Nijhout (1991a), © H. F. Nijhout, by permission. Courtesy of H. F. Nijhout.

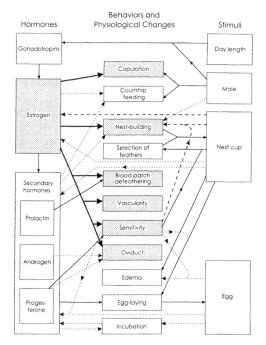

Fig. 4.11. Modularity of behavioral and physiological traits in the reproduction of female canaries. Dashed lines indicate negative effects. Bold lines show hormonal coordination of complex traits. Dotted lines indicate connections likely but not established. Shading indicates a trait complex whose coordinated expression is influenced by estrogen. Ability to be influenced by more than one hormone (as are brood patch defeathering and sensitivity) means that a degree of developmental overlap is possible and illustrates the limitations of the modularity concept. After Hinde (1970).

ders around the subunits we recognize as traits. There can be no doubt that there exist behavioral subroutines or subunits, for they are distinguishable from others in form, function, and discreteness, and sometimes in gene expression, as discussed later in this chapter (see also Raff, 1996). And they are dissociable, as shown by their repetition and reorganization in different performances. Bird song, for example, is elaborately modular and recombinatorial during performance:

> Notes are commonly 10 to 100 msec in duration. Groups of notes form syllables 100–200 msecs in duration assembled in turn into phrases, then grouped to form a song. In some bird songs every note is different, but more com-

monly, as in speech, notes and syllables are reused to produce trills and to create repertoires of different patterns, with rules for phonological syntax that vary from species to species. (Marler, 1984, p. 290)

Marler goes on to refer to the "developmental partitioning of song segments" (p. 296), with features characteristic of a local dialect concentrated in certain song segments while the structure of other

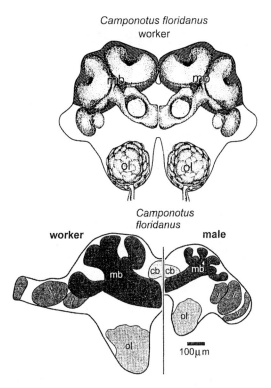

Fig. 4.12. Modular organization in the brain of an ant. The mushroom body (mb), the region of the insect brain concerned with behavioral integration, information processing, learning, and memory, is relatively large in workers compared with males, whereas the visual (v) regions are relatively large in the much larger-eyed males, indicating independent divergence of these somewhat independently used, modular brain regions in the two sexes. The olfactory (antennal) lobes (ol) are similar in size relative to the rest of the brain in the two sexes. The central body (cb) is a poorly known structure probably involved in leg movements (Gronenberg, personal communication; Gronenberg and Hölldobler, 1999). Courtesy of W. Gronenberg.

song segments varies individually. Members of a particular species tend to be consistent in the song features or segments to which imitations, improvisations, and inventions are assigned and where species differences are located (Marler, 1984, p. 296). At higher levels of behavioral organization, the movements and signals performed during courtship or territoriality usually can be distinguished from those performed during predation, and predatory behavior may take on different specialized forms depending on the nature of the prey (see Curio, 1976; Jackson and Wilcox, 1993) and the conditions of the hunt (Wilcox et al., 1993; chapter 23).

In terms of externally observed movements, the hierarchical aspect of modular organization is clearly manifested in behavior, as shown by the above examples of song (see also Fentress, 1983), for one can perceive that many complex stereotyped behaviors consist of subroutines, and these could be further analyzed into components down to the patterns of muscular contractions that perform them (Miller, 1988, p. 353). A spider's web is a material record of a complex behavior, web building, composed of several more-or-less stereotyped subcomponents (e.g., Eberhard, 1988). The subcomponents have a correlation structure, much as do the modular subunits of vertebrate morphology (e.g., Atchley and Hall, 1991).

Among the traits that characterize an individual phenotype, there can be varying degrees of subunit independence: the hand, both structurally and in its development, grades into the wrist and the arm, with which it shares muscles, nerves, blood vessels, and coordinated use. Still, the hand is identifiable as a semi-independent subunit of the phenotype in that it has specialized characteristics and uses distinguishable from those of an arm or a foot. Even the subunits of so-called modular organisms are only semi-independent. A branch of a tree, for example, is by its bifurcation point clearly recognizable as a morphological subunit, yet branches show autonomy in some aspects of their function and not in others. Old tree branches are autonomous in that when their respiration exceeds their ability to fix carbon, they do not draw carbon from the rest of the tree, and vascular constrictions at the base of a branch usually keep its water system somewhat independent of that of other branches. On the other hand, young branches are not autonomous regarding carbon supply, and branches under strong stress or forming a nutrient sink (e.g., when ripening fruit) may draw resources from the rest of the plant (Sprugel et al., 1991). Although an individual organism is clearly a discrete unit of biological organization, life cycles show connectedness between generations or continuity of the phenotype (see chapter 5). They also show modular discreteness in the boundaries between parents and offspring, the membranes between dividing cells, and degradation and reorganization that occurs between metamorphosed life stages (e.g., Nijhout, 1994a).

Comparative morphologists describe different degrees of modularity, or structural unity among parts, in terms of phenotypic covariance matrices (e.g., see Cheverud, 1982). These methods could perhaps profitably be extended to behavior and other fields, where some authors question the modularity of traits (above), in order to give discussions of modularity an operational basis linked to switches (a developmental basis of trait covariance). But in many cases it would be difficult to quantify the connectedness between subunits, say, between the caterpillar and adult stages of a butterfly. The variable nature of modularity and connectedness is illustrated well in a discussion of the mammalian skull:

in mammalian skulls, the individuality of single bones is sufficiently clear that phylogenetic homologies with other vertebrates are well-established. Yet . . . the shape of the skull as a unit is developmentally constrained to as great or a greater extent than are the size or shape of individual bones. . . . The positioning of boundaries between individual bones (i.e., sutures) can be quite variable at times when the overall shape of the skull . . . retains important phylogenetic information. When pieces of bone are excised in a region near a suture, bone growth will restore the shape of the skull, but which of the surrounding bones contributes most can be quite variable, and appears to be a function of minor local growth processes. . . . Thus, even at the level of gross phenotype, the individualized units we consider homologues may not be separable, and may overlap or intersect one another. (Roth, 1991, p. 175)

Similarly, the morphogenesis of the cochlea of the vertebrate ear is influenced by the growth of the temporal bone (Tilney et al., 1992, p. 271), and the brood cells, combs, and paper envelopes of social wasp nests, which form a coherent nest architecture, can, like the bones of the skull, be molded to fit together if damaged or irregular (Wenzel, 1993). The mechanism of flexibility in this case is the behavior of the building wasps.

Metamorphosis and
Life Cycle Modularity

All individuals of the "higher" insects—the butterflies and moths (Lepidoptera), flies and mosquitoes (Diptera), and wasps, ants, and bees (Hymenoptera)—along with many amphibians, invertebrates, and fish, undergo drastic change during their life cycles when they metamorphose from larva to adult. Insects transform from caterpillar to butterfly, from predatory swimming wriggler to blood-sucking flying mosquito, and from legless maggot to winged fly. Phenotypic remodeling during insect metamorphosis is so dramatic that the individual, as pupa, is temporarily inactive both ecologically and reproductively, even though, developmentally, it is in a hyperactive state.

Metamorphosis means extreme life-stage modularity. A single genome produces at least two highly distinctive morphological and behavioral phenotypes occupying very different niches. This allows extreme applicability of the rule of independent selection (West-Eberhard, 1996; see chapter 7), with extreme results in terms of phenotypic divergence of the cyclically alternating forms. Indeed, the evolutionary significance of complete metamorphosis is that it allows, via a profound switch, selection to produce extreme diversification of the morphological traits produced by a single genome. For a discussion on the evolutionary origin of this switch, see Truman and Riddiford (1999).

Another means of achieving life cycle modularity occurs in complex life cycles where multiple entire egg-to-reproductive cycles are carried out in association with different habitats or hosts, as in some parasites (Thompson, 1994) and in aphids. Aphids, like other insects lacking a larva-to-adult metamorphosis, have environmentally influenced cyclic dispersal polymorphisms. Many groups of aphids have complex life cycles in which successive parthenogenetic generations have dramatically different morphologies and occupy different ecological niches (especially, different host plants; for an evolutionary discussion of these facts without the bewildering jargon of traditional aphidology, see Moran, 1988). Parthenogenesis accomplishes for the aphids what complete metamorphosis does for the butterflies: it permits a drastic phenotypic reorganization, in this case beginning with an egg as a bridge to reorganization rather than a pupa. The primary evolutionary significance of cyclic parthenogenesis in aphids, then, may be that it allows the individual (genotype) to capitalize on an opportunity for altered gene expression, and selection thus produces widely divergent morphological and behavioral phenotypes from the same genotype—not only because the switch mechanism occurs early in development, which permits a distinctive phenotype to be constructed "from scratch" (with no pre-existing unmodified structure), but also because an early switch means that a correspondingly greater number of characters are brought under independent selection and can evolve together in a new direction.

These are two related but different factors in the evolution of intragenomic character divergence and polymorphism—one a developmental-mechanical one, a proximate cause of sharp divergence, and the other its evolutionary consequence, an ultimate cause of character divergence due to the independent selection of developmentally independent modular trait sets, an evolutionary consequence of modularity discussed further in chapter 7. For these reasons, in many examples of extreme polymorphisms (e.g., in aphids, social insects, and certain vertebrate examples involving heterochrony), the switch between alternatives is "pushed back" in development. Thus, as noted by Moran (1988), extreme polyphenism occurs along with cyclical parthenogenesis not only in aphids but also in rotifers and cladocerans, all characterized by a switch between forms that is in effect pushed back to the egg stage.

In some taxa, extreme polymorphism or polyphenism is an evolved response to extreme environmental contrasts (e.g., the aerial and aquatic leaf forms of some plants) or to extreme intraspecific competition (e.g., the sexual dimorphism of highly polygynous lekking species). However, given the extreme resetting of development offered by parthenogenesis, and the extreme redirection of selection offered by maternal choice of oviposition site in aphids, intraspecific divergence out of proportion to the seasonal environment change and/or intraspecific competition could be produced (genetic reasons for expecting rapid phenotypic evolution with cyclical parthenogenesis are given by Lynch and Gabriel, 1983). That is, it would be a mistake to think of these polymorphisms as extreme entirely because they are driven by extreme environmental or competitive circumstances: developmental plasticity and compartmentalization contribute to their exaggeration.

In clades like those of aphids, polymorphism is extreme in part because of a developmental device that is the opposite of a "constraint"—temporary, cyclic parthenogenetic reproduction yielding clearly compartmentalized developmental phases. A similar phenomenon occurs in the complex life cycles of plants and cnidarians (see Boero and

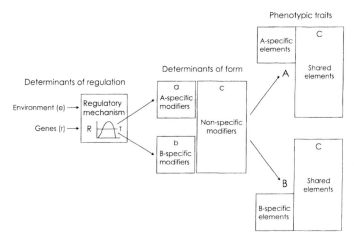

Fig. 4.13. Explanation of terms for components of phenotypic regulation and form. Phenotypes A and B may be morphological, behavioral, or physiological alternatives whose expression or use depends on the state of a condition-sensitive quantitatively variable regulatory mechanism (R) with threshold (T). The regulatory mechanism coordinates the expression and use of *specific modifier* gene products and environmental elements (a, b) that compose phenotype A or B but are not used in both. *Nonspecific modifiers* (C) are gene products and environmental elements shared (expressed or used) in both A and B. Genetic *modifiers of regulation* (r) affect the components of regulation (e.g., hormone system, neuromuscular morphology, sensory apparatus, metabolic rate), which, along with *environmental modifiers of regulation* (e), influence the value of T or the ability to pass T. Gene–product sets and environmental elements a–c and phenotypic traits A–C are referred to in the text as *elements of form*; gene set r and environmental factors (e) are termed *elements of regulation*. After West-Eberhard (1992a).

Bouillon, 1987, 1989a, b; Boero et al, 1992; Sarà, 1987). As noted by Buss (1987, pp. 161–162; see also Bonner, 1974, p. 14), by resetting the life cycle to a single-celled stage, the potential exists for diploid-phase–specific variants to arise and permit exploration of different morphological alternatives. Although the short generation times that accompany parthenogenesis in all these groups would accelerate evolution, as Moran (1992a) suggests for aphids, the developmental and selection effects of parthenogenesis just discussed seem fundamental to marked polyphenic divergence (Moran clearly points to parthenogenesis as the underlying cause, and raises the question, answered here in terms of developmental independence, of why this might be).

The Genetic Architecture of Modular Traits

A *switch point* refers to a point in time when some element of a phenotype changes from a default state, action, or pathway to an alternative one—it is activated, deactivated, altered, or moved. It is a

useful concept because it can apply to any phenotypic change at any level of organization. A switch point is the locus of operation of the mechanisms of responsiveness and the influence of the genetic and environmental factors that affect response thresholds. Switch-mediated developmental pathways occur at all levels of phenotype organization and in all forms of life, including viruses (see Ptashne, 1986). Switch points are the organizing points of development. They determine the patterns of gene expression and may therefore affect patterns of evolutionary change (see part II, and chapter 28).

Figure 4.13 diagrammatically defines terms that describe the genetic architecture of switch-controlled, modular traits. At any given, focal level of organization, this model distinguishes between regulation—the switch itself—and form—the properties of the phenotypic subunits (traits) whose expression or use is controlled by a switch. Viewed from a higher level of organization, the regulatory mechanism itself can be viewed as a modular trait with its own development (environmental and genetic inputs, modifiers, and switch-regulated phenotypic components). The distinction between regulation and form at a give focal level of organization is important, for each of the two aspects of the phenotype can change independently of the other. Changes in the inputs to regulation (e, r) can alter both R (the value of influential variables) and T (the threshold of their effects), thereby altering the frequency of occurrence of a trait in a population, and the degree of plasticity of its expression. Changes in form alter such qualities as the shape, effectiveness, and degree of elaboration or specialization of a trait.

Determinants of regulation include not only genomic and environmental inputs that contribute to a response, but also the phenotypic components of the responsive mechanism itself—any phenotypic factor that affects variation in the probability of trait expression (Falconer and Mackay, 1996; Zera, 1999). Determinants of regulation, then, include phenotypic traits that affect the threshold of a switch and the ability to pass the threshold, factors such as the structure and abundance of hormone molecules, efficiency of hormone-secreting organs, activity of inhibitors or substances that break down influential hormones, morphology of neurons, number and nature of responding tissues or organs, plus any aspects of morphology or physiology that modulate the sensitivity or response of such structures during the lifetime of the individual. For a nutrition-dependent switch, for example, such factors as food-getting ability or metabolic efficiency could be regulatory elements if they contribute to the likelihood of the nutritional threshold being passed. Here, the words "regulatory" and "regulation" refer to the control of expression of phenotypic elements, not, as the word "regulation" is sometimes used, to mean maintenance of a stable state. In general, given the number and complexity of the phenotypic "determinants of regulation," genetic influence on regulation is highly polygenic (see also Roff et al., 1997).

A switch implies some change in state, for example, between on and off, under certain conditions. If a process were constantly on or off regardless of conditions, there would be no operative switch. So condition sensitivity is an implicit quality of all switches. They mark developmental decision points that *depend on conditions*. Conditions in this case may refer to the internal environment, the social environment, or the external environment.

Phenotype *determination*, as in "sex determination" or "caste determination," is the choice made at a decision point. Thus, phenotype determination (as either A or B in figure 4.13, or the start or stop of a sequence or a process) is influenced by conditions as well as by the nature of the regulatory elements, products of both genotypic and environmental influence during their development. In a discussion of sex determination, Bull (1983) defined determination as recognized by "the earliest elements in ontogeny common to one sex that distinguishes it from the other sex, including environmental and genetic effects acting in parents or zygotes" (p. 8). The subsequent events, which are initiated by determination, Bull terms phenotype "development" or "differentiation." The initial determination event can be thought of as a master switch, with subsequent decision points acting as subordinant switches in a developmental sequence initiated, and ultimately controlled, by the master switch. It is often difficult to discern the earliest event in such a causal chain. Michener (1974) remarked on the difficulty of distinguishing between "determination" and "differentiation" with reference to the development of worker–queen caste differences in social insects:

> If a certain amount of food (greater than that necessary to produce a worker) is required to produce a gyne, does determination occur when workers receive a stimulus to assemble that much food in a single cell, or when they do so, or when the larva eats all the food, or when it eats the food beyond the quantity needed to make a worker, or when that food is assimilated? (p. 95)

Phenotype determination may be reversible (as in behavior) or irreversible (called "developmental" by some authors, e.g., "developmental polyphenism" in morphology: Lively, 1986; Smith-Gill, 1983).

Specific modifiers (a and b in figure 4.13) are any phenotypic subunits or genes that are expressed together due to control by the same switch and are not expressed in the alternative phenotype or state. In sex expression, for example, the specific modifiers are the traits that show sex-limited expression, in either male or female but not both.

Nonspecific modifiers (c in figure 4.13) are elements expressed or used in a phenotype subunit, or in both of two alternative phenotypes, but whose expression or use is not differentially affected by the switch. Morphology used in a particular behavior is often nonspecific: hand structure (and the underlying set of genes that influence hand structure), for example, would be a nonspecific modifier of both berry-picking and stone-throwing behaviors and would evolve under selection to be compatible with both. One source of connectedness and overlap in phenotype organization, and genetic correlations between traits, is represented by the nonspecific modifiers. Like the specific modifiers, they may be expressed or used in other phenotypic traits.

Shared-trait overlap can also occur vertically, between life stages, even in organisms with striking metamorphosis. Truman (1988) writes that in the moth *Manduca sexta* the larval CNS serves as the scaffolding for the construction of the adult CNS and many larval cells are remodeled during metamorphosis for their new adult functions.

Determinants of form are the components of an expressed trait, including both specific and nonspecific modifier gene products and elements of environmental origin. Each determinant may vary or be lost or gained independently of the others because each has its own subordinant switch, whose response to the master switch can vary independently of the others due to both genetic and environmental variation (the hierarchical organization of modular traits).

It is important to emphasize that assignment of a genetic locus to a box on the diagram of figure 4.13 does not imply exclusive expression there. A particular set of specific modifiers (a) controlled by a switch to influence a particular phenotype (A) may be expressed in other combinations that influence other traits—that is, may have pleiotropic effects not shown on the diagram. Extensive pleiotropy is expected under the combinatorial view of development and evolution that seems to be supported by many data (see especially part II). But there are also properties of developmental organization that reduce the importance of pleiotropy, such as tissue- and stage-specific gene expression (see below), and such properties can improve the prospects of directional evolutionary change (see chapter 7). Developmental recombination of expressed genes during development and evolution removes one objection I have heard to a modular view of the phenotype and the underlying gene sets, which is that there would not be enough genes to go around. That worry is based on the mistaken idea of strict trait-limited expression (one gene set, one trait).

Regulatory elements and modifiers of form do not correspond to so-called "regulatory genes" and "structural genes." Any gene that builds a sensory nerve, muscle fiber, or secretory cell may function as a regulatory element as defined here, in that its thresholds of sensitivity and response may influence the expression or use of some phenotypic subunit. At the same time, it is an element of form seen at a lower focal level of organization—the form of a nerve, a muscle, or a secretory cell. In eukaryotes, it is difficult to separate the regulatory and structural effects of gene products (see Raff and Kaufman, 1983; John and Miklos, 1988, p. 294). In a cascade of effects initiated by a switch, each successive effect functions as a regulator of the next and every phenotypic element of regulation has a structure or form. Different arrangements of master switches and subordinant switches are described in chapter 5, in a section on the comparative anatomy of switch-controlled traits.

Figure 4.13 may help to clarify thinking about how different genes are affected under selection on a trait. A gene may influence either regulation or form at the focal level of analysis shown in figure 4.13. If it influences regulation (left side of figure 4.13), it could affect the frequency of expression of a positively selected phenotype (A or B)—the frequency of a trait relative to others in a population. If a gene influences form (gene sets a, b, and c), it could affect the efficiency of the trait under selection. Which of these developmental roles a gene plays affects its response to selection on a focal trait (e.g., A or B). An allele that influences regulation, for example, may evolve to fixation and be expressed in all individuals of the population, and yet the trait whose regulation it influences need not be fixed: it can influence a threshold for turning the trait off and on that is advantageously set at a single level in all individuals. A specific modifier gene (one found in set a or b) that influences the form of a focal trait may be expressed in only a fraction of individuals even though it is present in all, or may be expressed for only a short and very specific time under the control of the switch in question.

Insofar as frequency or duration of expression affects response to selection (see chapter 7), such developmental-role-dependent difference in frequency of expression of a gene would affect the predictions of a single-locus model regarding the evolutionary fate of its alleles. An allele that is a nonspecific modifier (c) may be relatively slow to respond to selection compared to alleles (a or b) expressed in the same trait but controlled by the switch, because a nonspecific modifier may be subject to antagonistic selection, that is, selection in opposite directions, when expressed in contrasting trait A or B. Response to selection, then, depends not only on genetic variance but also on the developmental parameters diagrammed in figure 4.13, such as function (regulatory or regulated) with respect to a focal (selected) trait. In addition, there is feedback between selection on regulation and form: frequency of expression of A or B relative to its alternative affects the magnitude of its fitness effect, and therefore the direction of evolution of shared, nonspecific modifiers (c). This effect of frequency of expression on response to selection (see also Roff, 1996, p. 22) means that there must often be correlations between the evolution of regulation and the evolution of form, as further discussed in chapter 7.

The genetic architecture diagrammed in figure 4.13 implies several sources of genetic correlation between traits, and pleiotropic effects of genes.

Developmental switches, then, organize the phenotypic variation that is exposed to selection. Since evolution is ultimately concerned with genes, the following sections examine in more detail how the developmental architecture of the phenotype (figure 4.13) relates to genes.

Modular Traits as Subunits of Gene Expression

Molecular biologists have long recognized the modularity of gene expression as a product of coordinated regulation. The operon concept (Jacob and Monod, 1960), based on studies of gene expression in prokaryotes, denotes a cluster of functionally interacting genes whose expression is tightly coordinated. Earlier studies of gene expression in eukaryotes using electrophoresis also showed that genes are turned off as sets during development and that patterns of gene expression change during evolution (e.g., Markert and Moller, 1959; Whitt, 1980).

Gene-expression monitoring experiments using DNA microarrays have now greatly expanded this concept to include "synexpression groups" of coregulated genes in eukaryotes, including evidence that this mode of organization has facilitated evolution (Niehrs and Pollet, 1999). But that conclusion need not have awaited DNA microarray technology, for there is abundant older evidence for modular traits as gene-expression groups, as I show in this section. Coexpressed genes have been labeled with different names, including, in addition to *synexpression groups* (if they do not overlap with other sets of coexpressed genes; Niehrs and Pollet, 1999), *gene clusters* (an older term for coexpressed sets of genes that are chromosomally linked), and *syntagms* or *gene cassettes* (synonyms of synexpression group, used in *Drosophila* genetics; Niehrs and Pollet, 1999). There is a very tight relation between cofunction and coexpression (Niehrs and Pollet, 1999), for reasons discussed in chapter 7.

We can generalize about modular gene expression across all kinds of phenotypes and levels of organization as follows: all modular phenotypic structures, behaviors, and physiological traits diagrammatically represented in figure 4.13 are subunits of gene expression or gene-product use—they are subunits of gene action. In other words, the phenotypic traits that are linked together as a coexpressed set due to control by a common switch are underlain by genes whose expression is likewise linked by the switch—there is *developmental linkage* of coexpressed genes. Modular gene expression is not a new idea (Jacob and Monod, 1961; Britten and Davidson, 1969; Bonner, 1974, p. 60; West-Eberhard, 1992a; Raff, 1996). But the gene-expression modularity of phenotypes has not often been explicitly linked to the way evolutionary biologists think about fitness effects and gene-frequency change. Some molecular biologists have begun to connect modular gene expression with co-evolution of coexpressed genes (Niehrs and Pollet, 1999). Evolutionary consequences are further discussed in chapter 7.

In terms of evolutionary genetics, modular gene expression and gene-product use means that each phenotypic subunit is a somewhat independent system of genetic variation. This gives a mechanistic basis for the genetic treatment of single adaptive traits in semi-isolation from others. Atchley and Hall (1991; see also Hall, 1992, pp. 174ff) model the quantitative genetics of phenotypic traits with reference to the development of the mouse mandible (figures 4.14, 4.15) taken as a somewhat independent system within an individual organism. Modular gene expression helps to justify the genetic treatment of discrete morphological traits as evolutionary units semi-independent of others.

In terms of proximate mechanisms, modular gene action allows different cell types, body regions, and periods of time to be the scene of contrasting gene activities without interference from each other (see Gerhart and Kirschner, 1997). It also permits developmental recombination of genes during ontogeny—the use of the same gene product in different combinations with other gene products. Compartmentalized gene expression is therefore one means by which phenotypic organization achieves versatility in development. Due to *developmental recombination* of phenotypic subunits (see parts II and III), some genes or gene products are used in multiple, different combinations during ontogeny and in short-term reversible responses to environmental change. This is one basis for pleiotropic effects of genes, further discussed be-

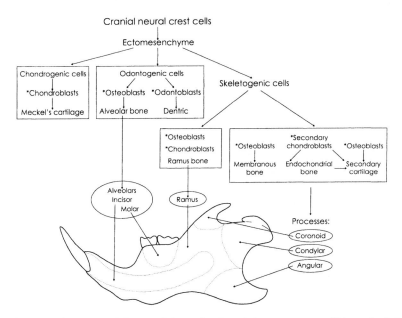

Fig. 4.14. Developmental origins of five modular subunits of the mouse mandible and of their quantitative variation. Five morphogenetic regions (circled labels) descend from neural crest cells via eight cell populations or condensations (asterisks) in four cell lines (boxes). The final size of each subunit varies semi-independently under the influence of the five factors described in the text. The mandible illustrates semi-independent developmental origin of trait subunits and their hierarchical organization: the mandible as a whole is a higher level subunit composed of smaller, lower level subunits whose independent variability depends on semi-independent developmental origins. Based on Atchley and Hall (1991).

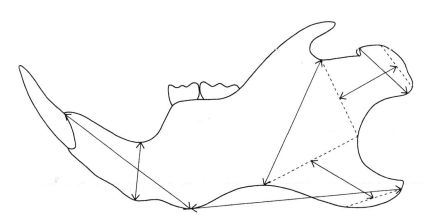

Fig. 4.15. The relationship between continuous and discrete variation in a morphological trait. The adult mouse mandible is a discrete subunit of skeletal morphology. Quantitative variation in its dimensions can be measured as shown. These dimensions are influenced by factors that determine the size of the four initial cell populations enclosed by boxes in figure 4.14, including (1) number of stem cells in the preskeletal condensation or aggregation of cells, (2) time of initiation of condensation, (3) rate of cell division, (4) fraction of cells involved in cell division, and (5) rate of cell loss (cell death) (Atchley and Hall, 1991). After Atchley et al. (1985).

low. In addition, the connectedness and integration of phenotypic subunits (above) suggest that there is likely to be gene-expression overlap, or genetic correlations, as well.

Many metabolic or "housekeeping" genes must be expressed in multiple or even all phenotype subunits as nonspecific modifiers not affected by the switches that alter form. The importance of overlap in subunit gene expression is dramatized by the finding that in *Drosophila* 75% of the genes that are essential for the normal functioning and development of the adult fly are likewise essential for the production of the "maternal dowry" provided to the embryo and supporting its early metabolism and organization (Anderson, 1989, p. 3).

Evidence for Modular Gene Expression

Evidence of modular gene expression is important to many of the claims made in later chapters. It is fundamental to the idea that modular traits, as subunits of natural selection, can show a response to selection and evolve semi independently of others, for this requires that they vary genetically somewhat independently of others. Modular gene expression is also fundamental to the idea of evolved developmental recombination of traits, which likewise depends on modular dissociability and reorganization of gene expression. The evidence of this section is necessary to make this model of developmental plasticity and evolution more than a mere assertion.

There is evidence for modular gene expression at many levels of phenotypic organization. In eukaryotes, protein domains are among the smallest subunits that can be recognized as developmentally and gene-expression-distinctive traits. Multidomain proteins are relatively rare in eubacterial and archaeal genomes but are common in eukaryotes, where domains can be associated with distinctive catalytic, adaptor, effector, and/or stimulator (signaling) functions as well as with specific aspects of transport, protein sorting, and cell-cycle regulation (Schultz et al., 1998). Protein domains are produced as units and are developmentally dissociable, being duplicated, deleted, and rearranged during both development and evolution, as discussed in chapter 17. Each domain corresponds to a set of chromosomal nucleotides (an exon) that is transcribed as a unit (figure 4.16). So each domain represents a phenotypic entity that is associated with a distinctive segment of DNA—it is a product of modular gene expression.

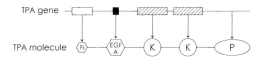

Fig. 4.16. Modular organization of a gene and its product. Different modular subunits (exons) of the tissue plasmagen activator (TPA) gene specify different functionally distinct protein domains: fibronectin I (FI); epidermal growth factor A (EGF A); Kringle (K); and a trypsin-like protease (P). Each binds to a particular, different type of macromolecule or macroscopic structure. Standard textbook representation (e.g., see Holland and Blake, 1990; Futuyma, 1998).

Different proteins are also clearly distinctive in terms of gene expression. Each corresponds to a consistent region or set of regions of DNA on a chromosome. In this case, the coordinating mechanism for subunit development is the condition-sensitive (often tissue-specific) process of gene transcription and protein assembly that takes place within a cell. Although it is now clear that the old rule of "one gene, one protein" does not hold, since the exon products of a single chromosomal gene can be spliced differently and somatically recombined to produce different proteins (see chapter 17), each protein produced is clearly genetically, or exonically, distinctive.

In multicellular organisms, different types of cells and tissues are well known to be characterized by differences in gene expression (Raff and Kaufman, 1983; Davidson, 1986; John and Miklos, 1988). Cell differentiation, induced by position effects and other interactions among cells, is associated with the regulation of cell- and tissue-specific genes. The imaginal discs of insects and other arthropods, sets of cells sequestered and latent in the larva, give rise to circumscribed sets of morphological structures in adults. The imaginal discs have been described as gene-expression "territories" (Dickinson, 1991, p. 132). Similarly, the distinctiveness of gene expression associated with local determinants of body form, especially allometric growth, is called "genetic autonomy" by Cock (1966, p. 177) in a still timely review of the genetics of growth and form. Hox genes are classic illustrations of modular gene expression (figure 4.17).

Local or subunit-specific gene expression can also be short-term and turned off and on rapidly. Suntanning—the rapid darkening of the skin on ex-

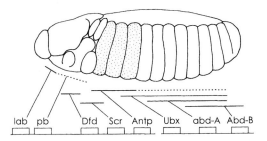

lab pb   Dfd Scr Antp Ubx abd-A Abd-B

Fig. 4.17. Modular gene expression at the body-segment level. Eight homeobox genes (boxes show chromosomal positions) specify eight protein homeodomains, expressed in different combinations in different body regions, the head (left), thorax (stippled), and abdomen (right) of a 10-hour *Drosophila* larva, where they regulate transcription of genes that determine development of segment-specific structures such as legs, mouthparts, and antennae. Homeobox genes shown are *labial* (lab), *proboscipedia* (pb), *Deformed* (Dfd), *Sex combs reduced* (Scr), *Antennapedia* (Antp), *Ultrabithorax* (Ubx), *abdominal-A* (abd-A), and *abdominal-B* (abd-B). Horizonal lines above the genes indicate approximate extent of epidermal expression of each gene. Horizontal lines beneath genes indicate the two clusters of related genes, bithorax (right) and antennipedia (left). After McGinnis and Krumlauf (1992).

posure to sunlight—involves rapid short-term activation of a pigmentation gene in already formed skin cells (mentioned in Ptashne, 1986, p. 3). Altered physiological states such as diapause and hibernation can involve switch-controlled gene expression, probably at multiple loci (Srere et al., 1992). Genes for pancreatic lipase (which liberates fatty acids from triglycerides at low temperatures) and pyruvate dehydrogenase kinase isozyme 4 (which inhibits carbohydrate oxidation and depresses metabolism) are up-regulated in the heart of the 13-lined ground squirrel (*Spermophilus tridecemlineatus*) during hibernation (Andrews et al., 1998). Day length–induced pupal diapause in the flesh fly *Sarcophaga crassipalpis* involves a partial shutdown in expression of genes as well as the expression of at least 14 brain proteins found only in diapausing pupae (Flannagan et al., 1998). A similar state of facultative inactivity in the nongrowing and nonfeeding dauer larva in *Caenorhabditis elegans* is induced by crowding and starvation and has been found by the study of mutants to involve

expression of several specific genes (see references in Flannagan et al., 1998).

Life-stage differences are also associated with gene-expression change during ontogeny. Metamorphosis is accompanied by massive regulation of large sets of genes. Yaoita et al. (1990) describe amphibian metamorphosis as "a cascade of changes in gene expression that occurs in every tissue of the tadpole as a response to thyroid hormone" (p. 7090). Many studies have documented quantitative and qualitative changes in specific gene products during amphibian ontogeny with and without metamorphosis (see brief review in Begun and Collins, 1992). Insect metamorphosis also involves "massive" changes in gene expression, effected by juvenile hormone (Nijhout and Wheeler, 1982; Nijhout, 1994a). Thus, not surprisingly, very distinctive life-stage phenotypes are also distinctive in terms of gene expression. Qualitative differences in allozymes of different life stages have been documented in lampreys (Potter, 1980, p. 1609), larvae versus adults in *Drosophila* (Carson and Kaneshiro, 1976, p. 335), and seedlings versus adults of the rain forest trees *Shorea leprosula* and *Xerospermum intermedium* (Gan and Robertson, 1981). Stage-specific quantitative differences are known in ants (Schmidt, 1973) and termites (Myles and Chang, 1984, p. 249). Strikingly compartmentalized stage-specific gene expression occurs in the development of all of the embryologically most studied animal species (*Drosophila*, *Caenorhabditis*, sea urchins, *Xenopus*, and mice), where a major fraction of the large set of genes that is transcribed in the embryo is not expressed in adult cells (Davidson, 1986, p. 172). These phenotypically distinctive life stages represent temporal partitioning of gene expression.

A spatial counterpart of metamorphosis-mediated change in gene expression is found in the spatially segmented body plans of arthropods. Here, the gene-expression distinctiveness of subunits has been demonstrated in experimental studies of segment-specific regulation by the "homeobox" genes of *Drosophila*. A homeobox is a base sequence that codes for a DNA-binding domain on a protein. That is, it is a genetic subunit that has a regulatory function. In *Drosophila*, the antennapedia and bithorax homeobox gene complexes orchestrate segment-specific gene expression responsible for the distinctive morphologies of the segments of the larval and adult head, thorax, and abdomen (see Raff and Kaufman, 1983, p. 249). "This control is effected at the time of [segmental cell fate] determination and apparently results from positive action of the loci in the complexes on batteries of other genes" (Raff and Kaufman, 1983, p. 251). Spatially restricted gene expression under the control of Hox genes has been demonstrated using mutants (Harding et al., 1985).

The queen and worker alternative phenotypes expressed in social insects, like sex differences in many organisms, are hormonally mediated and in some cases known to involve differential gene expression, either quantitative differences in gene product or qualitative differences (different sets of expressed genes; e.g., see Wigglesworth, 1961; Nijhout and Wheeler, 1982; Myles and Chang, 1984; Hartfelder et al., 1995; Evans and Wheeler, 1999). The earliest demonstrations of caste differences in gene expression known to me are those of Liu and Dixon (1965; see also Tripathi and Dixon, 1969), who found, using starch-gel electrophoresis, that the total protein concentration of third-day queen larvae is lower than in workers.

Brunnert (1967) was apparently the first to detect caste-specific differences in gene expression using electrophoresis in ants (*Formica polyctena*). This was followed by evidence on other ant species. Schmidt (1973) demonstrated both qualitative and quantitative differences in protein profiles between the castes in three *Formica* species, and Passera (1974, cited in Nijhout, 1994a, p. 183) and Craig and Crozier (1978) found such qualitative differences in *Pheidole pallidula* and in *Iridomyrmex* and *Brachyponera* species, respectively. Craig and Crozier (1978) cite studies by Hung et al. (1977) showing that the allele for malate dehydrogenase is expressed in workers but not in queens in a population of the ant *Solenopsis*. In *S. invicta*, the wingless workers and the queens that have shed their wings express α-glycerophosphate dehydrogenase at only one locus, whereas winged queens and males, which are always winged, have a second locus, possibly associated with increased metabolic efficiency for flight (Craig and Crozier, 1978).

Caste-specific protein profiles indicating both qualitative and quantitative caste differences in gene (protein) expression have also been found in bumblebees (Röseler, 1976), and termites (Wyss-Huber, 1981; Myles and Chang, 1984; Gruppioni and Sbrenna, 1992; Kubo et al., 1999). Severson et al. (1989) demonstrated nutrition-mediated caste-specific differences in transcriptional activity in honeybees, corresponding to periods of major morphogenetic differentiation between workers and queens during the larval and prepupal stages. They considered this "the first report of extrinsic regulation of bidirectional differentiation within a single organism" (p. 215), but the many previously documented caste differences in gene expression just listed would likewise qualify, since extrinsic influence on caste determination is the rule in these groups.

Especially interesting are the findings of Röseler and of Wyss-Huber that variations in protein-producing gene expression are associated with variations in hormone activity and morphology in social insects, since this begins to fill in the connections between environmental circumstances, behavior, hormones, and gene expression. Röseler (1976) showed that in bumblebees (*Bombus terrestris*) levels of a queen-specific protein are depressed by experimental application of juvenile hormone, which also affected the behavior and reproductive (ovarian) state of the queens, thus indicating the influence of the hormone on both gene expression and caste-specific behavior. Wyss-Huber (1981) demonstrated that in the termite *Macrotermes subhyalinus* activity of the esterases that bind juvenile hormone is highest in workers, followed by soldiers, and lowest in kings and queens. There were marked differences in hemolymph protein profiles between the reproductive versus the worker and soldier castes and between winged (virgin) and physogastric (mature) queens, indicating changes in gene expression during maturation. Worker protein profiles were similar regardless of sex, and there were qualitative differences between workers and soldiers.

More recently, mRNA differences have been demonstrated between mature soldiers and pseudergates in termites (Miura et al., 1999). These gene products have been identified with secretions of the mandibular glands of soldiers. The work by Evans and Wheeler (1999) is the first to identify proteins that are differentially expressed at the time of the developmental switch between workers and queens (Nijhout, 1999a, p. 5349). I predict that new techniques such as microarray DNA analysis (e.g., Eisen and Brown, 1999) will lead to an enormous proliferation of evidence for modular, phenotype-specific gene expression in the near future. But these important and heretofore overlooked (by evolutionary biologists) findings should not be forgotten.

Where molecular studies are lacking, there are several kinds of indirect evidence for differential gene expression in other complex alternative phenotypes. Juvenile hormone (JH), which is known to regulate gene expression in social insects as just shown (see also Nijhout and Wheeler, 1982), has been implicated in the control of many other insect polymorphisms and polyphenisms (reviewed in Nijhout, 1999c). Included are the migratory versus nonmigratory phases of locusts, the complex

Fig. 4.18. Polymorphism of *Papilio dardanus*. (A) Color patterns of males, nonmimetic females (*meri-ones* form), and palatable mimetic females, each a mimic of a different species of unpalatable butterfly: (B) *hippocoon* form, (C) *cenea* form, and (D) *planemoides* form. From Nijhout (1991a), © H. F. Nijhout, by permission. Courtesy of H. F. Nijhout.

polymorphisms of aphids, and various simpler polyphenisms, including color pattern polymorphisms of lepidopterous larvae, seasonal color polymorphisms of Orthoptera (especially grasshoppers), pupal and adult butterflies (Nijhout and Wheeler, 1982; Nijhout, 1991b), male earwig antennal and cercal polymorphisms, and the wing or dispersal polymorphisms of many orders of insects (reviewed in Matsuda, 1987). Nijhout and Wheeler (1982, p. 110) cite evidence that JH, like the steroids of vertebrates, acts directly on the genome after entering the nucleus by combining with a cytoplasmic receptor. Since some well-studied vertebrate polymorphisms, such as the normal and neotenic morphs of salamanders (*Ambystoma*; review in Matsuda, 1987), are known to involve failures of metamorphosis, it seems likely that differential gene expression is involved in vertebrate polyphenisms as well.

Genetically, among the best-known alternative phenotypes in insects are the mimicry polymorphisms of butterflies, thanks to decades of work by E. B. Ford, P. M. Sheppard, B. Clarke, J. R. G. Turner, and others (see Ford, 1971; Turner, 1977; for a synthetic overview, see Nijhout, 1991b). Studies of the color patterns of Batesian mimicry polymorphisms in nymphalid butterflies show that visually complex differences in the form of the mimic morph in geographic races having different models (e.g., in *Papilio dardanus*, figure 4.18) are likely caused by simple genetic changes involving only one or two loci plus their modifiers (Nijhout, 1991b). The wing of mimics (figure 4.18B–D) is tailless, a feature controlled by a single locus un-

linked to the genes for color pattern. Evidently, the geographically variable mimics track the variable form of their distasteful models via change in a small number of morph-specific modifiers: the nonmimetic malelike form of females (figure 4.18A) retains the tail and does not undergo evolutionary change in pattern like that of the mimics. Thus, there is morph-specific geographic variation in form, in accord with the distributions of their distasteful models (Turner, 1977, p. 184). These observations show how independent gene expression is associated with independent genetic variation and independent evolution of different alternative phenotypes.

A venerable source of indirect evidence that polyphenisms involve differences in gene *expression* rather than differences in genotype are bilateral mosaics that show the morphology of one alternative phenotype on the left side and the contrasting alternative on the right. This kind of evidence was used by Darwin (1868a [1875a]), who cited a bilateral mosaic in a rabbit to argue that complex morphological variants can arise without inherited (genotypic) differences (see chapter 16). Bilateral mosaics showing phenotypes of both alternatives are known in stag beetles (Lucanidae) with male weapon dimorphisms (Arrow, 1951 [drawn to my attention by L. Bartolozzi]) and in ants (Kinomura and Yamauchi, 1994; Heinze and Trenkle, 1997). Bilateral sex mosaics, or gynandromorphs, of insects have long been used to infer the architecture of sex-limited gene expression (see the brief modern summary in John and Miklos, 1988). This literature is recommended reading for anyone interested in the evolutionary significance of sporadic bilateral asymmetries (fluctuating asymmetry) in animals.

Some of the best examples of rapid change in gene expression associated with physiological traits come from studies of plants. The change from C_3 metabolism to crassulacean acid metabolism (CAM) metabolism in facultative CAM plants, discussed in chapter 17, involves a gene expression change (Cushman and Bohnert, 1996). Wound responses can also involve induced gene expression (Davis et al., 1991), including of proteinase inhibitors that function as defensive chemicals (Farmer and Ryan, 1990). In a remarkable case of this, defensive-substance gene activation is increased above the wound-stimulated level by an airborne volatile signal (methyl jasmonate) that passes from one leaf to another, and even between plants (e.g., from sagebrush to tomato). Since methyl jasmonate is widespread in plants, the fact that it can regulate gene expression in neighboring plants raises the interesting possibility that there could be interindividual and interspecific competition and manipulation using this signal (Farmer and Ryan, 1990, p. 7716).

Behavioral Subunits in Relation to Genes

Probably most biologists are more willing to grant the gene-expression distinctiveness of morphological or physiological traits than of behavioral ones. In an attempt to clarify the terminology of behavioral and morphological castes in social insects, for example, Peeters and Crozier (1988, p. 283) argue for a distinction between morphological "castes" and behavioral "roles," by arguing that the morphological differences are "produced by divergent developmental pathways coordinated by endocrine signals and [involving] . . . the expression of different sets of genes," as distinct from divisions of labor involving only behavior and/or reproductive physiology ("roles"). They reserve the terms "alternative phenotypes" and "forms" for *morphology* determined prior to adulthood. Yet behavioral and physiological alternatives, including those between social insect castes, can certainly evolve, and they are known to be genetically variable aspects of the phenotype (e.g., see Robinson and Page, 1988; Page and Robinson, 1991) whose expression is coordinated by endocrine signals (e.g., see Röseler, 1985). What is the relationship between behavior and genes? Obviously, every rapid movement or distinctive thought is not a phenotypic "subunit" underlain by massive instantaneous differences in gene expression. How, then, are these fleeting but presumably adaptive, evolved aspects of the phenotype to be related to genes?

The organization of behavior (movement) can be analyzed and related to genes in the same way as morphology can be. The regulation and form of a particular behavior depend on genetically influenced structure as much as does any other aspect of the phenotype. Particular animal behaviors are affected by particular arrangements of muscle, nerve, and bone or exoskeleton and by particular histories of practice and reinforcement, all known to leave physical traces in the structures of organisms that behave. The movements of plants obviously depend on specific structures as well. So behavior can be related to genes by asking which *gene products* are used together to produce the movement observed. The organization of behavior is also similar to that of morphology in being composed of elements coordinated by switches, or decision points (see chapter 3). The "specific modifier" *genes* in this case are the set of genes that influenced the structures coordinately used in a particular behavior pattern. The genes that influence regulation are those that affect the decision to move. They might include sensory systems, particular parts of the central nervous system, and hormonal systems that affect motivation. There is nothing mystical or vaporous about behavior that removes it from the material world of genetic influence: behavior is movement produced by evolved, genetically influenced structures that perform in certain ways because they are constructed in certain, genetically influenced and genetically variable ways. Because of the modular, switch-controlled nature of behaviors, it is not surprising that when ethologists and neurobiologists analyze the structure and development of behavior, they perceive a hierarchy of subunits, as discussed in chapter 3.

The sensitivity of highly conditional behavior to genetic influence is evident from the fact that some of the most rapid responses to artificial selection occur under selection on behavioral traits. Cairns et al. (1990) obtained exceptionally rapid responses, and reversibility of responses, to selection on such behaviors as degree of aggressiveness, "emotionality," and level of open-field activity in rats.

While not universal, the rapid expression of specific genes may be more common in the organization of behavior than is usually realized. Wingfield et al. (1987, p. 607) note that steroids in rats can affect gene transcription within 15 minutes of their release into the bloodstream and can affect the morphology of neuronal cell nuclei within 2 hours. In canaries (*Serinus canarius*) and zebra finches (*Tae-*

neopygea guttata), hearing a conspecific song for 45 minutes stimulates a detectable increase in mRNA levels of a regulator gene in a forebrain region believed to be involved in auditory processing (Mello et al., 1992). This may affect song learning or song control or general "arousal," and the precise effects are still being investigated. In vertebrates, the brain has been thought of as a target rather than a source of chemical (hormonal) signals. But in male zebra finches (*T. guttata*), enzymatic activity in the brain converts androgen to estrogen, which is released in large quantities into the blood and may affect both sexual differentiation of the brain and the capacity of males to learn song (Schlinger and Arnold, 1992).

The factors—stimuli and genetic-enzymatic pathways—that regulate brain synthesis of estrogen are not yet known, but it seems likely, given the results of Mello et al. (1992) on rapid gene expression in the brain of the same species, that local gene expression triggered by social or other environmental factors occurs in the brain and affects the development (including learning) of sexually dimorphic behavior. In these cases, the time delay between gene expression and gene product use is relatively short, whereas for other morphological components of behavior, this interval is often relatively long.

A complex, physiological and behavioral response known to involve environmentally stimulated gene expression occurs in the snail *Aplysia* (summarized in McAllister et al., 1986). Genitalic contact with the atrial gland during copulation induces local expression of genes producing A and B peptides, which in turn act on the bag-cell neurons, neurosecretory cells in the abdominal ganglion. There, genes of the same family as the peptide-producing genes are expressed, producing egg-laying hormone, which acts on the ovotestis to induce the smooth muscle contractions of egg-laying behavior. The same hormone also alters the firing patterns of nerve cells, affecting other behaviors. The result is a complex coordinated behavioral response involving egg extrusion, increased respiratory activity, and head waving.

We may exaggerate the degree to which behavior is less "genetic" or further removed from the genes than morphology because we do not so easily perceive the variability, flexibility, and condition sensitivity of morphological development, which in fact is quite similar in these respects to behavior. We may be deceived by the *durability* (nonreversibility) of structure into thinking that it is more fixedly determined by genes, when in fact its determination is always condition dependent.

What distinguishes behavior from morphological plasticity, then, is neither condition sensitivity nor freedom from genetic influence on component elements. Rather, it is the greater time delay between gene expression and gene-product use, and the number and reversibility of permutations or reorganizations of elements that can occur during the lifetime of an individual. *Phenotypic recombination*, or reorganization of the phenotype during development and evolution, resulting in the assembly of new combinations of traits, is common during the ontogeny of morphology, especially at the molecular level (see chapter 17). It is one form of pleiotropy, for the protein products of a single gene may be incorporated into several or many phenotypic traits at different levels of organization (for examples, see Mather, 1973, p. 15). But ontogenetic phenotypic recombination of behavioral subunits is far more extensive. This has been succinctly stated by Trewavas and Jennings (1986) in contemplating the differences between plants, which are noted for their physiological and morphological plasticity, and animals, noted for their behavioral plasticity: "The adaptiveness of animals lies in the brain, in the almost endless number of combinations in which the different tissues can be made to work together to produce different types of behavior" (p. 1).

Modular Gene Expression, Pleiotropy, and Genetic Correlations

The coexpressed traits whose expression or use is governed by a single switch are pleiotropic effects of the genes that influence the switch. As a result, they may show high phenotypic and genetic correlations due to their coordinated expression and selection as a functional set (Cheverud, 1982, 1984b, 1996; Levinton, 1988; Stearns et al., 1991; Windig, 1993). In effect, the coexpressed traits are *developmentally*, rather than chromosomally linked. In *Drosophila melanogaster*, for example, traits from the same imaginal disc, which form gene-expression subunits of the adult phenotype, have higher genetic correlations than do traits from different imaginal discs (Stearns et al., 1991, after Cowley and Atchley, 1990). *Developmental switches create genetic correlations within traits and break genetic correlations between traits.*

How genetic correlations are created by developmental mechanisms has been described in general terms by Cheverud (1996), who stresses that developmental integration through interactions of parts during ontogeny evolves first; then, the modular phenotypes thus formed structure genetic integration at the population level, leading to coevolution of developmentally related gene sets (see also West-Eberhard, 1992a). How correlation structure changes during ontogeny has been documented by Zelditch (1988) in studies of bone development in rats. Prior to weaning, jaw and cranial development are uncorrelated and developmentally independent. Later, when jaw and cranium grow together, they begin to influence each other's development and become genetically correlated as a result. The hormones and physiological mechanisms that coordinate development are among the "proximate causes" of genetic correlations (Ketterson and Nolan, 1992; see also Stearns, 1986, p. 39; Stearns et al., 1991). Since these mechanisms are highly condition sensitive, it is not surprising that the strength and the sign of genetic correlations can change depending on environmental conditions (see examples in Stearns et al., 1991).

Similarly, under sexual selection the genetic correlations of a Fisherian runaway process involving male signals and female preference (Fisher, 1930 [1958]) derive not from chromosomal linkage but from the highly coordinated and interdependent interactions of male and female. To the degree that they are genetically correlated, signal and response in effect develop as a single socially coordinated unit. Even though the subcomponents are expressed in separate individuals, both sexes come to carry and transmit the genes for both.

Hierarchical gene expression is an important means of developmental and phenotypic integration. But the potential independence of each node means that hierarchical organization is not firmly or irreversibly interconnected. The metaphor of a "gene net" (e.g., Bonner, 1988; McKinney and Mc-Namara, 1991) emphasizes the connections within, rather than the breaks between, subunits. A branching tree may better capture the two aspects, because it suggests a hierarchy of connections.

Hierarchical organization means that subordinate switches may somewhat disrupt formation of genetic correlations among usually coexpressed traits controlled by a higher level master switch. Hierarchical complexity is not readily brought to mind by a simplistic notion of pleiotropy. Pleiotropic effects occurring at several levels of organization may affect the genetic correlations among a single set of traits, if it is a mosaic composed of a subset of semi-independently controlled subordinant traits.

While a switch creates genetic correlations within subunits, it breaks genetic correlations between them. In mechanistic terms, it separates the expression of two sets of genes and phenotypic

traits, placing them on separate developmental trajectories and reducing the correlations between them. To the extent that two phenotypic subunits have shared, nonspecific modifiers (figure 4.13), they may remain to some degree genetically correlated even when (semi-)independently expressed. When traits show *antagonistic pleiotropy*—the traits have shared, nonspecific modifiers that are favorable to one of the pair of traits and unfavorable to the other—increase in the independence of their expression (e.g. increased divergent specialization such that some of the antagonistic shared traits are limited in expression to one of the alternatives) could have the effect of liberating them from genetic constraints, to more readily evolve in different directions (for a possible example, see Lenski, 1988a,b). Riedl (1978) called such overlap between traits, and antagonistic pleiotropy, "developmental burden," because it represents a potential impediment or cost to directional evolution, which Riedl compares to genetic load.

Arnold (1992, p. S102) discusses in theory how behavioral decision points can build and break genetic correlations. His arguments could be extended to development in general. Given the predominance of switch-controlled trait expression in phenotypic development, it seems likely that the making of genetic correlations via regulatory linkage of coexpressed traits, and the breaking of genetic correlations due to the evolution of bifurcating developmental pathways, is an exceedingly important factor in evolution. Since genetic correlations can cause correlated responses to selection, this result of modular genetic architecture must have widespread importance for rates and directions of evolutionary change (see chapter 6).

Predictions regarding the response to selection by correlated traits depend on the details of their regulatory architecture (see Stearns et al., 1991). For example, head size and ovary size are two aspects of genotype-influenced caste in the meliponine bees. Due to control by the same switch, these traits likely show a high degree of genetic correlation, as found in other traits under common developmental control (Stearns et al., 1991; Dingle et al., 1986). But the expected correlated response to selection may or may not occur, depending upon how the correlation is achieved developmentally. For example, selection for further decrease in head size could conceivably cause a change in JH titer throughout the body, since JH is involved in the master switch that underlies the dimorphism. In this case there would be a correlated change in ovary size. But selection on head size could also cause an effect via a change in the JH responsiveness of head cuticular cells, in which case a correlated change in the ovaries would not occur.

Chromosomal Linkage as an Alternative Model

Prior to evidence that coexpressed eukaryote gene sets can be developmentally linked and noncontiguous on chromosomes (see Britten and Davidson, 1969), research on prokaryotes, especially *Escherichia coli* and other bacteria, heavily influenced the molecular model of coordinated gene expression. It was a model that emphasized chromosomal linkage and coordination by regulatory genes (operons; e.g., see Herskowitz, 1962). In evolutionary biology, this idea was reinforced by widely appreciated (Felsenstein, 1985, p. 217) theoretical models showing that recombination is expected to be selected against as a result of its effect in breaking up coadapted genotypes (e.g., see Fisher, 1930; Kimura, 1956; Lewontin and Kojima, 1960; Feldman, 1972). As a result, alternative phenotypes like the Batesian mimicry morphs of butterflies came to be seen as models of alternative trait complexes controlled by alternative linked sets of genes, or *supergenes* (e.g., see Dobzhansky, 1951; Ford, 1964). This vision of the genetic architecture of complex adaptive traits may still exist in the minds of many evolutionary biologists. Goodnight (1988, p. 447, after Hedrick et al., 1978, and Wright, 1978) says that coadapted gene complexes are "generally believed" to be tightly linked, because this keeps favorably selected gene combinations from being "dissipated by recombination" (see also Dobzhansky, 1959, pp. 24–26). Nonetheless, some theoretical geneticists have discussed evidence (e.g., from *Drosophila* research) that natural populations are not composed of chromosomes carrying coadapted sets of genes in extreme linkage disequilibrium with each other (Maynard Smith, 1978, pp. 83–85).

The supergene view of regulatory structure was especially influential in ecological genetics (e.g., Ford, 1964; Clarke and Sheppard, 1971; Charlesworth and Charlesworth, 1975b, 1976a,b), and its influence continues into the present in the literature on complex polymorphisms. Indeed, not only is chromosomal linkage thought to underlie the cohesive coadapted nature of such traits and help to explain it, but given the lack of a widely accepted alternative model for the genetic basis of such traits, some consider a supergene genetic architecture to be actually *essential* for the evolution of complex adaptive polymorphisms and other coadapted complex traits (e.g., Templeton, 1982a, p. 23; Futuyma, 1986a, pp. 192–193; Mayr, 1988,

p. 428). Although aware of the difficulties of explaining the evolution of linked coadapted gene sets (Maynard Smith, 1978, p. 85), as recently as 1989 Maynard Smith wrote that " in a polymorphic population, a modifier mutation that improves one morph is likely to damage another, and so can spread only with tight linkage," and that "[Batesian] mimicry polymorphism is determined by a supergene" (p. 87).

The writings of E. B. Ford and others on Batesian mimicry butterfly polymorphisms were especially influential in establishing the linkage idea as the basis for complex polymorphisms, and Ford (1980) listed the supergene concept as one of the six fundamental insights that led to the evolutionary synthesis of the 1930s and 1940s. Ford and other ecological geneticists worked on the premise that alternative phenotypes correspond to alternative supergenes, and that, in general, linkage would evolve when gene "co-operation is needed" (Ford, 1961, p. 14). This vision of regulatory architecture was reinforced by theoretical models (e.g., Darlington, 1958; Turner, 1967) and by early work on regulation in prokaryotes, where true regulatory genes ("operons") are chromosomally linked to the loci they control (Jacob and Monod, 1961). So these empirical findings reinforced theoretical arguments and made chromosomal linkage a widely accepted idea. Models that consider the effects of linkage on divergence and speciation (e.g., Ford, 1971; Endler, 1977) may need to be reconsidered in the light of developmental mechanisms for correlations among coadapted traits.

While physically linked sets of genes may be an aspect of genetic architecture of some traits, as discussed below, this is certainly not an adequate model for the regulation of most switch-controlled traits, such as those influenced by complex regulatory mechanisms involving neuromuscular connections and hormones. The importance of understanding the genetic architecture of traits influenced by genes that are dispersed in the genome is illustrated by the kinds of mistakes that can be made if genetic architecture is misunderstood due to excessive faith in theoretical models, or is simply a black box. The supergene idea led to the idea that alternative phenotypes could be treated as if underlain by alternative alleles (or linked sets of alleles assorting as units; e.g., Ford, 1964, 1971). This was consistent with the idea that alternative phenotypes in general, like alternative alleles, require special equilibrium conditions of balanced or frequency dependent selection, in order to evolve (chapter 22). The erroneous one-allele, one-phenotype model of genetic architecture also led to the idea that the evolution of conditional alternatives expressed by different individuals in a population would begin with a genetic polymorphism, then require gene duplication of the polymorphic locus so that the potential to express both alternatives is present in the same individual (Bradshaw, 1973, p. 1258). In general, an overemphasis on the importance of gene duplication for the origin of phenotypic novelties is a product of the simplistic one-gene, one-phenotype model of genetic architecture.

The expectation of linkage among coexpressed and coevolved sets of alleles was so strong that it led to a needless controversy over the ultimate consequences of selection against recombination. By the chromosomal linkage hypothesis, selection should increasingly limit recombination that would break up coadapted sets of alleles, sometimes leading to the evolution of interacting linked polymorphisms (alternative superalleles or supergenes; Dobzhansky, 1951; Ford, 1961). But this led to the conundrum of why the genome does not congeal into a single optimal arrangement of genes (or two alternative supergenes) with no recombination at all (Turner, 1967; Lewontin, 1974; Maynard Smith, 1977). Lewontin (1974; see also Levin, 1975) argued that the genome does not congeal because continued adaptation in a changing environment requires recombination to generate new adaptive combinations.

Selection to maintain recombination is potentiated by the models of genetic architecture of eukaryotic traits suggested by Britten and Davidson (figure 2.5), which depicts the coexpression of coadapted traits controlled by a developmental coordinating mechanism or switch able to coordinate gene expression among loci including those on different chromosomes. This means that there need not be selection to limit recombination in the context of developmental mechanism, as there might be if coordinated transcription were dependent upon chromosomal contiguity. In eukaryotes transcription for even a single protein can involve transchromosomal loci developmentally linked by alternative RNA splicing (e.g., see Sharp, 1994).

The supergene concept is an extension of the idea of direct genotype-phenotype mapping, extrapolated to the evolution of complex coadapted traits. It was developed in the 1930s by C. D. Darlington (see Ford, 1980), well before the discovery of regulatory mechanisms that can coordinate the expression of genes that are dispersed in the genome (Britten and Davidson, 1969; Davidson, 1986; John and Miklos, 1988).

Although there are known highly conserved linkage groups (e.g., in *Drosophila* and other

Diptera) this is apparently not consistently related to linked function or expression (John and Miklos, 1988). Matthews and Munstermann (1990) document conserved linkage and consider three explanations in addition to functional association that might account for this poorly understood phenomenon. There are some eukaryote examples of linkage groups functioning together, such as the *bithorax* and *antennipedia* "homeobox" gene clusters of *Drosophila*, indicated by the underlining of the two major gene clusters in figure 4.17. Globin genes are another possible example, although their chromosomal contiguity may have been due to gene duplication rather than selection for linkage (see John and Miklos, 1988). Data on the alternative phenotypes of Batesian mimicry polymorphisms have been reanalyzed by Nijhout (1991b), who concludes that "the major differences between the forms can be readily explained as being due to variation in a single developmental parameter" (p. 158), not requiring the presence of multiple linked genes (a "supergene" or set of regulatory genes):

> The pattern diversity in the Batesian mimicry systems that have been studied provokes a strong suspicion that the systems are controlled by supergenes and that these supergenes consist of very few linked loci of ordinary structural genes. It must be emphasized, however, that the existence of supergenes for color pattern formation in Batesian mimicry has yet to be critically demonstrated in many of the species in which they are assumed to occur (p. 160).

Readers interested in this issue should consult the discussions by John and Miklos (1988) and Nijhout (1991b, 1994c) along with the earlier literature, where linkage is suspected or presumed.

These findings, and many others reviewed by John and Miklos (1988), show that the functional coadaptation of complex phenotype subunits in eukaryotes is usually based on regulatory linkage via co*expression*, under the coordinating influence of such factors as hormones or environmental inputs that influence more than one chromosomal locus at once, not genetic linkage. John and Miklos (1988, p. 95) provide much evidence for the "important principle" that the components of the genome can be rearranged and the genome fragmented without harming ability to function in development. Perhaps the most dramatic illustration of developmental regulation not involving chromosomal linkage occurs in plants, where there is coordinated expression of nuclear and chloroplast genes (Anderson et al., 1995, p.137).

The model of gene coexpression adopted here is consistent with the finding, reviewed in later chapters of part II, that evolution is extensively combinatorial, including at the molecular level (chapter 17). Regulatory linkage of coexpressed traits allows for flexibility of gene use (pleiotropy) as well as combinatorial evolutionary change—genetic reasons for why the genome does not congeal (Lewontin, 1974). It also explains coevolved coexpression without linkage. Extensive pleiotropy would oppose the tendency to link a particular allele to any one functional set, and might even lead selection to unlink genes coopted for multiple uses so as to reduce the bias toward any one (see chapter 7, on the parliament of the phenotype).

Limitations of the Modularity Concept: Connectedness

There are good reasons to moderate the concept of modularity, and good reasons not to abandon it. Dobzhansky (1970) wrote a tirade against modularity that is worth repeating as a caution:

Contrary to the views of early geneticists, the organism is not an aggregate of "unit" traits or characters or qualities. Traits, characters, and qualities are not biological units; they are abstractions, words, semantic devices that a student needs in order to describe and communicate the results of his observations. A trait has no adaptive significance in isolation from the whole developmental pattern that an organism exhibits at a certain stage of its life cycle; one may define a trait only as an aspect of the path of development of the organism. Talking about traits as though they were independent entities is responsible for much confusion in biological, and particularly in evolutionary, thought (p. 64).

While Dobzhansky's objection is well founded, the modular properties of phenotypes are as real as their connectedness. As concluded by Simon (1973), "everything is connected but some things are more connected than others' (p. 23).

The problem with emphasizing modularity is the risk of pushing a subdivided view of the phenotype once again to an extreme. This has already occurred for modularity in the literature on evolutionary psychology, which refers to behavioral traits and human abilities not as is if they have certain modular properties, such as localizability in particular brain regions, but as if they actually *are* separate modules, as in "the mind is a set of modules" in which

there "mental modules" (Pinker, 1997, p. 23), such as a "spatial relations module," a "rigid object mechanics module," a "tool-use module," a "fear module," a "friendship module," a "child-care module," and so forth (Tooby and Cosmides, 1992, p. 113). This use of a modularity concept is problematic not only because it risks an exaggerated view of compartmentalization but also because it seems to have no consistent operational basis for definition in terms of developmental or functional discreteness or uniformity. Does any human trait recognizable enough to have a name constitute a module? Does the "friendship module" include friendships among politicians as well as friendships between spouses? How do we know? The term "module" has also established itself within evolutionary developmental biology, even, as confessed by some of those who use it, without an agreed-upon definition (Raff and Raff, 2000).

Lewontin (2001) writes that "the price of metaphor is eternal vigilance," a quotation attributed to Arturo Rosenblueth and Norbert Weiner. The device that I propose for holding the modularity concept in check, in addition to a consistent developmental definition as a property of switch-controlled traits, is to eschew use of the word "module," a term that pushes the idea of physical discreteness beyond the more modest recognition of modular properties implied by the adjective "modular." Organisms do not consist of modules. They develop in ways that create modular behavior, such as switch-mediated semi-independence of expression, a degree of dissociability during development and evolution, and semi-independence, relative to independently expressed alternatives, under selection. The practice of thinking in terms of modularity, rather than modules, as a reminder of the limitations of the concept, will at least appeal to those who care about those limitations. In fact, use of older terms such as "trait" or "element" for a part of the phenotype could be considered synonyms of "module" that exaggerate discreteness just as badly. But these old words, unlike the banners of fashion, are less likely to mislead. For a discussion of artificial modularization in scientific thought, and a more ambitious solution, see Bohm (1980).

The omnipresent other side of modularity is the connectedness and integration of switch-controlled traits within the larger phenotype, as emphasized by Dobzhansky (see also Lewontin, 1992). Connectedness and overlap is represented by the nonspecific modifiers of figure 4.13. Connectedness among parts is one implication of hierarchical analyses of organization, where individuality and atomization is perceived in lower level analyses, sometimes called "reductionistic," and integration is perceived when the same structures are viewed at a higher level, sometimes called more "holistic." Even highly discrete functional subunits, such as molecules and cells, show *connectedness*—correlations and interactions among parts—when viewed in their associations at higher levels of organization. Epithelial cells, for example, have cytoskeletal connections that extend from cell to cell, and these connections may be important in certain coordinated, integrated responses of epithelium (Frankel, 1992). Plant thylakoid appressed modular regions (e.g., granal stacks) are physically interconnected by single membranes, the stroma thylakoids, and the interconnections facilitate coordination of light-condition acclimation that involves adjustment of the extent of the appressed membrane (see figure 3.7).

Behavioral subroutines are also functionally and structurally somewhat interdependent, being subunits of an integrated individual. The same appendages and sensory structures whose coordinated use in a behavior pattern makes them a subunit of morphology may be used in both courtship and predation, and the individual may not survive or reproduce if one of the subroutines is dropped. Fentress (1983) succinctly expressed the relation between modularity and connectedness in behavior: "If interactions were the only feature we would end up with so much homogenous soup, whereas if extreme compartmentalization were the rule there would be no way to obtain organized action" (p. 25).

Connectedness and interaction among modular traits are nicely illustrated by the bones of the vertebrate skull (Roth, 1991), as already described above (figure 4.5). Individuality of particular bones is sufficiently clear that homologies with other vertebrates are well established. Yet the positions of sutures that are the boundaries between bones can be quite variable. When pieces of bone are removed near a suture, bone growth restores the shape of the skull, but which of the surrounding bones contributes most is variable. This indicates variability in the importance of surrounding subunits for the construction of a particular modular trait—that is, a trait that has certain properties of modules. During morphogenesis, several different morphogenetic fields may affect the same trait, but their effects may not be absolutely congruent (Van Valen, 1962b; Roth, 1991).

Subindividual phenotypic components are only *semi-independent* because they share traits, or overlap, with other subunits, because they have physi-

cal connections with others, and because they cannot function, survive, and reproduce on their own. Differentiated subunits are bound to function as specialized parts of an integrated individual. They are not full-fledged units of structure or of selection because they cannot independently reproduce, though I will argue that they are subunits of selection with potentially quantifiable fitness effects. In some respects, individual organisms are not entirely reproductively independent either, insofar as their survival and reproduction depend upon interactions with other individuals such as parents, allies, and mates. While it is accurate to refer to the parts of organisms as having *modular properties*, it is potentially misleading to refer to them as "modules" because of the inevitable connectedness among them. I use the word *subunit*, rather than module, when in need of a noun for a switch-determined part of an organism. The prefix "sub-" is a reminder that (a) the subunit is a subordinate part of a larger entity and (b) the subunit is subindependent—less than a fully independent unit of organization or reproduction due to its connectedness with others and some degree of functional dependence on that connectedness.

Modularity does not imply rigidity, for plasticity and accommodation of parts are other universal properties of phenotypes. Riedl (1978, p. 112) uses nuts and bolts and jigsaw puzzles as metaphors for phenotypic subunits. This vividly represents their dissociability and somewhat stereotyped form, but it does not reflect the variation, plasticity, and connectedness that are also important for development and evolution. The same phenotypic subunits legitimately can be described either as modular or as connected to others, depending on the question at hand. A polyphenism, for example, can be considered two modular alternatives, in order to emphasize their independent, divergent nature (West-Eberhard, 1989). Or the alternatives can be described as connected ("one conditional strategy") in order to recognize their reciprocal frequencies and maintenance as a pair (e.g., Dawkins, 1980, p. 345). Some phenomena, such as cell-specific gene expression, are best described in terms of modularity. Others, such as integration and correlations between traits, are better described in terms of connectedness. Bonner (1988) discusses the paradox of integration and isolation of units of complex organisms and gives examples of both aspects at different levels of organization.

Modularity implies both discreteness, or evidence of boundaries around subunits, and connectedness, or integration within them. It is easy to think of cell or individual organisms as discrete entities, but in fact they are so actively tied to materials, resources, and other individuals in their environments and ontogenetic histories that the boundaries can be as misleading as they are useful. What is the material basis, or the *reality*, of the relation between discreteness and connectedness, component particles and infinite whole? To say that living modular entities are responsive and therefore inevitably connected to or influenced by other sources of movement and energy may be a beginning to a better understanding of biological organization but is certainly not a solution.

The most irksome limitation of the modularity concept is the worry that the apparent tangibility of biological units defined in terms of developmental switches will distract from their connectedness with each other and with inorganic nature. The only certainty I can see is that overemphasis on either modularity or connectedness is sure to distort reality, whatever reality may be. Emphasis on connectedness to the neglect of component modularity, as in the Gaia hypothesis or an extreme supraorganism view of groups or ecosystems, may misrepresent reality where modularity is important, as in the recognition of individuality in reproduction and selection (G. C. Williams, 1992; Hamilton and McNutt, 1997). Dobzhansky's statement, quoted above, that a "trait has no adaptive significance in isolation from the whole" may underestimate the power of semi-independent modular expression of gene sets to influence evolution, in partial isolation from the whole. At the same time, overemphasis on modularity to the exclusion of continuity is sure to mislead, as it does in the genetic view of development represented by the neo-Weismannian reduction, which regards individual phenotypes as discontinuous mortal units in the history of life, neglecting the importance of cross-generational continuity of the phenotype (see chapter 5).

Complementarity: Modularity and Connectedness as Complementary Aspects of Biological Organization

It would be difficult to overemphasize the importance of agility in being able to appreciate both the modularity and the connectedness of biological organization. Many patterns of evolution reflect reorganization of genome and phenotype that depends on their modular properties. And many phenomena, such as phenotypic accommodation of anomalies due to plasticity (see chapter 3) and the

evolution of different ways to produce the same phenotypic end point (see chapter 25), reflect connectedness that seems to transcend modularity (Greenspan, 2001). The tension between modularity and connectedness is not a new problem, as I show in chapter 24, on gradualism. Wake (1996c) refers to the extremes among biologists of the mid-twentieth century as the "mechanistic materialists" such as R. A. Fisher "and perhaps most western evolutionary biologists," and the more holistic Russian school represented by Schmalhausen and perhaps, surmises Wake, Sewell Wright. On the one hand, the reductionists focused on genes and atomistic explanations of organisms, with the whole a mosaic of separate, interacting but ultimately independent parts, and the whole a sum of the parts, with no emergent properties. On the other hand, the holists saw the parts so interconnected that they cannot be studied separately.

The continuous–discrete, gradualism–saltation question in biology was compared by Haldane (1958) to the wave–particle problem in the physics of light (see Holton, 1973). In an essay on William Bateson, Haldane remarked that "Bateson's discontinuities in evolution may be capable of description in other terms, as Newtonian particles can be described as wave packets" (p. 26). Just as the wave and particle theories of light have inspired controversy in physics (see Holton, 1973), the continuous and the discrete aspects of phenotypic variation have provoked deep philosophical divisions in biology that are still unresolved (see chapter 24). Some biologists have concluded that the analysis of continuous and discrete variation reflects different methodologies whose results cannot be compared: "The languages of systematics [where characters are discrete traits] and biometrics [which depends upon quantitative measures] must remain incommensurate" (Bookstein, 1994, p. 225). Johannsen (1911) credits de Vries with making a synthesis regarding continuous and discontinuous variation but believes that it was "too eclectic" for the stringent analytical tendencies of modern genetics then current, especially in American science (see also Hull, 1994, p. 493).

I propose that the solution to this problem is to accept a double standard for viewing development and phenotypic organization, and agility in jumping between discussion in terms of modularity and continuity, depending on the question at hand. This is the solution ultimately adopted in physics, where the particulate viewpoint sparked a bitter controversy when quantum mechanics (Heisenberg) challenged classical conceptions stressing continuity (most notably defended by Schroedinger), until it was pointed out by Bohr (1927 [1934]) that both describe different real qualities of the same phenomenon, depending on level of analysis or point of view (in that case, instrumentation). Similarly, in biology, by moving among levels of organization, we can see that continuous variation characterizes the dimensions of discrete traits. And by looking at mechanisms of plasticity, we can see that some, switch-controlled plasticity produces clearly modular and recurrent responses, while other, hypervariable mechanisms produce responses so finely adjusted to variable circumstances and so netlike in their pattern formation that their responses are sometimes labeled with the indeterministic sounding term "emergent."

Thus, when authors assert that "[d]iscrete polymorphisms are relatively rare and are not representative of the way morphological variation is distributed for most organisms," proposing instead that "we most often see phenotypic variation that is continuously distributed" (Reznick and Travis, 1996, p. 255), I would answer that the very ability to measure biologically meaningful continuously distributed variation often depends upon the occurrence of bounded phenotypic subunits that we recognize, and measure, as "traits." Similarly, when authors emphasize netlike pleiotropic interactiveness of gene expression, suggesting that this indicates the need for a new view of biological organization, which makes the treatment of subsets in isolation "fall by the wayside" (Greenspan, 2001, p. 386), I note that interactiveness presumes interacting entities, which are discrete things. Thus, the same author notes (p. 386) that evolutionary increase in the number of cell types (kinds of modular subunits) increases plasticity in the responses of a system: by changing level of analysis from the net to its components, he moves from an emphasis on connectedness to an emphasis on modularity, showing that both concepts are necessary. Bohr asked that physicists accept both wave and particle approaches, even though both could not be assessed in the same plane of focus at any given time, the choice depending on the theoretical or experimental questions at hand. Similarly, in biology one can emphasize the modular nature of a subindividual trait such as the vertebrate skull, or the fact that bony components of skulls are continuously variable and finely accommodated during interactive growth that cannot be described in terms of rigid isolated pathways for each part.

In biology, optimization theories presume that fine adjustments of quantitative variables are possible. Comparative studies of development, on the other hand, show that discrete variations can occur and that their dissociation plays a role in evolution (see part II). The discrete variants can be phenotypically accommodated by continuous plastic variation in the phenotype as well as genetically accommodated by selection on continuous, polygenic variation (see chapter 6). Both kinds of variation are complementary in the evolution of phenotypic novelties (discussed in parts II and III). Haldane's (1958) conclusion regarding discontinuous variation, written in an age of evolutionary thought dominated by Darwinian gradualism, was apt: " Bateson's fundamental notion of discontinuity in the evolutionary process, which he enunciated seven years before the rediscovery of Mendel's work, will remain, though doubtless with some modifications, a component of any theory of evolution (p. 26)."

Landmarks in the Evolution of Modularity

The origin of modularity during the evolution of life poses no special problem. Modular development is as old as organismic responsiveness itself. Modularity can be considered an automatic consequence of responsiveness: the ability to respond implies that a stimulus causes phenotypic change, or a developmental switch, and a developmental switch marks the differentation of a modular phenotypic trait. Compartmentalized phenotypes must be ancient properties of life, as old as branching decision points in ontogeny, which seem to characterize all forms of life including even viruses (see Ptashne, 1986, on phage λ) and bacterial molecules (Parkinson and Kofoid, 1992).

New modularization therefore does not depend on selection for modularity per se (cf. Wagner, 1996). But once a modular trait exists, it can undergo selection for *increased* modularity, or *consolidation* as a discrete trait. This can be achieved via increase in the two properties of modular traits—internal integration, and discreteness relative to other traits. These two aspects of increased modularity are termed by Wagner (1996, p. 38) increased *integration* ("increase in pleiotropic effects among primarily independent characters") and increased *parcellation* ("differential elimination of pleiotropic effects between members of different complexes"). In developmental terms, greater integration implies improved coordination by a switch. In functional terms, this means increased interdependence of the components of the modular subunit controlled by the switch (decreased dissociability of the components). Increased parcellation implies greater developmental independence and dissociability from other traits. The more highly consolidated a modular trait, the more independently it is subject to selection and evolution—the more legitimately it can be regarded as an independent subunit of selection (see chapter 7). All of the innovations heralded as "major transitions in the evolution of life" (Maynard Smith and Szathmary, 1995) are examples of increased modularization.

Consolidation beginning with increased parcellation produces new *subdivisions* of a phenotype (e.g., in the evolution of new cell types or organs), whereas consolidation beginning with increased integration occurs in the evolution of new higher level entities. An example of the latter is the gradual evolution of highly integrated insect societies, via decreased independence of the component individuals due to increases in the division of labor among them and consequently increased mutual dependence of the component individuals (West-Eberhard, 1979).

The origin of the cell nucleus was an increase in modularity that began with parcellation and marked a turning point in the evolution of life. The nuclear membrane separates the nucleoplasm from the cytoplasm of eukaryote cells, distinguishing them from prokaryotes such as bacteria and blue-green algae. E. O. Wilson (1975) calls the procaryote-eucaryote divergence "the deepest chasm in all of evolution" (p. 392). A nuclear membrane originally could have served functions having nothing to do with cellular physiology. It may have served as a barrier to the movements of molecular parasites and pathogens. As recognized by Gerhart and Kirschner (1997) in their general discussion of cell evolution, nuclear compartmentalization early could have been favored by ordinary natural selection for separation of enzymes that in a single compartment would interact and interfere with each other. In effect, the invention of the nucleus and other intracellular organelles creates a division of labor with separate compartments for different biochemical processes that occur within cells. This in turn makes possible the semi-independent selection and evolution of the enclosed processes. The rule of independent selection (see chapter 7) depends for its operation on modular structure.

In addition to possessing a nucleus, the eukaryotes are distinguished by the presence of mitochondria and a genome subdivided on multiple chromosomes, features not present in prokaryote cells. It is tempting to suppose that the precursors of the single prokaryote chromosome were selfish cytoplasmic molecules, suitable ancestors of selfish genes, wandering independently and out of control in the cytoplasm of unpartitioned anucleate cells. In a phenotype-centered view of evolution, the selfish bearers of manipulative instructions we think of as genes may have been a mixed blessing until they were brought under control in the interests of a cell. A circular unit chromosome like that of most prokaryotes is one way of restricting the independence of particulate genes. Another, adopted in the eukaryote line, is sequestration within a separate organelle, the nucleus. The usual explanation for the evolution of intracellular membranes is that just given—biochemical partitioning, either to keep reactants together (Maynard Smith and Szathmary, 1995, pp. 125–126) or to separate potentially interfering reactants (Gerhart and Kirschner, 1997). Whatever the original context of the evolution of the nuclear membrane (other hypotheses are considered by Margulis and Sagan, 1986), its ultimate, evolutionary significance is an increase in modularity and the consequent opportunity for divergent specialization between nuclear and cytoplasmic functions. Simultaneously, this internal specialization produces a division of labor and increased mutual dependence of parts, as in the social insects. This, in turn, reinforces internal integration, giving the evolution of modularity a self-reinforcing aspect.

As in the integrative pathway to the evolution of increased modularity in the social insects, the formation of the earliest cell membranes separated a modular unit of life from its molecular and physical environment, at the same time forming the first unit of selection above the molecular level. Then further integrative modularization occurred with the aggregation of cells to form multicellular organisms, followed by differentiation of different classes of cells to form separate tissues and organs. Again, the theme of internal division of labor, mutual dependence, and integration proceeded to consolidate a new modular structure and create a new, higher level of selection.

Many lesser landmarks of increased modularity some of them beginning with increased parcellation, and others with increased integration or complexity, are discussed in subsequent chapters. The reorganization of the jaw apparatus of cichlid fish enables a division of labor between the food gathering and the food-transfer functions of the mouth, with the result that striking dietary specializations became possible (see chapter 28, on adaptive radiation). The evolution of a unique hinge based on a novel division of the upper mandible of bolyerine reptiles (Frazzetta, 1975) had a similar effect on the predatory versatility of these snakes. And the mosaic of compartments of a butterfly's wing permit, and limit, the complexity of its color patterns (Nijhout, 1991b), while at the same time different color pattern elements, selected at a higher level, create coherent larger patterns. Compartmentalization of body segments and life history stages are other examples of modularity that underly the divergent complexity of development, always based on the flexibility afforded by a separation of activities or use.

Hypermodularity and Somatic Sequestration

Hypermodularity, or extreme compartmentalization, is sometimes associated with extreme divergent specialization of parts, providing that the modules are developmentally independent (this is not the case for self-organized hypermodularity, where different modules are only semi-independent in terms of regulation because they share the same rules of development or behavior). Examples of developmentally independent hypermodularity are the extreme life-stage compartments of holometabolous insects and many marine organisms (e.g., on hydroids, see Boero et al., 1992; and Sarà, 1987), the late versus early gene expression of metazoan embryology, and the somatic sequestration of insect imaginal discs and allied phenomena.

Weismann (1883), in his original discussion of germline sequestration, emphasized the "unalterable and inactive germ-plasm" as contrasted with the active idioplasm, or somatic material (p. 211). Similarly, in *somatic sequestration*, cells are set aside and temporarily inactive in development until a later stage. An example is the sequestration of insect imaginal discs, clusters of cells that are localized early in larval development and remain inactive until the pupal stage, when they participate in the development of adult morphology (Nijhout, 1994a). As I discuss below, one important adaptive developmental consequence of this may be that it assures that the initial cell phenotype that gives rise to localized development of specialized structures has not undergone irreversible specialization in the larva, and it also may guard against somatic mutation associated with repeated mitotic cell divisions (below). During development, this may have a kind of protective function. As in sequestration of the germline, somatic sequestration would help assure that any modifications of the genome that may be heightened during somatic activity and transmitted to descendents do not affect essential chromosomal regions or compromise totipotency of cell development. In effect, each imaginal disc

cell lineage in the adult starts with a relatively pristine genome, relatively free of any damage that may have resulted from activity during gene expression, and a uniform (dependably consistent), unspecialized cell morphology. The evolutionary result of hypermodularity like that promoted by cell sequestration is that larvae and adults can express and evolve extreme specializations semi-independently of each other.

The sequestration of the transmitted germline in an embryo has, as a kind of extension of sequestration, a division of labor between transmitted and untransmitted germline cells during the production of gametes in the adult. There are several mechanisms that seem designed to keep the expression of transmitted genes at a minimum. In some organisms (e.g., some insects; see Büning, 1994) the germline genes that are expressed during egg formation (e.g. in the nurse cells, granules, chromosomes or chromosomal segments that contribute to eggs and early embryogenesis) are not the same genes that are expressed in later development (the germ line is sequestered before these genes can be expressed by the germline cells). This, and the early versus late genes of embryogenesis (Davidson, 1986), permits extreme stage-specific functional specialization during development, because of the rule of independent selection and the consequent lack of genetic correlations among expressed gene sets.

Integration of a phenotypic subunit is achieved when all elements of a mosaic trait are well coordinated by a switch. This may be achieved by different arrangements of switches, as described in this chapter. How switch-controlled traits are arranged affects how they can be rearranged during evolution. The mosaic produced by a ganged switch in migratory locusts, for example, could be easily separated into its component parts if selection increased the discrepancy between the threshold responses to density of their switches. It would not be so easy to dissociate the elements of a self-organized mosaic because their switch mechanisms, while operationally independent, are not evolutionarily independent: they are duplicated subunits where all share the same switching rules.

General Evolutionary Consequences of Increased Modularity

The architecture of the phenotype affects, and in a sense "directs," evolution. But it would be a mistake to think that evolution is somehow limited by preexisting structure. A simple view of modularity might suggest that change is limited to shifting, deletion, and recall of unit traits, like moving furniture. But the very large number of subunits and their potential dissociability into smaller ones, in addition to their quantitatively variable dimensions, mean that modular traits are immensely modifiable. Modularity facilitates evolution both by use of already stable structural units and by the rapidity with which new complex structures can be assembled and quantitatively adjusted (Simon, 1973; see also Larsen, 1997).

There is plenty of evidence for evolutionary restructuring of development—for the creation of new phenotypic compartments, for increase in the number of phenotypic subunits, and for increase in their degree of individuation (increased modularity), as discussed in this chapter and in part II. Some authors have argued that increases in compartmentalization by decoupling (Lauder, 1981) can permit a burst of diversification involving the newly modularized structure, due to the liberating effect of a new division of labor (e.g., Liem, 1973, p. 425; Liem and Osse, 1975; Vermeij, 1973a,b; Lauder, 1981; Wake, 1982). New phenotypic subdivision allows novel specialization of parts that are newly independent subunits of development and evolution. Wake (1982) shows how stepwise increases in modularity of the feeding morphology of salamanders has led to diversification of feeding systems in that group. Increased modularity increases the number of independently varying quantitative elements. Conversely, fusions, or reductions in modularity, should produce reductions in evolutionary lability of the parts concerned.

Once a switch-regulated trait exists, modularity may increase in another way: by the evolution of further discreteness of the trait from others with which it shares nonspecific modifiers. This can occur, for example, if there is duplication of nonspecific modifier loci or shared phenotypic elements (see figure 4.13, c and C, respectively), followed by "capture" of different duplicates by different alternative phenotypes. This would reduce genetic correlations between the traits, increasing the independence of their expression and evolution (see chapter 7). Gene duplication is not required for the origin of discrete traits because they can be composed of different sets of preexisting traits and the underlying genes. But it would contribute to increased modularity providing that the duplicates become differentiated in their responses to the switch. Alternatively, modularity of alternatives would be increased (and genetic correlations reduced) if nonspecific modifiers come to be expressed in only one of two alternatives rather than

in both. Thus, alternative phenotypes (and all traits with shared modifiers) in a sense compete for shared elements under selection for increased specialization (chapter 7).

Several authors (e.g., Vermeij, 1973a; Lauder, 1981; Wenzel, 1993) have identified evolutionary versatility with the number of "decoupled elements" or independently varying traits or steps in a developmental sequence. Lauder (1981) presents evidence that phenotypic diversification in ray-finned fish (actinopterygians) is associated with "increase in the number of kinematic pathways governing movement of the jaws" (p. 436). This may have contributed to an associated increase in diversity of jaw structure in these fishes. Several examples from both plant and animal taxa are cited by Vermeij (1973a,b). One of them occurred in the evolution of actinopterygian bony fishes, where the freeing of the maxilla from the cheeks in the upper jaw of holosteans is associated with an increase in diversity of the shape and depth of the tooth row. A freeing of the maxilla from its ancestral connection with the cheek is associated with an unprecedented adaptive radiation of feeding types, including the origin of nonpredaceous diets. Conversely, reduced modularity is associated with reduced evolutionary diversity of a structure in frogs, in which fusion of the epicoracoid cartilage of the pectoral girdle is associated with convergence to a similar morphological shape (loss of morphological diversity; Emerson, 1988). Although these correlations are subject to the weaknesses of enumerative tests (see chapter 7), an increase in modularity or number of independently regulated switch points sometimes appears to have contributed to increased diversification of *that aspect of the phenotype* during the history of a taxon (whether or not this had anything to do with speciation rates, as sometimes implied by "key innovation" arguments, is another question).

Liem and others (see especially Galis, 1996a) propose that developmentally liberating compartmentalization has occurred in the evolution of feeding structures of cichlid fishes. Cichlid fish are remarkable for the diversification of their trophic morphology. They also differ from other fish in the presence of three small, functionally associated changes (Liem, 1973): a fusion of the lower pharyngeal jaws, accompanied by a shift of insertion of the fourth levator externi muscles, and the appearance of joints between the upper pharyngeal jaws and the basiocranium. This allows the pharyngeal jaws to specialize in masticating and swallowing food, freeing the premaxillary and mandibular (oral) jaws to specialize in food collection. The basic changes in the pharyngeal jaw apparatus allow for a stronger "bite" and is accompanied by evolutionary modifications of muscles (e.g., see Galis, 1993a) and nerves producing "an enormous range of possible functions" in processing different kinds of foods (Liem, 1973, p. 432). At the same time, they release the mouth structures to diversify in the food-collecting context. Following the initial reorganization, evolved specialization in many new directions was possible in both of the newly separated structures, in accord with the Rule of Independent Selection.

A similar connection between increased modularization and diversification is seen in sepsid flies. In some species the males, whose sternal sclerites are normally a single unbroken segment of exoskeleton, have a hinged or divided sternite associated with the evolution of exaggerated bristles (Hennig, 1949). The bristles are moved against the females during copulation, and in some species the new skeletal subunits are independently moved by powerful muscles and adorned with a diversity of species-specific brushes—behavioral and morphological innovations not observed in species with undivided sternites (Eberhard, 2001a, chapter 9).

The same correlation between increased modularity and increased diversification holds for the evolution of behavior. An increase in behavioral modularity occurred during the evolution of the worker–queen "castes" of social insects. Comparative study of Hymenoptera suggests that the distinctive worker and queen phenotypes originated by a separation or decoupling of cyclic female functions (see West-Eberhard, 1987b). Dominant, aggressive females monopolize egg-laying space in nests, suppressing oviposition but not brood care by less aggressive females, whose ovaries regress when they fail to oviposit. The result is a division of labor—a behavioral compartmentalization in which the reproductive tasks of a solitary female are divided between two classes of females, the reproductive queens and the nonreproductive workers who act as nurses and defenders of the nest. All of the elements of these two "social" phenotypes are present in nonsocial species. The behavioral subdivision or division of labor was likewise followed by radiating specializations in the two newly constituted phenotypic subsets (workers and queens), which have diversified independent of each other within the social insects (see Wilson, 1971a; Wheeler, 1991).

The fact that increased partitioning can result in an increased intra-individual (or intragroup) division of labor—increased complexity—possibly accompanied by increased organismal efficiency, raises the possibility that a purely organizational payoff may sometimes fuel the evolution of developmental or behavioral change. This would be a context in which modularity per se could be favored by selection, as suggested by Wagner (1995) and Gerhart and Kirschner (1997). Whenever a "trade-off" is involved—mutually compromised performance of tasks using shared phenotypic elements—phenotype partitioning per se (without or prior to increased modification of the new subunits) may increase efficiency enough to contribute to selection for a regulatory change such as that producing workers and queens in social insects and other examples of increased compartmentalization just discussed. Functional trade-offs characterize many alternative phenotypes, those associated with "antagonistic pleiotropy" (e.g., Moran, 1992b; chapter 16). Organizational advantage—the removal of functional conflict between complex traits having shared elements—may sometimes account for their origin.

Division of labor and mechanical efficiency is only one effect of increased modularity. Another is the effect on gene expression and selection. A new switch point breaks genetic correlations between formerly highly connected or correlated traits (e.g., those of a formerly unitary structure) and creates new ones within the new subunits created by the switch. This, along with increased mechanical versatility, would contribute to the diversification of forms following an increase in structural modularity. But the most striking evolutionary consequence of increased modularity is the creation of new subunits of gene expression and of natural selection.

The modularity concept contributes a new refinement to the debate over units of selection by showing that the relevant units are not always absolute—that there are degrees of independent selection, depending on the degree of consolidation and independence of the units involved.

5

Development

Development is phenotypic change in a responsive (plastic) phenotype due to inputs from the environment and the genome. Individual metazoan development begins with a ready-made, highly organized and reactive phenotype provided by a mother in the form of an egg. Cross-generational continuity of the phenotype makes it ultimately impossible to estimate the relative importance of environmental and genetic influence on development, for the inherited phenotype is a product of genomic and environmental influences intertwined and unbroken since the origin of reproducing phenotypes. Genomic and environmental factors have equivalent and potentially interchangeable developmental effects, effects that depend as much on the structure of the responding phenotype as they do on the specific inputs themselves. Each developmental response of the phenotype represents a new branch, or switch point, in development, associated with the production of a modular trait. While switches produce discrete traits, switch mechanisms are phenotypically complex and polygenically influenced, producing continuous variation in the dimensions of traits and quantitative genetic influence on regulation of trait expression. The quantitative adjustability of liability for switch-controlled responses is one aspect of gene–environment interchangeability and change in plasticity during evolution, since environmental and genetic influences on a threshold (or the ability to exceed it) are inversely related.

Introduction

Any comprehensive theory of adaptive evolution has to feature development. Development produces the phenotypic variation that is screened by selection. For a mutation to affect evolution, it must first affect development. In order to understand phenotypic change during evolution, one has to understand phenotypic change during development, as well as how to relate that change to selection and gene-frequency change (evolution). The evolution of the phenotype is synonymous with the evolution of development.

The genotype–phenotype problem addressed by the metaphors of chapter 2 is fundamentally a problem of development. Genetic programming, the canalized epigenetic landscape, and the recipes and blueprints contained in the genes—all are metaphors for development. Development is the missing link between genotype and phenotype, a place too often occupied by metaphors in the past. The task of this chapter is to outline a concept of development that connects it to mechanisms, on the one hand, and natural selection and evolution on the other, without a potentially misleading metaphorical crutch.

The portrait of development provided by developmental biology is not adequate to this task. Evolutionary developmental biology extensively treats the genomic correlates of gross morphological variation across phyla, with little or no discussion of behavior, physiology, life histories, and the kind of variation within populations that is required for natural selection to work. Some progress toward a population approach has been made in plant developmental biology (e.g., see Lawton-Rauh et al., 2000). But a strong emphasis on the genome means that environmental influence is systematically ignored. If you begin with DNA and view development as "hard-wired" (e.g., Davidson, 2000), you overlook the flexible phenotype and the causes of its variation that are the mainsprings of adaptive evolution.

I begin instead with the observation that DNA activity—gene expression—is universally condition sensitive and dependent upon materials from the environment. This implies connections between a DNA-centered approach and conventional insights about adaptive evolution in variable environments. The genome affects development at nearly every turn, so genes obviously play an important role in any theory of development and evolution. The problem of evolutionary developmental biology has been not too much of genetics but too little of the other factors of development: the organized flexible phenotype that is the setting for gene action, the influence of higher levels of organization on how genes are expressed, and the external environment as a source of developmental building blocks and information as crucial as genetic mutation for the origins of evolutionary novelties.

This chapter outlines a view of development designed especially to link data on developmental plasticity to principles of evolution by natural selection (see chapter 6). It proposes that individual development always starts with an organized phenotype inherited from previous generations. That phenotype is highly responsive to new inputs, which can be either genomic or environmental, as already discussed. These inputs cause the responses that correspond to new switch or decision points in development and give rise to new phenotypic traits. The result is development of a phenotype that is a mosaic of switch-determined modular traits. Beginning with this framework for relating genotype, environment, and phenotype during development, later chapters show how developmental plasticity facilitates evolution and leads to certain patterns of evolutionary change.

Continuity of the Phenotype

Continuity of the Germ Line and the Neo-Weismannian Reduction

Cross-generational continuity of the genetic material is a fundamental property of living things. Weismann's (1892) recognition of the continuity of the germ line was an important step toward the modern genetic theory of evolution and a crucial argument against Lamarckian inheritance, or somatic influence on the transmitted genes (for a brief history, see Mayr, 1982a). In many sexually re-

producing metazoans, the cells destined to give rise to gametes and gonads are set aside or sequestered early in development and thereby sheltered from participation in somatic activities. Even in organisms without early sequestration, such as plants and many invertebrate animals (see Buss, 1987), there is cross-generational continuity of the genetic material. There is an unbroken chain of genetic material uniting all branches of the tree of life since life began.

The idea of continuity of the genetic material is an indispensable part of any theory of biological organization. It is the basis for evolutionary improvement under natural selection. But the significance of this fact is exaggerated in what Frankel (1984, p. 1162) terms the neo-Weismannian reduction—the reduction of development to an expression of the genome. Although Weismann's term "germ plasm" meant gametes, or complete cells including cytoplasm and nucleus, it came to mean "what we call genes" (Dobzhansky, 1970, p. 31). In keeping with the problematic metaphors of chapter 2, the neo-Weismannian reduction treats continuity of the genetic material as if it means that a complete set of instructions for development is passed intact between generations except for changes that result from meiotic recombination and evolutionary change in gene frequencies due to selection or drift.

By this view, phenotypic organization is controlled by the packets of genes passed from parents to offspring; development begins with the new, zygotic genome; the genetic material persists while phenotypes come and go; and phenotypic organization in a new generation begins with genes. This sees the germ cells as immortal, whereas the soma or the phenotype "is the mortal part of the body of a multicellular organism that dies each generation" (Bonner, 1993, p. 118; see also Raff and Kaufman, 1983, p. 350). According to the germline view, which emphasizes cross-generational genetic continuity, the new individual begins when the male nucleus from the sperm fuses with the nucleus of the egg. Then "together they provide the genetic programme for development" (Wolpert, 1991, p. 137). A phenotype-centered view emphasizes connectedness between life cycles, where there is *phenotypic* continuity among generations (Bonner, 1974; Roth, 1994).

Continuity of the Phenotype

Continuity of the germ plasm is important for evolution, but it is not the place to begin thinking about individual development. Individual development al-

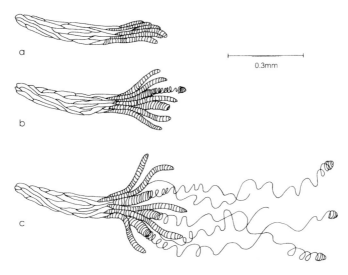

Fig. 5.1. Seed behavior in the orchid *Chiloschista lunifera*. (a) dry; (b) 10 seconds after moistening; (c) about 15 minutes after moistening. Threads fix the seed on moist bark. After Dressler (1981), based on Barthlott and Ziegler (1980).

ways begins with an *inherited bridging phenotype*—a responsive, organized cell provided by a parent in the form of an egg, a newly divided cell, or a set of cells that springs entirely from the previous generation, is adapted for survival and interaction in the gametic and embryonic environment, and is the active and organized field upon which the zygotic/offspring genome products and subsequent environments eventually act.

The bridging phenotype of a lichen is a spore containing both fungal and algal cells (Bonner, 1993, p. 29). That of a plant is a seed, whose inherited phenotype varies due to maternal genetic and environmental effects and may be "preconditioned" by the growth conditions of the parental plant to germinate and mature in a particular way (Baskin and Baskin, 1998; see chapter 8). An animal egg or the seed of a plant often has specialized physiological capacities and adaptive external morphology, including protective, respiratory, and attachment devices (see Zeh et al., 1989; Baskin and Baskin, 1998; figure 5.1). Cytoplasmic components of the parentally provided bridging phenotype can include organelles, ribosomes, proteins, and messenger RNAs (Gerhart and Kirschner, 1997, p. 382). The gametic cell—a phenotype—is "the minimum unit or link between one life cycle and the next" (Bonner, 1974, p. 5; see especially Roth, 1994, p. 311). As phrased by Rollo (1994), "organisms are lineage products, not discrete entities at all" (p. 205).

In many organisms, the inherited phenotype is evidently donated entirely by the female parent, but there are some important exceptions. The embryogenesis of plants differs from that of animals in that the products of plant meiosis in both sexes form multicellular structures—the embryo sac and the pollen grain (see a brief summary in Ammirato, 1999). Pollen grains contain from two cells (in many flowering plants) to as many as 40 cells in certain conifers (reviewed in Friedman, 1993, p. 16). In flowering plants, the pollen grain gives rise to a pollen tube, which delivers two sperm cells to an egg cell (called double fertilization, though only one of them contributes genes to the zygote). One of these combines with two maternally donated cells of the embryo sac to form the triploid endosperm tissue, which functions like a biparental placenta supportive of embryonic development, but set free of the parental soma. The participation of paternal cells in the provisioning of the embryo means that cross-generational paternal effects as well as maternal effects play a role in phenotype development and evolution in flowering plants (Westoby and Rice, 1982; Queller, 1983, 1984; Watson et al., 1995; Lacey, 1996). Gymnosperms are more like sexually reproducing animals in that the supplies to the embryo are maternal in orgin (from the gametophyte of the seed). In some primitive seed plants, the male gametophyte absorbs nutrients from the female ovule tissues (Friedman, 1993, p. 19), raising the possibility of negative pa-

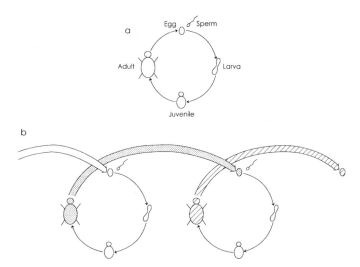

Fig. 5.2. Continuity of the phenotype. The life cycle of an animal is usually depicted as in (a), an endless cycle of temporary phenotypic forms. Cross-generational continuity is understood to be a function of inherited genes. In (b), individual life cycles are connected to show cross-generational phenotypic continuity (large arrows). Patterned arrows and eggs indicate inheritance of structure via the egg prior to fertilization. The same comparison applies in plants.

ternal effects on offspring number or nutritional status.

Paternal effects may be more widespread than heretofore appreciated. Paternal photoperiod affects the development time of progeny in *Drosophila simulans* (Giesel, 1988), but the mechanism by which this occurs is still obscure. Giesel reasons that such an effect can "only be due to alterations in the character of nuclear genomic information" (p. 1350) since sperm cytoplasmic elements are not known to be transferred to progeny. Other mechanisms are possible, however. Photoperiod effects on hormones or some other substance transferred in the seminal fluid, for example, might, via effects on the mated female, be transmitted as a maternal effect to the offspring. In the pteromalid wasp *Muscidifurax raptorellus*, the activation of oviposition, which occurs immediately after mating, evidently depends upon interaction with the male, and the reproductive response of females (number of eggs deposited per ovipositor insertion) is influenced by the *male* genome (Legner, 1988). Quantitative differences in oviposition behavior are heritable. Presumably factors transmitted in the male's seminal fluid modify the female's behavior (Legner, 1987), and female offspring are somehow affected. Cross-generational hormone effects are worth investigating in such cases.

Parental contributions to the inherited phenotype obviously present opportunities for manipula-

tion, and maternal manipulation of offspring via substances introduced into the egg may be a common phenomenon. Maternally derived substances can include hormones. Canary (*Serinus canarius*) mothers endow the eggs of later-produced chicks with higher doses of testosterone, which increases their begging vigor and may help to compensate the begging disadvantage of younger chicks, whereas cattle egret (*Bubulcus ibis*) mothers deposit more androgens in the first eggs, which may help older siblings eliminate younger ones during food shortages (Schwabl et al., 1997).

Structures surrounding the gametic cytoplasm of metazoans are products of parental somatic cells such as those of the ovarian follicles, whereas cytoplasmic materials such as nutrients and developmental instructions are provided by nongametic germline cells (e.g., nurse cells in insects; Büning, 1994). This division of labor has important consequences for evolution, as discussed in chapter 4. Parentally contributed materials can include organelles, ribosomes, proteins, and messenger RNAs (Gerhart and Kirschner, 1997, p. 382). So the life cycle of an individual does not begin with fertilization and end with the adult, as in textbook diagrams (figure 5.2a). Instead, it is just one loop in a continuous string of ontogenies linked by these gametic phenotypic bridges between generations (figure 5.2b). Parental effects endure into adulthood in organisms other than humans: in the goldenrod,

Solidago altissima, the soil on which the mother plant grew affects the germination, growth, and adult height of her offspring (Schmid and Dolt, 1994). There is no time in its life when an individual has a phenotype entirely of its own making or entirely dictated by its own genes. The hand-me-down bridging phenotype has lasting effects on the course of individual development, and it can include effects that have accumulated from uncountable generations in the past (Bonner, 1974; for a concrete example, see the discussion of locust maternal effects, below).

Continuity of the phenotype, or the unbroken and overlapping phentoypic connections between generations mediated by parentally constructed offspring phenotypes (e.g., eggs, spores, seeds, and effects on later development), is a fundamental fact of life. Development is a process where every change builds upon previous organization. Sometimes the offspring phenotype is manipulated by products of genes, and sometimes by factors in the environment. These factors nudge and join a pre-existing structure, a structure with some elements so responsive and internally active that they are sometimes said to organize themselves.

It would be a serious error to view the bridge between generations as entirely genetic, as if each new individual were a "return to the drawing board" (Dawkins, 1982, p. 259). This overlooks the insight from embryology that *all order proceeds from pre-existent order* (e.g., see Gilbert, 1991, p. 136). This insight had been articulated in the nineteenth century by E. B. Wilson (1894), citing earlier work of such embryologists as Nägeli, Hertwig, His, and Driesch, who were opposed to the the germ-plasm–centered view of Weismann and others:

We must not . . . lose sight of the cardinal fact that the organization of the idioplasm [the phenotype], which is at the bottom of every operation [during ontogeny], is *an inheritance from the past*. The idioplasm of every species has arisen through the modification of a preëxisting idioplasm, and every response that it gives to stimulus is an expression of its past history. (p. 123; italics in the original; brackets mine)

Britten (1998) has stated this view in modern molecular terms:

All living systems except a few parasites (viruses and DNA elements) have an unbroken lineage from cell to cell going back to some cell precursor, perhaps part of the origin of life, many billions of years ago. So the starting point can only be a complete cell. Thus in oogenesis the first units are bound to pre-existing structure and the pre-existing parental cell informs the early stages. (p. 9375).

As soon as proteins were synthesized by replicating molecules—the birth of the somatic phenotype—it was possible to endow replicates with substances that could be advantageously carried over and used in a subsequent generation (Bonner, 1974).

Phenotypes, including these cross-generational bridging phenotypes, are of course characterized by plasticity—the ability to react to stimuli with a change in form, state, movement, or rate of activity (chapter 3). This property of the bridging phenotype makes it susceptible to the new inputs that initiate ontogenetic change. Some inorganic structures also react to inputs from outside themselves, and such materials undoubtedly gave rise to the earliest life (Fox, 1980, 1984). The same kinds of built-in sensitivity that make phenotypes responsive to the environment can make them responsive to inputs specified by genes, to manipulation by parents and parasites, and to internal interactions among parts, a quality I call "interchangeability" (see below). Responsiveness to all of these different influences gives rise to development and to the variation that fuels selection and evolution.

Continuity of the phenotype implies that the individual's genome does not control its development: the zygotic genome is constrained to play upon the responsive structure that is in place when particular genes are expressed. Stent (1985) cites this as evidence that "development is a historical rather than programmatic phenomenon" (p. 1). Contrary to the impression given by genetic-control metaphors for development, the bare genes in isolation are among the most impotent and useless materials imaginable. Gene products are made from elements imported from the environment. Then, once a gene product is assembled, its biological significance depends on the reactivity *to it* of the phenotype. Exquisite precision in the timing of gene expression should not be taken as evidence for the genetic orchestration of development. Rather, it should be taken as evidence of the enslavement of the genome by the phenotype, a changing field for gene action that demands precise timing, because otherwise gene expression would be ineffective or misplaced.

In other words, the predictable effects of genes depend as much on the specific organized flexibil-

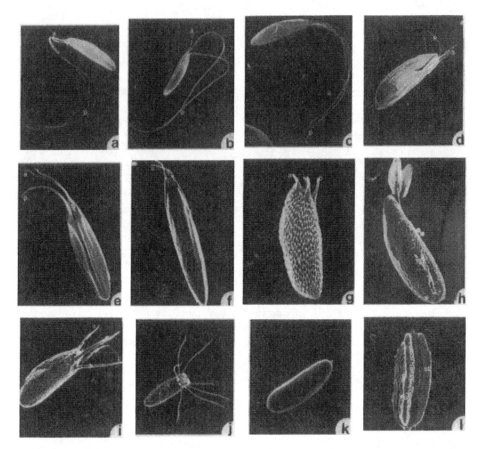

Fig. 5.3. Adaptive diversity in the eggs of drosophilid flies (Diptera, Drosophilidae; all but h–j are Hawaiian species). (a) *D. murphyi*; (b) *D. claytonae*; (c) *D. heteroneura*; (d) *D. formella*; (e) *D. truncipenna*; (f) *D. mulli*; (g) *D. longiperda*; (h) *D. willistoni*; (i) *D. virilis*; (j) *D. pattersoni*; (k) *Scaptomyza (Exalloscaptomyza) oahuensis*; (l) *Scaptomyza (Tantalia) albovittata*. Not to scale. From *Arthropod Structure and Development* (formerly *International Journal of Insect Morphology and Embryology*), Volume 22, Kambysellis, M. P., "Ultrastructural diversity in the egg chorion of Hawaiian *Drosophila* and *Scaptomyza*: ecological and phylogenetic considerations," pp. 417–446, © 1993, by permission of Elsevier Science.

ity, modular differentiation, and local conditions within a preexisting structure as they do on the specificity of the genes themselves. Roth (1982) stresses the importance of the nature of the phenotype—what Williams (1966) called the importance of the somatic environment—for specific responses to factors from the external environment. The same insight extends to the effects of genes (e.g., Milkman, 1961, pp. 35–36; see especially Larsen and McLaughlin, 1987). Without a responsive phenotype with a particular configuration, and specific environmental supplies, the selection-hewn codex written in the genes (*sensu* G. C. Williams,

1992) is meaningless. A particular gene or environmental factor of major effect may appear to be entirely responsible for the appearance of a particular trait, since in its absence the trait does not occur. But it has this decisive effect only because there is a structure that is poised by the peculiarities of its organization to respond in the observed way. The impact of gene expression at every stage of the life cycle depends on the presence of a structure susceptible to change. There is no stage of development in any organism without an organized phenotype capable in specific circumstances of some specific, active response.

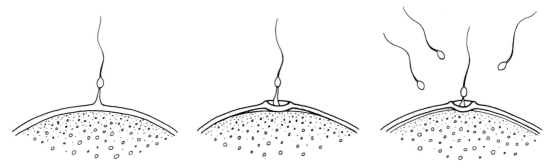

Fig. 5.4. Active sperm reception by the eggs of a starfish. A fertilization cone produced by the egg is extruded in response to a spermatozoan, which is then drawn into the egg. Thereafter, approaching spermatozoa elicit no response. After Fol (1879), reproduced in Deuchar (1975).

The Bridging Phenotype in Early Ontogeny

The inherited bridging phenotype is organized to respond to new inputs in specific ways that characterize early development. It is also organized for active survival in its distinctive environment (figure 5.3) and even for a kind of social life with gametes of the opposite sex. First, there are mating interactions between eggs and the spermatozoa (figures 5.4–5.6), which likewise have complexly organized phenotypes provided entirely by a parent. These interactions can be surprisingly complex and are a virtually unexplored frontier for evolutionary analysis (but see Pitnick and Karr, 1998). The pronucleus-visiting behavior of the egg pronucleus in a ctenophore (figure 5.5), for example, might be considered a mere curiosity if it were not for the fact that polyspermy, in which more than one spermatozoan enters the egg, occurs at some frequency in most insects (including the honeybee, where it is well studied; Milne, 1986) and in all elasmobranchs, urodeles, reptiles, and birds (Austin, 1965). This "physiological polyspermy" means that pairing behavior continues within the egg, until one male pronucleus unites with the female pronucleus. The other male pronuclei usually die. But in some birds, the rejected male pronuclei are relegated to the periphery of the blastodisc, where they induce cell division during early cleavage stages of embryonic development (Austin, 1965). Such events have been recorded with little attention to their possible adaptive significance. Developmental biologists seldom analyze this aspect of what they see, and evolutionary biologists seldom see these fascinating microscopic earliest

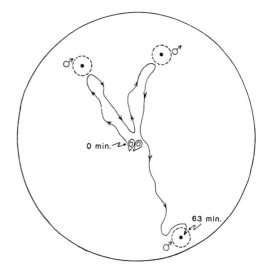

Fig. 5.5. Mate-selection behavior by the egg pronucleus of a ctenophore (*Beloe ovata*). Time-lapse video recorded movements of the pronucleus (center, at time = 0 minutes) toward three different sperm (black dots) that had entered the egg, finally fusing with one of them 63 minutes later (bottom). Areas within dotted lines, not crossed near the two bypassed sperm, are free of cortical granules. Sperm remain stationary while the female pronucleus moves. Courtesy of W. G. Eberhard, after Carré and Sardet (1984).

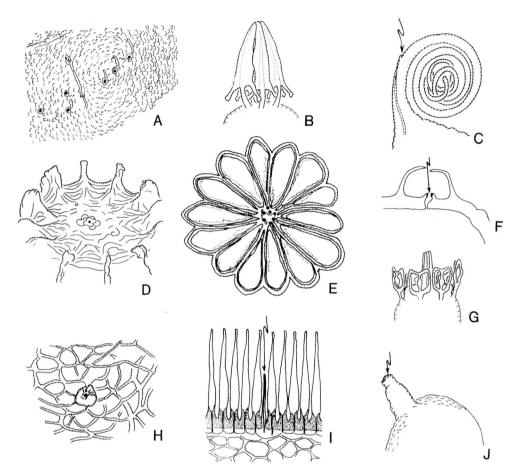

Fig. 5.6. Specialized morphology associated with micropyles (sites of sperm entry) of insect eggs. Micropyle indicated by black dot or arrow. (A) *Perlodes microcephala* (Plecoptera); (B) *Ortholomus* sp. (Hemiptera); (C) *Acanalonia bonducellae* (Homoptera); (D) *Hypena proboscidalis* (Lepidoptera); (E) *Habrosyne pyritoides* (Lepidoptera); (F) *Columbicola columbae* (Phthiraptera); (G) *Kleidocerys resedae* (Hemiptera); (H) *Fannia canicularis* (Diptera); (I) *Cylapofulvius* sp. (Hemiptera); (J) *Sialis lutaria* (Neuroptera). Compiled and redrawn by W. G. Eberhard, after Hinton (1981a,b).

stages of the life cycle as they pursue the implications of what Buss (1987) calls a theory of adults (see chapter 1). So this is an area ripe with possibilities for anyone willing to combine the two approaches (e.g., see Palumbi, 1999; Swanson and Vacquier, 2002).

Following fertilization, the egg proceeds, without input from its own genome, through the earliest, crucial phases of ontogeny—such events as establishment of polarity and axes of symmetry, and in some cases several cell divisions. In animals where early gene expression has been monitored, there is an initial stage in phenotypic development when the individual's genes are completely inactive (e.g., see Davidson, 1986; Buss, 1987). In some

species, such as the frog *Xenopus*, the zygotic genes are not detectably expressed until several hours after fertilization, when the blastula already contains 4,000 cells.

The embryonic phenotype is often characterized as a product of maternal gene transcripts. This draws attention to the maternal role in constructon of a cross-generationl bridging phenotype, but it errs in creating an impression of exclusive control by genes. The maternal environment is immensely influential as well (see especially Landman, 1991; Rossiter, 1996; Jablonka and Lamb, 1995, 1996; Sultan, 2000; chapter 26). In the insect egg, for example, the "informational" content of the egg—the gene products from the nurse cells, in-

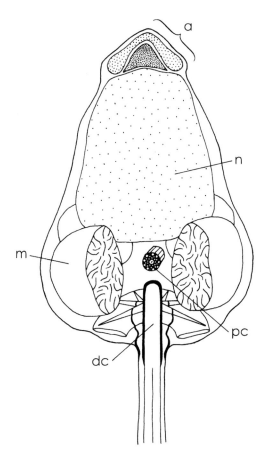

Fig. 5.7. Generalized diagram of an invertebrate spermatozoan, showing the acrosome (a), nucleus (n), mitochondria (m), proximal centriole (pc), and distal centriole (dc). After Baccetti and Afzelius (1976).

cluding messenger RNA, ribonucleoproteins, enzymatic proteins, ribosomes, and mitochondria—is only about 1% of the mature egg volume (Büning, 1994). The rest comes via the hemolymph of the maternal soma and can reflect environmentally influenced variables like diet, population density (crowding), photoperiod, and interactions with mates (chapter 5). Similarly, the zoomorphic phenotype of the spermatozoid is primarily a donation of the paternal phenotype, not a product of the sperm genome. Although a complex, motile sperm cell looks like an independent little individual organism (figure 5.7), the genes of the animal spermatozoan in most species are physically condensed and completely inert (Baccetti and Afzelius, 1976) during the gamete stage. The morphological, biochemical, and behavioral phenotype of the sper-

matozoan is a product of the paternal phenotype, not the genes within the sperm. The paternal phenotype is, in turn, a product of both its environments and its genes.

The egg phenotype qualifies as a full-fledged, organized and active individual. It may be morphologically equipped for dispersal, respiration, active transport of water and nutrients (figures 5.1, 5.3), and interactions with sperm (figures 5.4–5.6). The egg micropyle, or entryway for the sperm, is sometimes an elaborate and species-specific structure (Kambysellis, 1993; figure 5.6). The capacity of the egg for interactive behavior is revealed during its social debut, contact with a spermatozoan or pollen grain. Contrary to the image of a passive egg being penetrated by an active sperm, the animal egg can take an active and selective role in fertilization, engulfing (or not engulfing) a sperm (figure 5.4), and engaging in other activities that are necessary for fertilization to occur (reviewed in Eberhard, 1996). In a species with egg entry by multiple sperm, for example, the egg pronucleus was observed actively moving among male pronuclei before uniting with one (figure 5.5), in effect actively choosing its mate (*choice* being differential responsiveness to different alternatives). Similarly, in flowering plants, fertilization is potentially influenced by the maternal phenotype, though not directly by the egg phenotype itself, for the mother supplies substantial resources for the pollen-tube growth that enables pollen to reach eggs (Willson and Burley, 1983; Willson, 1994).

Maternal gene transcripts continue to be used for some functions after embryonic gene expression begins (Anderson, 1989; Davidson, 1986). This blurring of the line between generations underscores the importance of continuity of the phenotype. The eggshell that separates the chicken from the egg is not a convincing boundary, for the shell is a maternal secretion and it contains a phenotype built by the mother hen. Yet the new individual with the hand-me-down phenotype has its own genome and its own environment, and it will follow its own distinctive developmental path, lastingly influenced by its parentally structured beginning.

Parental effects continue long beyond embryogenesis in many organisms, especially in organisms with extended parental care (Cheverud and Moore, 1994). In collared flycatchers (*Ficedula albicollis*), for example, adding one egg to the mother's clutch affects the clutch size of her adult daughters, reducing it by an average of one fourth of an egg (Schluter and Gustafsson, 1993). In migratory locusts the maternally transmitted effects of crowding on the adult phenotype accumulate across generations, with four generations of crowding required to produce the full expression of the migratory phenotype (Kennedy, 1961). Similarly, underfed seed-beetle (*Callosobruchus maculatus*) larvae produce small eggs as adults, which leads to production of small eggs by their female offspring and to undersized grandprogeny (Fox and Savalli, 1998). Such phenomena provide at least a glimpse of the indefinitely extended consequences of continuity of the phenotype, in which events are "both the effect of earlier, and the cause of later developmental transactions" (Stent, 1985, p. 1).

Continuity of the phenotype has led to the idea that homology, or phenotypic similarity due to common ancestry, depends on cross-generational continuity of *information*, some of it transmitted via the organized phenotype, not just the germ line (Van Valen, 1982b; Roth, 1994). In some organisms, phenotypic inheritance includes environmental elements—nests, territories containing resources, and other individuals, such as parents and other group members (Wcislo, 1989), with which they communicate and from which they learn—and the social heritage is just as dependable and as necessary to their normal development as that passed between generations in the gametic nucleus and cytoplasm. This concept of inheritance predated the discovery of genes (West et al., 1994). The connecting thread of the extended phenotype bears and perpetuates the imprints of environments and genes past, transmitting those imprints between generations during the same phenotypic act of reproduction that transmits the genes. The uniqueness of the germ line compared with the phenotype line as a factor in evolution lies not in its continuity, a property shared with the phenotype, but in its relative immutability during development, the faithfulness of its replication, and the dependence of its rate of multiplication on the success of the phenotype as a whole. As a result of these three qualities, the population of genotypes reflects their history of selective screening across time. But the *continuity* of the germ line depends on the continuity of the phenotype that bears it.

A corollary of continuity of the phenotype is that *phenotypic* structures are the units of reproduction. A parental phenotype, whether a single cell, a multicellular individual, a group, or a species, gives rise to one or more offspring entities eventually capable of reproduction. It is this cycle of variable somatic activity and gametic reproduction that makes evolution by natural selection possible (Bonner, 1974). Genes replicate, but they cannot reproduce themselves across generations. Even in gametes they are packaged in phenotypes as eggs, spores, pollen, or sperm and sometimes vegetative fragments built by a parent. Over the long run, the survival and frequency of genes in a population depends on the packaging—the properties of phenotypes that reproduce. In meiotic drive a selfish gene's behavior may bias its probability of entering a gamete and being differentially transmitted to the next generation, but its transmission nonetheless depends on the success of the reproducing phenotype and its cross-generational continuity. Organismic genes are compelled to hitch a ride between generations encased in a phenotype. Brandon (1990), after Salmon (1984), calls this "screening off"—the genes are screened off from being the primary agents of cross-generational reproduction by the dependence of their reproduction on that of phenotypes. Because of screening off, *reproducing phenotypes are appropriately considered the units of selection*, even though evolution, a response to selection, is defined in terms of change in the frequency of genes (see the definitions in chapter 2).

Continuity of the phenotype is the thread that connects phenotypic development and evolution. Genetic and environmental factors that influence development always do so via effects on a preexistent phenotype.

The Dual Nature of All Regulation

Some people think of ontogeny in terms of how genes are "mapped on" to the phenotype, but this is too flat an image for the dynamic events of development. It would better evoke the events of development to ask, "How do environmental supplies, partially ordered by the genome, affect the highly reactive phenotype that exists before they arrive?"

Individual ontogeny begins with a responsive phenotype. *Development* is change in the phenotype as it is influenced by genomic and environmental inputs over time. The effects of these inputs depend crucially on the organization of the preexisting phenotype. Though Van Valen (1986b) suggests that the environmental information for a developing grass seed may closely resemble that for a developing lizard nearby, the operative environmental information—called the "ecological environment" (as contrasted with the larger "external environment") by Brandon (1992, p. 82)—is certainly quite different for a grass seed and a lizard. The concept of information is meaningful only with respect to a receiver phenotype that is organized to respond in a particular way (Bernays and Wcislo, 1994).

The preexisting phenotype is a transducer of both environmental and genomic information. The very specificity of gene action, and of environmental effects as well, depends as much on that structure as on the input that causes a response. Critical periods in development are periods of phenotypic readiness to respond to particular inputs. If those same inputs occur at another time, when the phenotype is *not* structured to respond, they have no effect. Tissue-specific gene expression is a product of local *phenotypic* divergence in the properties of cells. The genes are the same in the different cells but only act in those structured to interact with particular expressed genes. A phenotype poised to respond in one of these particular ways can then be stimulated by an "inappropriate" stimulus, and it may still respond. Nestlings can become imprinted during critical periods on people rather than on adult birds (Lorenz, 1952). Expression of particular genes can be induced by chemical treatments and heat shock (Lindquist and Craig, 1988). And the species-specific approach of a female to the distress call of her infant can be evoked by the courtship signal of an infant-mimicking male primate (Moynihan, 1970). Neither genes nor environmental conditions have any developmental significance without a phenotype already organized to respond.

With the responsive phenotype at the center of development, it becomes clear that the provenance of impinging stimuli—whether environmental or genomic—is of little developmental consequence. This is the death knell of the nature–nurture controversy, for it puts genes in perspective without detracting from their importance. Various critiques of the idea of genetic control of development emphasize the importance of other factors at the expense of genes. The "structuralist" or "structuralist heresy" (Frankel, 1984; see also Raff, 1996, pp. 131–132), for example, emphasizes the behavior of phenotypes rather than genes. But it relegates genetics and therefore the influence of evolutionary-genetic history to a minor role compared to certain physical and chemical processes, such as reaction–diffusion, that do not require extensive genetic instruction. Self-organization is another approach that explains many features of development that are not encoded by genes (e.g., Kauffman, 1993), demonstrating limits to the idea of genetic control. But self-organization sidesteps, as does the "pan-environmentalism" of writers like Matsuda (1987), the problem of relating genomic and environmental influence in development. The nature–nurture

dichotomy disappears with the realization that the developing phenotype responds to both internal and external stimuli in much the same way. As a result, genomic and environmental factors are interchangeable during evolution. If genetic and environmental influences are equivalent and interchangeable, they are not properly seen as opposed or even as complementary factors.

Despite the confusion engendered by the nature–nurture dichotomy, there is good reason to maintain the distinction between external environment and genome, at least in *evolutionary* discussions. Evolved improvement in the phenotype depends on the distinctive nature of the genome as faithful replicator and cross-generational transmitter of edited developmental information—what G. C. Williams (1992) calls the codex. Only the genetic inputs have been screened by selection. Selection may favor phenotypes that actively incorporate certain environmental elements, but it can only do this via gene frequency change.

Granted this one significant distinction between external environment and genes in evolution, it is important not to extrapolate and consider the genes more essential or specific than the environment in *development*. Note that I am not arguing here merely that the environment affects development. Everyone admits that. I am arguing that environmental effects on development can be as specific and essential as genetic effects, a far from universal belief. The metaphors for development reflect a belief in developmental inequality of genotypic and environmental inputs. There is a metaphorical genetic program, but no metaphorical environmental counterpart. The contours of Waddington's epigenetic landscape are specified entirely by genes, as discussed in chapter 1 (figure 1.5).

The dual nature of regulation is evident from the nature of phenotype determination at switch points. Phenotypic or trait *determination* means the adoption of one of two alternative states or pathways at a decision point (Bull, 1983). Determination can be distinguished from influence on regulation. A single genetic locus may *determine* whether or not a discrete trait is expressed if the alternative alleles at that locus inevitably cause a threshold to be passed or not passed, and the trait developed or not developed, under most circumstances. The switch may nonetheless be *influenced* by other genetic and environmental factors. For example, a particular range of temperatures may characterize the organism's usual environment, and in that range (and only in that range) the switch mechanism is poised so that the response is determined by the alleles in question. Possession of a particular trait rather than an alternative trait can be either genetically or environmentally determined, but regulation—the mechanism or the process—can never be determined by genes or environment alone, because the mechanism is an aspect of structure, and structure is *always* a product of both genetic and environmental influence. There is no exception to this universal law of *dual environmental–genetic influence.* There can be no exception because gene expression always requires raw materials or components ultimately imported from the external environment, as well as expressed genes themselves.

In *development*, genes and environment have complementary quantitative effects on switches, such that an increase in the influence of one implies a commensurate decrease in the influence of the other. There is a sliding continuum in their proportional influences on switch determination, such that environmental and genetic factors in a sense compete with each other for control of regulation (Bull, 1987). This relation is axiomatic. It follows from the fact that influence on trait determination can come from two and only two sources—genome and environment—and threshold effects are all-or-none in nature (a threshold is either passed or it is not).

The variation in heritability of a threshold trait, sex ratio in map turtles, is illustrated in figure 5.8. *Heritability* is the proportion of phenotypic variance of a population that is due to genetic variance (for methodology, see Falconer, 1981). It is a continuously variable measure of genetic influence on a trait in a population, whose observed value depends on the environmental conditions under which it is measured. In most conditions, heritability of sex ratio in these turtles is zero. That is, there is no measurable genetic influence on sex determination, which is completely dependent upon temperature. But in a narrow range of intermediate temperatures (28–30°C) the sex ratio is variable and heritability is very high (0.80; Cade, 1984, p. 359, after Bull et al., 1982a). The high heritability revealed by intermediate laboratory conditions may seldom occur in nature due to nest-site selection by females: in a natural population studied by Vogt and Bull (1984, cited in Bull, 1983, p. 121), nearly 80% of nests were all male or all female, with female-producing nests located in sunny places and male-producing nests located in the shade. So the high heritability artificially revealed in the laboratory in part may reflect lack of effect of selection on sex ratio in nature, where conditions exposing genetic variance in this trait may seldom occur. In other conditions, genetic variation

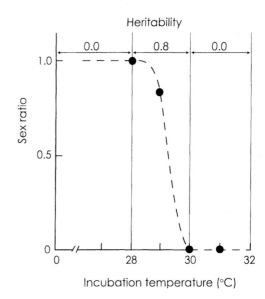

Fig. 5.8. The relationship between heritability and environmental conditions in sex ratio determination in map turtles. Curve after Bull (1983), based on Bull and Vogt (1979), Bull et al. (1982a,b), and Bulmer and Bull (1982). Heritability values from Bull et al. (1982a).

may be exposed, and the degree of genetic influence on sex determination and the sex ratio could be changed under selection (see also Rhen and Lang, 1998). That this has occurred repeatedly in turtles is discussed in a later section.

The difficulty of separating environmental and genetic causes of variation was established early in the history of modern genetics. In an article on "Environmental control of mutant expression," Sang (1961) illustrates this by describing the pioneer work of T. H. Morgan on the "eyeless" gene in *Drosophila.* Morgan found that the *eyeless* gene varies in penetrance (observed frequency) with the age of the culture in inbred strains. The expressivity of the gene varies as well, with some flies completely eyeless, and others showing various grades of eye reduction, even in laboratory strains known to be homozygous for the trait. The same kinds of variation were sometimes produced by mutants— environmental and genetic changes had the same effects. Subsequent refinements of culture methods, such as rearing larvae in axenic environments and manipulation of the nutrient medium, overcame these problems, but even environmental standardization and use of highly inbred stocks did not eliminate phenotypic variability. In a long list of well-known Drosophila mutants, including *vermillion, tetraptera, aristopedia, eyeless,* and *antennaless,* penetrance is affected by environmental factors such as temperature and diet (Wigglesworth, 1961). Such careful controls have to be imposed to isolate the genetic effects from the environmental ones that the difficulty of doing so should be considered a major datum in itself.

It is customary in biology to refer to some traits as "genetically determined" and others as "environmentally determined." But a close look always reveals the dual nature of determination and the imprecision of these terms. Most if not all "environmentally controlled" alternative phenotypes yield less than 100% of the expected morph under a given condition, suggesting genetic variation for morph determination. Environmental sex determination in the echiurid worm *Bonellia,* for example, is only 76% for larvae exposed to the determinant environmental factor, the female proboscis (Matsuda, 1987, after Leutert, 1974). Populations of mice and voles with photoperiod-influenced hibernation regularly show a small percentage of individuals breeding in winter (Gorman et al., 2001), and artificial selection in Siberian hamsters produced a line with 90% failure to respond to short day length (compared to 20–30% in unselected populations), demonstrating a genetic component to this "environmental" response.

Conversely, "genetic" traits invariably have demonstrable susceptibility to environmental influence if sought. The sickling defect in sickle cell anemia, a textbook example of a single-locus genetic disease, is always conditionally expressed, when oxygen is low (Herskowitz, 1962). Familial bipolar disorder (manic-depressive disease) correlates with day length (Goodwin and Jamison, 1990).

The highly variable relation between genotype and environment in phenotype expression is clarified by the concept of the *norm of reaction* of a genotype—"the array of phenotypes that will be developed by the genotype over an array of environments" (Gupta and Lewontin, 1982, p. 934, after Schmalhausen, 1949; the term apparently originated with Woltereck, 1909). As this concept implies, plasticity, or the set of responses provoked by environmental variation, is determined by genotype. If this is true, different genotypes may have different degrees and forms of plasticity; then, it follows that plasticity itself is subject to selection and evolutionary change.

An array of reaction norms associated with different genotypes were studied by Gupta (1978; see Gupta and Lewontin, 1982; figure 5.9a). A number of important points are illustrated by these graphs. First, different genotypes have different degrees of plasticity. The phenotype (bristle number) of genotype 8, for example, varied little over the range of environments (temperatures) studied, compared with the phenotype of genotype 2, whose range of variation was about three times as great. Furthermore, the pattern of plasticity varied, with some genotypes showing a rise, some a reduction in bristle number with rise in temperature during development. Still others had extreme phenotypic values (some high, some low) at intermediate temperatures. This clearly shows the deterministic role of genotype in plasticity. It also shows the deterministic role of the environment in setting not only the form of the phenotype but also the range of genetic variation expressed and, hence, exposed to selection.

The norms of reaction of a large number of genotypes were generally not parallel, so whether or not a difference in genotype gives rise to a difference in phenotype is heavily dependent on environmental conditions (figure 5.9b). Where there is overlap in reaction norms, the different genotypes produce overlapping distributions of phenotypes. In environmental conditions where the lines of genotypes intersect, the genotypic differences would be "invisible" to selection—there is less genetic variance in phenotype (region b in Figure 5.9b). On the other hand, in some environments the phenotypes expressed are markedly different. Genetic variance in those environments is high (region a in figure 5.9b), and the population is expected to evolve relatively rapidly due to the greater potential for a response to selection. When dealing with plastic traits, then, one cannot ignore the *dual role of the environment* in determining the strength of selection and the course of evolution: the environment is not only the agent of selection in the sense of being the arena where phenotypes are evaluated in a game of survival and reproductive success. It is also an agent of development, which by interacting differently with different available genotypes sets the phenotypes in the positions where they will be seen by selection.

There is no known method to estimate the proportional total influence of environment versus genes on individual or trait development. Measures like heritability and genetic versus environmental variance do not come close to a true assessment of the relative contributions of environment and genes to development. They refer only to causes of variation in particular quantitative traits, measured under specified environmental conditions and within a single generation. Heritability could be zero, for example, in an extreme environment or one that fluctuates so that negative and positive effects on

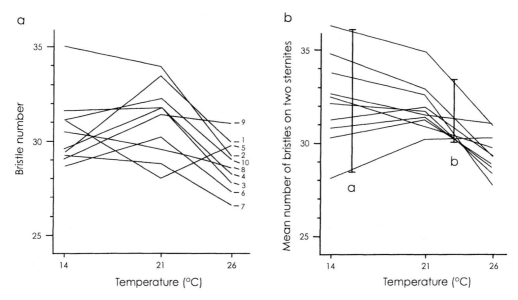

Fig. 5.9. (a) Genotypic differences in reaction norms and the condition dependence of results of selection: reaction norms of sternal bristle number in response to temperature variation in 10 female flies (*Drosophila pseudoobscura*). Genetically different individuals (compare 1 and 5) can have contrasting developmental responses to environmental conditions. Selection where their reaction norms cross (at about 16°C and about 25°C) would evaluate their phenotypes as identical, whereas selection at intermediate temperatures (about 21°C) would evaluate their phenotypes as opposite extremes. (b) Environmental influence on heritability and selectable variation: reaction norms of sternal bristle number in response to temperature variation in 10 female flies (*Drosophila pseudoobscura*). Selection on the variation exposed at low temperatures (a) would be more effective in modifying bristle number of flies developing at that temperature than would selection at higher temperatures (b), where there is little genotypic variation. After Gupta and Lewontin (1982).

heritability exactly cancel—yet obviously both genotype and environment (nutrients, water, permissive temperatures, etc.) would be necessary for the development of the trait. More important, given continuity of the phenotype, a comprehensive estimate of relative environmental and genotypic contributions to a trait would require disentangling these influences across generations, a truly impossible task, since phenotypic inheritance of both kinds of effect extends back to the beginning of life, with no natural break.

Four Meanings of "Genetically Determined"

By sheer repetition, the phrase "genetically determined" has become a source of confusion regarding the role of genes in development. So it is worth looking more closely at what these words actually mean. In common biological parlance, the term

"genetically determined" is used informally to mean several different things, all of them more accurately represented by the words "genetically *influenced*."

1. "Genetically determined" meaning *heritable*: Phenotypes are sometimes labeled genetically determined when parent–offspring or sibling correlations or selection experiments show a genotypic effect on their frequency of appearance. This confuses the technical with the everyday meaning of heritability (Levins and Lewontin, 1985), but it is quite common even among biologists. Results of a heritability study often appear with a title like "Genetic determination of phenotype X in species A," or "Genetic polymorphism in species A," meaning that heritability was detectable in laboratory conditions. But heritability could be 1.0 in one condition and near zero in another, for heritability is condition dependent, as already discussed. The most that can be said from a high value of heritability is

that phenotype determination is highly influenced by genotype *in the conditions observed*.

Caution is in order when laboratory measures of heritability are extrapolated to nature, for heritability measured under laboratory conditions may systematically exaggerate the importance of genetic influence. Two factors may contribute to this. Successful artificial rearing conditions may allow persistence of genetic variance that in natural conditions would have been removed by selection; and relatively constant laboratory conditions, designed to control environmental influence in order to measure the effects of genes, may reveal genetic variance not exposed in more variable natural conditions. Since gene expression and heritability are condition dependent (as discussed above), narrow conditions may stimulate the expression of one or a few loci while others are masked or unexpressed.

Pentatomid bugs (*Nezara viridula*) laboratory reared on diets of different quality have different heritabilities of size (McLain, 1987). In a well-fed population, individuals are significantly larger and heritability is high (mean, 0.591; standard error [S.E.] 0.147), whereas in a poorly nourished population heritability was low (0.304, S.E. 0.125). The underfed group resembled natural populations in body size. Similarly, James et al. (1984) found that gypsy moth (*Lymantria dispar*) and elm span worm (*Ennomos subsegnarius*) outbreaks in Connecticut corresponded to high densities of the wood thrushes (*Hylocichla mustelina*) that feed on them, indicating that these birds are usually (in nonoutbreak years) food limited. Such results are of particular interest for critical evaluation of laboratory heritability estimates. Many, perhaps most, heritability studies (e.g., those reviewed by Cade, 1984, and by Falconer, 1981) employ laboratory or domesticated organisms likely provided with ample food and rearing conditions in order to keep them healthy. And heritability estimates typically have large standard errors (> 10%; Markow and Clarke, 1997). Such estimates must be cautiously extrapolated to nature, where suboptimal and variable conditions may be the norm.

2. "Genetically determined" meaning *having a regulatory mechanism whose structure is influenced by one or more known genes*: Ptashne (1986) refers to the condition-sensitive regulation of alternative phenotypes in λ phage as governed by "a genetic switch" because it is a relatively simple system known to be underlain by a limited, known set of particular genes. All regulatory mechanisms presumably are influenced by particular genes, but in more complex organisms it would be difficult to

approach giving a complete list of them. Again, when referring to the role of specific genes in regulation, it is more accurate to use the term "genetically influenced," especially when, as in the case of λ phage, environmental influence is likewise known to occur.

3. "Genetically determined" meaning *genetically influenced form*: It is fairly common to refer to a phenotypic trait, such as family resemblance in the shape of the nose, as genetically determined, as if the inevitable condition sensitivity of shape can be ignored. It would be accurate to say that the shape of the nose is genetically influenced.

4. "Genetically determined" meaning *genotype specific or predictable on the basis of genotype*: Sometimes inheritance of a phenotype, even a complex one, occurs in simple Mendelian fashion, as if governed by alternative alleles at one or a few loci. In sex determination, sex expression may depend on a small number of genes inherited on a certain chromosome or chromosome set (see Bull, 1983). Even in these cases of especially strong genetic influence, particular circumstances may be required for genotype or karyotype to be decisive, and genotypic influence from other loci may indicate that a certain genetic background is also required for the influence of a particular genotypic input to be decisive in throwing the switch (Murfet, 1977). Therefore, such mechanisms are more accurately described as "genotype-specific" or "allele-specific" to indicate that in most natural circumstances certain genes are decisive for the development of a particular phenotype.

Genotype-specific phenotype determination invariably involves decisive influence of just one or a small number of loci, so individuals not expressing the phenotype possess the genetic and phenotypic potential to express it, save the few alleles that have a major effect on the switch. "Genetic" females bear the genes for male form but not the genes for male *determination*. This is easily shown by induced sex change under hormone treatments (see chapter 15). This potential to express both alternative forms, being present throughout the population, means that genotype-specific determination is an evolutionarily labile trait. It is because both genotype-specific and environment-specific alternative phenotypes have basically the same genetic-regulatory architecture (that shown in figure 4.13) that the sliding proportions of genotypic and environmental influence may easily change. Genotype-specific control means that one or a few loci have an overwhelming or controlling influence on trait determination (regulation). The genes that influence the form of the trait itself—the specific modifiers of form (figure 4.13)—could conceivably be identical while the elements of regulation vary.

Since this last meaning of "genetically determined" is the one most likely to foster the idea that genes can actually control development, it is worth examining closely the regulation of traits that appear to be under single-locus control.

Genes of Major Effect and Polygenic Regulation

In general, threshold responses are polygenic, as already mentioned (see especially Roff, 1996; Roff et al., 1997). They are subject to continuously variable genetic and environmental influence, which is not surprising since they are so often regulated by complex mechanisms, for example, involving hormones and sensory-motor equipment that are polygenic in nature. Early experiments on the genetics of extra digits in guinea pigs (Wright, 1934) and molars in mice (Grüneberg, 1947, 1963, cited in Hall, 1984) established the polygenic nature and environmental sensitivity of the inheritence of discrete threshold traits (reviewed in Hall, 1984, pp. 102–108).

Polygenic influence enables fine adjustment of degree of plasticity, in effect integrating or summing the effects of environment and genes on the threshold of a particular switch. Polygenic regulation also means that under certain circumstances, or because of an episode of natural selection, a single locus may prove to be a *gene of major effect*—meaning that one of the several polygenes that influence a switch mechanism may predominate over others in the size of its influence, giving the impression of single-gene (allelic) control of trait expression, even though several polygenes of minor or intermediate effect may still act (Falconer, 1981). The variable effects of polygenes are now better documented using molecular techniques of chromosomal mapping of quantitative trait loci (QTLs; Falconer and Mackay, 1996).

The dominant influence of a one or two loci on trait determination does not argue against a fundamentally polygenically influenced switch, even though in genetics threshold traits are treated under a polygenic model, and those controlled by genes of major effect under the classical Mendelian model as if they were fundamentally different in mechanism (e.g., see Roff, 1996, pp. 8–9). A major gene effect, however, such as that on caste determination in some social insects, obviously does not change the fact that caste regulation is highly polygenic, for hormone systems and hormone-

sensitive tissues and dietary differences are still involved in the switch (Buschinger, 1990a).

Either–or treatment of major gene versus polygenic regulation of adaptive traits (e.g., Orr and Coyne, 1992) may give the impression that two very different kinds of genetic systems are involved, when in fact a gene of major effect may occur at any one of many loci that influence polygenic traits. This has long been acknowledged in basic quantitative genetics texts (e.g., see Falconer, 1981, p. 95; Hatfield, 1997, p. 1025). Reports of major-gene effects are especially associated with the following events and conditions:

1. Mutation: Mutations sometimes cause large phenotypic effects on polygenic regulation, including pathologies that are inherited, such as those associated with familial disease. Mutations are also widely used in applied genetics to identify loci of polygenes involved in regulation of complex traits. A mutation of large effect overwhelms normal regulation at a switch point, overwhelmingly inhibits passing a threshold, or creates a new switch, to produce an unusual phenotype.

Mutations of major effect on polygenic regulatory mechanisms can give the false impression of single-locus genetic architecture and genetic control. As a result, the study of mutations in experimental genetics and medicine has been a powerful source of confusion about the importance of single genes, both within biology and in the popular press. When mutation in a single gene gives rise to a large phenotypic change, it is easy to forget that the phenotype is actually produced by a complex polygenic and environmentally susceptible mechanism. What has occurred due to a mutation, or when there is discrete expression of "Mendelian" alternatives, is that a single locus, one among many that influence the trait, has come to have a major effect on the polygenic threshold for producing the trait (Walker, 1979).

The eye-color mutations of *Drosophila*, the dwarf versus tall pea plants of Mendel, and other traits used in classic genetic experiments to dissect the structure of the genome, such as the mutations associated with familial inherited deafness or bipolar illness in humans, reveal a genetic component to phenotype determination and may enable geneticists to map the locations of particular influential gene loci on chromosomes. But they do not indicate that the traits they affect are normally influenced (certainly not controlled) by a single gene. Walker (1979) referred to the "asymmetry" of genetic effects between mutant and wild-type (normal) individuals (see also Sing et al., 1992, 1995):

[T]here are no specific [single] alleles for "wild type," *i.e.* for normal function. . . . In order to realize this function, myriads of alleles must be present in wild type form. Diverse mutations produce diverse effects . . . from which the breeder [or natural selection] can artificially select a specific mutant type, whilst he cannot select a specific wild type allele from a combination of wild type alleles. Thus, with regard to the selection process, mutant and wild type are not symmetrical options. (Walker, 1979, p. 250)

A mutation of large effect either overwhelms a threshold in most individuals or assures that a threshold will not be passed (it could inhibit the production of some trait). Other genes and environmental effects may still be important, for together they may poise the phenotype to respond to the mutant allele. Mutations of large effect abound in the popular press, where reports of the gene for obesity, alcoholism, or breast cancer raise the hope of an equally simple cure. Walker (1979, p. 250) regards failure to appreciate this asymmetry between normal (polygenic) and mutant effects on regulation a source of important confusion in evolutionary genetics.

In a polyphenism, an environmental factor, rather than a major gene, causes the large effect that overwhelms a switch. But the genetic structure of (polygenic) regulation in both cases is fundamentally the same (chapter 4, figure 4.13). In both, the underlying polygenic regulation allows for fine-tuning of the threshold of the switch under natural selection, making it possible to adaptively modify the effects of mutations and environmental factors.

In a polygenic mutationally mediated familial disease like bipolar (manic-depressive) illness, different families or populations showing the pathology may express the bipolar phenotype due to different mutations with equivalent effects. This is because regulation (expression or non-expression) of the bipolar phenotype is polygenic, and change in the pathway to its development can therefore be caused by mutations at different loci that cause the same developmental disturbance. Just as genetic and environmental effects are interchangeable, as discussed below, effects of different polygenes on polygenic regulation can be equivalent as well. It is for this reason that genetic studies, based on expectation that there is a single gene for bipolar illness, sometimes puzzle investigators when they do not localize "the gene" on the same chromosome or chromosomal region (Goodwin and Jamison, 1990). Such polygenic illnesses require an appropriately polygenic model of genetic architecture (Sing et al., 1992, 1995).

2. Controlled laboratory conditions: Genetic studies of peas (*Pisum sativum*) under conditions of constant day length and moderate temperature suggested single-locus control of early versus late flowering by alternative alleles of a locus (Sn; reviewed in Murfet, 1977). But the appearance of Mendelian segregation is reduced if the temperature is lowered or if the photoperiod is increased, and it disappears completely if seeds are exposed to continuous light from the start of germination. Flowering time in *Pisum* is now known to be a polygenic, environmentally influenced trait where the polygenes affect different aspects of regulation. Breeding experiments have identified four different loci that influence flowering time (reviewed by Murfet, 1977). They affect (1) sensitivity of the shoot apex to a flowering hormone, (2) production (or lack of production) of a substance inhibiting the initiation of flowering (the activity of this gene is sensitive to photoperiod), (3) the level of activity of the inhibitor produced by gene (2) above, and (4) the length of time the inhibitor-producing gene (2) is active. There are multiple alleles at all four of these loci. Furthermore, many environmental factors are known to influence flowering time in plants, including temperature, water, stress, and mineral nutrition (Murfet, 1977, p. 254f).

3. Recent strong directional artificial or natural selection: As discussed in chapter 6, selection is not expected to change allele frequencies at all influential loci simultaneously or at the same rate. For reasons discussed in chapter 6 (after Nijhout and Paulsen, 1997), under natural or artificial selection mutations or initially low-frequency alleles of relatively major effect are likely to be increased in frequency one at a time in order of their degree of influence on the phenotype, giving the impression, when high- and low-selected lines (or selected and unselected populations) are compared, that the trait in the selected populations is under single-locus control. Mutations of large effect may have both positive and negative fitness effects, and selection for modifiers that remove deleterious effects is expected to occur (Wright, 1941, 1977, p. 463). Over a large number of generations, more modifiers of the change would presumably accumulate, and the effects of multiple loci on trait expression would render the single-locus effect less predominant. By the model of Nijhout and Paulsen, this is the expected course of events during the evolution of polygenic regulation under selection "for" a particular trait (see chapter 6). Macnair (1991) made an additional interesting point, based on a simple

genetic model: under extreme conditions, in which survival depends upon rapid evolutionary adaptation to a contingency that is "a long way" from the mean prior to selection, such as toxic soil, then only populations with major genes that enable a large initial response to selection may be able to survive.

Taken together, these models may explain why major-gene differences are sometimes detected between recently isolated or artificially selected crop and pest populations that are subject to severe, new selection regimes in a new environment (Mcnair, 1991; Orr and Coyne, 1992; Ross and Shoemaker, 1997; Gottlieb, 1984). Local populations strongly selected for a particular trait, such as toxin resistence (e.g., see Antonovics et al., 1971), in nature or the laboratory, often show single-gene change. In the oligochaete worm *Limnodrilus hoffmeisteri*, resistence to environmental cadmium evolved in only a few generations both in nature and under artificial selection in the laboratory. Resistant populations synthesize high concentrations of a metal-binding protein, likely metallothionein, and crosses between resistent and nonresistent phenotypes indicated single-gene determination (Martinez and Levinton, 1996). Similarly, a recently evolved change in the social behavior of fire ants (*Solenopsis invicta*), which has evidently evolved since their introduction into the United States in the 1930s (Ross et al., 1996), has been traced to control by a gene of major effect. The gene, a protein-encoding allele Gp-9, mediates the switch between two forms of social organization (single- vs. multiple-queen colony formation) by affecting such traits as queen behavior, worker toleration of multiple queens, and worker tolerance of queens with alternative genotypes (Ross and Keller, 1998).

4. Major-gene effects on the switch between intraspecific alternative phenotypes in situations where conditional regulation is impossible or disadvantagous: Genotype-specific determination of complex alternative phenotypes in nature is relatively uncommon compared to environmental determination(see part III). But single-locus influence does evolve under certain conditions, especially when there is no environmental cue for an advantageous conditional switch between alternatives but there is a payoff for producing more than one form. Several authors note the scarcity of known cases of genotypic determination of phenotype expression in animals (see, e.g., Futuyma and Mayer, 1980, on multiple-niche polymorphisms; Shuster, 1989, on alternative mating tactics of males; Matsuda, 1979, on arthropod alternative phenotypes). Roff (1986b, p. 1011) lists known examples of one-locus determination of wing polymorphisms in insects, along with several species having polygenic control, and Gottlieb (1984) and Hilu (1983) list some examples in plants. Examples include the "pin" and "thrum" alternative forms of heterostylous species, and flowering time polyphenisms in some species (Hilu, 1983). Most of the genotype-specific "genetic polymorphisms" reviewed by Mayr (1963, pp. 152–154) involve relatively simple phenotypes, such as color morphs, blood groups, and Batesian mimicry morphs in butterflies (which, although phenotypically complex, appear to employ few phenotype-specific modifiers—see Nijhout, 1991).

A close look at putative genotype-specific control of complex phenotypic traits often reveals, as expected, an underlying polygenic, condition-sensitive developmental mechanism. So-called "genetic" wing-length polymorphisms in crickets, for example, are both polygenic and environmentally sensitive. Hormones are known to be involved in their regulation (Zera and Tiebel, 1988; Zera et al., 1992; Zera and Huang, 1999; Zera, 1999), and the thresholds for switching between morphs can be altered under artificial selection (e.g., Zera and Tiebel, 1988). Esterase titers involved in the regulation of this polymorphism are influenced by environmental factors such as population density (Zera and Tiebel, 1988).

In the marine isopod *Paracerceis sculpta*, three striking male morphs occur in Mendelian ratios in natural populations, and laboratory studies indicate not only influence by a single autosomal locus, but also an effect on growth and maturation rates (Shuster, 1989; Shuster and Wade, 1991a)—processes likely subject to genetically complex regulation. The well-analyzed genetically influenced polymorphism—often called a "genetic polymorphism" (Borowsky, 1987b; Ryan et al., 1992)—in size of male swordfish (*Xiphphorus* species) is another case in point. Genetic influence, measured under controlled conditions of ad libitum food, was found to be based on variation at a single locus on the Y chromosome. The same size differences are evident in collections from nature (Ryan et al., 1992). Large males mature later and court females, whereas small males mature relatively early and chase rather than court females (summary in Ryan et al., 1992). In *X. nigrensis* there are three alleles at the determinant P locus, corresponding to three modes in the size distribution of mature males (small, intermediate, and large; Ryan et al., 1992). The major gene associated with this polymorphism influences the onset of activation of the hypothalamic-pituitary-gonadal (HPG) system and the eventual secretion of androgens in males, which in poeceliid fish induces sexual maturation and stops or drastically reduces growth.

This example fits the predictions of the developmental-linkage hypothesis of genetic architecture in that the developmental mechanism of the switch between forms is obviously subject to polygenic influence, being a complex of structures composing the HPG system. Not surprisingly, given the large number of environmental factors that can influence size in these fish, including temperature (Borowsky, 1987a) and agonistic interactions with other males (Borowsky, 1987b), there is considerable overlap in male size classes, with genotypically small males sometimes larger than genotypically intermediate ones (Ryan et al., 1992). Agonistic interactions with other males are known to stimulate activity in the interrenal gland and increase production of corticosteroids (Borowsky, 1987a). Although behavior is better predicted by genotype than is size (Zimmerer and Kallman, 1989), behavior is also subject to alteration by interactions with other fish. Sneak-chase behavior, performed in the presence of large males, is a facultative alternative to courtship display. Given all of this condition-sensitive variation in both size and behavior, it would be misleading to conclude from the strong single-locus influence on morph determination under controlled conditions that the development of the alternative phenotypes is entirely under single-gene control.

The idea that strong single-locus influence represents exaggeration of the effect of one of many quantitative trait loci active in a particular switch seems to be supported by some findings of Buschinger and his associates on ants (Buschinger, 1975; Heinze and Buschinger, 1987, 1989; Winter and Buschinger, 1986; Buschinger and Francoeur, 1991; reviewed in Buschinger and Heinze, 1992). Most ants are typical of social insects in having nutritional determination of caste (for a concise review, see Nijhout and Wheeler, 1982), and there is little doubt that any departure from this (e.g., a strong genotypic component to determination) represents a secondary, derived state, as in some *Melipona* bees discussed above. In *Harpagoxenus sublaevis* and a *Leptothorax* species ("Sp. A"), Buschinger and his associates have shown that individuals homozygous for an allele "e" behave as the usual "wild type" and develop either queen or worker morphology, depending on conditions. In contrast, individuals bearing an "E" allele are intermorphs, intermediate in morphology between worker and queen. In *H. sublaevis*, the E allele affects one aspect of the complex social mechanism of caste determination in ants: it increases the susceptibility to the queen's inhibition of queen de-

velopment in larvae. EE larvae sometimes develop so slowly that they hibernate twice before molting to adulthood (Buschinger, 1990a, p. 51). This suggests that strong genotypic influence on the caste-determining switch evolved due to exaggeration of the influence of one of the polygenes involved in normal (ancestral) caste regulation in ants. Buschinger and Heinze (1992, p. 18) generalize these findings, comparing the worker–queen wing polymorphism of ants to those evolved as dispersal polymorphisms in other insects, and conclude, as emphasized here, that both environmental and genetic factors probably act via the same regulatory (hormonal) system. Their fascinating review of wingedness and intermorphs in ants is, for readers interested in development and evolution, a good modern-day supplement to Wheeler's (1937) "Mosaics and Other Anomalies Among Ants" (see also Wheeler, 1910b).

In sum, the regulatory architecture of complex alternative phenotypes that occur in Mendelian ratios or under single-locus control should prove to be fundamentally the same as that of traits showing polygenic influence on regulation of size, morphology, and behavior. Single-locus control often means that some element of regulation is affected by a gene of major effect on the response, relative to the many other genes and environmental factors that contribute to the regulatory mechanism.

Developmental Effects of Genes

If dual environmental-genomic influence on trait expression is the rule, and the genes are not the blueprint or program for development, then what, exactly, do the genes do?

Genes expressed in somatic cells do primarily two things during development: they replicate during cell divisions, with the result that as a rule every cell has a full complement of somatic chromosomes, and they act as templates for the production of particular molecules. Some gene products regulate the expression of others, and some specify proteins. All gene products are manufactured within cells. As a result, they are in a good position to affect the rates and kinds of interactions that occur there. This suggests a potential asymmetry between gene products and factors from the external environment in ability to influence development. Genes operate from within. Their effects project outward to every level of phenotypic organization. This helps to explain the enormous organizing influence of genes in many intracellular processes fundamental to life. But it does not mean that genes have the upper hand over the external environment in development, for there is a corresponding asymmetry that gives the external environment more power at higher levels of organization. Structures in contact with the external environment can respond in complex ways and can transmit environmental effects inward. Later I argue that this contributes to the ability of the external environment to take the lead over mutation in the origins of adaptive evolutionary change (chapter 26).

Gene products play two roles in development. Some function as phenotypic building blocks, and others function as signals or cues. A building-block function is served by such gene products as collagen, actin, tubulin, and hemoglobin proteins that form material phenotypic structures, often with highly specific characteristics. Cues are molecules such as enzymes and hormones that stimulate specific changes in state or in rates of interaction of other phenotypic elements. Of course, this distinction is not absolute, since a building-block component of a membrane, for example, may cue a specific process by its active transport of a particular substance into or out of a cell (see chapter 3), and an enzyme may act as a structural element by joining other molecules together, thus participating in the formation of an element of form. The point here is not to make airtight distinctions between gene–product effects but to describe the developmental functions of genes.

Developmental Effects of the Environment

The Genelike Nature of Environmental Effects
Genes are famous for being exquisitely well-designed, intricate structures, highly specific in their functions and predictable in their effects. Environmental factors, by contrast, have a reputation for being haphazard and incidental, more likely to be disruptive than dependable. By this line of reasoning, genes play a constructive role, whereas the environment plays only an unpredictable and potentially disruptive role. Indeed, many biologists (Rensch, 1960, p. 298, lists Sewertzoff, J. S. Huxley, and G. G. Simpson as having been among them) have defined evolutionary progress in terms of increasing *independence* of the environment—as if evolution were a triumph of genetically mandated order over environmental unpredictability. As long as this mistaken idea about genes versus environment prevails, it is easy to write or to read, without flinching, that the genome contains the recipe, blueprint, or program for development.

The environmental elements of phenotypes have many of the same properties that we ascribe to the products of genes. Consider carbon. As are many gene products, carbon is essential to normal development, recurrently and dependably present, and interactive in highly specific biochemical roles. In addition to being an essential building block, environmental carbon *level* can play an informational, or regulatory role, acting as a cue to switch physiological processes from one pathway to another. If carbon is absent or in low supply, a plant may switch into complex alternative metabolic pathways, become quiescent, or actively search for new sources of supply. Like many gene products, carbon is essential to normal growth and development. An environment without it would be developmentally equivalent to a dominant lethal gene.

The function of environmental elements as essential building blocks is so obvious that it may be easily forgotten. Carbon is one example. Others are oxygen, iron, calcium, water, incident light, and amino acids in food, as well as many other familiar essential dietary requirements (see any box of breakfast cereal).

Obvious kinds of developmental information from the environment include the factors we call environmental "cues," such as change in photoperiod, temperature, or rainfall, which commonly cause seasonal change in behavior, reproduction, or growth (e.g., see Tauber et al., 1986). Bonner (1974) gives examples of asymmetrical developmental responses of fungi that depend upon contrast between light and shade, some of them toward the light and others away from it. Knight et al. (1992) discuss wind as an "environmental signal" that modifies plant development. They point out that wind-exposed crops, like the wind battered trees of mountaintops, experience reduced growth and yields, and they experimentally examined how wind signals are perceived and transduced by plant cells via immediate increases in cytosolic calcium mobilized from organelles when plant tissues undergo wind-induced movements. Knight et al. suggest that calcium may be central in the control and generation of plant form mediated by mechanoperception and transduction of wind stimulation during morphogenesis.

We would not question the importance of information from the social environment for normal human development. Environmental effects are well recognized in development of other group-living organisms as well (e.g., see Tchernichovski and Nottebohm, 1998). Developmental information from the social environment with highly specific phenotypic effects includes stimuli from parents

Fig. 5.10. Stereotyped manipulations of development by organisms in the environment: 18 of the 48 kinds of leaf galls induced by gall midges of the genus *Caryomyia* (Diptera, Cecidomyiidae) on several hickory species (*Carya* spp., Juglandaceae). The gall is an extension of the phenotype of the inducing insect: it is specific to the midge species, not to the host species (Ananthakrishnan, 1984). The cavity within the gall, shown in cross-section (on the right in each pair of drawings), serves as home for a single midge larva. Letters refer to midge species numbers in Wells's drawing as follows: (a) 32; (c) 21; (d) 29; (e) 15; (f) 17; (g) 16a; (h) 27; (i) 33; (j) 5a; (k) 23; (l) 7; (m) 14; (n) 9; (o) 6; (p) 16; (q) 5; (r) 10. (b) is *C. holotricha* (after Gagné, 1989). Not to scale. Redrawn from Wells (1915) and Gagné (1989).

and siblings, the behavior of suitors (sexual signals), and the self-serving manipulations of hosts by parasites (e.g., see Le Moli and Mori, 1987).

Spectacular examples of stereotyped environmentally induced developmental manipulations occur in plants in response to gall-making insects (figure 5.10). Galls are produced by the plant, but their form and at least in some cases the chemical content of the gall tissues (Nyman and Julkunen-Tiitto, 2000) are determined by the inducing insect—the ovipositing female, the gall-inhabiting larva, or both (Nyman and Julkunen-Tiitto, 2000). Insect species specificity of gall phenotype applies to aphid, thrip, fly (Diptera), sawfly (Hymenoptera, Tenthredinidae), hemipteran, and cynipid-wasp galls (see, respectively, Stern, 1995; Crespi et al.,

1997; Southwood, 1973; Nyman and Julkunen-Tiitto, 2000; Hori, 1992; Stone and Cook, 1998), as well as to the midge galls (Diptera, Cecidomiidae) shown in figure 5.10. Galls have been called extended phenotypes of the inducing insects, for it is clearly the insect input that influences the plant's development in specific ways. Yet specific properties of the plant genome or environment must play a role too, because not all plant species bear galls, suggesting that not all are hospitable or responsive to galling insects. Gall midges, as common as they are on *Carya* species (hickories; figure 5.10; Gagné, 1989), are absent on the congeneric pecan (*Carya illinoensis*). And oak trees (*Quercus* species) seem to be particularly subject to galling insects of many genera of flies (Cecidomyidae) and wasps (Cynipidae; see examples in Metcalf et al., 1951). The fact that oaks also are among the boreal forest trees most subject to herbivorous insect damage in general (Metcalf et al., 1956, p. 763) suggests that their defenses are ineffective relative to the nutritional value of their tissues for insects.

The plant–insect interaction between plants and gall insects is carried to an extreme in the galls that involve a mutualism or symbiosis, as in the gall-making fig-wasp (Agaonidae) pollinators of fig trees. There, the plant produces specialized "gall flowers" that are the only substrate in which the female wasps can successfully oviposit (see summary in Southwood, 1973). The interdependencies are more complicated in some fig-wasp galls, which are induced not by the wasps but by their nematode companions (Southwood, 1973, after Currie, 1937). In gall production, the plant is a crucial and specific influential element of the insect's developmental environment, and the insect is a specific form-inducing element of the plant's developmental environment, whether the results be pathological or beneficial for the plant.

These interactions and their stereotyped results attest to the efficacy and specificity of environmental directives—environmental information—in development. How such environmental factors become developmentally entrenched in the course of evolution is the subject of chapters 6 and 26.

Van Valen (1986b) captured the informational role of the environment in development when he wrote:

> Consider development. The course of development can be different from what it would be otherwise in an environment which contains, say, thalidomide, lithium ions, a psychotic parent, hot plasma, too many littermates, or a scarcity of food. Differences in such environmental variables produce different phenotypes (or none at all) just as do differences in DNA sequences. Therefore developmental information resides in the environment as well as in the genome. (p. 67)

Environmental Information   An adequate review of environmental information in development would easily fill as many volumes as an adequate review of genetics; it would include the findings of entire fields such as sensory physiology, immunology, social psychology, environmental physiology, and epidemiology and large parts of neurobiology, endocrinology, parasitology, embryology, and ecology concerned with host preference and chemical interactions, to mention a few that come easily to mind.

Perhaps the most obvious demonstrations of the importance of environmental information in development are the sensory mechanisms that have evolved to gather and process it (Dusenbery, 1992; Stetson, 1989). Sensory structures for gathering the environmental cues that influence development include elaborate antennae, eyes, ears, proprioceptive organs, and chemosensory devices on skin, tongue, airways, and gut. Beyond them are the internal physiological devices that process information and connect stimuli to vital processes—the hormonal, conductance, circulatory, and neural systems that are familiar to all biologists. All of this apparatus bears witness to the fact that the environment provides crucial *information* for development and behavior. Like gene products, environmental factors activate, inhibit, modulate, and coordinate developmental events and physiological responses; like genes, they have pleiotropic (correlated) effects. Exercise during growth affects muscle size, and muscle action affects bone shape. Development of neural pathways depends on sensory inputs and nerve activity. These are familiar facts, and they mean that normal development depends on information from the environment, not just from genes.

It is now well established that the anatomy of the brain depends on how it is used, not only during infancy but during adulthood as well. Kittens experimentally raised in a visual environment filled with dots but not stripes develop cortical neurons sensitive to spots but not stripes (Pettigrew and Freeman, 1973, cited by Geist, 1978). Blindness in adult rats causes areas of the brain devoted to subsequently exaggerated use of nonvisual senses (the somatosensory cortex) to increase up to 40% in area (reviewed in Gerhart and Kirschner, 1997). Many examples of use effects in morphological development have been mentioned in the sections on somatic selection, and these effects are environmentally influenced because use depends in part on what is encountered by the developing individual. Thus, diet influences the mouth and head morphology of fish and of grasshoppers (chapter 3) and the claw size and feeding performance of crabs (Smith and Palmer, 1994).

The use of environmental cues or "information" in development, of course, does not necessarily imply central processing or intelligence (see chapter 23). In many cases, an environmental substance or situation acts as a signal that redirects development just as a gene product would. In many insects, chemical characteristics of larval host plants influence or determine food preferences so firmly that in some species individuals will go without food rather than accept diets that are acceptable to other individuals of the same species (Papaj and Prokopy, 1989).

Hormones and hormone receptors are ancient molecules, and many of them retain their interactivity across phyla. As a result, they are particularly influential aspects of the biotic environment. There is evidence for the presence of steroid hormone receptors in fungi and yeasts such as *Candida albicans*, and they apparently affect the severity of fungus and yeast infections that differ between the sexes in humans in accord with whether the hormone bound by the yeast is produced primarily by the host male or the female. Fungal pathogens of plants also use plant hormones as signals to time their germination and invasion of hosts. Flaishman and Kolattukudy (1994) speculate that the fungi acquire the response mechanism from the host. The possibility of a shared, conserved mechanism should be considered as well. It is known, for example, that a human cDNA clone can complement a yeast cell-cycle mutation (Lee and Nurse, 1987, cited in Shapiro, 1992), an indication of how regulatory fuctions, or at least crucial parts of them, can be conserved across taxonomic boundaries.

Environmental information in the form of thyroid hormone or hormone constituents ingested along with macroscopic prey causes the tadpoles of spadefood toads (*Scaphiopus multiplicatus*) to switch from an omnivore morphology to a carnivore morphology that enables them to more efficiently exploit such prey (Pfennig, 1992a). In this case, a hormone of the prey evidently has a hormonal function in the predator as well, altering its growth pattern in a way that happens to be, or has evolved to be, adaptive for carnivory. Use of female reproductive-tract hormones by courting males is discussed in chapter 15. It is quite common to find hormones of a manipulated organism produced by the manipulator, even when the two belong to different phyla. The saliva of some gall-forming insects contains plant hormone-like substances (auxins) that evidently manipulate the growth of the plant host in the specific ways characteristic of galls (figure 5.10; Shorthouse and Rohfritsch, 1992). Because gall form is specific to the gall-inducing species, not to the host (Ananthakrishnan, 1984), this is a demonstration of the high predictability of responses that can be induced by a stereotyped environmental input, in this case a substance produced by another organism (see also Batra and Batra, 1985). Ingested hormones or hormone components may quite commonly influence the reproduction of herbivorous animals. Many food plants contain phytoestrogens, and their concentration varies inversely with rate of plant growth (Lott, 1984). Lott (p. 279) speculates that the many known effects of food-plant condition on the reproduction of herbivorous vertebrates may be related to the hormone content of their food.

In other cases, nonhormone substances associated with certain dietary items have hormonelike effects. In rotifers (*Asplanchia*), for example, individuals that prey on relatively large herbivores ingest tocopherol, a substance derived from the algal diets of their prey, and this stimulates body size increase accompanied in some species by a progression of growth responses, including the formation of a distinctive campanulate cannibal form; in other individuals, hydraulically expandable cruciform morphologies protective against cannibalism are incapable of sexual reproduction (Gilbert, 1980). In all of these examples, external, environmental cues and interactions direct normal development as surely as do cues of internal, genetic origin.

My favorite demonstration of how thoroughly an environmental factor can be integrated into development is an ingenious mechanism that has evolved to mimic an environmental factor in fetal mammals. The development of precise neural connections requires nerve activity. In the case of the visual system, such activity is normally provided by an environmental factor—light. In mammalian fetuses, which develop in uterine darkness, there is a kind of head-start program that enables visual centers to develop precise connections without light. During a critical period of visual development, when the retinal cells are forming patterned connections in the lateral geniculate nucleus of the brain, retinal ganglion cells mimic the effect of light by producing spontaneous, synchronously generated bursts of action potentials. Not only do they stimulate the requisite nerve activity, but they do so in pulses followed by periods of inactivity, a pattern that optimizes coordinated connections and allows axons from the two eyes to sort out in a fashion approximating the topographically separate organization that characterizes the adult visual system (Shatz, 1996). By simulating the environmental information provided by light, this allows neural development to proceed prenatally, in the dark of the uterus.

Environmental Building Blocks In addition to the developmental information that comes from the environment, all gene expression requires substances from outside the organism. Humans have to eat 10 different amino acids to synthesize the proteins essential for growth and maintenance. Elements essential to plant growth are assimilated or ingested from the air or the soil. Cells require iron. One could say that iron is a component of life more widely essential than any single gene. Hemoglobins have evolved three times in insects alone (Locke and Nichol, 1992). Vitamin A functions as a hormone in the regulation of vertebrate gene expression, along with molecules of genomic origin such as steroid hormones and thyroxine. But vitamin A is a product of the environment, not of a gene. It is first ingested, then stored in the liver, which serves as its surrogate "endocrine organ," releasing it into the bloodstream. In the functionally active form of retinoic acid, vitamin A affects growth (by activating the gene for the production of growth hormone), metabolism (via effects on other genes, including thyroid and steroid hormone–responsive elements), and embryonic development in addition to its well-known effects on vision (Wolf, 1990).

For development to proceed, organisms also require what we call "favorable conditions" of temperature, humidity, and light. When these factors are unsuitable or extreme, aberrant phenotypes or death may result. Leigh (1991) discusses the extensive interdependence of organisms within ecological communities, for example, the dependence of plants upon fungi for uptake of nutrients and upon animals for pollination of flowers and dispersal of seeds. This is a direct result of the fact that normal development depends as much upon essential environmental materials and cues as it does upon particular genes.

These facts may seem almost too obvious to mention. But they are often forgotten in statements that imply that genes alone determine developmental regulation and phenotypic form, statements such as "the human genome contains the complete set of instructions for the construction of a human being." The idea that genes are in general more important for development than are environmental factors, or vice versa, is not scientifically justifiable (see especially chapter 26). The genetic and the environmental factors in development differ in the degree to which they incorporate *historical* information due to selection. But even on this score, inclusion of environmental factors reflects history of selection in a roundabout way, for they may come to be highly screened and specific due to evolution of specialized means to acquire and incorporate them, for example, through selection on habitat selection, food preference, and so forth. Phenotypes evolve so as to assure the acquisition of specific environmental elements that have become important for development and survival.

Inherited Environmental Effects *Inherited environmental influence* includes all factors or traits of nongenetic origin (materials, structures, behaviors, and cues, or their component parts) that are passed from one generation to the next and affect the phenotypic development of descendents. The inertness of the gametic genome has already been discussed. Its arousal from inertness, and the first progress of a zygote toward an organized response, is invariably environmentally induced, most notably by maternal factors, position effects, and cytoplasmic gradients that begin to structure the phenotype in the early embryo, as just described. The parental endowment contains nongenetic factors and has been structured by them as certainly as by genes. This fact is the basis for the assertion made earlier in this chapter that there can be no exception to the principle of dual environmental–genetic influence in the development of all phenotypes.

Maternal effects can be viewed in many different ways in relation to individual development. Some authors (e.g., Wcislo, 1989; Cheverud and Moore, 1994) focus on maternal effects as part of the environment provided by relatives. Others emphasize selection and genetics by regarding maternal effects in terms of kin selection (e.g., Cheverud, 1984b), and many concentrate on the genomic side of maternal effects, considering them a special class of "epigenetic" effects (e.g., Cowley and Atchley, 1992). *Epigenetic factors* have been defined as those heritable causal interactions between genes and their products during development that arise externally to a particular cell or group of cells within the same individual, and condition the expression of the cell's intrinsic genetic factors (i.e., genome) in an extrinsic manner (Cowley and Atchley, 1992, p. 496). In other words, epigenetic factors are the contributions to a cell's environment by genes in other cells of the same individual.

Cowley and Atchley object to calling epigenetic effects "environmental" with respect to the cell in question, which they believe may lead to confusion of heritable and nonheritable factors. This concept helps Cowley and Atchley to connect development with quantitative genetics by identifying the different sources of interacting gene products. But there is an unrecognized problem with this view: gene products contain, and cannot be produced without, the environmental elements that compose them, so the genetic and the environmental are always inextricably combined, and this combination must have evolved under selection. Cowley and Atchley (1992) refer to "random environmental effects" (p. 496) and do not consider environmental effects integral parts of development. Rather, they believe

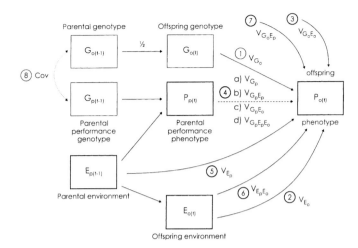

Fig. 5.11. Multiple sources of variation in the inherited phenotype. Seven routes of inheritance (numbered lines 1–7, with sources of variation listed) plus genetic covariance (cov) between genes expressed in both generations (8) contribute to the offspring phenotype. After Rossiter (1996; see also Cheverud, 1984a).

that environmental factors such as shifts in pH, oxygen tension, osmolarity, temperature, and diet "can have a large influence on phenotypic development; however they cannot contribute to permanent change brought about by selection" (p. 497). Their paper is a major step toward a synthesis of quantitative genetics and development (see also Atchley and Hall, 1991), but it gives no guidance to understanding how it is that evolution, under selection, has incorporated environmental elements such as oxygen and carbon dioxide into development, or how to deal with the fact that the environmental contributions to maternal phenotypes, always products of both environments and genes, must inevitably be involved when maternal effects guide the development of embryos and young. This problem is taken up again in chapters 6 (on adaptive evolution) and 26 (on environmental modifications).

The concept of "inherited environmental effects" (Rossiter, 1996) summarizes the many components of environmental effects on the cross-generational continuity of the phenotype, effects that are quite complex even when just one influential relative, a parent, is taken into account (figure 5.11). The literature on inherited environmental effects is enormous and growing rapidly (for reviews, see Jablonka and Lamb, 1995; Rossiter, 1996; Mousseau and Fox, 1998; Agrawal et al., 1999; Sultan, 2000). Although it has been widely recognized by evolutionary biologists only recently, environmental effects have long been known to be

passed from mother to egg, as in the cumulative influence of maternal rearing density on offspring characteristics in migratory locusts (Kennedy, 1961). In ectothermic vertebrates, yolk production is highly susceptible to environmental influence, with egg size affected by temperature in pupfish (*Cyprinodon nevadensis*) and by food availability in salamanders (*Plethodon cinereus*); in a frog (*Pseudacris triseriata*), "[n]on-genetic maternal effects are more important than genetic ones in explaining . . . observed variation in [offspring] larval growth rates (Travis, 1981)" (Kaplan and Cooper, 1984, p. 399; figure 5.12).

It could be objected that maternal endowments are energetic or nutritional and without influence on form, were it not known from experimental manipulations that egg size in sea urchins, for example, is responsible for species differences in larval shape, as well as development rate and growth rate (Sinervo and McEdward, 1988, ex Sinervo, 1990). These differences are species characteristic and might have been assumed to be due entirely to differences in *genetic* information were they not known from experiments to be environmental. Egg size also affects the growth rate and size of young salmon, where, as in many other polyphenisms, size determines or correlates with adult form (Skúlason et al., 1989a,b; Skúlason, 1990; Gross, 1985).

The charging of an insect egg with the elements of its new phenotype takes place in the ovary (reviewed in Büning, 1994). There, the egg receives a "euplasm," called by Büning the "informational"

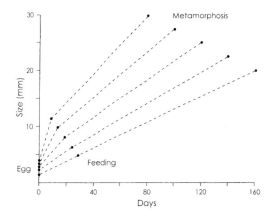

Fig. 5.12. Effect of egg size on size at and time to first feeding and metamorphosis in the California newt. After Kaplan and Cooper (1984).

component of the cytoplasm, which contains messenger RNA, ribonucleoproteins, enzymatic proteins, cytoskeletal proteins, ribosomes for protein synthesis, mitochondria, and "a basic stock of cellular membranes, vesicles, and vacuoles" (Büning, 1994, p. 107), most provided by nurse cells of germ-cell origin. To provide the gene transcripts of the euplasm, which is only about 1% (0.1–10%) of mature egg volume, nurse cells undergo chromosome duplications yielding polyploid genomes. A small portion of the euplasm comes from the oocyte itself in only a few taxa of insects (e.g., Hemiptera and Ephemeroptera). The heteroplasm, composed of nutrient macromolecules that fuel growth, is the product of somatic cell vitellogenesis, transported to the ovary via the hemolymph from the fat body and follicle cells. A fully provisioned insect egg is 1000 to 10,000 times the volume of the original germ cell.

In *Drosophila*, where early development is unusually well studied, maternal elements participate in embryonic nuclear division, nuclear migration, cellularization, gastrulation, definition of the polarity of the body axes, segmentation, neurogenesis, and segment identification (Anderson, 1989). In organisms such as mammals and flowering plants, where the egg phenotype is relatively poorly supplied when the egg is fertilized, cooperative parental structures such as the placenta and the endosperm charge the embryo after fertilization with phenotypic building blocks and developmental cues, not all of them genotypic, as I show below. In plants, the sperm makes a larger contribution to continuity of the phenotype than in the animals best studied in this regard (the fruitfly *Drosophila*, the

frog *Xenopus*, sea urchins, mice, and the nematode *Caenorhabditis elegans*—Davidson, 1986). Willson and Burley (1983) list numerous organisms in which paternal materials enter the egg (see also chapter 31). The transferred materials from the sperm include microtubules, mitochondria, plastids, starch grains, fat droplets, and other cytoplasmic materials. In some cases, female cytoplasm and inclusions are destroyed and the embryo gets its initial cytoplasm and organelles primarily, perhaps entirely, from the sperm.

There can be no question about the regulatory nature of some kinds of inherited environmental effects. Photoperiodic cues processed by the mother affect the prepubertal reproductive responses, growth rates, and activity rhythms of the young in various mammals (e.g., Pratt and Goldman, 1986; Stetson et al., 1986; Elliott and Goldman, 1989; Shaw and Goldman, 1995a,b; Gorman et al., 2001). A deficiency of a single environmental element, zinc, in the maternal diet of rats during gestation leads to offspring with reduced brain size, impaired DNA synthesis, decreased RNA polymerase activity in liver and brain, and behavioral abnormalities such as impaired avoidance learning and heightened aggressiveness (reviewed by Geist, 1978, p. 130). Many other examples, such as effects of intrauterine position and maternal social status, and hormonal endowment of the egg on sex expression and behavior of offspring, are mentioned in later chapters (see also Geist, 1978, for a review; Winkler, 1993). Lott (1984) gives many examples of prenatal hormonal effects in vertebrates caused by maternal stress, intrauterine position, and diet, all having long-term effects on behavior and social status of offspring.

The following offspring traits are influenced by parental environment: in plants, cold hardiness, seed weight, germination time, length of dormancy, resin production, amount of flower production, reproductive allocation, and adult size; in insects, coloration, degree of gregariousness, wingedness or lack of wings, hatching success, enzyme activity, survivorship, development time, weight, diapause propensity, heat resistance, body size, age of first reproduction, cold resistence, and fecundity (Rossiter, 1996).

Maternal effects, clearly influenced by the maternal environment as well as her genome, sometimes possibly adaptive and sometimes clearly not, include culturally transmitted traits and phenomena such as the inheritance of maternal social rank in nonhuman primates or fetal alcohol syndrome of humans. In other mammals, maternal "temperament" markedly affects the behavior of the young. When young bank voles (*Clethrionomys glareolus*) of the race *C. g. skomerensis*, which is docile when handled, are cross-fostered by mothers of a much more active race, *C. g. britannicus*, they adopt more active behavior than if reared by their own, docile mothers. The reverse cross-fostering produces the opposite result, but to obtain maximum docility pups must be both of docile genotype and reared by docile mothers (Alder, 1975). Intractable wild pacas (*Agouti paca*), large rodents hunted for food in Latin America, were successfully domesticated in a single generation by human handling from birth the offspring of captive mothers. Domesticated mothers then produced tame offspring without extensive handling (Smythe, 1991).

Maternal effects on behavior are not restricted to vertebrates. The change between solitary and gregarious, migratory morphs of locusts is accelerated (or retarded) by maternal effects. The progeny of crowded mothers not only are darker, are better endowed with fat reserves, and pass through fewer nymphal instars than those of solitaria females, but also are more vigorously marching, and in general more active (Hardie and Lees, 1985, p. 457). Maternal effects determine many aspects of insect life histories (review in Mousseau and Dingle, 1991). In some phytophagous insects (e.g., the treehopper *Enchenopa binotata* [Homoptera: Membracidae]), egg development is timed by information from the host plant. It is stopped in autumn, when plant dryness induces dehydration and dormancy, and initiated in spring, when plant sap flow induces hydration and initiation of embryonic development (Wood et al., 1990). The mother treehopper is involved because her choice of food plant determines the timing of these developmental events, which differ among populations of the same species on different hosts.

Nongenetic transmission of acquired phenotypes is most readily associated with culture and animal behavior, but it occurs even in the reproduction of single-celled organisms, where it has been termed "supramolecular inheritance" (Frankel, 1983). In fact, given continuity of the phenotype, transmission of acquired phenotypes can be considered a property of all ontogenies. Jablonka and Lamb (1995) give an extensive review of mechanisms of continuity during the life cycle, in what they call "epigenetic inheritance systems"—systems that enable the phenotypic expression of the information in a cell or individual to be transmitted to the next generation of cells or individuals. These mechanisms include the capacity of cells to make copies of themselves when they divide, an ability that provides

cross-generational continuity of the phenotype in unicellular organisms. Their discussion emphasizes Lamarkian (or quasi-Lamarkian) *qualities* of somatic inheritance, or the cross-generational inheritance of developmentally acquired characteristics, rather than the aspects that are not like Lamarkian inheritance such as the relatively transitory nature of traits inheritance in this way relative to those transmitted by genes (see Frankel, 1983).

The emphasis on Lamarkian aspects may be unnecessarily provocative, as was Waddington's (1942) paper on the canalization of development "and the inheritence of acquired characters," which no doubt contributed to the confused opposition of many biologists to the idea of genetic assimilation. All phenotypic characteristics are acquired during development, if not in the focal organism in a parent or, in a few cases, some more remote ancestor (Jablonka and Lamb, 1995, 1996). In this sense, anything other than DNA passed from parent to offspring is inheritance of an acquired characteristic. The unhappy fate of processes labeled Lamarkian is that they are likely to be dismissed as oddities even if common and important. Maynard Smith (1989) discusses many of the phenomena treated by Jablonka and Lamb, only to conclude not with a discussion of their special nature, but with the statement that "Lamarkian inheritance is rare" (p. 12) compared with ordinary inheritance mediated by DNA.

Continuity of the phenotype is not vulnerable to the rarity argument. It characterizes the development of all organisms. All phenotypes have a history of both environmental and genomic influence, and all development begins with a structured phenotype provided by a parent.

Rossiter (1996) discusses the transitory nature of acquired traits and points out that many environmental effects, such as seasonally recurrent change in temperature, photoperiod, or nutrition, are "sustainable" and therefore dependable factors in development. Among the dependable elements supplied by parents in addition to nutrients are a preregulated genome, defensive agents, symbionts, pathogens, toxins, hormones, enzymes, and cultural conditioning (Rossiter, 1996). Figure 5.11 shows how environmental factors and genomic ones are related in the cross-generational continuity of the phenotype.

In summary, environmental factors function in the same ways as do gene products during development. They serve as building blocks and cues, as specific and essential for development as the selection-honed products of genes, and their effects can be transmitted between generations. One result of the similarity between the *developmental* effects of environment and genome is that they sometimes prove to be interchangeable during evolution.

Gene–Environment Equivalence and Interchangeability

The developmental parallel between genotypic and environmental factors is perhaps best shown by their ability to replace each other in experiments and during evolution. The responsive phenotype acts as if it is indifferent to the provenance of the factors that cause it to respond, in that it may give the same response whether the effective stimulus is the product of a gene, a substance from the outside environment, or an activity of another organism. Given that gene products and environmental factors can function in parallel ways, it is not surprising that these sources sometimes show *interchangeability*—induction of the same response by either a particular genotype or an environmental factor (called "equivalence" by Zuckerkandl and Villet, 1988; the phenomenon was first extensively discussed by Goldschmidt, 1940; see also Van Valen, 1969, 1974, 1986a).

Naturalists and geneticists have often noted the occurrence of *phenocopies*—environmentally induced mimics of genetically specified traits (Goldschmidt, 1935). Extreme environmental conditions, such as abnormally high temperature during development, sometimes induce abnormal phenotypes that resemble the normal, inherited ones common in other populations or in other species. Phenocopies were extensively studied by Goldschmidt (1938), Waddington (1961), and others. As expected from the idea of interchangeability, the reverse of the phenocopy phenomenon also occurs: mutant genes sometimes mimic environmental effects to produce *genocopies* (Gause, 1942, attributes this term to Schmalhausen; see Scharloo, 1989, 1991, and Hall, 1992, for examples and discussions of phenocopies and genocopies).

Thoday (1964) indicated how polygenic variation, genes of large effect, and environmental factors may all interchangeably influence the same switch mechanism. In experiments using bristle number in *Drosophila*, Thoday and associates performed disruptive selection with negative assortative mating. In each generation, they selected the flies with the highest bristle number and mated them to those with the lowest bristle number. The result was a genotype-specific, high-low bristle number polymorphism, influenced by three loci of major effect. Thoday (1964) recognized the similarity between the genetic architecture of this example and that of familiar environmentally mediated polyphenisms:

Now while this example, as in sexual dimorphism in *Drosophila*, involves a *segregational* polymorphism, it makes clear that the essential genetic features of the system need not involve genetic segregation. If, as in the parallel systems that involve differentiation of cells and tissues, as in the determination of gregarious and solitary phases of locusts, or in some sexual dimorphisms like that of *Bonellia* [a marine worm having environmental sex determination], a sufficiently *regular* environmental switch is available, it could function equally well, in relation with the background genotype which determines the effects of the switch. (p. 110)

Thoday goes on to explain (p. 110) that any genetic locus influencing regulation might have served as the controlling "switch" locus, and he compares the control of polymorphism to a railway system with divergent lines (morphs, and their modifiers) connected by a single junction (regulator, or switch)—a model of regulatory architecture essentially the same as that of figure 4.13. Thoday, then, realized that segregational (Mendelian) control is potentially possible at any locus affecting genetically complex regulation, and also recognized the related phenomenon of interchangeability of environmental and genotypic control. Interchangeability of environmental and genetic factors, and of different genes, in regulation is a consequence of regulatory complexity and the universal condition sensitivity of phenotypes (I. Walker, 1983, pp. 810–811; Hall, 1992).

Zuckerkandl and Villet (1988) propose a biochemical mechanism for interchangeability. The effect of a gene product (e.g., an enzyme) depends in part on its rate of production (rate of gene expression). This rate can be affected by either a mutational change in structure that affects the biochemical affinities of the molecule, or a change in environment (e.g., amount of substrate, pH, presence of essential nutrients, temperature) that affects the relative concentration of the molecule (figure 5.13). This "principle of concentration-affinity equivalence" suggests that altered rate of expression or interaction of a gene product due to environmental factors can have the same effect as a structural mutation that alters its ability to interact. Zuckerkandl and Villet extend this principle to regulatory switches involving hormones and other complex mechanisms, and point out that the more

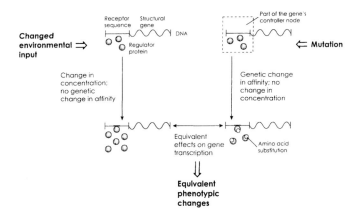

Fig. 5.13. Developmental equivalence of environmental and mutational inputs. Gene expression is a chemical process whose activation can be equally affected either by an environmental change (e.g., temperature or concentration of materials) or by a genetic change (e.g., affecting the reactivity of molecules or their affinities). Equivalence is the basis for gene–environment interchangeability during evolution. After Zuckerkandl and Villet (1988).

complex a regulatory node, the larger the number of different mutations or environmental factors that could have equivalent effects on the phenotype. They relate this equivalence to genetic assimilation and the commonness of reversibility and parallelism in evolution.

The complexity of regulation that permits gene-environment interchangeability is typical of switches that control behavior and development at higher levels of organization. The size-dependent switch between the large-horned and small-horned form of a male beetle, for example, depends upon variation in larval nutrition, as already discussed. Regulation therefore is potentially affected by numerous variable factors, including larval metabolic efficiency, food preferences, age of beginning to feed, hormone systems that affect the cessation of feeding prior to metamorphosis, or maternal choice of oviposition site. Therefore, changed frequency of the large-horned morph in a population, discussed in the following section, could conceivably be due to change in any one of these factors—they are interchangeable elements of regulation. A genetically mediated increase in metabolic efficiency could be equivalent to an environmental improvement in food availability.

Rensch (1960, p. 187) remarked that the often astonishing similarity of the effects of environmental modification and mutation may justify Lamarckian ideas in the minds of some biologists. But the phenocopy–genocopy phenomenon requires no Lamarkian inheritance of environmentally induced traits. As discussed by Zuckerkandl and Villet

(1988) and general considerations of regulatory complexity, it is predictable from the expected indifference of the responsive phenotype to the provenance of an effective stimulus. Genotypic and environmental effects on regulation are expected to be interchangeable, for it is the responsive preexisting phenotype that defines the precise form of the response once the threshold to produce it is passed, not the particular stimulus that elicits the response among the many possible.

The principle of concentration-affinity equivalence explains why the observation of interchangeability does not conflict with the fact that a gene may have a highly specific design evolved under selection to elicit a particular phenotypic response. But interchangeability does conflict with the habit of supposing that the specificity of the response comes entirely from the specificity of the gene. Interchangeability does not mean that every environmental element can be mimicked by a gene, or every gene by an environmental factor. The structural specificity of genes has to be seen in the context of the phenotype that has evolved to respond during development. Some of the best evidence for this comes from the fact that equivalent responses may be elicited by *different* environmental factors or different genes (Zuckerkandl and Villet, 1988). That is, not only are genetic and environmental inputs interchangeable, but different genetic inputs are interchangeable as well (see Tautz, 1992; Hall, 1992), as further discussed in the next section.

Interchangeability is a key to understanding both the organization of development and its mod-

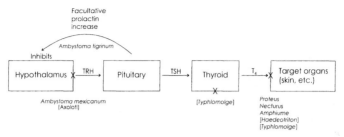

Fig. 5.14. Multiple pathways to neoteny in salamanders. Regulation of larva-to-adult metamorphosis in salamanders is by (left to right) hypothalamus secretion of thyrotropin-releasing hormone (TRH), pituitary secretion of thyroid-stimulating hormone (TSH), thyroid secretion of thyroxine (T_4), and tissue sensitivity to thyroxine. The pituitary hormone prolactin is an antagonist of thyroxine whose secretion is inhibited by TRH. Loss of metamorphosis (neoteny) occurs by (1) increased prolactin secretion by the pituitary (Norris et al., 1973; Norris, 1978), (2) blockage of TRH secretion by the hypothalamus (Norris and Platt, 1973), and (3) reduced sensitivity of target organs (Dent, 1968, on *Proteus* and *Amphiuma*; Noble and Richards, 1931 on *Necturus*; Dundee, 1961, on *Haedeotriton*). A fourth pathway, reduced thyroxine, may occur in *Typhlomolge*, where partial metamorphosis results from thyroxine application (Dundee, 1957). Brackets indicate partial blockage. Based on reviews in Matsuda (1982, 1987) and Levinton (1988) and D. O. Norris (personal communication).

ification during evolution. Interchangeability explains why there can be alternative pathways to the same phenotype, a common feature of the evolution of development (for examples, see especially Gerhart and Kirschner, 1997, and Raff, 1996). Degree of plasticity and degree of genotypic influence are reciprocal, interchangeable quantities. Both are subject to genetic variation and adjustable under selection (chapter 6). Interchangeability means that the evolutionary future of a favorable environmentally induced novel phenotype is as bright, in terms of evolutionary potential, as that of a mutationally induced novelty.

Interchangeability during evolution occurs because in a population, variation in trait expression (passing a threshold for expression of a trait) that is due to plasticity and variation that is due to genotype are continuously variable, complementary determinants of total variation—their sum equals 1.0: 100% of the total variation in the population. A decrease in one determinant implies a corresponding increase in the importance of the other. The sensitivity of phenotypes to inputs from the environment and the genome can be adjusted upward or downward under natural selection on response thresholds, on a sliding scale (see especially Bernays and Wcislo, 1994), for degree of environmental influence is the inverse of degree of genomic influence. The basis for this is change in the thresholds (or in the ability to pass them) of switch mechanisms that regulate the expression of traits (Hazel

et al., 1990; Roff, 1994b). Switches integrate inputs of these two kinds into a single threshold response of an individual.

Gene–Gene Interchangeability

Many authors have noted that change at different genetic loci may have equivalent phenotypic effects (e.g., see Dobzhansky, 1959; Wright, 1934; Avise, 1976; Gerhart and Kirschner, 1997)—there is genetic interchangeability as well as gene–environment interchangeability among loci that affect the same regulatory mechanism. Harris et al. (1990) tested for interpopulational differences in the genetic basis of the same flexible response. They made within-population and between-population crosses using a salamander (*Ambystoma talpoideum*) whose populations differ in propensity for metamorphic versus paedomorphic reproduction. In crosses between populations, hybrids showed clear intermediacy in proportions of the two life-history morphs, indicating that common genetic elements are involved in regulation. But in one pair of populations, hybrids showed a breakdown in hybrid intermediacy, indicating that regulation of the same polyphenism had a different genetic basis in the two populations. Why this can be so is shown in figure 5.14: because regulation of salamander metamorphosis is a genetically and phenotypically complex trait, loss of metamorphosis can be caused by genetic change at different points in the chain of events leading to metamorphosis. That is, different genes can be responsible for the same phenotypic result.

The polygenic nature of regulation in these salamanders is further evidenced by the ability of artificial selection to cause quantitative change in the threshold, or environmental sensitivity, of a switch. Semlitsch and Wilbur (1989, p. 111) artificially selected for altered thresholds of facultative metamorphosis in two populations of *A. talpoideum*. In both populations, four generations of selection for paedomorphosis resulted in a greater proportion of the population becoming paedomorphic when reared in controlled conditions. This indicates not only that regulation of the flexible response is genetically influenced (Semlitsch and Wilbur, 1989), but also that selection for a particular alternative can begin to produce its genetic assimilation—a lowered threshold to produce the trait that, if exaggerated, could lead to its fixation, as has occurred in some salamander species (Dent, 1968).

Interchangeability: Evidence

Phenocopies and Genocopies The ability of artificial selection to alter degree of phenotypic plasticity is evidence for the interchangeability of environmental and genotypic influence in the regulation of phenotype expression. The following examples are additional evidence of genotype–environment interchangeability:

- In *Drosophila*, for example, the crossveinless wing form can be produced by any of many mutant alleles, or as a phenocopy, by exposing pupae to a heat shock at a critical stage of development (Waddington, 1953b).
- In certain strains of mosquitoes (*Aedes aegypti*) with genetic sex determination, genetic males are feminized by environmental factors (e.g., if they develop above 30°C). Above 35°C they are strongly feminized, with genitalia containing reduced male elements (basimeres and telomeres), added female elements (cerci), and ovarian tissue and spermatheca (O'Meara and Van Handel, 1971).
- In *Drosophila melanogaster*, artificial selection produced two genetically distinct strains, "sitters" responding to food with exaggerated local food searching, and "rovers" with exaggerated distant searching. If sitters are starved before feeding and given low-quality food, they decrease local feeding and act increasingly like rovers. If rovers are only briefly starved and fed high-quality food, they in-

crease local searching, acting more like sitters. The two behavioral phenotypes are environmentally equalized despite genotypic differences (Bell and Tortorici, 1987).

- In butterflies (Lepidoptera), there are three morphologically distinct types of melanism, all of which can occur as intraspecific alternatives to light color forms. All three can occur as either genotype-specific or environmentally induced alternatives in different species of butterflies (Nijhout, 1991b, p. 138).
- In the nematode *Caenorhabditis elegans*, a facultatively expressed nonfeeding and nongrowing dauer larval stage is normally induced by crowding and starvation, but several "dauer-constituitive" mutations render the dauer larval pathway an obligatory stage (work by S. Gottlieb and G. Ruvkun, described in Flannagan et al., 1998).
- In *Drosophila*, most if not all homeotic mutations can be mimicked by environmental perturbations (Garcia-Bellido, 1977, ex Alberch, 1980, p. 657).
- In crickets (*Gryllus rubens*), wing length (whether short or long) can be influenced either by environmental factors (e.g., crowding or photoperiod) or almost exclusively by genotype in artificially selected short- or long-winged strains (Walker, 1987; Zera and Tiebel, 1989).
- In the carabid beetle *Calathus erythroderus*, a wing dimorphism is determined by a Mendelian single-locus, two-allele switch. Interchangeability of environmental and genotypic influence is indicated by the fact that in another species of the same genus, *C. melanocephalus*, the expression of the long-winged form is influenced by temperature and food supply (Aukema, 1986, cited by Wagner and Liebherr, 1992).
- In chickens, a mutation ("ptilopody") causes feathers to replace scales on the feet. The same phenotype can be induced by administration of retinoic acid (see references in Zuckerkandl and Villet, 1988).

Phenocopy–genocopy interchangeability occurs because both environmental and genetic factors act, in sliding or inverse proportion, on the same regulatory mechanism, or switch. In the crickets, where frequency of long- to short-wingedness can be increased by either environmental conditions or genetic change under artificial selection, involvement of a hormonal mechanism has been indicated by treatments with juvenile hormone (JH) (Zera and

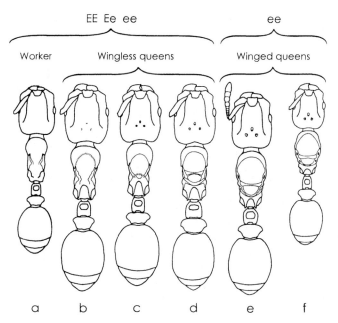

Fig. 5.15. Genetic and environmental influence on worker and queen morphology in ant (*Harpagoxenus sublaevis*) females. Workers (a) are non-egg-laying females with no sperm-storage organ, no ocelli, and reduced thoracic sculpture. Winged queens (e, f), usually an environmentally determined phenotype in ants, are homozygous recessive (ee). Queen versus worker determination for all genotypes (EE, Ee, ee) depends in part on environmental conditions: ee females may fail to develop wings under poor nutritional conditions or in the presence of a strong queen. Wingless queens (b–d) are egg-laying females with ocelli and thoracic sculpture that varies between workerlike and queenlike, as shown. Body size varies within all classes over a range similar to that shown here only for queens (e and f represent the size extremes among queens). After Buschinger and Winter (1975); genotypes as labeled in Hölldobler and Wilson (1990), after Passera (1974), in consultation with A. Buschinger.

Tiebel, 1989). During the last juvenile instar, when wing-morph is determined, individuals of a genetic strain selected for short-wingedness have a lower level of JH esterase in the plasma and have two isoforms of the esterase, whereas only one of them is present in the long-winged strain (Zera et al., 1992). JH esterase binds with JH, reducing its level in the plasma, and therefore could be an important element in the switch between morphs. This work suggests that the change in morph frequency, whether achieved by a genetic or by an environmental change, depends on change the same regulatory mechanism, in this case one that is sensitive to JH and JH esterase.

Interchange of genetic and environmental determination occurs frequently in the course of evolution. The carabid beetle *Calathus erythroderus*, for example, has a wing dimorphism determined by a Mendelian two-allele switch. In another species of the same genus, *C. melanocephalus*, the expression of the long-winged form is influenced by temperature and food supply (Aukema, 1986, cited by Wagner and Liebherr, 1992). When both modes of determination occur in different species of the same genus, or populations of the same species, interchange is indicated, although polarity (which one came first) may not be known. From other examples with known polarity, it is clear that genotypic and environmental determination is interchangeable in both directions. Caste determination, environmentally controlled in the vast majority of social insects, has a strong, secondarily evolved, genetic component in one genus of stingless bees (see below) and in a few species of ants (reviewed in Buschinger and Heinze, 1992; figure 5.15). In the ants, an allele (E) of major effect appears to suppress (raise the threshold for) wing production. Winged queens are very rare (< 1% of queens in more than a thousand colonies collected in nature—Buschinger and Winter, 1975;

Buschinger, 1978), suggesting that they are selected against in this slave-making species, which depends upon 2–3 years of worker production prior to investment in reproductive females and males (Buschinger, 1978).

Environmental Sex Determination   Environmental sex determination (ESD) has secondarily replaced allelic or chromosomal determinants of sex in a variety of organisms, and ESD may be ancestral in reptiles (Janzen and Paukstis, 1991a). The most commonly discussed examples are temperature-dependent sex determination in turtles and snakes (see Bull, 1983), but ESD is taxonomically more widespread than usually realized. At least 97 cases are known, most of them in invertebrates, and with a great diversity of environmental regulators, including not only temperature but also day length, ultraviolet light, metabolic products, parasite attack, exposure to the opposite sex of the same species, pH, salinity, light, water quality, nutrition, population density, oxygen pressure, chemical stimuli, and, in the case of parasitic or parasitoid species, host size, age, species, type, or parasite load (parasite density; Korpelainen, 1990). As in insect caste determination, sex determination in turtles has gone from primarily environmental to primarily genotypic in a few groups—the recurrent evolutionary establishment of a genocopy (figure 5.16). In lizards, sex determination also sometimes has evolved from genotypic to environmental—a phenocopy recurrently has been established as the species norm (figure 5.17).

The evolution of sex determination is a relatively well-studied phenomenon that illustrates how the evolutionary interchangeability of environmental and genetic determinants works upon a complex hormonal regulatory mechanism (Crews et al., 1994; figure 5.18). Bull (1983, p. 110) makes the important point that all ESD systems probably experience at least slight inherited effects on sex determination (see also Janzen, 1992; Rhen and Lang, 1998), and many genotypic sex-determination (GSD) systems experience rare environmental influences (Bull gives examples of the latter from birds, amphibians, and fish).

As expected in such systems, degree of genetic versus environmental influence on the sex ratio is geographically variable in some species, such as the Atlantic silverside (*Menidia menidia*), a fish in which sex is determined by joint effects of temperature and major sex-determining genes during a critical period of larval development (Conover and Van Voorhees, 1990). The sensitive period corre-

sponds to a hormone-sensitive period in other fishes (Conover and Fleisher, 1986), suggesting that the evolutionary lability of sex determination in this fish is due to the ease with which selection can adjust the environmental sensitivity of a polygenically influenced regulatory mechanism.

The potential for ESD to give rise to GSD and vice versa (figures 5.16, 5.17) is predicted by the interchangeability phenomenon. This lability may help explain the ability of some clades to survive periods of climate change. Janzen (1994a) discusses the possibility that climate change may endanger reptiles with ESD. A rise in global temperature of 4°C, within the range currently predicted to be in progress, could eliminate production of males. Janzen calculates that turtle populations would not be able to sufficiently rapidly evolve the necessary adjustments in their nest-selection behavior, which does not respond to annual or latitudinal variations in annual temperatures.

Janzen (1994a) also raises the question of how reptiles have survived similar global temperature changes in the past given the apparent lack of adaptive plasticity in response to temperature change. The history of interchangeability of ESD and GSD suggests the hypothesis that the adjustment may sometime occur not as a plastic response, but as a shift toward genetic sex determination, that is, by selection favoring individuals that produce both sexes regardless of temperature (figure 5.19). Given the lability of GSD–ESD and the principle of interchangeability, populations now showing ESD may have had GSD during the extremes of temperature fluctuations in the past. An ability to flip-flop rapidly between ESD and GSD under selection may explain why extant lineages with ESD are not extinct despite cycles of global climate change in the past (Korpelainen, 1990). In support of this hypothesis, Korpelainen (1990) provides data showing the ease with which sex ratio can be adaptively adjusted in organisms with ESD. Several organisms with ESD show adaptive geographic variations in sex ratio that compensate for variations in thermal environment and seasonality.

In terms of mechanism, ESD could be called maternally mediated and considered a highly polygenic aspect of continuity of the phenotype, which helps explain why it is so susceptible to interchange. Mothers select nest sites and can thereby affect the sex-determining environment of the eggs. In some ESD organisms, there is evidence that maternal hormone precursors added to the yolk bias the egg to develop as one sex or the other, and then maternal nest-site selection provides a temperature that acti-

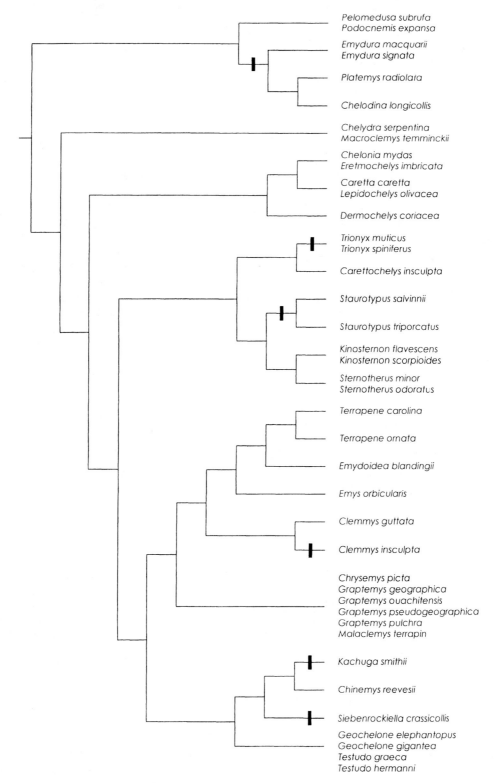

Fig. 5.16. Interchangeability of environmental and genotypic sex determination in turtles. Vertical bars indicate a change from environmental to genotypic sex determination. Based on Janzen and Paukstis (1991a).

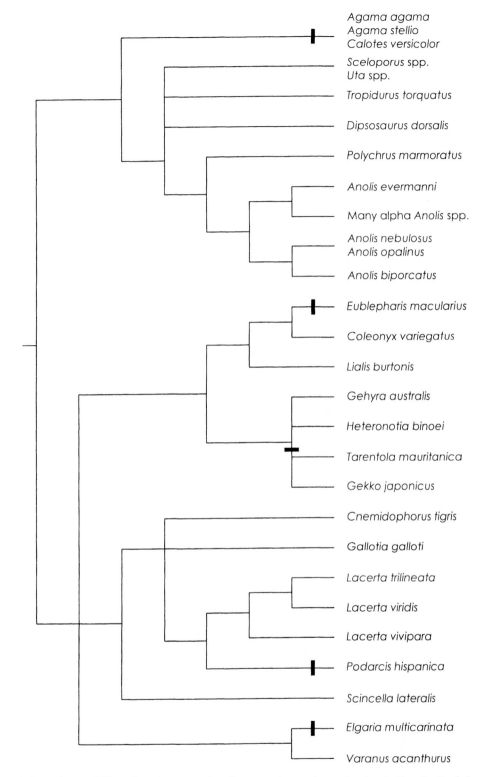

Fig. 5.17. Interchangeability of environmental and genotypic sex determination in lizards. Dark bars indicate a change from genotypic to environmental sex determination. After Janzen and Paukstis (1991b).

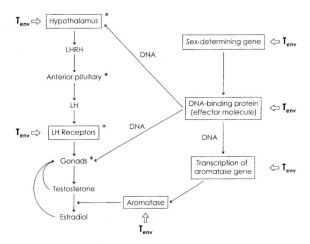

Fig. 5.18. Regulatory complexity as a basis for evolved interchangeability of environmental and genetic sex determination in reptiles. The diagram summarizes hypothesized regulation of temperature-dependent sex determination in reptiles. Boxes indicate factors and processes hypothesized or known to be temperature sensitive. Asterisks indicate genetically complex structures subject to polygenic influence on hormone sensitivity and output. Selection on quantitative genetic variation in these structures could change the effects of temperature-sensitive processes, either raising or lowering the degree of environmental influence (T_{env}) on sex determination. Lowered environmental influence would permit genetic sex determination. LH, luteinizing hormone: LHRH, LH-releasing hormone. After Janzen and Paukstis (1991b).

vates these precursors so that they produce hormones influencing the sex of embryonic gonads (Crews et al., 1989; Janzen, 1994b; Janzen and Paukstis, 1991b; Janzen et al., 1998). In effect, the mother manipulates the offspring hormonal system and thus bypasses or overwhelms offspring genotypic influence. At the same time, she may influence the aggressiveness and locomotory behavior of offspring, which has been demonstrated to correlate with egg steroid levels in at least some species (Janzen, 1995; Janzen et al., 1998).

Given maternal control, the ultimate cue in environmental sex determination is not the temperature of developing embryos but the conditions that affect the maternal decision to produce a male or a female clutch. Little is known about the natural history of nest-site selection in species with ESD (Bull, 1983, p. 122; see Janzen, 1994b). Crucial questions are not only how the mother chooses the appropriate temperature, but also what factors may influence her decision to invest in one sex or the other. Maternal hormonal state, known to influence that of offspring in at least some taxa (see Janzen et al., 1998; Buschinger, 1990a), may vary with maternal age, physical condition, or copulatory interactions (see Eberhard, 1997), and this may influence maternal ability or the desirability of producing successful offspring of a given sex. So

far, research on ESD has not focused on factors relating to the maternal decision, even though those factors, by Bull's (1983) definition of "determination" (adopted here; see above), are the earliest to define, or to determine, sex.

Horned Beetles: Hormonal Switches as Mediators of Interchangeability   Interchangeability and its regulatory basis have been demonstrated in some species of tropical horned beetles. A common type of horn-size dimorphism in males is controlled by a conditional, nutrition-dependent switch. There is a size-related developmental switch between forms, producing a bimodal distribution of horn sizes relative to body size (figure 5.20). Horn size is determined in the larval stage by quantity of food ingested by the larva (Emlen, 1994, 1997b), which correlates with a change in JH titer during a critical period (Emlen and Nijhout, 1999; Emlen, 2000). This channels development into either the large-horn (major) or the small-horn (minor) category. As in other facultative switches (e.g., see Schaffner, 1927, 1930), there is more variance in the determined character (e.g., horn size) near the neutral, or switch point between forms than at the extremes, as shown by the spread of points in the ascending part of the curve of figure 5.21a. For artificial selection experiments, Emlen (1996) mea-

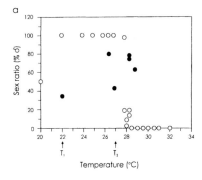

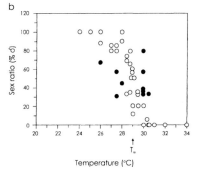

Fig. 5.19. Selectable variation and the potential for interchangeabilty of genotypic and environmental sex determination in turtles. (a) Temperature-dependent variation in offspring sex (male:female) ratio at different incubation temperatures within and among populations in *Chrysemys picta*. T_1 and T_2 are approximate switch threshold points for male and female production, respectively. (b) Temperature-dependent variation in sex ratio within and among populations in *Caretta caretta*. T_m is the approximate modal switch point between male and female production. Dark dots indicate females whose eggs are relatively insensitive or variable in their response to incubation temperature. If there is a genetic component to this variation, selection on such individuals could alter the degree of temperature sensitivity toward zero to produce GSD. After Janzen and Paukstis (1991b).

sured "residuals," or deviations from the mean at a given point on the curve of horn length and body size (figure 5.21a). I propose that one could use a plot of residuals to locate what Shaffner (1927, 1930) called the "neutral" or switch point between conditional alternatives, which is the point of least environmental influence, and therefore the point where variation in genotypic influence on the threshold of a switch (degree of plasticity, and morph ratio) is revealed. Emlen has kindly provided

a switch-point indicator plot based on his data for beetles (figure 5.21b).

Increased phenotypic variance near a switch point may be explained as follows: at extreme (high or low) values of the determining variable, in this case nutritional, there is a relatively clear signal to switch into one alternative pathway or the other— the switch threshold is either clearly passed or clearly not passed. At intermediate values, any genotypic differences in response threshold would be exposed (figure 5.21). Emlen (1996) took advantage of this by using estimates of residuals to exert bidirectional artificial selection. This altered the threshold for a response, demonstrating the presence of genetic variation for degree of plasticity and morph ratios (figure 5.22). Selection to change the switch point, or morph ratio, can effectively sample from the whole population by acting on the residuals in all environments, not just at the switch point. Polygenic threshold models (e.g., Hazel et al., 1990; Moran, 1992b; Roff, 1994b) predict a genetic correlation of 1.0 between the switch points (or morph ratios) found in two environments that differ in a single regulation-influencing factor such as temperature or body size (Roff, 1994b).

As long as genotypes with different switch points have parallel reaction norms, heritability in the two (above- and below-threshold) environments is not differentially affected by the environmental variable that governs the switch (Hazel et al., 1990). Thus, for example, selection to produce a higher ratio of horned beetles where the environmentally influenced variable is body size should

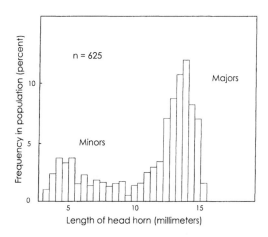

Fig. 5.20. Bimodal distribution of horn lengths in a dimorphic beetle (*Podischnus agenor*). After Eberhard (1980, 1982).

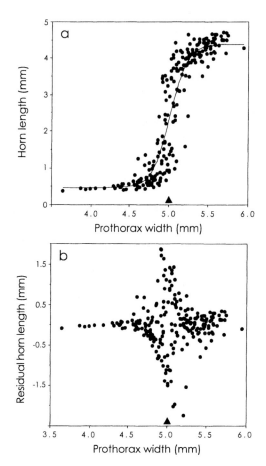

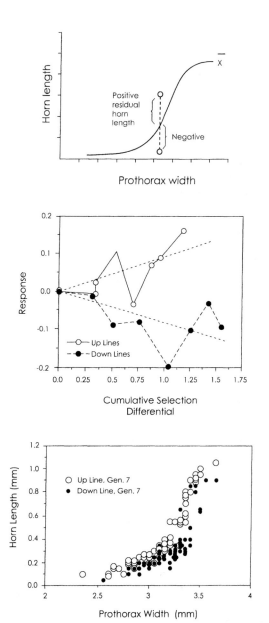

Fig. 5.21. Exaggeration of variation at the switch (neutral) point between alternatives. (a) Horn length in relation to body size (indexed by prothorax width) in a natural population of *Onthophagus acuminatus*. The switch between hornlessness and horn production occurs at about 5.0 mm (triangle). (b) Switchpoint-indicator plot. Residuals are measured as shown in figure 5.22 (top). The range of deviations from the mean is much greater near the switch point (triangle). In this region, the environmental signal (here, diet, reflected in body size) is weak and fails to overwhelm genetic and other causes of variation. Graph by D. J. Emlen.

Fig. 5.22. Phenotypic and developmental response to selection on a quantitative trait. (Top) Male beetles (*Onthophagus acuminatus*) with above-average horn lengths (in the positive residual range) were the upward-selected line; those with below-average horn lengths (negative residuals) were the downward-selected line. (Middle) The phenotypic response to selection: increased and decreased horn lengths. (Bottom) The developmental response to selection: divergence in the relationship between body size (indexed by prothorax width) and horn length, indicating change in the threshold of the switch between hornlessness and horn production. Based on Emlen (1996). Courtesy of D. J. Emlen.

be just as effective if it disfavors large individuals that fail to switch to horn production (delayed switchers) as it is if it favors small individuals that express horns (precocious switchers). The expectation of a high genetic correlation in morph ratio between environments is upheld by data on wing dimorphisms in insects (Roff, 1994b).

Some of the residual deviation from mean phenotype at a given body size could be due to varia-

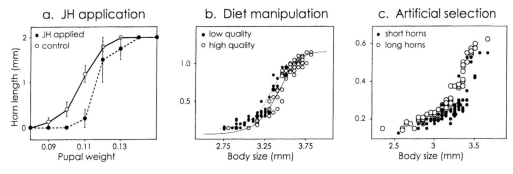

Fig. 5.23. Equivalence of environmental and genetic factors in the determination of alternative phenotypes. A shift in the size threshold for the switch to long versus short horn production in a beetle (*Onthophagus acuminatus*, Coleoptera: Scarabaeidae) can be produced either by (a) juvenile hormone (JH) application, which mediates the switch during the pupal stage; (b) dietary manipulation; or (c) artificial selection (Emlen, 1996, 1997b; Emlen and Nijhout, 1999). Courtesy of D. J. Emlen.

tion in *form*-influencing environmental and genotypic factors, not just variation in plasticity (regulatory factors). As discussed above (see figure 5.20), there is continuous variation in the dimensions of traits whose expression is controlled by a switch. Diverging selection that favors the largest and the smallest deviations from the mean, therefore, would simultaneously favor the extremes in plasticity and in within-morph variation, causing evolutionary change in both regulation and form.

Shifts in the response curve due to change in diet is similar to that obtained due to genetic change under selection (figure 5.23a,b). This supports the hypothesis of interchangeability of environmental and genetic factors in regulation by the switch. That this interchangeability is mediated by a hormonal switch mechanism is suggested, though not proven, by experiments that produced a similar shift using juvenile hormone treatments (figure 5.23c). Taken together, these findings are consistent with the hypothesis that the hormonal switch mechanism integrates the influence of genetic and environmental factors on the expression of horns, and mediates the interchangeability of environmental and genotypic influence (but see Zera and Denno, 1997, for a cautionary discussion of research needed to actually document hormonal control). A relatively clean hormonally mediated conditional switch, with little individual variation in response, is shown in figure 5.24. Such a switch is highly sensitive to environmental (size, and presumably dietary) variation.

Similarly, artificial selection lowered the threshold for switching to brown rather than green pupal color in butterflies (*Papilio polyxenes*; Hazel, 1977). There was a significant response to selection

after a single generation of selection for both raised and lowered sensitivity to a dark substrate. The regulatory mechanism evidently involves a hormone concentrated in secretory vesicles of ganglion cells and photoreceptors located in the larval head (reviewed by Hazel, 1995).

Moran (1992a) reviews evidence for genetic variation in the condition-sensitive regulation of alternative morphs in aphids. Clonal differences in morph ratio have been measured under uniform

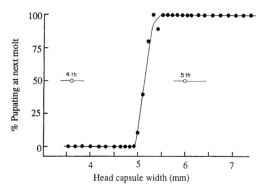

Fig. 5.24. A sharp nutritional switch to metamorphosis. Larval size was manipulated by varying larval nutrition, revealing a clean switch point (at about 5 mm head-capsule width, an index of body size), with little variation in response (metamorphosis). Still, some individuals (on the ascending portion of the curve) showed variation in response over a very small interval in size variation, and the number not responding was proportional to size. From Nijhout (1975), by permission of *The Biological Bulletin*. Courtesy of H. F. Nijhout.

conditions in seven species. Again, there is evidence for a role of juvenile hormone in regulating aphid wing polymorphisms, sexual versus partheno-genetic reproduction, host preference, and sex determination (reviewed in Moran, 1992a).

Interchangeability by Social Manipulation: Stingless Bees In social and symbiotic organisms, there is a level above the individual in the hierarchy of influences on trait expression: trait expression can be manipulated by other individuals. That is, interchangeability can be achieved by changes in the social environment. In stingless bees (*Melipona species*), the evolution of increased genetic influence on caste determination has occurred in a lineage with primarily nutritional determination. This was achieved via a social manipulation of development that in effect created a genocopy of the ancestral, conditional trait. In stingless bees (e.g., *Trigona* species), with nutrition-dependent caste determination, workers build dimorphic cells, with queen-producing cells much larger and more heavily provisioned than worker-producing cells (Michener, 1974, p. 105). In the *Melipona* species, with increased genotypic determination of caste (Kerr, 1950a,b, 1975), the brood manipulation performed by workers resembles that of laboratory scientists who wish to expose genotypic influence in a condition-sensitive trait. As if controlling environmental variables, workers make brood cells of nearly uniform size, and this is associated with increased uniformity in the distribution of provision among cells, thereby creating more nearly uniform rearing conditions. This increases the likelihood that genotypic differences among larvae are expressed (see also Emerson, 1958), and it is in species with such feeding regimes that queen–worker ratios appear to reflect Mendelian expectations.

Experimental manipulation of larval food supply that imposes extreme trophic differences between groups of larvae overcomes genotypic influence, restoring environmental control (Darchen and Delage-Darchen, 1975). Early work by Kerr et al. (1966) and Kerr and Nielsen (1966) showed that even with strong evidence of genotypic influence, there is still a nutritionally sensitive threshold. A certain trophic level is necessary for a queen genotype to produce a queen. Below this level queens are not produced, and above it the queen genotype is required or only a well-fed worker results. All of these observations accord with the expectations of interchangeability of genotypic and environmental factors, given a complex switch mechanism. As usual, when interchange of genetic and environmental influence occurs, caste in these bees involves

a complex switch involving worker behavior and hormones (Velthuis, 1976b; Velthuis and Sommeijer, 1990, 1991).

In stingless bees, the seemingly contradictory findings of different investigators, some revealing genetic determination and others supporting trophic determination, led to bitter controversy that sometimes even seemed to bring into question the integrity of the researchers concerned. The bees, however, have behaved well: they have never departed from a developmental pattern that takes both nature and nurture into account.

Does Interchangeability Confirm the Cybernetic Theory of Biological Order?

Kauffman (1993) cites the phenomena of genetic assimilation and the susceptibility of regulation to both environmental and genetic inputs as evidence for a far-reaching interpretation of development in terms of self-organization. He argues that "[g]enetic assimilation shows that exogenous perturbations can cause the same transformation of developmental pathways as those caused by selection on sets of modifier genes" and that this "suggests that the cell types of the genomic system are poised such that either minor external perturbations or alteration of internal components cause the same transformation" (p. 544). He goes on to interpret genetic assimilation as confirming a particular view of development. By this view, cells are "poised" to switch between a limited number of alternative states, or attractors. The poised condition leads to self-organization and the flexible bifurcating nature of development, with its sensitivity to slight perturbations by environmental or genomic inputs. Kauffman thus sees development as part of a more general class of phenomena governed by systems with their own "inherent" "spontaneous" order or self-organization.

Interchangeability of environmental and genomic inputs, in regulation controlled by condition-sensitive mechanisms, explains genetic assimilation and self-organization in conventional terms. Evolvable thresholds are easily related to quantitative genetic models on the one hand (see chapter 6), and self-organizing mechanisms of plasticity on the other (see chapter 3). The interchangeability of genetic and environmental influence is explainable in terms of the polygenic, poly-environmental nature of regulation involving response thresholds.

Some caution is in order regarding Kauffmann's conclusion that genetic assimilation supports the attractor concept. The fact that numerous genes or environmental factors can trigger the same transition when alternative phenotypes have previously evolved does not necessarily mean that only a few transitions are possible. Genetic assimilation is a product of selection, for a particular phenotype. It says nothing about others that might be produced by different stimuli, genetic or environmental. It does not, therefore, support the idea that only a few end points (implying attractors) are possible given a particular starting point. Living populations of organisms regularly show a diverse array of complex specializations referred to by biologists as "individual differences," as in the Cocos finches discussed in chapter 28.

Individual differences seem to be an expandable array of complex phenotypes, and they reflect the mosaic structure of the phenotype and its potential for combinatorial reorganization. In a study of individual foraging specializations in pigeons (Giraldeau and Lefebvre, 1985), the number of specializations could evidently be increased by increasing the complexity of the *environment*—in this case, the number of kinds of seeds offered to the birds. Only under certain circumstances, as in Galapagos and African finches (see chapter 28), is this indefinite array of possibilities forged, under environmental influence, into a limited number of alternative phenotypes that would fit the limited-end-points view of Kauffman. Most important, the phenotypic diversity observed depends as much on the environment as on the phenotype that is developmentally organized and reorganized. While phenotypic evolution is clearly influenced by preexisting structural organization and the genome (see part II), this is not the same as saying that it is severely limited or "constrained" by preexisting structure, since new inputs, including external ones, can intervene to multiply the number of variants in unexpected ways.

For these reasons, it is not convincing to argue that genetic assimilation indicates that organisms in general represent systems genomically poised for transitions to a limited number of alternative states or attractors.

I have argued that reorganization due to preexisting developmental plasticity, and subsequent genetic accommodation of regulation, is the basis for the emergence of new order in organic evolution. New phenotypic subunits begin and evolve as products of developmental plasticity. They originate when an environmental or genetic perturbation causes a shift in gene expression, and they are consolidated under selection for improved regulation and form. This is the *biological* basis for the evolution of complexity. It cannot be completely analogous to self-organizing evolution in the nonorganic world (cf. Kaufman and others), because multidimensional adaptively adjusted plasticity does not occur there.

Kaufman's book reaches beyond biology to show its links to more general physical processes that are the roots of organic nature. The idea of self-organization is a new perspective for understanding phenotypic organization and responsiveness, but it may tempt biologists once again toward an overly internalized view of phenotype determination. This could prove a complicating distraction from the familiar and powerfully explanatory facts of phenotypic plasticity, hierarchical modular structure, and quantitative genetics that seem sufficient to explain evolution, including the origin of novel phenotypes, the dispersed local decisions of self-organized systems, and the development of organized complexity.

The Organization of Development by Switches

Development has been described as a tree of binary decisions, or switch points (Garcia-Bellido et al., 1979; see also Weismann, 1893; John and Miklos, 1988; McKinney and McNamara, 1991; Raff, 1996; see figure 4.1). Switches are responsible for the universal properties of phenotypes discussed in other chapters. They determine the modularity of phenotypic traits and mediate the variation in phenotypic plasticity that permits the interchangeability of genetic and environmental influence on trait expression just discussed above.

Due to the influence of switches, the phenotype is a mosaic of semi-independent traits. The individual is still the fundamental unit of selection, for it is the individual that reproduces and thereby summarizes the effects of all subindividual traits on gene frequencies in the next generation. But the switch-organized modular subunits we call functional traits, each one brought into action under specific conditions, are semi-individually tested as contributors to individual fitness. Modular traits are subunits of switch-regulated gene expression and gene-product use, as described in chapter 4. So developmental switches determine which genes are exposed to selection and in which combinations. The ultimate justification for seeing how an individual phenotype is organized by switches is to understand the structure of the mosaic that is evaluated by selection. Here I describe some common patterns of switch coordination and how they can be analyzed.

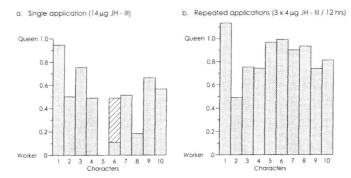

Fig. 5.25. The mosaic phenotype of the honeybee queen. Comparison of treatments (a) and (b) shows the semi-independent responses to juvenile hormone (JH-III) titer of (1) glossa (tongue) length; (2) mandibular tooth size; (3) ratio of head width to length; (4) size of mandibular gland; (5) development of corbicula (pollen baskets on hind legs); (6) pollen brush (dotted area = number of incomplete hair rows; hatched area = hairs evenly dispersed as in queens); (7) number of barbs on sting; (8) number of ovarioles per ovary; (9) diameter of spermatheca; and (10) ratio of length of medial and lateral venom duct, expressed as percent of the average value for 15 newly emerged queens (100%) compared with the value for workers (0). JH was applied to 3-day-old worker larvae. Response independence of these different parts illustrates the mosaic developmental structure of a highly integrated (queen) phenotype. After Beetsma (1979).

The Neutral Point as a Dissection Tool

When a strong signal coordinates the coexpression of a suite of traits, we perceive it as a single unit. But such units may be composed of many parts, as discussed in chapter 4. The mosaic nature of a trait can be exposed by adjusting the determinants of a switch so as to approach the *neutral point*—the equilibrium or threshold zone of responsiveness for a set of switches, called the "pivotal" point by Janzen et al. (1998; see also Ezhikov, 1934). This point has already been mentioned above as a range of conditions where individual differences in switch point may be revealed (figures 5.21, 5.24). In the neutral zone, all responses controlled by the switch are not clearly or completely triggered. The neutral zone was used by Schaffner (1930) to examine sex expression in corn (*Zea mays*). As noted by Schaffner, when a switch between two states takes place, the quantitative parameters that pass a threshold necessarily pass through a zero or neutral point in the transition from one state to the other, and "[t]he characters developed in tissues which have passed to the neutral condition are often of great interest" (p. 279). The neutral zone was discussed earlier because it can reveal individual differences in threshold that may reflect genetic variation susceptible to selection for altered sensi-

tivity. It also can reveal that what appears to be a single threshold is really composed of several that underlie a mosaic trait.

Experimental study of the neutral zone has demonstrated the mosaic nature of the worker and queen phenotypes of honeybees (*Apis mellifera*). Manipulations of larval diet produce a variety of intermorphs with some traits of workers and some of queens (Weaver, 1957a; Wirtz, 1973; Rachinsky and Hartfelder, 1990; Engels and Imperatriz-Fonseca, 1990). Such mosaics are rarely found in nature because worker-destined and queen-destined larvae usually receive sharply different diets clearly either passing, or failing to pass, the threshold distinguishing the castes (see Weaver, 1957a). Hormone manipulations reveal 10 phenotypic elements having somewhat different thresholds of expression (figure 5.25). As expected in a mosaic trait, each element is capable of independent quantitative variation in degree of development. The varied responses are evidently due to different sensitive periods of different imaginal discs that give rise to the separate traits. In normal queens, full development of all 10 traits is coordinated by adequate diet-induced JH titers during all critical periods.

This interpretation is supported by the results of repeated hormone applications, which caused the traits to approach the values found in normal queens (figure 5.25b). Variation among individuals

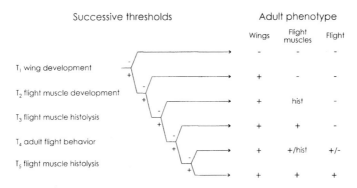

Fig. 5.26. Model of a serial ganged switch in insect dispersal polymorphisms. The loss of dispersal morphology and behavior often occurs in the same sequence, indicating different response thresholds (T) for different components of the mosaic trait. A minus sign indicates absence or reduction of the trait in the adult. Muscle formation may be followed by breakdown histolysis (hist). After Harrison's (1980) summary of Shaw (1970).

also occurs: intercastes subjected to the same feeding treatment showed different combinations of queenlike and workerlike traits (Weaver, 1957a). This variation could reflect genotypic differences potentially responsive to selection for modified form or altered frequencies of expression of the two morphs. The same phenomenon has been noted in ants: traits such as body size, ocellar size, type of thorax segmentation, number of ovarioles, and presence or absence of the spermatheca, which are strongly concordant in normal queens and workers, vary independently of each other in intercastes (Hölldobler and Wilson, 1990, after Plateaux, 1970). In the neutral zone, directional environmental influence is reduced, throwing any existing genetic component of variation into relief.

Comparison of the two treatments also shows that different elements respond independently of each other (have different regulatory pathways). Even though all are sensitive to a common regulatory element (juvenile hormone), they are a mosaic of traits that respond at different thresholds and rates. This study illustrates how complex phenotypes can be dissected using phenotypic engineering in the form of graded hormonal applications in the *neutral zone*, the zone of trait expression near the neutral, or switch point between alternative phenotypes or developmental pathways, where a variable (in this example, a hormone) that influences trait determination is at intermediate values.

The following sections describe some common arrangements of switches that organize the expression of complex traits.

Ganged Switches

Some compound switches consist of a temporal series of subordinate switches or sensitive periods that respond to the same environmental cue. Complete expression of the resultant mosaic trait requires that the environmental cue be sustained at sufficient intensity until all component switches are thrown. Mather (1955, p. 54) referred to this as a "ganged switch." He gave as an example a multilocus major gene (linked set), each locus of which controls a different phenotypic subunit. More commonly, a ganged switch might be a set of regulatory mechanisms subject to the same environmental or phenotypic factor at somewhat different times. The successive responses of a ganged switch require prolonged exposure to the same environmental factor. The stimulus may vary in intensity over time, with the result that some thresholds in the ganged switch may be surpassed in some individuals while others are not.

A relatively simple ganged switch is found in aphids (Shaw, 1970; figure 5.26). Winged migrant forms of *Aphis fabae* develop when alatiform nymphs are crowded throughout their development. If larvae of the winged form are isolated on successive days after hatching, different degrees of intermediates are produced, with traits dropping out in order depending on length of isolation (Shaw, 1970). Crowding for different amounts of time produces, in order of appearance, white abdominal markings, then full white markings, wing muscles, short wings, longer wings (without flight behavior), flight behavior, and, finally, the full migrant form. There is evidence (reviewed by Shaw, 1970) that development of all of these density-dependent traits is influenced by JH. Shaw speculates that successive responses depend on adequate JH being present during successive critical periods of development. Thus, as in locusts (see below), a full extreme migrant morph occurs only if the same stimulus (crowding) is continued long enough to induce the full set of complementary traits.

A well-studied ganged switch occurs in the determination of the migratory morph of migratory locusts (*Locusta migratoria*). Components of this complex mosaic phenotype include bright markings, small head, long wings, rapid development, high resting metabolic rate, lower oviposition rate and lifetime fecundity, and increased gregariousness and locomotory activity (Hardie and Lees, 1985; Kennedy, 1961). Each component is independently determined by a different regulatory mechanism. But the phenotype is structurally and functionally coordinated because all of the cues are correlated in some way with the same environmental factor—population density (see Kennedy, 1961; Nijhout and Wheeler, 1982; Hardie and Lees, 1985). Gregarious behavior (staying near others) results from tactile stimuli from other larvae followed by reduced movement. Development of *gregaria* pigmentation, on the other hand, requires multiple social stimuli (tactile, visual, auditory and olfactory, Kennedy, 1961, p. 85). The behavioral components of the alternative phenotypes are reversible in a matter of hours: a behaviorally solitarious individual becomes gregarious after only 4 hours of crowding and reverts to solitarious behavior after only 4 hours of isolation (reviewed in Collett et al., 1998), whereas other phase changes accrue across generations through maternal effects. The responses are hormonally mediated and cumulative. Crowding reduces corpora allata activity and causes earlier disappearance of the prothoracic glands (delayed maturation) in adult gregaria.

The development of phase differences under the ganged switch of locusts employs a compound switch whose elements are so obviously dissociated and whose response is so slow, with a full cumulative response requiring up to four generations of exposure to crowding, that it strains the concept of a single, switch-coordinated trait. It behaves in nature as if slowly passing from one morph, through the neutral zone, to the other. Yet the result, when all pathways are thrown in one direction or the other, is among the most striking coordinated alternative phenotypes known. For this reason, it illustrates clearly, as if photographed in slow mo-

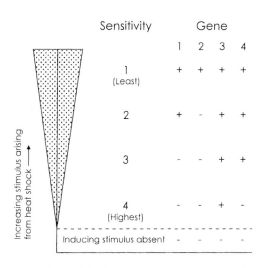

Fig. 5.27. Model of gene expression sequence in a serial ganged switch, based on study of heat-shock proteins. From Thomson (1987).

tion, how independent ontogenetic pathways ultimately converge to be expressed in unison as a functionally cohesive trait.

Serial Switches

In another kind of compound switch, subordinate switches have different thresholds for responding to the same cue. Under increasing values of the cue, the components of a mosaic develop in sequence, as quantitatively different thresholds are passed. Such a serial switch is hypothesized to explain the coordinated production of heat-shock proteins with different transcription thresholds (figure 5.27). Heat-shock proteins are transcribed as a set under stress (e.g., in response to strong or abrupt anoxia, temperature change, viral infections, harmful chemicals, etc.). They act as regulators that adaptively alter nuclear gene activity (see Thomson, 1987). Their coordinated transcription is governed by common affinity of their transcriptional control sequences for a particular binding protein that governs the transcription of all. The affinitiy of the binding protein for the different control sequences varies, and its affinity is condition sensitive—it rises as the environmental stress is increased. When the threshold affinity to all control regions is surpassed, the several heat-shock genes are expressed as a set, as shown in figure 5.27. The same model of switch coordination of a complex response is suggested by observations of responses at higher levels of organization (figure 5.26) and underscores the mosaic

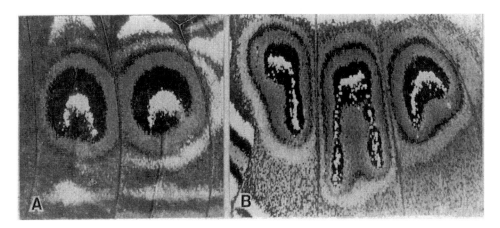

Fig. 5.28. Reaction-diffusion in pattern formation. Point and line sources interact to produce different patterns in the ventral forewing (A) and the ventral hind wing (B) of a butterfly (*Morpho hecuba*; Nymphalidae: Morphinae). From Nijhout (1991a), © H. F. Nijhout, by permission. Courtesy of H. F. Nijhout.

nature of switch-controlled traits, which is often revealed at intermediate levels of the indicing stimulus, near the "neutral point" or switch point for expression of the trait.

Another example of a serial switch occurs in the wing coloration of the cold-season phenotype of some butterflies, in which the melanin, pteridine, and red-pigment aspects of the color pattern are all cued by temperature. But each pigment has a somewhat different temperature threshold, so the colors appear sequentially in some populations having a long autumn emergence period (Shapiro, 1976).

Cascades

In a cascade, one switch-controlled component leads to, and necessarily precedes, another, such that the passing of an initial threshold automatically leads to the passing of a whole series of subordinate switches as a chain reaction. Once the initial switch is triggered, the subsequent ones follow, barring some alteration in their stimulus–response relations. An example of a cascade is the development of the sterile worker in *Polistes* and other social wasps having caste determination in the adult stage. An initially low degree of aggressive behavior relative to another female, perhaps reflecting hormonal differences, leads to subordinant rank in a dominance hierarchy, which leads in turn to a trophic disadvantage and lack of access to brood cells for oviposition, which leads in turn to ovarian resorption and associated behavioral differences (see references in West-Eberhard, 1996).

Switches Based on Spatial Relations, and Spatial Integration

All of the types of compound switches described so far involve temporal sequences of switches. The commonness of temporal mosaics is one reason that heterochrony, or temporal reorganization of the phenotype, is so often important in evolution (see chapter 13, on heterochrony). But compound switches can also depend upon spatial relationships among parts, especially in morphogenesis. Contact between growing tissues may alter their behavior, limiting growth, or inducing a switch to a new pattern of gene expression, and a set of such contacts in spatial contiguity may structure a mosaic trait. In the development of neural crest cells, for example, individual cells migrate until they contact epithelial cells, and depending on the type of cells contacted, they give rise to nerve cells, pigment cells, cartilage, or bone (McKinney and McNamara, 1991, p. 68).

Spatial diffusion boundaries of morphogens and other substances such as pigments also delimit phenotypic subunits, representing switch boundaries to spatial expansion, or determine a locus of deposition (Oster and Alberch, 1982; Nijhout, 1991b; Price and Pavelka, 1996; figure 5.28). In the determination of pigmentation pattern in birds, the sensitivity of melanocytes and the shape of the embryo at the time of pigment development seem to determine where pigment is deposited in a diffusion field (see Price and Pavelka, 1996). Changes in the sensitivity of all melanocytes mean that in a given epidermal area, the switch threshold for deposition of pigment is reached in some zones and before others. This leads to an evolutionary pattern in which pigmentation disappears in a certain order if the threshold is changed, presumably under selection, over time. In warblers, comparative study of species with different trait combinations shows that (*Phylloscopus* species) pigmentation is first lost from the wings, then wings and crown, a pattern that has evolved twice in the genus (Price and Pavelka, 1996; see chapter 14). Not surprisingly, color pattern is an evolutionarily labile trait in these birds, and intraspecific pigmentation anomalies that resemble patterns in related species are common.

Contiguous parts are often used together. By virtue of contiguity, they may form an integrated subunit. The color patterns of butterfly wings are an example where contiguous regions that represent semi-independent and bounded diffusion centers are displayed together as a single wing (Nijhout, 1991b). They therefore function together, forming an integrated unit pattern at the level of the wing.

The mouse mandible is another well-analyzed example of coordination involving spatial contiguity (summarized in Atchley and Hall, 1991; Hall, 1992). The ontogenetic origins of each component bony part of the mouse mandible have been described, as have their patterns of variation among species (Cheverud et al., 1991). The mouse mandible is typical of many morphological structures in that it forms a functional unit, and yet the subunits composing it are not controlled by a single switch. Rather, several elements of muscle and bone having different developmental pathways converge and interact in space to form an integrated structure (see figure 4.14, which shows the developmental pathways for the different elements of bone). Each subunit, or "fundamental developmental unit," is the product of a distinct cell lineage, whose final size (cell number) is subject to five general kinds of modulation, or "causal factors" (terminology from Hall, 1992). These are (1) number of stem cells in a preskeletal condensation or aggregation of cells, (2) time of initiation of condensation, (3) rate of cell division, (4) fraction of cells involved in cell division, and (5) rate of cell loss (cell death). These factors are subject to regulation by elements such as hormones and contact with adjacent tissues, which are known or suspected to be genetically variable (Atchley and Hall, 1991, p. 115). Patterns of evolutionary change in the shape of the mandible (Cheverud et al., 1991; Atchley and Hall, 1991) suggest hierarchical spatial organization where there is locally divergent

regulation *within* elements of bone, for example, due to interaction with muscle or to other kinds of local epigenetic factors (see Atchley and Hall, 1991, p. 121). The arrows in the "developmental tree" diagram of Atchley and Hall represent points of subunit determination and would be classified as switch points under the broad concept used here (see chapter 4).

The analysis of Atchley and Hall (1991) raises an important question regarding phenotype structure. What coordinates the higher unit of organization recognized as the (entire) mandible? Atchley and Hall refer to its coordinated development as being "choreographed" and discuss the importance of its covariance structure for correlated evolutionary change. But what *organizes* the choreography? Atchley and Hall list (their figure 15, p. 137) the "causal factors giving rise to morphological associations," for example, genetic, epigenetic, and environmental factors of maternal, intrinsic (progeny), and extrinsic origin. But why do we recognize these associations as a single, integrated structural unit?

The functional integration of the mouse mandible, or of any other such morphologically complex organ formed by the spatial convergence of different ontogenetic pathways, is based on its coordinated deployment or use and consequent selection as a unit (see chapter 7). This, in turn, depends importantly on spatial contiguity of parts, produced by spatially convergent cell movements and growth, which early in the evolutionary history of such a structure may have been coincidental rather than a product of selection. Neighboring parts have a relatively high probability of being moved or used together. Then selection molds the coexpressed or coused structures as a set. So the choreography begins with coincidental proximity and interaction during development. Integration may be increased both by phenotypic accommodation during use and by natural selection, yielding appropriate values of the five determinant, genetically variable parameters recognized by Atchley and Hall (1992) as influencing the composition of each component.

The importance of coordinated use for the evolution of integration of contiguous parts is shown by Zelditch (1988) in a study of cranial and jaw development in rats (*Rattus norvegicus*). At 21 days of age, the cranium and jaw vary independently of each other—they do not form an integrated structure. Later, in adult rats, they vary together as a single integrated unit. This is associated with an ontogenetic increase in coordinated use of the two structures. After weaning, the length of the dental diastema becomes increasingly correlated with the rest of the skull in association with the transition from sucking to grinding occlusion.

Simultaneous evolved loss or appearance of a morphological subunit, for example, as an anomaly or an established feature of a population or species, is evidence for common ontogenetic control. This fact could be used to help evaluate hypotheses such as that of Atchley and Hall (1991, p. 109) that the osteological components of the mammalian mandible develop from a single multipotential stem cell, rather than from different neural crest cells via four independent pathways. If all four components descend from a single cell, they would more easily be lost or modified as a set in natural or experimentally induced anomalies. Zelditch et al. (1992) discuss how spatiotemporal integration permits evolved heterotopy, or change in location of trait expression (see chapter 15).

Spatial proximity plays a role in developmental and functional integration at every level of organization. At the molecular level, coordinately transcribed segments of DNA occur together or are brought together by devices such as RNA splicing and transposons (John and Miklos, 1988). At the cellular level, migration and aggregation or "condensation" precede coordinated differentiation of cells (Atchley and Hall, 1991; Gerhart and Kirschner, 1997). At the organ level, contiguous tissues of different embryological origin form integrated structures such as the eye, the mandible, the stomach, or a limb (reviewed by Deuchar, 1975). At supraindividual levels of organization, in the evolution of social life, the first essential organizing step is spatial contiguity, or group formation, just as in the evolution of multicellular organisms the first step was likely cell aggregation (Bonner, 1988), either through migration and mutual attraction or by staying together following multiplication (population viscosity).

Local differentiation and alteration of contiguous elements may be a virtually inevitable result of proximity in living things. Proximity of differentially active, moving, growing, metabolizing, or aggressive elements often produces some kind of change in the interacting parts. It seems likely that prior to the evolution of neural and hormone systems, which permit coordination among noncontiguous parts, developmental coordination depended on proximity. The most primitive examples of distance signaling by hormonelike chemicals—the cell-attracting acrosins of amoebic cellular slime molds—function to promote cell aggregation (Bonner, 1988, p. 271), or proximity, the first step of organization into differentiated stalks and spores.

Dispersed Local Switches (Self-organization)

In self-organized development, shared decision rules of a large number of dispersed elements, each sensitive to local conditions, are the basis for emergence of a coordinated pattern. This was discussed in chapter 3, in a section on hypervariability and somatic selection. Establishment of patterned neural connections, for example, begins with production of a large number of processes, then selective survival of those used (Edelman, 1988; Gerhart and Kirschner, 1997). Patterned foraging trails of army ants begin with an exploratory front of individual ants searching for food, then adopts compact columns connecting food sources with brood, as individual ants react to local encounters with food and other ants to switch their behavior in accord with simple, shared rules (Deneubourg et al., 1989). This arrangement of switches has some similarity to a ganged switch in that different elements respond individually to the same type of stimulus. But dispersed subunits are a way of achieving more sensitivity to local conditions due to the extreme dispersion and independence of the responding subunits and their small size relative to the size of the field over which coordination can be achieved without direct communication among them. The independently responding elements of a self-organized system of local switches provide mechanisms of fine-tuned phenotypic accommodation (see chapter 3).

In summary, master switches or groups of coordinated switches define relatively large phenotypic subdivisions or modular traits, which may be a mosaic of subunits governed by subordinate switches (see also Thomson, 1988). The variable intermediates of a neutral zone of determination provide a window on the hierarchical developmental structure of mosaic phenotypes.

Complementarity, Continued: The Developmental Basis of Continuous and Discontinuous Variation

The problem of relating the continuous and discrete properties of biological organization was introduced in chapter 4. Some of the most divisive and persistent controversies of evolutionary biology revolve around different emphases on continuous and discrete variation. The debate over gradualism and saltation is such a controversy (see chapter 24). The problem of the continuous and the discrete comes up in any attempted synthesis of development and evolution because development, based on switches and modularity as already discussed, produces discrete qualitative traits. Yet important branches of evolutionary genetics are based on quantitative variation, and most studies of phenotypic plasticity deal with continuously variable reaction norms. A synthetic evolutionary theory has to be able to connect developmentally produced discrete variation with the continuous variation studied in these other fields. How, in terms of development and phenotypic structure, are discrete and continuous variation related?

This problem is discussed in chapter 4, where continuous and discrete variation are explained as complementary properties of modular traits. Continuously variable traits can be dimensions of discrete subunit traits, or of switch-controlled on–off processes, whose duration governs the size of the variable. Measurement of quantitative, continuous variation usually depends on the modular, or subdivided organization of phenotypes—on natural boundaries between measured traits and processes. Cock (1966) referred to "developmental autonomy" in the growth rates of organ rudiments once they are set on developmental trajectories at an early stage in development, as shown by *in vitro* culture experiments. This concept, and Cock's discussion, is of special interest because it clarifies the relationship between the local discontinuous nature (autonomy) of morphological subunits and physiological or behavioral processes, and the continuous nature of the variation of parts when a population of traits is compared. Natural boundaries between traits are defined by the developmental switch points that determine a phenotypic trait, or subunit. A phenotypic *subunit* is a component of the phenotype whose expression is governed by a switch or by a clearly localized, bounded process.

I have seen the objection to an emphasis on switches and discrete traits that most variation in organisms is quantitative, not qualitative or discontinuous variation. While it is true that "[c]ontinuous variation has been found in all characters of all organisms, where an adequate study has been made" (Mather, 1973, p. 17; see also Falconer, 1981), the word "character" refers to a discrete entity. In most cases, what is measured is a dimension of a discrete part. If there were not discernible boundaries and consistency among individuals in

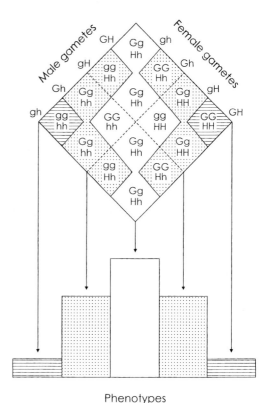

Fig. 5.29. The relation between continuous (polygenic) and discrete (allelic) genetic variation. With only two polymorphic loci, polygenic influence produces a continuously variable distribution of forms. The histogram shows the distribution of phenotypes obtained in an F_2 population (or a randomly breeding population with equal gene frequencies) with two diallelic genes (G and H) of equal and additive effect but without dominance, neglecting nonheritable variation. The quantitative expression is set equal to the number of capital letters in the genotype. After Mather (1973), based on Mather and Jinks (1971).

their definition, comparable measurements would be difficult or impossible. Measurement of continuous variation implies recognition of a discrete, measured entity, or bounded process. Even when a measurement requires the imposition of artificial boundaries, such as a defined interval of time or age in a continuous process such as growth, different measurements refer to some discrete natural unit such as an individual or a life stage with a distinctive rate.

The developmental causes of continuous variation fall into two classes, both of them based on

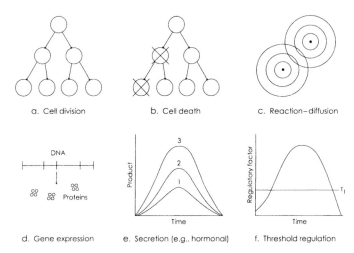

Fig. 5.30. Some developmental sources of continuous and discrete variation in the phenotype. Discrete repeated events, such as cell division (a) and cell death (b), that vary in rate and duration give rise continuous variation in trait dimensions; continuous processes, such as substance diffusion (c) and quantitative change in regulatory factors (d–f), can give rise to discrete events as borders are reached or reaction thresholds are passed.

the modular organization of phenotypes: (1) the value of a continuously variable trait may be the sum of several discrete or switch-controlled subunits, as in the continuous variation of polygenic traits (figure 5.29), in effect stepped or particulate gradation; or (2) it may be a dimension of a subunit whose value is determined by the stop–start action of switches on a continuously variable process such as growth, diffusion, or the secretion of a hormone (figure 5.30). The latter type may often prove to be finely particulate, as when growth involves cell multiplication or when diffusion is analyzed at the molecular level. Students of social insect queen–worker caste differences recognize these two kinds of continuous variation by distinguishing between *intermediates*, or individuals with intermediate values of a continuously variable trait, and *intercastes* (or intermorphs), individuals that combine discrete elements of two alternative mosaic alternatives.

Every discrete phenotypic trait has quantitatively variable dimensions, as in the length of a bone or a fingernail. Each discrete subelement of a well-studied polymorphism in migratory locusts, including degree of melanization, gregariousness, migratory restlessness, wing length, elytron length, head length, and fecundity, for example, shows somewhat independent quantitative variation if its dimensions are measured. When an entire population is examined, all grades of intermediates are found in all elements, leading Kennedy (1961) to describe it as a "continuous polymorphism." In this respect, the locust polymorphism is like many other switch-controlled morphological traits. Even when quantitative traits have a common regulatory mechanism, they can vary somethat independently due to having different thresholds of determination subject to somewhat different sources of variation (figure 5.31). Fighting male beetles have two discrete morphs, but within each there is continuous variation around the mean of each morphological class (see figure 5.20).

Any quantitative trait, such as bone length, is a dimension of a qualitative, discrete one, such as a bone (see figure 4.15). *Correlations* among quantitative traits can be produced when the same switch or set of off–on switches influences the expression of a complex phenotypic subunit—one composed of several quantitatively variable elements. Thus, we can state a principle of complementarity (see also chapter 4): *discrete and continuous variation are complementary, mutually dependent properties of every phenotypic subunit.* For these reasons, it may be misleading to think of a character exclusively as a "quantitative trait," "a misnomer since the same character can, even in a single set of observations, show continuous variation at the same time as the discontinuities in expression" (Mather, 1973, p. 17).

While it can be argued that all genetical evolution is quantitative change in gene frequencies, concepts such as species, homology, and Mendelian ge-

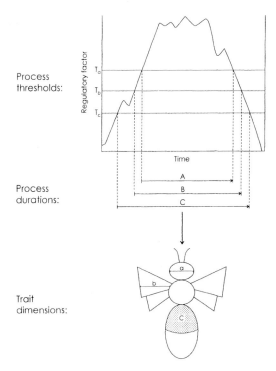

Process thresholds:

Process durations:

Trait dimensions:

Fig. 5.31. Developmental basis of continuous variation in the discrete components of a complex (mosaic) trait. A single quantitatively variable regulatory factor can determine the development of a set of correlated traits (head, wings, abdomen) whose underlying processes (a, b, c) have different thresholds (T_a, T_b, T_c) of sensitivity to the regulatory factor. The result is a coordinately produced mosaic of discrete traits whose quantitative dimensions (a, b, c, below) are highly but imperfectly correlated.

netics depend on the recognition and comparison of discrete variants (see Greene, 1994, regarding homology). So both approaches are clearly important to the study of evolution.

Seeing the relationship between the continuous variation and discreteness of phenotypic traits requires moving between levels of phenotypic organization. As one focuses downward on finer and finer elements of structure in a hierarchical arrangement of subunits, a trait, such as a bone, that appears to be a continuously variable trait when

viewed from a relatively high level of organization proves to be composed of smaller discrete subunits, for example, growth regions with differentiated cells (see Atchley and Newman, 1989; see also figure 4.14). As phrased by Mather (1973), "Each of the groups into which the individuals are divided by the major genic [or developmental] discontinuities then shows continuous variation among its members" (p. 17). Conversely, if you move upward in the hierarchy, the same change in viewpoint occurs, even above the individual level. There is continuous variation in population size, but separate populations are discrete entities (Van Valen, 1973, p. 93).

Consequences for Selection and Evolution

The organization of development by switches accounts for the fact that the phenotype of an individual is a mosaic of somewhat independent traits. Switches also account for the variable condition sensitivity of trait expression and use (phenotypic plasticity) and for continuous variation in the dimensions of modular traits.

Because switches are the focal points of environmental and genetic influence on development, they are the focal points of evolutionary change. As shown in this chapter, switches mediate the interchangeability of genetic and environmental influences that make it possible to adjust the condition sensitivity of regulation and thereby change the frequency of trait expression under selection. This enables adaptive evolution to proceed via quantitative genetic change, not as a shift in the frequency of alleles that stand in a one-to-one relationship to the favored trait, but as a shift in the frequency of alleles that regulate the *expression* of the trait. The potentially affected genes are those that influence response thresholds and the ability of organisms to pass them. In concrete terms, the genes that respond to selection are those that influence the responsiveness of mechanisms of regulation—neural and hormone systems, the sense organs that provide inputs to them, and the cells, tissues, and organs that either respond or fail to respond during development. Some of these mechanisms are described in chapter 3.

6

Adaptive Evolution

Evolution of a novel adaptive trait by natural selection begins with a recurrent developmental variant, a change in phenotypic organization that may be either genotypically or environmentally induced. Initial viability is increased by adaptive plasticity (phenotypic accommodation), followed by genetic accommodation due to selection, which may affect the regulation (frequency and conditions of expression) and form (efficiency and specialization) of the novel trait. Genetic accommodation of regulation may, but need not, lead to trait fixation (genetic assimilation). In evolution by genetic accommodation, quantitative genetic change in threshold of expression mediates the evolutionary establishment of a novel adaptive trait; genetic innovation need not await mutation; and environmental induction can initiate evolutionary–genetic change. Evolution by genetic accommodation includes genetic assimilation and the Baldwin effect but differs from those processes in that (1) innovation can begin with mutation, (2) condition sensitivity of expression can either increase or decrease under selection, and (3) genetic change can affect form as well as regulation (threshold and frequency of expression). Novel developmental responses are random with respect to future function, but nonrandom with respect to past function, since they involve reorganization of an adapted and adaptively plastic ancestral phenotype.

Introduction

Adaptive evolution—phenotypic improvement due to selection—is the central theme of Darwinian evolutionary biology. The concept of adaptive evolution by selection on heritable variation underlies every evolutionary analysis of form, function, and fitness. How biologists view adaptive evolution—how they are taught to view it—has a profound influence on thinking and research in virtually every area of biology.

Despite the importance of adaptive evolution, there is still controversy over how to relate development and selection in discussions of adaptive design (e.g. see Charlesworth, 1990; Amundson, 1996). It is a controversy that began in the late nineteenth century, when biologists unfortunately began to dichotomize development and selection, as if they were opposing factors in evolution (see the discussion of gradualism in chapter 24). The modern version of this debate dichotomizes selection and developmental constraints (e.g., see Maynard Smith et al., 1985; Bell, 1989; Ridley, 1993;

Amundson, 1996; see also chapter 1). Phylogenetic constraint is another version of developmental constraint, since it refers to limits imposed by ancestry (inherited developmental patterns) on current form as does the idea of developmental constraint.

The dichotomy between selection and development, as if they were opposed factors in adaptive evolution, is misconceived. Adaptive evolution is a two-step process: first the generation of variation by development, then the screening of that variation by selection (Mayr, 1962; Endler and McLellan, 1988, p. 395). If an established trait (one widespread in a population) persists through phylogenetic branching events, this is not evidence against selection as an explanation for the trait (cf. Coddington, 1988), as if speciation reduces the importance of selection. Rather, it could be taken as evidence that selection is important for maintenance of the trait in more than one lineage: trait persistence over long time spans certainly does not represent absence of continued selection, since it is known that traits no longer favored by selection, such as the eyes of cave animals and walking limbs

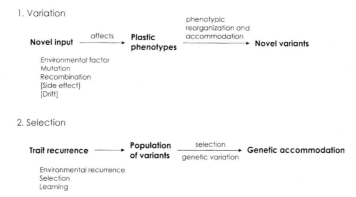

Fig. 6.1. Two steps in the evolution of novelty by genetic accommodation. (1) Initial stimulation of novel variants by mutation or environmental inputs to a plastic phenotype, which responds with phenotypic reorganization and phenotypic accommodation. (2) Recurrence of the initial stimulus (genetic or environmental) produces a population of variants on which selection can act to produce (given genetic variation in the variants) evolutionary genetic change (genetic accommodation).

in whales, are often lost (for other classic examples, see Rensch, 1960). Thus, trait persistence cannot be assumed to represent phylogenetic inertia or developmental constraints (for further discussion of this point, see Reeve and Sherman, 2001).

To treat development and selection as if they were competing explanations is to misunderstand evolution, like asking, "What makes a Nobel Prize, the originality of the winner or the Swedish selections committee?" Developmental variation and natural selection are two aspects of a single process, adaptive evolution. We may legitimately ask to what degree the form of an established (e.g., populationwide) trait is a developmental side effect of another, rather than a finely tuned result of selection; or whether the form of a trait originated as a single large developmental step (saltation) or because of a series of small steps favored by a long history of selection in the same direction (conventional gradualism). But a widely expressed trait is inevitably subject to selection, so the answers to such questions require attention to *both* development and selection. Both have inevitably affected the two-step process of variation and selection that produces the form we observe and causes it to persist in a population, or in the diversified descendents of a population.

Taking into account the two-step nature of evolutionary change (figure 6.1) and the developmental plasticity of organisms, I propose that adaptive evolution involves the following events:

1. *Trait origin*—the initial appearance as a qualitatively distinctive developmental variant, which occurs when some new input (e.g., a mutation or an environmental change) affects a preexisting responsive phenotype, causing a phenotypic change or reorganization.
2. *Phenotypic accommodation by individual phenotypes*—the immediate adjustment to a change, due to the multidimensional adaptive flexibility of the phenotype.
3. *Recurrence, or initial spread*—due to recurrence of the initiating factor, whether environmental or mutational. This produces a subpopulation of individuals that express the trait.
4. *Genetic accommodation*—gene-frequency (evolutionary) change due to selection on variation in the regulation, form, or side effects of the novel trait in the subpopulation of individuals that express the trait.

Somewhat similar ideas have been presented by Cairns et al. (1990), Johnston and Gottlieb (1990), Gottlieb (1992), and Sarà (1996), although with a different emphasis and in different terms. The Johnston and Gottlieb (1990) concept of *neophenogenesis*, or "persistent transgenerational change in the phenotypes of a population" due to both genetic and extragenetic change (p. 485), corresponds to steps 1 and 3 (trait origin and recurrence) above. These authors are primarily concerned with eliminating the nature–nurture dichotomy by showing, correctly, that neophenogenesis can occur with or without genetic change. "Our concern is with the process of neophenogenesis, with long-term change in the phenotypic makeup of populations, not with changes in their genetic makeup" (Johnston and Gottlieb, 1990, p. 481).

While I agree that phenotypic traits cannot be dichotomized as products of either nature or nurture (see chapter 5), and with the conviction of Johnston and Gottlieb (p. 483) that genetic change must follow, not precede populational phenotypic change (see Genes as Followers, below), I aim to show not only the nature of trait origins, but, beyond this, that novel phenotypes, whether mutationally or environmentally initiated, are virtually always subject to quantitative genetic variation and, if the variable novelty differentially affects fitness (undergoes selection), this can lead to changes in the genetic makeup of populations (evolution). The Johnston and Gottlieb discussion does not sufficiently emphasize the relationship between developmental and gene-frequency change and does not distinguish clearly between selection and evolution (both are defined as allele-frequency, or genetic, change; see Johnston and Gottlieb, 1990, figure 2 caption, p. 485).

Sarà (1989, 1996) also emphasizes the two-step nature of adaptive evolution: a responsive organism first responds to new inputs (he emphasizes environmentally induced responses), which are then filtered by natural selection. The new responses are "intelligent" in being related to environmental conditions from their initiation. In this way, environmental information gets incorporated into development.

The four fundamental aspects of adaptive evolution listed above are discussed in the remainder of this chapter and in part II (on phenotype transitions during evolution). Other aspects of adaptive evolution, or evolution by natural selection, are discussed extensively in later chapters. They include:

5. *Persistence as an alternative phenotype*—a new phenotype may be advantageous without ever becoming ubiquitous. It may begin at low frequency (as an anomaly), then persist indefinitely as an alternative phenotype, a variant that is common enough due to positive selection to be regarded as an adaptation rather than an anomaly, yet not expressed in all individuals of a population. In the alternative-phenotype phase, a phenotype is expressed as a switch-determined alternative to an established trait in the same stage of ontogeny.

The alternative phenotype could conceivably last for the entire evolutionary lifetime of a trait, with fluctuating frequencies of alternatives (see chapter 22, on maintenance of

alternative phenotypes). A new alternative can spread and persist even though relatively crude compared to a previously established alternative if it is more advantageous in some individuals or some circumstances (see chapter 22). Persistence and modification during this phase of evolution are discussed in detail in the chapters of part III, which show how persistence as an alternative phenotype facilitates the evolution of specialization and the acquisition of new adaptive modes within a species (see chapter 21).

6. *Modification*—modifier origin and its genetic accommodation. New rounds of evolutionary change (steps 1–5)—modification—may affect either the form or the regulation of a trait. Modification may occur in either the alternative phenotype phase or following fixation (see below). It implies the origin and establishment of new ontogenetic branches subordinate to the major switch that controls the trait modified (see chapter 21).

7. *Phenotypic fixation or deletion*—the alternative-phenotype phase ends if a trait either becomes the only alternative expressed (phenotype fixation) or fails to be expressed at all (deletion), due to change in the conditions or genetic elements that influence its regulation. Fixation can accelerate the evolution of increased specialization (West-Eberhard, 1986; see chapter 21), which may help to explain punctuated patterns of evolution (chapter 30).

The causal chain of adaptive evolution (figure 6.2) begins with development. *Development*, or ontogenetic change induced by genomic and environmental factors, causes phenotypic variation within populations. If the phenotypic variation caused by developmental variation in turn causes variation in survival and reproductive success, this constitutes *selection*. Then, if the phenotypic variation that causes selection has a genetic component, this causes *evolution*, or cross-generational change in phenotypic and genotypic frequencies (figure 6.2). Selection depends upon phenotypic variation and environmental contingencies only; it does not require genetic variation. But genetic variation is required for selection to have a cross-generational effect—an effect on evolution. From these causal relations, it is clear that development, not selection, is the first-order cause of design. Selection is a second-order cause that molds the distributions of traits in a *population* by screening the products of development and determining which ones persist and multiply across generations. It is evolution (de-

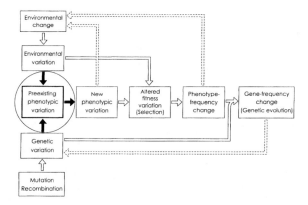

Fig. 6.2. Causal relations between environment, genes, phenotype, and selection in adaptive evolution. New phenotypic variation can begin with either environmental change (top) or genetic change (bottom). Dashed arrows indicate feedback among variables.

fined as cross-generational phenotypic change mediated by gene-frequency change), not selection, that depends on genetic variation. Therefore, selection (differential reproductive success) can continue due to phenotypic variation even after it depletes genetic variation, then have an immediate effect on evolution should any relevant genetic variation arise (West-Eberhard, 1979; Hamilton, 1987a, pp. 418–419).

The importance of genes for evolution is their inviolability: the relative immunity of the unexpressed transmitted germline genome to developmental contamination (chapter 17) means that the molecules whose copies were tested during one generation are closely identical to those inherited by the next generation. They are the most dependable currency of adaptive evolution due to selection, the most secure in that the next generation can depend on what it is getting following a round of selection. It is important to acknowledge this and not play with Lamarkian terms in discussions of development and evolution.

Prerequisites for Evolution by Natural Selection

Given the causal chain of events in adaptive evolution, all that is required for adaptive evolution to occur is intraspecific *recurrence* and *heritability* of a developmental novelty. Intraspecific *recurrence*, or the formation of a population of individuals that express the trait, enables the trait to be tested in

different genetic backgrounds (Rachootin and Thomson, 1981; Rossiter, 1996); heritability, due to genetic variation in the regulation or form of the trait, enables any variation in reproductive success due to the trait to cause evolutionary change. The degree to which a novelty satisfies these two conditions is a measure of its *evolutionary potential*— its ability to change frequency across generations under selection. *Genetic accommodation* is gene-frequency change due to selection on the regulation, form, or side effects of a novel trait.

It has been recognized that selection cannot act as efficiently on rare variants as on those at substantial frequencies in a population (Charlesworth et al., 1982; de Beer, 1958; Lande, 1978). Recurrence requires (a) shared capacity in a population of individuals to produce the trait in response to an inducing factor (e.g., a genetic or environmental input) and (b) recurrence of the inducing factor. Rossiter (1996) called recurrence the "sustainability" of a trait and noted that the evolutionary consequences of an inherited genetic or environmental effect depend on this. Recurrent development of a trait depends upon both (a) recurrence of the input (mutation or environmental change) that initiated the trait and (b) recurrence in the population of the ability to respond. It is due to the requirement of input recurrence (a) that environmental factors characterized by cyclic or seasonal recurrence, such as seasonal temperature, photoperiod, or nutritional effects, so often get incorporated into development.

The ability of preexisting phenotypes to respond consistently to a new input (b) is a hidden assumption of much evolutionary reasoning. We generally assume that when a mutation spreads in a population, its effect on the phenotype also spreads. This implies the additional assumption, seldom discussed, that all phenotypes respond in the same way to the mutant gene. I make the same assumption here, but for both genetic and environmental inputs: I assume that most members of a population of the same species share the ability to respond in a similar way to a given genetic or environmental factor. I stress, in addition, the importance of genetic variation in their ability to respond. The commonness of genetic variation in virtually every selectable (phenotypically variable) trait is well documented, as discussed below (e.g., see Lewontin, 1974, p. 92). I argue that gene-frequency change due to selection at loci that influence response capacity (genetic accommodation) is the usual basis for spread of an adaptive trait, whether genetically or environmentally induced. These assertions are justified with various lines of reasoning and various kinds of evidence in the present chapter.

For a novel phentoypic trait to evolve under selection, it has to recur in different individuals within a population. A unique, unreproduced event does not lead to selection and evolutionary–genetic change. Recurrent production of a phenotype allows a novelty to be screened by selection (Rachootin and Thomson, 1981, p. 185). Repeated expression is necessary to expose genetic variation, if it exists, to selection and produce evolutionary change.

Recurrence is not the same as replication. The distinction is important: phenotypic recurrence often involves a complex ontogenetic process with high susceptibility to variation within the bounds of normal development. The recurrence of a queen phenotype, for example, depends upon the state of a regulatory mechanism that is sensitive to nutritional level and social interactions. Queen ants may differ phenotypically and genotypically, but the queen phenotype recurs as a characteristic set of somewhat variable traits. Replication, by contrast, implies exact copies and highly restricted variation. Replicative accuracy would be reflected in trait-reproduction accuracy if there were a one-to-one relationship between phenotype and gene, but development complicates the genotype–phenotype relation. While this makes for reduced accuracy, the intervention of many genes actually may facilitate fine adjustments under selection: the effects of a precisely copied allele from one parent are combined, in the normal course of development, with other variables, such as different alleles from the parent's mate, environmental factors, and, most important, multiple modifiers of the allele's effect on the phenotype. Whereas replication implies precise multiplication, barring errors, trait reproduction, which involves many genes, must usually involve the regular production of selectable variation along with similarity of form. This variation facilitates phenotype evolution.

Mutationally induced phenotypic novelties are the responses of preexisting structure to a new genomic input. If a mutation is advantageous or neutral in its effect—it inflicts no net cost or disadvantage on its bearers—then it will spread in the population as its bearers reproduce, and its phenotypic effects (if any) will spread with it. Presence or absence of that gene satisfies the genetic-variation requirement for a response to selection, and meiotic replication of the gene satisfies the other, recurrence criterion. So mutationally induced novelties automatically satisfy both of the two prerequisites for evolutionary potential. The question is whether or not environmentally induced traits can follow a similar course.

The Origins of Novelty

Trait Initiation

If pressed to name the mechanism behind the origin of novel traits—the pivotal event that initiates change—most biologists would answer genetic mutation. Most would argue that mutation is ultimately the *only* legitimate source of evolutionary novelty, simply because evolution is defined as involving genetic change. The major cause of evolutionarily significant variation was summarized by Endler and McLellan (1988) as "mutation, in the broad sense"—base substitutions and deletions, changes in primary structure of gene products or regulatory networks, and hybridization or introgression between species. Even what Endler and McLellen term "phenotypic mutations"—changes in function at the cellular, tissue, or organism levels—are attributed by them to "molecular mutations," changes in DNA sequence. Mutation is seen as the prime motor of change even in the writings of leaders in the incorporation of evolutionary thought into developmental biology. Edelman (1988), for example, describes heterochrony as "alterations of tissues and form by mutations leading to changes in the relative rates of development of different body parts" (p. 51); Raff and Kaufman (1983) state that a "major characteristic of organismal evolution is that there is a genetically determined developmental program, and that evolutionary change occurs through genetic modification of this program" (p. 338). While they explicitly eschew "total genetic determinism," they see nongenetic flexibility as a buffering mechanism rather than an originator of novelty. Kauffman (1993) thinks that "mutations drive populations through neighborhood volumes of the ensemble of possible self-organized systems" (p. 24). These statements reflect assumptions about organization that see mutation as the ultimate and only source of evolutionarily important developmental novelty. As a result, while contributing to an evolutionary view of development, they fall short of integrating condition-sensitive development into the theory of evolution.

Mutation is undeniably responsible for the ultimate origins of all novel *genetic* material, and evolution without mutation would eventually come to a stop due to depletion of genetic variation. But the origin of a new phenotype is a different question. Does mutation explain the initiation and form of most of the phenotypic novelties that become established under natural selection? Phenotypic innovation depends on developmental innovation, and developmental innovation spans a broader field of possibilities than does mutation alone. In chapter 26 I argue that *the most important initiator of evolutionary novelties is environmental induction.* Here I briefly outline that argument. The causal chain of adaptive evolution (left to right in figure 6.2) is fundamentally the same whether it starts with genetic or environmental induction.

The initiation of a new phenotypic trait in theory could be caused by mutation, environmental change, or new genetic combinations due to drift (shifting balance, founder effects) or genetic recombination. Mutationally induced novelties, providing they are nonlethal, obviously satisfy the two requirements for evolutionary potential. Gene replication endows them with the capacity for recurrence, and they are eminently heritable.

Environmental induction can also fulfill the recurrence and heritability requirements for evolutionary potential. Environmentally induced novelties can spread without positive selection due to spread or recurrence of the conditions inducing them, without special restrictions as to population structure or developmental linkage to other traits. They can be heritable if there is genetic variation

in the developmental capacity to produce them in response to an environmental input. Both mutation and environmental induction are considered important modes of initiation of adaptive traits throughout this book.

Origin as a side effect can make a novel trait recurrent without positive selection on the trait itself, if it is a pleiotropic side effect of an advantageous trait (e.g., see Müller, 1990). Spread as a side effect can occur whether the primary trait is mutationally or environmentally induced. A side effect of an adaptive trait may persist and be recurrent indefinitely without itself becoming adaptive (Gould and Lewontin, 1979). But it should not be assumed to be *non*adaptive (as in Gould, 1984). The side effect may be neutral or even somewhat harmful in its effects on individual fitness, as when sterility of some females occurs as a side effect of nest-sharing in nonsocial wasps and bees (West-Eberhard, 1982; Sakagami, 1982). Then, by virtue of its recurrence, selection on any genetically variable aspects of the side effect could improve its adaptive value or alter the threshold of its occurrence.

Of the modes of initiation remaining, drift and genetic recombination are relatively low on the list of important sources of novelty in terms of evolutionary potential, despite their frequent mention in theoretical discussions. Wright's (1932) shifting balance theory of adaptive evolution proposes that drift initiates change: a chance combination of existing genes (e.g., due to drift or sampling error in a small population) happens to produce a novel developmental response. This response is locally favored by selection while some barrier to dispersal temporarily prevents the disappearance of the combination that would be caused by gene flow. Then later the adaptive trait spreads in its consolidated form to the population at large. There seem to be no well-recognized examples from nature of this difficult to document but conceivable mechanism (see Charlesworth, 1990; Coyne et al., 1997, 2000; Barton, 1998), and defenses of it (e.g., Peck et al., 1998) seem to be entirely on theoretical grounds. Wright's quest for documentation, a fascinating chronicle (Provine, 1986), was marked by his impressively uncompromising refusal to accept data thought by others, such as Dobzhansky, to support his theory. Wright's high standards were especially admirable in view of his strong belief in shifting balance on theoretical grounds. A paper (Lande, 1980c) cited by Provine (1989) as providing data in support of shifting balance contains only theoretical verification. Shifting balance should not be considered synonymous with "peak shifts" (shifts between high fitness peaks in an adaptive landscape), which can occur by other means (West-Eberhard 1986; see chapter 21; see also Coyne et al., 1997) or with founder effects that involve sampling error (see below).

Genetic recombination shares with shifting balance a genetic mechanism for the initiation and form of novelty that is based on a chance combination of genes. Genetic-recombinant anomalies are a kind of genetic novelty, a genetically induced developmental change with a transitory polygenic basis. Some "sports" that occur in nature as isolated variants may be of this type. They have no evolutionary potential unless the same gene combination can somehow be rendered recurrent (as in asexual reproductive phases of clonal organisms). I know of no evidence that genetic recombination is an important source of adaptive phenotypic novelties in sexually reproducing organisms, as important as recombination may be in the spread of alleles and their testing in different combinations.

By the reasoning of this section, the three initiators of novel traits most likely to be important for adaptive evolution are mutation, environmental induction, and pleiotropy. Initiation by pleiotropy ultimately must trace to one of the other two, which leaves us with mutation and environmental induction as the most likely major initiators of novelty. Of these two, environmental induction is superior in terms of evolutionary potential. Environmentally induced traits may be immediately recurrent due to the recurrence of the inducing factor. This means that the chances of occurrence in favorable genetic backgrounds is high relative to that of a mutation, which begins as a trait of a single individual and then spreads to its descendents. Environmental induction also increases the likelihood of consistent environmental matching and therefore consistent conditions for natural selection. Finally, environmentally induced traits are relatively immune to elimination by natural selection, in contrast to mutationally induced traits whose persistence and spread depend on selective advantage. These points are discussed more thoroughly in chapter 26.

The Sources of Novel Form

Mutation and environmental change may initiate a novel phenotypic response, constituting the expression of a new trait, but how does a new trait get its form? Novel phenotypes very often have their origins in the reorganization of old ones (see part II). Evolutionary change starts with a phenotype that is responsive to new inputs, and new in-

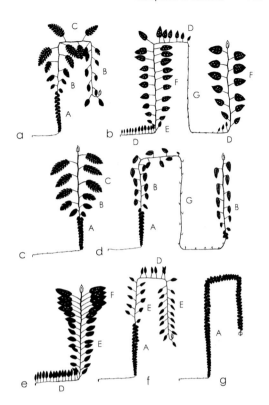

Fig. 6.3. Phenotypic recombination in the morphology of vines. Sequentially produced leaf forms separated by switches in tropical vines (*Monstera* species) are reorganized, deleted, and duplicated during evolution to form different patterns of morphology in different species. (a) *M. dubia*; (b) *M. siltepecana*; (c) *M. punctata*; (d) *M. acuminata*; (e) *M. lechleriana*; (f) *M. pittieri*; (g) *M. tuberculata*. After Madison (1977).

puts cause developmental change. When mutation induces phenotpic novelty, it is the developmental reorganization initiated by the mutation (e.g., due to its effect on the threshold for some response of the phenotype) that is responsible for the form of the phenotypic novelty, not the mutation itself, for all that a mutation can do directly is change the molecular product of a gene.

Reorganization of development can occur due to the action of new inputs on already existing developmental switch points. The vine *Monstera* provided an example above of switch-controlled metamporphosis in a plant (see figure 4.2). Figure 6.3 shows how switches between leaf forms have led to phenotypic recombinaton in the genus *Monstera*, producing a large number of variations on only a few leaf-shape themes. Reorganization of development can also result from the creation of a new switch, when a new input stimulates a novel response in a developmentally plastic organism. How developmental reorganization produces novel phenotypes beginning with preexisting ones is a major theme of this book, and the topic of the 11 chapters of part II. Those chapters describe some common pathways to novelty, via duplication of preexisting parts, deletions, reversions (reexpressions of traits that were developmentally deleted in the past), temporal change in trait expression (heterochrony), spatial change in expression (heterotopy), transfers of expression of sex-limited traits from one sex to the other, and correlated shifts in continously variable traits. Special attention is given to combinatorial change at the molecular level, and to the role of learning in the evolutionary reorganization of the phenotype.

The Simultaneous Origin of Regulation and Form

The reorganizational hypothesis for the induced origin of new traits solves an old problem in the evolutionary explanation of adaptive plasticity. The origin of a complex conditionally expressed trait has been considered a special problem in the past, since it would seem to require the simultaneous origin of both the trait and a regulatory mechanism capable of its condition-sensitive production (Williams, 1966, p. 81). Some biologists assume that a genetically fixed trait evolves first and that only after a developmental unit is complete can a condition-sensitive mechanism evolve to switch it off and on (Mayr, 1963, p. 220; Williams, 1966; Levinton, 1988, p. 251). Related to this is the idea that conditional alternative phenotypes are preceded in evolution by a genetic polymorphism (allelic switch between forms; e.g., Bradshaw, 1973; W. D. Hamilton, personal correspondence, 1990). This is simply a belief, for I know of no evidence or theory in support of the idea that constitutive expression must, or even is likely to, precede environmental induction or cues during evolution (none of the cited authors give any). I suspect that this belief is based on having overlooked the fact that a recurrent environmental factor can be as effective as a recurrent genetic one (e.g., a mutant allele) in producing a recurrent, genetically variable phenotype (further discussed in chapter 26).

Whether a trait is initially induced by recurrent genetic or environmental factors, it always has a genetic basis in that the responding phenotype has developed under genetic influence, and selection can favor modifiers of the initial population of somewhat variable responses to either increase or decrease environmental influence. The belief that genetic induction must come first also misses the point that all phenotypic traits are condition sensitive: they can only be expressed in some stage or circumstance during ontogeny and always involve a genetically influenced condition-sensitive switch activated under some particular developmental or environmental conditions.

Some authors (e.g., Levinton, 1986) visualize a separate origin of regulation and novel trait form. The order of events is presumed to be, first, the gradual evolution of a complex developmental unit (a novel alternative form), *followed by* the evolution of a "developmental switch gene" that "would arise only after the gradual evolution of the developmental unit is complete" (p. 265). This does not seem to me to be a reasonable hypothesis, for it is impossible to explain the presence of a coherent phenotypic unit trait without the presence of a mechanism for producing its coordinated development. If one visualizes trait origin as a novel response by a preexisting phenotype to a new genetic or environmental input, then this problem is solved.

An individual is composed of many kinds of responsive structures, some of them described under mechanisms of plasticity in chapter 3. An enormous repertoire of traits are turned off and on during development. The potential phenotypic repertoire of an individual is much greater than that actually expressed during its lifetime. This is shown when organisms are subjected to mutation or stress. Then they may produce a novel trait—a new subunit of form. The new subunit of form implies a new or reexpressed branch in development, and a new or reactivated switch mechanism—a new subunit of regulation, composed of the new input and whatever mechanisms are brought into play by its impact on the phenotype. Regulation and response are distinguishable but *inevitably* linked. In other words, there is no such thing as a new trait without a new developmental pathway.

The hypothesized evolution of the worker phenotype in social insects (Hymenoptera) illustrates the simultaneous origin of regulation and form (see supporting data in West-Eberhard, 1996). The worker phenotype is characterized by ovarian resorption and sterility, social subordinance relative to reproductive females (queens), brood-care behavior directed at the offspring of others, and nest defense. This contrasts with the phenotypes of solitary females which, like the ancestors of the social wasps and bees, lay eggs and rear their own young. The phenotypic change likely originated due to aggressive dominance within groups. Aggressive dominant females keep others from laying eggs by various means, including physically preventing their access to oviposition sites (empty cells), and they depress the energy budgets of subordinants by stimulating foraging while abstaining from foraging themselves and by aggressive solicitation of food from others (see Sakagami and Maeta, 1987a,b; West-Eberhard 1987b, 1996). The physiological basis for this dominance-mediated switch from reproduction to sterility can be observed in nonsocial species of wasps and bees, which undergo ovarian resorption like that of workers if they are protein-starved or prevented from laying eggs (see review in West-Eberhard, 1996). The hormones (especially juvenile hormone, and possibly others; see West-Eberhard, 1996) that regulate worker aggressiveness and age-related brood care activities in wasps (Röseler 1985, 1991) are common regulators of development and reproduction in nonsocial insects (Nijhout, 1994a). Dominance interactions have been observed in many normally solitary species when females are confined together in a group.

The simultaneous origin of regulation and form is especially evident when correlated changes are produced by a new input as in the two-legged goat (see chapter 3). The initial change, the reduction in the front legs, functions as a switch. It regulates a change in developmental pathway between the normal quadripedal goat phenotype and the bipedal one. All the components of the response—the altered musculature, the skeletal changes in thorax and legs, and the behavior of hopping on two legs—constitute a new phenotypic subunit or modular trait.

Phenotypic Accommodation

Darwin (1868b [1875b], p. 426) compared natural selection to a builder who uses uncut stones, the uncut stones being the variations produced by development. Baldwin (1902, p. 115) embellished this analogy by pointing out that individual accommodation—phenotypic accommodation due to preexisting adaptive plasticity—shapes and prepares the stones somewhat prior to the action of selection.

Phenotypic accommodation due to phenotypic plasticity is the immediate adaptive adjustment of the phenotype to the production of a novel trait or trait combination. It occurs in every generation and in every individual, for each individual is a somewhat different combination of independently varying parts. Phenotypic accommodation reduces the amount of functional disruption occasioned by a developmental novelty. It is exemplified by the two-legged goat (chapter 3) and its many counterparts in nature (chapter 16). Phenotypic accommodation is an automatic consequence of multidimensional adaptive plasticity in the face of a developmental change. Phenotypic accommodation may also contribute to the exaggeration of an induced change, as in the two-legged goat.

Genetic Accommodation

General Properties of Genetic Accommodation

In genetic accommodation, discrete, developmentally mediated changes in the phenotype are molded by quantitative genetic change to form adaptive traits. The skeletal genetic architecture of a novel trait is already in place when a trait originates, for a novelty is a response of individuals that depends on preexisting sets of genes organized in such a way that they can produce the response. Genetic accommodation improves a novel phenotype in at least three different ways: (a) by adjusting regulation, to change the frequency of expression of the trait or the conditions in which it is expressed; (b) by adjusting the form of the trait, improving its integration and efficiency; and (c) by reducing disadvantageous side effects. Genetic accommodation occurs whether a novel trait is mutationally or environmentally induced, for it depends on genetic variation at numerous loci brought under a new selective regime by the phenotypic change induced. A mutant allele may have been critical to the origin of the trait, so the spread of that allele would be critical to the spread of the phenotype, but not to its adaptive refinement by genetic accommodation. Modification of mutational side effects—a kind of genetic accommodation—is discussed by Charlesworth et al. (1982 p. 490).

Genetic accommodation of regulation is illustrated by the results of artificial selection for a novel trait in honeybees, namely, pollination of particular crop plants not previously part of the foraging repertoires of the bees. Nye and Mackensen (see Cale and Rothenbuhler, 1975) selected outbreeding lines increasingly specialized to forage on alfalfa. In one area of the midwestern United States, they increased the alfalfa foraging to 68% compared to 18% in unselected controls (see Cale and Rothenbuhler, 1975, p. 178). In effect, these experimental breeding programs selected for an increased preference for alfalfa. They demonstrate a potential for genetic accommodation of a novel flower preference if it proves advantageous. Other experiments have shown that the frequency of foraging for pollen or nectar can be increased under artifical selection in honeybees and that the change is due to increased frequency of particular alleles that affect the threshold of a response to sucrose (reviewed in Page 1997). Neuromuscular systems and hormonal systems that underlie the behavior and physiology of such traits as foraging in honeybees are phenotypically and genetically complex and therefore potentially subject to genetic variation at many loci. Such responses are poised not only for varied responses to new environmental stimuli or mutations, but also for their genetic accommodation, due to the involvement in any novel response of multiple loci where genetic variation can occur.

An example of genetic accommodation that reduces maladaptive side effects of a novel trait has been demonstrated in artificial selection experiments using bacteria. *Escherichia coli* mutants resistant to T-4 virus were favored under selection for resistance but showed deleterious side effects. After 400 generations in the absence of the virus, they showed a reduction in negative pleiotropic effects without losing their resistance to infection (Lenski, 1988a,b). This was due to spread of an allele that reduced their disadvantage, as measured in competitive growth experiments, compared with nonresistant populations. In this case, mutant resistance to viral infection was genetically accommodated by spread of an allele that modified its negative side effects. This is a good example of genetic accommodation as distinct from *genetic assimilation*—gene frequency change that fixes, or makes constitutive, the expression of a trait whose expression was formerly environmentally induced or strongly environmentally influenced. Since genetic accommodation applies to mutational as well as environmental inductions, somewhat deleterious mutations, or favorable mutations with deleterious side effects, instead of being eliminated or having their spread truncated by natural selection, may have their deleterious *effects* eliminated or neutralized by genetic accommodation, as in the example just given.

In theory, genetic accommodation would stop when the form and regulation of a trait have been maximized and negative side effects have been minimized by selection insofar as genetic variation permits. But genetic accommodation may continue indefinitely if selective conditions were to change or if new variation were to occur, due to mutation or gene flow, in the regulation of the focal trait or in the regulation of the lower level subunits that are elements of its form. Genetic accommodation

does not require mutation but may involve new (mutational) genetic variation in addition to pre-existing variation. Trait *modification*, by contrast, implies origin (by mutation or environmental induction) of a new phenotypic subunit, which alters (modifies) the form of a focal trait. Modification is origin of a trait at a lower level of organization and may be followed by a new round of phenotypic and genetic accommodation of that subunit.

Although he did not use the term "genetic accommodation," Wright (in a letter to V. McKusick, reprinted in Provine, 1989, p. 57) considered a process of genetic accommodation under selection as part of the shifting balance theory, where change is initiated by a chance favorable gene combination that is then reinforced and refined by selection on "multiple minor gene differences involved in quantitative variability." Frankel (1983) hypothesizes a process like genetic accommodation following mutational change in the evolution of protozoans.

Elimination by genetic accommodation of the harmful effects of mildly deleterious or "nearly neutral" mutations may be one means by which a neutral allele is established in the genome. This could allow accumulation of several alleles at the same locus that have equivalent effects on the phenotype (e.g. see Kimura, 1985; Ohta, 1992b). In altered circumstances, alleles neutralized in this fashion by suppression of negative phenotypic effects could, since they represent a somewhat distinctive form, contribute to selectable genetic variation in a new context.

The Meaning of Selection "for" a Trait

The concept of genetic accommodation shows how a novel trait can originate and spread without being tied to the spread of a particular mutant gene. In general, selection "for" a discrete trait means selection that favors the expression of one phenotypic alternative over another. For a novel trait, the alternative is the phenotypic state that preceded the origin of the novelty—the phenotype that is expressed in the absence of the novelty, when the novelty-inducing allele or condition is absent. Selection favors a particular alternative phenotype when it increases the frequencies of genes that influence the trait's regulation, for example, by lowering the threshold for its expression, thereby raising the frequency of the trait in the population. If selection for a trait is strong or persistent over a range of environmental conditions, the result may be evolved constituitive expression of one alternative alone or phenotypic fixation by genetic assimilation.

Change in the frequency of expression is often described as being due to a change in response thresholds and, if hormones are known to be involved, to be a change in the hormone system (e.g., Roff et al., 1997). But evolved change in phenotype frequency need not involve change in the threshold of a switch. It is possible for the response to selection to be a change in the ability to *pass* a threshold. This is included in the concept of *liability*—the continuously variable propensity of the individuals of a population to produce a particular threshold trait, being the result of variation in all phenotypic factors that affect the expression of the trait (Falconer and Mackay, 1996, p. 299).

A hormonally mediated switch, such as the density-dependent switch between short wings and long wings in grasshoppers (Pener, 1991), is actually a complex of associated traits that includes such things as sense organs used to translate population density into a developmental cue, and factors that affect the probability of passing the threshold, such as evolved change in behavior that causes individuals to move to more or less populated resting places, in effect altering the population density sensed by the individual. Changes in the brain or central processing where cues from the environment are integrated and/or passed to secretory organs (e.g., corpora allata and others) could alter the effectiveness of the incoming signal regarding population density. The secretory organs themselves could vary in their response to incoming cues, altering the amount of hormone output in response to a given central nervous system input. Then the sensitivity of wing anlage tissues to juvenile hormone titer could vary. In addition, there could be variations in the ability of wing-producing tissues to give a full response even when adequately stimulated, including limited resources to growth due to variation in metabolic efficiency and competing energetic needs such as that represented by ovary growth, which is frequently inversely proportional to wing development. In some insects, switches that correlate adaptively with body size are actually triggered by a stretch receptor in the gut that is sensitive, during a critical stage, to gut distention, which serves as an indirect indicator of trophic condition (Nijhout, 1979, 1984). All of these continuously variable phenotypic traits would contribute to continuous variation in liability to produce the long-winged or the short-winged morph.

An evolutionary change in liability can also be described as a change in phenotypic plasticity. The effect of selection on regulation may be to *broaden* the conditions under which a particular one of a pair of alternatives is produced, equivalent to making regulation increasingly insensitive to conditions (less plastic). When evolved fixation of a single alternative is achieved, regulation may not be condition sensitive at all, as in genetic assimilation; under all normally encountered situations, only one form is produced.

There is a potential for confusion when selection "for" a trait means selection for one extreme of a quantitative trait that is governed by a switch. For example, selection for large horns in a beetle that is dimorphic for large versus small horns could be expected to have two effects: it could affect regulation (the liability to produce large horns), lowering the threshold for the switch to large horn production or increasing the ability to pass it; and it could affect the form of large-horned individuals, for example by altering nutrient allocation so that horn length is increased within the population of large-horned beetles (see figure 5.20, which illustrates continuous variation in horn length within each of two discrete horn morphs).

The important point is that the effect of selection "for" a discrete trait acts on its regulation so as to increase its frequency of expression. Given the large number of factors that could influence thresholds and the ability to pass them, this must usually mean selection on a highly polygenic array of regulatory elements. The polygenic nature of change under selection for a particular trait needs to replace the image of gene-for-trait singe-locus change. In summary, *selection for a trait* does not mean selection for the spread of an allele that specifies the trait; it means selection for a change in regulation or behavior (including even habitat or diet selection) such that the trait is more readily, and therefore more commonly, produced. This may involve change in alleles at a variety of loci directly or indirectly affecting liability to produce the trait.

The initial response to selection may end up giving the impression that a single locus or a small set of loci is in control. As discussed in chapter 5, strong and recent selection for a particular trait is expected to favor the alleles of greatest effect on polygenic regulation (Nijhout and Paulsen, 1997). As a result, predominance of one or a few genetic loci in the regulation of phenotype expression following artificial selection may give the impression of simple genetic architecture. Following selection for high pollen collecting in honeybees, mentioned above as an example of genetic accommodation, two genomic regions had strong effects on the colony phenotype, explaining 59% of the total phenotypic variance observed in a backcross population, and these regions are known to contain loci that influence sensory responses of individuals to pollen and nectar (Page, 1997). Appropriately, these genomic regions are *quantitative trait loci* (QTLs), mapped loci that have been singled out from among a presumably large number of loci that influence a particular trait, in this case the pollen-foraging behavior of bees, a behavior that includes such complex and variable phenotypic traits as learning and ability to respond to nestmates' demands (Winston, 1987). Polygenic determination is probably general for behavioral traits (Plomin, 1990).

Because of the polygenic nature of the regulation of most adaptive phenotypic traits, quantitative genetic models of evolution should be broadly applicable to the evolution by genetic accommodation of traits governed by condition-sensitive developmental switches (e.g., see Cheverud, 1994; Cowley and Atchley, 1992; Atchley and Hall, 1991; Nijhout and Paulsen, 1997; Cheverud and Moore, 1994; Roff, 1996; Roff et al., 1997; and other references in chapter 22). Nijhout and Paulsen (1997) consider the effects of directional selection on a polygenically influenced quantitative trait such as a response threshold. They examine the genetic response to selection using a computer simulation of a randomly breeding sexual population where the quantitative trait under selection is affected by six different loci, each with two alleles with markedly different effects on the phenotypic value. The phenotypic response to selection and the genetic response at each of the six loci are shown in figure 6.4. Each of the six loci represents a different modifier of regulation. The simulations reveal that the gene-frequency shifts, under directional selection on a polygenic trait (e.g., during genetic accommodation), do not occur simultaneously at equal rates for all affected genes. Rather, the response to selection at a given locus depends on the strength of the correlation between the alleles at that locus and the phenotype selected: those most highly correlated with the phenotype, including those whose alleles have the most strikingly divergent effects, respond to selection first. Other loci, less influential when selection begins, at first show little or no response to selection—in effect, they are shielded by the greater phenotypic impact of other loci (Nijhout and Paulsen call this "pseudoneutrality"). As selection reduces genetic variance at the most influential locus, other loci take its place as most phenotypically correlated and begin to respond to selection until likewise driven toward fixation (figure 6.4B,C).

Nijhout and Paulsen (1997) point out that any parameter that affects the gene-phenotype correlation can be expected to affect the response to se-

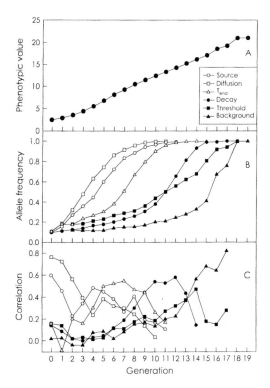

Fig. 6.4. Phenotypic, genetic, and correlated responses to selection. Truncating selection for a larger value of the phenotype was simulated by removal of individuals with phenotypic values below the mean minus half standard deviation in each generation. (A) Phenotypic response to selection. (B) Genetic response to selection. (C) Genetic correlations of the six developmental parameters (see symbol key) with the phenotype as selection proceeds. After Nijhout and Paulsen (1997). Courtesy of H. F. Nijhout.

lection. Thus, if a novelty undergoing genetic accommodation was initiated by a mutant allele of major effect on development, then that locus among all those influencing regulation of the novel phenotype will come under selection early during accommodation, and its status as an allele of major effect can be expected to decline as it goes to fixation unless, as sometimes though rarely occurs (see chapter 22), the allele is maintained in frequency-dependent or balanced equilibrium as a major "switch" allele determining the expression of adaptive alternatives. Loci whose expression is environmentally influenced would be relatively immune to change, especially early in the accommodation process, since their allelic differences would be relatively poorly correlated with the phenotype, a consequence not discussed by Nijhout and Paulsen but easily derived from their model. This may account for the fact that differences among recently diverged populations are often due to single-locus effects (e.g., see Orr and Coyne, 1992; see also chapter 4). Perhaps most important, the particular response to selection in the accommodation of regulation depends importantly on the genetic background, or the array of polygenic traits selected as a set when a particular phenotype is accommodated. This is one reason why identical or homologous phenotypes can have different developmental–genetic pathways. Because previously pseudoneutral loci are progressively brought under selection as the genetic background changes, heritability of the phenotype can remain high even while genetic variation is being eliminated under selection (Roff, 1994a). This may help to explain why artificial selection rarely fails to produce a result.

When does genetic accommodation end? In the simulations of Nijhout and Paulsen (1997) it ends with fixation of alleles at all the influential loci. Accommodation can be distinguished from *modification* or elaboration of a trait—increase in complexity—which implies subsequent episodes of developmental change producing novel subunits serving as phenotypic modifiers. These are products of subsidiary, newly originated developmental pathways subordinate to the original switch. Each modifier represents a new developmental branch point and would lead to a new round of genetic accommodation, this time at a lower level of organization.

Because selection on both continuous and discrete aspects of the phenotype affects genetic variation in regulation at switch points (see chapter 3), Scheiner (1993a, p. 38) was correct to conclude that a single evolutionary model can be used to describe the genetical evolution and accommodation of both continuous and discontinuous traits.

The Baldwin Effect and Genetic Assimilation

Genetic accommodation of flexible or learned traits was implicit in Baldwin's (1902) hypothesis, later known as the *Baldwin effect*—the idea that phenotypic accommodations to variable or extreme conditions can affect the direction of genetical evolution under natural selection, permitting survival in an unusual environment and allowing time for selection to favor adaptation to it (reviewed in Simpson 1953b; see also Wcislo 1989; Robinson and Dukas, 1999). Similar ideas were outlined at

about the same time by Baldwin (1896), Osborn (1897a,b), and Morgan (1896a,b), but the later treatment by Baldwin (1902) is by far the most carefully reasoned and detailed, and I treat it as the definitive version here. Readers interested in the Baldwin effect should read the meticulously written Baldwin (1902) book rather than rely on second- or third-hand accounts.

Gause (1942) seems to have been the first to make an experimental demonstration of the Baldwin effect. He produced genocopies of facultative adjustment to saline conditions in paramecia by artificially selecting for salt toleration. This increased ability of the paramecia to adapt to high concentrations of salt and demonstrated "parallelism of hereditary and non-hereditary variations" (p. 100). Gause proposed a "theory of hereditary fixation of adaptive modifications by means of natural selection of genocopies" (p. 101), an idea that clearly anticipates Waddington's genetic assimilation. Gause wrote that "[i]n the light of this conception adaptive . . . modifications pave the way for the subsequent evolutionary advance" (p. 100). Although this is similar to both the Baldwin effect and genetic assimilation, it was seldom mentioned in later literature on these subjects (but see Wcislo, 1989). Gause did not discuss the Baldwin effect, and Waddington (1942, 1952, 1953b, 1961) did not discuss Gause's theory of hereditary fixation.

The Baldwin effect has been unfortunately synonymized with genetic assimilation, for example, by Reid (1985) and by Dawkins (1982, p. 169), who refers to the "Baldwin/Waddington effect." *Genetic assimilation* is defined by Waddington (1961; reprinted in Waddington, 1975, p. 9) as "a process by which a phenotypic character, which initially is produced only in response to some environmental influence, becomes, through a process of selection, taken over by the genotype, so that it is found even in the absence of the environmental influence which had at first been necessary." The two ideas are similar in visualizing adaptive evolution beginning with plastic adaptability, followed by genetic change. But Baldwin emphasized that his hypothesis could produce increased, not decreased, phenotypic plasticity. Thus, the result may be polyphenism and movement toward "greater plasticity and intelligence" (Baldwin, 1902, p. 207), not fixation or increased genetic control. Although the writings of Lloyd Morgan, Osborn, Poulton, and Baldwin on "organic selection" and the Baldwin effect are often treated as essentially the same (e.g., by Simpson 1953b, Robinson and Dukas, 1999, and other authors), Baldwin explicitly objected to Lloyd Morgan's definition of organic selection as

the natural selection of coincident variations meaning genetic variation that coincides with, or is in
the same direction as, facultative responses. Baldwin (1902) recognized that facultative expression
or plasticity may be preserved "in a state of 'balance' between the organism and its environment"
(see also Cushing, 1941):

> In the state of balance, the [facultative] accom-
> modations would be made again and again in
> successive generations, and no further develop-
> ment of congenital endowment would take
> place. This flexibility of application which the
> principle of organic selection allows seems to be
> one of its great advantages (p. 210).

Both the Baldwin effect and genetic assimilation
have been confusingly discussed and grossly misunderstood in the past. As a result, they have been
summarily and unfairly dismissed as unimportant
for evolution (e.g., see Simpson, 1953b; Mayr, 1963,
p. 147). Simpson's (1953b) essay on the Baldwin
effect summarizes the early literature and concludes
that the idea is completely compatible with modern genetics, a point that Baldwin discussed with
remarkable accuracy 50 years earlier, as shown below. But Simpson considered the idea to be weakly
supported by evidence and concluded that "it is seldom assigned a major role in evolution" (p. 110).
This negative conclusion was more prominent and
frequently cited than Simpson's later (1958) more
favorable assessment (see Wcislo, 1989).

Waddington (1959, 1961) vigorously attacked
the Baldwin effect, which he erroneously described
as dependent upon the occurrence of new mutations in the direction of an environmentally induced
change. But the Baldwin effect as conceived by
Baldwin (1902) does not depend upon mutation:

> [o]rganisms which survive through individual
> modification [novel environmentally influenced
> traits] will hand on to the next generation any
> "coincident variations" (i.e., congenital varia-
> tions in the same direction as the individual
> modifications) which they may chance to have,
> and also allow further variations in the same di-
> rection (p. 149).

In other words, the Baldwin effect can utilize preexisting genetic variation; it does not depend upon
mutations but may utilize them. Further, "the mean
of the congenital variations will be shifted [by se-
lection] in the direction of the individual modification" (p. 150).

This misinterpretation regarding the importance
of mutation in the Baldwin effect has been perpetuated (Mayr, 1963, p. 147; Hall, 1992, p. 163) and
combined with others, such as the belief that "easy
modifiability of the phenotype has a retarding effect on evolution contrary to the claims of the ad-
herents of the 'Baldwin effect'" (Mayr, 1963, p.
147). Strong statements such as these from leading
evolutionary biologists were enough to discourage
serious consideration of this principle and that of
genetic assimilation for decades, with only a few
notable exceptions (e.g., Evans, 1966; Shapiro,
1976; Matsuda, 1981, 1987; Tierney, 1986; for a
review, see Wcislo, 1989). In fact, both objections
were unjust: under the Baldwin effect, modifiability or plasticity can either facilitate or retard evolution, depending upon the pattern of variation involved, and Baldwin (1902) carefully described
exactly the pattern of recurrent expression that, by
producing recurrent consistent responses, *can* accelerate evolution (see the section entitled Does
Plasticity Accelerate or Retard Evolution? in chapter 7). Elsewhere, Baldwin states that facultative responses, or "accommodations," would allow selection to produce "a shifting of the congenital
mean" (p. 210), and that "variations" (Baldwin's
consistent term for "congenital," or genetic, variations) used for ontogenetic accommodation would
be "utilized more widely in subsequent generations" (p. 97). This is a concept of evolution based
on preexisting genetic variation, not mutation,
though mutation is not necessarily excluded.

Schmalhausen's (1949) theory of "stabilizing selection" was a theory of genetic assimilation under
a different name. It visualized that "adaptive" (facultatively expressed) traits, for example, in cold
places or in time of drought, can eventually become
"stable" (expressed in all environments—genetically fixed; p. 83). Thus "forced adaptive modification" is a "transitional evolutionary stage" in the
origin of novel characters, and occasional "mistakes" (inappropriate triggered responses) select
against plasticity ("stabilizing selection"), producing "autonomous" (fixed) phenotypes (p. 86). In the
Neo-Darwinian Synthesis, stabilizing selection was
used to mean "normalizing selection"—limiting the
range of variability to near the norm favored by selection (Dobzhansky, 1977, p. 117) or the elimination of peripheral individuals (Mayr, 1963,
p. 282). The role of plasticity as a stage in evolution, emphasized by Schmalhausen, was passed
over in favor of the usual emphasis on stability and
homeostasis discussed in chapter 1 as a problematic aspect of modern evolutionary thought.

Waddington (1975), while recognizing that Schmalhausen and others "were following a line of thought almost parallel to that which . . . led to the idea of canalization and genetic assimilation" (p. 75), discounted this work as not being supported experimentally and also as relating not to genetic assimilation but to the Baldwin effect, an idea he considered untenable because he interpreted it as "a non-genetic phenomenon" with adaptation due to "a non-genetic plasticity of the phenotype" rather than an expression of hereditary potentialities (p. 89). Waddington seems not to have studied Baldwin's (1902) carefully worded text, which treats variation in plasticity as "congenetital [genetic] variation" (p. 98) and states that "the most plastic individuals will be preserved to do the advantageous things for which their variations show them to be most fit, and the next generation will show an emphasis of just this direction in its variation" (p. 98). Simpson's (1953b) discussion recognized that Baldwin used the word "variation" consistently to mean hereditary (genetic) variation. Unfortunately, rather than drawing attention to the importance of plasticity as a factor in evolution, Simpson's authoritative essay served instead to shelve the Baldwin effect as nothing more that conventional genetics (see the perceptive hisory of discussion of genetic assimilation in P. J. Taylor, 1987). Taken together, such discussions not surprisingly discouraged thinking along these lines for many years (but see Rau, 1933; Evans, 1966). As late as 1999, Orr wrote that "most evolutionists view assimilation as a curiosity of unclear relevance" (p. 343).

More recent discussions of genetic assimilation and the Baldwin effect occur in a climate more disposed to consider development and plasticity as factors in evolution (Matsuda, 1982, 1987; Tierney, 1986; P. J. Taylor, 1987; Wcislo, 1989; Robinson and Dukas, 1999), but misunderstanding persists. Some authors (e.g., Matsuda, 1987) use "genetic assimilation" to refer to any instance of fixation of an environmentally induced phenotype, or fixation of one phenotype of a polyphenism, under the assumption that fixation of an adaptive trait is due to genetic change. Change in frequency of an adaptive trait, however, can occur without genetic change, if an environmental change consistently induces (fixes) the trait. Unless genetic change has been demonstrated, I prefer the term *phenotype fixation* (West-Eberhard, 1986, 1989), which includes fixation by either environmental or genetic change.

Robinson and Dukas (1999, p. 583) portray the Baldwin effect as deficient in focusing on "very labile traits" such as behavior and physical and physiological flexibility, rather than on plasticity as it is currently conceived, namely, as "clear environment-dependent gene expression usually in a coarse grained environment," using an ANOVA approach that distinguishes between different sources of variation (environmental, genotypic, interaction effects, and developmental noise). This view of plasticity, excluding behavior and physical and physiological traits, seems far too narrow for a general evolutionary discussion, and this assessment of Baldwin's hypothesis, focused on its lack of relation to gene expression and ANOVA, seems unfairly dismissive, like faulting one's great grandmother for not having used an electric dishwasher.

How Genetic Accommodation Differs from the Baldwin Effect and Genetic Assimilation

Genetic accommodation can include the phenomena envisioned by the Baldwin effect and genetic assimilation, but it differs from those ideas in the following ways:

1. Genetic accommodation refers to adjustments of form as well as frequency of expression (regulation). That is, under genetic accommodation, selection may simultaneously improve the shape of a novel organ and change its frequency of occurrence or circumstances of expression. Genetic assimilation and the Baldwin effect refer only to change in regulation (toward increased or decreased genotypic influence on trait determination).

2. Genetic accommodation applies to any novel trait, whether it originated due to a genetic mutation or due to environmental induction. Genetic assimilation and the Baldwin effect apply only to phenotypes originally environmentally induced or whose expression is originally heavily environmentally influenced and then (in genetic assimilation) becomes increasingly genetically influenced.

3. Genetic accommodation may *increase* phenotypic susceptibility to environmental influence, rather than decrease it as in genetic assimilation. Genetic accommodation may *prevent* genetic assimilation or produce overwhelmingly environmental determination if conditional expression is favored by natural selection.

4. Genetic accommodation may alter or refine the particular circumstances in which an environmentally influenced phenotype is produced—its context of expression—whereas genetic assimilation decreases developmental sensitivity to the environment and thereby may *decrease* environmental specificity of expression.

5. Genetic accommodation can modify initially deleterious phenotypes, rendering them neutral or advantageous, whereas genetic assimilation and the Baldwin effect refer only to evolutionary change in the frequency of occurrence of positively selected traits.

6. Genetic accommodation does not apply only to increased frequency of expression or fixation relative to alternative traits, but applies as well to *reductions* in frequency of expression, including elimination of disadvantagous novelties.

Complexity of Regulation and Genetic Accommodation

Genetic accommodation may incorporate mutation, but it works best if there is a standing crop of genetic variation in the regulation of expression of a novel trait upon which selection can act. There are both theoretical and empirical reasons to expect genetic variation in the propensity to produce newly induced traits, without awaiting new mutation. One reason is the complexity of regulation, which means that genetic accommodation can be based on genetic variation at multiple loci. Most regulatory mechanisms are phenotypically, and therefore likely to be genetically, complex (Roff, 1996; see chapters 3–5), so there are multiple loci where selectable variation can occur to affect the frequency of trait expression. Even primary gene products, such as proteins, have multiple components that can vary and evolve under selection, by a kind of intramolecular genetic accommodation, as is shown below.

Polygenic regulation is discussed in chapter 5, but it is important to reemphasize it here because of its importance for genetic accommodation. No phenotypic response at supramolecular levels of organization could be produced by a single gene acting in isolation from the products of others. Even an apparently simple trait such as eye color in *Drosophila*, for example, is the product of complex epigenetic interactions where biochemical pathways incorporate environmental elements and many gene products to determine the formation of a particular eye pigment. Eye color is influenced by

at least 70 genes in *Drosophila* (Futuyma, 1998). Several mutations at different loci are known to affect pink eye color in *Drosophila* (Morgan et al. 1925) and red eye color in *Gammarus* (Sexton and Clark, 1936) (after Huxley, 1942, p. 511). This means that more than one, and usually many, potentially variable loci may influence the regulation (the rate, duration, and frequency of occurrence) of even an apparently simple trait. Indeed, often the appearance of simplicity is due to use of mutations of large effect in genetics, a technique that works in part because there are many mutable loci for a trait such as eye color that can be exploited.

The determinants of form are also genetically complex, and to the degree that this is true, it similarly facilitates the adaptive refinement of expressed traits under selection and genetic accommodation. The determinants of form (see figure 4.13) contain subsidiary branches with their own subsidiary epigenetic pathways of regulation, forming a mosaic of subunits, each subject to being eliminated or increased in frequency due to selection on polygenically influenced subsidiary switches.

The complexity of developmental pathways means that the accommodation of a threshold of expression upward or downward does not require fixation or even change in any single allele. Rather, it could be achieved by a shift in the frequencies of alleles at many different loci, or in the frequency of one of the polygenes with a relatively major effect. If anything has been learned during this century of evolutionary biology, it is that "a trait" of a population or species is really a population of variable traits, even after a history of strong stabilizing selection (e.g., Lewontin, 1974). Studies of genetic polymorphism and artificial selection indicate that there is sufficient genetic variation in conditionally expressed phenotypes to support the genetic accommodation of environmentally induced traits. Even in a presumably strongly selected trait, such as the photoperiod responsiveness of boreal mice and voles to induction of seasonal reproduction, will have genetic variation, as witnessed by the response to selection for unresponsiveness. Up to 90% of a selected line of Siberian hamsters failed to respond to short day length with inhibited reproduction, compared with only 20–30% nonresponders in the original population, and similar results have been obtained for other rodents (reviewed in Goldman and Nelson, 1993).

Developmental quantitative genetics is an emerging field of research in evolutionary biology (for example, see Atchley, 1987; Cheverud and Routman, 1995; Nijhout and Paulsen, 1997; Omholt et al., 2000). Already progress has been made in showing how the mechanisms of gene action help to explain, and validate, the predictive power of quantitative genetic theory (see especially Omholt et al., 2000). This is an important frontier which promises to unite the concepts of evolutionary genetics with knowledge of developmental mechanisms.

Evidence That Genetic Variation Is Widespread

Genetic accommodation depends on the presence of genetic variation for regulation of trait expression and of the expression of the subsidiary elements of form. The response to selection on degree of plasticity and frequency of conditionally expressed traits constitutes evidence that the kind of variation needed for genetic accommodation to occur is common in populations.

As expected from the complexity of regulatory mechanisms, and the many theoretical models for maintenance of genetic variation (see chapter 22), there is abundant experimental evidence for genetic variation in continuously variable traits of the kind that would support genetic accommodation. This evidence is of two primary kinds: electrophoretic studies of allozymes, and breeding experiments (e.g., artifical selection, sibling analysis, genetic strain comparisons, and common garden transplants of geographic variants).

Early summaries of results of artificial selection in domesticated plants and animals (e.g., Darwin, 1868a,b) are still cited by modern authors as important evidence for genetic variability in particular traits (e.g., see Briggs and Walters, 1988). Hill and Caballero (1991) list reviews of early literature on response to artificial selection on quantitative traits in plants. They conclude that the early plant work "was important in showing that almost any quantitative trait could be permanently altered, that responses (mostly) occurred as a consequence of changes in the frequencies of [pre-existing alleles] and not from mutations in the genes, and that, as *responses could continue well outside the range of the initial population*, many genes must be involved." (p. 287; emphasis mine) The potential to respond to selection beyond that usual in a population is especially important for the evolution of novelty, which may be induced by environmental and genetic extremes (see chapter 26 and part II). Known responses to artificial selection include discrete traits (particular aspects of form; emphasized by Darwin and other early authors), exaggeration of quantitative, continuously variable traits (see especially Hill and Caballero 1992), and regulation

per se (degree of plasticity, or environmental responsiveness independent of trait form; see Bradshaw, 1965; Scheiner, 1993b; see also examples given below). As concluded by Lewontin (1974),

> [I]f nearly any character can be selected for rather easily in nearly any population, then it is certain that a large number of genes of different function are segregating in natural populations, even if these genes collectively do not represent a large fraction of the total genome. (pp. 93–94)

Lewontin states that "[t]here appears to be no character—morphogenetic, behavioral, physiological, or cytological—that cannot be selected in *Drosophila*" (p. 92), and he cites only one exception, even that one successful in causing quantitative change. Maynard Smith (1989) stresses the same finding for the extemely well-studied eukaryote genetic system of these flies, noting that phenotypic traits that lack heritable variation are "exceptional" and adds that "there is no reason to think that *Drosophila* is peculiar in this respect" (p. 50).

The honeybee *Apis mellifera*, another well-studied animal species, also responds broadly to selective breeding, including in behavioral traits of interest to apiculturists (Rinderer, 1986). Artifical selection has produced lines with altered flight frequency of drones, orientation ability, degree of inhibition of dance communication following foraging, speed of flight and dance, wing-beat frequency, rate of dead brood removal, readiness to sting, responsiveness to alarm pheromone, and rate of hoarding behavior (Rinderer and Collins, 1986). Other breeding programs have produced bees specialized in citrus honey production, pollen hoarding, and syrup hoarding (Collins 1986); resistance to particular diseases (foulbrood, hairless black virus, mite infestations), oviposition rate, high and low alfalfa pollen gathering, longevity, and number of hamuli (wing hooks; Kulincević, 1986).

Some authors caution that artificial selection is a "false analogy" with selection in natural populations (e.g., Slatkin and Kirkpatrick, 1986, p. 291). They refer primarily to the abnormally high intensities of artificial selection and consequently high rates of change, due to the fact that the population has "no choice but to respond." This does not detract from the significance of artificial selection as an assay for genetic variation in natural populations. This means the potential for a response to selection. Though evolution may not be as rapid in natural populations, the time spans for selection to operate are much longer. Though the response to artificial selection often declines over time in managed and experimental populations (Falconer 1981), in nature it may be supplemented over longer time scales by mutation and gene flow (see G. C. Williams, 1992).

The presence of genetic variation in virtually all complex phenotypes is one of the best supported findings of twentieth-century genetics. Electrophoresis has documented high rates of protein polymorphism in a wide variety of organisms (e.g., see Lewontin 1974). Following the induction of a novel phenotype, a pool of previously neutral or slightly deleterious variation having a "latent potential for selection" (Ohta, 1992b, p. 278) may be brought into play, as long ago suggested by Schmalhausen (1949 [1986]). Ohta (1992b) cites data suggesting that this occurs.

Particularly important for genetic accommodation of environmentally induced change is evidence that conditional and developmental responses, such as some of those listed above for the honeybee, respond to artificial selection, at least under the conditions where this was tested. It has long been known that ratios of growth allocation to anatomical features in vertebrates show strong responses to selection, demonstrating that the allometric correlations of an environmentally sensitive process such as growth can undergo evolutionary change: realized heritabilities (detected by response to selection) are 0.60 for the ratio of cannon bone length to body weight in sheep; 0.57 for shank length/body weight in chickens; 0.56 for wing length/thorax length in *Drosophila* (older literature reviewed by Cock, 1966, p. 152). Degree of plasticity, or threshold of response to nutritional level, and therefore morph ratios in a dimorphic species, were altered by artifical selection in beetles (Emlen, 1994; Moczek, 1996) and earwig wing lengths (Briceño and Eberhard, 1987). There are enormous literatures on the genetics of behavioral responses (see the many references in Huettel 1986; Breed and Page, 1989; and Boake, 1994) and responses to selection affecting life history traits (Dingle et al., 1986; Roff, 1992; Stearns, 1992).

The masses of data on artificial selection in a very wide variety of plant and animal taxa that were collected by pre-Mendelian plant and animal breeders are of interest because they are based entirely on phenotypic data collected without precise information on genetics, beyond that deduced from a pedigree. As a result, environmental aspects of development were included in breeding programs to an extent that they might not be by a more sophisticated but purely genetic approach. Although this may come as a surprise to some, selection on

phenotypes that are environmentally exaggerated in the direction desired can be more effective than selection in a constant environment designed to minimize environmentally induced variation (e.g., see Via, 1986). This is because exaggeration of a phenotype in a particular direction includes exaggeration of its development, whatever environment-incorporating pathway that may include. One consequence of epigenetic complexity for evolution and genetic accommodation is that selection for a particular end point may encompass cryptic pathways for getting there.

This is shown by the following example from a manual for pheasant breeders (Allen, 1969), describing the origin of the "giant mutant" via artificial selection starting with small, normal ringneck pheasants:

> They were not crossed with Jumbo Ringnecks or any other pheasants to obtain the size. We achieved this by selective breeding, mating the largest and best specimens together from each generation; and also through the use of the best feed formulas for pheasants. A high protein feed, especially for the chicks, has been essential. The breeders received the best feed which produced large sized, quality eggs. These eggs produced large chicks which were fed high protein feed and given the best care. It is important to get chicks growing and keep them growing while they are young. . . . (p. 117)

While the lack of proper controls renders this description open to being considered ignorant Lamarkianism, with an emphasis on obtaining large size by means of a good diet, there is an important point to be made here. A proper geneticist, to maximize heritability, may standardize the diet at some moderate level in order to select individuals believed to be genetically large rather than large due to overfeeding. But in so doing, the sophisticated biologist would have fallen into an educated error of searching for genes for largeness independent of environmental interaction and maternal effects, in the belief that such genes are best exposed by eliminating environmental noise.

The breeder's success as an applied evolutionist comes from the realization, in practice if not in theory, that some of the genes for largeness may be those enhancing ability to digest high-protein food and convert it efficiently into weight and stature. Superiority in this aspect of the phenotype could not be screened for so effectively in birds on a low-quality diet. Put another way, efficient use of high-protein food is one epigenetic pathway contributing to large size. This example reveals how cryptic and circuitous developmental pathways contribute to the polygenicity of complex traits and their genetic accommodation.

Evidence That Genetic Accommodation Has Occurred

How can one substantiate the claim that it is the complex, polygenic regulator of a positively selected trait that responds to selection, as implied by the idea that quantitative genetic change in regulation underlies evolutionary change in trait frequency? This would be supported if a positive response to selection on a trait is accompanied by (1) a quantitative change in some aspect of its regulation, such as a hormone titer; or (2) there is correlated quantitative change in developmentally linked traits influenced by the same regulatory mechanism.

Artificial selection for high mating frequency in male quail (*Coturnix japonica*), for example, leads to significant correlated responses in aggressiveness and in the size of the cloacal gland, all testosterone-mediated traits (reviewed in Ketterson and Nolan, 1992). The fact that all were affected under selection on just one indicates that the effect of selection is to alter the underlying regulatory (hormonal) mechanism, a genetically complex system potentially subject to genetic variation at multiple loci affecting any one of its variable components (e.g., rate of production, use, inhibition, or breakdown of testosterone).

Roth (1992) gives many lines of indirect evidence for the genetic accommodation of initially environmentally influenced small size in the evolution of dwarf elephant species on islands. The evolution of dwarfism (small size, not pituitary dwarfism) likely began with stunting due to starvation in the face of food scarcity. Individuals genotypically predisposed to survive this would have a selective advantage generation after generation, leading to the gradual evolution of a dwarf form. There is fossil evidence of food stress in the dentition and skeletal proportions of dwarf individuals, and the extent of size change in the transition to small-size populations suggests genetic accommodation because it exceeds the size range that likely characterized the original population (Roth, 1992).

Genetic assimilation is genetic accommodation in regulation carried to its extreme end point—fixation of one of two alternative traits or phenotypic states. Geographically variable polyphenisms offer examples of this in nature. The skipper butterfly *Polites sabuleti sabuleti* (Lepidoptera, Heperiidae),

for example, produces many broods per year in lowland populations in western North America. Adults are polyphenic, with cool-weather (spring and fall) adults markedly smaller, darker, and hairier than summer adults. A closely related high-altitude subspecies, *P. s. tecumseh*, which is likely derived from the lowland populations (see Shapiro, 1975, p. 37), produces only one brood per year, and adults are very similar to the cool-weather phenotype of *P. s. sabuleti*. Individuals of these high-altitude populations reared under summerlike lowland conditions produce only the dark phenotype, indicating that phenotype fixation by genetic assimilation has occurred in the highland subspecies (Shapiro, 1975). Some possible examples of genetic accommodation are discussed in chapter 21 under a discussion of geographic variation and fixation of phenotype ratios (see also West-Eberhard, 1986, 1989).

Genes as Followers in Evolution

Certain conventional ideas about adaptive evolution have to change. First, genetic novelty is not neccesary for phenotypic novelty to evolve: phenotypic innovation does not await mutation (see also Johnston and Gottlieb, 1990). Contrary to the probably common belief that a phenotype of mutational origin has more evolutionary potential than does an environmentally induced trait, there are several compelling reasons to argue the reverse (see chapter 26). An environmental factor can immediately affect many individuals in a population at once, whereas a mutation initially affects only one individual. Widespread induction improves the likelihood (1) that the innovation will be selectively advantageous in at least some individuals and (2) that it will be tested in a variety of genetic backgrounds, beginning the process of genetic accommodation of the conditions of its expression and its form, and reducing deleterious side effects. Environmentally induced traits are relatively immune to elimination by natural selection compared with mutationally induced ones and therefore more likely to persist despite initial costs, raising the probability of favorable genetic accommodation under selection.

If the arguments of this chapter and of chapter 26 are correct, most phenotypic evolution begins with environmentally initiated phenotypic change, and even when the initiation of novelty is mutational, much of the subsequent gene-frequency change represents genetic accommodation, which does not depend upon mutation but can capitalize upon any influential mutations that occur. The leading event is a phenotypic change with particular, sometimes extensive, effects on development. Gene-frequency change follows, as a response to the developmental change. In this framework, most adaptive evolution is accommodation of developmental–phenotypic change. Genes are followers, not necessarily leaders, in phenotypic evolution.

This conclusion about the role of genetic change is pivotal for a synthesis of development and evolution. The idea that genes can directly code for complex structures has been one of the most remarkably persistent misconceptions in modern biology (Nijhout et al., 1986, pp. 445–446). The reason is that no substitute idea for the role of genes has been proposed that would consistently tie genes both to the visible phenotype and to selection. If not by coding for a selected structure, then how? The answer lies in the collective actions of multiple genes on the fine-tuning of devopmental change: complex structure emerges from complex chains of genetic and environmental interaction with the phenotype. The ability of genes to adjust complex structure comes from their large numbers, the precision of their effects, and their condition sensitivity in exerting those effects.

If genetic change in the early phases of an evolutionary transition involves genes of relatively large effect, as is known to sometimes occur (see above), then some of the genes of small effect in continuous polygenic variation may once have been major genes, whose effects have been modified or reinforced by selection at other loci, until their effects are not great relative to those of other polygenes that influence expression of the same trait. Of course, the opposite could conceivably occur: loci of small effect can have large effects in certain conditions, and this may be favored under certain selection regimes (see chapter 5). The polygenic nature of most regulation may additionally be a consequence of the fact that mutations of large effect are not necessary for the evolution of adaptive traits. This latter interpretation would follow if the environment, rather than mutation, is responsible for the initiation of much evolutionary change, as argued here and in chapter 26.

If genes are usually followers rather than leaders in evolution—that is, most gene-frequency change follows, rather than initiates, the evolution of adaptive traits—then the most important role of mutation in evolution may be to contribute not so much to the origin of phenotypic novelties as to the store of genetic variation available for long-term gradual genetic change under selection (Ohta, 1992b; see chapter 5). An important corollary of this view of evolution is that adaptation does not await a favorable mutation whose effects are random with respect to adaptation. Instead, adaptive innovation begins with reorganization of an already highly adapted phenotype (documented in part II), in which negative effects are ameliorated by adaptive developmental plasticity (phenotypic accommodation). Developmental variation, therefore, is not completely random with respect to adaptation, even though the factors that induce it (mutations or novel environmental inputs) may be. That genetically accommodated phenotypic novelties originate with novel inputs and developmental plasticity is a simple Darwinian explanation of adaptive design that depends on neither random mutation nor biblical miracles.

A Developmental Definition of Adaptive Evolution

We can now formulate a definition of adaptive evolution that can be used to relate selection, development, phenotypic traits, and genes: *adaptive evolution* is cross-generational change in phenotype frequency, accompanied by change in the frequency of expressed, phenotypically influential, genetic alleles, under selection that maximizes the positive fitness effects of the phenotypic traits whose development is influenced by these alleles. What selection maximizes, then, are the positive fitness effects of expressed genes. Selection favors the spread of genes that improve the phenotype, but the process need not begin with a mutation that directly affects form (the usual view of adaptive evolution). More commonly, selection adjusts the frequencies of genes that influence the regulation of gene expression. It is because adaptive evolution acts in this way—so as to maximize the fitness effects of expressed phenotypes and genes—that organisms develop and behave in ways that are beneficial to their inclusive fitness, even in variable conditions and different life stages. Genes or segments of DNA that are not expressed or active in regulation are not subject to selection and do not contribute to adaptive evolution. As emphasized in discussions of molecular evolution (e.g., Kimura, 1983), such segments of the genome evolve only by mutation and drift.

7

Principles of Development and Evolution

The conclusions of part I can be summarized as a set of principles that describe how developmental plasticity is expected to affect evolution. Adaptive phenotypic plasticity permits: phenotypic accommodation, and may exaggerate change; maintenance of alternative phenotypes that facilitate change without loss of adaptation; continuity of directional change under oscillating selection; and change without loss of the ability to change. Modular phenotypic organization determines semi-independent subunits of selection, accounts for the coevolution of coexpressed traits, permits the independent evolution of regulation and form, and allows dissociation and recombination in trait expression. Feedback between frequency of expression and rate of modification can cause mutual acceleration (or retardation) in the evolution of regulation and form. The hierarchical, or nested-traits nature of modular organization also promotes integration, a parliament of the phenotype in which selection at higher levels of organization to some degree screens off the effects of selection at lower levels. Regulatory complexity of switches permits interchangeability of environmental and genomic influences on development, multiple pathways to the same trait, and adaptive fine tuning of trait expression. The effect of plasticity and behavior on rates and directions of evolution depends on the patterns of variation produced. Developmental plasticity solves the cohesiveness problem, which sees present adaptation as an impediment to change. Previous solutions, including gene-for-trait gradualism, speciation, and cell-lineage competition, are shown to be unsatisfactory.

Introduction

So far, I have outlined the general properties of phenotypes, shown how they relate to development, and presented a model of adaptive evolution based on established principles of development and genetics. Now, using this general framework, I can summarize how developmental plasticity facilitates evolution.

Jacob (1977) characterized evolution as "tinkering." It shuffles and recombines what is already there. Frazzetta (1975), in another felicitous comparison with machines, wrote that evolution manages "the gradual improvement of a machine while it is running" (p. 20). Both of these qualities are possible due to characteristics of phenotypes that are *not* shared with most machines. Tinkering works because the phenotype is made of recombinable modular components that can be turned off

and on in different conditions and can function in more than one context, what Gerhart and Kirschner (1997; Kirschner and Gerhart, 1998) call "weak linkage" to any particular use. Improvement without disruption of function works because of the remarkable active flexibility, and redundancy, in the development of parts. As a result of these two qualities—modularity and plasticity—an organism has the unmachinelike ability to respond to a new situation or to a new gene with the production of a new trait, and then to multiply, through reproduction, the ability to produce this trait. Differential reproduction starts the cycle of variation, selection, and cross-generational change that we call evolution—the most unmachinelike process of all.

Many reasons have been given to believe that evolutionary change is difficult and even resisted in a well-adapted population (see chapter 1). The evolution of a novel specialization requires that a single lineage persist while undergoing extensive change. The conditions sometimes mentioned as favoring directional evolution, such as strong competition, very different or changing environments, small founder populations, or very long periods of time (see Mayr, 1982b), also favor population extinction. The idea of developmental cohesiveness, outlined in chapter 1, led to the further belief that major developmental change early in ontogeny would be disruptive. The cohesiveness theme persists even though it long has been clear that innovation does not occur exclusively by terminal addition (see chapter 1).

Adaptive evolution can work only if selection between character states and fitness has two characteristics: continuity and quasi-independence (Lewontin, 1978). Continuity, or cross-generational survival of integrated organization, requires that change in one part not disrupt others. Quasi-independence as described by Lewontin means that numerous potential pathways of change permit at least some to be used without being disruptive. "Continuity and quasi-independence are the most fundamental characteristics of the evolutionary process. Without them organisms as we know them could not exist because adaptive evolution would have been impossible" (p. 169). Lewontin abruptly ended his essay on adaptive evolution with that sentence. If you turn the page in the hope of seeing precisely how Lewontin sees the causes of continuity and quasi-independence, you find an article, by another author, on wave interference in soap bubbles.

In this chapter, I begin where Lewontin left off. Nondisruptive adaptive change is possible because of the subindividual modularity of phenotypic development, which partially atomizes *selection* so as to permit both integration and semi-independent evolutionary change of parts. The prospects for evolution are improved in several important ways by the universal modularity and plasticity of phenotypes. The point of this chapter is to show how specific consequences of developmental plasticity complement conventional Darwinian microevolutionary explanations of adaptive evolution. The principles outlined here resolve the controversies that have cast developmentalism against Darwinism and neo-Darwinism in the past (see especially chapter 24, on gradualism), showing how selection works *better* if developmental plasticity is taken into account.

This is not a balanced discussion of development and evolution, in that it concentrates primarily on the factors that promote change rather than those that may retard it. For this inbalance I offer no apology. This is a book about evolution, not lack of evolution. I also omit from this chapter consideration of biological consequences of plasticity that are not strictly speaking evolutionary, such as the likely contribution of plasticity to ecosystem stability and "resilience" (e.g., see Peterson et al., 1998).

Evolutionary Consequences of Plasticity

Phenotypic Accommodation of Change

The same mechanisms of phenotypic accommodation that permit integrated development (chapter 3) can facilitate evolutionary change (Goldschmidt, 1940; West-Eberhard, 1986, 1989, 1992a; Kirschner, 1992; Gerhart and Kirschner, 1997). To the extent that the accommodations are adaptive, having evolved under selection that favors normal development, they have an enhanced probability of contributing to a *functional* novelty, one that is likely to be viable and compatible with normal activities and ontogeny.

To appreciate the importance of phenotypic accommodation in evolution, one need only recall its importance in development. Individual ontogeny involves transitions between different morphologies and behaviors at different life stages, and in sexually reproducing organisms it involves the expression of two sets of genes that have been tested in two phenotypically different parents. Although genetic recombination is considered a positive factor in evolution because it permits varied, potentially favorable genes to spread within populations, it also promotes developmental variation and the union of dissimilar phenotypic traits. Genetic recombination would be useless and sometimes disasterous if individuals were unable to immediately accommodate the variations that result.

Slijper's two-legged goat, described in chapter 3 (see figure 3.13), is a clear example of phenotypic accommodation due to plasticity. All of the changes that occurred following the foreshortening of the goat's front limbs were phenotypic accommodations that compensated for the dysfunctional shortness of the front limbs. The initial adjustment was evidently behavioral, with the morphological ones due to the well-known impact of use on the highly plastic development of muscles and bones (see chapter 16).

As discussed in chapter 6, Baldwin (1896, 1902) and his contemporaries Morgan (1896a,b) and Osborn (1897a,b) were among the first to recognize the role of phenotypic accommodation for evolution. Baldwin (1902) called this phenomenon simply "accommodation" and considered it an adjunct to natural selection in the evolution of novelties:

> It may be said, indeed, quite truly, that this value of accommodation is implicit in the theory of natural selection; for, according to that theory, there is continued selection of certain fit individuals, and their fitness may consist in their being plastic or "accommodating." (p. 46)

Baldwin expanded on Darwin's analogy between developmental variants and uncut stones to point out that phenotypic accommodation prepares the rough innovations to function and persist, facilitating the evolution of integrated design (see chapter 6).

Other authors who have discussed phenotypic accommodation as a factor in the evolution of novelty include Frazzetta (1975), Rachootin and Thomson (1981), Alberch (1982), Kirschner (1992), Gerhart and Kirschner (1997), and West-Eberhard (1986, 1989, 1992a, 1998). Alberch (1982) cites a classical experimental demonstration of phenotypic accommodation described by Twitty (1932): when the eye of a large salamander species is transplanted into the embryonic skull of a smaller species, the eye grows to its genetically determined large size, and the host develops a proportionally larger cartilaginous optic cup to perfectly accommodate the larger eye, as well as a concordant change in the tectal neuron population of the midbrain corresponding to an increase in the number of retinal ganglion cells associated with the grafted eye.

Frazzetta (1975) calls phenotypic accommodation "compensation." He discusses how "animals experiencing sudden changes . . . are buffered through compensations from related systems" (p. 150). His examples are particularly compelling because his favorite subject is bone, a tissue we think of as among the most rigid in nature. In fact, bone growth involves hyperflexibility via somatic selection, as already discussed (chapter 3), and some of the most striking novelties created by plastic responses to environmental stress involve bone (see chapter 16).

Exaggeration of Change

Due to adaptive plasticity, a simple primary change, such as two-leggedness in the goat, induced by a small genetic or environmental change, can sometimes produce a large, or "saltatory" yet functionally integrated phenotypic change (Goldschmidt, 1940, pp. 296–297). Nonadaptive flexibility could exaggerate change but would not necessarily improve viability.

Exaggeration of change due to correlated shifts in adaptively plastic traits may help answer a question often raised by those who doubt the ability of Darwinian gradualism to explain the evolution of complex adaptive traits. Steele (1981) termed this the problem of "simultaneity": how can complex characters, whose imperfect intermediate or rudimentary states would seem to be of little use, evolve via gradual accumulation of essential features? Darwin, a thoroughgoing gradualist, envisioned functional advantage of intermediate forms (see Darwin, 1859 [1872], pp. 133–134, on the gradual evolution of the vertebrate eye). The idea that preexisting adaptive plasticity and correlated shifts could be involved in the origin of complex novelties, as proposed by Frazzetta (1975) for certain morphological traits in vertebrates, helps to solve two problems at once. Multidimensional adaptive plasticity accounts for coordinated change in several features at a time, and it explains why a sudden large change could be immediately adaptive or at least not disruptive. This plus the idea of genetic accommodation (chapter 6) solves the simultaneity problem without resorting to untenable Lamarkian solutions requiring transmission to the germ line of somatic change (e.g., as in Steele, 1981; for a critique of this hypothesis, see Dawkins, 1982, pp. 164–177).

Frazzetta (1975) criticized Goldschmidt's discussion of macromutations for "not always distinguishing between truly innovative evolutionary steps, those involving revision of the interaction responses among components, and the flexibility of integrated systems to provide an array of varied adaptive expressions" (p. 225). In fact, this is not an easy distinction to make, for the flexibility of systems may be a source of true innovation. The evolution of the worker phenotype in social insects is one example; part II describes many others. It is also difficult to draw a line between the development of a novel trait and its phenotypic accommodation, since phenotypic accommodation can involve a chain of correlated effects not unlike a developmental cascade, as in the two-legged goat.

Chapter 16 gives many examples of complex novel traits that are the result of correlated shifts in plastic traits.

The Buffering Effect of Alternatives and Phenotype–Environment Matching

A novel phenotypic subunit begins as a developmental alternative to another, preexisting trait. This is necessarily true, for a novelty could not usually become immediately ubiquitous in a population (see part III). This means that a novel phenotype can coexist alongside an established specialization, in effect allowing the individuals of a lineage to begin to exploit a new adaptation without abandoning an older, established one. The population is thereby buffered against decline or extinction as it acquires a new adaptation (West-Eberhard, 1986; see also J. W. Wilson, 1975). This can facilitate evolutionary change in a new direction. The buffering effect of alternatives was perhaps first widely recognized in discussions of gene duplication, where the redundancy of having two copies of a gene has long been recognized as a factor facilitating the evolution of new function (Ohno, 1970; Kimura, 1983; Ohta, 1989).

Why would an individual begin to adopt a new, possibly clumsy alternative when an old, relatively refined one is available? As shown in part III, new, secondarily evolved alternatives, and alternatives associated with a lower average fitness than other options (such as the lower payoff mating tactics of peripheral males compared with central or dominant ones) evolve in circumstances where some class of individuals would do even more poorly without them (see chapter 22). In facultative options, with condition-sensitive switches, individuals adopt a lower fitness option only when it is likely to be more profitable than others, and selection acts to refine the accuracy of the assessment process associated with the switch (see chapter 6).

The buffering effect of plasticity is most effective when the optional phenotypes are facultative, and performed only in circumstances or in individuals or times when likely to be advantageous. Genetic accommodation of regulation can organize the expression of facultative alternative phenotypes so that selection is always (or at least more likely to be) positive by matching the expression of the phenotype to environmental conditions. When this is accomplished, for example, via effective modes of assessment during development and behavior (see chapter 23), discontinuous plasticity becomes a device by which populations pass from one adaptive peak to another without passing through valleys of lesser adaptiveness between.

In the case of alternatives whose expression is determined by one or a few genes of major effect, phenotype–environment matching does not occur. But the polymorphic population or lineage may nonetheless be relatively buffered from extinction in a changing or fluctuating environment than would a monophenic population, by virtue of not having all its eggs in one specialized basket. Bet-hedging alternatives are evidently uncommon compared with conditionally expressed alternatives, both in theory and in fact (see chapter 22).

Adaptive conditional expression of alternatives is one of the properties of phenotypes that allows organisms to be modified while they are running.

Directional Change under Oscillating Selection: The Ratcheting Effect of Conditional Expression

In the few studies that have examined temporal variation in the fitness effects of particular traits, selection has acted in opposite directions at different stages of the life history (Schluter et al., 1991) and in different years (e.g., Grant, 1986; see also chapter 28). How can selection produce directional evolution if such reversals in direction are the rule? I suggest that directional change may often depend on the *ratcheting effect of conditional expression*—the fact that conditional expression matches a phenotype to an environment, making it possible for a phenotype to be exposed to positive selection when it is likely to be advantageous, and shielded from negative selection by nonexpression in conditions where it is inappropriate. This ratcheting effect could accumulate net directional change under episodic selection in temporally and spatially heterogeneous environments, even though the population may experience conditions that temporarily reverse the trend.

Panmixia and dispersal in a large population with a variable environment would be especially favorable to the ratcheted evolution of alternative phenotypes because the wide area sampled would increase the likelihood that both of two alternative phenotypes would be repeatedly expressed. It would therefore be an advantage to all individuals to be able to facultatively produce both. Under an oscillating selection regime, the phenotypic plasticity necessary for the ratchet would be best maintained. If one of the conditional phenotypes were to be genetically fixed in some localities due to selection on regulation, as occurs in some populations of buttercups (Cook and Johnson, 1968), then the ratchet would not apply to that population. Under panmixia and adaptive plasticity, gene flow would *promote* ratcheted divergence rather than prevent it. Local pockets of phenotypic specialization can occur in this way and may lead to speciation (Endler, 1977; see also chapter 27) but need not do so (cf. Maynard Smith, 1983a, p. 21).

In the conditional-ratchet model, evolving traits are developmentally, rather than geographically, partitioned. They are developmentally matched to their environments as they would be if adaptive habitat selection were occurring. As a result, the developmentally partitioned traits evolve semi-independently and divergently (the independent selection of independently expressed traits is discussed in a separate section below). This hypothesis simply projects the idea of a buffering effect of alternative phenotypes (West-Eberhard, 1986, 1989; see also above) into a spatially heterogeneous environment. The same argument would apply to continuous variation where extreme local environments conditionally produce an adaptive extreme or unusual form (see chapter 28).

Change without Loss of Ability to Change

One of the puzzles for purely genetic reasoning about evolution is to explain how change can go on indefinitely. Why do populations not evolve to an optimum, or exhaust their genetic variation, and then stop evolving? The usual answers are that continued evolution is driven by mutation, recombination, and environmental change, including evolutionary change in parasites and pathogens and the social behavior of relatives and mates. These are reasonable answers, but they may not be the whole story.

One consequence of phenotypic plasticity is that it can favor the evolution of specialized traits without loss of the ability to adopt others. The mechanisms of plasticity may be retained even when an alternative is fixed, since they are essential for its development. Then, recurrently stimulated by a new environmental input, they can foster a new response and evolution in a new direction.

Learning is a good example. Individual birds often have learned dietary specializations (see chapter 18). In some circumstances—if conditions repeatedly favor the development of a particular variant—a single learned specialization may predominate in a population (see chapter 28, on adaptive radiations). The ability to learn is not thereby diminished, for it may be indispensable for the normal development of the trait, as in language acquisition in humans and song acquisition in birds (Marler, 1998). Consequently, the population retains the ability to learn *new* specializations, and if one of them is recurrently learned, it may lead to new morphological, physiological, or behavioral specializations.

In many other kinds of highly plastic morphology and behavior, traits are easily molded by selection into different forms associated with small genetic changes, without losing their potential to take on a different form in the next generation should developmental conditions change. A morphological example is the pharyngeal jaw structure of cichlid fish: Liem and Kaufmann (1984) point out that the numerous differences between two morphs, one "papilliform" and feeding selectively on soft food items, the other "molariform" and consuming a greater proportion of snails, may, like comparable differences between species, involve correlated changes among multiple plastic features. "Head structural elements in fishes are coupled in specific patterns so that a perturbation in one element will elicit accommodating structural changes in many other component parts" and there is "well-documented evidence of regulatory interactions between muscle and bone during ontogeny" and "ability of bones to react to changes in stress regimes" (p. 210). Thus, small genetic changes can produce extensive restructuring of the trophic morphology, without a necessary loss in the basic plasticity of interacting muscle and bone. This means that even while specialized to a particular feeding morphology, species could retain the potential to give rise relatively quickly to other forms, a situation propitious to radiation because specialization is not always a dead end limiting further diversification.

Learning and other mechanisms of adaptive plasticity have probably contributed to unusually rapid radiations in nature (see chapter 28). By this "flexible stem" hypothesis, flexible morphology, behavior, or life-history patterns can give rise repeatedly to specialized traits without loss of the potential for a new round of change (chapter 28).

Evolutionary Consequences of Modularity

Chapter 4 showed that modularity is a property of phenotypic traits at all levels of organization. This includes not only morphology, at the molecular through the cell, tissue, organ, and social-individual level, but also physiological, behavioral, and learned traits, including reversible ones (traits that can be turned on and off during development). A broad application of the modularity concept to generalizations about selection and evolution is possible because of the underlying modularity of gene expression and gene-product use, discussed along with supporting data in chapter 4. Some authors (e.g., Robinson and Dukas, 1999, p. 583) see reversible behavioral alternatives and learned traits as falling into a different category for theoretical treatment from constitutively expressed traits. But the generalizations discussed in this section should apply to any phenotypic trait that has the general modular, switch-mediated form shown by traits A and B in figure 4.13, whether trait expression is reversible or irreversible, and whether its expression is influenced by learning or not.

Four features of modularity described and exemplified in chapter 4 facilitate adaptive evolution. (1) Modularity permits *compartmentalization of change*, which reduces developmental disruption during evolutionary change. Compartmentalized development, with one pathway or set of processes somewhat isolated from others by a switch or a physical boundary, such as a cell wall, can permit circumscribed change that is not disruptive to the larger whole insofar as it is confined within a subunit of development. A modular boundary to the effects of a change is one of the features, in addition to phenotypic accommodation (see chapter 3), that permits change while the organismic machine is running. (2) *Modular traits are subunits of selection.* The fact that a phenotypic subunit is a set of traits expressed or used together (chapter 4) means that it is selected as a unit. (3) Modular traits are *subunits of gene expression* (chapter 4), which, combined with their status as subunits of selection, permit a modularized response to selection, facilitating mosaic evolution.

Features (2) and (3) account for mosaic evolution, the independent specialization of individual traits expressed by the same genotype. Different limbs, segments, or juvenile morphologies can evolve in different directions and at different rates while others remain unchanged or evolve in other contexts. Finally, (4) modular traits are developmentally dissociable. This permits *combinatorial evolution*. Subunits of the phenotype can be reorganized and recombined—the tinkering aspect of evolution. The same subunit can be expressed in more than one trait or developmentally deleted somewhat independently of others (part II).

Several other authors have noted the importance of modularity for evolution. Most have been concerned with some particular aspect of the phenotype, such as morphology or cells. But the general implications are clear, because the modularity concept applies to all switch-mediated traits (chapter 4). Alberch (1982) recognized the importance of modularity for adaptive change: "Discontinuities are important to free the system from the contravening functional constraints that greatly limit the gradualistic action of directional selection" (p. 25; see also Cheverud, 1984b).

Simulation studies of networks designed to resemble genetic control and differentiation have confirmed the importance of modular organization for evolution. (Kauffman, 1983a,b). The required ability to change one control element ("gene") without altering the nature of other states was a feature of networks where, as in bifurcating developmental pathways, it is possible to pass directly from one state to only a small number of others. "Thus the observed branching nature of pathways of cellular differentiation may be a necessary and hence universal feature of systems capable of adaptive evolutionary change" (Maynard Smith et al., 1985, p. 268, based on Kauffman, 1983b). Although Kauffman (1986, 1993) sees gene interaction and "the complexity catastrophe" as a previously unexpected limitation of adaptation this problem has been discussed in population genetics in terms of genetic correlations and antagonistic pleiotropy between traits:

> The developmental constraints of Gould (1980), Alberch (1980), and Oster and Alberch (1982) are the source of the genetic constraints discussed [within a neo-Darwinian microevolutionary framework] by Charlesworth et al. (1982). . . . [t]he problem is not a microevolutionary theory incapable of accounting for developmental effects, but that those concerned with explanations of adaptation in terms of selection have ignored this connection between development, correlation, structure, and the evolution of cohesive adaptive traits. (Cheverud, 1984b, p. 169).

Wagner (1996) considers modularity responsible for the evolution of subindividual traits but adopts a view somewhat different from that presented here. For Wagner, a novel trait is not originally modular, but becomes modular as selection gradually eliminates pleiotropic effects between modules ("parcellation") or favors pleiotropic effects within modules ("integration"). I have given reasons to suppose that the regulation and form of a novel discrete trait necessarily originate simultaneously (see chapter 4), which implies that modularity exists from the beginning, though it may be subsequently improved or exaggerated by selection for increased parcellation and integration as suggested by Wagner.

Another line of argument suggests that modularity of gene action per se has evolved under natural selection, "since this feature [independence of genes' effects controlling different characters] of genomes is essential to adaptive evolution" (Leigh, 1987, p. 236). This argument implies selection among clades. Modularity per se could also evolve by selection on individuals, if it facilitates an efficient separation of biochemical processes, or division of labor, among different cells and other internal organs (Gerhart and Kirschner, 1997).

The four evolutionary consequences of modularity just listed suggest a set of general principles that relate modularity to selection and evolution.

Modular Traits as Subunits of Selection: The Rule of Independent Selection

In the past, evolutionary biologists have focused their discussions of fitness and of selection primarily on two levels of organization: the individual level (e.g., Williams, 1966), and the genic level (Williams, 1966; Dawkins, 1976), with some attention to selection at higher levels of organization (group, colony, species, or clade; Lewontin, 1970; Wilson, 1975, 1980; G. C. Williams, 1992; Gould, 1999). As a result, there is no conventional framework for dealing with selection at the level of the trait, which is intermediate between the level of the gene and the individual. There are several reasons for this. Like other subindividual levels, the trait level is screened off by selection on individuals (Brandon, 1990). That is, subindividual traits have no fitness independent of their effects on reproducing individuals. But the *fitness effect* of a trait—its contribution to individual reproductive success—in theory can be estimated separately from that of other traits of the same individual. Separable fitness effects, along with gene-expression differences between traits, account for the evolution of differences among them.

One impediment to understanding traits as subunits of selection on individuals has been opposition to considering the individual a unit of selection, reasoning that individuals are transient units not passed between generations as are genes (e.g., Williams, 1966; Dawkins, 1976). This led to the selfish-gene view of evolution, which confuses the issue by supposing that the units of selection have to be the cross-generational units of evolution,

rather than acknowledging that evolution by natural selection is based on phenotypic variation, differential fitness of different phenotypes, and heritability of traits relating to fitness (reproductive success; Lewontin, 1970). The selfish-gene view temporarily diverted attention from the evolutionary importance of higher units of selection (Lewontin, 1970). Similarly, discussions of selection above the individual level also fail to solve the problem of integrating subindividual developmental organization into natural selection theory (Smillie, 1995). Thus, when Mayr (1982c) listed evolutionary questions that remain without satisfactory answers, he included: "How can one explain mosaic evolution, that is more or less drastic change of part of the phenotype (and the portion of the genotype controlling it) without visible changes in the rest of the phenotype?" (p. 1131).

Modularity of phenotypic traits and underlying gene expression is a solution to the problem of explaining mosaic evolution. But modular traits do not qualify as full units of selection. One requirement of a unit of selection is that it be a reproducing entity, such as an individual (Lloyd, 1988). For this reason, a trait of an individual is only a *sub*unit of selection, for it does not reproduce, but only contributes to (or detracts from or has no effect on) the reproduction of an individual. As stated by Lloyd (1988), a unit of selection (interactor) "is an entity that has a trait; the interactor interacts with its environment through the trait, and the interactor's expected survival and reproductive success is determined (at least partly) by this interaction" (p. 69). In other words, there is a correlation between the trait and the individual's fitness. I have already referred to that correlation as the *fitness effect* of the trait—its contribution to individual fitness. Lloyd (1988, pp. 69ff) discusses various more formal and precise ways to express this relation. It is the correlation between variable fitness effects and variation in underlying genes (i.e., the modularity of gene expression), which allows subunits of selection to somewhat independently evolve (show a response to selection that is semi-independent of other, independently expressed subunit traits).

The modularity of trait expression suggests a rule of thumb for relating trait development to trait selection and evolution, which can be called for short the *rule of independent selection.*

To the degree that traits or life stages are independently expressed or used, they are independently subject to selection (West-Eberhard, 1992a). For example, the exposure to view of a butterfly's wings depends on wing position. During some kinds of display, wings are held open, exposing the

Fig. 7.1. Independent evolution of dorsal (top) and ventral (bottom) wing patterns of butterflies. (A) *Charaxes castor* (Nymphalidae: Charaxinae); (B) *Baeotus baeotus* (Nymphalidae: Limenitinae). From Nijhout (1991a), © H. F. Nijhout, by permission.

dorsal surfaces, whereas at rest the wings may be folded, exposing the ventral surfaces. As a result, the two surfaces have different uses and are independently subject to selection, causing them to diverge (figure 7.1). Development, the arbiter of trait coexpression and use, thus partitions the phenotype into semi-independent subunits or "blocks" with underlying gene combinations that are evaluated as sets under selection (each has its own independent fitness effect). Thus, not only do the dorsal and ventral wing patterns of butterflies vary and evolve somewhat independently of each other, but so do their larvae, genitalia, and use of food plants (Aiello, 1984). Since traits are never expressed completely independently, the rule of independent selection would be more accurately termed the rule of semi-independent selection.

The *developmental* partitioning of gene action and gene-product use, then, is the basis for subindividual diversification in all multicellular as well as single-celled organisms. The rule of independent selection accounts for the existence of trait independence that is assumed by many disciplines within biology even though seldom examined. The assumption of trait independence is fundamental to taxonomic classifications and cladistics; to adaptive interpretation of individual traits in terms of natural selection; and to population-genetic analyses that treat individual traits and the correlations among them.

The independent selection of compartmentalized traits greatly facilitates adaptive evolution because it implies release from an important class of evolutionary constraints, namely, those due to genetic correlations between traits (see "character release," below). The rule of independent selection means, for example, that a correlated response to selection between the sexes is not necessarily expected under sexual selection, to cite a much discussed example of supposed evolutionary constraints due to genetic correlations (e.g., Lande, 1980a; cf. Fisher's [1958] discussion of sex-limited expression). As emphasized by Darwin (1871 [1874]), sexually selected traits of males are *secondary* sexual traits commonly expressed only in adult males. That is, they are regulated by hormones and are often sex and age limited in expression. Variations in sex-limited adult male traits would never be expressed in females. So it is not surprising that sexual dimorphisms are so common in nature with adult males often diversifying independently of females (West-Eberhard, 1983, 1984; Amundson, 2000). There is no need to hypothesize special selection to eliminate male traits in females, any more than there need be selection to eliminate adult male traits from embryos or juvenile males. The only traits that would require such special selection would be non-specific modifiers of the sexually selected traits. This is a good example of how a developmental approach *supports* adaptationist arguments rather than being an enemy of Darwinism.

Schmalhausen (1949 [1986]) recognized the semi-independence of the evolution of subindividual traits with reference to the Batesian mimicry polymorphisms of butterflies (*Papilio dardanus*): "Every form has its own independent line of evolution, as indicated by the fact that . . . local geographic forms . . . mimic local subspecies [of the distasteful model]" (p. 77; see also Turner, 1977, on subspecies-specific modification of Batesian mimicry morphs). Divergent modification occurs between the phenotypically divergent subunits or morphs. Sewall Wright (1941) discussed, in somewhat different terms, the selective consequences of hierarchical, subunit organization of phenotypes in a review of a book by Goldschmidt:

> Within the organism as a more or less integrated reaction system, there is a hierarchy of subordinate reaction systems, each with considerable independence, as shown by capacities for self-differentiation. Thus there must be partially isolated reaction systems for each kind of organ and each kind of cell. It is difficult to see how any spatial pattern in the germ plasm can operate in determining these [Goldschmidt visualized chromosomal reorganization as the basis; linkage of functionally related traits is a related idea], but there is no theoretic difficulty with a branching hierarchic system of chain reactions in which genes are brought into effective action whenever presented with the proper substrates, irrespective of their locations within the cells. There is no limit to the number of reaction systems that can be based on the same set of genes, and *such systems may obviously evolve more or less independently of each other* (pp. 392–393; emphasis mine).

Nijhout (1991, p. 246) similarly noted that independent variation means independent developmental mechanism and independent potential for evolutionary change. He relates this to a model of butterfly wing pattern evolution where different sectors of the mosaic wing vary semi-independently while also under selection as a unit at the level of the wing.

Several authors have appreciated the independent evolution of alternative phenotypes without extrapolating to subindividual evolution in general. O. W. Richards (1961), in a discussion of polymorphisms, a major class of phenotypic subunits, noted: "One is perhaps led to the idea of a self-intensifying process. A particular type of environmental pressure leads to the evolution of two or more morphs, but *once the morphs exist they will be exposed to different pressures because they are different*" (p. 19; emphasis mine). Akimoto (1985) stresses the importance of developmental discreteness for the divergent evolution of aphid life-cycle morphs.

Character Release

Related to the rule of independent selection is the phenomenon of *developmental character release*, or simply "character release" (West-Eberhard, 1986)—increase in the relative freedom of a modular trait to evolve independently of other traits as the degree of its modularity, or independent expression, increases. In other words, release is the increase in evolutionary independence of a subindividual trait that corresponds to a decrease in its genetic correlations with other traits.

Character release begins with the origin of a switch, which permits the independent expression of two alternative modular traits, each characterized by its own set of specific modifiers. This permits the operation of the rule of independent selection, and the semi-independent evolution of the two alternative traits. The initiation of modularity, with the initiation of independent expression and selection, has been called "genetic release" by Gadagkar (1997b). I introduced the term "release" to refer to the increased freedom from genetic correlations that occurs when one of two alternative phenotypes increases in frequency relative to the other (West-Eberhard, 1986, 1989), a continuation of the release that occurs at the origin of a switch.

As a modular trait becomes common relative to other traits with which it shares quantitatively variable non-specific modifiers (figure 4.13), the optimum values of those modifiers come to conform increasingly to the optimum for the more frequently expressed (more strongly selected) trait. Thus, the compromise that is effected by antagonistic selection on shared modifiers is increasingly resolved in favor of the more common alternative. The expected result of an approach to fixation of one alternative trait is therefore a corresponding decrease in genetic correlations, a decrease in antagonistic pleiotropic effects, and a frequency-dependent acceleration of the rate of evolution of shared modifiers toward specialization of the more common or fixed form, discussed further later in this chapter under "Mutual acceleration in the evolution of regulation and form."

The expectation of character release is confirmed by models of phenotypic plasticity in continuously variable traits in spatially variable environments (Via and Lande, 1985) and, more explicitly, by models of mimicry polymorphisms in butterflies (Charlesworth and Charlesworth, 1976a; see also Turner, 1977, p. 185) and evolution in morph ratio clines (Clarke, 1966; chapter 27) (below).

Developmental character release should not be confused with "ecological character release," increased rate of trait evolution in an environment free of competitors or natural enemies, as on a newly colonized island (Lack, 1947; Losos, 1994b).

The Coevolution of
Coexpressed Traits

A corollary of the rule of independent selection is the *principle of coevolution of coexpressed traits*: traits that are expressed together are selected together and evolve together as a coadapted set. Coexpression of traits is a prerequisite for the evolution of integrated function under natural selection. Cheverud (1982, 1984b) discusses this relation between developmental coexpression, phenotypic correlation, and functional integration in quantitative genetic terms and gives supporting evidence, citing earlier insights to this effect by Olson and Miller (1958) and Riedl (1978). Wagner (1996) makes the same point in terms of selection for integration via increased pleiotropic effects *within* phenotypic subunits, and Niehrs and Pollet (1999) describe the functional integration of genes that are expressed together.

The independent coevolution of independently expressed coordinated traits in different life-stages has been recognized by students of hydromedusan Cnidaria (Hydrozoa), marine invertebrates with complex life cycles in which a zygote gives rise to a planula larva, which then becomes a polyp or hydroid, which produces other hydroids asexually,

which in turn produce sexually reproducing medusae (Boero et al., 1992). The hydroid and medusoid stages are so different that they have their own taxonomies, with those belonging to the same species sometimes impossible to associate (Boero and Bouillon, 1987). The contrasting patterns of evolution in the two stages has been called "inconsistent evolution" or "mosaic evolution" (see references in Boero and Bouillon, 1987), and the independent diversification of the two life stages has been explained in terms of selection in very different environments, the hydroids being asexually reproducing sessile benthic forms, and the medusae sexually reproducing forms that swim in the open water and have more complicated gastrovascular and nervous systems (Boero and Bouillon, 1987). But an important part of the explanation is their extreme modular developmental independence, for both hydroids and medusae can reproduce, and each begins its development from relatively undifferentiated cells as if it were a new individual, even though the hydroid is a phase of the life cycle that is like a juvenile stage interposed between the planula larva and the sexually reproducing medusa (Boero et al., 1992).

It is important to realize the cause-and-effect relationship between developmental organization, selection, and adaptive function: because selection cannot by itself give origin to organizational innovation, developmental organization (coexpression) should be regarded as the original source of coadapted functionality among traits, not the reverse. Coexpression has to occur first, before cofunctionality can be favored by selection.

The Independent Evolution of
Regulation and Form

A phenotypic trait and the switch mechanism that regulates the timing and frequency of its expression are different phenotypic subunits. Therefore, in accord with the rule of independent selection, they are somewhat independently subject to selection and evolution.

The fact that regulatory mechanisms are often recognizably different phenotypic subunits, or even organs, than the traits whose expression they control suggests that selection should be able to affect regulation without affecting form. This was noted by Raff and Kaufman (1983, p. 259), who discuss the independent evolution of "control" (regulatory) and "structural" genes (genes influencing form). It is also clear that the frequency and circumstances of expression of a trait can be altered independent

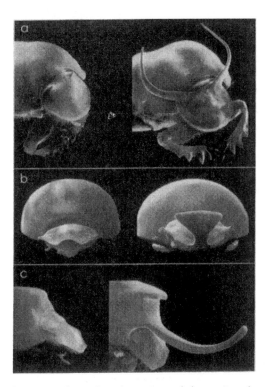

Fig. 7.2. Independent divergence of alternative phenotypes in male beetles (*Onthophagus* species). Hornless (left) and horned (right) males of (a) *Onthophagus taurus*; (b) *O. sharpi*; (c) *O. nigriventris*. Courtesy of D. J. Emlen.

of its form, as when morph frequencies change in polymorphic butterflies without change in the color phenotypes themselves (Turner, 1977). Selection on the threshold of a switch affects the frequency of expression of a trait, while selection on subordinate switches at lower levels of organization could modify the form of a trait by either omitting subunits or changing their dimensions.

Independent evolution of regulation and form is illustrated by studies of beetle horns and butterfly pupal coloration. In some beetle species, males are dimorphic, with a horned or hornless form depending on larval diet (Emlen, 1994). The threshold response, and consequently the ratio of horned to hornless morphs produced, can be shifted under artificial selection (Emlen, 1994, 1996; Moczek, 1996; see figure 5.22). In these artificial selection experiments, the shape of the horn is unaffected. Horn shape and form are species specific and vary among species of the same genus (Howden, 1979; figure 7.2), showing that morph form, as well as regulation of morph expression, evolves. Pupal

color polyphenism in butterfly pupae, which can match their background coloration by becoming either brown or green, is known to show genetic variation in both regulation (the threshold response sensitivity; Hazel, 1977; Hazel and West, 1979, 1982), and form (the degree of darkness of the pupa produced; Hazel et al., 1987) (review in Hazel, 1995). Partially independent modifiability of plasticity and form of a quantitative trait also was demonstrated using artifical selection on the mean and the temperature sensitivity of thorax size in *Drosophila* by Scheiner and Lyman (1991; see also Scheiner, 1993a).

The independent evolution of regulation and form of qualitative (threshold) traits requires that the two aspects be influenced by different sets of genes. This has been demonstrated in sticklebacks, which have three alternative morphological phenotypes (Bell, 1984). Intramorph variation in the form of morphs (lateral plate number in the low and partial morphs) is known to be influenced by modifier loci independent of the switch loci that influence plate number, which is a polygenic trait (Hagen and Gilbertson, 1972). Many other examples are given below in a section on the evolution of plasticity per se.

The independent evolution of regulation and form means that there can be separate genetic accommodation of regulation and form of novel phenotypes. Once a trait is widespread in a population, further rounds of induced change and accommodation may occur when new inputs cause *modifications* of the regulation and/or form of the trait.

Mutual Acceleration in the Evolution of Regulation and Form

It is obvious that a phenotypic trait must be expressed to be modified under selection. It is also obvious that a common trait stands a better chance of being modified than does a rare one. This suggests another corollary of the rule of independent selection: *the rate and degree of modification of a complex trait should be some positive function of its frequency of expression or use*, since this affects the extent of its exposure to selection relative to that of alternative phenotypic states. Rate of expression affects the strength of selection (size of fitness effect; Roff, 1996), which along with genetic variance determines the rate of evolution (Lande, 1979).

Increased frequency of expression is expected to increase the response to selection and rate of evolution in two ways: (1) by increase in the amount of genetic variation exposed to selection per generation—frequency of expression has an effect on rate of evolution analogous to that of shortened generation time, which likewise increases the sample of variants and variant combinations exposed to selection per unit time; and (2) by increased bias in the direction of selection on nonspecific modifiers of the trait—selection on loci that show antagonistic pleiotropy should favor the alleles that best suit the most frequently expressed phenotype among those they modify. Despite the semi-independent selection of regulation and form, the two interact under selection in this respect.

This effect is most clear in the case of modifiers of contrasting alternative phenotypes, because when one alternative increases, the other decreases in frequency. Alternative phenotypes share, and in a sense compete for, appropriate nonspecific modifiers. The nature of the nonspecific modifiers is expected to be determined by their functional compatibility with the most common alternative. This was pointed out by Slatkin (1979), who recognized "That the relative strength of selection on a modifier of small effect depends both on the incremental effect of the modifier on the fitness of each phenotypic class and on the frequencies of the different classes." Thus, a phenotypic parameter that affects both rare and common phenotypic classes would have to have a proportionately larger influence on the fitnesses of the rare classes for the selection not to be dominated by the effect on the common classes." (Slatkin, 1979, p. 396)

The same principle of mutual acceleration of regulation and form should apply to continuous variation. The modal value of a quantitative trait and any properties that happen to correlate with it are more often subject to selection than are less frequently expressed extreme values. Gupta and Lewontin (1982, p. 945) refer to the "myopia" of selection, which sees not the entire potential of a continuously variable response but only the range that happens to be produced, and produced most commonly. Selection therefore may accumulate modifiers of phenotypes such that design becomes most adaptive at or near the modal frequency, with extreme phenotypes eventually selected against. Not only are the relatively rare extreme phenotypes marginal with respect to the modal environment (presumably the prime evolutionary determinant of the mode), but they are also less favorably modified relative to modal phenotypes. Bimodal distributions of continuously varying traits could by this means set the stage for divergent modification of alternative phenotypes (see chapter 11, on deletion

of intermediates by developmental disruption of continuous variation).

The self-accelerating effect of change in morph ratio and morph fixation was noted by Clarke (1966) and used by Endler (1977) in his discussion of divergence and speciation in clines. The same principle—the importance of repeated expression for accumulation of modification—applies when rare phenotypes *fail* to become specialized in a particular direction. Slatkin (1979, p. 396) notes that this point was raised by Wright (1929) in arguing against Fisher's theory of the evolution of dominance: the frequency of affected individuals would be too small to allow selection on modifiers of dominance to overcome other possible pleiotropic effects of the modifiers. The enforced variability of immunoglobin development, described in chapter 3, seems designed to avoid repeating particular variants. It thereby scrambles the results of selection and maintains diversity by not allowing fixation of a few temporarily effective forms.

Because frequency of expression can be environmentally determined (*sensu* Bull, 1983), this is an area where "merely environmental" determinants of the phenotype can have a profound influence on the direction and rate of evolutionary change. For example, in some species where clinal variation in phenotype frequency may be largely driven by environmental factors, evolutionary (genetic) change is exaggerated at the clinal extremes (see chapter 27), suggesting an effect of the environmentally mediated clines in phenotype expression. This is an expected and observed effect of morph fixation, including when driven by environmental effects on development (see chapter 21). Fixation of alternatives removes any antagonistic pleiotropy between former alternatives and allows the formerly pleiotropic, nonspecific modifiers to specialize as specific modifiers of the fixed phenotype ("release"; West-Eberhard, 1989).

Dissociability and Combinatorial Evolution: A Head Start for Adaptive Evolution

Switch points in development divide the phenotype into semi-independent subunits. They are also control points where development can be turned off or on, accelerated or slowed—points of phenotype dissociability. Therefore, the modular subdivisions created by regulatory switch points are subunits of potential reorganization during evolution. Due to this capacity for dissociability, modularity is widely cited as a kind of developmental flexibility as well a basis for evolutionary lability at levels ranging from proteins to behavior patterns (e.g., Bonner, 1965, 1988, p. 174; McBride, 1971; Barlow, 1968; Gould, 1977; Gilbert, 1978; Raff and Kaufman, 1983; Smith-Gill, 1983; Wake and Roth, 1989; Lord and Hill, 1987; McKitrick, 1994; Edelman, 1988; Holland, 1992, in Wagner and Altenberg, 1996; Wagner and Altenberg, 1996). Dissociation occurs even at the level of a protein domain. The smallest dissociable phenotypic subunits may be the small functional folding units of proteins (DuBose and Hartl, 1989).

The principle of dissociability in development was first articulated by Needham (1933), who referred primarily to the dissociability of growth and differentiation and of those processes from normal embryonic functioning (e.g., respiration). Needham's description of dissociability is worth citing verbatim because it shows appreciation of both the connectedness and modularity of phenotypic subunits:

> In the development of an animal embryo, proceeding normally under optimum conditions, the fundamental processes are seen as constituting a perfectly integrated whole. They fit in with each other in such a way that the final product comes into being by means of a precise cooperation of reactions and events. But it seems to be a very important, if perhaps insufficiently appreciated, fact, that these fundamental processes are not separable only in thought; that on the contrary they can be dissociated experimentally or thrown out of gear with one another. This conception of out-of-gearishness still lacks a satisfactory name, but in the absence of better words, dissociability or disengagement will be used. (pp. 180–181)

Gould (1977) uses the dissociability concept in a discussion of heterochrony, gives a concise history of the concept, and calls it the "principle of modularity." Raff and Kaufman (1983) discuss the dissociability of morphogenesis and cytodifferentiation.

How dissociability at switch points contributes to patterns of evolution is shown in diagrammatic fashion in a genus of tropical vines (genus *Monstera*), discussed in chapter 6 (see figure 6.3), where seven distinctive leaf forms are expressed in different combinations in different species. This is a good example of phenotypic recombination—called "ontogenetic repatterning" by Wake and Roth (1989, p. 366; see also Roth and Wake, 1985). An example of phenotypic recombination in the evolution

of behavior is shown in figure 7.3, where five components of a signal are recombined to make a variety of species-specific calls (see below). Figure 7.4 shows a fossil example where a record of life-stage specializations is recorded in morphology.

Some examples of evolved phenotypic recombination are more complex. The recruitment dance of honeybees (*Apis mellifera*), for example, varies in complexity and form among different races of bees (von Frisch, 1950; Lindauer, 1961; Lindauer and Kerr, 1960; Dyer, 1985, 1991a; Dyer and Seeley, 1991) and has components that are widespread in insects and other organisms, recombining their expression to make a novel complex trait. The dancing bee translates a flight angle with respect to the sun into a deviation from vertical when running on the vertically hanging combs of the hive. A visually perceived angle is transformed into an angle with respect to gravity. This might seem a remarkable invention if it were not for the discovery of the "light-compass" reaction by Vowles (1954), who found that ants, when forced to fly at a known angle with respect to polarized light, and then to walk on a vertical surface, maintain the same angle in the field of gravity. Von Frisch and Lindauer (1956) list nine other arthropods which exhibit the light-compass reaction to polarized light, suggesting that this ability may be ancient in the group. Another characteristic of the stereotyped bee dance, alternation of right and left turns in the so-called "round dance," is also likely ancient in insects. Hemipteran bugs perform right-left alternating turns when tested in mazes, and this "correcting behavior" may be common in insects that walk on branching pathways such as plants (Dingle, 1961). In blow-flies (*Phorina regina*) individuals stimulated with a sugar solution perform repeated clockwise and counterclockwise turns, a behavior which resembles the bee dance in several ways: in continuing even when the fly is displaced to a different location; in having a turning rate that reflects the richness of the food (sugar) source; in having a dance duration that increases with the strength of the sugar solution; in having a consistent orientation with respect to both light and gravity (although no light-compass translation); and in being accompanied by regurgitation to other flies, which sometimes then perform excited gyrations of their own (Dethier, 1957). Such excited running and shaking is common in returning nectar foragers in social wasps (personal observation) although they do not communicate food location to nestmates, and has been observed in termites (Emerson, 1928) and ants (Schnierla, 1953). Even sun compensation, the ability of honeybees to estimate the sun's position at

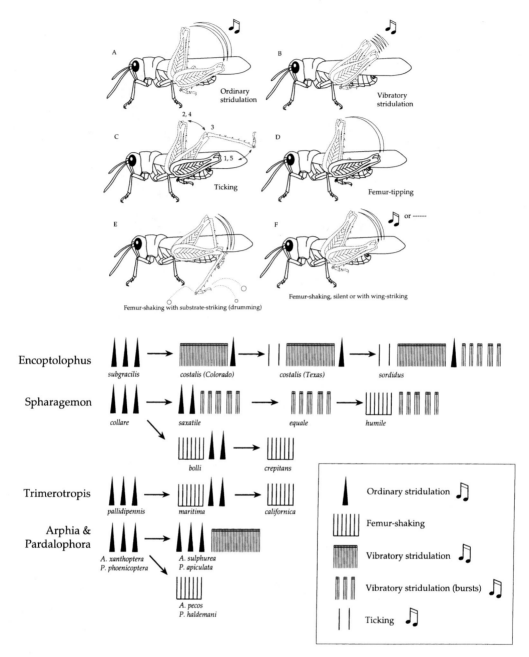

Fig. 7.3. Phenotypic recombination in the evolution of animal communication. Grasshopper (Orthoptera, Acrididae) courtship signals have five basic components (A–F). These are performed in different combinations in 16 different populations of five genera (Otte, 1970, 1974). Courtesy of D. Otte.

times of day they have not experienced, is an innate ability documented in a wide variety of animals (Dyer and Dickinson, 1994). Thus major components of the bee dance are likely to have been assembled by phenotypic recombination of preexisting traits.

Hundreds of other examples of dissociability and combinatorial evolution are discussed in part II (see also Matsuda, 1987). Unlike genetic recombination, in which most new combinations are lost during meiosis in every generation unless tightly chromosomally linked, novel combinations pro-

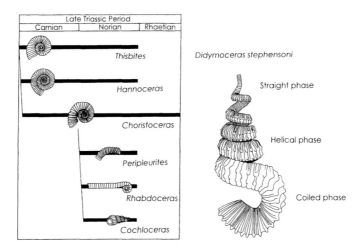

Fig. 7.4. Developmental switches and evolution of form in heteromorph ammonites. Ammonite ontogeny can include production of coiled, helical, and straight shell, sometimes in the same individual (right). These phases are expressed in different combinations in different ammonite lineages (left). *Didymoceras stephensoni* (right) after Raff and Kaufman (1983), based on Scott and Cobban (1965); phylogeny (left) after Raff and Kaufman (1983), based on Weidmann (1969), by permission of Cambridge University Press.

duced by phenotypic recombination can be preserved by *developmental linkage*—preservation and spread of novel phenotypic combinations due to selection on regulation that favors their coexpression or sequence. This—developmental rather than chromosomal linkage—is how new adaptive trait combinations are formed during evolution. Wright's shifting balance theory was designed to solve this problem (see especially Wright, 1980). There is far more evidence in favor of the developmental-linkage solution proposed here.

Both genetic and phenotypic recombination work because of the modularity of their units: genes and dissociable phenotypic traits can retain their functional integrity even though they are moved about. The nonblending discreteness or "particulate" quality of the recombined entities is important. Thus, the gene still functions in its new chromosomal setting, and an infant cry translocated into an adult courtship sequence does not get "confused" or blended with other adult vocalizations. This quality of dissociability of genetic and phenotypic units is so common that we now scarcely think about it as a property of special interest, although of course this was a signal contribution of Mendel to the study of inheritance.

In a discussion of heterochrony, Stearns (1986) remarks, "For changes in timing to evolve, processes that were previously integrated must become uncoupled in such a way that function is preserved

. . . we do not yet know in general what separates developmental processes that can be uncoupled from those that cannot" (p. 37). Degree of dissociability must depend not only on degree of developmental independence of traits, but also on their functional relations. Thus, a trait complex or organ indispensible for survival could not be dropped or transferred from an interdependent phenotype mosaic without disastrous results.

I will use the term *dependent* (phenotypes, alternatives, or morphs) to refer to the functional nondissociablity of such organs or alternative phenotypes. In a highly evolved social insect colony, for example, sterile worker and fertile queen phenotypes are "dependent" in that neither can survive or reproduce (have a selective or genetic effect) without the other (figure 7.5). The aerial and aquatic leaves of certain heterophyllous plants, by contrast, are relatively independent phenotypes: even though sometimes borne by the same individual, a single form can persist in the absence of the others (figure 7.6).

Mather (1973) discusses the distinction between dependent and independent morphs. Conditionally expressed alternative phenotypes are often highly independent and dissociable because not only are they switch contolled and therefore mechanistically dissociable, but they are adaptively dissociable as well. Due to the phenotype-environment matching of conditional expression, the less appropriate al-

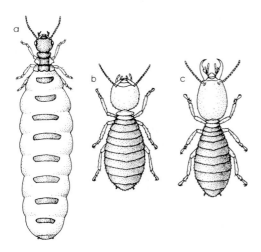

Fig. 7.5. Highly divergent dependent morphs: the castes of termites: (a) a physogastric queen, (b) worker, and (c) soldier of the black-mound termite *Amitermes hastatus*, a polyphenism in which none of the highly specialized kinds of individuals can reproduce without the others. After Skaife (1955).

ternative can be profitably deleted in a given circumstance. This is one reason for the evolutionary importance of alternative phenotypes (see part III).

Consequences of Hierarchical Organization: Integration, or the Parliament of the Phenotype

The branching nature of development produces a hierarchical arrangement of modular phenotypic traits within individuals (see chapter 4). Since modular traits can be considered subunits of selection, there is a hierarchical arrangement of subunits of selection as well.

Just as genes are screened off from being the primary agents of cross-generational reproduction and selection by their dependence on the reproduction of individuals (Brandon, 1985, 1990; Salmon, 1984), the cells and organs of a multicellular organism are screened off from being full units of selection since they do not live or multiply on their own but are only reproduced as developmental subunits of reproductive individuals. Each subunit of a phenotype can be regarded as having a *fitness effect*, defined as its lifetime contribution to individual reproductive success. A higher level subunit such as an organ would have a fitness effect that is a function of the fitness effects of its component tissues, the tissue a fitness effect that is a function of those of its component cells, and so on. Thus, there is a hierarchy of subunits of selection below the level of a full unit of selection, defined as a reproductive unit (see chapter 2).

The positive or negative effect of a lower level subunit may be canceled by the overall failure or success of higher subunits, with all of them ultimately screened off by the lifetime inclusive fitness of the individual. Tuomi and Vourisalo (1989) proposed such a hierarchy of functional and selective levels for modular and colonial organisms. It applies as well to the subdivisible developmental architecture of phenotypes in general. Because of this, any locally selfish behavior of a lower level subunit in a multicellular individual hurts itself by hurting the higher level subunits on which its ultimate success depends (Leigh, 1971). For this reason, the

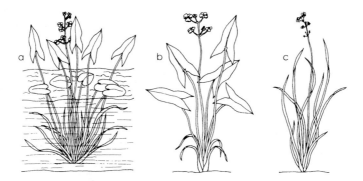

Fig. 7.6. Highly dissociable independent expression in the leaf form of a marsh plant (*Sagittaria sagittifolia*): (a) variations in a single individual showing submerged, water-surface, and aerial leaves on the same individual; (b) a completely terrestrial plant; (c) a completely aquatic (submerged) plant. The phenotype expressed depends on environmental circumstances, and one leaf type can be expressed without the other. After Schmalhausen (1949 [1986]).

"parliament of genes" (Leigh, 1983) is ultimately a *parliament of the phenotype*, the genes being the lowest level participants in a parliament whose fitness effects are ultimately measured at the level of the collective phenotype, not at the level of the collective genes.

As a result of this hierarchical screening off, integration is imposed on lower levels of organization by higher levels. This does not interfere with the fact that within levels of organization there is discreteness of gene expression, selection, and evolution of *alternative phenotypic traits*. It is for this reason that conditional alternative phenotypes— products of developmental decision points within a level of organization—are so important for adaptive evolution (see part III). Alternative adaptations permit diversification without loss of integration (West-Eberhard, 1986).

A subunit of the phenotype is a subunit of selection in the sense of making a semi-independent contribution to fitness, but it is not a subunit of selection in the sense of competing with other subindividual units for reproductive opportunities, as suggested by Buss (1987). Reproductive competition implies cross-generationally transmitted genetic differences. Intraindividual genetic competition may occur between cell lines, as in competitive growth of mutated cancer cells, but this is an evolutionary dead end since the cancer cell lineages and those with which they compete do not show cross-generational reproduction independent of the individual.

Consequences of Regulatory Complexity

Preceding chapters have discussed the importance of switches in development. The organization of the phenotype, including its flexibility, its modularity, and its continuously variable sensitivity to genomic and environmental inputs, all involve regulation by switches, broadly conceived to include all causes of pathway changes during development. For most decision points, switches are phenotypically and genetically complex, as already extensively discussed.

Regulatory complexity per se has important consequences for development and evolution. It means that a phenotypic subunit does not correspond to a single gene or a chromosomally linked set of genes. Instead, most subunit expression, especially at relatively high levels of organization, is subject to the effects of many genes and may also be influenced by the environment in a variety of ways.

Atchley and Hall (1991) make this point for regulation of protein production by a single gene, a complex process with considerable room for variation during the sequence of transcription, transcript processing (including RNA splicing), mRNA degradation, translation, polypeptide processing, and protein turnover and secretion (see Paigen, 1989; figure 7.7). Seeing the potential for variability at even this low level of organization, one can appreciate the very great potential for genetic and environmental variation in the much more complex developmental events at higher levels of organization, which may involve hormonal or neuromuscular responses produced by complex structures and subject to the influence of many genes and environmental factors.

Regulatory complexity has at least three important consequences for evolution: interchangeability, multiple pathways, and adaptive fine-tuning.

Interchangeability

The interchangeability of environmental and genetic factors in the regulation of trait expression is discussed in chapter 5. A single switch may be influenced by different complex systems, including hormonal, neural, and metabolic variables. As stated by Cheverud (1988), "Since environmentally caused variation must act through the same developmental pathways as genetically caused variations, one may expect the pattern of environmental correlation to also reflect developmental constraints, and be similar to the genetic correlation pattern, as in phenocopies" (p. 168). The more complex and numerous the regulatory mechanisms that influence a switch, the greater the number of genetic loci and environmental factors that can potentially affect the switch, and the more opportunities there are for equivalent effects. Interchangeability underlies genetic accommodation and its special case, genetic assimilation. It is responsible for the ability of selection to adjust the circumstances of trait expression, the degree of plasticity and its reciprocal, degree of genotypic influence, and frequency of trait expression.

It would be difficult to overemphasize the importance of interchangeability for development and evolution. It means that selection can alter the frequency of occurrence of a trait, for example, the sex ratio, not only by effecting genetic change in the nature of the regulatory apparatus (the usual way of thinking about evolutionary change in regulation), but also by manipulation of the environment, for example, indirectly or "epigenetically," by affecting, for example, oviposition behavior in a species such as turtles where regulation of sex expression is temperature dependent and oviposition

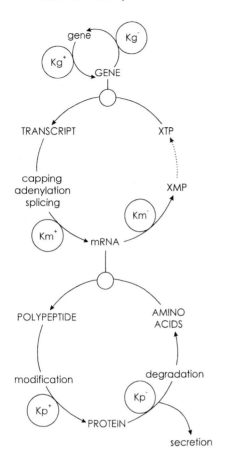

Protein concentration = $(Kg^+/Kg^-)(Km^+/Km^-)(Kp^+/Kp^-)$

Fig. 7.7. Regulatory complexity and interchangeability at the level of a single gene. Gene expression involves three interlinked metabolic cycles. Reaction-rate constants are circled. The equation relates the steady-state concentration of protein product to the six rate constants, three of them synthetic and three of them degradative. Each process is maintained by a balance of synthetic and degradative reactions, and catalytic linkage among cycles makes the level of protein production equally affected by a change in any of the six rate constants. Therefore, there are multiple potential sites for modulation of phenotype (protein) production, any one of which can produce an equivalent change in the level of gene expression. Equivalent results can be achieved by changes in DNA coding sequences; by changes in the metabolic cycles shown due to such factors as tissue-specific differences in protein degradation or use; or by hormone signals and presence of other interacting proteins (interchangeability of environmental and genotypic influence). So even at the gene-expression level, there can be (1) multiple pathways to the same phenotypic result and (2) interchangeability of mutational and environmental influence. After Paigen (1989).

sites vary in temperature. Maternal effects, for example, in embryogenesis, morphogenesis, and behavior, are well-known examples of this (see Anderson, 1989; Atchley and Hall, 1991; Wcislo, 1989). Similarly, if an alternative is size dependent, selection affecting nutrient acquisition or metabolism could alter switch points and morph ratios. Such "epigenetic" effects must be a common means of change at every level of organization in development, given the ubiquitous environmental–genotypic duality in determiation of all phenotypes.

Multiple Pathways, and Trait Conservation without Conservation of Developmental–Genetic Pathways

Multiple developmental pathways to the same trait are possible because with complex regulation the same switch can be determined or "thrown" by a number of different factors or subsystems (see

chapter 3). A nutritionally determined switch, for example, could be equivalently influenced by a change in food preference, a change in oviposition site preference of the mother, or a change in metabolic efficiency. These amount to different pathways to the same trait. Goldschmidt (1940 [1982], p. 270) noted that the same mutant phenotype can be caused by genetic mutations at different loci, since regulation is polygenic. Accordingly, Wright (1956) documented that there are at least five genotypically distinct ways to get the wild-type hair color in guinea pigs (Van Valen 1960).

Gerhart and Kirschner (1997) give several examples of multiple evolved pathways to conserved phenotypes in chordate embryology (see also Raff, 1996). The vertebrate neural tube and other phylotypic developmental stages are maintained across broad ranges of taxa even though the developmental sequences that produce them are different. Research on obesity illustrates the potential for multiple pathways in the control of a quantitative trait. Exercise and change of dietary habits are non-genetic routes to weight loss. Others could be manipulation of a gene that affects a brain region concerned with appetite, or ingestion of an enzyme known to affect fat metabolism (see also Sing et al., 1992, 1995 on multiple causes of cardiovascular disease). Any trait whose expression is governed by polygenic regulation that reflects multiple points of control has this potential to be altered via multiple developmental pathways. A major mutation at any one of the influential loci could conceivably fix the trait in a population under positive selection. Numerous environmental factors may accomplish the same thing due to the phenomenon of interchangeability (see chapter 4).

Multiple pathways can facilitate evolution by increasing functional *redundancy* in established traits (Walker, 1996; Gerhart and Kirschner, 1997). If one of the pathways is preempted or interrupted by a new switch, others may be brought into play. Redundancy was discussed as "duplication of function" by Darwin (1859), and the multiple pathways provided by gene duplication, allowing for the independent modification of duplicates, is now a well-known feature of genome evolution, and it is a common feature of phenotypic evolution as well (see chapter 10, on duplication).

The fact that multiple pathways are possible means that selection may seldom optimize development. Selection acts on end points (phenotypes) and may never have a chance to screen all possible pathways—it can only screen those that occur, in the order that they happen to occur as developmental variants. The best available may not be the best possible. As a result, a seemingly haphazard, inefficient developmental pathway that happens to produce a certain result may evolve. Once a pathway is established and improved, it may remain, even though at the outset some other pathway could conceivably have occurred and been, in the long run, a better solution.

It is in this sense that evolution can be irreversible. Complex traits may be lost and regained in blocks (see chapters 11, on deletion; 12, on reversions; and 19, on recurrent phenotypes). But the exact history of their evolution is unlikely to be replayed in reverse because the alternatives and conditions of each modifying step are highly unlikely to recur, least of all in the reverse order. As a result of the opportunistic nature of evolutionary change, development can be remarkably convoluted and inefficient. For example, the ovary of larval honeybee workers is as large as that of queens during an early stage of larval development, then regresses to end up relatively small (Snodgrass,

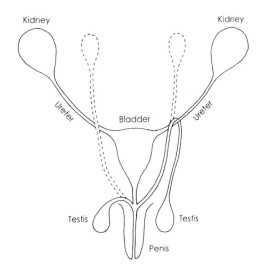

Fig. 7.8. Convoluted results of opportunistic evolution in the human male urogenital system. The long looping path of the vas deferens leading to the testis as it is (right, somewhat exaggerated) reflects its derivation from an ancestor in which the testes were internal and dorsal to the ureter, as shown by the dashed lines. As the testes became external in warm-blooded mammals, they moved dorsally to the ureter, while the vas deferens opened ventral to the ureter, producing a long vas deferens (right) rather than the short hypothetical one (left) that might have evolved had the testes originally moved ventral to the ureter. After G. C. Williams (1992).

1956). A classic example of this is the circuitous and long mammalian vas deferens of adult human males (figure 7.8).

Related to the potential for multiple pathways to the same trait is the fact that there can be conservation of a phenotype without conservation of its developmental genetic basis (see Gerhart and Kirschner, 1997; see also chapter 25, on homology). This means that developmental similarity may not be a reliable criterion of homology (see chapter 25). The converse may also occur: different, non-homologous traits may share homologous elements of regulation (Burian, 1997).

Adaptive Fine-tuning of Trait Expression

Polygenic influence on regulation of trait expression makes it possible for selection to fine-tune adaptive expression of traits. Phenotypically complex regulation means that many genes likely contribute to the level of a threshold and the frequency of expression of a trait. That this allows for fine-tuning of regulation is discussed under genetic accommodation in chapter 6.

Does Plasticity Accelerate or Retard Evolution?

The evolutionary effect of phenotypic plasticity is a subject of debate. Some argue that it can accelerate evolution due to environment matching of alternative forms (e.g., Baldwin, 1902; Rau, 1933; West-Eberhard, 1989; reviewed in Wcislo, 1989). Others maintain that plasticity retards evolution because it allows the phenotype to adjust nongenetically and therefore damps the genetic response to selection (e.g., Sultan, 1992; Schlichting, 1986). Still others have vacillated between these two views (e.g., Stearns, 1983; cf. Stearns, 1984, 1989) or discussed both effects, showing that either can occur, depending on circumstances (Wcislo, 1989; Papaj, 1994).

Phenotypic plasticity can have either result, depending on the effect of plasticity on the distribution of phenotypes, and on whether or not a quantitative trait or plasticity in switching between alternatives is involved. When a condition-sensitive switch produces recurrent expression of an alternative phenotype with environmental matching, plasticity can *accelerate* directional evolution of a recurrently expressed conditional alternative. When a hyperplastic mechanism such as learning, or a continuously variable reaction norm, produces a wide range of phenotypes, selection cannot act on a single recurrent trait or mode, and plasticity is expected to retard directional evolution. A hyperplastic device such as the immune system scrambles the results of selection and maintains a great diversity of phenotypic responses rather than a few recurrently. For this reason, researchers who deal primarily with reaction norms and continuous variation are likely to conclude that plasticity retards evolution, whereas those who focus more on discrete, conditionally expressed traits, where each phenotype is produced repeatedly, are likely to conclude that plasticity accelerates evolution. Some of the somatic-selection mechanisms discussed in chapter 3, such as the immune system, produce such a vast array of phenotypic variants under variable conditions that no single variant is sufficiently recurrent to respond to selection. In the immune system, the very genes that produce the randomly combined variants are made anew in individual B cells and are not transmitted as such to subsequent gen-

erations. It would be difficult to subvert such a system of somatic selection and permanently limit the phenotypic outcomes (immunoglobulins produced), for example, by persistent selection favoring fixation due to some particular pathogen. Such immunity to permanent effects of short-term specialization, for example, to suit the idiosyncratic needs of a single population or species, is an important quality of these highly plastic systems.

With hyperplasticity like that of the immune system and learning, which may permit switching among a multitude of alternatives, selection may come to act primarily on the mechanism itself, without producing directional specialization in any particular phenotypic result. Instead, evolution refines the decision rules and mechanisms of assessment to improve the ability for highly varied phenotype–environment matching. In a population with a diversity of learned specializations, for example, selection would favor the evolution of improved ability to learn and assess rewards for varied potential responses, an exploratory process where a multitude of potential pathways or switches are involved. The varied array of responses that may result could make it impossible for selection to focus in any particular direction. Where development or learning involves a recurrent binary switch, in contrast, alternative morphological specializations are possible within species because the behavioral alternatives are recurrent enough to be directionally modified.

The behavior of vertebrates and arthropods appears to represent different emphases in these two contrasting patterns of plasticity, with vertebrates specializing more often in learned alternatives, whose diversity and reversibility prevent recurrence that would foster the evolution of intraspecific morphological specializations. Arthropods, with a lesser development of learning, perform a more limited number of behavioral alternatives and are therefore more likely to evolve associated alternative morphological specializations. But it would be a mistake to overgeneralize about this, for both alternative morphological specializations, and well-developed mechanisms of plasticity such as learning and flexible physiological responses, can be found in both taxonomic groups.

Does Plasticity per se Evolve?

Plasticity presents special problems for studies of adaptation. The question "is plasticity adaptive?" can be interpreted in more than one way. It may be taken as a question of whether or not a particular response to the environment has a positive effect on fitness. Such a question poses special difficulties because adaptive plasticity is often a ubiquitous all-or-none response, so the environmental manipulation that would test for adaptation causes all organisms to change in the same way, leaving none for comparison to reveal how they would have performed had they not changed (Schmitt et al., 1999; see also chapter 23, below, on the adaptive significance of foundress helpers in social wasps).

Another way to interpret the question "is plasticity adaptive" is to ask whether plasticity itself can be considered an adaptive trait. Some authors maintain that the degree of plasticity of a trait can be affected only as a secondary result of selection on form (e.g., Via, 1993a; Hayes, 1997a; for discussion, see Via et al., 1995; Schlichting and Pigliucci, 1998). Others argue that regulation (e.g., degree of plasticity) and form (e.g., alternative phenotypic states expressed under environmental influence) can evolve independently of each other (Scheiner, 1993b). More commonly, authors *assume* that plasticity per se can evolve and proceed to discuss the costs and conditions of its evolution (e.g., DeWitt et al., 1998).

Both of the opposed opinions are correct, and both may imply limitations in their comprehension of the evolution of regulation and form. The view that plasticity is secondary is a truism, if it means merely that plasticity by itself, in isolation from its fitness effects on the expression of some (plastic) trait, has no meaning or independent existence. This is like saying that the heart per se does not evolve, except as a secondary product of selection on the circulatory system or, more fundamentally, selection on individuals (the ultimate conclusion of the secondary-result argument).

Quantitative-genetic models that take account of development and plasticity fall into two general categories, and this has sometimes led to unnecessary controversy about the evolution of plasticity. One, bileveled approach (e.g., Atchley 1987; Scheiner and Lyman, 1989) considers regulation and form to be distinct, though related and correlated, aspects of the phenotype taken at some convenient focal level of analysis. The second, unilevel approach (e.g., Via , 1993a; Slatkin, 1987, p. 801) assumes that plasticity or development per se does not evolve but that selection acts on some emergent aspect of form, and developmental properties are only secondarily changed.

The desireability of separate consideration of regulation and form is shown by the fact that an aspect of form can be deleted—completely unexpressed in an entire population—while its regulator (say, a hormone) continues to function in this and perhaps also other contexts (see chapters 11, on deletion; 19, on recurrence; 22, on maintenance of conditional alternatives). Elements of regulation and form represent different modular subunits of a phenotype. Nonetheless, if one considers which genes are actually treated by a quantitative genetic model, one inevitably concludes that the model describes the polygenically influenced regulatory elements underlying a phenotype being selected. Either the model treats the threshold of expression of the focal phenotype as a coordinated whole (the polygenic determinants of a continuously variable threshold response; e.g., Hazel et al 1990); or, if the quantitative trait is a single dimension or element of form, the genes in question are the polygenic determinants of its regulation—for example, stopping and starting growth, or affecting growth rate—seen at a lower level of organization (see chapter 3; Atchley and Hall 1991). Each subunit of the phenotype could be assigned its own fitness effect, but the fitness effects would be overlapping, not additive (see Nijhout and Paulsen, 1997).

The view that plasticity per se is a trait is based on the observation that selection on variation in plasticity can alter degree of plasticity (read: genetically variable traits that are mechanisms of plasticity, such as nerves, muscles, or the properties of a hormone system, are subject to selection and can evolve). By the rule of independent selection, this can occur without alteration in the form of the phenotypes whose expression is regulated (see Scheiner and Lyman, 1991, for experimental confirmation).

Thus, "selection on" a trait is always ambiguous in its meaning, for it may mean that selection affects either frequency or conditions of expression of a trait (regulation) or form (e.g., shape, elaboration, efficiency, etc.). This is a good example of a controversy that comes in part from lack of agility in looking at different levels of organization and the relations among them (see also the section entitled "The Evolutionary Importance of Mechanisms," chapter 3).

Thompson (1991) settles this question incisively: "Where genetic variation for plasticity exists, a population with a different mean plasticity can evolve" (p. 246). Sewell Wright (1931, cited by Stearns, 1982) called plasticity "itself perhaps the chief object of selection." Not surprisingly, degree of adaptive plasticity correlates with dependability of environmental cues in many organisms (e.g., see Scapini, 1988; Scapini and Fasinella, 1990; Scapini et al., 1988, 1993; Ugolini and Scapini, 1988, on orientation in sandhoppers). Degree of plasticity also correlates with degree of environmental variation in the parameter to which plasticity responds.

In the soapberry bug *Jadera haematoloma* (Insecta, Hemiptera), for example, only males from a natural population with a variable sex ratio in adult aggregations showed adaptive plasticity in guarding versus non-guarding mating behavior when tested with different degrees of female availability: males guarded females only when females were scarce. Males from a different, genetically isolated population with a less variable sex ratio showed less variation in the performance of copulatory guarding. This suggests that degree of plasticity has responded to selection in accord with variability in the supply of females (Carroll and Corneli, 1995). Similarly, female fig wasps (pollinators of *Ficus* species) adaptively adjust the sex ratio of their offspring in accord with the number of egg-layers per fig as predicted by local-mate-competition theory, but only within the range of foundress numbers that actually occurs in nature (Herre, 1987). For other tests of plasticity as an adaptation see Gotthard and Nylin (1995).

Did plasticity, as opposed to complete lack of environmental responsiveness, evolve? It is difficult to divorce this question from that of the evolution of degree of plasticity (see above), since even inorganic molecules, the undoubted predecessors of organic ones, show a range in degrees of responsiveness to environmental factors such as heat and contact with other substances (see also Stearns, 1982). Since the "reactions" of life's primal molecules to various kinds of environmental chemicals and events are central to explanations for the origin of life (e.g., Fox, 1980, 1984), implying that reactive molecules gave rise to living things, it is reasonable to conclude that phenotypic plasticity has been a property of living things since their origin.

Controversy over the nature and evolution of plasticity can be avoided by replacing abstract concepts of plasticity with discussions of concrete mechanisms, as pointed out in chapter 3. Regulatory mechanisms such as hormone systems are quantitatively variable traits with thresholds that underlie degree of plasticity. This means that plasticity is subject to selection as a quantitative genetic trait (see chapters 3 and 4). When theory is combined with knowledge of mechanisms, both are enriched (Bennett and Huey, 1990).

Does Behavior Take the Lead in Evolution?

The original meaning of "behavior takes the lead in evolution" is that behavioral plasticity can be the first step in evolutionary change, followed by morphology (e.g., Baldwin, 1896, 1902). The idea in modern terms can be paraphrased as follows: Given an array of potential responses (morphological, behavioral, plastic, and nonplastic), that with the highest sensitivity to environmental influence—the most plastic trait, usually behavior—is the first to change in a changing or extreme environment. The produced response subjects other attributes of the phenotype to an altered selective regime (e.g., a particular behavior may produce new physiological or morphological demands). This means that, given sufficient genetic variation in morphology, a recurrent behavioral response to the environment can affect the evolution of the structures affected or employed as a result. Thus, behavior being especially plastic, behavior must often take the lead in evolution.

It is difficult to challenge this statement, since it simply says that the most environmentally susceptible traits are the first to respond (a truism), and that the fitness-affecting variants thereby produced are subject to selection (a tautology). So it is not surprising that this seemingly obvious idea has been proposed repeatedly by students of animal behavior and evolution (e.g., Morgan, 1896a,b; Osborn, 1897a,b; Baldwin, 1902; Rau, 1933; Schmalhausen, 1949; G. Bateson, 1963; Evans, 1966; Mayr, 1958, 1959, 1963, 1970a,b, 1974a; P. Bateson, 1988; Wcislo, 1989). The same could be said for the physiological plasticity of plants: it is probably common that physiology is more plastic than morphology in plants (e.g., see Valladares et al., 2000) and therefore more responsive to environmental change.

What *is* surprising, however, is the fact that the idea that behavior can take the lead in evolution has been controversial, or considered to be of little importance in evolution (e.g., see discussion of the Baldwin effect, chapter 6). Indeed, some authors have argued the opposite, maintaining that morphological change occurs first, followed by affected behaviors. There is an inexorable logic behind this, too. As pointed out by Romer (1958), "For the development of a variety of responses it is necessary that an adequate series of sensory structures be evolved and that an efficient nervous system be present to enable appropriate . . . responses to be brought about" (p. 49). You cannot move an arm until you have one. Colbert (1958) reviewed the correlations between morphology and behavior in numerous vertebrates and concluded that "behavior patterns . . . can be completely understood only through knowledge of the morphological evolution of these animals" (p. 46). Emerson et al. (1990) discuss the complexity of morphological and behavioral interactions during adaptive evolution, giving examples that show behavior can remain unchanged even though the involved morphology changes, and that the adaptive significance of a behavioral (postural) change depends on the morphology of the individuals or taxa in which the change occurs.

Behavior and physiology are arguably the most highly responsive aspects of the phenotype in terms of speed of response and reversibility. While many features of morphology are environmentally responsive during their development (see chapter 3), they characteristically undergo relatively little change, compared with behavior, once they appear during ontogeny. Behavior and physiology, by contrast, are less often fixed by being expressed. Perhaps the distinguishing feature of behavior is that in many organisms it is highly, and reversibly, combinatorial and recombinatorial during a single life stage or lifetime. The same equipment is often used to produce different combinations of movement, so a very large and context-responsive repertoire of behavioral phenotypes can arise out of a single morphology. In fact, careful studies of movements show that for some behaviors an animal probably never repeats precisely in every detail any single movement (see references in Eberhard, 1990c).

Commonplace observations regarding the facultative adaptability of behavior have led to the belief that behavior takes the lead over morphology in adaptation to change (Wcislo, 1989). Other suggestive evidence came from precladistic comparative study indicating that certain morphological traits only occur in association with certain more widespread behavioral ones, giving the impression that behavioral specialization comes first. Mayr (1958) discusses several groups of birds where feather and head movements occur during courtship in many species, but only some of them develop crests and plumes that exaggerate these movements. This hypothesis is supported by the fact that there are no birds known that have developed a crest without the ability to raise it, suggesting that the behavior is a prerequisite for the morphological change (Mayr, 1963). In addition, there is an extremely consistent correlation, in hundreds of examples of sexual and social displays, between use of a body part in display, and structural modification of exactly the part used (e.g., see Brown, 1975; West-Eberhard, 1983, 1984; Eberhard, 1985, 1996). A typical example occurs in *Drosophila*. In 14 species of the *D. adiastola* species group, the male stands before the female, curves the abdomen upward and forward, and vibrates it while nearly touching the female's head. In 13 of the 14 species in this group, there is no obvious morphological modification, but in *D. clavisetae*, believed on the basis of chromosomal data to be phylogenetically highly derived relative to the others, the male abdomen is tipped with a brush of long clavate hairs that sweep over the female's head during the display (Carson, 1978). Such observations suggest a very strong correlation between recurrent behaviors and morphological trait exaggeration, but they only occasionally enable us to tell which came first during evolution.

Some recent critical discussions that take phylogeny into account (e.g., de Quieroz and Wimberger, 1993; Wenzel and Carpenter, 1994) interpret the idea that "behavior (or plasticity) takes the lead" to mean that behavior evolves more rapidly than morphology, not the same as asking whether behavior is the first step toward morphological innovation. For either question, phylogenetic analyses can help determine the sequence of changes in behavior and morphology, and the relative evolutionary lability of behavioral and morphological traits (e.g., see Gittleman et al., 1996, on quantitative traits; de Quieroz and Wimberger, 1993, on discrete traits). de Quieroz and Wimberger (1993) used traits selected by systematists for phylogenetic analyses, a preselection that could conceivably bias them toward low intraspecific variability, and it is uncertain whether such a bias would equally affect estimates of lability of the two kinds of traits.

These phylogenetic arguments address only the order-of-evolution side of the question. They do not assign or consider any causal role to intraspecific plasticity. Phylogenetic analysis shows that warning coloration (morphology) precedes gregarious behavior in butterfly larvae when both occur in the same lineage (Sillén-Tullberg, 1988). Gregarious behavior evolved independently 23 times, five of them without the prior evolution of warning coloration. But in the 18 lineages where both are present, gregarious behavior was accompanied by warning coloration in three and preceded by warning coloration in 15. Gregarious behavior did not evolve before warning coloration in any lineage. Marked evolutionary change in behavior often occurs without comparably obvious morphological change, as shown by the discovery of morphologically cryptic species using behavior (Evans, 1953; Alexander, 1962; Knowlton, 1993).

Morphology sometimes influences which individuals adopt which behaviors, and in this sense "lead" the generation of selectable variation in behavior. Finches with relatively large beaks, for example, specialize in hard-to-crack large seeds in times of drought (reviewed in Grant, 1986). The form of any behavior depends to a degree on the form of the morphology performing it (Evans, 1966). But the inevitable interdependence of behavior and morphology does not detract from the fact that phenotypic responsiveness focuses selection and evolution (Bateson, 1988).

Like plasticity in general (above), behavior may either slow or accelerate morphological evolution. If many behaviors are performed by the same morphology, then behavior can maintain morphological stasis. Then, if a single behavior pattern is fixed or predominant, this focuses selection on morphology in a single direction and leads to morphological change (see chapter 30, on punctuation). Perhaps the most convincing evidence that behavior precedes and leads morphological evolution comes from studies of alternative behavioral phenotypes showing accelerated morphological evolution following phenotype fixation (see chapter 21).

The idea that behavioral plasticity can promote stasis, and that fixation of a single behavior can promote morphological change, becomes important in later arguments regarding speciation, adaptive radiation, and punctuated patterns of evolution in animals (chapters 27, 28, and 30, respectively). It would be of interest to test the same principles with reference to plant physiology and plant morphological evolution.

As in other aspects of plasticity, the evolutionary role of behavior in relation to morphology depends on recurrence. If a particular movement is repeated in a particular context, and that context importantly determines the fitness effect of the morphology used, then the morphology may be modified under natural selection in that context. Hyperflexible behavior with few repeated elements is not expected to impose direction on morphological evolution.

Probably the best general approach to the relationship between behavior and morphology in evolution is to acknowledge that they will often evolve in concert, and that the aspect that takes the lead will be the one most flexible in producing a recurrent adaptive response. In grasshoppers (*Melanoplus femurrubrum*), for example, a behavior (consumption of relatively hard grasses) induces a morphological effect (larger head sizes), which in turn affects behavior (further increase in consumption of the hard grasses). This question is discussed further in chapter 28, on adaptive radiation.

Evolvability

One way to summarize the contents of this chapter is to say that the universal properties of phe-

notypes—modularity, flexibility, and the hierarchical organization of development by genetically complex switches—contribute to *evolvability*, the ability of organisms to evolve. The term "evolvability" is common in computational sciences and discussions of artificial life. Several evolutionary biologists have recently considered the evolution of evolvability per se, as if it were a character itself subject to selection. Most discuss evolvability with reference to genetic aspects (e.g., Vermeij, 1996; Wagner and Altenberg, 1996), defining it, for example, as "the genome's ability to produce adaptive variants when acted upon by the genetic system" (Wagner and Altenberg, 1996, p. 970). Such a definition returns us to a genetic view of evolution that is not explicit about the potential contributions of development and phenotypic plasticity to evolvability.

Kirschner (1992) and Gerhart and Kirschner (1997) define evolvability more broadly, as "the capacity to generate heritable, selectable phenotypic variation" (Kirschner and Gerhart, 1998, p. 8420). They emphasize that the mechanisms of flexibility and compartmentation, or modularity, that contribute to evolvability originate due to their advantages at the individual level (see chapter 3), then secondarily contribute to the ability of the clade to diversify and persist. Although Kirschner and Gerhart base this idea primarily on findings in animal cell biology and embryology, their conclusions regarding the contributions of flexibility, modularity, redundancy, and subunit versatility and reorganizability ("weak linkage" to any particular use) are likely to apply broadly to organisms, as suggested in the present chapter and documented in subsequent ones (see also West-Eberhard, 1998).

Kirschner and Gerhart (1998) suggest that the buffering effect of flexibility would have contributed to evolvability above the individual level by raising the number of nonlethal alleles that could be maintained in a population, thereby raising the genetic variation of the descendents of an individual and, at the population level, preventing extinction in the face of environmental change as well as permitting opportunistic use of new ecological opportunities. They point out that trends in some major groups of animals seem to have been toward increased modularity and flexibility, and that this may have contributed to their success. Precise tests of this idea will be difficult. Enumerative tests, for example, mapping increases in modularity on a cladogram and then comparing the diversification and survival of clades that contrast in this respect, are not usually satisfactory, because so many variables can intervene. One can often associate a spate of diversification, such as cichlid trophic versatility in the African lakes, with a particular phenotypic innovation, such as that of the cichlid pharyngeal jaw (Liem, 1973), but other factors, such as an unusual geologic history of lakes and strong sexual selection, have likely intervened, and it is virtually impossible to weigh their importance relative to the variable of interest. How often have such changes not led to diversification or enhanced survival? This is the usual problem with enumerative comparative tests, and in general they are unsatisfactory unless there is convincing information regarding effects of confounding variables, or the number of sampled clades is so large that confounding variables can be assumed likely to affect both sides of the comparison equally.

It seems likely that the universal properties of phenotypes have contributed to the survival and diversification of all known lineages of organisms. But the idea that evolvability per se evolves implies selection above the individual level (an individual cannot evolve), as well as the present or past existence of taxa with other, alternative phenotypic properties, (e.g., with less flexible or less modular phenotypes) that were less able to survive and diversify and therefore selected against at the lineage level. Woese (1998) imagines that some precellular organisms were organized as syncytia rather than internally modular. There seem to be no extant or fossil lineages known to completely *lack* the properties whose significance is to be tested, so a comparative test of the evolution of evolvability hypothesis presently cannot be done. Even fungi, whose mycelia are syncytia containing multiple nuclei, form partitions and undergo cellularization during reproduction (concisely reviewed in Buss, 1987), and their life histories are strikingly modular complex series of cyclically expressed distinctive stages. Either modularity and flexibility have been so strongly selected as features of evolvability per se that all competing modes of organization were eliminated prior to the evolution of fossilizable remains, or there is no alternative way to organize life as we recognize it.

Developmental Plasticity as a Solution to the Cohesiveness Problem

The principles of change introduced in this chapter contrast with others that have been invented by evolutionary biologists to explain subindividual diversification of the phenotype. Internal differentiation as a problem in *embryology* was raised and settled in terms of compartmentalized gene expression decades ago. Lillie (1927) and others raised the internal diversity question in developmental terms, asking how it can be that a single genome held in common by all somatic cells can account for differentated tissues and organs (called "Lillie's paradox" by R. Burian [personal communication], who attributes the term to Jane Maienschein). Burian regards the paradox as resolved by the end of the 1950s, citing a paper by Nanny (1958) showing how discrete gene expression can be the developmental cause of intraindividual differentiation. This idea has been extended from developmental biology to evolutionary biology in books by Bonner (1965, 1988) that refer to modular "gene nets" as explaining independent expression, dissociability, and independent selection and evolution of traits.

The cohesiveness problem has persisted, nonetheless, within evolutionary biology. The following three alternative solutions have been proposed, all of them flawed as general explanations of mosaic evolution: gradual evolution, speciation, and cell-lineage competition.

Gene-for-Trait Gradual Evolution

It is quite common to identify the diversity of traits with a diversity of the genes that underlie them— to consider that there is a gene "for" a trait. The idea of a gene for a trait assumes one-to-one genotype–phenotype mapping. This view may seem justified by mutation studies, where a particular phenotypic change is identified with a particular genetic one, and it is reinforced by a misconstrual of single-locus genetic models, which are theoretical constructs not intended as representations of development. Gene-for-trait thinking provides a simple and intuitively satisfactory image of intraindividual diversification and change, even if it is superficial and wrong.

The gene-for-trait idea is wrong because it treats a gene or mutation of major effect, for example, a single gene change that happens to decisively alter a response threshold, as if it were a complete control mechanism, which in most regulatory arrangements is a polygenically influenced complex trait (see chapter 6). Gene-for-trait thinking jumps past developmental mechanism and formative interactions among parts to see a trait as a direct product of a particular gene. This idea has been roundly criticized in the past (see chapter 1), and it is commonly acknowledged to be an oversimplification and a misrepresentation of the genetic architecture of traits (see especially Nijhout, 1990). The gene-for-trait view does not qualify as a satisfactory explanation of how multiple discrete traits can evolve over extended periods of time without disruption of harmonious integration.

Speciation

The speciational resolution of the cohesiveness problem (e.g., Mayr, 1963, 1988, pp. 423–438; Carson, 1985; Maynard Smith, 1983a; Eldredge and Gould, 1972; see chapter 30) sees all divergence as requiring a genetic revolution caused by a genetic splitting of a population into separate, isolated gene pools. By this hypothesis, phenotypic innovation requires speciation, and most phenotypic innovation occurs at the time of speciation, or branching of a phylogenetic tree.

According to the speciational hypothesis, a cohesive coadapted gene pool can give rise to change only via gene-pool splitting. This hypothesis is contradicted by occurrence of marked diversification within species that obviously does not involve speciation. Workers and queens of social insects are divergent phenotypes, but speciation has not occurred between them. Larvae often show species-specific specializations that are not reflected in adults, indicating that juveniles evolve semi-independently of adults without speciation. They occupy their own distinctive and specialized niches (Darwin, 1859 [1872], p. 67), but they do not constitute different species. The speciational hypothesis does not explain how subindividual diversification and mosaic evolution can occur. It explains divergence between populations, not divergence expressed within individuals.

The speciational hypothesis is also contradicted by the phenomenon of mosaic evolution, or different rates of evolution of different subindivudual traits. Mayr (1988) recognizes that mosaic evolution of different life-stage phenotypes is "the result of different selection forces" acting on each and that "different portions of the genotype are silent at each stage" (p. 491). This would seem to accord with the modularity hypothesis and contradict the idea that change requires speciation, unless one visualizes that the variation and selection responsible for adaptive evolution operate only during episodes of speciation, an untenable idea (see chapter 21).

Cell-Lineage Competition   The cell-lineage competition hypothesis (Buss, 1987) is the most carefully argued and explicit developmental solution proposed to explain the evolution of subindividual trait diversification. It sees specialization of parts as a product of competition among genetically different somatic cell lineages. Genetic innovations arise during ontogeny in somatic cells, then are transmitted across generations if those cells can gain access to the germ line. Buss shows convincingly that cohesive individual development ("individuality") is enforced in many animals by early germline sequestration ("Weismann's barrier") and maternal control of early embryogenesis, mechanisms that prevent somatic-cell access to the germ line. This section does not challenge that idea, but concerns only the cell-lineage competition hypothesis of diversification, which sees the specialization of subindividual phenotypic traits as a *departure* from individuality. By this hypothesis, developmental innovation evolves when a selfish mutation gives one cell lineage an advantage over others for incorporation in the germ line during a "window of heritability" between the end of maternal control and the onset of sequestration.

By the Buss hypothesis, all appreciable innovation preceded the evolution of early germline sequestration; the specializations of modular traits are a heritage of ancestral competition among cell lineages; developmental integration and differentiation are "opposing processes" (Buss, 1987, pp. 53–55); and organisms such as plants, fungi, and colonial invertebrates that lack early sequestration and maternal control of early development are the only ones where appreciable developmental innovation is possible. These "non-Weismannian" organisms therefore possess "enormous phenotypic plasticity" and "ecophenotypic variation," in contrast to metazoans with early sequestration and strong maternal control of early development (Buss, 1987, p. 24).

The following observations are inconsistent with the cell-lineage competition hypothesis:

1. Multiple specialized phenotypes have originated late in the phylogeny of many animals that are fully individual by Buss's criteria. Some of the most spectacular examples of specialized alternative phenotypes producible by the same genotype are in social Hymentoptera and Isoptera (termites), aphids, and hypermetamorphic insects (see figure 2.1), organisms where Weismann's barrier is well established. Such specialialized diversification, whether subindividual or social, is associated not with a *lack* of individuality, but instead with a high degree of individuality and integration, such that cooperation with minimal competition among parts is assured (West-Eberhard, 1979; Leigh, 1983). Many

mechanisms other than germline sequestration may act to selfishness among parts, including mechanisms assuring honest meiosis and, in social organisms, mechanisms that suppress competitive reproduction (Leigh and Rowell, 1995; Moritz and Southwick, 1992). Others reviewed by Buss (1987) are given in item 5, below.

2. Inherited genetic innovation specific to early ontogeny does not require a window of opportunity between the end of maternal control and the occurrence of germline sequestration, as pointed out by some previous authors in discussing the Buss hypothesis (Raff, 1988; Wolpert, 1990; Maynard Smith and Szathmáry, 1995). Instead, mutations have local, early or late effects due to cell-lineage-specific gene *expression* even though they have originated in a sequestered germline and are from the outset present in the nuclei of all somatic cells, including those where they are not expressed. As a rule, all genes are carried by all cells, a generalization thoroughly examined and confirmed for a wide variety of tissues and organisms by Davidson (1986). So traits that are specific to a particular cell lineage not due to lineage-specific mutation but due to lineage-specific expression (also extensively reviewed by Davidson, 1986; John and Miklos, 1988).

3. Competitive-sounding cell interactions such as differential cell death (apoptosis), induction, and competence to be induced, mentioned by Buss as evidence for a legacy of competition among cell lineages, do not resemble Darwinian reproductive competition. The cells concerned are genetically identical, so cell lineages are not expected to compete (Wolpert, 1990), and the interactions among cells of different lineages is not as expected if competition were involved. Differential cell death in the developing nervous system, for example, involves different cells of the same tissue (nerve cells), not differential mortality of cells from diferent cell lineages, and cells that fail to survive are those that *fail* to reach the tissue that they will innervate (Gerhart and Kirschner, 1997), not those that contact a competing lineage. Contact with another tissue promotes survival rather than death, the opposite of what one might expect if cell death were a legacy of competitive interaction between the two cell lineages involved. Cell death has been experimentally demonstrated to be a self-induced property of the

dying cell itself, regardless of its contact with other tissues, with contact in some cases *inhibiting* cell death (Raff and Kaufman, 1983). If anything this sort of differential mortality could be interpreted as altruistic, due to strong selection at the individual organism level (see Edelman, 1988). The altruism (kin-selection) interpretation is supported by the fact that apoptosis may have originated in bacteria as a defense mechanism against infection (Vaux and Strasser, 1996)— as suicide among kindred cells, not murder among competitors. In these and other examples of somatic selection (see chapter 3), there is no genetic difference between the entities that differentially survive or multiply.

4. Developmental innovations, including those in the adult stage, are not necessarily traceable to cell-lineage specialization in the early embryo, as implied by the "preformist development" of adults proposed by the cell-lineage competition hypothesis (Buss, 1987, p. 131). As extensively reviewed in part II of this book, novel traits can originate by reorganization of development within a life stage, including the adult stage, and they can involve, in addition to the origin of new cell types, duplications, deletions, and reversions at higher levels of organization. The Cnidaria show spectacular morphological and life-stage diversity (e.g., see Brusca and Brusca, 1990) despite being, along with the sponges, the "histologically least elaborate" of metazoans (Buss, 1987, p. 141).

5. Perhaps most telling for the cell-lineage competition hypothesis, I know of no case where somatic cell specialization brings enhanced access to the germ line, or where differently specialized somatic cells or their descendent cells produce gametes in competition with each other (Buss cites none). Specialized somatic cells are characteristically irreversibly nonreproductive. As a rule, previously undifferentiated, *totipotent* cells give rise to gametes, including in "non-Weismannian" organisms (Buss, 1987, p. 101). Examples include the I cells of hydroids, the neoblasts of platyhelminths, the archeocytes of sponges, and the meristems of plants (Buss, 1985, 1987). In colonial invertebrates and fungi, reproductive regions, are facultatively walled off from somatic regions so specialized somatic cells do not participate in reproduction. In fungi, all mycelia are evidently totipotent in that they can become repro-

ductive when contacted by foreign mycelia. These organisms illustrate a variety of alternative mechanisms for preventing reproductive competition among specialized somatic cells and assuring individuality without the kinds of maternal control and sequestration observed in "Weismannian" species. They do not illustrate the evolution of somatic diversification due to reproductive competition and access to gametes by specialized somatic cell lineages. So internal phenotypic diversification does not appear to be a product of cell-lineage competition and access to the germ line, even in organisms without germline sequestration.

The cell-lineage competition hypothesis predicts a greater degree of subindividual diversification in organisms where individuality is poorly developed. The present, modular-diversification hypothesis predicts the opposite: organisms and social groups showing the highest degree of individuality or multicellular integration should show the highest degree of specialized internal diversification, and the most complex internal divisions of labor. This is expected because the subindividual components, being highly subordinated to the whole, can increasingly specialize and depart from a totipotent condition that in the past enabled them to survive and reproduce as a performer of multiple functions, independent of the group or individual. This reasoning seems to apply to insect societies, where the most individuated "supraorganismic" genetically homogeneous societies such as those of army ants and some termites have the highest development of specialized morphological castes (e.g., see Jeanne, 1980). It also applies to multicellular organisms, where the greatest number of specialized cell types are found in vertebrates, not in colonial invertebrates or plants (figure 7.9).

Plants and other modular organisms can attain a high degree of cohesive individuality without germline sequestration to prevent propagation of selfish genes. One explanation for this is that a high degree of cooperation (individuality, or cohesiveness), can be achieved even with some genetic heterogeneity through mutualism, which does not depend upon relatedness although it has an extra genetic payoff among relatives (e.g., see West-Eberhard, 1978b; Michod, 1997a,b). In addition, many modular organisms reproduce clonally, keeping relatedness of cooperating cells high. But the lack of a division of labor between germline and somatic cell lineages in plants means that meristem cells must remain totipotent for both somatic and re-

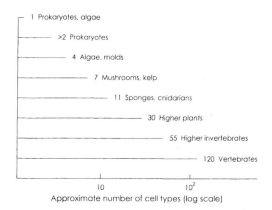

Fig. 7.9. Approximate log number of cell types in different types of organisms. After Raff and Kaufman (1983) and Bonner (1988).

productive functions. This may help account for the lesser proliferation of specialized cell types in plants compared to animals (Brink, 1962). Similarly, social insects such as *Polistes*, where adult females are totipotent for worker and queen development, have fewer specialized castes than do highly evolved social insects whose caste totipotency is lost in the larval stage.

I have discussed the cell-lineage hypothesis in some detail because it proposes a developmental solution to the cohesiveness problem and Lillie's paradox that contrasts with the one presented in this chapter, based on modular phenotypic organization in relation to gene expression, and selection. Buss (1987, p. 117) considers the complex epigenetic programs of metazoan development inexplicable within the neo-Darwinian tenet of the individual as the sole unit of selection. This chapter has shown how the cohesiveness problem is resolvable within the conventional paradigm, if one considers phenotypic traits subunits of gene expression and of selection, albeit restricted in their independent selection and evolution by phenotypic and genetic correlations among them (see also Cheverud, 1982, 1984b; West-Eberhard, 1996; chapters 4 and 5). The semi-independent expression of traits facilitates their independent specialization and permits diversification without departures from individuality.

Gene-expression modularity and semi-independent selection of independently expressed traits solve the internal diversity and cohesiveness problem in conventional neo-Darwinian terms, without contradicting what is known about genetics and development. It is a resolution of Lillie's paradox more in tune with the facts of development than the gene-for-trait, speciational, and cell-lineage-competition hypotheses proposed in the recent past. The importance of mosaic evolution was emphasized in debates over recapitulation, which implied evolution by terminal addition (e.g., de Beer, 1958). Mosaic evolution has been reemphasized in some recent evolutionary studies of development (e.g.,

Wray, 1992b). Here I stress the link between mosaic evolution and the genetic theory of natural selection: the principle of independent selection of independently expressed traits means that separate parts of the mosaic are semi-independently screened by selection. The modular gene expression of the parts means that they can respond to selection and evolve semi-independently as well.

8

Darwin's Theory of Development and Evolution

Everything new is the thoroughly forgotten old.
Russian saying

The only thing new in the world is the history you don't know.
Harry S Truman

Does the world progress? Or does it merely shuttle back and forth like a ferry?
Oates (1989, p. 12), after Julian Barnes

The concepts discussed in preceding chapters, and applied in the chapters that follow, capitalize on research in modern genetics, physiology, ethology, phylogenetics, and molecular biology, among other fields. It is impressive to realize, therefore, that Darwin reached similar conclusions more than a century ago, when he proposed a theory of the phenotype based on dissociable and independently evolving modular subunits; and a molecular theory of inheritence (pangenesis). Darwin avoided the false dichotomy between nature and nurture by insisting that inheritence includes both the transmission and the expression of traits. His views on the inheritance of acquired traits were influenced as much by false facts as by lack of progress in genetics. Seeing the achievements and shortcomings of Darwin's theory of development and evolution highlights the progress that makes a new synthesis possible. It is also a humbling indication of our collective amnesia regarding insights of the past when they do not flow immediately into the narrowed stream of progress at a particular time.

Biologists are fond of saying that Darwin was misled by an inadequate theory of inheritance—his theory of "pangenesis"—and that this was remedied by the rediscovery of Mendel's experiments in 1900:

As long as no coherent theory of heredity existed, the basis of natural selection could not be understood. Darwin's theory of pangenesis was an unfortunate anomaly. It was almost his only venture into the field of pure speculation. . . . One might speculate whether Darwin would have formulated his theory of pangenesis if he had been aware of Mendel's experiments. (J. L. Stebbins, 1977, p. 14; see also, e.g., Mayr, 1963, p. 10; 1966, p. xxvi; Selander, 1972, p. 185)

I have argued that some aspects of development can be related to natural selection by visualizing the phenotype as a mosaic of semi-independent subunits that are dissociated, recombined, lost, and recalled during evolution and that adaptive plasticity plays a role in evolutionary change. Darwin insisted on the same view in *The Variation of Animals and Plants under Domestication* (1868), where he gave the most complete exposition of his ideas about development and evolution. In a section titled "The Functional Independence of the Elements or Units of the Body," Darwin (1868a [1875a], p. 364) wrote that "the whole organism consists of a multitude of elemental parts, which are to a great extent independent of one another. Each organ . . .

has its proper life, its autonomy; it can develop and reproduce itself [in descendent individuals] independently of the adjoining tissues" (p. 364). And, "[T]he body consists of a multitude of organic units, all of which possess their own proper attributes, and are to a certain extent independent of all others. Hence it will be convenient to use indifferently the terms cells or organic units, or simply units" (p. 366). Furthermore, Darwin insisted that an acceptable theory of inheritance had to include development: "Two distinct elements are included under the term 'inheritance'—the transmission, and the development of characters; but as these generally go together, the distinction is often overlooked" (Darwin, 1871 [1874], p. 583). This was not a new idea: Aristotle had concluded nearly 2000 years before that what is inherited is "a potentiality to develop" (Simpson and Beck, 1965).

Far from being "purely speculative," Darwin's developmental view of the phenotype was supported by comparative and experimental data that was extensively cited and perceptively discussed by Darwin, especially in *Animals and Plants under Domestication*. That book contained so much fundamental data regarding inheritance that Haldane (1959) considered it "still indispensable to geneticists" (p. 103).

Darwin used several kinds of evidence to support his modular view of the phenotype, including broad comparisons within taxa to show that despite divergence and shuffling of traits, homologous subunits can still be identified, having maintained integration amidst change:

> According to the belief now generally accepted by our best naturalists, all the members of the same order or class, for instance, the Medusae or the Macrourous crustaceans, are descended from a common progenitor. During their descent they have diverged much in structure, but have retained much in common; and this has occurred, though they have passed through and still pass through marvellously different metamorphoses. This fact well illustrates how independent each structure is from that which precedes and that which follows it in the course of development. (Darwin, 1868b [1875b], p. 361)

The dissociability and independence of phenotypic subunits was evident to Darwin not only from observations on life-stage differences during development, but also in the patterns of expression of sexually dimorphic traits, polyphenisms, regulatory mutations ("transpositions"), and reversions, which revealed that complex traits could be lost, transferred, and recalled as sets. Clearly, the materials of their inheritance (which he called "gemmules") could be carried by an individual even though not expressed as a set, as shown by the following passages from Darwin (1868b [1875b]):

> Most, or perhaps all, of the secondary characters, which appertain to one sex, lie dormant in the other sex; that is, gemmules capable of development into the secondary male sexual characters are included within the female; and conversely female characters in the male: we have evidence of this in certain masculine characters, both corporeal and mental [behavioral], appearing in the female, when her ovaria are diseased or when they fail to act from old age. In like manner female characters appear in castrated males, as in the shape of the horns of the ox, and in the absence of horns in castrated stags. . . . The curious case formerly given of a hen which assumed the masculine characters, not of her own breed but of a remote progenitor, illustrates the close connection between latent sexual characters and ordinary reversion. (p. 393)

> With those animals and plants which habitually produce several forms, as with certain butterflies [Batesian mimics] described by Mr. Wallace, in which three female forms and one male form co-exist, or, as with the trimorphic species of Lythrum and Oxalis, gemmules capable of reproducing these different forms must be latent in each individual. (p. 394)

> The principle of the independent formation of each part . . . throws light on a widely different group of facts, on which on any ordinary view of development appears very strange. I allude to organs which are abnormally transposed or multiplied [in modern terms, homoeotic mutations. Many examples follow, including a chicken with the right leg articulated to the left side of the pelvis; supernumary fins in goldfish; limb doubling in amphibians and insects; and a crustacean whose eyestalk developed an antenna in place of an eye]. (p. 385)

Darwin attached special importance to explaining "reversions," by which he meant not only the sudden appearance of ancestral traits long lost (as we would apply the term today), but also patterns of transmission by which traits of grandparents appear in offspring without appearing in their parents—a phenomenon we would explain in terms of either recessive alleles (in the case of Mendelian traits), incomplete penetrance, or recombination with polygenic inheritance: "The principle of reversion, recently alluded to, is one of the most wonderful of the attributes of Inheritance. It proves to us that the transmission of a character and its development, which ordinarily go together and thus escape discrimination, are distinct powers" (Darwin, 1868b [1875b], p. 368).

Darwin wrote quite clearly about dual nature of regulation, and the extreme difficulty of disentangling environmental and genetic or "constitutional" causes of variation:

[A]lthough every variation is either directly or indirectly caused by some change in the surrounding conditions, we must never forget that the nature of the organisation which is acted on, is by far the more important factor in the result. We see this in different organisms [species], which when placed under similar conditions vary in a different manner, whilst closely-allied organisms under dissimilar conditions often vary in nearly the same manner. . . . [Conditions] will cause the individuals of the same species either to vary in the same manner, or differently in accordance with slight differences in their consitution . . . the whole organisation becomes slightly plastic. Although each modification must have its own exciting cause, and though each is subjected to law, yet we can so rarely trace the precise relation between cause and effect, that we are tempted to speak of variations as if they arose spontaneously. (Darwin, 1868b [1875b], pp. 415–416)

Darwin's ideas on development became more sophisticated as he got older, especially as regards the relation between transmission and development, or expression of traits. For example, in the section of the *Origin* discussing the morphological and behavioral distinctiveness of social insect castes, he initially wrote, in the first edition (1859):

As ants work by inherited instincts and by inherited tools or weapons, and not by acquired knowledge and manufactured instruments, *a perfect division of labour could be effected with them only by the workers being sterile; for had they been fertile, they would have intercrossed, and their instincts and structure would have become blended.* (1859, p. 242; emphasis mine)

In the fourth edition (1866) this was amended in such a way that the italicized, erroneous portion was omitted (see Peckham, 1959, p. 421). As written, the omitted sentence contradicted the fact, acknowledged elsewhere by Darwin (see above), that polymorphic divergence (e.g., between the two sexes and in the mimicry morphs of butterflies) does not require genetic isolation (barriers to interbreeding).

In retrospect, there were three main defects of Darwin's "pangenesis" theory of inheritance: (1) the presumption that there was a gemmule corresponding to every trait and specific to one and only one trait; (2) the idea that copies of all gemmules are transmitted to the next generation; and (3) the idea that somatic changes occurring during the lifetime of an individual could locally alter gemmules proceeding from the affected organ and then be passed to subsequent generations via gemmule replicates contained in the gametes. Inheritance of acquired traits was the best explanation Darwin could think of for the origin of adaptive variants, and it was supported by some experiments and contradicted by others, as Darwin discussed. For example, he noted the "at first sight fatal" objection that repeated mutilations, such as cutting off the tails of dogs or male circumcision by generations of Jews, failed to alter inherited traits (1868b [1875b], p. 391). But he considered (p. 392) that perhaps the great accumulation of gemmules from many previous generations was still able to overwhelm the fewer altered gemmules. There was seemingly careful experimental work on guinea pigs, which, following amputation and disease of toes, in 13 cases gave birth to young with defective feet lacking the corresponding toes. And there were reports of cows that, after loss of a horn, developed a severe infection at the side and then gave birth to calves lacking the corresponding horn.

How, then, in accordance with our hypothesis can we account for mutilations being sometimes strongly inherited, if they are followed by diseased action? The answer probably is that all the gemmules of the mutilated or amputated part are gradually attracted to the diseased surface during the reparative process, and are there destroyed by the morbid action. (p. 392)

It was a heroic attempt to push a hypothesis to explain contradictory evidence, some of it erroneous, which he had done so successfully with the idea of natural selection. In this case, Darwin was seriously misled by not always being able to distinguish between reliable and unreliable reports at a time when the causes of variation were still a mystery. As he once remarked,

> False facts are highly injurious to the progress of science, for they often long endure; but false views, if supported by some evidence, do little harm, as everyone takes a salutary pleasure in proving their falseness; and when this is done, one path towards error is closed and the road to truth is often at the same time opened. (1871 [1874], p. 909)

Perhaps the insight of Mendel that most would have helped Darwin was the realization that each parental gamete contains hereditary material for only one of two or more alternative character states (now known as alleles at the same locus). Thus, although Darwin saw that traits could be "latent" (transmitted but not expressed), he did not realize that sometimes lack of expression means that not all traits present in ancestors are transmitted to all descendents. But Mendel's discoveries did not completely address the problem of inheritance, as conceived by Darwin. Mendel's results did not explain blending inheritance, documented by Darwin in his own experiments on inheritance in peas (e.g., Darwin, 1868a,b [1875a,b]; see also several papers by Darwin reprinted in Barrett, 1977a,b), and they did not treat the developmental side of inheritance so important to Darwin as he endeavored to explain patterns (e.g., sex-limited and stage-specific inheritance) that we would term aspects of gene *expression.*

Darwin knew that his theory of inheritance was tentative and vulnerable. He presented it (1868b [1875b], pp. 349–350) as the "provisional hypothesis of pangenesis" with a quote from Whewell, an historian of science: "'Hypotheses may often be of service to science, when they involve a certain portion of incompleteness, and even of error.' Under this point of view I venture to advance the hypothesis of Pangenesis." It was designed to explain phenomena of transmission and development for which "no other hypothesis . . . has as yet been advanced." But Darwin's pangenesis (for a history of earlier versions, see Zirkle, 1946) failed to explain even some of Darwin's own insights regarding inheritance, such as his explanation of the worker phenotype in social insects. Darwin realized that the evolution of specialized behavior and morphology in the sterile worker caste could be explained by selection at the level of the family or the fertile parents, where the reproducing members of a colony pass traits to their worker offspring even though not expressed in themselves.

Although not mentioned by Darwin, this explanation conflicts with the pangenesis hypothesis, which requires that trait-specific gemmules accumulate in specialized body parts (in this case, in the workers), then get passed to the gametes and the next generation, a step that is impossible in the sterile workers. Under pangenesis, there was no mechanism for factors accumulated in the workers to get passed to the reproductive queens.

As a subsidiary of the theory of natural selection, pangenesis (including the developmental aspects not explained by Mendelian genetics until decades later) was discarded without seriously threatening the selection hypothesis itself. One wonders what might have happened if Darwin had made the theory of inheritance preeminent, and selection subsidiary, for scientific "progress" seems so often to involve a high mortality of correct and well-supported subsidiary ideas each time some even trifling reason can be found to discard the whole. As Darwin (1859) noted, theories are readily rejected by "[a]ny one whose disposition leads him to attach more weight to unexplained difficulties than to the explanation of a certain number of facts" (p. 482). Thus, for example, pangenesis was "disproved" by Francis Galton in 1871 (see response by Darwin in Barrett, 1977b, p. 165) by showing that blood transfusions would not transmit characters from one individual to another, showing that gemmules were not carried by the blood. Darwin replied that "I have not said one word about the blood . . . which can form no necessary part of my hypothesis." But, he continued, "from presenting so many vulnerable points [such as having no known medium of gemmule transport], its life is always in jeopardy; and this is my excuse for having said a few words in its defence." For discussions of pangenesis and its critics, see Ghiselin (1969, 1975) and Mayr (1982a).

A serious blow to pangenesis was dealt by Weismann's (1892) demonstration (see Mayr, 1982a) that germ cells descend directly from the germ cells of the previous generation, in many animals being separated early in development. Somatic cells are derived from germ cells but in organisms with germline sequestration do not give rise to them (for a discussion of exceptions, and alternative ways of isolating sexually reproductive cells from specialized somatic functions, see Buss, 1987). So the mechanism of somatic development has to be seen as separated and independent of transmitted materials, and somatic processes cannot usually influence the gemmules (genes) as visualized by Darwin.

Modern biologists think that the fate of the "Lamarkian" idea of pangenesis was settled by

Weismann's insight, and they seldom look back to see the broader question that Darwin was trying to solve. Weismann's barrier and the demise of the "transportation aspect" of pangenesis (see Mayr, 1982a, p. 693f) left unsolved the problem of how to explain the observation that environmentally altered phenotypes, being subject to selection, are preserved and multiplied across generations. Twentieth-century evolutionary biology passed over this difficulty and relegated the environment to a nondevelopmental, purely selective role (e.g., see Ghiselin, 1975, p. 52).

Much of this book is dedicated to solving what might be called the "pangenesis problem"—the problem of showing how environmentally induced variation *does* affect the developmental side of inheritence and evolution, even though there can be no direct adaptive environmental effect on the materials of inheritance, the genes. We have rejected this aspect of Darwin's theory, but a rereading of Darwin shows that we rejected far too much, not only misinterpreting many of Darwin's words, but also failing to explain many of the major phenomena he recognized as significant. His sections on "direct and definite action of the external conditions of life" and on "use and disuse" (1868b [1875b]), for example, refer not to Lamarkian inheritance but to what we would now call phenotypic plasticity. His use of "inheritance," as in "sex-limited inheritance" and "inheritance at corresponding periods of life," meant what we would now call sex-limited and life-stage specific *expression*. Darwin was trying to solve Lillie's paradox and the cohesiveness problem using molecular (gemmule) genetics a century before the discovery of DNA.

That Darwin was far from being a Lamarkian is shown by his discussion of the origin of selectable variation: "Variability is the necessary basis for the action of selection, and is wholly independent of it" (Darwin, 1871 [1874], p. 916).

We know not what produces the numberless slight differences between the individuals of each species. . . . With respect to the exciting causes we can only say, as when speaking of so-called spontaneous variation, that they relate much more closely to the constitution of the varying organism than to the nature of the conditions to which it has been subjected. (p. 442)

In other words, adaptive evolution involves first variation and then selection—it is not the result of acquired traits that both originate and spread due to uses and disuse; and the variations that become adaptations more likely arise from within the organism than as a result of the conditions of life. Elsewhere, Darwin argued specifically against Lamarkian inheritance (and, incidentally, against pangenesis, as already noted):

peculiar habits confined to the workers or sterile females [of social insects], however long they may be followed, could not possibly affect the males and fertile females, which alone leave descendents. I am surprised that no one has hitherto advanced this demonstrative case of neuter insects, against the well-known doctrine of inherited habit, as advanced by Lamarck. (Darwin, 1859 [1872], p. 207)

In development and evolution, the environment is a source of building blocks and cues, not just an agent of selection. So environment does, as Darwin perceived, take an active role in the "programming," informational side of change. These matters were central to pangenesis. They were bypassed by Mendelian genetics and its descendent, twentieth-century evolutionary biology.

It would be foolish not to acknowledge that Weismann's insight, combined with the Mendelian focus on transmission genetics, greatly clarified thinking about inheritance beyond what Darwin and his contemporaries were able to achieve. Lewontin (1992) has identified the "causal rupture between the processes of inheritance [transmission] and the processes of development" as the essential feature of Mendel's contribution. But Darwin had already recognized this distinction and was struggling to explain both. It is the "rupture" between the two that makes a reunification necessary now.

Perhaps pangenesis contributed to the progress of genetics by focusing Weismann's attention on this problem (e.g., Weismann, 1892 [1893] indicates this). If so, Darwin clearly would have been pleased at some utility from this idea, which he had thought about so much that he "had lost all power of judging it" (1867 letter to Lyell, in F. Darwin, 1958, p. 281).

With the demise of pangenesis and the rise of modern genetics, considerations of development seemed to fade from the forefront as a complement to studies of inheritance and selection. In 1930, J. H. Woodger (cited by Buss, 1987) noted, "There are books in plenty on experimental embryology but none on theoretical embryology," nor, as Woodger went on to consider, on embryology in relation to natural selection. Books relating ontogeny and phylogeny do not really fulfill this role, for they focus on pattern of diversification, not selection. How could two such exquisitely intertwined subjects as development and selection become so thoroughly divorced in the minds of biologists? Mayr (1982a) has offered a clue:

[A]ll geneticists from Nageli and Weismann to Bateson failed to develop successful theories of heredity because they attempted to explain simultaneously inheritance (transmission of the genetic material from generation to generation) and development . . . It was Morgan's genius to put aside all developmental-physiological considerations . . . and concentrate strictly on the problems of transmission. His pioneering discoveries from 1910 to 1915 were entirely due to this wise restriction. (p. 832).

Like Lewontin (see also Buss, 1987, p. 10), Mayr sees the divorcement of transmission genetics and embryology as indispensible for the progress of understanding of inheritance. The fact remains that what was wise for genetics was not a completely happy solution for the study of evolution. It has taken more than 120 years to come back to where Darwin left off, in constructing a theory of evolution that gives attention to both the transmission and the expression of genes.

A major synthesis is possible now due to convergent progress in disparate fields: the molecular biology of development and gene expression makes it possible to pinpoint the genes actually expressed in particular phenotypes. Studies at the cell and cell interaction level show how flexible developmental processes are achieved, revealing the mechanisms of production and accommodation of environmentally influenced variation. Physiological and hormone studies link environmental cues to expressed genes and show how gene expression is coordinated within phenotypes. Behavioral ecology of both plants and animals, and sociobiological studies demonstrate the adaptive plasticity of individual organisms and its relation to fitness and reproductive success (natural selection). Population and evolutionary genetics draw on decades of progress in experimental and theoretical genetics to show how the fitness consequences of adaptability are connected to natural selection and evolutionary–genetic change. And cladistic analysis makes it feasible for the first time to use consistent methods to trace phenotype origins and see with increasing accuracy the broad pattern of phylogenetic change. Meanwhile, the boost to population genetics and classical systematics given by molecular methods enables finer resolution of intra- and interspecific variation and the structure of evolving populations. Only in the last few decades have these fields grown to the point where their interconnections are obvious and their findings detailed and confident enough to be put together on a broad scale like that attempted in this book.

Darwin was a close correspondent of Weismann and wrote a preface to one of his books, published the year Darwin died. There, Darwin makes it clear that he regarded the evolutionary role of environmentally mediated variation as the greatest unsolved problem of evolutionary biology:

Several distinguished naturalists maintain with much confidence that organic beings tend to vary and to rise in the scale, independently of the conditions to which they and their progenitors have been exposed; whilst others maintain that all variation is due to such exposure, though the manner in which the environment acts is as yet quite unknown. At the present time there is hardly any question in biology of more importance than this of the nature and causes of variability. . . . (Darwin, 1882, p. vi)

These were among Darwin's last published words. They are a prophetic link to a new age of biology, just beginning, where the genomic and environmental causes of variation can be related in concrete terms.

PART II

THE ORIGINS OF NOVELTY

9

The Nature and Analysis of Phenotypic Transitions

The next 10 chapters show how novel phenotypes can originate by developmental reorganization of preexisting phenotypes. Tracing the evolutionary transitions between phenotypic states requires evidence of homology between novel traits and their ancestral components, and phylogenetic evidence for the direction of change (polarity). Discussions of evolutionary transitions need to distinguish between the factors that initiate change, such as mutation or environmental factors, and the sources of novelty in the ancestral phenotype; between development and selection as agents of change; between the origin of novel phenotypes (developmental branching) and the origin of new species (lineage branching); and between development and saltation as explanations for novelty. Complex novelties can be stimulated by a simple new environmental or genetic input such as a mutation. But regulation by a single switch does not demonstrate one-step, saltatory origin. Rare phenotypes (anomalies) are valuable and unjustly neglected indicators of trait origins, for recurrent anomalies are material for selection and evolution in new directions.

Introduction

Part II is about origins: how do new traits arise from old phenotypes? People of all ages are fascinated by the question of origins. Origins are the common concern of evolutionists and creationists, of ethnic historians, of Mormon geneologists and the Daughters of the American Revolution, of adopted children searching for their biological parents—indeed, of all who have wondered where Johnny got his patience, his sense of humor, or his big nose. Darwin was a clever publicist when he titled his most famous book *The Origin of Species*. He touched deep human chords by discussing not only the origin of species but the origin of marvellously complex morphological and psychological traits—specialized limbs, sexual behavior, intelligence, heroism, and the vertebrate eye, to mention just a few.

Research on selection and adaptation may tell us why a trait persisted and spread, but it will not tell us where a trait came from. This is why evolutionary biology inevitably intersects with developmental biology, and why satisfactory explanations of ultimate (evolutionary) causation must always include both proximate causes and the study of selection. Novel traits originate via the transformation of ancestral phenotypes during development.

This transformational aspect of evolutionary change has been oddly neglected in modern evolutionary biology, even though it is an integral part of human curiosity about origins in other fields. From classical mythology to modern-day childrens' books, origins are explained in terms of transformations of the phenotype, alongside attention to developmental mechanisms and adaptive functions. Consider this excerpt from *The Apeman's Secret* (Dixon, 1980), a Hardy Boys adventure book:

[T]he Apeman hated cruelty of any kind. Whenever he saw crooks or villians do something nasty to a helpless victim, he would fly into a rage. This would change his body chemistry and cause him to revert to the savage state. Then, with bulging muscles and fearsome growls, he would beat up the villians and wreck their criminal plot. . . . (p. 7)

The Apeman story contains all three elements that are essential to an evolutionary explanation—what Huxley (1942, p. 40) called "the three aspects of biological fact": *origin* (the initial induction and phenotypic source of a novel phenotype, given a presumed ancestral state), *polarity* (the temporal order of appearance of alternative states), and *spread* (the selective context, fitness effect, or adaptive function that causes the trait to increase in frequency or be maintained).

Of these three elements of evolutionary explanation, transformational origin is the most neglected. There is a large recent literature that examines the polarity, or the phylogenetic order of change in traits, and on the selective basis of change. The recent neglect of phenotype origins contrasts with the extensive treatment given this topic in the older literature, where the situation was reversed: in the early twentieth century, theories of selection and phylogenetics were relatively primitive or unattended, and embryology was at the forefront of biology. Research on the relationship of ontogeny to phylogeny has a long history, and most of the fundamental embryological concepts (e.g., von Baer's laws) relevant to this relationship predate Darwin's *Origin of Species* (Wake and Roth, 1989; reviews in Gould, 1977; Rieppel, 1994). But the recent neglect of this aspect of biological fact has made it necessary to propose anew that "transition morphology," taken broadly to mean the origins of phenotypic forms, be adopted as an important area of contemporary research (Galis, 1996a).

Chapter 7 introduced the idea that modular dissociability can give a head start to adaptive evolution. The chapters of this section show how a phenotypic *novelty*—a trait that is new in composition or context of expression relative to established ancestral traits—can grow out of previous organization, and why a complex evolutionary change in the phenotype need not depend on a long series of genetic mutations. Because evolution can occur by reorganization, the phenotypic divergence between populations or species need not be proportional to the genetic distance between them. The examples also suggest that development, while it affects the form of novel traits, is not a severe constraint for evolution. Modular traits are shuffled, deleted, retrieved, and recombined, in so many ways and at so many levels of organization that the possibilities for reorganization are enormous, and each new arrangement makes further permutations of change possible.

These chapters describe the nuts and bolts of evolutionary change—the preexisting phenotypic organization and developmental mechanisms that permit new adaptive traits to be pieced together from old ones. Explaining transitions between phenotypic states is the traditional old core of evolutionary biology. Turning to it now, with modern analyses of development and phylogenetics in hand, makes it possible to see origins with unprecedented clarity, since in many cases the hormonal and genetic mechanisms are known in some detail and the effects of environmental factors and genetic change can actually be traced.

Missing Chapters: Hybridization, Polyploidy, Horizontal Gene Transfer, and Phenotypic Fusion and Fission of Modular Traits

The chapters of part II grew out of research on alternative phenotypes. Originally I intended to examine the different ways that new alternative phenotypes can originate by making a collection of examples based on comparative study. It was soon clear that most examples fall into just a few categories and that each category, represented by a chapter of this section, is a common mode of phenotype reorganization during evolution.

I have not attempted to make rigid, nonoverlapping classification of modes of origin, and many if not all of the examples could be classified under

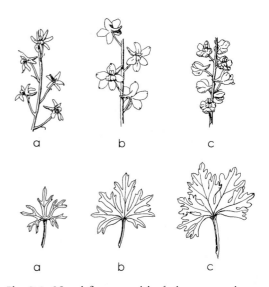

Fig. 9.1. Novel flower and leaf phenotypes due to hybridization. (a) *Delphinium recurvatum*; (b) *D. gypsophilum*, hybrid between a and c; (c) *D. hesperium pallescens*. From Lewis and Epling (1959), by permission of *Evolution*.

two or more of the categories discerned. Nor are the chapter titles of this section an exhaustive list of modes of origin. Among the important categories of origins of novelty not included are five that especially deserve mention: hybridization, polyploidy (chromosomal copy increase), transfer of genes between organisms, and the coupling and decoupling of modular traits.

Hybridization can be a source of phenotypic novelty in both plants (figure 9.1) and animals (Grant and Grant, 1996). Origins of novel phenotypes due to hybridization between species, and changes in number of chromosomes (polyploidy), may be especially important in plants (e.g., see Grant, 1963). Horizontal gene transfer, including possibly massive transfers during the evolution of prokaryotes giving rise to the major types of genes found in eukaryotes, has evidently played an important role in evolution of genomes (e.g., see Rivera et al., 1998) and would have profound consequences for the evolution of phenotypic novelty.

Well-known examples of novelty due to the coupling or fusion of traits occur in the evolution of bird limbs (figure 9.2) and in the evolution of the insect head, in which several ancestral body segments are fused (see chapter 10). All of these modes of origin can be regarded as reorganizational, like those more thoroughly discussed in this section.

The addition of modular structures by duplication followed by diversification of duplicates, as in the evolution of gene families, body segments, and novel cell types (chapter 10), is an important aspect of the evolution of complexity. But increased

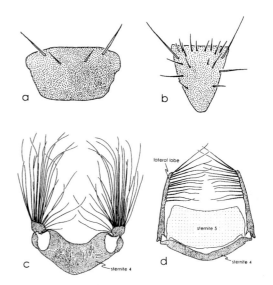

Fig. 9.3. The birth of an appendage—Increased modularity and plasticity by subdivision of a sternite (ventral segment of the abdomen) in flies (Diptera, Sepsidae): (a) unmodified sternite 4 of female *Archisepsis diversiformis*; (b) sternite 4 of male *A. diversiformis*, modified by bristles rubbed against female during courtship; (c) sternite 4 *Paleosepsis* sp. males with exaggerated bristles attached to semi-articulated lateral lobes capable of movement posteriad (upward in the figure) and ventrad (toward the viewer) by associated muscles; (d) male of *Pseudopalaeosepsis nigricoxa*, with specialized setae on highly articulated sternal lobes capable of limblike posteriad and ventrad movements. After Eberhard (2001a). Courtesy of W. G. Eberhard.

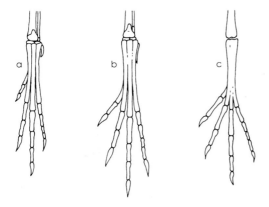

Fig. 9.2. Novelty by fusion. The five metatarsal bones in the feet of reptiles were reduced to three in bipedal archosaurs (a), partially fused in *Archaeopteryx* (b), and highly fused in modern birds (c). Based on Frazzetta (1975), Ostrom (1976), and Hall (1984).

modularization need not involve duplication. It can involve subdivision (decoupling) of traits as well, as in the evolution of multisegmented appendages (for examples, see Kukalova-Peck, 1997; Coen, 1991). Recent comparative studies of the functional morphology of flies (Diptera, Sepsidae) show how a new moveable abdominal appendage can originate by subdivision and muscularization of a segment of the abdominal exoskeleton, in this example under sexual selection (Eberhard, 2001a; figure 9.3). The modified segments occur in a group of closely related species and genera where the sexually dimorphic abdominal segment contacts and is moved against the female during courtship and copulation. There is a variety of sternal modifications among the males of related species and genera, with muscles of the abdominal wall evidently recruited to form specialized musculature that moves the modified lobes (Eberhard, 2001a). Modification of novel phenotypic subunits by muscle recruitment has been noted before, for example, in the evolution of accessory hearts in insects, shown by comparative study to incorporate muscles by splitting fibers from other muscles and shifting their attachment sites, or by displacement of a muscle portion (Pass, 2000, p. 512).

Developmental Recombination as a Complex Response to a Simple Input

Part I describes phenotypic traits as products of developmental switches and showed how a novel trait and its regulatory switch mechanism can originate at the same time. Developmentally, the origin of a phenotypic novelty is the response of a preexisting phenotype to some new genetic or environmental input.

The fact that novelty is the response of a preexisting structure has far-reaching consequences for evolution. As documented in this section, evolution is often, and perhaps always, combinatorial. Many new phenotypes are literally rearrangements of old ones, with subunits deleted, duplicated, recalled, and moved—shifted in their expression and recombined in myriad ways—the ontogenetic repatterning of Wake and Roth (1989).

Combinatorial evolution, or evolution by reorganization of preexisting traits, is possible because of the modular quality of phenotypic organization and because the switches that control development are subject to influence by numerous genetic and environmental cues. A new input can evoke correlated change in several elements at once, causing them to be expressed as a new combination (see chapter 16). Switch-controlled traits can also be turned off and on in new sequences, as illustrated in chapter 6 by patterns of evolution in tropical vines (see figure 6.3), where the switch points between leaf forms during ontogeny mark reorganization points in the development of different forms in different species (Ray, 1990).

Phenotypic recombination can produce a new trait without loss of an old one, through repeated gene *expression* and novel epigenetic interactions. Components of parental behavior, for example, can be transferred between the sexes without being lost in the sex of origin (see chapter 15). Combinatorial evolution is not just moving the furniture. It can increase phenotypic complexity, multiplying the

potential for further variation and evolutionary change. So increasing the phenotype repertoire of the genome, by increasing the potential for further phenotypic recombination, is a self-accelerating process that greatly augments the production of selectable variation.

Genetic recombination is important for passing genes around in a population, but it is difficult to see how it could be important in the origins of novelty given that chromosomal linkage of adaptive gene combinations is not usual (see chapter 4). Genetic recombinants are generally ephemeral, being broken up by crossing over and new chromosomal combinations in every generation. But developmental recombinants can be reproduced, providing that the environmental or genetic factors that induced them recur. It is this property of reproducibility that makes origins via phenotypic recombination important for genetical evolution. I will use the terms phenotypic recombination and developmental recombination as synonyms, to refer to the origin of new traits via altered patterns of expression of old ones. Although the word "recombination" in evolutionary biology is most closely identified with genetics, in developmental biology it has been used to describe novel phenotypes derived via experimental manipulation, as in "chimeric recombinations" and "blastomere recombinations," which are *in vitro* manipulations of cells (Davidson, 1989). Novel phenotypes obtained in Garcia-Bellido's experiments on the effects of cellular position in the development of *Drosophila* have been termed the results of "somatic recombination" (Raff and Kaufman, 1983, p. 250).

Important Distinctions

The Initiation versus the Sources of Novel Traits

In the origin of a novel phenotype, there are two related events. First, there is a new input in the form of a genetic mutation or an environmental change. Second, there is a developmental response that produces a new phenotype. For a novelty to have evolutionary potential, both of these events must be recurrent in a population and across generations (see chapter 6).

The role of environmental change in the origins of novelty is treated in chapter 26. This section (part II) deals with the developmental-response aspect of origins, to focus on the sources of new phenotypes—where do they come from? How does the preexisting, ancestral phenotype get transformed into the descendent one?

Development versus Selection as Agents of Change

It is quite common to use the word "origin" to refer to the selective context in which a trait evolved, rather than its source in the ancestral phenotype, for example, in such statements as, "Increased range of browsing accounts for the origin of the giraffe's long neck" or "global drying was a factor in the origin of life on land." But selection is powerless to change form. It can only change the frequency or range variations produced by development. Lack of clarity and disagreement about the relative importance of these two contributions to novelty underlie the perennial debate over gradualism and saltation, and its modern manifestation, the division between those who see selection as paramount and those who focus on "developmental constraints" (see chapter 24). Such controversies often reduce to a debate over how much of an observed phenotype is traceable to a one-step developmental change (origin) and how much is due to small developmental modifications that have accumulated under directional selection. In other cases it reduces to a failure to recognize that no established, adaptive trait gets its form from selection. Form always has a developmental origin, even though the spread and persistence of an adaptive variant are usually (though not necessarily) due to selection (e.g., see the exchange comprising Wake, 1991; Reeve and Sherman, 1993; Wake, 1996b).

Developmental trait origin necessarily precedes, and is independent of, trait selection. A phenotype has to appear before it can affect fitness (selection): "Natural selection can do nothing until favourable individual differences or variations occur" (Darwin, 1859 [1872], p. 127). Nonetheless, developmental origins and their proximate causes are often passed over in discussions of functional significance, adaptation, and selection. This was pointed out by Jamieson (1986) and Gould and Lewontin (1979), but their discussions unfortunately confused the issue in other respects (e.g., see chapter 15).

The Origin of Novelty versus the Origin of Species

Origins and speciation get confused in at least two ways. Some authors synonymize the origin of novelty with the origin of species. Cracraft (1990), for example, defines "evolutionary innovations" as "singular phenotypic changes that, subsequent to arising in individual organisms [as developmental variants, or "prospective innovations], spread through a population and become fixed, thus characterizing that population *as a new differentiated evolutionary taxon*" (p. 27, italics mine). While it is true that diagnostic characters of closely related taxa may represent innovations (Cracraft, 1990, p. 29), this of course does not mean that every innovation causes a speciation event and establishes a new taxon. Such an approach implies that every new branching point in *development* necessarily involves the origin of reproductive isolation. If this were true, as claimed by some authors (e.g., Eldredge and Gould, 1972, 1997; Eldredge, 1995), then one is led to the absurd conclusion that all phenotype-diversifying evolution, including that leading to differences between larvae and adults, the two sexes, and alternative phenotypes within populations, depends on speciation.

Another permutation of the innovation–speciation confusion is to think that individuals with markedly novel traits have no evolutionary future because such deviant individuals would be unable to find phenotypically similar mates (e.g., Mayr, 1963, p. 438; Frazzetta, 1975, p. 86), as if selective breeding with others of the same unusual phenotype is required for novelty to persist within an interbreeding population. Again, the implication is that evolutionary change requires speciation (reproductive isolation between very different forms) for phenotypic divergence to be preserved. Beneath this kind of reasoning runs the one-genotype, one-phenotype idea, which ignores within-genotype developmental diversification and therefore supposes that separation of gene pools is needed to keep novel phenotypes from getting swamped by gene flow. One need only think of the striking sexual dimorphisms that exist in many species to realize that phenotypic divergence does not depend upon breeding isolation between unlike phenotypes.

Developmental Origin versus Evolutionary Saltation

The assertion that all evolutionary novelties are developmental in origin does not imply a belief in saltation or macromutation (see chapter 24). Even small phenotypic changes are developmental in origin.

A synthesis of development and evolution has repeatedly foundered upon its identification with saltatory change, and therefore with antigradualism, anti-Darwininism, and critiques of selection theory, for example, in the provocative writings of Bateson (1894) and Goldschmidt (e.g., 1938, 1940; see chapter 24). It is important to acknowledge that there are small as well as large developmental variants and that all selectable phenotypic variation, whether large or small, continuous or discontinuous, necessarily reflects developmental variation. This is one reason for the fundamental evolutionary importance of development.

Problems in Interpretation

One-Step Regulation Does Not Prove One-Step Origin

The fact that an entire complex phenotype has been shifted, lost, or regained during evolution, and that the expression of all of its components are now controlled by a single switch, does not necessarily mean that the shift occurred in a single sudden step. Complex traits can be consolidated by gradual evolution, then turned off and on by a single switch. The recurrent evolution of flightlessness in insects, sometimes but not always proceeding to the point of wing loss, is a good example of a complex change that may occur gradually, with loss of flight behavior, flight muscles, and wings occurring at different times. This is a good example of a complex change that may occur gradually (see chapter 11).

It is not a simple matter to deduce from comparative study whether the evolution of a *complex phenotypic trait*—one composed of several modular elements whose coordinated expression or use is regulated by the same switch—originated via one step or many. In the following chapters, many phenotypic transitions are described, ranging in complexity from those involving a single molecule to those involving the transfer of immense behavioral and morphological trait complexes from one context of expression to another. Many of these transitions may have involved multistep reorganizations of ancestral phenotypes, and in most cases it is impossible to tell on the basis of comparative study whether the change occurred gradually or in a single initial step. The chapters of this section are less concerned with the question of gradualism than with the attempt to trace the sources of new traits in ancestral phenotypes.

When a novel complex trait, like the social insect worker phenotype or the mouse mandible, is a mosaic of several elements, it may have a com-

plex origin, involving the gradual (stepwise) accumulation of large or "saltatory" developmental changes as well as small ones. "The" origin of such a trait is difficult to pinpoint. Does the worker phenotype originate with the advent of the earliest evolved distinctive component (say, defensive stinging, or resorption of disused ovaries)? Or does it originate with a final defining step, say, the advent of lifetime sterility? Each time we define a trait of interest at a different level of organization, we alter the question of mode of origin. This need not be a paralyzing difficulty. It can be avoided by focusing on one developmentally defined trait and level of organization at a time.

Concepts and Criteria of Homology

Tracing origins using comparative study means searching for homologies, or similarities due to common descent. Problems and definitions of homology are discussed in detail in chapter 25, and readers especially concerned with the thorny problem of homology may wish to read that chapter now rather than this abbreviated statement. In one sense, homology is an all-or-none concept: either two traits are homologous, or they are not. But combinatorial evolution makes a concept of partial and mixed homology more accurately descriptive of the nature of homology. By the cladistic definition of homology, as similarity with unbroken expression during descent from a common ancestor, the all-or-nothing concept of homology, allowing for some modification during descent, seems straightforward and clear. But homologous traits, even of sister populations, can be mixed—simultaneously homologous with more than one ancestral trait. The ear of modern corn (*Zea mays*), for example, is a composite of the ancestral male and female structures and is therefore homologous with both (see chapter 15). Furthermore, similarity can represent a continuum between similarity entirely due to common descent and similarity entirely due to convergent evolution. Here I adopt a broad definition of homology (the broad-sense homology of chapter 25), which includes similarity due to common ancestry of characters that do not necessarily show unbroken descent from a common ancestor but may be widely dispersed on a cladogram (see chapter 19). The criteria versus the definitions of homology are discussed in detail in chapter 25. Several criteria of homology are used in the chapters of part II, not all of them applicable to all examples. The greater the number that apply to a given example, the stronger the hypothesis of homology.

Detailed phenotypic similarity is a commonly used criterion of homology between traits. In order to test the hypothesis that female parasitic bees (*Paralictus asteris*) have traits that are homologous to those of males in the host species from which they are derived, Wcislo (1999) used a principle-components analysis to show that females of the parasite are morphologically more similar to the males of their own and their host species than they are to host females (see chapter 15). Similarly, Wray (1996) adapted cladistic techniques, which offer refined methods for estimating degree of phenotypic similarity, to assess degree of parallelism, or homology, among recurrent traits in echinoderms. The same "phylogram" technique could be applied in other kinds of phenotypic transition.

Developmental similarity is another type of resemblance that supports a homology hypothesis. This may take the form of inducibility by the same hormonal or environmental stimuli, or similarity of expressed genes. Sometimes discussions of homology are confused, however, by the expectation that homologous traits should be genetically identical or develop via the same ontogenetic steps (see chapter 25). Given the regulatory complexity of switches between alternative states, there can be numerous routes to the loss or gain of a trait (see chapter 7). The developmental basis for the recurrent loss of metamorphosis in salamanders, for example, varies, as indicated by the fact that different genetic loci and different events in the control of metamorphosis are involved in different episodes of change (Shaffer and Voss, 1996; see chapter 25). The hormonal regulation of metamorphosis is the same, but it is genetically complex (polygenic), so genetic diversity in disjunct episodes of change is to be expected (for an early discussion of this point see Haldane, 1958, p. 21). In fact, the possibility of multiple pathways for evolutionary change in the regulation of alternative traits has been cited as one factor that facilitates recurrent *parallel evolution* (Walker and Williams, 1976), the evolution of similar traits in different lineages, beginning with similar or identical ancestral traits.

Intraspecific alternative phenotypes or *distinctive life-stage phenotypes* can illuminate homology by showing that the same or similar phenotypes can be turned off and on during development, providing reason to expect the phylogenetically disjunct expression of a trait. In some groups, such as seasonally polyphenic butterflies, the developmental basis of a recurrent phenotype, such as the dark-colored vernal form, has been studied in polyphenic species, and phenocopies are inducible in related species normally lacking the phenotype (Shapiro,

1976). This combination of comparative and experimental evidence has led to hypotheses regarding the selective circumstances that favor development of the dark form, and predicting the populations where it is expected to be expressed (Shapiro, 1976; Kingsolver 1995; Kingsolver and Wiernasz, 1991). Similarly, the existence of a striking polyphenism for the recurrent migratory form of grasshoppers makes it possible to demonstrate that the basis of recurrence in different grasshopper species is the recurrence of the environmental circumstance—prolonged high population density across several generations—which leads to the full expression of the migratory form (Kennedy, 1961). Evolutionary pattern often reflects patterns of intraspecific developmental plasticity (Higgins and Rankin, 1996), and this fact can be used to support homology hypotheses.

Even without a cladogram it is sometimes possible to see an association between intra- and interspecific alternatives suggestive of homology between the latter. For example, in six cricket subfamilies containing wing-polymorphic species there are also short-winged and long-winged *species*, and the intraspecific alternatives appear as interspecific alternatives within genera (e.g., in *Scapteriscus*, *Gryllus*, and *Anurogryllus*; Masaki and Walker, 1987). Providing that the genera are monophyletic groups, this suggests a phylogenetic association between patterns of intra- and interspecific variation.

How would one proceed to find the developmental basis for a recurrent phenotype not known to occur or be experimentally inducible as an intraspecific alternative or low-frequency anomaly? Other kinds of developmental flexibility, such as differences between the sexes, can offer clues and support a recurrence (homology) hypothesis (see chapter 15).

The homology problem is not simple. An exceptionally good fossil series of sticklebacks (*Gasterosteus doryssus*) discussed in chapter 19 shows how difficult it is to distinguish between homology and convergence, even in very closely related populations. The different morphological series seen in fossils and different extant populations suggest that the pelvic ontogeny that was the starting point for pelvic reduction in the fossil *G. doryssus* was not the same as that leading to pelvic reduction in more recent populations (Bell, 1988). If this is true, then the two different pathways of pelvic reduction may produce convergent evolution. But the remnant "ovoid vestiges," apparently so similar as an end point in both sequences (Bell, 1987), conceivably could still owe their similarity to common ancestry and legitimately be considered homologous in the traditional broad sense of the term. The homology hypothesis is supported by the fact that paedomorphic intermediate pelvic vestiges sometimes occur in extant populations, which usually show pelvic reduction via distal truncation. This suggests a broad propensity for paedomorphosis (Bell, 1987, p. 373) and lends support to a hypothesis of similarity due to common ancestry (homology), albeit due to development so flexible that different pathways to the same end point are possible. While some cladists are disturbed by such tentativeness regarding assignment of homology under the broad-sense definition (e.g., see Rieppel, 1994), it does not serve understanding of evolution to hide these complexities under the operationally facile but insufficiently informative label of homoplasy. The homology problem is more thoroughly discussed in chapter 25.

Polarity

Description of a unidirectional evolutionary transition requires some evidence of the polarity, or direction of change: which of two traits hypothesized to be homologous is likely to be more similar to the ancestral state, and which is more likely derived? Sometimes this can be examined by optimization of the two traits on a cladogram (for methods and examples, see Mickelvich and Weller, 1990; Brooks and McLennan, 1991; Emerson et al., 1997).

For many of the transitions hypothesized in these chapters, a cladogram was not available, but the polarity hypothesized for the transition is strongly supported by biological information. For example, female hyenas have an erectile penis that is virtually indistinguishable from that of a male and is associated with malelike hormonal conditions during development (see chapter 15). Given the well-established male function of intromittent genitalia in vertebrates, the most parsimonious polarity hypothesis is that the transition was from femalelike to malelike genital morphology, with the malelike genitalia of female hyenas derived, rather than the reverse. The biological evidence for this, based on comparison with sex hormones and genital morphology in the vast majority of vertebrates, is so strong that cladistic evidence is not required to support it. If cladistic evidence were to suggest the opposite polarity, the cladogram would have to be considered questionable on biological grounds. Similarly, courtship displays in many birds incorporate feeding of potential mates, or movements that resemble those of feeding in adults of both

sexes and in juveniles as well (chapter 13). It would not be parsimonious, on biological evidence, to hypothesize that the feeding movements originated as courtship behavior and then were secondarily expressed in the trophic context in descendents. Other kinds of situations where a phylogeny is either unnecessary or inconclusive are discussed by Losos (1994a; see also Losos and Miles, 1994; Losos, 1999).

Often, the derived form of a reorganized phenotype is less perfect in its functional details than is the primary, ancestorlike form (S. Emerson, personal communication). A femalelike "beta" male morph of an isopod crustacean with three male forms, for example (Shuster, 1987), is highly likely to be derived by the transfer of expression of female morphology and behavior to males (see chapter 15). The derived, feminized male, however, is a less-than-perfect version of a female. It retains some malelike traits, such as greater persistence during courtship, and differs from females in various other behavioral and morphological details (Shuster, 1990, 1992b; S. M. Shuster, personal communication). Such lack of perfect resemblance is expected because during phenotypic reorganization the reexpression of ancestral traits can be incomplete, producing what some authors have called "partial" homologies (see chapter 25). A derived trait could conceivably evolve under a long period of selection to be more elaborate and specialized in the new context than an ancestral trait from which it is derived, so this is only an approximate criterion of polarity that depends on the derived trait being newer and therefore less closely adaptive in the context where it is expressed.

In other examples, such as the derivation of gold color from black coloration in a cichlid fish, the transformation and its mechanism can be directly observed within individuals and is irreversible (see chapter 11), so the polarity is unmistakable. Even if the character states were reversible within individuals, the events involved in the phenotypic transtition between states—a major question for this chapter—would be evident from individual ontogenies without a cladogram. If phylogenetic analysis were to suggest that the gold color is older in cichlid fishes, we would still have to conclude from the study of ontogeny that a pathway for reversion to gold from black has evolved.

In some examples, the ability to switch back and forth between character states within individuals or populations is so marked that I have not attempted to discuss polarity and have focused instead on how the phenotype is reorganized to produce the bidirectional transitions observed. In some of these examples, one or the other of the alternative states is fixed as a species or population characteristic in a related taxon (Michener, 1985; West-Eberhard, 1986, 1989). This suggests the hypothesis that one of the developmental transitions observed within the polymorphic population has been involved in an evolutionary transition. It does not indicate which of the states was the first to evolve, but that question is of secondary interest for this section. Given that the transition is known to occur in both directions (e.g., in a polymorphic population), the question addressed here is how, in terms of phenotypic reorganization and development, the transitions occur.

In Praise of Anomalies

Anomalies, or low-frequency discrete phenotypes, along with seemingly unpatterned variation called imprecision or noise, are often passed over in studies of variation, as if unusual variation were the enemy of insight. There is a prejudice against variation that is nonadaptive in studies of adaptive evolution, against variation that is nongenetic in studies of genetics, and against virtually all variation in studies of *the* genome in *the* worm (*Caenorhabditis elegans*), *the* fly (*Drosophila*), and *the* frog (*Xenopus*). Unusual variation is abnormal, at least in the sense of being rare, and sometimes even grotesque. But anomalies represent new options for evolution if they are recurrent enough to be modified by selection or if they happen to prove advantageous in conditions that were previously absent or rare. This quality of anomalies is dramatized in computational studies of evolution and the Baldwin effect (Mitchell and Taylor, 1999). It is also confirmed by the fact that phenotypes that appear as anomalies in one species sometimes appear as widespread functional traits in others (Eble, 2001; Shubin et al., 1995).

Experimentally induced anomalies in the form of mutations have long been used in genetics to study development. Mutant lines created for modern studies of evolution have been baptized "designer organisms" (Dworkin et al., 2001). Environmental or hormonal manipulations have been used to induce developmental anomalies, called the "pseudomutant technique" by Wilson (1981) and "*phenotypic engineering*" by Ketterson and Nolan (1992). It will be especially interesting to apply an experimental approach to the study of naturally occurring anomalies in order to investigate their causes and evolutionary prospects. This is already occurring in the study of "developmental instabil-

ity," a respectable-sounding name for the causes of anomalies which may finally bring them the attention they deserve (see Polak, 2002).

Use of anomalies to illuminate common ancestry has a long history in evolutionary biology. Experimentally induced anomalies, revealing underlying developmental capacities, have long been used as clues to phylogenetic relationship. "Inferring phylogeny from concealed variation is a technique which has been rediscovered almost as often as penicillin" (A. M. Shapiro, 1984c, p. 186). In addition, phenotypes seen as intraspecific and/or interspecific alternatives often occur as very low-frequency "mutants" or environmentally induced anomalies in natural populations of related species. If one of a pair of alternatives is present in a species, it is often possible to artificially induce the missing alternative as an anomaly that is strikingly similar in both form and development to an observed established trait. Here are some illustrative examples:

In crickets, a long-winged form can be induced by crowding in species such as *Gryllodes supplicans* and *Pteronemobius nitidus*, which are usually 100% short-winged in the field (Masaki and Walker, 1987). Phenocopies of the wing morphs can also be induced by hormone manipulations (Zera and Tiebel, 1988). The feet of guinea pigs, as in other representatives of the family Caviidae, lack the thumb, big toe, and little fourth toe. But they commonly show as an anomaly a fourth toe that is exactly like that of related rodents (Wright, 1968). In protists, microfilaments that are used in the motility of amoeboid species form the flagella of flagellates, and an individual of these one-celled organisms is either amoeboid or flagellar but cannot be both. Some protist species are flagellates, and some are ameboid (these forms are interspecific, or intertaxic alternatives); in the rhizopod amoebae of the genus *Naegleria*, a flagellate form sometimes occurs as an anomaly (Stebbins and Basile, 1986, after Kühn, 1971). Similarly, nematodes of the genus *Caenorhabditis* usually have double-armed ovaries; in some other nematodes, the ovary is single armed, and the single-armed condition can be experimentally produced as an induced anomaly (by laser elimination of a single cell during development) in *Caenorhabditis* (Raff and Kaufmann, 1983), showing that an apparantly major change (elimination of an entire arm of the ovary) is developmentally feasible. Such observations, of frequent reversion within a species, whether as an anomaly or an alternative phenotype, is evidence that rapid reversal and recurrent fixation of homologous phenotypes may have occurred in the history of the group.

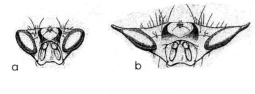

Fig. 9.4. Developmental plasticity in male head morphology of *Zygothrica dispar* (Diptera, Drosophilidae) as shown by an anomaly: (a) head of male at the extreme non-hypercephalic end of the range for this species; (b) extreme hypercephalic male; (c) rare asymmetrical variant (1/5000 specimens examined) showing both extremes in a single individual. Drawings (a, b) from Grimaldi (1987); electron micrograph (c) from Grimaldi and Fenster (1989), by permission of the authors.

Extensive studies of variation in salamanders (Shubin et al., 1995; see chapter 19) and of anomalies in bird limbs (Raikow et al., 1979) and ants (Buschinger and Stoewesand, 1971) showed that bilateral asymmetry of anomalies, or "fluctuating asymmetry" (Van Valen, 1962), sometimes occurs. Sometimes asymmetries show the normal alternative (or a plesiomorphic state) on one side and the anomalous one (or a relatively derived state) on the other, as in the bilaterally asymmetrical male drosophilid fly (*Zygothrica dispar*) shown in figure 9.4c. Such a grotesque aberrancy is unlikely to be important for evolution, but it can be an important tool for evolutionary analyses. Males of some species related to *Z. dispar*, for example, have heads like the left side of the anomaly as a fixed characteristic, and other species have heads similar to elongate right side. Finding the same extremes in a single individual, as well as commonly within the normal range of variation of *Z. dispar* (figure 9.4a,b), shows that a transition between the extremes is developmentally easy. In general, finding alternative states producible by the same individual demonstrates the potential for lability of switching from one to the other during the course of evolution. Especially, it suggests susceptibility to environmental influence, since, barring somatic mutation, both are produced by the same genotype.

Recurrent, viable anomalies seem to appear wherever they are systematically sought. Yablokov (1966 [1974], pp. 161–162), cites Berry (1963, 1964) as having examined the skeletons of various mammals and found 53 different types of qualitatively distinguishable recurrent deviations from the "normal," including such things as a curved nasal cavity, presence of interfrontal bones, fusion of certain bones of the skull, presence of abnormal processes on certain skull bones, absence of certain molars, fusions of vertebrae, fusion of pelvic bones, and so on. Not a single skeleton was completely free of discrete deviations. The number of "deviating traits" per individual varied from 12 to 29. Raikow et al. (1990) examined hindlimb muscles in two passerine birds, the northern cardinal (*Cardinalis cardinalis*) and the wood thrush (*Hylocichla mustelina*). Most leg muscles were invariable in the 23 and 25 specimens of the two species, respectively, that were examined, but some showed quite striking variations. For example, in the cardinals, one muscle (the M. fibularis longus) has a small, ovate area of ossification in the tendon where it bifurcates to insert onto the tibial cartilage (on one side) and to join the tendon of another muscle (on the other side). The ossification showed variation in size not associated with sex or developmental stage, being about 2 mm long in both legs of 11 specimens (48% of the sample); about 1 mm long in both legs of 9 specimens (39%); 1 mm long on one side versus 2 mm long on the other side in 2 specimens (9%); and completely absent on both sides in one specimen (4%). In the wood thrush, only two specimens (8%) showed such ossifications. Thus, a trait that is a variable but typical feature of one species is a low-frequency trait in the other and can be completely lost in individuals of a species where it is usually present. Another muscle (the M. extensor digitorum longus) showed extra attachments in a small percentage of cardinals (Raikow et al., 1990, their figure 4, p. 367).

In studies of variation like these, "the meaning of 'normal' loses its meaning because there are no 'normals' as such in nature. There is only a more or less usual group of different deviations from some mean values" (Yablokov 1966 [1974], p. 161). While this may make the description of species more difficult, it renders the explanation of evolution by natural selection easier by showing the enormous standing crop of variation available for the action of natural selection. The great diversity of developmental variation is the basis for the impression that selection can produce adaptive design seemingly of any kind.

Atavisms, as recurrent variants in a population, are material for the evolutionary establishment of a reversion. But evidence that this has actually occurred in nature requires an exceptional knowledge of phylogenetic relationships as well as attention to sporadic anomalies that indicate a capacity for coordinated reappearance of the trait (e.g., see Buschinger and Stoewesand, 1971). These conditions are satisfied in a few recent studies, which demonstrate anomalies resembling the established derived traits of related species. Studies of the hindlimb musculature of birds (McKitrick, 1992; Raikow et al., 1979) indicate that the iliofemoralis externus muscle is absent in most passeriforms, regularly present in species of related groups of birds, and present as an anomaly in the drongo (*Dicrurus*) and some starlings (*Sturnus* species; McKitrick, 1992). It has also been found as an anomaly in the fox sparrow (*Passerella iliaca*; Fringillidae; Raikow, 1975) and in *Aechmorhynchus cancellatus* (Scolopacidae; Zusi and Jehl, 1970, cited by Raikow, 1975). The same muscle is typically present in bowerbirds and their relatives (Paradisaeidae, Ptilonorhyncidae, Callaeidae; McKitrick, 1992). This pattern of occurrence suggests that the iliofemoralis externus was suppressed in most passerines but recurs sporadically as an atavism in some groups, a phenomenon that would have allowed it to be reestablished in the bowerbirds (see Raikow et al., 1979; McKitrick, 1992). Given the absence of this muscle as an established trait in all of the diverse passerines that have been studied, Raikow (1975) estimated that it was lost "at least several million years" ago. He speculates that the genes required have been maintained in other selective contexts (by pleiotropic effects) and that only the coordinated production of this particular phenotypic aspect was lost, not the underlying genes.

As would be expected if anomalies are in fact material for selection, anomalies of one species have been recorded to appear, as in the case of the iliofemoralis esternus muscle, as fixed traits in other species (see also chapter 19). Induced major mutant anomalies of *Volvox carteri*, for example, resemble established phenotypes in other species of *Volvox* and, in some cases, more distantly related species (Koufopanou and Bell, 1991). In one of the few systematic studies of recurrent anomalies, the salamander-limb survey of Shubin et al. (1995), the fact that so many (about 40%) of the patterns considered anomalous within *Taricha granulosa* turned out to be found as frequently expressed or fixed traits in other taxa is evidence of the significance

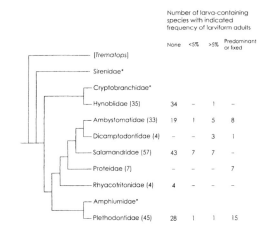

| | Number of larva-containing species with indicated frequency of larviform adults | | | |
	None	<5%	>5%	Predominant or fixed
[Trematops]				
Sirenidae*				
Cryptobranchidae*				
Hynobiidae (35)	34	–	1	–
Ambystomatidae (33)	19	1	5	8
Dicamptodontidae (4)	–	–	3	1
Salamandridae (57)	43	7	7	–
Proteidae (7)	–	–	–	7
Rhyacotritonidae (4)	4	–	–	–
Amphiumidae*				
Plethodontidae (45)	28	1	1	15

Fig. 9.5. Recurrence of the larviform (branchiate, paedomorphic) adult morph in salamanders. Total number of larva-containing species per family is in parentheses. Families characterized by a partial metamorphosis (adult retention of some larval traits rather than a complete larva-adult metamorphosis) are indicated with an asterisk (*) and are omitted from the analysis. The other families usually have complete metamorphosis. The temnospondyl fossil genus *Trematops* serves as an outgroup. Phylogeny from Shubin et al. (1995). Data on occurrence of larviform morph from Collins and Sokol (1980), kindly updated (July 2000) by D. B. Wake (personal communication).

of developmental anomalies as material for evolutionary change in the phenotype (see chapter 19).

The picture of anomalies as material for selection is completed by another comparative study of salamanders, where a major morphological and life history novelty—larviform, brachiate adults that mature sexually without metamorphosis—varies greatly in frequency of expression in different species of the 10 families of salamanders (figure 9.5). In some species this recurrent morph is unrecorded; in some it is rare (<5% of population); in some it is common and regarded as part of a polyphenism; and in others it is predominant or fixed as the characteristic adult form (review in Matsuda, 1987; see also Collins et al., 1993; Shaffer and Voss, 1996). In some species where larviform adults are fixed, metamorphosis occasionally occurs in nature or can be induced by hormone treatments. But a few species have lost all ability to metamorphose even under hormonal application (reviewed in Matsuda, 1987).

From the distribution of the larviform adult morph within families, Collins and Sokol (1980) estimate that this recurrent form arose independently at least 16–18 times in salamanders. Obligate paedomorphosis has evolved independently a minimum of four times in *Ambystoma* (Shaffer and Voss, 1996; based on the molecular phylogeny of Shaffer and McKnight, 1996). Even within a single population of *A. gracile*, there are some individuals that metamorphose regardless of environmental conditions, some that are neotenous regardless of rearing conditions, and some that metamorphose or not, depending on conditions (Sprules, 1974, cited by Matsuda, 1987). This indicates a high potential for evolutionary lability in adopting or omitting metamorphosis in that lineage. The fact that the paedomorphic adult recurs as an intraspecific alternative indicates a potential for saltatory, or unitary, gain and loss of this complex trait, wherein reproducing larvae omit metamorphosis and complex postmetamorphic developmental stages. But gradual parallel evolution is also possible, as shown by the fact that some families of salamanders (asterisks in figure 9.5) have partial metamorphosis, or retention by the adult of some larval traits, indicating that mosaic intermediates are possible.

The ready-made quality of recurrent developmental anomalies as material for selection, and the fact that recurrence can occur at large phylogenetic distances (see chapter 19), may help to explain the mimicry polymorphisms of butterflies that have challenged ecological genetics research for decades, making these polymorphisms among the best studied in nature. The central puzzle of the evolution of mimicry was how the detailed mimicry of a distantly related model could evolve gradually from an originally poor mimic that would have little selective advantage as a mimic until it was already highly evolved toward perfection. Ford (1971, pp. 272ff) reviews the major theories on this point, including those of Punnett (1915), who suggested parallel mutations in both mimic and model; of Goldschmidt (e.g., 1940), who proposed developmental limitation of the number of patterns possible in both groups, and therefore a possibility of initial close general resemblance—a view roundly criticized by Ford; and of Ford's own hypothesis (credited to Fisher, 1927), of a major switch gene subject to some genetic variation. Goldschmidt's much maligned developmental-constraints view seems closest to the modern one. For the case of *Papilio dardanus*, one of the best-studied examples, Nijhout (1991b, pp. 147–148) concludes that the color patterns of closely related [monomorphic species of Papilio (e.g., *P. phorcas*, *P. hesperus*, and *P. echerioides*) resemble the derivative mimetic patterns of *P. dardanus* much more closely than they do the malelike (and presumably primitive) nonmimetic pattern of the nonmimetic female morph. This suggests that the derivative mimic patterns of *P. dardanus* could have originated as atavisms.

The quantitative counterpart of discrete variants, sometimes called noise or imprecision (e.g., Eberhard, 1990b, 2000), may also serve as material for selection and evolution. How much of continuous variation is due to developmental imprecision and noise is even less often measured and compared among populations and species than are discrete anomalies, undoubtedly because it is difficult to completely eliminate the possibility that measurement of some unappreciated genetic or environmental factor would reveal pattern or function in the variation. But imprecision and noise are evidently common, at least in behavior, where even simple patterns are rarely exactly repeated, even by the same individual in the same conditions (Gallistel, 1980). A relatively simple locomotory silk-spinning behavior of fly larvae (*Leptomorphus* sp.; Mycetophilidae), for example, showed variation in several kinds and sequences of movement even when the larvae moved across a clean glass substrate designed to eliminate the kind of stimuli that normally cause variation in their spinning behavior (Eberhard, 1990b). The idea that such variability in behavior may be due to neuromuscular imprecision that could be subject to natural selection and give rise to novel behaviors (Eberhard, 1990b) was also suggested by the observation that unusual evolutionary lability in web design occurs in a spider genus (*Wendilgarda*) that also has unusual intraspecific variability and imprecision in its web building (Eberhard, 2000).

To see pattern and extremes in variation is to glimpse the world as seen by natural selection. It is not a world of uniformly tiny, mutationally based, or exclusively quantitative variants. Rather, it is one full of recurrent developmental anomalies that vary in accord with the genetic makeup of individuals and also with their environmental circumstances.

The remaining chapters of part II show how novelty can result from different kinds of reorganization of ancestral phenotypes. This mode of origin of qualitatively distinct novel phenotypes is so common that the burden of proof now lies with anyone who would argue that such novelties can arise due to genetic gradualism—the gradual, step-by-step accumulation of small innovations that were not present in ancestral populations and are the results of a series of genetic mutations.

10

Duplication

Gene duplication is widely recognized as a first step in genetic diversification. Phenotypic duplication, or repetition of parts, can be a first step in the evolution of novel morphology and behavior as well. Duplication leads to divergence only if duplicated parts are independently expressed. Only then can traits become developmentally independent of other copies and independently subject to variation and selection. Concerted evolution, or change without divergence of duplicated parts, is a feature of multigene families. Concerted evolution occurs in multiphenotype families, as well, and is most likely to characterize iterated, simultaneously developed parts with a common regulatory mechanism.

Introduction

Duplication occurs at all levels of phenotypic organization. It was among the first developmental phenomena to attract the attention of biologists interested in phenotypic transitions during evolution. Duplication and repetition of similar segments, or metamerism, appeared in the writings of Aristotle, Cuvier, Owen, and Darwin, and many early twentieth-century authors discussed it (reviewed in Lauder, 1981). More recently it has become a major theme in discussions of genetic diversification (e.g., Ohno, 1970; Weiss, 1990).

Nijhout (1991a) considers iteration or redundancy of parts followed by divergence of the replicates "the major principle of morphological evolution." But it is not easy to rank modes of origin in order of their importance, because it is so easy to reclassify examples from different points of view. Just as heterochrony can be emphasized by focusing on the timing aspect of any regulatory change (as discussed in chapter 13), duplication can be emphasized by focusing on the iterative aspect of repeated expression at different times, places, or contexts. Phenotypic recombinations classified in these chapters as heterotopy, heterochrony, cross-sex transfers, or correlated shifts could be reclassified as duplications, since many are reexpressions of the same trait in a new place or context without loss of the original trait. Similarly, pleiotropy can be de-

scribed as the duplicated expression of genes (Hamburger, 1980).

This chapter focuses mainly on adjacent duplications that result in serial repetitions—serial homologues, or "iteration" (Nijhout, 1991a). Heterotopy, or nonadjacent duplication within the same individual and life stage, is discussed in chapter 14, even though the separation is somewhat arbitrary.

The differences in position and surroundings that result in differences in phenotypic pattern are, in their effects, regulatory differences, and it is just such local differences that cause developmental processes to diverge. It is a difference in regulation, then, that causes duplicates to diverge phenotypically and ultimately renders them subject to diverging selection and divergent evolution. One could reason from this that divergence between duplicates is virtually inevitable given enough time, because the local developmental environments on all sides of duplicate parts would rarely be absolutely identical over long periods. For this reason, I predict that concerted evolution will be found to most often characterize the internal members (rather than the terminal members) of series of serial duplications, or iterative parts, which are simultaneously produced during ontogeny. In this situation spatial heterogeneity is minimized because each duplicate contributes to the similarity of developmental environment for those adjacent

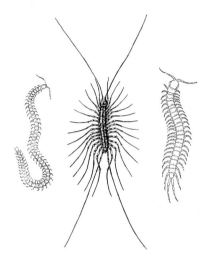

Fig. 10.1. Concerted evolution in a multiple-phenotype family. The segmental and appendage duplication of centipedes (Myriapoda, subclass Chilopoda) creates a multiple-phenotype family of appendages that vary together in different taxa. Left, a gelomorph centipede; middle, *Scutigera coleoptrata* (order Scutigeromorpha); right, *Otocryptops sexspinnosa* (order Scolopendromorpha). After Snodgrass (1952).

to it, and temporal variation in developmental conditions is minimized by simultaneous development. That is, not only the parts are duplicated, but many factors that might influence their development and activity are duplicated as well. The segmental appendages of arthropods (figure 10.1) are possible illustrations, where the appendages of internal segments are more similar than those of either the head or the posterier regions, whose developmental environments and uses are relatively distinctive, as are their (consequently) divergently evolved morphologies.

Duplication and the Rule of Independent Selection

The divergence of duplicated parts illustrates the significance of compartmentalization for both flexibility and evolution, since bounded structures such as cells and body segments frequently have different functional specializations. A duplicated structure or repeated behavior can be expressed in a new context and then, without disturbance of original function, can diverge due to variation and selection in the new circumstances. Duplication produces a new subunit that, to the degree that it is independently expressed, is subject to the rule of independent selection. Bateson (1894, p. 67), perhaps the first to realize this, referred to the independence of duplicates as "discontinuity in variation." Later, geneticists, beginning with Haldane (1932) and more recently especially Ohno (1970), became interested in gene duplication as a means of genetic diversification having the same properties of buffered evolution in a new direction as that characterizing morphological duplications. Haldane (1932) also anticipated the rule of independent selection in a discussion of gene duplication in polyploid organisms, noting that polyploids could be favored because they "possess several pairs of sets of genes, so that one gene may be altered without disadvantage, provided its functions can be performed by a gene in one of the other sets of chromosomes" (p. 194). It is now known that divergence between pairs of genes under polyploidy is common (e.g., in salmonid fish—Allendorf and Thorgaard, 1984; Forbes et al., 1994).

Duplication of phenotype subunits is probably a very common mode of origin of novelties at all levels of organization. It is known to be responsible for the evolution of new kinds of cells (Gerhart and Kirschner, 1997), new genes and proteins (John and Miklos, 1988), and higher level units of morphological and behavioral organization (see below).

Sometimes duplication is cited as a prelude to divergence as if the two phenomena were automatically linked. But duplication need not lead to divergence, as acknowledged by Bateson (1894; see also Lauder, 1981). In a discussion of gene duplication, applicable as well to phenotypic evolution, John and Miklos (1988, p. 280) note that duplicates can have three different fates: they may be used as duplicates, as in multigene or multiphenotype families, where multiple copies increase product quantity; they may diverge, as in the case of embryonic and fetal globin genes; or they may become nonfunctional, for example, yielding the so-called pseudogenes.

Duplication itself sometimes may initiate divergence evolution. As with identical twins, the close presence of the other twin becomes an important part of the environment of each, and there may be asymmetries between members of the pair in surroundings and position. In animal morphology, the environmental difference between the anterior and the posterior segments of a moving body must be responsible for the fact that sensory organs (eyes, antennae) are commonly concentrated on the anterior segments, and once such anterior–posterior divergence had evolved, it would be further reinforced by selection to move about anterior first due to the protection afforded by those structures. Riedl (1978) notes that the duplication—he calls it "doubling"—of complex morphological traits is rarely perfect, so for this reason, too, there must often be immediate material for selection and divergence.

Any difference in the environment, phenotype, location, or use of duplicates sets the stage for different regimes of selection and the divergent evolution of duplicate traits.

Gene Duplication

Gene duplication has become a textbook example of how genetic novelties originate. Mayo (1983, p. 61) cites Bridges (1935) and Stephens (1951) as among the early Mendelians who, like Haldane (1932), considered the role of gene duplication in the origin of functional genetic novelties (others are given by Lauder, 1981, p. 438). The best known and most extensive treatment is by Ohno (1970).

A few well-studied examples can serve to illustrate how differences in the timing or location of duplicate genes are associated with their divergence. This suggests that the same principles proposed to underlie phenotypic divergence apply to genomic divergence as well, namely, that independent expression is fundamental to the independent selection and evolutionary divergence of initially identical traits. The human beta globin gene family, for example, has five functional genes (Ohta, 1993, based on Goodman, 1976), all of them apparently produced by tandem duplication (Nei, 1987, p. 114; for a concise review, see Meireles et al., 1995). One gene (epsilon) functions only in the early embryo, two (G-gamma and A-gamma) produce globin polypeptides only in the fetus, and two (delta and beta) function in adults. In this example, the different times of expression of the duplicated genes during the human life cycle would account for their divergence. Similarly, the alcohol dehydrogenase (adh) gene is duplicated in some species of *Drosophila*, with one form expressed in larvae or (in some species) larvae and ovaries and the other in adults. As expected given the different times of expression, functional divergence has been inferred from different relative activity levels of the two genes in larvae and adults (Dickinson, 1991). Other examples where adaptive divergence between duplicate genes has been studied are the classic case of the lactate dehydrogenase (LDH) isozymes (e.g., Markert and Faulhaber, 1969; reviewed in Almeida-Val and Val, 1993) and the chal-

cone synthase (CHS) gene family, which influences flower color in morning glories (*Ipmoea*; Durbin et al., 1995). In some genes, divergent duplicates show different tissue-specific functional interactions that could contribute to maintenance by selection of particular variants within multigene families (Chourey and Taliercio, 1994).

Homeobox genes in at least some vertebrates and some insects are evidently products of gene duplication followed by modification by addition of gene-specific secondary expression domains, and then participation in different segment-specific regulatory complexes where they play key roles as DNA binding sites (Holland, 1992; Galis, 1996b). The divergence of homeobox genes exemplifies divergence promoted by spatial differences in expression. Other means of novel gene origin, most notably exon shuffling, are discussed in chapter 17. There also, duplication is involved, but of gene segments (exons) rather than entire genes (Ohta, 1991).

The term "gene duplication" includes a broad range of phenomena, all of them involving evolution via incorporation and/or modification of extra copies of one or more preexisting DNA fragments. For example, the fibronectin gene—strictly speaking, the protein fibronectin—is said to have "evolved by gene duplication" (Holland and Blake, 1990, p. 35) because two of its domains (FnI and FnII) were likely derived via the duplication and modification of a common ancestral domain possibly accounting for their architectural similarity to each other and to various other domains (kringle and epidermal growth factor). Intragenic duplication also occurs in the evolution of tetracycline resistence in bacteria. Two complementary gene segments specify two different domains of a membrane-spanning protein that mediates tetracycline efflux. The pair of fragments are formed by duplication and then diverge, forming a plasmid product that confers resistence (see Johnson and Adams, 1992).

The importance of duplication for functional genome evolution could hardly be overemphasized. In the well-studied genomes of the plant genera *Arabidopsis* and *Zea* (maize), mutational inactivation of a gene *invariably* uncovers another nonallelic duplicate or homologous gene encoding a similar function (Chourey and Taliercio, 1994). Duplication followed by combinatorial evolution of initially identical useful genes permits reuse of already functional genes while allowing them to evolve in new contexts. This is a better arrangement than would be possible with combinatorial evolution alone: reutilization by reexpression in multiple contexts without duplication would foster conservatism rather than change, for change in any one context is likely to compromise function in others. This is thought to be one basis for the growing numbers of examples of conservatism in the active regions of molecules, including proteins, cellular signals, and hormones—molecules notably functional in multiple contexts (Gerhart and Kirschner, 1997). Gene duplication permits an escape from this kind of conservatism without disturbing function (Dickinson, 1991).

Although gene duplication is now a familiar phenomenon to students of general biology, the details of how duplicate copies diverge still seems to be largely a matter of speculation (see Holland and Blake, 1990; Smith et al., 1990; Zuckerkandl, 1994). Smith et al. (1990) suggest a role for mechanisms that participate in alternative splicing such as the small nuclear ribonuclear protein particles (snRNP), and they acknowledge that "the ability to place copies of the same gene under separate regulatory pathways" is an important factor (p. 184). The problem is more difficult than for more complex phenotypes, where multiple epigenetic influences provide multiple potential pathways for divergent change.

Duplication in the Origin of Novel Morphologies

The evolution of body segments is a clear case of the role of duplication as a step in the origin of morphological novelty. In annelids, arthropods, and chordates, ancestral forms are believed to have evolved through a stage of segment multiplication (Barnes, 1987), followed by specialization of different segments to such a degree that in some cases their serial homology is barely recognizable in modern derived forms. The insect head, for example, is derived from four highly fused segments, each corresponding to a different specialized structure or the adult head and mouthparts. But the homologous nature of the segments is revealed in both anomalies (e.g., antennapedia in Drosophila and other insects—see Beeman, 1987) and experimental embryology (e.g., on the homeobox genes of *Drosophila*, see Garcia-Bellido, 1977, p. 625; on vertebrates, see Holland, 1992; see also Gerhart and Kirschner, 1997). The homeobox genes themselves appear to be products of duplication, as already mentioned.

A possible example of a macroevolutionary innovation in plant morphology initiated by a dupli-

cation is the evolution of endosperm, the embryo-nourishing tissue in angiosperms (flowering plants). Recent comparative studies of reproductive biology suggest that endosperm evolved when a redundant, duplicate embryo formed by double fertilization, then specialized in nourishment of its genetically identical sibling embryo, replacing the nutritive function of the maternal gametophyte found in nonflowering plants (Friedman, 1995). The triploid endosperm characteristic of angiosperms was apparently a later modification.

Duplication in the Origin of Novel Behaviors

There are many examples of duplication—repetition in a new context—followed by modification of behavior, especially in the origin and evolution of social displays (figure 10.2). Comparative studies suggest that ritualized displays of many kinds (e.g., those used in courtship and "appeasement" or subordinance) have originated via duplicated expression, in a new context, of a variety of behaviors, including movements used in locomotion, heat regulation, "comfort" or settling down, and prey capture, in addition to feeding behaviors and infantile vocalizations (Blest, 1961; Eibl-Eibesfeldt, 1970; Wickler, 1967; Moynihan, 1970). For example, courting roosters duplicate in a new context the pecking movements used in feeding. This is an effective mate-attracting signal because foraging chicks and adults use such movements as cues in their search for food and commonly move toward individuals seen pecking the ground. Male pecking movements are subsequently elaborated into some of the most complex courtship displays known in animals (Eibl-Eibesfeldt, 1970, p. 103). Another example of such a "sensory trap" used in courtship is the imitation of infant cries by males, who thereby capitalize on the strong tendancy of females to respond (see West-Eberhard, 1983, after Moynihan, 1970). "Sensory exploitation" (Ryan, 1990) is a similar phenomenon but refers to manipulation by signals tuned to match the sensitivity of a particular receptor system. The peak sensitivity of a receptor, however, is not necessarily linked to evolved strength or compulsion of response, so sensory exploitation may or may not constitute a sensory trap. Because examples of so-called ritualized behavior have been treated in the ethological literature as changes in timing rather than duplications, I discuss them more thoroughly in chapter 13, on heterochrony.

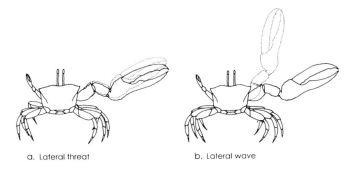

a. Lateral threat   b. Lateral wave

Lateral wave patterns

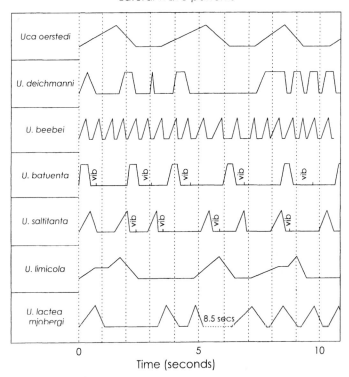

Fig. 10.2. Duplication in the origin and evolution of displays in fiddler crabs (*Uca* species). The lateral threat posture of alarmed crabs (a), is duplicated and exaggerated as a lateral wave (b), a movement that is in turn serially duplicated in different stereotyped patterns in different species of the subgenus *Celuca* (below) and performed by territorial males toward other males and toward females prior to mating attempts. Lines indicate claw movements; a plateau indicates that the claw is held in place at a wave's peak; vib indicates vibrations more rapid than 1 per frame on film exposed at 24 frames per second. After Crane (1975).

The evolution of social interactions resembles the evolution of development in its use of preexisting elements. Strongly incorporated essential responses, evolved in nonsocial contexts, are played upon by other individuals to manipulate the phenotypes of companions to their own advantage. Hundreds of proposed examples of manipulative behavioral duplication are found in the details of courtship and copulation in animals (Eberhard, 1985, 1996). Males produce a fantastic array of intricate movements and stimulatory chemicals that appear to increase the probability that females will accept, retain, and utilize their sperm rather than those of other males. Among the tactics that may involve duplication are the production by the males of some species of femalelike hormonal substances known to affect key processes in the female reproductive tract. It remains to be seen whether this chemical mimicry involves duplication (male expression of developmental pathways originally evolved in females—see chapter 15, cross-sex transfer) or convergent evolution of a functional analog without reuse of preexisting traits.

Simple serial duplication, or repetition, is exceedingly common in social displays (figure 10.2; see also figure 7.3).

Concerted Evolution and Diversification in Multigene and Multiphenotype Families

Iterative duplication creates families of traits of two general kinds: sets of identical traits and sets of divergent serially homologous traits. Multigene families of duplicated DNA are well-known examples. *Concerted evolution*, or a synchronous series of changes in nucleotide sequence among the members of a multigene family, can occur by various mechanisms such as gene conversion, the transfer of DNA between nonallelic genes (see below; John and Miklos, 1988). Hughes (2000) applies the term "concerted evolution" to the formation of a divergent set of duplicates within a multigene family, by multiple duplications of one divergent member, a phenomenon that might better be called secondary duplication (duplication of a diverged duplicate).

Concerted phenotypic evolution can occur in multi*phenotype* families as well, when there are serially homologous traits ("iteration"; Nijhout, 1991a) whose expression is controlled by the same regulatory mechanism. Then, distinctive subfamilies within families may evolve when a subset comes under independent control. Segmented bodies or appendages, for example, often have a series of identical duplicated segments alongside others that have diverged (figures 10.3 and 10.4). The homologous compartments of butterfly wings sometimes evolve in concert and sometimes diverge, with some genes expressed in a whole set of compartments and others expressed in only one (Nijhout et al., 1990). Bateson (1894), in a passage with a decidedly modern tone, long ago recognized that the duplicated structures of a meristic series may vary "simultaneously and collectively" and "yet it is true that in variation single members of such series may vary independently and behave as though they pos-

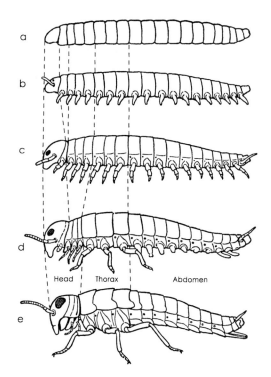

Fig. 10.3. Duplication, followed by diversification, fusion, and deletion in arthropod body segments: (a) hypothetical wormlike ancestor; (b) onychophoran, with simple appendages serially duplicated; (c) myriapod and (d) wingless (apterygote) insect, with divergent specializations among appendages; (e) winged (pterygote) insect, with fused head segments, further specialization of appendages of head and thorax, and deletion of abdominal appendages. After Snodgrass (1935 [2000]).

sessed an 'individuality' of their own" (p. 67). This pattern of parallel yet divergent iterative variation in meristic structures has been compared to the theme and variations of a Bach fugue (Wilson, 1986). It depends upon the pattern of regulation within multiphenotype, or iterative, families of traits.

How can duplication and regulation of phenotypic parts sometimes lead to correlated change and sometimes to divergence? To visualize this at the level of the phenotype, Nijhout (1991b) depicts the border of a developmental compartment as a ridge defined by contour lines like those of a topographic map, with development proceeding as waves in a fluid across an undulating landscape—a reasonable depiction of development since it is in fact believed to be governed by waves of chemicals turning genes

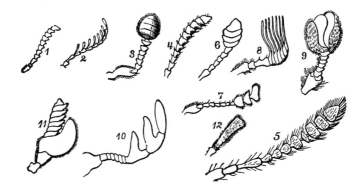

Fig. 10.4. Duplication and diversification in a multiple-phenotype family, the antennal segments of beetles (Coleoptera). Modifications include deletion of segments, as in the two-segmented antennae of *Adranes caecus* (12). From Kellogg (1904).

off and on according to patterns influenced by compartmental borders. Duplication produces a series of such ridges. Then, starting with a series of duplicated patterns, one can visualize that some changes would be propagated as uniform waves over the ridges, whereas others would show local alterations or divergences confined to certain ones, due to differences in position and surroundings—aspects that could change under selection. The result is both concordant and divergent change. Nijhout (1991a,b) illustrates this with the evolution of color patterns in butterfly wings, an elaborately compartmentalized iterative two-dimensional system. Homologies occur between front and hind wing, as well as between dorsal and ventral surfaces and within the wings, whose veins mark out smaller serially homologous compartments. Fore and hind wings show both correlations due to overlap in gene expression (pleiotropy) between these two homologous subunits of the phenotype, as well as independent variation due to expression of independent sets of genes (figure 10.5). This interpretation is supported when artificial selection on a single spot produces correlated effects on others (Monteiro et al., 1994; figure 10.6).

Selection on multigene families may sometimes prevent divergent modification. Massive duplication sometimes functions to increase amount of gene product and is associated with concerted evolution of the duplicated genes. Interactions during crossing over and gene conversion are thought by some to be mechanisms for concerted change within multiple gene families (Ohta, 1991). *Gene conversion*, a term that, surprisingly, is more than 50 years old (Klein, 1986, attributes it to Linde-

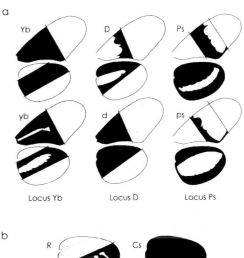

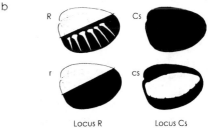

Fig. 10.5. Correlation and independence in serially homologous (duplicated) traits: the fore- and hindwings of butterflies. (a) Color-pattern phenotypes produced by alternative alleles at three loci that affect pattern in both fore and hind wings in *Heliconius* butterflies. Even though these patterns are genetically correlated, the effects of the alleles are different in the two backgrounds. (b) Phenotypes produced by two loci that affect only the hindwings. After Nijhout et al. (1990).

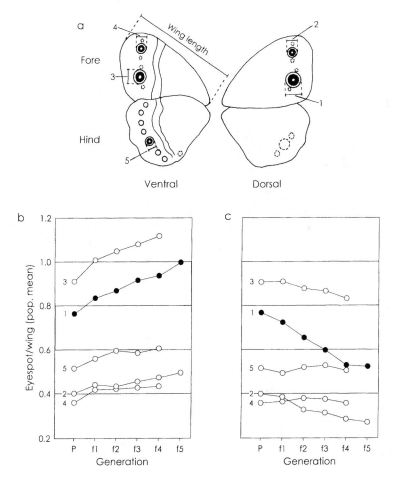

Fig. 10.6. Concerted response to selection in the wing spots of a butterfly (*Bicyclus anynana*): (a) spots (1–5) measured, (b and c) change in spots 1–5 during successive generations (f1–f5) of selection for (b) larger size of spot 1 and (c) smaller size of spot 1. Response to selection of spots 2–5 (white circles) was a correlated result of selection on spot 1 (dark circles). The result is concerted evolution of the homologous spots. After Monteiro et al. (1994).

gren, 1949), is transfer of portions of DNA from one gene to another. It was first hypothesized to explain abnormal segregation in fungi. The original hypothesis pictured a switching of the replication mechanism from one allele to its homolog (Lindegren, 1949; see Klein, 1986). As the term *concerted evolution* is sometimes applied to eukaryotes, it refers to transfer of DNA sequences by some mechanism between nonallelic genes. There is some evidence for the involvement of transposable elements (Thompson-Stewart et al., 1994). Divergence would interfere with these kinds of interactions. Ohta (1991) cites Walsh (1987) as suggesting that, when twice the mutation rate is larger than one tenth the conversion rate, the two

duplicate genes will be sufficiently different to evolve independently (see also Clark, 1994, on the maintenance of newly duplicated genes).

The pleiotropic participation in diverse traits established by duplication must often limit the divergent evolution of duplicates. But concerted phenotypic change can sometimes be adaptive, as it is in the concerted evolution of gene families, when there is a premium on quantitative increase in the same product. As pointed out by Nijhout (1991b), "the eyespot on a peacock's tail feather is nice, but a whole fan of them is spectacular" (p. 61). Indeed, animal calling and courtship signals are often repetitious, and the precision and rate of repetitions are known to influence their effectiveness (reviewed in Andersson, 1994, p. 353). When such correlations among parts have adaptive consequences, selection would reinforce concerted change in the duplicates, forming multiple phenotype families. For example, the serial uniformity of a centipede's legs must be important for coordination, as well as a cause of exaggeration by multiplication of adaptive change (figure 10.1).

It would be interesting to know if there is ever a connection between massive gene duplication during gene expression, of the kind that occurs, for example, in the ovarian nurse cells of insects (Büning, 1994), and the ease (frequency) of duplication of germline genes of the kind that contributes to genome evolution. The production of large amounts of yolk proteins is preceded by many-fold duplications of the expressed genes. Does a history of such "programmed" duplication ever promote structural changes in DNA that might increase the susceptibility of a germline gene for duplication? If certain genes evolve under selection for increased facility of duplication, it is conceivable that the same properties could facilitate their duplication in the germ line.

An important aspect of concerted variation in a phenotype set is that it can contribute to concerted facultative responses. Butterfly wing patterns are good examples of iteration and concerted as well as divergent change (Nijhout, 1991a,b). In the seasonally polyphenic butterfly *Bicyclus anynana* (Nymphalidae), one form of the wing, with multiple eyespots, is expressed in the wet season, when the butterflies are especially subject to predation (eyespots are believed to distract predators from seizing the butterfly's body); a cryptic form lacking eyespots is expressed in the dry season (Brakefield and Larson, 1984; Brakefield, 1987). Both temperature and larval development times influence the expression of the eyespots (Brakefield and Reitsma, 1991; Windig, 1993). In this situation, the concerted regulation of the multiple eyespots would be an advantage, as may their concerted evolution. The correlations should be maintained by selection, not broken as they are in the case of some duplicated structures. Consistent with this, Monteiro et al. (1994) demonstrated highly correlated responses to selection on the size of these developmentally homologous characters (figure 10.6).

Elder and Turner (1994) documented geographically variable concerted evolution of a satellite DNA array in pupfish (*Cyprinodon variegatus*), one of few demonstrations of intraspecific concerted gene evolution in nature. Sequences were consistent within populations and individuals and divergent between populations.

11

Deletion

Deletions give the impression of regressive evolution, yet they are reponsible for some major evolutionary innovations, such as the origin of direct development in echinoderms and body-plan simplification in the most derived arthropods. Deletions can lead to loss of life stages and loss of sexual reproduction in both plants and animals. Disruptive selection against intermediate phenotypes can lead to their developmental deletion, producing a clean switch between extremes, as in the evolution of workers and queens from continuously variable female phenotypes in social insects. Developmental deletion of phenotypes shows that lost traits need not involve lost genes, through evolved or induced changes in gene expression.

Introduction

Deletion, or trait loss, may seem a step backward rather than a step toward something new. Goldschmidt (1940) emphasized the regressive aspect of deletion by calling it "rudimentation." But trait deletions create novelties by subtraction, in at least four different ways. First, a complex trait lacking an element may immediately have an altered function. A worker honeybee, for example, is a mature brood-tending female minus the ability to lay eggs, and a queen is a solitary female minus the ability to care for the brood. These reciprocal, complementary deletions, reinforced by kinship, make the two kinds of female into mutually dependent collaborators. Second, the loss of a trait may have correlated developmental effects that force the remaining phenotypic elements into new configurations, as with the virtually deleted forelegs of the two-legged goat (see chapter 3). Third, a deletion, if it is repeatedly produced, makes the resultant phenotype subject to divergent evolution under selection, simply because it is different. Finally, deletion of a phenotypic subunit can release other, genetically and developmentally correlated traits from the evolutionary constraints represented by these correlations, freeing the remaining traits to evolve more rapidly and independently.

Deletions can evolve gradually, by change in regulation to reduce the frequency of expression of a trait, as in loss of an alternative phenotype, or by change in form such that elements of the phenotype are gradually lost, as in flight reduction in insects beginning with loss of flight behavior, then wing musculature, then wings (Shaw, 1970; see figure 5.26). Classical gradualism refers to the latter type of change—gradual loss of elements of form—as suggested by the occurrence of mosaic intermediates (e.g., a flightless population that possesses wings). A deletion occurs every time an alternative phenotype evolves to fixation, which means that its former alternative is no longer expressed. Since this is a step in the evolution of many constitutive qualitative trait (see part III), regulatory deletions of alternative phenotypes must be common events. As with other kinds of phenotypic change, deletions occur at different levels of organization, from pieces of genes to elements of behavior and whole life stages of individual development.

Melanophore Deletion in the Midas Cichlid: Novelty by Subtraction

A simple deletion with a striking effect, and immediate consequences for direction of selection, occurs in the color metamorphosis of the midas cichlid, *Cichlasoma citrinellum* (Dickman et al., 1988). In clear lakes, all individuals are gray with black markings. In murky lakes, about 8% of the adults have undergone a color change that involves the death of superficial melanophores and the consumption of their remains by macrophagelike cells, revealing bright gold or yellow-to-red coloration beneath. Variation in the capacity to undergo this change is inherited, but the triggering and regulatory mechanism is unknown. Cell death may occur due to lack of neural stimulation (Dickman et al., 1988, p. 12), suggesting that environmental conditions could play an important role. Once color metamorphosis is complete, it is irreversible: individuals cannot alter their hue toward the more cryptic dark colors and may therefore be at a disadvantage under predation. On the other hand, gold coloration is socially advantageous, for gold fish dominate dark-colored and white ones and have better mating success. Dickman et al. note that many kinds of fish show a similar color metamorphosis, indicating that melanophore deletion is readily achieved developmentally. This example shows how a phenotypically simple, and possibly environmentally initiated, deletion can produce an adaptively important novel trait.

Deletion in the Evolution of the Arthropod Body Plan

Just as segmentation and compartmentalization have contributed, via duplication and modification, to the evolution of arthropod diversity, they have also provided the conditions for segment-specific deletions. Arthur (1984, p. 200) points out (after Lewis, 1978) that the evolution of insects has apparently involved deletion of legs from the abdominal segments of millipedelike ancestors. Most elements of the hind wings have been deleted during the evolution of flies (Diptera), leaving only wing rudiments (halteres) intact. If it is correct that appendage loss occurs due to changes at switch points, one expects to sometimes see variants (atavisms) where the deleted appendage is restored. Both leg-restoring and wing-restoring mutants occur in *Drosophila*, and they involve only a small cluster of pseudoallelic genes (the bithorax complex; John and Miklos, 1988). These patterns suggest a role for deletion in macroevolutionary change.

Life-Stage Deletions

Most organisms change form as they grow and mature. These changes can be products of continuous,

correlated variation in traits whose dimensions depend on the rate and duration of their growth or development (e.g., Atchley and Hall, 1991). In addition, growth and maturation can involve switches marking relatively sharp changes in behavior and morphology, as in the hormonal revolutions of metamorphoses. Many organisms have complex life cycles or alternation of generations, with drastically different cyclic reproductive morphs such as gametophyte–sporophyte, or contrasting host-specific specializations. A single genotype produces unusually profound periodic alterations of the phenotype in these life-stage polyphenisms. Unlike the life stages of individual ontogeny, each stage is independently capable of reproduction and therefore may be potentially highly dissociable from the others should selection favor its independent existence. In migratory organisms such as salmon, and the many seasonally polyphenic animals and plants (see part III), extensive switching occurs between different extreme states of physiology, behavior, and sometimes morphology that is similar to life-stage switching but may occur repeatedly or cyclically during the lifetime of a single individual. These cyclic phenotypes are subject to massive deletions involving very large portions of the phenotypic repertoire of a genome, because of their highly compartmentalized structure and dissociability. Massive deletions are illustrated by the following examples.

Recurrent Deletion of the Medusa in Hydromedusae (Cnidaria)

The hydroids and medusae of hydromedusae (Cnidaria, Hydrozoa) were introduced in chapter 7 as examples of highly modular, independently evolving life stages. Rather than being dependent forms, like the larva and adult of a butterfly, neither of which can reproduce independently, the hydroids and medusae are independent morphs in that both can reproduce, the hydroid polyps asexually and the medusae sexually. As is generally true of highly modular features, these two life-stage forms are highly dissociable, and many hydrozoan lineages, such as the thecate families Aglaopheniidae, Plumulariidae, and Sertulariidae, have lost, or greatly reduced, the medusa stage (Boero and Bouillon, 1987, 1989a). In such lineages, the hydroids become the "adults" and sexually reproduce (for a review of paedomorphism in hydroids, see Gould, 1977). In at least one species, *Laodicea indica* (Cnidaria, Leptomedusae, Laodiceidae) in the Bismarck Sea near Papua New Guinea, a normal al-

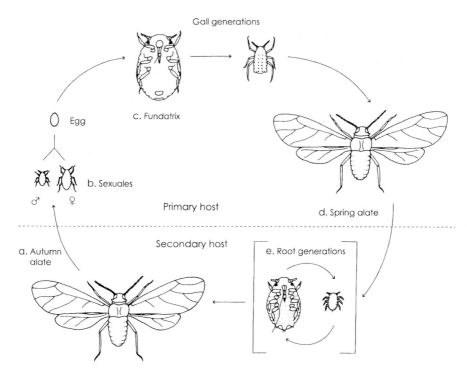

Fig. 11.1. Life-stage deletion in aphids. The diagram shows a typical life cycle of aphids of the family Pemphigidae. In *Pemphigus betae*, the autumn alate and the gall generations of the primary host are deleted, leaving only the root generations (e) found on the secondary host (enclosed in brackets; Moran, 1991). After Akimoto (1985).

ternation of hydroid and medusoid generations occurs during the wet season, but in the dry season the planula larvae give rise directly to medusae— the hydroid stage is facultatively deleted (Bouillon et al., 1991). True to a pattern seen in many examples of recurrent transitions (see chapter 19), facultative reduction of the medusa is seen seasonally in some species (e.g., *Clytia linearis*; Campanulariidae; Boero and Sará, 1987).

Multistage Deletions in Aphids

In aphids, deletion of one or more host-plant phases from complex life cycles has occurred repeatedly in many genera of the several families showing host-plant alternation (Moran and Whitham, 1988; Moran, 1991, 1992a). Some of these deletions involve enormous ontogenetic sequences including several discrete morphs.

A complete life cycle in a typical species (*Pemphigus betae*) of the family Pemphigidae contains five morphologically and behaviorally distinct morphs (Moran, 1991; Akimoto, 1985; figure 11.1): (a) a winged migratory form that produces sexual offspring of both sexes by ovipositing under the bark of trees; (b) a nonfeeding, wingless sexual form that produces one female-producing egg per female under bark; (c) a parthenogenetic wingless female (fundatrix) that induces galls on leaves, feeds, and produces a large number of winged migrants; (d) a winged migrant that leaves the galls to locate secondary hosts, where they produce a brood of females; (e) females that crawl into the soil and live and reproduce on roots. The most common type of life-cycle reduction is "anholocycly"— omission of the winged migrants that in the ancestral alternating cycle colonized the primary host. In some populations of *Pemphigus betae* where cottonwood trees (the "primary" host) are unfavorable to aphids due to lack of crossing with other species, clones remain on secondary hosts, where they reproduce without the winged phase: stages a–d are deleted, fixing the wingless secondary-host-inhabiting parthenogenetic form (Moran, 1991). This may occur in a single step, for sometimes alternating and anholocyclic lineages coexist within the same population, and there are no intermediates, since deletion of the winged migrant (e.g., by a failure of metamorphosis) automatically deletes the other primary-host-inhabiting generations to which she would give rise. Matsuda (1987, pp. 189ff) noted the susceptibility of aphid metamorphosis to the influence of temperature, photoperiod, and crowding, and implicated these environmental factors in the initiation of evolutionary change.

Recurrent deletions of host-plant-specific generations in aphids could occur via loss of the primary host and transfer of its resident morphs to the secondary host (Moran, 1992a), a suggestion similar to the "generation packing" hypothesis of Aoki and Kurosu (1986, 1988). This amounts to the deletion of an environment, only secondarily of a morph insofar as the phenotype was distinctive on the deleted host. In terms of the phenotype itself, a relatively minor behavioral deletion might be involved, deleting only the ability of the winged migrant to disperse between hosts.

Heteromorph Ammonites

Major morphological deletions occur in the fossil heteromorphic ammonites (Cephalopoda; Raff and Kaufman, 1983; see figure 7.4). Ammonite shells record the ontogeny and developmental recombination of three growth patterns: coiled, uncoiled, and helical. The ancestors of the heteromorphs were coiled, and their fossil record shows slow and gradual evolution of continuously variable details such as size, tightness of enrollment, complexity of suture patterns, and coarseness of ribbing (Raff and Kaufman, 1983, p. 40), with no major change in overall form. Then, in the Late Triassic, there ensued an era of profound and rapid changes in shell configuration (figure 7.4), producing a variety of forms with some coiled, some uncoiled, some spiral, and some with discrete serial growth stages showing two of these modes or even all three. Raff and Kaufman (1983) underline the contribution of switch-defined modularity to these patterns, characterized by sudden change: "The rapid evolution of heteromorphs suggests that the flexibility provided by switches in growth mode was exploited in ammonite evolution" (p. 286).

The Jack Male of Salmon

Deletion can be associated with acceleration of reproductive development. Some developmental changes usually classified as heterochrony, such as progenesis, or precocious gonadal maturity, cause failure to express certain life stages. For example, the "jack" phenotype of Pacific salmon (e.g., the coho, *Onchorynchus kisutch*) is a precociously mature male with juvenile body form. It develops mature gonads after only one year of oceanic life rather than the three years required for development of the large hook-nosed-male phenotype (Gross, 1984, 1985). Precocious maturation of the jack male deletes a substantial growth phase and the complex specialized morphological and behavioral phenotype characteristic of the hook-nosed form with its distinctive head, dorsal hump, and coloration. The large hooked snout and enlarged teeth are used in grappling and biting opponents as well as influencing female choice (Fleming and Gross, 1994). The jack males obtain mates by sneaking, so behaviors of fighting and courtship are unexpressed. Intermediate-sized males are at a mating disadvantage (Gross, 1984; Fleming and Gross, 1994). There is evidence that the precocious maturity of the jack male is associated with a size differential among fry (Gross, 1991). The result is a striking polyphenism among breeding males.

Larval-Stage Deletions in Echinoids and Other Marine Invertebrates

Sea urchins (Echinodermata) are among the "Big Five" taxa of cellular and molecular embryology, along with fruitflies (*Drosophila*), frogs (*Xenopus*), nematodes (*Caenorhabditis*), and mice (*Mus*). The echinoids are the premier group when it comes to *comparative* studies of development. Information on development is available on at least 100 species in 10 different orders of echinoids (Raff et al., 1990, based on Emlet, 1986; see also Emlet and Hoegh-Guldberg, 1997). This situation has been exploited to achieve a truly comparative evolutionary and experimental study of development among related species (see discussions by Raff et al., 1990; Strathmann et al., 1992; Sinervo and Basolo, 1996; Emlet and Hoegh-Guldberg, 1997).

Sea urchins typically have a free-living, plankton-feeding pluteus larva. This larva in echinoderms is so similar to the larvae of hemichordates that direct development via a pluteus stage was likely ancestral in both groups (Strathmann, 1978). Deletion of the pluteus stage, or "direct development," has occurred independently in 6 of 10 echinoid orders (Raff et al., 1990) at least 27 times (Emlet and Hoegh-Guldberg, 1997). As expected, evolutionary lability can be related to developmental plasticity: direct development sometimes occurs as an alternative phenotype. In the genus

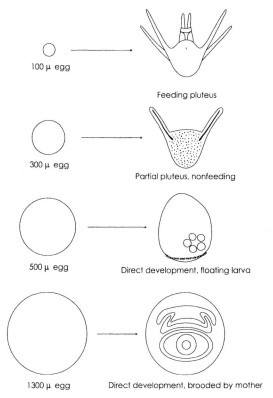

100 μ egg

Feeding pluteus

300 μ egg

Partial pluteus, nonfeeding

500 μ egg

Direct development, floating larva

1300 μ egg   Direct development, brooded by mother

Fig. 11.2. Egg sizes and modes of development in sea urchins: (a) a typical species; (b) *Peronella japonica*, which lacks a gut and has a reduced pluteus skeleton; (c) *Heliocidaris erythrogramma*, lacking all pluteus structures except the vestibule; and (d) *Abatus cordatus*, which develops juvenile structures (coelom, ectoderm, intestine, mesenchyme, and tube-foot rudiment), with little vestige of pluteus morphology. After Emlet et al. (1987), Raff (1987), and Raff et al. (1990).

Clypeaster, there are both facultatively feeding and obligatory feeding species (Emlet, 1986; see also Emlet, 1990). Some genes in sea urchins are expressed only in the feeding pluteus larva (Davidson, 1986, p. 232). So the evolution of direct development implies the silencing or heterochronic expression of those genes (Wray and McClay, 1989; see chapter 13).

The hierarchical, or internal subunit, structure of the pluteus larva is reflected in the fact that its deletion can be partial or complete (figure 11.2). Subunit organization is also shown by comparative study of larval ontogeny. Pluteus larvae typically develop in the sequence archenteron → skeleton → gut → coelom → hydrocoel → vestibule (Raff et

al., 1990). In two different direct-developing species, the larval gut and skeleton are deleted, giving the sequences archenteron → coelom → hydrocoel → vestibule in *Heliocidaris erythrograma* and archenteron → skeleton → vestibule → coelom → hydrocoel in *Peronella japonica*. This shows that different cell lineages within the larval stage are dissociable from one another and develop relatively autonomously. So there is no reason to suppose that all cell lineages undergo the same heterochronies or deletions (Raff et al., 1990). In *P. japonica*, the *sequence* of principal event is altered, with the vestibule forming before rather than after the coelom and the hydrocoel. This shows that ontogenetic events do not necessarily have to occur in a particular order but can be shuffled.

Changes of this sort are in fact quite common in the evolution of development, with similar end points produced by very different pathways, as mentioned in chapter 9 (see also chapter 25, on homology; Gerhart and Kirschner, 1997). The sea-urchin findings contradict the common belief, discussed in chapter 2 (see also Raff et al., 1990, p. 75), that evolutionary change by terminal addition is expected because change in early development is virtually impossible due to the netlike organization of development and the strict necessity for a particular order of events. The evolution of direct development and deletion of the feeding pluteus are further discussed in chapter 26.

Deletion in the Evolution of Behavior

Behavioral sequences, like developmental ones, often involve orderly sequences of cues that trigger specific events in a particular order. In some sequences, a preceding activity creates the cue for performance the next act. For example, in wasps (*Polistes*), a mature ovarian egg or its hormonal correlate is associated with the initiation of a new cell on the nest (see West-Eberhard, 1969); then the presence of an empty cell stimulates oviposition. But the cell-construction step can be deleted if a female encounters an empty cell built by another female or vacated by an emerged offspring adult, and cell reuse is quite common even in solitary and primitively social species (e.g., West-Eberhard, 1978a, 1986; Kurczewski and Spofford, 1998). This is an important preadaptation for social life, since the reproductive division of labor of highly social species concentrates building behavior in workers who do not themselves lay eggs, and all oviposition is by queens who do not themselves build cells. Deletion of construction behavior also occurs in brood parasitic wasps, bees, and birds, which place eggs in nests built by others without nest building themselves.

Similarly, brood care by worker females that have not themselves laid eggs may be stimulated out of sequence when a subordinated female encounters a hungry larva, even though it is not her own offspring, thus causing her to skip ahead in the normal reproductive cycle, deleting the oviposition phase. If such behavior happens to be advantageous (e.g., when the hungry larvae are genetic relatives), selection may favor maintaining such altered behavioral sequences in the new context (West-Eberhard, 1988).

Helping behavior by at least temporarily nonreproducing adults occurs in a variety of birds and mammals (J. L. Brown, 1987; Stacey and Koenig, 1990; Solomon and French, 1997) and is often interpreted as originating as misplaced parental care (i.e., parental care stimulated, e.g., by the presence of young, in a new situation; see references in Craig and Jamieson, 1990), which persists if it is favored by selection. In canids, social suppression of subordinate adults prevents their mating but is not strong enough to drive them away or prevent estrus and false pregnancies, which are accompanied by maternal solicitude toward the young of dominants—possibly the selective basis for tolerance (Asa, 1997). The evolution of helping based on finely modulated suppression by dominants would involve the adjustment, under selection, of the socioendocrine interaction: the "genes for helping" are the many modifiers of endocrine and behavioral systems that respond to selection on reproductive and dominance levels. From a developmental point of view, alloparental care can be described as behavioral heterochrony—a shift in the timing of brood care so that it precedes reproduction—or, if permanent as in the workers of eusocial Hymenoptera, a deletion of the oviposition phase of reproductive behavior.

Other kinds of alternative behavioral tactics can originate via deletion of serial behaviors when an individual physiologically primed for a particular behavior happens to encounter the cue eliciting it, even though out of sequence. This may explain the occurrence of entering preexisting holes rather than digging new ones in the digger wasp *Sphex ichneumoneus* (Brockmann, 1976, 1979). Opportunistic use of ready-built nests rather than nest digging can be expected to occur in proportion to the number of nests available to be entered, especially

if new nests are costly to dig. Brockmann's (1976) data support these expectations: use of abandoned holes depends on number available per searching wasp in a nesting area as well as on the time cost of digging new holes. Whatever the mechanism or function of this enter-rather-than-dig behavior, females that enter ready-made holes perform a nesting sequence in which digging is deleted from the usual sequence dig → provision → oviposit → close.

The nests of social wasps record the behavior of their builders and provide a structural record of behavioral ontogeny and evolution. This has been exploited by Wenzel (1991, 1993, 1996). His comparisons of nest architecture with a cladogram based on other characters (Carpenter, 1991) indicate the recurrent deletion of the pedunculate support structure and, in some species of one lineage (genus *Agelaia*), deletion of the nest envelope.

In vertebrates, threat displays can represent truncated aggressive sequences, where some elements of aggressive behavior have been retained and others deleted. For example, gaping or baring the teeth may occur without pecking or biting; claw raising and waving without striking or grasping in crabs; and jumping forward briefly without actually attacking in badgers and squirrels (Eibl-Eibesfeldt, 1970, p. 95). I have observed threat displays in *Polistes* wasps where dominant females adopt a stinging posture without stinging (West-Eberhard, 1982). Approaching nestmates withdraw rather than continuing to approach displaying females.

In behavioral deletions, as in morphological ones, partial omissions are possible, with some elements deleted while others remain. For example, the red-headed finch (*Amadina erythrocephala*) uses nests of other birds yet performs nest-building movements while sitting on the nest, pulling in nonexistent materials as if it were forming a nest (Eibl-Eibesfeldt, 1970, p. 193). Unless this behavior accomplishes some unknown function, it argues either for a recent deletion of nest building or for persistence of "dispensable phenotypes," even though presumably costing some energy, without selection to maintain them.

Deletion of the Male Phenotype in Unisexual Flowers and Fishes

In the evolution of flowering plants, unisexual flowers have apparently been derived from bisexual flowers by the gradual reduction and deletion ei-

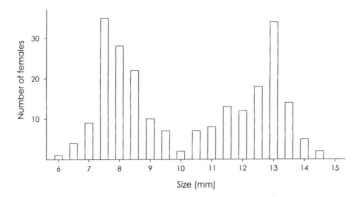

Fig. 11.3. Environmental deletion of intermediates in the parasitoid *Dasymutilla bioculata* (Hymenoptera, Mutillidae). Adult body size depends on host size: large adults (above 10 mm in body length) feed on cocoons of the sand wasp *Bembix pruinosa*, whereas small adults (10 mm and below in body length) feed on cocoons of *Microbembix monodonta*. After Mickel (1924).

ther of stamens to produce female flowers, or of carpels to produce male flowers (Takhtajan, 1969). Many different genera of plants with unisexual flowers have vestigial organs or rudiments indicative of a former bisexuality and suggesting a gradual rather than one-step transition. In some species, bisexual flowers occur as an atavistic anomaly (Takhtajan, 1969, pp. 14–15), showing that one-step regulation is developmentally possible.

Unisexuality by deletion also occurs in the Amazon molly, *Poecilia formosa* (Poeciliidae), a clonal all-female fish that originated via hybridization of *P. latipinna* and *P. mexicana*, a mode of initiation that suggests a saltatory origin. Sperm from one of the ancestral male-containing species is required to initiate embryogenesis, but the diploid eggs contain only the maternal genome. When newborn *P. formosa* were treated with androgens until mature, they developed a full set of male traits resembling those of the ancestral species, including male body proportions, pigmentation, complex insemination apparatus, sexual behavior, and spermatogenesis (Schartl et al., 1991). This superb example of phenotypic engineering indicates that the very large, complex subunit of the phenotype corresponding to male-specific morphology and behavior can be deleted in a coordinated manner via hormonal change, then reversed with an apparently simple manipulation. The impression of potential evolutionary lability is supported by the fact that phenotypic males of *P. formosa* have been collected, albeit rarely, in natural habitats and sometimes occurs spontaneously in laboratory stocks (Schartl et al., 1991, p. 8761).

Deletion of Intermediates: Disruptive Development Due to Disruptive Selection

A common kind of variation in natural populations is a bimodal frequency distribution of some continuously variable trait (see figure 5.20). Sex differences are common examples of bimodality caused by a developmental switch. Bimodality can also occur as a side effect of other processes, such as seasonally pulsed reproduction or different prey species, which can delete the intermediates from a fundamentally continuous distribution. Thus, for example, a bimodal distribution of adult sizes occurs in a parasitoid mutillid wasp that preys on two discrete sizes of host (Mickel, 1924; figure 11.3). Each parasitoid larva consumes an entire, single host pupa, and the discontinuous size distribution is the result of oviposition on two discrete-sized hosts: a large sand wasp, *Bembix pruinosa* (adult length, 16–19 mm) and a small species, *B. bioculata* (adult length, 8–14 mm). There were no intermediates in the total sample of 440 females and 206 males from which were drawn the 232 females represented in the graph (the approximately equal numbers of large and small females may not be representative of the population as a whole). Host-size association was confirmed by laboratory rearing from pupae of known host species (Mickel, 1924). Because of their discrete size differences, the two size classes were formerly classified as different species, but males and females of the two sizes were observed to cross in both field and laboratory

(Mickel, 1924). No morphological differences, including in male genitalia, between size classes could be found. Males are wingless, and perhaps as a consequence of a greater investment in growth rather than wing musculature, are somewhat larger than females. Their sizes (not shown) are likewise bimodally distributed with no overlap between size classes (Mickel, 1924).

Host-dependent size variation also occurs in the egg parasite *Trichogramma evanescens* (Hymenoptera, Chalcidoidea), whose trimodal distribution of adult sizes depends on the egg size of three different hosts (Salt, 1940, 1941). Progeny of a single female 0.40 mm in length measured 0.46 mm when emerged from eggs of *Ephestia kuehniella*, only 0.34 mm when reared from an egg of *Sitatroga cerealella*, and up to 0.57 mm when reared from an egg of *Agrotis c-nigrum*). When females oviposited on eggs of a single species (*Sitotroga cerealella*) selected for contrasting large and small size, the size of the offspring was influenced by the size of host individual and departed from that of the mother (Salt, 1940, 1941).

Bimodal distributions can also result from disruptive selection against intermediates, or from frequency-dependent selection favoring different discrete forms (Mather, 1973). But theoretical models of disruptive and frequency-dependent selection do not directly address the mechanisms producing the bimodal distribution of forms, and they invite potentially misleading views of the genetic architecture of development and evolutionary change. One-locus and game-theoretic models of frequency-dependent selection suggest the presence of an allelic switch, or of genes of major effect on a developmental switch (e.g., see Maynard Smith, 1989), and bimodality results from the existence of two genotypic classes of individuals, each with one or the other of the alternative alleles. Models that begin with polygenic determination of continuous variation and see disruptive selection as causing a shift in allele frequencies involving many loci give the impression that each genotype has a particular phenotype, and that intermediate genotypes are eliminated leaving two continuous genotype distributions having different, extreme modes. Both models encourage thinking of disruptive selection as producing a bimodal distribution of genotypes underlying the bimodal distribution of phenotypes observed: "Natural selection diversifies the gene pool in such a way that two or several classes of genotypes having optimal adaptiveness in different subenvironments are produced" (Dobzhansky, 1977, p. 117). Disruptive selection is seen as increasing genetic variance (Thoday, 1972; Falconer,

1981), and increased genetic variance is sometimes taken to indicate that disruptive selection has occurred (e.g., Endler, 1986, p. 211; Grant, 1985).

Another kind of evolutionary response to disruptive selection is possible, and the study of alternative phenotypes indicates that it may be the most common one (see chapter 22). Disruptive selection can lead to the evolution of polymorphisms, polyphenisms, and alternative behavioral and physiological phenotypes, with the alternative adopted by an individual dependent on conditions or on a small number of loci influencing regulation. That is, the result of disruptive selection can be the evolution of *disruptive development*—alternative developmental pathways. This produces a bimodal distribution of *phenotypes* in the population without a bimodal distribution of genotypes. Intermediates are developmentally deleted by selection that favors ability to switch between extremes, rather than by elimination of intermediate genotypes from a continuously variable array.

The potential for disruptive selection to produce a developmental polymorphism has long been acknowledged by theoretical and experimental geneticists, as in Fisher's (1930) discussion of the nature of sexual dimorphisms and in the writings of Mather (1955, 1973) and Thoday (1972). As pointed out by Bradshaw (1965), "[T]he ease with which plants are able to respond to disruptive selection by plasticity is perhaps itself the reason why genetic polymorphisms are not conspicuous" (p. 126). But the idea that disruptive selection could lead to a plastic response was controversial. Surprisingly, Waddington, while a leader in relating development to genetics and evolution, was among those who most clearly expressed doubt, as revealed in an exchange of letters with P. M. Sheppard (Waddington, 1975, pp. 161–167). He wrote: "[D]oes anything yet show that if a population is subject to two different selection pressures, and yet the whole population mates completely at random in every generation, you achieve a bimodal distribution of phenotypes?" (p. 162). Waddington found polymorphism based on disruptive selection "as yet unproved, and pretty difficult to believe," and insisted that "the production of an environmental switch mechanism controlling two distinct phenotypes is a relatively rare outcome" (p. 162). This opinion is echoed in more recent discussions (e.g., Futuyma and Moreno, 1988) that continue to treat phenotypically complex alternative phenotypes as behaving as if they were allelic polymorphisms.

In the following examples of disruptive selection, the developmental deletion of intermediates has resulted in populations of genotypes capable of alternative developmental pathways. These examples are particularly instructive because they give insight into how switch mechanisms, like other novel aspects of form, are constructed during evolution by tinkering with preexisting responses, sometimes capitalizing on different pathways to the same end.

Disruptive Selection
Experiments in Drosophila

In experiments using disruptive selection for the extremes of a continuously variable trait, several investigators demonstrated the emergence of variants with a developmental switch mechanism eliminating intermediates (Scharloo, 1970; Thoday, 1972; Mather, 1973). A pioneer in this area was Thoday, who began by outcrossing two selected lines of *Drosophila melanogaster*, one with a high sternopleural chaeta number and the other with a low chaeta number. He then subjected the population to disruptive selection with negative assortative mating (high-chaeta-number females with low-chaeta-number males and vice versa). Both extremes of each sex were always used, permitting no consistent correlation between chaeta-number extremes and sex.

The result was "a reasonably stable polymorphism" for chaeta number, with a complex genetic basis: there were three switch-influencing alleles located together on chromosome III, and modifiers enhancing their effects, corresponding to the specific modifiers of figure 4.13, elsewhere in the genome. Thoday (1964) clearly saw the relationship of these findings to the broad class of developmental switches:

> [W]hile this example, as in sexual dimorphism . . . involves a *segregational* polymorphism [the switch is thrown by allelic differences between individuals], it makes it clear that the essential genetic features of the system need not involve genetic segregation. If, as in the parallel systems that involve differentiation of cells and tissues, as in the determination of gregarious and solitary phases of locusts, or in some sexual dimorphisms like that of Bonellia, a sufficiently *regular* environmental switch is available, it could function equally well, in relation with the background genotype which determines the effects of the switch. (p. 110)

In later work, Scharloo (1970, cited in Thoday, 1972) demonstrated the phenomenon predicted by

these remarks—the origin of an environmentally influenced polyphenism under disruptive selection. Using a negative assortative mating system and selecting on cubitus interruptus expression in *Drosophila*, he obtained a bimodal distribution that was sensitive to environmental variation.

Indirect Evidence That
Disruptive Selection Affects
Bimodal Distributions

According to Endler's (1986) criteria, there are few direct demonstrations of disruptive selection in the wild. He lists examples (p. 214) in dog whelks, *Nucella lapillus*, newts *Triturus vulgaris*, and lizards *Uta stansburiana* (after Bell, 1974, 1978), and the Galapagos finch *Geospiza conirostris* (Grant, 1985). Another listed example in *G. fortis* must be subtracted following a reanalysis of the data (Schluter, 1988, p. 859).

In contrast, indirect evidence for disruptive selection can often be provided in a dimorphic species when (1) the extremes within each morph perform better or incur a higher benefit–cost ratio than do intermediates showing a less extreme development of one of the alternative traits; and (2) there are clear performance differences between morphs with respect to contrasting tasks or situations. Classic examples are provided by the Batesian mimicry polymorphisms of butterflies, where a cryptic morph and a mimic morph resembling a distasteful model are produced, and intermediates showing imperfect mimicry have been shown to be more susceptible to predation (reviewed by Turner, 1977, p. 175).

Selection against intermediates has been demonstrated in studies of alternative phenotypes in males. Intermediate-sized males of Pacific coho salmon (*Oncorhynchus kisutch*) are at a disadvantage in mating competition with both larger, hook-nosed fighting males and smaller jack males that use a sneak mating tactic (Gross, 1985). Moczec (1996) showed that in aggressive interactions among horned males of the dimorphic beetle *Onthophagus taurus*, males with larger horns won more fights than intermediate males with small horns, independent of body size. Experimental horn removals showed that horns are a disadvantage in running through tunnels where males compete and must pass each other to pursue females. So males with intermediate-sized horns would pay a price in terms of mobility relative to hornless males, while also being poor fighters relative to large-horned males. There may also be developmental costs to the possession of horns that may weigh against intermediates bearing part of the cost without reaping the benefits of large horns: in some horned beetles, there is a negative correlation between size of head horns and size of eyes or wings (Nijhout and Emlen, 1998). Such trade-offs are quite common in morphological development (see chapter 16) and must sometimes affect the evolution of alternatives under disruptive selection, both limiting the extremes and affecting the costs and benefits that determine the position of a switch point between forms.

While it is clear from the experiments of Thoday and Scharloo that disruptive selection can convert a continuously variable trait into a discontinuous one governed by a switch, it is not always clear which came first in a given dimorphic species in nature, the disruptive selection or the developmental change. If a threshold effect were to appear as a developmental variant in a species, say, a beetle with continuously distributed horn sizes, preexisting plasticity in the use of the ancestrally variable trait could, by immediately capitalizing on the extremes, lead to the occurrence of disruptive selection against intermediates less successful in either realm. It is of course possible to use a cladogram to show the distribution of continuous versus discontinuous variation within a taxon, but this by itself would not reveal whether disruptive selection or disruptive development came first in a transition from continuous to discontinuous variation unless there were some additional clue regarding developmental and selective causes of the change. In this regard, it would be of interest to see if there are any bimodal distributions *not* accompanied by disruptive selection. A possible example in a mutillid wasp is discussed above (figure 11.3).

Mechanisms of Disruptive Development

What kinds of developmental mechanisms could evolve under disruptive selection to give rise to bimodal size distributions and diphasic allometries such as those seen in some beetles and social insects? It is clear that a developmental threshold effect is sometimes involved because in at least some species the two modes show different allometric relationships when plotted against size, even though size itself is continuously variable (on beetles: Eberhard, 1979; Eberhard and Gutierrez, 1991; Emlen, 1994; Moczec and Emlen, 1999; on wasps; Jeanne and Fagen, 1974). This indicates a change in epigenetic trajectory governed by a switch. The switch is subject to environmental influence, as shown by the effect of diet on individual phenotype (e.g., see

Emlen, 1997b, on beetles; Wilson, 1971a, on social insects). And it is susceptible to genetic influence as shown by artificial selection experiments that shift the critical size for the switch point (Emlen, 1996).

But the developmental–genetic architecture of the bimodal distribution is most clearly revealed by the finding of a hormonal basis for the switch: in the horned beetle *Onthophagus taurus*, there is a brief juvenile hormone (JH)-sensitive period at the end of the final larval instar, and the level of JH present at that time determines the horn morphology of males. Addition of JH during the critical period causes small males to develop horns (Emlen and Nijhout, 1999). So a genetically complex hormonal switch mechanism converts a continuous distribution of male sizes into a bimodal distribution of investments in horn production at a threshold larval size. How larval size is translated into a hormonal signal is not known in this species, but in some insects abdominal stretch receptors respond to gut distension in well-fed individuals and thereby trigger a hormonally mediated molt (Nijhout, 1979; see also Nijhout, 1984). Such a mechanism could serve as a cue reflecting trophic status and relative size. Correlated with the morphological switch are changes in behavior that affect the access of males to mates (Eberhard, 1979; Conner, 1988; Moczek, 1996; Moczek and Emlen, 2000; Emlen, 1997a). The hormonal basis of social insect polymorphisms is well known (reviewed in Nijhout and Wheeler, 1982; Engels, 1990). Condition-sensitive deletion of wings in crowded crickets is associated with JH action (Zera and Tiebel, 1988, 1989; Zera et al., 1992).

The following additional examples indicate the circuitous pathways by which environmentally sensitive switches can evolve beginning with continuous variation.

Discrete Size Dimorphisms in Social Insects: A Diversity of Regulators

In wasps and bees with discrete size dimorphism of large queens and small workers, but lacking external structural differences between the castes, the following mechanisms can cause the deletion of intermediate-sized females:

1. Worker construction of two discrete size classes of brood cells, accompanied (known in Vespinae) by quantitatively increased provisioning of the larger (queen-producing)

cells (Vespinae: *Paravespula vulgaris*, Sprad-bery 1973; *Vespa crabro*, Potter, 1965, ex Spradbery 1973, pp. 222–223; Polistinae: *Ropalidia ignobilis*, Wenzel, 1992b; Halicti-nae: *Lasioglossum* [*Dialictus*] *umbripenne* [one population], see Michener, 1974, p. 97)—males reared in the two size classes of cells are also size dimorphic (Potter, 1965), as are social parasites that feed on the same food (Michener, 1974, p. 96), indicating that discrete food supplies are adequate to explain the size differences without the action of a developmental switch intrinsic to the larva itself.

2. Discrete or bimodal seasonal broods (voltin-ism) in species with seasonal increase in size of larval provisions (due to increase in number of provisioning workers or food avail-ability to foragers; halictine bees: *Evylaeus cinctipes*, Knerer and Atwood, 1966; *Polistes* wasps, West-Eberhard, 1969).

3. Sudden seasonal increase in the worker-to-larva ratio (and therefore food per larva) due to egg-eating by workers at the time of in-ception of queen-rearing (bees: *Bombus* pollen-storing species, Michener, 1974, p. 101).

4. The same effect as in item 3, but caused by a decrease in oviposition rate by the queen (bees: *Bombus* species, Cumber, 1949, cited by Michener, 1974, p. 101]

The cues triggering these changes are incom-pletely known. Röseler (1970) provided evidence for queen regulation of worker behavior via a pheromonal cue that changes with queen age in a bumblebee (*Bombus terrestris*; Michener, 1974, p. 100). Since all of the examples given occur in temperate zone populations, photoperiodic or other environmental cues may be involved (see Kamm, 1974, for tentative experiments examining these factors and cell size in *Lasioglossum zephyrum*).

A further variation on the manipulative larval-provisioning theme occurs in solitary and commu-nal bees of the genus *Perdita* (Hymenoptera: An-drenidae; Danforth, 1991). The males of *P. portalis* are strikingly dimorphic: fewer than 50% of them are "normal" winged males that mate with females at flowers, and the remainder are large, flightless, macrocephalic fighters that remain in the natal nest, where they fight to the death with competitors. The winners then mate with females about to oviposit. The large-headed males have elongate genal pro-jections and highly developed hypostomal carinae apparently functioning to protect the vulnerable membranous head and neck regions during fights; enlarged mandibular adductor muscles attached to a greatly enlarged head region dorsal to the ocelli (Danforth, 1991); and complete lack of dorsoven-tral and longitudinal flight muscles. A similar di-morphism occurs in *P. mellea*, another species in the same subgenus (*Macroteropsis*).

This striking dimorphism may be largely a prod-uct of disruptive selection on a preexisting allome-try (Danforth and Desjardins, 1999). Male head al-lometry occurs in six different subgenera of *Perdita*, including *Macroteropsis* (Norden et al., 1991), and phylogenetic analysis indicates that dimorphism is derived relative to monomorphism (Danforth, 1991). Small males resemble the small-headed males of *P. portalis*, and aspects of head size and shape change with increased male size so that the larger males of a continuous distribution resemble the fighting males of *P. portalis* (for a detailed anal-ysis of such an allometry, see Norden et al., 1991, on *P.* [*Hexaperdita*] *graenicheri*).

The nearly discrete size-class dimorphism of *P. portalis* may have originated by developmental deletion of intermediates when ovipositing females placed haploid male-producing eggs on female-size brood balls, initially by "mistake," as has been sug-gested to explain origins of male dimorphisms in halictine bees (Knerer and Schwarz, 1976, 1978; Kukuk and Schwarz, 1988). This hypothesis is sup-ported by the fact that the size distribution of large males is similar to that of females and different from that of small males, which resemble monomorphic males of other species (Danforth, 1991, p. 244). Subsequent selection could explain other aspects of the dimorphism, such as the fact that in many body dimensions the large- and small-headed morphs have different allometries, indicat-ing that they have diverged in form under some-what independent selection.

These examples show that the same result—dis-crete size dimorphism—can be achieved via differ-ent mechanisms of producing increased adult size due to increased larval food supply. All of these ex-amples show that the same basic pattern obtains: manipulation of larval food supply due to discrete changes in the behavior of adults is followed by al-tered adult body size. These examples show not only that the same result (discrete size dimorphism) can be achieved via different pathways for manip-ulating diet, but also how regulatory mechanisms can originate using preexisting elements of the an-cestral phenotype. Even in the most specialized so-cial bees, discrete female size classes are associated with discrete size classes of the brood cells in which they are raised (Michener, 1974).

The recurrent evolution of brood-cell dimorphism as a mechanism of disruptive regulation of size is a pattern that invites experimental manipulation. It would be interesting to investigate the possible tendency in size-monomorphic wasp and bee species (e.g., *Polistes* and some halictine bees) to differentially dispense food in accord with size of brood cell, delivering more food to larvae in large cells. If this occurs, it may be possible to artificially induce dimorphism by providing provisioning females with larvae in artificial cells of two sizes. This would permit investigation of the ancestral cues that have formed the basis for parallel evolution of size dimorphisms in several more specialized social species.

Diverging Selection Following Environmental and Incidental Deletion of Intermediates

The possible effect of dimorphic brood-cell size on the dimorphism of male bees points to the environmental deletion of intermediates as a factor in the evolution of a bimodal distribution of forms. Alternative developmental pathways can arise when some environmental circumstance deletes intermediates, leaving only the extremes of what had been a continuous distribution. This creates two phenotypically different subpopulations that, following the rule of independent selection, are independently selected and may evolve in different directions once there are gene-expression differences between them.

An example of environmental deletion has already been described (above) in the velvet ant *Dasymutilla bioculata* (Mutillidae), where two nonoverlapping size categories of adults correspond to two different discrete sizes of the hosts available to them as prey (figure 11.3; see Mickel, 1924; Brothers, 1972). Similarly, cleptoparasitic bees (*Coelioxys funeraria*; Megachilidae) with two host species with nonoverlapping sizes produce a bimodal distribution of offspring sizes without overlap (Packer et al., 1995).

Another situation in which environmental discontinuities can impose phenotypic discontinuities is in populations that show continuously variable adaptive plasticity. Heterophyllic amphibious plants such as water buttercups (*Ranunculus* species) show variation in leaf shape (heterophylly) depending on conditions of growth. In populations and species where this variation is most marked (see Cook, 1968; Bradshaw, 1965; Cook and Johnson, 1968), an individual can produce one type of leaf when submerged and another in the air. Although a gradual change in growing conditions provokes a series of intermediates during the transition between the extremes (Cook, 1968), if individuals of the same genotype are grown either in purely terrestrial conditions or completely submerged, two distinct phenotypes (terrestrial vs. aquatic) are produced (Cook and Johnson, 1968); intermediates are deleted due to a consistent contrast in growing conditions.

In the tropical shrub *Psychotria marginata* (Rubiaceae), there are indications that disruptive development producing a bimodal distribution of forms comes about through pulsed leaf development superimposed on basically continuous variation in physiological acclimation (Mulkey et al., 1992). Leaf production occurs in seasonal "flushes," with intermediate states deleted by intervening periods of little or no leaf production, much in the way that intermediate-sized wasps and bees are deleted in some species by brood production in temporally separated bursts (voltinism). In this plant, whatever mechanism regulates the production of leaves in seasonal flushes has become a regulator of bimodal development of adaptive leaf characteristics. Bursts of leaves produced just prior to the dry season have relatively marked drought-resistant characteristics, including higher specific mass, lower stomatal conductances, and higher water-use efficiencies in response to dryness than do wet-season leaves (Mulkey et al., 1992). The cyclic change in characteristics persisted despite experimental irrigation during the dry season. Dry-season characteristics are developed during periods of high water availability prior to onset of drought, indicating that a specialized anticipatory timing mechanism has evolved in addition to the deletion of intermediates that otherwise would be continuously distributed over time with respect to some environmental factor(s).

Discrete phenotypes sometimes occur in bivoltine insects, species that produce two seasonally separated, synchronized broods in a year (Shapiro, 1976; Windig, 1993). Some of these may constitute a further example of deletion of intermediates due to temporally separated flushes of reproduction. Bivoltine populations of the butterfly *Pieris occidentalis*, for example, have a bimodal distribution of color forms, whereas univoltine populations are monophenic (Shapiro, 1976). The "highly multivoltine" pierid *Colias eurytheme* shows effectively continuous (or highly multimodal) variation. In such populations, discontinuous brood production places broods in discrete segments of a continuously varying environment, and discrete phenotypes are the result. Bimodality could originate by environmental effects, independent of disruptive selection, and then be refined or exaggerated by disruptive selection if discrete forms happen to be advantageous.

What Happens to the Genes?
Dispensable Phenotypes versus
Dispensable Genes

Extensive phenotypic deletions have raised the question of the fate of the underlying genes (Schartl et al., 1991). Are they immediately lost? Are they eventually modified by unselected mutations until they become dysfunctional? Are they carried intact although simply unexpressed? If so, how long can an unexpressed gene persist and still maintain the potential for functional expression (Ohno, 1985)? Or is it usually only the phenotype that is unexpressed, not the associated genes, which may be functional in other contexts or in other expressed gene sets? There has long been abundant evidence from classical and molecular genetics for deletion of chromosomal segments and pieces of DNA (Herskowitz, 1962). Deletion, along with substitution, insertion, duplication, and inversion, is a major kind of genetic mutation (Glass, 1982). So it is not unreasonable to think that genetic deletions may correspond to phenotypic ones. Alternatively, given the evidence that evolutionary change is reorganizational, with extensive reexpression of preexisting genes, it is also resonable to hypothesize that pleiotropic effects and use in other contexts could maintain the genes that are unexpressed as a set when a phenotypic trait is lost. Lost traits do not necessarily mean lost genes.

Many authors assume that lost phenotypes imply lost genes. Schartl et al. (1991, p. 8759), adopting a term used by Ohno (1985), regard the genes that had been associated with the deleted male phenotype in *P. formosa* as "dispensable." Similarly, Gould (1977) considered the genes formerly expressed in the adult ancestors of progenetic forms to be "extra" genetic material, "unemployed" in the precociously mature progenetic larva, but his main point was that lack of expression in the adult may, as in gene duplication, provide genetic material "freed from the need to function in . . . one way and therefore available for experimental [evolutionary] change" in another context (p. 339). Raff et al. (1990) concluded, regarding the loss of the echinoid pluteus larva, that with the provision of a large yolk "feeding structures can be omitted and the genetic programs that produced them lost"

(p. 90). The idea of a lost "program" of organized expression is somewhat different from the idea of a "lost gene"; I discuss this further below.

Sometimes lost phenotypes *do* correspond to lost genes. In Hawaiian *Drosophila* of the *grimshawi* group, larval patterns of alcohol dehydrogenase (ADH) expression correspond to deletions of particular segments of DNA (Rabinow and Dickinson, 1986). Genes coding for photosynthetic enzymes have been lost from the chloroplast genomes of some nonphotosynthetic parasitic plants (dePamphilis and Palmer, 1990). But the occurrence of atavisms and reversions and their frequently *repeated* occurrence (see chapter 12) suggest that the deletion of a phenotypic subunit often involves a regulatory change that affects gene expression while leaving the unexpressed DNA intact. The question remains: is that DNA used in another context? If not, what is the fate of unexpressed genes when a phenotype is deleted for a long period of time?

There is ample evidence of pleiotropic gene effects from the genetic correlations observed under artificial selection (Falconer and Mackay, 1996) and from classical work in Mendelian genetics, which led Wright (1968) to declare a "principle of universal pleiotropy" (p. 60). It is probably fair to place the burden of proof with those who would argue that any gene or set of genes *lacks* pleiotropic effects. With reference to the particular examples of "dispensable" and "unemployed" genes underlying sex-limited or life-stage-limited traits discussed by Schartl et al. (1991) and Gould (1977), some genes affect traits expressed in only one sex as well as traits expressed in both. For example, the white eye-color genes of *Drosophila* affect not only eye color (a trait of both sexes) but also the shape of the spermatheca (an effect that would be expressed only in females; Huxley, 1942, p. 62). A classic example of expression of a single gene in different combinations is that of the ADH gene of *Drosophila*, which is expressed in one set of tissues in the larva and in a different, partially overlapping set in the adult. Two different promoters direct transcription; both are active in the embryo, only one is active in the larva, and the other is active in the adult (Ptashne, 1986, p. 66; see also Rabinow and Dickinson, 1986; Dickinson, 1991). Deletion of the adult male phenotype, as in a unisexual or progenetic descendent, would not necessarily render this gene "dispensable." The persistence, due to pleiotropic effects, of genes associated with deleted phenotypes is discussed by Price and Pavelka (1996; they cite Wright, 1968, and Lande, 1978, for theoretical discussions).

The "dispensable" and "unemployed" genes idea is another example of genotype–phenotype confusion: it presumes that there is a set of genes corresponding to a phenotype, and if the phenotype is dispensable, unemployed, or lost, so are the genes. As Ohno (1985) states, "The indispensability of a gene can only be judged by the performance of homozygous deficiency individuals" (p. 160), and tests for the effects of complete absence of the gene would have to include performance of young and adults, males and females, and a range of different environmental situations in nature, not just in the laboratory. Although Ohno discusses "dispensable" to mean that the gene can be "discarded" without detriment to fitness—it is "ignored by natural selection," to use Ohno's phrase (p. 160)—his discussion is confusing because some of the examples given imply that "dispensable" means merely "not essential for survival"—not the same as ignored by natural selection (not affecting fitness). For example, he cites "healthy" humans and laboratory populations of rats and fish lacking an intact "dispensable" gene, including some lacking serum albumin, where other substances substitute for its function (p. 163). But healthy survival, especially in the laboratory, is not an adequate test of the fitness effect of a gene, omitting as it does effects on social, mating, and other untested biotic interactions. Ohno gives no indication that the fitnesses of normal and homozygous individuals lacking a given "dispensable" gene was compared.

How long unexpressed or unselected genes persist evidently depends importantly on the accuracy with which they are replicated and repaired. Various authors (e.g., Ohno, 1985; Schartl et al., 1991) have attempted to estimate how long unimpaired functionality can be expected to persist in unexpressed alleles given known deleterious mutation rates. Ohno (1985) suspects that taxa like vertebrates, with efficient DNA polymerases and repair mechanisms, may have a large proportion of uneliminated "dispensable" genes or "junk DNA" compared with groups like viruses and bacteria, where repair mechanisms are less efficient and dispensable genes are more readily lost due to mutation. The correlation he notes, however, between low mutation rate and large genome size in vertebrates may not be due to the greater accumulation of dispensable genes, as he hypothesizes. It may be due to the greater phenotypic complexity of the vertebrates compared with viruses and bacteria, and the correspondingly greater conservation of genes due to multiple uses of each (pleiotropic expression). Factors in addition to mutation rate that may influence the rate of loss of dispensable genes (e.g.,

in alloploids) are discussed by Kimura (1983, p. 108).

Marshall et al. (1994) used empirical data to estimate that "silenced" (unexpressed) genes and developmental pathways have a significant probability over 0.5–6.0 million years of being reactivated, and concluded that this is unlikely to occur after >10 million years without having function maintained by selection in some context. It is interesting to compare this estimate with the findings of dePamphilis and Palmer (1990), who estimated gene loss from the chloroplasts of the nonphotosynthesizing parasitic plant *Epiphagus virginiana* to have occurred within 5–50 million years, an unusually rapid change in plastid DNA content. A related parasitic species (*Striga asiatic*) has an intact complement of these genes, and the time of separation between the lineages was estimated from fossil data. In *Epiphagus*, one photosynthetic gene was detected using DNA probes but did not map to the plastid genome. The authors suspect that it may be incorporated in the mitochondrian genome, which, like the nuclear genome, they consider "more prone . . . to the retention of nonfunctional endogenous sequences" in plants (p. 338). It would be of interest to know whether plastid genes are less subject to pleiotropic or intermittent use than are nuclear genes, a factor that may explain these results.

These estimates suggest that individual alleles may sometimes be maintained over long periods of time due to multiple phenotypic effects. But what are the chances of the capacity for expression as an organized set being maintained? Is it true, as concluded by Raff et al. (1990, p. 90) that when a trait is unexpressed its *genetic program*, or capacity for coordinated expression, is lost? The loss of expression of a phenotype obviously involves some change in development, but what does it mean to lose a developmental program? In one, trivial sense, the program is lost—some essential step or interaction is missing—whenever a phenotype is unexpressed. But such a deletion can occur due to lack of some crucial environmental cue or a genetic change slightly affecting a response threshold, readily reversible changes that would seem to leave the information essential for development largely intact. On the other hand, deletion could conceivably involve some more profound or extensive change in developmental pathway that would be less easily reversible. The many examples of atavisms and reversions having remarkably detailed resemblance to those expressed in related taxa (see chapter 12) show that both form and regulation can recur intact, sometimes after millions of years of subunit loss. This is the subject of chapter 12.

12

Reversion

Reversions and atavisms due to apparent switching on of lost traits are more common than would be expected if they required step-by-step reversal of evolutionary history (one interpretation of Dollo's Law). Some reversions are environmentally induced, and some are evidently resemblances to ancestral states due to gradual loss of complex modifications or life stages. Recurrent reversion is facilitated by the complexity of regulatory mechanisms that provide multiple pathways for reexpression.

Introduction

Deletion sets the stage for *reversion*, the reappearance during evolution of traits lost earlier in the evolution of a lineage. Reversion can also refer to a return to an ancestral phenotypic state, as when a lineage of multicellular organisms gives rise to a lineage of unicellular ones, even though the single cell that results may not closely resemble the ancestral single cell. Deletions can occur due to changes in regulation that are small relative to their phenotypic effects. So some deleted traits may be subject to atavistic recall and reversion with little environmental or genetic change. An *atavism* is a low-frequency or sporadic reversion, the reappearance of a lost character of remote ancestors not seen in the parents or recent ancestors of the individuals that express it (Hall, 1984).

Like heterochrony and heterotopy (see chapters 13 and 14), reversion, as a category of evolutionary transition, overlaps on all sides with other categories. The reestablishment of a lost, ancestral trait that had evolved via heterochrony, for example, may occur if the heterochrony, or change in timing of expression, is reversed. Such a reversion could itself accurately be called a heterochrony. Many reversions could be classified as deletions, if they involve the loss of a recently evolved trait. One of the examples discussed in this chapter, the reversion to solitary reproduction in lineages of social bees, could be termed a deletion because it involves the loss of worker behavior.

Examples

Atavisms

Atavisms occur in a wide variety of organisms. An often-cited example of a revealing and useless atavism is the case of one humpback whale (*Megaptera nodosa*) with two hind limb-like appendages reflecting its terrestrial ancestry. Each limb was over a meter long and contained a nearly complete femur, tibia and vestiges of tarsal and metatarsal bones (Andrews, 1921, cited by Hall, 1984, as *M. novaeangliae*; Levinton, 1988; John and Miklos, 1988). Skeletal atavisms are relatively well studied in whales because Russian researchers (cited in Yablokov, 1966) have taken advantage of large numbers of specimens available in a whale factory in the Central Kuril Islands. Normally, the only vestige of ancestral hind limbs in the whale is the presence of much reduced pelvic bones, and these have been called "vestigial organs" by many authors (see references in Yablokov, 1966 [1974]). But the pelvic bones have acquired functions in the urogenital systems of some modern whale species, where they are sexually dimorphic in size and shape and essential for penis erection in males and for anovaginal contraction in females (Yablokov, 1966 [1974]; see also Hall, 1984). The fact that selection in a new context maintains part of the skeletomuscular structure formerly associated with terrestrial limbs may contribute to the ability to reexpress rudiments of other bones as atavisms in some

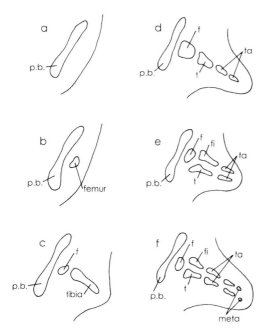

Fig. 12.1. Atavistic expression of posterior appendages in the pelvic girdle of the sperm whale (*Physeter catodon*). Labels indicate the vestigial pelvic bone (p.b.) and atavistic rudiments of femur (f), tibia (t), tarsal (ta), fibula (fi) and metatarsal (meta) bones. (a) Usual structure; (b–f) variants with various combinations of atavistic rudiments. After Yablokov (1966) and Hall (1984).

whales (figure 12.1): well-developed atavistic hind limbs have been seen only in whale species that possess rudimentary hind limb skeletons (Hall, 1984). Other recurrent atavisms (e.g., in the wing musculature of birds; Raikow, 1975) are described in chapter 21.

Other examples of atavisms include the reappearance of polydactyly in horses. In modern horses, the hoof is a single highly developed toe (the third metacarpal), the other toes having been lost (see MacFadden, 1992). Supernumary toes sometimes occur in modern horses (figure 12.2) and presumably represent atavistic expression of ancestral metacarpals. Other well-studied examples of atavistic polydactyly occur in guinea pigs (Wright, 1934) and some other organisms (see Rensch, 1960; Hall, 1984). In *Drosophila*, certain mutations such as bithorax and tetrapter produce individuals with two pairs of wings, restoring the ancestral condition that preceded the evolution of the Diptera where the hind pair of wings are normally

reduced to rudimentary halteres (Goldschmidt, 1940).

Two celebrated atavisms have recently been challenged. First, the experimental induction of an *Archeopterix*-like limb in a chick (Hampé, 1959) has been termed a "misinterpretation" (Marshall et al., 1994, p. 12283). Attempts to replicate the experiments (Müller, 1989), while they produced changes in relative bone length, showed that this result was due to a shortening of the tibia rather than a lengthening of the fibula, with no atavistic recovery of the ancestral fibular articulation that would suggest reactivation of that particular ancient developmental pathway. Nonetheless, the replication experiments did produce an atavistic, reptilelike muscle arrangement associated with the differently constituted but reptilelike bones. Three major muscle configurations shifted to a typically reptilian pattern (see Thomson, 1988). This illustrates a common feature of recurrence and conservation of phenotypic traits, namely, that different

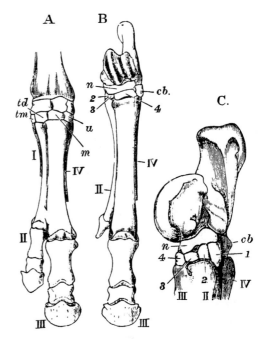

Fig. 12.2. Atavistic polydactyly in a modern horse: (A) left fore foot; (B) left hind foot; (C) tarsus of right hind foot from the inside. I–IV, bones homologous to metacarpals I–IV. Other labels indicate navicular (n), cuboid (cb.), trapezoid (td), trapezium (tm), unciform (u), magnum (m), ectocuneiform (4) and cuneiform (1–3) bones. From Bateson (1894).

developmental pathways can produce similar results. Evidently, the change in bone structure affected muscle development. So this is also an example of the two-legged goat phenomenon (see chapter 3), where correlated effects magnify change and produce a functional arrangement that mimics an evolved one (see also chapter 16).

The other classic atavism to be challenged by some recent work is the atavistic expression of teeth in a chick embryo (Kollar and Fisher, 1980). This was another spectacular experimental atavism, one that seemed to require an enamalin gene or a close homolog to persist unexpressed or used in some other context for 80 million years. This oft-cited example is now "clouded by the possibility that the mouse dermis [used to induce tooth production in the chick tissue] may have been contaminated by mouse epithelial cells" (Marshall et al., 1994, p. 12286). Furthermore, no homolog of the mammalian enamelin gene has been found in birds. Background on these two classic examples of experimental atavisms is summarized in Hall (1984). Although these particular examples have been challenged, the resilience of a flexible modular genome to reconstitute "lost" subunits from fragments still maintained in unexpected ways has been documented in other examples (see chapter 17).

Atavisms sometimes serve in evolutionary biology and taxonomy to indicate common ancestry. For example, the discovery of an anomalous winged male in a wingless species of parasitoid wasp suddenly made it possible to associate the females with males in collections. The taxonomic confusion had led to placement of the unusual wingless females in a succession of taxonomic families, including Tiphiidae, Torymidae, and Braconidae. The atavistic wings of the anomalous males revealed them to be closely related to a winged species of Braconidae in Brazil (Ramirez and Marsh, 1996). Similarly, Goldschmidt (1940 [1982], p. 257) cites Pantel (1917) as having found an atavistic winged specimen in a normally wingless species of earwig (*Anisolabis annulapis*). The atavistic wings had "exactly the structure" found in a related, winged genus (*Psalis*), suggesting a secondary and "saltatory" reacquisition of wings as an anomaly. In some of the cases discussed below, atavisms have been experimentally induced as phenocopies, sometimes by artificial stimuli such as heat shock. This illustrates the fact that atavistic recall need not involve precise duplication of ancestral conditions (A. M. Shapiro, 1984c). Rensch (1960), Rachootin and Thompson (1981), Shapiro (1984c), and Hall (1984) list other atavisms.

Atavisms are instructive not just as curiosities or rare taxonomic tools, but also, as Bateson (1894) long ago emphasized, in revealing the modular developmental structure of the phenotype and showing that that traits can be lost and regained as units. Hall (1984) emphasizes the importance of atavisms as a demonstration that developmental pathways may persist for long periods of evolutionary time. The next step, if such persistence is to have evolutionary significance, is to show that atavisms sometimes provide material for positive selection and the reestablishment of adaptive traits in populations. Atavisms are commonly viewed as oddities that have little evolutionary significance, "hopeless monsters rather than hopeful change" (Levinton, 1988, p. 252; Thomson, 1988, p. 90).

It is undoubtedly true that most atavisms, like other qualitative developmental variants, never multiply to the point of becoming regularly occurring, established traits. But many do evolve to high frequencies. When this happens, an atavism is converted into a reversion, and the "ontogenetic memory" of a lineage may have accelerated evolutionary change (Fabian, 1985).

Reversions

Many examples of reversions, or atavisms that have become established traits, are discussed in chapter 19, on the phenomenon of phylogenetically recurrent phenotypes. The vast majority of reversions known to me turn out to exemplify recurrence as well. They consistute a subclass of the phenomenon of homoplasy that has proved so common in phylogenetic studies. Here are some examples of reversions not known to occur independently more than once:

- Isodont teeth in toothed whales: Primeval whales (Archaeoceti) possessed triangular heterodont molars and premolars, but their descendants, the toothed whales, have numerous isodont teeth resembling those of the reptilian ancestors of mammals and therefore regarded as a reversion to a more ancient type of dentition (Rensch, 1960).
- The second molar in the lynx: Primitive carnivora possessed several molars. Most modern species of the cat family (Felidae), however, have only the first set of molars. But in the lynx (*Lynx lynx*), an additional group of structures have reappeared, sometimes including the second molar, otherwise missing

since the Miocene, a minimum of 5 millions years (Kürten, 1963).

- Lower jaw teeth in a frog: Frogs have lacked teeth in the lower jaw since the Jurassic, 140 million years ago or more. Lower jaw teeth can be experimentally induced, however, and they regularly occur in a South American tree frog (*Amphignathodon* species; see references in Futuyma, 1986a, p. 298).

- Restoration of eyes in oonopid spiders: Most spiders have eight eyes, but those of the family Oonopidae have a reduced number, usually six. In one as yet unidentified British oonopid, perhaps of the genus *Orchestina*, numerous individuals (but not all) have eight eyes as do the rest of the spiders and presumably, therefore, as did the ancestors of Oonopidae (Ruffell and Kovoor, 1993).

A frequently produced kind of experimental reversion is the autonomously reproducing single-cell lineage derived from multicellular organisms in laboratory cell cultures. In the man-made, global environment of biomedical research laboratories, artificial selection has allowed these artificially created atavistic protists—unicellular eukaryotes derived from multicellular ones—to not only propagate in large numbers but also in some cases to spread to other continents and to compete with other protist lineages that have originated in the same way (Strathmann, 1991b). Some of these laboratory cell lineages, derived from cancer tissues, have invaded and displaced others in tissue culture despite efforts by researchers to keep them separate (see references in Strathmann, 1991b). They resemble natural species in this and other ways, being isolated from other protists and their metazoan progenitors by their asexual reproduction, and being able to gain new genetic material by cell fusion and through viral infection. The experimental manipulations have produced not only a reversion to a long-lost condition in the evolution of complex organisms, but also a favorable environment for its survival, propagation, and dispersal as well. Is such a long phenotypic jump backward possible only in the laboratory? Some reversions observed in nature seem just as improbable and yet are recurrent. My favorite example is the reversion to the panoistic type of ovary considered ancestral in the insects (see chapter 19).

Another class of reversions of macroevolutionary proportions are the reversions to solitary life in highly social insects—species that live in colonies with reproductively differentiated reproductive queens and sterile workers. Early popular writers on the social insects (e.g., Maeterlinck, 1901 [1958]) considered insect sociality a stable evolutionary end point, a pinnacle of evolution too profitable and successful to permit a return to the "bottom of the scale" where the female bee is found "working alone, in wretchedness" (p. 22). But later authors (e.g., Michener, 1969; Wilson, 1971a) predicted that reversions to solitary nesting may sometimes occur if the benefits of group life were decreased (e.g., under lessened parasite or predatory pressure), so that they no longer outweighed the automatic costs, such as competition with other group members for crucial resources (Alexander, 1974). Using modern phylogenetic analysis, four independent reversions to solitary nesting have now been confirmed in the sweat bees (Halictidae) alone (Richards, 1994; figure 12.3; see also Eickwort et al., 1996; Packer, 1991; Packer et al., 1994; Plateaux-Quenu et al., 1997; Michener, 1990; Danforth et al., 1999). Recent more extensive phylogenetic analysis indicates that social-to-solitary reversion may have ocurred as many as 12 times in halictid bees (Danforth, 2002). Others are recorded in allodapine bees and in numerous other groups of social organisms (other insects, and spiders, birds, and ascidians; reviewed by Wcislo and Danforth, 1997).

Wcislo and Danforth (1997) point out that some or all of the reversions to solitary life in bees may be environmentally induced, for example, by climatic factors (below) with the ability to express social nesting still retained. It is a moot point whether the expression of solitary nesting, and other (e.g., morphological) reversions, need be obligatory (fixed by genetic assimilation) to be considered true reversions, as implied if reversion is defined as the loss of a *capability* to express a formerly plastic trait (Wcislo and Danforth, 1997). Loss-of-capability evidence is not usually required for morphological reversions, though a developmentally plastic morphological or life-history trait, like a behavioral one, can be fixed (expressed in all members of a population) due to a widespread persistent environmental change. Lost traits never expressed in present-day natural populations sometimes can be induced by extreme environments (Shapiro, 1976, 1978b, 1980, 1981), showing that the capacity to produce them has not been lost. The phenomenon of recurrence (see chapter 19) shows that loss of capability to express a trait is an elusive concept.

If a trait is found as a facultative alternative phenotype in a population whose ancestors lacked

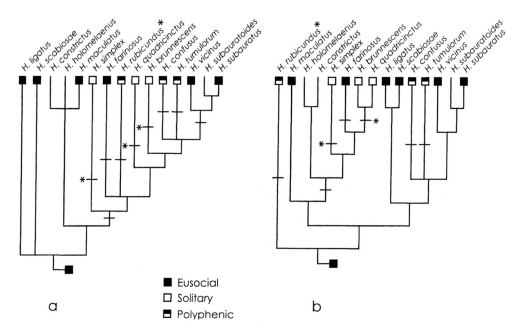

Fig. 12.3. Recurrent reversions to solitary nesting in halictid bees. Reversions to solitary nesting have occurred at least three or four times in the subdivisions *Halictus* and *Seladonia* of the bee genus *Halictus*, for which two equally parsimonious trees are shown. Both trees support a social or socially polymorphic ancestor. Asterisks (*) denote reversions to solitary behavior; horizontal bars indicate changes of character state (loss or gain of social nesting). *H. rubicundus* (asterisk, above) is labeled polyphenic, but some montane populations are exclusively solitary (Eickwort et al., 1996). Phylogenies after Richards (1994).

the trait, as in the paedomorphic larviform adult of some salamander populations (chapter 19), I would regard the return of that morph as a reversion, even though the capacity to express the alternative morph has not been lost. If environmental change is sufficient to fix an advantageous variant, as may be the case in some halictid bees, then genetic change is superfluous for adaptive change to occur. This would be an example of adaptive phenotypic change favored and maintained by natural selection (against other variants) without genetic (evolutionary) change. In such changes, environmental determinants are incorporated into normal development as surely as a genetic change would be (see chapter 26), and a long-lasting consistent environmental change is as effective for maintenance under selection as a genetic change would be.

This is not to say that genetic change has no effect on the permanence of a trait. Retention of behavioral flexibility in female bees probably facilitates reversion, and genetic fixation or modification evidenced by certain morphological specializations could act as "preventing mechanisms" (Gadagkar,

1997a) to reduce the probability of reversals, creating what Wilson (1971a) called a point of no return to solitary life. In boreal populations of social halictid bees, queens act like solitary females during the rearing of their first brood, so they retain the brood-care ability that is essential to solitary life. In *Halictus*, some species are polymorphic for social and solitary behavior (figure 12.3). Reversion invariably occurs in montane or temperate regions where the nesting season is too short to permit rearing of two offspring broods, rendering production of offspring likely to be unprofitable for the worker generation. The connection between such a polymorphism and a reversion to solitary life is shown in some populations (e.g., of *Halictus rubicundus*) where there is a solitary-social polymorphism in the first brood (Yanega, 1988, 1989) and a geographic cline in morph ratios ending in fixation of solitary behavior at one extreme (Eickwort et al., 1996). These bees have probably undergone a double reversion, flip-flopping from social to solitary and back to social again, due to climatic change and range extension during their evolutionary history.

The Developmental and Genetic Basis of Atavisms and Reversions

Deletions, as discussed in chapter 11, raise the question of lost genes, and reversions seem to answer it. Reversions are evidence that the materials of inheritance, or, more accurately, the capacity to produce certain phenotypes, can remain latent and be transmitted between generations without a particular visible effect. This realization crops up repeatedly in the writings of evolutionists, like a kind of intellectual reversion itself:

1868

That a being should be born resembling in certain characters an ancestor removed by two or three, and in some cases by hundreds or even thousands of generations, is assuredly a wonderful fact. . . . If . . . we suppose . . . that many characters lie latent or dormant in both parents during a long succession of generations, the foregoing facts are intelligible. (Darwin, 1868b [1875b], p. 25)

1933

there is increasing proof that the genes for many characters lie latent for long periods, so that similar structures appear in different branches of the same family or order. . . . the evidence for germinal continuity and latency seems rather convincing. (Cockerell and Ireland, 1933, p. 972)

1979

[T]here is reason to believe that the loss of a structure in the phenotype does not necessarily mean that the genetic information controlling its development has also been eliminated from the genome. This hypothesis is based on the existence of reversional anomalies and the asymmetry of structures. It is consistent with current ideas of developmental genetics, especially . . . the temporary inactivation of regulatory genes. (Raikow et al., 1979, p. 206)

Even in an age of relative ignorance regarding the nature of heredity, Darwin clearly recognized that atavisms demonstrated transmission of the potential to express lost phenotypes (see also chapter 8):

When a character which has been lost in a breed, reappears after a great number of generations, the most probable hypothesis is, not that one individual suddenly takes after an ancestor removed by some hundred generations, but that in each successive generation the character in question has been lying latent, and at last, under unknown favourable conditions, is developed. (Darwin 1859 [1872], p. 118)

Darwin attached particular importance to atavisms because they argue against the separate creation of each species, the most parsimonious explanation for similar variations being inheritance of the same tendency to vary from a common ancestor.

A reversion need not involve a step-by-step reversal or repetition of precisely the same environmental or genetic elements that caused an earlier deletion, so it should not be considered a form of "regressive evolution" or a violation of a general evolutionary irreversibility. Evolution, like any other historical process, is irreversible in the sense that the precise details and circumstances of a particular change could not possibly be reproduced, so historical events, like individuals and the traits they express, are, strictly speaking, unique (Gould, 1970). *Dollo's Law*, or the principle of phylogenetical irreversibility, is nonetheless a subject of perennial debate because it has been interpreted in so many different ways (see Rensch, 1960; Gould, 1970). In Dollo's original statement (1893, quoted by Rensch, 1960, p. 123), organisms cannot return *even partially* (*ne peut retourner meme partiellement*) to a former state. By that criterion, many of the reversions described in the present chapter, some of them apparently involving reexpression at the same genetic loci, do violate the original sense of Dollo's Law. But a strict version of Dollo's Law may lead us to call some things different or convergent that we should consider the same in view of the flexible and variable nature of development in the production of species-specific traits. All normal humans, for example, learn to walk upright, but different individuals accomplish this in different ways—via different developmental pathways. Should we therefore consider their walking behavior convergent rather than the same behavioral trait? In practice, we do not, and it is important to acknowledge that homologous traits, including reversions, may develop and evolve via different pathways. The difficult problem of homology is discussed further in chapter 25.

Simpson's "principle of the irrevocability of evolution" is perhaps more generally relevant than is Dollo's Law as an explanation of recurrent phenotypes in evolution. As a complement to Dollo's Law, Simpson (1953a) proposed that "influence of a different ancestral condition is not wholly lost in a descendant group" (p. 310). Once a phenotype has evolved, the involvement of multiple genetic and environmental determinants in its expression means that a reversion can occur due to any one of many effects on regulatory thresholds. Altered regulation can cause reexpression of an integrated phenotype without detailed reconsition of the regulatory mechanism that existed before.

Once a phenotypic trait is established as a relatively invariable feature of a species, it is more easily turned off and on as a unit later in evolution (Nijhout, 1991b; Kavanau, 1990; Price and Pavelka, 1996). Well-defined melanic wing patterns in birds, for example, have a tendency to reappear following loss (Price and Pavelka, 1996). As a genetically accommodated subunit, such a trait is susceptible to regulation, dissociation, and recall as a coordinated set, whereas initially it may have been more crudely or erratically produced within the population by certain genotypes or under certain conditions, for example, as an occasional and somewhat coincidental variant.

The widespread occurrence of viable atavisms and reversions (Hall, 1984) shows that phenotypic and regulatory integration are not immediately lost, even though the expression of a particular phenotype or gene combination is suspended, sometimes for millions of years. Fabian (1985) calls this the "ontogenetic memory" of the lineage and discusses consequences for reversions and acceleration of diversification. In effect, evolutionary progress toward integration can be retained and "stored" as a raised frequency of certain influential alleles in the affected population of genomes. The genotypes able to express the phenotype in a functional, accommodated form under favorable environmental conditions are more widespread in the population than they were when the trait first originated. This ratchet, or improvement–storage effect, along with an even more ancient ground plan tendency to respond in a given way, which had to be present to give origin to the phenotype in the first place, would explain reversions. The same factors would help to explain the common pattern of phylogenetic recurrence, the subject of chapter 19.

Pleiotropy and Silent Genes

It cannot be overemphasized that the capacity to reexpress a lost phenotype does not depend on preservation of a particular intact ancestral set of genes. Rather, it depends on the preservation of a capacity to develop a particular phenotype whose developmental pathways may vary in their details and show the interchangeability of environmental

and polygenic factors that is typical of complex phenotypic regulation (see chapter 5; see also Wright, 1934; Hall, 1984). Levinton (1988, p. 608) suggested the term "epigenetic pleiotropy" rather than genetic pleiotropy to describe effects of a given developmental event on disparate parts of the phenotype, and associated this with the long-term maintenance of unexpressed developmental capacities whose components may be maintained by expression and selection in other contexts.

The hypothesis of pleiotropic maintenance of essential genetic components of lost responses is more often assumed or asserted than tested. A promising approach for research in this area is with comparative studies of animal communication. Ethologists have studied the origins of social and sexual displays and know that they often come from preexisting essential movements originally expressed in noncommunicative contexts (e.g., see Hinde, 1970; Eibl-Eibesfeldt, 1970). It is reasonable to expect, therefore, that if the communicative function were to be deleted, the capacity to respond may nonetheless be maintained by the same pleiotropic effects that were in place before the communication use evolved. In most populations of the three-spine sticklebacks (*Gasterosteus aculeatus*), for example, males display a red abdomen during courtship. In one 8,000-year-old population, however, males are black, yet females prefer red-abdomened males when experimentally exposed to them (McPhail, 1969; see also Mayr, 1974a, n. 19; Bonner, 1980, p. 151). Two different explanations have been proposed to explain this experimentally induced reversion in the behavior of females. One is the unselected-silent-phenotype hypothesis, which suggests that the capacity to produce the trait persists, without selection, as a long-dormant female response, returning as an atavism after "an extraordinarily long retention of an innate releasing mechanism in the absence of any reinforcement by natural selection" (Mayr, 1974a, p. 653). Curio (1973, p. 1049) calls such behaviors "cryptotypes"—hidden potentials, no longer expressed and therefore not under selection. Alternatively, the response may be retained due to pleiotropy or use in another context.

The cryptotype interpretation may apply to some atavisms. For example, ostriches perform flight movements under artificial stimulation even though they belong to a lineage that has not shown flight behavior in millions of years (Eibl-Eibesfeldt, 1970, p. 192). But in sticklebacks, the pleiotropy hypothesis for maintenance of a female response to red signals from courting males is supported by several observations (reviewed by Hart and Gill,

1994). During feeding, sticklebacks prefer red prey. They have eyes whose optical pigments and structure are especially suited for a good response to red wavelengths, and they forage in freshwater habitats where red is especially visible against the generally green background coloration. Several common stickleback prey types, such as chironomids and copepods, often have red pigments. Thus, pleiotropic effects of a response to red could help explain the ready reversion to use of red spots in courtship. The reexpression of red spots in the males could stimulate the female sensory response apparatus that had been maintained in the context of feeding behavior.

Recurrent genetic mutation has been suggested to explain recurrent reversions. Hall and Betts (1987), for example, hypothesize that mutations of "cryptic" (unexpressed, or silent) genetic alleles may explain recurrent acquisition of ability to utilize β-glucoside in cultures of *Escherichia coli*, which spontaneously and repeatedly mutate at a particular site so as to be able to utilize cellobiose. They hypothesize that the silent inactive state of the gene (the "cryptic allele") is maintained in the population not because it is functional in another context, but because of occasional selection for its mutated derivative. That is, it is retained as a mutatable allele potentially able to utilize cellobiose once mutated. A fluctuating or heterogeneous environment may sometimes select against the alternative, normal allele. Presumably, the functional mutant form is favored often enough to keep the cryptic (reverse-mutated) form from being lost by drift or due to other disabling or negatively selected mutations (Hall and Betts, 1987, with reference to a model by Li, 1984), for an allele cannot be selected due to its *potential*, only due to its actual effects on the phenotype.

The extent to which such a mechanism of reversion could work in prokaryotes is not yet clear. In the genetically complex regulatory systems of eukaryotes, the setting of most of the examples discussed here, many loci affect trait expression. A recurrent mutation in a particular gene is less likely to explain recurrent reversion in such a situation than is regulatory threshold change due to change at different loci. I mention the cryptic-allele hypothesis especially to warn that the term "cryptic genes" carries this special implication of advantageous mutability, not mere silence. These are not the "neutral" or "nearly neutral" genes of Kimura (1983) and Ohta (1992b) that are expressed but neutral compared to other expressed alleles in terms of fitness effect. Nor are they the "cryptotypes" or hidden potentials of G. Osche (see Curio, 1973,

p. 1049), or the "silent genes," which are potentially functional alleles that exist unexpressed in the genome but then can be expressed due to environmental or evolutionary change affecting their regulation. The "cryptic gene" hypothesis of Hall and Betts (1987) would be better called the "adaptive mutability" hypothesis to avoid confusion with these other phenomena.

The question of how long genes usually remain intact if they are truly silent—unexpressed for many generations—was discussed in chapter 11. To summarize the conclusions of Marshall et al. (1994), using data on genes that specify particular proteins, "there is a significant probability over evolutionary time scales of 0.5–6 million years for successful reactivation of silenced genes or "lost developmental programs." They also related this finding to the likelihood of recurrent phenotypes: "[F]or groups undergoing adaptive radiations, lost features may 'flicker' on and off, resulting in a distribution of character states that does not reflect the phylogeny of the group" (Marshall et al., 1994, p. 12283). Marshall et al. also conclude that reactivation may occur over longer time scales (<10 million years) if function is maintained in other contexts, that is, by pleiotropic effects as discussed here.

One-Step and Gradual Reversions

Conditionally expressed alternative phenotypes (chapter 22) make both rapid deletion and rapid reversion possible if there is an abrupt local change in conditions. In addition to the example of the social bees described above, documented examples of environmental phenotype fixation as well as rapid reversion include

- Contrasting aerial and aquatic leaf forms in buttercups in areas lacking water versus in lakes (Cook and Johnson, 1968)
- The dietary reversion of the Great Basin kangaroo rat, whose specialized feeding behavior on salt-bush leaves was fixed without genetic change due to loss of alternative food plants, but yet alternative feeding behaviors are immediately regained when individuals are presented with the missing food plants (Csuti, 1979; see chapter 26)
- Fixation of a seasonal color morph in butterflies in areas lacking seasonality, without loss of the ability to produce the deleted phenotype when environmental conditions inducing it are restored (Shapiro, 1976).

In all of these examples, the lost phenotypes are reexpressed under experimental conditions designed to evoke them. In some populations of buttercups and butterflies where one alternative has been fixed, a switch to the other is no longer inducible (Cook and Johnson, 1968; Shapiro, 1976, 1980, 1981), suggesting that genetic accommodation of regulation has proceeded to the point of genetic assimilation.

There is, then, no clear line between conditional expression (short-term failure of an alternative to be expressed, followed by reexpression) and "atavistic" recall of unexpressed traits. A polyphenism can be regarded as a system of alternating reversions on a short time scale. Each individual in the population bears the genes necessary for the production of two different mutually exclusive phenotypes depending upon conditions. If phenotype-influencing conditions fail to change during one or more generations, one of the alternatives may be fixed for an indefinite time period, yet the genome may retain the ability to conditionally produce the unexpressed phenotype should its developmental conditions recur. Behaviors expressed in low-oxygen environments in shallow-water mollusks can be unexpressed for hundreds of generations in deep-water populations, then reappear when snails are kept in the laboratory (Russell-Hunter, 1978). There are many other such examples, where an alternative phenotype is unexpressed for multiple generations due to environmental conditions, yet can be fully expressed if conditions inducing it return (Rollo, 1994; see also chapter 6 and part III). We can assume this to be a general property of polyphenisms. There is a continuum in regularity of expression of "lost" phenotypes between polyphenisms, where the less frequent alternative may be expressed at low frequency in most generations, and reversions, where the less frequent alternative has fallen to zero for a period of generations and then revives and increases in frequency to fixation.

The potential for rapid reversion of complex behaviors due to environmental change in nature is further illustrated by the following cases: The Galapagos dove (*Nasopelia galapagoensis*) shows an antipredator distraction display near the nest, even though mammalian predators, against which such displays are presumably effective, have been absent from the islands until very recently (Eibl-Eibesfeldt, 1970, p. 194). Similarly, Galapagos finches from predator-free islands respond appropriately when experimentally presented with snakes and raptors (Curio, 1965, ex Eibl-Eibesfeldt, 1970, p. 194). There is abundant experimental and comparative evidence using field populations of bees that the transitions between solitary and social nesting (discussed above) can be environmentally induced in both directions (see especially Sakagami and Maeta, 1987a,b; Stark, 1992a,b; Plateaux-Quenu et al., 1997; reviewed in Wcislo and Danforth, 1997).

In general, evidence that a reversion can be environmentally induced, in nature or in the laboratory, shows that such a reversion could conceivably evolve by a one-step developmental change (Plateaux-Quenu et al., 1997). Of course, environmental induction does not demonstrate that this is what in fact occurred during evolution, and gradual genetic accommodation or assimilation of the developmental change may have been involved.

Reversions do not necessarily occur via selection on one-step developmental variants; they may also evolve gradually. As an example of this, Rensch (1960) describes vertebral evolution in snakes. During the course of evolution from amphibians to reptiles, the vertebral column became increasingly differentiated, developing distinctive neck, breast, pelvic, and tail subregions. This differentiation was lost in snakes and slow-worms (Anguidae, Amphisbaenidae) when limbs were gradually reduced, which led to the gradual reappearance of an undifferentiated vertebral column. As in the case of deletions, change between discrete-appearing states cannot be assumed to occur via saltatory or one-step change. Another example of gradual reversion has occurred in echinoids with lecitrophic (nonfeeding, yolk-nourished) larvae, where loss of complex structures of the planktonic pluteus larva has been accompanied by reversion to the pleisiomorphic (ancestral) state of nonfeeding larvae that existed in the Paleozoic (chapter 24). Several mechanistically distinct developmental processes are associated with this change, and mosaic intermediates are evident both within and between species (Wray, 1994, 1996).

13

Heterochrony

Heterochrony is historically the most important category of evolutionary change in development. If defined as evolutionary change in regulatory timing, it can be construed to include virtually all evolutionary transitions between phenotypic states. Heterochrony has been discussed primarily as a feature of the evolution of animal morphology. Here I define heterochrony as evolutionary change in the life stage of expression of a trait, and discuss examples showing that it occurs during behavioral, physiological, molecular, and life-history evolution as well as during morphological evolution, and in plants as well as in animals. Early authors associated heterochrony with saltatory change, but it can involve either gradual or single-step developmental change in the phenotype.

Introduction: Definitions and Perspectives

Heterochrony is evolutionary change in the timing of expression of a phenotype trait, that transfers expression of the trait from one life stage or behavioral or physiological phase to another—"the shifting of characters from one part of an ontogeny to another" (Valentine, 1977b, p. 260) or simply "the displacement of characters in time" (Gould, 1977, p. 225). A clear fossil example occurs in monograptids, where there is a temporal shift in the onset of a modified morphology, a life-history modification that is recorded in the structure of the body (figure 13.1). Heterochrony can occur at any level of organization, including the molecular level, where evolutionary changes in the timing of gene expression have been demonstrated, for example, in echinoids (figure 13.2; see also Ferkowicz and Raff, 2001) and a fibronectin gene of amphibians (Collazo, 1994).

Some recent authors (e.g., McKinney and McNamara, 1991; Reilly et al., 1997) define heterochrony as any change in the timing of regulatory events. Most novel traits qualify as heterochrony by this regulatory-timing definition, since virtually all evolutionary change involves change in the timing of developmental events. This regulatory-timing definition of heterochrony leads to a classification of evolutionary change termed panhete-

rochrony by McKinney and McNamara (1991)—a classification of evolutionary transitions that lumps all evolutionary change under the heading of heterochrony. "Because all developmental events occur along a time line, any significant change is likely to result in a heterochrony at some level" (Raff, 1992, p. 211). As expressed by Barbara McClintock in a staff meeting at Cold Spring Harbor Laboratories, "If I could control the time of gene action, I could cause a fertilized snail egg to develop into an elephant. Their biochemistries are not all that different; it's simply a matter of timing" (quoted by Wallace, 1989, p. 38).

I will adhere to a classification that divides the effects of regulatory change into their different kinds of effects on phenotypes, in keeping with the general emphasis on phenotypes in this book. Beneath this phenotype-centered classification is the question, "How do phenotypes change during evolution, relative to their ancestral states?" If one lumps all innovation as if it were the same in terms of development, the diversity of phenotypic results is less evident, and some of the convoluted pathways that may underlie phenotypic reorganization and complexity fade into the background. More important for understanding adaptive evolution, panheterochrony may lose sight of the fact, underscored by a phenotypic classification, that traits evolved in a particular selective context can be switched to another, as when trait expression is

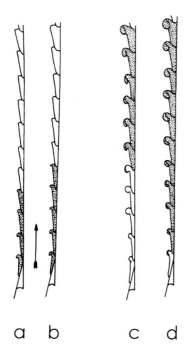

a b   c d

Fig. 13.1. Heterochrony in fossil monograptid colonies. Arrow indicates direction of colony growth, beginning at the base (below). (a) Ancestral colony morphology of (b), a derived morphology with more distal expression of subunit modification; (c) ancestral morphology of (d), a derived morphology with more proximal expression of a subunit modification. After Gould (1977), based on Urbanek (1973).

transferred from one part of the body to another (heterotopy) or from one sex to the other (cross-sexual transfer).

Gould's *Ontogeny and Phylogeny* (1977) disentangled the confusing terms that formerly characterized discussions of heterochrony and of developmental evolutionary biology in general. Gould clarified the central issues and made the key concepts accessible to modern biologists, opening the way for a growing interest in development and evolution to be connected to long-standing controversies. To that 1977 book we owe our freedom to discuss heterochrony without first memorizing the definitions such words as arthallaxis, cenogenesis, hypermorphosis, metathetely, prothetely, and tachygenesis, to mention just a few of the classical subcategories of morphological heterochrony alone. This is not to say that all terminological discussion has stopped, for new terms for distinguishing different kinds of heterochrony continue to be invented (e.g., see Reilly et al., 1997).

Gould (1977) took a somewhat different view of the evolutionary significance of heterochrony than the one taken here or by him in later writings (e.g., 1982a, 1988). In the 1977 book he saw it not as a way of transferring particular coevolved trait complexes from one context to another, but (1) as a way of altering life history parameters (timing of reproduction, investment in growth vs. dispersal, K vs. r selection) and (2) as a *simplification* of the phenotype allowing *de novo* evolution and specialization in new directions—by "unbinding selection from morphology," heterochrony was seen as providing a "laboratory for morphological experimentation" (Gould, 1977, p. 339; for similar views, see also Hardy, 1954; Takhtajan, 1976). Later (e.g., 1982a, p. 341), Gould gave more emphasis to traits as covariant sets, trait dissociability, and reexpression of traits "as blocks" to "produce a new and interesting mixture of ancestral juvenile and adult features" (e.g., in progenesis)—what I would call phenotypic recombination of modular traits. In agreement with this latter view, Raff and Kauffman (1983 p. 341) describe heterochronies as non-disruptive modifications in a developmental path. "Existing integrated processes are shifted with respect to each other, but overall functional integrity is maintained; a reproductively mature organism with a larval morphology retains a set of environmental adaptations and a working body structure" (p. 341). The interpretation of neoteny as a simplification is problematic, for juvenile forms are not necessarily less specialized in terms of their behavior, physiology, and morphology than are later forms. It is also wrong to suppose, as in Gould's 1977 discussion, that despecialization must precede specialization in a new direction. Novel specialization without despecialization occurs commonly in alternative phenotypes (see part III).

Treatment of heterochrony as a temporal shift of traits in "blocks" does not mean that heterochrony cannot give rise to true novelties Gould (1982a) gives examples of how phenotypic recombination via progenesis forms new trait mosaics and is "not just an integral juvenile stage suddenly able to reproduce" (p. 342). Some authors refer to such evolutionary recombinations as "partial" homologies, and one could in this spirit refer to partial heterochronies (see chapter 25 for references and further discussion of this point).

The word "heterochrony" is the most prominent modern-day key word for access to discussions of the developmental origins of new traits. Two recent books on heterochrony (McKinney, 1988; McKinney and McNamara, 1991) that adopt the

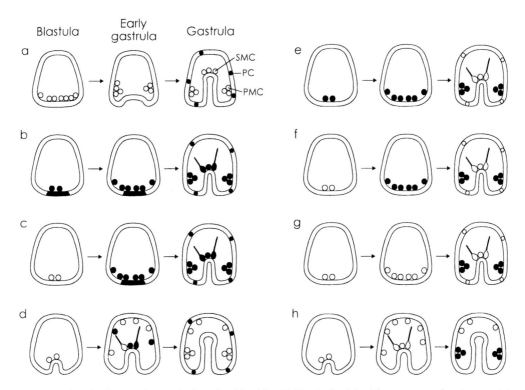

Fig. 13.2. Molecular heterochrony in larval echinoids. (a) Typical echinoid ontogeny, showing positions of primary mesenchyme cells (PMC) and secondary mesenchyme cells (SMC), some of which differentiate into pigment cells (PC). (b–d) Timing and location of expression of the Meso 1 protein (black dots) in (b) *Lytechinus variegatus*, *L. pictus* (Euechinoidea, Camaradonta) and *Arbacia punctulata* (Euechinoidea, Stirodonta), (c) *Mellita quinquiesperforata* (Euechinoidea, Clypeasteroida), and (d) *Eucidaris tribuloides* (Cidaroidea, Cidaroida). (e–h) Timing and location of expression of the msp130 protein (black dots) in (e) three Camaradonta species and *A. punctulata* (Stirodonta), (f) *M. quinquiesperforata*, (g) *M. quinquiesperforata*, and (h) *E. tribuloides*. After Wray and McClay (1989).

regulatory-timing definition are really very general books on development and evolution, much broader than their titles may suggest to a reader using the narrower definition adopted here. Guided by the regulatory-timing definition, McKinney and McNamara (1991) used heterochrony as the central theme for a synthetic treatment of development and the origins of novelty. Although they deny embracing "panheterochrony," the only kinds of phenotypic innovation I can think of that are not included under their definition are what could be called "true" or "ultimate" molecular novelties: changes in primary gene products due to mutation, and incorporation of new environmental elements, for example, the incorporation of iron into a heme group—changes that could not be called a change in timing or rate. Even cross-sexual transfers (see chapter 15) can be included by slightly stretching the idea of timing to include more than one gener-

ation for traits that are borne but not expressed by every individual, and McKinney and McNamara include sex transfer as a kind of heterochrony in their book.

There is another problem with the regulatory definition. I think it makes it more difficult for people to remember that *heterochrony is not a "mechanism" of evolution*. Heterochrony, like duplication, deletion, and reversion, describes a *kind* of variation that has become established during a cross-generational transition, a change in the phenotypic composition of a population or species, not a cause (like recurrent environmental induction, or natural selection) of such change (see figures 6.1, 6.2). Because regulatory timing refers to a mechanistic aspect of development, and heterochrony is an evolutionary phenomenon, the two levels of analysis may get confused. In a heterochronic transition, phenotypic variation is caused by some mechanism (e.g., an altered hormone or an environmental stimulus) causing a shift in timing, and the implied evolutionary change is caused by selection or drift. "Heterochrony" has a potentially misleading developmental-mechanism ring to it, and with the regulatory definition, that sometimes confusing overtone is amplified.

Raff et al. (1990) adopt a definition of heterochrony that refers to both regulation and phenotypic reorganization. They describe heterochrony as "dissociation [of phenotype elements] in which relative timing of two developmental processes undergoes an evolutionary shift" (p. 87). *Dissociation* implies that a dissociable (i.e., switch-controlled), modular subunit of the phenotype is displaced during ontogeny relative to others. This phenotypic definition preserves the traditional meaning of heterochrony as a temporal change in phenotype pattern while visualizing a regulatory process beneath it. For Raff et al. (1990), as here, some traits, such as the large egg cytoplasm and wrinkled blastula in direct-developing sea urchins, are "nonheterochronic" in origin, although there may be shifts in the timing of onset, duration, or rate of gene expression that are considered "heterochronic" by a strictly regulatory definition of heterochrony. There is no reason to regard a phenotypic-transition definition as a rule against looking at lower levels of organization to examine the ultimate intracellular mechanisms of temporal change.

Heterchrony, even under the relatively narrow phenotype-transition definition, includes an enormous class of examples. Gould (1977), emphasizing phenotype reorganization, referred to "thousands of documented heterochronies," concluding that "this process has dominated the evolution of many important lineages" (p. 234). But, as noted by Wray and McClay (1989), there is a bias in the kinds of cases discussed in the literature, toward morphology and toward events late in ontogeny—and, they might have added, toward animals. Here I highlight the relatively neglected categories of behavioral, physiological, molecular, and plant heterochrony, without attempting a comprehensive review that would include numerous already familiar morphological cases, such as paedomorphosis in salamanders and neoteny in humans (Gould, 1977; McKinney, 1988; McKinney and McNamara, 1991).

Behavioral Heterochrony

Phenotypic recombination via heterochrony is common and conspicuous in the evolution of be-

havior (for many examples, see Cairns et al., 1990, p. 55). Classical ethologists approached comparative ethology as if it were comparative morphology, extensively documenting evolutionary transitions in behavior and seeking the origins and mechanisms of novel behavior patterns. Many of their findings were subsequently neglected as interest focused more on adaptation and selection than on development and evolution (but see Lawton and Lawton, 1986; chapter 7 in McKinney and McNamara, 1991). The ethological literature of the mid 1900s gives a rich source of examples of combinatorial behavioral heterochrony. Except when two behavior patterns are performed at once or a single behavior functions simultaneously in two ways, all evolutionary change in the context of performance of behavior is heterochronic. So I sometimes use "change in timing" and "change in context" interchangeably to describe heterochronic behavioral change. I also consider examples where an established behavior comes to be repeated at a different time. Such examples could just as well be classified as duplications.

Animal Communication and "Ritualization"

Barlow (1977, p. 118) states that most behavioral patterns in animal communication have been derived from basic patterns of coordination, such as locomotion, respiration, feeding, elimination, and protective reflexes—movements transferred from one context to another by a change in timing of expression (heterochrony).

The traditional ethological term for behavioral heterochrony in communication is "ritualization," the evolutionary process by which a noncommunicative process becomes a social display or part of a display, via a change in the context or timing of its performance (Tinbergen, 1951, 1952; for reviews; see Tinbergen, 1951; Hailman, 1977). Examples of heterochronic behavioral recombination are very common in the evolution of ritualized courtship behavior. Movements performed during copulation, nest building, feeding, brood-care, and times of social tension or frustration ("thwarting" and "displacement" movements) often appear as elements of visual courtship displays in birds and mammals (reviewed for birds by Andrew, 1961; see the many references in Hailman, 1977, pp. 201–202). One reason for the commonness of transfer of signals from other contexts in the evolution of animal communication may be the fact that the receiving individuals are already equipped and evolved to pay attention to these behaviors,

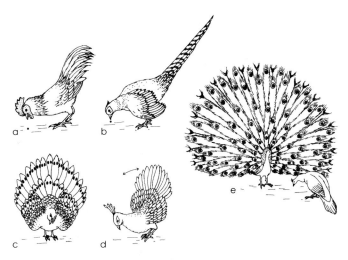

Fig. 13.3. Behavioral heterochrony in courtship displays of birds. Courting males of the family Phasianidae scratch and peck the ground in the presence of a hen, a temporal displacement of feeding behavior that attracts females and stimulates them to seek food near the male. Domestic roosters (*Gallus*) (a) and ring-necked pheasant males (*Phasianus colchicus*) (b) scratch and peck the ground while calling, even if no food is present. Impeyan pheasants (*Lophophorus impejanus*) (c) bow low before the hen, then peck vigorously. Peacock pheasants (*Polyplectron bicalcaratum*) (d) and peacocks (e) spread and vibrate elaborate tail and wing plumage. After Schenkel (1956) and Eibl-Eibesfeldt (1970).

and therefore they are preadapted for attention-getting as components of courtship—the "sensory trap" idea, further discussed below.

Many birds, such as woodpecker finch males (*Cactospiza pallida*), use food-begging behavior typical of young birds as part of courtship display, and this elicits feeding behavior by females, which keeps them near the males (Armstrong, 1965; Eibl-Eibesfeldt, 1970). A well-studied example is the incorporation of feeding movements into the courtship displays of pheasants and peacocks (reviewed by Eibl-Eibesfeldt, 1970). In wild pheasants and related birds, including domestic chickens (*Gallus*), hens attract their chicks to food by scratching the ground and pecking as well as making special sounds. Courting roosters and male ring-necked pheasants (*Phasianus colchicus*) behave similarly. Even in the absence of food, roosters scratch at the ground with the feet, then step back, and peck at the ground while calling, sometimes picking up small pebbles as if they were food. Hens respond by running toward the rooster, and if food is present they eat it. It is possible that partial reinforcement by sometimes obtaining food plays a role in the persistence of the response early in its evolution as a courtship element.

In other pheasants, false-feeding movements are variously amplified and modified ("ritualized") independent of the presence of food, and in some species the movements have become embellished with special movements of the head and tail, which come to be greatly exaggerated and adorned with elaborately decorated plumage (figure 13.3). In false-feeding courtship signals, the response of the female is heterochronically shifted, as is the behavior of the male: the henlike behavior of the male stimulates a shift in the timing and context of the female's feeding response, a response most consistently observed in chicks rather than adults. Because the feeding movements are performed in a new context, they become modified and elaborated independent of feeding, in accord with the rule of independent selection (see chapter 7).

Behavioral heterochrony in animal communication may often begin as a *sensory trap* (West-Eberhard, 1979, 1984; see also Baerends, 1950; Dawkins and Krebs, 1978; Wickler, 1967; Markl, 1985). In the example just discussed, it was selection on the observer's preexisting *responses* that initiated the evolution of a communication system. No change in the reflex signal itself was initially required. So origin of a communication system does not require simultaneous origin of signal and response as sometimes believed (e.g., Manning, 1979), although coevolution of signals and responses may subsequently occur. Courtship communication in phaisanids began with a preexisting behavior of adults and a response common in juveniles, and it was the male signal that was subsequently elaborated.

Behavioral heterochrony sometimes involves mimicry of established signals, which act to stimulate a shift in timing and context of a response. For example, courtship and alarm signals sometimes resemble infant cries, for example, in the night monkey (*Aotus trivergatus*), where juvenilelike squeaks are produced by males during sexual displays, effectively attracting females' attention presumably by playing upon deeply ingrained maternal responses (Moyhihan, 1970). This acts to cue maternal-phase behaviors at the time of mating. Geist (1978, p. 54) identifies "infant mimicry" as a common occurrence in birds and mammals (see also Wickler, 1967). In courtship he refers to it as "parasitizing the maternal emotions of the female" (p. 54). Parental responses of males also sometimes can be used by their mates, as in monk parakeets (*Myiopsitta monachus*), where breeding females imitate the begging behavior of young birds and are fed by males (Eberhard, 1998a). "In essence, communication systems take advantage of existing response mechanisms" (Geist, 1978, p. 54).

A chemical example of sensory-trap heterochrony occurs in the foraging behavior of ants (*Aphaenogaster* species), where the poison-gland secretions serve as an alarm signal, which draws colony mates to an alarmed worker. In these species, when a forager discovers a prey object too large to be carried by a single ant, she releases poison-gland secretions into the air and thereby attracts nestmates toward the source. When a sufficient number have assembled, they together carry the prey swiftly to the nest (Hölldobler, 1995). Defensive responses are thereby shifted in time and context toward foraging.

Modulation and Increased Complexity of Signals

Social communication is famous for its complexity. This has been explained in terms of selection as an unending race among competitors and their groups (e.g., see West-Eberhard, 1979, 1983, 1984). There is also a developmental contribution to the unending race, in the form of a virtually unending supply of neural and sensory gimmicks built into responsive organisms through selection in other contexts of life, selection that continually enriches the array of stimuli and responses that can be coopted to function as signals.

How social communication builds upon itself through temporal recombination of signals is vividly portrayed in a short essay by Hölldobler

(1995) on social communication in ants. An individual ant is a veritable Pentagon of communication devices in the form of pheromone-producing glands, sensory cells and bristles, and intricate behaviors, used in contests within groups, colony territoriality, kinship discrimination, food getting, and sex. Hölldobler emphasizes the multicomponent nature of signals and their origins. Ants alone provide many examples of signal origins by heterochronic (context) change. The poison gland, primarily a hunting or defensive organ, produces a venom spray that contains metabolic by-products of venom production in addition to the functional venom. Venom spray can be activated at different times. In some ant species, poison gland products serve as nondefensive signals, including (1) forager recruitment signals, (2) trail pheromones, (3) a calling signal for aid in carrying large prey, and, combined with Dufour's gland substances, (4) a long-lasting homing orientation pheromone, and (5) a home-range marker. These effects of the Dufour's gland chemicals may involve heterochronic transfer of ancient hymenopteran (wasp, ant, and bee) signals, since the Dufour's gland produces nest-building secretions in bees (Michener, 1974) and nest-marking substances in wasps (briefly reviewed in West-Eberhard, 1996, p. 312). So it is striking that in the ants, the Dufour's gland substances sometimes become associated with nest-oriented homing and marking the area near the nest.

In addition to the signals just mentioned, Dufour's gland secretions are combined in some species with formic acid, a substance that enhances the spread and penetration of the spray. In some signals, the chemical components are modulated not by addition of other chemicals but by heterochronic transfer of signals of other kinds. For example, chemically signaling *Aphaenogaster* foragers usually stridulate as well. The substrate-borne sounds accelerate the responses of their nestmates to the chemical signal and stimulate them to release chemical signals of their own. Similarly, leaf-cutting ants (*Atta*) augment chemical recruitment signals with substrate vibrations like those that occur during the activity of leaf cutting. The same stridulatory sounds also double as context-dependent stress and rescue signals (Markl, 1965, cited in Hölldobler, 1995, p. 21). In some species, they additionally function in a sexual context (Hölldobler, 1995).

One interesting aspect of the signal heterochronies of ants is the use of originally neutral traits—what Hölldobler calls "biochemical noise"—such as the venom-synthesis by-products used as trail pheromones in the myrmecine ants.

Like intention movements and displacement activities, these contain information by association with other states and events. They are used not because they were especially designed for a signal function, but because they are there and they happen to work (West-Eberhard, 1979; G. C. Williams, 1992). This is another illustration of the convoluted, tinkering nature of phenotype evolution, which seems to characterize every level of organization where selection and evolution occurs. The long run of social evolution is likely to be a tangled accumulation of manipulative exploitations of preexisting responses.

There is a striking resemblance between social signaling and the signaling that occurs between cells during the development of a multicellular organism (e.g., see Robertson, 1977; Gerhart and Kirschner, 1997). As in, for example, infant vocalizations, multiple, pleiotropic effects are characteristic of hormones such as vertebrate steroids and invertebrate juvenile hormones and other ancient internal signaling molecules where heterochronic transfers of function are rampant. Juvenile hormone, for example, regulates functions as diverse as aggressiveness, molting, and diapause in the same species of insect (see Nijhout, 1994a). The resemblance between internal signaling and social communication does not end there. Communication in ants and other organisms is likewise characterized by feedback, threshold, and regulatory effects. And, like these molecules (see Gerhart and Kirschner, 1997), multifunctional social-signal elements should likewise prove to be evolutionarily conservative, even though the complex displays they compose, like signal sequences in morphological development, are not. As in regulation by hormones, the stridulatory signal of *Atta* is the same whether in alarm or recruitment; it is the context that determines the distinctive response. At the same time, there is variation on a common theme in the responses observed; both recruitment and defense involve moving toward a nestmate—just as in multiuse internal regulatory molecules whose design suits them to similar roles (e.g., transport or ligation) but in different contexts (see chapter 19).

Learning as a Mechanism of Behavioral Heterochrony

Given the familiar abilities of animals to learn to associate novel cues with repeated, rewarding or punitive events and to employ trial-and-error learning (reviewed in Mackintosh, 1974), it seems likely that ritualization and other behavioral heterochronies sometimes involve learning as a mecha-

nism producing a recurrent response (see Hinde, 1970, on the role of learning in the development of stereotyped stimulus-response chains; see Jeanne, 1996b, on associative learning and the evolution of insect defensive behavior). Since phenotype recurrence facilitates selection (see chapter 6), any learned recurrent pattern could be followed by genetic accommodation based on genetic variation in responsiveness, including ability to learn or pay attention to appropriate cues.

Even without genetic accommodation of regulation altering the frequency of the trait, the form of the learned recurrent trait itself would be subject to evolved modification by genetic accommodation. During play fighting, for example, juveniles practice the movements involved in adult fighting and predation (e.g., see Eibl-Eibesfeldt, 1970), and their improved skill likely includes learning to pay attention to certain cues anticipating the moods and movements of other individuals. For example, autonomic responses such as pilomotor actions— feather or fur erection ("goose bumps" in humans)— are often associated with aggressiveness and fear (see Darwin, 1872; Hailman, 1977), and being automatic or involuntary, such responses are potentially dependable warning signals of impending attack. In species capable of learning to associate the reflex with attack behavior, such autonomic responses have repeatedly become ritualized as aggressive threat signals performed without usually being followed by attack (Morris, 1956; Hailman, 1977). Ritual modification suggests genetic accommodation of both the signal and the response of heeding and assessing the variable-intensity pilomotor movement as a warning signal. Initial heterochronic shifts in context of performance, whether due to learning or driven by natural selection, are expected to be followed by coevolutionary change in signal and response, leading to further elaboration or ritualization. The role of learning in evolutionary transitions is further discussed in chapter 18.

Behavioral Heterochrony Induced by Other Species

Heterochronic reorganization of behavior can take place within life stages as well as between them. Stimulation of behaviors in new or initially inappropriate contexts can be regarded as a kind of heterochrony, for it involves a change in timing relative to the usual stimuli that served to time the performance of the behavior. Ethologists have long recognized the role of context or timing change in the evolution of manipulative interspecific interac-

Fig. 13.4. Behavioral heterochrony induced by aggressive signal mimicry. A female firefly (*Photuris* species "B"; Coleoptera) feeds on a male of *Pyractomena angulata* that she has attracted by precisely imitating the calling flash pattern produced by females of the male's own species, displacing the male's sexual response in time and context. Courtesy of J. Lloyd.

tions, variously called "aggressive mimicry" or "false signals" (Lloyd, 1977) and "sign stimuli parasitic on existing releasing mechanisms" (Baerends, 1950, p. 353). The head adornments of anglerfish, for example, serve to change the context and timing of attack behavior in their prey. They lure small predatory fish to within gulping distance by stimulating approach behavior that is inappropriately timed relative to its original adaptive context of attraction to food. Other predators capitalize on sex attractants to attract prey. Female fireflies (*Photuris* species) mimic the flash signals used by the females of other species to attract males and then prey upon the attracted males (Lloyd, 1975, 1977; figure 13.4); bolas spiders mimic the sex attractant of female moths in order to draw and capture the males (Eberhard, 1977); and some orchid flowers mimic the sex attractants of their hymenopteran pollinators, attracting males which pollinate them without receiving nectar rewards (Dressler, 1981).

Many examples of aggressive mimicry are given by Baerends (1950), who points out (p. 355) that these and other sensory traps are especially effective if playing upon essential responses of the manipulated individuals, using some mechanism that has so much biological importance that it will be retained rather than readily selected against even though somewhat costly in its temporally displaced context (see also West-Eberhard, 1984). For many examples of behavioral heterochrony induced by interspecific mimicry of evocative stimuli, see Wickler (1968).

Socially Induced Heterochrony in the Evolution of Termites

Termites (Dictyoptera, Isoptera) are ancient insects that originated, along with their closest living relatives, the cockroaches (Dictyoptera, Blattaria) and mantids (Dictyoptera, Mantoidea), in the Mesozoic (reviewed in Nalepa and Bandi, 2000). It has long been recognized that termites are likely derived via heterochrony from their cockroachlike ancestors (e.g., see Matsuda, 1979). Recent analyses of the behavior and natural history of developmental plasticity and paedomorphosis in cockroaches (Nalepa and Bandi, 2000) show how this could have occurred. Developmental rate in cockroaches is modified by some of the same environmental factors known to mediate caste differentiation in termites, such as crowding, minor injuries, and food quantity and quality. Crowding can induce a syndrome of juvenilized morphological traits such as winglessness and juvenilized tracheal system, eyes, ocelli, cerci, and integument. All stages of extant termites exhibit behaviors characteristic of juvenile (but not adult) cockroaches, including gregariousness, social grooming, and consumption of conspecific feces, a behavior relating to the acquisition of intestinal symbionts that aid in digestion of wood and other difficult-to-digest foods.

Nalepa and Bandi (2000) argue that juvenile cockroaches depend more than do adults on use of gut symbionts, as evidenced by their relatively greater sensitivity to experimental elimination of gut microbiota, and their inability to range widely in search of food. Termites depend on the symbionts throughout their lives. Increased social stimuli (crowding) with increased duration and importance of life in groups would have intensified the paedomorphosis-inducing "group effect" seen in cockroaches. Then these widespread responses of cockroaches could have been further refined as caste-determining mechanisms in termites.

Life-History Heterochrony in Vertebrates

Delayed maturity in external morphology of birds is discussed as both behavioral and morphological heterochrony by Lawton and Lawton (1986), who give a large number of interesting and well-studied examples as well as a review of the literature (see also Grant, 1990, for a subsequent well-documented example in Darwin's finches; McDonald, 1989, on manakins). In many species, delayed maturity in plumage, external morphology, and behavior gives males a femalelike appearance that reduces the frequency of aggressive attacks by other males and increases their success in obtaining surreptitious copulations. In other species, femalelike or subadult plumage is associated with residence and helping behavior in a group of relatives. In manakins, extremely delayed plumage maturation is hypothesized to facilitate cooperation among age cohorts of males in cooperatively displaying groups (McDonald, 1989). Lawton and Lawton (1986) show that this type of heterochrony is often associated with the evolution of complex social behavior in birds, especially with the occurrence of communal breeding and helpers at the nest.

A similar delayed-maturity phenomenon appears to occur in mammals, also under social pressure or manipulation. Prepubertal mammalian males often resemble females more than they do mature males (e.g., see Geist, 1978, p. 99). In various Northern canids (e.g., the arctic fox, *Alopex*, and red fox, *Vulpes vulpes fulva*), there is a "silver" or black pelage phase that is an alternative to red pelage and appears to be an "outward manifestation of a taxonomically recurrent syndrome of behavioral (less aggressive), physiological and morphological adaptations" whose frequency changes over time and location, perhaps correlating with high density phases of population cycles (Guthrie, 1975, p. 414). The less aggressive behavior and silver or black coloration are regarded as neotenic (newborns are chocolate-black) and may function to reduce aggression by conspecifics.

The evolution of the hormonally mediated worker phenotype in social insects has sometimes been labeled "neoteny" (Kennedy, 1961) because the sexual development of socially dominated individuals is arrested or juvenilized relative to that of dominant reproductives. The same label could be applied to the worker adults of naked mole rats (Sherman et al., 1991). Behavioral heterochrony has evolved recurrently in social spiders. A review of the systematics and biology of the social-spider genus *Stegodyphus* suggests that the neotenous origin of adult sociality has occurred independently in three different lineages (Kraus and Kraus, 1988). Each of three social species has a nonsocial sibling species with gregarious juveniles. Sexual maturity can be displaced backward to instars 7, 8, and 9, producing three size classes of mature males in

some species; and postmaturation molts occur in females of some species. This developmental flexibility, and the gregariousness of juveniles in the nonsocial sibling species, has led Kraus and Kraus (1988) and others (see references in Kraus and Kraus, 1988) to hypothesize that "periodically social forms may have been 'phylogenetic precursors' of permanently social species" (p. 242; see also Seibt and Wickler, 1988).

As in morphological heterochrony (see Gould, 1977; Matsuda, 1987), behavioral heterochronies similar to natural ones can be induced experimentally by hormones. Armstrong (1965, pp. 329–340) lists many examples from early studies of avian endocrinology. Administration of testosterone propionate in black-crowned night herons causes month-old chicks to behave like adults, occupying territories, building nests, and exhibiting all adult display ceremonies including copulation and brooding, and undergoing changes in vocalizations and plumage (Armstrong, 1965, after Noble and Wurm, 1943). Hormonal involvement has also been demonstrated in socially mediated behavioral heterochronies (delayed maturation of aggressive behavior and correlated reproductive-system traits) in fish (*Xiphophorus*; see references in Borowsky, 1987a, p. 795) and mammals (Faulkes et al., 1991).

These behavioral parallels to morphological evolution underline the essential similarity of behavioral and morphological development. Both classes of heterochrony depend on the sensitivity of preexisting structure to environmental stimuli that evoke specific responses at specific times. This sensitivity makes the structure subject to reorganization by temporal shifts in stimuli.

Some classical examples of morphological heterochrony, such as larviform adults in salamanders and interspecific variation in leg length in frogs, may be a product of selection on life-history traits such as timing of reproduction rather than the morphological variants that result (e.g., see Emerson, 1986; Whiteman, 1994). This is a possibility because sexual maturity without metamorphosis, and the timing of metamorphosis that affects body proportions such as relative limb length, also affects the timing of reproduction.

In at least some examples of accelerated metamorphosis, it is the timing of metamorphosis per se that is the object of selection. In red-eyed tree frogs (*Agalychnis callidryas*), behavioral heterochrony—acceleration of hatching behavior relative to morphological development—occurs in response to predation. Under attack by snakes or wasps, some individuals from the aerial, leaf-borne egg masses escape predation by hatching preco-

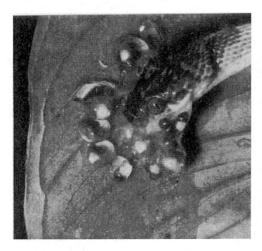

Fig. 13.5. Predation of red-eyed tree-frog eggs by (left) a snake and (right) a wasp (*Polybia rejecta*). Courtesy of K. M. Warkentin.

ciously and dropping into the water below. They face a trade-off between almost certain death in the egg mass and hatching at a relatively young stage when they are less likely than more mature hatchlings to survive in the water (Warkentin, 1995, 1999a,b, 2000; figure 13.5).

Heterochrony in Plants

Morphology

The modular morphology of plants, like the modular life cycles of animals with metamorphosis, lends itself to compartmentalized evolutionary events and also makes them easier to see. In plants, the stages of growth are often preserved in linear series along a stem. Morphological heterochrony is heterotopic change as well, so when mature individuals of related species such as vines are compared, heterochronic variation is sometimes diagrammatically clear (see figure 6.3) even though the evolutionary polarity of change may be unknown. The iterative modular structure of plants and the development of all of the divergent parts of stem, branches, and flowers from the same type of undifferentiated meristem make all of these parts homologous.

Morphological heterochrony is a well-recognized phenomenon in plants (Takhtajan, 1969, 1976; Bradshaw, 1965, p. 144; Guerrant, 1982; Lord and Hill, 1987). But only recently have the models and concepts invented by zoologists to analyze heterochrony been extensively applied to plants (e.g., Guerrant, 1982; Lord and Hill, 1987; Ray, 1987, 1990; Wilson, 1986). Heterochrony has been invoked to explain such major features of plant evolution as the recurrent derivation of herbaceous stems from woody plants having nonwoody juvenile stems (Cronquist, 1968; Takhtajan, 1969) and the origin of cleistogamous flowers (Lord and Hill, 1987). Still, the heterochrony approach, and other development-based hypotheses emphasizing switch-controlled reorganization of ontogeny (see especially chapter 15), remain relatively unexploited as explanations of evolutionary change in plants, compared with the continuous-plasticity reaction-norm approach adopted in many studies of plant plasticity (e.g., Via et al., 1995; Schlichting and Pigliucci, 1998; Scheiner, 1993a,b, 1998).

Interest in plant heterochrony was more prominent in the older literature. For example, Corner (1964) discusses several phenomena as "neoteny," including the simplification of life cycles and structures in marine algae; the origin of the pollen tube by suspended development (juvenilization) of the ancestral male prothallus; delayed maturation of female seed leaves (sporophylls) giving rise to the carpels of flowers that mature only after pollination has occurred (in more primitive plants, ovules grow to full size before being pollinated); and certain grasses that resemble bamboo seedlings and can be regarded as neotenic or precociously matured bamboos, with slender stems, small inflorescences, and small seeds. Like Gould (1977), Corner attributed abbreviated life cycles to r-selection, which he called "quick returns." In a remarkable and seldom cited paper, "Phase Change in Higher Plants and Somatic Cell Heredity," Brink (1962) reviewed heterochrony in plants as a reversible change in "phase," or abrupt morphological change during ontogeny, aptly termed "metamorphosis" by Ray (1990). Doorenbos (1965) reviewed reports on experimental and anomalous heterochronic change in woody plants, noting that in contrast to the situation with herbaceous plants, most such work is undertaken by applied researchers and is unknown to basic scientists in botany and evolutionary biology. This fascinating account of the older applied literature is recommended reading for present-day evolutionary biologists interested in plant heterochrony. Efforts by horticulturists to accelerate seedling growth and force early flowering have led to many findings of potential evolutionary significance.

Some trees, for example understory representatives of the dipterocarps *Hopea*, *Stemonoporus*, and *Vatica* flower while still juveniles. This neotenous event is obvious when such trees are compared with their "normal" and presumably ancestral canopy relatives (Ashton, 1988). The life cycles of some *Clusia* (Magnoliopsida) species (Lüttge, 1996) resemble those of direct developing animal species such as sea urchins and salamanders (see chapter 11) that sometimes develop via a complex juvenile phase and sometimes omit it. Both an epiphytic and a terrestrial form can occur in the same species or progeny of a single individual. Seeds that germinate on the ground develop directly into independent trees, whereas those germinating on trees first form aerial roots, then become stranglers that eventually kill their host, and finally free-standing trees. As in animals, the evolution of direct development could be considered either heterochrony (accelerated development of the tree form) or deletion (loss of the juvenile epiphytic or strangler phase).

Roots also show heterochrony. Some cacti, for example, are epiphytic, and some are desert inhabitants. Roots of epiphytic species show retarded thickening and accelerated growth of ensheathing fibers compared with desert species of comparable age, differences evidently reflecting specialization of the epiphytes to attachment in the canopies of trees (Nobel and North, 1996).

Heterochrony probably explains the origin of numerous examples of inducible defenses in modular organisms, including plants and some invertebrates (Harvell and Padilla, 1990). In some plants, spines are a juvenile trait that is usually lost in later-developing tissue, but spines are sometimes produced again when mature tissue is damaged and regeneration occurs (Harvell, 1991). Other similar examples include resinous adventitious shoots of willow, sweeper tentacles of corals and octocorals, and defensive zooid production in Bryozoa. Many clonal

organisms, (e.g. fungi, herbaceous plants, and colonial invertebrates) show heterochronic growth forms (Blackstone and Buss, 1992). The complex life cycle from which they are thought to be derived has two distinctive phases of growth: an early, lineal runner, in which primary polyps give rise to lineal extensions of stolons; and a sheetlike phase in which an encrusting sheet forms by iterative initiation of new stolons with numerous feeding polyps forming along them. Some heterochronically derived species have a paedomorphic or juvenile, runnerlike morphology resembling the early lineal-growth phase, while others have a sheetlike colony form with more closely packed polyps, a kind of direct development that omits the lineal runner phase. Runner and sheet dichotomies are common interspecific alternatives in colonial invertebrates (Blackstone and Buss, 1992; Jackson and McKinny, 1990).

Physiology

There are probably many undiscovered, cryptic examples of physiological heterochrony in plants. The physiological and growth responses of plants are the vegetable equivalent of animal behavior, being the primary means for a sedentary organism to adapt to a changing environment. Morphological heterochrony of leaf development (see preceding section) is probably often accompanied by biochemical or physiological heterochrony (e.g., see Kluge and Brulfert, 1996, p. 326; Winter and Smith, 1996a; Raven and Spicer, 1996, p. 365). Comparative study of physiological heterochrony and its relation to patterns of evolution in morphology is a promising frontier for plant physiologists interested in adaptation and evolution.

Bromeliads illustrate a pattern suggestive of physiological and morphological heterochrony. Most epiphytic bromeliads are of two morphological types. One type is "atmospheric," with succulent leaves, a dense cover of elaborate trichomes, and high leaf reflectivity, with no provision for the collection or "impoundment" of water. The other is "tank forming," with water-collecting architecture, large flattened leaves, less elaborate trichomes, and low surface reflectivity (Adams and Martin, 1986). Corresponding to these morphological differences are physiological ones: the atmospheric morphotype can maintain positive rates of net CO_2 exchange under desiccation much longer than can a tank-type adult. In the subfamily Tillandsioideae both types occur, and in a few species (e.g., of the genera *Tillandsia*, *Vriesia*, and *Guzmania*) individual development is heterophyllic, with juvenile leaves atmospheric and adult leaves tank forming.

Studies of *Tillandsia deppeana* have confirmed that there are major physiological differences between the two morphologically distinct life stages (Adams and Martin, 1986). This raises the possibility that atmospheric species may sometimes have evolved via heterochrony from heterophyllic ancestors via sexual maturation without metamorphosis to the tank-forming phenotype (suggested by several previous authors cited in Adams and Martin, 1986, p. 124). Similarly, the tank-forming species may have evolved via heterochrony toward direct development, or maturation without an atmospheric juvenile phase. Adams and Martin review findings bearing on the possible polarity of change in these traits. The phylogeny is as yet insufficiently known, as are the juvenile stages of many key species believed to be basal within the subfamily.

Some seasonal responses of crassulacean acid metabolism (CAM) in plants could be regarded as heterochronic derivatives of leaf ontogeny. In *Clusia uvitana* in Panama, CAM leaves revert largely to C_3 photosynthesis during the wet season. In this response they resemble juvenile leaves: each young leaf (less than two months of age) shows C_3 photosynthesis regardless of conditions, then becomes capable of facultative CAM activity during its "adult" life, which extends up to about 12 months of age (Zotz and Winter, 1996). The seasonal switch is a rapid brand of environmentally induced intraspecific heterochrony. That it is reversible in response to conditions was shown when a heavy rain during the dry season caused a return to C_3 photosynthesis and abolished the nocturnal CO_2 uptake that characterizes CAM. The importance of developmental plasticity in the evolution of plant physiology is further discussed in chapter 26.

Gradual versus One-Step Heterochrony

In Darwin's *Origin* (1859 [1882]) heterochrony was identified with saltatory transitions and accorded little attention (see chapter 24). Different modes of heterochronic change were extensively documented and discussed during what might be called the Golden Age of classical evolutionary embryology in the early 1900s. But in mainstream evolutionary biology, developmental change, and especially heterochrony, continues to be identified by many with macromutation. For this reason, it is instructive to examine the gradualism-saltation controversy briefly with special reference to heterochrony (for a more detailed discussion see McKinney and McNamara, 1991, on "incremental" heterochrony; see also chapter 24, on gradualism).

Gradual Evolution of Behavioral Heterochrony under Artificial Selection

Artificial selection can produce a rapid heterochronic shift in the ontogeny of aggressive behavior in populations of mice (Cairns et al., 1990). Even though the change is rapid, it is gradual. Laboratory mice were selected for subordinance according to several criteria of degree of aggressiveness, including reactivity to tactile stimulation; behavior toward conspecifics, including attack by slashing and biting; freezing upon social contact; and "tunneling" or burrowing beneath the body of the other animal (Cairns et al., 1990). Selection was continued on a high- and a low-aggressive line for 18 generations. The result, in the low-aggression line, was a gradual change in the shape of the curve describing the development of aggressiveness with age, with juvenile subordinant behavior retained in progressively older age classes as selection proceeded—a change the authors interpreted as "neotenous." The high-aggressiveness line responded to selection with increased aggressiveness but no change in the timing of aggressive development (see figure 5 in Cairns et al., 1990). In this example of behavioral neoteny, a complex set of traits that characterize juvenile subordinant behavior was shifted to a later developmental stage by gradually prolonged expression. While it is possible that not all of the behavior patterns responded to selection at the same rate, the result was a correlated evolutionary change in a set of traits expressed in a coordinated fashion under certain circumstances, namely, confrontation with conspecifics.

The Origin of Soldier Morphs in Aphids: Heterochrony and Saltation Revisited

A debate regarding the origin of fighting larvae (soldiers) in aphids (Insecta, Homoptera) revives the age-old controversy over heterochrony as saltation and shows the importance of a case-by-case evaluation of competing hypotheses for origins, even within the same taxon. Aphids, with their complex life cycles and numerous discrete morphological phenotypes (see figure 11.1), are especially interesting subjects for evolutionary studies of development and life-cycle evolution. In many species of the families Hormaphididae and Pemphigidae, there are morphologically and behaviorally specialized "soldier" larvae that defend the plant galls that serve as their homes against insect predators (Aoki, 1987; Stern and Foster, 1996; figure 13.6).

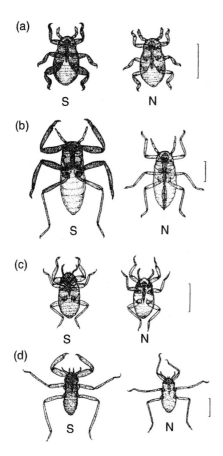

Fig. 13.6. Soldier (S) and normal (N) first-instar larvae of aphids: (a) *Pemphigus spyrothecae*; (b) *Colophina monstrifica*; (c) *Pseudoregma bambucicola* from primary host, *Styrax suberfoliae*; (d) *Pseudoregma bambucicola* from secondary host, bamboo. Scale bars = 0.5 mm. From Stern and Foster (1996), by permission of Cambridge University Press. Drawings by Kristina Thalia Grant. Courtesy of D. Stern.

The soldier larvae do not molt to adulthood and are therefore sterile. Soldiers have evolved separately at least nine times in the aphids (Stern et al., 1997). They have been gained at least five times and lost at least once in the family Hormaphididae alone (Stern, 1994; Stern and Foster, 1996). In some pemphigid genera (e.g., *Colophina* and *Eriosoma*), the soldiers have mouthparts that are too short for feeding and are used instead for defensively piercing predators (figure 13.6a, left). "Normal" secondary-host first-instar larvae, on the other hand, have long mouthparts that penetrate the stems of the host plant (figure 13.6a, right). They are an unaggressive morph specialized for feeding on the secondary host.

The origin of the aphid soldier morphology could conceivably have occurred in at least three different ways: (1) gradually, by directional or disruptive selection on continuously variable polygenic traits; (2) gradually, via numerous reorganizational changes in development, leading to the gradual transfer of a mosaic trait originally evolved in another life stage (gradual heterochrony); or (3) via single-step heterochrony, with key elements of the soldier phenotype shifted as a set from another lifestage.

Akimoto (1985) proposes a one-step-heterochrony hypothesis (hypothesis 3, above) for the origin of the short-stylet soldier morphology in Eriosomatinae (Pemphigidae). He argues that a regulatory mutation caused a secondary-host-inhabiting female to produce short-stylet larvae like those on the primary host. The production of the short-stylet form alongside normal long-styled feeding larvae by the same females would yield the larval dimorphism characteristic of soldier-producing species. As evidence for the feasibility of one-step change, Akimoto cites the work of Aoki (1980) showing that in one genus (*Colophina*) the soldiers on secondary-host plants resemble closely the primary-host larvae, which in that genus are specialized to attack predators. In addition, Akimoto (1985) observed another species (*Hemipodaphis persimilis*) of the subfamily Eriosomatinae with five anomalous larvae in the galls of the primary host whose morphology bears a detailed resemblance to secondary-host larvae. These anomalies demonstrate that a one-step heterochronic shift in morphology from secondary to primary host generations can occur. They also provide some support for a gradual heterochronic shift (hypothesis 2, above) because variation in the degree of expression of soldierlike morphology among the five anomalous individuals reveals that the soldierlike morphology is a mosaic of dissociable traits. One anomalous larva was distinctive in only one trait (wax plates); two showed an intermediate degree of resemblance involving several traits like those of secondary-host larvae; and two others resembled the secondary-host larvae almost completely. Such a mosaic set of traits could be subject to either a gradual heterochronic shift or a shift in a single block.

Stern (1994, p. 208) briefly presents a heterochrony hypothesis similar to that of Akimoto for the origin of soldiers in species of Cerataphidini (Hormaphididae), whose soldiers have head horns (figure 13.6d) rather than piercing mouthpart stylets. Heterochrony in this group is suggested by the fact that secondary-host soldiers are found only in species with primary-host soldiers, suggesting that the secondary-host-soldier traits originate in the primary-host larvae.

The Akimoto one-step hypothesis has been criticized by Kurosu and Aoki (1988) and Aoki and Kurosu (1988), who hypothesize a gradual evolution of the soldier morph. *Colophina clematicola* produces structurally monomorphic aggressive larvae in the secondary host. These larvae attack predators using their mouthparts, which are intermediate in length between those of sterile soldiers and normal larvae of sterile-soldier-containing species. The *C. clematicola* soldier larvae are phenotypically intermediate in that they feed, molt, and reproduce. Aoki and Kurosu (1988, p. 870) conclude that these observations render the Akimoto one-step hypothesis "less compelling." But aggressive behavior in structurally monomorphic larvae could lead to either the gradual or the sudden origin of soldier morphology—it does not by itself support either hypothesis more than the other.

In an earlier paper Aoki and Kurosu (1986, p. 102) considered the possibility that larval dimorphism and soldier production could have originated in another species (*Pemphigus spyrothecae*) via what they termed a "generation slip" or heterochronic mutation, as proposed by Akimoto (1985). They observed sporadically produced abnormal females (virginosexuparae) that produced two different kinds of larvae as envisioned by this hypothesis. But they were skeptical regarding its evolutionary future, since "[i]f a virginosexupara flies to either the primary host or the secondary one . . . parts of its progeny will not grow," some being specialized for one host and some for the other. "Therefore, this morph is a maladaptive one, and has not been created gradually by natural selection" (p. 102). This implies that they consider gradualism a requirement of adaptation, a view that might predispose them to doubt any one-step hypothesis, and it also implies that heterochrony involves an entire individual rather than a trait-by-trait change in expression, a common but potentially confusing view, as discussed by Reilly et al. (1997). In the aphids, soldier morphology could be subject to heterochronic change independent of features that permit survival on a given host.

In some species of the aphid subfamily Cerataphidinae, soldiers occur with no evidence of a het-

erochronic origin. Possible transitional states are represented in related species of the same subfamily, and Aoki (1987) plausibly hypothesizes gradual evolution of sterile soldiers in these species, starting with monomorphic larvae that are aggressive toward conspecifics, then toward predators, and finally a dimorphism with soldiers morphologically specialized to attack predators. Heterochrony is not supported in these species, for soldier traits produced in the two life stages are not homologous: soldiers on the primary host attack predators with the stylets, whereas those on the secondary host use horns (Aoki, 1987). Stern (1994) has suggested, however, that the soldier behavior of attacking rather than fleeing may have originated in larvae on the primary host and then was transferred as a behavioral heterochrony, or "generation slip" (to use Aoki and Koruso's term) to the secondary-host morphs (Stern, 1994).

The aphid-soldier controversy is of special interest because it represents in the microcosm of modern aphidology the age-old debates over gradualism, illustrating the plausibility of both sides and illuminated by a rich collection of comparative data (for a recent review, see Stern and Foster, 1996). The debate reveals the impossibility of generalization about the origin of a heterochronic change even within a single family of insects: the evidence in some groups supports a one-step or gradual heterochrony or "generation slip" hypothesis, while in other groups heterochrony is not plausible and there is evidence for gradual evolution of soldier traits. A phylogenetic analysis of the tribe Cerataphidini suggests that strong dimorphism arose first, followed by the gradual loss of soldier structural dimorphism, and, finally, of soldier behavior (Stern, 1998).

In this debate, developmental anomalies are cited as critical evidence on both sides. Perhaps a high frequency of developmental anomalies has contributed to the remarkable phenotypic diversity of the aphids. But this frequency of anomalies may not be unusual (see chapter 9), and the documentation of anomalies may say more about the acuity and industry of aphidologists than anything unusual about aphids: Akimoto (1985) inspected a total of 4,764 minute larvae of *Hemipodaphis persimilis*, and only five of them showed the anomaly he discussed.

14

Heterotopy

Heterotopy, or evolutionary change in the location of trait expression, is the spatial counterpart of heterochrony. It transfers traits between segments, from flowers to leaves, and even between structures within cells. It changes the patterns of colors on mammalian fur and on butterfly wings. Some heterotopies are classic illustrations of the role of environmental induction and of underlying developmental processes in the patterns of adaptive evolutionary change.

Heterotopy is the spatial analogue of heterochrony: it is evolutionary change in the site of expression of a phenotypic trait. Gould (1977) attributes the word "heterotopy" to Haeckel, who used it in a more specialized way, to mean evolutionary change in the germ layer from which an organ differentiates. Wray and McClay (1989, p. 810) list several examples of heterotopy, including the origin of muscles in tetrapod forelimbs from different somites, the origin of vertebrate primordial germ cells from different germ layers, and homeotic "heterotopic" mutations that transfer appendages from one body segment to another. A broad definition of heterotopy extends the concept to include spatial patterning, not only transposition from one location to another, but spatial organization of quantitative processes such as growth (Zelldditch et al., 1992) or the location of precursors during the development of homologous traits (Wray and McClay, 1988, p. 313).

As in other categorizations of transitions, heterotopies could as well be classified in other ways, such as duplication. Severtzoff (paraphrased by Goldschmidt, 1940, p. 388) signaled a general relationship between heterochrony and heterotopy when he wrote that "heterochrony in development is a means of topographic coordination; i.e., new adaptation of the parts to each other." Many morphological heterochronies in plants produce heterotopic change, since the morphological ontogeny of a plant is recorded in its adult architecture. Thus, changes in timing of expression of juvenile and adult leaf forms result in heterotopic change in architecture of the mature plant, with the juvenile leaves appearing high on the stem, rather than only basally as before.

A clear and oft-described example of environmentally mediated heterotopic change was demonstrated in early experiments on melanization in the Himalayan rabbit (Sturtevant, 1913; Iljin, 1927; Iljin and Iljin, 1930; see discussions in Schmalhausen, 1949 [1986]; Huxley, 1942; Levinton, 1988). In the Himalayan rabbit, as in the Siamese cat, pigment normally develops only in the extremities (figure 14.1b), where skin temperature is below the general body temperature. Individuals raised at temperatures above 30°C develop white extremities (figure 14.1a). If the hair is shaved off another part of the body and the rabbit kept in the cold, pigmented hair grows in the shaved areas (figure 14.1c). This shows that a heterotopic change in pigment pattern can be initiated by an environmental change, or as a secondary effect of change in another trait, such as reduced pelage in some body region. The temperature threshold for melanization is known to be alterable by a genetic mutation in an allele of the albino series (Huxley, 1942). So this is an illustration of gene–environment equivalence in development (see chapter 5), as well as heterotopy. Under artificial selection on domesticated breeds of rabbits and cats, the threshold is kept at a level just sensitive enough to produce localized melanization of the relatively cool extremities. This is a good model of how heterotopic change works (see also Price and Pavelka, 1996): a heritable potential for a particular threshold response, susceptible to both environmental and genotypic alteration, responds to phenotypi-

Fig. 14.1. Environmental influence and interchangeability in determination of pigmentation pattern in the Himalayan rabbit: individuals raised (a) above 30°C; (b) at about 25°C (the normal pattern); (c) with left flank hairless (shaved) (artificially cooled below 25°C); temperatures indicated are critical body-surface temperatures above which no pigment is formed (Levinton, 1988, after Stern, 1968). Thus, development of the normal pattern is a result of relatively low temperature in the extremities. Only strains homozygous for a Himalayan-specific allele at one gene of the albino gene series are temperature sensitive in this way (Stern, 1968), indicating that an albino strain could be created under selection against that allele (interchangeability of genotypic and environmental determination of color pattern). Based on Iljin (1927) cited in Huxley (1942), and Levinton (1988) after Stern (1968).

cally local developmental conditions in a new way. Then, providing that there is genetic variation in the response, and recurrence of the new conditions that initiated it, selection and genetic accommodation can propagate the change.

Another classic example of heterotopy occurs in geckos of the genus *Lygodactylus*, where the feet have hairy scales that facilitate climbing on vertical surfaces (Underwood, 1954; figure 14.2). In addition, some species have similarly modified scales on the underside of the tail (Loveridge, 1947). This example has been used to illustrate the importance of heterotopy in evolution, and the general point that phenotypic transitions can involve reexpression of preexisting traits: "To argue that this caudal pad appeared de novo without relation to the prior existence of digital pads would be unconvincing in the extreme" (Underwood, 1954, p. 366). A textbook example of genetic assimilation, the evolution of calluses in the ostrich at skin areas exposed to contact with the ground in the unusual avian anatomy of that species (e.g., Waddington, 1975), also illustrates a heterotopic shift in the expression of a specialized tissue. Examples such as these from the older literature would be good subjects for comparative research using modern phylogenetic techniques and information on development of specialized tissues.

Homeosis, the assumption by one member of a meristic series, of the form or characters proper to other members of the series (Garcia-Bellido, 1977), was among the heterotopic anomalies recognized by Bateson (1894). The myriad modification of appendages, following segment duplication in arthropods, into mouthparts, antennae, genitalia, and so

forth, are prime material for subsequent homeotic mutation (Garcia-Bellido, 1977) and homeotic anomalies can also be induced in vertebrates (Weiss, 1990; Mohanty-Hejmadi et al., 1992). Are homeotic mutations important in evolution? Goldschmidt (1940 p. 333) argued that they are, giving as possible examples a rudimentary halterelike wing form of a fly (*Termitoxenia*) resembling homeotic wings of *Drosophila* mutants, as well as halterelike wings in various wing-reduced species of Hymenoptera, Neuroptera, and Strepsiptera, but these are perhaps more accurately described as examples of truncated wing development rather than homeotic heterotopies. Raff and Kauffman (1983) found no evidence that homeotic mutations have themselves led to evolutionary change.

A good place to search for evolved homeosis is in the complex diversity of butterfly wings, where homeotic anomalies are common and readily visible (figure 14.3). Homeotic transformation between forewing and hind wing, and between dorsal and ventral surfaces of wings, occurs in a broad diversity of lepidopteran species (Nijhout and Rountree, 1995). In the buckeye butterfly (*Precis coenia*), eyespot organizing centers for the forewing can induce eyespots in hind wing tissue and vice versa, indicating that identical processes occur in both locations (Nijhout and Rountree, 1995; Weatherbee et al., 1999; figure 14.3). Many well-documented examples of heterotopic shifts are provided by the evolution of wing patterns in butterflies where, for example, long wing stripes may be assembled by the evolutionary realignment of many short stripe fragments originating in different subdivisions (cells of the wing). This "pierellization" (Nijhout, 1991b,

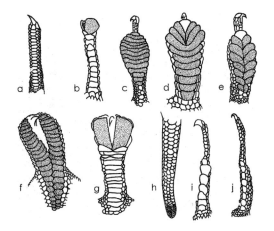

Fig. 14.2. Heterotopic variation in the expression of epidermal pads on the feet and tail in different species of geckos. Epidermal pads (shaded) are specialized scales with a velvety covering of fine hairs that facilitate climbing. (a) Simple fourth toe of *Coleonyx variegatus* without pads. (b–g) Patterns of enlarged, padded scales in the fourth toe of different species. (h) Padded scales at the tip of the tail of *Lygodactylus picturatus*, which has toe pads similar to those in (e). (i) Enlarged, unpadded proximal digital scales in fourth toe of *Gonatodes humeralis*. (j) Small, unpadded proximal scales in fourth toe of *Gonatode fuscus*. After Underwood (1954).

but also which elements of regulation and form are likely genetically variable and subject to selection and evolutionary change. They focused their attention on three white (unmelanized) patches in the plumage of a genus of warblers (*Phylloscopus*), located on the wing, crown, and rump (figure 14.4). The pattern of occurrence of these patches among different species indicates a developmental hierarchy: white wing bars can occur alone, while if only two patch types are present they are always wing bars and crown stripe; the white rump patch occurs only in the presence of the other two.

Price and Pavelka (1996) propose a developmental explanation for this evolutionary pattern in the topographic location of white patches on the bodies of warblers. They hypothesize that addition of successive patches may originate when selection favoring one (the wing bars) leads to a generally raised melanization threshold so that a white patch appears in the next most susceptible location—the crown; and if selection raises the threshold further (e.g., due to advantage of increased size or number of white patches), then the third (rump) patch may appear. Selection affecting the melanization threshold could likewise reduce patch number, which should occur in the reverse order (loss of rump patch followed by loss of crown stripe and finally loss of wing bar. Cladistic analysis tracing the gains and losses of patches during phylogeny confirmed this prediction of the threshold hypothesis.

p. 54) amounts to phenotypic recombination by spatial shifts in the expression of preexisting pigment patterns. Nijhout (1991b, pp. 183–185) reviews homeotic anomalies in the wing patterns of butterflies. Those between forewing and hind wing are most common, but they also occur between dorsal and ventral surfaces and between the sexes (in gynandromorphs). Although he does not relate these changes directly to evolutionary transitions, clearly this is the stuff of which adaptive transitions are made.

Spatial patterns of gene expression in the coloration of animals such as the Himalayan rabbit of figure 14.1 help us visualize how the interaction between development and selection can produce heterotopic shifts that, like some of the butterfly examples, are not duplications of pattern but utilize pathways and genes expressed elsewhere. Because some information is available on the mechanisms of melanization in the plumage of birds, Price and Pavelka (1996) were able to begin to explain not only why certain combinations of markings repeatedly occur together in many groups of birds,

Fig. 14.3. Heterotopic (homeotic) expression of eyespots in the hind wings of a butterfly (*Precis coenia*). The large hind-wing eyespots replace the small eyespots usual on the hind wings and resemble those of the forewings with which the eyespots are homologous (Nijhout and Rountree, 1995). Courtesy of H. F. Nijhout.

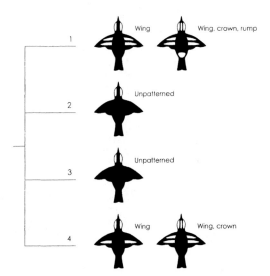

Fig. 14.4. Plumage-color heterotopy in four clades of warblers (*Phylloscopus* species). Unmelanized patches (white areas) occur on different body regions in clades 1 and 4 and are absent in clades 2 and 3. After Price and Pavelka (1996).

The mechanism of melanization in birds (summarized in Price and Pavelka, 1996) is particularly interesting because it involves somatic selection (see chapter 3) in the form of differential cell death. Melanocytes are organismlike cells that have a life cycle, behavior, and population biology or cellular ecology of their own. They begin as melanoblasts that divide and proliferate at the base of the feather germ, migrating into the feather as it grows. Under some conditions, they differentiate into melanocytes, produce melanin under stimulation from local epidermal signals, and die. If melanoblast-epidermis conditions are unfavorable to melanocyte differentiation, then the cells die prematurely and melanin synthesis does not occur. This can be caused by spatial variations in epidermal signaling early in development, or by a genetically variable life span of the melanocytes, whose durations show intraspecific and species differences. Thus, genetic variation in both of these qualities can affect the threshold for melanization and the pattern of white patches.

Warbler melanization patterns illustrate how developmental potentials characteristic of a taxon influence the variation available to selection and therefore affect patterns of social and adaptive evolution. The color patterns of birds are known to be frequently important in species-specific communication and display (e.g., see West-Eberhard, 1983),

and the basis for their diversity is found in a developmental heritage that provides both recurrent pattern and malleability, in the form of genetically variable responses of cells. This example shows clearly why adaptive traits are not properly understood as entirely molded by selection; nor is selection irrelevant, for it determines the variants that persist and multiply. Although this may seem obvious, misunderstanding of this point has been the crux of perennial controversy in evolutionary biology over the relative importance of selection and development in the determination of form (see chapter 24).

Wray and McClay (1989) describe an example of molecular heterotopy: a change in the location of two proteins within the gastrula of a sea urchin via shifts in expression from one mesodermal cell type to another in different echinoderm families (figure 14.5). Chitin, a component of the insect cuticle, seems particularly prone to heterotopic leaps during phylogeny if we count as heterotopy changes in context of expression that bring changes in location of expression. Chitin synthetase appears not only in arthropods but also in yeasts, *Xenopus laevis*, and a fish (Wagner et al., 1993). Its widespread occurrence and sporadic reactivation and functionality may owe to maintenance of the chitin synthetase gene or fragments of it in some context

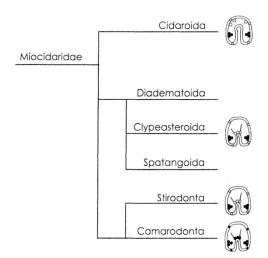

Fig. 14.5. Molecular heterotopy in orders of echinoids. Cell types are as in figure 13.2. Black dots indicate expression of S-12 antigen in gastrulae of representative species (see figure 13.2). Order Cidaroida is in the subclass Cidaroidea; the others are in the subclass Euechinoidea. After Wray and McClay (1989).

other than chitin production. Wagner et al. speculate that it may function to produce short chitin molecules used in lipid-membrane intercellular signaling, and thereby may have been conserved as a widespread functional gene by being expressed in locations other than the skeleton.

It may at first seem that heterotopies, being topographical, could only be morphological. But behavior and physiology also can have a topographical dimension. For example, abilities learned by practice with the right hand, if transferred to the left hand, represent in some respects a learned behavioral heterotopy. Behavioral heterotopies are recorded in the webs of spiders, where comparative study indicates phylogenetic change in the location on webs of such traits as the application of sticky silk (from the flagelliform gland), the location (central or peripheral) of the hub of radial webs, and the location of the resting place or retreat (Eberhard, 1990a).

In some plants (*Dalechampia* vines, Euphorbiaceae), resin secretion by leaves in some species protects them from herbivory. In other species of the same genus, only the flowers produce resins,

and they are chemically similar to those of leaves. Phylogenetic study has shown that resin production originated as a floral reward, then became defensive in flowers, and then came to be expressed in leaves (Armbruster et al., 1997). Microheterotopic change in location of gene expression within cells may have contributed to the evolution of CAM (crassalusean acid metabolism) in plants, by associating in new ways different metabolic processes. A number of changes in the main sites of intracellular expression of enzymes (e.g. carbonic anhydrase) and in regulation (e.g., of vacuolar H^+, K^+, $malate^{2-}$, and of phosphoenolpyruvate carboxylase (PEPC) have occurred in the evolution of CAM from C3 metabolism (Raven and Spicer, 1996), suggesting that shifts in the intracellular locations of preexisting processes have been instrumental in producing a novel physiological result in plants.

These examples show that heterotopic trait expression occurs at many levels of organization, from behavior to the transfer of appendages from one body segment to another, to molecular heterotopy within cells and early embryos.

15

Cross-sexual Transfer

Sexually dimorphic organisms can show extreme morphological and behavioral divergence between the sexes, and yet every individual bears the genetic potential to express both male and female traits, which often share an ancient homology. Hormonal systems that regulate sex-trait expression are highly sensitive to both genetic and environmental variation. This renders developmental reorganization of the sexual phenotype common and relatively easy to study using endocrinological techniques. Cross-sexual transfer, in which traits originally expressed in only one sex are secondarily expressed in the other, is a common source of evolutionary novelties, especially in the evolution of parental care, courtship communication, and alternative reproductive tactics in animals. In plants, combinations of male and female traits may lead to the evolution of striking novelties, such as the ear of domestic corn.

Introduction

Distinctive male and female traits are perhaps the most familiar of all divergent specializations within species. In *cross-sexual transfer*, discrete traits that are expressed exclusively in one sex in an ancestral species appear in the opposite sex of descendants. An example is the expression of brood care by males in a lineage where ancestral females are the exclusive caretakers of the young, as in some voles (Thomas and Birney, 1979).

Despite the prominence of sexual dimorphism and sex reversals in nature, and an early explicit treatment by Darwin, discussed in the next section, cross-sexual transfer is not often recognized as a major factor in the evolution of novelty (but see, on animals, Mayr, 1963, pp. 435–439; Mayr, 1970, p. 254; on plants, Iltis, 1983). When more widely investigated, cross-sexual transfer may prove to rival heterochrony and duplication as an important source of novelties in sexually dimorphic lineages. For this reason, I devote more attention here to cross-sexual transfer than to these other, well-established general patterns of change.

The Organization and Reorganization of Sex Expression

The male and female of a sexually dimorphic species may be so different that it is easy to forget that each individual carries most or all of the genes necessary to produce the phenotype of the opposite sex. Sex determination, like caste determination and other switches between alternative phenotypes, depends on only a few genetic loci or, in many species, environmental factors (Bull, 1983). There is considerable flexibility in sex determination and facultative reversal in some taxa. Among fish, for example, there is even a species wherein sex is determined by juvenile size at a critical age (Francis and Barlow, 1993). The sex determination mechanism, whatever its nature, leads to a series of sex-limited responses, often coordinated by hormones and not necessarily all occurring at once.

A distinguishing aspect of sexually dimorphic traits in adults is that there is often a close homology between the secondary sexual traits that are differently modified in the two sexes. In plants, male and female flowers, like the external genitalia of mammals (but not of all animals, e.g., insects; Snodgrass, 1935 [2000]), are closely homologous: the organs of both sexes, even though strikingly different in adults, begin development as identical, totipotent structures that are modified by differential abortion or divergent modification of certain parts during ontogeny (e.g., see Sundberg, 1987, on the inflorescences of maize, *Zea mays*; Frank et al., 1990, on the spotted hyena, *Crocuta crocuta*). When homology applies, similarities (e.g., tissue, positional, and other structural) sometimes may facilitate the evolutionary tranfer of expression between the sexes.

This is not to say that homology is required for cross-sexual transfer, because the mosaic, hierarchical organization of the phenotype means that subunits, including of the sexual phenotype, are developmentally dissociable and their components reexpressible at many levels of organization (see chapter 4). Sexual differentiation is a good illustration of the hierarchical nature of development. In humans, for example, sex is "determined" (*sensu* Bull, 1983) by the presence (in the male) or absence (in the female) of a Y chromosome. This sets the *chromosomal*, or genotypic, sex of the individual (see Crews, 1989, for this and the subsequent terms). But the rest of sex expression is spread out over ontogeny and so can be modified piecemeal: first there is *gonadal* sexual differentiation, leading to *anatomical* and *behavioral* sex differences. Most secondary sexual characteristics occur in sexually mature adults. Examples are sex differences in body size, hair growth, and distribution of subcutaneous fat in humans. These adult characteristics develop asynchronously and so are to some degree developmentally dissociable, with the result that a range of variants occurs within a normal distribution, with only the extremes dysfunctional. This means that even when sex is "determined" genotypically in the zygote, altered expression of sex-limited traits is still possible by rearrangement of the subsequently regulated subunits, due especially to hormonal variants that may result from environmental or genetic effects.

Because the sexual phenotype of any species is a mosaic of subunits, there are many kinds of "intersexes" that amount to continuous variation between the masculine and feminine extremes. In humans there are quantitatively variable aspects such as size, hairiness, and development of certain areas of the brain (e.g., see Goy and McEwen, 1980). One evolutionary consequence of this is that many degrees of change are possible, from partial sex transformations, involving a single small trait, to

Fig. 15.1. A young female spotted hyena, showing her malelike genitalia. By permission of *Bioscience*. Courtesy of K. E. Holekamp.

complete role reversals with extensive morphological change.

The potential for evolutionary lability and cross-sexual transfer is easy to appreciate from the fact that hormones and hormonelike substances are so often involved in sex differentiation in both animals and plants. Hormonal mediation is notoriously sensitive to both environmental and genetic influence. Hormones can coordinate the expression of complex arrays of traits, and they are known to directly mediate gene expression (for references, see Arnemann et al., 1993; Hayes, 1997a; Parker, 1993). So there is an enormous potential for combinatorial evolution of the sexual phenotype via hormonal change, and for selection to affect the frequencies of the affected genes. The genes for novel phenotypic recombinants are present in both sexes, so relatively small genetic changes may result in a large cross-sexual transfer of trait expression. Some changes assumed to be the result of a long evolutionary history of genetic change may in fact be largely or entirely environmentally mediated such that large phenotypic changes represent little initial change in gene frequencies.

Cross-sexual recombinants are often easy to recognize because they may have distinctively male and female elements mixed together, as in the clitoral penis of the female spotted hyena (figure 15.1), which is traversed by a functional birth canal even though virtually indistinguishable in outward appearance from the penis of a male (Kruuk, 1972; figure 15.2). In some species (e.g., of hermaphroditic vertebrates and monoecious plants), both sexes are expressed in the same individual, separated either in time (in sequential hermaphrodites)

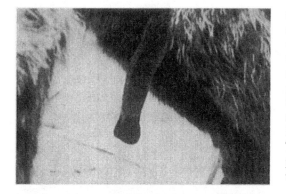

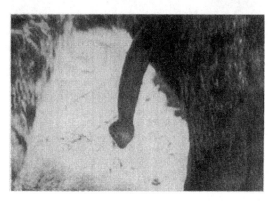

Fig. 15.2. Female pseudopenis (top) and male penis (bottom) at ages 8 and 9 months, respectively, in the spotted hyena. Successful mating requires that the male maneuver his penis into the female's pseudopenis, a task facilitated by the female being able to retract the phallus upon itself "much as one pushes up a shirt sleeve" (Frank, 1997, p. 58). From Frank et al. (1990), by permission of Cambridge University Press. Courtesy of L. Frank.

or in space (e.g., in monoecious plants, with male and female flowers at different locations on an individual). Even in these cases of intraindividual flexibility in sex determination, the coordinated expression of sex-limited traits is regulated by hormones, often the same ones that are involved in the sexual dimorphism of separate individuals (e.g., see Chailakhyan and Khryanin, 1980, on plants; Crews, 1989, on animals). As shown in this chapter, cross-sexual phenotypic resemblance often involves hormonal resemblance as well.

Most previous discussions of evolutionary change in sex expression have concentrated on its adaptive significance (e.g., for reviews, see Ridley, 1978; Charnov, 1982; Knowlton, 1982; Lloyd and Bawa, 1984; Reeve, 1993). Of course, adaptive

change is associated with positive fitness effects, but it also depends on the production of selectable variation. There is perhaps no class of evolutionary transition where the origins of novelty are so clearly seen to have arisen from preexisting adaptive complexity, for the two poles of a cross-sexual transfer can often be found within the same species, along with the developmental mechanisms that have permitted and perhaps promoted the change. This neglected factor in the evolution of sex expression—selectable, hormonally mediated phenotypic modification—is the focus of this chapter.

Darwin's Theory of Cross-sexual Transfer

Darwin (1859) considered monomorphism in secondary sexual traits to be a result of characters favored in the sexually selected sex (usually the male) being automatically expressed in the female. He thought that this would usually occur unless opposed by selection on females not to express the traits. In such *primary sexual monomorphism*, both sexes express similar traits and presumably similar sets of genes, and there is no evidence of a history of sex-limited expression in sister taxa. In *secondary sexual monomorphism*, by contrast, both sexes express a trait that historically was produced in only one sex, and the resemblance may be due to cross-sexual transfer of expression.

Darwin gave examples of cross-sexual transfer of traits, using this as evidence that all individuals inherit and transmit the capacity to produce the characteristics of both sexes, even if only one is expressed:

[C]haracters are occasionally transferred from the male to the female, as when, in certain breeds of the fowl, spurs regularly appear in the young and healthy females. But in truth they are simply developed in the female; for in every breed each detail in the structure of the spur is transmitted through the female to her male offspring. Many cases will hereafter be given, where the female exhibits, more or less perfectly, characters proper to the male, in whom they must have been first developed, and then transferred to the female. (Darwin, 1871 [1874], p. 584)

Darwin believed that transfer from female to male is less frequent than the reverse but gave one example of female–male transfer: the occurrence of a "perfectly developed" pollen-collecting apparatus in bumble bee (*Bombus*) males. He paid attention to the importance of ascertaining the polarity of the transition, reasoning in this case that "we have no grounds for supposing that male bees primordially collected pollen as well as the females" (p. 584) because the bumble bee male pollen baskets are unique among wasps and bees.

Darwin also noted that cross-sexual transfers in animals frequently occur as anomalies—variants on which selection could act—citing examples that we would now recognize as due to regulatory (hormonal) change:

[C]haracters proper to the male are occasionally developed in the female, as when she grows old or becomes diseased, as, for instance when the common hen assumes the flowing tail-feathers, hackles, comb, spurs, voice and even pugnacity of the cock. Conversely, the same thing is evident . . . with castrated males. (Darwin, 1871 [1874], p. 584)

With those animals and plants which habitually produce several forms, as with certain butterflies described by Mr. Wallace, in which three female forms and one male form coexist, or, as with the trimorphic forms of Lythreum and Oxalis, gemmules [genes] producing these different forms must be latent in each individual. (Darwin, 1868b [1875b], p. 394)

This would explain the "curious case of a hen which assumed the masculine characters, not of her own breed but of a remote progenitor illustrates the close connection between latent sexual characters and ordinary reversion" (Darwin, 1868b [1875b], p. 394). In his work on plants, Darwin (1868b [1875b]) remarked that "stamens are so frequently converted into petals, more or less completely, that such cases are passed over as not deserving notice" (p. 386). He also gives an extensive review of cross-sexual transfer in plants with references to early reviews.

Kinds of Evidence for Cross-sexual Transfer

The following kinds of observations support the hypothesis that resemblance between the sexes has originated by cross-sexual transfer, and indicate the direction of transfer:

1. The trait(s) resemble characteristics clearly more appropriate to, and usually exclusive to (in the same or related species), the opposite sex. There can be little reasonable doubt, for example, that the penile morphology and displays of female hyenas, discussed earlier in this chapter, were originally characteristic of males rather than females.

2. Detailed resemblance of a subset of one sex to individuals of the other sex within a species primarily dimorphic for the trait; or between the sexes in a larger taxon where phylogenetic information indicates that sexual dimorphism for the trait is ancestral. Emerson (1996) discusses the use of phylogenies in analyses of the evolution of sexually dimorphic traits and sex transfer.

3. A sexually monomorphic trait hypothesized to be a product of sex transfer in one of the sexes is more complete or elaborate in the hypothesized sex of origin than in the sex where it is hypothesized to be secondarily expressed. This is a weak criterion, since in both sexes traits can evolve to be either elaborated or simplified subsequent to transfer.

4. A trait suspected of cross-sexual transfer has maladaptive features in the hypothesized sex of secondary expression that are not seen, or would be adaptive, in the other sex. The malelike genitalia of female hyenas, for example, greatly complicate parturition by constricting the birth canal (Frank et al., 1995).

5. Hormone studies indicate hormonal feminization or masculinization of the phenotype.

6. Lack of evidence for the alternative explanations listed below.

Alternative Explanations for Sexual Monomorphism

The following transition hypotheses are alternatives to cross-sexual transfer as explanations for similarity of males and females.

Primary Sexual Monomorphism

As already explained, resemblance between the sexes may be due to a history of selection acting similarly on both sexes so that monomorphism is ancestral and male–female divergence is limited. Or, as Darwin argued, traits evolved under selection in one sex may be expressed in the other if not costly. Either situation could maintain a primary sexual monomorphism.

Resemblance between the sexes may sometimes be maintained by selection on one sex to resemble or "mimic" the other, but the resemblance may have a common origin not involving cross-sexual transfer. Young or malnourished male rove beetles (*Aleochara curtula*, Coleoptera, Staphylinidae), for example, produce the female sex pheromone, which enables them to feed on carrion without being attacked by mature males (Peschke, 1987). But the sex pheromone is produced by newly emerged adults of both sexes (Peschke, 1986) and may have originated as an incidental product of some physiological process common to immature individuals regardless of sex. The same young-adult sexually monomorphic trait, then, may have been opportunistically used by males in two ways: to locate young females who exude the odor when young (a common origin for female sex pheromones in insects, where males simply take advantage of a cue incidentally produced by females; West-Eberhard, 1984; see also G. C. Williams, 1992); and, due to their own pheromonal resemblance to females, to gain access to food with less probability of attack by aggressive males. In such a sequence, the selective advantage of female mimicry may help to maintain the male–female resemblance, but cross-sexual transfer of the phenotypic trait need not have been involved. The absence of cross-sexual transfer is also suggested by the lack of femalelike behavior or morphology in the pheromone-producing males (as discussed in a later section, many femalelike alternative male phenotypes suggestive of sex transfer involve expression in males of more than one femalelike trait, with very close female–male similarity).

These contrasting interpretations could be examined in various ways. A phylogenetic analysis may illuminate the sequence of evolutionary changes if sufficient variation is present in behavior of different lineages: if common ancestors can be deduced to have produced the pheromone in both sexes prior to the evolution of a sex-attractant function, this would support the primary-monomorphism hypothesis (no cross-sexual transfer). If common ancestors likely produced the pheromone in only one sex, cross-sexual transfer is supported. If adult femalelike behaviors not present in juveniles accompany the expression of femalelike pheromones, this would support a sex-transfer hypothesis by suggesting that the female traits are expressed as a unit. Some examples of femalelike plumage in adult male birds may be derived, via heterochrony, from monomorphically dull juvenile plumage rather than by cross-sexual transfer.

Convergence

In convergence, resemblance is due to similarity of nonhomologous traits. An example of convergent sexual monomorphism (or pseudomonomorphism) is the bright anal spots of some primates (*Cercopithecus*), which in males are formed by bright red hairs and in females by red-pigmented skin (Wickler, 1967). Even if such resemblance is secondary to an ancestral dimorphism in appearance, it is unlikely to have involved cross-sexual transfer.

If different hormonal backgrounds are involved in cases of secondary monomorphism, this may be an indication that the resemblance is the result of convergent evolution achieved via different developmental pathways. If a secondary resemblance between the sexes has the same or a similar hormonal background, this is positive but not decisive evidence that the similarity is due to homology rather than convergence.

Deletion of Sex-Limited Trait(s)

Increased resemblance between the sexes could be due to deletion (or reduction) of distinctive sex-limited traits—monomorphism by subtraction (see discussion of the garter snake "she-males," below).

An example of this sort may occur in females of some socially parasitic bees, where females are more similar to males than they are in the nonparasitic groups from which they derive (Wcislo, 1990, 1999). Perez (1884) described this as due to "hereditary transport" from one sex to the other (cross-sexual transfer, or masculinization; see also Wcislo, 1990), and in one parasitic species, *Paralictus asteris*, the females are phenetically closer to males than they are to the females of their host *Lasioglossum imitatum* (Wcislo, 1999), suggesting sex transfer. But in some other parasitic bees, the similarity of females to males may represent deletion of female specializations, or defeminization, as cautioned by Michener (1978). In parasitic halictid bees, females do not build or provision nests and have lost the pollen combs and nest-building structures found in nonparasitic species, so resemblance could at least sometimes represent convergence due to loss of female traits rather than feminization due to expression in males of female traits. Feminizing deletions of male traits can be produced in some mosquitoes by high-temperature rearing of male larvae. This causes the male-distinctive traits to be unexpressed, leaving the femalelike traits unchanged (O'Meara and Van Handel, 1971), implying a sexual dimorphism in which males always express female traits, plus some distinctive male-limited ones, the opposite of the situation in bees.

Sex-Expression Pathologies Induced by Parasites and Pathogens

Normal sexual development often involves the manipulation of one individual's phenotype by another individual, such as a parent or a mate, the most dramatic examples being sex change due to interactions during courtship (discussed below). So it is not surprising that pathological cross-sexual variants sometimes occur due to manipulations by individuals of other species.

Mermithid nematodes, for example, are famous for the ability to induce feminization or masculinization in their invertebrate hosts (review in Vance, 1996). This sometimes creates intersexes so highly modified that their original sex is difficult to discern. Male mayflies (*Baetis bicaudatus*) infected by the nematode *Gasteromermis* sp. are dramatically feminized (Vance, 1996) and may sometimes be completely sex-reversed. They adopt an array of female behaviors, including upstream dispersal and oviposition behavior. Cross-sexual transfer due to parasitization may begin as an incidental side effect of infection, but in the mayflies it appears to be an extended phenotype of the host that is adaptive for the parasite: female behavior ensures a return to water and upstream transport for the completion of the parasitic life cycle, and it preserves host resources for use of the parasite, since femalelike dispersal uses less energy and fat reserves than would male swarming behavior (Vance, 1996).

Wheeler (1910b) long ago cited many examples of altered sex expression in parasitized insects, crustaceans (spider crabs, hermit crabs), and plants. Some otherwise mysterious cases of cross-sexual transfer involve parasites. For example, genetic females of the sexually dimorphic crustacean *Armadillium vulgare* (Isopoda, suborder Oniscoidea) frequently show partial or complete masculinization of the external morphology, which is inherited in non-Mendelian fashion by a variable proportion of their offspring. The cause proved to be a vertically transmitted cytoplasmic virus (Juchault et al., 1991). Males of some species of the same isopod genus are feminized when infected by symbiotic bacteria (Juchault et al., 1991).

Cross-sexual Transfer in Plants

The following sections describe examples of cross-sexual transfer in a large number of plants and animals. One purpose of these examples is to show the importance of this relatively neglected phenomenon. Another is to illustrate problems as well as approaches in evolutionary studies of sex expression, and to suggest future avenues of research. Some, such as sex transfer in the evolution of maize, are discussed in considerable detail in order to show the complexities of a well-studied example and the questions raised in its interpretation.

Sex expression in plants is an exceedingly complex topic due to the numerous arrangements of male and female structures in plants, and the specialized terminology that has accumulated to describe them. Charnov (1982) and Lloyd and Bawa (1984) provide synthetic discussions of gender modification in plants and of the possible adaptive significance of different patterns of sex expression (sex allocation) in response to environmental conditions. Here, a detailed discussion of a single much-debated transition, the evolution of maize, can serve as an introduction to cross-sexual transfer as a mode of evolutionary change.

The Origin and Evolution of Maize

The flower head of *Zea mays* has been called "the most striking morphological novelty among the higher plants" (Maynard Smith, 1981, quoting a remark by G. L. Stebbins—see Gottlieb, 1984, p. 693). Modern corn (maize, *Zea mays mays*) descended from a Central American teosinte, probably Balsas teosinte (*Z. mays parviglumis*; Doebley, 1983, 1990; Wang et al., 1999; see also Iltis, 1983; Walbot, 1983; Doebley et al., 1990; cf. Manglesdorf, 1974, 1986, who proposed that the distinctive characteristics of maize originated prior to domestication: the "wild maize" hypothesis; see Doebley, 1990). The ears of maize—female structures—occupy positions on the stem homologous to the lateral branches of teosinte that bear the terminal male inflorescences (tassels). There has long been agreement that the pivotal change in the origin of maize from teosinte was branch shortening, and the production of a female inflorescence (ear) in the position, at the end of a branch, formerly occupied by a tassel (figure 15.3). Thus, as early as 1937, J. H. Kempton (cited by Iltis, 1983 p. 887) wrote, "All authorities recognize that the ear of corn is a transformed terminal inflorescence of a

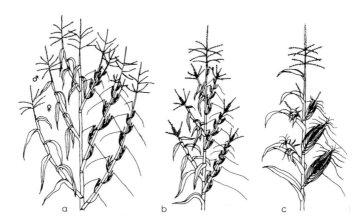

Fig. 15.3. The origin of the maize ear: (a) a teosinte (*Zea mays mexicana*), with long lateral branches tipped with male inflorescences (tassels); (b) a maize × teosinte hybrid, with shortened branches and feminized tassels; (c) maize (*Z. mays mays*), in which lateral branches are condensed into the cob and lack the terminal male inflorescences (tassels). After Iltis (1983).

lateral branch and that its covering of husks came about through a shortening of the internodes."

Branch shortening and sex transformation of the terminal inflorescence from male to female lead it to suppress, through apical dominance, the later-blooming ears that develop along the branches of teosinte. This diverts most branch nutrients to a single large ear (Iltis and Doebley, 1984, pp. 607ff). The shortened maize branches thus produce one large, husked cob close to the main stem, with the internodes not reduced in number but compressed into the "shank" of the cob. Iltis and Doebley (1984) credit Kellerman (1895) and Montgomery (1906, 1913) with being the first to invoke cross-sexual transfer in the origin of maize.

Several kinds of evidence support the cross-sex transfer hypothesis for the origin of maize from teosinte:

(a) Sex-transfer anomalies (e.g., mixed-sex inflorescences, feminized tassels, and masculinized cobs) are common in both maize and teosinte. Both modern and archeological maize ears often have staminate "tails" (attached tassels), suggesting ease of transformation between tassels and ears.

(b) Plant parts seen as homologous under the sex-transfer hypothesis show detailed morphological similarity. Maize grains, like spikelets of the teosinte tassel, are always paired (unlike the solitary grains of the teosinte ear), and each of the two florets of each tassel spikelet in teosinte has a vestigial ovary, the upper one of which can be regarded as hormonally reactivated in the maize ear (Iltis, 1983, 1987; Sundberg and Orr, 1990, p. 150). The shank of the maize ear is a lateral branch off the

main stalk, clearly showing the internodes corresponding to those of a lateral branch. The ear occupies the terminal position of a tassel (see Iltis, 1987, figure 19.9, p. 209)—it is not in the position of a secondary branchlet like a teosinte ear. In some "primitive races" of maize, rudimentary cobs sometimes appear as secondary lateral appendages to the maize cob shank, in the position where a teosinte ear would be (see Doebley et al., 1990, their figure 1, p. 9888), a kind of atavism that supports the sex-transfer hypothesis. The leaves of the condensed branch (shank) correspond to the layered leaves of the cob husk, further reinforcing the hypothesis that the maize ear occupied the apical position of the ancestral tassel.

(c) Cross-sexual transfers (e.g., between tassel and cob) are inducible by treatment with sex hormones (gibberellins) in maize (reviewed in Dellaporta and Calderon-Urrea, 1994), and hormonal changes accompany environmentally induced sex reversals (Rood et al., 1980). Reduced ambient light intensity causes the apical inflorescence of maize to change from male to female, and this is accompanied by a rise in gibberellin levels (Rood et al., 1980, p. 795).

(d) The hypothesized cross-sexual transformations can be experimentally influenced either via genetic change under selective breeding or via environmental induction. In maize (*Z. mays mays*), the primordia for producing both sexes are present in both male and female inflorescences (Richey and Sprague, 1932), and in greenhouse specimens of another subspecies (*Z. mays mexicana*) close to the hypothesized ancestral teosinte subspecies, *Zea mays parviglumis* (see Doebley et al., 1990, p. 9888), sexual transformations (feminized tassels) are easily induced (Iltis, 1983, n. 65). Environmentally induced feminization also occurs in up to 100% of maize tassels under short day lengths, under lowered temperature (Richey and Sprague, 1932; Schaffner, 1927, 1930), and under reduced light intensity (Rood et al., 1980). Polygenic variation in environmental susceptibility is indicated by response differences of different genetic strains and in crosses between strains (Richey and Sprague, 1932).

Recent research has shown that the salient differences between teosinte (*Z. mays parviglumis*) and modern maize are largely influenced by only four or five genetic loci found at four chromosomal locations (Walbot, 1983; see summaries in Doebley and Stec, 1991, 1993; Doebley et al., 1995; Szabo and Burr, 1996). These include a locus (*teosinte branched 1*, tb1) with pleiotropic effects on branch length and inflorescence feminization (see especially Doebley et al., 1997; Wang et al., 1999) that has an effect twice as strong as its allele in teosinte due to evolutionary change in its promoter region (not its protein-coding region); a locus (*teosinte glume architecture 1*, tba1) that strongly influences seed case (glume) softness (Dorweiler et al., 1993); and other loci that show simple Mendelian inheritance in *maize–parviglumis* hybrids (Szabo and Burr, 1996). These loci influence traits of the female inflorescence (ears): two-ranked versus many-ranked ears, inclination of the kernels toward the rachis, distichous versus polystichous central spikes, and paired versus single spikelets (the latter apparently influenced by two loci; Szabo and Burr, 1996).

Unresolved Questions   These observations on environmentally influenced sex expression in maize and its relatives, and the genes responsible for sex-limited (female) traits, could serve as a model demonstration of gene–environment equivalence in development, and interchangeability in evolution (see chapter 5). They also imply the capacity for genetic accommodation of the sex-determining switch that would permit evolution of a major novelty, even if it had been originally environmentally induced.

In keeping with this capacity for interchangeability, there are indications that developmental plasticity of the ancestral teosinte may have played a role in the origin of maize. As in many other monopodial annuals, teosinte (*Z. mays parviglumis* and *Z. luxurians*) produces branches in uncrowded conditions and unbranched plants in thickets and maize fields and in dense stands of teosinte (Iltis, 1983, p. 892). Doebley et al. (1995, after unpublished observations of Doebley) also note that both slender unbranched and robust highly branched teosinte plants occur under natural conditions in Mexico. Both Iltis (1983, pp. 890–892) and Doebley et al. (1995, p. 342) hypothesize that the evolution of maize began with environmental induction of branch shortening, followed by genetic assimilation of the short-branched morphology.

The possibility also needs to be considered that environmental stress could have produced a transposon-mediated rearrangement of the genome with phenotypic consequences, since stress-related transposon-mediated novelties have been documented in maize (Walbot et al., 1987; chapter 3). Whether the response to stress was initially phenotypic (due to developmental plasticity) or genomic (due to transposon activity), genetic accommodation would be expected to follow, for example, under selection by humans.

There is general agreement that cross-sexual transfer accounts for the origin of maize, that developmental plasticity and genetic assimilation likely played an important role, and that a few genetic loci are highly influential in producing the major morphological differences between teosinte and modern corn. But there are still questions to be resolved regarding the importance of a partial rather than a complete sex-transfer early in the transition to maize and regarding the precise role of genetic loci of large effect.

As pointed out by Sundberg and Orr (1990) and Sundberg (1990) after Sattler (1988), given the close homology between the teosinte tassel and ear, and their sex-expression totipotency during early development (Sundberg and Orr, 1990), the debate over whether maize is derived from a tassel or an ear is a pseudoquestion. The real question for debate concerns whether the evolution of maize began with a partial or a complete sex transformation at the morphological position usually occupied by a teosinte tassel. The catastrophic sex-transmutation theory of Iltis (1983) proposed that some characteristics of the maize ear, such as increased softness of the seed cases or glumes, were remnant male (tassel) traits. That is, the theory visualized a transformation that combined characteristics of the female and the male inflorescence. One mixed aspect of the transition to maize was the change in developmental position of the ear to the location on the plant formerly occupied by the tassel, with the corresponding change in apical dominance and resource allocation, to a single ear per branch rather than, potentially, several. The result is a relatively well-nourished, large ear (Iltis, 1983). Another hypothesized change was partial feminization of the entire inflorescence (a sex mosaic), rather than a complete feminization, or a mixed zoned inflorescence like those common in teosinte and maize, where the inflorescence has a female (ear) zone near the stalk and a male (tassel) zone distally (e.g., Sundberg and Orr, 1986). The possibility of an initial partial feminization has not been sufficiently investigated. Are there tassel traits, such as soft glumes and perhaps others, that could have made a partially feminized sexual mosaic attractive for human use? And are such traits evident in sex mosaics in teosinte?

Early work by Schaffner (1930), described below, showed that unzoned sex mosaics are possible in maize. The conclusion of Doebley et al. (1995) that a complete feminization was involved, with all evolutionary genetic change in the inflorescence beginning with an entirely female substrate (teosinte ear), seems inadequately justified, for it employs an advanced product of maize evolution, the highly derived, enhanced-function maize alleles at the tb1 locus, which when transferred into a teosinte background produce a complete feminization (Doebley et al., 1995). An enhanced function allele present in modern maize, with double the phenotypic effect of the teosinte allele (Wang et al., 1999), cannot be assumed to represent the earliest steps in maize evolution. Similarly, the tga1 allele of modern maize that has a strong influence on softness of glumes (Dorweiler et al., 1993) may be a relatively recent improvement, and it could mimic and enhance, by an entirely new pathway, an originally male-derived soft-glume phenotype. Selection sees only phenotypes and can preserve and exaggerate them without conserving their original developmental basis (see chapter 25).

The genetic data of Doebley and others certainly establish the evolutionary importance of several loci of major effect, contrary to the expectations of Iltis (1983). But it would be premature to conclude that mutations at a few key loci explain the origin of maize (e.g., following Beadle, 1939). As discussed in chapter 6, genetic accommodation of a recently evolved trait may often be characterized by fixation of alleles of relatively large effect on quantitative traits, giving the impression of "control" by single quantitative trait loci. Findings of Lukens and Doebley (1999) show that the tb1 locus in teosinte and maize has greater phenotypic plasticity of expression (in response to plant density) in teosinte than in maize, and that its effects on maize branch architecture may have evolved under selection for a gene complex, not a single

Mendelian locus. These conclusions are consistent with a vision of evolutionary change (see chapter 6) in which novel phenotypes originate due to environmental or mutational effects on a developmentally plastic phenotype, then are genetically accommodated by change in quantitatively variable traits, such as hormone systems, that affect gene expression.

Solutions Using Developmental Plasticity   The story of the evolution of maize is fundamentally a story of evolutionary change in sex expression whose mechanisms and archaeology are still incompletely understood (Piperno and Pearsall, 1998). Early experimental and morphological studies of sex expression in plants by Schaffner (1927, 1930) show how developmental plasticity can be exploited to examine the possibility that partial sex transfers played a role in the evolution of maize, if combined with modern genetic, microscopic, and physiological techniques, and applied to teosinte. Schaffner (1930) investigated the sex reversal transition in flowering plants, particularly maize, by scrutinizing the "neutral zone"—the zone of environmental conditions where sex determination is developmentally poised at the switch point between male and female pathways (see chapter 5). By manipulating environmental conditions, especially day length, he was able to extend the transitional neutral zone between female and male expression in maize tassels. In some plants, when the neutral point was reached just as reproduction and sex determination were beginning, he obtained entirely "neuter" tassel spikes with neither female nor male phenotypes expressed. In others he obtained interdigitated sex mosaics—mixed inflorescences, but not zoned ones like those studied developmentally in the past (e.g., Sundberg and Orr, 1986). What is going on, in terms of gene expression and hormone action, in that attenuated transitional zone, and how would it look in teosinte?

Schaffner (1930) was emphatic in pointing out that this effect is dependent on the "various physiological gradients induced" by the environment. "Any theory of sex which would explain the diversity of sexual expressions by an appeal to a diversity of gene constitutions is entirely beside the mark" (p. 285). Schaffner also provided evidence for individual differences in response, showing "the interaction of the hereditary constitution of the individual with the environment" (p. 289). Genetic variation is clearly only one factor in the evolution of developmentally plastic cultivars exploited under different environmental conditions by mobile human populations with varied culinary tastes.

There is evidence that a marked change in glume softness, for example, originated not in the Balsas valley populations of Mexico, which is the home of *Zea mays parviglumis*, but in Panama (Piperno and Pearsall, 1998). The evidence for gene–environment interchangeability (see above) means that environmentally induced anomalies could have been crucial at different stages in the evolution of maize, not only in its origin but also in the appearance of selectable variation in other traits, as teosinte cultivation spread under different conditions of soil, climate, and culture in the Americas.

What is missing from the story of the evolution of maize is more detailed genetic and morphological study of developmental plasticity and genetic variation in natural populations of teosinte, with special attention to the feminized inflorescences and gene–hormone regulation of short-branched plants. Variation in natural populations, and effects of extreme conditions that might have affected their evolution and domestication, have scarcely been investigated (Iltis, 2000). Sachs (1988b) presents a path-breaking discussion of plant evolutionary endocrinology, recognizing phytohormones as sources and indicators of evolutionary lability in plants. The evolutionary study of *Zea mays* is ripe for a combined genetic, phytohormonal, morphological, and archaeological study of variation in native and experimental populations of teosinte, for no single approach can adequately solve the mystery of the origin and evolution of maize.

Cross-sexual Transfer in Other Plants

Anomalies   Meyer (1966) reviews the older literature on developmental anomalies in plants, many of them products of cross-sexual transfers. As usual for anomalies, most examples are in papers dealing primarily with other subjects. Anomalies are virtually ignored in publications after the 1800s: "A non-heritable abnormality is unlikely to be reported in modern journals. . . . heritable abnormalities are responsible for the majority of the reports of abnormal differentiation" since the turn of the century (Meyer, 1966, p. 178). Wheeler (1910b, p. 417) also lists various environmentally induced sexual transformations in plants caused by fungal, nematode, and aphid infections.

The many indications that environmental induction is important for sex-transfer transitions in plants (Meyer, 1966; Wheeler, 1910b; and the data on maize, above) would seem to contradict the belief of Hilu (1983) that "[r]eproductive structures generally show more stability under different environmental conditions than do vegetative structures" (p. 98). If this is so, then environmental induction must be very important indeed in the evolution of vegetative novelties in plants, since it is so prominently associated with floral anomalies.

Without counting reductions in sexuality (e.g., transformation of a stamen into a petal), and without counting the results of hybridization, Meyer's (1966) review lists a total of 263 species with partial and 82 with complete cross-sexual anomalies in flower structure (for additional examples see Bawa, 1980). "Partial" transfers involve either transitions to or from nonsexual flower parts such as sepals, petals, leaves or bracts, or sex mosaics (partial feminizations or masculinizations of a flower; production of hermaphrodite flowers). "Complete" transfers convert stamens to carpels or vice versa. Not only are sex-transfer anomalies common, but they are also taxonomically widespread, observed in 67 of the 99 families of flowering plants listed by Meyer (1966, his table I). Most of the 32 taxa with no known examples are small, little-studied families. Hilu (1983) lists many cross-sexual transfers caused by induced mutations in plants. Taken together, the data on mutations, anomalies, and regularly occurring sex reversals suggest that cross-sexual transfer is likely to be an important mode of evolutionary transition in flowering plants.

Not all sex reversals in plants are anomalies. Many plants naturally undergo inflorescence sex reversals either as a temporary, reversible phase or as an irreversible decision in their life cycle, and this at least in some cases seems to be part of a "sex allocation strategy" that enhances the reproductive success of the plant (Lloyd and Bawa, 1984). In some cases of regularly occurring condition-sensitive sex change, it is difficult to tell whether the switch is adaptive. In maize and teosinte, for example, both complete and partial (zoned) sex mosaics are exceedingly common in response to numerous environmental conditions, including temperature extremes, infection, soil conditions, and change in day length (see references in Iltis, 1983), but these variants are not yet known to be adaptive in natural populations.

Homeoses Homeosis, the assumption by one member of a meristic series of the form or characters proper to other members of the series, was discussed in chapter 14, as a type of heterotopy. Goethe (1790) is usually credited with being the first to realize that all appendices of the leafy shoot of a plant—leaves, bracts, sepals, petals, stamens, and pistils—are homologous (Meyer, 1966; Coen,

1991). On the basis of an extensive review, Meyer (1966) concluded that all of the organs of the flower are capable, in some plant or other, of developing in the form of any of the other organs present in the normal flower. Plant ontogeny is a progressive series of metamorphoses in the form of the organs produced, starting with the leaf and culminating in the parts of the flower (for modern treatments, see Coen, 1991; Ray, 1990; Freeling et al., 1992). So plants are constructed of a series of homologous organs and are eminently subject to homeotic developmental and evolutionary change.

The homeosis concept is relevant to evolutionary sex transfer in plants because the male and female structures are often arranged in meristic series, as in teosinte, with its male flowers (tassles) distal to the female ones (ears), or as in *Arabidopsis*, with male structures (stamens and carpels) adjacent, and convertible, to female ones (petals and sepals; see Meyerowitz, 1994). Cross-sexual transfers, then, as in the evolution of maize, may involve homeotic transfers from one segment or module to another. Some recent discussions of cross-sexual expression appear in the literature as homeoses (e.g., see Coen, 1991; Lehmann and Sattler, 1993; Meyerowitz, 1994) and involve regulatory mutants, such as AGAMOUS, that invite comparison with animal hox (homeobox) genes, and like those genes have highly conserved homologous sequences in plants (see Meyerowitz, 1994).

Developmental genetic studies of homeotic change, primarily in *Antirrhinum majus* and *Arabidopsis thaliana*, are reviewed by Coen (1991). In the woodland poppy *Sanguinaria canadensis* (Papaveraceae), multiple petals develop from stamen primordia and occupy positions taken by stamens in related plants (Lehmann and Sattler, 1993), also indicating a cross-sexual transfer of expression. Lehmann and Sattler give reasons to believe that the polarity of the change was from male (stamen) to female (petal) and relate their findings to the molecular biology of gene expression in other plants.

In summary, research at many levels of organization and on many kinds of plants indicates that sex transfer is an important kind of evolutionary transition in flowering plants.

Cross-sexual Transfer in Animals

As in plants, sex transfer of traits seems to occur in animals at low frequency in many species, and sex-transfer anomalies are occasionally described even though not often a subject of special study, except in the older literature. W. M. Wheeler

(1937), for example, devoted an entire book to *Mosaics and Other Anomalies Among Ants*. And Beach (1961, chap. 2), in a classic book on hormones and behavior, lists many interesting sex-transfer anomalies and experimental results from early endocrinological literature on both invertebrates and vertebrates.

Some of the first evolutionary studies of cross-sexual transfer where by Goldschmidt (e.g., 1940) using the gypsy moth (*Lymantria dispar*), where partial to complete cross-sexual transfers in genitalia and antennae occur when pupae are subjected to extreme temperatures during critical developmental periods (Goldschmidt, 1940 [1982], pp. 301–304). Comparisons with other species reveal that such variants may have played an evolutionary role: in a closely related genus (*Orgyia*), the male antennae are like those of a partially feminized *Lymantria* male, having one long and one short row of lateral branches, suggesting the possibility of their origin via cross-sexual transfer. There is also evidence for evolutionary change in male genitalia by cross-sexual transfer of homologous parts in moths, and differences between species can be simulated by temperature treatments producing intersexes (Goldschmidt, 1940, p. 302).

Despite the commonness of sex-transfer anomalies in animals, sex transfer as source of evolutionary novelties comparable in importance to duplications, deletions, reversions, heterochrony, and heterotopy has seldom been discussed. In this regard, a paper by Schultz (1987) on the origin of the silk-producing apparatus (spinnerets) of both male and female spiders is exceptional in its recognition of sex transfer as a likely widespread kind of evolutionary transition. The two leading hypotheses for silk-and-spinneret origin both imply cross-sexual transfer. One idea is that they began in males, as modified glands and secretions originally involved in the construction of the spermatophore, and were later adapted for use in construction of the sperm web, then for the webs used in catching prey. The other hypothesis is that they originated in females, as glands and secretions for construction of the egg sac. The hypothesis that silk production originated in a reproductive context in one or the other of the two sexes is supported by (1) the likely location of the spinnerets near the genital opening in ancestral spiders, suggesting that they represent modified gonopods; (2) the fact that the most primitive types of silk glands in spiders are used in construction of sperm webs or egg sacs; (3) the fact that silk in other adult arthropods is used largely in manipulation or protection of sex cells; and (4) the fact that when silk has multiple uses, one of them usually has to do with reproduction (Schultz, 1987). Schultz reviews comparative data on spiders and related arachnids (Amblypygi, Uropygi) and terrestrial arthropods (myriapods, hexapods) and suggests ways to discriminate between the female and male origins in future research.

Animal Communication and Signaling Morphology

Insects   Insect taxonomists sometimes note patterns of variation that suggest cross-sexual transfer. For example, Richards (1961, p. 6, after Balfour-Browne, 1940) wrote, "A number of Dysticids [Coleoptera] have a peculiar dimorphism in the female, one form having the elytra sculptured as in the male and the other much duller and more reticulate, as in the female of most other species." Malelike color morphs or "androchrome morphs" are common in the females of some insects, for example, in damselflies and dragonflies (Odonata; Fincke, 1994; Cordero and Andrés, 1996; Corbet, 1999) and butterflies ("transvestitism"; Clarke et al., 1985), and femalelike behaviors occur in male scorpionflies (Thornhill, 1979). Such morphs, when they occur regularly within a few species and closely resemble the opposite sex, are good candidate examples of origin by cross-sexual transfer.

The Odonata are especially promising for research on the role of developmental plasticity in the evolution of sexual monomorphism and dimorphism in coloration. Ratios of female morphs, including a malelike bright-colored form hypothesized on the basis of genetic experiments to be under single-locus control in controlled conditions in the laboratory (Cordero, 1990, on *Ischnura graellsii*), are also known to vary seasonally in natural populations (Cordero, 1990), and coloration of males and of the three female morphs changes with age (Cordero, 1987). In some species of this genus, the bright coloration is temperature dependent and reversible: a single individual of dull (gynochromatypic) coloration can change to blue (androchromatypic) and back to dull again, a type of reversible temperature-induced color change that occurs in 12 genera of seven families of Odonata (Corbet, 1999, p. 280). In large samples of laboratory and natural populations, a femalelike male was once observed as an anomaly, as well as an atypical female whose coloration resembled that of related species in the same genus (Cordero, 1992). In some groups, species closely related to the reversibly color-plastic species have fixed coloration (Corbet, 1999); and there is often geographic variation in

morph ratios (Andrés and Cordero, 1999). Such data, and the masterly synthetic review of the biology of the Odonata by Corbet (1999), poises this group for evolutionary studies of development and evolution possibly involving cross-sexual transfer.

Anomalous cross-sexual transfer of behavior is associated with aggressive circumstances in crickets. Although females of most crickets (Gryllidae) lack the file-and-scraper stridulatory apparatus on the wings and therefore cannot sing, during aggressive encounters they sometimes raise the wings slightly and move them as males do when they produce aggressive songs (Kutsch and Huber, 1989). Whether this capacity is adaptive or not, the neuromuscular circuits for singing develop early in ontogeny, for late juvenile instars of *Gryllus campestris* already fight, and during aggressive encounters they display typical adult behaviors except sound production. The corresponding thoracic muscles are activated, as during aggressive song, indicating that a juvenile sings "internally" even though it lacks external singing morphology (wings; Kutsch and Huber, 1989, p. 306). Therefore, both flight and stridulatory motor patterns are completed before the wings have developed fully and become functional.

Given this hint of a developmental propensity for sex transfer of song, it could be supposed that if song were favored in female orthopterans, it would likely evolve via transfer from the male. But in at least one group where both males and females sing, the bush crickets (Tettigonoidea: Phaneropteridae), the morphological apparatus for sound production has evolved independently in the two sexes. In some species, males rub a toothed file on the left wing over the inner edge of the right wing, producing sound when the wings are closed beginning from a widely open position, whereas females rub a thickened vein of the left wing over specialized spines on the dorsal surface of the right wing and produce sound without first widely opening the wings (Heller and von Helversen, 1986). Examples of this kind raise questions that evolutionary biologists have not begun to answer: Why, in the bush crickets, did cross-sexual transfer not occur? Are sex-expression anomalies absent or uncommon in related species having male-only song? If so, what is distinctive about sex expression in bush crickets that could account for this? Perhaps the answers lie in preexisting variations in wing structure that inclined the males and females toward different stridulatory mechanisms.

In army ants, the shape, size, and even the pheromone-gland array of the male resemble those of the queens, and males shed their wings when en-

tering a colony as do queens (Franks and Hölldobler, 1987). There is an unusually detailed similarity of male and female pheromone-producing
glands along the dorsal surface of the abdomen,
which the workers constantly lick, in both males
and queens. The army ant male must be allowed
by aggressive workers to enter the ant colony in order to mate. But this is not really queen mimicry,
for the workers would reject additional queens, as
presumably would the queen herself—masquerad-
ing as a competitor would not be the ideal courtship
device. Franks and Hölldobler (1987) propose that
worker choice of queen is based on pheromonal criteria, and that the same criteria have come to be
used to screen males. Evidently, the pheromones
themselves have not been analyzed, however, so it
is not known whether the striking morphological
similarity is biochemical as well, as suggested by
this hypothesis. Franks and Hölldobler term the resemblance between the sexes "convergence," but
Gottwald (1995, p. 155) suggests that these highly
similar features may be homologous, which would
imply origin of the femalelike male traits by cross-
sexual transfer. Cross-sexual transfer likely occurs
in the mating behavior of a number of other ants,
as discussed below in the section on alternative mating tactics in insects.

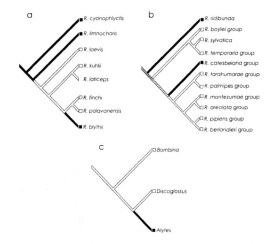

Fig. 15.4. Evolution of female vocalization in three
clades of anuran amphibians: (a) Southeast Asian
fanged frogs (genus *Rana*; based on Emerson and
Berrigan, 1993; Emerson and Ward, 1998); (b)
North American ranids (based on Hillis and Davis,
1986); (c) midwife toads and relatives (based on
Hay et al., 1995). Black lines indicate species in
which females vocalize, which is usually a male activity. After Emerson and Boyd (1999).

Frogs   Vocalization in frogs is universally sexually
dimorphic (Emerson and Boyd, 1999). In most
species, only the males give advertisement calls.
Usually female calling is limited to a simple release
call given by an unreceptive female when clasped
by a male. In a few species of frogs, however, the
females call during mate location (figure 15.4), and
there is good evidence that female calling has
evolved via cross-sexual transfer from males. Female mating calls are more similar to male advertisement calls than they are to the female release
calls, and the majority of differences between the
male and female calls can be explained in terms of
sexual differences in laryngeal and oblique muscle
morphology. Female mating calls, moreover, are
produced only during courtship around the time of
ovulation, when androgen levels in females are significantly higher even than in males. Experimental
work has shown that in *X. laevis* and *Hyla cinerea*
females produce a malelike mating call when
treated with androgens (Emerson and Boyd, 1999).
Emerson and Boyd suggest that female calls have
evolved by co-opting the preexisting advertisement
calling pathway of males, in other words, by cross-
sexual transfer. A correlation in several lineages between the evolution of female calling (figure 15.4)
and male territoriality further supports the hy-

pothesis that female calls function in the territorial
context, either to identify the caller as a female
rather than a male competitor, or to distinguish,
via the male response, between territorial and satellite males.

Birds   In a pioneer study of sex transfer in birds,
Mayr (1933, 1934) described extraordinary variation in the plumage of specimens from various
South Sea islands. In the Australian robin (*Petroica
multicolor*), for example, Australian mainland populations have a standard type of sexual dimorphism, with a bright male and a dull, cryptically
colored female (see also Mayr, 1982a, p. 838). On
some islands with populations derived from the
Australian birds, both sexes are cryptically brown-
ish; and on still other islands the two sexes are
monomorphically bright, the females having acquired malelike brilliant black, white, and red coloration suggestive of cross-sexual transfer. Mayr
discusses findings on bird hormonal effects to support this conjecture.

It has long been established that in sexually dimorphic birds, at least those studied in the temperate zone and at least some in the tropics, such
as some parrots (see below), the "default" plumage
coloration is usually male: the "neutral plumage"

of castrated adults of both sexes resembles the male plumage, and injection of female hormones (estrogens) or transplantation of ovaries at the critical period for feather development induces acquisition of female coloration by males (Mayr, 1933, p. 9; see also Welty, 1962; Winterbottom, 1929, 1932). This is true even for the strikingly dimorphically bright eclectic parrot, discussed below: castrated males retain their bright green plumage, whereas castrated females lose their bright red colors and express vivid green, the default, male plumage (Welty, 1962, p. 47).

The house sparrow *Passer domesticus* is cited by Mayr (1933) as an exception to the "neutral plumage" rule: castrated individuals maintain their dimorphism (females do not develop the male plumage). In house sparrows, males play an important role in incubation of the eggs (Oring et al., 1989). In some other species, such as the grey- and red-necked phalaropes (*Phalaropus fulicarius* and *P. lobatus*), femalelike male role is more exaggerated, and sexual dimorphism is reversed compared to that in most birds: the female is more brightly colored than the male, and the development of nuptial plumage in both sexes is induced by testosterone (Johns, 1964, in O'Donald, 1980, p. 32), which occurs in greater amounts in ovaries than in testes in these species. Thus, in these species, sex role reversal in plumage coloration seems to be accompanied by hormone role reversal as well: the female plumage is the default pattern, male rather than female hormones play the predominant role in determination of plumage coloration, and the hormones have a stimulatory rather than an inhibitory effect. The regulation of adult plumage development in lineages where bright plumage is of male origin, then, depends on the balance between the hormones usually associated with maleness (e.g., testosterone) and femaleness (e.g., estrogens): the "neutral point" (absence of steroids) allows expression of brightness of male origin in both sexes; female-hormone dominance suppresses traits of male origin leading to sexual dimorphism; and male-hormone dominance promotes exaggeration of coloration of male origin even in the female (as in phalaropes, above).

This pattern of regulation suggests that in the absence of selection on females to suppress expression of male coloration, sexual plumage monomorphism would be the rule in most such birds, with the traits selected for in males automatically expressed in females, as suggested by Darwin (discussed above). This should not be understood as arguing for developmental determination as opposed to selection: when avian males and females are monomorphically bright or morphologically extreme, as in some hummingbirds, toucans, ovenbirds, and parrots, there is evidence that the bright plumage is positively, socially selected in both males and females (West-Eberhard, 1983; see also Irwin, 1994).

In many birds song is the acoustical equivalent of plumage display, and it is usually the male that sings complex or conspicuous territorial and courtship songs. In many tropical species, both males and females are territorial and aggressive (Farabaugh, 1982). In over 200 such species, male–female pairs perform complex antiphonal duets (e.g., R.N. Levin, 1988; Levin and Wingfield, 1992; Levin, 1996). Unexpectedly, early work suggested that there are differences between the endocrinology of aggressiveness seen in many temperate-zone birds, which is thought to be typically mediated by testosterone, and that of some tropical species, (e.g., the white-browed sparrow weaver *Plocepasser mahali* in Africa and the bay wren *Thryothorus nigricapillus* in Panama), where year-round aggressiveness occurs but testosterone levels are usually low. In two of these species, the white-browed sparrow weaver and the bay wren, further studies suggest that aggression is likely independent of hormonal control or, instead, mediated by some hormone other than testosterone, such as luteinizing hormone (LH; Levin and Wingfield, 1992). In another species, the spotted antbird (*Hylophylax* sp. n. *naevioides*, in Panama), testosterone is associated with aggression only during extremely prolonged social challenges (Wikelski et al., 1999).

Observations like these raise challenging questions about the evolutionary history of hormone function in birds, and invite development of a field of *evolutionary endocrinology* (Moore, 1991) that would help to explain species differences in hormonal mechanisms via comparative research to test explicitly evolutionary hypotheses, treating both adaptive significance and phylogenetic history. A major function of LH, which is "remarkably conserved" throughout the vertebrate subphylum (Cheng et al., 1998, p. 5477), is to trigger the cascade of endocrine responses that culminate in ovulation. On that basis, LH, like estrogen, can perhaps be considered a primarily female-function-mediating hormone, even though, like estrogens, it can be produced by males. It seems reasonable to very tentatively offer a *sex-transfer-history hypothesis*—that hormonal mechanism may sometimes offer a clue as to sex of origin of a sexually monomorphic trait. LH-mediated and estrogen-stimulated traits would by this hypothesis be predicted to have originated, perhaps in some phylogenetically distant ancestor, as female morphologies or behaviors and then, if expressed in males under LH influence, to have been transferred cross-sexually to the males from females.

The corresponding hypothesis for traits of male origin would be that they can be recognized by their susceptibility to stimulation by maleness-associated hormones (e.g., testosterone) or inhibition by female hormones (e.g., estrogen), as in the plumage of galliform birds (Kimball and Ligon, 1999, their table 1). Adult canaries (*Serinus canarius*) can be induced to sing a simplified version of male song by testosterone treatment, which triggers growth of vocal control nuclei in the brain (brief review in Nottebohm, 1980).

Thus, the primary explanation for the LH mediation of aggressiveness in some tropical birds perhaps should be sought in the evolutionary history of the groups concerned, and may not always correlate with their tropical versus temperate-zone locations. Alternatively, separation of functions may underlie the difference: the advantage of separating control of year-long territorial song from a testosterone-mediated reproductive cycle may have favored use of an independent, LH-mediated regulatory mechanism for male song, facilitating the expression of song in females.

Whatever the fate of these tentative ideas, the utility of a comparative approach that considers both adaptive significance and a phylogeny of mechanisms is clear. In at least some taxa, regulation of sexual dimorphism is phylogenetically conservative. Regulation of plumage dimorphism in birds is an example (Kimball and Ligon, 1999): all galliform and anseriform species for which data are available have the male coloration as the default plumage, and female coloration requires estrogen; in all charadriiformes (e.g., gulls, phalaropes, ruffs), the female coloration is the default plumage and male coloration requires testosterone; and passeriformes (songbird) plumage determination is LH dependent in all seven species where hormones are known to have an effect (Kimball and Ligon, 1999). Other sexually dimorphic (nonplumage) traits often have a different hormone profile; for example, the wattles and behavior of chickens are influenced by testosterone in both sexes (Owens and Short, 1995).

Internal as well as external morphology can be involved in cross-sexual transfers. Extreme sexual dimorphism in bird song, (e.g., with males singing complex songs and females silent, as in the zebra finch, *Poephila guttata*, or in canaries) is accompanied by striking differences in the neural control

regions of the brain. These develop under hormonal influence early in ontogeny (Arnold et al., 1986). Estradiol produces sex differences in cell number and in the proportion of the cells of the brain that are steroid targets in adults. Arnold et al. (1986, p. 29) compared two closely related wrens (*Thryothorus* species) having distinctly different degrees of song dimorphism (p. 29). They noted that such differences could originate via reorganization of sex expression, hypothesizing that a sexually monomorphic species could evolve from a dimorphic species via a "simple mutation" leading to greater estradiol secretion during development in females, which would masculinize many attributes of the song system.

Arnold et al. (1986) then raise a question: "If a masculinizing hormone is secreted in females during development, how are other potentially sexually dimorphic functions (e.g., ovulatory physiology, nest building, courtship) spared from the effects of masculinization? How can these other female attributes remain feminine?" (p. 29). The answer was suggested earlier in this chapter: due to the hierarchical nature of sex expression, and the fact that regulatory evolution involving hormones typically involves change in receptors or responding tissues (Bolander, 1994), like the brain cells studied by Arnold et al., cross-sexual transfers can be partial, creating mosaic male–female phenotypic recombinants, as could occur if there were change in the steroid-target brain cells without change in other steroid-influenced targets, such as those concerned with plumage coloration. Species that are phenotypically intermediate in degree of dimorphism, with females showing intermediate degrees of development of most of the neural centers concerned with song (Arnold et al., 1986), indicate a potential for gradual and partial sex transfer of vocal behavior. Only one region—the robust nucleus of the archistriatum (RA)—is without intermediates, being clearly dimorphic in somal size in the zebra finch and monomorphic in somal size in all species examined with any singing by females (reviewed in Arnold et al., 1986). This shows not only that the size of the RA (or some correlate or determinant of RA size) may be a critical threshold trait for song production, but also that singing is developmentally a somewhat dissectable mosaic. Since differentiation of ovaries and testes occurs first in development, followed by secondary sex attributes (Schwabl, 1993), it is not surprising that it is developmentally possible to decouple the expression of different secondary sexual traits.

Partial cross-sexual transfers have been achieved in other hormone experiments on adult birds: female canaries (*Serinus canarius*) normally do not sing, but when treated with androgens they sing, albeit a song less complex than that of males (reviewed by Arnold, 1982). This is an additional example of partial sex transfer made possible by the mosaic organization of sex expression.

A study of avian sexual differentiation in zebra finches (*P. guttata*) and canaries (*S. canaria*) illustrates the convoluted pathways of sex expression that can evolve. The egg yolks of these species contain testosterone, even when the eggs are freshly laid by unmated females, which shows that the testosterone is of maternal origin (Schwabl, 1993). The amount of testosterone per newly laid egg varies both between species and within the same clutch and seems to affect the social rank of both male and female hatchlings. Evidently, the hormone dosage bestowed upon their young by female songbirds modifies the behavior of offspring (Schwabl, 1993), a phenomenon reminiscent of the intrauterine effects observed in mammals (see chapter 5). Schwabl (1993, p. 11449) points out that such manipulations could have the potential to influence the differentiation of secondary sex characteristics, including sex-specific brain pathways and behavior, for example, by changing the sensitivities of target tissues to the embryo's own gonadal secretions later in life. It is just such variation that can provide the material for selection in behavioral evolution. So the origin of a cross-sexual transfer in bird song could conceivably begin with some environmental or genotypic variation in mothers causing them to produce female offspring inclined to sing.

At first glance, the data on the neuroanatomy of song may suggest a powerful constraint on the evolution of vocalization by females: females of nonsinging species lack a very complex array of necessary equipment in the brain, a highly evolved difference between the sexes. But cross-sexual transfer of these complex traits can circumvent these presumed constraints. It is therefore potentially misleading to identify a particular arrangement of "canalized" or complex development as a developmental constraint for evolution. For this reason, the idea of "developmental constraints" should be used with caution (or not used at all; see Antonovics and van Tienderen, 1991).

Complex, little understood patterns of variation in degree of sexual monomorphism and dimorphism occur in other groups of birds. Most parrot species, for example, are sexually monomorphic (Forshaw, 1977), with adults of both sexes often bearing bright markings used in social display (see references in West-Eberhard, 1983). In most families of parrots, this is presumably primary sexual

monomorphism, since the parrots as a whole (Psittaciformes) are sexually monomorphic for plumage (Kimball and Ligon, 1999). But within one genus, the African lovebirds (*Agapornis*), there is considerable variation among species in degree of sexual dimorphism (see Forshaw, 1977, pp. 299–314). Phylogenetic analysis (Eberhard, 1998b) indicates that *A. roseicollis* is a secondarily sexually monomorphic descendent of species with bright males and relatively dull females. This suggests cross-sexual transfer of bright plumage from male to female, and it would be of interest to do detailed studies of resemblance and/or hormone manipulations to ascertain whether homology (cross-sexual transfer) or convergence is involved. Females in some parrot species (e.g., the eclectic parrot, *Eclectus roratus*) have bright plumage in both sexes that is obviously not a product of sex transfer because the female plumage is of a strikingly different color and pattern from that of males (see Forshaw, 1977). The eclectic parrot is a reminder that independent evolution of bright plumage in the two sexes is possible.

There are many examples suggestive of sex transfer that, like the Australian robin variants discussed at the beginning of this section (after Mayr, 1933), call out for further investigation of their pattern and adaptive significance. Sex-expression anomalies occur in a hummingbird, the white-necked jacobin (*Florisuga* species), in Panama (Angehr, 1980; T. and D. Robinson, personal communication). This species is largely sexually dimorphic, the adult males being brightly colored with blue head, green back and mostly white tail, and the majority of females drab. But many of the females examined by Angehr (1980) in a museum collection had "some degree of male-type plumage," and some of them "would have been indistinguishable from males in the field" (p. 54). Again, the presence of intermediates indicates the possibility of a partial cross-sexual transfer of plumage coloration. Individuals of *F. mellivora* with male-type plumage (perhaps including some females) were observed to engage in aggressive encounters with conspecifics at feeding sites (Angehr, 1980). Further study would be required to determine whether the plumage variation is functionless developmental variation or a correlate of social role.

Sex-expression intermediates in birds, as in the Mayr (1933, 1934) and Angehr (1980) studies, shows that cross-sexual transfers can occur gradually. What are the behavioral, breeding-system, and genetic correlates of complicated patterns of variation in degree of dimorphism within a genus like

Petroica (see Mayr, 1963, p. 318, for other examples), sparrows (*Passer* species; see Winterbottom, 1929), weaver birds (e.g., *Ploceus* species; Crook, 1964), or lovebirds (*Agapornis*; see Dilger, 1960; Eberhard, 1997, 1998b)? These groups, with social systems including solitary, familial, and colonial breeding, would reward comparative study that combines ethology, endocrinology, and phylogenetic analysis.

Mammals   Cross-sexual changes in expression, such as the "male uterus" in whales, are common anomalies in mammals (Yablokov, 1966, p. 243). Yablokov likened these anomalies to atavisms and pointed out that such variants are material for selection.

The mammalian sex-transfer example that perhaps best matches the evolution of maize in terms of both scale and ability to stir controversy is the spectacular penile morphology and display of female spotted hyenas (*Crocuta crocuta*). Females have a penislike clitoris or "pseudopenis" capable of erection and used in displays (Wickler, 1965; Kruuk, 1972). The resemblance to the male penis is so close that it is "almost impossible even for a trained observer to distinguish males and females by external signs" (Eibl-Eibesfeldt, 1970, p. 153). This phenomenon has given rise to a debate that exemplifies perfectly the needlessly dichotomous views of biologists regarding the role of development and selection in evolution (e.g., see Gould, 1981; Alcock, 1987; Sherman, 1988, 1989; Jamieson, 1989a,b; Frank, 1997), the same debate that has characterized evolutionary biology since Darwin (see chapter 24). Neither side clearly acknowledges that all traits that become widespread in a population must be products of both developmental innovation, which is necessarily responsible for the origin of any selectable form, and natural selection, which is responsible for its persistence, spread, or reduced occurrence. The only trait that would be immune to screening by selection would be one with no fitness effect whatsoever. Here I am concerned not so much with the fitness effects of the trait (see East et al., 1993; Frank, 1997) as with the evidence that both the morphology and the display represent a masculinization, or cross-sexual transfer to females of the expression of a complex trait originally evolved in males.

In hyenas, the female clitoris is enlarged to form a phallus with the urogenital tract inside. The homology of female with male morphology in this case is evidenced by close detailed resemblance of the male and female external genitalia. While the internal female reproductive anatomy is "unremarkable," the external genitalia of the two sexes are nearly indistinguishable when prepubertal females are compared with males (figure 15.2) (following puberty, changes occur in the internal structure of the clitoris, associated with intromission and delivery of young). The only differences between the sexes are more enlarged retractor muscles and shorter penis length in females (Hamilton et al., 1986, give background and references). Hormone studies suggest that prenatal testosterone exposure is involved in the genital masculinization in females (Racey and Skinner, 1979; Lindeque and Skinner, 1982, cited by Hamilton et al., 1986). There is no significant difference in androgen levels between prepubertal males and females (Hamilton et al., 1986), in contrast to the differences usually present in other mammals at this stage. These findings support the hypothesis that the malelike genitalia of females are the result of a transfer of expression from male to female, with the genital structures developmentally homologous in the two sexes.

By one hypothesis (East et al., 1993; Frank, 1997), the evolutionary transition may have begun when fetal androgen exposure was heightened as a side effect of the extreme aggressiveness of adult females, then may have been further exaggerated by change in the placental regulation of the enzyme aromatase under selection for aggressiveness in the young, which compete fiercely for food and social status. Aromatase keeps testosterone from being converted to estradiol (Wibbels and Crews, 1994). Fetal hormonal conditions are well known to cause cross-sexual changes in expression of sex-limited traits in mammals, as in the masculinization of female fetuses located next to male fetuses in utero, due to higher androgen levels (e.g., in Mongolian gerbils, Clark et al., 1991; human twins, McFadden, 1993). Changes in maternal androgen levels also affect fetuses in gerbils (Clark et al., 1991).

Female stumptail macaques (*Macaca arctoides*) engage in elaborate malelike homosexual mounting interactions that also have the earmarks of origin by cross-sexual transfer. Male macaques have distinctive copulatory behaviors, including teeth chattering, mounting, thrusting movements, puckered lips, lip smacking, square-mouthed expression, and (in the ejaculatory phase) body tenseness and rigidity, with a round-mouthed facial expression accompanied by vocal expirations (Hrdy, 1981, after Chevalier-Skolnikoff, 1974). Mounting females perform all of these behaviors, including even "the behavioral homolog of the entire species-typical male ejaculatory pattern" without seminal emission

(Goldfoot et al., 1980, p. 1477). While mounted they also showed intense uterine contractions and sudden increases in heart rate while performing malelike behaviors during a pseudoejaculatory phase (Goldfoot et al., 1980). This behavior is different from that accompanying female responses during copulation with males (Mitchell, 1979, after Chevalier-Skolnikoff, 1974). Hormones appear to influence this striking behavioral cross-sexual transfer. Mounting frequency by females, at least in a captive colony of *M. nemestrina*, correlates positively with dominance rank (Messeri and Giacoma, 1986), suggesting that hormones could be involved in regulation of both traits.

Many other potential examples of cross-sexual transfer occur in the displays of primates under the heading of "sexual mimicry" (reviewed in Wickler, 1967). Females of some primate species have conspicuous enlarged and sometimes brightly colored "sexual skins" that they "present" by displaying them toward males. Most commonly this is a female trait, but in some species with female displays, the males and even juveniles develop similar markings and behaviors, and observations indicate that they, like the penile displays of hyenas, function as appeasement (aggression-reducing) displays within social groups. At least some of these examples of mimicry are products of intraspecific convergent evolution rather than homology: in the vervet monkey (*Cercopithecus aethiops*), for example, females in estrus have a deep red perineum, swollen bright blue vulvar margins, and a conspicuous protruding red clitoris. They also have a bright red patch of skin around the anus, as already mentioned. The anogenital region of the adult males has the same appearance when seen from behind, but it is achieved by different structures: the perineum is scarlet, the scrotum (for which there is no homology in these monkeys) is bright blue, the penis red; and the anal region has a red appearance due not to the skin but to a fringe of long, vivid red hairs that are absent in the female (Wickler, 1967).

Female Hormones and Neurotransmitter Substances in Male Semen and Accessory Glands

Courtship communication includes not only precopulatory interaction and display, but also internal signaling and manipulation during copulation, especially by the male (Eberhard, 1985, 1996).

Some of these interactions seem to stimulate females to respond in ways that enhance the fertilization and reproductive success of males, for example, by triggering physiological responses that alter processing of the male's sperm, induce oviposition whose timing favors the performer, or resist further mating attempts (Eberhard, 1996).

Cross-sexual transfer may be intimately involved in the evolution of some of these sexual manipulations. Males of several species of insects and mammals produce femalelike hormones and neurotransmitters that are passed to the female during mating and likely affect, or even are essential for, her reproductive responses (Eberhard, 1996). That these substances are homologous to female molecules rather than independently evolved is indicated by their detailed structural and functional similarity to substances produced by females. For example, in the crickets *Acheta domesticus* and *Teleogryllus commodus*, mating males transfer to females an enzyme that acts in the female spermatheca to produce prostaglandin, a hormone known to occur in females and to cause their increased oviposition in these and other species (Destephano and Brady, 1977; Loher et al., 1981, cited in Eberhard, 1996). Prostaglandin also occurs in females and in the seminal fluid of the silkworm moth *Bombyx mori* (Setty and Ramaiah, 1980) and in many ticks (review in Eberhard, 1996). An impressive array of hormone molecules known to influence female reproductive behavior are found in the seminal plasma of male mammals (reviewed in Eberhard, 1996). They include follicle-stimulating hormone (FSH), luteinizing hormone (LH), chorionic gonadotrophin, prolactin, estradiole-17β, estradiole, estriol, and estrone (see references in Eberhard, 1996). Some of these, such as LH, function in vertebrate males in other contexts (e.g., in regulating aggressive and parental behavior; see sections on signals and parental behavior below). So their presence in male mammals cannot be assumed to be a product of cross-sexual transfer in the female-manipulation context without further comparative study.

Alternative Reproductive Tactics

Insects

Small males of the rove beetle (*Leistotrophus versicolor*; Coleoptera: Staphylinidae), rather than interacting aggressively with larger, more aggressive males, behave like females in being unaggressive and remaining at dung where both sexes prey on flies (Forsyth and Alcock, 1990). When challenged by another male, they exhibit behavior that is indistinguishable from that of courted females, turning and presenting the elevated abdomen tip, then responding to the antennal courtship behavior of the male by waving the abdomen from side to side and walking forward like a courted female while the courting male follows. Large males sometimes adopt this femalelike behavior when approached by another male if their weapons (mandibles) are occupied by feeding or courtship activity (figure 15.5). The alternative tactics are rapidly reversible: the same male sometimes employs both female mimicry and fighting in an interaction with the same opponent. In another rove beetle species (*Aleochara curtula*) discussed above, young, starved, or multiply mated adult males release a female sex pheromone found in the cuticle but evidently do not engage in femalelike behavior (Peschke, 1986, 1987). Although the pheromonal trait could have been originally sexually monomorphic (see above), its usefulness in the female-mimicry context could have set the stage for advantageous sex transfer of traits that improve male resemblance to females.

In the order Odonata (dragonflies and damselflies), especially in the damselfly family Coenagrionidae, there is sometimes a female dimorphism in sexually dimorphic species, as mentioned above. In addition to the normal, cryptic female there is commonly a malelike brightly colored morph. In *Ischnura ramburi* the malelike adult females are identical to males in all details of coloration, being distinguishable only by minute differences in external genitalia and in a uniform coloration of the pterostigma, a small sector of the wing (Robertson, 1985). They also behave like males in heterosexual encounters, hovering and circling face to face. Males do not attempt to copulate with them during these malelike interactions and eventually move away. Malelike females mated only half has many times as did normal females. Robertson hypothesizes that the malelike features may be advantageous in reducing mating attempts in this species, which has very long copulations (an average of three hours) in proportion to their adult life span (only a few days), with consequent increased time and energy expenditure and exposure to predation during mating. Another hypothesis (D. R. Paulson, cited by Robertson, 1985) is that the malelike appearance allows females to avoid male interference with oviposition. Robertson suggests that the disadvantage of bright coloration under predation could prevent the adoption of malelike coloration

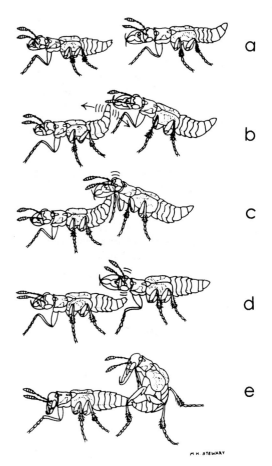

Fig. 15.5. Femalelike sexual behavior of male rove beetles (*Leistotrophus versicolor*): (a) when challenged by a larger, aggressive male rival (right, with large mandibles) a noncombative male turns to face away from the rival; (b) the smaller male presents his abdomen tip as does a courted female; (c) the larger male antennates and taps the abdomen tip with his head as in heterosexual courtship; (d) the pseudofemale walks slowly forward, waving his abdomen from side to side while the larger male follows; as the pair gets farther from the mating site, the follower eventually leaves; (e) Heterosexual copulation, the only stage of courtship not performed in male–male interactions. Behaviors a–d could not be distinguished from those observed during heterosexual courtship. From Forsyth and Alcock (1990), Figure 1, p. 326, by permission of © Springer-Verlag GmbH & Co.

throughout the population. This polymorphism is unusual among complex alternatives in being inherited in a simple Mendelian fashion, with the malelike form a homozygote recessive (Robertson,

(Robertson, 1985). So frequency dependence or bet-hedging could explain its maintenance (see chapter 22). The detailed resemblance between males and malelike females suggests a cross-sexual transfer of traits.

The detailed specialized resemblance of these femalelike behaviors and morphologies to those of normal females suggests origin by cross-sexual transfer. But comparative endocrinology cannot be used to help evaluate the hypothesis, as in vertebrates, because the mechanisms of the switches between tactics are unknown (see Forsyth and Alcock, 1990). Expression of reproductive behaviors in adult insects is hormonally mediated (Nijhout, 1994a), but morphological sex differentiation is cell autonomous rather than globally coordinated by hormones (John and Miklos, 1988; Nijhout, 1994a, with some possible specialized exceptions—see p. 159). Nonetheless, parasitism can cause global or phenotypically extensive morphological changes, as discussed above in the section on alternative explanations. And genetically male but femalelike "ergatoid" ants and stingless bee males have many morphological features of females alongside those of males (see discussion of parental care, below). This indicates that even when cells are genetically one sex, their responses can be altered so that features of the other sex are expressed. So in the developmental cascade of sex expression in insects, local regulatory responses permit phenotypic sex mosaics to occur.

An Isopod Crustacean

The marine isopod *Paracerceis sculpta* (Sphaeromatidae) has three discrete male morphs, called, from largest to smallest, alpha, beta, and gamma (Shuster, 1987, 1989; figure 15.6). The alpha males are hypothesized to be the primary male form in a basically sexually dimorphic species, since their enlarged telsons and elongated uropods are like those of males of many other dimorphic sphaeromatids (Shuster, 1987), and their behavior and interactions with females set the stage for alternative tactics of the other two male morphs. Even if anatomical sexual monomorphism were pleisiomorphic in these isopods, the functionality of the femalelike *behavior* of beta males depends on the prior existence of the alpha male behavior that they manipulate.

The specialized morphology of alpha males is used in courtship interactions with females. They are aggressive toward other males and sequester a harem of females within a spongocoel. Beta males are likely derived via cross-sexual transfer, judging by their extremely close morphological and behav-

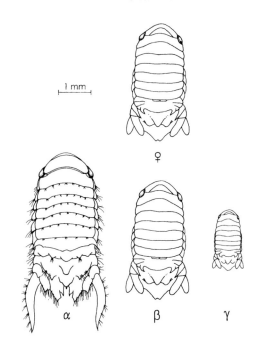

Fig. 15.6. Female and three male phenotypes in the marine isopod *Paracerceis sculpta* (Crustacea, Isopoda). The behavior and external morphology of the beta males are virtually indistinguishable from those of females. After Shuster (1992b).

ioral resemblance to females. They lack ornamentation and are the same size and morphology as females, and they exploit this remarkably close resemblance by deceptively entering spongocoels where alpha males maintain harems. Beta males mimic female behavior "to the closest detail" (Shuster, 1989, p. 1687; Shuster and Wade, 1991b). They initiate oral contact with alpha-male uropods and solicit bouts of shaking as do females that approach a spongocoele held by an alpha male. They then enter the spongocoel just as do females, without their male identity being discovered by the resident male. Gamma males, by contrast, resemble immatures and gain entry by stealth, which capitalizes on their small size. If discovered by an alpha male, they are vigorously pursued and attacked.

The developmental basis of this polymorphism has not been investigated. Shuster and Wade (1991a) provided evidence suggesting that morph determination is strongly influenced by one or a few genetic loci of major effect. Subsequent studies (Shuster and Sassaman, 1997) have shown that the female-mimic beta males originate via two different developmental pathways: some are genetic males, the homogametic sex, and some are genetic females that have changed sex to become beta males! Both sexes change sex in this species. Transformation from female to beta male is evidently due to the interaction of certain autosomal genes that influence morph determination, and primary sex determination loci (Shuster and Sassaman, 1997). A suggestion that sex reversal in males may be influenced by extrachromosomal factors produced by bacteria or virions (Shuster and Sassaman, 1997) has not been verified (Shuster, personal communication, 1999; but see discussion of parasites below). These findings show the developmental versatility of sex expression in *P. sculpta*.

Given the complexity of morph determination in *P. sculpta*, it is worth noting that, as mentioned above, cross-sexual transfer of phenotypic traits (feminization and masculinization) occurs in many crustaceans, insects, and plants due to infection by parasites (reviewed in Wheeler, 1910b). Infection is usually accompanied by sterility, but in some crustaceans, for example, the spider crab *Inachus mauritanicus* parasitized by the cirriped *Sacculina neglecta*, male genitalia are retained, and if the parasite drops off after morphological feminization has occurred, males regenerate their male gonads and produce spermatozoa (Wheeler, 1910b, pp. 411–412, after Smith, 1909). This raises the possibility that feminized morphs such as the beta males of *P. sculpta* could sometimes by induced by parasites. The examples reviewed by Wheeler show the readiness with which cross-sexual transfer of trait expression can occur facultatively in many organisms.

Fish

There is an extraordinarily large number of examples of femalelike alternative male phenotypes in fish (reviewed in Magurran, 1987; Taborsky, 1994), variously called "pseudofemales," "female mimics," and "transvestite males" (Taborsky, 1994). This is perhaps not surprising in view of the great flexibility of sex expression in fish, among the few vertebrates in which some individuals facultatively and reversibly change sex (Policansky, 1982).

One of the best-studied examples, and one that is in many ways typical, occurs in bluegills (*Lepomis macrochirus*). Large males compete aggressively for territorial positions and build nests in breeding colonies. Small males, on the other hand, adopt behavior "essentially the same as that of functional females" (Dominey, 1980). The femalelike males are not aggressive, do not build nests, stay in the same areas as females, and solicit nesting males for courtship, following males in their courtship turning in the nest and even adopting the

dark barred coloration characteristic of females that is known to reduce male aggression in a related species (Colgan and Gross, 1977, cited by Dominey, 1980). They remain in the male's nest as if paired with him, then evidently spawn simultaneously with him when a female releases eggs in the nest, and subsequently leaves the resident male to care for the eggs (Dominey, 1980).

The very detailed resemblance of the femalelike male behavior and coloration to that of females lends support to the hypothesis of origin by cross-sexual transfer. This interpretation is also supported by hormone studies showing male resemblance to females in another fish, the sequentially hermaphroditic stoplight parrotfish (*Sparisoma viride*). In this species, all individuals begin life as females, then transform into territorial males, sometimes via a femalelike phase. The femalelike males have high estrogen levels (reviewed in Moore, 1991), suggesting hormonally mediated cross-sexual transfer.

Similar female "mimicry" is reported in other fish, including three species of sticklebacks (*Pungitius pungitius*, *Gasterosteus aculeatus*, *Apeltes quadracus*—Gasterosteidae), leaf fish (*Polycentrus schomburgki*—Nandidae), corkwings (*Drenilabrus melops*—Labridae; reviewed by Gross, 1984), and a Mediterranean species of *Tripterygion* (Dominey, 1980). Taborsky (1994) provides a review and critical discussion of interpretations regarding adaptive significance.

Lizards

Adult male tree lizards (*Urosaurus ornatus*) comprise two distinctive phenotypes: a relatively small and territorial form, with the dewlap, a throat fan used in signaling, orange and blue (OB males); and a relatively large, relatively unaggressive form with an orange dewlap lacking blue (O males), which is more femalelike in color and engages in sneaking or satellite behavior when pursuing mates (Thompson et al., 1993). The two classes of males have different growth trajectories from an early age, the large femalelike male morph growing faster than the small morph.

Hormone studies suggest to me that the O morph is a feminized male. Males castrated early in development (day one after hatching) develop into the O phenotype (Zucker and Boecklen, 1990; Thompson et al., 1993, p. 139). In contrast, male hatchlings treated with testosterone developed dewlaps with more blue yielding, an orange–blue intermediate (Hews et al., 1994). In females, dewlap color is deeper orange in gravid and near-gravid females, indicating a hormonal influence as is known to occur in other iguanids (Zucker and Boecklen, 1990). And progesterone, which in all-female parthenogenetic *Cnemidophorus* lizards promotes male-like behavior, also promotes expression of the male-like morph when applied to *U. ornatus* male hatchlings (reviewed in Moore et al., 1998).

The regulation of sex expression in *U. ornatus* is further illuminated by experiments with females. Females treated with dihydrotestosterone (DHT) are masculinized, and express color phenotypes in the same proportions as observed in the male of their natural, parent population (see Moore et al., 1998). This is consistent with a genetic architecture in which (1) capacity to express the alternative color morphs is present in both sexes, and (2) during male development (following sex determination, when sex-limited male differentiation is under the influence of masculinizing hormonal conditions) individual color is determined either by a single polymorphic locus, or by a switch (between alternative pathways of color production) that is strongly influenced by a locus of major effect.

The body size relations of the two morphs at first seem to contradict the feminization hypothesis: in these lizards males are larger than females, yet the femalelike O morph has been described as the larger of the two male morphs, and experimental application of androgens does not consistently increase body size (Hews and Moore, 1995). In these studies, however, size refers to body length (Moore et al., 1998). More extensive morphometric analysis shows that the body shape of the aggressive OB male morph is, as described by Moore et al. (1998; after Hews and Moore, 1994) more massive, or more short-and-stocky (although not differing in fat-body mass), than that of the less aggressive O morph, which is described as more slender. The more massive shape of the aggressive OB morph could be considered fighter morphology, while the slender shape of the O morph could be an adaptation to, or an ontogenetic result of, its nomadic behavior (described by Moore et al., 1998). Moore et al. state that the OB males are aggressively territorial early in the breeding season and that food is scarce at that time. This may explain the seemingly paradoxical smaller size of the more aggressive morph, as well as its more "massive" shape, perhaps due in part to muscular activity during growth.

A similar color polymorphism occurs in side-blotched lizards (*Uta stansburiana*; Iguanidae), where there are three throat-color morphs with different mating tactics. One of them is a femalelike morph with yellow-striped throat that obtains mates by sneaking and often mimics a "female rejection display" in encounters with dominant males (Sinervo et al., 2000). The more aggressive, territorial orange-throated males have higher testosterone levels (summarized in Sinervo and Lively, 1996). This is consistent with the interpretation of the femalelike morph being hormonally less masculine, which would provide a mechanism for sex-transfer of behavior and throat color. A comparison of male and female hormone profiles would help in testing the sex transfer interpretation. Weight differences between the orange- and heavier yellow-throated males suggest that morph divergence occurs early in development (Sinervo et al., 2000), so hormone comparisons may need to be done on juveniles. But hormonally mediated changes in throat color and behavior do occur in adults: some yellow-throated males lose their yellow throat coloration (which changes to blue) and femalelike behavior when dominant males disappear, and this defeminization is accompanied by a rise in testosterone level (Sinervo et al., 2000).

Snakes

Male red-sided garter snakes (*Thamnophis sirtalis parietalis*) compete for females in large aggregations to which they are attracted by a pheromone produced by the female (Mason and Crews, 1985, ex Moore, 1991). A small number of femalelike "she-males" are courted as if they were females, and a hexane wash of their skin elicits chin rubbing courtship behavior from males, suggesting that they produce femalelike pheromones. The she-males also have femalelike levels of aromatase enzyme in the brain and skin (Krohmer, Mason and Crews, unpublished observations, cited by Moore, 1991, p. 1168). Furthermore, experimental application of estrogen feminizes males and makes them attractive to males (Crews, 1985). All of these findings would seem to support a cross-sexual transfer interpretation for the origin of the "she-male" phenotype. But some other facts indicate that the she-males are not simply products of cross-sexual transfer or feminization but that "demasculinization," or developmental deletion of male traits by failure to express them, may also be involved. The she-males lack certain methyl ketones used in pheromonal male recognition. And their skin lipids lack squalene, a characteristic of normal males (Mason et al., 1989). Thus, this novel phenotype may have evolved toward female resemblance by developmental deletion, as well as cross-sexual transfer involving change in levels of aromatase production.

A Mammal

A common testosterone-mediated tradeoff between sexual and parental behavior (see Ketterson and Nolan, 1992, and below) has led to an alternative reproductive tactic of an unusual kind in Mongolian gerbils (*Meriones unguiculatus*)(Clark et al., 1997). Males that were gestated between two male siblings have relatively high testosterone as adults, and better mating success when paired with unfamiliar females. Males gestated between two female sibs, in contrast, have relatively low testosterone as adults, and lower success in impregnating females. This handicap turns out to be a blessing in disguise, however, for the low-testosterone males, which dedicate more attention to pups in the absence of the female, have increased reproductive success in terms of pups delivered by females following copulations during postpartum estrus. The authors interpret this difference as due to better condition of females following their lesser investment in the first litter, thanks to the help of the low-testosterone males, leaving them in better condition for rearing a second litter. An alternative hypothesis is that females facultatively adjust litter size due to cryptic female choice following cohabitation with males, investing more when mated by a male of proven paternal ability (see Eberhard, 1996). Whatever the explanation of altered litter size, the greater paternal investment associated with the more feminized parental investment of low-testosterone males did not increase the probability of survival or rate of growth of pups (Brown et al., 1997), so the payoff to the more paternal male was not a direct result of paternal care, at least not in these captive populations.

Parthenogenetic Species

Some vertebrates and insects lack males (e.g., see White, 1978; Pardi, 1987; Crews, 1989). Sometimes this is the result of repeated hybridization between closely related species. In other species, those with both sexual and asexual modes of reproduction, selection has presumably favored elimination of the cost to females of sexual reproduction (Williams, 1975; Maynard Smith, 1978). In some unisexual species (e.g., in some fish), sperm from a male of one of the parental species must penetrate the egg for development to begin, even though the male genes are not incorporated into the zygote (see chapter 31). In other taxa, pseudomale and pseudocopulatory behavior is performed by females toward other females (Werner, 1980). This behavior may or may not be necessary for reproduction.

There is evidence in some species that the male-like behavior of parthenogenetic females is a product of cross-sexual transfer. In the all-female whiptail lizard *Cnemidophorus uniparens*, malelike courtship behavior performed by females is indistinguishable from that of males of the known sexual ancestral species (Crews and Bull, 1987). The malelike behavioral morph depends not on genotype but on an environmentally conditioned hormone surge in the postovulatory phase of the female reproductive cycle. Pairs of females housed together develop complementary oppositeness in the phasing of their reproductive cycles, alternating gender roles as do some hermaphroditic fish and polychaete worms (Sella, 1990). Reversion to production of behaviorally, morphologically, and physiologically intact males is inducible in this all-female species by treatment of embryos with an inhibitor of aromatase, which promotes masculinization by inhibiting conversion of testosterone to estradiol (Wibbels and Crews, 1994). Obviously, the male sex differentiation genes are not lost in the unmanipulated unisexual species, only unexpressed.

Malelike behavior of females of the parthenogenetic curculionid beetle *Otiorrhynchus pupillatus cyclophtalmus* (Pardi, 1987) is highly suggestive of origin by cross-sexual transfer. Females approach, antennate, mount, and then abdominally stroke the abdomens of other females in the same manner as do males of the same species. Often the mounted female then extrudes an egg, but pseudocopulation is not necessary for oviposition to occur. Pseudomale behavior is most often performed by females whose ovaries contain mature eggs and by females that are the more active of a pair of interacting females. This indicates that it is associated with reproductive readiness and perhaps level of an associated hormone. There are many other examples of pseudocopulatory behavior in insects and vertebrates (reviewed, respectively, by Pardi, 1987; Crews, 1987, 1989). Because of their detailed resemblance to male behavior, these examples suggest cross-sexual transfer and invite more specific studies of the pathways leading to the origin and evolution of the traits.

Cross-sexual Transfer of Parental Care

Social Insects

Males of nest-building species of insects may have more contact with the brood and with the conditions that stimulate brood care than do non-nest-building species. It is perhaps not surprising, therefore, that males of nesting insect species, like those of nesting mammals and birds, sometimes show femalelike parental behavior. Males occasionally attend the brood, dispensing solid food to larvae and collecting larval saliva as do workers, in species of primitively eusocial wasps such as *Mischocyttarus* (Jeanne, 1972), *Polistes* (West-Eberhard, 1969, 1975; Hunt and Noonan, 1979; Cameron, 1986), and *Ropalidia* (Kojima, 1993a). *P. instabilis* males forage for nectar, which they sometimes share with workers and larvae; dispense worker food loads to larvae; and fan their wings to cool the nest, as do workers (O'Donnell, 1995). In *R. plebeiana* (Kojima, 1993a) and *M. collarellus* (Smith, 1999; O'Donnell, 1999), males frequently attend larvae and participate in dominance–subordinance interactions with females. They resemble young adult females in behaving as nurses, while not building or bringing foraged materials to the nest.

Femalelike behavior by male ants is unusual, but several cases have been recorded. Males of the carpenter ant *Camponotus herculeanus* store food in their crops and regurgitate it to workers and other males (Hölldobler and Wilson, 1990). And the ergatoid, or workerlike males of *Cardiocondyla nuda* have occasionally been seen to carry brood, behavior considered "remarkable" in ants, as such behavior had previously not been observed in ants (Heinze et al., 1993), although workerlike transport of larvae by males of this species had earlier been observed by Santschi (1907; cited by Wheeler, 1937, p. 60). Male *C. nuda* also lick larvae and groom each other, as do workers (Heinze et al., 1993). Similarly, ergatoid males of a male-polymorphic *Hyperponera* species differed from alate (normal) males of the same species in their workerlike behavior of soliciting and receiving regurgitated food from workers, and in regurgitation to young queens (Hashimoto et al., 1995).

These examples of workerlike behavior in males are interesting exceptions to the rule of no paternal care in ants because the ergatoid males are morphologically workerlike in their antennal structure, lack of wings, reduced eyes, and general body form (figure 15.7). These examples are especially striking since the males of ants usually resemble queens more closely than they do workers (Wheeler, 1910a). Wheeler (1910a, p. 60, citing work by previous authors) points out that workerlike males are, morphologically, anteroposterior gynandromorphs, since their posterior genitalia and gonads are unfeminized. In this connection, it is interesting to note that a rare gynandromorph, the first

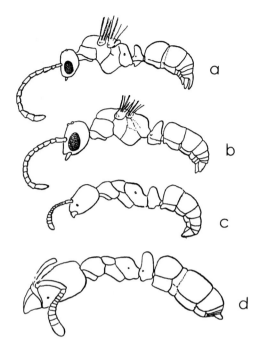

Fig. 15.7. Workerlike males in ants (*Ponera* species): (a) normal male of *P. coarctata*; (b) normal male of *P. eduardi*; (c) workerlike male of *P. eduardi*; (d) workerlike male of *P. punctatissima*. Workerlike males resemble workers in lacking wings, in having small eyes and workerlike antennae and thorax, and in sometimes performing workerlike behaviors. After Wheeler (1910a).

recorded for the genus *Myrmecia*, proved to be not a male–queen mosaic but a male–worker mosaic ("ergatandromorph"; Crosland et al., 1988).

A congener of *Cardiocondyla nuda*, *C. wroughtonii*, has both winged (normal) and ergatoid males (Heinze et al., 1995). Some workerless, parasitic ants (e.g., *Anergates* and *Epoecus*) have queenlike males (Wheeler, 1910a), and those males, true to their queenlike morphology and parasitic status, do not to my knowledge work, though behavioral observations of parasites are rare. Evidence that the ergatoid male morphology is in fact derived via a cross-sexual transfer from the ancestral worker morph rather than by convergent evolution comes from a recent study of gynandromorphs, or anomalous bilateral male–female mosaics. *C. emeryi* gynandromorphs with the male side winged have a queenlike (winged) female side, whereas those with an ergatoid male side have a workerlike female side (Heinze and Trenkle, 1997).

Male bumblebees (*Bombus griseocollis* and *B. pennsylvanicus*) perform brooding behavior. They incubate pupae by vibrating the wing muscles while standing over cells, but they do not feed larvae (Cameron, 1985). Maeta et al. (1992) believe that the greatest diversity of femalelike behaviors performed by males of any species of bee occurs in the Taiwanese xylocopine *Braunsapis hewitti*, where males occasionally perform nest maintenance behaviors, pollen handling, moving immatures, food solicitation, withdrawal of honey from larvae, and feeding honey to adults. If this is the leader in feminized male behavior in bees, the males of stingless bees (Meliponini) represent a close second. Meliponine males during the first third of their lives perform many femalelike tasks, including waste dumping and nectar dehydration (Sommeijer et al., 1990; van Veen et al., 1997), incubation of larvae, wax production and manipulation, nest defense, feeding the queen (see references in Kerr, 1997), and following of odor trails made by females (Kerr, 1990; the latter perhaps allowing males to locate nests and virgin females (Roubik, 1990), so the polarity of sex transfer in this trait is uncertain).

The most remarkable of these femalelike tasks is wax secretion and handling, an exclusively female activity in all other bees and one that requires specialized morphology and behavior: in colonies of *Melipona marginata* and *Plebeia droryana*, hundreds of males produce wax abundantly (Kerr, 1997). Males of another species, *M. compressipes*, also behave like workers, taking the wax scales from their tergites and putting them in the wax deposits around the brood nests. They also build small wax columns, pots, and involucrum sheets, but not brood cells.

Worker females in many species of stingless bees are in several traits morphologically more like males than like queens (Bonnetti and Kerr, 1985; Hartfelder and Engels, 1992), although they are closer to queens than to males in the dimensions of the antennal scape, a trait used during mating by males (Kerr , 1997). Kerr and Cunha (1990) suggest that because of this similarity, mutations that would affect females may also affect males, in effect suggesting cross-sexual transfer due to broad overlap in gene expression. Kerr (1987) and Kerr and Cunha (1990) consider stingless bee workers to be masculinized females that have escaped the feminizing influence of juvenile hormone that governs expression of female traits of queens in well-nourished last-instar larvae (Kerr, 1987; see also Kerr, 1997). They see traits selected in workers as also expressed in males because both are under a

similar "masculine" hormonal regime during the last larval instar. Brief reviews of femalelike worker behavior in male Hymenoptera are given by West-Eberhard (1975) and Kerr (1997).

Fish and Amphibians

Comparative study of fish indicates that male-only care is usually derived from no care; biparental care is derived from male-only care; and female-only care is usually derived from biparental care (Clutton-Brock, 1991a, p. 114). If cross-sexual transfer of parental care occurs in fish, therefore, it is most likely to be from male to female.

Parental care is unusual in anurans (frogs and toads). In poison dart frogs (Dendrobatidae), parental care is performed by females in some species and by males in others, with some species having biparental care (Wells, 1981; Caldwell, 1997; Summers et al., 1999). The most elaborate parental behavior occurs in the genus *Dendrobates*, where it can include egg guarding, moving tadpoles to water, and production of trophic eggs that are consumed by the tadpoles (Summers et al., 1999). Mapping parental care onto a phylogeny of *Dendrobates* species shows that male parental care is likely ancestral and that maternal care evolved via a stage with biparental care (Summers et al., 1999; Summers and Earn, 1999; Clough and Summers, 2000). This is of interest because in vertebrates in general, monogamy correlates with increased hormonal and behavioral sensitivity to social conditions (Wingfield et al., 1990), presumably favored by selection. The evolution of biparental association, therefore, could set the stage for a hormonally mediated sex transfer of parental care behavior to females.

In keeping with the increased parental investment by males, females appear to compete for mates in *D. auratus*: numerous females sometimes follow single calling males, fight among themselves for access to males, and appear to court them (Wells, 1981; Summers, 1989, 1990). But there is no evidence for sex-role reversal of the type that occurs in some birds, where access to males limits the reproductive success of individual females (Summers, 1989). Male parental care by males who continue to attract mates may involve little investment in terms of reproductive opportunities lost (K. D. Wells, personal communication). To further examine the sex-transfer hypothesis for the origin of female parental care in *Dendrobates*, it would be of interest to investigate the possibility that in this lineage, as in the fanged frogs (see below), and

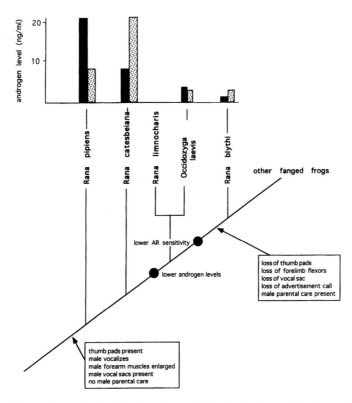

Fig. 15.8. Hormonal correlates of behavioral and morphological changes in the evolution of feminization in fanged frogs: the ancestral phenotypic state of males (lower box); the derived, feminized state
(upper box), which is accompanied by lowered androgen levels and androgen-receptor sensitivity (AR)
as shown. Black bars indicate mean testosterone levels; stippled bars indicate levels for breeding males.
From Emerson et al. (1997), by permission of © Wiley-Liss, Inc., a subsidiary of John Wiley & Sons,
Inc. Courtesy of S. B. Emerson.

birds and mammals (discussed in subsequent sections), hormonal change sets the stage for or accompanies change toward more parental activity by
females, and to compare the regulation and details
of parental behavior to assess the degree of its homology in opposite sexes. Tadpole transport,
whether by males (*D. leucomelas*) or by females
(*D. histrionicus*), involves allowing tadpoles to
wriggle onto the adult's back and then carrying
them to water (Summers, 1992), so this is a potential candidate for cross-sexual transfer of behavior. In *Eleutherodactylus* frogs, when males show
parental behavior (egg guarding), their plasma levels of testosterone decline (Wingfield et al., 1990,
after Townsend and Moger, 1987). Hormone data
are not available for *Dendrobates* (K. Summers and
K. D. Wells, personal communication).

The origin of male parental care in Southeast
Asian fanged frogs, *Rana blythii*, was preceded during evolution by lowered androgen levels and other
signs of feminization such as loss of thumb pads,
forelimb flexors, vocal sac, and advertisement call
(Emerson et al., 1997; figure 15.8). There is no indication of female parental care in the relatives of
these frogs among the ranids. Rather than being a
case of hormonally mediated cross-sexual transfer,
parental care in the fanged frogs evidently evolved
secondarily to, and was perhaps permitted by, the
hormonal change rather than resulting directly or
immediately from it (Emerson, 1996; Emerson et
al., 1997).

Birds

Taxa with parental care of young often show
species differences in the sex that performs the care
behavior. This raises the possibility that cross-
sexual transfer has occurred. In birds, the female is

the sex that invests the most effort in care of the young in the vast majority of species, and there is a correspondingly greater dedication to courtship activity by males, as predicted by sexual selection theory (Trivers, 1972). Comparison of both role-normal avian species (e.g., chickens and turkeys), where females brood the young and males engage in courtship of mates, and role-reversed species, where males brood the young (e.g., sandpipers, pied flycatchers, and house sparrows), suggests that parental care can be potentiated in either sex by selection that promotes decreased gonadal steroid levels. In birds where females incubate, there is a decrease in estradiol and progesterone at the time of heavy incubation. In species with male incubation, on the other hand, there is a decrease in testosterone, and increasing it experimentally suppresses paternal care (Oring et al., 1989; see also Ketterson et al., 1992, on juncos, *Junco hyemalis*; Wingfield et al., 1990, on house sparrows, *Passer domesticus*, and pied flycatchers, *Ficedula hypoleuca*). The same seems to be true in pipefishes (Syngnathidae), mainly marine teleost fishes where the males brood the young and females compete for mates. In three species of pipefish, testosterone, the primary androgen in breeding males of most teleosts, was absent or at very low levels in brooding males (Mayer et al., 1993).

Such observations have led to the hypothesis that there is a testosterone-mediated tradeoff between investment in sexual and parental behavior in male vertebrates (Ketterson and Nolan, 1992). As discussed above, in mammals this may sometimes lead to the evolution of alternative male reproductive tactics, with high-testosterone males investing more in pre-copulatory mate-acquisition activies, and low-testosterone males engaging relatively more in parental care which keeps them in contact with females during postpartum estrus (Clark et al., 1997).

The potential for sex reversal in avian parental care is dramatically shown by sex-reversal anomalies in domestic fowl. In an Orpingon race of chickens, a hen was observed to change sex after more than three years as an egg-layer. She developed a roosterlike appearance and sired two chicks. An autopsy revealed functioning testes and a shriveled left ovary, apparently destroyed by a tumor (Crew, 1923, cited by Welty, 1962). I cannot resist mentioning a less fortunate intersex, an egg-laying hen with male plumage, that was reportedly burned for witchcraft in the Middle Ages (recounted by Winterbottom, 1929), confirming my grandmother's admonition that "whistling girls and crowing hens will come to no good ends."

Sex-role anomalies have also been observed in a few natural populations of birds, including western gulls (*Larus occidentalis*; Hunt and Hunt, 1977) and hooded warblers (*Wilsonia citrina*), in which a young male with normal adult plumage but an undeveloped cloacal protuberance nested and incubated in the territory of an aggressive male who fed the incubating male as he would have a female (Niven, 1993). Well-documented cases of sex role reversal in nature are probably rare not because the phenomenon is unusual, but because it may be most common in sexually monomorphic birds where homosexual pairs are mistaken for heterosexual ones by observers (Dilger, 1960; Huber and Martys, 1993).

Morris (1955) noted that four conditions are associated with pseudofemale and pseudomale behavior in the courtship behavior of vertebrates: (1) hormonal and/or structural abnormalities that influence the sexual system, (2) submissiveness or subordinate behavior of males or the reverse (aggressiveness or dominance in females), (3) the arousal and subsequent thwarting of highly motivated sexual behavior (which often leads to immediate performance of sexual behavior like that of the opposite sex as a kind of redirected activity), and (4) the presence of the stimuli that release sexual behavior of the opposite sex.

In birds, whether the female or the male is the care-giving sex, incubation is accompanied by a sharp rise in the pituitary hormone prolactin (Beach, 1961, on females and artificially injected males; Oring et al., 1986a,b, on males). The occurrence of regulation of parental care by this characteristically "maternal" hormone in males of role-reversed species supports the hypothesis that cross-sexual transfer of hormonally regulated gene expression may be involved—that the evolution of male parental care involves expression of gene sets organized under selection on females rather than convergent evolution of a new pathway in males.

Phylogenetic analysis supports the hypothesis that biparental care, widespread in birds, is also primitive in the group, with the sexual division of labor for egg brooding, for posthatching and postfledging care, or for male–female similarity in these tasks being evolutionarily labile (McKitrick, 1992).

Mammals

Until recently, paternal care in mammals received little attention in general discussions of parental role reversal, since finding extensive participation of males in the care of young was considered unlikely in a taxon where dependence on maternal lactation requires exclusively female feeding of infants (Knowlton, 1982; Yamamoto, 1987). Several striking examples of mammalian paternal care in natural populations are known, however, especially in rodents, carnivores, and primates. In addition, male parental behavior in species not normally showing it sometimes can be induced either by repeated exposure of males to infants or by feminizing hormonal manipulation (Gubernick and Nelson, 1989). Most studies of hormonal mechanisms of male parenting in mammals have involved artificially induced examples of this sort. Such studies provide valuable clues as to the circumstances and preexisting response capacities that could contribute to an evolved sex transfer of parental care.

In the raccoon dog (*Nyctereutes procyonoides*), males and females take turns caring for the young during early rearing, including at parturition (Yamamoto, 1987). The male's behavior appears to be identical to that of the female in every respect except for the absence of lactation: like the female, the male licks the newborn, eats the umbilical cord and placenta, huddles with the neonate during the birth of subsequent littermates, licks the anogenital region to induce defecation, retrieves pups, and places pups in contact with his venter in nursing position. As pups age, the male's time with them increases from 60% of the day to 80% of the day at 16 days old, while the time spent by the female decreases from 85% to 25% of the day (Yamamoto, 1987).

Such close detailed resemblance makes origin by cross-sexual transfer seem likely, though a primary monomorphism is not impossible given the fact that paternal care, albeit often less elaborate, is common in canids (Kleiman, 1977).

Rodents: Mice, Voles, Hamsters, and Mole Rats   Like the raccoon dog, male California mice (*Peromyscus californicus*) show parental behavior remarkably similar in detail to that of females. The two sexes spend equivalent amounts of time in the nest in contact with the young from birth until weaning (Gubernick and Alberts, 1987). Fathers display all of the parental behaviors shown by mothers except lactation, even responding to ventral probing by pups with an arched nursing posture and immobility. Gubernick and Nelson (1989) found that male plasma prolactin levels rose soon after parturition of pups and were similar to those of mothers. In dwarf hamsters (*Phodopus* species), males of *P. campbelli* that participate in care of the young show a rise in prolactin and a reduction in testosterone during their mate's lactation, whereas males of a nonpaternal species (*P. sungorus*) do not (Re-

burn and Wynne-Edwards, 1999). All of these observations suggest a hormonally mediated cross-sexual transfer of parental behavior, as do endocrine studies of humans (see Storey et al., 2000, and Berg and Wynne-Edwards, 2001), though Gubernick and Nelson (1989) caution that correlational studies do not always determine whether a hormonal change is a cause or a consequence of paternal behavior.

The prairie vole (*Microtus ochrogaster*), like the California mouse, is a monogamous rodent in which both males and females care for the young (Thomas and Birney, 1979). Comparison with a congeneric species, the meadow vole (*M. pennsylvanicus*), which does not show male parental care, suggests that the neural mechanism of parental care may involve the Arg-8 vasopressin-immunoreactive (AVP-ir) fibers of the lateral septum of the brain. Characteristic paternal behavior can be induced in young males by injection of AVP before pairing, and the pattern of change in density of the AVP-ir nerve fibers correlates with the parental-care cycle in unmanipulated males (Wang et al., 1994). Injections of AVP also increase maternal responsiveness in female rats (Pedersen et al., 1982, cited by Wang et al., 1994), so the same regulatory pathway may be involved in parental behavior in both sexes, suggesting evolution by cross-sexual transfer. Female prairie voles differ from unmanipulated males, however, in not showing change in the innervation pattern following sexual experience. This indicates that the developmental pathway for parental behavior may not be identical in the two sexes.

A transition to paternal care may occur with little genetic change in voles once selection has favored male cohabitation with lactating females. Small rodents are noted for variability in both social organization and behavior (Wang and Novak, 1992, on voles), and some of this variation is in response to variation in social conditions, population density, and intraspecific competition (Lott, 1984, 1988). Various mate-acquisition and mate-guarding tactics as well as offspring benefits could lead to a rodent male staying with a female through gestation and birth (Elwood, 1983). Mate acquisition activities could bring males in contact with the young and incidentally induce femalelike parental responses in individuals with a low threshold to express them, for example, when testosterone is low during the nonmating season. Prairie vole males begin licking their pups as soon as they are born. In other rodents, this results in ingestion of salty urine, and salt injections are known to increase the release of the parental-care-associated substance AVP from the lateral septum in rats (Wang et al., 1994). Laboratory observations show that stimulation by infant rats and mice can produce components of parental care (Noirot, 1972, cited by Lott, 1984) even in species not showing paternal care in nature (Elwood, 1983).

The voles (*Microtus*) promise to be a key group for research on the evolution of mammalian parental care. Voles show both intraspecific variation (e.g., individual differences; Lott, 1984, 1988) and interspecific (Elwood, 1983) variation in the degree of parental care by males. Some species, such as the biparental pine vole, *Microtus pinetorum* (Powell and Fried, 1992), and the meadow vole, *M. pennsylvanicus* (Wang and Novak, 1992), also have helping behavior by juveniles. They should lend themselves to comparative study of the developmental mechanisms involved in evolutionary transitions, and of the possibility of cross-sexual transfer and heterochrony in the evolution of complex novel behavior patterns. The voles even have a fossil record, allowing a relatively precise determination of phylogeny at the species level, leading Chaline (1987, p. 240) to call them the "*Drosophila* of paleomammalogists."

The naked mole rats (*Heterocephalus glaber*) are unusual among mammals in that permanently non-reproductive adults of both sexes participate in parental behavior, in colonies that have been compared to those of social insects (Sherman et al., 1991). Males, females, and young cohabit in subterranean burrows where they pass their entire lives. Breeding males and females do most of the handling of the young, grooming them and carrying and pushing them (Lacey and Sherman, 1991). These behaviors seem to be qualitatively sexually monomorphic except for nursing, but breeding males do significantly more handling of young than do breeding females. Nonbreeding males and females carry and provision food, carry nesting material, dig and maintain tunnels, and defend the nest against intruders. In most of these activities, with the exception of handling nesting material and removing obstructions, males and small adult females are more active than large females (see table 10-5 in Lacey and Sherman, 1991). Urinary progesterone is high in breeding females but undetectable in nonbreeding females. Breeding females also have significantly higher concentrations of plasma luteinizing hormone (LH). Similarly, breeding, care-giving males also have higher urinary testosterone and plasma LH than do nonbreeding males (Faulkes et al., 1991). Thus, similar hormonal profiles are associated with variations in parental care in the two sexes, as in other vertebrates having

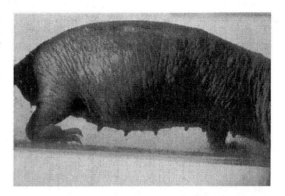

Fig. 15.9. Nipple development in a nonbreeding naked mole rat (*Heterocephalus glaber*). A female is shown. Nipple development is identical in nonbreeding males. Photo by Nicola Kountoupes. Courtesy of P. W. Sherman.

parental care by both sexes. This hormonal similarity (an approach to hormonal sexual monomorphism) is consistent with a hypothesis of cross-sexual transfer of parental-care behavior.

In mole rats, the approach to sexual monomorphism of breeding characteristics extends to morphology. As the breeding female nears parturition, both male and female nonbreeders develop nipples (figure 15.9), although they do not lactate (Jarvis, 1991; I thank J. U. M. Jarvis for an extensive series of photographs documenting this). The nipple development suggests a rise in brood-care-associated hormone levels in both sexes at the time of breeding. I hypothesize that nipple development may be a sign, or physiological side effect, of adaptive permissiveness of partial reproductive development in nonbreeders, that is, only partial suppression of reproductive traits, which allows parental care beneficial to dominants, without parenthood that would be detrimental to them, as suggested for canids by Asa (1997; see chapter 11). When young are born, nonbreeders of both sexes groom and clean them, huddle with them in the nest, carry them to safety if a disturbance occurs, bring small bulbs and tubers to the nest for mother and young, and supply partially digested fecal pellets (caecotrophes) laden with nutrients as well as intestinal endosymbionts that aid in digestion. This considerably relieves the lactational load on a reproductive female, which enables her to gestate unusually large litters compared with most other rodents (see the concise summary in Sherman et al., 1999). Similarly, the adaptive-permissiveness hypothesis may explain a common pattern of queen

Fig. 15.10. Male solicitude toward infants in baboons. A male holds a black infant. The nearby female (left) is not the infant's mother. Courtesy of S. Alberts.

control in social wasps, in which queens allow ovary-developed competitors to engage in oviposition and associated parental activities (especially construction of brood cells), then eat their eggs (West-Eberhard, 1969, 1978b), permitting parental behavior while prohibiting reproduction. If reproductive physiology were to be completely suppressed in subordinates, parental activities beneficial to the dominant reproductives may also disappear.

Primates

Males of different primate species show vastly different degrees of parental solicitude, ranging from performance of all behaviors shown by mothers except lactation (in marmosets, tamarinds, and titis), to tolerance and only occasional defense (in young male chimpanzees; reviewed in Mitchell, 1979; Whitten, 1987). Within species, there are also marked individual differences among adult males in the amount of parental care expressed (e.g., on baboons, *Papio species*, see Altmann, 1980; Collins, 1986; figure 15.10).

One might think that this would make primates the ideal group for comparative study of the causes and origins of parental solicitude, including by cross-sexual transfer. But primate societies are a complicated setting in which to search for behavioral homologies. First, female roles themselves vary both between and within species, in complex ways that are sometimes difficult to unravel but that offer ample opportunity for comparison. At one extreme, the mother tree shrew leaves her babies alone in the nest immediately after birth for one or two days, whereupon she returns and squirts

milk into their mouths during a perfunctory 10-minute visit before leaving them again for two days. If she does not find them within two minutes of returning to the nest, she abandons them forever (Jolly, 1972). At the other extreme, normal mother chimpanzees are highly attached to their young and greatly distressed if they are ill, lost, or threatened (Goodall, 1986). In many species, social relations are complex, and solicitousness toward others can be affected by learning, mimicry, kinship, and alliances involving nonkin (e.g., see especially Goodall, 1986, on chimpanzees). Quality of mothering is known to be affected by individual experiences during infancy (Mitchell, 1979; Goodall, 1986) and by differences of unknown cause in hormone levels of adults (Rosenblatt, 1991).

The importance of learning is emphasized by the fact that in some primate species there are even indications of culturally transmitted variations in gender roles. Males of some troops of Japanese macaques (*Macaca fuscata fuscata*), for example, are solicitous toward infants, while in other troops they are not (Jolly, 1972, after Itani, 1959). In 1940, the Barbary macaques of Gibraltar showed participation by subadult females in the handling and socialization of infants, whereas in later observations at the same locality (Burton, 1972), adolescent males were prominent caretakers, females being rebuffed and isolated from the group. This led Mitchell (1979) to suspect "tradition drift" in sex assignment of infant care. In this population a dominant adult male helps direct infant care by orienting small infants away from their mothers and permitting certain subadults to seize and carry them (Mitchell, 1979). This raises the possibility of a flexible regulatory pathway, which can be socially manipulated: subadults selected by a dominant to be those in contact with infants then respond to them with possessive and solicitous behavior. Direct contact with an infant is known to stimulate a surge of prolactin production and paternal care in male marmosets (Rosenblatt, 1991) as in other mammals (above).

Primate parental solicitude toward infants also is known to be exploited in various ways, as a way of forming pair bonds; by using an infant as a shield in self-defense, as in "agonistic buffering" (Deag and Crook, 1971), where a subordinant male seizes an infant to deflect the attack of a dominant male; and in "aunt" behavior by females who seek contact with offspring not their own (see especially Hrdy, 1977), a kind of solicitousness that may or may not be homologous with maternal behavior and could likewise be a source of evolved behavior potentially transferred to males.

With so many selective contexts and developmental pathways, including learned and cultural ones, leading to infant care and handling, it is sometimes difficult to decide on the basis of behavioral observations alone which cases may be homologous with parental care in the opposite sex. But this very difficulty makes attention to development and endocrinology of special interest. Functional interpretations of parental-appearing behaviors such as "aunt" behavior and agonistic buffering depend on context and behavioral details to deduce the "motivation" or function of the behavior, and, by implication, the adaptive context in which it may have evolved. Sometimes such observations yield convincing evidence of the likely motivation for handling infants: if a young male pursued by a ferocious dominant male five times its size runs to seize an infant and uses it as a shield, thereby stopping the onslaught while putting the infant at risk, there is little doubt as to the nonparental motivation and function of the behavior. Such a response seems unlikely to have originated via cross-sexual transfer involving hormones of solicitous behavior evolved in females. But suppose it is learned, due to the observation that females holding infants are relatively immune to aggression. Then agonistic buffering would be an example of a convergently produced resemblance between the sexes, with a physiologically distinct pathway. Hormone studies can help illuminate whether parental behavior in a particular age, cyclic (e.g., seasonal), or social class of male is likely to be parentally, sexually, or defensively motivated, if particular hormones are associated with these different classes of behavior. Information on motivation indicates the adaptive context in which the behavior likely evolved, not necessarily its developmental pathway.

Primates show hormonally mediated patterns of male parental care quite similar to those found in birds and small mammals. As in birds and many other animals, in primates there is an association between sexual size monomorphism, monogamy, and parental care of the young by males [Compare the body weights of the two sexes (Table 16-1 in Harvey et al., 1987, p. 183), with the list of species with intensive male caretaking, most (nine of ten) of which are monogamous (Whitten, 1987, p. 355, Table 28-3).] (see also Alexander et al., 1979; Selander, 1972, on birds). In one of the few hormone studies of biparental behavior comparable to those on birds, male parental behavior in the marmoset *Callithrix jacchus* proved to correlate with elevated blood levels of prolactin that were induced by carrying an infant (Dixson and George, 1982; cited in Rosenblatt, 1991). This is the same pattern that occurs in some other female mammals (see above), including some other primate females (Mitchell, 1979, on the pigtail macaque, *Macaca nemestrina*), associated with the onset of maternal care (Rosenblatt, 1991). *C. jacchus* males are the major caretakers of infants, performing all parental duties except nursing, including carrying, protecting, food sharing, grooming, playing, and remaining in close contact and proximity (Whitten, 1987). Taken together, these hormonal and behavioral studies indicate a high degree of resemblance between males and females suggesting a hormonally mediated cross-sexual identity of parental traits.

In Japanese macaques, males that adopt yearling offspring do so seasonally, when females are giving birth to new infants (Jolly, 1972, after Itani, 1959). Not only are the infants temporarily needy and perhaps solicitous toward males, but the males are not engaging in the same intensity of mating and male–male aggressive activities as during the mating season in these highly seasonal temperate-zone (30–41°N latitude) monkeys (Kurland, 1977). This raises the possibility that in some primates, as in birds (see above), high steroid titers are incompatible with a high degree of parental-care behavior. In keeping with this pattern, parental care by male Barbery macaques ceases with the onset of the mating season (Whitten, 1987).

In some species (e.g., Barbery macaques and Hamadryas baboons), immature or nonbreeding males handle infants especially frequently (Walters, 1987), a situation reminiscent of helpers at the nest in birds (e.g., see J. L. Brown, 1987). The occasional examples in chimpanzees of extended care by males of younger sibling orphans also involved young, prepubertal rather than adult males (Goodall, 1986). Testosterone levels, which are positively correlated with degree of aggressive behavior and which may interfere with the expression of parental care when high, are lower in such males than in fully mature adults (Mitchell, 1979). The possible importance of low testosterone is further suggested by the fact that in rhesus monkeys, where males usually avoid infants, both castrated and elderly males show more infant care than do other males. Differences between female and male parental care in these animals appear to be quantitative rather than qualitative (Mitchell, 1979). Taken together, these observations suggest an association between low testosterone and feminized responses toward infants.

There is also a strong positive correlation between the degree of male–male social competition (life in multi-male groups) and sperm competition (testis size). Males in one-male groups (e.g., harems) and monogamous males have relatively reduced testes (Hrdy and Whitten, 1987; Clutton-Brock, 1991b). When testis size correlates with hormonal condition, this would help explain, in terms of mechanism, the fact that the highly sexually dimorphic gorilla nonetheless shows a higher degree of male parental care than predicted from the usual correlation between sexual monomorphism and paternal care (above): gorillas live in groups where typically only one male breeds, and they have relatively small testes (only 0.02% of body weight, compared, with, e.g., savanna baboons, whose testis weight is 0.26% of body weight; Hrdy and Whitten, 1987, after Harcourt et al., 1981). If small testis size is associated with low testosterone, this may facilitate the evolution of paternal behavior in gorilla males. While probability of paternity would be a factor contributing to the *adaptive* significance of paternal solicitude in one-male groups, the mechanism permitting it to occur as a selectable variant and to evolve under selection is of equal interest.

Little attention has been paid to the hormonal and developmental aspects of primate social behavior and evolution in major compendia on primate behavior (e.g., Mitchell, 1979; Hinde, 1983; Smuts et al., 1987), even in those dealing extensively with maternal behavior, ontogeny, and proximate factors (e.g., Hinde, 1983; Else and Lee, 1986; but see Bateson, 1991; Hrdy, 1999). Manipulative studies are more difficult than in some other vertebrates for reasons discussed by Rosenblatt (1991; but see Mitchell, 1979). Perhaps more important, sociality is considered by most of the authors in exclusively adaptive terms without attention to the mechanistic causes of selectable variation. The possibility needs to be examined that sexual monomorphism per se, or the possibly more femalelike hormonal profile of young, seasonally breeding, or nonfighting males, contributes to the likelihood of their showing femalelike infant care, due to the greater ease of cross-sexual transfer of originally maternal traits in those classes of individuals. A hormonally mediated trade-off between aggressiveness and parental solicitude (Ketterson and Nolan, 1992) may be the mechanism underlying the common association between large harem size (male–male competition) and lack of male parental solicitude.

Cross-sexual Transfer of Switch Mechanisms

It is common for the expression of alternative phenotypes to be confined to a single sex. Thus, males may show alternative tactics of mate acquisition not expressed in females, and likewise females may have female-specific sets of alternatives not seen in males, such as the worker and queen "castes" of female social Hymenoptera. This implies that there are sex-limited regulatory mechanisms. Are these, like the dissociable phenotypes they regulate, also sometimes transferred from one sex to the other? And is the sex-determining switch ever preempted for use in another regulatory context? The answer to both of these questions is yes, and this lends another level of complexity to the potential for sex transfer in phenotypic reorganization.

The dimorphic males of some ants are discussed above. In some *Cardiochondyla* species, there are both normal winged males and wingless, fighting, workerlike ("ergatoid") males. From data on emergence order of the two forms, Yamauchi and Kinomura (1993) hypothesize that the differentiation of these two classes of males may be based on a mechanism similar to that governing the difference between queen and worker. This suggestion is reinforced by the studies of gynandromorphs, or sex-mosaic anomalies, in *C. emeryi*, where some individuals combined winged male and winged queen traits whereas others combined wingless ergatoid male and wingless worker traits (Heinze and Trenkle, 1997). Since these associations were more frequent than random in the sample of gynandromorphs known, Heinze and Trenkle conclude that factors influencing wing polymorphism in the two sexes are the same. This implies sex transfer of a female caste-determining mechanism to determination of a polymorphism among males.

In many bees, both social and solitary-nesting species, the size of a brood cell in the nest determines the amount of food given the larva it contains (see the many examples in Michener, 1974). Cell size therefore sometimes becomes part of a mechanism for determination of size dimorphisms between the sexes or between the female castes. In the stingless bees *Nannotrigona* (*Scaptotrigona*) *postica* (Meliponinae), for example, relatively large (and therefore more heavily provisioned) cells produce queens. In at least 10 such species of stingless bees, colonies regularly produce "giant" males along with males of normal size (Bego and de Camargo, 1984; Engels and Imperatriz-Fonseca, 1990). The result is a striking size dimorphism among males, with virtually nonoverlapping distributions in values of various traits correlated with size (see Bego and de Camargo, 1984, on *N. postica*). It has been hypothesized that the giant males are the result of "accidental" ovipositions, especially by unmated (male-producing) workers, in the large (and therefore more heavily provisioned) cells destined to produce queens. In support of this hypothesis, the size of giant males is similar to that of queens, and in one species (*Paratrigona subnuda*) giant males are known to be produced in up to 30% of "royal" (queen) cells (Imperatriz, 1970). In a species (*N. postica*) where worker oviposition is rare, giant males are also rare (Bego and de Camargo, 1984). This appears to be an example of sex transfer of the mechanism of determination of a female size dimorphism to males.

In this example, the giant males have no distinctive morphological traits and are not known to have any specialized behaviors or functions. But the phenomenon shows how a novel and potentially functional phenotypic variant can be repeatedly produced as a developmental accident due to cross-sexual transfer of a switch mechanism (in this case, large cells associated with large provisions) evolved for phenotypic manipulation of the opposite sex.

Such recurrent developmental accidents that transfer a body size dimorphism to the opposite sex may have been the initial stage for the evolution of a morphologically and behaviorally specialized large male "fighter" morph in *Perdita portalis* (Hymenoptera: Andrenidae), a solitary-nesting species with two male morphs. One is a small male that resembles the males of related species in head size, wingedness, aggressiveness toward other males, and mating with females foraging on flowers. The other is a large wing-reduced male that has a disproportionately large head, reduced compound eyes, enlarged facial foveae, and fully atrophied flight muscles that fights, sometimes to the death, within the subterranean nests where the bees mate.

According to a phylogeny of *Perdita* and related genera, the distinctive morphological characters of the large males are almost certainly derived (Danforth, 1991). The high frequency of occurrence of the large-male morph (50% of all males) as well as their specialized mating tactic leads Danforth (1991) to believe that this is not an accidental form. But it is possible that the size-dependent switch between the morphs originated, as suggested by Kukuk and Schwarz (1988) for a similar dimorphism in males of *Lasioglossum erythrurum*, when females mistakenly placed a male egg on a pollen ball that is large like those usually destined for female larvae. The large male morph could then become specialized, under selection capitalizing on its large size, to mate within the nest. The sex-transfer hypothesis for the origin of the size-dependent switch is supported by the fact that the large males are similar in size to females; by a strong positive head-size/body-size allometry in closely related

species not showing dimorphism (even though thousands of other species of bees show "more or less isometric size variation" among males—Kukuk and Schwarz, 1988); and by the fact that provisioning mass is known to correlate with offspring body size in various hymenopteran species (reviewed in Danforth, 1991).

The large flightless fighting males of *L. erythrurum* not only mate within the nest but also work in ways that are unusual for hymenopteran males except as anomalies (see the discussion of social insect males, above). They do a small amount of nest maintenance in the form of tunnel repair and moving of dirt toward the nest entrance, and defend the nest by attacking invading ants. These males, then, appear to be "feminized" with regard to nesting behavior as well as size—some *behavioral* cross-sexual transfer therefore may have occurred, perhaps as an automatic result of the feminized diet or induction by contact with stimuli usually only experienced by females, or, alternatively, evolved under subsequent selection.

In some termites, the genetic switch mechanism governing sex determination is transferred to a quite different context, the determination not of sexual phenotypes but of neuter (worker and soldier) subcastes (Noirot, 1989, 1990). Although the sex-switch mechanism is used as a caste-switch mechanism in many species of termites, the caste corresponding to male or female is not always the same. Almost every permutation of sex-dependent and sex-independent differentiation of workers and soldiers has been observed: in some genera, workers are female and soldiers are male (*Trinervitermes*); in some, small workers are female, large workers male, and soldiers are male; in some, small workers and soldiers are female and large workers are male (*Macrotermes*); in some all neuters (both soldiers and workers) are female (*Schedorhinotermes*); and in some all neuters are male (*Acanthotermes, Cornitermes*; Noirot, 1990). In these latter species, where all nonreproductives are of the same sex, aberrant sex ratios are observed, and production of neuters is unusual in continuing even in the season when reproductive males and females emerge (Noirot, 1990). Thus, a switch that originally promoted sexual divergence serves as the basis for behavioral and morphological divergence between nonreproductive phenotypes and regulation of the relative numbers produced within colonies (Noirot, 1990).

Despite the strong influence of genotype on the determination of soldier and worker castes, environmental conditions can reverse the switch. For example, in unisexual lab cultures of some species, the soldier phenotype is expressed in the "wrong" sex, and the same result can be produced by application of a juvenile hormone analog (Noirot, 1989, p. 11). In some species where both soldiers and workers are of the same sex, the sex ratio among these nonreproductive individuals is strongly skewed, with production of only that sex during most of the year, suggesting a seasonal cycle. In one species (*Schedorhinotermes lamanianus*), there is evidence for seasonal modification of sex-determination chromosomes (Noirot, 1989, after Renoux, 1976), a suggestion rendered less preposterous by the extreme complexity, plasticity, and lability of sex-determination mechanisms, which in the termite families Kalotermitidae and Rhinotermitidae sometimes undergo Y-linked autosomal translocations with massive inactivation of involved genes (see Syren and Luykx, 1977; Bull, 1983; Lacy, 1980; Crozier and Luykx, 1985; Myles and Nutting, 1988). The effects, if any, of the complexity of chromosomal evolution on the evolution of sex and caste determination remain to be understood. It would be of special interest to determine if the inactivated genes are those expressed in the ancestral sexual (female) phenotype.

It is difficult to imagine how such a diversity of genetic caste-switching mechanisms has evolved from sex determination, since it is not obvious from these patterns that anything like feminization or masculinization is involved. But the transfer of a switching mechanism from one phenotypic setting to another illustrates that the essential quality of a switch—the determination of developmental divergence via coordinated gene expression—obtains, even if the selective context, and therefore the alternative forms regulated, has changed.

The Social Environment as an Inducer of Cross-sexual Transfer

All of the phenotypes discussed in this chapter affect social interactions and are sensitive to environmental influence. How far can these changes in sex-trait expression be pushed by change in the social environment alone? How far toward sexual monomorphism can a dimorphic population be moved by simply increasing the competition among females or by putting individuals in extreme social situations? In species without paternal care of young, how far toward parenting can a solitary male be pushed if he is forced, under sexual competition, to stay with a female and her begging, crying, pheromonally perfumed young?

The answers to these questions undoubtedly depend on the species in question. Polygynous male vertebrates, for example, appear to be less responsive to social environmental cues that influence reproductive behavior (courtship and aggression) than are monogamous males (Wingfield et al., 1990). But there are also individual differences in responsiveness within species in hormonally mediated behaviors (Wingfield et al., 1990)—material for change in degree of social responsiveness under selection.

Homosexual courtship behavior, including courtship feeding and attempted copulation, sometimes occurs in homosexual pairs of female birds, with one of the females adopting the male role (e.g., see Hunt and Hunt, 1977, on western gulls). Male zebra finches (*Poephila guttata*) perform the female invitation-to-copulation display on many occasions when they are thwarted in their attempts to court an unresponsive female; females sometimes respond to this "pseudofemale" behavior by mounting the displaying male and performing the full male copulatory pattern (beating their wings, lowering and twisting the tail, etc.; Morris, 1955). Similar observations on the behavior of mammals are reviewed by Dagg (1984). Although the pseudofemale behavior of zebra finches appears to be an anomaly, such environmentally stimulated responses to thwarted mating could be the prelude to the evolution of the femalelike tactics that characterize some adaptive alternative male reproductive patterns that are expressed in subordinate males (see above).

The behavior of fish in the laboratory, and under unusual conditions in nature (e.g., see Sauer, 1972), has revealed conditions that could conceivably lead to the origin of adaptive cross-sexual expression of behavior in natural populations. When female 10-spined sticklebacks (*Pygosteus pungitius*) are abundantly fed, their egg development is accelerated, and if such females are kept together without males in the same tank, females occasionally perform the male courtship dance (a series of head-down jumps) toward other females. If the approached female responds, the first stages of the usual courtship sequence are sometimes performed by the female in the male role (Morris, 1955). Greenberg (1961) reviews many such examples in different fish species kept crowded in unisexual groups, including some where females paired to spawn. Some of these pairs spawned alternately, in an interaction reminiscent of that between simultaneous hermaphrodites in fish (Fischer, 1988; see chapter 23), raising the possibility that little genetic change would be needed to establish such mating systems in species where they are adaptive. Morris (1955) summarizes factors associated with such sex-transfer behavioral anomalies, including hormonal or structural abnormality of the sexual system, low social rank or subordinance of males and aggressiveness or dominance in females (for examples in cockroaches, see Rocha, 1991), high sexual motivation followed by thwarting, and presence of stimuli that promote the sexual behavior of the opposite sex.

It is often difficult to know how much influence is imposed by social environment and how much is due to genetic change under selection in the diversification of sexual behavior and morphology. This is illustrated by patterns of variation in the sexual dimorphism of salmon. One of Darwin's (1871) featured examples of a sex-limited male trait was the kype of the male Atlantic salmon, *Salmo salar*. The kype is an exaggerated extension of the lower jaw wielded in ferocious battles among hyperaggressive males. Unlike the Pacific salmon, which reproduces once and then dies, the Atlantic salmon reproduces in more than one year. The seasonally reversible polyphenism of the Atlantic salmon is the most remarkable known to me in any vertebrate (described in Tchernavin, 1938, 1944; Fleming, 1996; notes on the Tchernavin papers kindly provided by I. Fleming): the kype of the male develops only during the spawning season and then is lost during the nonbreeding season. Seasonal development of the breeding male phenotype involves an increase in absolute size and change in shape of the skull, elongation and change in shape of the cartilaginous front end of the jaws to form the kype (which may be 1.5 times its former length in large males), elongation of all jaw bones, and production of a distinctive set of breeding teeth that replace the feeding teeth. Breeding teeth are long, broad at the base, curved, and anchored to the bones, with lengthened premaxillaries that reach beyond the cartilaginous kype.

Following the breeding season, the breeding teeth degenerate and are replaced by newly grown feeding teeth, which are long and narrow especially at their bases where fastened to the jaw bones, and the premaxillary bones are short and wide posteriorly. Jaw bones and kype shrink, and the skull slowly returns to its former, marine form. All of this is accompanied by changes in scales, coloration, and muscles and thickening of the skin. Changes in the skull also occur in females, but to a much lesser degree, and females do not show the increased relative growth of kype and premaxillary bones compared to body size seen in males. Breeding females become darker in color, whereas males

become reddish and show an increase in adipose fin size not seen in females (Tchernavin, 1938, 1944).

In Pacific salmon, such as the Pacific coho, *Oncorhynchus kisutch*, and the other five North American Pacific *Oncorhynchus* species, females as well as males fight, with females defending breeding sites not only at the time of acquisition (as in Atlantic salmon) but also after spawning, when they defend the nest sites until they die (Quinn and Foote, 1994; Fleming, 1996). In the Pacific salmon, females develop the kype to variable degrees (van den Berghe and Gross, 1989). The kype and head hump in females are less developed than in males (Fleming and Gross, 1989; Quinn and Foote, 1994), and female kype development varies geographically in accord with the local degree of female–female competition, being markedly reduced in hatchery populations where there is little fighting among females (Fleming and Gross, 1989). This suggests that kype development may be condition sensitive and rapidly lost or gained under changed social circumstances.

Enhanced kype expression in female Pacific salmon is a good candidate for origin via environmentally mediated sex transfer of expression from males. As noted by Fleming and Gross (1989), the populational differences in kype expression are likely adaptive, evolved products of strong selection in the different competitive situations described. But the seasonal flexibility in development and loss of the kype in *S. salar*, variable female kype development in populations with different levels of female aggressiveness, and rapid kype loss in fishery females of Pacific salmon all suggest the hypothesis that rapid adaptive change in kype expression may get a boost from developmental plasticity. Hormonal mediation is likely involved, as suggested by the fact that the kype is absent in the precociously mature "jack" or parr males, which have relatively low androgen levels compared with the kyped males (Mayer et al., 1990a,b,c), but there are few other endocrinological studies of behavioral and morphological change in salmon (I. Fleming, personal communication; see Leatherland et al., 1982; see also references in Mayer et al., 1990a,b).

As just discussed, sex expression in fish, including cross-sexual transfer of expression of malelike traits to females, is often affected by social interactions including aggression (reviewed in Baerends, 1975). It is therefore reasonable to hypothesize that the observed divergence among salmon populations represents change initiated by the social environment itself, possibly mediated by hormones, followed by genetic accommodation of the change.

Kype development occurs during the migration from the sea to the freshwater spawning grounds and therefore could not be a influenced by plasticity during the spawning interactions where the kype, modified teeth, and other modifications are used (Fleming and Gross, 1989). But it would not be surprising if migrating salmon interact during migration, at least to the extent of being sensitive to population density of migrants. Kype expression is the kind of phenomenon that would reward a comparative study that includes behavior during migration, endocrinology, morphology, and genetics of variation in populations with different degrees of female aggressiveness such as those described by Fleming and Gross (1989).

Social competition for mates and other resources often leads to the evolution of adaptive alternative phenotypes expressed by the losers in social interactions (see part II), and the examples of the present chapter suggest that hormonally mediated developmental plasticity may often contribute to the origin of such alternatives via cross-sexual transfer (see also Moore, 1991). For this to occur requires phenotypic plasticity during the adult stage. Such flexibility is found in many vertebrates, as recognized by the *organization-activation theory* of hormone action, which describes the ontogeny of sexual dimorphism in terms of both organizational (relatively permanent) and activational effects of sex hormones (see Moore, 1991). Activational effects occur relatively late in development and may turn sexually dimorphic traits off and on seasonally or facultatively (e.g., under particular social conditions). The social environment could influence both effects, including organizational influence via maternal effects, but activational hormonal effects are likely to be especially important in the evolution of facultative alternative phenotypes and sex transfer (Moore, 1991; see also Emerson, 1996, p. 280).

16

Quantitative Shifts and Correlated Change

Much of evolution involves developmental recombination of modular parts, but novel phenotypes can also originate due to quantitative change, facilitated and exaggerated by developmental plasticity and by developmental correlations among traits (the two-legged-goat effect). Some quantitative novelties are responses to environmental extremes followed by genetic accommodation. Fitness trade-offs sometimes involve antagonistic developmental processes, such as internal resource competition, temporal interference between activities, and proportional but opposite responses to the same regulatory cue. Allometries and correlations are subject to change under selection and should not be regarded as inescapable constraints in evolution.

Introduction

Preceding chapters have discussed evolutionary transitions as changes in the expression of discrete, modular traits. This chapter discusses transitions that are due to shifts in the magnitude, rather than the time, place, or repetition, of trait expression. Especially, it considers examples where environmental extremes induce quantitative change in the expression of continuously variable plastic traits. Quantitative shifts can produce novel extremes, novel combinations of extremes, or simultaneously opposite shifts due to negative correlations between traits, as in trade-offs. As pointed out by Brien (1969; see also chapter 7), correlated adaptive shifts can produce major changes in which large steps are not lethal because the usual adaptive plasticity of the organism accommodates the kind of change that occurs when "a new type of organization is born" (p. 731).

The same developmental plasticity that is responsible for phenotypic accommodation and homeostatic stability (see chapter 3) can produce correlated change as well. Correlations among the environmental responses of plastic traits (figure 16.1) mean that several quantitative traits can change at once, if they respond simultaneously to the same mutation or environmental factor. The two-legged goat described in chapter 3 (see figure 3.13) shows how the multidimensional plasticity of

the phenotype can produce a strikingly novel form that, lacking intermediates, appears to be a qualitative change—a change in kind, not merely degree. The fact that the same plasticity-mediated changes would occur whether the cause of shortened front legs were due to a mutation or to an environmental effect early in development illustrates the interchangeability of genetic and environmental factors in inducing correlated change.

Raff and Kaufman (1983, p. 202) called correlated effects due to developmental relationships among continuously variable traits "relational pleiotropy." Positive relational pleiotropy can result when numerous positively correlated traits respond in unison to a single stimulus or condition, such as variation in size. Negative relational pleiotropy can give rise to *trade-offs*, or negative fitness effects among traits such that an increase in the magnitude of one means a decrease in the magnitude of one or more others. Fitness trade-offs are a major subject of life-history theory and related fields, including those dealing with optimal foraging, sex allocation, parental investment, and morphometric study of allometries.

Previous discussions of quantum evolution and of fitness trade-offs have focused primarily on their adaptive nature and evolution in response to natural selection. This chapter focuses, instead, on the much less discussed developmental underpinnings of quantum phenotypic novelties and fitness trade-

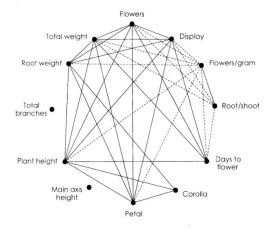

Fig. 16.1. Correlated plastic responses in a plant (*Phlox cuspidata*). Lines connecting different traits indicate their degree and direction of correlation (plasticity correlations) when grown under different environmental conditions. Solid lines indicate positive Spearman rank correlations (similar responses across treatments); dashed lines indicate negative correlations. In this plant, an environmental condition that reduces the weight of the plant, for example, would produce a correlated shift toward shorter stature, smaller and fewer flowers, and smaller roots and shoots. At the same time there would be an increase in the relative investment in flowers and in roots versus shoots per unit weight. After Schlichting (1986).

offs, in order to understand the possible contributions of developmental plasticity to the origins of these phenomena.

Correlated Extremes: Two-Legged Goats and Developmental Constraints

Given the commonness of plasticity in quantitative traits, the two-legged goat should be the standard bearer for a major category of selectable variation and evolutionary change. Yet, because plasticity has been associated with stability and developmental noise, the two-legged goat has hobbled through the literature as an oddity not obviously relevant to any specific example of evolution.

It is not always obvious how the congenitally deformed could be the leaders of invention. Frazzetta (1975) remarked that Goldschmidt's discussion of trait origins "leaves me with the feeling that he often did not distinguish between truly in-

novative evolutionary steps, those involving revision of the interaction responses among components, and the flexibility of integrated systems to provide an array of varied adaptive expressions" (p. 225). There is, in fact, no clear line between a truly innovative step and the extremes producible by ordinary adaptive plasticity. Innovation can spring from routine plasticity in a new or extreme environment. Frazzetta (1975) gives testimony to this, in his pioneer discussions of "the use of existing integrative processes in evolutionary changes" (p. 225) and the accommodation of change due to preexisting "integrated relationships" (p. 227), or phenotypic accommodation. Phenotypic plasticity achieves both adaptive accommodation, the aspect emphasized by Frazzetta, and evolutionary change, when there is genetic variation and selection for correlated responses induced by recurrent conditions.

Mrosovsky (1990) shows how physiological plasticity can promote both stability and change. He shows how the homeostasis concept has served physiology while at the same time distracting attention from condition-sensitive adaptive *change* in the physiological phenotypes of animals. As an antidote, he proposes the concept of "rheostasis," or "the physiology of change." Under extreme conditions, homeostatic mechanisms may reach new "set points" for adaptive regulation.

The inherent flexibility of physiological adjustment puts homeostasis in a new light. Homeostatic mechanisms function within the limits of environmental fluctuation normally experienced by a population, but they are not thereby "frozen" or constrained to respond only within those limits. Similarly, Waddington's metaphorical canalization of development, also a kind of restorative flexibility, is accurate only if one considers that the retraining walls of the valleys are deformable, given an unusual genomic or environmental nudge. Developmental homeostasis is just plasticity held in check under a limited set of circumstances—those historically presented for screening by selection. Beyond that, under the insult of an unscreened input from the genome or the environment, occasional or unprecedented departures from the norm can occur. Two-legged goats and other flexible "monsters" indicate that the outer limits of adaptive plasticity—the location of constraints—are less severe than may be suggested by the stereotypy of normal development and physiological homeostasis.

Charlesworth (1990) listed some doubts about two-legged goats that are probably shared by many biologists: Slijper's two-legged goat "did not produce offspring, and so its potential as a 'hopeful monster' is purely speculative. The evidence from genetics is overwhelmingly unfavorable to the idea that single Mendelian genes with major morphological effects on a complex of characters can become established in evolution" (p. 53). As Charlesworth says, for a trait to be established in evolution, individuals with a developmental novelty would have to reproduce—the novelty would have to be recurrent in more than one individual for selection to occur (see chapter 6). But his next point, that genetic research is unfavorable to the idea that a mutant allele of major effect can be established in evolution, is questionable. As already discussed in chapter 6, the spread under selection of a complex advantageous novelty need not depend exclusively on the increase in frequency of a particular mutant allele, even if such an allele initiates the production the novel trait. Whatever the initial cause—mutation, environmental induction, or pleiotropic effect of other change—genetic accommodation due to selection on genes of small effect can increase the capacity to express the favored response, tailor the conditions of its expression, and reduce negative side effects that the novel trait may have.

The hypothesis that quantum shifts in plastic quantitative traits—the two-legged-goat effect—can contribute to the origin of evolutionary novelties is supported if (1) complex correlated shifts are found as one-step anomalies in natural, experimental, or domesticated populations and (2) the same or similar traits are found in related populations or species as established, widespread traits. The following section discusses some possible examples.

The Two-Legged-Goat Effect in Domestic and Natural Populations

Vertebrates

One of Darwin's examples of correlated change was based on his own study of the skulls of lop-eared rabbits, large domesticated breeds with "prodigiously" large ears. The heavy ears droop down at the sides of the face, and this has "influenced in a small degree the form of the whole skull" (Darwin, 1868a [1875b], p. 120). Especially, the lop ears correlate with a notable change in the size, form, and direction of the bony meatus. How could he be sure, you may ask, that this was due to the exaggerated ears, and not some poorly understood

side effect of generations of artificial selection on this odd domesticated breed? Darwin showed that the altered skull is developmentally associated with the large pendant ears by exploiting a natural anomaly with a built-in control. "Half-lop" rabbits have one ear (the "control") that is short and upright, as in normal rabbits, while the other is long and pendant, as in the lop-eared breed (figure 16.2). These specimens reveal a clear difference in the form and direction of the bony meatus on the two sides, and this "slightly affected on the same side the structure, of the whole skull" (p. 124; figure 16.2, left). These observations show a complex change produced in a single step by the same genotype; extensive, gradual evolutionary change was not involved.

The correlated effects of lop ears are probably even more extensive than Darwin realized. It is now known that the upright ears of a hare, which weigh one third as much as the entire head, function in a shock-absorption mechanism during saltatory running (Bramble, 1989, summarized in Galis, 1996a). Therefore, lop ears may have correlated effects on locomotory behavior and, if locomotion were inhibited due to impaired ear function, body musculature, in addition to the cranial changes noted by Darwin.

Numerous real-life examples of the two-legged goat effect occur in normal ontogeny and evolution of vertebrate morphology (Hanken, 1983; Frazzetta, 1975). In rats, for example, anomalous early fusion of two bones in the base of the skull can lead (via plasticity) to a host of correlated skull changes (DuBrul and Laskin, 1961, cited by Frazzetta, 1975). Skulls similar to the anomalous rat skull recur in various mammalian species having an upright stance. This suggests that more or less parallel variants of the mammalian skull could thus be derived in different species via incorporation of plastic responses from a basic growth pattern present in a common ancestral species.

Atchley (1990) and Atchley and Hall (1991) have expanded this kind of observation into a model study of correlated change in the mouse mandible. According to Atchley (1990, p. 290) the craniomandibular region is a mosaic of individual components whose patterns of growth and morphogenesis must be highly coordinated during ontogeny. The osteo-, chondro-, and viscerocranium each possesses its own distinctive pattern of growth, morphogenesis, maturation and function (see figure 4.14). However, each is a phenotypic subunit whose development is tightly integrated with the others. A failure of integration can cause serious malformation. At the same time, small

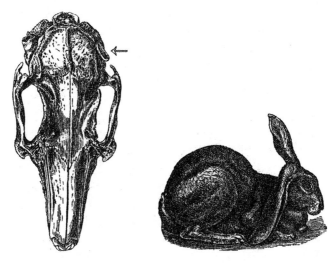

Fig. 16.2. Correlated developmental shifts in skull morphology under selection for large ears. The half-lop rabbit (right) with the large drooping ear of the lop-eared breed on one side and the ear of a normal domestic rabbit on the other, has a highly asymmetric skull (left), especially in the development of the auditory meatus (arrow). This shows that the morphological differences between artificially selected breeds, expressed together in this anomalous individual, are likely due to a simple genetic change amplified by correlated effects. After Darwin (1868a [1875a]).

changes in the genetic and epigenetic controlling factors studied by Atchley and Hall can result in extensive correlated morphological changes, as exemplified in the skull and jaw of various breeds of domestic dog (Wayne, 1986). The correlated changes are both extensive and accommodated to give a functional result. These are examples of evolutionary changes rendered by artificial selection. Work by Atchley and Hall (1991) and Hanken (1983) shows that the changes observed are due largely to developmentally correlated effects, not step-by-step genetic changes, favored by selection.

There is evidence that the two-legged-goat effect has influenced the evolution of morphology and behavior of fish. Cichlid fishes are well known for the flexibility of their pharyngeal jaw apparatus in response to diet. When individuals of *Astatoreochromis alluaudi*, an African species that is highly variable in terms of jaw development in nature, are reared on contrasting diets of snails and soft food in aquaria, they show complex correlated differences in morphology (Greenwood, 1965; see chapter 28). The aquarium fish develop a pharyngeal mill that is similar to that of related species with soft (insect) diets, including reduced bone mass, fewer enlarged pharyngeal teeth, less flattened and molariform shape of teeth, reduced size of neurocranial apophysis, and smaller and differently shaped basioccipital facets. The changes owe to

both the mechanical effect of reduced muscles (associated with processing soft rather than hard food) and differences in calcium intake, which affect bone development (Greenwood, 1965). So dietary differences have both exercise and nutritional effects. More recently, Meyer (1989a, 1990a,b) demonstrated a diet-dependent polyphenism in the jaw morphology of the neotropical cichlid *Cichlasoma citrinellum*, with the primarily snail-feeding versus softer-diet phenotype showing a different body shape as well (Meyer, 1990a). Associated with this intraspecific morphological plasticity is a notable evolutionary lability of the cichlid pharyngeal jaw apparatus (Liem, 1973, 1984; Liem and Kaufman, 1984). This suggests that plasticity and correlated change have been important in the evolution of cichlid jaw diversity (see chapter 28, on adaptive radiations).

Diet-induced behavioral and morphological variation appears to be common in fishes (Ivlev, 1961; Hart and Gill, 1994). When the diets of a "benthic" and a "limnetic" species of threespine stickleback (*Gasterosteus*) are reversed, the morphological differences between them are reduced to degrees ranging from 1% to 58%, depending on the trait (Day et al., 1994). Diet-related mechanical stress affects cartilage and bone, which in turn is associated with change in several otherwise uncorrelated traits, especially gill raker length, head depth, and snout length. Patterns of correlated intraspecific variation that parallel interspecific variation are also found within and between species of cyprinid fish in the genus *Barbus* (Banister, 1973), and it would be interesting to test the hypothesis that these are correlated developmental effects of diet. They could be correlated either due to direct developmental interactions among them (as in some fitness trade-offs; see below) or due to independent susceptibility of each to influence by the same factor (diet).

There are correlated changes in morphology and behavior in male guppies (*Poecilia reticulata*) depending on whether they are reared in fast- or slow-moving water (Nicoletto, 1996). Males reared in fast-moving water have increased muscle development and, consequently, wider caudal peduncles. They also swim faster, display longer during courtship, and spend more time displaying relative to other activities. Comparable differences are observed in some natural populations of guppies: males in highland populations in fast-moving streams are larger and display more vigorously than do males in lowland, slow-moving streams (Nicoletto, 1996). These observations suggest that stream velocity may be among the factors (in addition to predation rates) that contribute to phenotypic divergence among populations (reviewed in Nicoletto, 1996). Correlated responses to a single variable (ambient water velocity) appear to exaggerate change.

Comparative morphological studies by vertebrate ecologists and systematists sometimes reveal patterns of variation that suggest a widespread role for quantum shifts due to plasticity and point to areas where research on the evolutionary role of plasticity may be rewarding. Frog tadpoles (*Rana chiricauensis*), for example, show patterns of behavioral and morphological variation that are correlated with degree of water movement as in the poeceliid fishes just described (Jennings and Scott, 1993). And tadpoles exposed to predators engage in burst swimming and facultatively develop large, brightly colored tails and short bodies; this morphotype experiences only about half the mortality under predation as do controls not reared with predators (Van Buskirk and McCollum, 2000). In the characid fish genus *Paracheirodon*, recurrent losses of particular cranial parts of the laterosensory system and associated bones often result in a characteristic change in the morphology of neighboring elements, and this pattern of correlated shifts occurs in several species and genera of the family Characidae (Weitzman and Fink, 1985). Such recurrent patterns suggest that correlations

due to developmental plasticity may be involved in evolutionary change.

Caterpillars, Grasshoppers, and Locusts

Morphological flexibility in response to diet is not limited to vertebrates, although this could be assumed to be true due to the familiar responses of vertebrate muscle and bone to use and disuse, and the seeming contrast of vertebrate tissues with the presumed inflexibility of the chitinous exoskeletons of such organisms as insects. Yet it is now known that individual caterpillars that specialize in feeding on relatively hard leaves (e.g., grasses) develop relatively larger heads than do those that feed on soft material such as the leaves of herbaceous dicots, and that this intraspecific difference parallels variation among species (Bernays, 1986). Cohorts of caterpillars (*Pseudaletia unipuncta*) fed on tough grasses have greater positive allometric growth of the head in subsequent instars than when fed on soft food. Thus, in an insect, as in vertebrates, muscular effort is associated with effects on skeletal morphogenesis (head size).

Phase polyphenism in locusts (see chapter 5) is an example of a set of continuously variable traits whose extremes are expressed in unison under high population densities due to common, independent correlations with that single factor. A full response, which occurs only at especially high population densities, includes extreme expressions of several continuously variable, density-dependent traits found in numerous other Orthoptera (though not all are universal; e.g., see Zera and Tiebel, 1989, on *Gryllus rubens*). For example, in *Locusta migratoria* and *Schistocerca* species, the the extreme gregarious phase expresses the following traits (summarized by Kennedy, 1961): (1) long wings (also associated with crowding in some crickets—Alexander, 1961; Masaki and Walker, 1987; and in many other insects—Matsuda, 1987, p. 21), (2) low fecundity (frequently correlated with long vs. short or absent wings in insects in general—Matsuda, 1987), (3) bold black and yellow or orange coloration (also found in exceptionally dense populations of other grasshoppers not normally considered liable to phase change—Rowell, 1971, p. 175), (4) reduced developmental time (also associated with crowding in some crickets—Alexander, 1968), and (5) distinctive behaviors, such as aggregation rather than aggressiveness or retreat from others (also inducible by high density in crickets—Alexander, 1961). The combined set of extremes constitutes a density-induced "gregarious form," a coordinated character complex seen only in extreme conditions. At moderate densities, each element of this complex trait has its own independent rate and pathway of development (see chapter 5).

When each component of a complex response is found as a widespread facultative trait in related species, as are the elements of the migratory phenotype of locusts, the complex novelty could conceivably be initiated by environmental change (in this case, high population density) rather than mutational change. Indeed, environmental induction in such a group is a more parsimonious explanation, given the information on related species, than mutation or a series of mutational changes, although these possibilities are not excluded.

Social Insect Workers

The origin of the worker phenotype in the social Hymenoptera is discussed in chapter 11, on deletions. It can also be seen as a set of correlated effects that stem from a drastic change in the social environment. A worker is a behavioral and physiological two-legged goat in being a complex novelty with a severe handicap—complete sterility—yet under kin selection or family-group selection, it has become a major adaptive innovation in several speciose clades of ants, wasps, and bees.

Group life imposes a set of correlated behavioral and physiological changes on primitively social hymenopteran females (summarized from West-Eberhard, 1996; see also Sakagami and Maeta, 1987a,b). On the one hand, females with elevated ovarian development, high juvenile hormone titers, and a trophic advantage due to social dominance (a set of consistently correlated traits) are able to maintain reproductive and social dominance over other females, who are consequently pushed to the opposite extremes in these traits. The result is to reinforce initial differences and cause a developmental bifurcation between the extremes, such that the two ends of the continuum become more different and separate than in solitary populations. Social dominants, for example, exaggerate their status by securing more food via aggressive solicitation from subordinants, which in turn likely further exaggerates the differences in their ovarian development. When dominant females can physically control access to brood cells and oviposit in them, they effectively exclude others from egg-laying. Then, those that lay eggs develop larger ovaries while those unable to lay eggs experience ovarian regression.

These correlated and mutually reinforcing responses are known to be derived from responses widespread in nonsocial hymenopteran females (West-Eberhard, 1996). That is, the worker phenotype is based on ancient characteristics of this group of insects. Solitary wasps, for example, respond to territorial aggressiveness, starvation, and failure to oviposit in the same ways that social wasps do (see references in West-Eberhard, 1996). Each major element of the worker phenotype, like each distinctive trait in the two-legged goat or the migratory locust, is a result of environmental influence (in this case, social interaction) pushing a preexisting plasticity (in this case, in degree of ovarian development) to an extreme state (in workers, atrophy of all oocytes; in queens, hyperdevelopment).

Many authors concerned with the evolution of the worker phenotype have written about "genes for altruism." Yet, as Rachootin and Thomson (1981) point out regarding the novel features of the two-legged goat, "No one would maintain that goats have genes for developing an S-shaped spine" (p. 18). If there are genes for altruism, many of them must be acknowledged to have been present in the remote ancestors of worker females. Then, under the altered conditions of life in groups, they produced a novel phenotype that paid off in a new way. Subsequent selection and genetic accommodation would involve other "genes for altruism," those that adjust the switch between worker and queen at levels favored by natural selection and increase the efficiency of performance of their respective tasks.

Buttresses in Trees

Striking morphological variants involving correlated effects are induced by mechanical stress in trees. Pines growing in strong prevailing winds, for example, have a distinctive form and structural traits different from those of trees growing in sheltered conditions. These traits can be duplicated within a group of genetically homogeneous plants by subjecting seedlings to mechanical perturbation (Vogel, 1988, based on Telewski and Jaffe, 1986).

Reaction wood is a kind of secondary xylem produced by trees in response to mechanical stress on branches and trunks (Wainwright et al., 1976). Reaction wood is an environmentally induced growth device that plays an important role in the normal life of trees, enabling them to actively resist tension, compression, bending, and impact, under constant stresses (e.g., in response to wind) or in response to emergencies. In softwoods (gymnosperms) this compensatory growth occurs on the side where increased length will restore normal position and is called "compression wood." In hardwoods (dicot angiosperms), it occurs on the side

Fig. 16.3. Buttress roots of a tropical tree. Photo by M. Guerra.

where shortening acts to restore the original position and is called "tension wood." Compensatory, posture-restoring growth of this kind can be experimentally induced by application of indoleacetic acid, a plant growth hormone. The cells of these two kinds of reaction wood have a number of special properties that may enable these tissues to exert position-restoring force (discussed by Wainwright et al., 1976; see also Scurfield, 1973, on the special phenotypic properties of reaction wood).

Reaction wood is responsible for a number of familiar responses of trees, such as the restoration of upright posture in trees that have been knocked over and continued to grow. It reinforces heavy branches at their joints with the trunk. The force exerted by reaction wood is demonstrated by allowing reaction-wood-producing roots to grow into potted soil: when aerial roots of a fig tree (*Ficus*) penetrate the soil, they produce tension wood and then contract, with force sufficient to lift the container off the ground.

It has been hypothesized that the buttress roots characteristic of many tropical trees owe their existence to the effects of mechanical stress on the growth of superficial lateral roots (Ennos, 1993). That buttresses are a response to stress is supported by the following observations. (a) Buttresses are absent in young trees. They grow rapidly in forest trees only as individuals reach the canopy. (b) They are most common in trees growing in shallow soil (e.g., where there is a high water table or shallow humus over rock or subsoil). In such conditions, roots are superficial and are dominated by lateral supporting root branches.

Buttresses (figure 16.3) are more common in tropical trees, which lack deep tap roots or sinkers close to the trunk, and which have relatively thin bark (there is a strong negative correlation between buttressing and bark thickness; Smith, 1979, cited by Hartshorn, 1983). Buttresses occasionally occur on temperate-zone poplars and spruces growing on shallow, water-logged soils, a situation that resembles tropical conditions associated with buttresses. Finally, (c) as expected if they are derived from a reaction-wood response to stress, buttresses grow asymmetrically, and their orientation is related to the direction of stress (e.g., prevailing wind direction) or failure of support (e.g., on the banks of a stream). The stress-origin hypothesis also is supported by the correspondence between buttress morphology and a computer analysis of the geometry of stress (Mattheck, 1993). In sum, much evidence points to the origin of buttresses as a response to the "mechanical environment" of a tree (Ennos, 1993, p. 350).

Trade-offs

A *fitness trade-off* refers to a negative correlation between the fitness effects of two traits such that both cannot be optimized at the same time. Fitness trade-offs are often associated with "adaptive syndromes," such as a high activity, high metabolic rate, and small body size syndrome (as in hummingbirds) versus a low activity, low metabolic rate, and large body size syndrome (as in cows). An activity-rate versus body-size trade-off may underlie some contrasting adaptive strategies, such as the search-and-chase versus the sit-and-wait ambush tactics of some predators (Rollo, 1994). A similar trade-off, between photosynthetic gains and transpirational losses under high sunlight, has marked the physiological evolution of plants, producing alternative adaptive syndromes and conditional switches between them (see chapter 17, on CAM in plants). And allocation of meristematic tissue in a growing plant can be to either grow or reproduce (Watkinson and White, 1985), creating a trade-off similar to that seen in animals. Fitness trade-offs have been found even at the level of protein structure, where structural configurations that favor improved catalytic function may reduce molecular stability (Shoichet et al., 1995).

Although the term "trade-off" has a modern ring, the concept traces back to the old concept of "material compensation," or the idea that "in order to spend on one side, nature is forced to economise on the other side" (Darwin, 1859 [1872], p. 110, after Goethe). Darwin (1868b [1875b]) gave many examples of novelties due to negatively correlated changes under the heading of "Compensation of Growth, or Balancement." An extensive treatment of this subject, written in 1857, appears in the posthumously published draft that was the basis for the *Origin of Species*, the so-called "big species book" (Darwin, 1975, pp. 304ff). Darwin (1859 [1872], p. 110) also discussed the problem of distinguishing between proximate and evolutionary causation in analyses of allocation trade-offs, a problem that has since been approached with some ingenious experimental and comparative methods (e.g., see Strauss, 1984; Atchley and Hall, 1991; Ketterson and Nolan, 1992; Sinervo, 1990; Mole and Zera, 1993).

Not every fitness trade-off is based on a *developmental trade-off*—a negative correlation between traits due to their dependence on the same developmental mechanism, process, or limited bodily resource. Predation-induced accelerated hatching in frogs (Warkentin, 1995), for example, is a fitness trade-off that is not based on antagonistic developmental processes. The presence of a predator causes precocious hatching of eggs by accelerating a developmental switch. This potentially increases fitness because it allows hatchlings to drop into the water and escape from aerial predators such as snakes. But there is a fitness trade-off because precocious hatching compromises ability to survive in the water, where aquatic predators more easily capture the unusually small, precociously hatched tadpoles (Warkentin, 1995).

By stretching this definition of a developmental trade-off, most traits could be considered to involve a trade-off with every other trait, since the development or performance of every trait involves some cost in terms of time, material, or energy and therefore may detract from others. So one could organize a broad theory of adaptive evolutionary change around trade-offs, as Gould (1977) did by associating r- and K-selection trade-offs with heterochrony.

Here I discuss trade-offs that have a developmental basis. Developmental trade-offs seem to be primarily of two kinds.

1. *Resource-competition antagonisms*, characterized by internal competition for limited materials or energy for different developmental processes or traits. Such antagonisms would influence all simultaneously developing traits, but they would be most obvious for especially costly traits. An example is the competition between reproductive activities such as egg production and somatic activities such as growth or dispersal.

2. *Temporal-interference antagonisms*, in which shared mechanisms or morphological equipment used for two activities means that the activities cannot occur at the same time (Holling, 1959). They therefore require "time sharing" (McFarland, 1974, 1983) or asynchronous performance. Speaking and swallowing, or approach and escape involve temporal interference in that both cannot be performed at the same time.

Internal resource competition is analogous to ecological "scramble" or "consumptive" competition (Schoener, 1983), where the fastest to acquire or consume a resource deprives others of it. Simultaneously growing or adjacent tissues are examples whose resource-competitive nature may be revealed if the two activities are prioritized when resources are scarce, for example, in the preferential oxygenation of the brain versus the extremities under metabolic stress in the cold. Temporal-interference antagonisms and time sharing, on the other hand, resemble "interference" (Park, 1962) or "preemptive" (Schoener, 1983) competition, where the simple presence of one contestant blocks access of another. These distinctions are not always clear-cut, and I present them only to organize the discussion of examples.

Hormones are sometimes seen as the developmental cause of trade-offs, for they often affect different processes simultaneously in opposite directions. In vertebrates, for example, steroid hormones increase aggressiveness and decrease parental behavior (see chapter 15), and prolactin production, which is high in a nursing mother, inhibits ovulation and further reproduction. These processes are fundamentally different from the other two developmental antagonisms listed above, however, in that they represent different independently evolved responses to the same signal, not a trade-off where the negatively correlated variables depend upon each other. Negative correlations mediated by a third factor (such as a signal) can independently be eliminated by selection for change in their relation to the third factor, such as change in signal-response relations. The "challenge hypothesis" of behavioral regulation (Wingfield et al., 1990), for example, suggests that compartmentalization of responses has alleviated the antagonistic effects of steriods on aggressiveness and parental behavior: androgen levels rise in parental individuals only during the "challenge" of mating or aggression. This is borne out in many birds and some lizards (briefly reviewed in Creel et al., 1993). This restriction in the expression of androgen peaks can be interpreted as a device that permits a behavioral response to challenges while circumventing a regulatory conflict with parenting. Some data on mammals indicate that the conflict may be resolved via temporal changes in sensitivity of different behaviors (aggressive, sexual, paternal behavior) to androgens (Creel et al., 1993).

Although hormones may sometimes mediate the level at which a developmental trade-off occurs (see below) and may therefore be mistaken for the cause of a trade-off, response antagonisms like those sometimes seen in response to hormones are perhaps best seen as negative phenotypic correlations due to common control by a third factor. All such third-factor correlations differ from developmental trade-offs in their potential for independent responses to selection, as just explained. Exposure to the sun, for example, can cause both decreased skin whiteness and increased sweating. The negative correlation between skin whiteness and sweating is not due to a developmental antagonisms because the two variables do not influence each other, but are related only through their common relation to sunlight. Leroi et al. (1994) discovered a pseudo-trade-off of this sort when they attempted to test trade-off predictions in *Drosophila* with artifical selection on starvation resistence, expecting to find a correlated effect on fecundity, since the flies show a negative phenotypic correlation between the two traits. The expected effect on fecundity was not obtained, because the phenotypic correlation is due to dependence of both traits on a third factor—the amount of yeast in the diet—rather than to any direct physiological or developmental interaction.

The developmental basis of a trade-off may go unnoticed when attention centers on adaptive significance to the exclusion of mechanisms. Nonetheless, many authors acknowledge that internal allocation mechanisms underlie some fitness trade-offs, especially in life-history traits, and discuss the difficulties of demonstrating this (see especially Roff, 1992; Stearns, 1992; Ketterson and Nolan, 1994; Zera, 1999).

The remainder of this chapter discusses examples in which developmental trade-offs may contribute to the origin and evolution of fitness trade-offs. A focus on developmental aspects should not be interpreted, however, as an argument against the importance of selection. Selection is expected to adjust the parameters that govern the level at which the trade-off will occur, that is, which of two negatively correlated variables will predominate, and to what degree. And selection can change developmental parameters, and with them the correlation structure of the phenotype (Riska, 1986). Developmentally mediated correlations can be adjusted and even decoupled under selection (see the section on allometry, below, and chapter 10, on duplication and concerted evolution). Individuals subject to developmental trade-offs can sometimes facultatively alter their allocation ratios under extreme conditions (Mrosovsky, 1990), and such compensatory responses differ among species, indicating a genetic, evolved component in the organization of trade-offs.

The widespread trade-off between growth and reproduction, while it may be governed by internal resource competition, is well known to be adjustable. When two species of freshwater snails, for example, were starved by diluting their nutrients by 75% with indigestible cellulose, one of them increased growth more than reproduction (90% vs. 59% compared to controls), whereas the other did the opposite, disproportionately increasing reproduction (116% compared to 38% for growth; Rollo and Hawryluk, 1988). A large literature on the adaptive evolution of parental investment syndromes (e.g., see Clutton-Brock and Godfray, 1991) is testimony to the power of selection to adjust parameters of an offspring-size versus offspring-numbers trade-off that is widely acknowledged to involve resource allocation antagonism that has profoundly affected the evolution of male and female reproductive physiology.

Internal Resource-Competition Antagonisms: Material Compensation and Allometry

Trade-offs due to internal resource-competition antagonisms are those in which two or more simultaneously developing body parts or processes have their growth or development limited by a limited supply of resources required by both. Those that affect energy budgets and biochemical processes are studied by physiologists. Those that affect growth and form are most often studied by developmental biologists and morphologists. Those that affect life histories are studied by ecologists. This is an enormous and important class of phenomena for understanding the internal origins of variation in adaptive syndromes. The best I can do here is give a few examples that show the parallels, in terms of underlying cause, among adaptive syndromes seen in these different fields.

Material compensation, sometimes called "regulation" (Goldschmidt, 1940), compensation of body material, or "pleiotropism in a wider sense" (Rensch, 1960), describes a negative correlation in the size of body parts presumed to reflect dependence on the same resources for their growth. Extended to energy- or material-requiring activities other than growth, it results in negative correlation in degree of development or rate of performance of different organismic activities or processes. Because this seemed to indicate a reciprocally "balanced" relation between parts, this phenomenon was at first described as governed by a "loi de balance" (Geoffroy St. Hilaire, 1822, cited by Darwin, 1868b [1875b]; Rensch, 1960). Even earlier in the nineteenth century, Goethe considered body form to be the result of trade-offs among investments of matter and force in different body parts, with the elongated bodies of snakes ("gleichsam unendlich"), for example, due to the fact that they do not need to invest in extremities (Nordenskiold, 1928).

A large number of examples of negative correlation thought to result from material compensation are listed in reviews by Rensch (1960) and Matsuda (1979). They include enlargement of legs with wing loss in insects, enlargement of hind legs with reduced lumbar ribs in vertebrates, increased ovarian development with wing reduction or loss in insects, and increase in size of canine teeth accompanied by a loss of premolars in some mammalian dentitions. Many such examples continue to be described, such as that discovered in thrips (*Oncothrips tepperi*) in which short-winged nymphal (immature) morphs have larger forelegs,

stouter fore tibiae, and stouter front tarsal claws than do the long-winged morphs (Crespi, 1992a).

Material compensation is sometimes treated as an unexplained phenomenon and therefore one of uncertain significance (see Cock, 1966, p. 173). Here I define *material compensation* as a negative correlation among parts that is due to resource competition during simultaneous or interactive growth. This is an operational definition that is open to experimental examination.

Holometabolous insect larvae just prior to metamorphosis to the pupal stage are especially appropriate for experimental tests of the resource-competition hypothesis as they cannot respond to changing resource consumption with altered intake (Nijhout and Wheeler, 1996; Nijhout and Emlen, 1998). They are an energetically closed system, for they do not feed, and undergo a striking metamorphosis fueled entirely by a limited supply of resources sequestered during earlier larval stages. Furthermore, these insects have identifiable competing growth centers in the form of imaginal discs, set aside in early larvae and activated in the pupal stage. Each disc differentiates into a circumscribed body region. The imaginal discs grow explosively in pupae and are sufficiently discrete that they can be selectively extirpated.

Models of competitive growth predict that imaginal structures compete for larval reserves, and, when resources are limiting, diminished growth of one trait should be compensated by excessive growth in another (Nijhout and Wheeler, 1996). These predictions held in tests using butterfly wings and beetle horns (Nijhout and Emlen, 1998). When one or two hind-wing discs are removed from butterflies early in the final larval instar, the forewings and some other thoracic (forelegs, overall thorax size) of metamorphosed adults were disproportionately large for their body size. Features of the head and abdomen, being more widely separated during development, were unchanged. This latter finding is reminiscent of a rule, attributed to Karl Pearson (in a 1900 article with Whitely) by Rensch (1960, p. 179), which states that the neighboring parts of an organism are more closely correlated than are more distant parts.

A different kind of test of the resource-competition hypothesis was performed using beetles (*Onthophagus taurus* and *O. acuminatus*; Coleoptera: Scarabaeidae) that are dimorphic for possession of horns on the head. Large males are horned, and small males are hornless. Intermediate-sized beetle larvae were manipulated with juvenile hormone to raise the body-size threshold for horn production (Emlen and Nijhout, 1999). This caused them to

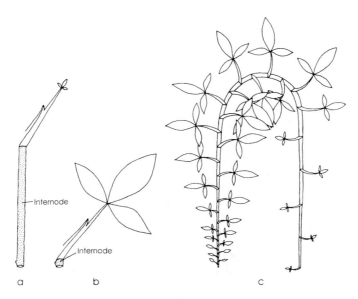

Fig. 16.4. Trade-off between growth of internode and leaf within segments in vines (Araceae): (a) a structural segment with long internode and small leaf; (b) the other extreme, a large shade leaf with short internode; (c) variation in leaf-internode proportions during the ontogeny of an individual plant (*Syngonium standleyanum*), with climbing stem on the left, descending stem on the right. A strong correlation between the diameter of the internode and the weight of the entire segment results in a trade-off between size of leaf and internode; total dry weight of a segment of a given diameter is fixed. After Ray (1986).

invest less than usual for their body size in the development of horns. The result was a compensatory increase in the size of the nearby compound eyes (Nijhout and Emlen, 1998).

As in the butterflies, the strongest compensatory response was in a structure closely adjacent to the one manipulated, suggesting material compensation, but the possibility that the hormone acts independently on the two structures to produce the correlation needs to be tested. In beetles with thoracic rather than head horns, there is a negative relationship between horn size and wing size—both thoracic structures (Kawano, 1995, cited in Emlen and Nijhout, 2000). Artificial selection for increased head-horn size relative to body size, and for decreased relative horn size, produces, as a correlated effect, the opposite direction of change in the size of the compound eye (Nijhout and Emlen, 1998). All of these results are consistent with the hypothesis that the trade-off observed is based on developmental resource-competition antagonism. The possibility of third-factor (hormonal) control of the correlation is not yet eliminated. But the understanding of mechanism in beetles is approaching a level where further evolutionary hypotheses can be considered using comparative study. The possiblity might be examined, for example, that the

observed trade-offs began as as material compensation and then, under selection to regulate investment ratios, competitive growth came to be manipulated by hormones to which the sensitivity of competing tissues could be readily adjusted under selection.

Comparative studies of plant morphology reveal negative phenotypic correlations that could conceivably be produced by resource-competition trade-offs. In palms (*Geonoma* species), for example, single ("solitary") stems are associated with thicker axes, and multiple stems (branched axes) are associated with thinner axes (Chazdon, 1991). Resource competition between simultaneously growing stems could explain the negative correlation, but so could selection to bolster the strength of single stems have produced some other mechanism that facultatively causes greater thickening of single stems. In vines (Araceae), growth patterns suggest that there is a trade-off between making a large leaf or a long internode that could be mediated by resource competition between parts. The total dry weight of a segment is fixed, and apportionment between leaf and stem can vary widely, from allocation of 95% dry weight to a large leaf, to allocation of 98% to a long internode (Ray, 1986; figure 16.4). The allocation ratio can change dramatically during individual ontogeny as well as among species, and this profoundly influences the form of the plant (figure 16.4c; see also figure 6.3).

The fixed weight of segments suggests a developmental rule that limits the growth resources per segment, a circumstance that could impose resource competition between leaf and stem. But leaf and stem develop sequentially, so noncompetitive mechanisms of allocation are conceivable, such as a condition-sensitive allocation to stem, followed by leaf development using whatever remains of the fixed resource ration per segment. Such a mechanism, if it exists, could be construed to have evolved via a manipulation of internal resource competition (material compensation) that prioritized resource access of one tissue or the other. The possibility of prioritized trade-offs in plants is discussed in connection with the currently debated "growth-differentiation balance" hypothesis of plant resource allocation (e.g., see Tuomi, 1991; Lerdau et al., 1994). Resource allocation adjustments under selection are suggested by the highly variable relationship between resource availability, growth, and defensive-compound production in different species of plants (see Lerdau et al., 1994, p. 59). The literature on this controversy contains many examples of investment trade-offs where developmental antagonisms may conceivably occur. Such examples are a reminder of the importance of research on mechanisms for understanding precisely which aspects of the phenotype are affected by selection and how they may change during evolution to achieve a particular adaptive pattern of growth.

It is sometimes difficult to distinguish between internal resource competition, regulatory antagonism, and third-factor regulation as causes of negative correlations in morphology. This is illustrated by a study of the carnivore and omnivore morphs of spadefoot toad tadpoles (*Scaphiopus multiplicatus*; Pfennig, 1992a,b) (now called *Spea multiplicata*—D. Pfennig, personal communication). Carnivorous morphs have diet-induced enlarged jaw (orbitohyoideus) muscles and short intestines compared with omnivore morphs. In these toads, muscle use (processing harder and larger prey) could conceivably cause hypertrophy of jaw muscles independently of the shortened gut length, which is known to be alterable by diet in anurans (citations in Pfennig, 1992a, p. 172). Experimental administration of thyroxine with the food, however, induces both the muscle hypertrophy and the correlated decrease in gut length without change in ingested food type (Pfennig, 1992a), showing that the correlation is not due to independent direct responses to diet as a third factor. But an *indirect*

third-factor regulation could still be involved, if the diet contains or induces a growth regulator such as thyroxine, which then affects both traits independent of each other (Pfennig, 1992a, notes that tadpole prey contain thyroid hormone and its constituent iodine). Alternatively, the negative size relation between the two organs could be due to internal resource competition between them (material compensation). Deuchar (1975) and Buss (1987) discuss various possible examples of competitive growth during early development, but the examples need to be subjected to experimental tests to eliminate alternative explanations.

Confirmation of material compensation requires some evidence that energy or material consumed by growth or performance of one trait diminishes the level available to the other. Material compensation is thrown in doubt even in an energetically closed system if the negatively correlated traits are sequentially developed (Mole and Zera, 1993), as in the stems and leaves of vines discussed above, or if the correlated traits are subject to influence by the same hormone(s), as in the beetles, butterflies, and toads, raising the possibility of correlation due to control by a third factor.

Gould (1977) discusses intraspecific alternative phenotypes with contrasting life-history tactics. Many originate via heterochrony and involve trade-offs between precocious gonad development and elaborate adult morphology, trade-offs that may be a result of material or energy-investment antagonism when either resources or growing time is limited. Following this line of reasoning, Matsuda (1979) sees even metamorphosis (larva–pupa–adult) as involving material compensation—the reorganization of material resources during metamorphosis to construct morphs having the same size but with exaggeration of different structures. Others have depicted metamorphosis as a change in the investment of limited resources, as in this colorful statement by Carrol Williams (1958):

> The earth-bound early stages built enormous digestive tracts and hauled them around on caterpillar treads. Later in the life-history these assets could be liquidated and reinvested in the construction of an essentially new organism—a flying-machine devoted to sex. (cited by Angelo and Slansky, 1984)

But one cannot conclude, without considering alternatives, that the trade-offs involved are a direct result of resource competition (material compensation), as assumed by Matsuda (1979, 1987) in discussing this and other negative correlations such as those between wing and ovarian development in insects. Matsuda's (1987) critique of neo-Darwinism as a doctrine that neglects development in favor of selection is weakened by this assumption.

Allometry is a disproportionately large or small change in a body part or activity relative to change in another or in the dimensions of the body as a whole. An allometric change contrasts with an isometric, or proportional, change. Allometry implies disproportionate investment, or allocation of resources, to particular parts or activities during development. An allometric curve reflects an investment ratio for one trait, such as an appendage or a horn, relative to another, such as body length. If the curve changes in slope during the course of development, this means, in effect, that a resource allocation or investment decision has been made. If resources for growth are limited, as they must usually be, change in one allometric ratio may affect another, where the affected structures compete for internal resources during development. So allometries may be affected by internal resource competition and material compensation. Not surprisingly, allometries are sometimes discussed along with life-history traits as subject to fitness trade-offs during their evolution (e.g., see Gould's [1977] discussion of heterochrony).

A resource-based explanation for allometric relations, involving limited growth resources, was indicated by Huxley (1932 [1972]) in his discussion of heterogony (allometric growth):

> [I]t is not necessarily the actual rate of growth which is regulated in accordance with our formula, but the *limitation of the total amount of growth achieved*. . . . [A]t any given body-size the total amount of material which can be incorporated in the organ is proportional to the body-size raised to a power. . . . The mechanism of this relation is at present obscure . . . experiment alone can decide the point. (p. 49; italics in the original)

Huxley (1932 [1972], p. 80) gave an impressive illustration of the resource-dependent nature of allometry, showing the reduction in the degree of exaggeration of the male dorsal crest of a male newt (*Triton cristatus*) under starvation.

There is some support for the idea that internal resource competition affects the evolution of adaptive allometries in holometabolous insects, where marked exaggeration of one structure is sometimes associated with marked reduction in another, as already discussed. In addition to the examples examined experimentally in butterflies and beetles, so-cial insects show many negative allometries that may have similar explanations. In army ants (*Eciton*) and leaf-cutter ants (*Atta*), there is a spectacular hypertrophy of the head in some individuals. As predicted by the resource-competition hypothesis, as head size increases relative to body size, there is a corresponding decrease in the slope of allometry of leg length with body size, and this is more exaggerated in army ants (*Eciton*), where head size is proportionately larger, than in *Atta*, where head hypertrophy is less (Wheeler, 1991). All of these allometries may be influenced, as already discussed, by a third factor, such as a hormone, that could affect both overall growth rate and allocation to particular traits.

In vertebrates, and in arthropods with incomplete metamorphosis, development is *relatively* continuous and more open to resource acquisition during growth, but not completely so. The semicompartmentalization of growth stages, by such phenomena as molting (in birds and arthropods) and growth spurts (in mammals; see Atchley et al., 1985, 1990), may influence the correlations between different structures, since structures growing during a particular phase are more likely to compete with each other for growth resources, as well as formatively interacting in other ways as described by Atchley and Hall (1991). Different correlation structures are known to occur at different ages, and correlations can even change sign, as in the correlations between certain mandibular dimensions and body size at certain stages in some vertebrates (Atchley et al., 1985, 1990). But the contribution, if any, of internal resource competition to the negative correlations observed is not to my knowledge demonstrated, although it has been listed as a potentially important factor (Cowley and Atchley, 1990).

Since artificially selected changes in fatness have marked effects on correlation structure at different ages (Atchley et al., 1990), it may be of interest to see if lean strains, or individuals poor in growth resources, show competitive effects in growth of different structures growing at the same time. Hanken and Wake (1993) review the sometimes spectacular correlated effects of size reduction that have occurred independently in various miniaturized lineages of animals. Some of the morphological novelties that result are clearly due to correlated effects that could be due to resource limitation and competitive growth of parts. In reptiles and amphibians, for example, there is a negative allometry between inner-ear size and head and body size. In extremely small heads of miniaturized species, the inner ear is so enormously enlarged relative to other structures that there is a drastic rearrangement of the skull, especially the jaw. Since many morphogenetic mechanisms are size dependent, reduced size can lead to radical structural rearrangements with no change in developmental mechanisms themselves (Hanken and Wake, 1993). Size-correlated changes are associated with the origin of important novelties, including some major bauplan changes (e.g., in early reptiles, snakes, lizards, and bivalves—reviewed by Hanken and Wake, 1993, p. 508). Are such dramatic evolutionary changes virtually automatic given ancestral allometries? This could be expected if negative allometries have as their basis a conserved and seemingly simple mechanism like resource competition during growth. It is of some interest to examine the resource-competition hypothesis and alternative explanations, such as common influence by a third factor (e.g., a hormone).

Allometry is sometimes considered a conservative force in evolution, one that constrains or limits body proportions and produces certain morphological trends (e.g., see Hersh, 1934; Rensch, 1960; Simpson, 1944), even to the extreme of proposing that extinction has been caused by the rigidity of allometries in horn growth and jaw growth leading to difficult births and poor dental occlusion. The assumption of fixity of allometric relationships leads to the view that phenotypic novelties or exaggerated traits may be consequences of altered body size independent of selection, even producing maladaptive exaggeration (see references in Gould, 1974, 1977; Levinton, 1988, p. 314). But there is abundant evidence that allometry itself is an evolutionarily labile trait that is subject to change under selection if an allometric relationship were to prove maladaptive. Altered allometries have been produced under artificial selection (e.g., see Wilkinson, 1993), and allometries clearly evolve in nature: variants having altered allometries or relative-growth investment ratios ("allometric acceleration"—McKinney and McNamara, 1991, p. 266) sometimes have been favored during the evolution of extreme polymorphism in social insects (see Wheeler, 1991), in fighting male beetles (Eberhard and Gutierrez, 1991), and during phyletic increase in the size of mammalian (titanothere) horns (McKinney and Schoch, 1985; see McKinney and McNamara, 1991).

In ants, where the developmental basis of altered allometry has been investigated, one can examine the diverging effects of three kinds of size-related developmental change: first, a reprogramming of size-at-metamorphosis against a background of isometric head growth (e.g., in *Solenopsis invicta*),

which produces two size classes of workers (majors and minors); second, a similar reprogramming of size at metamorphosis, but against a background of allometric head growth (in *S. geminata*), producing a class of macrocephalic soldiers; and a third class of species (e.g., *Camponotus festinatus*) where growth parameters (allometric relations) are altered to produce a "diphasic allometry," with workers and soldiers differing in proportions at comparable size (reviewed by Wheeler, 1991). A "triphasic allometry" was described in *Oecophylla smaragdina* by Wilson (1953), but it "appears to be uncommon" and the developmental basis of the intermediate third phase "cannot be assessed without further study" (Wheeler, 1991, p. 1228). It is possible that this represents not a third allometric phase but true intermediates near the switch point of a diphasic allometry, where (as in Wilson's graph) individuals of intermediate size fall into neither the small- nor the large-female distribution. Similar distributions of intermediates occur in other dimorphic species such as the beetles *Onthophagus incensus* (Eberhard and Gutierrez, 1991), *Podischnus agenor* (Eberhard, 1980), and *Onthophagus acuminatus* (Emlen, 1996). Another complicating factor is the fact that the inflection point of a diphasic allometry is plastic and can shift with diet (Emlen, 1997b).

These examples are important because they show that allometry should not be regarded as a synonym of constraint, or limitation on the evolution of form, with size representing the only evolutionarily labile parameter to effect change. Although this may appear to be true in some taxa (e.g., see Gould, 1989a, on *Cerion* snails), it may be only temporarily true. For an interesting history and refutation of the idea of allometry as a constraint, see Cock (1966), who gives examples of altered allometry in plants and animals based on small (singe-allele) genetic differences, and Eberhard and Gutierrez (1991) on beetle horns. Frazzetta (1975) emphasizes the importance of allometry in the evolution of diversity, as well as the importance of altered allometries.

What happens, genetically and developmentally, to change an allometry? This question is analyzed in models by Nijhout and Wheeler (1996) for the allometries of holometabolous insects such as butterflies, beetles, and bees, having complete metamorphosis—a discrete larval, pupal, and adult stage. This is a favorable potential setting for competitive growth because the prepupa and pupa constitute a relatively closed system where resources stored during the larval stage are apportioned to different structures. The imaginal discs are specialized tissues set aside in the embryo that give rise to different structures of the adult insect (e.g., a particular set of appendages, head structures, genitalia, or portions of the thorax and abdomen). The models indicate that allometries are influenced by three parameters: body size at time of metamorphosis, which defines the amount of resources available for growth; initial size, growth rate, and growth duration of potentially competing imaginal discs; and hormonal or environmental factors that alter the first two parameters, affecting the timing of pupation and/or the growth responses of cells during metamorphosis. Simulations based on these resource-competition models explain common patterns of allometry seen, for example, in beetles (Eberhard and Gutierrez, 1991) and social insects (Wilson, 1953; Wheeler, 1991). For example, they show that a discontinuity in some feature between morphs, without a change in allometric slope, can be caused by a change in the critical size for metamorphosis; and that parallel or intersecting allometries differ due to differences in growth constants of different imaginal discs. These models set the stage for a comparative, mechanistic study of allometry as competitive growth. "A comparative understanding of the control of growth and development of imaginal disks remains one of the great gaps in our understanding of insect development" (Nijhout, 1994a, p. 84).

How often does the hypertrophy of an appendage or organ provoke, or even require, the reduction of some other organ or activity? To cite an almost grotesquely exaggerated example of morphological hypertrophy, minor workers of the weaver ant Camponotus saundersi have mandibular glands that show such spectacular enlargement that, rather than being confined to one small region of the head as in most ants, they occupy virtually the entire head, thorax, and abdomen. These kamikaze workers explode when provoked, splattering their enemies with a sticky mandibular-gland secretion (Maschwitz and Maschwitz, 1974, their figure 2). What, if anything, is compensatorily sacrificed in the anatomy and behavior of such females during their development? Comparison with minor workers in related species, and a detailed comparative study of mandibular-gland ontogeny, may tell. Clues may lie in the relationships among imaginal discs that compete for limited resources during development in pupae (Nijhout and Wheeler, 1996). The specialized second-instar soldiers of some aphids are known not to feed or reproduce. But why, or by what kind of a developmental transition, have these functions been lost? It could be assumed that unused characters are lost due to se-

lection on that trait alone, to reduce its size or complexity in the interests of overall economic use of developmental resources. But internal resource competition—loss of particular structures due to the hypertrophy of others dependent on the same growth resources—is another, alternative hypothesis. The many unexplained examples of apparent material compensation (e.g., in Matsuda, 1979, 1987) are a menu for research on the developmental basis of exaggerated traits in evolution.

Developmental Antagonisms and Life-History Evolution

Investment decisions in the expression and evolution of life-history patterns often depend on materially competing developmental processes, especially those that affect investment in growth versus reproduction (Matsuda, 1979; Charnov, 1982; Partridge and Sibly, 1991; Stearns, 1992; Roff, 1992; Finch and Rose, 1995; Sinervo and Svensson, 1998). Such trade-offs are known to occur in a wide variety of organisms, including barnacles, fishes, and polychaete worms (reviewed by Berglund, 1991; reviewed in plants by Roff, 1992; Stearns, 1992; Charnov, 1982). The fitness trade-off between reproduction and growth is evidently an ancient and recurrent feature of evolution. It characterizes the gamete dimorphism of the vast majority of sexually reproducing organisms, where the trade-off between somatic growth (investment of resources in the egg) and purely genetic reproduction (investment in large number of gametes, or sperm) presumably has led to the phenotypic divergence between the sexes (see Parker et al., 1972; West-Eberhard, 1979). Cavalier-Smith (1980, p. 49) argues that the trade-off between reproduction and growth is a universal characteristic of life, having its origin in unicellular organisms and is even inescapable for viruses, which must divide their resources and activities between rapid intracellular replication and dispersal and invasion of hosts. In this section, the issue is to examine the extent to which life-history trade-offs may be influenced by developmental antagonisms.

Mole and Zera (1993; see also Zera and Denno, 1997) studied the developmental basis of a life-history trade-off in a wing-dimorphic cricket (Gryllus rubens). They determined that the short-winged, flightless, large-ovaried morph and the long-winged, flying, small-ovaried morph of the cricket G. rubens are equivalent in their food-intake and digestive efficiency. During the first two weeks of adulthood, the flight muscles, already dimorphic (vestigial in the short-winged morph), do not change, but the ovaries of the two morphs develop divergently. The ovaries of the flightless morph attain an average mass 1.6 times that of the long-winged morph, whose flight-muscle mass averages 1.8 times that of the flightless morph. Based on these findings, Mole and Zera (see further support in Zera and Denno, 1997) hypothesize that if there is a direct energetic trade-off, it is between flight-muscle *maintenance* and egg production in the adult stage. Mole and Zera considered this "the first evidence for a life-history trade-off that has a physiological basis which is limited to the allocation of acquired and assimilated nutrients within the organism" (p. 121).

The dispersal-reproduction trade-off in insects is mediated by hormones, whose role can be regarded as setting the level of allocation to each activity. Hormonal influence in a cricketlike dispersal dimorphism in grasshoppers (migratory locusts) has long been known (e.g., Kennedy, 1961). In crickets, juvenile hormone (JH) negatively affects muscle and wing development but stimulates ovarian development (see references in Mole and Zera, 1993, p. 125), a pattern seen in numerous other insects (Matsuda, 1987, p. 21). JH levels are also known to differ during the nymphal stage of the two morphs (Zera and Tiebel, 1989; Roff and Fairbairn, 1991), raising the possibility that JH-modulated resource competition, occurs during juvenile stages (e.g., between development of wing-muscle primordia and fat body development). The Mole and Zera (1993) study is important progress toward understanding the developmental, and therefore the evolutionary, basis for the common "oogenesis-flight syndrome," about which so much has been written in the literature on insect dispersal polymorphisms and life-history strategies (reviewed in Gould, 1977; Harrison, 1980; Matsuda, 1987; Roff, 1984, 1986a,b, 1992; Rankin and Burchsted, 1992; Zera and Denno, 1997). JH is not the only hormone involved in regulation of this polymorphism in crickets and probably other insects. Ecdysteroids have also been implicated (Zera and Bottsford, 2001).

Hormones are a major regulatory intermediary for life-history trade-offs (see Finch and Rose, 1995). But life-history studies of fitness trade-offs seldom mention endocrinology. "The dearth of hormonal studies is surprising because hormones can produce correlated effects with potentially antagonistic fitness consequences, precisely the situation we seek when attempting to demonstrate the existence of trade-offs" (Ketterson and Nolan, 1992, p. 37). The trade-off between egg number

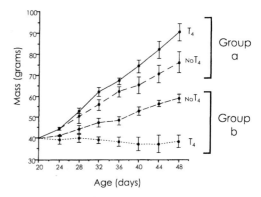

Fig. 16.5. Experimental evidence for a resource-allocation trade-off in cotton rats (*Sigmodon hispidus*). With free access to food (group a) treatment with thyroid hormones (T_4), which increases basal metabolic rate and therefore energy expenditure in bodily maintenance, does not completely inhibit growth. With restricted access to food (group b), thyroid treatment inhibits growth. This suggests that there is a resource-allocation trade-off between growth and maintenance. It is only evident when food resources are limited. After Ketterson and Nolan (1994); based on Derting (1989).

and size examined by Sinervo (1990; see above), for example, can be experimentally manipulated using follicle-stimulating hormone (Sinervo and Licht, 1991a,b). Similarly, thyroid hormone mediates a trade-off between growth and maintenance (metabolic rate) in cotton rats (Derting, 1989; figure 16.5). But, as emphasized by Zera and Denno (1997), documenting hormone involvement does not completely explain the physiological basis of a developmental trade-off, for their are various conceivable ways that an apparent trade-off could be affected. A hormone, for example, could have opposite (e.g., inhibitory vs. stimulatory) effects on two processes not in any other way connected with each other; or it could stimulate (or inhibit) only one, which competes with another for nutrients; or it could simultaneously stimulate two processes that then compete for resources.

Wing-muscle loss and ovarian hypertrophy are often accompanied by hypertrophy of appendages (see the many examples in Matsuda, 1987; see also Rankin and Burchsted, 1992). Evidence for a resource-competition basis was provided by Tanaka (1976): when hind wings of macropterous females in a wing dimorphic cricket (*Pteronemobius taprobanensis*) were removed immediately after emergence, the ovaries developed more quickly, as they do in the micropterous morph, which also lays more eggs. Related to these observations is the fact that the winged queens of ants lose their wings as soon as they have dispersed and have entered or excavated the nest where they will lay eggs. Wing muscles are resorbed, and this is presumably an important source of resources for egg production in the so-called claustral, or solitary, queens that reproduce for a time without workers to provide them with food (Hölldobler and Wilson, 1990). If there is a direct developmental antagonism between wings (or wing-muscle maintenance) and ovaries, as suggested (but not demonstrated) by the commonness of the oogenisis-flight syndrome, then the developmental trade-off may explain the evolution of wing-shedding and muscle resorption in reproductive ants.

Matsuda (1987, p. 28) reviewed the reallocation processes within organisms associated with a life-history specialization called "embryonization" (growth of larvae within eggs), representing an evolutionary change from production of many small to fewer large eggs. In *Hydra*, embryonization leads to the production of a single large egg via adsorption of neighboring oocytes. In the archiannelid polychaete *Diurodrilus*, it occurs via oocyte fusion. In these two organisms, evolutionary change in the investment trade-off between many-small and few-large eggs occurs via a mechanism that directly affects the organization of egg production within the gonads. In some other organisms, a similar change involves change in the allocation of resources between gonads and other tissues (see Sinervo, 1990).

The different pathways to embryonization are a reminder that even though the reproduction versus growth trade-off is common in nature, it cannot be assumed to be an automatic consequence of particular kinds of developmental antagonisms. Fitness trade-offs probably evolve due to many kinds of negative pleiotropy, and they can be modified under selection, so the evolutionary history of trade-offs may traverse paths as twisted as those of morphology. In guppies (*Poecilia reticula*), for example, females experimentally prevented from reproducing did not grow larger, but instead developed fat reserves that could be used in future reproduction (Reznick, 1983)—resource limitation did not provoke a trade-off. In wing-dimorphic gerrid bugs (*Gerris* species), total fecundity between the two wing morphs did not differ, but the flightless females had shorter preoviposition periods, leading to greater egg production during the first 20 days of reproduction—a trade-off occurs, but rather than being between dispersal and fecundity it is between dispersal and speed of reproduction. In higher plants, growth and reproduction are meristematic, and if a meristem produces a flower, it cannot undergo further vegetative growth, indicating a developmental antagonism between reproduction and growth. The reciprocal relationship does not always apply, however, for an early commitment to vegetative growth can lead to increased fecundity later on (discussed by Roff, 1992, p. 157). Other phenotypic correlations follow the common pattern of a reciprocally negative relation between growth and reproduction and may be presumed due to resource competition during development, yet may prove to have a very different cause: in barnacles (*Balanus balanoides*), weight declines during the reproductive period, but this is in part due to a failure to feed (Barnes, 1962, cited by Roff, 1992, p. 151). This is a reminder that a common type of fitness trade-off may have an uncommon type of developmental cause.

The principal challenge for a developmental evolutionary approach to life-history trade-offs is to understand the relationship between widespread sources of developmental antagonism, and how selection operates on developmental mechanisms to adaptively modify them. Sinervo (1990) invented an experimental approach to distinguish between the ancestral developmental groundplan and its adaptive modification. He mimicked population differences in egg size in the lizard *Sceloporus occidentalis* by experimentally removing egg yolk, and thus determined the degree to which yolk size itself affects incubation time, hatchling size, growth rate, and behavior (sprint or running speed) and to what degree population divergence in these features is likely due to evolved genetic differentiation between the populations in the physiology of growth. Manipulations within clutches showed that there is both a maternal effect (of yolk-size endowment) on a size versus growth-rate trade-off, and a genotypic (variable among families) component. The genotypic component may reflect heritable differences in growth rate correlated with thermal physiology (preferred body temperature and growth rate are positively correlated). This indicates a pathway by which selection could alter growth rate independent of a fundamental developmental relation between size (yolk endowment) and growth rate. Even though offspring size is correlated with "adaptive" behavior and life-history traits in a very great diversity of animals, including fish, amphibians, reptiles, mammals, and insects (partially reviewed by Sinervo, 1990, p. 291), it is possible to begin to test the degree to which these correlations are ancestral or specially evolved in an observed adaptive context.

Reznick (1983) showed that allocation to reproduction vs. growth in guppies is subject to genetic variation and evolution. Populations of guppies from different localities have different ratios of allocation to reproduction and growth even though their total energy budget (somatic plus reproductive) differ little among populations. Reznik's work shows how this question can be examined via comparative study of populations varying in the relevant parameters. One of the mechanisms by which the developmental trade-off between growth and reproduction is altered during evolution is via heterochrony. This was noted by Gould (1977) who went on to propose that patterns of adaptation in life-history strategy—especially, the association of resource limitation with low fecundity, large eggs and slow growth ("K selection") versus resource abundance with high fecundity, small propagules and rapid growth ("r selection")—are achieved by evolutionary changes in the timing of development (retarded reproduction and large investment in growth in the first case, and precocious maturity in the second). So, Gould argued (1977), what was at that time "the most widely discussed topic in studies of life history strategies, the theory of r and K selection, offers particular promise as a framework for understanding the immediate significance of heterochrony in ecological terms" (p. 290). Conversely, this means that what we perceive as contrasting means of "optimizing" life history tactics have their origins in a continuum of developmental trade-offs, where adjustments in the timing of sexual maturity can affect how limited resources are allocated between reproduction and growth.

The point here is to emphasize that common kinds of life-history trade-offs frequently have a basis in internal resource competition that precedes their evolution and serves as a basis for the form that they take. Antagonistic processes of relative growth, energy investment, physiology, and behavior (e.g., inability to do two things at once) may determine the form of many tactics selected for their divergence per se.

The importance of antagonistic evolution in tactics borne by single individuals (though not always expressed in all individuals) has been recognized for social interactions and termed "intergenomic conflict" (Rice and Holland, 1997). But the phenomenon may better be termed intragenomic (since the genes for the performance of both antagonistic tactics are borne, even though not necessarily expressed, by the same individual). It is a phenomenon that characterizes many alternative phenotypes, not just those evolving under social competition.

Other kinds of developmentally influenced life-history trade-offs are possible, such as those between reproduction and longevity (Partridge, 1992) or number and quality of offspring (including number vs. size of eggs). Trade-offs that involve internal resource competition are easily extrapolated to external competition and behavioral decisions (e.g., compare Lessells, 1991, and Milinski and Parker, 1991).

Temporal-Interference Conflicts in the Evolution of Behavior

Temporal interference between mutually exclusive actions has long been recognized by ethologists as an important source of novel patterns of animal behavior. Courtship and aggressive displays often include behaviors symptomatic of internal motivational and possible neural (regulatory interference) conflict over which of two temporally antagonistic activities to pursue, (e.g., between approach and avoidance, attack and flee; see Baerends, 1975). A famous illustration is the courtship behavior of spiders, where the male engages in complicated tense- and tentative-looking movements believed to be based on conflict between sexual approach and defensive fear, since the courted female is a potential predator (Manning, 1979, p. 187). Often the movements symptomatic of temporal interference, such as grooming or trembling, become incorporated into displays. The zig-zag dance of courting sticklebacks, for example, has been interpreted as a product of conflict between aggressiveness and cooperative leading to the nest, and the "zig" (approach) part of the dance sometimes actually provokes an attack, while a "zag" (away) movement may end with the male leading the female toward the nest (Manning, 1979).

In stingless bees (Meliponidae), conflict-derived display is evident in the behavior of workers, many of them developed with ovaries and capable of oviposition, when they participate in the "oviposition ritual" that precedes egg-laying by the dominant queen (summarized in Sakagami, 1982). The queen in some species violently taps with her antennae on the bodies of workers, who insert their bodies into cells; and workers exhibit varied species-typical behaviors that can be interpreted as inhibited aggressiveness toward the queen, including violent darting with forebody raised, forelegs extended, mandibles opened, the base of the tongue exposed, and wings asymmetrically fluttered (in *Scaptotrigona*); darting with peculiar up-and-down shaking of the head (in *Scaura*); or passive sitting over the cell (in *Schwarziana*).

Zucchi (1994) employed a phylogenetic analysis of ritualized dominance signals associated with the oviposition ceremony to show that complex, intensively displayed ritualized signals are more primitive (plesiomorphic) than are simple, more abstract or symbolic traits. This is predicted if the signals originate in, and have as their basis, regulatory conflict, then become simplified as more efficient signals. With apologies for anthropomorphism, Zucchi describes the behavior of the workers as indicating conflict between fear of the aggressive queen and competitive threat behavior.

Zucchi's reticence to use words reflecting "inner conflict" as a basis for observed behaviors is typical of care taken by professional ethologists to avoid anthropomorphisms, a taboo that has been both a healthy caution and a curse. Anti-anthropomorphism has contributed to the neglect and even denial in evolutionary biology of internal mechanisms of behavior, and modern resistance to use of such terms as "motivation," "appetitive behavior," or "drive" (see discussions in Hinde, 1970). Yet the study of motivation, or response potentiation, has prospered in experimental physiological psychology and illuminated the mechanisms of evolved behaviors (e.g., Stellar, 1987; Stellar and Stellar, 1985; Gallistel, 1980). These studies show that hormonal and neural states potentiate certain responses and affect decisions among mutually antagonistic activities. Steroid hormones, for example, prime neural systems to mediate specific behaviors such as courtship and mating (Arnold and Breedlove, 1985).

That animals show "goal-directed" behavior underlain by specific motivations can be shown by behavioral observations, as in studies of social wasps indicating that individuals go in search of particular items (e.g., food, water, pulp or nectar) when they leave the nest on foraging trips, in accord with colony needs (see references in Calderone and Page, 1992, on bees; Jeanne, 1996a, on wasps). The fact that worker bees search for nectar sources using information transferred during the famous bee-dance communication in the hive (von Frisch, 1950) is evidence of highly specific goal-directed behavior in an insect. Zoologists have engaged in such extreme denial of motivation and goal-directed behavior, not to mention animal consciousness and complex intellectual abilities, that until very recently mechanisms for them are not widely sought or even hypothesized (but see Griffin, 1976, 1984; Crook 1980). At present, this is perhaps the greatest conceptual void in evolutionary ethology.

Behavioral fitness trade-offs raise many of the same questions regarding the nature of underlying regulatory conflict as those raised for life-history trade-offs (for critical discussions, see Hinde, 1970; Manning, 1979; Mackintosh, 1974). Bernays and Wcislo (1994) discuss a kind of neural interference or "attention" competition as a component of behavioral trade-offs. They review evidence for neurological limitations in the ability of insects and vertebrates to simultaneously focus attention on two tasks, such as feeding and self-defense against predators. This kind of neural (or sensory) focusing, which they argue may have evolved due to selection for the resulting increase in decision efficiency, sets the stage for a trade-off among breeding, feeding and defense, which must inevitably be evaluated under selection and may affect the tactics adopted in circuitous ways.

Bernays and Wcislo (1994) cite experiments showing that salmon feeding in the presence of a predator make more mistakes in a food selection task than they do if not "distracted" by a predator. As the authors point out, parasitism and predation can be almost instantaneous events, so even brief periods of inattention are significant. Although the cost of competing attention is probably most familiar in vertebrates, it is also clear in observations of insects: social wasps (*Polistes*) guard their brood against parasitoids (Ichneumonidae) via visual alertness at the margin of the nest, and when guards become distracted by other tasks, the parasitoids, which have remarkable stealthy patience, remaining immobile for many minutes, rapidly oviposit in the formerly defended cells (West-Eberhard, 1969).

The sensory-competition or attention hypothesis also receives indirect support from observations showing elaborate mechanisms functioning to close specific pathways when they are not useful. For example, Bernays and Wcislo (1994) describe a shutdown in food-specific sensory devices that occurs in grasshoppers after feeding to satiation. It includes moving away from the stimulus (food); hormone (corpora cardiaca)-mediated closure of the chemoreceptor pores on the tips of the palps used to sense food (Bernays and Mordue, 1973); and a rise in the threshold of responsiveness to particular nutrients, associated with presence of increase of those particular nutrients in the blood (Simpson and Simpson, 1992).

Just as material investment ratios and allometries can change under selection, so can behavioral trade-offs based on temporal interference. Galis (1996a) points out that in lizards, as in the early tetrapods that were the ancestors of terrestrial vertebrates, there is a regulatory antagonism (and fitness trade-off) between running and breathing because both use muscles attached to the ribs, but in antagonistic ways. Birds and mammals have been liberated from this by the evolution of relatively long vertebral transverse processes to which the locomotory muscles are attached, while respiration remains associated with movements of the ribs. This is an example of decoupling, or increased compartmentalization, allowing functional divergence between formerly correlated phenotype subunits.

Quantum Shifts and Environmental Extremes

The factors that initiate correlated shifts can be either internal or external in origin. Internal initiation can begin with a relatively small developmental anomaly, a genotypic or epigenetic quirk, subsequently exaggerated by phenotypic accommodation during ontogeny. Examples of this type include the shifts in vertebrate skull morphology following a small change in growth rate of an associated organ like the eye, or deletions of particular bones. In the teosinte-maize transition discussed in chapter 15, a relatively simple shortening of teosinte branch length may have caused the formation of the complexly distinctive maize cob. It matters little, for explaining the phenotypically complex transition, whether the original anomalies were due to mutation or to environmental effects: the coordinated complexity of the change is due to preexisting phenotypic plasticity. But it is notable that several of the examples discussed in this chapter, such as the altered jaws of fish reared on hard foods, the increased musculature of those reared in fast-flowing streams, and the buttress roots of trees strongly suggest a role for environmental factors in initiating recurrent correlated shifts that could then become established, via genetic accommodation or genetic assimilation, as evolved traits.

A skeptic could argue that the evolved examples discussed here only coincidentally or convergently resemble facultative responses sometimes seen in nature in related species; or that the facultative responses followed the fixed ones in the course of evolution. The question cannot be convincingly settled by a cladogram, because, as discussed in chapter 19 on recurrence, traits known to be facultatively expressed may change state (including between fixed, facultatively expressed, and not expressed at all) faster than can be tracked by a phylogeny based on extant species.

The most convincing evidence for the evolutionary importance of extreme conditions in quantum change comes from taxa where universal, or

very common, physiological responses produce, under extreme conditions, a set of responses that is found as part of a specialized established adaptation in some species of the same taxon, suggesting that the widespread traits were the basis for the specialized one under conditions of prolonged or cyclic extremes. Vertebrate hibernation, for example, may have evolved as an elaboration of starvation responses that are common in vertebrates. Starvation and hibernation share numerous features (reviewed by Mrosovsky, 1990). In starved adults, gonadal hormones, reproduction, and parental solicitude decline; thyroid hormone levels fall, lowering general metabolic rate, so that the individual's characteristic oxygen consumption level relative to body weight is regulated at a lower "set point"; and body temperature is regulated at a lower level. Lowered oxygen consumption and lowered body temperature are both energy-saving responses that together result in a kind of torper that resembles hibernation and may have given rise to it in extreme environments during evolution. Mrosovsky (1990) calls starvation a minor version of hibernation.

A counterpart of these correlated responses to starvation occurs in plants, and similarly may have served as the basis for a specialized adaptive syndrome. Many plants, when subjected to low-resource environments (e.g. places that are dry, flooded, cold, saline, shaded, or with poor soils; Chapin et al., 1993), facultatively grow slowly, a response that involves a hormonally coordinated change in water, nutrient, and carbon balance. They show low rates of tissue turnover and high concentrations of secondary metabolites when compared with individuals from high-resource environments. This "stress resistence syndrome" has become a permanent specialization of some species that live in low-resource environments, a specialization that likely originated as a set of correlated, facultative responses that evolved to fixation via genetic change in the thresholds of expression of the component traits (Chapin et al., 1993). In this example the environmental condition itself becomes a switching cue that coordinates the expression of the component variables as a set. Chapin et al. (1993) review many examples in both plants and animals where correlated responses to extreme conditions may have been genetically accommodated in this way to form regularly produced complex traits.

Environmental extremes may be important in the origin and evolution of fitness trade-offs, since trade-offs are often only evident in extreme environments (Derting, 1989; Hoffman and Parsons,

1991). Then "sustainable scope" (Peterson et al., 1990; see Rollo, 1994) or the limits of energy-consuming physiological processes become evident, and trade-offs appear (Ketterson and Nolan, 1994; figure 16.5). Lessells (1991) gives a good illustration of the resource dependence of trade-offs: in conditions of wealth there may be a positive correlation between house ownership and car ownership, whereas poverty may reveal a negative relationship (trade-off) in resource allocation between owning a house and owning a car. This illustrates not only the resource-dependent nature of trade-offs, but also the distinction between a phenotypic correlation (e.g., between house ownership and car ownership, present in both examples) and a trade-off (the negative correlation that occurs only when resources are limited; see also Arnold, 1992; van Noordwijk and de Jong, 1986). Trade-offs may be difficult to discern in organisms living under favorable conditions and then appear under stress (Rollo, 1994, pp. 297–8).

Since extreme environments both induce phenotypic changes and may select for them, often in the same direction, environmental extremes may be disproportionately important in evolution even if rare. Starvation often translates to reduced body size, as may have occurred in the evolution of miniaturized elephants (see chapter 26), which in turn often correlates strongly with developmental rates, lifeways, and competitive ability (Peters, 1983; Schmidt-Nielsen, 1984), which in turn are important components of fitness (for a concise review of the evolutionary importance of size, see McKinney and McNamara, 1991). Not surprisingly, measurements of natural selection in nature often involve populations under extreme conditions (Grant, 1986; Hoffmann and Parsons, 1991). Many of the examples of natural selection in the wild compiled by Endler (1986, table 5.1) occurred under environmental duress such as high toxicity, predation, thermal extremes, or epidemics of disease.

How internal resource competition can set the stage for the evolution of morphological alternatives, which are then modified under selection, is illustrated by the evolution of shell shape in snails (DeWitt et al., 2000). A common morphological trade-off in snails is that between a shell with wide aperture and rotund shape and a shell with narrow aperture and elongate shape. Calcium-limited freshwater snails lack sufficient calcium to simultaneously strengthen the shell by making it thicker, and narrow the aperture by adding special structures such as barricading teeth. Therefore, freshwater populations are stuck with a developmentally inescapable dilemma: either make a wide aperture and invest more in a robust shell, making the snail relatively vulnerable to entering crayfish predators while relatively protected from shell-crushing fish; or make a narrow aperture and a more delicate shell, and be better protected from crayfish while vulnerable to shell-crushing fish. DeWitt (1995, 1998) showed that natural populations of snails (*Physa gyrina* and *P. heterostropha*) where crayfish are the primary predators have a higher frequency of narrow-apertured, elongate shells; and those more exposed to fish predation have wider apertured, more rotund shells. Snails reared for about a month in water containing active predators, either crayfish or fish, facultatively modified their morphology, with shells in fish treatments 6% more rotund than in those with crayfish and controls (water alone). Shells in the crayfish treatment showed a less marked but likewise significant response (DeWitt, 1995).

This study could be cited as an example of developmental constraints. But it also shows how the same continuously variable correlations that constrain directional evolution under conflicting selective factors can serve as a basis for adaptive plasticity, and how the range of phenotypes produced can lead, under selection, to divergent evolved specializations.

17

Combinatorial Evolution
at the Molecular Level

Genomic introns, like developmental switches at higher levels of organization, are points of dissociation and reorganizational flexibility in eukaryote development (gene transcription) and evolution. Alternative RNA splicing produces alternative molecular phenotypes (proteins) that may contribute to the origin of new genes (segments of DNA that function together to produce a protein). Recombinatorial use of the same sequences in different genes and proteins helps explain both conserved gene sequences and the C-value paradox (lack of correlation between degree of specialization and genome size). Combinatorial gene expression and molecular evolution challenge the conventional definitions of evolution and the gene and suggest a new hypothesis for germline sequestration in animals.

Introduction

Some of the best evidence for combinatorial evolution comes from studies of molecular evolution. This chapter discusses combinatorial molecular evolution and shows that it is facilitated by the same properties of the molecular phenotype—modularity and flexibility—that facilitate combinatorial evolution at higher levels of organization.

This is not a review of molecular or genomic evolution, and I am aware that by the time it is published it will lack the latest references even on the few topics discussed. I suspect that continued progress will only make the main point of this chapter more obvious: in many respects, evolution at the molecular level follows the same pattern as that seen at higher levels of organization, for it involves modular reorganization and developmental plasticity as architects of evolutionary change.

A combinatorial view of structural change has long been commonplace in chemistry, since all of the materials of the organic and inorganic world come from different combinations of only 112 elements listed in the periodic table. Since biochemistry and molecular biology focus on the fundamentally modular structure and behavior of biological molecules, it is perhaps not surprising that they arrived early at a combinatorial view of

evolution, and that it was a molecular biologist (Jacob, 1977) who described evolution as "tinkering" with preexisting pieces.

The lowest level of combinatorial evolution is based on the "changeability" of the genetic code—its ability to undergo rearrangement without loss of functionality (Maeshiro and Kimura, 1998). A reorganizational basis for some kinds of mutation was also proposed by premolecular geneticists like H. J. Muller (see discussion of this work in Huxley, 1942, p. 92), who saw minute rearrangements as a kind of mutation distinguishable from "substantive" change of the chromosomal-damage type caused by ultraviolet radiation. More recently, Dickinson (1988) refers to a "combinatorial" model for the evolution of gene regulation. And genetic engineering makes extensive use of combinatorial principles in creating novel substances and genes (figure 17.1). Molecular biologists, focused on a single level of organization, and doing comparative morphology on a grand scale with the help of sequencers and DNA libraries, have been among the first to extensively document a reorganizational model of change.

Dorit et al. (1990; cited by Keese and Gibbs, 1992) state that all proteins may be composed of a set of no more than 7,000 distinctive types of exons. Since more than 2,000 enzymes (just one cat-

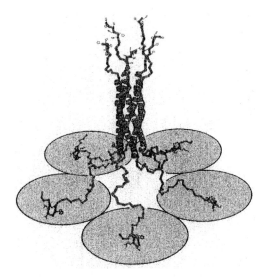

Fig. 17.1. Phenotypic recombination in protein engineering. A ribbon model of the three-dimensional structure of the artificially created peptabody (pabs) protein showing five receptor regions represented by shaded circles (radius, 40 Å). Association of multiple duplicate binding regions allows multivalent high avidity binding of selected peptide ligands. The modular building-block structure of protein molecules facilitates the construction of novel desirable combinations of parts. From Terskikh et al. (1997). Courtesy of J.-P. Mach.

egory of protein) have been identified on the basis of the chemical reactions they catalyze (John and Miklos, 1988, p. 325) and a single protein commonly involves the expression of several or many exons, reorganization must have been important in protein evolution if the claim of Dorit et al. is true. A widespread role for combinatorial evolution is also suggested by the mechanisms believed responsible for the evolution of molecular diversity. Divergence of the molecular phenotype is commonly ascribed to genomic changes such as DNA duplications and rearrangements, exon shuffling, transposition, and alterations in regulatory pathways (Keese and Gibbs, 1992; Consortium, 2001). Furthermore, there is some evidence for a "standard modular structure" of proteins (Berman et al., 1994). Virtually all polypeptide chains with more than 200 amino acid residues are organized into two or more autonomous structural regions or domains (Holland and Blake, 1990).

Brenner (1988) is credited by DuBose and Hartl (1989) as having proposed a "principle of local functionality," which summarizes a view of protein evolution that involves modularity, plasticity, and adaptive structure. By this hypothesis, "small local folding units assemble into larger units (or domains) having specific properties or functions, and these can be shuffled among evolving proteins; their sequences can then be refined for different functions by gradual replacement of amino acids" (p. 9966). The idea of postorigin fine-tuning under selection corresponds to genetic accommodation, discussed in chapter 6 as characterizing phenotype evolution in general. The principle of local functionality also assumes that protein folding results mainly from local *interactions* among physically proximate amino acids (DuBose and Hartl, 1989, p. 9966). This implies that phenotypic plasticity, or responsiveness to elements of the environment, is expected to act as an organizing factor even at the submolecular level.

In findings consistent with the principle of local or modular functionality, DuBose and Hartl (1989), using site-specific mutagenesis in *Escherichia coli*, found that some short stretches of α-helix could be replaced with other orderly fragments without destroying enzyme function, but a random, nonhelical array in the same region produced a nonfunctional enzyme. They cite other evidence, on lysozyme, hemoglobin, and the low-density lipoprotein receptor, which they interpret as showing that "random" reorganization (vs. intact subunit reorganization) is more likely to produce a nonfunctional structure. Such experiments show how novelty can originate via subunit recombination. Preexisting subunits of proteins are already functional to some degree. Their functionality and stability may be transferred to a new context and then improved under selection. Preexisting structure must often give a head start toward functionality, or at least toward integration with fewer negative effects.

Combinatorial Evolution in Regulatory Molecules

The control of eukaryote gene transcription has been described as the result of a multitude of combinations of a finite number of transcription factors, where any imaginable combination is likely to be found to exist at some gene locus (O'Malley and Tsai, 1993). A classic example of combinatorial regulatory evolution involves the homeobox sequences of *Drosophila* and other organisms (see especially McGinnis and Krumlauf, 1992). Eight homeobox sequences present in *Drosophila* are ex-

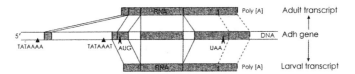

Fig. 17.2. Life-stage-specific phenotypes at the molecular level. Life-stage-specific gene products in adults and larvae are due to alternative patterns of transcription of the *Drosophila affinidisjuncta* Adh gene. Shaded regions of the DNA molecule represent exons, connected by lines to their RNA products. Triangles indicate locations of TATA-box sequences; arrowpoints indicate points of initiation (AUG) and termination (UAA) codons for RNA translation. Introns (spaces between shaded bars), exons, and the two transcript maps are drawn to the same scale. After Rowan et al. (1986).

pressed in different combinations in different tissues (see figure 4.17). Four of these are closely similar to sequences employed in analogous functions in mice, where they are expressed along with others *not* found in *Drosophila*. The homeobox sequences are expressed in different combinations both within and between the two species (summarized in McGinnis and Krumlauf, 1992). The conserved protein-domain products of homeobox expression whose functions have been tested function in DNA binding and are parts of larger proteins that act as transcriptional regulators (summarized in Levine and Hoey, 1988; McGinnis and Krumlauf, 1992). That is, they have similar, widely used functions that are important in a variety of circumstances and tissues.

Much research in molecular developmental biology focuses on only a few species, and as a result, most "evolutionary" or cross-taxon comparisons involve enormous phylogenetic leaps, for example, from viruses, to bacteria, to nematodes, thence *Drosophila* flies, frogs (*Xenopus*), mice (*Mus*), and humans. Comparisons of this sort do not often illuminate the nature of evolutionary transitions, best studied by comparison of closely related species. For this reason, research on the molecular evolution of regulation in Hawaiian *Drosophila* is of exceptional interest (reviewed in Dickinson, 1991). This work traces the evolution of *cis*-acting control regions, that is, control regions located adjacent to, or on the same chromosome with, the regions they control (compare *trans*-acting control by regions located on different chromosomes). The regions examined influence the expression of six enzymes in 14 different tissues and stages among nearly 30 different species.

These *Drosophila* studies support a combinatorial model for the evolution of novel patterns of regulation. The Adh locus, for example, has two TATA-box-containing promoters or transcription initiators, the proximal one generally active in lar-

vae and the distal one generally active in adults (figure 17.2). But there are species differences in their identity and use (reviewed in Dickinson, 1991, p. 129). For example, in *D. prostopalpis*, both promoters contribute about equally to transcript present in all larval tissue except midgut; in *D. formella*, both promoters act in all adult tissues; and in adult *D. affinidisjuncta*, both transcripts appear in Malpighian tubules, only proximal transcript occurs in midgut, and only distal transcript is found in other tissues. *Trans*-acting factors have also been implicated in a few species, and some complex differences result from variation in both kinds of elements (Dickinson, 1991). Dickinson (1988, 1991) concludes that combinatorial principles apply to gene regulation, and suggests how both complexity and precision can be achieved with a small number of regulatory molecules. New patterns "need not involve de novo evolution of regulatory components or even modification of the properties of existing ones other than as a consequence of assembling new combinations of modules" like the TATA-box promoter sequences shown in figure 17.2 (Dickinson, 1991, p. 160).

As in comparative studies of morphological or behavioral phenotypes, species differences in molecular phenotype can be found as low-frequency variants in related species. Molecular anomalies are sometimes systematically sought in broad surveys designed to uncover variants for in experimental work. These surveys often reveal that distinct variants are recurrent, and frequencies exceeding 5–10% for some patterns are probably common (Dickinson, 1991, p. 135).

Comparative endocrinology is a rich source of evidence for regulatory molecular diversification based on modular organization (Bolander, 1994). The diversification of peptide hormones, for example, has involved gene duplication, exon shuffling, gene-family branch jumping, and possibly reverse transcription. In branch jumping, two divergent branches of the same duplicated gene family have binding sites with distinctive receptors that overlap somewhat in structure and are bound by different hormones. Then, a single hormone either evolves the ability to bind both kinds of receptors or changes from the ability to bind only one to binding only the other. The transition, or evolutionary "jump" from binding members of one branch of a gene family to binding members of another branch, is facilitated by the similarity between the two kinds of receptor, based on common ancestry of duplicated genes. Combinatorial evolution of hormone molecules is facilitated by the fact that many hormones (e.g., steroids, prostaglandins, and amines) are simple compounds that can be synthesized in only a few steps from common precursors, so they are readily produced molecules (Bolander, 1994, p. 496). Some of these molecules are found in organisms where they do not function as hormones, but are breakdown products of proteins, or molecules with nonsignal functions. As in the evolution of animal communication, a hormone's ability to function as a signal depends on the evolution of receptors. Most hormone molecules by themselves have no intrinsic activity (very few act as enzymes, transport, or structural molecules; Bolander, 1994).

Combinatorial Evolution in the Genome

The idea that the evolution of functional genes themselves may be combinatorial, or characterized by reorganization of preexisting functional subunits of DNA, began by extrapolation from the study of the domain structure of proteins (for short histories, see Rossmann, 1990; Stone and Schwartz, 1990b). This idea gained momentum with the discovery of the intron–exon structure of eukaryote genes by Breathnach et al. (1977; see Holland and Blake, 1990).

In a subsequent series of three almost simultaneous papers in *Nature* by three different authors, Gilbert (1978) proposed that introns, or genes in pieces, could facilitate nondisruptive crossing over and recombination, including in the production of alternative protein products from the same gene; Doolittle (1978) suggested that "split genes" and RNA splicing of exon products following transcription had likely occurred in the earliest organisms, the "progenote" ancestors of both prokaryotes and eukaryotes; and Blake (1978) suggested that the intron–exon structure of genes corre-

sponded with coding-sequence borders to the functional domains of proteins.

Subsequent findings, summarized in Stone and Schwartz (1990a), support a fundamentally modular and combinatorial view of eukaryote gene structure evolution, and the exon–intron organization of eukaryote genes (summarized in Caporale, 1999a). There is still controversy, however, over whether introns evolved very early in the history of life, or first appeared in eukaryotes (introns-early vs. introns-late; W. F. Doolittle, 1987). Does the lack of introns in modern prokaryotes represent a secondary loss, or the ancestral state (e.g., see Hurst and McVean, 1996; Rzhetsky et al., 1997; de Souza et al., 1998; Roy et al., 1999)?

The "exon theory" as proposed by Gilbert includes both the idea of exon–intron organization with exon shuffling, and the idea of an early origin of introns. It sees the primordial genes ("urgenes") as having encoded short polypeptides that were folding units or functional domains. Clusters of such protogenes, held together and separated by introns, may have specified the first enzymes. Later genes were assembled from ur-genes by the fusion of introns. This theory sees introns as "residual glue": "A gene today has introns because at some time in its history it was assembled by recombination within those [original] introns" (Gilbert and Glynias, 1993, p. 138). The further evolution of genes, by this theory, is seen as having involved the shuffling and reassortment of exons and the creation of more complicated exons from simple ones (details in Gilbert and Glynias, 1993 see also Long et al., 1995).

Supporters of this introns-early theory argue that structural comparisons of the ancient enzyme, triosephosphate isomerase (TPI), in vertebrates (chickens) and plants (maize) demonstrates that the exon–intron structure of the TPI gene predated the separation of plants and animals (Gilbert and Glynias, 1993). This would imply that the TPI gene had an exon–intron subunit structure in some very early, possibly single-cell, organism. Another line of evidence cited in support of the intron-early view is the correlation between the modular organization of proteins with the exon–intron structure of the genes encoding them (Go and Nosaka, 1987; Go, 1994).

Other authors argue that the intron–exon structure characteristic of eukaryote genomes, in contrast to prokaryotes, which lack introns, may be an extremely ancient trait, preceding the prokaryote-eukaryote split in evolution, with the lack of introns in prokaryotes being a derived trait associated with the advantage of a compact genome (summarized in Doolittle, 1990; Gilbert and Glynias, 1993). By this view, "the current structure of the prokaryotes is not an image of the primitive world but is a highly evolved structure that has undergone a great deal of evolutionary modeling" (Gilbert and Glynias, 1993, p. 138). Among these are the authors who see the introns as added elements, possibly originally transposons or "selfish elements" having nothing to do with protein structure and function, with the original genome a continuous coding region (e.g., Cavalier-Smith, 1991). The introns-early theory has been modified in recent papers of Gilbert and associates as the "mixed theory," which holds that some introns are very old and were used for exon shuffling in the progenote, while many introns have been lost and added since (Roy et al., 1999).

Recent studies provide introns-late interpretations of the evolution of the TPI gene and challenge the intron-early interpretations of eukaryote genomes by Gilbert and associates (Logsdon, 1998, summarized in Hurst and McVean, 1996, which gives a valuable critique of flawed evolutionary reasoning on both sides). There is reason to think that intron insertion (introns late) could have evolved, including in ancient genes (conserved in both prokaryotes and eukaryotes), under selection for increased splicing efficiency, rather than being interpreted as a relict of progenote organization (de Souza et al., 1998) or something as yet unexplainable in terms of selection (Hurst and McVean, 1996). The properties and insertions of introns should be regarded as selectable variables. The base-compositional contrast in G, C, and U content between exon and intron sequences in the vicinity of splice sites in maize, for example, correlates with splicing efficiency (Carle-Urioste et al., 1997).

Given the involvement of introns in splicing, and the importance of splicing in eukaryote gene expression, it is to be expected that intron properties and insertions favorable to splicing efficiency have been molded by selection. Some comparative studies suggest that exon shuffling, or intronic recombination, is best demonstrated in higher eukaryotes (i.e., metazoa) and may have become significant only after the evolution in these organisms of spliceosomal introns, which, in contrast to the self-splicing introns of fungi, protists, and plants (which contain structures that catalize their own splicing reaction), guide transcription without playing a role in their own excision (they are excised from RNA by special mechanisms; Patthy, 1996, 1999; for a concise discussion of the possible origins of spliceosomal introns from mobile introns found in bacteria, mitochondria, and chloroplasts, see Eickbush, 2000).

Whatever the outcome of the introns-early–introns-late debate, the important fact for the combinatorial view of evolution presented in this chapter is that "if any finding is accepted in this contentious field, it is that exon shuffling does occur in eukaryotes" (Hurst and McVean, 1996, p. 535). Even without introns, prokaryote genome evolution sometimes involves "intragenomic recombination," a class of mutations wherein preexisting subunits of DNA are reorganized to produce novel genes (Thaler, 1994; de Chateau and Bjorck, 1994, cited by Patthy, 1996). Whether facilitated by introns or not, modular recombination occurs during genome evolution in a very wide spectrum of forms of life.

Divergence between duplicated genes may sometimes involve exon shuffling in the duplicated DNA. Present-day proteins evidently have descended from a small number of ancestral proteins, and this has frequently involved gene duplication (Holland and Blake, 1990; Consortium, 2001; see chapter 10), but it is not always clear how the duplicates diverge. One hypothesis is that duplicates diverge simply due to selection on random mutations that differ in different copies (e.g., see Li, 1983). Divergence by exon shuffling "would allow significant evolution of proteins to proceed in major steps rather than through a series of intermediates reflecting minor incremental changes" (Hunkapiller et al., 1982, p. 167; see also Holland and Blake, 1990). Here we see contrasting mutational-gradualism and developmental-reorganization approaches pursued on the molecular level.

Sometimes gene fragments that contain several exons and introns are incorporated into a new gene, as indicated by the study of domain homologies in related proteins. In other innovations, there is a serial, tandem repeat indicating multiple duplications of the same genome section, which are then used together to code a single new protein. Fibronectin, for example, contains multiple copies of each of three domain types (figure 17.3). One of them corresponds to a "Kringle" domain found in four other proteins, and the "core" Kringle unit has taken on different specifications and functions in the different proteins by the addition of variable polypeptide loops on the exterior of the structure, possibly for use in anchoring the molecule at the sites of action (Holland and Blake, 1990, p. 17). This suggests evolutionary modification at the exon/domain

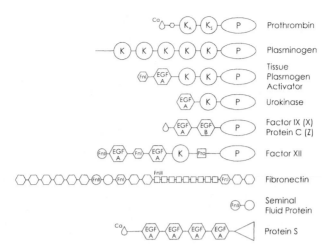

Fig. 17.3. Combinatorial rearrangements of nine domains in the evolution of a family of proteins (blood coagulation, fibrinolytic, and related proteins). Domains include calcium binding (Ca), Kringle (K), epidermal growth factor A (EGF A), epidermal growth factor B (EGF B), fibronectin type I (FnI), fibronectin type II (FnII), fibronectin type III (FnIII), serine protease catalytic (P), and a proline-rich region (Pro). After Holland and Blake (1990).

level following or accompanying duplication and use in a new context (see section on genetic accommodation, below). Some Kringle-containing units (e.g., the Kringle/FnII and Kringle/FnI units in plasma proteins) may have undergone a unification or fusing of domains. Deletions, reversions, and novel combinations of exons occur via RNA splicing in both development and evolution (McKeown, 1992; Sharp, 1994), with intron deletion likely important in the evolution of some genes (Holland and Blake, 1990, pp. 28–29). Intron *capture* has been suggested to occur in the origin of novel genes (Gilbert, 1978). The phosphoglycerate kinase gene in a trypanosome, for example, has unusual sequences that correspond to the positions of introns in that gene in some other organisms (e.g., some ciliates). Introns may have been captured by the protein and used as part of the coding sequence (Golding et al., 1994).

Another combinatorial means of functional-gene evolution, in addition to exon shuffling and intron capture or deletion, is by *overprinting*, or the translation of portions of existing coding and noncoding material of adjacent genes without altering their sequences (Grassé, 1977; Keese and Gibbs, 1992). Overprinting gives rise to overlapping genes that share coding regions translated in different reading frames to produce distinctive proteins. In some organisms, the phylogenetic history of reorganization can be traced, as in the case of certain steroid-related receptor genes and viral

genomes (Keese and Gibbs, 1992). The transcript of the thyroid hormone receptor (TR) alpha gene, for example, undergoes alternative spicing as described in the next section to produce a novel hormone-binding region unique in its family of steroid receptor genes and containing a nucleotide sequence that partially overlaps that of a related thyroid receptor (Keese and Gibbs, 1992, p. 9489), suggesting that the novel splicing product evolved via overprinting. In addition, hundreds of human genes appear likely to have resulted from horizontal transfer from bacteria, and 47 human genes appear to be derived from transposable elements, all but four of them from DNA transposons (Consortium, 2001).

Many of the reorganizational events that characterize genome evolution are of uncertain functional significance. But genome studies are proceeding so rapidly that this situation is changing, and there is increasing caution in use of the term "junk DNA" (see especially Shapiro, 1999; Consortium, 2001, discussion of the human-genome results). Less than 5% of the human genome is estimated to consist of coding sequences. About 50% consists of repeat sequences. The majority of these (about 45% of the human genome) are transposon-derived "interspersed repeats." The remainder are "processed pseudogenes" (mostly inactive retroposed copies of cellular genes derived from protein-coding genes and small structural RNAs), and simple sequence repeats, tandem repeats, and segmental duplications copied from one region to another (Consortium, 2001). If nothing else, these findings attest to the importance of reorganizational change in the evolution of the human genome.

Phenotypic Recombination by RNA Splicing

One way that the same set of exons can produce diverse products during gene expression is by differences in their transcription from different promoters (see figure 17.2). Another is by differences in how the transcripts are joined together, or spliced (see figure 3.9). Alternative splicing was introduced in chapter 3 as an example of phenotypic plasticity at the molecular level. RNA splicing cuts and then joins together (splices) sections of transcribed pre-RNA into the mRNA that will form the functional gene product (e.g., a particular protein). Alternative splicing produces alternative molecular phenotypes, and as might be predicted, it is involved in the origins of molecular novelties.

In alternative splicing, the primary transcripts, or pre-RNAs, of different sets of genomic sequences are joined so that they produce more than one kind of messenger RNA (mRNA) and thus can participate in the expression of more than one protein phenotype. For example, human fibronectin-gene DNA gives rise to pre-RNAs that are alternatively spliced into mRNAs that specify more than 20 different fibronectin proteins, having different phenotypic characteristics, locations, and functions. Some of them are shown in figure 17.3. One form of fibronectin, produced in liver cells, is specialized, by deletion of some protein domains, to more readily circulate in the bloodstream because it has fewer domains "recognized" by cell-surface receptors than do other forms of fibronectin. Secreted fibronectin, produced by other cells, includes the domains missing in the liver cells and is more often found in the extracellular matrix of solid tissue (Sharp, 1994). Other examples of tissue-specific alternative splicing occur in neurohormonal peptides of the calcitonin and tachykinin families, where different peptides are produced from the same gene due to alternative splicing in neurons versus endocrine tissues (Thorndyke, 1988; figure 3.9). Variation in cellular adhesion is associated with alternative splicing in other parts of the mRNA. Thus, through alternative splicing, the same set of gene sequences can be utilized for different functions (Sharp, 1994, p. 806). Although most spliced exon products are from contiguous exons, proteins need

not be specified by coding regions at the same locus or even on the same chromosome. Some are formed via RNA splicing of exon transcripts from different loci both within and between chromosomes. Such "*trans*-splicing" occurs in trypanosomes and nematodes (references in Sharp, 1994).

The products of alternative splicing are miniature polyphenisms, or conditionally expressed alternative phenotypes. Alternative splicing results from sensitivity of the splicing mechanism to differences in intracellular environments. Factors hypothesized to account for the tissue- or stage-specific splicing include snRNPs (small nuclear ribonuclear proteins), particles essential for pre-mRNA splicing and possibly varying in form among cell types (Smith et al., 1990; Sharp, 1994). Alternative splicing can be sex specific as well as tissue specific. In *Drosophila*, alternative splicing determines sex and influences the gene products responsible for the sexually dimorphic, divergent characteristics of adults (Baker, 1989). For example, the "doublesex" gene is borne, and expressed, by both sexes, but the pre-RNA undergoes sex-specific splicing: in females the product of the third exon is spliced to the fourth exon product and that of the fifth exon product is excised, while in males the third exon product is spliced to the fifth, omitting the fourth (Inoue et al., 1992). So the condition-sensitive operation of alternative splicing early in development produces a self-sustaining molecular polyphenism that is the basis for male and female alternative phenotypes in adults.

As with alternative phenotypes at higher levels of organization (see part III), molecular alternative phenotypes produced by alternative splicing are sometimes reflected in the pattern of molecular evolution. The fibronectin gene, for example, contains three types of exons, each encoding one of the protein domains that compose the 20 or more proteins formed from products of those exons by alternative splicing. Homologous domains are found in various other proteins, indicating that they have been duplicated, recombined, and somewhat divergently specialized in new contexts during the course of evolution. Evolution by duplication and recombination of eight different protein domains, three of them the fibronectin domains, is shown in figure 17.3. Sharp (1994, p. 806) regards the fibronectin proteins as having evolved via tandem and dispersed duplication of exon units using breakage and joining within the intron sequences. These exon-specified units are assembled by RNA splicing. Sharp regards this theme of duplication and utilization of an exon-specified domain in a new context via alternative splicing a common mechanism in the origin of genes (i.e., innovation in which segments function together to produce a protein) encoding many cell surface receptors and other types of proteins in vertebrates. In other words, not only is RNA splicing an important means of phenotypically recombining exon products during ontogeny, it is also associated with combinatorial origins of novel proteins during evolution.

The functionality of the novel molecules is enhanced by the preexisting functionality of their component domains. Holland and Blake (1990) argue that without the folding pathways already built into preexisting exon products, creation of novel proteins using novel untested exons "would reduce to almost zero the chances of a folded protein being created by linking such units together" (p. 13).

Alternative splicing, then, may complement duplication in the generation of novel molecular phenotypes. As discussed in chapter 10, duplication by itself does not increase phenotypic diversity, although it is often cited alone as a mechanism of divergence. The duplicate must be independently expressed or used to contribute to divergence. Splicing provides one mechanism for the divergent use of duplicated exons, capitalizing on the buffering effect of alternative phenotypes to preserve old functions while new ones evolve (see chapter 7).

During transcription, introns serve as organizing points. The noncoding sections of DNA (including introns) are organizational analogues of the regulatory switch points of higher levels of organization. These noncoding regions are often bordered by highly conserved consensus sequences that act as regulatory (e.g., splicing) markers. Splice sites are the stopping and starting places for structural subunits, and so are points of regulation and dissociation, as are switch points that regulate development, morphology, and behavior at higher levels of organization. Therefore, the same ideas regarding modularity, flexibility and evolution (see part I) should apply at all levels of organization.

Previous discussions of the adaptive significance of introns have emphasized their genomic-replication and cohesiveness roles, for example, in facilitating recombination during meiosis by providing larger "targets" for recombinational splitting of DNA without splitting regions that are functional for producing proteins (Gilbert, 1978); or in serving as "residual glue" to hold together the hypothetical proto-genes (or "ur-genes"), proposed ancestors of present-day genes in the hypothesis of Gilbert and Glynias (1993). The role of introns as switch points in molecular development raises the possibility that introns, whenever they originated

(whether early or late in the ancestors of eurkaryotes) originally functioned as developmental devices, then later became important, in addition, because of their significance in facilitating recombination, exon shuffling, and the increased evolutionary lability of clades (for discussion see Holland and Blake, 1990; Sharp, 1994).

The *developmental-function hypothesis*, that a primary function of introns is developmental, with recombination functions during sexual reproduction and exon shuffling during evolution sometimes a secondary result, is perhaps supported by the phylogenetically limited role of exon shuffling in the evolutionary diversification of proteins. Even though introns are ubiquitous in eukaryotes, a recent survey shows that modular multidomain proteins produced by exon shuffling are largely restricted to metazoa (multicellular animals), where they occur in all major groups from sponges to chordates. Metazoan multidomain proteins have functions linked to multicellular integration. For example, they include contituents of the extracellular matrix and body fluids, proteases involved in tissue remodeling, and membrane-associated mediators of cell–cell interactions (Patthy, 1999). Complex proteins formed by exon shuffling are most common in the most complex metazoans, whereas they are absent in protists, and (despite the presence of introns) there is little evidence for such proteins in plants and fungi (Patthy, 1999). This indicates a possible relation between the prominence of exon shuffling in protein evolution, and the prominence of sexual reproduction (accompanied, of course, by genetic recombination). Sexual reproduction is obligatory primarily in "higher" metazoa (Williams, 1975). In metazoa, introns are important in both the developmental and the evolutionary reorganization of the phenotype. Further evolutionary speculations regarding introns should include study of their relation to the evolution of sexual reproduction and genetic recombination, including mechanisms of chromosomal behavior.

Even if the details of their evolution may be subject to debate, the evidence for intron-mediated split-gene organization of gene expression, and for combinatorial gene evolution by exon shuffling in eukaryotes (especially metazoa) is now considered "overwhelming" (Stone and Schwartz, 1990c, p. 65).

Genetic Accommodation at the Molecular Level

Once domains have been shifted from expression in one set to expression in another, by exon shuf-

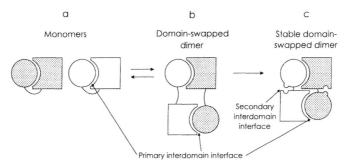

Fig. 17.4. Domain-swapping during development and evolution: (a) the monomeric form of a protein has two domains (circle and square), with a primary interdomain interface between them, in each of two polypeptide chains (stippled and unstippled); (b) formation of a domain-swapped dimer, with interchange of domains between polypeptide chains facilitated by the preexisting shared primary interdomain interface, and likely not requiring genetic change; (c) evolved formation of a secondary interdomain interface, favored by selection for dimer stability (genetic accommodation of a novel molecular morphology). After Bennett et al. (1994).

fling or by alternative splicing or overprinting, their sequences can then be refined for different functions by gradual replacement of amino acids (Du-Bose and Hartl, 1989), in other words, by genetic accommodation at the molecular level. As in genetic accommodation at higher levels of organization, the process suggested by DuBose and Hartl acts on phenotypic form to fine-tune the functional structure of the new variant, for example, modifying particular aspects of enzymatic function (Du-Bose and Hartl, 1989, p. 9970). This is genetic accommodation of form, not genetic assimilation (increased frequency of expression of the new trait).

How genetic accommodation of a combinatorial change could occur at the molecular level is illustrated by the evolution of complex proteins, such as the diptheria toxin, believed to have evolved from smaller molecules via domain swapping. In *domain swapping*, two or more monomers of an ancestral monomeric protein are joined (Bennett et al., 1994; figure 17.4). The novel structure is then stabilized by small evolutionary changes. As in combinatorial evolution at higher levels of organization, a developmental reorganization of the phenotype occurs first, then is genetically accommodated by modifications not present in the ancestral phenotype, such as secondary interfaces between the newly joined monomers (figure 17.4c).

The combinatorial, domain-swapping hypothesis is supported by the fact that many complex proteins, such as the diphtheria toxin, have monomeric counterparts that can be environmentally induced (e.g., by manipulating the pH of their medium) to form the dimeric form. The complex structure is

adaptive in that each of the domains performs a particular function. In the diphtheria toxin, for example, each of three functions—receptor binding, membrane translocation, and catalysis—is associated with a different folding domain of this complex, dimeric protein. Bennett et al. (1994) consider domain swapping followed by addition of small structural details not accounted for by the initial combinatorial event more feasible than other hypotheses, which postulate a gradual accumulation of complementary mutations, with several occurring at a time, an unlikely event. The hypothesis of evolution of oligomers by domain swapping may also account for the otherwise puzzling observation that some oligomeric enzymes have their active sites at the junction of two subunits (Bennett et al., 1994, p. 3131).

Combinatorial Evolution and DNA Sequence Conservation

Combinatorial evolution helps explain the extreme conservatism of many genomic sequences in the face of extensive phenotypic evolution (see also Gerhart and Kirschner, 1997). Dissociation and recombination can occur at all levels of organization where there are developmental switch points. The most conserved subunits over the long span of evolutionary time should be the most elemental modular traits—those that cannot be divided without loss of function, a criterion that likely applies to some extensively shuffled coding domains of DNA; and those whose functions are extremely versatile in being able to function in many combinations, such as the small noncoding consensus sequences that serve as tags during gene transcription. Such regulatory sequences could be preserved by selection even though they are not expressed. Dickinson (1991) remarks that "it has become common to use comparison of sequences in non-coding regions [often containing regulatory signal sequences] to locate [conserved] elements that might be involved in gene regulation" (p. 140).

In examples of genetic conservation, what is actually conserved is not usually the entire DNA sequence that specifies a protein—what we would refer to as a gene. Instead, conserved sequences are often fragments of genes corresponding to exons, and small or fundamental regulatory units—used in phenotype (e.g., protein) production. The term "conserved gene" is used rather loosely to refer to conserved parts of genes. For example, a "highly conserved gene" studied in nematodes (*Caenorhabditis*), flies (*Drosophila*), and humans (*Homo*) corresponds to a protein, or, more accurately, to a family of proteins, found to be only 49% identical in sequence similarity between nematode and fly, and only 58% identical between human and fly (Campbell et al., 1993). The famous homeobox genes, noted for their "conserved" presence in a phylogenetically wide range of organisms (e.g., mice, flies, perhaps nematodes) really specify "homeodomains" or parts of large protein molecules that function as transcriptional regulators (see McGinnis and Krumlauf, 1992). Their structures resemble DNA-binding proteins found in prokaryotes, indicating that both their structure and their function may be very old. The selection for conservatism in relatively small sequences involved in regulation of gene expression may be very strong, since a change in tissue specificity of expression can be caused by a change in only two base pairs within a binding site (Elsholtz et al., 1990, ex Dickinson, 1991, p. 161).

Two opposite-sounding properties of conserved sequences may favor their conservation. One is single-function precision, or utility in a particular task; another is multiple-function versatility, or the ability to function in multiple contexts. In combinatorial evolution, the product of a conserved sequence may be expressed in numerous higher level functional subunits. It becomes a fundamental building block or signal under selection in multiple structural and regulatory contexts, perhaps performing the same specialized function in each. Sequence conservation due to specialized function in multiple contexts is analogous to the conservation of nonspecific modifiers of alternative phenotypes, which are one basis for conservative genetic corre-

lations between traits (see chapter 7). Riedl (1978) refers to the "burden" of multiple functions and states that the greater the number of developmental events depending on a particular trait, the higher it "burden" or resistence to evolutionary change. He sees the "freedom" of standard parts as decreasing at lower levels of organization (p. 106) due to their involvement in multiple higher level functions. In still other conceptualizations, "mature" established genes (or subunits) are seen as not readily mutable (see Zuckerkandl, 1994, for a discussion of "mature" and "infant" proteins in relation to evolutionary potential and conservation of form). Multiple functions have been seen as leading to "generative entrenchment" (Wimsatt and Schank, 1988).

These concepts are relevant to a discussion of DNA sequence conservation because they suggest that specialized function in multiple contexts can in a sense protect integrated functional sequences even when one or more contexts of expression have been reduced or lost. Expression in a similar context then may be regained as a reversion or a component of a novel trait. Chitin synthetase, for example, contributes to chitin production in the exoskeleton of arthropods and in the cell wall of fungi. It was formerly believed to be limited to nonvertebrates (Wagner, 1994). But similar genes now have been found in the amphibian *Xenopus* and in the bacterium *Rhyzobium*, where it plays a crucial role in the symbiotic interaction with plants. In fish, a chitin synthetase also produces chitin, although it is not used in this way in other vertebrates. Chitin synthetase-like genes may be maintained in vertebrates that do not produce chitin because they synthesize short molecules used in lipid membrane signal substances. So this is an ancient gene, evidently conserved by multiple but shifting functions (Wagner et al., 1993; Wagner, 1994).

Examples of conserved sequences, presumably conserved due to consistency of function in multiple contexts, are

1. The consensus sequences common to vertebrate, plant, and yeast cells, which are small relative to introns and exons and function to tag the points of excision and splicing as already discussed (see Sharp, 1994)

2. Tubulin and actin genes: β-tubulin is 86% identical between *Giardia* (a protozoan) and humans, despite some diversification, following duplications, as a multigene family. Actin, which binds to many other proteins, has diverged only about 10% in eukaryotes in the past billion years. Both have conserved

functions as assemblers of polymers within cells (see summary and discussion in Gerhart and Kirschner, 1997)

3. Factors that influence widespread growth phenomena such as dorsal–ventral patterning (e.g., see Padgett et al., 1993), axial patterning (McGinnis and Krumlauf, 1992), or the condition-responsive expression of certain proteins in tissues as disparate as mammalian brains and various tissues of plants (Lu et al., 1992)

4. The so-called "housekeeping genes" that mediate widespread essential tasks, such as signal-transduction pathways, occurring in all cells (e.g., see Bennett and Scheller, 1993).

Some other sequences, such as the exons that specify protein domains, which are often shuffled among different molecules, are probably conserved because of participation as standard parts, or building blocks, in multiple combinations. Finer analysis may indicate that these shuffled building-block sequences serve similar functions in the different proteins in which they occur.

The conservation of fundamental subunits and their use and reuse in different kinds of structures mean that the expression of a particular gene or gene product in more than one species, or higher taxon, does not necessarily indicate close phylogenetic relationship between them. The problem of phenotypic recombination and homology is taken up in chapter 25.

Molecular Terminology and the Definitions of Evolution and the Gene

Discussions of molecular evolution frequently intermix or conflate genome and phenotype. For example, Gilbert and Glynias (1993, p. 140) discuss "introns" in the secondary structure of an enzyme (triosephosphate isomerase), as if the introns of the genome were an aspect of the phenotype, when it would be more accurate to state that certain aspects of protein structure *correspond to* certain genomic introns. Similarly, Sharp (1994) uses the words "intron," "exon," and "gene" to describe aspects of the genome and the soma or posttranscriptional events, interchangeably, as in "exons . . . are joined by RNA splicing" (p. 808; read: exon-specified products are joined by RNA splicing). As a result, some of the discussions of combinatorial phenomena that may *seem* to apply to the genome

in fact refer to the development and evolution of the phenotype.

While there is a close and fundamental relationship between flexible phenotype and evolved pattern, it is still crucial not to equate phenotype and genes, even—or especially—at the molecular level. Terminological imprecision seems not to have hampered molecular biology. But the genotype–phenotype border has to be respected in evolutionary biology, in order clarify cause and effect in discussions of selection and evolution. It is possible, as in some of the sections of this chapter, to view the genome as phenotype and to see many of its antics as if it were a miniature organism with its parasitic invaders and active responses. Many of those interactions have far-reaching consequences for evolution. But until they affect fitness at the level of the individual, genomic properties, like phenotypic traits, have no evolutionary significance. Selection sees only the phenotype; and only the genetic material that correlates with phenotypic fitness is affected, increasing or decreasing in frequency at the population level as a response to selection. This relationship is the essence of Darwinian evolution.

Clarity regarding cause and effect in *evolutionary* biology absolutely requires keeping the phenotype–genotype distinction clear. For this reason, the definition of evolution as cross-generational phenotypic change involving gene frequency change, adopted by most evolutionary biologists, is strictly followed in this book, and the genotype–phenotype distinction is conserved. What a molecular biologist may regard as pure pedantry or excessive timidity in use of terms instead reflects the fact that progress in evolutionary biology depends on precision in seeing the cause and effect relations between phenotypic variation, selection, genotype, and evolutionary change.

The findings on combinatorial gene expression strain the conventional distinctions between genotype and phenotype and challenge the usual definitions of evolution and the gene. The findings described above show, for example, that evolution can involve change in the timing or tissues of *expression* of a gene, without change in its frequency. By the strict definition of evolution as gene-frequency change, the gene-expression divergence observed in Hawaiian *Drosophila* species (Dickinson, 1988) would not necessarily be regarded as evolved even if favored and therefore maintained by natural selection because, strictly speaking, it may not involve gene-frequency or genomic change. In molecular biology, reorganizational phenotypic change is unhesitatingly called "evolutionary."

Dickinson's (1991) title, for example, is "The Evolution of Regulatory Genes and Patterns in *Drosophila*," not because a gene-frequency change is documented or assumed, but because a populational change in gene-expression pattern has occurred. The change could have involved gene-frequency change in the regulation of gene expression, fitting the conventional definition of evolution, or it could have involved some environmental change without any gene frequency change at all.

Another challenge to a sacred definition is the finding of widespread phenotypic recombination via RNA splicing. This strains the definition of the gene as a linear segment of the genome. Phillip Sharp, in the published version of his Nobel lecture (Sharp, 1994, p. 808), gave as his "current working definition of a gene" "a linear collection of exons that are joined by RNA splicing." This definition mixes units that are genomic (exons) with a process that is phenotypic (RNA splicing). To be consistent, it would have to be redone, to define either a somatic unit ("a linear collection of *exon products* that are joined by RNA splicing") or a genomic unit ("a linear collection of *exons whose products* are joined by RNA splicing"). The second, genomic definition would be consistent with current usage in evolutionary biology, since it is a genomic definition of the gene. Even with this distinction clear, some phenomena are left unclassified, since RNA splicing does not always join products of a linear set of exons. Some proteins are formed by *trans*-splicing, or RNA splicing of exon transcripts from different loci both within and between chromosomes (e.g., in trypanosomes and nematodes; Sharp, 1994). Like Dickinson's view of evolutionary change, Sharp's definition of the gene begins to overstep, at the molecular level, that thin but jealously guarded line in evolutionary biology between the phenotype and the gene.

Other proposals for a modern definition of the gene similarly involve development, or gene expression. Singer and Berg (1991, p. 622), for example, define the eukaryotic gene as "[a] combination of DNA segments that together constitute an expressible unit, expression leading to the formation of one or more specific functional gene products that may be either RNA molecules or polypeptides." This definition takes into account the growing conviction that gene expression must be part of the gene concept without violating the conventional idea that the material of the gene resides in the genome. Portin (1993) concludes a history of concepts of the gene with the statement that "We are currently left with a rather abstract, open, and generalized concept of the gene, even though

our comprehension of the structure and organization of the genetic material has greatly increased" (p. 173).

Sharp (1994, p. 808) regards the "unit of inheritance"—the conceptual descendent of the Mendelian notion of the gene—to be a single exon. Is this to be regarded as the fundamental unit of evolutionary change—that which changes in frequency when populations evolve? If so, it would not include the highly conserved splice sites that influence regulation of splicing; and the introns, which are sometimes deleted and otherwise altered during genomic "evolution," including the formation of new "genes" (Gilbert and Glynias, 1993). It is also worth noting that sometimes only part of an exon product is used, as in the expression of rat α-Tropomyosin in brain tissue, where part of exon 13a "is treated as an untranslated sequence" (it is completely used in various other tissues; Smith et al., 1990, p. 171). So even at the subexon level, not all that is inherited as a unit is used as a unit. Furthermore, even the smallest recognized functional subunits of proteins—domains—in some cases may be polygenic (or polyexonic), with each of several exons encoding a different folded substructure (Holland and Blake, 1990, p. 23).

While this conflation of the genetic and the somatic may sometimes be confusing, it is also gratifying as evidence of the importance of the phenotype for interpretations of the gene. The sometimes unconscious somatization of the concept of the gene in molecular biology confirms the significance of *phenotypic* gene products as the ultimate subject of interest. Although it is still possible to distinguish between the genome and its environment, the distinction gets difficult as soon as the genome begins to act. This is exactly the same moment when the genome becomes of biological interest.

The prominence of differential, environmentally mediated responsiveness at the molecular level challenges evolutionary biologists to keep an open mind regarding the relationship between genomic and phenotypic evolution. Do we want to rigidly consider gene-frequency change the only true evolution? Or can reorganization of gene expression by a persistent environmental factor be included? As I discuss further in the following section and in chapter 26, this implies keeping an open mind regarding the informational role of the environment in development and evolution (Bernays and Wcislo, 1994; Wcislo, 1997). The decisive role of the cellular environment in *directing* the use of genes, even determining how their products will be assembled by splicing them into usable molecules, gives a clear demonstration of the informational role of the already constructed phenotype, whose inducible and inducing structure incorporates, always, elements from both environments and genes.

Any evolution that takes place on the cytoplasmic side of the line between genome and phenotype may involve information from the environment. Thus, for example, there are more than 20 proteins produced by different combinations of exons from the fibronectin-influencing zone of the genome. The protein produced in the liver environment lacks two exon-specified domains present in fibroblast proteins—molecules assembled from the same "gene" (exon set) but in a different cellular environment (see Sharp, 1994, and above). Whenever a catalyst of genomic origin fails to switch on a reaction or cascading pathway due to the absence of some element or signal of external environmental origin (sunlight, oxygen, nitrogen, water), or is initiated due to the *presence* of such an element, environmental information has influenced the course of development—as surely as information of genomic provenance has. The fact that the the phenotype is "preset" to respond in a particular way, under the influence of its genome, should not be taken as a sign of genomic control, because the entire responsive apparatus has involved particular, and sometimes variable, environmental inputs. When condition-dependent RNA splicing is an agent of development, the unadulterated effect of the transmitted DNA may be lost even before it has made a messenger RNA. The influence of environmental factors is so prompt and ubiquitous that the attempt to classify any character as "genetically determined"—even one as close to the genes as a protein—is rendered imprecise even before its transcript leaves the nucleus of the cell.

By somaticizing the concept of the gene, readily interchanging genomic and somatic terms in common usage, molecular biology—that most reductionistic and most genomically focused field of biology—has unwittingly dramatized for evolutionary biologists the inseparability of genotype and environment, nature and nurture, as agents of development. The distinctions are easily blurred, even at the finest, molecular level. But the distinctions must be maintained to understand selection and evolution. Selection reflects properties of the soma. Through those properties, it affects the differential reproduction of genes.

Speculations

Laboratory biologists, looking for ways to connect development and evolution, sometimes ask evolu-

tionary biologists what kinds of questions are of interest for future work. Certain very broad questions are raised by the observations of this chapter.

Genomic Behavior and Genomic Evolution

Gene expression, by tissue-specific transcription and alternative splicing, is combinatorial, and so is genome evolution. Could there be any connections between the two? Does the dynamic behavior of the genome during gene expression dispose it to evolve along certain lines, the way patterns of phenotypic flexibility in a sense direct the selection and evolution of the phenotype? In general, the switch points of development coincide with the dissociation points of evolution. If a comparable relation holds at the level of the genome, it would imply that the activities and mechanisms of that regulate patterns of gene expression and manipulate the genome during ontogeny can presage and influence restructuring of germline DNA rather than only the reverse. This would occur not due to direct effects of ontogeny on the germline genome, but due to evolved predispositions of the genetic material that stem from its organization to behave in certain ways during development.

The expectation that gene-expression behavior may influence genome evolution has been expressed by some molecular biologists (e.g., see Shapiro, 1992, 1999), and seems borne out by various observed phenomena. Introns, for example, function as both gene-expression markers and markers for gene duplication, some transposable-element-mediated repeats and rearrangements, and exon shuffling during evolution (e.g., see Kidwell and Lisch, 2001; Consortium, 2001). Since introns presumably evolved primarily and originally as devices that facilitate gene expression during development, this is consistent with the hypothesis that a genomic, developmental device has secondarily facilitated evolution by exon shuffling. A corollary of this is that alternative splicing, including *trans*-splicing of chromosomally noncontiguous exons, may sometimes precede the evolution of the lineal sets of contiguous exons that characterize many genes. The challenge is to discover precisely how properties of genomes that are important in gene expression may affect the origin of permanent alterations in the transmitted genome. The evolution, under selection, of an exon-associated splice site, for example, may influence genome evolution because it predisposes that site to serve as an insertion point for transposable elements active in the germ line. This could conceivably facilitate exon shuffling and the origin of new genes (e.g., see Consortium, 2001; Kidwell and Lisch, 2001).

"Programmed rearrangements" in DNA sequence are sometimes, though not usually, prerequisites for gene expression (Haselkorn, 1992; Davidson, 1986, p. 13). Sequence rearrangement occurs in a yeast mating-type locus and some trypanosome coat proteins. In vertebrates, gene segments that are separate in germline cells are active only after they are rearranged in somatic cells to make a functional relationship (John and Miklos, 1988, p. 98). The stupendous diversity of molecules generated by the vertebrate immune system is a product of genes assembled anew *during development* via translocation of genomic fragments in somatic cells. In certain bacteria and fungi, gene expression involves moving unexpressed segments of DNA to homologous expression sites where they are inserted and then expressed (Thaler, 1994). Facultative somatic gene rearrangements are extremely common in bacteria (Haselkorn, 1992). Are the sites involved in such somatic events more likely to undergo rearrangement during evolution?

Gene duplication is another process that can characterize both gene expression and genome evolution. Are genes that are massively duplicated during gene expression (e.g., in the chorion genes of insect nurse cells—Büning, 1994; the salivary glands of the fly *Rhynchosciara anglae*, insect ribosomal genes in many species, various genes that provide toxicity resistence; and lampbrush chromosomes in insects and amphibians—reviewed in Davidson, 1986) more readily duplicated during evolution? Are some exons more frequently shuffled during evolution, and if so, are they also more dissociable and shuffled during gene expression? Since phenotypic behavior so often leads, via genetic variation and selection, to the genetic accommodation of particular morphologies, it is germane to ask if this kind of somatic gene responsiveness and behavior, in the immune system or elsewhere, ever leads to generation of selectable variation and evolved change in genome morphologies as well.

Some previous authors have suggested a relationship between somatic gene behavior and genomic evolution (McClintock, 1984; Thaler, 1994). McClintock titled her Nobel lecture (1984) "The Significance of Responses of the Genome to Challenge." She cited evidence, based on genetic studies of maize, that genome evolution is influenced by preexisting genomic behavior, or responsiveness. She argued that the changes that occur in the genome, including those induced by such experimental "shocks" as x-irradiation, are not really "mu-

tations" but the responses of living chromosomes. The reorderings are not yet predictable in advance, but nonetheless are repeatable if the same reorganizing "challenge" is repeated (see also Wessler, 1988). McClintock's suggestion implies that certain genomic alterations are more likely than others (see also Caporale, 1999a) and that the likelihood of alteration has to do with environmental responsiveness. It is only a small step from that idea to the possibility that evolved sensitivity to environmental influence during gene expression could influence susceptibility to certain kinds of structural change during evolution. Caporale (1984) reviewed kinds of DNA behavior during gene expression that could select for certain properties of the DNA itself, features that may then predispose the genome to certain kinds of variation and evolution (see also Shapiro, 1999).

To appreciate the argument that DNA behavior and manipulations during development may influence DNA evolution, it is only necessary to consider that the DNA itself has a selectable phenotype. DNA structure and behavior are subject to selection, for example, in the context of its interactions with RNA during transcription, and during DNA repair. Thus, for example, the "redundancy" of the genetic code—the fact that there are synonymous codons that can vary without affecting protein structure—may reflect, not that those codons are "neutral" or without functional effects, but that sequences evolve in the context of variable DNA behavior as well as in the context of protein specification (Caporale, 1984).

A very different view of how molecular interactions could affect genome evolution comes from the realization that cell nuclei contain a pool of unused and varied pre-RNAs that, like the noise or imprecision in the behavior of a fly larva described in chapter 9, could be seized upon opportunistically if useful (Herbert and Rich, 1999). Only about 5% (by mass) of pre-mRNA leaves the nuclei as mRNA (Lewin, 1980, cited by Herbert and Rich, 1999). Evidently, these are incidental transcripts, not part of the functional gene expression of a particular cell. As pointed out by Dickinson (1991), genes may respond to changing regulatory environments at times and places where they make no active contribution to the functions of a cell. In the nervous system of *Drosophila*, for example, there is patterned expression of enhancer detectors that "jumped at random into many genomic sites" (see references in Dickinson, 1991, p. 156). If such transcripts are not processed, they either are destroyed within the nucleus or remain untranslated. But if sporadic connections among these potentially functional gene products were to prove beneficial, like the recurrent anomalies of a lizard's foot (see chapter 19), they could become novel established traits.

Proteins that show domain sharing, or the appearance in different proteins of the same exon-specified domain, could conceivably originate in this way, from opportunistic use of anomalous but coherent fragments from a pre-RNA pool. Domain sharing is very common in the human genome and also occurs at only slightly lower frequencies in *Drosophila*, the nematode *Caenorhabditis elegans*, and at least some yeasts (Li et al., 2001). Similarly, novel defensive chemicals and signal pheromones may, more often than is usually realized, originate by "combinatorial chemistry" from pools or "libraries" of related compounds whose "random" or unordered flexibility facilitates rapid changes that elude the countermoves of enemies or competitors (Denton, 1998; Schröder et al., 1998). Evolutionary hypotheses for molecular evolution that begin with arrays of developmental and phenotypic anomalies or noise as the material for selection (e.g., Herbert and Rich, 1999) are closer to the general pattern seen in phenotypic evolution, as described in these chapters of part II, than is a view based on gradual accumulation of novel genomic mutations. Evolution at all levels of organization seems to capitalize on preexisting parts in new arrangements or contexts of expression.

Developmental plasticity may also influence the origin of genes by exon shuffling. In one view of the evolution of a new gene via exon shuffling, the rearrangement could occur first as a chance event and then be preserved because a functional combination happens to result (see Federoff, 1999). Alternatively, shuffled positions in the genome could follow, and depend upon, a phase of evolution in which gene products are developmentally linked (e.g., by RNA splicing) before being consolidated as a series of contiguous chromosomal exons. Such a phenotype-first hypothesis for the origin of new genes was suggested by Stone and Schwartz (1990c) in a discussion of the origin of introns. By their hypothesis, the production of a novel protein, by domain association in the cytoplasm, precedes the consolidation of a new gene. Consolidation occurs by crossing over and linkage followed by selection for the combination that produces the functional protein. Could unequal crossing over (a linkage hypothesis) account for the evolutionary consolidation of new genes in eukaryotes? *Trans*-splicing—joining of mRNAs from distant exons—occurs (see above), but the vast majority of proteins are encoded by linked or relatively contiguous DNA (E. Davidson, personal communication). This suggests that chromosomal contiguity is a more stable or efficient configuration than is noncontiguity for exons that are expressed together.

The Lamarkian-sounding suggestion that processes involved in gene use (expression) could influence the evolution of the genome may seem close to heresy. But the hypotheses just listed are not Lamarkian. They are standard Darwinian hypotheses of natural selection for fitness-enhancing structure and behavior, as applied to the structure and behavior of DNA and its products within cells.

Weismann's Barrier Revisited

The dual nature of inheritance, implying both the transmission and the development of traits (Darwin, 1859 [1872]; see chapter 8), is reflected in the dual functions of the genome as the agent of both the transmission and the expression of genes. The findings described in this chapter indicate that these two functions may sometimes be at odds. Gene expression requires active participation in a world of "chemical hostility" within somatic cells where DNA, as a polymer, is subject to being "replicated, degraded, nicked, modified, circularized, moved around and generally tinkered with" (John and Miklos, 1988, p. 15). Gene expression itself sometimes, though not always (see Davidson, 1986; John and Miklos, 1988), involves rearrangements of DNA that are passed from one cell generation to the next during individual development, as discussed in preceding sections. Gene transmission, by contrast, is expected to be selected for replicative accuracy. Although there may be some allowance for mutability permitting change, there is a premium on fidelity and resistence to the unscreened changes that can be inflicted, even at low frequency, during participation in somatic activity. In short, there is a potential conflict between the information-transmission and the information-expression functions of the genome.

Weismann (1893) noted that sequestration of the germ-cell lineage early in development serves as a barrier to inheritance of somatic effects on transmitted genes: the cells of the soma participate in growth and differentiation, but then they die, while the germline cells, set aside early in development, serve as an uncontaminated bridge to the next generation. "Weismann's barrier" to somatic influence on the germ line was a prime argument against Lamarkian inheritance of somatically acquired characters. Germline sequestration also circumvents intraindividual competition among cell lineages (and their mutable genomes) for access to the transmitted line, thus promoting the cooperation of somatic cells and the integrity of the individual (see Buss, 1987).

Here I propose that a beneficial side effect of germline sequestration may be protection of the germline genome from gene expression. This *genome-protection hypothesisis* is suggested by the fact that (1) gene activity probably increases the probability of genomic damage, rearrangement, or change, which would be to the detriment of transmission fidelity; and (2) certain patterns of gene expression suggest minimization of germline gene activity.

There is evidence in mammals that active regions of chromosomes are more susceptible than inactive regions to damage (reviewed in Bernstein and Bernstein, 1991, p. 28). In addition, the insertions and sequence duplications mediated by tranposable elements, which account for a large proportion of plant and animal genomes (up to 50% in human and maize; see Consortium, 2001; Kidwell and Lisch, 2001), depend to some extent on activity of the host genome in order to occur: transposition activities require host enzymes. It would be of interest to know whether silenced genomes (e.g., in the very condensed nucleus of animal sperm, or in silenced X chromosomes) have silenced transposons as well. That would strengthen the hypothesis that limited gene expression by the transmitted genome means limited rearrangement by transposons, for transposon activities, while sometimes leading to beneficial mutations, are known to reduce fitness (reviewed and discussed in Kidwell and Lisch, 2001).

Various observations suggest that suppression of expression has been an important mechanism by which host genomes combat parasitic transposon activity. Indeed, some authors have suggested that parasite silencing evolved first and was the basis for the later evolution of gene regulation that allowed differential gene expression during development (see references in Kidwell and Lisch, 2001, p. 10). McDonald (1998) briefly reviews various mechanisms associated with the silencing of transposable elements. The mechanisms (e.g., methylation and chromatin formation) for suppression of transposon activity suggest that temporary inactivation of the genome as a whole would likewise protect the transmitted genome from transposon activity that could alter the sequence of germline DNA. Thus, temporary germline sequestration and inactivity could perhaps be added to the list of

"global silencing" mechanisms (McDonald, 1998) that serve to protect the genome from parasitic rearrangements.

In many organisms, described below, germline cells or genes are inactive, or relatively inactive, compared with the somatic cells and genomes. The possible protective functions of sequestration do not conflict with those suggested by Weismann and Buss. Nor does the protection hypothesis conflict with the fact that embryos can sometimes be cloned from evidently undamaged adult somatic cells, since the hypothesis does not imply that the genomes of all somatic cells are damaged during somatic activity, only that genomic activity raises the probability of genomic damage or irreversible alteration.

Several observations are consistent with the genome-protection hypothesis:

1. The pattern of gene expression and quiescence in the germline genome: Diploid germline cells express many genes during cell divisions and migration within embryos prior to meiosis. Recombination during meiosis involves DNA repair and realignment of functional sequences (Bernstein and Bernstein, 1991). As a result, the postmeiotic germ cells contain a freshly cleaned up and orderly genome. There are many indications that the relatively pristine postmeiotic germline genome is then protected from further potentially damaging activity until again ensconced in the embryo: gene expression is reduced by various special devices. The chromosomes of mature gametes in insects, for example, are extraordinarily compacted rather than extended, a configuration characteristic of chromosomal inactivity (Büning, 1994). The morphologies of both sperm and egg are products of parental cells, not the germ-cell genome, and when germline cells contribute, the cells that are active in provisioning and structuring the egg or sperm are separate from those that become gametes.

This division of labor between germline and somatic products in provisioning of gametes seems to be universal in animals (Davidson, 1986): in general, the maternal transcripts used in early embryogenesis (called the "euplasm" in Büning's discussion of insect oocytes) come from the maternal germline genome, whereas the macromolecules that are nutritional rather than informational come from somatic and germline chromosomes (e.g., in insect nurse cells) that are separate from the transmitted germ line. Chromosomes of cells that are actively provisioning eggs sometimes undergo dramatic increases in ploidy and transcriptional activity, as exemplified by lampbrush chromosomes (Davidson, 1986; Büning, 1994). Büning (1994) regards this division of labor between somatic and germline as necessary to allow increased ploidy for the rapid provisioning of oocytes that are limited to a haploid state. Genomic protection could be an additional or secondary result. Although a few gametic genes are expressed postmeiotically during the period of sperm maturation in animals (especially mammals), only half of those are expressed somatically as well (Erickson, 1990). Such devices suggest a minimization of expression of transmitted genes that are to be used in the zygote, and a restriction of genes expressed during sperm maturation to genes used primarily in that context.

By these various mechanisms, postmeiotic genomic activity is reduced and the meiotically repaired germline genome is relatively protected during its transmission phase.

2. The extensive control of early embryonic development by maternal gene transcripts could contribute to protection from somatic gene-expression activity in cells destined for the germline, in addition to suppression of presequestration cell-lineage competition, as hypothesized by Buss (1987), who also noted restriction of mitotic activity as an important aspect of sequestration.

3. In some nematodes and chironomid flies (Diptera), there is "genomic reduction" or elimination of some chromosomes from somatic cells, but not from future germ cells (see references in Deuchar, 1975; John and Miklos, 1988). Since experimental excision of these chromosomes in *pregametic* germline cells prevents the formation of normal gametes (John and Miklos, 1988), these chromosomes evidently contain genes that are expressed only in the pregametic phase of development. That is, there is evidently some division of labor between genes active in the germ line and those that construct embryos. By this device, genes essential to gamete formation would be protected from any changes wrought during somatic activity.

4. *Hydra*, one of the animals discussed by Buss (1987) as lacking germline sequestration, nonetheless evidently has a kind of transmission-expression division of labor that may limit the gene expression of pregerm cells. The I cells of *Hydra* are like germline cells in being totipotent and in going through many rounds of mitosis before giving rise to gametes. They are also ameboid and relatively featureless, or limited in their specializations (Buss, 1987), suggesting that their totipotency is associated with limited gene expression compared with that of other cells.

5. All unicellular ciliates except one asexual marine group have dimorphic nuclei, with the germ line sequestered in a small micronucleus whose genome is completely silent (reviewed in John and Miklos, 1988, pp. 104ff). The micronucleus is used only in cell mating and is the only one capable of meiosis; the macronuclei make the RNA for the cell (Prescott, 1999). When ciliates mate, they exchange haploid micronuclei, and the resultant diploid micronucleus then divides without cell division, giving rise to a new micronucleus and a new macronucleus, while the old macronuclei and unused haploid micronuclei are degraded (Prescott, 1999). Somatic (vegetative) development and growth involves activity of the DNA of one or more large macronuclei in the cytoplasm. A peculiarity of the micronuclear genome is that some of its genes are "scrambled": the segments destined for macronucli are disordered. Then some unknown mechanism unscrambles them during macronuclear development so that they are in precise order for recombination (Prescott, 1999). Prescott speculates (p. 313) that this "fluidity" of the micronuclear genome, tolerated due to its transcriptional silence, may facilitate shuffling of segments of DNA that could be important in the orgin of novel genes.

Since plants lack germline sequestration, somatic mutations and rearrangements (e.g., due to transposon activity) can be inherited (e.g., McClintock, 1984; Wessler, 1988). In view of this, under the protection hypothesis one could expect special mechanisms in plants to deal with the greater potential for somatic damage to the transmitted genome. Postmeiotic gene expression in plant germ cells is restricted to one male pronucleus (Erickson, 1990). Transposable elements are relatively silent during normal development in plants and are activated by stresses such as wounding, pathogen attack, and cell culture (Wessler, 1996, cited in Kidwell and Lisch, 2001). This has been interpreted as a mechanism, due to selection on both host and parasitic transposable elements, to limit damage, with increased activity under stress interpreted as an adaptive response of the transposable element to increase likelihood of transmission to at least one offspring (Kidwell and Lisch, 2001). It is perhaps worth noting, however, that wounding, pathogen attack, and perhaps other "stresses" are associated with increased gene expression at loci involved in various kinds of defensive responses (e.g., Ryan, 1994; Ryan and Jagendorf, 1995; Lindquist and Craig, 1988). It would be of interest to know if the increased activity of transposable elements under stress in plants has as its proximate cause a general increase in genomic activity. This would be consistent with the protection hypothesis, which associates increased genomic activity with increased likelihood of genomic alteration.

Various other peculiarities of plants may reduce transmission of somatic mutations, whether or not they originally evolved in this context. Selfing, or self-fertilization, for example, is commonly interspersed with cross-breeding sexual reproduction in plants, and it quickly results in the segregation of lethal or sublethal types as homozygous recessives among offspring. Such homozygotes may die early in their growth, permitting somatically altered genotypes to be eliminated and well-adapted ones replicated (Briggs and Walters, 1988). Among other advantages (e.g., see Williams, 1975), the alternation of selfing with outcrossing may permit plants to control somatic damage to the genome while still having the benefits (e.g., variation and heterosis, or hybrid vigor) of sex. Although few comparative data are available, the supposed vulnerablility of the unsequestered plant genome to somatic mutation does not seem to have put the maize genome much behind the sequestered human genome in terms of cumulative mutation by transposable elements, which stands at about 50% of the contemporary genome for both (see above).

The interactions between transposable elements and organism genome can be viewed as products of a coevolutionary race between genomic parasites and host genome (Kidwell and Lisch, 2001). Many observations are consistent with this. In maize, for example, one of two genes expressed by the MuDR transposon is down-regulated in terminally differentiated somatic tissue and up-regulated in floral tissues (Donlin et al., 1995, cited by Kidwell and Lisch, 2001, p. 8). This would enhance likelihood of cross-generational transmission of the transposon. Some transposable elements are active only in germline cells, and in some species somatic activity of transposable elements is inhibited by proteins encoded by host genes (Kidwell and Lisch, 2001). These examples are evidence of the coevolutionary race between transposable elements and their hosts, in which the transposable elements win if they are able to parasitize germline cells. In such interactions, inactivation could be an effective defense of genomic integrity.

Whatever the fate of the very tentative protection-by-inactivity hypothesis to explain differences in the behavior of the somatic and transmitted genomes, it is clear that genome studies are pushing evolutionary biology to a new level of understanding of the nature of mutation. We need to face the growing evidence that there is feedback between genome activity and evolved genomic change.

Does Genetic Information Limit Evolution?

The common idea that evolutionary innovation depends on mutation implies that new phenotypes cannot arise without genomic change—that the information in the genome limits evolution. One effect of combinatorial evolution at the molecular level is to multiply manyfold the information potential of the genome. But the information explosion does not stop there, for it is again multiplied by the many kinds of combinatorial evolution at higher levels of organization described in the other chapters of part II. By analogy with human language, biological evolution proceeds not only by making many words (molecules) from a single set of letters (the genetic code). By taking a limited vocabulary of words, it can rearrange those into an even more enormous number of distinctive sentences, and then can continually and simultaneously reorganize paragraphs and pages and chapters and books (the phenotype at different levels of organization) without increasing the volume of the basic vocabulary at all. Chapter 18 shows, in addition, that reorganization of behavior, the phenotypic level most removed from the genes, can occur virtually instantly (in evolutionary time) due to learning.

Is there is a limit to the amount of information product that can be generated in this way, by reorganization, without changing the size of the genomic code? In theory, the number of permutations achievable by multiple recombinations at multiple levels of organization beginning with some 7,000 exons (the number estimated to compose all known proteins) is extremely large. The coded instructions are augmented at every level by the environmental and structural information built into the phenotype as it develops. That type of information is also affected by generations of selection on phenotype form and the devices that filter and direct organismic interactions with the environment. Evolved phenotypic diversification is therefore not likely to be limited by the variety of genomic information—by the mutation rate. If there is a limit imposed by information, it is more likely comparable to the limits to the information explosion when libraries fill until their walls burst, their floors collapse, their paper supply dwindles, and their attendants drop from exhaustion. That is, the limits to organic diversity may be physical-material-energetic, not informational.

Stated in other terms: developmental plasticity, more often than genomic expansion, increases the number of cell types and phenotypic specializations of multicellular organisms, enabling them to get more adaptive specialization per codon. This is self-evident from the condition sensitivity and combinatorial nature of gene expression and alternative splicing. This is an aspect of evolution that is systematically excluded by the genetic-informational approach. Thanks to plasticity, environmental materials and information are captured and used by a living organism and then are recombined along with the genotypic ones. This reduces the expectation of a proportionality between genome size and organismic (e.g., cellular) complexity. All of the major transitions to increased level's of complexity in the evolution of life, such as the transition to intracellular specializations, to multicellular organization, to sexual reproduction, and the origin of complex social organization (see chapter 29), depend upon the capacity for developmental plasticity in the use of a single genome.

The *informational fallacy*, the belief in a necessary, quantitative correspondence between informational (genomic) and phenotypic evolution, is inherent in the now refuted expectation that "genetic distance," or the degree of genetic difference (e.g., measured in base pairs, or allele-frequency changes) between populations would correspond to phenotypic distance. This expectation was strikingly contradicted by the very small genetic difference between humans and chimpanzees, estimated to be only about 1.1% of their total base pairs (King and Wilson, 1975).

The informational fallacy is carried to an extreme in the blueprint metaphor for development and the idea that "the problem of the complexity of organisms can thus be reduced to that of the complexity of their blueprints" (Eigen, 1992, p. 9). This has led to the expectation of a proportionality between increased DNA content (sometimes expressed in terms of C value) and the complexity of life, for example, measured in terms of number of specialized cell types (Kauffman, 1993), which has undergone a general increase over evolutionary time (see figure 7.9). The expected correlation generally does not hold, a fact known as the "C-value paradox" (e.g., see John and Miklos, 1988; Jockusch, 1997; Roth et al., 1997). Genome size, measured in picograms of DNA (1 pg = 10^{-12} g) or number of base pairs, does not increase with cell-type number, and even very closely related taxa sometimes have vastly different amounts of DNA. *Drosophila virilis*, for example, has a C value (0.34 pg) nearly double that of *D. melanogaster* (0.18 pg;

from a review in John and Miklos, 1988). Given the ubiquity of combinatorial evolution and the fact that phenotypic complexity depends on selective DNA expression, not total DNA amount, these are not surprising facts. The C-value paradox could well be recast as the C-value prediction: due to combinatorial evolution and developmental plasticity in gene expression, organismic complexity is not expected to be proportional to the DNA content of cells.

In Defense of Speculation

The interface between developmental molecular biology and Darwinian thought is the Wild West of evolutionary biology. Genome evolution and gene expression represent rapidly advancing frontiers that will continue to bombard evolutionary biology with unexplained facts about development. It is at such frontiers that speculation is most risky and most useful.

There are important cultural differences as well as similarities between these converging fields, molecular and evolutionary biology (see Raff, 1992; Wake, 1996b). Molecular biology has made spectacular advances based on experimental ingenuity, technical precision, and the generation of enormous databases (as in the human genome project). Alongside technical and interpretive precision in the things that matter within molecular biology is a conceptual sloppiness in things that matter in evolutionary and organismic biology. There is teleological language that refers to evolution in response to "needs"; disregard for the boundary between phenotype (protein) and genotype; and high-profile papers on "flies and worms" as if all flies and all worms were the same, or "the fly and the worm" as if there were only one of each. "Vertebrates" may mean humans and mice or, among extreme generalists, may include "frogs," meaning the African aquatic fanged frog *Xenopus laevis* (Anura, Pipidae). On the other side of the turbulent stream are the evolutionary organismal biologists, for whom a worm could be an annelid rather than a nematode, or a zinc finger part of a robot. While placing a high value on conceptual originality and terminological precision, this breed of biologist will sometimes construct elaborate theoretical edifices on a set of data smaller than those generated in a molecular lab in less than an hour. Among the similarities between the fields is an abundance of testable speculations regarding functions and origins of unexplained patterns. Informed just-so stories have long distinguished the writings of the acknowledged leaders in both fields, for they draw pattern out of an otherwise confused mass of facts.

Molecular, developmental, and evolutionary biology have existed side by side for decades with little overlap beyond the use of borrowed molecular techniques as tools in evolutionary biology, and the occasional use of borrowed concepts (homology, phylogeny, selection, and drift) in discussions of molecular evolution. But there is a growing convergence of interests, and this is finally driving a more profound synthesis (see especially Raff, 1996; Gerhart and Kirschner, 1997; and Caporale, 1999b). Molecular developmental biology has opened the black box between the phenotype and the genes, bridging the gap between what is selected and transmitted and what is expressed; and modern evolutionary biology can help to make sense out of a humbling mass of poorly understood facts about mechanisms of development and the behavior of the genome. A collaborative synthesis is anticipated from both sides:

> As our predecessors in embryology were well aware a century ago, comparative considerations can be immensely illuminating, and, even in the advanced state to which we have been carried by our superb molecular technology, it might still be said that we need all the help that we can get in trying to understand embryonic development. (Davidson, 1991, p. 1).

> While Huxley and de Beer, and later, Waddington were concerned with the development-evolution relationship and made important contributions, the promise of a synthesis has yet to be attained. . . . [T]echnical and conceptual advances have at last made such a synthesis an achievable goal. (Wake, 1996b, p. 97)

18

Phenotypic Recombination Due to Learning

Contrary to common belief, learned traits can have high evolutionary potential. Learning, like other condition-sensitive regulatory mechanisms, can produce close environment–phenotype matching and the recurrent expression of behavioral traits. It exposes behavior and associated morphology to context-specific selection and, if a particular environment prevails, directional evolution. Learning itself can be fine-tuned under selection and then can mimic selection to create, and rapidly spread, novel adaptive traits. The other side of learning—forgetting—is also a specialized trait with its own set of distinctive properties.

Introduction

Learning, like consciousness, is something that everybody can recognize and no one can define without provoking controversy. Perhaps this is why some important books dedicate hundreds of pages to learning without defining it (e.g., Mackintosh, 1974; Marler and Terrace, 1984). In one unusually candid book, the indexed page that promised a definition of learning proved to be completely blank (Bell, 1991, p. 301). That stimulated me to make my own definition, something that is easier for a person who is not an expert in the field: *learning is a change in the nervous system manifested as altered behavior due to experience* (based on discussions in Marler and Terrace, 1984; Bell, 1991; Mackintosh, 1974, 1983; Papaj, 1994).

Most people, including most biologists, probably underestimate the importance of learning in the biology of nonhuman animals. But there have been important exceptions, for example, in the writings of Baldwin (1902), Hinde (1959), Partridge (1983), Roper (1983a,b), Slater (1983, 1986), Shettleworth (1984), Davey (1989), Wcislo (1989), Real (1993, 1994), Dyer (1994), Morse (1980), Marler (1998), and others (see Marler and Terrace, 1984). Some form of learning, whether habituation, associative learning (Pavlovian conditioning, in which a reward or punishment is associated with some cue such a color, odor, or sound), aversive learning, or trial and error learning (operant conditioning, in which a rewarded behavior is repeated or a pun-

ished one stopped), seems to occur in all animal groups where there is enough versatility in movement to allow it to be recognized. The venerable animal psychology text by Maier and Schneirla (1935 [1964]) gives many interesting examples from a time when researchers sought to demonstrate learning in a wide variety of organisms. They found it even in protists. In more recent research in areas such as foraging behavior and kin recognition (e.g., see Heinrich, 1979; Fletcher and Michener, 1987), learning has proven to be important but is a sidelight to research more concerned with optimization and adaptation. So learning itself has not always received the attention it deserves as a phenomenon of general evolutionary interest.

This chapter does not attempt to review the natural history of learning or even its evolutionary significance. In keeping with the other chapters of part II, I focus instead on the importance of learning for the origin of novel adaptive phenotypes. Learning, like other kinds of regulatory mechanisms, influences the frequency and conditions in which traits are expressed. It can therefore affect the rate and the direction of evolution. This effect of learning obtains regardless of the precise nature or kind of learning that is involved. For the present discussion, then, we can sidestep certain controversies, such as that between the "cognition" (internal representation) and the "releasing-mechanism" schools of thought (see Marler and Terrace, 1984), and move on to the question of their phenotypic results. As long as there is a behavioral decision mediated by learning, learning potentially affects the recurrent expression of particular behaviors, the action of selection, and the course of evolution.

In development, a continuum of decisions occur at different levels of organization, from the intron-mediated expression of genomic exons to learned decisions between complex behavioral alternatives. Learning can involve integrated assessment of multiple factors and may employ numerous perceptual and internal abilities, including memory. On this continuum of complexity, learning is among the most highly condition-sensitive and also highly polygenic developmental mechanisms. As such, by the arguments of chapters 6 and 7, learned traits and learning mechanisms can have relatively great evolutionary potential. Learning, like other genomically and environmentally responsive regulatory mechanisms, can also produce maladaptive novelties, and this is expected to affect the evolution of learning itself.

Learning in Relation to Selection and Evolution

How can *learning* lead to genetical evolutionary change? It would seem a poor agent of genetical evolution. Learned correlations form anew in each individual. Except in species with cultural transmission of particular traits (e.g., in birds that imitate song and in primates that imitate food cleaning behaviors, dominance status, or language), the learned correlations, however adaptive they may be, seem destined to die with the individual who expresses them. The answer to skepticism regarding the evolutionary importance of learned traits is the same as for any other conditionally expressed trait (i.e., all traits): recurrent phenotypes, whatever the mechanism of their production, can be subject to genetic variation, selection, and genetical evolution. Learning, as shown in chapter 3, is just one among many environmentally responsive regulatory mechanisms that coordinate trait expression and determine the circumstances in which they are exposed to selection.

Learning as a Developmental Source of Evolved Correlations

Learning is a developmental mechanism that, like hormones and positional interactions, acts to form correlations among traits. In the learned dietary specializations of some birds, for example, the form and the size of the beak render some dietary choices more quickly rewarding than others, producing *morphology-biased learning*. This sets up a correlation between beak morphology and behavior that may in turn affect intestinal morphology and physiology (discussed below). The integrating mechanism is learning: it produces an association between beak morphology, diet, food-handling behavior, and gut physiology. Then this coordinated, complex expressed phenotype is subject as a unit to selection.

Recall that a conditional phenotype need not be *fixed* to be modified as a unit by selection, but it does need to be recurrent. Learning causes recurrence of particular, reinforced behaviors or phenotype sets. Recurrence of a learned trait complex such as the beak–behavior–dietary–gut complex just described exposes it to selection that can favor supporting modifications in genetically variable morphology (e.g., of mandibular muscles, gut characteristics, and bone structure during growth). This can become a self-reinforcing process, for the evolved phenotypic changes can augment the learned preference as individuals become better equipped to obtain a particular reward, which leads to further recurrence of trait expression and further selection toward specialization at the expense of other options. The specialized trait is still learned, but it has been refined and elaborated—genetically accommodated—by selection. The importance of learning for adaptive radiations is discussed in chapter 28. Like hormones (Ketterson and Nolan, 1992) and other developmental coordinating mechanisms, learning creates phenotypic and (due to selection) genetic correlations. Correlation building by simple learning must be immensely important for the evolution of novel adaptive traits, more important than some other mechanisms that have received far more theoretical attention in evolutionary biology, such as chromosomal linkage (see chapter 4).

The potential for the evolutionary modification of learned traits without making them genotypically determined is a key to understanding the evolutionary biology of learning. In the past, evolutionary discussions of learning and evolution have tended to emphasize either the idea that learning ability is the thing that evolves, or the idea that learned traits can be genetically assimilated—the idea that a "genetic" or constitutively expressed trait can begin as a learned one (Baldwin, 1902; Papaj, 1994). Both of these may occur, but the more important general relationships between learning and genetical evolution are the fitness-screening aspect (see below), and the potential for self-reinforcing adaptive modification just described.

The Genetic Accommodation of Learned Traits

A common line of reasoning sees learned traits as immune to selection and divorced from evolution as a result of being "nongenetic" in nature. Contrary to this belief, learning is a prodigious source of adaptive novelties. As already explained, learning is a mechanism for producing trait recurrence, one of the prerequisites for adaptive evolution (see chapter 6). In this respect, learned traits have a greater evolutionary potential than do mutational ones, since many individuals of a population may simultaneously and suddenly learn the same things in the same circumstances, or due to mimicry of other individuals (see chapter 26). At the same time, learning can match trait to conditions, thereby bringing a learning-linked trait complex under consistent selection in the environment where it is likely to be adaptive.

How rapidly genetic accommodation of an advantageous learned trait can occur is illustrated by observations on honeybees. Individual honeybees are well known for their ability to learn to return to rich food sources. Weaver (1957b) noted the marked individuality of different bees in their methods and approach to foraging on certain flowers, especially the variable amount of time spent in tripping blossoms and foraging from them (Weaver, 1956; see also Heinrich, 1979, on bumblebees). Such variability with respect to exploitation of particular plants can have both a selectable, genetic component and fitness consequences for bees in different ecological circumstances.

The genetic component was verified in pioneer efforts to breed honeybees (*Apis mellifera*) specialized to pollinate particular domesticated crop plants that would not have been part of the environment of ancestral bees. Nye and Mackensen, in studies reviewed by Cale and Rothenbuhler (1975), recorded differences among colonies in foraging behavior and were able to select outbreeding lines increasingly specialized to foraging on alfalfa (*Medicago sativa*). In one area of the midwestern United States, they increased the alfalfa foraging to 68% compared to 18% in controls (see Cale and Rothenbuhler, 1975, p. 178). In effect, these experimental breeding programs selected for an increased preference for alfalfa, showing that plant specialization

is heritable and that there is a potential for genetic accommodation of a novel flower preference if it proves advantageous. This demonstrates the potential for a shift in ecological niche as proposed by Baldwin (1902), first by learned or flexible behavioral responses, then followed by genetic accommodation increasing the frequency of a particular specialization, in this case one among several in the repertoire of a genetically variable, generalist forager.

The propensity for genetic accommodation of a trait such as flower or host preference could be increased by the complexity of its regulation. A preference for alfalfa, for example, could involve genetic variation in such diverse characteristics as sensory acuity in distinguishing among plants, learning ability, morphological ability to work a particular type of flower, and physiological abilities of both adults and larvae to process particular kinds of pollen. The switch to a particular specialization such as alfalfa foraging is undoubtedly subject to polygenic influence.

In bumblebees, flower preference correlates with size and tongue length of individual bees (Heinrich, 1979). Those with long tongues specialize in long-corolla flowers. Evidently, this is due to reinforcement learning to prefer them over short-corolla flowers, where their long tongues can be a liability (Heinrich, 1979, p. 152). Clearly, there is a kind of circular interaction between morphology and learning in evolution, learning being influenced by morphology and the resulting learned specialization in turn influencing the direction of selection on morphology itself (see Grant, 1986; Wcislo, 1989; see also chapter 28).

Mimicry of Natural Selection by Learning: Learning as a Fitness-Effect Screening Device

Learning has a special quality that sets it apart from other mechanisms of plasticity and makes it unusually important for adaptive evolution. When learning involves evolved motivational cues and reward criteria, it imposes a fitness-effect assay on behavioral decisions. Motivations and rewards molded by natural selection provide a built-in means of trial-and-error screening for those behaviors that are likely to have positive fitness effects, leading to the repetition of those likely to be beneficial to the organism. In the past, learning and natural selection have been compared within the framework of genetics. The learned counterparts of genes have been called "memes" (Dawkins, 1976) and the transmission aspect of learning emphasized: learned traits spread within populations, then may be transmitted culturally to the next generation. When the evolutionary significance of learning is considered in this framework, the analogy between learning and selection in the evaluation of the fitness effects of traits is overlooked.

If the effect of learning were confined to making certain behaviors recur, then learning would be developmentally comparable to a mutagenic device that forms novel phenotypes of variable effects, most of them negative. But a learned novel phenotype influenced by evolved motivation and reward criteria is not random with respect to fitness in the same way a mutant genetic mutant novelty is. A learned novelty is a product of the assessment of alternatives, which are screened by a system of differential rewards. If a honeybee, for example, discovers a novel flower patch with high nectar yield, it learns to recognize and return to that place. The bee is motivated to forage for food, and nectar rewards train it to exploit rich sources (summarized in Winston, 1987). Because bees vary genetically in their sensory acuity and responsiveness to sugar concentration (e.g., see Page et al., 1998; Pankiw and Page, 1999), selection can lead to the increase of the frequency of a particular foraging specialization in a population if it were to prove overwhelmingly profitable compared with other specializations. This has been demonstrated by successful artificial selection on honeybees specialized to forage on alfalfa, as discussed above. Reward-based learning also leads to improved efficiency in the manipulation and exploitation of particular kinds of flowers by individual bees (e.g., see Heinrich, 1979, on bumblebees). Young honeybees foraging for pollen have to learn to "trip" flowers in order to reach the pollen, and inexperienced bees often get trapped, but experienced bees avoid the trap and forage more efficiently (Gary, 1975). Efficiency, or the ease of obtaining a reward, depends in part upon morphology (e.g., tongue length), and individual bees develop learned specializations (Heinrich, 1979; Weaver, 1956, 1957b).

Rewarded behavioral routines and behavior–morphology combinations are facultatively repeated; unrewarded ones are not. Most important, the criteria for a positive reward can be defined by selection, such that the rewarded behavior has an increased likelihood of a positive effect on fitness (e.g., it satisfies an internally specified need). This built-in, evolved fitness assay means that with learning there is a rough prescreening of adaptive correlations. The behavioral and morphology–behavior (use) combinations that are repeated by individuals of a population are those that are subject to selection. And since the reward and punishment structure of simple learning can be a good predictor of fitness payoff (e.g., success in obtaining some resource, or avoidance of harmful stimuli), this can lead to the construction of correlations with likely fitness benefits, giving a head start to adaptive evolutionary change.

It is easy to think of disadvantageous learned traits in humans, such as a love of alcohol or speeding on highways. It is not so easy to think of examples in other animals, presumably because selection would adjust such parameters as reward criteria and motivation thresholds so as to decrease the frequency of maladaptive learned traits.

Learning, then, is a fitness-enhancing mechanism par excellence, a mechanism formed by natural selection in such a way that it mimics natural selection as a multiplier of adaptive traits. Context-specific motivation defines the context for selection, and positive and negative reinforcement carries out selection during ontogeny, retaining (in accord with evolved context-specific reward criteria) the responses that are likely to be adaptive. It is this property of learning that makes evolutionary or sociobiological theories based on natural selection theory good predictors of pattern in behavior, even when the patterns themselves are learned, not evolved.

This is one reason why some genetic models have been successfully applied to learned (including culturally transmitted) traits, even though the (genetic) parameters of the models are not actually in play during the ontogeny of the traits. Cues that indicate degree of genetic kinship with other group members, for example, are learned by individuals in many kinds of social animals, from isopods to humans, and individuals perform beneficent behavior in accord with their relatedness to beneficiaries of the acts. They behave as predicted by quantitative genetic models of kin selection (see chapter 23), which are stated in terms of genes for altruistic behavioral traits. Genetic models may accurately predict the patterns of adaptive behavior. But, as pointed out by Papaj (1994), the impression that behavior is therefore genetically determined is false, and this has been a serious source of confusion in discussions of behavioral evolution. Instead, learning may have enabled fulfillment of one or more of the genetic parameters of the model (in the example just given, it enabled estimation of degree of genetic relatedness).

When a foraging bee specializes on the most profitable available flower type, it behaves as predicted by natural selection theory, but it does so on the basis of learned rewards; it behaves, due to learning, as a population of similar bees would if natural selection had screened behavior in the precise situation of flower availability experienced by the bee for a period generations. But the bees are able, through learning based on an evolved system motivations and rewards, to evaluate adaptive options immediately, without waiting for natural selection to do so. It is evident that there can be an evolved analogy between the operation of natural selection and the operation of learning to produce adaptive situation-appropriate behaviors (e.g., see also Lorenz, 1965). The exact nature of this analogy has long been a subject of debate (Boyd and Richerson, 1983).

Here I discuss how motivation and rewards are likely to evolve so as to reflect and predict the fitness effects of particular learned behaviors, thus allowing learning to anticipate selection in screening for adaptive traits. This mimicry of selection by learning is likely to accelerate the evolutionary improvement of learned behaviors, as demonstrated by simulations that compare the rates of evolution of learned and constituitive traits (Papaj, 1994).

Evolved Components of Learning

It is often said of learned traits that what evolves is not the trait itself but the capacity to learn. Here I regard both as somewhat independent potentially genetically variable aspects of the phenotype: with reference to figure 4.13, learning is a regulatory mechanism that influences the expression of certain behavioral traits and the use of certain morphological ones. This implies that both the traits themselves and the capacity to learn can evolve.

What is "the capacity to learn"? It could be assumed to reside entirely in the central nervous system and to have to do with memory. But this is an incomplete idea of learning and would lead to incorrect conclusions, such as the idea that evolved improvement in learning requires improvement in the brain, or that all context-specific learning is due to the evolution of specialized brain centers dedicated to particular tasks. But enhanced learning ability can derive from other components of learning, such as the specificity of motivation, sensory acuity in the identification of rewards, or boldness in exploration.

The kind of learning that is most often documented in the natural history of animals is improvement in the ability to find and acquire re-

sources, such as food, water, shelter, or mates. Learning in these contexts seems always to involve the following four evolved components.

1. Resource-specific motivation: *Motivation* can be operationally defined in terms of the intensity or magnitude of an animal's response to stimuli associated with acquisition of a particular resource, for example, the vigor of mating behavior or the rate or amount of eating by a food-deprived animal (Stellar, 1987). Frank (1996, p. 468, after McFarland and Bösser, 1993) aptly calls motivation an "internal value system." The response to appropriate stimuli reflects an internal state, involving such things as hormones, reception of external stimuli, and sensory devices that indicate internal scarcity or external availability of a resource. It can include thresholds for activity of specific kinds, and what Posner (1994) and Bernays and Wcislo (1994) call "attention" to specific resource-related stimuli, reflecting the absence of a resource essential to performance of a particular task (metabolism, reproduction, escape, nutrients, etc.). Motivation is resource specific. An symptom of motivation is the sensation of hunger in a food-deprived individual.

2. A repertoire of motivation-specific *exploratory behaviors* that have proven successful in the evolutionary or experiential past: Sufficiently high motivation activates a repertoire of evolved, context-specific unlearned or previously learned "trial" movements. This is the bag of tricks that have been successful in the past and have therefore been accumulated due to either natural selection or previous learning. These might include such behaviors as searching in a particular pattern, handling particular kinds of objects, investigating particular kinds of places, alertness to a particular search image, or simply increased general activity, alertness, or investigative boldness. Because these maneuvers are motivation specific—induced by a motivational state connected with a particular resource either by evolutionary history or associative learning during ontogeny—they are context appropriate. An example of the exploratory aspect of learning is, in a hungry animal, sniffing the air, looking for particular kinds of prey-associated movements, or flying around the leaves of particular types of plants (Bell, 1991).

3. *Reinforcement* by context-specific sensations of reward or punishment: Reinforcement, or somatic selection, of successful solutions results from evolved, and potentially learned, pleasurable feelings associated with acquisition of a particular kind of resource, and avoidance of inappropriate solutions. An example of a reward, or positive rein-

forcement, in the food-searching context, is agreeable rather than disagreeable taste; negative reinforcement may be accomplished by distasteful bitterness, or illness immediately after ingestion.

4. *Discernment and memory* of rewarded or punished trials: Part of central processing during learning is discernment regarding which tentative solutions were successful, and ability to remember them in association with a particular quest and reward. We might describe this as part of "analytical ability"—perceptual acuity, or being "observant" as well as having a good memory.

It is easy to see room for genetic variation in all of these evolvable components of the ability to learn. Individual organisms vary genetically in their motivational levels whenever there are genetic differences in such things as hormone systems, or sensory acuity or ability to see relevant stimuli in the environment. They may vary genetically in aspects of morphology (e.g., muscle size, beak length) that make certain exploratory maneuvers easier, or more effective than others. They may vary genetically in their "tastes" or precise sensations of what is delicious or disgusting, and they may vary in their ability to observe the details of successful maneuvers and remember them or match them to particular tasks. In other words, differences in learning ability have to do with motivation, including hormones and social interactions, motivation-maneuver matching, rewards, and sensory apparatus—not just with the memory centers of the brain. So it is not surprising that artificial selection for learning ability, or "conditionability" in flies ended up producing increased sensory acuity (Hirsch and Holliday, 1986). This is not a failure to select for learning ability. It could be interpreted as successful selection for an important component of the ability to learn. Learned behavior is the category of phenotypic variation most likely to be polygenic.

Because these elements of the mechanism of learning are subject to natural selection, there can be species-specific specialization in efficiency at particular kinds or contexts of learning. Marler (1998) calls this "innate learning predisposition." Inexperienced forager bees, for example, learn most quickly when trained using colors that are most likely to be food signals (violet, blue, and purple), whereas background colors (e.g., green) are learned more slowly (see references in Waddington, 1983, p. 219). This could conceivably be due to species-specific exploratory behavior limited to the colors more likely to be successful (component 2, above), sometimes referred to as "phylogenetic prelearning" (term attributed to R. Menzel by Waddington, 1983, p. 218).

There has been a general reluctance in biology to use the term "motivation". But some remarkable devices associated with adaptive flexibility show how the vague-sounding phenomenon of motivation manifests in concrete terms (see references in Bernays and Wcislo, 1994). Following a full meal, locusts (*Schistocerca americana*) shut down their attention to food-getting activity. They move away from the food item they have been eating, thus terminating taste input from the tarsal chemoreceptors; hormone secretions from the corpora cardiaca cause closure of chemoreceptive pores on the palps; and increased nutrients in the blood cause elevated chemosensory thresholds to the ingested nutrients. All of these changes are concrete and presumably evolved manifestations of a lowered motivation to seek food, and they would reduce exploration and, temporarily, ability to learn in this particular context.

I give this rather elemental discussion of the evolvable rudimentary components of learning because most biologists do not think about learning, and most learning theorists do not think about learning in evolutionary terms. Discussions of learning, in both biology and psychology, are incomplete as a result.

Of course, there are other evolved traits and kinds of phenotypic variation that can affect learning ability. Tropical species of wrens, for example, vary in their degree of flexibility in foraging behavior (Greenberg, 1985). While laboratory tests of foliage-type learning showed no difference in learning ability between the species, the one observed to be most flexible in the field was less fearful of novel objects. This boldness, or what we might call "active curiosity," could affect likelihood of learned flexibility. To learn by trial and error in a particular context requires that an individual dare to perform the appropriate trials.

Learned Components of Evolved Traits

Learning is probably far more important as a mechanism of development of adaptive traits in nature than is usually realized. It is well known to be essential for such behaviors as feeding preferences, orientation and homing, and individual trophic differences, in many birds, insects, mammals, and fish (see references in Scapini, 1988; Magurran, 1987; Giraldeau, 1984). Complex learning does not require an unusually large brain. Multifactorial learned discrimination occurs in the desert isopod

Fig. 18.1. Desert isopods (*Hemilepsis reaumuri*) at the entrance to their burrow. From Linsenmair (1987). © John Wiley & Sons Limited. Reproduced with permission. Courtesy of K. E. Linsenmaier.

Hemilepsis reaumuri, a relative of the familiar "pill-bugs" found under rocks in North America and Europe (figure 18.1). This is a group not expected on the basis of brain size to be especially brilliant, since an individual has a brain of only about 10,000 neurons (the human cerebral cortex contains about 20 billion neurons—Calvin, 1983). But even six-week-old babies of *H. reaumuri* learn to distinguish kin by memorizing a composite identification odor composed of a complex blend of substances representing the odors of all family members (reviewed in Linsenmair, 1987). Research on learned kin recognition indicates that most if not all of the many species of diverse vertebrates and invertebrates that live in family groups and defend their group or residence against conspecific intruders probably resemble the desert isopods in being able to learn a family identification badge (reviewed in Fletcher and Michener, 1987; Hepper, 1991).

The discrimination and learning capacities of isopods perhaps should not be surprising, since crustaceans are well known to employ learning in other contexts. Individual mantis shrimps (stomatopods) "show considerable plasticity in learning how to open various kinds of shell," some breaking the apex of the shell first, others beginning with the lip, and others shearing off the whorls on one side, improving with practice when given novel food items. Each animal developed its own style of attacking shells and used a favorite rock as an anvil (Caldwell and Dingle, 1976). Talitrid shrimp (amphipod crustaceans) use evolved visual orientation cues for locating the direction of the sea but are able to adjust them through learning when resident

on shifting seashores (Ugolini and Scapini, 1988). The reward for success is hypothesized to be positive reinforcement of branchiae rehydration on or in the damp sand (Scapini, 1988).

Return migrations and homing orientation are contexts, along with kin recognition, in which normal development inevitably and necessarily involves learning in a very great diversity of animals. Scapini (1988) concisely summarizes experimental knowledge of learning in relation to heritability in the orientation of crabs, turtles, fish, and birds. Learned nest-site orientation also occurs, of course, in all of the many species of nest-building insects and mammals. Individuals of many of these species occupy a series of different nest sites during their lifetimes and so are equipped to facultatively forget as well as to learn.

Learning can affect species-specific traits usually regarded as stereotyped and unaffected by learning. Insect mating behavior is such a trait. But in fruit-flies of the genus *Drosophila*, learning and memory affect courtship (see especially Siegel and Hall, 1979; Koref-Santibañez, 1986). Males exhibit less active courtship behavior toward receptive females after spending time with unreceptive (mated) females. This appears to involve memory because mutations such as "amnesiac," known to interfere specifically with retention of learned behaviors, interfere with the depressed-courtship response. In chapter 28 I discuss how learning may have contributed to the adaptive radiation of *Drosophila* species in Hawaii by influencing the oviposition sites of females (Jaenike, 1982, 1985, 1988). Learned character sets seen as individual differences in one species sometimes parallel those seen as established traits in a closely related species (see chapter 28). This accords with the hypothesis that learned novelties, like other kinds of flexible alternative phenotypes, can play a role in the origin of fixed traits (see part III).

Individually learned behavior patterns could contribute to niche shifts of the type associated with ecological specialization and species differences if they were to be learned with increasing frequency by the individuals of a population. Particular individual blue butterflies (*Glaucopsyche lygdamus*) repeatedly oviposit on one species of legume, even though others used by different butterflies of the same species are available. Once a particular legume species is used, it is reused, and landing on others does not result in oviposition (Bernays and Wcislo, 1994). Similarly, female sphecid wasps (*Stictia heros*) normally hunt and capture free-living prey, but when nesting in aggregations they sometimes steal prey from other females or other nests. Once this occurs, it appears to be perpetuated in a particular individual (Villalobos and Shelly, 1996). "Runs" of at least temporary specialization on an encountered food are very commonly observed (e.g., see Partridge, 1976) not only in birds and insects but also in a mollusk and several rodents (Partridge and Green, 1984, p. 211) and the lizard *Anolis lineatopus* (Curio, 1976). This kind of behavioral recurrence may or may not be experience or success mediated, but it can at least provisionally be assumed to be due to reinforcement, for it seems unlikely to occur if unrewarded (fruitless). Even though the mechanism (e.g., the nature of a reward, and the conditions of extinction or switching to another pattern) may be unknown, such examples produce recurrent associations of particular movements and morphological and physiological traits with particular environmental conditions—the property of learning that is of interest here because it allows selection to act on fitness-affecting variations in performance.

These kinds of relatively simple learning should not be taken to indicate unusual "higher faculties." Belief that an effect of experience must involve humanlike abilities may contribute to the neglect of learning in evolutionary biology. Darwin's ideas on sexual selection were ignored for decades because female choice was interpreted to imply a sophisticated aesthetic sense and memory, when all that is required to make it work is differential responsiveness to phenotypically different males. It would be unfortunate if a similar misconception were to detract from the study of learning as a taxonomically widespread cause of correlations among traits.

Learning and Individual Differences in the Evolution of Specialization

A common result of simple learning is to create individual differences in behavior. Learned individual differences are the result of multidimensional plasticity in continuously variable traits whose magnitudes (e.g., frequency or intensity of performance) can be influenced by experience. This establishes idiosyncratic combinations or interrelations of traits such that different individuals have different complex phenotypes with each one at a low frequency in the population (hence the *individuality* of the differences). The greater the number of variable traits involved, the more highly idiosyncratic individual differences can be.

In contrast with some other authors (e.g., Magurran, 1987; Clark and Ehlinger, 1987) I dis-

tinguish between low-frequency individual differences, which are the behavioral counterparts of complex morphological anomalies (low-frequency variants), and alternative phenotypes (*sensu* West-Eberhard, 1986), where particular trait sets are at a relatively high frequency and governed by a single switch. Individual differences that increase in frequency could become alternative phenotypes, so the distinction is quantitative and somewhat arbitrary. The solitaria and gregaria "morphs" of migratory locusts illustrate an intermediate condition where a ganged switch, when not completely thrown, produces intermediate forms that could be called individual differences (see chapter 5).

Not surprisingly, given the commonness of learning in the animal kingdom, learned intraspecific individual differences in behavior are common in a wide taxonomic range of species in nature, including insects, fish, birds, and mammals (reviewed in Giraldeau, 1984; Magurran, 1987; Clark and Ehlinger, 1987) as well as crustaceans (Caldwell and Dingle, 1976)—to mention only those where individual differences have been singled out for study or special mention. Armitage (1986) describes individuality in the social traits of marmots (*Marmota flaviventris*), including in such continuously variable characteristics as dispersal versus residentiality, dominance and subordinance, and approach versus avoidance of strangers. Learning is likely to influence this individuality, since these are traits known to be experience mediated in other mammals (see references in Armitage, 1986).

Learned individual specializations in diet are very common in birds (Giraldeau, 1984). For example, after four days of feeding on a mixture of seven seed types, examination of the stomach contents of 57 feral rock doves (*Columba livia*) showed extreme individual specializations, with every one showing significant departures from the seed-type proportions present in the mixture (Giraldeau and Lefebvre, 1985; figure 18.2). Similar data are given by Bryan and Larkin (1972) on trout.

The digger wasp *Sphex ichneumoneus* could be described as a generalized predator that provisions with diverse species of three families of Orthoptera (Brockmann, 1985a). Certain females, however, show marked individual specializations. In a population where most individuals (58%) used prey species of from three to five orthopteran genera, one eccentric female accounted for 65% of all the cone-headed katydids captured by that population (Brockmann, 1985a). Most revealing was the behavior of some injured females, which suggests that the physical phenotype of individuals affects which behavioral trait is learned. Two individuals spe-

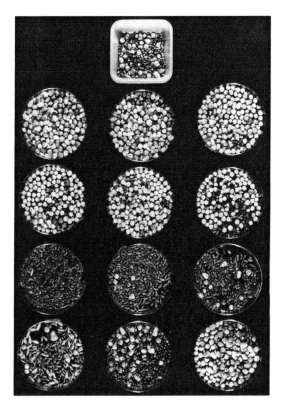

Fig. 18.2. Individual feeding specializations in feral rock doves (*Columba livia*). A commercial mixture of seven seed types (square dish, top) offered to feeding birds, and 12 of the 57 crop-content samples (round dishes). Individuals represented on the left specialized in a single seed type (top to bottom): corn, white peas, wheat, and oats. Those represented by the middle column specialized in a mixture of two or three seeds (top to bottom): corn, white peas, and maple peas; white peas and wheat; wheat and a few white peas; corn, wheat, and milo. The right column gives other examples of crop content (top to bottom): white peas and corn; white peas, vetch, wheat, and corn; white peas, corn, maple peas, and wheat. Individual differences are more striking in color. From Giraldeau and Lefebvre (1985), by permission of NRC Research Press. Courtesy of L. Giraldeau.

cialized in tree crickets (*Oecanthus*), relatively small and distinctively located arboreal prey that were not captured by any other females; these two wasps had defective wings and could fly only with difficulty. Predatory wasps with heavy prey often carry them to high places from which they can more easily take flight with them (personal observation),

and the two handicapped females may have been successful in predation and retrieval of tree crickets due to their location in high places suitable for ease of takeoff while bearing a load. Since those females repeatedly predated tree crickets, they may have learned to do so, a possibility that could be experimentally investigated in this species by altering or weighting wings. Brockmann's observations indicate that individual differences are not haphazard variations but may show pattern dictated by morphological variation or other phenotypic peculiarities of individuals—a two-legged goat effect that could, if a morphological change were recurrent, become an established trait.

It may not be obvious how the unpatterned, complex variation of individual differences could have evolutionary importance. The conversion of idiosyncratic learned or compensatory variants into recurrent, evolved genetical ones can have many causes, but the first critical step, upon which all others will depend (given genetic variation and a fitness effect), is for the trait to become recurrent. This is where learning could play a role, as already discussed.

In the remaining sections, I focus on three sets of studies that show how learning could act to form complex phenotypic correlations or "individual differences" with evolutionary potential. The examples used represent two contrasting major groups of animals—vertebrates and insects—and emphasize the role of learning in its natural setting. In all cases, field observations of behavior and its ecological correlates have suggested learned reorganization of behavior, and this was subsequently documented with experimental or observational research on learning per se.

Feeding Specializations in Birds

Individual learned specializations in nature are particularly well documented in birds, especially the "Darwin's finches" (Emberizidae: Geospiza) of the Galapagos (Ecuador) and Cocos Island (Costa Rica). The Cocos Island finch (*Pinarolaxias inornata*) shows a great range of foraging specializations (Werner and Sherry, 1987; T. K. Werner, 1988). Eight specializations that involve complex differences in behavior could be distinguished in the birds observed: hunting on branch surfaces, probing branches, gleaning from leaf surfaces, extracting leaf-miner (Lepidoptera) caterpillars, hunting by probing dead leaf clusters, nectar feeding from extrafloral nectaries, nectar feeding from flowers, and ground feeding on seeds and insects. Long-term observations of the 89 individually identified finches most often observed feeding showed that specialization was independent of age, season, sex, time, place observed within the study area, and morphology. Measurements of body size, beak size and shape, and toe length showed the birds to be remarkably uniform morphologically. Individuals were not territorial or restricted in their movements within the study area, a *Hibiscus* thicket described as homogeneous in terms of feeding locations for the birds. So individual specialization was not due to restricted feeding opportunities. Werner and Sherry present evidence that the individual specializations are learned early in life and influenced by the feeding activities of adults and even sometimes birds of other species. Juveniles followed, watched, and imitated the feeding behavior of adult birds, including those of other species as diverse as warblers and sandpipers. They also exhibit apparent exploratory behavior toward potential prey objects (Sherry, 1990).

In related species of Darwin's finches in the Galapagos, the ontogeny of individual feeding specializations has been studied in more detail (summarized in Grant, 1986; Grant and Grant, 1989). Juveniles sometimes take up to a year to attain the foraging efficiency of adults (Grant, 1986, p. 128), and young fledglings peck "apparently haphazardly" at leaves, bark, flowers, and the ground in an apparently exploratory phase of learning. Later (12–25 days after fledging), fledglings of different species associate with each other and appear to copy each other's feeding behaviors, sometimes attempting to feed in ways that adults of the same species do not. But eventually "they drop these heterospecific behaviors from their repertoires, probably as a result of inefficiency and poor rewards" (Grant, 1986, p. 128). Without experimental tests, it is difficult to know for sure whether such changes represent learning ("practice") or maturation (see Mackintosh, 1983, p. 150). But the presence of apparent long-term individual differences in food preferences and associated behaviors suggests that simple maturation toward a species-typical routine is not an adequate description of behavioral development in these finches.

Observations by the Grants and their associates indicate that beak size and shape in combination with local food availability can influence intraspecific variations in adult diet. For example, in *Geospiza difficilis*, *Opuntia* cactus is used on one island (Genovesa) only by relatively short-billed individuals. In *G. fortis*, as in the Cocos Island finch, different individuals have different feeding specializations within a population. Habitat and diet selection were influenced by bill characteristics: indi-

viduals foraging in woodlands where larger seeds were more common had slightly but significantly deeper and blunter bills than did those foraging in a nearby region with smaller seeds. Small bill size caused individuals to fail in attempts to crack hard seeds of *Bursura* that larger billed individuals could crack, and the largest billed birds were quickest at cracking hard seeds of *Opuntia*.

In addition, there were differences in behavior even among individuals of the same bill size feeding on the same food:

[d]ifferent individuals of *G. fortis* used slightly different maneuvers in attempting to crack a *Tribulus* mericarp and extract the seeds. Peculiarities of individuals stand out after repeated observation. In addition some individuals, but only some, fed parasitically from others, and from *G. magnirostris*, by rushing in to seize a seed or mericarp fragment when a large piece was split or shattered, and then rapidly hopping away to consume it. Thus bill size is an important determinant of diet, but so too is behavior that varies to some extent independently. (Grant, 1986, p. 138)

Giraldeau (1984, p. 73; see also Dill, 1983) reviews a large body of experimental evidence showing that learning is a "major component" in the development of foraging behavior in many birds (see Giraldeau, 1984, for numerous references). In many cases, learning magnifies the effects of small individual foraging differences into pronounced individual differences. In other cases, small chance differences in individual experience result in considerable individual variation. "Differences in learning propensity for particular tasks lead to dramatic changes in individual performances when even minute changes are made to the task to be learned. (Giraldeau, 1984, p. 73).

Hinde (1959, p. 87) cites comparisons that illustrate the developmental role of learning particularly well: great tits consistently use their feet for holding pieces of food, whereas cardueline finches show individual as well as species differences in ability and frequency in use of the feet. When the foot is used, learning plays an important part in the development of the behavior:

In the great tit this learning is almost bound to occur because of the way in which young birds hold the food down near their feet. . . . Thus the inter-specific differences may not be due to neural mechanisms which are directly determined genetically, but to other factors, such as body structures and postures, which govern the course of development by determining the learning that occurs. (p. 87).

Hinde notes that individuals with grossly deformed beaks sometimes adopt scooping movements and other bizarre behaviors foreign to the normal repertoire of the species to obtain food (see also references in Wcislo, 1989, on learned adjustments of birds with deformed beaks). This same two-legged goat effect would operate within a normal range of beak variation to exaggerate differences in behavior based on individual differences in beak size and shape. This would affect the nature of the learned specialization, as appears to occur in the Darwin's finches.

Hinde (1959) further cites evidence for size differences in certain muscles due to foraging and dietary differences in birds. These can be added to the skull differences sometimes associated with facultative dietary specialization already discussed, and to the changes in gut morphology and digestive efficiency that can occur as a consequence of diet (e.g., in birds, mammals, insects, and bacteria; reviewed by Partridge and Green, 1984; see also Willson and Comet, 1993, and references therein). Thus, morphology affects (via learning) behavior, which in turn affects morphology (e.g., muscle, and in some insects and vertebrates skeletal and intestinal development). This kind of self-reinforcing cycle of learning-facilitated change can occur when increased efficiency leads to increased frequency of performance, leading to further increased efficiency, and so on. The cycle could, of course, be broken if preferred items were in short supply or if other factors, such as the need for a balanced diet, intervene to put a brake on specialization. Learning is the integrating mechanism for all of these developmentally linked traits that combine morphological and behavioral specializations.

Of course, dietary differences such as those observed in the Galapagos and Cocos Island finches are likely to involve different skills—different search behaviors and uses of beak and feet, and competitive opportunism (including stealing)—like those studied experimentally and comparatively by ethologists. Taken together with the ethological studies, the quantitative field studies of behavior and natural selection (Grant, 1986; Grant and Grant, 1989) provide strong evidence that a learned dietary bias can contribute, within an individual, to the establishment of a complex correlated shift involving behavior, biased use of particular aspects of morphology (sensory, trophic, locomotory) and

digestive physiology, and habitat selection in the search for preferred food. This can be recurrent throughout life, and so there is ample chance for fitness effects such as those documented (when exaggerated under seasonal or occasional stress) in finches (Grant, 1986; Smith, 1990c). The legendary "boldness" and "curiosity" of island organisms, such as the Galapagos finches (figure 18.3), has sometimes been attributed to the absence of humans and other natural enemies on islands (e.g., Darwin, 1845). While this may help to explain the persistence of such behavior, boldness and curiosity would be highly positively selected if learning is especially important to these birds, for boldness in approaching new things (lack of "neophobia") is an important component of learning in birds (Greenberg, 1983, 1985, 1990). As Brown and Godin (1999) expressed it, "Who dares, learns."

As discussed in chapter 28, on adaptive radiations, learning is a neglected aspect of adaptive evolution and one that demands more attention from evolutionary biologists. Different bird species, for example, even quite closely related ones, show different degrees of learned modification in specific aspects of natural history (Marler, 1998). Such differences may affect rates of adaptive evolution in different clades. Greenberg (1983, 1990) has shown that different species of warblers (*Dendroica*) and sparrows (*Melospiza*) differ in their learning-mediated switching among different classes of food, and this is associated with the degree to which they specialize in the field. By testing both species of warblers in learning experiments, Greenberg (1985) showed that they did not differ significantly in standardized tests of conditioned learning ability. Both species reached the criterion response for training to a new foliage type in a nearly identical number of trials and committed a similar number of errors while learning. Their differences in rates of learning in nature were due instead to what in humans could be called "personality" differences: chestnut-sided warblers (*D. pennsylvanica*) were relatively timid or neophobic, reluctant to sample new foods or feed on accustomed foods in new situations; bay-breasted warblers (*D. castanea*), on the other hand, fed with less hesitancy at unfamiliar food sources. Such a divergence in approach to novelty makes the difference, in this case, between a narrow dietary and habitat specialist, and a "generalist" species in the same winter habitats.

Another neotropical migrant warbler (*Helmitheros vermivorus*) preferentially searches dead leaves as a juvenile prior to self-feeding. Then, at an age when individuals would be arriving in their neotropical wintering grounds, they show instead

Fig. 18.3. Boldness and curiosity: an adult male large cactus finch (*Geospiza conirostris*) peers into a camera on Española Island, Galapagos, Ecuador. From Grant (1986). Photograph by Cleveland P. Hickman, Jr., courtesy of the photographer.

a persistent exploratory behavior manipulating substrates, particularly foliage, whether reinforced with food or not. This enables them to track changes in abundance of dead-leaf and live-leaf arthropods, leading them to a species-specific but learned specialization in dead curled leaves in the tropics during the winter and a specialization in live leaves during the breeding season in the temperate zone (Greenberg, 1987a,b).

Wilson et al. (1994) discuss the "shy–bold" continuum in various other species as a product of natural selection and attempt to relate it to fitness differences in fish. They conclude that the "evolutionary implications are unknown" but would merit consideration. The work by Greenberg summarized above makes it possible to relate the shy–bold continuum to learning and, via that link, to variation in such things as habitat selection and diet that is material for natural selection (see chapter 28).

Kleptoparasitism of food is very common in some families of birds, especially in opportunistic feeders and predators of large prey that associate with other species at feeding sites (Brockmann and Barnard, 1979). In some species, such as frigatebirds, kleptoparasitism is accompanied by evolved specializations (unwebbed feet, loss of oily feathers) that make independent feeding more difficult. It has been suggested that such behavior may originate by a combination of preexisting opportunistic responses to the sight of prey and reinforcement learning. For example, in a region where kleptoparasitism had not previously been observed, Bengston (1966, cited by Brockmann and Barnard, 1979, p. 494) observed an individual tern (*Sterna paradisaea*) in a feeding association of terns and horned grebes (*Podiceps auritus*) suddenly retrieve a fish dropped by a grebe. The next time the grebe appeared with a fish, the tern swooped down on it, causing it to drop its prey. Bengston suggested that kleptoparasitic behavior, common in some populations of terns, may originate via such "accidents."

Bengston's insight provides another example of a rare observation that is valuable without large sample sizes: a behavioral accident is analogous to a morphological anomaly, except that in the case of rewarded behavior accidental phenotypes can be quickly reproduced to become individual specializations or (through imitation) established traits within a population. Given the propensity among birds for observational and imitative learning of feeding behavior (see above), it would be interesting to investigate the possibility that the innovation of a single individual could spread in a population and even become propogated through generations as a learned tradition. This may have happened when titmice learned to open milk bottle caps in England, a behavior that began in one region and then spread rapidly and widely by imitative learning (Hinde and Fisher, 1951). Thus, learning could be both a mechanism of trait origin and a means of rapid spread.

Foraging Specializations in Bees

Bees are the Darwin's finches of the insect world, in terms of known abilities for learned foraging specialization influenced by trophic morphology. The apine counterpart of the avian beak is the tongue, or proboscis. As in birds, the feeding apparatus of bees is often taxonomically distinctive at fairly low levels of classification (generic, or subgeneric; Michener et al., 1994), suggesting evolutionary lability and rapid diversification due to selection. Like the beak of birds, the bee tongue is involved in learned correlations among traits. Detailed studies of the learned individual foraging differences among bumblebees (summarized in Heinrich, 1979) show that novice foragers begin clumsily to extract nectar from a diversity of flower morphologies encountered, then eventually specialize in long runs on a particular species with a dramatic increase in efficiency of working the flower. For example, naive bees visiting monkshood—a plant pollinated exclusively by bumblebees—have difficulty managing these morphologically complex flowers, where the nectar is hidden in the tips of modified petals under a hood. Inexperienced individuals sometimes approach buds as well as flowers and often fail to reach the nectar at all, whereas monkshood specialists move rapidly from one flower to the next, entering via the easiest route (over the anthers) and reaching the nectar in a single smooth motion. Inexperienced bees on complex flowers took about an hour to collect a load of nectar, while experienced bees accomplished the same task in about six minutes. Improvement in foraging efficiency with age is known in other social insects, including wasps (O'Donnell and Jeanne, 1992) and ants (Schmid-Hempel and Schmid-Hempel, 1984).

Some individual bumblebees learn to shake the flower to remove pollen, or to bite a hole in its base to go directly to the nectar. By engaging in "major" and "minor" individual specializations, and by adjusting their behavior in accord with changing rewards, bumblebees track changing food supplies via flexible tactics of foraging. Tongue length correlates with body size, and large-bodied bee species specialize on nectar sources unreachable by short-tongued bees. The same correlation between tongue length and flower structure influences individual specializations within species (Morse, 1978; Inouye, 1980). So, as in finches, learning in bumblebees sets up a correlation involving complex searching and handling behaviors (such as shaking vs. biting and different flower entry routes) as well as tongue length and body size. And, as is common in evolution in general (see part III), a pattern gener-

ated within species by adaptive phenotypic flexibility parallels patterns of divergence between species. In the bees, it is also known that there are individual differences in "innate preferences [for different flower colors] that determined rates of learning, rates of forgetting, and ability to switch to more rewarding food sources" (Heinrich, 1979, p. 138), indicating that genetically variable individual color preferences can be added to the complex of traits that become correlated in their expression due to learning. This correlation would be subject to reinforcement by natural selection.

Although honeybees (*Apis mellifera*) and stingless bees (Meliponidae) have less size variation within colonies than do bumblebees (Waddington, 1988), their tongue length also correlates with body size (Waddington and Herbst, 1987) and individuals of different sizes feed at different flower species, presumably because of the effect of efficiency differences on learned specialization (Morse, 1978; Inouye, 1980). In the honeybees this has an interesting consequence for social behavior: when successful foragers return to the hive, they perform the famous recruitment "dance" that informs nestmates of food locations (von Frisch, 1950). The groups following the dance of a particular forager prove to approximate her in body size (Waddington, 1981, 1989). This would prevent recruitment of size classes to food sources where they would be relatively inefficient. Whether learning plays a role in the choice of dancing partners is not known.

The learning abilities and related sensory capacities, neurobiological characteristics, and genetics of bees are extensively documented (e.g. see Menzel et al., 1973; Menzel, 1985; Raveret Richter and Waddington, 1993; Menzel and Müller, 1996; Fahrbach and Robinson, 1995; Page et al., 1995a,b, 1998; Scheiner et al., 1999). These and a rich literature on the comparative natural history of learning, only touched upon in this section, promise to make the social Hymenoptera and their solitary relatives model groups for studies of learning in relation to evolution.

Social Competition and Learning

Research on learning in comparative psychology (e.g., see Mackintosh, 1974) focuses on the precise measurement of individual learning abilities under controlled conditions, and there is a tendency, therefore, to consider the individual apart from its environment—alone in a testing situation with a stimulus and a reward. In nature, learning may occur within sight of competitors and in surroundings rich with potentially informative cues. Individuals are often alert to the resource-acquisition activities of conspecifics, and stealing (kleptoparasitism) and "socially facilitated" flocking to newly discovered supplies (both food and mates) are common in insects and vertebrates (Brockmann and Barnard, 1979; Curio, 1976; Barnard, 1984). In birds, individuals sometimes learn by mimicking other individuals, as, for example, in the acquisition of song or foraging behavior (Grant, 1986). But more commonly, and in a greater range of taxa, individuals may be literally forced or channeled into particular specializations by early dominance relations or just simple accidents of association, wherein certain resources are preempted by siblings or neighbors that happen to have a preference or superior ability to acquire them, thus leaving other options relatively open.

The effects of social competition on learned dietary specialization were investigated experimentally by Milinski (1982). He observed sticklebacks (*Gasterosteus aculeatus*) two at a time and classified them as good or poor feeding competitors by measuring their relative rates of predation on *Daphnia*. He then presented each pair with two size classes of prey having equal handling times and found that the "poor" competitor of each pair specialized on small prey. This entrainment persisted even when they were able to feed on a mixture of prey sizes independent of the superior competitor. In some other species, the same relation between competitive ability and size of prey obtains, but inferior competitors switch to larger prey when on their own (Magurran, 1987, p. 346). Parasitized sticklebacks specialized in small prey, thereby avoiding direct confrontation with stronger competitors and maximizing their food intake (Milinski, 1984b, cited by Magurran, 1987).

The competitive aspect of learning and experience-mediated specialization has been documented in the wild by several investigators, who have taken advantage of natural experiments in resource scarcity to observe the effects on foraging behavior. In Galapagos finches, severe drought caused relatively large-beaked individuals to rapidly (within a season) specialize in large hard-to-crack seeds, whereas they normally fed on both large and small seeds along with smaller beaked conspecifics (summarized in Grant, 1986). Similarly, in a population of African weaver finches (*Pyrenestes ostrinus*; Estrildidae), dietary differences between a large- and a small-billed morph were most marked during seasonal food shortage. When their food plants (two species of sedges with different seed sizes and hardnesses) were scarce, the large-billed

morph, which fed along with the other morph on soft seeds during the breeding season when food was abundant, specialized on large hard-to-crack seeds, and the small-billed morph (which showed superior ability at handling soft seeds—Smith, 1987) ate soft seeds of the sedge and other small seeds (Smith, 1990b,c).

Learning takes on an entirely new aspect when competition is for social status and socially defined rewards, rather than other kinds of resources, especially in organisms, such as humans and perhaps other animals, capable of manipulating and learning socially defined rewards and punishments. Then social development (socialization) can come to occur within a framework where successful behaviors change from group to group, and generation to generation, for success no longer is defined by some fixed or slowly evolving criterion, such as ability to find and handle particular kinds of food (which may be sufficiently recurrent to produce genetic correlations between phenotypic traits). Under socially defined competition, success has relatively flexible and changeable criteria invented and imposed by parents and influential dominant leaders, reinforced by mimicry and strong rewards and punishments that enforce both conformity and change. This is one reason why it can be extremely misleading to suppose that particular human learned behavioral traits are evolved. The role of genetic variation and evolution should be sought in any traits that would recurrently influence success despite rapid social change, such as generalized ability to communicate, observe and adapt to shifting criteria of success, and so forth. This follows from the principle that to be molded by natural selection, a trait not only must be heritable (show genetic variation correlated with phenotypic variation) and affect reproductive success, but also must be sufficiently recurrent in a population across generations for natural selection to act (see chapter 6).

The Importance of Forgetting

Both birds and bees show sensory bias in what is most easily learned, and in both cases the bias can be overcome by circumstances (e.g., resource shortages, or competing stimuli) that influence them to adopt alternative behaviors (Greenberg, 1987b, on birds; Heinrich, 1979, on color bias in bees). This implies that learning can be adjusted by selection to favor rapid changes in the adoption of certain behavioral traits. But at the same time, flexibility is not lost, and adaptively favorable departures from the usual learned behaviors can occur under extreme conditions, in emergencies, or when a particular learned pattern ceases to be rewarding. Under changing conditions, a learned specialization that ceases to be rewarded can be dropped or forgotten and a new one learned.

A cycle of learning followed by forgetting as conditions change occurs in bees and wasps that learn to return repeatedly to a rich food location or nest site, then cease to return there and resume searching when the food supply at a particular site is exhausted or the colony is moved. Forgetting of specific depleted sites is particularly important in wasps, such as *Agelaia* species, that utilize small, isolated sources of carrion. *Agelaia* workers hunt using odors, then return repeatedly to a carcass, sometimes a small caterpillar or adult insect. When nothing is left, visits to the site completely cease (personal observation). Presumably, there would be selection on such wasps to cease returning to learned sites, due to the very short duration of the resource and its lack of association with any consistent habitat or cue other than odor. These observations also suggest the hypothesis that associative learning could have played a role in the evolution of use of rotting-meat odor in food location, which predominates as a cue in the genus *Agelaia* compared with most other neotropical social wasps (Vespidae; see review of foraging cues in Raveret Richter, 2000). Wasps use characteristics of surrounding vegetation and other landmarks as orientation cues when they learn the locations of their nests and prey (Rau and Rau, 1918; Baerends, 1941; Raveret Richter, 2000). But they are not known to generalize between habitat characteristics and prey presence as a result of experience. It would be of interest to investigate this possibility.

Social wasps and bees can learn to associate colors and odors with food rewards (reviewed in Menzel, 1985, 1990; Raveret Richter, 2000). Associative learning may be widely important in the evolution of foraging insects, for the recurrent use of consistent food-location cues under associative learning is the type of recurrence that could lead to phenotypic fixation via genetic assimilation. Rate of site forgetting may also be affected by patterns of distribution of food sources, with isolated sources associated with faster forgetting in short-term memory, whereas somewhat clumped or durable sources (which would consistently reward continued searching in the general area of a success) may be associated with a slower extinction of a response. This is confirmed for honeybees by the finding that training to a first color is rapidly ex-

tinguished if a reward associated with a second color comes soon after the first (Menzel, 1985). But there is also plasticity in forgetting behavior: bees that have learned the location of a new hive can still find the old one if the new hive is removed (Robinson and Dyer, 1993). So cues are not really forgotten even when not used. They can also use the old cues in new contexts (Dyer, 1991b).

The ability to selectively forget is a hallmark of a continuing ability to learn, as distinct from imprinting or irreversible learning during a critical period. The mechanisms of context-specific forgetting seem as interesting as those of learning (on extinction of tactile learning in relation to sucrose rewards in honeybees, see Menzel, 1985; Scheiner et al., 1999; on the effect of rate of forgetting on the effectiveness of Müllerian and Batesian mimicry in butterflies, see Speed and Turner, 1999). The ability to forget or reverse an acquired trajectory distinguishes a durable ability to learn from some other kinds of condition-sensitive but relatively irrevocable decisions made by plants, nerve fibers, or mitotic-spindle microtubules following "exploratory" movements or growth and eventual "fixation" of morphology in some other kinds of somatic selection (see chapter 3). Some of these processes also have features analogous to forgetting, for example, cell death in the exploratory proliferation of a cell lineage. So it would seem that one of the essential features of even a simple durable learning capacity is that, under pressure, a learned pattern can be revoked. Learning may, due to sensory and other evolved biases, channel behavioral development in a stereotyped and species-specific manner (Menzel, 1985). But the capacity for reversibility is usually not lost.

The evolutionary consequences of such a developmental device are profound. The ability to forget means that learned traits can be adaptively modified to the same degree as are other recurrently developed phenotypic traits and yet, under pressure of competition or resource scarcity, increased motivation (e.g., in level of aggressiveness, hunger, thirst, or hormonal state) may overcome the usual threshold, bringing the mechanisms of learning (e.g., searching and reinforcement) back into play, and with them the establishment of new correlations and new avenues of selection. Mechanisms of flexibility may generate fixed, specialized phenotypes, but in so doing they do not lose the capacity to generate new variation. For this reason, learning and plasticity are powerful and continuing agents of biological diversification.

Forgetting should probably be considered as specialized an adaptive trait as is learning. Rate of forgetting can presumably be modified under selection to correlate to different degrees with strength of the learned reward or punishment, and with the frequency with which it has to be repeated to prevent forgetting (maintained a learned behavior). The properties of forgetting have important effects on other organisms in coevolutionary relationships. In butterfly mimicry complexes, for example, the protection from predation afforded by a warning color pattern associated with distastefulness depends in part on the forgetting curve of a predator that learns to associate a color pattern with distastefulness. Simulations of these effects have shown that the mode of forgetting has a more significant effect on mimetic relationships and frequency-dependent Batesian mimicry polymorphisms than does the rate of learning (Speed and Turner, 1999).

19

Recurrence

Phylogenetically discontinuous recurrent traits are exceedingly common and phylogenetically widespread, as witnessed by the commonness of homoplasy in cladistic analyses. Comparative study with attention to intraspecific variation often reveals that the same traits that are recurrent between species are expressed as variants within species, suggesting that the interspecific similarity is homologous, due to common descent from a developmentally plastic ancestor. In some groups, the same traits recur at different frequencies in different species, as anomalies, intraspecific alternatives, and fixed characters. Recurrent reversion, or flip-flopping between alternative traits over time, gives an impression that some traits blink on and off during evolution. Sometimes, but not always, expression of a particular trait is associated with particular environmental conditions. Recurrent phenotypes are so common in some taxa that they render cladistic analysis difficult, but they represent a pattern of evolution that is predicted by a combinatorial, reorganizational theory of evolution.

Introduction

Recurrent phenotypes are similar or identical phenotypic traits with discontinuous phylogenetic distributions, which owe their similarity to common ancestry (homology). A recurrent trait may be found as a fixed trait, as an alternative phenotype (one morph of a polymorphism or polyphenism), or as a low-frequency developmental anomaly. *Recurrence*, then, is the phyletically disjunct appearance of homologous traits. An example is the repeated evolution of larviform (paedomorphic) adults in salamanders (figure 19.1). The larviform morph is characterized by retention in the reproductive stage of homologous larval traits such as external gills and a tail. This has involved changes at various points in the hormonal mechanism that controls metamorphosis in all salamanders (chapter 25), perhaps under selection for accelerated reproduction in stressful environments (Whiteman, 1994). As is characteristic of recurrent phenotypes, the occurrence of the reproductive larviform adult morph varies in frequency from one species of salamander to another: it can be absent, an anomaly (<5% of population), a common (>5%) alterna-

tive to complete metamorphosis, or a predominant or fixed form (see figure 9.5). Even within the genus *Ambystoma*, the unmetamorphosed larviform adult occurs as an occasional anomaly in some populations, as a facultatively expressed alternative phenotype in others (e.g., *A. tigrinum*) and as a fixed form in others (e.g., *A. dumerilii*; Collins et al., 1993). All atavisms and reversions (see chapter 12) are examples of recurrence.

Discontinuity of expression is expected in combinatorial evolution, where traits are turned off and on and expressed in different combinations due to regulatory change. The growing evidence of homoplasy in phylogenetic studies is important evidence that combinatorial evolution occurs and that homoplasy itself is worthy of study, not just a source of "noise" in cladistics (Wake, 1996a). *Homoplasy* has been defined as "possession by two or more taxa of a character derived not from the nearest common ancestor but through convergence, parallelism, or reversal" (Mayr and Ashlock, 1991, p. 418; figure 19.2). More simply, homoplasy is the recurrence of similarity in evolution (Sanderson and Hufford, 1996). Homoplasy is often due to recurrence, either the repeated switching on of a lost trait

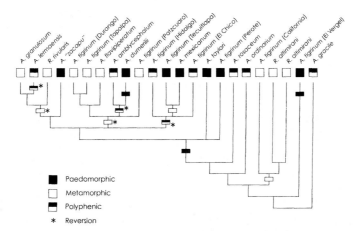

Fig. 19.1. Recurrent evolution of paedomorphic (nonmetamorphic) populations of salamanders (*Ambystoma* and *Rhyacosiredon* species) in Mexico. Some lineages show reversions, and some (*A. lermaensis, A. dumerilii*) show flip-flopping between states. After Shaffer (1984).

or, as in salamanders, a repeated phenotypic change based on lability in a common ancestral developmental mechanism.

Historical Discussions of Recurrence

It is a tribute to the prominence and commonness of recurrence that it has been a recurrently designated Law of Nature, albeit under different names. Darwin (1859 [1872]) called recurrent phenotypes "analogous variations" and mentioned earlier work by B. D. Walsh (see also Darwin, 1868b [1875b], p. 344) who named the phenomenon the "Law of Equable Variability" and reviewed cases in insects. Darwin (1859) listed many examples of recurrence in both cultivated plants and domestic animals as well as in nature. The very distinct transverse bars on the legs in zebras, for example, sometimes appear in the domestic ass. The shoulder stripe, lacking in the white ass, and in some dark-colored asses is present as an occasional trait in pallas koulans, hemionus (whose foals usually have leg stripes), and in some dun- and bay-colored horses, and it is present as a regular feature of the Kattywar breed of horses in India. This pattern—occurrence as an occasional anomaly in some populations, as a regular but not ubiquitous alternative phenotype in others, and as an established characteristic in some—is common for recurrent phenotypes and evidence of its basis in developmental plasticity. Darwin notes the variability in expression of these

stripes and also makes the astute observation that their recurrence in related species is due to "having inherited the same constitution and tendency to variation, when acted on by similar unknown influences (Darwin, 1859 [1872], p. 117). This statement leaves open the possibility of both environmental and genomic effects among the "unknown influences" (Darwin considered both, as shown in chapter 8). It also leaves open the possibility that the response is not a reversion to a precisely identical ancestral phenotype, but could be due to "constitutional" responsiveness (inherited ability to respond to environmental variation) that could assemble anew the same or similar phenotypes under analogous circumstances.

Recurrence was rediscovered later by the Russian school of developmental geneticists in studies of cultivated plants, who described recurrence as "phenotypic parallelism" (attributed to Timofeeff-Ressovsky by Yablokov, 1986, p. 38) or "homologous variation" (Vavilov, cited by Van Valen, 1974, p. 7). Vavilov summarized the "Law of Homologous Series in Hereditary Variability," more recently called the "Law of Directed Series" (Levinton, 1988) or simply "Vavilov's law," as follows:

1. Species and genera, genetically close, are characterized by a similar order of hereditary variability with such a regularity that, knowing a number of forms within the limits of one species, it is possible to predict the finding of parallel forms in other species and genera. The closer genera and linneons [lineages] are genet-

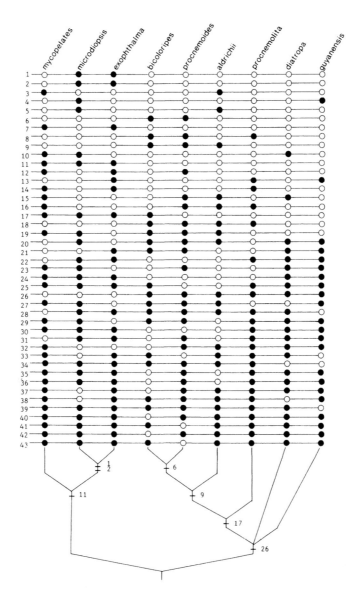

Fig. 19.2. Rampant homoplasy in flies of the *Chymomyza aldrichii* species group (Diptera, Drosophilidae). Presence (black dots) and absence (open dots) of 43 morphological characters of adults of nine species (top). From Grimaldi (1986), by permission of the New York Entomological Society.

ically in the whole system, the greater is the similarity in the order of their variability.

2. Whole families of plants in general are characterized by a specific cycle of variability passing through all genera and species comprising the family. (Yablokov, 1986, p. 35)

The classic illustrations of this law were recurrent discrete alternative traits in wheat (*Triticum vulgare*) and other cereals, including barley, oats,

wheatgrass, rye, maize, and millet. These traits included awned and non-awned, serrated versus smooth awns, open versus closed flowers, and alternative forms of the spikelets. These variants sometimes occur as intraspecific forms as well as interspecific alternatives. A whole school of Russian pheneticists active in the first half of the twentieth century focused on the fact that comparative study of *qualitative* discrete traits often revealed a "remarkable parallelism" in alternative character

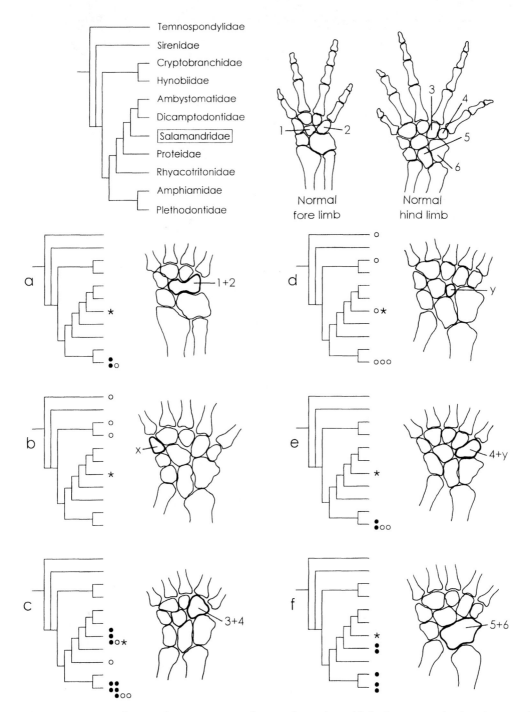

Fig. 19.3. Recurrence of anomalies as intraspecific morphs and established traits in the foot bones of salamanders. Top left: a cladogram of the 10 families of urodele amphibians. Top right: normal limb bones of the newt *Taricha granulosa* (Salamandridae), numbered for comparison with six anomalies (a–f below) found in a large sample (*n* = 452) of this species. Recurrence in other families is shown on a copy of the cladogram (unlabeled) at the left of each anomaly (a–f): open dots indicate species showing the variant at low frequency; black dots indicate species showing the variant as an alternative phenotype or fixed trait; asterisks (∗) show the phylogenetic position of *T. granulosa*, where the variant occurs at low frequency (<6% if unilateral, 1% or less if bilateral). Dots in a horizontal row indicate independent recurrences; dots in a vertical arrangement indicate occurrences in different genera of the same family that may or may not be independently evolved. Phylogeny, diagrams, and data after Shubin et al. (1995).

states (Yablokov, 1986, p. 35; see also Schmal-hausen, 1949 [1986]).

Rensch (1960, p. 198) also emphasized the importance of recurrent parallel evolution, which he called "iteration"—"the repeated evolution of parallel structures originating from a generally conservative type"—not to be confused with the more specialized current usage "iterative homology" or serial homology—traits repeated in different body segments. Rensch gave several examples, including albinism, melanism, and pituitary dwarfism in vertebrates. He considered these to be attributable to "homologous genes" or "identical genes," called "paripotency" by Haecker (1925, ex Rensch, 1960). His most elegant illustrations came from his own studies of snails (see Rensch, 1960, p. 195). They show disjunct recurrence of three distinctive alternative shell forms (flat, crested, and ribbed) found in different races and species in 12 genera "well-differentiated" on the basis of soft-part morphology (especially genitalia). In some cases, the morphs occur as "quite frequent" intrapopulational alternatives rather than fixed forms. Rensch emphasized the difficulties posed by such marked recurrent variants for the taxonomy of the snails.

Goldschmidt (1940 [1982]) paid more attention to the developmental side of recurrence. He thought that mimicry in butterflies could be based partly on the fact that a limited number of developmental patterns recur in even different families of butterflies. This view was subjected to harsh criticism (e.g., see Ford, 1971, p. 273) but has been extensively confirmed by recent research showing recurrence in developmental characteristics of butterfly wings (e.g., see Nijhout, 1991b).

More recently, systematists have uncovered very large numbers of recurrent traits among the characters used in classification and cladistics. They have been called by many names, such as apomorphic tendencies, unique inside parallelism, indicators of latent homology (Wake, 1996a) underlying synapomorphy (see below), homiology, and non-universal derived character states (see references and many examples in Sluys, 1989; additional synonyms in Wake, 1996a). Recurrence has been discussed as a product of *underlying synapomorphy*—"agreement in capacity to develop parallel similarity" (Tuomikoski, 1967), or "parallelism as a result of common inherited genetic factors including parallel mutations" (Saether, 1979, p. 305; Saether, 1983; see also Alberch, 1980; Garcia-Bellido, 1977). Association of recurrence with intraspecific developmental plasticity means that the pattern of occurrence of a trait need not coincide with the pattern of speciation or branching of a phylogenetic

tree. Traits may blink off and on both within and between populations and species, as in the limb bones of salamanders (figure 19.3) and other examples discussed in this chapter.

Given the developmental reasons for *expecting* recurrence, it should not be surprising that recurrence has so often been recognized as a major feature of evolution. As phrased by the Italian hydroidologist F. Boero (personal communication), given the combinatorial nature of evolution, discovering recurrent phenotypes is "like finding humidity in wells." Only recently, with the improvement in phylogenetic study represented by cladistics has the full extent of the incongruence between branching pattern and phenotypic transition pattern become evident. Rampant homoplasy is perhaps the most revealing and revolutionary finding to emerge from modern phylogenetics for ideas about how evolution works. By an ironic twist, the homoplasy that can only be discovered with cladistic methods has turned out to be so common that it ends up being a challenge for cladistic methods themselves (Wake, 1991; see Consequences for Systematics, below).

Problems in the Interpretation of Recurrent Similarity

The Homology Problem Again

Recurrent traits are homologous, or similar due to common descent, not products of convergent evolution. As discussed briefly in chapter 9 and more thoroughly discussed in Chapter 25, on homology, parallel traits are somewhat confusingly termed "convergent" under the cladistic homology concept even though their similarities may be due to common descent. The recurrent identical wing patterns of different races of *Heliconius* butterflies, for example (Brower, 1994; figure 19.4), are "convergent" in cladistic terms, even though they are found as geographic variants in the same species, but they are parallelisms in the broader sense of being based on shared physiological determinants of wing patterns found in those butterflies (see Nijhout, 1991b). This enabled the somewhat isolated populations of the same species to undergo the same type of change independently in different descendent sublineages.

Homoplasy is a term that blurs distinctions that are important in evolutionary biology, for it lumps both recurrence (parallelism and reversal), where similarity is due to common ancestry, and conver-

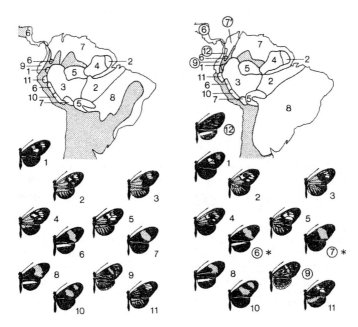

Fig. 19.4. Recurrent wing-coloration phenotypes in *Heliconius* butterflies: geographic distribution of color patterns (numbered, below) in *H. melpomeme* (left) and *H. erato* (right). Parallel, independently evolved (recurrent) patterns in the two species are numbered the same. Circled numbers in *H. erato* correspond to a Western clade; uncircled numbers, to an Eastern clade; and asterisks indicate two patterns (6, 7) that have evolved independently in both *erato* clades (Brower, 1994). Thus, patterns 6 and 7 evolved independently three times in these two closely related species, both of them distasteful Müllerian mimics of each other (hence, the close geographic overlap of similar color patterns in the two species). After Turner (1981), modified with data from Brower (1994).

gence, where similarity is not due to common ancestry. I will call the subclass of homoplasy that is due to common ancestry rather than convergent evolution *recurrence homoplasy*.

Often, recurrent traits are lumped under the term "parallelism" or "homoplasy" without distinction as to their origin, whether it be by gradual reconstruction based on common developmental propensities, or the repeated expression of a unit trait. In this chapter, *parallel evolution* means that a trait began with the same or similar initial trait or ground plan and independently underwent the same type of evolutionary change due in part to the common starting conditions (similar selective context may also be involved). In theory, parallelism could be either gradual—a shift in a quantitative trait or a series of small qualitative steps—or single step, if it is caused by a switch to an alternative state. Parallelism is distinguished from *convergence*, or similarity due to independent evolution that begins with different initial conditions (or a different aspect of the phenotype), with the similarity having either a different or only a superficial

underlying structural, developmental, or genetic resemblance (after Gosliner and Ghiselin, 1984, p. 258). A good example of convergent recurrence is the repeated evolution of reduced numbers of parietal (lateral wall) plates in barnacles, which occurs differently in different lineages, via either the fusion or the loss of different combinations of plates (Palmer, 1982). Repeated convergent similarity does not represent recurrence, for recurrence implies broad-sense homology between the recurrent traits (see chapters 9 and 25).

Homology between recurrent traits can be relatively easy to identify if the trait is turned off and on as a unit. So phenotypic plasticity can be used as a research tool for testing a hypothesis of homology between recurrent traits. In general, if a recurrent trait occurs as an intrapopulational alternative phenotype, as a stage-specific phenotype, as a developmental anomaly, as a stress-induced extreme of continuous variation, or as an experimentally inducible trait in a species normally lacking it, expression of the trait can be easily turned of or on as a unit in the taxon concerned. This supports a hypothesis of homology between disjunct occurrences by showing that disjunct expression of the trait is developmentally feasible. Negative results, or failure to induce the trait, would not constitute decisive evidence against homology, however, because genetic accommodation and modification of phenotypes secondarily lacking the trait may reduce or eliminate the ability to express it.

Recurrence per se supports a homology hypothesis if the populations showing recurrence are known to be monophyletic, especially if they are very closely related. Lake populations of three-spine sticklebacks (*Gasterosteus aculeatus*), all recently descended from a common marine ancestor, have recurrently evolved slender limnetic and broad-bodied benthic forms in different lakes (see chapter 28). It is more parsimonious to suppose that such repeated similarities are homologous (recurrence) than to suppose that they represent repeated convergent evolution or even repeated independent parallel derivations. The recurrent-homology hypothesis is further supported for the sticklebacks by the presence in marine populations of ontogenetic stages that resemble the two forms (Andrews, 1999; see chapters 27 and 28).

When the developmental mechanism of recurrence is known, it often involves mechanisms such as hormones that respond to environmental conditions and influence gene expression. For instance, Curio (1973, p. 1049) noted that hormones bring out "hidden potentials" in morphology and behavior. He cited research that employed steroids (methyltestosterone) to induce sword formation in males of some swordtail species that lack swords in nature but belong to a genus (*Xiphophorus*) where the sword is a recurrent phenotype (Meyer et al., 1994; Schluter et al., 1997; figure 19.5). This work suggests that hormonally based developmental plasticity mediates the evolutionary gain and loss of the sword. This in turn supports the hypothesis of homology for the different recurrences of the sword. Of the five treated species normally lacking a sword, only two responded to the treatment by producing one, so the potential for a reversion would require more change in some populations than in others.

One-Step or Gradual Recurrence?

Alternative phenotypes are exceedingly common and found in many kinds of organisms (see part III), and often their expression is reversible even within individuals. The vast majority are facultatively expressed. This means that such complex coordinated phenotypes are highly labile evolution-

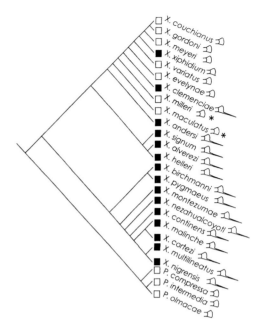

Fig. 19.5. Recurrent gain and loss of the sword in male swordfish and platies (*Xiphophorus* species). Dark boxes indicate presence of the sword (*X. xiphidium* has a small sword); open boxes indicate absence of the sword. Asterisks indicate hormonal inducibility of the sword in species where it is usually absent. Phylogeny after Meyer et al. (1994).

arily. If a trait complex can be reversibly expressed within an individual, or be either present or absent in the members of a single brood, then it may be rapidly turned off and on as a unit during evolution, should that be favored by selection or induced by environmental circumstances. Rampant recurrence, where a trait seems to blink on and off within a taxon as an anomaly, an alternative, or an established trait on different branches of a phylogenetic tree, is evidence for one-step recurrence. It suggests that recurrence is not repeated gradual parallel evolution but rather is reversion to an ancestrally consolidated unit trait whose switch-controlled development has been activated and deactivated repeatedly in the course of evolution, even though it may initially have been gradually evolved.

Alternatively, similar phenotypes may be recurrently assembled gradually from a common ancestral ground plan (*parallel evolution*). In recurrent gradual parallelism, one could expect to find qualitative mosaic intermediates in related species or populations, rather than well-consolidated alternatives controlled by a single switch. But the uncer-

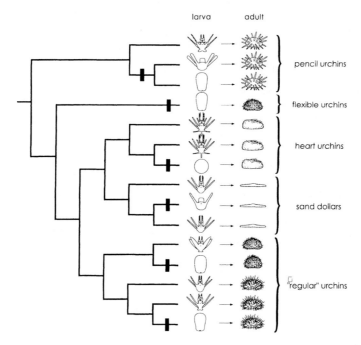

Fig. 19.6. Recurrent evolution of direct development in sea urchins. Although adult morphology is conservative in most urchin orders (listed by their common names on the right), and most retain the ancestral long-armed, feeding pluteus larva, direct development has arisen independently an estimated 27 times, in the six families indicated by the dark vertical bars. From Raff (1996), by permission of University of Chicago Press. Courtesy of G. A. Wray.

tainties of the distinction are indicated by experiments with echinoderm larvae. In echinoids, direct development, or replacement of the feeding planktonic larval stage by a nonfeeding larva without an elaborate skeleton, has evolved at least 20 times (figure 19.6). It is reasonable to cite the occurrence of intermediates as evidence of gradual evolution (e.g., Wray, 1996). But this is not a decisive criterion for distinguishing between parallel evolution and single-step recurrence, because a mutation or environmental factor of major effect conceivably could alter the expression of a large or small portion of the phenotype in extant species, secondarily creating a graded series that may or may not have occurred during the evolutionary history of the lineage. In echinoderms, large variants in larval morphology, approaching those between orders, are found in natural populations of some nonfeeding species in which individuals usually possess a full skeleton (Raff, 1987) and extreme skeletal reduction and loss can be environmentally induced in the laboratory (Wilson, 1894; see chapter 26). This indicates a potential for sudden change. It may often be difficult, therefore, to distinguish with cer-

tainty between one-step and gradual recurrence or parallelism.

Phylogenetic Mapping

Even if homology is well established, multiple occurrences alone do not indicate recurrence. Some phylogenetic evidence of disjunct expression is always required. For example, Jannett (1975, cited by Raikow et al., 1970, p. 206) found that "hip glands" of voles (*Microtus* species) are present in the mature males of some species but are absent in others. Hip glands also can be hormonally induced in species lacking them, suggesting developmental lability of a kind that could lead to recurrence. But it leaves open the possibility that all the natural occurrences represent the unbroken expression of the trait in all descendent species starting with its origin in a common ancestor.

The only satisfactory way to document the independence of disjunct origins using extant populations is to map homologous phenotypes onto a cladogram, in order to see if repeated occurrence of homology in different species is likely disjunct (recurrence), rather than an unbroken product of a single origin (synapomorphy). This can provide an estimate of the minimum number of times that a trait has recurred but may underestimate losses and gains too rapid to be tracked by a cladogram based only on presence and absence in extant populations.

In some groups (e.g., salamanders), recurrence and parallelism have rendered phylogenetic analysis difficult (Wake, 1991). In such taxa, disjunct expression of structurally homologous traits sometimes has been hypothesized on the basis of their isolated or sporadic occurrence "highly nested" (as a specialized or unusual form) in different genera of a conventional taxonomy (Wake, 1991; see also Ross and Carpenter, 1991). This is best justified if there is some evidence that the separate taxa represent well-circumscribed and speciose natural groups, within which the homologous characters of interest are found only sporadically or are taxonomically clumped. In that situation, it is more parsimonious to suppose that the trait is apomorphic (derived) or disjunct from its other occurrences, than to suppose that it is a synapomorphy that has been lost in the majority of species. But this procedure can be misleading, and in some such groups the attempt to say how *much* recurrence or parallelism is present is virtually hopeless (Eickwort et al., 1996). All that can be said is that recurrence exists, as indicated by an extreme degree of independence in the evolution of the repeated alternative states observed.

Unfortunately, the taxa of most interest for a study of recurrent phenotypes are just those where dependable cladistic analyses based on morphology are most difficult (e.g., see Gosliner and Ghiselin, 1984; Sluys, 1989; Wake, 1991). Molecular phylogenies are not necessarily immune to this problem, since molecular evolution is similarly combinatorial (see chapter 17) and subject to homoplasy (Hillis, 1994), but the level of molecular homoplasy is presumably independent of that in the nonmolecular phenotype.

Fossils

Fossils are another source of evidence for recurrence. Recurrence has been hypothesized based on the fossil record of mollusks (Rensch, 1960; Gould, 1969; Stanley, 1972; Gosliner and Ghiselin, 1984), ammonite shell form (Raff and Kaufmann, 1983), fish skeletal features (McCune et al., 1984), and bryozoan colony form (Cheetham, 1986b; Cheetham and Jackson, 1996; Anstey and Pachut, 1995).

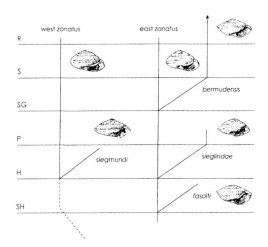

Fig. 19.7. Recurrent paedomorphosis in fossil and recent snails (*Poecilozonites bermudensis*). Paedomorphic populations (branches, flattened shells) have repeatedly descended from two allopatric nonpaedomorphic populations (west zonatus and east zonatus) of the land snail *P. bermudensis* in Bermuda. Fossil strata are Shore Hills (SH); Harrington (H); Pembroke (P); St. George's (SG); Southampton (S); and recent (R). After Gould (1969).

Fossil evidence of recurrence requires evidence that the populations considered to represent repeated appearances of similarity were disjunct in time, or descended from ancestors separated in space. Bermuda land-snail populations (*Poecilozonites bermudensis*) are hypothesized to show recurrent production of populations with paedomorphic shell forms (Gould, 1969; figure 19.7). Extensive samples from six time zones and numerous geographic localities provide strong evidence for the allopatry of the two stem species and of their paedomorphic offshoots, which are never found in the same samples as the stem (zonatus) populations. In the eastern lineage, the paedomorphic populations are disjunct in both space and time and the temporally overlapping paedomorphs are disjunct in space (allopatric).

While the similar *P. bermudensis* populations are convincingly disjunct, they illustrate a problem of fossil material. It is often difficult to distinguish between allopatric morphs that develop under different conditions in a spatially heterogeneous environment, and genetically diverged traits of reproductively isolated populations. This problem is thoroughly discussed by Gould (1969), who cites literature showing that calcium-poor soils can have a direct effect on the thickness of shells. The distinctive morphology of the paedomorphic forms could be a result their distinctive habitat, for they consistently occur in red calcium-poor soils. Gould attributes their thin shells to this fact and speculates that paedomorphosis may facilitate the production of thin-shelled adults and therefore be favored in the calcium-poor environment.

One argument given by Gould against the idea that paedomorphosis is a facultative response to soil quality is the allopatry of the normal and paedomorphic forms, which he considers evidence for their genetic isolation. But allopatry is not sufficient evidence of genetic isolation. Facultatively induced alternatives could occur in a freely interbreeding population and still show allopatric separation of forms due to a combination of local environmental induction and habitat selection at habitat boundaries. This is especially possible if the separation between morphotypes is very small or if the organism has widely broadcast propagules. The paedomorphic population of *sieglindae* is estimated by Gould to have occupied a linear distance along the outcrop where it was found not exceeding 200 meters. This lends some support to the hypothesis that it may have been a morph associated with a peculiar local environment rather than a widespread, genetically distinctive form, expressed (genetically fixed) despite heterogeneous conditions more likely in a larger range.

The morph versus species problem, difficult in taxonomy of extant forms, is of course especially difficult in paleontology (see chapter 30). But the morph versus species hypotheses may be testable in *P. bermudensis*, since Gould (1969) describes extant populations of both the paedomorphic and nonpaedomorphic form. Environmental effects on the shell morphology of snails are reviewed by Matsuda (1987; see also DeWitt, 1998; Rensch, 1960, discussed above). Even if the two forms of *P. bermudensis* were to prove genetically distinct, it is of interest to test the hypothesis that intraspecific developmental plasticity accounts for the recurrent origin of similar forms and their predominance in different parallel environments, a hypothesis more parsimonious than the supposition that they originated independently due to genetic divergence in allopatry (see chapter 28).

Biochemical recurrence in the chemical composition of shells has occurred in brachiopods, where skeletons composed of scleroproteins and calcium carbonate have originated independently a minimum of four times: in the Obelellida, Craniacea,

Craniopsidae, and Trimerellacea (Williams and Hurst, 1977). All of these studies used a fossil record evidently sufficiently complete to rule out artifactual disjunct occurrences due to lack of preservation of connecting forms. Some, such as the evolution of ammonite shell form, involve recurrence of discrete traits apparently governed by switches (Raff and Kaufman, 1983; see figure 7.4). Observation of different recurrent morphs as lifestage morphotypes within different species supports a hypothesis of homology among like forms. Ontogenetic origins of recurrent encrusting and arborescent colony forms in the Bryozoa is suggested by the presence in every arborescent colony of an encrusting base of variable size (sometimes very reduced), as well as by colony form polymorphism in living species of the same genera (see chapter 30). In this example, a continuously variable trait—the size of the encrusting juvenile phase of colony growth—produces a recurrent trait-encrusting colony form, by gradual quantitative change.

Some fossil series show recurrence by gradual parallel evolution, or "chronoclines" (Simpson, 1953a), for example, in some ammonites (Kennedy, 1977, p. 283), Bryozoa (Schopf, 1977), barnacles (Palmer, 1982), and mammals (Simpson, 1953a; Gingerich, 1977; Roth, 1992). Here the problem of cryptic morphs is avoided by the presence of continuous variation among forms. A beautiful Miocene section of lacustrine diatomite in Nevada yielded such a series for a fossil stickleback, *Gasterosteus doryssus*, that revealed first the gradual evolution of the pelvic girdle, then its gradual reduction (Bell, 1987, 1988). Pelvic reduction is a recurrent phenotype in lacustrine sticklebacks, which are descended from *anadromous* forms—forms that are primarily marine but migrate to fresh water to breed. Anadromous sticklebacks have a well-developed pelvic girdle. Comparative study of variation in the fossil populations and in populations of extant sticklebacks (*Gasterosteus*; Bell, 1987, 1988) shows several of the salient features and complications of recurrence homoplasy (figure 19.8). First, the expected parallel between intrapopulation variation, here in the form of ontogenetic change, and evolved end points is observed in populations in two environmental extremes. Second, the comparison of fossil and extant stickleback populations illustrates the principle of multiple pathways to the same recurrent phenotypic endpoint. Most of the morphological series in extant populations show pelvic reduction via distal truncation (figure 19.8b), whereas pelvic reduction in the fossils was paedomorphic (figure 19.8c), a reversal of the sequence seen during ontogeny (figure 19.8d).

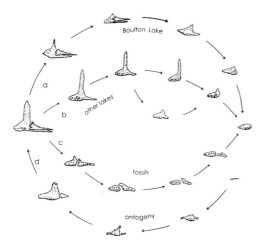

Fig. 19.8. Recurrent pelvic reduction by different pathways in freshwater sticklebacks (*Gasterosteus* species complex). Beginning with complete pelvic development in the marine ancestor (left): (a) pattern of reduction seen in Boulton Lake only; (b) usual pattern of reduction in all other lake populations; (c) pattern in a fossil series (*G. doryssus*), which is the reverse of (d), the ontogenetic sequence of the marine form. After Bell (1987), by permission of Academic Press Ltd., London.

Taken together, the data on fossil and extant populations suggest that full development of the pelvic girdle involves a developmental cascade, since the usual sequence of reduction and loss is the reverse of the ontogenetic sequence, namely (in order of loss), (1) pelvic spine; (2) posterior process, (3) ascending branch, and (4) anterior process (Bell, 1988). This means that a single switch that affects the earliest steps of the cascade could cause rapid pelvic loss, and this may have contributed to the sudden fixation of the extremes in the fossil record.

Patterns of Recurrence

Intraspecific Alternative Phenotypes and Fixed Derived Traits

As discussed above, interspecific patterns of recurrence often reflect intraspecific developmental plasticity found in the ancestral population, in the form of either alternative phenotypes, life-stage forms, or extremes of continuous variation within the same sex or life stage. In sweat bees (Haltictidae), for ex-

ample, both social and solitary nesting are recurrent behavioral phenotypes of females (Richards, 1994; see also figure 12.3). Both nesting modes also are found as alternatives to each other within some species (Wcislo and Danforth, 1997). The Richards (1994) phylogeny has been confirmed by Danforth et al. (1999), who also found that the ancestral state for the genus *Halictus* is eusociality and that reversal to solitary behavior has occurred at least four times. The phylogenetic finding and the discussion of Eickwort et al. (1996) indicate that the lineage of *H. rubicundus* may have flip-flopped between solitary and social behavior four times, beginning with (1) solitary nesting, which is the plesiomorphic (ancestral) state for all social bees, followed by (2) social nesting, the plesiomorhic state for *Halictus*), (3) polyphenic and solitary (during migration from Eurasia to North America), to (4) eusocial (most U.S. populations), to (5) solitary (extant derived montane populations) (figure 12.3).

Socially parasitic reproduction is also a recurrent species trait, and, like social nesting, occurs as an intrapopulation alternative (Wcislo, 1987; Wcislo and Cane, 1996). This pattern of parallel variation within and between species has often been noted in studies of vertebrate morphology (e.g., Alberch, 1983b; McCune et al., 1984, p. 38; Shubin et al., 1995), suggesting that traits capable of expression as alternatives within populations may have been separately fixed in different branches of a phylogenetic tree.

Size-associated phenotypes are especially common among examples of recurrence that involve both intra- and interspecific alternatives. Rensch (1960) pointed out that evolutionary change in body size and its correlated effects can be an important factor in recurrence: "[I]f there is a parallel increase or decrease in body size, causing correlative changes of numerous proportions of the body, animal types will arise showing marked similarity of many characters" (pp. 97–98).

Size-associated alternative phenotypes are conspicuous among polyphenisms (see part III). Body size can be a particularly important evolutionary factor because of its influence on the outcome of social and ecological competition, as well as on correlated change in growth and shape during development and on life-history tactics (see chapter 16), where size correlates with developmental stage. Such size-correlated changes during ontogeny predispose to heterochrony during evolution (Gould, 1977; McKinney and McNamara, 1991). Miniaturization in fish, which has evolved at least 34 times in tropical freshwater species (Weitzman and Vari, 1988), involves paedomorphosis (sexual mat-

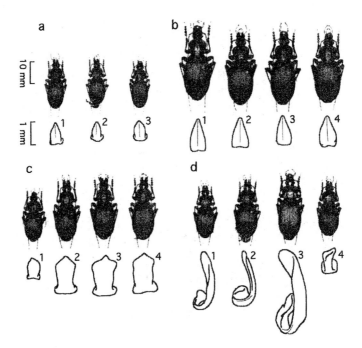

Fig. 19.9. Four recurrent suites of traits in male carabid beetles (*Ohomopterus* species). Each set (a–d) contains a representative male of different species with similar size, genitalic structure, and other morphology. The copulatory piece (shown below each specimen) is a chitinous part of the male genitalia (a–c in dorsal view, d in lateral view). From Su et al. (1996). Courtesy of Z.-H. Su.

uration of individuals with juvenile body forms) accompanied by a recurrent suite of morphological and ecological traits in addition to change in size. Even without heterochrony, size-correlated niche shifts in diet, enemies, and social role may be virtually automatic given the effects of size on ability of individuals to handle food and prey, run and hide, and win in threat displays and fights (see chapter 16).

The physical effects of the environment are also size dependent, for the effects of factors such as gravity, inertia, viscosity, and surface tension change with respect to size (LaBarbera, 1989). Dwarf forms (e.g., in fishes, *Tilapia*) sometimes occur in taxa where stunting is seen as a facultative intraspecific variant associated with adverse conditions (Fryer and Iles, 1972; see also Roth, 1992, on the evolution of dwarf species in elephants). So it is not surprising that size-associated novelties are recurrent as well (Roth, 1992; Weitzman and Vari, 1988). Discrete size classes (dwarfism and gigantism) are sometimes involved in the evolution of parallel species pairs, so the significance of size for evolution will be taken up further in the chapter on speciation (chapter 27).

Recurrent Interspecific Alternatives

Two or more phenotypes may be recurrently fixed in different related species. This pattern is called "type-switching" by Su et al. (1996), who describe recurrences in multiple characters in species of a genus of ground beetles (*Ohomopterus*, Carabidae) in Japan. The genus contains four distinctive sets of coexpressed traits, each characterized by a distinctive and discrete form of male and female genitalia, body size, and in some cases antennal length (figure 19.9) plus some other less conspicuous traits. Every population of this intensively collected group is characterized by one of the four phenotypes, and no intermediate phenotypes have been found—the complete syndrome of traits is always expressed as a set, even in geographically separated local populations. A phylogenetic analysis (figure 19.10) suggests that there were at least 14 switches between types, with type a independently derived at least three times, type c six times, and type d five times. Identical specimens of types d and c were found in both of the major two lineages.

A classic example of recurrent interspecific alternatives occurs in the shape of the ejaculatory

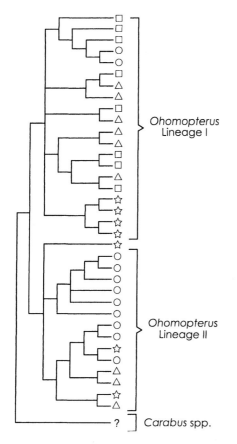

Fig. 19.10. "Type switching" in male carabid beetles (*Ohomopterus* species). The four suites of traits (a–d) shown in figure 19.9 recur independently many times, as shown on a molecular phylogeny of two major lineages of the genus *Ohomopterus*. Symbols correspond to groups of figure 19.9: a = squares; b = triangles; c = stars; d = circles. Phylogeny after Su et al. (1996).

bulb in the *repleta* group of *Drosophila* (Throckmorton, 1965; figure 10 in Futuyma, 1986a). A map of the five discretely distinctive forms of the bulb onto a chromosome-based phylogeny shows a minimum of six instances of recurrence. One of the variants independently evolved at least three times, and another twice. In addition there was one reversion to the apparently ancestral elongate form following its loss, an example of flip-flopping recurrence (see below).

Recurrent interpopulational and interspecific trophic alternatives are exceedingly common in fishes. Examples include the miniaturized paedomorphic species of neotropical freshwater fishes (Weitzman and Vari, 1988), several trophic forms

in cichlids (Liem and Kaufman, 1984); benthic and pelagic forms of European whitefish (*Coregonus lavaretus*), a dwarf pelagic phenotype in different races of eastern lake whitefish (*C. clupeaformis*), trophic ecotypes in rainbow smelt (*Osmerus mordax*), and habitat-specific life-history traits in arctic charr (*Salvelinus alpinus*), threespine stickleback (*Gasterosteus aculeatus*), sockeye salmon (*Oncorhynchus nerka*), Atlantic salmon (*Salmo salar*), and brown trout (*S. trutta*; see references in Bernatchez et al., 1996, p. 632; see also chapters 27 and 28).

These interspecific alternatives are sometimes but not always associated with intraspecific polymorphism even in groups where this possibility has been rigorously tested. In whitefish (*Coregonus*), for example, the alternatives occur sympatrically but represent genetically divergent populations (species or incipient species; Bernatchez et al., 1996). The pattern has been explained as "evidence for the potential of rapid, parallel evolution, which may reflect population response to local selection in the face of similar ecological opportunities" (Bernatchez et al., 1996, p. 632), that is, as a product of parallel selection pressures (see also Schluter and Nagel, 1995; Rundle et al., 2000). This explanation could apply to either convergence or recurrence of traits already established in the ancestral population. Such rapid and frequently repeated similarity suggests the latter. The recurrence hypothesis is supported for sticklebacks by evidence of ontogenetic stages in the ancestral populations that resemble the two morphs (Andrews, 1999; see chapter 27).

Most examples of recurrence seem to involve nonsocial traits, but an unusual example of recurrence in the evolution of courtship signals occurs in lacewings (Chrysopidae), insects that produce sounds by vibrating their bodies against a substrate. Courtship communication is usually exceedingly diverse, even (or especially) among members of the same genus (West-Eberhard, 1983, 1984). The *carnea* group of green lacewings is no exception, in that several very distinctive songs have evolved, but they are unusual in that one of them has evolved independently four times, cladistically interspersed among other varied songs (Henry et al., 1999).

Parallel Evolution

As mentioned above, in some taxa such as the ground beetles (*Ohomopterus*) there are recurrent traits with no evidence of intrapopulational alternatives, ontogenetic progressions, or anomalies that would indicate a potential to switch developmentally to the novel form in a single step. Nonetheless, there is evidence in some for homology based on the presence of intermediates indicating gradual evolution starting with a common ground plan. In insects, hearing organs have evolved more than once in distantly related taxa (different families) in both Diptera (flies) and Orthoptera (crickets, grasshoppers, katydids, and related groups; reviewed in Gwynne, 1995). In the Diptera, parasitoids in two different families (Tachinidae and Sarcophagidae) that locate their hosts by sound have independently evolved hearing organs on the front of the thorax. In Orthoptera, singing insects have independently evolved hearing organs on the front legs in both grylloids (crickets and their relatives) and tettiginiioids (katydids and their relatives). Both occurrences are so similar that taxonomists have been reluctant to consider the derivations independent, but a phylogeny suggests that they are examples of parallel evolution in these orthopterans, both likely derived starting from the same structure—the tibial receptor cells—considered to have been present in the ancestors of both groups (Gwynne, 1995). In tettigoniids there are tibial sensory organs on all six legs, and there is embryological evidence for their homology with the hearing organ (see references in Gwynne, 1995).

It seems likely that most, if not all, examples of parallel evolution depend on some ancestral developmental plasticity. In the cheilostome Bryozoa, colonial marine invertebrates that grow by the multiplication of their component structures (zooids), there is a recurrently evolved alternative growth pattern, zooidal budding, which produces a continuously extending tube with calcareous structure developed in the proximal, more mature part of the tube. This contrasts with the pleisiomorphic growth form by intrazooidal budding, or intermittent expansion by outward growth of an uncalcified portion of a zooid body wall (Lidgard, 1986). Zooidal budding apparently originated in the Late Cretaceous, then recurred independently in several major bryozoan stocks, sometimes co-occurring as an interspecific alternative to intrazooidal budding even within the same genus (Lidgard, 1985, cited by Lidgard, 1986, p. 231). What in the ancestral developmental ground plan of this group of organisms could account for the recurrent origin of this trait?

Terminal defensive stolons are a candidate phenomenon that is widespread and inducible, and therefore could have recurrently given rise to zooidal budding. When approaching or in contact with neighboring colonies or colonial ascidians, some cheilostome bryozoans produce fingerlike "terminal stolons" at the colony periphery, apparently to redirect growth of competitors and prevent overgrowth (Osborne, 1984). Terminal stolons appear to be widespread in cheilostomes: five of eight of the Australian cheilostomes studied produced them, and similar outgrowths have been noted by other authors (e.g., Stebbing, 1973). They lack calcified walls, so they would not be preserved as fossils. But they satisfy some of the criteria of a protozooidal bud in being widespread, inducible intraspecific variants, long, continuous extensions of a preexisting zooid, whose continued extension does not depend on prior partition formation, and (like zooidal buds) development of a calcified interior wall and calcified skeleton inside uncalcified cuticular and tissue layers (figure 1 in Lidgard, 1986).

Recurrent Reversals (Flip-flopping)

As expected if recurrence depends on developmental switches between alternative phenotypes or ontogenetic stages, some traits show recurrent reversions, or flip-flopping between alternative states both within and between populations. Flip-flopping between alternatives during phylogeny has been called "switchback evolution" by Van Valen (1979) and "flickering" by Marshall et al. (1994, p. 12286). An example has already been described in the recurrently recombined leaf forms in vines of the genus *Monstera* (see figure 6.3).

Recurrent flip-flopping evolution sometimes involves surprisingly complex traits, features of macroevolutionary dimensions, that is, traits usually characterizing higher taxonomic categories such as families or orders. An example is the flip-flopping evolution of the insect ovary. In insects, there are three major types of ovary: the *panoistic* ovary, which lacks nurse cells (oogonia are transformed into oocytes), and two types of *meroistic* ovary (polytrophic and telotrophic), in which oogonia generate both oocytes and nurse cells and show various arrangements for transport between them (figure 19.11). In meroistic ovaries, the nurse cells shrink as the eggs develop (King and Büning, 1985). The evolutionary transitions among ovary types are shown in figure 19.12. The panoistic ovary (a) is plesiomorphic (ancestral), and the transition to a polytrophic meroistic type (b) occurred twice, once in a line ancestral to both the Paraneoptera and the Holometabola and once (b) in the earwig (Dermaptera) line, which has a distinctive meroistic ovary indicating convergent evolution. The polytrophic meroistic type then gave rise to the telotrophic meroistic type independently three times (figure 19.12).

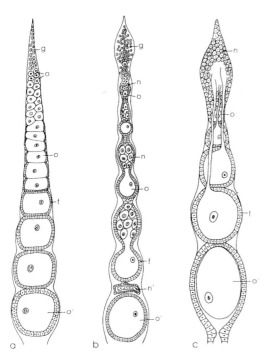

Fig. 19.11. The three major types of insect ovaries: (a) panoistic, (b) polytrophic, and (c) telotrophic, showing germaria (g), immature oocytes (o), mature oocytes (o'), nutritive (nurse) cells (n), remains of depleted nurse cells (n'), and follicular cells (f). Diagrammatic after Imms (1964, and King and Büning (1985).

Reversions to panoistic ovaries occurred at least four times, in the Thysanoptera, some Megaloptera, some Neuroptera, and some Mecoptera (King and Büning, 1985). Most remarkable for such a major character, usually specific to whole orders and even subclasses of insects, there was flip-flopping between the panoistic ovary, typical of the oldest orders of insects, and the polytrophic meroistic form (b), so that in one lineage of fleas (Siphonaptera, Hystrichopsyllinae) the ovary evolved from panoistic, to meroistic, to panoistic, and then back to meroistic again (figure 19.12). This implies an enormous change in the female reproductive system, egg phenotype (e.g., germ-band organization; see figure 3.42 in Büning, 1994) and life history of the insects. The internal nature of mother-offspring relationships is altered in that in the panoistic ovary the oocytes are surrounded only by maternal somatic (follicle) cells, whereas in meroistic polytrophic ovaries the oocyte is nurtured by a cluster of nurse cells that are sister germline cells of the oocyte itself. Meroistic telotrophic ovaries

are similar except that the nurse cell cluster is located anteriorly but is still attached to the sister oocyte by a nutritive cord (figure 19.11c).

Some authors have been incredulous that macroevolutionary flip-flopping of such major proportions could occur (e.g., see Büning, 1994, p. 318, for challenges and replies). But the panoistic ovary can be derived from a polytrophic or telotrophic meroistic ovary by a single step—the reduction of cytoblast divisions to zero (Büning, 1994, pp. 324, 317ff). The reversion to the polytrophic state from secondary panoism is evidently facilitated by the fact that a crucial step—cell-cluster formation during germ cell development—persists in the *males* of panoistic species, even while the genes influencing this step are silenced in females (Büning, 1994, p. 316). This cross-sexual pleiotropy may be responsible for the long-term evolutionary lability of the switch from panoistic to polytropic ovaries.

Although I know of no detailed discussion of the selective context for these major changes in the ovarian styles of insects, different ovary types have contrasting functional characteristics that profoundly influence the life histories of the offspring. Panoistic ovaries, which are typical of the insect order Orthoptera (crickets, grasshoppers, and katydids), produce eggs with relatively small amounts of euplasm, the informational materials provided by the maternal germ line. This is associated with slow embryogenesis, taking up to several weeks.

Polytrophic meroistic ovaries like those of the Hymenoptera (wasps, ants, bees) and Diptera (flies, including *Drosophila*), by contrast, have large euplasms provided by the nurse cells, and relatively rapid embryogenesis occupying only a few days (Büning, 1994). It is easy to imagine that such differences may influence evolutionary outcomes when selection favors an evolutionary switch between ovary types, for example, in situations that favor a change in rate of offspring production.

Flip-flopping recurrence of macroevolutionary proportions is also seen in the alternation between solitary and social life in some bees. Eusociality, or life in colonies containing queens aided by sterile workers, has evolved at least four times in the bees (Michener, 1974), undoubtedly an underestimate if all reversions following losses were to be counted. At least six independent reversions to solitary nesting have occurred in one family (Halictidae) alone (see chapter 12).

A possible example of macroevolutionary flip-flopping occurs in the evolutionary history of snakes (reviewed by Greene and Cundall, 2000). The three major lineages of snakes, blindsnakes,

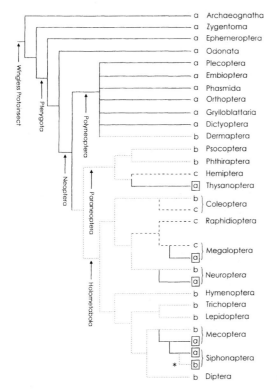

Fig. 19.12. Recurrence and flip-flopping in the evolution of insects. Letters show the phylogenetic distribution of ovary types on a tree of insect orders, with ovary types as in figure 19.11: (a) and solid lines: panoistic; (b) and dotted lines: polytrophic; and (c) and dashed lines: telotrophic. Letters in boxes indicate reversions; the asterisk (*) indicates a lineage (the Siphonaptera) with the inferred flip-flopping evolutionary sequence a–b–a–b. The panoistic ovary (b) and the telotrophic ovary (c) have each evolved "independently" three times. Phylogeny after Kristensen (1981). Mapping on of ovarian types after King and Büning (1985).

pipesnakes and shieldtails, and macrostomatid snakes (including pythons, boas, and advanced snakes), are probably descended from a legless common ancestor (since all lack legs, the unparsimonious alternative is to suppose that legs were independently lost repeatedly by descendants). Some fossil macrostomate snakes, *Haasiophis* and *Pachyrhachis*, have regained well-developed limbs. The potential to regain hindlimbs is suggested by the commonness in living snake species of external hind limbs, which are sometimes present as claw-

like vestiges. There is also evidence for the regulation of such vestigial leg development by the hox family of genes (Cohn and Tickle, 1999, cited in Greene and Cundall, 2000).

All of the manifestations of recurrence—repetition of the same traits as a low-frequency anomaly, as an alternative, and as a widespread fixed trait—are sometimes seen in the same taxon. A striking example of this occurs in the limb morphology of salamanders (*Taricha glanulosa*, Salamandridae) studied by Shubin et al. (1995), who examined a large sample (452 individuals) of a California population. Six limb-bone anomalies occur at low frequency (<6% unilateral occurrence; 1% or less bilateral occurrence; Figure 19.3). About 75% of the patterns found as anomalies in *T. granulosa* are found in other taxa either as fixed traits or alternative phenotypes (40%), or as anomalies (Shubin et al., 1995). The *T. granulosa* study employed "the largest known series . . . from a single urodele population" (p. 874), raising the possibility that the patterns revealed may be found in other organisms if comparable samples were to be examined. A similar pattern of recurrent intrapopulational and interspecific variation occurs in the neotropical salamander genus *Bolitoglossa* (Alberch, 1983b). Although it is unusual to *document* this pattern so thoroughly, it is a pattern familiar to taxonomists, as Shubin et al. discuss (see also Consequences of Recurrence for Systematics and Phylogenetics, below).

Not surprisingly, given the flexibility of behavior, behavior can show the same patterns of frequent recurrence seen in morphology. For example, in the social wasps (*Polistes*), parasitic oviposition in nests of other females is rare in some populations (West-Eberhard, 1969), a common alternative of variable frequency in others (Klahn, 1988), and fixed in still other species that are obligatory social parasites (Cervo and Dani, 1996). The behavior of obligatory social parasite species shows clear homology with that of facultative social parasites and is believed to be derived from that of usurpers in primarily nonparasitic species (reviewed in Cervo and Dani, 1996).

Environmentally Correlated Recurrence

Recurrent transitions sometimes correlate with recurrent environmental conditions, as in the high-altitude halictine bees discussed in the preceding section (see also chapter 28, on parallel species pairs). This could reflect the role of the environment as an inducer of traits, as an agent of natural selection, or both. In bees, a derived, conditionally expressed alternative phenotype was recurrently deleted, producing a reversion to the ancestral state that lacked it. Recurrent phenotypes can also reflect recurrent environmental opportunities against a background of large standing crop of developmental variation, with sporadically produced low-frequency anomalies occasionally advantageous, as suggested by the patterns of variation in the foot bones of salamanders (see figure 19.3). A recurrent variant could remain a sporadic atavism for generations following its loss, then reappear at high frequency if favored by selection.

The recurrent evolution of a reduced pluteus larva in echinoderms (Emlet, 1986; Raff, 1996; see figure 19.6), along with the recurrent evolution of the paedomorphic shell form in snails (see figure 19.7), the recurrent evolution of parallel species pairs in fish (see chapters 27, 28), and the recurrence over time, measured on a paleontological time scale, of dietary specializations in kangaroo rats (see chapter 26) may all be examples of recurrence importantly facilitated by parallel recurrent environmental change on both development and selection.

The role of recurrent environments in the recurrent induction of novelties is perhaps best seen in animals capable of learning. Kleptoparasitism of nests or food is a recurrently evolved behavior of birds. It often occurs as an individual opportunistic variant within species (Seger and Brockmann, 1987; Field, 1992a). In some species (e.g., frigate birds), it is an obligatory specialization. Similarly, cannibalism is quite common among predators, which occasionally prey on conspecifics: under prolonged trophic stress, cannibalism may be established as a regularly occurring alternative specialization (Polis, 1981; Elgar and Crespi, 1992). Initially, cannibalism may require no special phenotypic change beyond the behaviors and morphology of ordinary predation. But species where cannibalism is frequent sometimes have morphologically specialized cannibal morphs, for example, in rotifers (Gilbert, 1977), salamanders (Pfennig and Collins, 1993), and spadefoot toad tadpoles (Pfennig, 1990a,b, 1992a,b). In these examples, a new direction of evolution begins not with the advent of a novel phenotypic variant but with the advent of a novel environment (e.g., scarcity of ancestral prey and proximity to conspecifics) that favors the cannibalism-adept variants already present in the population.

Learning has not been a popular subject among evolutionary ecologists. Learning was proposed to be involved in the evolution of habitat selection in mice, in a classic paper (Wecker, 1963) prominent as much for the loneliness of its interpretation as for the excellence of its insights regarding the potential importance of learning in evolution. Doubts cast on the importance of learning due to the failure of Wecker and others to induce a learned habitat change by forcing mice to inhabit woodlands rather then fields (e.g., Morse, 1980) seem to misconstrue the nature of adaptive learning: woodlands may not reward a field mouse, so such experiments are not a fair test of adaptive learning but instead require mice to learn stupidly. Adaptive learning should reinforce rewarded (appropriate) habitat choice, not produce choice of inappropriate, unrewarding habitats.

The role of environmental induction and developmental plasticity in the recurrent evolution of novelties is more thoroughly discussed in chapters 26, 27, and 28. But the question of environmental influence bears mention here because it is so often the object of benign neglect by students of parallel evolution. Recurrence is often assumed to be entirely a result of selection in similar environments (e.g., see Gould, 1969; Schluter and Nagel, 1995; Rundle et al., 2000). This may be largely because it has never been sufficiently clear how environmental induction can be important for the genetic change that is of most interest—a problem that is largely resolved by gene–environment interchangeability and genetic accommodation (see chapters 5 and 6). Recurrence and parallelism should always be regarded as cues for attention to developmental plasticity, including environmental effects. Recurrence virtually requires pursuit of the working hypothesis that recurrent similarity is due to ancestral developmental traits and therefore not entirely due to independent evolutionary events.

Consequences of Recurrence for Systematics and Phylogenetics

Recurrence contradicts the impression of concordant phenotypic evolution sometimes given by the taxonomic and phylogenetic practice of arranging groups of organisms in lineal and hierarchical series to approximate a geneology. The construction of branching phylogenetic trees may systematically downplay intrataxonomic variation because it is considered a source of noise for analyses not specially designed to utilize such variation (Wiens, 1995, 1999). This may unwittingly encourage a lineal view of evolution, when in fact the phenotype often evolves in mosaic, combinatorial and intermittent fashion, as shown in the previous ten chapters (part II).

On one hand, recurrent alternative phenotypes are an important and sometimes underestimated source of homoplasy (e.g., see Rachootin and Thomson, 1981, p. 189; Gosliner and Ghiselin, 1984; Matsuda, 1987, p. 51; Wake, 1991; Frumhoff and Reeve, 1994; Schluter et al., 1997; Losos, 1999). Recurrence violates common assumptions of parsimony analyses, such as character-state irreversibility, single-time trait origins, and intraspecific constancy. If recurrence homoplasy is common, including change within species in the form of alternative phenotypes, existing cladistic methods based on parsimony may fail (Sluys, 1989; Wake, 1991) or be subject to a high degree of uncertainty (Schluter et al., 1997). In groups with rampant recurrence homoplasy, the accuracy of parsimony in construction of a cladogram may actually decrease as the number of characters included is increased (Bull et al., 1993, cited by Wiens and Servedio, 1997). For a recent discussion of this problem and the potential uses of information on polymorphism in cladistics, see Hillis (1994) and Wiens (1999).

Recurrence and phenotypic polymorphisms also raise difficult questions regarding character weighting in systematics and phylogenetics (see Mayr and Ashlock, 1991, for a general discussion; see Wiens, 1995, 1999, for detailed discussions). Some cladists (e.g., Allard and Carpenter, 1996) argue that character weighting is never justified. Others believe that it may be desirable to weight polymorphic characters in accord with their phylogenetic signal (Mayr and Ashlock, 1991, p. 218; Wiens, 1995, 1999). The findings of this chapter urge caution in the use of the following four assumptions sometimes used in character weighting:

1. Complex characters are less subject to homoplasy, "because the probability of evolving independently two or more times is considerably lower for a complex than it is for a simple feature." (Mayr and Ashlock, 1991, p. 218). This assumption must be treated with caution because very complex characters are known to be recurrent. There is recurrent evolution of such complex traits as worker production and social cooperation in insects (Wilson, 1971a; Michener, 1974, Crespi, 1996); direct development in echinoderms and salamanders; and ovary type in insects, with the panoistic and telotrophic ovaries each having evolved in-

dependently three times (figure 19.11). The assumption that complex characters are unlikely to show homoplasy is also contradicted by the complexity of many conditionally expressed polyphenic traits, which are subject to rapid fixation and repeated appearance (Shaffer and Voss, 1996; see part III). In theory, likelihood of repeated independent evolution depends less on degree of complexity than on such factors as degree of developmental dissociability (the extent of control by a single switch) and degree to which the components of form and regulation of a lost phenotype are maintained by selection in other contexts even while not expressed as a coherent set (see chapter 12).

2. Novel traits are less liable to recurrent gains than to deletions, "because the probability of a new feature evolving independently two or more times is low, while reduction or loss of a feature may occur several times independently" (Mayr and Ashlock, 1991, p. 218). Recurrent reversions and parallel origins (above) are common enough to throw reasonable doubt on this conclusion as a generally dependable rule. It needs to be tested in a wide variety of taxa, but this is presently difficult because such an assumption is often incorporated into cladistic methodology and so becomes a self-fulfilling prophesy that biases the available information and affects the opinions of comparative biologists regarding likelihood of recurrent gains. Some well-documented recurrent gains of complex trophic phenotypes in fishes are discussed in the section on binary radiations in chapter 28.

3. Traits with multiple functions are not likely to undergo reversion and parallel origins: "[T]he probability of a feature serving a wide range of roles evolving independently two or more times is less than that of a feature with a single biological role" (Mayr and Ashlock, 1991, p. 218). This reasoning is flawed, because functions can be added following the origin of a trait, and multiple origins (reversions) following deletion may be *more* likely for a trait capable of functioning in a variety of contexts, because such a trait has a greater likelihood of functioning in some context upon reexpression. Thus, multiple diverse roles may *favor* homoplasy rather than militate against it. Multiple functions are expected to be an impediment to deletions but not to recurrent origins.

4. Complexity of developmental interconnections is an impediment to recurrence, because multiple origins are less likely in "a feature that has an ontogenetic development that depends on a complex pattern of embryological modifications and interconnections with other developmental features

rather than one that has a simple ontogeny" (Mayr and Ashlock, 1991, p. 218). This is based on the cohesiveness view of development that is contradicted by mosaic evolution and modular dissociability of phenotypic traits, including complex ones (see chapter 4). Complexity per se is not an impediment to dissociability and recurrence.

Patterns of recurrence can provide clues as to phylogenetic relationship when none was suspected. In characid fishes, certain characters are evolutionarily labile when analyzed at lower taxonomic levels, causing some authors to discount their utility for phylogenetic analysis. But they may be indicative of monophyly at some higher taxonomic level (Weitzman and Fink, 1983, p. 342). These authors use evolutionary lability as a character of a taxon above the species level. They point out that certain kinds of osteological lability distinguish the subfamily Glandulacaudinae as a whole from other characiform fishes. Similarly, the unique occurrence of dioecy in two species of planarians, a taxon with no other dioecious species, has been used to support a hypothesis of common ancestry of these two species (Sluys, 1989).

Some authors (e.g., Saether, 1979, 1990; Cronquist, 1987; Sluys, 1987) have suggested that recurrence and parallelism can be used as part of a formal method to detect latent characters or underlying synapomorphies within a group of related species, and they suggest using this information, along with established characters, as positive indicators of relationship (see Wiens, 1999), especially in taxa with recurrence homoplasy in many traits, where parsimony methods are otherwise weak. This approach is based on the frequently expressed notion (e.g., see Mayr and Ashlock, 1991; Sluys, 1989) that parallelisms are more likely in closely related groups than in unrelated taxa.

While concealed and inducible variation sometimes may be useful in systematics, the phylogenetic level at which it is significant is difficult to evaluate. Numerous abnormalities can be induced by hormone treatment in plants of the family Caryophyllaceae, for example, and many of the same anomalies are seen as sporadic variants in natural populations (Meyer, 1966). Such anomalies may illuminate phylogenetic relationship, but at what taxonomic level? In this family there are anomalies, especially in flower structure, corresponding to most of the taxonomic differences by which different genera and subfamilies are recognized (Briggs and Walters, 1988).

In the gypsy moth (*Porthetra dispar*; formerly *Lymantria dispar*), normal females with short-branched antennae that are subjected to extreme temperatures during development, or the offspring of interracial crosses, develop anomalous platelike lamellae instead of short hairs on the antennae. Such lamellate antennae are normal in a distant family of moths, the Cossidae (Goldschmidt, 1940 [1982], pp. 305–306). Should a species with a sporadic variant resembling members of another genus or family have its status reexamined? Recurrent traits can indicate membership in a particular taxon—that group where the trait sometimes occurs, as opposed to all other groups where the trait is never seen (strict "inside parallelism"—Sluys, 1989). Sluys (1989, p. 364) draws an arbitrary line between "inside parallelisms," those within two outgroup nodes of each other, and more distant "outside parallelisms." But the arbitrariness of this is troublesome. Although some authors believe that recurrence and parallelism are more common in more closely related species (e.g., see Simpson, 1951, p. 67), this is not known to be a generally applicable principle. Extremely disjunct recurrences, such as that of the insect ovary after millions of years of absence, and other atavisms discussed in chapter 12, suggest that likelihood of reversion is not be proportional to phylogenetic nearness of the last occurrence of the trait. Adequate estimates of this would require taxonomically wider and locally larger samples than those usually possible. So it is difficult to disprove the hypothesis that a recurrent trait is anything but an *underlying plesiomorphy* (a term attributed to Kristensen, 1984, by Sluys, 1989, p. 358)—a developmental potential shared with less derived groups.

What, then, should be done with recurrent intrapopulation variation and rapid flip-flopping between character states in systematics and phylogenetics? Using various criteria of phylogenetic utility, Wiens (1995, 1999) finds that polymorphic traits do contain phylogenetic information, and that the best way to take advantage of that information is by weighting methods that take morph frequency into account. This conclusion has been confirmed by simulation studies (Wiens and Servedio, 1997).

The construction of cladograms is one problem. Tracing the evolutionary history of particular developmentally plastic traits is another. First, the apparent pattern of gains and losses of a plastic trait depends on how the analysis codes for intrataxic variation (Wiens, 1999; see chapter 20), and on how completely each data point samples the represented taxon. Carpenter (1989) proposed a phylogenetic test of a hypothesis regarding the evolutionary transition from solitary to social life in the vespid wasps, which is widely regarded among cladists as an example of "rigorous inspection" of an adaptation hypothesis (e.g., Novacek, 1996, p. 333). This test seems to contradict the feasibility of one predicted stage (nest sharing without workers) but overtypifies the ancestral group (the eumenine wasps) as solitary (Carpenter, 1989, p. 137), rendering invisible the social species of that taxon that were crucial to the evolutionary hypothesis. Tracing the evolutionary origin of a trait may require inclusion of more intrataxon, including intrapopulational, variation than is usual in a cladistic analysis.

Second, and more difficult, is the problem of ancestor-state reconstruction. Polymorphic traits (alternative phenotypes) are switch controlled and therefore can be assumed to be developmentally labile in their presence or absence and frequency of expression. It may at first seem paradoxical that a cladogram coding for polymorphism would reconstruct the ancestral states of two polymorphic branches as absent rather than polymorphic, as occurs under some coding procedures (Wiens, 1999). But it is equally accurate to call the ancestral state of a sometimes polymorphic trait absent, polymorphic, or present, because there is no way to estimate the evolutionary history of such a trait from data on extant populations. Given control by a switch, the state of the trait at any given time depends primarily on selection, which is known to be variable in time and space if the trait varies in frequency within and between species. To the degree that past selection is unpredictable from present selection, *the ancestral state is nonreconstructible.* Assumptions that apply to DNA polymorphisms, such as unlikelihood of a gain following a loss, do not apply to the phenotypic polymorphisms, where gains and losses depend on a polygenic switch mechanism that is adjustable by selection and where a lost trait can be regained via an environmental or regulatory change.

The potential for evolutionary lability applies even to switch-regulated traits that are at extremely high or low frequency or genetically fixed in some populations, given gene–environment interchangeability and genetic accommodation (see chapters 5 and 6). Detailed studies of recurrence in such traits as the salamander bones and bird muscles discussed in this chapter suggest that low-frequency anomalies occur wherever they are sought. In such traits, change can be expected virtually whenever it is favored by natural selection. In other words, developmentally labile traits should be regarded as virtually free of so-called phylogenetic inertia, or phylogenetic constraints on their gain and loss. Frequency-based cladistic reconstructions of their ancestral condition, which weight changes based on extant trait frequencies, are meaningless.

As expected from the observations of this chapter, there is sometimes a correlation between the level of homoplasy of a trait and its intraspecific variability (Wiens, 1995). Wiens gave several tentative explanations for this which are unlikely to apply to adaptively flexible traits, namely, that intraspecific variants may be nongenetically determined and therefore not indicative of relationship; that fixation may be the result of random drift in the sampled isolates during geographic speciation, such that a particular morph happens to get fixed, obscuring its close relationship to the parent population; or (after Farris, 1966) that intraspecifically variable traits are less constrained by selection and therefore haphazardly fixed. Considerations of development and evolution suggest a different and more commonly applicable explanation, at least for nonmolecular traits. Switch-determined polymorphisms and low-frequency anomalies are associated with recurrence homoplasy because numerous genetic and environmental factors can trigger their production (see chapter 5; Shaffer and Voss, 1995). Whenever selection happens to favor one of these variants, it may evolve to fixation, creating recurrent fixations (homoplasy). The selective context need not be always the same, as shown in the recurrent evolution of paedomorphosis as an alternative developmental mode in salamanders, where it evidently can serve either as a means of remaining in the aquatic habitat, or as a way of salvaging reproduction in individuals too small to metamorphose to the normal adult form (Whiteman, 1994).

Schluter et al. (1997) give a quantitative error-rate estimation for accuracy of ancestral state reconstruction based on maximum likelihood. They conclude (cf. Schultz et al., 1996) that if change is estimated to have been frequent, then ancestor reconstruction based on extant species is subject to a high degree of uncertainty. For developmentally plastic traits, this conclusion can be emphasized—as a rule of thumb, fixed-trait and frequency data on extant populations cannot be used to deduce the ancestral state of a developmentally plastic phenotypic trait that is known to switch between alternative states within an individual or (in a population) change frequency between generations.

For developmentally plastic characters, application of parsimony is misleading and estimates of degree of uncertainty need to be applied (see Schluter et al., 1997; Crespi and Sandoval, 2000) for reasons discussed by Frumhoff and Reeve (1994) and Wiens (1999). In summary, this conclusion is based on (1) the association between intrapopulation and interspecific variation within clades (discussed in this chapter), which shows that developmental plasticity is indeed reflected in evolutionary flexibility of the same traits; (2) the ease with which the expression of a switch-regulated trait can be turned off and on due to environmental factors and to selection for change at any one of multiple loci (see chapter 5); and (3) the inability to predict past selection (morph frequencies) from present selection.

Large samples and character weighting may ameliorate the polymorphism problem for cladogram construction (e.g., Hillis, 1994; Wiens, 1999) but not for ancestor reconstruction. As discussed in this chapter, cladistic mapping (optimization of traits on a cladogram) can suggest hypotheses regarding the point of first appearance, minimum number of recurrences, degree of lability, and derivation by phenotypic reorganization of a flexible trait. But it cannot determine precisely how many times a switch-controlled trait has been lost or gained in that clade. In such lineages, application of parsimony to determine ancestor states (e.g., Carpenter, 1997) is inappropriate. The occurrence of the trait as an alternative within species means that it is regulated by a switch, so gains and losses can occur repeatedly *between* branching points. As a result, two sister groups may possess (or lack) a switch-controlled trait that their common ancestor did not possess (or lack).

The Evolutionary Significance of Recurrence

Recurrence, like reversion (see chapter 12), shows that phenotypic components lost in the distant past can be resurrected to participate in the origins of novelty without being passed in an unbroken line of expression to immediate descendants. Dissociable phenotypic traits may be lost, gained, and repeatedly used in different combinations at different times over the course of evolution. Since this implies inheritance of developmental capacities that are not visible in the phenotypes of every generation, or in every species of a monophyletic group, it demonstrates the importance of keeping a broad view of the meaning of similarity due to common descent in studies of evolutionary transitions and the origins of novelty.

Interspecific variation often reflects intraspecific variation within a clade, as shown by the studies of variation in skeletal morphology of salamanders discussed in this and in the other chapters of part II. The similarity between intra- and interspecific variation is not surprising, since evolved differences between related species must reflect developmental

processes that generate the selectable variation within them. But the failure to recognize this pattern has had profound consequences in the history of evolutionary biology. The fact that developmental variation governs evolved variation is the reverse of the causal relations between ontogeny and phylogeny implied by Haeckel's Biogenetic Law, as it was summarized by the famous phrase "ontogeny recapitulates phylogeny," as if phylogeny were the causal agent and ontogeny the result. Instead, there is a sense in which phylogeny reflects ontogeny: development gives rise to the variation that underlies phylogenetic diversification (Gottlieb, 1992).

The explanation for patterns of trait diversification and recurrence should be sought in patterns of development rather than the reverse. The practice of looking to development to explain phylogenetic pattern is especially fruitful in studies of adaptive radiation, where diversification in well-studied taxa clearly reflects patterns of developmental plasticity within species (chapter 28). In general, recurrent patterns of evolution reflect recurrent or widespread patterns of developmental plasticity within the same taxa, and rapid diversification reflects adaptive evolution that has capitalized on some shared capacity of related species to respond rapidly to environmental variation, competition, or stress (see chapter 28). If you see pattern in phylogeny, look for its causes in development and ontogeny.

Why are unit phenotypes so often recurrent? One reason is that regulation capable of turning modular traits off and on during evolution is complex. This means that the same off–on switch mechanism can be influenced by any one of a number of individual genes or environmental factors or combinations of factors. So expression of a trait does not depend upon any particular mutation or environmental stimulus. A transition from wingedness to winglessness in insects, for example, is reversible, and wing polymorphisms are common (Harrison, 1980; Roff, 1986a,b). In crickets (*Gryllus rubens*), wing loss can be environmentally induced by crowding, or by artificial selection, implying genetic change (Zera and Tiebel, 1988). In every insect species where the mechanism of the transition between wingedness and winglessness has been studied, it involves juvenile hormone (JH) and enzymes involved in JH-mediated regulation (e.g., Zera and Tiebel, 1988, 1989; Zera et al., 1992; reviewed in Nijhout and Wheeler, 1982; Matsuda, 1987; Nijhout, 1994a; Roff et al., 1997). So there is a widespread polygenic developmental arrangement that is responsible for wing loss and recurrence in these insects.

Evolutionary lability is also favored by the fact that the gain and loss of wings are highly subject to changing contexts of selection for dispersal ability and certain life-history traits, such as a trade-off between wing production and fecundity (Mole and Zera, 1993; Zera and Bottsford, 2001). Not surprisingly, a similarly recurrent wing loss in water striders (Hemiptera, Gerridae) proved upon close examination to give rise to two different classes of short-winged morphs, one emerging from fifth-instar nymphs with distinct wing pads, and the other from nymphs with reduced wing pads (Andersen, 1993), in keeping with the expectation that different developmental modes of wing-loss should be possible in a complexly regulated trait. So the wing-loss transition seems to be recurrent due to both the readiness of its utility under selection and the complexity of its regulation (which makes it developmentally susceptible to change by various new inputs).

It is not certain how often recurrence is due to discrete reversions based on evolved change in switch mechanisms controlling traits lost in the past, and how often it is due to repeated de novo assembly via gradual parallel evolution. There is some reason to expect that once a phenotypic trait has been consolidated by genetic accommodation and modification, it may have an increased probability of recurrence following loss. Consolidation means establishment as an independently regulated trait, via an evolved breaking of genetic correlations between alternatives and other traits, and a building of within-trait correlations (see Berg, 1960, on the evolution of correlation pleiades). In effect, the consolidation of a trait increases modularity by exaggerating the developmental–genetic boundaries around traits, enhancing between-trait dissociability, and increasing within-trait integration. Both aspects would contribute to the likelihood of unit recurrence. In addition, there is the prospect for adaptive advantage of a ready-made and previously adaptive trait, should it recur during evolution following its loss. If this reasoning is correct, it is reasonable to hypothesize that the first occurrence of a recurrent trait involves more gradual genetic change than do its subsequent occurrences (reversions).

Given the potential for incongruence between patterns of speciation and patterns of phenotype evolution, phylogenies may give an erroneous impression of the orderliness of events during evolution. Phenotypic divergence obviously is facilitated by the gene-flow barriers of speciation. But development does not respect historic barriers in that evolutionary change receives much of its direction from developmental responses that, within the *genetic* boundaries of the species, allow populations to developmentally skip over ancient phenotypic changes to recall long-lost traits. It cannot be overemphasized that the evolution of phenotypes does not necessarily show boundaries that correspond to species boundaries and phylogenetic branch points, or map neatly onto phylogenetic pattern. The data of this chapter are one kind of evidence of that.

The nature of variation within populations contradicts certain assumptions about selection and evolution that I had long imagined to be reasonable. Phenotypically complex anomalies are not rare, and the same ones recur over and over again. Nor are they always random with respect to selection, for recurrent phenotypes represent traits that have been assembled and favored or at least tolerated by selection in the past, so they are not products of casual chance, even when induced by mutations. Selectable variation is not small or exclusively quantitative. And large discrete variants are not always devastating to survival and reproduction, but can be bearable handicaps that eventually prove advantageous and become widespread established traits. Thanks to the modular nature of phenotypic development, innovative variants are semi-isolated from other functioning traits, and therefore not so disruptive as they could be if all development were interconnected.

Phenotypic variants are to a degree products of history, but they are not bound by the past history recorded in the genome, for the numbers of potential new developmental rearrangements are enormous, and they can be either mutationally or environmentally induced (see chapter 6). As concluded by Wake (1991), recurrence homoplasy complicates phylogenetic analysis enormously, but at the same time it enriches our understanding of evolution as a process that combines and recombines discrete traits at all levels of organization.

PART III

ALTERNATIVE PHENOTYPES

20

Alternative Phenotypes as a Phase of Evolution

A novel trait expressed as an alternative phenotype can undergo extensive evolution in a population buffered from the disruptive effects of change by the presence of more than one adaptive option in a given functional context. The evolutionary importance of alternative phenotypes has been underestimated due to misunderstandings regarding their nature and commonness. Alternatives are developmentally and functionally semi-independent of each other. Not only are they independently subject to selection, but they also can become independently expressed in different individuals or lineages. Consequently, they can be an important phase in the evolution of major adaptive novelties that characterize species and higher taxa.

Introduction

So far I have discussed two aspects of the origins of novel traits, their initiation and spread under selection (chapter 6), and their sources in preexisting, ancestral phenotypes (part II). This chapter addresses another aspect of origins, namely, that qualitative novelties must pass through an alternative-phenotype phase during their evolution. If new traits first arise in single individuals and then increase in frequency, they automatically render their population polymorphic (Roth, 1991, p. 185). The four chapters of this section discuss the far-reaching evolutionary consequences of this.

The branching nature of development implies that a novel trait originates as a new branch in a previously existing developmental pathway (Oster and Alberch, 1982)—a new modular alternative to another, previously established one. The new trait may then evolve to fixation and persist alongside preexisting traits as a simultaneously expressed complementary form, representing an increase in complexity. Or it may be maintained indefinitely as a polymorphism or polyphenism, representing an increase in both complexity and developmental plasticity. Both kinds of new traits are modular subunits with the properties of dissociability and independent selection.

Polyphenisms, polymorphisms, and alternative behavioral and life-history or physiological traits, collectively called alternative phenotypes (West-Eberhard, 1986), are of special interest for evolution because they are adaptive options of potentially great complexity that can come to characterize different phylogenetic lines. Alternatives permit the elaboration of a new trait without elimination of an established one, thereby facilitating the evolution of new adaptive specializations. They can represent a crucial phase in the evolution of discrete adaptive novelties, allowing macroevolutionary changes to occur via microevolutionary processes within species, and without disruption of preexisting adaptations (West-Eberhard, 1986, 1989; see chapter 29). They also set the stage for punctuated rates of change in quantitative traits (see chapter 30).

The four chapters of part III describe the nature and evolution of alternative phenotypes themselves, as a prelude to showing their broad evolutionary significance.

Terms and Distinctions

Alternative phenotypes are different traits expressed in the same life stage and population, more frequently expressed than traits considered anomalies or mutations, and not simultaneously expressed in the same individual (West-Eberhard, 1986). Although some authors (e.g., Austad, 1984) include only adaptive alternatives in the definition of alternative phenotypes, I employ a fitness-neutral definition to include alternatives even when their fitness effects are negative, negligible, or unknown, as may be true in the early stages of their evolution. The general term "alternative phenotypes" refers to alternatives of all kinds, whether in behavior, morphology, physiology, or life history, whether reversible or irreversibly determined, and whether regulated primarily by genotypic or environmental factors.

Alternatives were recognized as a special phenomenon in the eighteenth century by Rousseau, who, according to Sang (discussion following Richards, 1961), invented the word "polymorphism" to refer to several forms at the same stage of development. Alternative phenotypes played an important role in the birth of modern genetics. The phenotypes of peas used in classic genetic studies by Darwin (1868a,b [1875a,b]), Mendel (1863), and the early Mendelians (de Vries, Tschernak, Correns, and others; see Yablokov, 1986, pp. 24ff) included discrete characters expressed as alternative phenotypes. Indeed, the word "mutation," attributed by Yablokov (1986) to Waagen (1869), originally referred not to genes but to alternative phenotypes—discrete new morphological traits in the fossil record of Jurassic ammonites. Shull (1915) was among the first to recognize that what we perceive as "traits" exist as alternatives to others, and as discrete entities are subject to combinatorial evolution. He defined a "unit character" as "alternative differences of any kind which are either absent or present as a whole in each individual and which have the potential to be united in new combinations with other traits" (Yablokov, 1986, p. 29). Later, the terms "gene," "genotype," "phenotype," and "allele" were introduced by Johannsen (1909) to distinguish among these now-familiar kinds of alternative forms.

Following a strong focus on genotype-specific polymorphisms in ecological genetics (e.g., see Ford, 1940, 1961, 1964, 1971), it was left for ethology and behavioral ecology, with their emphasis on adaptive phenotypic flexibility, to produce the recent renaissance of interest in condition-sensitive "strategies" and "tactics" of resource acquisition, bringing alternative phenotypes to the forefront as a common pattern of adaptive evolution.

The terms that describe alternative phenotypes give an idea of their nature and diversity as well as a feeling for their commonness. *Polyphenisms* are irreversible environment-specific alternative phenotypes (in contrast to genotype-specific "polymorphisms"), most commonly morphological ones (e.g., Michener, 1961; Mayr, 1963; Moran, 1992b; Evans and Wheeler, 1999). Examples are the horned or hornless male phenotypes of some beetle species (Eberhard, 1979) and the seasonal color forms of butterflies (Shapiro, 1976). *Reversible alternatives* are those where more than one alternative can be expressed by a single individual, as in the winter and summer plumages of some birds, the shade and sun leaves of some plants, and many behavioral alternatives. Behavioral alternatives are sometimes called *polyethisms*.

For a phenomenon long believed to be rare, alternative phenotypes have inspired a wondrous array of specialized terms. My favorites are those of Wheeler (1910a) denoting the morphotypes of ants, which include dinergates (soldiers), phthisaners (abortive pupoid males), and dichthadiigynes (externally workerlike females with enormously developed ovaries). Wheeler's terms are extravagant without being controversial. "Polymorphism," by contrast, is simple but so burdened with contrasting definitions that it can scarcely be used with impunity (see especially the audience discussion following Richards, 1961). Every subdiscipline of biology seems to have a proprietary idea of what polymorphism should mean. In Mendelian genetics, it refers to different alleles at a single locus or, more recently, allozymes (e.g., Lewontin, 1974; Kimura, 1983), whereas in ecological genetics (e.g., Ford, 1971) it refers to different complex genotype-specific structural phenotypes maintained in the same population. Some insist that it applies only to morphology, whereas others prefer a broader interpretation and include behavior and physiology. Some (e.g., Ford, 1971; Curio, 1973) insist that the alternative forms must be "genetically determined" (allelically switched), while others allow the inclusion of environmentally cued forms (as in caste "polymorphisms" of social insects; Michener, 1961).

In evolutionary ecology, "polymorphism" sometimes is used with a modifier that refers to context of selection (e.g., multiple-niche polymorphism, density-dependent polymorphism), not mode of regulation, and so could include both genotype-specific alternatives and environmentally switched alternatives (polyphenisms). The term "polymorphism" also has been applied to geographic variation, as in the "polymorphic" (or polytypic—Mayr, 1963) species of Müllerian mimicry complexes in butterflies, and even to mean simply "variable." As a result, the attractive word "polymorphism" is now of limited usefulness in a general discussion.

Some authors (e.g., Schmalhausen, 1949; Smith-Gill, 1983) have devised elaborate classifications to distinguish among the different general properties of alternative phenotypes. I am not much concerned with those distinctions here. Instead, I focus on several shared properties of alternative phenotypes that are of special evolutionary interest: all are switch-controlled, more or less discrete, modular phenotypic traits. All are subunits of gene expression or gene-product use, potentially subject, semi-independently of each other, to natural selection (see chapter 7). Due to regulation by evolutionarily adjustable thresholds or switch mechanisms, all are potentially dissociable during development and evolution. That is, regulatory changes can cause them to be shifted in time or place of expression, or in some cases deleted, without alteration of those aspects of form that are determined by specific modifiers controlled by the switch. In these respects, alternative phenotypes are like other modular traits. But because they are alternatives, they can serve as options in a single functional context. It is this property that enables them to serve as bridges between adaptive peaks, for two solutions to the same problem (e.g., food getting, predator escape, or mate acquisition) can be elaborated at the same time if they are expressed as alternatives within a population.

Alternative Phenotypes as Models for Relating Development and Evolution

Alternative phenotypes epitomize the switch-controlled, environmentally sensitive, adaptive traits that characterize all of ontogeny (see chapter 6; see also figure 4.13). λ Phage, a viral parasite of bacteria (*Escherichia coli*), is a real-life illustration of that structure, well understood from decades of research by molecular geneticists. The alternative phage phenotypes resemble the polyphenisms of higher organisms in having a condition-sensitive switch mechanism, specific modifier genes whose expression is controlled by the switch, and divergence in the behavior and morphology of two alternative phenotypes (figure 20.1). The two alternative forms of λ phage are the lytic form, in which the phage particles are multiplied and then burst out of the bacterium to infect others; and the lysogenic form, in which the phage DNA inserts itself

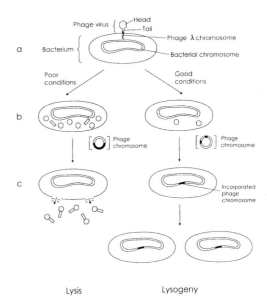

Fig. 20.1. Conditional alternative phenotypes in a virus: (a) λ Phage virus attaches to a bacterium and injects its chromosome; (b) under poor conditions (the lytic cycle, left), signaled by enzymes of the bacterial host, phage genes (brackets), produce viral head proteins (hexagons) and tail proteins (bars), and the chromosome is replicated (circles); (c) the bacterial cell then lyses and releases phage viruses. Under good conditions (the lysogenic cycle, right), only two viral genes are expressed (brackets): one a facilitator of insertion into the host chromosome, the other a repressor of protein production. The lysogenic cell then divides, reproducing its chromosome and the incorporated phage chromosome. The lysogenic cycle is repeated indefinitely under favorable conditions. Other phage genes are expressed during phage growth prior to the conditional switch. After Ptashne (1986).

into the bacterial chromosome and is passively replicated along with the bacterial DNA. The differences between the alternatives is clearly a gene-expression difference controlled by a switch. In the lysogenic (dormant) state, the activity of the phage genome is greatly reduced. A repressor protein encoded by a single active gene (cI) turns off all other phage genes, including both those on its own chromosome and those of any other phage particles that happen to penetrate the host cell. One other gene product (int) inserts the phage chromosome into that of the bacterium. In effect, the lysogenic form parasitizes the bacterium by inserting itself into the host genome and being replicated there.

The phenotype of the alternative, lytic form is more complex. It consists of a coat made of 15 proteins, a head, and a tail (figure 20.1). This form is capable of fairly complex behavior. It attaches to a bacterium, makes a hole in the cell wall, and injects its DNA contents into the cell. Production of the active lytic phage involves the coordinated expression of two sets of phage genes that are controlled by the switch. One set is contiguous on the phage chromosome and expressed during growth. The other set is expressed following the switch to the lytic phase of development (Ptashne, 1986). This regulatory structure reinforced the early idea that chromosomal linkage would characterize coexpressed genes in all organisms. As discussed in chapter 4, chromosomal linkage sometimes occurs but is not a general characteristic of coexpressed gene sets in eukaryotes.

Ptashne titled his 1986 classic summary of research on λ phage *A Genetic Switch*, and the condition sensitivity of the bacteriophage switch has never been particularly emphasized. But this was the first example of conditional adaptive alternatives to be thoroughly analyzed at the molecular level. External environmental conditions such as the nutritional state of the bacterial host cell stimulate switching between the two alternative phenotypes. As in all alternative phenotypes, the divergent alternatives are genetically distinctive in that they are underlain by gene-*expression* differences, rather than by gene-frequency divergence as in genetically divergent populations. It is the gene-expression difference between alternatives that gives them the potential to evolve extreme differences in form (see chapter 21).

Alternatives illuminate development in general because switching between alternative developmental pathways has been called "an almost universal part of ontogeny" (Raff and Kaufman, 1983, p. 282). For this reason, the studies of alternative phenotypes in the λ phage virus have been called "a miniature problem in development" (Ptashne, 1986, p. 63). They are, like other alternative phenotypes, a miniature problem in evolution, as well. Switch-mediated alternatives are a convenient place to examine the links between development and evolution. They occur within individuals at all levels of phenotypic organization from alternatively spliced molecules to alternative behavioral tactics, so they can be observed by any biologist in any field from molecular biology to ethology. They provide opportunities to document the connections between gene expression and selection, for context of expression can be an indicator of adaptive significance. Switches between alternatives are subject to manipulation by experimental rearing and artificial selection, and this facilitates study of the effects of genotypic and environmental influence on developmental decisions. For these reasons, several preceding chapters used alternative phenotypes as examples to illustrate general principles of development and evolution.

Novel discrete traits often begin as novel alternative phenotypes. In general, the environmental conditions that favor the origins of novel traits are associated with the occurrence of novel alternative phenotypes as well. "Stress," "competition," "defense," and "heterogeneity" are good key words in a search for examples of both trait origins and alternative phenotypes. Switches between alternative phenotypes are often sensitive to stress in the biotic and physical environment, status in social or sexual competition, presence of predators and parasites, and seasonal change and spatial environmental heterogeneity that encourage the cyclical or opportunistic adoption of different modes of life.

To consider alternative phenotypes characteristic of a phase of evolution does not imply progressive evolution favoring the more recently derived of a set of intraspecific alternatives. For this reason, and to avoid any teleological implications, it is best to call the alternative-phenotype state a "phase" rather than a "stage" or a "step" in evolution. An alternative may either increase or decrease in frequency, and it may persist indefinitely while fluctuating in frequency relative to others. Meanwhile, more than one alternative can be undergoing evolutionary change in form.

Following a period of evolutionary accommodation and modification, one of two alternative phenotypes may increase in frequency and be fixed. This is suggested by the fact that innovations that are fixed in one species quite commonly occur as alternative phenotypes in other, related species (see part II). Gould (1977) and McKinney and McNamara (1991) note in their synthetic reviews of heterochrony that heterochronic novelties are often found as intrapopulational polymorphisms ("heterochronic morphotypes"). When such a pattern of trait distribution occurs within and between populations, it suggests a role for alternatives in trait origins and a history of polymorphism (Dickinson, 1991; Wiens 1995).

Phenomena Easily Confused with Alternative Phenotypes

Alternative phenotypes are easily confused with other phenomena, including the following examples:

1. Variants of geographic origin, and species: Alternative phenotypes may be confused with characteristics of full species or populations that are divergent due to reduced gene flow, as in the subspecific or "racial" geographic variation within so-called polytypic species. Divergent forms of geographic origin are the result of genetic divergence between populations, not developmental switching between traits within a population.

In Darwin's time, when Wallace first described the Batesian mimicry morphs of *Papilio* butterflies, most alternative phenotypes were unwittingly classed as distinct species. Trimen (1874, p. 140) quotes the well-known nineteenth-century entomologist W. C. Hewitson as having objected that recognition of such forms as intraspecific variants or polymorphisms "would require a stretch of the imagination, of which I am incapable" and that they "shock one's notions of propriety." Victorian prudishness regarding polymorphism, at least in butterflies, has largely disappeared, but alternative phenotypes are still mistaken for species. This may be especially problematic in fields such as paleontology when small morphological differences, known to characterize some genetically distinct populations, are argued to be valid general criteria for recognition of species (e.g., Jackson and Cheetham, 1990, on Bryozoa) even though intraspecific morphs of a species-level degree of morphological divergence are known in the same taxa and sometimes mistaken for species (e.g., Winston, 1979; Anstey, 1987; Anstey and Pachut, 1995).

Sometimes geographically variable species, such as the butterflies of Müllerian mimicry complexes, are referred to as "polymorphic" even though any given population is monomorphically and genetically specialized to a single form. It would be more accurate to call such species polytypic, with different populations adapted to different, geographically variable conditions (in the Müllerian-mimic butterflies, local differences in distasteful models). Other potentially confusing terms are "ecophenotype" and "ecotype," as used in the writings of plant genecologists (e.g., Turesson, 1922; Clausen et al., 1940). Sometimes these terms refer to geographic variation, and sometimes they refer to switch-regulated alternative phenotypes.

There are several standard methods for distinguishing between alternative phenotypes due to phenotypic plasticity, and genetically diverged forms. One is the "common garden" experiment, where representatives of one morphological ecotype are transplanted to the habitat characteristic of another, to ascertain the degree to which the contrast in form is influenced by genotypic divergence and to what degree it is associated with environmental differences per se. Other methods include laboratory rearing and molecular genetic tests (see Hoagland and Robertson, 1988; Hewitt et al., 1991; Hillis and Moritz, 1990). If two or more complex alternatives can be reared within the brood of a single female, this is good evidence that they do not represent genetic divergence of geographic origin. Similarly, if genetic tests (e.g., using electrophoresis or microsatellites) have indicated that alternative phenotypes are part of a single gene pool, this supports the hypothesis that their divergence is developmental rather than due to breeding isolation. Sandoval (1994a) discusses an unusual case in which polymorphism is reinforced by gene flow (hybridization) from a contrasting environment. As a rule, however, breeding isolation and developmental switches are operationally distinguishable sources of divergent phenotypes.

2. Bimodal continuous distributions: Switch-controlled alternative phenotypes have quantitative dimensions, and so may show continuous variation centered about two or more different modes. This may be difficult to distinguish from a bimodal distribution that is not caused by a switch, as illustrated by the mutillid wasp *Dasymutilla bioculata* (see chapter 3), which preys on two discrete sizes of host and therefore rears two discrete size classes of offspring (Mickel, 1924; see figure 11.3). The two phenotypic classes observed were caused not by a developmental switch but by an environmental (dietary) discontinuity. Eberhard and Gutierrez (1991) discuss sampling problems and apply statistical tests that distinguish true developmental polyphenisms governed by switches from bimodal continuous distributions.

Some polyphenisms, like that of migratory locusts, have been labeled with the seemingly self-contradictory term "continuous polymorphisms" (Kennedy, 1961; Pener, 1991). In locusts, the alternatives are mosaics of several sets of alternative traits, each with its own threshold response to population density and, in some cases, other environmental factors (e.g., temperature and humidity; see the section on the ganged switch in chapter 5). If all switches of such a mosaic are thrown simultaneously (e.g., by a strong cue sufficient to pass all thresholds at once, as in the larval diet of worker vs. queen honeybees), we perceive a well-coordinated polymorphism controlled by a single factor, even though under intermediate conditions, when not all switches are thrown, the mosaic nature is evident in a more-or-less continuous-appearing spectrum of intermediates.

3. Extremes of continuous variation: Biased sampling (e.g., at separate times of year or in environmental or habitat extremes) may give an impression of polymorphism or polyphenism. Paper wasps (*Polistes fuscatus*), for example, show variable amounts of brown versus yellow body coloration and continuous variation in body size if the sample examined includes individuals collected throughout the temperate-zone summer. Small, primarily yellow females emerge in early summer, and large, primarily brown females emerge in late summer. Since early-summer colonies contain both large, brown females from the previous summer, and small, yellow females that are newly emerged, the species then appears to have a queen–worker polymorphism, but it is only an artifact of seeing two extremes of the seasonal variation together in colonies lacking intermediates (West-Eberhard, 1969).

4. Generalists versus polyspecialists: Sometimes authors (e.g., Futuyma and Moreno, 1988) distinguish between *specialists*, meaning species or individuals that show a narrow range of traits highly suited to a particular host or diet, and *generalists*, species or individuals that adopt a wide range of solutions (e.g., hosts or dietary items). Although this is not always recognized, the generalist category is actually composed of a continuum between two importantly different patterns—true generalists and polyspecialists (figure 20.2). At one extreme, a true *generalist* individual or species performs a broad and highly variable range of tasks, often with little distinction in the morphology, sensory capacities, and behaviors used to accomplish each. At the other extreme, a *polyspecialist* performs a limited number of distinctive alternative tasks using alternative sets of behaviors, morphological equipment, search images, and so forth.

Organisms that appear to be generalists may turn out to be polyspecialists. Casual observation suggests that rock doves, for example, are generalist seedeaters. A flock of doves will quickly consume a wide variety of seeds. As discussed in chapter 18, however, inspection of stomach contents indicates that they are in fact polyspecialists, with different individuals selecting particular kinds of seeds from among a wide variety available (Giraldeau and Lefebvre, 1985; see figure 18.2).

True generalists and polyspecialists are importantly different in terms of both development and natural selection. Polyspecialists switch or choose among distinctive developmental and behavioral pathways, and the involvement of somewhat discrete alternative phenotypes means that selection can act to effect specialization in different direc-

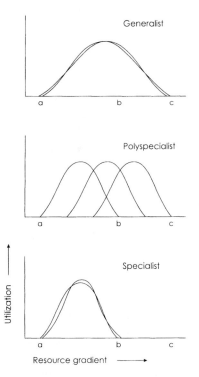

Fig. 20.2. The distinctions between a generalist (top), polyspecialist (middle), and specialist (bottom) individual or population in use of variable resources. Points a, b, and c refer to different quantitative aspects, or proportions of kinds, of the resource concerned. After Grant (1986).

tions within the species or population. In a true generalist, by contrast, the variety of tasks performed by a single versatile morphology or behavior means that selection is expected to produce continuous variation around a single modal morphology or behavior and a compromise or continuously variable plasticity capable of performing a multitude of tasks (Klopfer, 1962; Van Valen, 1965; Curio, 1976; Grant, 1986; Wilson, 1989).

5. Cyclic or life-stage variants: Sometimes multiple forms in the same population are the result of maturational, size-specific, or age-specific changes passed through in sequence by most or all individuals. Femalelike plumage and behavior of young adult males in some species of birds are examples (see chapter 15). Again, stage-specific and alternative phenotypes show some overlap in categories: stage-specific phenotypes are sometimes conditionally expressed out of sequence, if the stage-specific conditions occur out of sequence. Plants, for example, commonly produce shade-adapted leaves

when young and sun-adapted leaves when they grow into sunlight at a later age. In some but not all plants, such as some species of the tropical vine *Monstera*, older plants reexpress a juvenile morphology if they grow downward or reenter shade (see figures 4.2, 6.3). In such an organism, it is difficult and unproductive to attempt to distinguish between parts representing an intra-individual division of labor, life-stage phenotypes, and alternative phenotypes.

Historical Misconceptions about Alternative Phenotypes

Historically, some insistent prejudices have stigmatized alternative phenotypes and kept them from getting the attention they deserve in evolutionary biology. Some of these ideas are discussed in preceding chapters, but I list them here because of their historical interest and bearing on evolutionary concepts.

The Idea That Conditionally Expressed Alternatives Are Nongenetic Traits

Environmentally induced variants, in a genetically deterministic framework, appear as vaporous ghosts of elusive outline, awkward to model, of marginal interest, or even (for some) difficult to believe. Yet biologists routinely use condition-sensitive expression to test adaptation hypotheses, under the assumption that if a trait is facultatively expressed in a given condition, it is likely adaptive in that context (for many examples, see Lott, 1988). Shapiro (1976) gives a brief history of skepticism regarding environmentally cued polyphenisms. As mentioned in chapter 1, environmentally induced phenotypes have been described as "nongenetic" (Hoffman and Parsons, 1991, cited by Shuster, 1992a, p. 1249); likely rare (Waddington, 1975, p. 167; Wade, 1984; Stearns, 1982, p. 243); unimportant for evolution (or lacking "evolutionary potential"—Svardson, 1961); "only an embarrassment" (see Bradshaw, 1965, p. 148); examples of elaborate machinery gone wrong (R. A. Fisher, cited by Wigglesworth, 1961, p. 107); and "special cases" best forgotten in a general evolutionary model (O'Donald, 1982, p. 67). Skepticism persists regarding the evolutionary importance of environmentally influenced traits, despite immense progress in studies of conditional regulation of gene expression, and the well-known condition sensitivity of all development. One can only conclude that

findings in developmental biology and behavior have not been sufficiently connected to genetics in the minds of many evolutionary biologists.

Curiously, the connection has not been made because of two, opposite attitudes, underestimating the importance of genetics in some situations and overestimating its importance in others. An example of underestimation is the idea that since all individuals of a polyphenic population are capable of multiple forms, they are all genetically equal regarding polyphenism. Accordingly, Shapiro (1984a) considered polyphenism to be "the production of two or more phenotypes in individuals which do not differ in their genetic makeup" (p. 178); Shapiro (1984b) also considered that "the genetics of polyphenism has not been looked at before simply because one cannot do genetics in a genetically uniform population" (p. 306); Shapiro used interpopulation hybrids to expose genetic variation in thresholds). Similarly, Crozier (1992, p. 219) interpreted the potentiality of all females of a social insect species to adopt either the worker or the queen phenotype to mean that "queenness strikes at random" with respect to genetic makeup. Although correct in one sense—the phenotype adopted is due primarily or entirely to environmental, not genotypic differences in many species—these assertions are incorrect in another, important sense: even if the same individual switches back and forth between alternative forms during its lifetime (obviously, without changing genotype), its threshold for doing so may differ—genetically—from that of other individuals. Therefore, it would be incorrect to describe the population as genetically uniform in its propensity to adopt one phenotype or the other. The phenotype adopted does not strike completely at random with regard to genotype.

Genetic variation in the switching thresholds of polyphenisms and reversible condition-sensitive alternatives has been revealed by genetic experiments in many organisms (e.g., see Hazel, 1977; Hoy, 1977; Jain, 1979; Shapiro, 1984a,b; Dingle, 1984, 1991, 1994, 1996; Dingle et al., 1986; Briceño and Eberhard, 1987; Parker and Gatehouse, 1985, cited in Dingle, 1994; Messina, 1987, cited in Dingle, 1994; Walker, 1987; Semlitsch and Wilbur, 1989; Knülle, 1987, 1991; Page and Robinson, 1991; Page et al., 1989, 1995a,b; Dingle, 1984; Janzen, 1992; Frumhoff and Baker, 1988; Harris et al., 1990; Zera et al., 1992; Thompson et al., 1993; Roff, 1996; Roff et al., 1997; Emlen, 1996).

The other, opposite attitude is to overestimate the importance of genetics, taking evidence of genetic influence on the switch between forms to mean that alternatives can be treated as if they were genetic (allelic) polymorphisms. This misconception is discussed in chapter 5, where the meanings of "genetically determined" are discussed, and is further discussed in chapter 22.

The Supergene Concept

The evolutionary study of complex polymorphisms became important in the mid-twentieth century, when ecological genetics, a field most closely identified with E. B. Ford and his associates, began to examine mimicry polymorphisms in butterflies and other insects (see Sheppard, 1959; Ford, 1971). While this field greatly advanced the evolutionary genetic study of complex alternative phenotypes, especially the Batesian mimicry polymorphisms of butterflies, it also promoted some misconceptions regarding the genetic architecture and maintenance of complex alternative phenotypes and complex phenotypes in general. One of these was the idea that alternative phenotypes in general must be "balanced" (due to either heterozygote superiority or frequency-dependent selection) in order to persist, a mistaken idea that I return to in chapter 22. Another misconception was the idea that complex polymorphisms and phenotypes in general are underlain by alternative linked sets of genes, or "supergenes" (see chapter 5), and the associated expectation that coadapted sets of alleles would come to be linked on the same chromosome and thereby have their integrity conserved through reduction in the frequency of crossing over. The ability to acquire and process information about relatives, for example, has been described as "expected to become tightly linked into a supergene" (Alexander and Borgia, 1978, p. 466). This vision of genetic architecture has continued to influence thought about coexpressed traits into the recent past (e.g., see Felsenstein, 1981; Zimmerer and Kallman, 1989, p. 1305), even though known from genomic and developmental studies to be wrong as a general description of the genetic architecture of such traits (see chapter 5).

The Stability Problem

There has been some reluctance to believe that alternative phenotypes, or complex polymorphisms, could be stable enough to be important in evolution. This stems from identification of alternative phenotypes with the much narrower phenomenon of "genetic" or allelic polymorphism. If alternative genetic alleles are unequal in their fitness effects, which must usually be the case, one of them should

evolve to fixation, and so the same expectation accrues to alternative phenotypic traits.

For many years, most theoretical work relating to alternative phenotypes addressed the problem of maintenance of *genetic* polymorphisms (including the complex ones thought to be underlain by "supergenes" behaving as alternative alleles) and has concluded that this could only occur under special conditions (e.g., of frequency-dependent equilibrium or heterozygote advantage; see below). Even models designed to apply to behavior (e.g., some game theoretic "evolutionarily stable strategy" (ESS) models—see chapter 22) may parallel models of genetic (or genotype-specific) polymorphisms in their conclusion that maintenance requires equal payoff conditions. Similarly, models of polymorphism in relation to sympatric speciation often depict exploitation of multiple hosts and multiple niches as if it were a product of genetic polymorphism (e.g., see reviews and discussions by Futuyma and Moreno, 1988; Grant and Grant, 1989).

The stability problem has been so important in evolutionary thinking about polymorphisms and has led to such a tangle of misconceptions that I devote a whole chapter to it (chapter 22).

The Idea That Conditional Alternatives Are Difficult to Evolve

Conditional expression solves the stability problem but raises another difficulty. In the abstract, it seems difficult to imagine the simultaneous origin of two complex adaptive alternatives as well as their adaptive, condition-sensitive regulation. How could these two qualities of polyphenisms get started at once? This is probably a common worry:

> The origin of a fixed adaptation is simple. The population merely needs to have or to acquire some genetic variation in the right general direction. The origin of a facultative response is a problem of much greater magnitude. Such an adaptation implies the possession of instructions for two or more alternative somatic states or at least for adaptively controlled variability of expression. It also implies sensing and control mechanisms whereby the nature of the response can be adaptively adjusted to the ecological environment. A facultative response would require much more delicate genotypic adjustments than a comparable fixed response. (Williams, 1966, pp. 81–82)

How can this difficulty be resolved?

The evolution of condition-sensitive adaptation appears difficult because of the habit of thinking of evolution in terms of genetic changes alone. It becomes easier with the realization that all phenotypes are condition sensitive. If this were not true, they would be continuously expressed in every life stage from egg to adult. Universal condition sensitivity means that a change in conditions or context alone can signal a switch between states or traits given a preexisting capacity to respond. A trait that originates as a response to environmental induction (see chapter 26) from the start has an environmental component to its expression. Genetic accommodation of the response threshold can come later (see chapter 6). Williams's (1966) separation of response and trait origins is therefore potentially misleading: whatever induces a novel response, whether a mutation or some environmental factor, provokes a new developmental pathway—a new branch point in ontogeny, behavior, or physiology and a new modular trait. The new phenotype and its regulatory mechanism are simultaneous creations (see chapter 6, on the simultaneous origin of regulation and form).

Some authors (e.g., Bradshaw, 1973; W. D. Hamilton, personal correspondence, 1990) assume that polyphenisms arise secondarily from genetic polymorphisms, under conditions favoring the evolution of condition-sensitive regulation. Shuster and Wade (1991a) made a similar argument. While they acknowledged the predominance of conditional alternatives in nature, they concluded that "the presumed ancestral condition for male alternative strategies" is one of being "genetically distinct" (p. 610). Other authors (e.g., Levinton, 1986, pp. 259, 265) believe that the regulation of a novel traits is initially polygenic or condition sensitive and that a "developmental switch gene" may come to control it "only after the gradual evolution of the developmental unit is complete." Here I maintain that a novelty can be initiated by either mutation or environmental induction.

Given the condition-sensitive expression of phenotypes mentioned above, there is no reason to suppose that alternatives always start with determination by a genetic polymorphism, and I know of no data or convincing theory suggesting that conditional alternative phenotypes are usually descended from genotype-specific ones. Nor is there any reason to think that regulation is always originally environmental. Given that the overwhelming majority of alternatives are conditional (see below), the burden of evidence falls on those who claim that regulation by genetic determination usually evolves first. Evolution both toward and away from

greater plasticity is of course possible (see chapter 6). Bradshaw's 1973 discussion was a pioneer step toward a formalization of the conditions under which degree of plasticity can evolve in nature. Both sides of this controversy regarding the evolution of genetic and environmental control of sex determination are summarized by Korpelainen (1990).

The belief that conditional alternatives are difficult to evolve is probably a result of the habit of thinking about trait origins in terms of genetic mutation. The supposed difficulty of evolving conditional regulation has greatly influenced biologists' thinking about alternatives in general, most especially regarding the expectation of rarity, discussed in the following section. Theory is sometimes more powerful than fact in determining what people are willing to believe about alternative phenotypes. When Thornhill and Alcock (1983) gave many well-documented examples of conditional male mating strategies and suggested, based on a massive survey of empirical studies, that "conditional strategies are probably the dominant source of variation in mating systems within a species" (p. 291), Wade (1984, p. 707) nonetheless declared this suggestion "unwarranted" on *theoretical* grounds.

The Rarity Problem

The idea that alternative phenotypes are rare has a long history and many causes. An old and still important cause of this impression, and one that is particularly relevant for evolutionary discussions, is the practice in systematics and phylogenetics of typological classification and overtypification of taxa, which can make intraspecific variants such as alternative phenotypes appear to be more rare than they really are. This problem is discussed in chapter 19, on recurrence. The false impression of rarity comes up again in the discussion of punctuated patterns of morphological evolution in paleontology (chapter 30), where faith in morphospecies can make alternative phenotypes appear to be rare by causing us to treat them as species. How one chooses to code non-ubiquitous traits in a phylogenetic analysis can make alternative phenotypes appear to be either common, rare, or completely absent, with drastic consequences for the portrayal of evolutionary transitions (figure 20.3).

Theoretical preconceptions about the instability and difficult-to-evolve nature of alternative phenotypes have reinforced the impression that they are too rare to be important for evolution (e.g., O'Donald, 1982). Below I discuss the dubious implication that rare phenomena are unimportant for

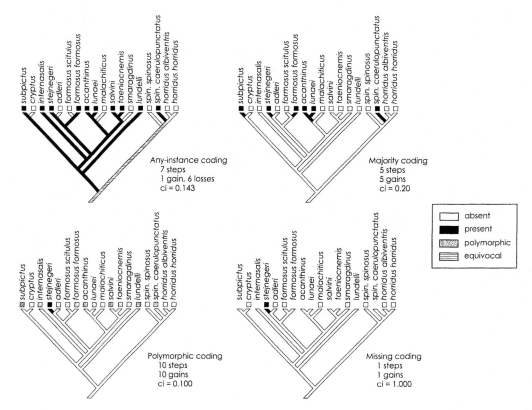

Fig. 20.3. The effect of coding method on the interpretation of intraspecific variation in evolutionary-phylogenetic analyses. From Wiens (1999). ci is the consistency index. With permission, from the *Annual Review of Ecology and Systematics*, Volume 30 © 1999 by Annual Reviews www.AnnualReviews.org. Courtesy of J. J. Wiens.

evolution. First I ask only whether it is reasonable to consider alternative phenotypes rare.

Contrary to the expectation of rarity, systematic searches often reveal alternative phenotypes to be common. The use of electrophoresis to reveal alternative molecular phenotypes is a good example: protein polymorphism turned out to be much more common than expected on theoretical grounds (Lewontin, 1974; Nevo et al., 1984; cf. Ford's [1971] citation of H. J. Muller and others as arguing for the rarity of genetic polymorphism, pp. 124ff). More recent methods have yielded even more impressive evidence of genetic (DNA) polymorphism (e.g., Golding, 1993).

A similar explosion of polymorphism discovery occurred at the whole-organism level with a flurry of interest in alternative tactics of foraging, mating, and life-history tactics during the 1980s. Multitudes of heretofore undescribed and *unseen* alternative behavior patterns suddenly came to light in

sociobiology and behavioral ecology when ESS and optimal foraging theories focused attention on alternative phenotypes (for partial reviews, see Krebs and Davies, 1991).

The same correlation between research focus and case abundance is witnessed at the individual researcher level. When a particular biologist sets out to document the presence of alternative phenotypes, large numbers come to light; for example, see Bradshaw (1965) on plants; Ford (1971) on "genetic" polymorphisms, especially in insects; Shapiro (1976) and Moran (1992a) on seasonal and cyclic polyphenisms; Richards (1961) on anatomical polymorphisms in insects; Jackson (1986) and Jackson and Wilcox (1993) on salticid spiders; Gross (1984) on alternative mating tactics in fish; Roff (1986a,b) on dispersal polyphenisms in insects; and West-Eberhard (1986, 1989) and Roff (1996) on alternative phenotypes in general. Shapiro (1976) concluded that "the literature on polyphenisms is enormous. . . . A comprehensive survey even of seasonal polyphenisms would have to be a classic Teutonic tome" (p. 262). McLeod (1984) found them to be "nearly universal" in tropical pierid butterflies. Sometimes a specialized focus gives the impression that, because polymorphism is believed to be rare elsewhere, it is concentrated in the investigator's field of interest. Dominey (1984), for example, pursuing research on male alternative tactics, concluded that "[p]erhaps the largest class of condition-dependent tactics can be traced to male–male competition" (p. 389).

The importance of polyphenism-awareness for the abundance of documented cases was brought to light when I naively attempted to settle the rarity question by counting the alternative phenotypes in a particular group of organisms (spiders) by sampling the literature on a single taxon using a large collection of reprints on spider behavior and natural history, assembled "blind" with respect to the question of alternative phenotypes. Proceeding through it in alphabetical order, I found only a few examples, until I encountered the publications of Robert Jackson (e.g., Jackson, 1984; Jackson and Wilcox, 1993) and his associates. There I found a very large number of well-documented examples of alternative tactics, one of Jackson's principal interests. Similarly, the work of B. Crespi (see Crespi and Mound, 1997; Crespi et al., 1997), inspired by modern concepts and methods of behavioral ecology and, like Jackson's work, emphasizing detailed observation of living individuals rather than just collection and morphology, recently produced a sudden increase in reported cases of polymorphism in thrips (Thysanoptera), a taxon formerly thought (Richards, 1961) to be completely devoid of examples.

In aphids, a "soldier" morph was described for the first time by Aoki (1977). Since then, the number of examples has increased steadily: 10 years later there were 10 times as many known examples, soldiers having been found in 10 species of three tribes in one family (Pemphigidae; Itô, 1989, p. 71). Twenty years later, the number had risen to 50 species of six tribes in two families (Pemphigidae and Hormaphididae; Stern and Foster, 1997). Sage and Selander (1975) state that the first consideration of nonsexual trophic polymorphism in fishes was quite recent, with the work of Roberts (1974) on *Saccodon*. Many examples have been documented since (e.g., see Sage and Selander, 1975; Echelle and Kornfield, 1984; Meyer, 1989a,b, on cichlids; Noakes et al., 1989, on salmonids). Similarly, until recently, studies of alternative reproductive tactics in the social insects were limited to the reproductive division of labor between queens and workers, but recent research has revealed an extraordinary diversity of tactics in sexual adults. In ants alone, this includes such phenomena as wingless morphs, reproduction by mated workers, thelytokous parthenogenesis, and complete workerlessness (reviewed by Heinze and Tsuji, 1995).

The references just cited indicate that alternative phenotypes are taxonomically widespread. They also occur in every life stage, including sperm (Snook, 1997, 1998) and eggs (Crespi, 1992b); larvae (for insects, see review in Greene, 1996; for vertebrates, see Hoffman and Pfennig, 1999); and pupae (see the many examples in Greene, 1996) in addition to adults. But enumerative attempts to demonstrate quantitatively that any biological phenomenon is "common" or "rare" or "more common in some groups than in others" are always unsatisfactory due to quirks of sampling and history such as those just discussed. It is now clear that there is a great variety of kinds of alternative phenotypes, and that they are not limited in their occurrence to just a few animal or plant taxa.

However common or rare they may be, the argument that alternative phenotypes are important in evolutionary transitions does not depend on their commonness. Relatively rare events, such as favorable genetic mutations or transitions between levels of organization (Buss, 1987; Maynard Smith and Szathmary, 1995), may be disproportionately important in evolution. Even if alternative phenotypes were to prove uncommon in a particular group of organisms, they are likely to prove uncommonly important as agents of change, for reasons discussed below.

Although alternative phenotypes have to be recurrent to be genetically accommodated and modified, and rate of modification depends on frequency of expression see chapter 7), rarity is not an impediment to their persistence and functionality: a conditional alternative can be completely absent for an entire generation if the conditions for its expression are gone, and then reappear if those conditions return. As soon as an alternative increases in frequency, it gains in evolutionary potential for genetic accommodation of both regulation and form (see chapter 7).

Imagined Barriers to Evolution within Species

The belief in an adaptive optimum phenotype underlain by a cohesive interacting set of genes has sometimes led to a conviction that the evolution of divergent new phenotypes is *necessarily* linked to speciation (see chapter 7). Speciation, or increased isolation between populations, permits divergence in sexual species by reducing or eliminating gene flow between populations under different selective regimes, allowing the accumulation of genetic differences between them. Trait divergence within populations has been considered possible only if there is positive assortative mating between males and females of the same phenotype (and, by implication, genotype; Maynard Smith, 1966). Interbreeding would be a barrier to divergence, so divergence is strongly dependent upon speciation.

This role for speciation as the exclusive agent of divergence fits well into a general scheme where alternative traits are identified with alternative genes, and where phenotype divergence therefore depends on genetic novelty and reproductive isolation. In extreme versions of this view (e.g., in at least some versions of the punctuated equilibrium hypothesis such as the original proposal by Eldredge and Gould, 1972), divergence is so strongly tied to speciation that appreciable change is considered impossible except during a small window of time when speciation is occurring. Then, if it proves genetically distinctive due either to a founder effect at multiple loci (e.g., Mayr, 1954; Carson, 1985) or to a few genes of major effect (Templeton, 1981), the new population may evolve in a new direction (for a history and assessment of this idea, see Provine, 1989). The idea that diversification depends upon speciation is related to the "fallacy of adaptive uniformity" of species (G. C. Williams, 1992), an idea that leads not only to an erroneous theory of phenotypic evolution but also to misclassification of alternative phenotypes as species in taxonomy and paleontology, fields where research would be easier if intrapopulational diversification did not occur.

The erroneous arguments in favor of species-dependent divergence have created a widespread impression that adaptive evolution within species is not feasible. This is demonstrably not true, as shown by such data as a lack of correlation between morphological divergence and number of speciation events separating taxa (e.g., see Emerson and Ward, 1998). The existence of complex alternative phenotypes is further evidence, for they imply that divergence in contrasting directions in a single context not only can but does occur without speciation. This is important evidence against the all-importance of speciation in the evolutionary diversification of phenotypes.

Major Categories of Conditional Alternatives

Here I list some major categories of alternatives. Their divergence and maintenance within populations is discussed in later chapters.

Status-Related Phenotype-Dependent Alternatives

Some alternatives are *phenotype dependent* in that the alternative adopted depends upon relative ability of individuals having different characteristics to acquire resources (e.g., food, shelter, water, or mates) in competition with others having other phenotypic characteristics (figure 20.4). When alternatives are *status dependent*—a function of variable ability in social competition—a phenotype-dependent trait can depend on such variables as size, degree of aggressiveness, social display characteristics, motivation, social relations with other individuals, and fighting ability (Hamilton and McNutt, 1997). Consequently, these have been called phenotype-limited alternatives (Parker, 1982), alternatives that "make the best of a bad job" (Dawkins, 1980), or status-dependent alternatives (Gross, 1996). Male horn dimorphisms are examples of phenotype-dependent alternatives in that they depend on a phenotypic variable, body size, that affects individual ability to win in fights with other males.

Secondary status-dependent alternatives can be regarded as fitness-salvaging tactics by individuals who lose out in social competition with individuals superior at performance of a primary alternative tactic, such as fighting (West-Eberhard, 1979; Eberhard, 1982; Parker, 1982). Secondary alternatives such as sneaking or parasitism are often associated with small body size or other traits that are handicaps in the primary, high-payoff competitive mode. Large size is a common correlate of success in sexual and social competition, and large individuals can often virtually monopolize critical resources unless losers adopt alternative tactics to obtain them.

The complexity of status-dependent behavior is illustrated by observations of the spider *Agelenopsis aperta* (Riechert (1982). Females fight over already constructed webs, and losers resume searching and may have to build new webs. Decisions between fighting and searching are phenotype dependent in that the larger individual wins 91.2% of the time. Decisions also depend on resource abundance or "energy needs," as fighting is stronger in conditions of site scarcity. And decisions are affected by severity of the habitat, which affects the distance within which a spider will in-

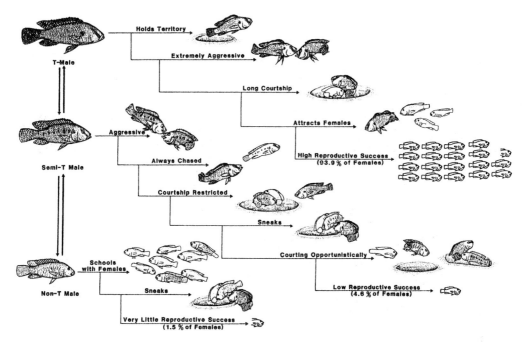

Fig. 20.4. Status-dependent male reproductive alternatives in the African cichlid fish *Pseudocrenilabrus philander*. Adoption of one of three tactics (left) depends on relative male size and competitiveness. T-males are aggressively territorial at a spawning site. Semi-T males perform aggressively opportunistic courtship and parasitic spawning when a territorial male spawns. Non-T males are female mimics in color and behavior and opportunistic parasitic spawners when a territorial male spawns. Individuals sometimes switch between tactics. Reproductive success (shown by number of small fish at end of each arrow) was estimated from eggs spawned per spawning circumstance, divided by the number of males that spawned, not genetic studies of paternity. Higher profit tactics are estimated to have higher energy expenditure and risk of injury. Drawing from Chan (1987). Courtesy of T.-Y. Chan. Description based on Taborsky (1994) and T.-Y. Chan (personal communication).

teract agonistically with a conspecific web owner (Riechert, 1982, p. 288). Such factors are not adequately considered by purely frequency-dependent models, so it is not surprising that the quantitative predictions of those models are seldom satisfied (see Riechert, 1982, pp. 312–313). A game theoretic analysis tailored to data on these spiders shows the multiplicity of factors that affect competition-dependent selection (Hammerstein and Riechert, 1988; see also Hamilton and McNutt, 1997).

Phenotype-dependent social tactics can be maintained even when resources are abundant and the original reasons for the evolution of status-based decisions have disappeared (West-Eberhard, 1979, 1986). When evolved status-based relations have supplanted scramble competition as a means of access to resources, their performance may become insensitive to short-term fluctuations in resources due to selection that favors establishment of status

early in ontogeny and competition for status throughout life as if it were a commodity in itself. This confers a kind of stability on secondary social tactics that resembles that of a highly specialized trophic structure in a monomorphically or polyphenically specialized species, where specialized morphology obliges an individual to pursue a particular strategy even though it is disadvantageous in some places or times. Social alternatives may additionally be stabilized if they become complementary and mutually interdependent, as in the divergent alternatives that produce a division of labor within social groups, or mutualisms between species (further discussed below).

The advantageous maintenance of phenotype-dependent secondary alternatives depends on the maintenance of variation in resource-acquisition ability that corresponds to variation in phenotype. Status-dependent alternatives are not stable if contested resources become abundant, or if some other phenotype-correlated trait were to prove more effective. Game theoretic (ESS) models describe the conditions for stability of phenotype-limited alternatives and their switch mechanisms, where stability is defined in terms of the conditions where the strategy, when adopted by most members of a population, cannot be invaded by the spread of any alternative strategy (Parker, 1984). They are sometimes described as frequency dependent, but as shown by Repka and Gross (1995), frequency dependence is only one factor that influences a competition-dependent switch. If an individual of superior phenotype completely monopolizes the access to a resource by individuals engaging in the same tactic, as in social insect colonies containing a dominant queen, then variable frequency (variation above one individual using the primary tactic) is of limited importance.

Status-dependent alternatives may correlate with certain phenotypic variables, such as size, while in fact having polyfactorial determination by such additional factors as maternal social rank, variable hormonal condition, and signalling ability. Models need to be tailored to consider the multiple idiosyncratic determinants of status in a species if they are to make realistic predictions of advantageous status-dependent switching.

Multiple-Niche Phenotype-Dependent Alternatives Not all phenotype-dependent alternatives are status dependent. Those such as foraging tactics adopted by finches with different-sized beaks (Grant, 1986) involve indirect, ecological competition (e.g., under food scarcity) rather than direct interaction or social competition. In nonsocial phenotype-dependent competition, individuals may sort themselves into behavioral alternatives or dietary specializations in accord with phenotypic variation in ability to pursue different tactics of acquisition of limited resources. In Galapagos finches, trophic preferences and, consequently, food-handling tactics are influenced by beak size, especially during times of food shortage (Grant, 1986; see chapter 28). In many other species, morphology influences the behavioral alternative assumed (Galis, 1993a). In humans, career decisions can be strongly influenced by body size, as shown by the size differences between the average racing jockey or college professor and the average professional basketball player.

Seasonal Polyphenisms Seasonal change in temperature, rainfall, and biotic circumstances (abundance of food, predators and parasites, mates, etc.)

Fig. 20.5. Seasonal polyphenism in a butterfly (*Precis almana*): (A) dorsal and (B) ventral sides of the wet-season form; (C) dorsal and (D) ventral sides of dry-season form. From Nijhout (1991a), © H. F. Nijhout, by permission. Courtesy of H. F. Nijhout.

can oblige or favor a change in morphology or behavior. Thus, many Lepidoptera adopt dark, heat-retaining coloration and sun-basking behavior in cool weather, and lighter coloration with shade-seeking behavior in warmer seasons (Shapiro, 1976), and the cryptic pelage and plumage of some arctic mammals and birds are dark in summer and white in the snowbound winter. Seasonal polyphenism in tropical butterflies often involves a more cryptic alternative during the dry season (Brakefield and Larson, 1984; Nijhout, 1991a; figure 20.5). Seasonal change can also force a switch to migratory or dispersal phenotypes or to dramatic alternative physiological states such as estivation and hibernation. Migration, including long-distance migration, is a seasonal alternative commonly associated with annual change in local conditions (Beebe, 1949; Haber, 1993; Dingle, 1996).

Stress-Induced Alternatives Many kinds of secondary alternatives are associated with nonseasonal environmental extremes in climate or resource availability. In some cases the alternatives involve escape, for example, by dispersal, estivation, or migration.

In plants, growth may move an individual through unavoidable environmental extremes (e.g., of shade and light, or moisture and dryness), leading to such alternatives as shade- versus light-adapted physiology and morphology and submerged versus aerial leaf morphologies (Cook and Johnson, 1968; see figure 7.6). Some of these, such as crassulacean acid metabolism, are costly but enable the plant to continue photosynthesis in circumstances where metabolic activity could otherwise approach zero. Some stress-related conditional alternatives are enforced by catastrophic environmental heterogeneity in space or in time, environmental variation so extreme that, as in the case of extreme social monopolies, they leave no choice but to switch or lose out severely or even completely. Examples are some seasonal alternatives, such as physiological altered states of overwintering organisms (see Tauber et al., 1986, on hibernation or diapause in insects); costly defensive responses to predation (Lively, 1986; Harvell and Padilla, 1990; Harvell, 1991); responses to submergence or drought (e.g., the submerged vs. aerial leaves of heteromorphic plants; Cook and Johnson, 1968); or air breathing and land locomotion by fish during seasonal drought or stagnation of waterways (Graham et al., 1978). Drastic forced seasonal change in diet causes change in the gut morphology and digestive physiology of the rock ptarmigan (*Lagopus*; Gasaway, 1976). Mammalian hibernation has the characteristics of an extreme response to starvation (Srere et al., 1992).

Less drastic food shortages produce switches in diet, hosts, and prey, which may be associated with discrete or bimodal variation in morphology. Alternative soft versus hard foods in cichlid fish and grasshoppers can cause morphological change in the mouth and head morphology (see chapter 28; Bernays, 1986), indicating that seasonal, geographic, or competitive variation that forces changes in diet can have discrete morphological effects.

Social stress, or losing out in social competition, can also lead to dispersal (e.g., independent nesting—West-Eberhard, 1987) or to adopting secondary alternatives of extremely low profitability, such as worker behavior in social insects. In Africanized honeybees, a worker has an estimated inclusive fitness of only 0.04 male offspring equivalents, whereas a queen's inclusive fitness in the same colony is nearly 1000 (see appendix 1, West-Eberhard, 1981).

Forced Opportunistic Alternatives Probably most organisms are much more exploratory and resourceful than we think when we observe them doing the same things in the same places for long periods of time. Leks, or non-resource-associated mating aggregations, are famous for taking the same form in the same localities year after year. Mating systems in general are often species specific or even genus wide traits, for example, in bower birds, where most or all males court in individual constructed arenas. Drastic switches in mating system can occur, nonetheless, as opportunistic forced responses to sudden extreme environmental change. For two consecutive years, the males of an Italian social wasp (*Polistes biglumis bimaculatus*) formed a lek of tiny territories clustered together in a white shoreline of a stream that served as a conspicuous landmark among green fields (Beani and Lorenzi, 1992; Beani, 1996). Late in a very dry summer, when nectar was scarce, a few males formed territories at a nearby flowering plant that was visited by many insects, including females of *P. biglumis*. Then, during another dry summer, a flood drastically reduced the conspicuousness of the landmark lekking site, and covered with mud the sets of stones that had been defended by individual males. In this situation, the males shifted their territories onto low green bushes that were conspicuous, sheltered from the wind, and laden with honeydew-secreting aphids attractive to females. These males, then, opportunistically shifted from a classical landmark lek mating system to resource-defense territoriality, using two different resources attractive to females, when forced to do so by two different episodes of environmental change.

Opportunistic secondary alternatives sometimes evolve because of recurrent encounters with relatively unprofitable or high-cost resources that are frequently available. The specialized scorpion-catching behavior of the generally insect-eating grasshopper mouse, *Onychomys torridus* (Horner et al., 1965), may be an example of this. Hunting and foraging animals must frequently take and process food of secondary value or high cost simply because it is there (the principle of "a bird in hand"), and if recurrent, this could lead to the evolution of specializations for taking and processing a secondary resource. Learning sometimes turns an opportunistic secondary alternative into a practiced and therefore less costly and even primary one. This is suggested by some individual differences in the feeding behavior of birds, such as the famous habit of opening milk bottles in a population of titmice in Britain (Hinde and Fisher, 1951).

Beneficent Secondary Alternatives among Kin In groups of relatives, there are reproductive options that reduce individual fitness while increasing the fitness of genetically related individuals. Examples include helping behavior by sterile females (workers) in social insects, and helpers at the nest in many birds. Such alternatives can be favored by natural ("kin") selection if they increase the performer's *inclusive fitness* (Hamilton, 1964a,b), defined as individual fitness (e.g., number of own offspring) plus

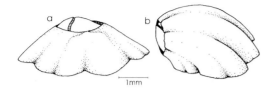

Fig. 20.6. Facultative defensive morphology in the acorn barnacle (*Chthamalus anisopoma*). The typical conic form (a) predominates in predator-free areas. The bent form (b) occurs only in the presence of the carnivorous snail *Adanthina angelica*. After Lively (1986).

the sum of increases in the fitnesses of relatives that are due to effects of the performer, devalued by a relatedness factor, and minus the cost of aid. "Hamilton's rule"—$K > 1/r$—describes the advantageous switch point between selfish and self-costing beneficent social behavior, where K is the ratio of benefit to cost in fitness, and r is the relatedness (fraction of genes identical by descent) of beneficiary and performer of a social act (Hamilton, 1964a,b).

There are many examples of beneficent aid among relatives, and there is considerable evidence that this is a conditional secondary reproductive option whose switch point depends on the costs and benefits of aid as well as degree of relatedness of performer and benefitted individuals (West-Eberhard, 1969, 1975; chapter 23).

Defensive Alternatives   There are many examples of predator-induced phenotypes in both plants and animals (Lively, 1986, Karban and Myers, 1989; Haukioja, 1990; Adler and Harvell, 1990; Harvell, 1991; Brönmark and Pettersson, 1994; Ryan and Jagendorf, 1995; Warkentin, 1995, 2000; McCollum and Van Buskirk, 1996; Agrawal, 1998; Agrawal et al., 1999; DeWitt, 1998; Tollrian and Harvell, 1999). Predator-induced phenotypes are evidently secondary alternatives, in that they are expressed only when induced by the presence or attack of a predator. Some defensive alternatives, such as the bent shells of barnacles (Lively, 1986; figure 20.6), involve morphological change that develops over a period of time. Others, such as precocious hatching in frog tadpoles, are rapid responses taking only a few seconds (Warkentin, 1995, 2000; see figure 13.5). Some defensive alternatives, such as the defensive chemical responses of plants to herbivore attack (Ryan and Jagendorf, 1995), are known to involve rapid changes in gene expression (see chapter 4).

How Alternatives Facilitate Evolution

That alternative phenotypes can be bridges for innovation is not a new idea. Darwin (1859 [1872], p. 136) used alternative phenotypes to illustrate evolutionary transitions. He mentioned heterophyllic plants in a discussion of the evolution of novel leaf forms, and discussed the ability of some fishes to switch between gill and swim-bladder respiration during the transition from aquatic to terrestrial life. Chapter 7 lists some of the ways in which alternative phenotypes facilitate evolution. Here in part III I consider their contribution in more detail.

Alternative phenotypes are modular subunits of the phenotype with all of the properties of modularity that facilitate their semi-independent evolution and dissociability (chapter 7). But alternative phenotypes have three additional properties that give them a special evolutionary significance.

First, the rule of independent selection applies in the extreme to the evolution of alternative phenotypes. Alternative phenotypes represent contrasting options in a particular functional context. They are always divergent from each other due to selection under the different conditions of their expression (see chapter 7). In addition, some are selected for divergence from each other per se. Competition-dependent alternatives evolved under social or sexual selection, for example, may favor options that alleviate competition between unlike individuals or may opportunistically exploit contrasting conditions, such as large or small size or wet versus dry habitats. This means that alternative phenotypes may be unusually divergent from each other and sources of especially innovative change (see chapter 21).

Second, the buffering effect of alternatives—the availability of options in the same functional context—provides an opportunity for profound change. Novel traits can evolve alongside established adaptive traits, allowing populations to move into new adaptive zones without abandoning old ones (West-Eberhard, 1986, 1989; see chapter 7). Alternative phenotypes enable condition-sensitive evolutionary experimentation within populations.

Third, alternative phenotypes show extreme dissociability during evolution. The degree of evolutionary dissociability of a trait depends on its degree of developmental and functional independence. Alternative phenotypes do not have to be produced in sequence or simultaneously, so they are ontogenetically independent. Their developmental independence of each other further depends on their number of nonspecific modifiers (see figure 4.13): the fewer the number of nonspecific (shared) modifiers, the more developmentally independent they are.

Alternative phenotypes are functionally independent of each other by virtue of their mutually exclusive expression. This means that one of them could be lost and the other would still function, because they do not have to be present at the same time in order to work. The aerial and aquatic leaf forms of aquatic plants (see figure 7.6) are examples of functionally independent alternatives. Although loss of one or the other form would be detrimental to a plant or a population growing in a variable environment that includes both aquatic and terrestrial conditions, either form can be fixed in environments that are uniformly aquatic or terrestrial (Cook and Johnson, 1968). A plant on land can completely dispense with the aquatic leaf form. Other alternatives are functionally *dependent* in that they have complementary functions in all conditions. Workers and queens in social insects and alternative tissue forms in multicellular organisms are examples of dependent forms, for neither can reproduce without the other. Dependent alternatives can be developmentally dissociable, but they are not functionally dissociable.

Alternative phenotypes, then, differ from other modular traits in the potential to be expressed independently of each other, as mutually exclusive traits. While the regulation of their expression may be frequency dependent, and the fitness effect of one may affect the relative desirability and frequency of expression of the other (see chapter 21), they are not simultaneously expressed. This means that they are extremely dissociable compared with other modular traits: each can stand alone, independent of its alternatives, as a trait of an individual organism, population, or lineage, giving rise to lineage-specific adaptations on different branches of a phylogenetic tree. That is, the developmental isolation between alternatives, which has promoted their distinctiveness of form, can be converted into population and species differences through phenotypic fixation.

As a result of these three special properties—independent selection, condition-sensitive buffering effect, and extreme evolutionary dissociability—alternative phenotypes play a special role in two types of major evolutionary change: (1) major lineage-specific adaptive innovations, those that characterize entry into new adaptive zones of species and higher taxa; and (2) transitions to higher levels of organization. These consequences of alternatives are discussed in the chapters that follow.

21

Divergence without Speciation

[E]volution, however formally defined . . . is quintessentially a matter of speciation. . . . There simply is
no evidence (nor any convincing theoretical model) that substantial "genomic" change occurs in the
absence of speciation.

—N. Eldredge

A developmental or behavioral switch divides a population into two phenotypic classes,
each involving somewhat different expressed genes or gene-product usage, and each
semi-independently subject to selection and evolution. Intraspecific diverging selection
increases the phenotypic diversity that can be produced by a single genotype. It can
produce evolved change in the dimensions as well as the complexity of alternative
traits. Compared with divergence between reproductively isolated populations, divergence between alternative phenotypes within populations can be especially marked,
due to selection for antagonistic specializations. Condition-sensitive expression can
ratchet directional evolution even in large populations, in heterogeneous conditions,
and through episodes of negative selection. Phenotype fixation and developmental character release accelerate evolution, with selective feedback in the evolution of regulation and form an important driving force for evolution within populations.

Introduction

Part II discussed the developmental origins of novelty in terms of how the phenotype is reorganized
during evolution. It did not deal extensively with
the problem of adaptedness during evolutionary
transitions. How are we to explain transitions from
one well-adapted state to another? Many still-in-
fluential discussions of adaptive shifts, such as
Simpson's (1944) treatment of quantum evolution
and Wright's (1932) discussion of shifting balance,
associate change with fitness cost. Speciational theories of change depict change as dependent upon
reproductively isolated populations in new environments. This chapter discusses divergence without reproductive isolation of novel forms, where
the presumed cost of change is sidestepped because
of the presence of adaptive options in the population undergoing change.

Darwin's solution to the problem of maladaptation during change was strict gradualism in
monomorphically adapted populations. Darwin
(1859 [1966]) reasoned that transitions between
specialized adaptive states need not be disruptive if
they were to occur by a series of small steps.

Wright's (1932) shifting balance is another solution to the same problem, but in Wright's theory,
change is initiated by a chance combination of
genes that happens to suit a population to a new
adaptive mode. Without a gradual adaptive change
or a lucky gene combination, a shift between two
peaks on Wright's adaptive landscape would imply
passing through a valley of inferior adaptedness.
Alternative phenotypes offer a third kind of solution, one that requires neither strict gradualism in
a monomorphic population nor chance genetic
events. In species with alternative phenotypes, a recurrent novelty that happens to prove advantageous to some individuals or in some circumstances
can be refined via gradual genetic accommodation
as an optional trait. Since this involves developmental diversification, not transformation or loss
of existing traits, the new option develops as a new
specialization alongside old ones.

Shapiro (1984a, p. 28; Shapiro, 1984c) notes
that conditional expression of alternative phenotypes is a way of having two adaptive specializations "without carrying a genetic load," or a cost
of genotypes that oblige expression of phenotypes
less favorable than the fittest one. Mather (1969)

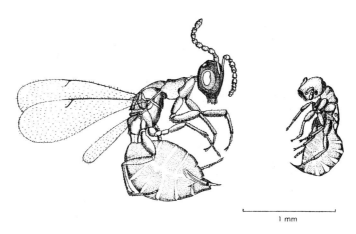

Fig. 21.1. Winged and wingless males of the fig wasp *Pseudidarnes minerva*. Scale = 1 mm. From Cook et al. 1997, figure 1, p. 748), by permission of the Royal Society of London.

also showed that the fitness of a particular phenotype/genotype can be relative and situation dependent (specifically, competition dependent) and argued against the concept of genetic load, which is based on the assumption that there is a single best genotype and that all others represent a fitness burden (load) in the population. There is, however, a kind of burden, or retarding effect on the adaptive evolution of alternatives, in the form of genetic correlations between alternative traits (see chapter 4). The independent expression of alternative traits permits their divergent evolution, but genetic correlations between them, due to shared nonspecific modifiers of form, limit the degree to which they can divergently evolve. The retarding effect of genetic correlations may explain why there are so often only two alternative phenotypes rather than three or four. Although some trimorphisms and quatrimorphisms are known (e.g., Shuster, 1987, on isopods; Gilbert, 1980, on rotifers; Wilson, 1986, on worker castes in ants; Eberhard et al., 2000, on a horned weevil), *di*morphisms are the rule. The answer may lie with selection rather than with any intrinsic developmental constraint, for binary switch points can easily give rise to three or more forms via a sequence of two or more developmental bifurcations, as in social insects with multiple nonreproductive castes (Wheeler, 1986). But multiple sources of genetic correlations could severely limit adaptive specialization, making dimorphisms more stable and advantageous than larger numbers of alternative forms.

Several previous authors have recognized the buffering effect of alternative adaptations in populations undergoing change (see references in West-Eberhard, 1986, 1989; see also chapter 7). The idea of buffered change implies that alternative phenotypes continue to evolve and accumulate differences beyond those that may have been induced at the time of their origin, when a mutation or environmental change caused a developmental change that led to the expression of a novel alternative trait. This claim requires evidence of postorigin divergence between alternatives, a question taken up later in this chapter.

Specieslike Aspects of Alternatives

Marked Adaptive Divergence

A characteristic of species or variation between reproductively isolated populations is their phenotypic divergence in a particular context due to selection in different local environments. Analogous divergence occurs between alternative phenotypes, due to selection on traits *expressed* in different environments in response to different social conditions.

A few examples can serve to illustrate the extreme divergence that can occur between alternative phenotypes in ways that are usually identified with differences between species. Hamilton (1979) remarked on the "very deep divergence of morphs" in male fig wasps (Torymidae, e.g., *Idarnes* species; figure 21.1), where one morph is winged, unaggressive, and femalelike while the other is wingless and in some species highly modified for fighting. The fighting morph has a "totally different form and behavior" that includes enlarged head and mandibles, shieldlike modifications of head and anterior pronotum in some species, and extreme aggressiveness in mortal combat where opponents are sometimes bitten in two. Numerous differences in morphology, physiology, and behavior also distinguish the statary and migratory morphs of locusts (see chapter 5), and the genetic underpinnings of divergence between alternative phenotypes were documented for several examples in chapter 4.

Some alternative phenotypes have specializations so divergent that in separate populations they would usually be regarded as due to a long history of genetic isolation. The complex polyphenisms of economically important insects such as locusts, aphids, honeybees, and parasitoid wasps are well-studied examples, but it would be a mistake to presume from the biased attention afforded pests and pollinators that insects are exceptional in the complexity of their alternative phenotypes. Probably many other polyphenisms would prove comparably complex were they subjected to comparably intense scrutiny. In the spadefoot toad (*Scaphiopus multiplicatus*; now called *Spea multiplicata*), for example, larval diet determines a thyroxine-sensitive switch between a carnivorous and an omnivorous morph (Pfennig, 1990b, 1992a,b; Pfennig et al., 1993). The carnivorous form is large, light in color, rapidly developing, solitary, and predatory. It employs a specialized mouth, with a notched, keratinized oral beak and distinctive buccal musculature, to feed on small animals such as shrimp and tadpoles. This morphology enables the carnivorous morph to "generate a powerful suction to despatch macroscopic prey quickly" (Pfennig, 1992a, p. 168). The omnivorous morph, by contrast, is small, slowly developing, and gregarious. It feeds on detritus and algae and has larger fat reserves and better postmetamorphic survival. The two morphs forage in different locations using different behaviors and have different gut morphologies. Carnivores are more efficient at capturing shrimp, and compared with similar-aged omnivores, they have shorter guts. The gut morphology and more concentrated skin melanophores of the carnivorous morph are characteristics associated with accelerated development: the carnivores resemble larvae that are near metamorphosis (Pfennig, 1992a).

In a related (congeneric) species, *S. bombifrons*, with a similar polyphenism, the two morphs have evolved a divergent specialized ability usually associated with sociality, namely, kin recognition. But the contrasting morphs use this ability in opposite ways: carnivores, which are sometimes cannibalistic, preferentially associate with nonkin and sometimes eat them. Omnivores, on the other hand, preferentially associate with kin (siblings). This reduces their probability of being eaten (Pfennig et al., 1993, 1994, describe a similar example in polyphenic salamanders). To cap this story of developmental virtuosity, the polyphenism is reversible: the same individual can change from one morph to another if subjected to a dietary change (Pfennig, 1992b). Comparable complex divergence between alternatives occurs in plants, where contrast in form can be morphological, as in the aquatic and terrestrial leaf morphology of heterophyllic plants, or physiological, as in the switch between C_3 and crassulacean acid metabolism (see chapter 17).

These examples and many others featured in these chapters of part III show that, contrary to Eldredge's (1976) claim, the epigraph at the head of this chapter, extensive phenotypic divergence can occur without speciation. This is also evident in life-stage specializations and sexual dimorphisms. But the divergence of alternative phenotypes is especially important for evolution because of their greater potential for dissociation to characterize different lineages. Although divergence between alternatives does not depend upon reproductive isolation, it may become associated with speciation (chapter 27) and macroevolution (chapter 29).

Selection for Divergence per se: Intraspecific Character Displacement, Mutual Antagonism, and Complementary Alternatives

Some alternative tactics are "phenotype dependent," functioning to "make the best of a bad job" or salvage fitness of individuals that are at a competitive disadvantage relative to others. Their handicap may be small size, large size (see Halliday and Tejedo, 1995), emergence relatively late in the breeding season, or some other factor causing them to be losers in competition with conspecifics (West-Eberhard, 1979; Pfennig and Murphy, 2000). Such individuals often adopt antagonistic or opposite alternatives that especially suit their handicap, while being costly or unprofitable for their competitors, and the two sets of individuals are therefore placed on different competitive fields. Large, horned males in beetles, for example, have a clear advantage in fights, whereas small, hornless males are faster at running through tunnels where they sneak copulations (Emlen, 1997a; Moczek and Emlen, 2000).

Antagonistic, or mutually exclusive, specializations are common in the evolution of alternative phenotypes. The large-beaked morph of the African finch *Pyrenestes ostrinus ostrinus* is more efficient than the small-beaked morph at cracking hard seeds, whereas the small-beaked morph is more efficient at cracking soft seeds, and intermediate beak forms are selected against (Smith, 1987, 1990a,b,c, 1993; see chapter 28). The molariform morph of the cichlid fish *Cichlasoma citrinellum* can crack snail shells immune to attack by the papilliform morph, which feeds more efficiently than the molariform morph on soft prey (Meyer, 1989a). The carnivorous morph of spadefoot toad tadpoles survives better in highly ephemeral ponds with high densities of animal prey (shrimps and other tadpoles), whereas the omnivore morph survives better in long-lived ponds with abundant detritus (Pfennig, 1992a). In these antagonistic alternatives, the *principle of divergence*, or the advantage of adopting opposite extreme phenotypes or tactics (Darwin, 1859 [1872]), at least partially enables individuals to escape competition with conspecifics that would otherwise be their closest competitors: selection especially favors tactics that, while suited to the individual's phenotype or circumstances (say, its small size, morphological peculiarities, or late seasonal emergence), at the same time are *unsuited* to competitors having contrasting phenotypes (e.g., large or early-emerging individuals). The result is the evolution of alternative phenotypes that not only are different but also are especially divergent.

By analogy with character displacement between sympatric species (Brown and Wilson, 1956), this is *intraspecific character displacement*, divergence in form between contrasting intraspecific alternative phenotypes under competition between subpopulations, each phenotypically better suited to perform one alternative than the other(s). There are many examples in fish where intraspecific character displacement involving phenotypic plasticity (e.g., in lakes where conspecifics are the primary competitors) parallels that between species in multispecies lakes (Robinson and Wilson, 1994).

Mutually antagonistic adaptations, or fitness trade-offs between alternative phenotypes, have been demonstrated in a large variety of organisms in addition to the examples just described, including plants (Cook and Johnson, 1968), bryozoans (Harvell, 1986), barnacles (Lively, 1986), aphids (Moran, 1992b), and butterflies (Kingsolver, 1995) (for references on fitness trade-offs for plasticity in general, see Van Buskirk et al., 1997). Wilson (1989) discusses a model of how multiple-niche

alternatives can evolve without genetic isolation between morphs, and how morphology and learning interact to produce contrasting trophic specializations between benthic and water-column foragers in bluegills (*Lepomis machrochirus*) and many other fishes (further discussed in chapters 27 and 28).

The evolution of antagonistic divergent specializations (West-Eberhard, 1979) obtains especially under social and sexual competition (West-Eberhard, 1979; Gadagkar, 1997b). The same idea has been applied to explain the evolution of anisogamy, or divergence between large, material-filled, relatively immobile eggs and small, material-depleted, motile sperm (Parker et al., 1972). Social insect worker and queen phenotypes are similarly "opposite" in that the workers are usually relatively small, motile foragers with little weight in the form of oocytes and fat bodies, whereas queens are large and sometimes (in highly specialized species) so heavily burdened with ovarian eggs and stored nutrients that they are unable to walk or fly (see figure 7.5). A comparable phenomenon occurs under strong ecological (natural) selection, for example, when seed-eating birds with different beak sizes and seed-cracking abilities experience food shortages seasonally or during times of drought (see chapter 28; see also above). Under food scarcity there is a premium on individual feeding efficiency, but there may be an additional premium on divergence and food partitioning per se, due to the alleviation of direct competition among individuals for food.

The developmental plasticity of alternatives representing intraspecific character displacement has proven useful in the study of interspecific character displacement (Brown and Wilson, 1956), a hypothesis that has been controversial in the past (Walker, 1974; Grant, 1972, 1975; see also other references in Pfennig and Murphy, 2000). Tadpoles in both of two species of spadefoot toad (*Spea bombifrons* and *S. multiplicata*) have facultatively expressed antagonistic trophic alternatives (figure 21.2). Tadpoles of these two species were reared together in the laboratory and censused in ponds where they occur together in nature. The carnivore morph of *S. bombifrons* is superior to the carnivore morph of *S. multiplicata* in competition for shrimp, and the omnivore morph of *S. multiplicata* is superior to that of *S. bombifrons* in competition for detritus. When the two species occur together, carnivore morph frequency increases in *S. bombifrons* and decreases in *S. multiplicata*—a kind of character displacement, manifested as divergence in morph frequencies rather than phenotypic form.

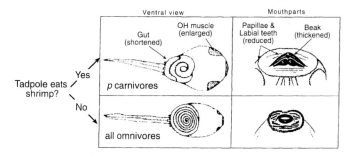

Fig. 21.2. Diet-dependent morphological alternatives in tadpoles of spadefoot toads (*Spea multiplicata*). Ingestion of shrimp, in nature or laboratory, causes tadpoles to adopt a carnivore morphology (top), with different body shape (due to enlargement of mouth-associated muscles), and a short intestine. Ingestion of detritus causes an alternative, omnivore morphology (bottom). Alternatives are reversible if diet is switched. After Pfennig (1992b). Courtesy of D. Pfennig.

Since this is a parameter that can change under competition within a generation, it is an especially sensitive assay for detection of interspecific character displacement.

When divergent antagonistic specializations involve complementary functions, such as somatic and reproductive functions in workers and queens, or material provisions versus dispersal and egg activation in dimorphic gametes, divergence reaches unusual heights. The result may be a division of labor, mutual dependence of the two alternatives, or the emergence of a new level of organization. This has occurred in some highly evolved social insects, where individual workers and queens have complementary phenotypes and are unable to survive and reproduce on their own. The honeybee queen morphotype differs from that of workers by its rounder head, shorter mouthparts, smaller eyes, sharper and more pointed mandibles, lack of pollen baskets (corbiculae) on the legs, recurved and unbarbed sting, smaller brain, absent or vestigial wax, hypopharyngeal and Nasonov glands, larger number of ovarioles in the ovary, and larger spermatheca (Michener, 1974; Winston, 1987). The queen's sharp mouthparts and recurved sting are used in vicious fights among queens, and her enlarged reproductive organs reflect selection for a prodigious output of eggs.

Queens also have distinctive behavioral and physiological traits. Young queens perform "piping" sounds, a highly specialized signal unknown in workers or in any other bee species, and queens produce elaborate pheromonal signals of dominance (Velthuis, 1977). In contrast, workers alone secrete wax and manufacture royal jelly, communicate via the highly specialized dances that help guide fellow workers to food sources and nest sites, build the perfect hexagonal cells of the comb, and perform other specialized tasks of brood care and nest defense seen in honeybees (Winston, 1987). Workers have pollen baskets, barbed stings, and large brains, all specializations for worker-specific tasks such as foraging, nest defense, and sophisticated feats of memory and complex integration, including nest-site orientation, kin recognition, and memory of food sources.

It is no surprise, given these profound phenotypic differences between workers and queens, that there are marked gene-expression differences between them during their larval development (Severson et al., 1989; Evans and Wheeler, 1999). Like other highly specialized social insects, the mutual dependence of the two castes not only makes them obligatory group members but also liberates them for unusually extreme divergence. The division of labor means that morphology as well as behavior and physiology can specialize to a single set of tasks with the assurance that complementary functions will still be performed; and a clean and ontogenetically early switch between castes minimizes correlations between alternative traits, allowing them to be expressed and selected unusually independently of each other. The divergent specializations contribute to fitness at an emergent new level of organization—the integrated colony.

Similarly, divergence between the sexes can lead to their obligatory cooperation in reproduction (see chapter 31). Mutual dependence and a division of labor between cells, tissues, and organs obviously characterizes the organization of multicellular organisms. Such mutual developmental and reproductive dependence implies not only the emergence of a new level of phenotypic organization but also the emergence of a new level of selection (Maynard Smith and Szathmary, 1995).

Developmental Hybrids

The adaptively divergent phenotypes of honeybee queens and workers described above are examples of extremely divergent alternative phenotypes. Like characters of different populations, the morphological traits of queens and workers "vary independently around their own means" (Weaver, 1957a, p. 291). The two alternative phenotypes are also specieslike in that developmental intermediates, or intercastes, show developmental defects comparable to those of hybrids, or intermediates due to crossing between individuals from different genetically diverged populations. This was revealed by laboratory feeding experiments. Larval diets can be manipulated so that larvae receive an intermediate amount of food, near the neutral point for switching between queen and worker (Weaver, 1957a), revealing the mosaic nature of complex alternatives (see chapter 5). The mosaic worker–queen intermediates produced near the neutral point proved to experience higher mortality rates than did normal larvae. If larvae were grafted into queen cells when more than 3.5 days old (just after the critical period for caste determination), almost half died before adulthood, and a few that matured appeared workerlike but failed to emerge and died in their cells. Some of these early fatalities were anatomically intermediate between the two castes. Weaver (1957a) suggested that "development of normal queens, or normal workers, is a well integrated process . . . slight departures from the normal are not especially deleterious, but . . . a state intermediate between the castes is somewhat lethal" (p. 294). This hybridlike consequence of developmental intermediacy between extreme alternatives means that there would be strong selection on colonies to avoid the trophic conditions at the neutral point—strong selection in favor of clearly distinct feeding regimes for workers and queens that would assure the production of clearly discrete forms. Disruptive selection against the handicapped developmental hybrids, or intercastes, and in favor of a clean switch between extremes would lead to the evolution of disruptive development (discrete developmental pathways) rather than to genetic polymorphism, as discussed in chapter 11. A clean switch between alternatives is analogous, in terms of favoring divergence, to a reproductive isolating mechanism between species.

Consolidation by Selection against Intermediates

Some discussions of speciation (e.g., Thoday and Gibson, 1962; Maynard Smith, 1966; Dobzhansky, 1970) emphasize the importance of selection against intermediates in the consolidation of species differences in areas where two forms overlap. This led to the idea of "character displacement," or increased phenotypic divergence between sympatric populations in a given selective context (Brown and Wilson, 1956). In analogous fashion, disruptive selection that affects regulation helps to consolidate the development of an intrapopulation alternative as a discrete subunit of the phenotype. Selection against intermediates improves coordinated expression to produce a clean switch between traits. The regulatory consolidation of a trait increases its phenotypic discreteness and its evolutionary dissociability as a unit. In effect, consolidation adds a discrete building block to the phenotypic repertoire of a genome and increases the scope for evolution by developmental reorganization through regulatory change. The phenotypic divergence observed is similar, but the mechanisms are completely different: in populational character displacement, divergence is due to genetic differences between noninterbreeding populations, whereas in regulatory consolidation, divergence is due to improvement in a developmental switch.

Chapter 4 describes examples of switch-regulated phenotypes with different degrees of regulatory consolidation. The "ganged switch" of migratory locusts exemplifies alternatives where the switch is poorly consolidated, and a complete switch between extreme forms occurs over a long period of time. The different elements of the mosaic phenotype controlled by the switch are poorly synchronized in their expression. By contrast, the complex mosaic honeybee worker and queen phenotypes are highly coordinated in their expression by a strong nutritionally influenced switch during larval development, and the mosaic nature of the phenotypes is revealed only by experimental manipulation of the switch (see figure 5.25).

Effective regulatory coordination of trait expression is sometimes called "canalization" (Waddington, 1942). The channels or valleys of Waddington's epigenetic landscape (see figure 1.3) refer to a restricted range of expression of a continuously variable trait. I prefer the term "consolidation" for the evolution of well-coordinated trait expression, for it implies unification of a set of components that retain their separate identities even though expressed as a set. Canalization does not evoke so well the idea of a set of coordinated subelements that, while expressed as a set, are still potentially dissociable and independently variable at lower levels of organization and therefore potentially dissociable and subject to somewhat independent evolutionary change. Consolidated alternatives may be repeatedly lost, gained, and recombined as units over the long course of evolution, in many different ways, as shown in part II. Under some conditions, subunit components of traits may also vary quantitatively, individually, or as a correlated set, as shown in the variation of plate number and size within plate morphs of sticklebacks (discussed below).

Evidence of Postorigin Divergence

Complex alternative phenotypes show that intraspecific variants can have highly distinctive forms, but what is the evidence that alternative phenotypes diverge after they have originated, as expected under the rule of independent selection (chapter 7)? Geographic variation in alternatives can provide evidence of such divergence. If homologous alternatives found in different populations or species vary in form, and there is no evidence for recurrent independent origins (i.e., a polymorphism is a shared trait, or synapomorphy, of the taxa where it is found), then the differences likely evolved subsequent to the origin of the developmental bifurcation that produced the alternatives. Taxonomic data can often be used to demonstrate postorigin divergence between alternative forms. In highly polymorphic species of social insects such as many ants, for example, taxonomy is based on the morphology of workers, and workers often cannot be associated with queens of the same species or identified using the same keys (e.g., see Wheeler, 1910a, pp. 129ff). This shows that the workers have diversified independently of the queens. Their diversification must have occurred after the origin of the castes, because all ants have worker and queen alternative phenotypes, so both alternatives were likely present in the common ancestor of extant species.

In general, lack of concordance between alternatives in taxonomic and phylogenetic analyses supports a hypothesis of their independent evolution. Key characters are chosen because of their lack of correlation with each other, evidence that they are evolving independently. Lack of correlation between alternative morphologies is also evidenced by

different allometrical relationships (e.g., see Eberhard and Gutierrez, 1991, on beetle polymorphisms; Wheeler, 1991, on caste polymorphisms in ants). In many male-dimorphic horned beetles, with large males horned and small males hornless, the horns of the horned males are sufficiently diverse to be used as taxonomic characters in some genera that would be difficult to distinguish on the basis of morphology of the hornless morph (Howden, 1979). This is evidence that the two morphs are evolving independently of each other. In the Australian genus *Blackburnium* (Scarabaeidae: Geotrupinae), for example, horns are species specific in form, as well as geographically variable within species (Howden, 1979). The abundant and well-studied species *B. angulicorne* shows geographic variation in horns with little or no corresponding variation in male genitalia or female morphology. Behavioral observations suggest that sexual selection is responsible for the diversification of male horns, which are used in fighting in all beetles where horn function has been closely observed (e.g., Eberhard, 1979, 1981; Moczek, 1996; Emlen, 1997a).

Similarly, in ants the morph-specific divergence of the worker caste involves both behavioral and morphological traits. In relatively unspecialized species, behavioral differences are determined in the adult stage and workers and queens are morphologically similar or identical. In more specialized taxa, caste is determined in the larval stage and morphological modifications are more extreme (D. Wheeler, 1986, 1991; Nijhout and Wheeler, 1982, 1996). Ant workers may have specialized more than sexual morphs (queens and males) during their evolution, for the queens "differ in no essential respect from the corresponding sexual phase of the solitary species" (W. M. Wheeler, 1910a, p. 104) or differ mainly due to deletions of ancestral female traits rather than in large numbers of novel additions (Michener, 1974; Evans and Wheeler, 1999). Queens are "the normal females of the species and bear the same morphological relations to their males quite irrespective of the nature of their larval food" (Wheeler, 1910a, p. 104). Wheeler concluded that the dietary determination of caste mainly affects the worker caste, and the worker morphology and behavior that appear to diverge most from ancestral (solitary) forms (see also Wilde and Beetsma, 1982). It would be interesting to do a phylogram analysis of caste diversification in social insects (as in figure 21.9, discussed below) to evaluate the relative divergence of different morphs. Whichever caste proves to be most derived, it is clear that one or both castes have diverged independently of each other, and they represent different sets of expressed genes as required if divergence is to occur (Evans and Wheeler, 1999; see chapter 4).

The Batesian mimic polymorphisms of butterflies are well-studied illustrations of extreme adaptive morph-specific divergence in form. Individual morphs show independent geographic variation in *Papilio dardanus* (Ford, 1971, pp. 266–268), and the genetic divergence of mimics in different regional populations in this species is indicated by the breakdown of mimicry pattern in interregional hybrids (Turner, 1977). The quality or precision of mimicry varies geographically in accord with the frequency of particular distasteful models, and independently of variation or lack of it in the nonmimic morph. In *P. dardanus*, for example, the mimic morph that resembles the model species *Amauris echeria* is a "better mimic" in Southern Africa, where *A. echeria* is common, than it is in central Africa, where the model is uncommon. Similarly, the "hippocoonides" form of the race *tibullus* is a "beautiful mimic" of *A. niavius dominicanus* north of Delagoa Bay, where that model is very common and the mimic compromises 85–90% of the female population. This contrasts with the west coast of Africa and in central Africa, where the mimic morph (there called "hypocoon") is modified to resemble a darker race of the distasteful model (*A. niavius niavius*). In the mountains of central Kenya and Tanganyika, mimicry breaks down. There butterflies show variable and more malelike patterns, because the model is rare or absent. Another morph ("trophonius") of *Papilio dardanus* simultaneously mimics another distasteful butterfly, *Danaus chrysippus*, and its variation independently parallels that of its model's geographic variation. This morphspecific environmental tracking is evidently produced by genetically simple quantitative change in pattern elements that are homologous in models and mimics (Nijhout, 1991b, pp. 182–183).

Heterophyllic buttercups (*Ranunculus* species) have contrasting aquatic and aerial leaf forms on the same individual plant. The two leaf forms vary independently within and between species. Rearing experiments have ruled out environmental influence as the source of the divergence and shown it to be genetic, presumably evolved under natural selection (Cook and Johnson, 1968).

These examples show that alternative phenotypes evolve independent of each other by genetic modification following their origin, as claimed by the rule of independent expression, selection, and evolution (see chapter 7). Morph-ratio stability or morph fixation is not required for morph-specific evolution to proceed. The dark morph of the classic industrial-melanism polymorphism of the moth *Biston betularia* studied by Kettlewell (summarized in Ford, 1964) is a "unifactorial" or allelically determined color morph alternative to the "normal" light form. The light form compised about 2% of the population in the Manchester area by 1895 (Ford, 1964). This polymorphism has been studied for a long period of time and is known to fluctuate in relative morph frequencies both geographically and temporally (Sheppard, 1975). There is clear evidence for modification of the dark morph associated with its increased commonness in parts of England, where it has undergone intensification of the black coloration, since older collected specimens of melanics have a thin white line on the forewings and hind wings that had disappeared in later specimens (Ford, 1964, p. 308).

Why Alternatives May Foster Divergence More Effectively Than Speciation

Reproductively isolated sister populations may or may not diverge in isolation. Barring founder effects in small populations, they begin with phenotypically and genotypically similar distributions of traits, and they would be equally likely to evolve in parallel or divergently. So their adaptive divergence depends primarily on the magnitude of environmental differences between them. Because sister populations may be subject to only small differences in selection regimes, drift and founder effects in small populations have sometimes been considered important for the evolutionary divergence of populations (Wright, 1932; Carson, 1985; Mayr, 1963). But alternative phenotypes are necessarily subject to different selective regimes, even without environmental differences or genetic biases, simply because they are different. Their phenotypic differences mean that they interact differently with the environments they may share. Beyond this, and also by contrast with species, divergence of alternatives can be accentuated by selection for divergence per se, as explained above.

These considerations challenge the claim that "speciation is orders of magnitude more important than phyletic evolution as a mode of evolutionary change" (Gould and Eldredge, 1977, p. 116). The impression that speciation is the prime mover of diversification is especially prominent in paleontology (e.g., Eldredge and Gould, 1972, 1997; Eldredge, 1976; Stanley, 1989; Jackson and Cheetham, 1990), where speciation sometimes may be credited with diversity that is in fact a consequence of complex polymorphisms, polyethisms, or phenotype fixation of traits. Sudden and brief episodes of morphological change, thought by some paleontologists to characterize morphological evolution, follow exactly the pattern of change that would be expected to accompany developmental character release following fixation of a behavioral alternative, an event that would usually be invisible in the fossil record (see chapter 30).

Diverging Selection

The examples described in this chapter show that alternative phenotypes can become more specialized by the addition of new modifiers of form as well as by quantitative changes in dimensions, and that the phenotypic distance between them increases during evolution. Recent efforts to document the qualitative aspect of divergence have applied methods derived from cladistics (e.g., Wray, 1996; figure 21.9). These efforts are important, for they make it possible to discuss in concrete terms the changes in complexity of evolved traits, whereas before only their dimensions were being compared.

As discussed in chapter 1, general discussions of selection and evolution often list three kinds of selection—stabilizing, directional, and disruptive selection—that affect the evolution of quantitative traits within species (e.g., Mather, 1973; Endler, 1986; Futuyma, 1986a, 1998). The three types of selection (see figure 1.1) depict the response to selection in terms of quantitative change, not an erroneous portrayal but one that omits the possibility of change in complexity of the phenotype. They also sometimes omit the possibility of increase in the range of a quantitative distribution when there is selection for extremes and against intermediates (figure 1.1, right).

The study of divergence between alternatives suggests that a fourth category of selection needs to be added to the customary list of three: *diverging selection* is selection that increases the difference between alternative phenotypes, through change in either their quantitative traits (dimensions) or their complexity (by addition or subtraction of elements), or both. Diverging selection differs from the classical idea of disruptive selection (figure 1.1, right; compare with some later usages, e.g., in Futuyma, 1986a) in that it can extend the *range* of variation of a quantitative trait as well as reduce the frequency of its intermediate values, and in the inclusion of changes in trait complexity. In its effects on quantitative traits, diverging selection is like directional selection that acts in two directions rather than one, whereas disruptive selection is like stabilizing selection that acts on two modes rather than one (see figure 1.1).

Evolution under diverging selection has been proposed to explain the evolution of anisogamy, or the dimorphism of eggs and sperm (Parker et al., 1972). The evolution of eggs and sperm is a good illustration of diverging selection driven by a fitness trade-off, in this case between the benefits of large size (in eggs) and small size (in sperm). The result is antagonistic polymorphism between the male and female gametes and, ultimately, between the two sexes. Diverging selection may preserve either large or small qualitative or quantitative variants.

"Diversifying selection" has been applied to some aspects of the phenomenon I am calling diverging selection, but it has been used by previous authors to mean somewhat different things. Dobzhansky (1970) and Mayr (1974b) used "diversifying selection" as a synonym of disruptive selection, a term that (Dobzhansky, 1970, p. 167) considered suggestive of disorder or disarray. Dobzhansky (see also Cavalier-Smith, 1980) applied the term "diversifying selection" to any increase in the range of phenotypes producible by a single genotype, thus including both discrete variants and quantitative shifts. He used it to explain the maintenance of genetic polymorphism, including under frequency-dependent selection and in different niches. Mayr (1974b) later proposed using Dobzhansky's term diversifying selection in place of disruptive selection, to counteract confusion he felt was caused by Thoday's (1972) use of the term "disruptive selection" to include selection against hybrids at species borders. Neither Dobzhansky's nor Mayr's objections widely influenced usage, and disruptive selection is still the more common term for selection against intermediates and favoring the extremes of a continuous distribution (as in Mather, 1953; see figure 1.1).

The Ratcheting Effect of Conditional Expression

Directional evolution is widely considered to be difficult in large panmictic populations, where the environment is highly heterogeneous and individuals disperse widely to mate. Gene flow is believed, on theoretical grounds, to cause a spatial averaging of selection pressures (Kirkpatrick, 1996, p. 129). Some authors consequently believe that change occurs only if a small offshoot population becomes geographically isolated (Giddings et al., 1989). Others argue that small populations are not necessary for evolutionary change, and some maintain that large populations can be even more favorable to change than small ones (reviewed in Coyne and Charlesworth, 1997).

Condition-dependent expression adds support to the case for evolution in large populations. In chapter 7 I hypothesize that episodic directional evolution could be facilitated by conditional expression of alternatives in a spatially or temporally heterogeneous environment. Condition-sensitive expression confers directionality in the face of fluctuating or episodic selection. For convenience, I will call this effect the *conditional ratchet*: because conditional alternatives are expressed primarily when they are subject to positive selection, they experience directional selection even under fluctuating conditions.

How the conditional ratchet works is illustrated by a morphologically polyphenic species such as heteromorphic buttercups (*Ranunculus* species; Cook and Johnson, 1968). Suppose that the buttercups compose a genetically variable, panmictic, sexually reproducing population distributed in space without regard to genotype. In populations that span both wet and dry areas, there is variation in phenotype, with aquatic leaves in wet localities or periods and terrestrial leaves in dry sites or periods (Cook and Johnson, 1968). Selection in wet conditions favors genotypes best able to produce the aquatic leaves. The aquatic phenotype would be selected against in dry areas if it were expressed there, but it is not. Therefore, both specialized phenotypes are protected from negative selection by environment matching due to facultative expression.

If such a population were to have readily dispersed gametes and panmixis, individuals in the dry areas or eras would acquire wet-morph-specific modifier alleles, and individuals in wet conditions would obtain dry-specialist alleles. Each episode of selection on the aquatic or the terrestrial leaf form would lead to improvement in that form, and barring loss by genetic drift, this improvement would persist until the phenotype is expressed again, when it would be exposed to another round of selection and potentially improved. Lack of expression in contrasting conditions would protect the improved morph from negative selection, causing a ratchet effect that moves change in each of the specialized phenotypes in a single direction.

Grant and Grant (1995b) emphasize the importance of episodic selection and changes in sign of fitness effects of particular traits, and the fact that selection nonetheless often produces a directional trend in the long run. They give examples of trends furthered by episodic selection (p. 249). The conditional ratchet would promote such change. Var-

ious other examples are discussed in chapter 28. Learned phenotype-environment matching can generate a directional trend in morphological evolution despite changing directions of selection. In island birds, for example, individuals in times of plenty may take the easiest obtainable food regardless of beak morphology. But in times or localities of food scarcity, individuals (e.g., those with relatively strong beaks) morphologically predisposed to learn to specialize in a food less available to other individuals may be favored, leading to episodes of change in beak size and shape. During periods of food abundance, selection on beak form is relaxed, so evolved specialization could persist through a period of facultative generalized foraging. Periods of negative selection such as those documented by Grant and Grant (1995b) could affect such a trend, producing slippage in the conditional ratchet, but would not reverse it completely if variation in beak size persists and episodes of positive selection recur.

The ratchet effect of phenotypic plasticity may be especially important in marine organisms with planktonic larvae, where both gene-flow and habitat heterogeneity are potentially high. It could facilitate phenotypic divergence in some of the marine invertebrates used in paleontology such as mollusks and plantlike hard-bodied colonial organisms (bryozoans and corals). Hoagland (1979), in a monograph on the snail genus *Crepidula*, discusses the problem of evolution and plasticity in a mollusk (see also Piaget, 1949) and gives examples of the resultant taxonomic confusion and its importance for paleontology. She notes how the substrate where a snail happens to grow can affect shell shape, and how environmental responses are exaggerated by phenotypic accommodation among correlated traits. Due to the ratcheting effect, condition-sensitive expression of phenotypic extremes in such an organism could give rise to geographic variation without reproductive isolation (see chapter 30, on punctuation). The diverse and variably plastic living species of *Crepidula*, where there is much parallel evolution in morphology and species differences in key developmental traits (e.g., planktonic vs. brooded larvae) may be suitable material for research on the role of plasticity in the evolution of a group with a paleontological record.

Oscillating selection may favor maintenance of alternative phenotypes even when they do not respond to every reversal of conditions. In carpenter bees (*Xylocopa pubescens*), dominant females experience both costs and benefits by tolerating the presence of subordinates, who guard the nest against robbers and usurpers but may usurp the dominant position themselves. The subordinate alternative is maintained in populations by episodes of intense competition for pollen and nests, when the benefits of tolerance by the dominant clearly outweigh the costs (Hogendoorn and Velthuis, 1993).

Phenotype Fixation and Developmental Character Release

Rate of Divergence as a Function of the Ratio of Alternatives

Evolutionary change in the regulation of alternatives can have at least three different effects on the evolution of phenotypic form. First, it can affect the degree of discreteness of alternatives, thereby reducing genetic correlations between traits and permitting their independent evolution (see chapter 7). Second, it can affect the precision of environmental matching, or assessment of when to switch, thereby influencing the consistency of selection and evolution in a particular direction. Finally, it can change the threshold setting of the switch or the ability to overcome the threshold, and thereby alter the ratio of alternatives produced, which affects the evolution of form because it affects the frequency with which variation in each form is exposed to selection. As one alternative increases in frequency relative to another, it is therefore expected to improve at an accelerating rate (see chapter 7, on mutual acceleration in the evolution of regulation and form). The expectation of accelerated improvement, or developmental character release, with increased frequency of expression is the subject of this section.

The more frequent the expression of a phenotype due to change in its regulation or environment, the more rapidly it should evolve toward increased specialization, other things (e.g., strength of selection, amount of genetic variation) being equal (chapter 7). The extreme dissociability of independent alternatives permits the fixation of single alternatives and the loss of others. Phenotype fixation may be due to evolutionary change in the frequency of expression (genetic assimilation), or it may occur without genetic change in regulation, due to consistent environmental induction of single form.

Phenotype fixation by genetic assimilation is discussed in chapters 6 and 7. Selection on the threshold for production of one alternative relative to another can lead to an increase in frequency of the favored phenotype, and the same change can be caused by an environmental change that increases the frequency with which the threshold for that phenotype is passed (see especially figure 5.23). Carried to its extreme end point, such change results in phenotype fixation.

Phenotype fixation by environmental change is much less discussed than is fixation by genetic assimilation, but environmentally mediated fixation is known to occur and should be investigated as a possibility whenever phenotype fixation occurs. Several examples are discussed by Wcislo and Danforth (1997) in their review of reversals in transitions between solitary and social behavior in insects. In several lineages of bees, solitary and social behaviors occur as alternatives within populations in some environments, whereas social nesting is absent in others. Montane populations of *Exoneura bicolor* are polyphenic, but heathland populations are fixed for solitary nesting. The environmental mediation of phenotype fixation was demonstrated by experimentally mimicking the montane distribution of favorable nesting conditions, and the result was an increase in social nesting to montane levels in the heathland bees (Hurst et al., 1997). In *Halictus rubicundus*, there is a reduction in the frequency of social nesting with increased altitude in this polyphenic solitary–social species. At very high altitudes solitary nesting is fixed (Eickwort et al., 1996; Yanega, 1997).

When fixation of a single alternative occurs, it is expected to be accompanied by accelerated evolution of the fixed trait, or developmental character release (see chapter 7). As the more frequently expressed alternative is liberated from genetic correlations with other traits, formerly shared, nonspecific modifiers are more free to specialize in the direction of the fixed trait (West-Eberhard, 1986, 1989). Expression of only one alternative means that selection on the remaining alternative(s) is increased (more frequent), which further drives specialization in a single direction. Release in both its genetic and selection aspects can be regarded as a continuum beginning with the origin of an alternative (see chapter 7) and continuing with its increasing frequency of expression, sometimes accentuated by improvement of clean switching between alternatives (as in advanced social insects), and sometimes culminating in fixation of a single one (see chapter 7). Any event that reduces genetic correlations between traits and antagonistic pleiotropy, or increases the frequency or independence of expression of a single alternative (e.g., see Rice, 1998), contributes to developmental character release.

It has been recognized that nonspecific modifiers favorable to only one alternative morph should not become established unless that morph is at high frequency (Charlesworth and Charlesworth, 1976a; see also Turner, 1977, p. 185). Turner (1977 p. 184) cites Pilecki and O'Donald (1971) as saying, "[T]he commoner a mimetic form is, the stronger will be the selection pressure on the modifying loci, and hence the better the mimicry." Then both the high-frequency morph and its modifiers should increase to fixation. Because of this change in rate of evolution due to phenotype fixation, plasticity has sometimes been seen as promoting "accumulation of concealed reserves" of variation, which then can produce accelerated evolution in particular, less variable environments (Schmalhausen, 1949, p. 274).

Phenotype fixation is similar to increased modularization in releasing traits from genetic correlations. The term "coupling" has been used to refer to the performance of two functions by a single structure (Maynard Smith, 1989); in other words, the structure is a shared or nonspecific modifier of the two phenotypic components. "Decoupling," or increase in number of components, has been associated with the evolution of diversity in derived forms (Lauder, 1981; Wake, 1982). Thus, accelerated evolution can be associated with release from genetic correlations, whether due to separation (modularization) or fixation of parts. In either change, there is an acceleration of modification with increased independence of expression and decreased dependence on shared, nonspecific modifiers or traits.

Since developmental character release is frequency dependent, it may occur in polyphenic populations whenever there is skew in the proportions of morphs produced, even though morph fixation has not occurred. But release should be most rapid in formerly polymorphic populations where a single alternative has been recently or rapidly fixed. As selection reduces the genetic variation in the nonspecific modifiers of former alternatives, the effect of character release should decline. Subsequent directional change would then occur due to innovation and genetic accommodation as described in chapter 6.

Kinds of Evidence That Developmental Character Release Occurs

The release hypothesis predicts change in particular kinds of traits, namely, those held in check when two somewhat antagonistic alternatives are expressed. The character-release hypothesis gains support if one or more of the following four kinds of change occur in a population when one alternative phenotype approaches or reaches fixation:

1. A trait with multiple functions becomes more specialized in a single function when that function becomes predominant or fixed. In a pioneer discussion of developmental character release, Darwin (1859 [1872]) noted that increased specialization due to simplification of function is a common feature of morphological evolution. He cited an example in barnacles. The ovigerous folds in the skin of pedunculated barnacles perform the double function of egg retention and respiration. In sessile barnacles (Galanidae), the egg-holding function is dropped and the ovigerous folds become exaggerated to form respiratory branchiae. Similarly, the mouth anatomy of salamanders (Plethodontidae) participates in both feeding and pulmonary respiration. Lung loss in six separate lineages of salamanders (Plethodontidae) has been accompanied by the evolution of tongue projection (Wake, 1982). Lung loss "frees salamanders from one extremely important and very restrictive functional design constraint—filling the lungs by buccal pump mechanisms" (p. 56). This permits developmental rearrangements and specialization of the tongue apparatus.

2. A fixed trait that is expressed as an alternative in some species or populations shows improvement that would have been detrimental if expressed in the lost or declining alternative. In bimodal-breathing fish (facultatively air and aquatic breathing individuals), there is "interference" between respiratory modes due to the presence of gills (Kramer, 1983, pp. 149–150): during air breathing, gills are a liability because they allow diffusion of oxygen from the blood into the water, due to the very properties that make the gills effective in water breathing. Obligate air-breathing species have reduced gills (Kramer, 1983 p. 149). This suggests character release accelerating gill reduction as the frequency of air versus water breathing increases. Character release of this type is likely to have accompanied the vertebrate colonization of land, a major transition in the evolution of terrestrial vertebrates.

3. A morph-ratio cline coincides with a trait-specialization cline, with the phenotype that is increasing in frequency increasing in specialization as well. Female pitcher-plant mosquitoes (*Wyeomyia smithii*) have a north–south morph-ratio cline in two modes of reproductive development in the eastern United States. In the North, most are "autogenous" and produce mature eggs without a blood meal, using nutrients stored in a well-nourished lar-

val stage; in the South most require a blood meal to reproduce (Lounibos et al., 1982). The cline in nutritional mode corresponds to a cline of increasingly adequate nutrition afforded by decreasing density of larvae in pitcher plants (Bradshaw and Lounibos, 1977) with increasing latitude. North of 40°N, females are unable to feed on blood even if reared on an inferior diet (Lounibos et al., 1982). These obligatorily autogenous northern females, designated as a separate "race" by Bradshaw and Lounibos (1977), emerge with more precocious ovarian development than do southern females, and they also mate earlier than do southern (Florida) autogenous females (O'Meara and Lounibos, 1981). The comparison of populations at the two ends of a phenotype-ratio cline indicates increased genetic specialization accompanying morph fixation, in keeping with the release hypothesis.

4. Trait specialization is greater in populations or species where a trait is fixed than in related populations or species where the same trait occurs as an alternative phenotype. This is a common pattern that may indicate developmental character release, especially if the increased specialization is likely to have been opposed under selection for the omitted alternative.

Several possible examples are summarized below. Many other studies could be cited where observations suggest that character release may occur if evidence for it were to be sought. In *Timema* walking-sticks (Phasmatodeae: Timemidae), for example, predation has played a role in the evolution of polymorphism, as well as radiation onto different hosts (Crespi and Sandoval, 2000). Phylogenetic analysis shows that the ancestral populations of these insects were likely host-plant generalists with color polymorphisms of three or four color forms maintained due to crypticity on their host plant leaves and stems (Sandoval, 1993, 1994a,b). Polymorphic ancestors have evidently given rise to monomorphic species via morph fixation multiple times. This is the kind of taxon where it would be of interest to look for signs of character release, for example, increased specialization in cryptic coloration or behavior, in the descendent specialists.

Possible Examples of Developmental Character Release

Aphids   The aphid *Pemphigus betae* has a complex life cycle with a root-inhabiting morph in herbaceous plants and a gall-inhabiting morph on the leaves of cottonwood trees—two very different ecological niches (Moran, 1991). In some localities the gall morph is deleted. When performance (size/ fecundity, developmental rate, and mortality) was compared among different clones on root hosts, root-morph populations that had deleted the gall morph performed better on roots, with faster development, lower mortality, and larger size. These results suggest that when both morphs are expressed, the evolution of improvement in the root-inhabiting morph is limited by genetic correlations between morphs, or antagonistic pleiotropy, and that character release allowing specialization to the root environment has occurred due to deletion of the gall morph (Moran, 1991). Geographic variation in morph ratios, including phase deletion, is common in aphids that have complex life cycles (Moran and Whitham, 1988).

Crickets   In   the   Japanese   ground   crickets *Pteronemobius nigrofasciatus* and *P. mikado* (Orthoptera, Gryllidae), bivoltine (double-brooded) southern populations are seasonally polyphenic for length of the egg stage: the eggs of an early-summer brood pass only a few weeks in the soil, and the eggs of a late-summer brood overwinter in diapause for several months in the soil (Masaki, 1986). In univoltine (single-brooded) northern populations, by contrast, the long egg stage is fixed: all eggs pass the winter in diapause. In the polyphenic populations there is alternation between selection for short and long ovipositors. Short-duration eggs need not be placed deep in the soil, and there is selection against long ovipositors, which implies increased oviposition effort, susceptibility to breakage, and greater cost of offspring emergence (Masaki, 1986). Long-duration eggs select for long ovipositors that place eggs deeper in the soil, affording greater protection from desiccation and better survivorship of eggs that remain for many months in the soil. The overwintering-egg phenotype is fixed in northern populations, and, as predicted by the character-release hypothesis, the ovipositors of northern females are longer in relation to body size than in southern populations, where there is egg-duration polyphenism (figure 21.3). Whether polyphenism or monophenism is ancestral, these data confirm the expectation of increased specialization in fixed-phenotype populations. As pointed out by Masaki (1986), ovipositor length in crickets is evolutionarily labile and adjusted either upward or downward in accord with local conditions. In populations or species with obligatory egg diapause, or resident in areas with dry or unstable soils, ovipositor length is longer than in closely related populations or species with nymphal rather than egg diapause, with short-duration egg diapause, or inhabiting moist stable soils (Masaki, 1986). When contrasting conditions are fixed for two noninterbreeding, closely related populations, ovipositor divergence is more marked (figure 21.4) than between populations interbreeding along a cline of conditions with phenotype fixation in only one (figure 21.3). This is expected because population divergence can be favored by both phenotype fixation and reproductive isolation.

Solitary Predatory Wasps   Most solitary predatory wasps (Sphecidae) build nests, but females of some species (e.g., *Anoplius marginalis*) show an intraspecific alternative behavior of occupying the burrows of their prey rather than building their own nests (Evans, 1953). Still other species (e.g., of *Aporus* and *Psorthapsis*), use of host burrows is fixed. In such species, specialized digging morphology such as the tarsal comb is sometimes lost, suggesting that developmental character release has affected the evolution of the appendages.

Developmental character release is also suggested in the evolution of prey-carrying behavior and morphology. Females of most solitary species use the legs to carry their prey (Evans, 1962). But females of several species of crabronine wasps (Crabroninae, Sphecidae) impale the prey on the sting, leaving the legs free when the wasp lands and enters the nest (see references in Evans, 1962). In some species, impalement of the prey occurs only under special circumstances: in *Oxybelus bipunctatus* (Sphecidae, Crabroninae) and some females of *O. uniglumis quadrinotatum*, impalement reportedly occurs only while the wasp is opening and entering the nest (Peckham et al., 1973). In *Lindenius armaticeps*, it occurs only if the nest entrance has been obstructed (Miller and Kurczewski, 1975). In *Crossocerus elongatus* and *C. maculiclypeus*, it has been observed only when the wasp has been captured and held in a vial (Nielsen, 1933; Miller and Kurczewski, 1975). And in *O. strandi*, impalement depends on prey size (figure 21.5; Tanaka, 1985). Prey impalement has also been observed in *O. subulatus* (Peckham et al., 1973). In the widespread and commonly observed species *O. uniglumis*, fe-

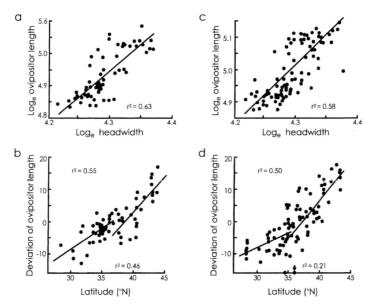

Fig. 21.3. Morphological specialization, phenotype fixation, and accelerated change in cricket ovipositors. In *Pteronemobius nigrofasciatus* (left) and *P. mikado* (right), ovipositor length increases with body size (head width; a, c) and latitude (b, d). Ovipositor length determines the depth of egg placement in the soil. Bivoltine (double-brooded) southern populations (south of 35–37°N) are seasonally polyphenic for overwintering versus not overwintering in the egg stage, with the intensity of selection on deep placement of overwintering eggs increasing (along with mean ovipositor length) with latitude. Univoltine northern populations (north of 35–37°N) have the overwintering egg stage fixed. In keeping with the character-release hypothesis, ovipositor length in populations where egg overwintering is fixed is relatively longer than in the polyphenic populations when the regressions (r) of their deviations on latitude are compared (b, d), and the slope of the regression lines changes at the latitude corresponding to morph fixation (arrow in d). For all morphological measures, 1.0 = 1/40 mm. After Masaki (1986).

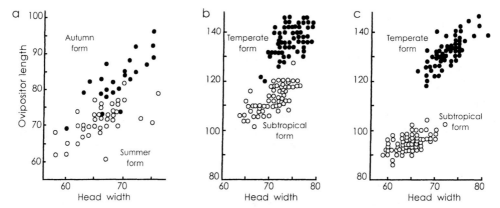

Fig. 21.4. Divergent ovipositor length in closely related noninterbreeding cricket populations with different overwintering stages. Females of populations with the long (overwintering) egg stage fixed or at high frequency (solid circles) have longer ovipositors relative to their body size than do those in populations that where egg overwintering is absent (individuals overwinter as nymphs; open circles). (a) Sympatric noninterbreeding seasonal forms of *Velarifictorus micado*; (b) the lawn ground cricket, *Pteronemobius*, in temperate-zone (*P. mikado*) and subtropical (*P. taprobanensis*) populations; (c) the band-legged ground crickets *Pteronemobius*, in temperate-zone (*P. nigrofasciatus*) and subtropical (*P. fascipes*) populations. The increased divergence compared with that in figure 21.4 could be due to phenotype fixation (here, in both populations), reproductive isolation, or both. After Masaki (1986).

males usually impale the prey on the sting and have a barbed sting believed to secure the prey during transport. Barbs are absent or very small in at least three related species (*C. elongatus, O. sericium, Crabro argus*) that do not impale the prey, and in the facultative prey-impaling population of *O. strandi*. These observations tentatively support the hypothesis that frequent impalement, or fixation of the sting-transport alternative, is associated with morphological character release in the form of a barbed sting.

The behaviorally and morphologically variable genus *Oxybelus* is the kind of taxon that invites a more thorough test of the release hypothesis. It contains 215 recognized species worldwide, the vast majority of them undescribed in terms of prey-transport behavior and sting morphology (Bohart and Menke, 1976).

Wasp Social Parasites   Social wasps (Hymenoptera, Vespidae) sometimes engage in intraspecific social parasitism, where females sneak opportunities to lay eggs in the nests of others, or aggressively usurp nests built by other females and utilize the ready-made structures and workers to raise their own young (reviewed in Cervo and Dani, 1996). Fixation of parasitism in two different subfamilies, Vespinae and Polistinae, has been accompanied by the evolution of morphological specializations to

parasitism that are absent in facultatively parasitic species. Morphological specializations include enlarged mandibles, a thick cuticle, and (in the vespines) a recurved sting (West-Eberhard 1989, 1996; Cervo and Dani, 1996), which has evolved independently in the vespines at least twice (Figure 21.6). *Polistes* social parasites also perform greatly exaggerated versions of certain behaviors seen especially in reproductively active females of nonparasitic species, such as gastral rubbing of the nest (Turillazzi, 1992). Vespine parasites have enlarged Dufours glands (Jeanne, 1977), traits seen primarily in queens of nonparasitic species but in less exaggerated form. These observations suggest that fixation of the parasitic alternative is accompanied by release of morphological traits and certain behavioral traits to specialize toward parasitism. The fact that such similar parasitic specializations have evolved recurrently suggests that the parasitic species have inherited a similar capacity for morphological variation but that selection toward the observed specializations is checked in nonparasitic species where other alternative reproductive modes (worker phenotypes and nonparasitic queen) persist.

The release hypothesis could be tested by investigation of the possibility that some of the parasitic specializations may be disadvantageous if expressed in females that are only facultatively parasitic.

Hyperdevelopment of the Dufours glands in vespines, and exaggerated nest-rubbing behavior in polistines, may be selected against in nonparasites. Added to the energetic cost of producing exaggerated secretions and behaviors may be a social cost, since glandular and behavioral overactivity may subject females to attack by dominant nestmates, whereas in an aggressive social parasite these traits are evidently favored as tools of usurpation. Similarly, the conspicuous, white uncamouflaged pupal caps of the social parasite *Polistes atrimandibularis*, hypothesized to function as a warning signal to other potential usurpers of the presence of a hyperaggressive parasite female (Cervo et al., 1990), may be selected against in nonparasitic wasps where it would make the nest more conspicuous to predators and parasites.

Spiders   Populations of the spider genus *Portia* (Araneae: Salticidae) contain both web makers and cursorial (terrestrial) predators in varying proportions (Jackson and Hallas, 1986). *P. fimbriata* is sometimes a terrestrial predator of salticid spiders and sometimes invades alien webs to feed on their builders. It uses special vibratory movements to lure spiders toward them, a kind of "aggressive mimicry" of the vibrations made by entrapped insects. In Queensland, where terrestrial salticid spiders are especially common, *P. fimbriata* has a specialized "cryptic stalking" behavior that enables them to capture the highly visual salticids without causing them to flee. In a habitat of the Northern Territory of Australia where terrestrial prey are scarce and web-building spiders are abundant, *P. fibriata* lacks this specialized stalking behavior. When Queensland and Northern territory *P. fimbriata* were tested for capture efficiency in controlled laboratory observations, Queensland spiders were far more efficient at capturing salticids than were those from the Northern territory, indicating that the high frequency of expression of the cryptic stalking behavior has led to evolved improvement in its efficiency. Spiders from both regions employed vibratory signals and were effective at capturing spider prey on webs (Jackson and Hallas, 1986). These data are consistent with the idea that the degree of evolved specialization and effectiveness of a behavioral alternative depends in part on the frequency with which it is employed, a correlation fundamental to the release hypothesis.

The data on *Portia* document a pattern noted elsewhere in the evolution of specialized predation, namely, that the evolution of specialized mechanisms for dealing with dangerous prey is often associated with specialization on those prey (Horner

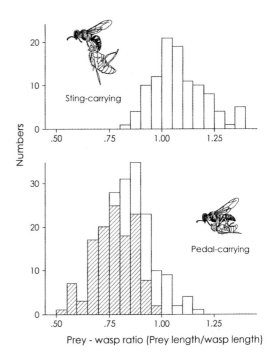

Fig. 21.5. Alternative prey-carriage behaviors in the wasp *Oxybelus strandi* (Sphecidae). Sting-carrying (top) occurs with relatively large prey, pedal-carrying (bottom) with relatively small prey. Shaded bars: direct returns without temporary prey deposition at nest entrance. After Tanaka (1985).

et al., 1965; Li and Jackson, 1996). Most salticid spiders catch relatively innocuous insects, but some species catch ants and spiders, which are dangerous prey. Ants can sting and bite and attack predators in large numbers, and of course spiders are themselves highly effective killers capable of rapid attack, silk-wrapping of victims, and a deadly bite. Each of nine species of myrmicophagic salticids uses an ant-specific behavior for catching ants and a different behavior for catching other insects. All nine species with the ant-specific behavior also showed a preference for ants when offered a choice of prey in three different kinds of tests (Li and Jackson, 1996). The arenophagic salticids *P. fimbriata* and *P. labiata*, which have specialized spider-catching behaviors, as already described, have a pronounced preference for spider prey, and those specialized to prey on other salticids prefer them (Li and Jackson, 1996).

Why should there be a pattern of increased preference with increased specialization to dangerous prey? Several explanations are possible, and one or

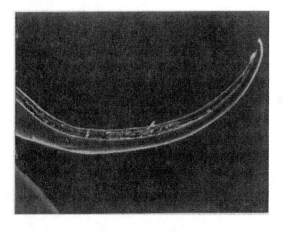
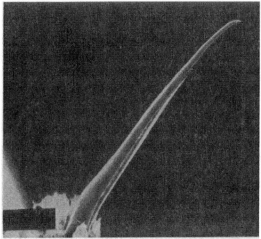
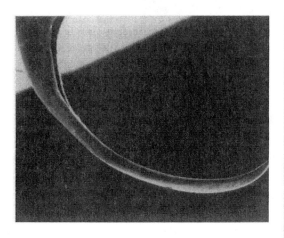
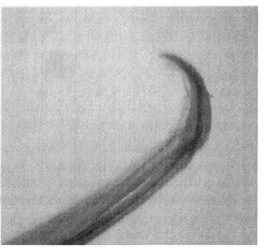

Fig. 21.6. Exaggeration of sting modification in obligate social parasites in wasps (Vespidae, Vespinae). Top left: the highly modified (curved, bent-tipped) sting shaft of the social parasite *Vespula austriaca*. Top right: the unmodified (straight) sting shaft of its nonparasitic host *V. acadica*. Bottom left: the curved shaft of the social parasite *Dolichovespula arctica*. Bottom right: the slightly modified (bent-tipped) sting shaft of the facultative social parasite *V. squamosa*. After Reed and Akre (1982), with permission of Psyche. Courtesy of H. C. Reed.

more may obtain to produce the association. (1) Selective feedback between regulation and form is expected in the evolution of specialized traits (see chapter 7): the more efficient a trait, the more frequently it should be performed relative to alternatives. Thus, selection for improvement leads to selection to lower the threshold for performance, and more frequent performance leads to more frequent exposure to selection and accelerated improvement of the trait. (2) Dangerous prey may be less wary and otherwise not as well defended once their dangerous qualities are overcome, for they are usually relatively immune to predation and therefore have relatively few kinds of evolved defenses. (3) Relative immunity to predation means that there is little competition among predators for dangerous prey. In effect, once dangerous prey can be handled, they are a trophic bonanza. But the correlation between handling and specialization on dangerous prey is expected to obtain only for prey that are locally abundant and a dependable source of food. Whatever the explanation or combination of explanations, selective feedback in the evolution of regulation and form would contribute to develop-

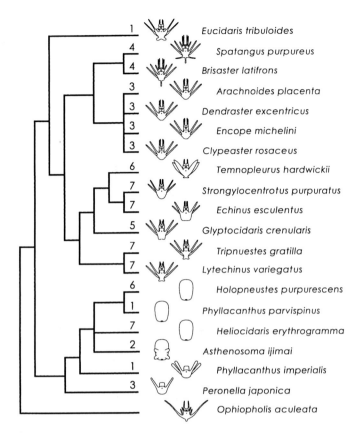

Fig. 21.7. Echinoid phylogeny based on larval morphology. Numbers indicate Order as in figure 21.8. One major clade (above) comprises feeding pluteus larvae with long arms; another major clade (below) comprises armless nonfeeding larvae, and facultative feeders and nonfeeders. After Wray (1996). Copyright 1996 from "Parallel evolution of nonfeeding larvae in echinoids," by G. A. Wray. Reproduced by permission of Taylor & Francis, Inc. Routledge, Inc., http//www.routledge-ny.com. Courtesy of G. A. Wray. Numbers added.

mental character release in populations with increasing preference or performance of one alternative and an approach to phenotype fixation.

Larval Echinoids   In most echinoids (sea urchins, starfish, and their relatives) larvae are of one of two types, either free living and feeding on plankton (plankotrophic) or direct developing and nonfeeding. A few species have facultatively planktotrophic larvae (Wray, 1996). A phylogenetic analysis based on larval morphology shows that nonfeeding and feeding larva comprise two morphologically distinct groups (figure 21.7) that are evolving independently of the adult morphology on which the "true" phylogeny of the echinoids is usually based (figure 21.8).

If the facultatively feeding larvae are included in a phenogram of larvae, they prove to be morphologically similar to the feeding ones (see figure 21.9a). When larval morphology is optimized on a phylogeny of adults like that of figure 21.8, the branches leading to nonfeeding larvae are very long relative to those where there has been no change in mode of larval feeding, and branches leading to facultatively feeding larvae are similar in length to those of feeding larvae, indicating little divergence in their morphology (Figure 21.9b). This accords with the prediction of the release hypothesis, that fixation of a novel alternative (here, nonfeeding in larvae) is associated with increased morphological specialization (measured here as a significant increase in morphological distance on a phenogram; Wray, 1996).

The Wray phenogram is a technique that could be applied to other taxa with morph-ratio variation among populations or species, as a quanti-

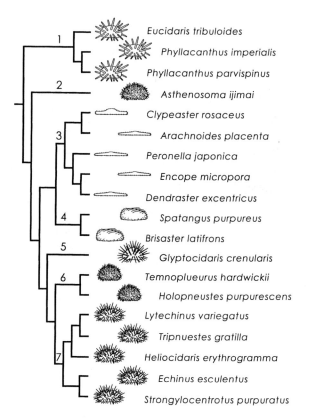

Fig. 21.8. Echinoid phylogeny based on adult morphology, with representative species names. Numbers indicate Order: (1) Cidaroida, (2) Echinothurioida, (3) Clypeasteroida (sand dollars), (4) Spatangoida, (5) Phymosomatoida, (6) Temnopleuroida, (7) Echinoida (sea urchins). After Wray (1996). Copyright 1996 from "Parallel evolution of nonfeeding larvae in echinoids," by G. A. Wray. Phylogeny and drawings reproduced by permission of Taylor & Francis, Inc./Routledge, Inc., http//www.routledge-ny.com. Courtesy of G. A. Wray. Numbers added.

tative indicator of developmental character release.

Sticklebacks Threespine sticklebacks (*Gasterosteus aculeatus*) have three discontinuous forms of adult lateral-plate morphology (Bell, 1981; figure 21.10):

1. The "complete" morph (figure 21.10c), in which a continuous row of lateral plates extends from behind the head onto the caudal peduncle.
2. The "partial" morph (figure 21.10b), in which the abdominal lateral plates end anterior to an unplated area that is followed by a short row of plates on the caudal peduncle.
3. The "low" morph (figure 21.10a), in which the abdominal plates are absent on the caudal peduncle as is the caudal keel which is

present in the other two morphs. Some extreme forms of the low morph are plateless (Bell, 1976, 1987).

Although these three forms are highly variable and could be said to intergrade with each other, they are treated as alternative morphs because of strong associations with alternative habitats and behaviors, and because the members of a population can usually be classified as belonging to one of the three morphs.

Plate morphs vary locally and geographically in both their frequency and degree of modification (see reviews in Bell, 1976, 1984; Bell and Foster, 1994b). Anadromous (marine, with freshwater spawning) populations of the Pacific coast of North America are virtually monomorphic for the complete morph, a situation that probably has been stable since the Pliocene (Bell, 1981) and is considered

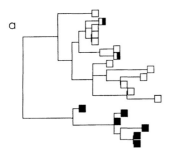

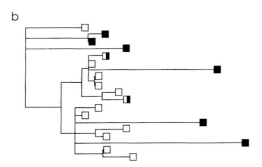

Fig. 21.9. Phenotype fixation and accelerated morphological change in echinoid larvae. (a) A phylogram with degree of changes in larval morphology optimized on the "false" tree based on larval morphology (see figure 21.7). The longest branch is that supporting the nonfeeding larvae (dark squares). Facultatively feeding larvae (half-dark squares) fall within the group of feeding-larvae (open squares). (b) Optimization of the same larval data set over the "true" phylogeny based on adult morphology (see figure 21.8). Branches with fixation of nonfeeding larvae (dark squares) are all long relative to those with facultatively feeding larvae or feeding larvae. Such exaggerated divergence of fixed relative to facultatively expressed phenotypes is consistent with the character-release hypothesis. After Wray (1996).

ancestral relative to the variants found in fresh water (Walker and Bell, 2000). Freshwater populations, by contrast, may be trimorphic, dimorphic for any two of the three morphs, or monomorphic for any one of the three morphs (reviewed in Bell, 1984).

Plate number correlates with local predation regime (reviewed in Bell, 1984; Reimchen, 1994), and morph specialization occurs against a background of geographically fluctuating morph ratios. All lake populations are recently derived from marine ancestors (Walker and Bell, 2000). In lakes where the complete morph is absent and the low morph is common or fixed, plate, spine, and pelvic morphology is most reduced, sometimes even beyond the range of the low plate morph of figure 21.10a. In five samples from a single creek drainage in California, the plate number of the complete morph was higher in populations monomorphic for that morph than in polymorphic populations (Bell, 1982, cited in Bell, 1984, p. 456). That is, the armor is more exaggerated in the direction of the complete-morph specialization when that morph is fixed, as expected under the release hypothesis.

Patterns of geographic variation in plate number observed in European *G. aculeatus* are compatible with the expectations of the release hypothesis. In Europe, northern populations are monomorphic for the complete morph, which extends from the White Sea south to the Atlantic and North Sea coasts of Sweden and central Britain to about 55°N, and along the coast of Iceland and in the southern Baltic and Black Seas. Trimorphic marine populations occupy the more southern portion of the North Sea, and the frequency of the complete morph declines toward the south along the coast of western Europe (reviewed by Bell, 1984). In keeping with the prediction of the release hypothesis, there is a corresponding decrease in the number of lateral plates in the complete morph from north to south in Europe (see references in Bell, 1984, p. 455). Bell (1984), unable to see an obvious explanation for this, tentatively speculated that the pattern may be an unselected developmental response to change in temperature. The release hypothesis offers an evolutionary explanation: plate-morph number is known to be subject to selection even over short geographic and time scales (Bell, 1984), and this, combined with the expected change in plate count associated with change in plate-morph frequency, could explain the clinal change in number of lateral plates.

Variance in plate number has been found to be greater in partial morphs when that morph is at low frequency (presumably, under weaker selection; reviewed in Bell, 1984, p. 457). The association between variance and low frequency of expression at first seems contradicted by the finding that populations monomorphic (at high frequency) for the low-plate morph have large variance in plate count compared with polymorphic populations (Bell, 1976, after Hagen, 1973). But there is evidence that selection on plate morphology likewise may be different in streams, for a different reason: stream sticklebacks are less heavily armored presumably in response to low vertebrate predation and increased insect predation (Reimchen, 1994).

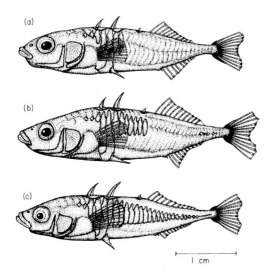

(a)

(b)

(c)

| 1 cm

Fig. 21.10. Lateral plate morphs of the stickleback *Gasterosteus aculeatus* from a single population (Sports Lake, Alaska): (a) low; (b) partial; (c) complete. From Bell and Foster (1994b), in *The Evolutionary Biology of the Threespine Stickleback*, by permission of Oxford University Press. Courtesy of M. A. Bell.

The Laysan Finch Rapid morphological specialization following behavioral fixation is perhaps best documented in a vertebrate, the Laysan finch (*Telespyza cantans*). This bird is omnivorous on Laysan, where it feeds on insects, carrion, birds' eggs, and a variety of seeds. After only 20 years of isolation on islands, 78% of the birds specialized on a narrower diet of large, hard *Tribulus* seeds, which comprise only about 4% of the diet of the known ancestral population on Laysan. The new specialization was associated with a measurable increase in bill depth. Founder effects were considered and eliminated as a cause of the divergence in behavior and bill depth (Pimm, 1988).

Geographic and temporal variation in intensity of selection is expected to cause variation in rates of evolution independent of change in the ratios of alternatives (i.e., selection may cause variation in rates of evolution even if morph ratios are stable). And reproductive isolation and consequent gene-pool distinctiveness can also contribute to divergence, independent of any effect of character release. In divergent, reproductively isolated populations, the past contribution of these other factors may be difficult to distinguish from that of release. But when traits expressed in more than one morph of a formerly polymorphic population repeatedly

show exaggeration in the expected direction when a particular morph is fixed or at high frequency, as in sticklebacks, then it is reasonable to hypothesize that developmental character release has played some role.

Signs of phenotype fixation and character release have not been sought systematically in research on divergence and speciation. In the future, it would be desirable to design further ways to document the role of developmental character release in evolutionary transitions, and to distinguish between divergence before and after morph fixation, for example, by study of change in clines.

Buttercups Heterophyllic buttercups *(Ranunculus flammula)*, like many other aquatic plants, have two types of leaves (see figure 7.6), an aerial leaf form that is relatively broad and lanceolate and an aquatic leaf form that is relatively linear and facilitates gas exchange and persistence in the face of the mechanical action of turbulent water (Cook and Johnson, 1968). Individual plants in lakes with seasonally fluctuating water levels regularly produce both leaf forms as a facultative response to conditions. But populations from wet meadows contain only individuals with aerial leaves, and those under constant immersion in lakes contain individuals with exclusively aquatic leaves. Under environmental conditions where only one morph is expressed for many generations, aquatic and terrestrial populations show some morphological specialization in form, toward narrower and broader leaves, respectively. Hybrids between such populations develop intermediate morphology, and in common garden experiments, survival is reduced when a form is reared in the conditions that characterize the other phenotype, indicating a genetic, selectively favored basis for the increased specialization in form (Cook and Johnson, 1968). In buttercups, phenotypes originally diverged entirely because of differences in gene *expression* in populations where all individuals were part of the same gene pool. Subsequently, their phenotypic divergence was exaggerated due to genetic divergence in separate populations in which different alternatives are fixed. The exaggeration is in the same traits—leaf width and survival ability in different exposures to terrestrial or aquatic habitats—that characterize differences between alternative leaf forms. This suggests that in heterophyllic populations, specialization in both directions is checked when both leaf forms are expressed by the same genotype, and that these characters are released for increased divergence when one of the alternatives is fixed.

An Experimental Demonstration in *Drosophila* In an experiment using *Drosophila melanogaster*, Rice (1998) dramatically reduced genetic correlations between the sexes by causing 99% of the haploid set of genes to be transmitted from father to son as a giant synthetic Y chromosome composed of chromosomes X, II, and III, excluding the small dot chromosome IV. Most of the genome was thereby released from counterselection due to expression in females. The result, after 41 generations of breeding with a particular "clone-generator" line of females, was a dramatic increase in male fitness, as predicted by the release hypothesis, and a sharp drop in the fitness of females mated to males from the experimental line. Male fitness components measured were mating speed, remating rate, and rate of preferential use of the male's sperm by multiply mated females.

Genetic Assimilation Revisited

The phenomenon of fixation and release draws attention to the potential importance of genetic assimilation in evolution. Genetic assimilation, or trait fixation due to genetic change in regulation, is among the most maligned and poorly understood concepts in evolutionary biology, and yet it has probably been involved in the evolution of most if not all traits that go through an alternative-phenotype phase of evolution to become obligatory or constitutive traits of populations. Blind prejudice against genetic assimilation rivals similar past prejudices against group selection, macromutation, and sympatric speciation, subjects almost taboo during certain eras in mainstream evolutionary biology. Yet genetic assimilation is simply genetic accommodation carried to an extreme, and probably most adaptive evolution involves genetic accommodation of regulation and form (see chapter 6). If the objections to genetic assimilation cannot be overcome, there is little hope for establishing the links between environmentally influenced development and genetically mediated evolution.

Phenotypic fixation need not involve genetic assimilation, for fixation can occur due to altered frequency of induction by environmental factors. The terrestrial form of a water buttercup (*Ranunculus*), for example, can be fixed in populations where there is a lack of water, and this can occur without genetic change due to the facultative response of the plants (Cook and Johnson, 1968). So the mere observation of a monomorphic population derived from a polymorphic one does not imply that genetic assimilation has occurred. But, given the

likely commonness of environmentally induced novelty (see chapter 26), and the fact that all adaptations, whether environmentally or mutationally induced, become established via genetic accommodation of regulation (see chapter 9), most *evolved* fixation of adaptive traits must involve quantitative genetic change in regulation, or genetic assimilation.

Opinions about genetic assimilation and the Baldwin effect, with which it is often synonymized (e.g., Simpson, 1953a), could serve as indices of the status, high or low, granted development and plasticity as factors in evolution. With scattered exceptions (e.g., Cushing, 1941, 1944; Cushing and Ramsey, 1949; Huxley, 1942; Grant, 1963; Evans, 1966; Shapiro, 1976; Matsuda, 1982, 1987; Tierney, 1986; Wcislo, 1989), genetic assimilation and the Baldwin effect have been considered exceptional processes, at best "subsidiary factors" (Simpson, 1953b) in evolution. Simpson took pains to show how the Baldwin effect is completely consistent with the genetical theory of natural selection. Then, rather than enthusiastically endorsing the reasonableness and importance of the Baldwin effect, Simpson dismissed it as not worthy of special attention! Dobzhansky (1970, p. 211) accomplished the same dismissal with a polygenic explanation in less than a paragraph. Mayr (1963) considered that genetic assimilation is an example of "the normal process of selection of a polygenic character and requires no special terminology" (p. 612). Genetic assimilation and the Baldwin effect became like the native Fuegians imported to Victorian England by Captain Fitzroy (see Darwin, 1845): once dressed in normal clothes, the most profound differences in character could be overlooked. The naked implications of environmental influence, dressed in conventional genetic terms, were virtually lost from view. Even today, many evolutionary biologists probably share the view of Wake and Wake (1988) that "[g]enetic assimilation has been demonstrated experimentally, but it has never been seen to be a dominant evolutionary force" (p. 238).

Waddington's (1953b) experiments on genetic assimilation may have captured the imaginations of biologists had they involved a threshold trait such as horn production in beetles, whose ratio of expression is environmentally influenced, whose adaptive significance is known, and whose expression responds to artificial selection (Emlen, 1996). This would have clarified how selection for a trait changes the genetically influenced threshold for its production, and thereby increases trait frequency in a population. Such an example would have helped explain how the response to selection on environmentally induced traits such as crossveinless, extra wing veins, and thoracic anomalies (bithorax) in *Drosophila* can relate to evolutionary transitions in discrete adaptive traits.

There is abundant evidence that phenotype fixation by genetic assimilation is feasible. Some of this evidence is reviewed in chapter 6 (on genetic accommodation). The interchangeability of genetic and environmental determinants of a switch (see chapter 5) is further support for the feasibility of genetic assimilation, which amounts to replacement of environmental influence with genotypic influence on a switch. Beyond this evidence of feasibility, how can one show that fixation by genetic assimilation actually occurs? Contrary to the conclusion of Simpson (1953b), it is quite easy to find evidence of genetic assimilation in nature. One need only find instances where an alternative phenotype reaches fixation in some populations, along with the environmental circumstances that favor it, and where there is evidence that this has involved genetic change.

Buttercups express only the aerial leaf form in constantly dry areas, and rearing experiments have shown that regulatory fixation is in some cases genetically assimilated because the alternative, aquatic leaf form can no longer be induced (Cook and Johnson, 1968). The same has been shown in experiments with seasonal polyphenisms of butterflies (Shapiro, 1976), and in the autogenous versus blood-feeding life-history form of pitcher-plant mosquitoes (reviewed in West-Eberhard, 1986), as discussed above. Other examples are suggested by phylogenetic study of recurrent phenotypes (see chapter 19). These studies provide evidence of polarity, that is, that there has been a change from polymorphism to monomorphism (fixation). Beyond this, support for genetic assimilation requires evidence that a genetic change is involved. This can be documented by using common garden or hybridization experiments (Cook and Johnson, 1968; Shapiro, 1976).

Other examples of genetic assimilation involve traits whose regulation is mediated by manipulation, or alteration of a flexible phenotype by the activities of another individual of the same or a different species. The evolution of phenotypic manipulation may involve selection on either the manipulator or the manipulated individual, or both, and it may engender conflict between them, as in conflict between parent and offspring over parental investment in young (Trivers, 1974). An example of evolved phenotypic manipulation is parental manipulation of offspring phenotypes (Alexander, 1974), especially maternal manipulation of embryonic development (chapter 5), maternal habitat selection (Wade, 1998), and other populationwide maternal effects (Mousseau and Fox, 1998; see chapter 5). The species-specific nature of early metazoan embryology—the characteristic behavior of maternally supplied cytoplasm, manifested after fertilization but prior to the expression of the zygotic genome (see chapter 5)—may be an example of genetic assimilation via the evolution of maternal manipulation, where maternal inputs to the structure of an egg stabilize a highly condition-sensitive developmental trajectory that otherwise may vary because of development in an unpredictably changing external environment. In an odd twist on the manipulation and genetic assimilation theme, manipulation by maternal colonies to standardize larval feeding regimes evidently has contributed to the genetic assimilation of caste determination in stingless bees (*Melipona species*; see chapter 5). Genetic assimilation via maternal effects can occur even though the regulation-affecting genes are expressed in a different (adult) stage from their ultimate phenotypic effects on offspring.

Conclusions

Extensive divergence between alternative phenotypes within populations challenges the idea that "speciation is orders of magnitude more important than phyletic evolution as a mode of evolutionary change" (Gould and Eldredge, 1977, p. 116). This does not mean that all divergence is via alternative phenotypes. Nor does it detract from the well-established fact that reproductive isolation (speciation) promotes evolutionary divergence. But it does mean that we can confidently negate the idea that there is "no such thing as evolution without speciation" (Eldredge, 1976). Intraspecific divergence between alternatives must be ranked along with speciation as an important setting for major innovation in the history of life.

22

Maintenance without Equilibrium

The maintenance of alternative phenotypes has been treated as an equilibrium problem analogous to that of explaining the maintenance of genetic polymorphisms. But the applicability of a model depends in part on the mode of regulation it assumes. Some widely discussed models assume regulatory mechanisms that are rare or unknown in nature. Most putative examples of genotype-specific complex alternatives have proven to have condition-sensitive regulation. Equal-payoff genetic models become applicable to conditional alternatives if theoretical equilibrium points are interpreted as advantageous switching points. Conditional alternatives have properties that promote long-term stability without equal fitness payoffs, and switching between alternatives may contribute to the still-enigmatic maintenance of genetic polymorphisms.

Introduction

The old question of the maintenance of alternative phenotypes appears in a new light when seen from a developmental point of view. In the past, alternatives were treated as genetic polymorphisms and explained in terms of genetic equilibria. During the synthesis of genetics and evolutionary theory in the early twentieth century, single-locus and linkage-group genetic polymorphisms were seen as maintained in three different ways: by heterosis (heterozygote advantage), by frequency-dependent selection, and by balancing selection in a spatially heterogeneous environment (summarized in Dobzhansky, 1970). Later models showed how alternative behavioral and morphological tactics in theory could be maintained as "evolutionary stable strategies" (ESSs; see summary and history in Parker, 1984a). ESS models treat the maintenance of alternatives not as a problem in genetics but as a problem in game theory.

This chapter views the maintenance of alternative phenotypes as a problem in the evolution of regulation, where the genes responsible for the maintenance of alternatives influence regulation at condition-sensitive, polygenic developmental decision points whose thresholds can be adjusted upward and downward under selection. These adjustments affect the ratio of alternatives observed.

Regulation may sometimes be determined by a small number of alleles, but even then, insofar as most readily observed alternative phenotypes are concerned, they are alleles that influence a polygenic switch (see chapter 6). Rephrased in developmental terms, the question of maintenance of alternative phenotypes, whether genotype specific or conditional, becomes: What selective factors can influence the evolution of regulatory mechanisms so as to preserve the expression of two or more alternative traits rather than just one?

Along with the rest of this book, this chapter focuses on the importance of developmental plasticity for evolutionary interpretations. It describes three types of regulation that have been assumed by theoretical models, and argues that two of them—genotype-specific and stochastic regulation—are relatively uncommon in nature, even though common in theoretical research. I then discuss reasons for the superior stability of the third, conditional type of regulation and argue that the properties of complex adaptive alternative phenotypes may help explain the maintenance of genetic polymorphisms. This chapter is not a review of models. Rather, it is a commentary on the uses and misuses of certain kinds of models and on selected aspects of the evolutionary maintenance of alternative phenotypes that especially relate to developmental plasticity.

Matching Models to
Modes of Regulation

Modes of regulation hypothesized to account for switches between alternatives fall into three categories: genotype-specific regulation, stochastic regulation, and condition-sensitive regulation. In *genotype-specific regulation* (sometimes called genetic determination or genetic polymorphism), the alternative expressed depends on which allele among a small number of alleles of major effect on a switch is present at one or a few loci. Determination is irreversible (an individual genotype is capable of only one option). In *condition-sensitive* or *conditional regulation*, the alternative adopted by a given individual or by a particular individual at a given time depends primarily on conditions rather than primarily on genotype. Conditional regulation can be either reversible, as in many behavioral or physiological alternatives, or irreversible, as in polyphenisms. In *stochastic regulation*, the alternative adopted is governed by stochastic environmental or genetic processes in such a way that the alternatives yield equal mean payoffs in an unpredictably varying selective environment over the long run. Bet hedging by producing eggs that emerge at different times of year is a kind of alternative sometimes thought to involve stochastic regulation.

For a model to apply to a particular organism or class of alternative phenotypes, the mode of regulation assumed by the model and that observed in the organism have to coincide. This is not a trivial consideration for the ultimate, evolutionary explanation, for it is the regulatory mechanism that governs the ratio of alternatives and their maintenance in a population. If the terms of a model cannot possibly apply to the mechanism observed, or if the mechanism required by the model is not operative in the observed switch between alternatives, then some other adaptive explanation must be sought for the maintenance of the alternatives observed. In the application of models for the maintenance of alternatives to real organisms, the nature of the switch mechanism is an important part of the data and is essential to the evolutionary explanation. It is possible to ignore mechanisms in some kinds of research on adaptation and natural selection, but this is not one of them.

Correct identification of the mode of regulation can cause theoretical difficulties to dissolve. The genetical (kin-selection) theory for the evolution of social behavior applies only if genes for beneficent behavior are already present in the population (Hamilton, 1964a, p. 14), for the principle does not work unless beneficiaries and donors have a sufficiently high probability of sharing the alleles that favor social aid. The problem of initial spread is solved with reference to a developmental view of evolutionary transitions. The origin of worker behavior in the social Hymenoptera or in birds, for example, can occur in group-living species when some individuals aggressively prevent oviposition by others (Brown and Brown, 1980, 1990; see chapter 11). If the suppressed individuals retain brood-care behaviors even though their own reproduction is suppressed, then all would possess the polygenically variable responses that give rise to kin-benefiting aid. If their nestmates are relatives, all of the genetic conditions necessary for Hamilton's model to work obtain prior to the advent of a dominance-induced division of labor. The preexisting responses would presumably be refined and adjusted under selection for advantageous thresholds of switching between donor and beneficiary social roles. Thus, a developmental view of this evolutionary transition helps to solve what is otherwise a problematic aspect of the theory.

Genotype-Specific Regulation

An equilibrium model for the maintenance of genetic polymorphism is probably the most frequently reinvented wheel in theoretical evolutionary biology. In an early review, Maynard Smith (1970) listed six models proposed between 1953 and 1970. During the 1970s and 1980s, there was a great proliferation of genetic-polymorphism models, perhaps due in part to the interest in explaining the profusion of genetic polymorphisms uncovered by allozyme data (summarized in Lewontin, 1974). I easily garnered a list of 30 models proposed between 1970 and 1986 (for partial reviews, see Grant and Grant, 1989; Hedrick et al., 1976; Hedrick, 1986; Seger and Brockmann, 1987; Futuyma and Moreno, 1988). All of these genetic-polymorphism models, with the exception of those that depend upon heterosis (heterozygote superiority), are equal-payoff models. That is, they predict that polymorphisms will evolve toward an equilibrium state where the average fitness payoffs for the alternative phenotypes are equal.

Equal-payoff genetic-polymorphism models fall into several general categories in terms of the conditions that give rise to polymorphism, for example, adaptation to multiple niches (e.g., Levene, 1953; Maynard Smith, 1966); frequency-dependent advantage (e.g., Mather, 1973; O'Donald, 1980); density-dependent advantage (Clarke, 1972); competition-dependent advantage (Mather, 1969);

adaptation in spatially heterogeneous environments (e.g., Levins, 1963, 1968); and adaptation in cyclic, temporally heterogeneous environments (e.g., Haldane and Jayakar, 1963).

The idea that polymorphisms must have balanced fitness payoffs is usually attributed to Fisher (1958), although Haldane (1930) has also been cited as the first to invent such a model (Seger and Brockmann, 1987), with a frequency-dependent model by Wright and Dobzhansky (1946) not far behind. Subsequent major influential landmarks are the two-page paper on multiple-niche polymorphisms by Levene (1953) and the comprehensive book by Levins (1968). But equal-payoff maintenance of polymorphism is a concept that dates back to Darwin, who explained male dimorphisms in the amphipod crustacean *Orchestia* with a verbal model of a fitness equilibrium: "[T]he two male forms probably originated by some [individuals] having varied in one way and some in another; both forms having derived certain special, but nearly equal advantages, from their differently shaped organs" (Darwin, 1871 [1874], p. 619). For these equal-payoff models of maintenance of genetic polymorphism to apply directly to complex alternative phenotypes, the phenotypes must have genotype-specific regulation, that is, regulation in which the alternative expressed depends on the alleles present at one or a few genetic loci.

Some game-theory ESS models for the maintenance of alternatives, such as models of the mixed ESS, also require alternatives with genotype-specific regulation. In a *mixed ESS*, alternative phenotypes are expressed with fixed probabilities (e.g., in condition c, adopt phenotype A with probability P_A, adopt phenotype B with probability P_B), and the probabilities evolve toward equilibrium values where the fitness effects of the alternatives are equal (see Parker, 1984a). There are two regulatory types of mixed ESS: stochastic and genotype-specific. In the *stochastic mixed ESS*, discussed further below, the probabilities for each alternative are set by selection, and each individual's phenotype, or each individual performance of a phenotype, is assigned by some developmental randomizing process. In a *genotype-specific mixed ESS*, the population is genetically polymorphic, and ESS individual-strategy frequencies correspond to genotype frequencies. In a genotype-specific mixed ESS, an individual adopts a lifetime irreversible alternative phenotype that corresponds to its genotype, and equal payoffs for different alternatives are expected.

The term "mixed ESS" can potentially be confused with the term "mixed strategy" used briefly by Trivers (1972) and followed by some other au-

thors (e.g., Lyon, 1993) to describe reversible conditional alternative reproductive tactics of males and females, such as opportunistic brood parasitism combined with independent nesting. Pfennig (1992b, p. 1417) calls these "environmentally determined mixed ESSs," but there is a potential for confusion because, in the game theory ESS literature, conditional behaviors are classified as pure strategies (a single phenotype per condition), not mixed strategies (more than one phenotype per condition; Maynard Smith, 1982; Parker, 1984a).

Genetic and game-theory models often have been inspired by observations of complex alternative phenotypes, such as beetle-horn polymorphisms (Gadgil, 1972) or insect wing polymorphisms (Hedrick et al., 1976) now known to have a strong condition sensitivity. By treating a phenotype as a gene or a tactic, the properties of complex alternatives (e.g., adaptation to environmental heterogeneity or multiple niches, and frequency-dependent advantage) have been translated into plausible explanations of stable alternative alleles. Sometimes this translation has created the mistaken expectation that flexible phenotypes will behave like genes and conform to the same model predictions, or a belief that long-term maintenance depends upon genotypic control. I have seen a complex set of alternatives described as genetically controlled and, hence, potentially maintained by selection, as if a polyphenism with an environmentally controlled switch could not possibly persist. In fact, conditional alternatives may be among the most stable of the three classes of regulation, for reasons discussed below.

The landmark discussion of multiple-niche genetic polymorphisms by Levene (1953) ended with a prophetic statement: "The model here proposed is obviously not realistic; however, if it is modified by supposing that individuals move preferentially to niches they are better fitted for . . . conditions will be more favorable for equilibrium" (p. 332–333). Here, Levene anticipates more recent work on the phenotype-dependent, conditional alternatives that are likely important in evolution under sexual selection, ecological competition, and predation (e.g., see especially chapters 27 and 28). Levene's idea of relaxing the conditions for genetic polymorphism by introduction of condition sensitivity was formally developed by Levins (1963; see also Lively, 1999).

Conditional alternatives do not require equal payoffs to be maintained in a population, as discussed below. Therefore, evidence for conditional regulation can be used to eliminate a large class of (equal-payoff) genetic and genotype-specific models from consideration as explanations of particular cases. But not all of them can be eliminated, for some equal-payoff explanations, such as the ideal free distribution (Fretwell and Lucas, 1970), apply to conditional alternatives, and the equilibrium point of an equal-payoff model can be interpreted as the switch point for conditional alternatives (see below). Demonstrations of equal payoffs are sometimes used to support equal-payoff models that specify or tacitly assume genotype-specific or stochastic regulation (e.g., Brockmann and Dawkins, 1979; Gross, 1985; Ryan et al., 1992; Shuster and Wade, 1991a), but this should be considered weak support. Conditional alternatives can yield equal payoffs without satisfying the regulatory requirements of equal-payoff models, for example, if they represent an ideal free distribution or are observed in conditions near the switch point between alternatives where their fitness payoffs approach equality (below).

Demonstrations of Genotype-Specific Regulation
Given the condition sensitivity of all phenotypic development, the developmental ease of interchangeability of genetic and environmental influence (see chapter 5), and the advantage of adaptively appropriate conditional rather than genetic determination if an adequate environmental cue can be used, one would expect genotype-specific determination that is immune to environmental effects to be unusual. Numerous well-documented cases of genotype-specific alternative phenotypes are known (Hedrick, 1986). Some of them are relatively simple phenotypes, such as the color morphs of *Cepea* snails (Ford, 1971); the color morphs of butterflies, moths, and flowers; and leaf form in plants (e.g., Ford, 1964; Kettlewell, 1961).

Seemingly simple genotype-specific alternatives sometimes, of course, may be more complex than meets the eye. A color polymorphism can be developmentally linked to complementary behaviors, such as different behavioral tactics, as in the trimorphic color forms of side-blotched lizards (*Uta stansburiana*; Sinervo and Lively, 1996), where heritability of color form in nature is very high ($h^2 = 0.96$) and is associated with morph-specific behaviors. Habitat selection or philopatry can match color to background, as in some walking sticks described below (Sandoval, 1994a). In lazuli buntings (*Passerina amoena*), plumage variation puts dull-colored yearling males at a disadvantage in maintaining territories, but they adopt a behavioral alternative tactic to obtain good territories by settling near aggressive older males that are less aggressive toward them than toward brightly colored, more adultlike yearling males (Greene et al., 2000). In all of these examples, a relatively simple trait, body color, becomes developmentally linked to complexly divergent alternatives.

In vertebrates and other organisms where learning is involved in the development of many advantageous traits, relatively simple morphological alternatives, such as the genotype-specific beak polymorphism of African finches (*Pyrenestes*; Smith, 1987), indirectly can foster complex differences in foraging and feeding behavior due to beak-dependent differences in the rewards for using different types of seeds (see chapters 16 and 28). Clearly, genotype can determine alternatives without the expression of alternatives being genotypically specified. Genotypic influence on development must often act in this way, playing a decisive role in determination while the details of expression are heavily environmentally mediated. Genotype, for example, may influence body size or vigor decisive in social contests, but then the social environment of competitors, rather than the genotype, would mediate access to resources. For this reason, it is more instructive to think in terms of genotypic influence on particular turning points in development, rather than genetic programming of the unfolding of development.

A major well-studied class of genotype-specific regulation of complex polymorphisms is sex determination, which, when genotype specific, is also under the control of one or a few linked genes (Bull, 1983). Is there evidence for genotype-specific determination of other complex alternative phenotypes? If such regulation is not very common, then genotype-specific models for the maintenance of complex alternatives are correspondingly unimportant relative to other kinds of evolutionary models.

The answer seems to be that there is some evidence for genotype-specific complex alternatives, but they are relatively rare. Genotype-specific determination is suspected in the dimorphism of fig wasp males on the basis of indirect evidence and theoretical reasons to expect it. Dimorphic males such as those of figure 21.1 have complex alternative phenotypes that are determined in the preadult stage, without known anticipatory cues that would permit conditional development of fighting morphology only when competitors are present or likely to be present (Hamilton, 1979; Cook et al., 1997). The morph ratio in at least one species, *Pseudidarnes minerva*, is said to be consistent with a one-locus, two-allele model of morph determination, but there has been no experimental study of morph determination in any dimorphic fig wasp population (Cook et al., 1997). There is, however,

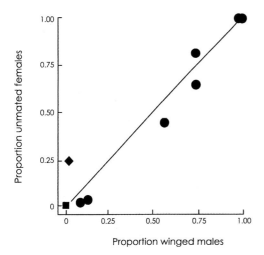

Fig. 22.1. Proportion of unmated females and winged males in different species of fig wasps. Square: seven species with the same value: three studied by Godfray (1988) and four from Hamilton (1979). Diamond: genus *Apocrypta* (from Godfray (1988). Circles: species studied by Hamilton (1979). After Godfray (1988).

good agreement with the prediction (Hamilton, 1979) that the proportion of winged males in a genotype-specific dimorphic species should correlate with the average population abundance of unmated females trapped with the males in developing figs (Godfray, 1988; Cook et al., 1997; figure 22.1). Other examples of alternative male mating tactics for which genotype-specific determination is suspected but not definitely shown are ruffs and a bethylid wasp (Kearns, 1934, ex Cook et al., 1997).

An apparent example of genotype-specific regulation that fits the detailed predictions of a genetic model occurs in a walking stick, *Timema cristinae* (Phasmatodeae: Timemidae), which has cryptically colored striped and unstriped alternative morphs (Sandoval, 1994a,b). Morph frequencies approximate those expected if they are determined by a one-locus, two-allele genetic mechanism. These insects also show variable patchy spatial distributions with "grain" like that discussed by Levene (1953) and Levins (1968) as favorable to the maintenance of genetic polymorphisms, but heretofore virtually untested (Sandoval, 1994a). Individuals are wingless and stay within a variety of different-sized patches of their two distinctively colored host plants, where they are differentially subject to predation due to the color polymorphism. The striped morph is more cryptic on chamise (*Adenostoma fasciculatum*, Rosaceae), whereas the unstriped

morph is cryptic on ceanothus (*Ceanothus spinosus*, Rhamnaceae). The frequency of a given morph is high when the host plant patch on which it was best protected is large, as predicted by the Levene model.

Perhaps the reason why models like that of Levene are rarely tested, despite their theoretical interest, is that the assumptions of genotype-specific regulation and particular environmental parameters are rarely met, and difficult to show, as illustrated in the following section.

False Positives in the Search for Genotypic Control

Several kinds of data are regularly mistaken for genotype-specific determination of alternatives and treated as if environmental influence is absent or negligible.

1. Mendelian ratios of alternative phenotypes under controlled environmental conditions: Since laboratory genetic studies often, if not always, strive to control environmental variation and maintain organisms in good condition and as free of environmental stress as possible, the appearance of Mendelian ratios in such studies may sometimes be an artifact of constant conditions, or the elimination of environmental variables that would importantly influence phenotype determination in nature. A male dimorphism in courtship behavior between independently displaying and satellite ruffs (*Philomachus pugnax*) is among the few alternative mating tactics still considered controlled by a genetic polymorphism (Lank et al., 1995). Available data show genetic influence, but as the authors point out, their captive conditions, which included ad libidum food availability and benign conditions, could have damped the kinds of environmentally mediated variation in size and rearing conditions that commonly influence switches between alternative male tactics. The proportion of captive males adopting satellite behavior did not differ greatly from that observed in nature (Lank et al., 1995). This may indicate genetic influence, but social contests may give consistent proportions of winner and loser tactics whether rank is determined by genotype or environment. Even social insects, where worker and queen determination is strongly influenced by larval diet in all known species, can produce Mendelian ratios of queens and workers in species where laboratorylike uniform feeding conditions have evolved (Kerr, 1950a; see chapter 5).

2. High heritability, or any evidence of significant genotypic *influence* on determination taken to indicate unconditional genetic determination: The terms "genetic polymorphism" or regulation by

"genetic control" (Roff, 1984, p. 34) are sometimes applied to alternatives that are not genotype specific, for example, when data from breeding experiments (Roff, 1984) or a response to artificial selection indicates that genetic influence is involved. Some examples of this type are discussed critically by Gross (1996).

Male mating tactics in swordtail fishes (*Xiphophorus* species; Poeciliidae) illustrate well the distinction between genotypic *influence* and unconditional genotypic *determination* of complex alternatives. In *Xiphophorus*, well-documented genetic influence is often mistaken for genotype-specific determination, and as a result, this genus has been cited as one of the few examples of a genetic switch between alternative mating behaviors (e.g., Shuster and Wade, 1991a; Ryan et al., 1992; Cook et al., 1997).

X. multilineatus (*X. nigrensis* of Zimmerer and Kallman, 1989; Ryan et al., 1992) has four male color forms that correspond to four genotypes at a locus (P) on the Y chromosome that affects pituitary function. When reared in the laboratory, presumably under relatively constant conditions compared with those experienced in nature, the four color forms have broadly overlapping size distributions with different mean values (Zimmerer and Kallman, 1989). The size variation is apparently influenced by effects of the P locus on the age, and therefore the size, at which individuals mature (Morris and Ryan, 1990). Thus, there is evidence for a genotype-specific determination of color and for genotypic influence of the same loci on adult size and age at maturation, which (at least in one species, *X. maculatus*) are clearly associated with genotype even under environmentally variable conditions (see Policansky, 1983).

The sneaker male mating tactic is performed only by the s-allele-bearing morph, which has the smallest mean size. The other three genotypes always court rather than sneak. Phenotype determination in *X. multilineatus* correlates better with color (and evidently hormonal class) than with size, for when equal-sized individuals known by their color to bear different alleles (s vs. I) compete with large males, only the s-bearing males switch occasionally to the sneaking tactic (Kallman, 1984, 1989; Zimmerer and Kallman, 1989; Ryan and Wagner, 1987). These findings mean that behavior is strongly genotypically influenced, but not that it is genotype specific. Particular social conditions (competition with a large male) are always required to induce expression of sneaking behavior. Otherwise, the small males court rather than sneak (Ryan et al., 1992).

Another species, *X. nigrensis*, shows a variation on the same theme where the conditional nature of the behavioral alternatives is clear. It has three rather than four alternative genotypes distinguished by color and size, and alternative mating tactics that correlate better with size than with genotype (Ryan and Causey, 1989, p. 343). All genotypes perform both mating tactics, though at different, size-related frequencies. This indicates the existence of size-dependent conditional alternative mating tactics with strong genetic influence on the conditional factor, size. These findings are consistent with the hypothesis that the *X. nigrensis* behavioral switch has the same basic polygenic regulatory mechanism (including P-locus participation via size effects on competition) as does that of *X. multilineatus*, but that the P-locus genetic polymorphism does not have a major-gene effect on the switch as it does in *X. multilineatus*. The direct effect of the alternative color-associated alleles in *Xiphophorus* species appears to be an effect on size and therefore, secondarily, on the switch between competition-dependent alternatives, as originally suggested by Borowsky (1987b). In sum, research on *Xiphophorus* demonstrates genetic influence on the phenotypic correlates of alternative mating tactics, but it does not demonstrate unconditional genetic determination of alternatives.

3. Confusion of continuous genetic variation in regulation (including genotypic determination of alternatives at the extremes) with genotype-specific switching between forms: Cade (1981) demonstrated heritability of calling time of male crickets (*Gryllus integer*) using artificial selection. Since this species has alternative calling and noncalling tactics, the results have been misinterpreted as evidence of genotypic determination of alternatives (Thornhill and Alcock, 1983 p. 293; Maynard Smith, 1982, pp. 86–88; Ryan et al., 1992, pp. 27–28) even though Cade (1981) explicitly regarded the alternatives as "separate genetic strategies whose expression is conditional on extrinsic variables" (p. 564) and observed that individual males sometimes switch mating tactics (Cade, 1979, pp. 367–369, 374). Presumably, a male with zero calling duration—at the lower extreme of continuous variation in calling duration—may permanently adopt the alternative tactic of noncalling males, which is to seek mates near calling males (satellite behavior). Satellite behavior would be genotypically determined in such a male, but not in the population as a whole, where duration of calling and male tactics are known to be influenced by environmental factors such as population density, male aggressive interactions, calling intensity

of neighbors, and time of night (Cade, 1979), in addition to genotype. Cade (1984) reviews evidence for genetic variation in switching between alternative mating tactics. There is considerable evidence that alternative male mating strategies in these and other crickets is conditional rather than genotype specific (Zuk and Simmons, 1997, pp. 94–95).

4. Geographic variation in morph ratios that persists in common-garden rearing experiments: Sometimes there is a mistaken expectation that a (single) common garden experiment must produce phenotypic convergence between genotypes from contrasting environments in order to demonstrate plasticity. The opposite result—persistence of phenotypic differences despite identical conditions—is then taken to indicate genetic determination (e.g., see Thompson et al., 1993). A common garden experiment can show whether or not a particular environmental difference influences a particular phenotypic difference, but it cannot demonstrate unconditional genetic determination (lack of plasticity). What is needed to reveal plasticity, if it is present, is not a common garden experiment but a series of *multiple contrasting garden* experiments, or rearing of individuals of identical or similar genotype in opposite extremes of a environmental variables hypothesized to possibly affect a switch. Contrasting garden tests may fail to be informative if the environmental factors that influence a response are poorly chosen, but candidate factors often emerge from observations of natural history. Behavioral observations, for example, suggested that the presence of a dominant queen is the primary environmental determinant of caste in certain social wasps, since queen removal caused developing female sibs to switch from the worker to the queen phenotype (Strassmann and Meyer, 1983; West-Eberhard, 1978b). In these insects, rearing in different nutritional conditions may not affect queen determination.

Sometimes the attempt to test the assumption of genotype-specific control leads to the discovery of unexpected complexity in morph determination and to new insights regarding development and evolution. This is true of research on alternative male phenotypes in the marine isopod *Paracerceis sculpta* (Shuster and Wade, 1991a,b) As discussed in chapter 15, on cross-sexual transfer (see figure 15.6). *P. sculpta* has three male morphs that differ in size and behavior. Alpha males are largest and defend harems within sponges using elongated posterior appendages; beta males have femalelike morphology and invade harems by mimicking female behavior; small, gamma males invade harems by sneaking, secretive behavior. Male size, as affected

by number of molts and maturation rate, at first appeared to be determined by a single locus with three alleles (Shuster and Wade, 1991a). Both laboratory crosses and extensive sampling of natural populations showed a good approximation to expected Mendelian ratios of the three morphs, which persist at Hardy-Weinberg frequencies and seem to have equal mating success (Shuster and Wade, 1991a).

As mentioned above, equal-payoff data are relatively weak support for genotype-specific regulation. Shuster and Wade (1991a) based mating-success estimates on paternity experiments using genetic markers and two-male harems, from which they extrapolated to assign male mating success to 14 types of differently composed groups of up to six males and a variety of harem sizes observed within spongocoels in nature. As pointed out by Gross (1996), more direct measures of paternity would be desirable, and using mating success alone omits potential fitness differences between morphs due to their differences in maturation rate or dispersal mortality.

Subsequent data on morph determination (Shuster and Sassaman, 1997; Shuster and Guthrie, 1999) have weakened the case for simple genotype-specific control but have revealed some bizarre and interesting developmental details. As discussed in chapter 15, some of the female-mimicking beta males are actually genetic females that have changed sex, evidently due to the interaction of certain autosomal genes, genes that influence morph determination, and primary sex determination loci (Shuster and Sassaman, 1997). Records of food, temperature, and phenotypic variation in field and laboratory populations of *P. sculpta* have shown that temperature correlates negatively with body size within morphs but does not affect morph determination. Seasonal change in algal food abundance at the levels measured had no effect on growth rate, adult size, or morph determination (Shuster and Guthrie, 1999). Experiments designed to further test the influence of diet are still in progress (Shuster and Guthrie, 1999). While the switch mechanism in *P. sculpta* may not be a simple example of genotype-specific regulation, it has permitted the identification of a gene (phosphoglucomutase, or Pgm) closely linked to the major gene locus most influential in the switch between alternatives, and has illuminated the evolutionary origins of a novel femalelike male phenotype.

Claims that different alternative forms are "genetically determined" should always be viewed with caution. In general, a developmental view of adaptive evolution requires more precise discussion of regulation than has been customary in the past. Confusion of genetic *influence* (effect on a reaction norm) with genetic *determination* (genotype-specific expression) is exceedingly common. Continuous variation in reaction norms, including continuous variation in threshold responses, is best described as influenced, not determined, by genes. "Determined" should be reserved for cases where a genotypic or an environmental factor decides which of two alternatives is adopted in all normally encountered environments and genetic backgrounds, as in genotypic or environmental sex determination (chapter 15). "Conditional" is more accurate than "phenotypic" to distinguish a response that is primarily environmentally determined from one that is genotypically influenced. Thus, the sentence, "Wing dimorphisms could be purely phenotypic (polyphenism), but the available evidence suggests a genetic basis in most species" (Roff and Fairbairn, 1991, p. 249) would be more precisely reworded as, "Wing dimorphism could be primarily conditional in regulation, but the available evidence suggests strong genotypic influence in most species."

Stochastic Regulation

Several influential hypotheses for the maintenance of alternative phenotypes require stochastic regulation of the switch between alternatives. They include bet hedging (Seger and Brockmann, 1987), coin-flipping hypotheses (Cooper and Kaplan, 1982; Kaplan and Cooper, 1984; Fagan, 1987; see the classification of Seger and Brockmann, 1987), and the stochastic mixed ESS (type A of Maynard Smith, 1982, p. 78).

In adaptive *bet hedging*, the mean individual fitness in a population or brood is lowered by the adoption of alternatives that reduce the variance in fitness. Rather than optimizing fitness, selection might favor a bet-hedging alternative that reduces the chances of complete failure (Seger and Brockmann, 1987; Philippi and Seger, 1989; Moran, 1992b). Bet hedging randomly assigns phenotypes to recurrent but unpredictable environments, on the chance that some will fall into an environment where they are adapted and thus save the lineage from the decline or extinction that would occur if no individuals were able to survive and reproduce. Alternatives that prevent "putting all eggs in one basket" are of this category. In the stochastic *mixed ESS*, individuals adopt alternatives at a frequency that is set by selection, but individual decisions are determined randomly, not by particular genotypes or conditions. Instead, the organism acts or devel-

ops as if controlled by a roulette wheel (Maynard Smith, 1982, p. 76) to produce the alternatives at equilibrium frequencies.

An example of stochastic regulation is the sex ratio of offspring, seen as a trait of sexually reproducing parents with genotypic sex determination (e.g., birds, mammals, and many insects, such as *Drosophila*). Supposing that frequency-dependent selection sets the populationwide sex ratio at 1:1 male and female (Fisher, 1930 [1958]), the sex ratio of the offspring zygotes produced by an individual female, barring sex-ratio distortion, is determined stochastically, due to random segregation of genetic sex-determining factors during meiosis in the heterogametic sex (for a more thorough discussion, see Maynard Smith, 1982). So sex ratio in such organisms is a stochastic mixed ESS.

Bet hedging is considered likely to evolve in an environment with unpredictable temporal variation on a scale longer than a single generation, where monomorphic specialization could be fatal to a lineage that does not hedge its bets by producing a few individuals able to survive in some other environment, even though this handicaps them or incurs extra costs in the usual environment (Seger and Brockmann, 1987). Bet hedging is considered more likely as a response to temporal environmental fluctuations than as a response to spatial environmental heterogeneity. In keeping with expectations, the most convincing examples of bet hedging, such as the hypopus stage of mites (see below), involve temporal environmental variation, as do examples that involve diapause and hibernation versus reproduction in unpredictable, extreme, temporally separate environments (Seger and Brockmann, 1987; see also Goldman and Nelson, 1993, p. 242, on variation in the photoperiodic responses of mammals). Tuljapurkar (1990) discusses a stochastic model for delayed reproduction in a temporally unpredictably variable environment.

Alternatives maintained by *spatial* environmental variation seldom represent adaptive bet hedging, but instead favor the evolution of genetic polymorphisms and polyphenisms (Seger and Brockmann, 1987). However, bet-hedging polyphenism that takes advantage of spatial environmental heterogeneity may evolve in a group of relatives, because kin selection may favor a reduction in competition that would hurt relatives. This could be achieved via diversification of tactics for resource acquisition or habitat use. In that situation, a bet-hedging assignment of alternative phenotypes to an alternative environment or a set of alternative behaviors may be favored (Moran, 1992b). Spatial bet hedging has been hypothesized to explain the pattern of tissue development in the shade leaves of some tropical understory plants (Givnish et al., 1994) in which the juvenile, shaded leaves are highly dissected—dentate, lobed, or perforated with large holes. In a partially shaded place, this could function to sample a broader space than would a solid leaf of equal photosynthetic area, hedging the plant's bets in tissue investment by not putting the entire leaf in the shade. Although this idea is unsubstantiated in the plants discussed, it illustrates how the bet-hedging concept can work as an adaptation to irregular, small-scale spatial environmental heterogeneity such as the little patches of sunlight that fall upon a partially shaded leaf.

Variation in egg size and clutch size in some cold-blooded vertebrates may sometimes represent "coin flipping," where individual phenotype is determined by stochastic environmental variation (see discussion of mechanisms, below; Kaplan and Cooper, 1984). The environmental determinants of phenotypic variation are known, and there are good reasons, such as unpredictable environments at the time of determination, to expect this sort of developmental variation to be advantageous.

ESS game theory models give the conditions for stability of a trait or switching strategy that, if followed by the members of a population, cannot be invaded by the spread of any rare alternative strategy (Maynard Smith, 1982; Parker, 1984a). A *stochastic mixed ESS* describes alternative phenotypes in which each developmental or behavioral decision chooses randomly from a set of possible actions (H and D) according to the evolved rule 'play H with probability P, and D with probability $(1 - P)$.' This could apply to a genotype-specific switch mechanism if the probability P were to depend on the frequency of a particular allele relative to another at the same locus. At equilibrium, the two alternatives have equal fitness payoffs (Maynard Smith, 1982, p. 15). The equal-payoff ratio of alternatives is adjusted by selection.

Adaptive stochastic regulation is required by the mixed ESS. An example is genotypic sex determination in organisms such as birds and mammals, with little ability to facultatively influence the sex of offspring (but see Lott, 1984, on conditional influence in mammals). While the alternative (male or female) phenotype is determined by the genotype of the fertilized egg, the genotype is stochastically determined by segregation during meiosis. Support for various other putative examples of a stochastic mixed ESS, presented by Maynard Smith (1982) and others (Brockmann and Dawkins, 1979; Gross, 1985), is in the form of (1) evidence for equal payoffs, as required under the mixed ESS hy-

pothesis, and (2) evidence *against* other types of regulation—evidence or arguments that the alternatives are not genotype specific, and not conditional, and therefore may be randomly determined or stochastic.

All examples of the stochastic mixed ESS proposed in these early discussions are now suspected or known to have conditional regulation (Gross, 1996). One of these, cited as a possible mixed ESS by Maynard Smith (1982), is the subordinate and dominant behavioral alternatives in male Harris sparrows (Rohwer et al., 1981; Rohwer and Ewald, 1981; Rohwer, 1982). This example has the earmarks of being phenotype dependent and conditional. Individuals with relatively high testosterone titers and dark plumage are more aggressive and adopt dominant behavior, whereas low-testosterone, light-colored males are subordinate. Dominance relations can be reversed by painting light males dark (Rohwer, 1977) and injecting them with testosterone (Rohwer and Rohwer, 1978). Interpretation of this case as an example of a stochastic mixed ESS is evidently based on Maynard Smith's (1982) doubt that subordinate males could be victims of circumstances (making the best of a bad job), for he wrote: "If, by the miniscule expenditure of producing more testosterone and more melanin, a bird could become dominant, the advantages of being a dominant cannot be great" (p. 82). This overlooks the possibility of unavoidable phenotypic (e.g., hormonal and dominance) differences among competitors, for example, due to social or nutritional differences during development. Variation in the dominance status of individuals may be advantageously signaled by plumage color, as suggested by Rohwer (1982), to avoid the costs of having to back up a signaled dominant position when physically ill-equipped to win fights. This example, then, seems best interpreted as condition dependent rather than stochastic.

Another proposed example of a mixed stochastic behavioral ESS is the dig versus enter hole-nesting alternatives of the digger wasp *Sphex ichneumoneus* (Sphecidae; Brockmann and Dawkins, 1979; Brockmann et al., 1979). The digger-wasp example has been called the only well-documented case of a stochastic mixed ESS (Austad, 1984). It is based upon extensive field data (Brockmann, 1976) that permit tests of numerous alternative explanations (e.g., see Brockmann and Dawkins, 1979; Field, 1989; see also below). Brockmann and Dawkins (1979) examined numerous possible conditional bases for the individual decisions to enter rather than dig a nest, and found all of them to be contradicted by the evidence. One hypothesis they overlooked was that wasps may conditionally enter rather than dig when digging is costly (Field, 1989), a feature of conditional alternative tactics in some other sphecid wasps such as *Ammophila*, which is kleptoparasitic primarily at the end of the day or season, when prey are relatively scarce (Kurczewski and Spofford, 1998).

This conditional hypothesis receives some support from observations of Brockmann (1976), which show that entering rate was greater when the cost of digging was greater: in successive years (1973–1975) in the Michigan population, mean digging time per hole increased annually from 85 to 95 to 110 minutes, and rate of use of abandoned holes (percentage of available holes per wasp actually occupied per wasp) increased accordingly, from 29% to 63% to 100% (based on figure 3-8 and table 3-2 in Brockmann, 1976). There also is a significant negative correlation between the number of abandoned holes occupied and the number of parasites observed in the nesting area in a given year (Brockmann, 1976, p. 281), further supporting a hypothesis of conditional decisions. The finding of equal egg-production rates for the dig and enter nesting decisions (Brockmann and Dawkins, 1979) may reflect not equal fitness payoffs but equal average times for egg production in both nesting alternatives. *S. ichneumoneus* show remarkable abilities for condition-appropriate decisions in related activities such as nest excavation and provisioning behavior (Brockmann, 1980b, 1985a), so conditional digging and entering would not be surprising in this species.

With the digger-wasp example showing some signs of conditional regulation that invite further investigation, there is no remaining undisputed within-individual (non-genotype-specific) example of the stochastic mixed ESS. As concluded by Parker, "Mixed ESSs commonly collapse into conditional ESS if we allow that there will be phenotypic differences between competitors" (Parker, 1984a, p. 56).

Alternative searching tactics are another large category of alternatives sometimes interpreted as governed by stochastic rules (see Bell, 1991). Observations and experiments indicate, however, that "the overwhelming majority of examples of alternative tactics are derived from environmental cues initiating different sets of events in an equipotent individual" (p. 23).

Hypothetical Mechanisms of Stochastic Regulation
The various hypotheses for stochastic alternative phenotypes would gain support if a mechanism of stochastic regulation could be discovered. To my knowledge, aside from sex-ratio determination (see above), no specific regulatory mechanism for a particular stochastic mixed ESS has ever been described in the literature on alternative phenotypes. For coin flipping and bet hedging, two classes of purely hypothetical stochastic-determination mechanisms have been proposed: primarily environmental and primarily genotypic mechanisms.

Stochastic Conditional Regulation   Stochastic regulation could be achieved if the alternative adopted could be randomized by a randomly varying environment to which development was sensitive (Cooper and Kaplan, 1982; Kaplan and Cooper, 1984). The degree of sensitivity of development, and therefore the phenotype frequencies, could be subject to selection. This would provide a conditional switch mechanism without environmental matching. Another variant of the conditional-regulation idea is the developmental noise hypothesis of Spudich and Koshland (1976; see also Soulé, 1982; Cooper and Kaplan, 1982), which invokes variation in the internal rather than the external environment. By this hypothesis, randomization of phenotype production is inherent in the organism's development and requires no external environmental input. Random switching could be produced either by fluctuations in the internal environment or in the development of the threshold for switching.

Some of the mechanisms of plasticity described as involving somatic selection (see chapter 3) could provide material for a stochastic switch. Various aspects of development have a random component, as surmised from comparing left and right sides of the human body: the coronary blood vessel arrangement differs on the right and left sides of the heart (Gerhart and Kirschner, 1997), for example, and nerve connections involve initial overconnection and later pruning back (see chapter 3) but differ in detail. If a small but indeterminate number of the regularly produced entities involved in hypervariable mechanisms of development determine a property at some key developmental stage, then in theory, individual variations can be generated even in a genetically and environmentally uniform population. As suggested by Spudich and Koshland (1976), because of small numbers at the critical stages, they can be subject to chance diversity at the extremes of a Poisson distribution describing variation in their effects, which would not appear as significant variation if a larger number of molecular determinants were involved—a kind of sampling error or *developmental drift*—determination of a switch by stochastic events. (This idea differs from the "developmental system drift" of True and Haag [2001], which, unlike the idea of drift usual in evolutionary biology, refers to developmentally divergent underpinnings of similar phenotypes that may be due to selection, not sampling error or random events.)

Spudich and Koshland (1976) were attempting to explain stochastic "individuality" in the behavior of bacteria, but the same type of explanation perhaps could apply to bet hedging in organisms such as insects and mammals, where the hormonal mechanisms of regulation of switches are only beginning to be understood (e.g., see Nijhout and Wheeler, 1982; Nijhout, 1999c; Goldman and Nelson, 1993). Phenotype determination near the neutral point of a switch (see chapter 5) could sometimes allow small numbers of molecules, or sampling error at critical points, to determine life-history decisions (such as the decision to remain in diapause or not).

Another hypothesis for conditional stochastic regulation envisions maternal manipulation of the developing phenotype. Alternatives such as dispersal polymorphisms could be important even in uniform and predictable environments if they enable an organism to retain a foothold at home as well as to occasionally colonize empty sites free of competition (Hamilton and May, 1977; evidently this idea assumes that the environment is not uniform as far as distribution of competitors is concerned). Selection would then favor parents who impose dispersal on a certain fraction of offspring so that some compete locally and others do so away from their natal site. To accomplish this, the mother's genotype may specify some maternal influence on each ovum (or testa or fruit) that determines the fraction of dispersing offspring (Hamilton and May, 1977). In an endearing finale, Hamilton and May (1977) wrote, "It would be satisfying to conclude by listing a few biological illustrations of principles suggested by the models. Unfortunately most of the examples which have come to mind are more in the region of jokes than of reality" (p. 580).

Still, their suggestion of maternal manipulation of the offspring phenotype (e.g., via differential endowment of ovules due to fluctuations in timing of oviposition, or differential provisioning of successive offspring) is a reasonable place to look for a mechanism of stochastic polymorphism. It would be neither genotypic (cf. Van Valen, 1971) nor adaptively conditional, and would be independent of offspring genotype and environment. Possible examples occur in aphids, where morph determination shows both maternal and grandmaternal effects, and where stochastic processes have been pro-

posed for determination of wing polymorphisms (Blackman, 1979, pp. 273–274). Various presumably stochastic developmental processes, such as gene translocation and effects of transposable elements showing cyclic phase change (e.g., in maize), have been mentioned as candidate determinants of stochastic timing (Blackman, 1979).

The eggs of mosquitos seem to show a stochastic regulation of egg dormancy. Eggs remain dormant in the egg stage for long and irregular periods of time. Flooding causes some but not all eggs to hatch, in effect leaving a reserve to be triggered at a later time (Breeland and Pickard, 1967). Many similar examples occur in seed dormancy of plants (reviewed in Baskin and Baskin, 1998). Again, maternal manipulation of egg properties seems a reasonable regulatory candidate.

Stochastic Genotypic Regulation Nonconditional polygenic regulation of stochastic switches is a hypothetical possibility discussed by Roff (1975) and Roff and Fairbairn (1991). By this hypothesis, genetic variation that is random with respect to the individual condition or situation at hand affects the probability of a switch between forms. By analogy with sex-determination theory (see Bull, 1983), it should be possible to produce nonconditional alternatives under multilocus genetic control, with the alternative expressed determined by additive effects of alleles at numerous variable loci—called "polyfactorial determination" by Bull (1983). Bull gives three criteria used as evidence for such determination: (a) large between-family morph-ratio variance, (b) paternal and maternal effects on brood morph ratio, and (c) a morph-ratio response to selection (p. 94). It would be difficult to distinguish between polyfactorial determination and two-factor determination with environmental effects (Bull, 1983). None of these authors give examples where stochastic genotypic regulation has been shown.

Recent studies of gene expression indicate that normal gene activity has a greater stochastic element than previously recognized (for a concise review, see Klingenberg, in press). When and how often a particular gene is transcribed depend upon molecular signals in complex cellular regulatory networks, and the protein product encoded by one gene often regulates expression of other genes. Simulations based on studies of gene expression in prokaryotes show that once a promoter is activated within such a network, proteins are expected to be produced in short bursts at random time intervals, and this may result in large differences in the timing of events in different cells of the same clonal population (McAdams and Arkin, 1997). McAdams and Arkin note that random patterns of expression of competitive effectors can affect switching mechanisms and channel development, randomly, between alternative pathways. They point out that organisms can exploit such inherent fluctuations "to achieve nongenetic diversity where this makes the population more capable of surviving in a wide range of environments"(p. 819)— precisely the suggestion of bet-hedging arguments that depend upon stochastic mechanisms. Then, when a deterministic outcome is advantageous, selection would favor "reaction parameters" that produce a dependable switch (McAdams and Arkin, 1997). That is, the stochastic mechanisms may be a constant, with the familiar switches observed in development superimposed. Extrapolating from the model of Klingenberg and Nijhout (1999), genetic variation in sensitivity to random developmental noise could convert developmental instability into adaptation if the random factors were to become patterned. While such discussions are still far from providing a mechanism for any particular stochastic switch in the whole-organism variants discussed in this chapter, they suggest a reorientation of how we think about stochastic processes and regulation that may put the evolution of stochastic alternatives in a new light.

Examples attributed to stochastic regulation often prove, upon further examination, to be conditional or genotype-specific alternatives (Seger and Brockmann, 1987; Gross, 1996). It is difficult to confidently eliminate the possibility that regulation is stochastic rather than influenced by some unknown environmental polygenic factors, and the problem is compounded by the fact that no mechanism of stochastic regulation has yet been demonstrated for any of the purported examples other than the sex ratio. Until concrete mechanisms for stochastic regulation are known, coin-flipping, bet-hedging, and stochastic mixed ESS theories will remain tentative and subject to doubt, especially since so many of them have proven under close study to possibly have conditional or genotype-specific regulation.

The stochastic-regulation question relates to a broader theme in debates over development and evolution. I have avoided talking about developmental "constraints" because it is so much more fruitful to look at developmental possibilities, given the enormous potential for developmental reorganization in the evolution of all organisms. But the virtual absence of evidence for specific stochastic devices in alternative phenotypes where they are expected in selection-based theories challenges the conviction of many biologists that any trait favored by selection is likely to appear. Maynard Smith (1982) voiced this conviction when he wrote, of stochastic regulation, "If it were selectively advantageous, a randomizing device could surely evolve" (p. 76). If, as presently seems likely, stochastic regulation has its origins in a very widespread stochasticity of intracellular gene expression, then switching between stochastic alternatives may be the result not of natural selection in the context of those alternatives, but rather, of the absence of selection for deterministic switching, leaving the switch between the alternatives in question to a stochastic background that is always present.

Like the question of gradualism versus saltation (see chapter 24), the question of mechanisms in relation to adaptive alternatives brings out the different attitudes of two philosophical camps, one that insists on the power of development to influence design, and one that, like Maynard Smith in the passage just cited, claims that selection can produce virtually any advantageous design, implying that selection is likely responsible for any adaptive design observed. Ironically, developmental biology may have to come to the aid of selection extremists if this claim is to be substantiated, because a long history of controversy within evolutionary biology shows that blind faith in all-powerful selection is academically linked to a professed lack of interest in doing research on mechanisms (for justifications and commentaries, see Mayr, 1961; Hamburger, 1980, pp. 99–100 see also other essays in Mayr and Provine, 1980; Sherman, 1988; Dewsbury, 1994). In the matter of stochastic alternatives, research on mechanisms may be on the road toward answering an important ecological and evolutionary question, unaware that it had been asked.

Condition-Sensitive Regulation

In general, equal-payoff models do not apply to conditional alternatives. Conditional alternatives, sometimes called *conditional ESSs* (Parker, 1984a), are selectively advantageous as long as an alternative is expressed primarily when it is more advantageous than other alternatives, even if equal fitness contributions do not obtain. The superiority of a secondary, relatively low-overall-payoff conditional alternative may be temporary and occasional, such that it never attains an average fitness effect that equals those of more frequently expressed, higher-payoff alternatives.

Models of ideal free distributions are exceptions to the rule of inapplicability of equal-payoff models to conditional alternatives. Models of *ideal free distributions* predict equal payoffs for conditional decisions (e.g., to stay or move among alternative resource patches; Fretwell, 1972; Maynard Smith, 1982; Pulliam and Caraco, 1984; Parker, 1984a; and Milinski and Parker, 1991). In an ideal free distribution, individuals switch between alternatives when diminishing returns for alternative A (e.g., staying at a patch) equal the payoff for performance of alternative B (e.g., moving to another patch). A classic example is the stay versus leave behavior of male dungflies (see Parker, 1970), which colonize fresh cow dung that is attractive to females, then leave as its quality declines with time. Leaving is conditional in that selection sets a time to leave based on a combination of average values of declining resource quality and numbers of competitors present at a site (see summary and review in Milinski and Parker, 1991). Other examples of alternatives evidently maintained as ideal free distributions occur in stickleback fish (Milinski, 1979), water striders (Rubenstein, 1984), and dragonflies (Rowe, 1988). T. J. Walker (1983) describes a temporal ideal free distribution of calling times in male crickets.

Maintenance of Conditional Alternatives

Several factors contribute to the maintenance of conditional alternatives and undoubtedly help account for their evidently great commonness in nature compared with genotype-specific or stochastic alternatives. With conditional regulation, alternatives can be maintained under changing conditions, and no particular equilibrium ratio is approached, only conformity to the developmental switching rule: benefit/cost tactic B > benefit/cost tactic A. Equal payoffs for conditional alternatives are *possible*, but not *necessary* for the maintenance of conditional alternatives. An example of equal payoffs for conditional reproductive alternative tactics was demonstrated in coho salmon (*Oncorhynchus kisutch*; Gross, 1985), and the fit of this example to a conditional model was shown by Hazel et al. (1990). Because there is no requirement for equal payoffs or cross-generational equilibria of phenotype ratios, these features may or may not indicate an evolutionarily stable state.

Immunity to the equal-payoff condition means that the evolved improvement in one alternative can outdistance that of another, thereby increasing its fitness effect and its relative frequency in the population, without leading to fixation of the most profitable alternative and loss of polymorphism. Conditional alternatives are highly stable in the sense of being immune to extinction because the mechanisms for switching can remain intact even if one of the alternatives is unexpressed for more than a generation. A lost conditional alternative can be immediately restored by the return of favorable, inducing conditions, provided that the mechanisms for producing it have not degenerated during a period of disuse (see chapters 11 and 12). In this respect, conditional alternatives are more immune to elimination by selection than are genetic polymorphisms. Many seasonally expressed phenotypes disappear completely every year. The saltbush feeding specialization of the Great Basin kangaroo rat remained unexpressed for millenia, then was reexpressed when climatic change made it once again useful (Csuti, 1979; see chapter 26). How long an unexpressed alternative may persist intact without being expressed is unknown and undoubtedly varies from case to case (see chapter 11; Shapiro, 1976, 1980, reviews experimental studies in the induction of lost alternatives). But one point is certain: stable equilibrium is not required to explain the long-term persistence of alternatives, especially of conditionally expressed ones. As long as a secondary alternative is superior in some environment, and that circumstance recurs, a capacity to facultatively produce or perform the alternative may be advantageously maintained as long as the cost of assessment is not too high. In general, the circumstances for maintenance of conditional alternatives are much broader than those for genetic polymorphism (Lively, 1999).

Another source of stability comes into play when alternatives become complementary. Contrasting alternatives, especially those that are phenotype dependent (e.g., size, morphology, or rank dependent) in expression, may become complementary and mutually dependent because the strongly divergent forms can lose the ability to perform the opposite specialization (West-Eberhard, 1979). The evolution of new *mutually dependent, complementary traits* typifies the emergence of new levels of organization, divisions of labor, increased complexity, and new levels of selection (Bonner, 1974, 1993; West-Eberhard, 1979; Maynard Smith and Szathmary, 1995). Mutual dependence implies stabilizing selection for maintenance of alternatives. As long as an individual or group depends on the presence of both of two alternative forms, neither is likely to be dropped.

While the evolution of mutually dependent complementary specializations has often been noted as a factor in the emergence of new levels of organization (e.g., Maynard Smith and Szathmary, 1995), it plays a part in what could be called day-to-day evolutionary change, as well. Within organisms, functional differentiation of parts may begin due to selection on alternative phenotypes. For example, the intestine of the shrimp *Gammarus fossorum* (Crustacea, Amphipoda) has two functionally differentiated zones (Barlocher, 1982, 1983, cited by Anderson and Cargill, 1987). A slightly acid anterior zone digests carbohydrates, and a more alkaline posterior zone digests microbial and leaf proteins. If different individuals use different behaviors and habitats to feed on carbohydrates or proteins, these two specialized regions of the gut could be regarded as parts of two alternative trophic tactics. But if both kinds of food are ingested simultaneously and with little or no difference in searching and handling, the gut specializations would constitute interdependent and complementary parts of a single feeding regime with an internal division of labor between sections of the gut.

Similarly, two conditional alternative phenotypes can be regarded as parts of a single strategy (e.g., Dawkins, 1980), in that individuals that lack either one would be at a disadvantage. Even so, control by a switch means that the two alternatives are potentially developmentally dissociable, should selection favor expression of only one. In effect, a single conditional strategy has three modular components, two alternative tactics and a switch mechanism, all of them potentially dissociable for expression in other contexts (see part II).

General Requirements for the Maintenance of Conditional Alternatives

It takes little reflection to realize that not all conditional alternative phenotypes rely on equal fitness payoffs to persist. Queen honeybees produce thousands of offspring, and workers none; even the indirect contribution of a worker to the reproduction of the queen—its inclusive fitness (Hamilton, 1964a)—is miniscule compared with that of the queen (West-Eberhard, 1981). Dominant male horses monopolize whole harems of mates, while subordinates sneak only a few furtive copulations. How are these unequally rewarding styles of life maintained in the same population, if not in an equal-payoff equilibrium? It is easy to explain the persistence of a relatively high-payoff trait. The challenge is to explain the persistence of *secondary alternatives*—traits whose fitness effect is lower than that of a *primary alternative*, a trait with a large per-occurrence fitness effect relative to that of

the others in a set of alternative phenotypes. The answer, for all advantageous conditional alternatives, is that primary alternatives are high-payoff traits only in some circumstances; in other conditions, a secondary alternative is more profitable. The key to the maintenance of conditional alternatives is in their timely expression. It depends on matching trait expression to the conditions where they are expressed. Therefore, adaptively condition-sensitive mechanisms of regulation are crucial in the maintenance of conditional alternatives. If the switch mechanism evolves so that each alternative is expressed when it is likely to be more advantageous than others, and the cost of assessment is not excessive, then the alternatives can be maintained (see Lively, 1999). The use of adequate and affordable environmental cues is the key to such a mechanism and is the subject of chapter 23.

Several models treat the maintenance of conditional secondary alternatives in terms of the adjustment, under selection, of polygenically influenced thresholds of expression (e.g., see Hazel et al., 1990; Moran, 1992b; Gross and Repka, 1997, 1998). Since phenotype ratios depend on polygenic regulatory mechanisms (see chapter 5), these polygenic threshold models represent an advance toward linking development and evolutionary genetics. Genetic models that describe alternatives as if underlain by homozygous alternative alleles (e.g., Levins, 1963; Lloyd, 1984; Lively, 1986) successfully explain many features of alternative traits and their distributions in organisms (see especially Lloyd, 1984). They can be seen as analyses of alleles of major effect on polygenic regulation (chapter 5). Nijhout and Paulsen (1997) discuss the limitations of single-locus models for analyses of development.

It is not the purpose of this chapter to exhaustively review models for the maintenance of alternatives. Suffice it to say that conditional alternatives persist as long as switching between them is favored—as long as a secondary alternative is advantageous relative to its cost and the cost of switching mechanisms in at least some individuals or conditions. In general, environmentally cued polymorphisms are evolutionarily stable over a wider range of parameters than are allelically switched alternatives (Lively, 1999).

Costs of Conditional Regulation

Switching between alternatives may have various kinds of costs (Padilla and Adolph, 1996; DeWitt et al., 1998; Agrawal, 2001), and there may be fitness trade-offs associated with the development or performance of different alternatives. This is shown by research on the costs and benefits of predator-induced polyphenism in amphibian tadpoles (McCollum and Van Buskirk, 1996), hatching decisions by anuran eggs (Warkentin, 1995), and switching between morphs in toad tadpoles (Pfennig, 1992b). A simple comparison of contrasting alternatives themselves is often enough to indicate the costs of adopting the opposite alternative (West-Eberhard, 1979; Emlen, 1997a). For example, in the defensive polyphenism of barnacles studied by Lively (1986) the predator-induced "bent" shells grow more slowly and are less fecund than are conics (the normal form developed in the absence of predator attack; see figure 20.6). So the bent form evidently incurs a cost in terms of growth and fecundity, and the conics pay a cost under predation. This fitness cost of being bent contributes to the maintenance of the conical alternative even though it is less protected from predation.

Not all alternatives involve fitness trade-offs of this kind. Under social and sexual competition, the primary alternative (e.g., the position of queen in a social insect colony, or of dominant male in a lek) may be a profitable option for only a few individuals who exclude all others from pursuing it. For a born loser or subordinate individual, the value of the primary tactic therefore may be extremely low or even zero, so virtually nothing is lost by not pursuing it, and secondary alternatives with very low payoffs (such as worker behavior, or patrolling the foraging sites of relatively unreceptive females) may be stable ways to salvage at least some reproductive success (West-Eberhard, 1979; Dawkins, 1980). Similarly, a switch to a food source of secondary value could be forced by periodic scarcity of the primary resource rather than by a trade-off in the usual sense. Or a switch could be driven by lowered cost of the secondary tactic, for example, if there were an opportunistic bonanza of an easy to catch, even though less nutritious, food.

Developmental costs of plasticity (Pfennig, 1992b; DeWitt, 1998) can include such things as (1) resources expended in the production and maintenance of regulatory devices (assessment and production devices such as sensory morphology, hormonal systems, or a complex nervous system), (2) costs of converting from one state to another (e.g., tissue reorganization, lower fitness of transitional intermediate forms, or time required to learn a new skill or develop a new morphology in a rapidly changing or stochastically fluctuating environment; see especially Padilla and Adolph, 1996), (3) developmental instability or inefficiency (e.g., if plastic development is more variable than development

strictly selected to produce a single form), (4) the unreliability of a switching cue in predicting the results of a switch, as in a fine-grained environment.

The idea of a cost of plasticity is potentially misleading if it is taken to imply that there are inevitable or *net* costs to flexible phenotypes. Switch points should evolve toward optima where benefits more than compensate costs, even for very costly alternatives (see Zeh and Zeh, 1988, on the circumstances that favor conditionally developed sex ornaments). With a switch point at the optimum, conditional switching is always beneficial, barring errors of assessment, generational lag in the evolution of advantageous switching thresholds, or such disadvantages of plasticity as susceptibility to phenotypic manipulation by others (Fagen, 1987).

In a sense, alternatives compete with each other for developmental priority. Improved efficiency of performance could conceivably convert a secondary alternative into a highly profitable primary one. This may have occurred in the jack male of salmon. Like the secondary mating tactic of many species, the jack male is smaller, more femalelike, less aggressive, and less modified for combat than is the alternative aggressive, large, hook-nose male. Yet in the Pacific coho salmon *Oncorhynchus kisutch*, the jack males come from the larger juveniles in the population and have mating success rates that equal or even exceed those of hook-nose males (Gross, 1985; Charnov, 1993). It is conceivable that the jack male tactic began as a secondary alternative of small males in competition with large ones, and that small size associated with accelerated maturity and other modifications enabled the sneaking tactic to become so effective that it became primary. The hypothesis that the jack male was originally a secondary alternative may be testable using comparative and phylogenetic study of the life histories and mating success of related species. If such studies were to reveal that the coho salmon situation is derived relative to others, that is, that the ancestral state was for jack males to be small as juveniles and to pursue a clearly secondary (relatively unprofitable) sneaking tactic, this would indicate that a secondary alternative has evolved toward being a primary alternative.

Stable maintenance of conditional alternatives depends crucially on the existence of an adequate cue, or mechanism of assessment that can match phenotype to conditions in a variable environment (Levins, 1968; Moran, 1992b). In the past, one stumbling block to acceptance of the idea of conditional alternatives has been doubt regarding the feasibility of adequate evolved cues. Chapter 23 treats the nature and evolution of *assessment*, the ability of organisms to evaluate and respond appropriately to environmental or phenotypic variables used as decision cues. Despite remarkable capacities for assessment, there may sometimes be imprecision in the degree to which the switching threshold reflects the true relative values of the alternatives controlled, especially when alternatives are evolving at different rates under selection. Switching imprecision would be an additional cost of developmental plasticity (D. Pfennig, personal correspondence).

Equilibrium Points as the Switch Points of Conditional Alternatives

Some genetic models can be applied to conditional alternative phenotypes if the equilibrium point of a model for the maintenance of genetic polymorphism is taken to be the switch point, or equal-payoff point between conditional alternatives (Warner et al., 1975; West-Eberhard, 1979; Parker, 1982; Repka and Gross, 1995). As a result, many insights regarding quantitative variation in parameters treated by genetic models can be used to illuminate the evolution of condition-sensitive regulation.

Given the overwhelming commonness of conditional alternatives relative to genotype-specific ones, the most useful genetic models for understanding the maintenance of alternatives are those that can easily be applied to conditional phenotypes. Hamilton's (1964a,b) formulation of kin-selection theory illustrates this. As a model of genetical evolution, Hamilton's hypothesis describes the gain:loss equilibrium point, or least required gain, for selection to favor an allele that promotes altruism (Hamilton, 1964b, p. 18). Translating from this to a behavioral decision process, Hamilton wrote, "The social behaviour of a species evolves in such a way that in each distinct behaviour-evoking situation the individual will seem to value his neighbours' fitness against his own according to the coefficients of relationship appropriate to that situation" (p. 19).

Hamilton's application of inclusive fitness theory as both a genetic model and an optimal decision rule was one of the reasons for its outstanding success. Most empirical tests of kin-selection theory have used or reformulated Hamilton's rule as a decision rule. In that form, the required relation between costs, benefits, and relatedness for a switch from independent reproduction to a helping tactic has proven robust to field tests (e.g., see West, 1967; West-Eberhard, 1969, 1975; Noonan, 1981; see also chapter 11). Parker (1989) compares conditional and unconditional gene expression in a series of kin-selection models. Charlesworth (1980) and Seger (1981) conclude that genes that cause sterility can only spread if they have low penetrance or are conditionally expressed (see also Queller and Strassmann, 1998).

Frequency-Dependent Selection

Some frequency-dependent models for the maintenance of alternative phenotypes require genotype-specific regulation, and some imply conditional switches. With genotype-specific regulation, average payoffs are equal. With conditional regulation, frequency dependence can be important but equal payoffs need not obtain. Well-known examples of frequency-dependent models are among the game-theory ESS models developed to analyze strategies with alternative behavioral and ecological tactics, stability meaning that if most members of the population have adopted the strategy, it cannot be invaded by any other strategy (Maynard Smith and Price, 1973; reviewed in Parker 1984a). In a frequency-dependent ESS, the fitness effect of a given alternative depends on the numbers of the alternative phenotypes present in the population; under negative frequency dependence, an alternative has a fitness advantage when rare.

In classical genotype-specific frequency dependence, the phenotype adopted is irreversible and selection adjusts the ratio of alternative genotypes in the population at an equilibrium level where increase in either would be selected against. Standard examples are sex ratio in organisms with genotypic sex determination (Bull, 1983) and the Batesian mimicry, polymorphisms of butterflies, where phenotype is determined by a small number of linked loci (Ford, 1964). In Batesian mimicry one morph is protected from predation by color-pattern mimicry of a distasteful model species, and the other is protected by cryptic coloration. Frequency dependence obtains because the mimic morph is palatable and is selected against if it begins to outnumber the distasteful models.

One example of genotype-specific (or highly heritable, $h^2 = 0.96$) frequency dependence without stable equilibrium payoffs occurs in the trimorphic males of side-blotched lizards (*Uta stansburiana*), where there is cyclic change in morph ratios due to circular frequency-dependent selection: aggressive territorial orange-throated males, when they are common, are defeated by sneaker yellow-throated males, which are in turn defeated when common by mate-guarding blue-throated males, which in turn are defeated when common by the territorial orange males. This produces an endless cycle of change in morph ratios that maintains genetic variation in the switch threshold that controls morph ratios, while preventing a stable equilibrium (Sinervo and Lively, 1996). Paternity data indicate that the morph ratio change is due to gene frequency change under this unusual circular frequency-dependent selection regime (Sinervo and Lively, 1996).

Conditional frequency dependence requires some mechanism for individual assessment of phenotype frequency. Then frequency-dependent selection adjusts the threshold frequency for a facultative switch between alternatives. A well-documented example of evolutionary adjustment in the threshold for a conditional frequency-dependent switch occurs in spadefoot toad (*Scaphiopus multiplicatus*) tadpoles, which have a reversible diet-induced omnivorous morph, which is specialized to feed on detritus, and a carnivorous morph, specialized to feed on shrimp (Pfennig, 1992b; see figure 21.2). In short-lived, rapidly drying ponds, carnivore morphs fare better than do omnivores because carnivores grow more rapidly and metamorphose at a larger size, then survive better as juveniles and adults. In long-lasting ponds, omnivores do better, evidently because their larger fat body enhances postmetamorphic survival (Pfennig, 1992b). As expected, short-lived ponds contain toad populations with relatively high proportions of carnivorous tadpoles, and long-lived ponds contain more omnivores. Is this due to local differences in the strength or frequency of the morph-determining cue, or to evolutionary change in the threshold for switching between morphs, or to both? Available evidence suggests both: tadpoles develop into carnivores only if they ingest shrimp (Pfennig, 1990b), and shrimp are more common in ephemeral ponds (Pfennig, 1992a). But there is also a frequency-dependent component to the switch: morph determination is reversible if diet changes (tadpoles must continue to feed on shrimp to maintain the cannibal morphology), and increased competition for food among the more common morph makes the less common morph do better. Some individuals switch diets and consequently switch to the opposite morph. Rearing experiments using individuals from different ponds show that there is frequency-dependent equilibrium morph frequency in each pond, implying that the threshold for morph reversal is different for different ponds; individuals reared at a frequency above the pond's equilibrium frequency metamorphosed at suboptimal sizes, and this affects adult survivorship (Pfennig, 1992b). This is strong evidence that the local switch point

for morph determination has been adjusted by selection to produce evolved differences between ponds.

Adaptive *facultative* changes in the "set point" or threshold for switches is also quite common (see Mrosovsky, 1990). In a natural population of horned beetles studied in Panama, the switch point between male morphs appears to change seasonally in response to seasonal change in diet and body size distribution (Emlen, 1997b). Other examples of shifted switch points (Gross, 1996) occur in response to changes in population density in the size-dependent alternative tactics of male mites (*Caloglyphus berlesei*; Radwan, 1993) and in response to predation in guppies (*Poecilia reticulata*; Godin, 1995).

Another example of conditional frequency-dependent regulation is the switch between water foraging and other tasks in social wasps (Jeanne, 1996a). As the number of water foragers goes up, the need for them goes down. Individual foragers evidently assess the change by the waiting time for unloading water to recipient nestmates. Water foraging is affected not only by the frequency of water foragers but also by the water usage rate and demand in the colony (Jeanne, 1996a, 1999). Water foraging thus has a frequency-dependent component without being strictly frequency dependent in its determination and evolution. This is probably true of most of the polyethisms that have been used to illustrate frequency-dependent ESSs in social behaviors, such as alternative tactics of combat and courtship, or tit-for-tat egg trading and fertilization in simultaneous hermaphrodites (Fischer, 1988). Most if not all of these alternatives are competition dependent, not strictly frequency dependent. Parker (1982) acknowledged this in a discussion of *phenotype-limited* alternatives, in which individuals switch between alternatives depending on such factors as their size in order to "make the best of a bad job" (Dawkins, 1980; Eberhard, 1982), but where frequency dependence is probably important as well. All types of intraspecific parasitism and cooperation obviously are frequency dependent to some degree.

Competition-Dependent Selection

The problem of multiple determinants, with frequency dependence one of them, was solved for a large category of social and ecological alternatives by Mather's (1969) model of competitive selection, which has not received the attention it deserves. Mather's competitive selection model is a genetic model similar in purpose (but not form) to the game-theory ESS models that were developed at about the same time (e.g., see Maynard Smith and Price, 1972) but taking account of phenotype dependence and frequency dependence in the same formulation. Like Maynard Smith, Mather (1969) recognizes that "the fitness of an individual . . . depends not just on itself but on the constitution of the population as a whole" (p. 536) and emphasizes alternatives that function in social competition. Mather refers to alternatives as competition dependent rather than frequency dependent, although they may sometimes have a frequency- or density-dependent component, as Mather points out.

The Mather model is more satisfactory than only frequency dependence as a description of socially selected, and some ecological, alternatives, such as the trophically competitive behaviors and morphology of fish in lakes (e.g., see especially Robinson and Wilson, 1994). Many social and ecological alternatives that depend upon what others are doing are only minimally frequency dependent: they may depend much more on how *well* others are doing things than on how *many* others are involved. In the competition-dependent alternatives of a social insect colony or a lek, the frequency of dominant individuals need only be one to make subordinate alternatives more advantageous, and the frequency of dominant behavior may only vary between one (when the superdominant position of queen or harem chief is filled) and zero (when the position becomes vacant). In ecological competition-dependent alternatives of spadefoot toads, the advantage to an individual of switching between the omnivorous and the carnivorous phenotypes would depend not only on the number of each present but also on relative individual success in competing with others for food and in handling and metabolizing the food obtained. These variables would also affect rate of ingestion and metabolization of shrimp, the proximate mechanism for the switch.

Competition-dependent selection may be affected, in addition to frequency, by relative size, vigor, age, sensory acuity, hormonal status, and learning ability. Social contests are also affected by psychological factors, such as individual history of wins and losses in fights or associations with high-ranking individuals (e.g., Alexander, 1961; Goodall, 1986) and by density of interactants and opportunities (Eadie and Fryxell, 1992). Thus, competition-dependent alternatives can be at once frequency dependent, density dependent, phenotype dependent, and opportunistic; each of these factors has inspired a name for a kind of model, tactic, or strategy, for example, frequency dependent (Maynard Smith, 1982), density dependent (Kennedy, 1961; Semlitsch and Wilbur, 1989), phenotype dependent (Parker, 1982), opportunistic (Barnard and Sibly, 1981), and status dependent (Gross, 1996), but none is likely to apply in isolation. Eadie and Fryxell (1992) found brood parasitism in goldeneye ducks (*Bucephala islandica*) to be an opportunistic, density-dependent, and frequency-dependent alternative to independent nesting. Even a model incorporating all three of these factors is not completely satisfactory, as these authors point out, since other influential variables such as individual age and social status are not included (Eadie and Fryxell, 1992).

All mutually dependent alternatives, such as male and female and worker and queen, are to some degree frequency dependent, for a group or population would not persist if all members are of one alternative phenotype or the other. Selection must therefore favor the ratio of alternatives most favorable to their mutual collaboration and division of labor.

Alternative Phenotypes and Maintenance of Genetic Polymorphism

High levels of genetic polymorphism characterize many populations of organisms, and attempts to explain this have been a perennial subject of debate in evolutionary biology (Lewontin, 1974; Gillespie, 1991). Several properties of complex alternative phenotypes suggest an explanation in terms of natural selection. One is the capacity for alternatives to evolve in contrasting directions in the same context. This is likely to increase the number of genetic loci subject to antagonistic pleiotropy, or opposite selective effects in different conditions. Another is the polygenic nature of regulatory mechanisms, which means that an adaptive alternative phenotype can approach fixation without fixation of alleles that influence the frequency of its expression.

Antagonistic Pleiotropy

In *antagonistic pleiotropy*, different pleiotropic fitness effects of a trait or a gene are opposite in sign, positive in one context of expression and negative in another. In theory, this could maintain a genetic polymorphism, and significant antagonistic pleiotropy has been demonstrated between fitness components in *Drosophila* (Rose, 1982). A related concept is *structured pleiotropy*, or covariance between the effects of a gene or gene substitution on two phenotypic traits, based upon correspondence in expression (functional constraint) regarding the two traits (de Jong, 1990). Structured pleiotropy may either be synergistic (positive covariance) or represent a trade-off (negative covariance of effects). In the latter case, as a kind of antagonistic pleiotropy, it could be expected to help maintain genetic polymorphism.

Antagonistic pleiotropy and structured pleiotropy in general must often characterize the shared modifiers of alternative phenotypes (Moran, 1992b)—the nonspecific modifiers of figure 4.13. Recall that nonspecific modifiers are gene products or traits that are shared by alternatives—whose presence is necessary for both alternatives but whose expression either is not controlled by the switch between them or is consistently turned on in both (see chapter 4). An example is the legs of horned beetles, which are essential for both fighting and sneaking but whose development is not affected by the switch between morphs. Since alternatives often show opposite or antagonistic functions (see chapter 21) and always have different uses, genetic variation in nonspecific modifiers must often be subject to selection of opposite sign. This is one reason to expect character release when one alternative approaches fixation, freeing nonspecific modifiers from antagonistic pleiotropy and permitting a unidirectional response to selection (West-Eberhard, 1986, 1989; Moran, 1991, 1992a; see chapter 21). Nonspecific modifiers of facultatively expressed alternative defenses to pathogens or diseases, or cyclic escape mechanisms such as host alternation, may be subject to cyclic reversals in direction of selection such as those discussed by Hamilton et al. (1990) for pathogens and disease, but in the many more selective contexts where alternative phenotypes occur. The fact that contrasting alternative phenotypes are common and frequently reversible even within individuals, and that relative frequencies of contrasting alternatives fluctuate both between generations and geographically (chapter 21), may make them a major factor in maintenance of genetic polymorphism. Alternatives produce not just graded environmental variation but also flip-flopping between adaptive states on a very short time scale within individuals and populations.

Many alternative genetic alleles have no known functional differences and appear to be evolving due to drift rather than selection. These functionally neutral (or adaptively equivalent) alleles are commonly assumed to evolve to fixation more readily than do those under selection, though there are reasons to doubt that the facile assumption of adaptive neutrality for rapid gene substitutions in molecular biology is always justified (Gillespie, 1991). Studies of enzyme kinetics and relative temperature stability have demonstrated functional differences between numerous allozymes (e.g., see review in Gillespie, 1991; Powers et al., 1993), and their allele ratios have been shown to vary geographically in accord with predictions based on adaptive properties and clinal change in relevant environmental variables. These studies support the hypothesis that quantitative differences in the effects of the alleles in question have evolved under natural selection, and that they show the kinds of fitness trade-offs under different conditions characteristic of antagonistic pleiotropy. Some proportion of gene loci with rapid substitution rates may be specific modifiers of alternative phenotypes whose frequency is changing over time under selection, a common phenomenon that gives rise to intra- and interspecific recurrent phenotypes and flip-flopping evolution (see chapter 19).

Fixation of Phenotypes without Fixation of Genes

A single alternative phenotype can evolve to fixation without fixation of the genes that influence its polygenic regulatory mechanism. This may help to explain the persistence of genetic variation in populations, even under strong selection (Roff, 1994a). When a single alternative is favored by selection, it at first increases rapidly in frequency due to genetic change in the threshold for its production. Then, as selection proceeds and the trait approaches fixation, the rate of genetic change in regulation declines (figure 22.2a). One possible explanation for this decline is that the phenotypic variation in response threshold that allows selection to proceed decreases as the frequency of the trait approaches fixation. But simulations show that the heritability of the selected trait can actually rise after many generations of selection, especially if regulation is highly polygenic (figure 22.2b). As a result, a phenotypic trait can approach fixation in a population without exhausting genetic variation for its production (figure 22.2b), and a phenotype can be fixed without the fixation of any particular genetic allele (see chapter 6). This implies the persistence of genetic variation even under strong selection. It is a result that could be overlooked if it were forgotten that selection acts on phenotypes and not genes and that selection intensity (as distinct from the response to selection) depends upon *phenotypic* variation and not on genetic variation.

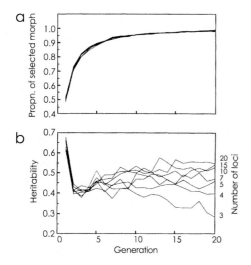

Fig. 22.2. Phenotype fixation without loss of heritability: (a) simulated selection to increase the frequency of one morph of a polymorphic trait with a polygenic regulatory mechanism causes the frequency of the morph to approach fixation; (b) change in heritability with selection. Each line is the mean of 20 replicates per number of loci influencing frequency of trait expression (right). Heritability of the trait at fixation is higher when a higher number of loci (right) influence trait expression. After Roff (1994a).

There are several possible explanations for continued genetic variation in switch points and phenotype ratios, in addition to reduced intensity of selection when a favored alternative approaches fixation:

1. When more than one alternative is favored, the best switch point may vary in time or in space, such that no single value is consistently selected.
2. Neurohormonal system components (e.g., juvenile hormone in insects, steroids in vertebrates) frequently are involved in regulation of several traits at once and are therefore subject to multiple and variable selective regimes, comparable to environmental heterogeneity in their effects on the response to selection.
3. Even under stable conditions and without conflicting selection, genetic variation in regulatory elements may be self-perpetuating, due to extensive interaction among elements. For example, variation in sensory acuity in

perception of environmental factors could lead to over- or underproduction of a hormone, which could in turn produce variable selection on effector responsiveness to return output to a favorable level. Due to such interactions, a stable genetically specified configuration of regulatory components may be difficult to achieve as long as there is genetic variation in any one element.

4. At intermediate values of conditions that influence phenotype determination—near the neutral zone for phenotype determination—there is no consistent advantage of one alternative over another. Yet it is in this zone of conditions that fine-tuning of a switch point must occur, since extreme conditions are more likely to obscure genetic variation in response. For this reason, small genetically based variations in response may be able to persist.
5. The switch point for alternatives such as wing-length morphs in an insect could be altered by genetic effects on numerous traits, including such things as sensory acuity, hormone output, tissue sensitivity to hormones, and ovarian development. Selection is blind to which of these traits is responsible for a particular change in switch point. As a result, selection for what appears to be "the same" trait (e.g., a higher response threshold) can in fact select for numerous underlying traits, with different ones effective in different individuals. It follows that the response to selection may involve numerous loci, each contributing a small effect to a statistical shift in the threshold, with no one of them markedly or consistently more important than any other (compare the model of Nijhout and Paulsen, 1997, where alleles of relatively major effect go to fixation).

All of these considerations support the idea that directional selection can lead to phenotypic fixation while genetic polymorphism is maintained.

Strong environmental influence on trait expression may mask any genotypic variation in switch point by exceeding the threshold of all genotypic variants. Thus protected, genetic variation can later be exposed when environmental influence is relaxed. This means that phenotypic plasticity can shield genetic variation from selection and allow it to accumulate across generations. The protective effect of a strong environmental switch is shown by the neutral-point experiments discussed in chapter 5, where intermediate larval diets resulted in variable queen–worker mosaics. The common occurrence of increased scatter near the neutral point of a conditional switch (see figure 5.21) is likely due at least in part to genetic variation that is hidden at the extremes, where the environmental influence is strong. In a species like the honeybee, with a very effective nutritional switch, variable intermediates showing genetic variation in response rarely occur in nature. The expected result should be an accumulation of genetic variation in switch point due to generations of lack of exposure to selection of "subliminal" (unexpressed) genetic variation in this trait.

In general, the more consistently effective the *environmental* determination of a switch, the greater should be the accumulation of unexpressed genetic variation in response. This hypothesis can be tested using the neutral-zone technique—environmental manipulation to expose variation near a conditional switch point, followed by artificial selection on that variation. The response to selection on the switch point should be greater in a species like the honeybee, with strong environmental determination and few intermediates, than in a species with weak or variable environmental determination and many intermediates in nature.

If genetic variation in switch points is maintained by variation in the *optimum* switch point, such that no single threshold is consistently favored by selection (see item 1, above), this leads to various predictions. There should be an association between (a) high environmental variation, (b) high variability in switch point, and (c) high genetic variation for switch point. All three phenomena have been documented in studies of the stored-product mite *Lepidoglyphus destructor* (Glycyphagidae; Knülle, 1987, 1991). *L. destructor* is a tiny and extremely abundant arthropod, a component of the "dust" that sparkles in rays of light filtering through the cracks of old haylofts. This organism has an extremely variable life-history switch. In favorable conditions, development is uninterrupted (direct) from egg to adult. But in unfavorable dietary or drought conditions, a preadult stage, the protonymph, forms a specialized morph (deutonymph or hypopus) encased in a modified protonymphal skin. The hypopus is a dispersal form, capable of escaping on air currents to other places. It is also highly resistant to the high temperatures and low humidities of drought fatal to all other stages of the mite's life cycle. The hypopus remains in diapause until favorable conditions induce it to molt and resume development.

The unfavorable temperature and humidity conditions that induce the hypopus are frequent, sporadic, and highly variable in duration. As expected, the threshold for hypopus induction is highly genetically variable. There is a rapid response to artificial selection for either readiness to form a hypopus, or direct development (omission of the hypopus stage; Knülle, 1987; Athias-Binche, 1995). There is also extreme variation in duration of the hypopus stage, which may last from one week to as long as a year. During that time, the mite undergoes a complex developmental process called "physiogenesis," then molts when stimulated to do so by high humidity and moderate temperature. If unfavorable conditions persist, it remains quiescent.

Also in accord with expectations, wild-caught mites show extreme variation in both hypopus induction and hypopus duration, with a strong genetic component to the variation (Knülle, 1987, 1991). Genetic variation in hypopus inducibility and duration was revealed not only by a rapid response to artificial selection (see above), but also by breeding experiments with populations having different average hypopus durations in the wild and by the fact that laboratory populations showed evolutionary change in hypopus duration (Knülle, 1991). Under a standardized rearing regime where the hypopus is induced and then given conditions favorable to molting, a stable hypopus duration evolved. In cultures maintained for over 30 years under favorable nutritional and humidity conditions, the ability to form the hypopus was nearly lost, although it can be induced at low frequency by a markedly inferior diet. This indicates a greatly reduced threshold for hypopus expression and an approach to genetic assimilation of direct development (Knülle, 1987). Genetic assimilation of hypopus loss has occurred in populations of the *Acarus siro* that have invaded stored agricultural products, where the hypopus cannot be induced even under conditions that usually stimulate hypopus formation (Athias-Binche, 1995). Other possible examples of "storage" of genetic variation due to variable estivation or dormancy are the so-called seed banks of plants in the soil (Ellner, 1984, cited by Warner and Chesson, 1985) and variable diapause duration of copepod eggs in lake bottoms (Hairston and Dillon, 1990). Similar phenomena in marine invertebrates are hypothesized to give rise to some plankton "blooms" (Boero, 1994).

In summary, environmental variability, because of extreme and irregular changes in direction of selection on the threshold of a conditional switch, can maintain genetic variation in a population. This suggests that environmental responsiveness, a universal quality of living things, may contribute to the maintenance of genetic variation, also a widespread if not universal quality of natural populations.

Maintenance of Genetic Variation: Artifact or Adaptation?

Some authors argue that rampant variation in switch point, such as that of the hypopus stage of *L. destructor*, which protects genetic variation from selection, could be selected for as mechanisms for maintenance of genetic variation per se. Hairston and Dillon (1990) take this view of the exaggerated egg diapause in copepods, which they interpret as multigeneration "storage" of genotypes. They propose that individual selection favors production of broods with variable timing in the egg diapause duration of copepods because the pool of inactive individuals would be protected from selection and therefore represent a reserve of genetic variation. That pool would then be capable of a rapid response to selection with no depletion of genetic variation, since there would always be a protected inactive pool in reserve. Knülle (1991) emphasizes the selective advantage *to the population* of the enormous variability produced by this combination of environmental sensitivity and genetic variation in response. Both arguments propose that protection of genetic variation has evolved as an adaptive trait. A population-level advantage is not necessary, as shown by the individual-selection argument of Hairston and Dillon (1990).

Selection at the colony level for maintenance of intracolony genetic variation per se has been hypothesized for social insects. Honeybee queens, for example, mate multiple times, and this increases the genetic variation within the colony. One hypothesis suggests that maintenance of intracolony genetic diversity is adaptive in social insects as a hedge against uniformity that could favor explosive multiplication of specialized pathogens and parasites within colonies (Hamilton, 1987b; Sherman et al., 1988, 1998; Oldroyd et al., 1997; Schmid-Hempel, 1998). Another is the adaptive task-threshold diversity hypothesis (e.g., Crozier and Consul, 1976; Crozier and Page, 1985; Moritz and Southwick, 1987; Page et al., 1989, 1995a; Page and Robinson, 1991; Keller and Reeve, 1994; Sherman et al., 1998; Gadau et al., 2000). Colonies depend for their survival and reproductive success on the performance by workers of a full set of specialized tasks. There is a genetic component to task determination, as shown by the fact that workers of different "subfamilies" (having the same mother but a different father) are biased toward the performance of certain tasks (Calderone and Page, 1988; see also Frumhoff and Baker, 1988; see reviews in Robinson, 1992; Moritz and Southwick, 1987; Page et al., 1995b). This has led to the hypothesis that multiple mating by queens "is favored because it increases the genotypic variation within colonies and this variation is a basis for the division of labor among workers" (Page et al., 1995a, p. 387).

Page et al. (1995a) tested this adaptive task-threshold diversity hypothesis experimentally. They constructed four contrasting sets of genetically homogeneous colonies by using sperm of drones from four different queens to artificially inseminate large numbers of queens, each with sperm from a single male. For comparison, a fifth set of colonies were made genetically heterogeneous by insemination of each introduced queen with a mixture of sperm from all four sources combined. All colonies were placed in the same area, grown during the same time of year, and observed well after the introduced queen's progeny were the only ones present among the workers. Colonies were then compared using 19 characteristics, including frequency of performance of certain tasks (e.g., defense, hygiene, corpse removal, foraging for pollen and nonpollen, nest repair) and overall condition (e.g., drone production, pollen and honey stores, forager production, weight of combs). Genetically uniform colonies showed greater deviation from the average in task performance. The genetically heterogeneous colonies, by contrast, approached the average for the study population as a whole. Page et al. (1995a) interpret these results as supporting the adaptive task-threshold diversity hypothesis: colonies that show the average distribution of tasks are less likely to fail because of inappropriate behavior or a restricted range of responses to changing conditions.

These results under experimetally equalized colony conditions show an effect of genetic variation on task performance, predictably nearer the population mean in genetically heterogeneous colonies. But if task performance is highly condition sensitive, as it is known to be in social insects (see below), then in diverse and changing conditions the contribution of this genetic variation would be small compared to influential environmental signals (see chapter 5), reducing the importance of genetic variation per se. Then selection should improve mechanisms of plasticity, not mechanisms to maintain genetic diversity.

Many other studies show that genetic variation within colonies is not required to maintain adaptive task diversity in social insect colonies. Social insect workers, including not only honeybees (see Winston, 1987) but also in groups such as termites with long-term monogomy of queens, are phenotypically flexible and capable of performing a diversity of tasks, varying the frequency of performance in response to colony conditions. Workers are well known to adjust conditional task performance ratios in response to colony needs in natural populations of social insects, including honeybees (Jeanne, 1996a; Gordon, 1995a,b; Seeley, 1986, 1989). No genetic variation in task thresholds is required for such task diversity, though of course genetic variation in task thresholds like that shown by Page et al. (1995a) is required for adaptive levels to evolve.

Conclusions

Understanding the maintenance of alternative phenotypes is clarified by realizing that it always involves the evolution of regulation at a polygenic switch or decision point that is potentially subject to both genotypic and environmental influence. The findings discussed in this chapter underline the importance of the distinction between genetic *influence* and genetic *control*. Future modeling efforts should be directed at understanding the evolution of conditional alternative phenotypes. Unsolved problems include (1) the quantitative relations during evolution between the frequency of alternatives and the improvement of their form, (2) the importance of complex regulation of phenotype expression for the protection of genetic polymorphism, and (3) the degree to which maintenance of genetic polymorphism is a product of selection for genetic variation per se, or an artifact of fluctuations in the selective value of alternatives. It may be fruitful to abandon the search for stable states of complex alternative phenotypes in nature in favor of a new focus on the importance of environmental fluctuations and dynamic instabilities—the factors that drive the evolution of the polygenic regulatory mechanisms that control the expression of adaptive alternative phenotypes.

23

Assessment

Adaptive switching between alternative phenotypes requires assessment of environmental conditions sometimes considered beyond the capacities of nonhuman organisms. But remarkable feats of assessment occur, both in the regulation of alternative phenotypes and in many other contexts, such as selective mating, selective feeding, and discriminatory social aid, where organisms exercise choice. Like other aspects of the phenotype, adaptive mechanisms of assessment capitalize on preexisting phenotypic elements such as hormone systems, which are sensitive to environmental conditions and are versatile in their effects.

Introduction

A book on developmental plasticity needs a chapter on assessment, if only to show that adaptive environmental assessment occurs. Skepticism regarding the ability of nonhuman organisms to assess conditions well enough to make adaptive decisions has a long history in evolutionary biology, and it has been an important barrier to understanding the evolution of adaptive developmental plasticity. It is worth briefly reviewing this history in order to understand certain preconceptions about assessment that still persist.

In the nineteenth century, critics of Darwin's theory of sexual selection (Darwin, 1871) balked at the idea of an "aesthetic sense" in lowly creatures that would enable female choice of mates (representative papers are reprinted and discussed in Bajema, 1984). Later, the barrier persisted for other reasons. Even though naturalists routinely used the condition-appropriate expression of phenotypic traits to support adaptation hypotheses—a practice that assumes adaptive assessment of conditions as it is defined here—theoretically inclined biologists paid little attention to the question of facultatively expressed traits. Part of the difficulty lay in the problem of explaining how adaptive assessment could evolve within the framework of conventional genetics.

Theodosius Dobzhansky, one of the twentieth century's leading evolutionary biologists, acknowledged this unresolved problem in remarks following a lecture by J. S. Kennedy on the phase polyphenisms of migratory locusts (Kennedy, 1961). Dobzhansky referred to the "challenge to a geneticist" of explaining the adaptive switch between the sedentary and the migratory phenotypes of the locusts, which had been shown to be largely independent of genotype. He suggested that an extrachromosomal factor may be involved, a symbiotic microorganism that acts as a "plasmagene" whose multiplication would eventually stimulate phase change (Kennedy, 1961, p. 88). Although Dobzhansky's proposal was no more preposterous than some of the regulatory devices that have actually been discovered, Kennedy (1961) minced no words in his reply to this suggestion:

[W]e need not feel obliged to invoke a second organism to explain [phase polymorphism] unless we are reluctant to concede an important part to the environment as well as to heredity in moulding development. That is the only challenge to geneticists I can see. (p. 88)

Reluctance to attribute adaptive assessment capacities to nonhuman organisms continues to be a leading reason for theoretical neglect of conditional alternatives in favor of stochastic and genotype-specific ones that seem to offer simpler explanations (e.g., see Riechert, 1982, p. 313; Maynard Smith, 1982; Rosenzweig, 1990). This was especially common in the 1980s, before the predominance and feasibility of adaptive conditional tactics was well established, and the mixed evolutionarily stable strategy (ESS) was attracting attention to alternative strategies. When Rohwer (1975, 1977) demonstrated that dark plumage markings are used by male Harris sparrows to assess degree of dominance of individual males, who adjust their aggressive tactics accordingly, this "status-signaling badge" interpretation was criticized due to doubts that both assessment and adaptive response could arise simultaneously (each depends upon the other) and not be readily invaded by evolved mimicry of the dominant badges by subordinate weaklings (Maynard Smith, 1982). Rohwer (1982) responded to this criticism within the theoretical framework in which he was criticized. He hypothesized that a mutation that produces a dependable or "honest" badge could conceivably spread by several genetic means, for example, (1) if linked to a gene for superior resource-holding power, (2) if a single gene provided both the badge and (as a pleiotropic effect) improved fighting ability, or (3) if a gene were facultatively expressed so as to "code for a badge when in a dominant individual" (p. 532). The third, developmental-plasticity solution requires some mechanism that links plumage coloration and aggressiveness. Until such a mechanism is evident, the conditional explanation is less convincing than are those based on more familiar genetic mechanisms such as chromosomal linkage and pleiotropic effects.

In fact, a mechanism supportive of the developmental-plasticity explanation is readily available in the form of a developmental linkage known to occur in birds in nature, namely, a hormonally mediated correlation between fighting ability (aggressiveness) and badge (plumage coloration). Hormones have long been known to influence both the aggressiveness and plumage variation of birds (e.g., Beach, 1961; Mayr, 1933; Wingfield et al., 1987). A correlation between the two has been hypothesized to underlie other male plumage and behavioral polyphenisms, such as that of male ruffs (*Philomachus pugnax*; Andersson, 1994, pp. 389–390). The hormonally mediated correlation would make the badge an honest indicator of aggressive dominance, and both correlated traits would vary in their degree of expression due to genotype, age, sex, and environmental variation, as documented in Rohwer's publications, allowing status-dependent conditional alternatives to be adaptively cued by relative plumage coloration as an indicator of relative aggressiveness. Maynard Smith's worry that nonaggressive individuals could mimic the dominance-badge coloration, and thereby undermine the system, would presumably

be selected against: since the plumage badge is an honest signal of phenotypic condition, subordinates should benefit by *not* mimicking the dominants' badge, thereby avoiding costly and futile fights.

In retrospect, it is remarkable that discussions of adaptively flexible social behavior have contained so little mention of developmentally (especially, hormonally) mediated correlations of this kind. In a treatise on sexual competition, Darwin (1871 [1874]) noted that breeding plumage of male birds, antlers of male deer, and kypes of male salmon develop as alternatives to less exaggerated features only in the breeding season, a phenomenon that readily suggests mediation by sex hormones. In the first half of the twentieth century, the hormonal basis of adaptive changes in vertebrate behavior was extensively discussed by comparative psychologists such as Kuo, Beech, and Lehrman. But their interest in the proximate causes of behavior did not cross the barrier between comparative psychology, with its emphasis on learning in nonhuman animals, and classical ethology, with its greater emphasis on so-called instinctive, or "natively determined" (Maier and Schneirla, 1964) behavior. Ethology, personified by such figures as Tinbergen, Lorenz, and von Frisch, was more evolutionary in its emphasis on field studies of species-typical and adaptive behavior. But tests of explicitly evolutionary hypotheses were limited to ethological observations of adaptive function in ethology, or to gross comparisons of organismic "levels" in comparative psychology (e.g., Maier and Schneirla, 1964) without reference to systematic phylogenetic analysis. The landmark text of Hinde (1970) was a synthesis of ethology and comparative psychology, but Hinde's discussion of evolution is brief and largely limited to questions of speciation and phylogeny. So Hinde's book did not serve as a bridge between mechanism-oriented comparative psychology and the emerging branch of animal behavior now called behavioral ecology, which has focused heavily on the formulation and testing of hypotheses concerned with adaptation and natural selection. At stake during the history of animal behavior studies were differences of opinion about proximate and ultimate causation, nature and nurture, and the usual differences in personalities and politics that divide fields (for perceptive accounts, see Hinde, 1970, pp. 425–434; Dewsbury, 1990; Barlow, 1991). It is important to realize that embryology was not the only discipline from which a developmental approach could have reached evolutionary biology and yet failed.

Given this history of estrangement among fields, perhaps many biologists still share the opinion of

Wade (1984), that since "facultative expression . . . depends on the prior existence in individuals of an ability to gather a great deal of information about the environment . . . the very liberal use of conditional strategies in adaptive explanations is unwarranted" (p. 707). Nonetheless, adaptive conditional alternatives have proven exceedingly common, and Wade later acknowledged this (Shuster and Wade, 1991a), citing 17 different references and concluding that "[m]ost polymorphisms in male mating behaviour seem to be condition dependent" (p. 610).

Given the resilience of conditional alternative phenotypes in the face of environmental change, and their apparent commonness in nature, it is odd that they have so persistently been considered more difficult to evolve and maintain than genetic polymorphisms or stochastic mixed tactics. Explanation of assessment—of how seemingly intelligent adaptive decisions get incorporated into development—is an important frontier, as important for the explanation of adaptive development as equilibrium models have been for the explanation of genetic polymorphisms.

In this chapter, I extend the discussion of assessment beyond switches between alternative phenotypes to give a broader view of assessment in behavioral and developmental decisions. Assessment of mates, for example, is not part of a switch between alternative phenotypes, but it is similar to other kinds of social assessment, such as the assessment of dominance in social insects, that does influence the switch between alternative worker and queen phenotypes. So it is useful to discuss these two examples of assessment together. If adaptive assessment is understood as a widespread phenomenon that has reached a high degree of refinement in a variety of biological settings, then skepticism regarding assessment in general should diminish.

Terminology: Assessment and Other Anthropomorphisms

Assessment—evaluation of environmental circumstances, competitors, or mates—occurs whenever a particular response correlates consistently with some environmental variable. *Choice* between two or more actions, pathways, objects, or individuals occurs when there is a differential response to stimulus differences associated with the alternatives, that is, if an organism responds differentially to different stimuli (West-Eberhard, 1983; Bateson, 1990, p. 149). Obviously, these are very broad concepts of assessment and choice. They would apply

even to the very simple discriminations implied when an organism moves toward or away from some stimulus, indicating that it has assessed stimulus gradients or environmental heterogeneities of specific kinds and chosen to respond differentially to them. Some of the controversy over assessment and choice has been the result of anthropocentric concepts of choice that require evidence that organisms "actually know of the alternatives . . . that the animal's central nervous system contains information about both alternatives simultaneously" (Rosenzweig, 1990, p. 164). I prefer the more readily operational definition just given, especially since it is open to question whether human mate choice and other vital decisions in fact involve the simultaneous comparisons implied by Rosenzweig's definition of choice. While introspection suggests that humans can make direct comparisons using simultaneously available information, this does not mean that they actually do so rather than responding differentially to other factors known from introspection to be important, such as "emotional reactions" or "physical attraction" that need not involve our capabilities for simultaneous comparison.

The mechanisms of assessment discussed in this chapter could be included, along with those of chapter 3, among the mechanisms of flexibility. Chapter 3 emphasizes the continuously variable adjustments that permit phenotypic accommodation. This chapter emphasizes instead switch-mediated plasticity, or decisions among alternatives, including alternative behaviors and other developmental pathways as well as alternative mates, beneficiaries of aid, or nesting sites. There is no absolute difference between continuous and discrete plasticity, for graded accommodation often involves switchlike behavior by the responding elements, as when a variable number of microtubules either depolymerize or switch to the polymerized state, or a discrete behavior is repeated a variable number of times. The important quality of assessment is that some sort of environmental evaluation, indicated by differential responses in different conditions, is involved.

No doubt some readers will object to use of the purposeful-sounding word "assessment" for the processes that influence the behaviors and morphologies of such nonhuman organisms as toads, insects, plants, and viruses. In the words of Ohno (1985), a theoretical molecular geneticist well known for discussions of gene duplication (see chapter 10), the "ingenious rationalizations" of research on adaptive social behavior have "led to discussions of evolutionary strategies giving ants and bees foresight worthy of Prometheus" (p. 160).

Darwin had a good answer for such worries when he faced accusations of anthropomorphism and teleology for use of the word "selection" to mean the differential survival of variable individuals:

> In the literal sense of the word, no doubt, natural selection is a false term; but who ever objected to chemists speaking of the elective affinities of the various elements?—and yet an acid cannot strictly be said to elect the base with which it in preference combines. It has been said that I speak of natural selection as an active power or Deity; but who objects to an author speaking of the attraction of gravity as ruling the movements of the planets? . . . With a little familiarity such superficial objections will be forgotten. (Darwin 1859 [1872], p. 64])

For some authors, a rigorous test of choice involves sequential encounters, and memory is required (e.g., Bateson, 1990). Memory is known to be involved in some of very common kinds of decisions I classify as choice, for example, in behaviors that selectively distinguish kin from nonkin, where memorized identity badges of kinship are borne by nestmates (see below); and in orientation choices by displaced homing organisms (Scapini, 1988), where individuals memorize cues that indicate home. But memory need not be involved in choice, for example, when a female preferentially mates with ("chooses") the one of a succession of suitors that stimulates her effectively enough to elicit a positive response.

Critics of anthropomorphism worry that biologists will attribute uniquely human capacities to nonhuman organisms. I have the opposite worry, that obsessions with human specialness will prevent us from seeing the remarkable assessment capacities of nonhuman organisms, and how these capabilities relate to our own sometimes less-than-rational decision processes. The fact that nonhuman organisms are known to *adaptively* switch between conditional alternatives implies that they somehow use criteria of advantageousness to assess the conditions surrounding their decisions. That many are unlikely to do this using conscious thought or a college education makes their behavior all the more intriguing.

Selected Examples

Every adaptively flexible aspect of development and behavior reflects some means of assessment of the environment. This chapter contains an admittedly haphazard sample of an enormous field, just to illustrate some contexts in which assessment can occur and to discuss how it can evolve.

Assessment during Feeding and Foraging

European ethologists pioneered experimental research on assessment with elegant and incisive experiments on adaptive flexibility in predation behavior and assessment of provisions (Curio, 1976; Bell, 1991). For example, classic work by Baerends (1941) showed that female wasps (*Ammophila campestris*; now *A. pubescens*) inspect their brood cells before hunting. Then, if the larva within the cell is small, the female brings from one to three prey caterpillars; if it is medium sized, she brings from three to seven; if it is fully grown, she seals the cell and starts another one. In a recent extension of this work, Field (1992b) showed that *A. sabulosa* females assess the relative sizes of prey, not just prey numbers. If the first caterpillar caught is large, the female may close the cell, but if it is small she brings additional, small (rarely large) prey. Birds, such as the loon *Gavia stellata*, also adjust prey size to the size of the young destined to consume it (Reimchen and Douglas, 1984).

Bean weevil females (*Callosobruchus maculatus*; Insecta, Coleoptera) assess the food that will be available for their larvae before ovipositing in a bean. They judge the size of beans and whether or not they already contain an egg. They then lay eggs in the largest available beans first and avoid laying a second egg if another is present, even on very large beans, as long as there are unused beans available (Mitchell, 1975). I return to these beetles later in this chapter to discuss how the larvae in turn assess the quality of competitors within the beans in life-or-death contests with other larvae.

Optimal foraging theory (reviewed by Stephens and Krebs, 1986) makes detailed predictions regarding decisions based on food quality, abundance, and foraging risk that would seem to demand near-human capacities for assessment by humble organisms. Research stimulated by these models and the older observations of ethologists (e.g., see Curio, 1976; Bell, 1991) have demonstrated the commonness of adaptive foraging decisions that require fine-tuned assessment of the costs and benefits of resource acquisition.

In 125 tests of optimal foraging theories that used a variety of organisms, including mammals, birds, fish, crustaceans, and insects, animals proved capable of the required assessments of many parameters (Stephens and Krebs, 1986, their table

9.1). Although the mechanisms of assessment are usually not discussed, and the precise parameters assessed are therefore open to debate, these studies demonstrate refined assessment of a wide variety of kinds: of prey value with preference for more profitable prey (44 studies); of prey abundance and quality, with selectivity increasing when prey is abundant (25 studies); of foraging conditions, with individuals remaining in a particular food patch longer when foraging success is poor (22 studies); of relative value of patches, with individuals moving to a new patch when its value declines, such that all visited patches are reduced to the some "marginal value" (four studies); and of prey size relative to foraging distance, with larger prey being brought from greater distances (nine studies).

Organisms "should and really do make their foraging decisions not only according to the expected net benefits but also according to the associated variances" in rewards (Regelmann, 1986, p. 321). Some organisms conditionally choose constant over variable rewards when they are running below their energy requirements, a complex feat of assessment called risk sensitive foraging observed in birds (juncos and white-crowned sparrows; Caraco, 1983) and shrews (*Sorex araneus*; Barnard and Brown, 1985). Such conditional adjustment did not occur in sticklebacks (*Gasterosteus aculeatus*), perhaps because cold-blooded vertebrates can more readily survive periods of starvation and therefore survive without high-risk foraging (Milinski and Regelmann, 1985; Regelmann, 1986). Risk-sensitive foraging does occur in other fish, such as juvenile creek chubs (*Semotilus atromaculatus*), which shift from a safe to a hazardous site at the level of food shortage that accords with theoretical predictions (Gilliam and Fraser, 1987).

Learning is so common in the foraging decisions of insects that the *absence* of learning is sometimes considered noteworthy (Parmesan et al., 1995). The butterfly *Euphydryas editha* has been nominated as a model "nonlearning" insect whose behavior contrasts with that observed in most studies of insect foraging, "all of which have shown that learning is an important component of foraging behaviour, causing search efficiency to improve with experience" (Parmesan et al., 1995, p. 161). The larvae of the dragonfly *Hemianax papuensis* (Odonata: Aeshnidae) show unlearned specialized alternative behaviors in handling different kinds and sizes of prey: when a naive larva is offered a medium-sized snail for the first time, for example, it immediately performs a specialized routine of stalking, maneuvering around the snail, grasping it with the labium, and peeling the snail out of the shell, a sequence that contrasts with that used on small snails, which are lifted to the mouthparts and attacked by mandibles and maxillae while the labium presses the snail toward the mouth (Rowe, 1987).

Learning plays an important role not only in improvement of food location but also in food selection and achievement of a balanced diet in insects and other animals (Bernays, 1993; Bright et al., 1994). Aversion learning occurs in rats and other mammals, flies, grasshoppers, and slugs. Slugs develop aversions to foods that lack essential nutrients, and preferences for other, complementary foods. They can discriminate, and develop an aversion for, a food lacking only one essential amino acid (Delaney and Gelperin, 1986, cited by Bernays, 1993).

Kangaroo rats (*Dipodomys spectabilis*) not only make fine assessments of food quality but also manage their food stores in a manner that improves their diet. Kangaroo rats store large quantities of seeds in underground caches. They prefer slightly moldy seeds, rather than nonmoldy or very moldy ones. Experimental manipulations using laboratory chambers show that they move nonmoldy seeds to the most humid locations available and that they place "ripe" seeds with the preferred level of mold in relatively dry places where further mold growth will be inhibited (Reichman et al., 1986).

Honeybees also engage in mass management of their food industry, by allocating (through individual foraging decisions) more foragers to rich nectar sources (Seeley et al., 1991). A bee that has visited a rich source performs a recruitment dance at the hive that is up to 100 times longer and contains many more turns or "circuits" than the ones it performs for a weak source (Seeley and Towne, 1992). Individual bees do not systematically sample different sources available in the neighborhood and compare them. Rather, they react to concentration and nearness to the hive. But they do compare nectar richness with that sampled in their recent experience, and they are especially sensitive to *declining* nectar richness, which causes a proportionately greater change in dance rate than does *increasing* quality (Raveret Richter and Waddington, 1993). Bees also do not systematically sample and compare dances of other foragers before making a decision. Rather, a worker that has abandoned a poor source may follow several dancers but then concentrate on one, apparently at random with respect to dance intensity (but see below). Recruitment is proportional to the number of circuits in dances associated with particular sources, and not necessarily skewed toward the richest source (Seeley and Towne, 1992).

Accuracy of assessment may be improved by the simple mechanism of size uniformity within dancing groups: the subsets of colony members that associate in transfers of information contain bees of similar size relative to the range of sizes present in the colony as a whole. Since the desirability of a food source depends partly on ease of nectar collection, handling time, and transport, and because large bees, for example, can more easily extract nectar from flowers of certain structure and corolla length, a "good" source for a large bee is not the same as a good source for a small bee. So size similarity within dancing groups as well as within colonies as a whole could increase the accuracy of communication (Waddington, 1989). In keeping with this, Waddington (1989) also found a weak negative correlation between size variance within colonies and foraging efficiency, and notes that bees (Apidae, Meliponini) having food-source communication also have uniform worker sizes compared with bees lacking such communication.

Honeybees assess the colony nectar supply (intake rate and stored supplies) by the time required to dispense foraged nectar to nestmates, who are slower to take in nectar when the colony is well supplied (Seeley, 1986). When demand is low, foragers preferentially abandon weak-nectar sources in favor of richer sources (Seeley, 1989).

Many organisms adjust their predatory tactics to suit the prey at hand (Curio, 1976; Bell, 1991). The specialized behavior adopted by a mouse in catching scorpions is an example (Horner et al., 1965; see chapter 19). Another is the choice of tactic used by the tiny hawk, *Accipiter suprciliosus*, to catch hummingbirds, either waiting in ambush near the territorial perch of one individual prey, or flying rapidly between the perches of several (Stiles, 1978). The tactic used appears to depend on assessment of the local density of territorial hummingbirds available as prey.

Tactical judgment of this kind is not limited to vertebrates. Some jumping spiders (Salticidae) of a genus aptly called *Portia* prey on web-building spiders by mimicking their prey. *Portia* makes prey-like vibratory signals on the victim's web until the owner approaches to make the kill and then is itself killed and eaten (Jackson and Wilcox, 1993). When first stalking its spider prey, *Portia* broadcasts a variety of tentative vibratory signals, then, as a consequence of feedback from the prey spider's movements and web vibrations, it narrows its performance to particularly effective signals. It thus uses trial-and-error experimentation and learning to select the most effective luring behavior for a particular prey. By this means, *Portia* is able to ex-

ploit a variety of types of web-building spiders, in contrast with most such mimics, which have a much narrower repertoire of mimicry signals (Stowe, 1986; Jackson and Wilcox, 1993). In a further flourish of tactical assessment, *Portia* spiders exploit episodes of windiness to move forward quickly while the vibratory "noise" of the wind disguises the vibrations of their approach to prey (Wilcox et al.,1996; Jackson and Wilcox, 1998). Similarly, ichneumonid parasitoids of *Polistes* wasps assess the alertness of host guards at the periphery of their hosts' nest. They remain immobile for as long as twenty minutes when faced by a vigilant wasp, then jump onto the nest and lay their eggs the moment a guard turns away (West-Eberhard, 1969).

Assessment and planning ahead also occur when spiders are obliged to construct webs with a limited supply of silk. Orb building starts with a series of radiating straight lines of nonsticky silk, followed by an overlay of spiral sticky-silk loops. When a spider's supply of nonsticky silk has been depleted due to previous construction activity, it behaves as if it is aware that it has a limited supply. Without reducing the area of the web or the regular spacing of lines, it reduces the number of radii and sticky loops constructed (Eberhard, 1988).

Many observations indicate that social animals evaluate not only the resource itself but also the consequences of communicating about it. Early observations of chimpanzees in the wild, for example, showed that when food is abundant, rich sources of fruit cause noisy excitement, which attracts others to the site, whereas in times of food scarcity chimps search in small groups and exploit their finds quietly (Reynolds, 1970). So they assess both food quality and the social expediency of advertisement. Cockerels in domestic chickens tailor their vocalizations to the nature of the food they discover as well as to the nature of the conspecifics that are present and likely to respond (Marler et al., 1986a,b, cited by Curio, 1988).

Assessment of Where to
Live and What to Wear

Many insects are host specific, especially in the larval feeding stage, and in some of them host-specific adaptations have evolved. Not surprisingly, alternative hosts of contrasting colors are sometimes associated with alternative color phenotypes, and their study has revealed varied mechanisms of assessment. While determination of color morph is sometimes influenced by ambient temperature or humidity, and some insect larvae respond to the wavelength of incident and reflected light in adjusting larval or pupal body color to match that of their hosts (see references in Greene, 1996). A remarkable seasonal polyphenism in caterpillars of the moth *Nemoria arizonaria* (Lepidoptera: Geometridae) is governed by larval diet. Caterpillars hatched in spring feed on oak catkins and are virtually indistinguishable from them in appearance, developing yellow and brown splotches, densely rugose skin, two rows of green dots that resemble oak stamens, and large processes on both sides of the body, giving them a catkinlike, knobby appearance. The same morphology can be induced by feeding later-hatched caterpillars on oak catkins. Caterpillars hatched later in the summer and feeding on oak leaves closely resemble twigs, with grayish-green smooth skin and less pronounced lateral projections (Greene, 1989, 1996). Temperature, day length, and color of light have been ruled out as influential factors.

Hermit crabs are decapod crustaceans that protect their bodies in empty snail shells with only their legs exposed, switching to larger shells as they grow (Hazlett, 1981). When two crabs exchange shells, they first engage in stereotyped rocking, shaking, and rapping movements. Exchanges of experimentally modified shells suggest that this "fighting" behavior serves to communicate information about the quality of shells and to influence whether or not an exchange will take place (Hazlett, 1981, p. 12). The highly developed assessment of shells reflects the fact that changing shells is costly, partly because suitable shells have specialized characteristics that are often in short supply.

In keeping with a suggestion of Childress (1972) that hermit crabs choose shells that allow room for growth, there is evidence that at least *Pagurus longicarpus* adjusts its standards of choice depending on assessment of the local *supply* of shells. When individuals from a shell-limited location are given free access to shells, they choose larger ones than do individuals from a region with excess shells (Scully, 1979). Behavioral observations indicate that this is due to experience rather than an evolved change in preference.

The house-hunting expertise of the hermit crab is impressive, but that of European honeybees is even more so because it involves both assessment and mass politics. Individual scout bees evaluate the suitability of potential nesting cavities primarily on foot, sometimes walking a total of about 50 meters around the inside of a cavity during a single inspection (Seeley, 1977, 1982). Upon return-

ing to the hive, they perform propaganda displays, in the form of excited dances with zig-zag runs and buzzing of wings, and they also follow the dances of numerous other scouts and eventually reach a consensus by successive votes (Seeley, 1982). They are able to distinguish between cavity volumes differing by 15.0 liters, and given a choice between 10-, 40-, and 100-liter cavities, they prefer the 40-liter ones. Offered a series of choices, they prefer a cavity 5 meters in height over one that is 1 meter in height. They also assess the size of the entrance, and prefer an entrance whose area is less than 50 square centimeters in size, at least 2 meters above the ground, at the bottom rather than the top of the cavity, and facing southward rather than northward. They choose a previously inhabited site over a previously uninhabited one, and a site distant from their former hive (> 300 meters away) over nearer ones (Seeley et al., 1978; Seeley, 1982).

Assessment is complicated by the fact that a few hundred scouts from a single swarm sometimes inspect many potential sites over a period of several days before coming to a decision. Scouts communicate with each other by dancing on the surface of the swarm, vigorously for a high-quality site and "unenthusiastically" for a low-quality one. They also follow the dances of other scouts, which contain information on the quality and location of sites. Using this information, they visit sites discovered by others, eventually reaching a consensus when all scouts visit and recruit to the preferred site, to which they then guide the waiting swarm (Seeley et al., 1979; Seeley, 1982). Because scouts follow dances in proportion to the amount of dancing by others for particular sites, the scouts can reach a decision without direct comparison of sites by individual scouts, a finding confirmed by an experiment that prevented scouts from visiting more than one site (Visscher and Camazine, 1999). But the finding that prevention of direct comparisons did not retard the decision between equal sites does not justify the conclusion (Visscher and Camazine, 1999; Camazine et al., 1999) that direct comparison of sites plays no role in decisions. Indeed, contrary to the authors' expectations, one could predict that prevention of direct comparison of *equal* sites, as in these experiments, would speed the decision process by removing checks on the snowballing effect of following dances in proportion to those being performed. A more incisive test would be to measure the effect on decision speed of preventing direct comparisons in decisions between *unequal* sites.

Assessment of Relatedness

The birth of sociobiology (E. O. Wilson, 1975) led to serious concerns, especially in the human social sciences, that a biological theory of social behavior would promote rampant genetic determinism. Despite these worries, much of the research inspired by kin selection theory documented not genetic determination of behavior, but adaptive developmental and behavioral *flexibility*, especially in the expression of alternative social traits that leads to the division of labor between helpers and helped within social groups. Patterns in the division of labor seemed to reflect evaluation of costs, benefits, and relatedness, even by insects (e.g., see West-Eberhard, 1969, on wasps). Although experimental demonstrations of nestmate recognition in social insects date back at least to the nineteenth century (Lubbock, 1884), the new theoretical framework led to far deeper research, not only on the fitness consequences but also on the mechanisms by which "foresight worthy of Prometheus," formerly beyond the imaginations of most of us, is actually achieved in nonhuman animals.

Some of the most preposterous-sounding stories invented in what was seen by some as a euphoria of adaptationism turned out to be true, to a degree far beyond what a cautious observer would have predicted (for a concise review on the social insects, see Queller and Strassmann, 1998). Without the convincing fitness-oriented logic of kin selection theory (Hamilton, 1964a,b), most workers in these fields would not have noticed the remarkable capacities for assessment of group-member recognition and relatedness that have been shown to exist. The conceptual and factual progress stimulated by Hamilton's work is comparable, as a scientific bonanza within evolutionary biology, to the modern progress in molecular biology that was stimulated by the work of Watson and Crick on DNA.

Hamilton's rule ($K > 1/r$) gives the conditions for advantageous conditional aid and implies assessment of both the costs and benefits to the performer (K), discussed in the following section, and degree of genetic relatedness to the beneficiary (r, their fractional degree of relatedness by descent). Early kin-selection discussions took a conservative view of the assessment abilities of insects, adopting a suggestion of Hamilton (1964b) that kin associations could be fostered by *population viscosity*, a simple failure to disperse from an organism's birthplace. Although many social organisms do show population viscosity (e.g., see West-Eberhard, 1969, on the social wasp *Polistes*; Johannsen and Lubin, 1999, on social spiders), they do not neces-

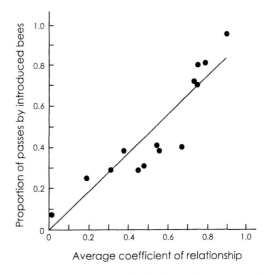

Fig. 23.1. Assessment of relatedness by nest-guarding bees (*Lasioglossum zephyrum*, Halictidae). Acceptance of bees introduced to the nest entrance was indicated if a guard bee allowed the introduced bee to pass into the nest. Bees ranged in degree of relatedness from those collected at distant sites, through family members (cousins, aunts and nieces, sisters of guards), to highly related individuals obtained by sib mating (inbred lines). Points were weighted by the number of interactions for that degree of relationship, using a total of 593 bees to obtain 1,586 interactions. After Greenberg (1979).

sarily depend on it, and stopping with that (e.g., in West-Eberhard, 1969) turned out to underestimate the assessment capacities of the social insects. Even relatively primitive societies of social wasps and bees soon were shown to be capable of distinguishing among various degrees of kin (Greenberg, 1979; figure 23.1). *Polistes* wasps learn and remember the identity of nestmates and distinguish them from nonnestmates, using chemical cues acquired from the nest (reviewed by Gamboa, 1996; see also Panek and Gamboa, 2000).

Some specialized social insects, such as honeybees, could be expected to show the most highly refined abilities for kin recognition. In honeybees, workers belong to different patrilines due to the multiple mating of the queen. But so far, they have not proven capable of differentiating between full and half siblings in their treatment of larvae, although the data on this are equivocal (for critical discussions, see Oldroyd et al., 1994; Visscher, 1998). During the fissioning of a swarming insect colony with more than one queen, when females could be expected to divide into groups on the basis of kinship, kin discrimination does not seem to occur (see Queller and Strassmann, 1998, Solís et al., 1998, on wasps; Visscher, 1998, on honeybees). Nor could within-colony kin discrimination be demonstrated in the multiple-queen tropical social wasps *Parachartergus colobopterus* (Strassmann et al., 1997) or in cooperating carpenter bees (*Xylocopa pubescens*) when nonrelatives supersede relatives as dominant reproductives (Hogendoorn and Leys, 1993). There may be colony (queen)-level selection against intracolony discrimination in social insects. But the possibility of intracolony kin discrimination under some heretofore untested conditions promises to perpetuate research on this question (Mohammedi and Le Conte, 2000). In the stingless bee *Melipona panamica*, for example, discriminatory aggressiveness toward intruders increases when there is a disturbance at the nest, and also depends on the age of the intruder (Inoue et al., 1999). And in some ants, colony fission occurs only when relatedness is high, possibly indicating some mechanism for assessment of intracolony variation in relatedness (Heinze et al., 1997b).

Researchers who have sought refined kin discrimination in birds have not been so disappointed. An intricately social African bird, the white-fronted bee-eater (*Merops bullockoides*), proved a favorable subject for demonstrating such discrimination, as they live in multigenerational but fairly small extended families of 3–17 individuals, with two or three mated pairs plus a coterie of single adults, all closely related (summarized in Emlen et al., 1995). Bee-eaters are cooperative breeders in which helpers other than a breeding pair participate in nearly every breeding activity except copulation, digging nest chambers, bringing food to breeding females, incubating the clutch, and defending the young. When a helper has a choice of breeding pairs to help, it aids the most closely related pair in more than 90% of observed cases. Ironically, this loyalty is repaid by harassment of close relatives when help is needed: parents sometimes interfere with the courtship and nesting attempts of their own sons in order to recruit them as helpers. This dramatically increases the provisioning per nestling and raises the number and condition of offspring fledged (Emlen and Wrege, 1991).

There have been many surprises in the study of kinship assessment. Kin aggregation by tadpoles (Waldman and Adler, 1979) was at first greeted with skepticism because there seemed to be no adaptive reason for this to occur. Subsequent work (reviewed by Blaustein et al., 1986) showed a similar phenomenon in various other amphibian species. Now it is known that tadpoles of at least some amphibians are often cannibalistic, and that they preferentially dine on companions who are not their kin, a preference that is most marked in the individuals, sibships, and species that are most frequently cannibalistic (Pfennig, 1997, 1999). The cannibal morph of spadefoot toad tadpoles avoids eating kin, whereas when the same individuals revert to the noncannibalistic morph they do not avoid eating kin (they are still occasionally cannibalistic; Pfennig, 1999). This is a key observation because it indicates that kin discrimination is an evolved specialization of the cannibalistic morph, and is morph limited in expression (controlled by the switch between morphs).

Tadpoles of the salamander *Ambystoma tigrinum nebulosum* seem to use the same olfactory signals that influence their fussy eating as a morphogenetic cue: when in water conditioned by the presence of kin, they are less likely to transform into a cannibalistic morph. The timing of the larval transformation varies in accord with the degree of kinship sensed, indicating that "small differences in kinship environment influenced morphogenesis" (Pfennig and Collins, 1993, p. 837; see also Pfennig and Frankino, 1997, for similar findings on spadefoot toad tadpoles, *Spea bombifrons*). This example nicely illustrates the similarity between behavioral assessment and regulation of morphological development. Kin selection has been shown to be the likely selective context responsible for the kin-recognition abilities of these salamanders (Pfennig et al., 1999).

The prize for most surprising assessment of relatedness in invertebrates should perhaps go to an isopod. Desert isopods (*Hemilepsis reaumuri*) live in groups of up to 80 relatives within fiercely defended burrows located in crowded aggregations of frequently interacting but closed family groups (Linsenmair, 1987; see figure 18.1). They use a composite identification odor or badge composed of the memorized individual odors of all family members combined. In thousands of field observations, there was not a single case of mistaken identity, and this is accomplished with a brain consisting of only 10,000 neurons, fewer than 6,000 of which are concerned with central processing of chemical stimuli—a minuscule structure compared, for example, with the 20 billion neurons of the human cerebral cortex (Calvin, 1983).

There is a close similarity between kin recognition in social organisms and self-recognition in fungi and other multicellular organisms that permits avoidance of infection and parasitism by alien genotypes (Rayner and Franks, 1987). This again draws attention to the similarity between regulation of internal developmental processes and that of behavioral ones. As in the cannibalistic tadpoles, where kin recognition affects aspects of morphogenesis, it is not easy to draw a line between internal developmental processes and behavioral ones.

Despite this progress in understanding nestmate recognition and assessment relatedness, it was still evidently true, as late as 1998, that "the actual biochemical mechanisms of recognition have never been determined for any species" (Breed, 1998, p. 463). There is considerable evidence that in social wasps (*Polistes*) and honeybees, chemicals (hydrocarbons) produced by nest builders and incorporated into the nest are then transferred to the inhabitants of the nest, learned by nestmates, and used as recognition cues (Espelie et al, 1994; Gamboa et al., 1996; review in Breed, 1998). Similarly, human mothers can identify their own babies by smell and by cry: a mother can pick out her own baby's T-shirt from among others within a day of birth (references in Hrdy, 1999, p. 158). But the ingredients of the babies' chemical bouquet have not been analyzed.

Dominance and Aggressiveness in the Assessment of Costs and Benefits of Aid

Some social organisms behave as if they evaluate the costs and benefits of social aid, in keeping with Hamilton's Rule. In the bee-eaters, whose kinship discrimination was mentioned in the preceding section, harassment of older sons and their subjection as helpers decline as their potential for independent success increases (Emlen et al., 1995). This suggests that parents cease to exact a price from their offspring as the cost, in terms of lost reproduction by sons, rises. How can there be assessment of such a changing cost? In many social animals, individuals engage in ritualized aggressive interactions and fighting within groups of potential reproductives during periods when reproductive dominance is being established (e.g., soon after nest founding by a group of females, or if a primary reproductive disappears). The most dominant individuals reproduce, while others have reduced reproduction or none at all (Pardi, 1948). In social wasps, for example, relative dominance correlates with relative ovarian development (Pardi, 1948; reviewed in Reeve, 1991; West-Eberhard, 1996; see also Field et al., 1998), due to the influence of juvenile hormone on both (Röseler, 1991; Robinson and Vargo, 1997).

These facts have led several authors to suggest an *assessment hypothesis* for the function of dominance within groups of relatives, the idea that relative aggressiveness or dominance rank could serve as an indicator of relative reproductive capacity, thereby identifying the most promising beneficiaries of aid. Costs and benefits of aid to kin could be adjusted in the right direction if subordinates (with relatively little to lose) selectively help reproductively superior relatives capable of augmented reproduction (West, 1967; West-Eberhard, 1969, 1975; Hrdy and Hrdy, 1976; Forsyth, 1980; Brown and Pimm, 1985). Dominance and social status, including territorial defense and song in birds, appear to serve as measuring sticks (or honest indicators) of physical quality or resource-capture ability in mate choice in some species (e.g., Kodric-Brown and Nicoletto, 1993; discussed in Andersson, 1994, pp. 203–205). The assessment hypothesis applies the same principle to the acquisition of social aid as a resource, rather than mates, and implies that worker choice of queens (like female choice of mates) is sometimes influenced by social status as an honest signal of quality.

Subordinates with relatively high reproductive value in populations with high queen turnover sometimes adopt an idle waiting tactic that raises the probability of future direct reproduction on a nest built by others (e.g., Litte, 1977; West-Eberhard, 1981; Starks, 1998). Whatever an individual's probability of direct and indirect reproduction, with or without residence in a group, an adaptive decision to work or not to work as a subordinate helper of relatives should in theory take into account the relative reproductive capacity of the individuals concerned. Potential workers should use signs of relative reproductive capacity of nestmates in the decision whether or not to work.

In the assessment hypothesis proposed by Hrdy and Hrdy (1976), based on dominance hierarchies in female primates, dominance rank is seen as an indicator of *relative reproductive value*—the individual's prospect for future reproduction (see also West-Eberhard, 1981). Forsyth (1980) applied the assessment model to the special case of worker execution of inferior supernumary queens. Various authors (West-Eberhard, 1969, 1975; Craig, 1983; Brown and Pimm, 1985) have developed kin-selection-based quantitative models treating aid by subfertile individuals, and Craig (1983) tested the idea with a simulation. Recently, an assessment hypothesis, in which subordinates use honest signals of queen capacity to make decisions that increase their fitness, has been reinvented and has gained support in the literature on ants (e.g., Keller and

Nonacs, 1993, p. 790; concisely reviewed in Liebig et al., 2000). Peeters et al. (1999) refer to "fertility cues" underpinning aggressive interactions with potential reproductives in insect societies, especially ants. Crewe (1999) presents evidence that dominance pheromones function as honest signals of reproductive capacity in cape honeybees (*Apis mellifera capensis*).

All of these discussions support an assessment hypothesis for the function of dominance interactions and pheromonally signaled dominance (e.g., Velthuis, 1976a, 1977) in social animals, the proposal that dominance rank reflects relative reproductive value and can be used to assess the costs and benefits of aid to particular individuals, or in evaluation of aid as an alternative to other reproductive tactics. As pointed out by Keller and Nonacs (1993; cf. West-Eberhard, 1981), if signals are used as cues in reproductive decisions (rather than as simply threat displays), they need not be backed up by real fights. Instead, the evolution of signaled reproductive superiority may be subject to departures from honesty due to a runaway process, as befalls all contests involving "choice": successful signals benefit both signaler and (genetically related) chooser, since those producing them win in social competition and may pass this signaling ability (whether backed up by phenotypic superiority or not) to their offspring, thereby increasing the inclusive fitness of themselves and their helpers.

The runaway evolution that results from a genetic correlation between competitively successful signal and criteria of choice could contribute to the elaborateness of signals of dominance such as the chemically complex queen pheromones of honeybees (Velthuis, 1977). As in sexual selection (Fisher, 1958), such a runaway process would ultimately be limited by natural selection if the cost in terms of true reproductive value comes to outweigh the advantage of a superior signal; thus, it does not contradict this hypothesis that there is still a correlation between ovarian development and the range of signal compounds produced by honeybees (Velthuis et al., 1990). New signals of true superiority would be expected to feed new variants into a runaway process, alongside signal components that are valuable as signals per se, producing an unending increase in the complexity of signals used in social choice.

The assessment hypothesis was first tested in newly founded colonies of social wasps (*Polistes* and *Mischocyttarus*), where degree of kinship has been estimated by using direct observation of marked progeny of known queens (West, 1967; West-Eberhard, 1969; Litte, 1977). In colonies where dominance relations determined the queen, the conditions of Hamilton's Rule seemed to be met. Later, electrophoresis and microsatellite techniques improved estimates of relatedness (reviewed in Queller and Strassmann, 1989; Crozier and Pamilo, 1996; Arévalo et al., 1998). In tests of the assessment hypothesis using social wasps and bees, the inclusive fitness of subordinates in groups of nest-founding females ("foundresses") has been estimated, to see if it is sufficient under Hamilton's Rule to compensate for their sterility as subordinates. How subordinates and queens would do on their own is estimated either by comparison with solitary foundresses or with established dominants and subordinates forced to nest alone (Queller and Strassmann, 1989; Shakarad and Gadagkar, 1997).

Experimental tests of the assessment hypothesis, like measurements of the speed of light, are beset with an indeterminacy problem. What needs to be measured—lifetime inclusive fitness of the same female in two different situations—is inevitably altered by the manipulations involved in the measurements, for the same female cannot be evaluated as both a lifetime reproductive and as a lifetime nonreproductive, to see how much she loses by being a sterile helper or gains by being a dominant queen or a temporarily idle reproductive (see Queller and Strassmann, 1989). Furthermore, lumping females of similar rank from different nests to compare their average characteristics is likely to blur the distinctions made by the insects within colonies. A high-ranking female in one group can be a subordinate in another, as shown by the fact that when group composition is changed by successive queen removals, an initially very low-ranking female can eventually become a dominant queen (Strassmann and Meyer, 1983; West-Eberhard, 1986). This may be one reason why comparisons using lumped samples from different colonies have sometimes failed to show significant differences in solitary reproductive success of dominants and subordinates, throwing the assessment hypothesis into doubt (Röseler, 1985; Queller and Strassmann, 1989; Reeve, 1991).

Other studies of wasps and bees, despite use of the conservative measures provided by across-colony averages, have detected differences large enough to support the hypothesis, in *Polistes* (West-Eberhard, 1969; Metcalf and Whitt, 1977; Noonan, 1981), *Ropalidia* (Shakarad and Gadagkar, 1995, 1997), and a primitively eusocial bee (*Halictus ligatus*; Packer, 1986). In all of these studies, colonies in which dominance relations reflect ovarian condition, and determine which females serve as workers and queens, Hamilton's Rule for advantageous gain versus loss in helper fitness is satisfied. These results are consistent with the hypothesis that dominance relations can mediate a decision to help that favors the helper's inclusive fitness.

In the past, discussions of the assessment hypothesis have considered individual differences in what we can call the *basal reproductive value* of an individual—its reproductive value independent of social interactions, called "inherent" differences by Röseler (1991, p. 332). Basal, or solitary reproductive value includes egg-production capacity as well as such factors as nutrient legacy from the larval stage, early maturity, general health and vigor, and freedom from parasites and disease. But the reproductive value of a social individual also depends on its *social reproductive value*—individual ability to function effectively as a reproductive or a helper within a social group.

The social reproductive value of a dominant individual includes, for example, ability to obtain resources (nutrients, egg-laying space) from other group members; to suppress conflict that interferes with efficient foraging, nest construction, and brood rearing (see especially Cole, 1986); and to engage in various games of aggressiveness and tolerance treated by models of "skew" or "transaction theory" (e.g., Vehrencamp, 1983a,b; Reeve, 2000). The social reproductive value of a subordinate helper (called indirect reproductive value by West-Eberhard, 1981) includes its efficiency in contributing to the reproduction of others, as in the workers of social insects and grandparents among primates (Hrdy and Hrdy, 1976; West-Eberhard, 1981). The *total reproductive value* of a socially reproducing individual is the sum of its basal reproductive value and its social reproductive value (which may have both a direct and an indirect component). Reeve and Nonacs (1997) call this simply "value," defined as the ratio of expected long-term group productivity with the individual present, to the group productivity with the individual absent.

Dominance can reflect more than one of the components of total reproductive value. Dominance itself contributes to social reproductive value, so assessment of relative dominance automatically reflects an important component of relative social reproductive value. At the same time, it can indicate relative basal value, because it is often highly correlated (hormonally linked) with components of basal reproductive value, such as superior ovarian development (Pardi, 1948; reviewed in Röseler, 1991) or body size. Relative body size sometimes, but not dependably, correlates with dominance rank in primitively eusocial insects such as *Polistes*,

Belonogaster, and bumblebees (e.g., Turillazzi and Pardi, 1977; Sullivan and Strassmann, 1984; Stark, 1992a; Keeping, 2000; reviewed in Reeve, 1991, p. 128) and at least sometimes reflects possible aspects of reproductive value, such as nutritional condition (e.g., body fat content: Sullivan and Strassmann, 1984). But queens are not always the largest females on a nest (Sullivan and Strassmann, 1984; Stark, 1992a; Field et al., 1998), and size differentials between dominant and subordinant individuals do not predict reproductive differentials or skew (Field et al., 1998).

Ovarian development is a more dependable correlate of rank than is body size (Röseler, 1985; Hughes and Strassmann, 1988). In some species, however, this correlation breaks down among subordinates in established colonies, perhaps due to the intervention of multiple factors such as variation in access to food and in energetic costs (e.g., of foraging) that can affect ovarian development independently of hormones (for a good discussion of multiple factors, see Keeping, 2000, pp. 152–153). This breakdown in the earlier, stronger correlation between dominance and ovarian size is accompanied by a growing breech between the dominance–ovarian status of queen and subordinates. It therefore supports, rather than contradicts, the assessment hypothesis for the function of dominance, which, as mentioned above, involves a self-fulfilling prophecy: a dominant female becomes increasingly dominant and increasingly well-endowed for reproduction, due to the social-nutritional advantages of dominance itself.

Age can also predict dominance rank in wasps (Pardi, 1948; Hughes and Strassmann, 1988) but only in relatively young females during the ascending phase of ovarian development, when age correlates positively with ovarian development (see Pardi, 1948; West-Eberhard, 1996; Hughes and Strassmann, 1988, sampled females of that age class). The length of the ascending ovarian, or maturation, phase may vary among species (Hughes and Strassmann, 1988); in only one study has worker age been shown to correlate with dominance independent of its correlation with ovarian development, with which age is highly correlated (Arathi et al., 1997). So age, if it usually reflects ovarian or hormonal condition, may be more than a "conventional cue" serving to arbitrarily resolve conflicts over queen succession driven by differences in relatedness alone (cf. Queller et al., 1997, p.3): it could reflect relative reproductive value, including the developmental (ovarian) head start of older females in species with seasonal reproduction (see discussion below of Darwin's similar hypothesis for mate choice in temperate-zone monogamous birds).

Despite the difficulties in measuring fitness consequences, which include variable kinship relations (Queller et al., 1997) as well as variable reproductive values, there is considerable evidence that dominance reflects the reproductive value side and that it is used by organisms in decisions between alternative social and reproductive roles. Dominance rank correlates with ovarian development and with access to resources within colonies in many social insects, in addition to social wasps. In ants (*Leptothorax* species), workers in colonies where the queen is removed engage in dominance interactions, and high-ranking workers lay eggs (reviewed in Heinze et al., 1997a). Monnin and Peeters (1999) estimate that dominance determines reproductive success in about 100 species of queenless ponerine ants where mated workers (gamergates) are the reproductive females. Bumblebees (*Bombus terrestris*), like social wasps, show a hormone-mediated correlation between dominance and ovarian condition (briefly reviewed in Bloch et al., 2000).

In large colonies of social insects where dominance relations have been studied, pheromonal signals of dominance and reproductive status appear to replace direct aggression in the maintenance of reproductive dominance by the queen (West-Eberhard, 1977, 1978b; Keller and Nonacs, 1993; Liebig et al., 2000). In the ant *Dinoponera quadriceps* (Ponerinae), one of the most common aggressive behaviors of dominant females is to seize the antenna of a nestmate and rub it against the dorsal surface of her cuticle, which requires a special contortion of the dominant ant (figure 23.2). The rubbed area of cuticle is coated with distinctive hydrocarbons characteristic of reproductive females and contrasting with sterile ones (Peeters et al., 1999). Keller and Nonacs (1993) review evidence that signaled dominance is more likely than direct chemical suppression of reproduction (sometimes called "queen control") as the function of queen pheromones in social insects, a conclusion reached also by Velthuis (1977) for honeybees and West-Eberhard (1977) for wasps. Swarms of wasps (*Metapolybia aztecoides*) contain numerous queens, and workers in newly founded colonies appear to test the relative dominance of queens in a ritualized "dance" that sometimes turns to a vigorous attack and ends with some but not all queens being forced to leave the nest or to become workers (West-Eberhard, 1978b). This suggests worker choice of queens using relative dominance, possibly indicated by pheromonal signals, as a criterion

Fig. 23.2. Antenna rubbing in the ant *Dinoponera quadriceps* (Ponerinae). A dominant egg-laying female (right) rubs the antenna of a nestmate against the dorsal surface of her cuticle, which is coated with hydrocarbons characteristic of reproductive females. After Monnin and Peeters (1999). Courtesy of C. Peeters.

(West-Eberhard, 1978b; see model in Forsyth, 1980), sometimes called "worker policing" (Ratnieks, 1988). Keller and Nonacs (1993, p. 792) propose the term "pheromonal queen signal" for situations where workers or subordinate queens react to queen pheromones in ways that increase their (and possibly the queen's) inclusive fitness, thus reviving and supporting the assessment hypothesis for the special case of pheromonally signaled dominance.

In species where threat and pheromonal displays replace or supplement direct dominance interactions, occasional violent fights indicate that ritualized assessment mechanisms such as pheromonal or behavioral displays are backed up by true strength (e.g., Forsyth, 1975, on *Metapolybia*). Honeybee workers can assess the reproductive status (developed ovaries) of both queens and workers (Visscher and Dukas, 1995) by their pheromonal signals. Numerous laying workers are pheromonally equivalent to a queen in suppressing oviposition by nestmates (Free, 1987). The pheromonal signal of dominance that identifies the honeybee queen is so effective that synthetic queen mandibular gland pheromone chemicals can replace her as the inhibitor of nestmate reproductive development (Pettis et al., 1995).

In some species, the suppression of reproduction by nestmates is aided by worker policing (e.g., in the wasp *Metapolybia aztecoides*—West-Eberhard, 1978b; for other possible examples, see Ratnieks, 1988; Monnin and Peeters, 1999). But in all cases, the ultimate determinant of the queen's status and discriminatory worker behavior toward other females is some indicator of relative aggressive dominance that correlates with ovarian or hormonal status, such as behavior and pheromones, for example, the queen mandibular gland secretion in honeybees (Velthuis, 1977) and cuticular hydrocarbons in some ants (Monnin and Peeters, 1999; Peeters et al., 1999), whose quantities or qualities correlate with both social status and ovarian development. The evidence that social dominance and ovarian development are hormonally linked is also strong, especially for juvenile hormone (reviewed in Röseler, 1991; West-Eberhard, 1986; for other ecdysteroids, see Bloch et al., 2000). All of these data support the assessment hypothesis by showing that dominance is associated with some component of reproductive value (rank and access to resources, ovarian condition), and that it is the mechanism by which nestmates determine which female shall be the primary recipient of social aid.

Relative dominance and subordinance may be widely used as an assessment mechanism in social animals when there is an association between aggressiveness and superior physical condition or reproductive value that indirectly or directly benefits the assessing individual (West, 1967). Dominance often determines skewed reproductive success in cooperatively breeding birds and mammals, and cooperative groups are often composed of relatives (see Stacey and Koenig, 1990, on birds; Solomon and French, 1997, on mammals). To the degree that dominance indicates relative phenotypic reproductive capacity in such groups, the assessment hypothesis may apply. In some primates (e.g., chimpanzees), there is a positive correlation between dominance rank and reproductive success in males (see references in Mitani and Nishida, 1993) or in females (Hrdy and Hrdy, 1976). While there are many complicating factors, including the potential to form mutualistic alliances among nonrelatives, this correlation is more often positive than negative for male and female primates (Silk, 1987). If such groups contain relatives, it could be advantageous to defer to or aid a dominant relative when resources are limited, and kin selection may set the threshold dominance differential for advantageous deference. Mutualistic alliances might also involve use of dominance rank to assess fighting ability or effectiveness of social display. In competitive interactions where cooperative roles are not at stake, such as between males competing for mates, honest assessment or status-signaling badges can of course also be advantageous, for example, as a means of curtailing costly fights (Lyon and Montgomerie, 1986).

Similar assessment using aggressiveness as a measure of quality occurs during mate choice in some sexually reproducing species, as mentioned below. Fighting or competitive display among males, for example, in a lek, is followed by preferential mating by females with the winner(s), called passive choice by Thornhill and Alcock (1983, p. 406). In effect, females are using dominance rank to screen for the quality of their mates and offspring, the recipients of their social (parental) aid.

Darwin (1872, pp. 572–573) proposed that another mechanism of mate-quality assessment could occur in temperate-zone monogamous birds, a hypothesis that may apply as well to the choice of a queen in temperate-zone wasps. Darwin's idea was that early nest initiation per se could serve as an indicator (to males) of phenotypically superior females because it would indirectly select for superior nutritional condition if the two variables (early nest initiation and superior female phenotype) were phenotypically correlated. Darwin emphasized that the correlation had to be phenotypic, not inherited, for it would not be advantageous to select for an earlier and earlier initiation of the breeding season in temperate-zone birds (see also Fisher, 1930). Trivers (1972, p. 172) pointed out that in birds "aggressiveness may act as a sieve," admitting only those females whose high motivation to form early pairs correlates with early egg laying and high reproductive potential. In the social wasp *Polistes fuscatus*, all nests are founded during a short period in springtime, and all subsequently active females become subordinate helpers on the nests of the earlier-nesting females (West-Eberhard, 1969). This could indirectly result in relatively poor reproductive females helping relatively high-quality relatives with superior reproductive and survival prospects, with or without the use of dominance relations as an estimator of reproductive quality.

In summary, social dominance can reflect in a single, detectable and quantitatively variable trait— relative aggressiveness—several components of basal and social reproductive value. This makes dominance rank a feasible means by which individuals can assess their own reproductive value compared with that of others and make decisions between alternative reproductive behaviors within groups. The utility of dominance rank as a graded indicator of relative reproductive value is probably one reason why ritualized and pheromonal dominance interactions so often determine reproductive rank and direction of resource flow in social insects and other group-living organisms, where the inclusive fitness of individuals with socially reduced fertility depends either temporarily or entirely on the reproductive capacity of one or a few dominant relatives.

Assessment of Aggressiveness and Territoriality Not Involving Aid

In group-living animals, dominance and threat behavior is often the primary determinant of access to resources, whether food, shelter, or mates. Organisms live and seek mates in groups because they derive individually selfish benefits from being near others, sometimes called mutualism (see Alexander, 1974). At the same time, there are costs of living in groups. One of the obvious costs is competition with other group members. If there is no alternative means for subordinates to obtain contested resources, then individuals are expected to fight to the death (West-Eberhard, 1979). But the evolution of alternative tactics of resource acquisition, and the advantage of establishing aggressive dominance rank via threat display rather than costly fighting make ritualized, or signaled, dominance and subordinance behavior common in obligatorily social species.

When ritualized resolution of dominance disputes is advantageous, clear signals of rank are expected to evolve (West-Eberhard, 1979). Signal *antithesis*, or sharp contrast between signals of dominance and signals of subordinance achieved by adoption of opposite postures or movements, occurs in a wide variety of animals (Darwin, 1872; figures 23.3, 23.4). Darwin considered the "principle of antithesis" to be a fundamental rule of animal communication and proposed that its function is to facilitate unambiguous assessment. A similar principle is a general property of signal amplification in physical and biological systems (such as neural nets and logical thought), where polar opposites, although they sacrifice information, facilitate decisions (Platt, 1956). A recent study of selection on plumage coloration in yearling males of a bird (lazuli buntings, *Passerina amoena*) demonstrated disruptive selection against intermediate-colored plumage, with the dullest (unattacked) and the brightest individuals (most successful in competition) obtaining the highest-quality nesting territories in interactions with adult males, and therefore the most mates (Greene et al., 2000).

Antithesis can take many forms, including not only postural opposites, such as those of figure 23.4, but also interactional opposites, such as the offering of food or regurgitated fluid by a social subordinate and receiving by a dominant (e.g., Pardi, 1948; West-Eberhard, 1969; Jeanne, 1972), or offering and receiving between males and females during courtship, as in some birds where the male signals lowered aggressiveness by offering, which facilitates his approach to a potential mate

Fig. 23.3. Darwin's principle of antithesis. The clarity of signals with opposite effects is increased by oppositeness of form: threat posture of a domestic dog (top); submissive posture (bottom). From Darwin (1872).

(Thornhill and Alcock, 1983). Human communication and logical thought also employ antithesis, with arguments stated in dichotomous "black-and-white" terms that eschew the middle ground, as in nature–nurture, learning and instinct, gradualism and saltation, art and science (see Shearer and Gould, 1999). It has been argued, by a Chinese correspondent (Chianmg Cheng-Chung, personal communication, 1976), that polarized reasoning and argument, like the Cartesian coordinates, are peculiarities of Western culture. But the evidence for the very wide occurrence of antithesis in the communication of nonhuman animals (figure 23.4), and its association with dichotomous decisions and dimorphisms in behavior and development, suggests a more ancient origin.

The assessment of aggressive dominance takes many forms. Stallions assess relative dominance by the length of a squeal given upon meeting a competitor. Sonographs show that squeals of dominants are longer and spectrally broader than those of subordinates (figure 23.5), and playback experiments demonstrate that individuals respond appropriately to different squeals (Rubenstein and Hack, 1992). Horses also assess dominance indirectly by sniffing pheromone-impregnated dung,

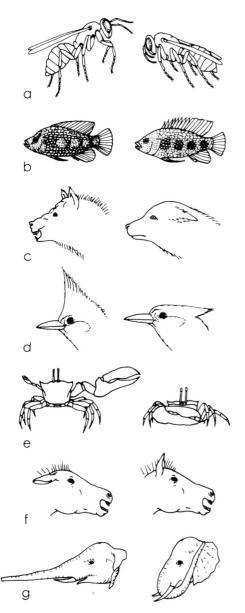

which carries information about individual identity that they can likely associate with dominance rank on the basis of past encounters (Rubenstein and Hack, 1992). Pheromonal assessment of dominance rank occurs in insects as well. It is well studied in social insects (Free, 1987; Liebig et al., 2000; see also section on costs and benefits of aid, above) and in cockroaches (Moore, 1997). In the cockroach *Nauphoeta cinerea*, dominant males secrete a larger and less variable amount of a multicomponent pheromone in which two of the components vary with dominance and inversely with each other (Moore et al., 1995; Moore, 1997). This inverse relationship between dominance-related components of the pheromone blend would facilitate discrimination of relative dominance. The pheromonal signal affects both male–male dominance hierarchies and female choice of mates (Moore et al., 1995; Moore, 1997).

Other dominance assessment devices include the intensity and duration of roaring in red deer stags (Clutton-Brock and Albon, 1979), the breast stripe

Fig. 23.4. Widespread applicability of the principle of antithesis in animal communication: Dominant, threat, or aggressive signals (left) and contrasting subordinate or appeasement signals (right) of (a) *Polistes* wasps (Vespidae), after Pardi (1946); (b) cichlid fishes (*Hemichromis fasciatus*), after Wickler (1965); (c) dogs, after Lorenz (1953, in Eibl-Eibesfeldt, 1970); (d) Steller's jays, after Brown (1964); (e) fiddler crabs (*Uca maracoani maracoani*), after Crane (1975); (f) zebras, after Trumler (1959); and (g) African elephants (*Loxodonta africana*), after Kühme (1963).

a. Dominant

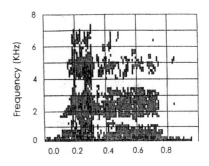

b. Subordinate

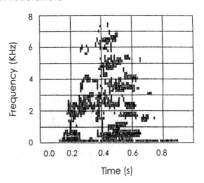

Fig. 23.5. Sonograms of dominant and subordinate squeals of horses. After Rubenstein and Hack (1992).

of great tits (Jarvi and Bakken, 1984, cited by Rubenstein and Hack, 1992), and the degree of melanization of feathering on the throats and crowns of Harris sparrows (*Zonotrichia querula*; Rohwer and Rohwer, 1978).

Territoriality, or defense of a bounded area, is a variation on the theme of aggressive dominance. Animals make cost–benefit assessments regarding territories just as they do about the value of dominance or subordinance (Davies and Houston, 1984). Individuals of some species, such as the pied wagtail, *Motocilla alba*, and the pygmy sunfish, switch from nonterritorial behavior to territorial defense when resources are scarce, especially rich, or clumped (Zahavi, 1971; Rubenstein, 1981). Others adjust territory size in accord with resource abundance (e.g., flower density and/or nectar content), as in rufous hummingbirds, *Selasphorus rufus* (Kodric-Brown and Brown, 1978; Gass et al., 1976), golden-winged sunbirds, *Nectarinia reichenowi* (Gill and Wolf, 1975, 1977), and *Sceloporus* lizards (Simon, 1975). The territory size of some birds, such as the sanderling *Calidris alba*, correlates better with intruder pressure than with resource content (Myers et al., 1979). Intruder pressure and resource value have proven to have complex but predictable interactions in various species (see Eberhard and Ewald, 1994, and references therein).

Social assessment, including assessment of mates (see following section) often depends upon size. The mechanism of interindividual size assessment is unknown in most species, but a few examples indicate that there is a great variety of mechanisms for the assessment of relative size of opponents and mates, giving the impression that virtually any trait that happens to correlate with size in socially competing species can become a signal. Toads (*Bufo bufo*) judge the size of males by the deepness of their croaks (Davies and Halliday, 1978; see also Arak, 1983, on *Bufo calamila*). Female frogs (*Physalaemus pustulosus*) use the frequency of the "chuck" component of the male call, which is lower in large males (Ryan, 1985). Female tree frogs (*Hyla chrysoscelis*) also use the frequency of the call, likewise lower in larger males (Morris and Yoon, 1989). In some crickets (*Gryllus bimaculatus*), conspecifics judge the size of males by the repetition rate of chirps and pulses (Simmons, 1988; see also Shuvalov and Popov, 1973) whereas in others (e.g., *Oecanthus nigricornis*) low frequency songs are used by females to choose large males as mates (Brown et al., (1996). In a very diverse set of organisms, relative size is assessed during the performance of "parallel bouts" of fighting in which males line up side to side or head-to-head and strike or push each other or raise their bodies until the smaller individual withdraws. This occurs in thrips (*Elaphrothrips tuberculatus*) and is suggested by observations of ants (Hölldobler, 1999), broad-headed flies (Moulds, 1977), red deer, elk, giraffes, reptiles, and fish (see references in Crespi, 1986, p. 1332). In thrips, the contesting males strike each other with their abdomens, a measuring episode that influences their subsequent behavior toward each other (Crespi, 1986).

Chimpanzees have spectacular social assessment behaviors that include signals that seem designed for assessment by members of other groups, or foraging subgroups when separated in space. Communicative drumming by chimpanzees on buttress roots of trees is a possible example (Reynolds and Reynolds, 1965; Goodall, 1986; Arcadi et al., 1998). Drumming signals can be heard more than three kilometers away (Goodall, 1986), at least as far and perhaps farther than the loudest vocalizations. They are occasionally performed by females and juveniles, when their group passes a drumming tree, or in an outburst of excited display upon return to familiar ground after socially enforced silence during a walk through unfamiliar territory. But drumming is performed primarily by males, often by groups arriving together at a rich new food site, and is frequently accompanied by loud vocalizations called pant-hoots. Some authors hypothesize that drumming serves to coordinate spacing and travel among subgroups of a larger stable group (reviewed in Arcadi et al., 1998). But various observations suggest that, alternatively or additionally, it may be a long-distance dominance or threat display that allows assessment of strength of neighboring groups or males. Individual differences in drumming pattern occur that could permit identification of individual males; coalitions among specific males are used to establish rank and access to mates and other resources within groups; and there is sometimes lethal fighting between neighboring males of different subgroups (see data and references in Arcadi et al., 1998). Behavioral observations also suggest threat and assessment of relative strength:

> When males hear pant-hoots and drumming from an obviously *larger* number of unhabituated adult males, they sometimes stare toward the sounds in silence, then hastily retreat. When the number of males in the two parties appears to be similar, members of both sides usually engage in vigorous displays with much drumming. . . . (Goodall, 1986, p. 491)

And, as is typical of socially competitive displays (West-Eberhard, 1983), drumming displays vary among populations.

The fact that particular trees are favored as "drumming trees" (Goodall, 1986) suggests that there is care in choice of instruments. Yet apparently nothing is known about individual differences in drumming expertise, buttress assessment, or learned improvement in technique via practice or mimicry. Innovation in use of instrumental media for mass communication has received far less attention than has tool use (e.g., see Goodall, 1986), undoubtedly due to the long-time emphasis on the invention of tools as a milestone of human prehistory and the longtime emphasis in evolutionary biology on survival and ecology rather than social factors (West-Eberhard, 1983). The alternative hypotheses of cooperative and competitive assessment as functions of chimpanzee drumming remain to be investigated.

Noteworthy feats of social assessment occur in other primates. In at least some monkeys, individuals distinguish not only relative dominance rank but also kinds of relationships (e.g., distinguishing between mother–offspring pairs and others). Vervet monkeys, for example, that have observed fights between a member of their own family with a member of another family are more likely to act aggressively toward the "enemy" family in the future (Cheney and Seyfarth, 1990).

Ritualized combat that rivals any found in primates, with apparent group assessment of strength, is known to occur in ants. This phenomenon is best analyzed in the honey ants *Myrmecocystus mimicus*, which conduct "tournaments" sometimes involving hundreds of ants (summary in Hölldobler, 1999). Individuals perform highly stereotyped aggressive displays toward a series of opponents in defense of their colony territories, adopting a stilt-like raised posture and sometimes even displaying down on the opponent while standing on a stone, both behaviors that increase the apparent size of the ant. Experiments indicate that opposing colonies assess each others' strength through assessment of the number of relatively large individuals among the opponents, responding with retreat if individuals encounter a large proportion of major workers. Relatively small "reconnaissance ants" move through the crowd of displaying ants and recruit reinforcements by laying chemical trails between the nest and the display arena. Tournaments result in adjustment of territorial boundaries and, sometimes, enslavement of the weaker colony.

A more ambitious review would undoubtedly uncover many more examples of discriminatory ag-gression in other kinds of animals. Even in insects, such as the cicada-killer wasp *Sphecius speciosus* (Hymenoptera, Sphecidae), nesting females are less aggressive toward females that are neighbors (females nesting within 1 meter of their own nest) than they are toward nonneighbors, and they also adjust their aggressiveness depending on the size of an intruder and whether or not there is an exposed prey near their nest (Pfennig and Reeve, 1989). Although solitarily nesting, these females can assess neighbor identity and body size of competitors, and degree of aggressiveness is proportional to ovarian development, as in social wasps, all preadaptations for social life, as the authors point out.

Assessment of Mates

Among the better known kinds of assessment by nonhuman organisms are the often subtle distinctions used in female choice of mates (reviewed in Andersson, 1994; Eberhard, 1996). In reproductive decisions, the connections between behavioral interactions and developmental decisions are particularly clear. The stimuli exchanged between the sexes during courtship and mating have been studied at various levels of organization, from the behavioral (Thornhill and Alcock, 1983; Andersson, 1994) to the internal (Eberhard, 1996), and the physiological and hormonal consequences of many events are known. There is still controversy over the adaptive significance of male–female interactions—whether they most often function to screen mate quality, to determine species identity, or mainly to coordinate the transfer and activation of gametes. But it is clear that these interactions involve assessment and discrimination on the part of the "choosy" sex (usually females) and competitive, manipulative stimulation, especially by males. The various gadgets and communicative devices that we call sexual signals are very much like the internal signaling events of development. Interactions during courtship stimulate sequential events: first external attractive stimuli, then internal interactions during copulation, with many circuitous and complex physiological responses eventually leading to fertilization. In one sense, sexual interaction is an aspect of adult reproductive development, one where essential environmental input comes from a conspecific individual of the opposite sex.

In species where both sexes care for the young, the social effects of mate choice do not end with mating. Experimental studies of mate assessment in captive zebra finches (Burley, 1988) show that relatively attractive individuals get more help from their mates. Unattractive birds invest more in parental effort per offspring and have shorter lives and lower long-term reproductive success than do more attractive individuals of the same sex. Female white-fronted bee-eaters, *Merops bulockoides*, assess the family situation of a male before deciding to join his group as a mate instead of staying in the maternal group as a helper of breeding family members. Males with social characteristics associated with low reproductive success are rejected by females, whereas socially attractive males, that is, males accompanied by helpers, are usually paired (Wrege and Emlen, 1994). Thus, females of these birds make complex comparative assessments of their own family situation and that of potential mates, and they behave as predicted by a computer simulation using measurements of fitness-affecting parameters from field studies (Wrege and Emlen, 1994).

Sexual behavior is also interesting from a developmental point of view because it draws attention to the *manipulability* of environmentally sensitive responses. The susceptibility of behavior and development to environmental influence means that they are eminently subject to influence by other organisms if natural selection favors behaviors that cause them to pervert development for their own selfish ends. In this respect, sexual manipulations of female reproductive physiology by male conspecifics (Eberhard, 1985, 1996) are just one of a very large category of developmental manipulations by outsiders, such as the manipulation of caste by adult social insects in their interactions with larvae, the induction of galls by a multitude of plant feeding insects (see figure 5.10), the selfish deformation of host phenotypes by parasites (Eberhard, 2000), the mimicry by social parasites in some insects (Wilson, 1971a,b) and birds (Brown, 1975; Payne and Payne, 1994, 1995) of host stimuli known or likely to affect acceptance and resource acquisition (e.g., Lyon et al., 1994), and the acculturation and education of human infants, children, and adults.

Assessment of When to Change Sex

Sex reversal occurs in a number of coral-reef fishes (Shapiro, 1984). In the bluehead wrasse, *Thalassoma bifasciatum*, an inhabitant of tropical coral reefs, some individuals are sequential hermaphrodites. These individuals are born female and at first are variable in color. When they are large enough to compete as territory holders in leks, they change into a bright-colored territorial male (summarized in Warner et al., 1975). In populations where leks are monopolized by a very small number of males, there is an additional secondary, "initial phase"

male morph, which is born a male but resembles the females in being of variable and changeable color. Initial phase males, rather than spawning with one female at a time like territorial males, are nonterritorial and unaggressive. They spawn in groups with a single female and intrude on the territories of large males. In their "terminal phase" when older and larger, these males finally change to the bright male coloration and become territorial.

Both sex change and morph change among males are highly condition sensitive. If all terminal phase males are experimentally removed, the largest remaining fish changes to terminal phase coloration. If that fish happens to be a female, she changes sex as well. This leads to different proportions of sexes and morphs on different reefs, depending on population size and competition. In accord with theoretical predictions of when to switch, Warner et al. (1975) found a positive correlation between relative abundance of a species on a particular reef and the frequency of initial phase males, indicating that selection has favored finely tuned assessment of conditions that permits an adaptive switch.

Some polychaete worms (*Ophryotrocha puerilis*) undergo facultative sex-role change in circumstances quite parallel to those of the blueheaded wrasse (summarized in Sella, 1990; see also Charnov, 1982, for earlier work). But in the polychaete, individuals hatch as males and then change to females when large, and small rather than large males have a mating advantage. Females inhibit sex reversal in large males. Sex change is highly condition sensitive. If two females are confined together without males, the smaller one resorbs it oocytes and becomes a male. If the two females are similar in size, they sometimes undergo repeated simultaneous sex reversals by both individuals. If reciprocal sex change proceeds for more than a month, both worms become simultaneous hermaphrodites and reciprocally exchange eggs and sperm. This is of special interest because it suggests an origin in phenotypic flexibility (facultative simultaneous hermaphroditism) for the egg-trading simultaneous hermaphroditism that characterizes a related species (*O. diadema*). Egg-trading has also evolved in fishes (Serranidae; summarized in Fischer, 1988) but apparently not in wrasses (Labridae).

Assessment of When to Reproduce

Many plants and animals reproduce seasonally, and this can mean changing cyclically between reproductive and nonreproductive phenotypes, including in morphology such as antlers, plumage, and color

(Darwin, 1871 [1874]). Individuals of many species depend on anticipatory cues to begin their phenotypic transformation prior to the breeding season. Photoperiod, or day length, frequently cues the initiation of reproductive development (Tauber and Tauber, 1981; Tauber et al., 1986; Nelson et al., 1990; Goldman and Nelson, 1993; Baskin and Baskin, 1998). Day length is also involved in timing of migrations, hibernation, mating, nesting, and the germination of seeds.

How can organisms assess day length, and how can natural selection fine-tune so many kinds of responses to the same cues? The apparent use of day length as a cue for regulation of seasonal switches is so common in such a diversity of organisms, including plants, insects, mammals, and birds (Baskin and Baskin, 1998; Tauber et al., 1986; Underwood and Goldman, 1987; Nelson et al., 1990; Farner, 1980), that biologists may take it for granted as an obvious mechanism for assessment of time of year. Yet it is not immediately obvious how such an abstract cue, which has no direct effect on the phenotype, could become associated with so many kinds of switches in so many kinds of organisms.

Evidently the answer lies in the ubiquity of metabolically entrained intracellular circadian (daily) cycles. Virtually all organisms have *circadian rhythms*, or daily 24–hour physiological cycles, that can entrain to light (B. Goldman, personal communication). One of the first discoveries of a circadian rhythm was in a cyanobacterium, where the daily rhythm was explained in terms of the advantage of temporal compartmentalization of the incompatible metabolic functions of photosynthesis and nitrogen fixation. A nitrogenase rhythm is mediated by daily fluctuations in the amount present, which is controlled in turn by a daily rhythm of the enzyme coding mRNA (see references in Hall and Rosbash, 1993).

Although circadian rhythms are virtually ubiquitous, not all organisms show *photoperiodism*, or *photoperiodic time measurement*—the ability to extract information from day length, as evidenced by seasonal changes regulated by changes in day length. In fact, in most photoperiodic organisms, it is not day (or night) length itself that is used but rather the timing of light exposure, which evidently interacts with the circadian rhythm to entrain it to the light–dark cycle (Elliott and Goldman, 1981; by another hypothesis, there are two internal oscillators whose phase relations change as a result of day length; Gorman et al., 2001). Photoperiodic time measurement requires that connections be made between an internal circadian clock and neurohormonal systems that affect behavior and reproduction. Not all organisms time critical events such as reproduction or seed germination using day length, and of those that do, not all do it in the same way.

How the circadian signal of mammalian brain cells is translated into a seasonal assessment mechanism that governs a phenotypic switch is perhaps best known for mammals (figure 23.6). Not surprisingly, assessment of day length begins in the eye, which in mammals are the only photoreceptor organs. The daily cycle of retinal stimuli entrains rhythmic oscillations of neural activity in the brain, specifically, in the suprachiasmatic nuclei of the hypothalamus. The precise mechanism of response to day length is still controversial, but as mentioned above, it evidently involves either entrainment of a critical photoinductive daily period, or two entrained oscillators whose phase relations produce critical signals (Gorman et al., 2001). The suprachiasmatic nuclei then stimulate, in rhythmic fashion, the pineal gland, which adopts a corresponding rhythm in its secretion of melatonin. It is not day length that matters to the the pineal gland, but night length: the pineal gland is nocturnally active, and the longer the period of darkness, the longer the period of daily secretion of melatonin. The pineal gland, then, translates a neural signal into a hormonal signal. Melatonin is released into the blood and influences the pituitary gland to alter its secretion of hormones (FSH, LH, and PRL; see figure 23.6). These in turn regulate the various changes in behavior, morphology, and physiology that an adult mammal undergoes in the field. Of course, the multiple effects of hormones, and of their ability to control different events at different time, owe to the fact that different tissues (gonads, hair follicles, melanophores, and neuromuscular tissues) respond in different ways and sometimes have different thresholds of response to the same hormones.

Each of the phenotypic traits enclosed in boxes in figure 23.6 can be presumed to be a semi-independent, genetically variable subunit of the phenotype whose responses can evolve somewhat independently of the others. In some mammals, the mother is able to transmit day-length information to her fetuses, and these signals can affect the offspring responses to postnatal photoperiod experience (Shaw and Goldman, 1995a). This highly polygenic, highly environmentally sensitive system is typical of the hormone-mediated assessment mechanisms that regulate seasonal reproductive phenotypes. It has the potential for adaptive adjustments under natural selection at many levels of organization, for each of the different responses involved.

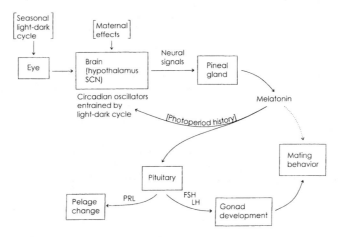

Fig. 23.6. Assessment of day length and the regulation of seasonal phenotypes in mammals. Influential environmental factors are in brackets; phenotypic subunits are in boxes. Visual signals entrain a circadian rhythm in cells of the brain, especially the suprachiasmatic nuclei (SCN) of the hypothalamus. The SCN, via neural connections, drive a rhythm of melatonin secretion by the pineal gland. The night-length-dependent melatonin pulse regulates seasonal variations in the secretion of pituitary hormones. These hormones in turn stimulate or inhibit responsive tissues (hair cells, gonadal cells, and neuromuscular systems involved in behavior). Pituitary hormones influenced by melatonin include follicle-stimulating hormone (FSH), luteinizing hormone (LH), and prolactin (PRL). Melatonin may also act directly on some parts of the reproductive axis in ways that do not involve changes in pituitary hormone secretion (dashed arrow). (Based primarily on studies of the Siberian hamster, *Phodopus sungorus*, as summarized in Gorman et al., 2001)

When seasonal resources limit breeding success and are not dependably correlated with day length, more direct cues may trigger reproduction, and multiple cues are sometimes used. In the montane vole *Microtus montanus*, for example, long photoperiod is essential for reproductive activity, but so is ingestion of the compound 6–methoxybenzoxlazonlinone (6–MBOA), a component of fresh vegetation (Berger et al., 1977, cited by Goldman and Nelson, 1993). The effects of 6–MBOA are transmitted as a maternal effect to the offspring, which show significantly increased growth and gonadal size as a result (Frandsen et al., 1993), so offspring quality may be as important in food-plant cuing of reproduction as is favorable seasonal timing. A number of boreal mammals, including male prairie dogs (*Cynomys ludovicianus*) and Norway rats (*Rattus norvegicus*), are regarded as reproductively nonphotoperiodic, but this may reflect inadequate investigation of their responses to photoperiod (Goldman and Nelson, 1993). Some high-arctic tundra species of lemmings (e.g., *Dicrostonyx groenlandicus*) are considered photoperiodic, and some (e.g., *Lemmus sibiricus*) are not (Negus and Berger, 1998). Breeding in *L. sibiricus* coincides with the appearance of the first sprouts of its pre-

ferred food plants, monocots such as *Dupontia fisheri* and *Carex stans*, which, like the plants consumed by *M. montanus*, are high in 6–MBOA. Laboratory studies show that 6–MBOA causes increased uterine mass in *L. sibiricus* females, but not in *D. groenlandicus* females. Marked yearly variation in the 6–MBOA content of the plant *D. fisheri* coincided with population density fluctuations in *L. sibiricus* (Negus and Berger, 1998).

In plants, the decision regarding when to germinate is in many ways comparable to the animal decision regarding when to reproduce. Seeds characteristically undergo a period of dormancy prior to germination. Germination may be stimulated by such environmental stimuli as a change in temperature or soil moisture, and plants are sensitive to variation in various properties of light, including light:dark ratio, photoperiod, and light quality (Baskin and Baskin, 1998). As in some animals, the reproductive response can be altered by maternal effects, called "preconditioning" in plants (Baskin and Baskin, 1998, p. 191). Preconditioning or maternal effects on a seed's decision about when to germinate can be caused by numerous factors in the maternal environment, including carbon dioxide levels, competition with other plant species, day length, fungal infection, herbicides, hormone treatments, length of growing season, spectral quality of light, mineral nutrition, maternal age, position on the mother plant, defoliation, soil moisture, amount of sunlight, temperature, and seasonal time of seed production (Baskin and Baskin, 1998). All of these factors are known to affect the timing of seed germination, but not all are known to lead to adaptive (fitness-enhancing) changes.

These observations illustrate the genetic complexity of day-length measuring systems. There are individual differences and polymorphisms in responsiveness to photoperiod in field populations of both animals and plants, and a heritable component to these differences has been demonstrated (on reproductive timing in mammals, see Blank and Freeman, 1991; Freeman and Goldman, 1997; on germination in plants, see Baskin and Baskin, 1998).

In organisms with juvenile-to-adult metamorphosis the timing of reproduction is often determined by the timing of metamorphosis to adulthood. Timing of metamorphosis in such organisms may correlate with favorable conditions for reproduction, including not only environmental conditions such as temperature or rainfall, but also phenotypic condition, such as body size (Werner, 1986; 1988).

Assessment in the Development of Morphological Alternatives

Many morphological alternative phenotypes, in addition to seasonal ones, reflect adaptive assessment of environmental conditions. A number of morphological alternative phenotypes correlate adaptively with adult size, and so they imply some means of anticipatory assessment of adult size during the period of growth and phenotype determination. Size-correlated social and reproductive alternative tactics are especially common, since size is a common correlate of success in courtship, reproduction, and battle (for reviews of examples, see Thornhill and Alcock, 1983, on insects; Payne, 1984, on birds; Andersson, 1994). Size is also a common determinant of alternative ecological and mating strategies and morphologies in animals, and the size–tactic relationship is known in some organisms to be genetically variable and subject to selection (Travis, 1994).

Juvenile size is often a good predictor of adult size. In insects such as the social wasps, ants, bees, and termites, larval size is a frequent correlate of early developmental switches toward queen- and worker-specific morphology that will benefit adults of a certain size class, and the same is true in some

male polyphenisms such as the horn in large-bodied beetles that engage in fights. In the social Hymenoptera and beetles, the switch between alternative morphologies necessarily occurs prior to adulthood since these are holometabolous insects and their exoskeletons do not change size or shape during adulthood. In ants and honeybees a juvenile-hormone-mediated nutritional switch between specialized worker and queen morphology occurs early in the larval stage (summarized in Nijhout, 1994a, 1999c), whereas in the beetle *Onthophagus taurus*, the switch between horned and hornless morphology, also mediated by juvenile hormone, occurs after larvae have stopped feeding, and adult body size is most predictable from larval size (Emlen and Nijhout, 1999; Emlen, 2000). The earlier switch in honeybees accords with expectations, because honeybees impose a highly predictable, relatively uniform bimodal feeding regime on their larvae that determines the larva's caste (Engels and Imperatriz-Fonseca, 1990). The food supply of dung beetle larvae, by contrast, is continuously variable and depends on the size of a dung supply provided in the egg stage (Emlen, 1994), so the total larval supply may not be internally assessable by the larva until the entire supply has been consumed.

In the social Hymenoptera, the environmental parameter that correlates with adult phenotype is larval diet during certain sensitive periods (reviewed in Wheeler, 1986, 1991, 1994; Engels, 1990; Nijhout, 1994a). While this can be regarded as a kind of developmental assessment cue, in most species it is regarded as a product of manipulation of the phenotype by the adults who feed the larvae. In fact, the larvae themselves may play a more active role in caste differentiation than is usually acknowledged. Larval hunger cues direct adult feeding behavior in some ants (Cassill and Tschinkel, 1995, 1996), raising the possibility that begging signal effectiveness can affect caste determination. This is an insufficiently explored aspect of social insect biology, where the study of behavior and natural history has lagged behind studies of physiology and relatedness.

Even in these relatively well-studied examples, the social insects and the horned beetles, the detailed mechanism by which the organism detects quantitative differences in diet that cue the hormonal changes and size-related morphological development is not known; only the hormonal response has been measured. How are dietary differences actually sensed? A clue was provided in studies of size-dependent metamorphosis in milkweed bugs (*Oncopeltus fasciatus*), which are sexually dimorphic in size. Different critical weights trigger the final molt in the two sexes, and body size is sensed by stretch receptors in the abdominal wall (Nijhout, 1979).

In abdominal stretch reception, neurons associated with certain abdominal muscles sense gut size in blood-sucking bugs, such as *Dipetalogaster maximus* (Nijhout, 1984) and *Rhodnius prolixus* (Hemiptera: Reduviidae), and probably many other insects (reviewed in Nijhout, 1984; see also Chapman, 1998). Release of hormones that stimulate metamorphosis can be inhibited by cutting abdominal nerves, and undersized individuals can be induced to molt by saline injections that simulate expanded size. In *Rhodnius*, the chain of events governing the decision to molt is extraordinarily well known, thanks to the work of Wigglesworth and others (briefly reviewed in Nijhout, 1994a). It involves the following complex series of potentially variable structures and events: a blood meal causes the abdomen to expand, which stimulates stretch receptor nerves in the abdomen, which in turn stimulate specialized cells in the brain to secrete prothoracicotropic hormone, which is released into the blood and stimulates the prothoracic gland to secrete the molting hormone, ecdysone, which stimulates molting (Wigglesworth, 1936). Evolutionary change in any of the involved structures, tissues, or responses (blood, size of egg and subsequent developmental stages, nerves, muscles, secretory cells, tissue sensitivity to hormones, or stretchability of gut) would have the effect of altering the molting threshold and adjusting the assessment of size. Most organisms have a species-specific modal individual body size that does not vary much from one generation to the next. Nijhout (1999b) regards the problem of how developing animals in general assess their body size as "one of the oldest still unanswered questions in developmental biology" (p. 239).

The remarkable assessment of shells by hermit crabs is discussed above. Hermit crabs also have the ability to alter their size if shell assessment indicates that the size of their otherwise suitable shell is too small to accommodate growth (Markham, 1968). When that occurs, the crab actually decreases its body size at the next molt.

Various animals respond to predation and cannibalism with adaptive morphological change. Females of the rotifer *Asplanchna sieboldi* occur in three distinctive morphotypes with intermediates: a small, saccate morphotype; an intermediate-sized cruciform morphotype; and a large cruciform morphotype with four humped outgrowths of the body wall (summarized in Gilbert, 1973). The latter, campanulate females, are induced when saccate females ingest a large prey or a conspecific individual along with α-tocopherol, present in food organisms such as *Euglena* (Gilbert, 1977). In the absence of dietary α-tocopherol, they revert to the saccate morphology after several generations. Various lines of indirect evidence show that the campanulate morph evolved as protection against predation by cannibalistic females (Gilbert, 1973, 1980).

In spadefoot toads (*Scaphiopus multiplicatus*), the switch between the carnivorous and the omnivorous gut, eye, and mouth morphology (see figure 21.2) is also known to be cued by the nature of the prey: a carnivorous diet (shrimp ingestion) causes a switch to a carnivorous morphology, and thyroid hormone or its constituents present in the shrimp prey may directly act on the tadpole's development to stimulate the morphological (and, to an unknown degree, the behavioral) change (Pfennig, 1992a). In salamanders (*Ambystoma tigrinum*), a large-headed cannibal morph is induced by crowding, and tactile cues for other larvae are required, with larvae of other species effective at an earlier age (Hoffman and Pfennig, 1999).

Learning and Assessment

Trial and error learning with negative or positive reinforcement of responses implies assessment of punishments and rewards. As discussed in chapter 18, the mechanism for this involves context-specific motivation determined by internal physiological states that underlie such measurable parameters as hunger, thirst, and sexual interest (Gallistel, 1980). Variations in these parameters often help to explain why a given stimulus (from outside the organism) is effective in eliciting a response on some occasions and not others. Variation in stimulus effectiveness is not random noise but is a result of internal variation in the mechanisms that underlie stimulus assessment. Changes in motivation alter thresholds for switches between alternative behavioral states.

Trial and error exploratory behavior with assessment of conditions occurs even in bacteria and in plants. Moving bacteria stop periodically and then renew movement randomly in a new direction (Koshland, 1977, cited by Sachs, 1988a, p. 551). Movement that results in improved conditions is continued longer. Motion in the appropriate direction is therefore not programmed but rather selected from many random possibilities that are assessed by the moving organism. Trial-and-error searchlike behavior ending in a directed response in plants was studied by Darwin and described in

The Power of Movement in Plants (1880; see also chapter 3, and Bell, 1959). During these searchlike movements, the plant in effect inspects and assesses its environment, responding when particular conditions of light, gravitation, and support are satisfied.

Learning simplifies the proximate process of assessment and decisions by making an integrated evaluation of numerous environmental and phenotypic variables that may affect the success of a tactic, ending with one overreaching criterion: whether or not, taken together, they resulted in a reward. The complex set of factors that are thereby collapsed into one could include variable morphology and behavior involved in searching, distinguishing, and handling some resource, and variable environmental features and cues that are encountered during the search. If the combination of selective searching, handling, idiosyncratic morphology, and reinforced cue is successful in obtaining a reward, the whole combination, having been rewarded, will be repeated under a regime where learning governs behavior. If something *works* (attains a reward), it is repeated. Each component of the successful maneuver need not be separately assessed by the organism. But at the population level, each will be assessed, due to effects on fitness, by selection.

Foraging decisions by the medium Galapagos ground finch *Geospiza fortis*, described in chapter 18, illustrate how learning can provide integrated assessment of multiple variables to produce an adaptive decision. Relatively large-bodied, large-beaked individuals of a population feed selectively on large seeds, especially the hard seeds of *Bursura*, and this likely reflects the differential rewards for doing so in large- versus small-beaked birds (Grant, 1986; see chapter 18). The large-beaked birds also show habitat selection, frequenting woodland (vs. park land), which is relatively richer in larger seeds (Grant, 1986). Boag (1983) reported high heritabilities (0.65–0.91) of these bill characteristics and of three indices of body size in a natural population of *G. fortis* (Grant, 1986, p. 181). In this species, environmentally influenced and experience-modulated decisions regarding diet and habitat choice would have a strong heritable component through the influence of genetically variable morphology. Learning, by differentially reinforcing different behaviors in morphologically different individuals, makes a feeding decision that integrates the effects of a complex combination of inherited morphology and ecological opportunity.

It should not be supposed that learning as a developmental mechanism necessarily implies an extraordinarily high degree of flexibility or diversity in switching between alternatives. Like any other condition-sensitive developmental mechanism, learning connects environmental conditions to some developmental pathway, but once the connection is made, it may not be easily reversible, as in imprinting and other kinds of critical-period learning in birds (see Gill, 1995, p. 430). Learned responses are, like other conditional responses, flexible in that they are subject to manipulation or to novel inputs during the period when a learned decision is being made. This is evident from research on learned individual feeding specializations. In stickleback fish, for example, individuals that lose out in competition for large prey begin to take smaller prey and then develop a *preference* for smaller prey that persists even when competitors are removed (Milinsky, 1982). This is evidently advantageous, and reinforced, because individuals become experts in handling the secondarily adopted prey (Milinsky, 1982) and perhaps in recognizing it as desirable food (see Bernays and Wcislo, 1994). Like career choice in humans, the initial decision may be phenotype limited and then becomes a highly developed skill where expertise can raise payoffs, conceivably above those attained in some initially more attractive option.

Learning may well play an important role in the evolution of alternative male mating tactics where blind instinct is presumed to rule. Observed cases of switching between mating system tactics, such as that observed in Italian social wasps, in which males change the location and type of mate-searching behavior from one year to the next if the ecological situation changes (Beani, 1996), may be mediated by trial and error learning where the reward is presence of conspecific females or some other indication of a suitable mating site, a possibility that needs to be investigated. Such possibilities are not usually raised in studies of adaptive mating systems, because of divergent research traditions that separate fields. Rigorous study of animal learning has been largely confined to experimental psychology of a few species in the laboratory (see Mackintosh, 1974, p. 2, for a history and justification; see also Marler and Terrace, 1984, for a critique), whereas rigorous study of mating systems is now largely confined to behavioral ecology, where learning and other mechanisms that produce the observed behaviors are often ignored (but see Wcislo, 1989, 1992; Real, 1994).

Learning does sometimes create a large number of individualized alternatives, as in the highly individualized foraging tactics in some populations of birds (Giraldeau and Lefebvre, 1985; T. K. Werner, 1988; Werner and Sherry, 1987). Multiple alter-

natives in a single functional context mean that selection is less able to improve the form of particular alternatives. But with multiple alternatives, selection on assessment is strong, and assessment may
be more complex when multiple variable options
are involved. The expected result is accelerated evolution in plasticity and assessment per se. This may
give rise to a self-accelerating process, where multiple alternatives bring improved assessment ability, and improved assessment ability further multi-
ples alternatives in a mutually reinforcing spiral of
change.

The pivotal event at the point of upward inflection in the evolution of human brain size may have
been some breakthrough in the evolution of flexibility, causing a self-accelerating process such as
that just described, where multiple learned alternatives switch the focus of selection from a small
number of evolved specializations, to a large number of learned alternatives, and where environmentally (in this case, socially) complex variables
increase selection on plasticity and assessment ability per se. In highly social organisms, social competition screens access to virtually all crucial resources (food, space, protection, and mates;
West-Eberhard, 1979). Dominance at feeding sites,
for example, is a good predictor of winter survival
in song sparrows (Arcese and Smith, 1985), and socially dominant social insect females are the only
ones that lay eggs, to the exclusion of hundreds and
sometimes thousands of potential competitors.
Probably as a result of this, traits used in social
competition are notable for their exaggeration
(Darwin, 1871 [1874]; West-Eberhard, 1983). An
exaggerated trait like the human brain in such an
eminently social (and socially competitive) animal
seems likely explained at least in part by feats of
social maneuvering and assessment (see also Jolly,
1966; Humphrey, 1976; Barkow et al., 1992; Byrne
and Whiten, 1988; for discussions of the relationship between nutritional and social factors in the
evolution of human intelligence, see also Aiello and
Wheeler, 1995; Milton, 1988). Humans engage in
fine-tuned assessment of relatedness, status, and
reciprocity in alliances and exchange, where they
make precise quantitative assessments and remember them for long periods of time.

For these reasons, hypotheses for the evolutionary increase in the size of the human brain seem to
me most convincing when they deal with social aspects of judgment and intelligence, such as use of
language (for recent references, see Tattersall,
2000) or the expansion and assessment of social alliances (Alexander, 1979), and least convincing
when they address ecological aspects, such as tool

making or throwing ability of hunters (Calvin,
1983). Throwing ability of warriors would be more
credible, but not as convincing as assessment of allies and tactics on the battlefield, where an unending, runaway process of evolution under social selection would apply (West-Eberhard, 1983).

How Complex Mechanisms of Assessment Originate and Evolve

An introduction to the evolution of assessment is
provided in preceding chapters, which discuss the
simultaneous origin of regulation and form, and
how new phenotypes begin with the developmental reorganizations of older ones. Mechanisms of
assessment are no exception to this pattern. In
many of the examples described in this chapter, the
makings of an assessment mechanism were present
before a new trait emerged from an ancestral phenotype. The mammalian response to photoperiod,
for example, apparently capitalizes on circadian
rhythms present in all cells, as well as hormonal
systems employed in other contexts, whose communication and gene-expression control systems
can be exploited in new ways by linking them (e.g.,
via a new receptor) to a new task. Birds and lizards
have pineal-gland melatonin rhythms but, unlike
mammals (figure 23.6), do not use them in photoperiodic time measurement (B. Goldman, personal communication, 1999).

The opportunistic evolution of assessment
mechanisms from preexisting phenotypes explains
other qualities that assessment shares with development in general, namely, seemingly indirect or
convoluted means of attaining an adaptive result.
Adaptive condition-sensitive expression of alternatives implies environmental sensitivity that can be
molded under selection, so indirect cues characterize many examples of assessment. Any happen-
stance connection that may arise between a suitable environmental cue and the original condition
sensitivity of a trait could become part of an adaptive regulatory or assessment device, the way
pineal-gland rhythms have come to be used by
mammals in the assessment of day length. Photoperiod, for example, is an environmental variable
that is often used as a seasonal cue. This is not only
because day length is a widespread and dependable
environmental correlate of time of year, but also
because it is associated with very widespread intracellular metabolic rhythms that can become entrained by daylight-associated "circadian" activities
of organisms and then connected (e.g., via hor-

mones) to gene expression, metabolism, and growth. Similarly, spadefoot toad tadpoles associate preferentially with kin, not because they can recognize related individuals directly by some inherited cue, but because they can recognize those that have been raised on the same type of food as themselves. They prefer unfamiliar nonsiblings reared on the same food, or a habitat containing water strained through the food on which they were reared, to siblings reared on a different, unfamiliar food (Pfennig, 1990b). So chemicals that are associated with siblings have proven dependable enough to serve, under selection for their recognition, as cues of relatedness.

Certain hormones (e.g., juvenile hormone in arthropods, steroids in vertebrates) connect a diversity of environmental cues to a diversity of expressed genes. Hormones can participate simultaneously in many kinds of decisions because the level of production of a hormone can respond to environmental inputs through various pathways, as illustrated in the hormonal response to stretch receptors in blood-feeding insects; and because the same hormone can stimulate or inhibit responses in different parts of the body where appropriate receptors are found. Thus, juvenile hormone can influence such disparate activities as metamorphosis, aggressive behavior, and ovarian development in the same insect. Given the environmental sensitivity and integrative power of hormones and other sensory–response systems, it is easier to understand how adaptive assessment can evolve via the evolutionary adjustment of thresholds of response-specific receptors. Thus, potential mechanisms for adaptive assessment, in the form of versatile hormone and neural systems, *do* often precede conditional expression in the evolution of conditional alternatives, as seemed improbable to Wade (1984; see above). As in the badge of aggressiveness discussed at the beginning of this chapter, originally fortuitous developmental correlations, such as a hormonally mediated relation between aggressiveness and plumage color in birds, can serve as assessment cues, and selection can then improve the consistency of the correlation as well as the threshold of a response.

In the insect order Hymenoptera (wasps, ants, and bees), sterile workers have repeatedly evolved in nonsocial lineages where larval nutrition could be markedly affected by environmental variation in the quality or availability of larval food, and where larval nutrition affects the ovarian development of independently reproducing adults (reviewed in West-Eberhard, 1996). This helps to take the mystery out of the multiple origins of the food-regulated caste determination system found in the specialized social Hymenoptera, where assessment of whether to switch between reproductive queen and sterile worker depends on larval diet (Wilson, 1971a).

The environment is an inexorable inducer of phenotypic variation and can be more persistent under negative selection than a gene (chapter 6). An environmentally induced novelty—that is, a novel conditional alternative—may be recurrent in a population even though it is disadvantageous. Genetic accommodation can then gradually adjust the form of the trait and the responsiveness that produces it, and the result would be adaptive assessment and regulation. Small size is a good example of a recurrent trait that is often disadvantageous. Environmentally influenced size variation is common, and extremely small size is often accompanied by certain correlated, automatic morphological and behavioral peculiarities. Selection can mold such peculiarities into adaptive behaviors and morphological features that are in effect cued by small size. Selection can also fine-tune the size threshold for their expression, by adjusting whatever mechanism caused the original correlation with size. We would then say that a size-dependent alternative with an adaptive threshold of expression had evolved, and that development had incorporated some mechanism for assessing size.

The evolution of refined capacities for assessment in nonhuman organisms becomes further understandable if one considers the high premium on assessment accuracy. As pointed out by Bell (1991), the penalty for inappropriate decisions is severe. This is dramatically illustrated by the plight of larval bean weevils (*Callosobruchus maculatus*), which feed and develop enclosed within a bean where they have been placed by their mothers (Mitchell, 1975; Thanthianga and Mitchell, 1987; figure 23.7). Most beans contain sufficient nutrients for the development of only one larva, but when the majority of beans are occupied, females sometimes oviposit in large beans that already contain a larva. Females do not discriminate beans containing their own eggs; thus, larvae that share a bean are usually not sibs (Thanthianga and Mitchell, 1987), so competition between them is not expected to be softened by kinship. A larva assesses the presence and strength of another larva by its chewing vibrations. If a larva is relatively weak, as indicated by its weaker vibrations, its burrowing is inhibited (Thanthianga and Mitchell, 1987). If burrows intersect, the two larvae engage in mandible-gnashing aggressive behavior, and the weaker larva dies, presumably killed by the larger

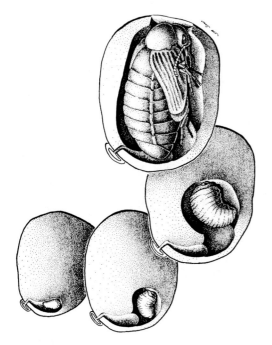

Fig. 23.7. Larvae of the bruchid bean weevil *Callosobruchus maculatus* at different stages of development (pupa at top). From Thanthianga and Mitchell (1987), figure 1, p. 17, © Kluwer Academic Publishers, with kind permission from Kluwer Academic Publishers. Drawing by Suzanne M. Sasso. Courtesy of R. Mitchell.

and older competitor. But if the dominant's vibrations cease, due either to its death or pupation, the weaker larva resumes chewing and its development proceeds (Thanthianga and Mitchell, 1987). Since it is the larger beans that contain two larvae, the smaller one has a good chance of benefitting from avoiding a contest. But the outcome of a contest is decided before the contestants meet, by the vibration-induced inhibition of the growth of one. So there is a very high premium on correct assessment of relative strength of chewing, and resumption of burrowing when a competitor's signals have ceased.

The Evolution of Assessment Involving Choice

Discriminatory choice, or differential responsiveness to signals produced by socially competing individuals, occurs in several of the kinds of assessment discussed in this chapter, including female choice of mates, parental choice or favoritism of

offspring, and worker choice of queens in social insects (West-Eberhard, 1983). Like mate choice, where one sex (usually the male) engages in attention-getting displays, parental choice sometimes involves specialized displays by offspring and differential responsiveness by parents. Some colobine monkey babies have flamboyant natal coats of strikingly different color than those of adults, especially in species where alloparental behavior, or caregiving by nonparents, plays an important role in infant care. Contrasting coat color is hypothesized to attract the attention of alloparents toward the young (Hrdy, 1999). Other examples occur in birds. The outlandish and colorful facial markings of nestling coots, which are most exaggerated in the youngest nestlings, have been experimentally shown to attract the adults, which feed the colorful chicks in preference to experimentally modified dull-colored ones (Lyon et al., 1994). In other birds, such as great tits (*Parus major*), begging calls of nestlings attract the attention of mothers, and the size of the maternal response is positively related to the intensity of the call (Kölliker et al., 2000). In these birds, there is also an analogy to indirect mate choice based on male dominance (above): parents appear to exploit competitive interactions among siblings to exercise indirect parental favoritism: parents establish feeding positions, and the dominant offspring that monopolize those positions are the ones best fed (Kölliker et al., 1998).

The evolution of assessment has a special dimension when choice is involved, because of the potential importance of manipulative cues. *Manipulative signals* (Dawkins and Krebs, 1978) trigger a response that benefits the signaler, without necessarily being indicators of any underlying quality other than the ability to produce the signal and evoke a favorable response. Good-genes arguments have come to dominate discussions of sexual selection so much that the social functions of signals sometimes seem all but forgotten, even though this was the aspect emphasized by Darwin (1871 [1874]). Such signals can be distinguished from *indicator signals*—signals whose variations reflect variation in some underlying trait that would enhance the fitness of the chooser or the chooser's descendants if favored, or that otherwise adjusts the costs and benefits of choice in favor of the chooser. Large size, for example, can serve as an indicator signal if it correlates with ability to survive. Hunger signals may indicate need in an offspring, which would therefore especially benefit from being favored in an episode of parental choice. Threat displays are good examples of indicator signals, since they evolve to be accurate indicators of fighting ability. A threat signal is only effective if backed up by true strength, and opponents are expected to evolve so as to test the truth of display, and not give in to false indications of strength, or bluffing (West-Eberhard, 1979).

Since manipulative signals evolve unrestrained by their correlation with any underlying quality, they can be elaborated and improved without limits other than their cost under natural selection. Choice, then, may screen for three aspects of quality:

1. Phenotypic quality, which rewards accurate assessment due to superior performance of the chosen individual during the chooser's lifetime (e.g., as a helpful mate or a productive queen)
2. Genetic quality under natural selection, which rewards accurate assessment due to superior quality of descendants or other relatives in the nonsocial environment (e.g., in food getting, predator escape, and resistence to parasites and pathogens)
3. Genetic quality under social selection per se, which includes ability to excel in the social environment through effective signals.

In the first two aspects, selection favors close correlation between the indicator signal and some underlying trait, and choice that detects the truth of the indicator. The signal itself has no value except as an indicator. In the third, selection favors the best manipulative signalers, and choice that distinguishes the best signals, due to the advantage of signaling superiority of descendants. It is the latter type of choice that can lead to a genetic correlation between signal ability and discrimination ability, and so-called runaway change (Fisher, 1930). For example, in the begging signals of nestling great tits mentioned above, begging-call intensity is genetically variable, and cross-fostering experiments show that it correlates better with the discrimination ability of the true mothers than with that of the foster mothers, suggesting a genetic correlation between traits characteristic of a runaway process under parental choice (favoritism) of offspring (Kölliker et al., 2000).

Much of the current research on choice focuses exclusively on "good genes," indicator signals (Andersson, 1994). So in analyses of social assessment, it is worth remembering the following neglected points:

1. So-called good genes can include both survival genes and social genes—genes that favor social or sexual success as a signaler.

2. Choice can involve multiple criteria. Truth in advertising can be combined with manipulation.
3. Under choice, all indicator signals, since they are used in social competition, are potentially manipulative and so are susceptible to "contamination" (departures from truth) in the social context. To the extent that social selection is stronger than the cost under natural selection of signal contamination by attractive but deceptive or dishonest improvements, the indicator signal will evolve as a purely manipulative signal and be subject to runaway change. If the indicator signal is subject to very strong natural selection as a true indicator and cannot (like threat signals) be verified during choice but only retrospectively by cross-generational selection, then social selection is expected to produce episodes of evolved departures from truth in the signaling context per se, followed by episodes of corrective discrimination, or use of additional indicator signals. Alternating episodes of honesty enforcement and runaway change may account for the great complexity of some signals of social dominance, such as the queen pheromone of the honeybee.
4. Even without a runaway process, manipulative signals are less constrained in their evolution than are indicator signals, since there is no ceiling of truth to their change, other than their costs under natural selection (Fisher, 1930; West-Eberhard, 1979). The manipulative signals least limited in their specialized elaboration are those least costly to the signaler. This may help account for the taxonomically widespread explosive evolution of genitalic structures used to stimulate females internally during copulation (e.g., Eberhard, 1985, 1996). The elaborate but minute structural traits that divergently evolve are unlikely to be very costly to develop or maintain. Nor are genitalic stimuli likely to be indicator signals. Therefore, male genitalia are especially likely to be subject to runaway selection as well (Eberhard, 1985, 1996).

Under choice, there are in theory at least two ways to avoid manipulation and impose honesty on socially competitive displays. One is to use competition among the contestants themselves as a screening process, since threat and fighting must be backed up by true strength. The winner has then demonstrated superior rank in whatever is required to win a contest. Another way to impose honesty is to pay attention only to signals that have some cost, so that purely selfish manipulative display is limited (see Zahavi, 1975, on the handicap principle; Godfray, 1991, on begging displays of offspring). As discussed below, dominance contests are used in mate choice and in worker choice of social insect queens. There is also evidence that sibling rivalry is used as a screening mechanism by parents attending their young (Hrdy, 1999; Roulin et al., 2000).

One may expect that honest signaling would always be favored when choice occurs in groups of kin, or in status contests among allies, where it would seem to benefit the inclusive fitness of all individuals, even the losers, for the highest quality individual to be chosen. But as long as social choice is involved, a runaway process is possible. Runaway evolution is expected to especially affect social signals in the following situations, among others:

1. When social competitive ability per se is a major determinant of access to resources required for survival and reproductive success, as within social groups, or under strong sexual competition for mates.

2. When social signals or stimuli influence choice by manipulating essential processes selected strongly in other contexts, such as maternal or physiological reflexes of females, or startle responses or shelter cues used in predator escape (Christy, 1995)—responses that cannot readily be eliminated by selection (called a sensory trap by West-Eberhard, 1979). Some genitalic and cryptic female choice signals may be in this category (Eberhard, 1985, 1996).

3. When indicator signals are so nearly equal among competitors that they do not permit choice, as when competition on nonsocial grounds is highly escalated and top competitors are closely matched. Then, as in a lek, or in the assessment of candidates for highly contested academic jobs with large numbers of superb applicants, the indicator criteria of choice may become virtually arbitrary with respect to performance in nonsocial contexts, and social signals per se become decisive, and in fact enhance the (social performance) value of the individuals under choice. This may explain why birds, such as coot, with so-called programmed overproduction and routine brood reduction, are among those that have have extreme development bright marks and competitive signaling in nestlings.

PART IV

DEVELOPMENTAL PLASTICITY AND THE MAJOR THEMES OF EVOLUTIONARY BIOLOGY

24

Gradualism

If it could be demonstrated that any complex organ existed, which could not possibly have been formed by numerous, successive, slight modifications, my theory would absolutely break down.

Charles Darwin

Darwin's adamant defense of gradualism, or small-step evolution by natural selection, set the battle lines for anti-Darwin arguments that portray developmental variation, rather than selection, as the primary architect of adaptive form. The eventual result was to caricature development as an enemy of Darwinian theory. Gradualism is still the target of critics of Darwinism, one so prominent that it is sometimes falsely linked to other issues, such as the debate over punctuation (really a controversy over changing rates of evolution, not whether or not it is gradual); the idea that developmental change implies saltatory speciation (not a requirement or a consequence of large-step phenotypic change); and the question of proximate versus ultimate causation, which draws another type of artificial dichotomy between development and selection as causes of evolutionary change. The genetic synthesis of the mid-twentieth century, following R.A. Fisher, endorsed gradualism and the primacy of selection by invoking an unsatisfactory argument that overlooked the role of development: because all selectable variation is genetic, or ultimately mutational, in origin, and mutation is random with respect to adaptation, then selection (not developmental variation) must be completely responsible for adaptive design. Nine erroneous beliefs, a legacy of this view, are discussed in the light of developmental plasticity and genetics. The main point is that development is an ally, not an enemy, of Darwinian evolution by natural selection, and should be re-integrated into the Darwinian synthesis.

Introduction

Ever since Darwin, there has been a tension between selectionists and developmentalists over the question of gradualism versus saltation and selection versus variation. Does form evolve by a series of small modifications, each one mediated by selection? Or does complex change in form originate suddenly due to a developmental change? Why should there have been such an enduring controversy over these questions? There would seem to be no intrinsic conflict between a belief that development produces variations, some of them discrete and phenotypically complex, and a belief that selection chooses among them.

The significance of gradualism for Darwin's argument regarding natural selection has often been lost from view. The gradualism versus saltation question is not just a debate over whether or not large variants can occur and be selected, although this might seem to be the issue from the dichotomy "gradual versus saltatory." One may get the impression from Bateson's (1894) compendium of complex developmental anomalies (figure 24.1), and Goldschmidt's (1940 [1982]) discussion of hopeful monsters, that Darwin overlooked the evolutionary importance of large developmental variants. In fact, Darwin (1868b [1875b]) extensively reviewed developmental anomalies, including meristic freaks such as those emphasized by Bateson, and considered large qualitative variants likely important in artificial selection producing certain breeds of dogs.

In essence, the gradualism–saltation debate is a debate over the causes of adaptive design. Is adaptive design primarily the result of selection, which

Fig. 24.1. Discontinuity in variation: a representative sample of the developmental variants compiled by Bateson (1894) to show that variation in nature need not be gradual and shaped by selection, as argued by Darwin. Compiled from Bateson (1894).

molds the phenotype step by small step, as Darwin argued? Or is it mainly due to the nature of developmentally mediated variation, with selection playing only a minor, if any, role in the creation of form, as argued by Bateson? Bateson (1894) was among the first to articulate the variationist position:

> the crude belief that living beings are plastic conglomerates of miscellaneous attributes, and that order of form . . . has been impressed upon this medley by Selection alone; and that by Variation any of these attributes may be subtracted or any other attribute added in indefinite proportion, is a fancy which the Study of Variation does not support. (p. 80)

If the critical aspects of design are already present in variation when selection acts, then selection plays only a minor role in evolution relative to development; if, on the other hand, selected variants are small and change is gradual, selection must be the principal cause of adaptive design. Gradualism gives selection more power, relative to development, in the origin of form and the "direction" of evolution. This is why Darwin placed such emphasis on *gradual* evolution, involving selection of small, even "insensible" variations; it was the cornerstone of his defense of selection. This role of gradualism in the Darwinian argument also helps to explain why the gradualism debate is revived each time there is a resurgence of interest in development as a determinant of form. Darwin (1868b [1875b]) came to regard "selection as the paramount power" (p. 426), whereas Bateson (1894) decided that "Variation . . . is the essential phenomenon of Evolution. Variation, in fact, *is* Evolution" (p. 6, emphasis in the original).

There are indications in some of Darwin's earlier writings that he was at first undecided over which factor—variation or environmentally mediated selection—to emphasize. In the unpublished manuscript of his "Big Species Book," of which the *Origin of Species* was an abstract, the chapter devoted to natural selection, completed in March 1857 (see Darwin, [1975], p. 254), portrays selection as secondary to variation in the evolution of form: it describes selection as "*second* only *to variability*, the basis on which the power of selection rests" (emphasis added). Then later in the same chapter Darwin wrote:

> in regard to the several contingencies favorable to natural selection, I am inclined to rank changed relations or associations between the inhabitants of a country . . . as the most important element . . . The amount of variability, which is largely contingent on the number of individuals, as of secondary importance. . . . But the subject is far too much involved in doubt for us to be enabled to weigh ⟨or to strike any balance between⟩ these several contingencies. (p. 273)

Later, however, he ended the two-volume work on *The Variation of Animals and Plants under Domestication* (1868a,b) with a resounding defense of "selection as the paramount power, whether applied by man to the formation of domestic breeds, or by nature to the production of species." And he used an architectural metaphor to illustrate the relationship between variation and selection in the evolution of design:

> If an architect were to rear a noble and commodious edifice, without the use of cut stone, by selecting from the fragments at the base of a precipice wedge-formed stones for his arches, elongated stones for his lintels, and flat stones for his roof, we should admire his skill and regard him as the paramount power. Now, the fragments of stone, though indispensable to the architect, bear to the edifice built by him the same relation which the fluctuating variations of organic beings bear to the varied and admirable structures ultimately acquired by their modified descendants. (Darwin 1868b [1875b], p. 426)

As for the "macromutations," or saltatory variants produced by such developmental processes as heterochrony, which could serve as a basis for nongradual transitions in evolution, Darwin invented an ingenious answer:

There is another possible mode of transition, namely, through the acceleration or retardation of the period of reproduction. . . . Whether species have often or ever been modified through this comparatively sudden mode of transition, I can form no opinion; but if this has occurred, it is probable that the differences between the young and the mature, and between the mature and the old, were primordially acquired by graduated steps. (Darwin, 1859 [1872], pp. 137–138)

With this concept of *retrospective gradualism*—the idea that large change involves traits previously evolved gradually via small steps—Darwin in effect proposed that portions of the phenotype could be suddenly shifted in their expression once they have been gradually consolidated by natural selection. In this paragraph, Darwin showed that he would not deny that phenotypically large developmental variants could occur, survive, and be selected. Sudden transitions between phenotypes is not the issue. The issue is whether the phenotypes that are shifted in timing or location of expression were originally products of gradual evolution under natural selection.

Darwinian gradualism, including retrospective gradualism, thus differs somewhat from the more simplistic version common in modern discussions where gradualism is restricted to mean evolution by selection on small (e.g., single-locus) variants, or continuous variation. Darwinian gradualism included large reorganizational change once complex traits had gradually evolved. It also included the gradual evolution of large or macroevolutionary differences between higher taxa by species or clade selection (see chapter 29).

While Mendelian genetics showed how genetic change could accumulate within populations due to selection, it did not resolve the gradualism controversy, because selectable variation is a developmental, not just a genetic phenomenon. Mendelism has been portrayed as "completing" Darwinian theory by supplying the mechanism for heredity that had eluded Darwin. But it was also a distraction from the more comprehensive view of the origins of variation that had enabled Darwin to effectively defend his theory.

Modern Permutations of the Gradualism Controversy

Why, now that selection is such a well-established tenet of evolutionary theory, do evolutionists still debate this old issue? The gradualism debate is far from dead. The either–or attitude toward variation and selection persists, for Bell (1996) writes, more than a century after Bateson, that there are two theories of evolution, one based on variation and the other on selection. Reeve and Sherman (1993) challenge us to document the influence of developmental constraints "rather than" selection in the evolution of form. And Johnston and Gottlieb (1990) find it necessary to point out the inadequacy of selection as the sole explanation of phenotypic evolution even while they grant it an important role.

We are just emerging from a productive but one-sided emphasis on selection in adaptive evolution, with a corresponding neglect of development. So, with the rise of evolutionary interest in developmental biology, the variation versus selection question, the basis of the gradualism debate, has been revived. The modern descendants of Bateson and Goldschmidt no longer speak in terms of saltation, macromutation, and instant speciation. Instead, the discussion is in terms of "developmental constraints," major change such as abnormal metamorphosis, and "epigenetic" approaches to the origins of traits, as opposed to "panselectionism" (Matsuda, 1987), "ultra-Darwinism" (a term attributed by Johnston and Gottlieb, 1990, to Romanes, 1897, but current in some recent debates; e.g., see Dennett, 1995) and what are seen as the excesses of the "adaptationist programme" in overemphasizing selection as explanations of form (e.g., Lewontin, 1978; Gould and Lewontin, 1979; Jamieson, 1986, 1989b; Matsuda, 1986; Ho, 1988; Kauffman, 1993, p. 24; John and Miklos, 1988; for opposing views, see Emlen et al., 1991; Reeve and Sherman, 1993).

Aside from the introduction of these new terms, present-day critiques of selectionism are remarkably similar to those of a century ago. Compare, for example, the phraseology of Bateson (1894) with that of Gould and Lewontin (1979):

If any one is curious on these questions of Adaptation, he may easily thus exercise his imagination. In any case of Variation there are a hundred ways in which it may be beneficial, or detrimental. . . . on this class of speculation the only limitations are those of the ingenuity of the author. (Bateson, 1894, p. 79)

Since the range of adaptive stories is as wide as our minds are fertile, new stories can always be postulated. And if a story is not immediately available, one can always plead temporary ig-

norance and trust that it will be forthcoming. . . . (Gould and Lewontin, 1979, p. 587)

Although Gould and Lewontin advocate "Darwin's pluralism" on the question of variation versus selection as contributors to form, and state in one passage, in explicit agreement with Darwin, that selection is paramount ["Darwin regarded selection as the most important of evolutionary mechanisms (as do we)"—p. 589], they are more Batesonian than Darwinian in their summary statement regarding variation as predominant over selection as a determinant of form:

An adaptationist programme . . . is based on faith in the power of natural selection as an optimizing agent. . . . We criticize this approach and attempt to reassert a competing notion . . . that organisms must be analysed as integrated wholes, with Baupläne so constrained by phyletic heritage, pathways of development and general architecture that the constraints themselves become more interesting and more important in delimiting pathways of change than the selective force that may mediate change when it occurs. (p. 581)

A host of other recent discussions both of the role of developmental "constraints" (e.g., Maynard Smith et al., 1985; Levinton, 1986; Stearns, 1986; Eberhard and Gutierrez, 1991; Galis, 1993b; Reeve and Sherman, 1993) and of Goldschmidt's work (e.g., Bush, 1982; Charlesworth, 1982; Gould, 1982; Harrison, 1982; Matsuda, 1987; Templeton, 1982b) indicate that the fundamental issue raised by the gradualism debate—the relative importance of selection versus developmental variation in the evolution of adaptive form—is still alive. Before discussing this debate in the light of preceding chapters, it is best to eliminate some peripheral questions that may confuse the issues.

What the Gradualism Controversy Is Not

Not a Controversy over Rates of Evolution (Punctuated Equilibria)

The controversy over punctuated equilibria is sometimes mistakenly taken to be the modern descendant of the gradualism–saltation debate. Punctuated equilibrium is a hypothesis regarding rates of phenotypic evolution and does not challenge gradualism. The patterns and causes of change in evolutionary rates are at issue, not the relative importance of selection versus development, for both can be involved in either stasis or rapid change. Gradual evolution can be either fast or slow. Punctuation need not imply saltation. Rather, it can be accelerated gradual evolution following a period of relatively slow gradual evolution (stasis). Large gaps, under punctuation theory, are hypothesized to be due not to saltation but to selection among species.

Both the gradualism/saltation controversy and the punctuation debate have engendered a discussion of whether or not microevolutionary, intrapopulational processes are sufficient to account for macroevolutionary phenomena, such as the phenotypic gaps between species and higher taxa. Punctuationists are on the negative side in this question because they propose that gaps originate due to (gradual) species selection rather than gradual divergence between lineages that are independently undergoing phyletic change. Again, selective gradualism is not at issue, but its units are debated. In fact, Darwin (1859) made the same species- and clade-selection argument in discussing his principle of divergence, in which gaps can originate due to gradualism driven by selection at various phylogenetic levels (see chapter 29).

Not a Controversy over Modes of Speciation (Macrogenesis)

The idea of an instant mutational origin of species (macrogenesis) originated with H. DeVries (1910) at the turn of the century. Later it was associated with Goldschmidt's (1940 [1982]) critique of neo-Darwinism and therefore was much debated in the twentieth century (e.g., see Mayr, 1963; Lande, 1980b). The term "macrogenesis" to describe saltatory speciation originated with Mayr (1963). Wright (1930) succinctly summarized the macrogenesis question as follows:

During the latter part of the nineteenth century, increasing difficulty was felt in accepting Darwin's conception of the evolutionary process as one in which variation merely plays the subordinate (though necessary) role of providing a field of potentialities, through which the actual direction of advance is determined by natural selection. Theories were developed according to which the "origin of species" was to be sought more directly in the "origin of variation." . . .

The rediscovery of Mendelian heredity was a direct consequence of the mutation theory of the origin of species and was naturally seized upon as supporting this view. (p. 349)

Although Goldschmidt discussed macrogenesis and other ideas regarding the physiology of mutation that were anathema to contemporary evolutionists, he emphasized the importance of phenotypic flexibility for the accommodation of novelty and recognized the evolutionary significance of alternative phenotypes, especially the seasonal polyphenisms of Lepidoptera, as well as the interchangeability of genetic and environmental stimuli (especially with relation to the production of phenocopies). These insights proved of less interest at the time than the heretical ones that seemed to threaten gradualism in explanations of adaptation speciation. Gould (1982b) makes these points with reference to specific passages in Goldschmidt's *The Material Basis of Evolution* (1940 [1982]).

Not a Controversy over Proximate versus Ultimate Causation

Beginning with the dichotomy between development as a proximate, or mechanistic cause of phenotypic form, and selection as an ultimate, evolutionary cause, the debate over development versus selection sometimes gets shelved by labeling development and variation as a proximate process not relevant to evolution (ultimate causation), or a simple difference in levels of analysis. As pointed out by Jamieson (1989b) in an exchange of opinions with Sherman (1988, 1989; see also Alcock and Sherman, 1994), this sidesteps the issue of what determines observed form. Different levels of analysis may be involved in the study of selection (relative fitness) and development as determinants of form. But the assertion that a developmental phenomenon rather than selection is responsible for some aspect of form is a statement about *ultimate* causation (origin during the course of evolution), not proximate causation (how it is presently produced during individual ontogeny).

Alcock and Sherman (1994) come close to saying this when they state that "constraints" hypotheses can be considered aspects of ultimate causation—"forces other than natural selection that might affect the historical pathway leading to a currently existing trait" (p. 59). But this falls short of recognition that all selectable variation is developmental in origin. Other authors have reached a conciliatory position that also falls short of the mark.

Reeve and Sherman (1993; paraphrased in Reeve and Sherman, 2001) conclude that "developmental constraints represent a special form of, not an alternative to, selection and can be studied as fitness costs arising from perturbations to existing developmental programs" (p. 64). But developmental constraints are not a *form* of selection; they are a cause of the variation *screened by* selection, and fitness costs are the dark side of variation that, seen in another light, is the cause, under selection, of all evolutionary change. Unfortunately neither side in this exchange pinpointed the underlying issue: development and selection are seen as competing explanations for evolved form. Until they are seen as complementary rather than competing explanations, the selection–variation controversy beneath the gradualism debate will not be resolved.

Fisher's Solution, or Why the Neo-Darwinian Resolution of the Gradualism Controversy Was Unsatisfactory

Fisher (1930 [1958]) promoted belief in the primacy of selection over variation as the architect of design beginning with the premise that all selectable variation is genetic, or mutational in origin. If all new selectable variation is mutational, and mutation is random with respect to adaptation, then selection is completely responsible for the direction of evolution and the origin of adaptive design, the only possible exceptions being the rare examples of environmentally influenced recurrent adaptive mutation (Jablonka and Lamb, 1995) and the cases when the range of available mutations may be said to limit the scope of selection (Leigh, 1987). Since these two exceptions are usually regarded as negligible, selection is still regarded by many as being entirely responsible for adaptive design (e.g., Bell, 1996). As argued throughout this book, new selectable variation can originate due either to mutation or to environmental change. Variation originating in both ways can have evolutionary consequences and can lead to adaptive evolution due to genetic accommodation under selection (see especially chapters 6 and 26).

This means that both variation and selection always contribute to the evolution of adaptive design. Neither can be assigned a dominant role, because development is the source of all selectable variation, and selection determines which variations among those produced by development spread and persist (see chapter 6). If a trait can be

considered a product of "design"—if it is an adaptation in the sense of contributing positively to function and fitness due to its phenotypic form—then neither development nor selection can be eliminated as historically involved.

The concept of genetic accommodation (chapter 6) reconciles the "saltatory" developmental production of discrete, complex variation with the gradual process of evolutionary genetic change. Again, the two occurrences are complementary, not contradictory, in the evolution of phenotypic novelties. Any major, or macroevolutionary adaptive change is followed by microevolutionary quantitative genetic accommodation (see chapter 29). Qualitative novelties such as those reviewed in part II define new zones of adaptation within which quantitative genetic change can accommodate functionally innovative traits. Further developmental variations at lower levels of organization can alter the new form and in turn be gradually genetically accommodated by standard microevolution (gene frequency change under natural selection). This can add up to major adaptive phenotypic change. Genetic accommodation, then, helps resolve the gradualism controversy as well as the question of whether microevolutionary processes can account for macroevolutionary change (Charlesworth et al., 1982; see also chapter 29).

Nine Modern Beliefs about Gradualism Reexamined in the Light of Developmental Plasticity

Having suggested a developmental resolution of the gradualism controversy, and the tension between variation and selection in explanations of adaptation, I now review several related ideas that need to be revised when developmental plasticity is taken into account.

The Expectation That Degree of Adaptation Attained Is Proportional to the Difficulty of Further Adaptive Change

Fisher (1930, pp. 41ff) gave a formally elegant rationalization of small-step gradualism. He described the probability of improvement by a given phenotypic change in terms of an n-dimensional hypervolume with perfect adaptation at its center, O, and n the number of dimensions, or complexity, of the trait. Under perfect adaptation, current adap-

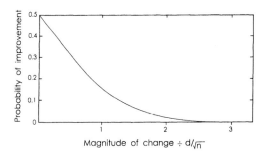

Fig. 24.2. The relation between the magnitude of an undirected change (e.g., one caused by a mutation) and the probability of improving adaptation. See the text for explanation. After Fisher (1958).

tation (*A*) would coincide with *O*. In the relatively simple three-dimensional case, *O* is at the center of a sphere whose radius, *r*, is the deviation of current adaptation (*A*) from perfect adaptation, *O*. For a new trait to represent an improvement, it would have to fall within the sphere (closer to *O* than the established adaptive state, *A*), and the greater the complexity of the trait (*n*) the more restricted the possibilities for improvement by a change of a given size. In this model, the larger the change, even if in the right direction, the greater the likelihood that the trait will exceed *r* and fall in maladaptive space. The larger the change and the more complex the trait, the smaller the probability of improving adaptation (figure 24.2).

A corollary of Fisher's model is that the better the initial adaptation, the more difficult it is for a large change to improve adaptation. Fisher (1930, p. 44) compared the requirements for adaptive improvement with the difficulty of improving an instrument such as a microscope, once it has been adjusted for distinct vision. He argued that any intentional change in the transparency, curvature, position, or polish of a lens is unlikely to improve the carefully engineered adjustment of the instrument as a whole.

The Fisher model assumes that all *n* dimensions are equally interconnected so that an effect on one has an effect on all others, and that there is no phenotypic ability for adaptive accommodation among parts. These assumptions are violated by the hierarchical and modular organization of development, and the common occurrence of adaptive phenotypic plasticity. Fisher's model does not anticipate alternative phenotypes and two-legged goats. Unlike precisely machined optical instruments, complex phenotypes have multidimensional plasticity, not just multidimensional constraints. Fisher's metaphor

fits well with the idea of the phenotype as a stable, coadapted trait complex, with netlike interconnections among parts, and with his view of conditional alternatives as elaborate machinery gone wrong (Wigglesworth, 1961; see chapter 1).

It is interesting to contrast Fisher's description of a rigidly constructed optical instrument with the discussion of flexible alternatives in the optical design of organisms given by Darwin (1859 [1872]), after Wallace:

> As Mr. Wallace has remarked, "if a lens has too short or too long a focus, it may be amended either by an alteration of curvature, or an alteration of density; if the curvature be irregular, and the rays do not converge to a point, then any increased regularity of curvature will be an improvement. So the contraction of the iris and the muscular movements of the eye are neither of them essential to vision, but only improvements which might have been added and perfected at any stage of the construction of the instrument." (p. 134)

This still timely paragraph, elegant in its way, emphasizes the properties of complex adaptation that are favorable to change, such as the possibility of different developmental routes to the same improvement, and the role of flexibility in accommodating change. Thus, the addition of the iris, or muscular movement, need not modify and thereby detract from the adaptedness of existing features. Contrary to Fisher's argument, Wallace's example shows that change is not necessarily destructive to adaptation in proportion to the degree of adaptation already achieved.

The Dichotomy between Development and Selection as Alternative Causes of Form

The classical gradualism debate concerns the contributions of selection and developmental variation to *adaptive* phenotypic form. The resolution of that debate is simple: *every* adaptive trait requires *both* an origin in developmental variation, and spread within a population due to selection (see especially Johnson, 1987). Selection cannot generate variation, and variation by itself cannot explain spread in a population.

Adaptive evolution, by definition, means that the spread in a population of the trait in question occurred due to selection (e.g., see Williams, 1966; West-Eberhard, 1992c). An adaptive trait could conceivably spread as a hitch-hiking side effect of another, more strongly positively selected trait (Gould and Lewontin, 1979; Müller, 1990) or due to the recurrence of an environmental factor inducing it (West-Eberhard, 1998; see chapter 19). Under the side-effect and the environmental-induction hypotheses, selection gets an assist from ancillary factors, but the trait(s) in question would have to be either neutral or positive in its net fitness effects in order to spread. Gould and Lewontin (1979) used the spandrels of cathedral arches as an illustration of a complex trait that has no architectural function, evidently overlooking the fact that spandrels are often modified in ways that show that they have come to function as adornments. Their cultural spread as an architectural trait may well have depended in part upon their aesthetic and not their purely engineering effects (see Mark, 1996). Nonetheless, the spandrel example illustrates how phenotypic novelties can originate and may spread as developmental side effects of structures invented and selected in other contexts.

At another extreme of the modern selection versus development controversy, sometimes labeled "panselectionism" by its critics (e.g., Matsuda, 1987), selection is given full credit for form: every feature of the phenotype is considered adaptive—each detail added by an episode of selection and therefore of functional significance. This extreme could as well be labeled "pangradualism," for the possibility of selection preserving a relatively large developmental variant with net positive fitness value despite some irrelevant or costly details is not considered. It is sometimes even asserted that if a trait is known to be adaptive, it must not be developmental in origin. For example, Ewer (1960, p. 169) supported an argument against the claim of a developmental origin (neoteny) for the larvalike lateral line canals of sexually mature *Xenopus* by pointing out that the allegedly neotenous traits are adaptive, implying by this that they must therefore not be developmental in origin—the exact converse of Gould's (1984) likewise mistaken claim that if a trait is known to have a developmental, condition-sensitive origin, this is evidence of *non*adaptation.

One reason for the predominance of extreme adaptationism—but not an adequate justification for its extreme versions—is the remarkable payoff, in terms of insight into function, of an uncompromising adaptationist approach (see Alcock, 1987; Lewontin, 1978, p. 125; West-Eberhard, 1992c). But the cost of extreme adaptationism is also high. As pointed out by Gould and Lewontin (1979) and many others (Thompson, 1961 [1992]; Strauss,

1984; Matsuda, 1987; Ho, 1988; Goodwin, 1994), it has led to a serious neglect of such form-influencing and "self-organizing" processes as allometry, correlation, heterochrony, and phenotypic plasticity in general. This amounts to a neglect of nonadaptive variation, even though nonadaptive "anomalous" variation has to be regarded as the material for selection that gives rise to adaptive novelty.

The antidote to both extreme selection–adaptationism and extreme developmentalism–nonadaptationism is the two-step vision of adaptation, as *always* involving, first, developmental variation to explain origin, then selection to explain spread and maintenance within a population. Thus, a statement like "the evolutionary origin of the enlarged brain in [geomyoid rodents] may have been developmentally mandated and not originally the direct object of selection" (Hafner and Hafner, 1988, p. 1097) applies to every adaptive phenotypic trait, once the sentence is changed to read *"must have"* rather than *"may* have been developmentally mandated."

Genetic Gradualism

Some biologists argue that quantitative genetics establishes gradualism as the primary mode of evolution by showing that most traits are polygenic in nature and that gradual change in frequency in genes of small effect at many loci is likely sufficient to explain the evolution of phenotypically complex traits. By this reasoning, "there is usually no reason to appeal to macromutations [mutations of large phenotypic effect] to explain the production of a given phenotype" (Charlesworth et al., 1982, p. 489). But the issue is not one of any *necessity* for large-step change; rather, it is to determine whether large phenotypic variants in fact *do* play a role in evolution (e.g., see Orr and Coyne, 1992). Complex selectable variation is known, making large-step phenotypic change feasible; the well-established fact that evolution involves (gradual) change in gene frequencies does not settle the issue, for such change could represent genetic accommodation of complex traits. As pointed out by Wright (1949) "Acceptance of any hypothesis of evolution by statistical transformation of populations permits differences of opinion as to whether it is the fixation of rare major adaptive mutations or the accumulation of a large number of minor ones" that is responsible for change (p. 366).

Charlesworth et al. (1982) end up suggesting a reconciliation of gradualism and saltation similar to that discussed here, but still incomplete:

The evolution of certain types of characters may occur (or begin) with a gene having major effects, if minor modifiers are selected to remove deleterious pleiotropic effects of the major gene during its evolution (Wright, 1977, p. 463), or if selection on the main effect is strong enough to overcome deleterious side-effects. (p. 490)

This hypothesis attributes saltatory change to genes of major effect. As a result, the examples given are mostly relatively unusual kinds of "balanced" polymorphisms maintained by heterozygote advantage, such as sickle cell anemia in humans and warfarin resistance in rats. When complex phenotypic change due to environmental effects is allowed, the proposal is similar to that of chapter 6 in that it combines the possibility of a phenotypically large mutational effect with gradual genetic change (genetic accommodation). This visualizes evolution in two steps: (1) production of a recurrent, relatively large developmental variant, followed by (2) genetic accommodation. The applicability of this proposal is greatly widened, as discussed in chapter 26, by including environmental induction alongside mutation as a source of selectable variation.

The Idea That Continuous Variation Implies Gradual Evolution

Virtually all characters of organisms that have been examined have been found to show quantitative variation in their dimensions, and a genetic component of such variation is also common if not virtually universal (Falconer, 1981; see chapter 6). This could be taken as an argument in favor of gradualism, since selection on such variation can result in gradual directional change.

Although it is likely that most if not all evolutionary change in the phenotype involves quantitative genetic change in the form of genetic accommodation (chapter 6), this "genetic gradualism" does not eliminate the possibility that novelty originates via a saltatory or large mutationally or environmentally induced phenotypic change, subsequently gradually accommodated as discussed in the preceding section. Whether a phenotypic variant is seen as continuous or discrete depends on the grain, or scale, of analysis. In a hierarchically organized phenotype, quantitative variation simply refers to the dimensions of a qualitative trait. Focusing downward, one moves alternately from a qualitative, discrete component to quantitative continuously variable dimension. Since even the ability to measure quantitative variation implies the recognition of boundaries (discreteness), the ubiquity of continuous variation cannot be regarded as an argument for gradualism, or the absence of discrete variation and discontinuities in change.

The Idea That Gradualism Is Enforced by the Interconnectedness of Traits

The vision of phenotype structure as a highly integrated net has led to the idea that small-step gradualism is necessary to avoid disruption of intricately interconnected development. This idea is related to Fisher's argument (discussed above) that the more complex and highly adaptive a phenotype or organism, the more difficult is adaptive evolutionary change.

The interconnectedness argument can be overcome in two ways. Phenotype structure is compartmentalized and therefore composed of dissociable parts—the net idea exaggerates the disruptiveness of change. Since phenotype subunits can be either small or large, modular reorganization can occur by either small or large steps. The phenotype is also flexible and able to accommodate variations, large and small. So netlike interconnectedness is flexible rather than rigid, and may accommodate rather than resist change.

The Impression That a Large Phenotypic Change Is More Difficult Than a Small Change

It is frequently claimed (e.g., Fisher, 1930 [1958]) that small evolutionary steps are easier than large ones because they cause less disruption of function. It is easy to show that this is not necessarily true: in various primates and hyenas, the sex transmutation of male genitalia, coloration, specialized musculature, and display to females must be regarded as a phenotypically large change and is of a kind that conceivably could be accomplished hormonally and without disruption in a single step, yet such a change has occurred repeatedly in mammals (see chapter 15 for this and other examples). The many examples of recurrent phenotypes (see chapter 19) represent complex traits that are evidently lost and gained single steps as integrated units. Nonetheless, some unit deletions, such as of a single kind of molecule—hemoglobin—from the blood, would be fatal, and many small (molecular) phenotypic building blocks are highly conserved (Gerhart and Kirschner, 1997), showing that phenotypic integration may sometimes limit the reorganization of the phenotype. Modular structure is necessary, but not sufficient, for such change to occur.

It could be argued, exactly contrary to the Fisherian view, that major complex phenotypes are more, not less, susceptible to evolved change in the time and place of their expression than are small ones. Complex phenotypes may have complex regulatory mechanisms subject to polygenic and polyenvironmental influence and therefore offer multiple pathways for regulatory change. Altered expression of a particular single molecule, on the other hand, may require a highly specific inducer or internal environmental circumstance, with no alternative route to change, and therefore evolutionary gain and loss of such relatively simple traits may be more difficult to achieve.

In any case, the evidence reviewed in preceding chapters shows that selectable variants, in the form of recurrent anomalies, occur in both large and small sizes (see part II), and that such variants can become established traits of populations (see chapters 21 and 27). This means that both gradual (small-step) and saltatory (large-variant) evolution occurs. Those who doubt the viability and evolutionary potential of large anomalies probably underestimate the ability of phenotypic plasticity to accommodate them (see chapters 3 and 16).

The Claim That Complex Developmental Anomalies Rarely Become Adaptive Traits in Nature

The collections of anomalies amassed by Bateson (1894) and others are often regarded as no more than monstrous freaks of little relevance to evolution: "There are few if any genetically well established cases of morphological macromutations which have been fixed in natural populations of animals" (Lande, 1980b, p. 234).

Some of the examples of part II contradict this assertion, and they probably represent only a small sample of those that could be listed if biologists paid more attention to the inducibility of evolved novelties, anomalies, and recurrent traits (Matsuda, 1987). There is experimental evidence, for example, that large steps toward the loss of pluteus skeletal morphology, like that recurrently fixed in echinoids, can be induced by experimental manipulations of diet simulating changes that are not unthinkable in nature (see chapter 26). The same is true of neotenous morphology in salamanders (Gould, 1977). Similar changes have become fixed in natural populations of these taxa, indicating that morphological macromutations could conceivably have been important in the evolutionary history of these groups in at least some of the multiple times these traits have evolved to fixation (see chapter 19 and figures 19.1 and 19.6). Homeotic mutations occur in a great variety of both invertebrates and vertebrates, suggesting that macromutations that affect iterative and meristic traits could have been important in evolution. In some plants bisexual flowers occur as atavistic anomalies in taxa where they have been fixed in the past (Takhtajan, 1969) showing that one-step change is developmentally possible in a major morphological trait that is sometimes fixed. I can only conclude from this very incomplete list that skeptics regarding macromutational morphological change did not search very hard for evidence that such change is eminently feasible and likely sometimes important in evolution.

The Idea That Presence of Intermediates in Related Species Demonstrates Gradual Transitions between States

Bateson (1894) was quick to point out that a complete series of intergrades among related species does not necessarily support a hypothesis of gradual origin. Bateson used the example of intersexes as an illustration:

> [T]hough these intermediate forms perhaps exist in gradations sufficiently fine to supply all the steps between male and female, it cannot be supposed that the one sex has been derived from the other, and still less that the various degrees of hermaphroditism have been passed through in such Descent. (p. 67)

Just because intermediates are developmentally possible and viable does not mean that they were passed through during evolution, though it does indicate developmental feasibility of such a sequence and the ability of intermediates to survive. In some examples of recurrence, there is evidence that the recurrent event could conceivably evolve by a combination of small, gradual changes and rather sudden ones. In the evolution of direct development in echinoids, for example, there is evidence for gradual change via a series of intermediate steps (figure 24.3), and there is also evidence that large changes in larval skeletal morphology can occur within populations (see chapter 26).

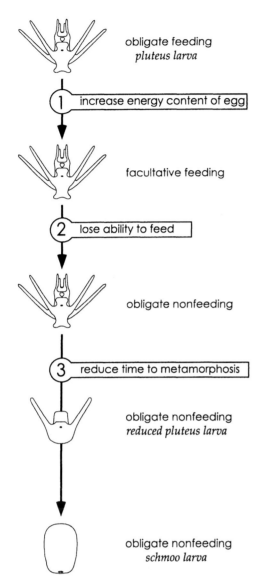

obligate feeding
pluteus larva

① increase energy content of egg

facultative feeding

② lose ability to feed

obligate nonfeeding

③ reduce time to metamorphosis

obligate nonfeeding
reduced pluteus larva

obligate nonfeeding
schmoo larva

Fig. 24.3. Possible steps in a gradual evolution of direct development in echinoid larvae. Modified from Wray (1996). Copyright 1996 from "Parallel evolution of nonfeeding larvae in echinoids," by G. A. Wray. Reproduced by permission of Taylor & Francis, Inc. Routledge, Inc., http//www.routledge-ny.com. Courtesy of G. Wray.

*Mating Problems of Hopeful
Monsters as a Barrier to
Saltatory Change*

It has been repeatedly argued that a large phenotypic change would be an evolutionary dead end, even if individually viable, because of the impossi-

bility of an individual with an extremely deviant phenotype of finding a mate (e.g., Mayr, 1963, p. 438; Frazzetta, 1975, p. 86). This objection arises in discussions of instant speciation (macrogenesis), but sometimes it is given as a reason to doubt saltatory phyletic evolution because the chance of two identical rare mutant individuals arising in sufficient propinquity to produce offspring seems too small to consider as a significant evolutionary event (e.g., see Erwin and Valentine, 1984).

Matsuda (1979, p. 223) invented a solution to this supposed problem by hypothesizing that the original change could be environmentally induced and therefore could affect a large population simultaneously.

It has never been clear to me what the monster-mating problem is supposed to involve. Is the monster sexually unattractive? Is it that by being obliged to mate with a normal individual the monster's distinctiveness would be diluted by interbreeding with normal individuals? Or would the variant simply be so different that it would be incompatible with any other individual in terms of ability to pair and/or produce viable offspring? None of these objections is convincing. Intraspecific phenotypic divergence is common in the form of alternative phenotypes (polymorphisms and polyphenisms) and sexual dimorphisms in nature. Contrasting morphs of extreme diversity freely breed within the same species without fatal detriment to viability or reproductive success and without loss of distinctiveness. In sexually dimorphic species, the truly monstrous phenotypic differences between the sexes are sometimes so great that the two sexes were initially classified as separate species, genera, or even families. Yet the differences are no impediment to mating. And in species having dimorphic males, females of a single (unimodally variable) type regularly breed with two monstrously different classes of males. There is no evidence that "saltatory" anomalies such as some of those described in chapter 19 lead to breeding and viability problems, since they are recurrently established (and lost) as the norm within lineages. Clearly, phenotypic deviation from the norm does not inevitably cause mating difficulties, loss of distinctiveness, or inviability.

Is Darwinian Gradualism
Falsified by a Developmental
Evolutionary Biology?

None of the modern defenses of gradualism stand up to scrutiny. Does this mean that Darwin's theory "absolutely breaks down?" The answer to that question is a resounding "No." Darwin's own defense of natural selection theory was in some ways stronger than the later, neo-Darwinian one. Darwin, while arguing for the power of selection, never lost sight of the developmental causes of variation and therefore managed to defend his theory with explicit reference to them.

It is a peculiar and striking fact that, with the great diversity of origins of novel phenotypes gathered together under different headings in part II, none clearly violates Darwin's standards of gradualism. Many of them violate gradualism as it was later conceived, to exclude the origin and spread of any relatively complex developmental variant whose form could not be attributed to the action of a single allele.

Darwin's retrospective-gradualism argument—that the adaptive organization of heterochronic novelties should be attributed to previous gradual evolution—was an intellectual tour de force. Discussions of heterochrony were new at the time Darwin mentioned them in the *Origin of Species* (1859). He could not have known how important heterochrony would later prove to be as a class of evolutionarily significant developmental change (see Gould, 1977, for a history). His thorough treatment anticipated a potentially important class of objections to his theory.

The retrospective gradualism argument applies to other classes of developmental reorganization involving prefabricated, dissociable phenotype compartments and preexisting responses. There are many examples of phenotypically major deletions, sex transfers, heterochronies, and reversions—transitions that are phenotypically saltatory in their evolutionary effects. But the original consolidation of the subunits concerned usually, if not always, can be attributed to previous gradual change. As noted by Levinton (1986), "The quantal nature of development is thus somewhat deceiving. A long process of gradual evolution may be behind the current presence of a quantally determined developmental program" (p. 265). It would be exceedingly difficult to show that this is not true. In many of the examples discussed in the chapters of part II, the sources of novelty are identified as preexisting elements, but it is not always known whether they were gradually or suddenly assembled. Indeed, previous consolidation is one reason why phenotypic recombination works: it is the reordering of components that are already accommodated as complex functional subunits of the phenotype. Perhaps the best claims of true saltation could be made for corre-

lated effects, where traits previously unexpressed together form a complex novelty with striking consequences, such as the buttress roots of tropical trees, the two-legged goat, or the worker phenotype of social wasps and bees (see chapter 16). But even in those cases, Darwin could argue that the responses giving rise to novelty were preexisting ones gradually evolved under natural selection.

Broad surveys of phenotypic variation (e.g., West-Eberhard, 1983, 1984, 1989) show that there are two fundamentally different patterns of phyletic evolution under selection. Type 1, exemplified by sexually selected traits, resembles classical gradualism in being long-continued, progressive, and seemingly continuous modification in a particular direction. Type 2 shows quantum jumps identified with adaptive shifts, especially via complex alternative phenotypes, and sometimes involves major developmental change. The type 1, continuous pattern is often found in coevolved competitive races, for example, under social or sexual selection or host–parasite or host–pathogen coevolution. The type 2, discrete pattern is more often associated with ecological adaptive evolution, niche change, and escape from extreme social or ecological stress (part III). Continuous and discrete patterns can occur, of course, in both contexts, so this is not an absolute categorization. But it is no accident that the classic illustrations of gradual trends are traits, such as mammalian horns and body size, that are likely important in social competition (figure 24.4). Clinal and local geographic variation in bird plumage and other display traits (Mayr, 1963) have been used to illustrate gradual divergence that increases with time and geographic distance and that shows discrete change only when there are gaps in geographic distribution or other barriers to interbreeding. These socially wielded traits have become prominent illustrations of Darwinian gradualism because, in contrast to adaptation to nonbiological (nonevolving) aspects of the environment, they may continue to evolve in social arms races over long periods of time without reaching an optimum that would mean an end to change (West-Eberhard, 1979, 1983; Eberhard, 1985).

Conclusions

Perhaps the greatest contribution of a developmental-plasticity view to our understanding of evolution is that it shows development to be an ally, rather than an enemy, of Darwinian selection theory. Far from detracting from the role of selection,

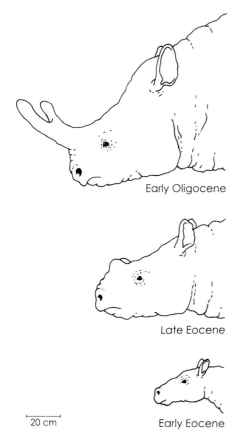

Early Oligocene

Late Eocene

20 cm

Early Eocene

Fig. 24.4. Gradual increase in size and horn development in fossil titanotheres. After Osborn (1929), used by Simpson (1949), Futuyma (1986a), Ridley (1993) and others to illustrate a directional evolutionary trend.

developmental recombination and plasticity greatly multiply the number and variety of selectable variants produced, and flexible developmental mechanisms accommodate variants in ways that increase their likelihood of persistence and modification under natural selection, in effect giving selection room to act gradually. Gradual evolution by natural selection is thus rendered more viable as an explanatory theory than without consideration of development.

Selectable variants undeniably come in many sizes, but—and probably few would deny this—selection determines which ones are multiplied and therefore is primarily responsible for determining which designs spread and persist, as Darwin insisted from the start. It does not detract from this conclusion to admit that development influences the variants presented to selection and in that sense, in modern terminology, must inevitably contribute to or "constrain" design. This is shown to be a trivial limitation when one considers the many levels at which development can be dissected and rearranged to produce distinctive variants. Subunits can be dissociated and recombined, and they can vary slightly in form, due to both genotypic and environmental variation, so the number of developmental variants that can be produced is enormous—vastly larger than the number that could be produced by mutation and genetic variation alone.

An awareness of the immense capacity for variability of development returns us approximately to where Darwin was in 1868 when he published the architect analogy with its insightful recognition of the role of variation in relation to selection in adaptive evolution. While selectable variants—the stone fragments of his analogy—may be somewhat limited in their variety, they nonetheless play a minor role in design compared with the "paramount" influence of selection in determining which ones will be used and in which position of the many feasible. Developmental anomalies of many sizes and shapes occur wherever they are sought, and many of them even recur quite frequently *without* being favored by selection (see chapter 19). But until they are positively selected, obviously they have no influence on the direction of evolution.

Nijhout (1991b) ends a synthetic developmental and evolutionary study of butterfly wing patterns with a similar conclusion regarding the causes of form:

> Perhaps the most important point to note in conclusion is that all the data presented . . . (concerning development, comparative morphology, genetics, and so on) speak with a single voice: It is possible (and straightforward) to have both gradualistic change and saltational change within this system. In fact, numerous kinds of saltational changes in phenotype emerge from gradual changes in genotype, according to the model. In addition, the ability of different kinds of parameter changes to produce similar phenotypic effects argues strongly that even if one kind of parameter change is constrained from causing a particular morphology, some other kind of parameter change could do so . . . . Thus, although there may be a short-term constraint on the evolution of certain morphologies, this system, if we have interpreted everything correctly, is virtually free of constraint in the long run. (p. 248)

This conclusion has profound meaning for the gradualism (selection vs. development) controversy. It means that the variants producible by development as material for selection may not be limiting factors in evolution. Far from having to await suitable mutations (or being limited by those that may newly arise), selection is presented with the enormous diversity of variants produced by development's alternative pathways and accommodating responses. Each time a new phenotypic combination is established in a population, with or without a genetic novelty, development has a new element of recombination and a new multiplier of selectable variation. If development is "virtually free of constraint," then selection is virtually responsible for evolved form.

25

Homology

Homology, or similarity due to common descent, is a difficult concept that is sometimes illuminated, but not simplified, by considerations of developmental plasticity. There are two homology concepts in general use: cladistic homology, or similarity due to unbroken descent from a common ancestor, and broad-sense homology, or similarity due to descent from a common ancestor whether unbroken or (as in recurrent phenotypes) not. Other, more specialized definitions result from the confusing practice of including one or more of the many criteria of homology (developmental pathway similarity, genetic similarity, etc.) in the definition. Because inherited traits are seldom identical, virtually all phenotypic homology is "mixed," or partial, a combination of similarity by descent and dissimilarity. Often the degree of homologous resemblance decreases as one focuses downward toward lower levels of organization to examine developmental and genetic pathways, for phenotypic similarity can be preserved even though genetic and developmental causes of similarity change. Multiple developmental pathways for the same trait, and combinatorial evolution, make homology concepts difficult to apply and greatly complicate the study of homology. Nonetheless, attention to developmental plasticity reveals fascinating stories of how similarity with descent can be preserved despite developmental change.

Introduction

Homology—similarity due to common descent—is the cornerstone of comparative evolutionary research. Wake (1994) calls it "the central concept for *all* of biology" (p. 269). Yet homology, like "fitness" or "species," is an elusive concept (Wagner, 1989b, p. 51; Hall, 1994, p. 2). There is unceasing debate within evolutionary biology regarding its meaning and use.

Combinatorial evolution and the extensive recurrence of similar traits revealed by modern phylogenetic study (see chapter 19) compel biologists to reconsider many ideas about homology. This is already becoming apparent in recent discussions of homology in relation to developmental variation (e.g., Wagner, 1989a,b; Hall, 1994). The traditional idea of homology visualizes a linear series of changes whereby an ancestral trait has been transformed into a descendent one (see discussion by Cartmill, 1994). By this idea different homologues may appear differently modified on different phylogenetic branches, but each descendant homolo-

gous trait has at its core a single ancestral trait. If two characters are homologous, that means that each is the modified descendant of a single ancestral trait of a shared common ancestor, and characters can be homologized in simple two-member pairs.

Combinatorial evolution raises the possibility that derived traits may often contain elements of more than one ancestral trait, and that what was formerly seen as a de novo modification actually involves the recombined expression of preexisting traits. Homology may involve not just different degrees of similarity, due to divergent modification, but may be "mixed." The maize ear, for example, combines features of both male and female ancestral inflorescences (see chapter 15), and the insect head evolved by the fusion of six ancestral body segments (Kukalova-Peck, 1997). Tracing mixed homologies requires separately homologizing pairs or series of ancestral and derived states for different elements of the same descendent trait, not just lineal comparisons focusing on the modifications of a single ancestral form.

Other problems are raised for the homology concept by the fact that "the same" feature may be conserved via different developmental pathways, and by the frequent and evolutionarily important occurrence of duplication and modification (see chapter 8, on duplication), producing "serial" or "iterative" homology ("homonomy") and positional shifts (see chapter 12, on heterotopy) of similar and historically related structures (for discussions of these problems, see Roth, 1984, 1988). Examples of serial homology include such things as metamerism (arthropod body segments; vertebrae), gene families produced by gene duplication, and "identical mass individualities" (Riedl, 1978) such as hairs, cells of the same tissue, and so forth. Although some authors exclude such repetitions from being called "serial homology" (Leschen, unpublished ms), it is not easy to justify where to draw the line. Such duplicated compartments, no matter how small or how frequently repeated, are expected to manifest the same principles of origin and potentially independent transformation as any other phenotype subunit.

This chapter considers how a developmental view of evolution affects the homology concept, both challenging its traditional meaning and clarifying certain controversies regarding its use.

Cladistic and Broad-Sense Homology

Some of the controversy about homology stems from dual usage of the word. *Cladistic homology*, the homology concept used in cladistics and generally in modern phylogenetic analyses, means similarity due to unbroken descent from an immediate common ancestor, or synapomorphy—"the relation which characterizes [traits of] monophyletic groups" (Patterson, 1982). Cladistic homology is recognized by the occurrence of similar traits in a set of taxa adjacent on a cladogram. In contrast, what I will call *broad-sense homology*—sometimes called "biological" or "transformational" homology (see Roth, 1991)—admits homology due to more distant common ancestry, and would include (1) the products of independent parallel evolution, where similarity could spring in different related branches from possession of a shared ancestral ground plan or developmental propensities, indicating "latent homology" (de Beer, 1971) in the group as a whole; and (2) recurrence—phylogenetically disjunct expression of phenotypes possessed by a common ancestor and subsequently lost, then repeatedly regained and perhaps modified. To distinguish this usage from the cladistic one, I will call it "broad-sense" homology, or simply "homology," since it includes cladistic homology as a subset. Both usages agree that homology means similarity due to common descent, and both can involve stable underlying developmental patterns (Young, 1993).

Although cladists claim that homology is an entirely phylogenetic relationship (monophyly), as if phylogenetic analysis were the only criterion of cladistic homology (e.g., see Nelson, 1994, p. 104, and references therein), in fact both the cladistic and the broad-sense approaches have criteria of similarity and phylogenetic analysis as their first two steps. Both approaches begin with (1) formulation of a homology hypothesis based on various criteria of similarity (see following section) and (2) a test of the homology hypothesis by mapping comparative data on a phylogeny. There the similarity ends, for cladistic homology requires unbroken expression of the traits within a monophyletic group, whereas broad-sense homology identifies the relevant monophyletic group as all the descendants of the nearest common ancestor of traits being compared, even if the traits are irregularly distributed (e.g., recurrent) in a cladogram, and even if it is necessary to go back to the origin of life to find the most recent common ancestor (the inclusive monophyletic group) of those traits.

Patterson (1982) alleges that the broad-sense definition of homology in terms of common ancestry leads to a circular procedure, with common ancestry used to define homology, yet the evidence of common ancestry is homology. This misconstrues the use of common ancestor identification in studies of broad-sense homology: the use of common ancestry to identify a monophyletic group is not *evidence* of homology, as in cladistics; it only serves to define the taxa within which the transitions of interest (e.g., as evidenced by intermediate forms) are to be evaluated. The cladistic concept has another kind of problem, which I have not seen discussed, namely, that it is inconsistent in its application of the requirement of unbroken expression of traits. If a trait is broken between generations within a population, is cladistic homology lost? Eye color is not consistently inherited between generations. Should blue eyes in the offspring of a brown-eyed father be considered "convergent" with those of the blue-eyed grandfather? Similarly, in polyphenic populations, parents can be of one morph and offspring of another. To be consistent, the cladistic homology concept should consider the regained morph "convergent" rather than homologous with that of its predecessor in previous generations. The cladistic concept is inconsistent in ignoring these breaks in within-population expression, while recognizing the same break in expression if it occurs between species (e.g., eye color or morph becomes fixed). The broad-sense definition is more consistent in its recognition that homology can occur even though there are breaks in expression during descent.

One potentially confusing consequence of these two homology concepts is that some disjunct similarities that are considered homologous under a broad-sense definition are considered nonhomologous "convergence" under a cladistic definition. Under the cladistic definition of homology, all homoplasy (convergence, parallelism, and recurrence) is seen as contrasted with, not due to, common ancestry. Thus, while an evolutionist may take pains to distinguish between convergence and parallel evolution, a phylogeneticist would sometimes lump them, or deny that homology is involved, even when discussing (disjunct) similarities clearly due to common descent.

In keeping with the cladistic homology concept, homoplasy has been defined as structural resemblance due to parallelism or convergence "rather than" (immediate) common ancestry (e.g., Lincoln et al., 1982; discussed critically by Rieppel, 1994, p. 66), or as "mistaken homology" (de Queiroz and Wimberger, 1993, p. 46). That is, parallel evolution and recurrence based on a common ancestral ground plan are not regarded as giving rise to homologous traits even if, as sometimes occurs, there is evidence that a recurrence is due to a change in a regulatory pathway known to be shared by all members of the clade concerned, as in the case of recurrent neoteny in salamanders (see chapter 19). As stated by Haas and Simpson (1946), homology expresses an opinion as to how similarity arose. Homoplasy expresses an opinion as to how the similarity *did not* arise, that is, that it *did not* arise by homology, but it does not express an opinion as to how the similarity *did* arise.

When rigidly applied, the cladistic homology concept can even assert that genetically and developmentally identical traits—"the same traits"—found in sister species, unbroken in their expression (expressed in some individuals of every generation) and undeniably similar due to their common ancestry, are nonhomologous simply because they are independently fixed in different branches of a cladogram:

[t]wo species showing the same ancestral polymorphism [may have] experienced similar selection pressures leading to fixation of the same trait. Because the fixed trait arose [was fixed] more than once evolutionarily, its various manifestations among different species are not evolutionarily homologous. (Brooks and McLennan, 1991, p. 13)

Here, the designation of traits as nonhomologous extends not only for phylogenetic purposes, but for "evolutionary" homology as well. Furthermore, this paragraph implies that a trait does not exist as a homologue until it is fixed. As phrased by Meyer (1999), cladistically homologous traits are "those that are evolutionarily always expressed" (p. 141). But for evolutionists interested in the sources of novel traits and aware of a potential for reversions, it is important to recognize similarities due to common descent in the "convergences" of the cladistic concept (e.g., see Edwards and Naeem, 1993, on the recurrence of cooperative breeding in birds).

It does not make sense, in an evolutionary discussion, to regard the event of fixation as being required to establish homology, and some have argued the same for cladistics (see Wiens, 1995). Nor does divergent modification ("various manifestations among different species") destroy homology, for if modification in different species were to invalidate homology, even narrow-sense cladistic homology would disappear from use, since it invariably refers to "various manifestations among different species" of the same trait. This is a good example of the kinds of confusion that can arise in discussions of homology, and it shows the importance of retaining a broad-sense usage alongside the cladistic one, in order to make sense of evolutionary relationships among phenotypes. Both usages can occur side by side without conflict and without losing the essence of homology—similarity due to common descent. There is no confusion if it is specified which meaning is in use.

Description of evolutionary transitions and broad-sense homology also demands a different usage of the word "arose." A complex, adaptive trait expressed as a polyphenism must be regarded as having "arisen"—occurred in a population—before it has been fixed. While overtypification of a polymorphic population may sometimes be convenient for purposes of taxonomy and cladistics (but see Wiens, 1995), it is not permissible in evolutionary studies where alternative phenotypes, even at low frequencies, may be crucial in an evolutionary transition. As discussed in part III, alternative phenotypes such as the aquatic leaf form of buttercups may be unexpressed for an entire generation in a particular time or place (e.g., in a dry zone or a dry year). Yet when they reappear under wet conditions, it would be absurd to consider them convergent, rather than homologous, with those of previous generation. Many such examples are given in part III, and they reveal the inadvisability of making continuity of expression part of the definition of true homology for anyone interested in tracing the origins of traits. A trait can disappear from an entire lineage for many generations and then reappear in some descendent species and not others. It would not be any less homologous by virtue of its long period of nonexpression, or its sporadic expression in only one or a few descendant populations, or even extreme modification with time.

It has been suggested to me that the cladistic concept of homology is so important and predominant in current usage that broad-sense homology will only cause confusion. But the broad-sense homology concept is a reminder that the cladistic concept of homoplasy concatenates and therefore confounds such evolutionarily important, though intergrading, phenomena as parallelism, convergence, and reversion. If the cladistic homology concept is the only one, distinctions among these phenomena, crucial for evolutionary comparative biology, could easily be neglected. Rigid adherence to the cladistic homology concept is too narrow and restrictive for evolutionary research and is likely to lead to errors in understanding evolutionary transitions.

I believe that such errors have already occurred. I recently saw a spirited article showing the value of phylogenetics in the study of evolution where it was suggested that if we had a phylogenetic tree and a detailed set of biological data on the great apes, we would be able to ascertain, among all the traits humans display, which are historical legacies and which are uniquely evolved in *Homo sapiens*. The possibility that some human traits could be reversions reflecting more ancient ancestry was not considered. This can only mean that the extensive parallelism and recurrence homoplasy uncovered by cladistic analyses, and the evidence for its developmental basis (see chapter 19), have not been sufficient to bring home the obvious evolutionary conclusion: parallelism and recurrence often reflect common ancestry, and from an evolutionary transition perspective, a broad view of phylogeny and inheritance is imperative.

A belief that the historical legacy of a species can be entirely represented by the close extant relatives known to compose a monophyletic group implies that the cladistic concepts of homology and convergence are being used in an evolutionary discussion to the exclusion of the broader evolutionary concepts, which allow for the possibility of inheritance from distant ancestors without unbroken expression and for parallel changes based on similarity of developmental ground plan. While the broad-sense definition would cause confusion if applied in cladistics, so the cladistic concepts can cause confusion when applied to evolutionary questions. So both definitions are essential, each in its own domain. Comparative biologists—cladists and evolutionary biologists—should perhaps follow the example of quantitative genetics in its distinction between "broad-sense" and "narrow-sense" heritability (see Falconer, 1981) and tolerate two definitions while taking care to note which is being used. The two usages of homology represent two complementary approaches to understanding the evolution and stability of structure (Shubin, 1994, p. 252).

The Criteria versus the Definition of Homology

The cladistic versus broad-sense definition problem is easily solved by context-dependent switching between definitions. Much of the remaining controversy regarding homology grows out of another problem, namely, the attempt to incorporate criteria for recognizing homology into its definition, a problem noted by many previous authors (see below). In the first phase of homology research, when similarity is used to hypothesize homology, different criteria of similarity may be used—genetic similarity, developmental similarity, detailed structural similarity, position relative to other structures, and experimentally induced transformations, for example, can all be used as criteria of similarity. When de Pinna (1991) writes, "In its most basic form, homology means equivalence of parts" (p. 368), he refers to criteria of similarity or homology—how you recognize it, not what it means (whether similarity due to common descent, or synapomorphy). At this point, as de Pinna observes, "consensus ends" because parts can be shown "equivalent" in different ways that may indicate common descent, confirming a homology hypothesis.

Much of a recent and authoritative book on homology (Hall, 1994) is dedicated to debating the relative merits of different criteria of homology, some of them in the form of definitions. But it is a mistake to *define* homology in terms of a single criterion (see also Raff, 1996), since not all indicators of similarity are usable in every case. When genetic criteria, for example, are criticized by saying that they need to be viewed in terms of developmental pathways (e.g., Galis, 1996b; Müller and Wagner, 1996), there is a risk of putting too much confidence, in turn, in developmental pathways, known in some cases to vary in producing the same phenotypic trait (e.g., see Gerhart and Kirschner, 1997; see also chapter 5, and below).

There have been repeated well-reasoned appeals to keep the definition separate from the criteria of homology (e.g., Simpson, 1975, p. 17; Mayr, 1982a, p. 45). As pointed out by Donoghue (1992) "The choice of a [specialized] definition is, at least in part, a means of forcing other scientists to pay closer attention to whatever one thinks is most important" (p. 174). It also invites endless argument over what the "correct" definition should be. The choice of criteria is partly a pragmatic matter. The most powerful procedure is to recognize that there are numerous criteria, and given the difficulty of tracing homologies, as many as possible should be used. The approach of Gosliner and Ghiselin (1984) to phylogenetic reconstruction applies to homology reconstruction as well: "In this endeavor, any valid canons of evidence that reveal the truth about genealogy are legitimate. . . . We ask what occurred and why, and attempt to discover the answer" (p. 256).

New homology criteria due to recent progress in molecular and developmental biology add to the available arsenal of tools. But the new findings do not settle forever the homology problem and in some respects make it seem more difficult, as discussed below. Many criteria of homology are possible, and all that are useful in a particular taxon should be used. This seems less confusing than the alternative solution, which is to admit multiple operational definitions. One can legitimately debate the relative usefulness of different criteria without discarding any of them, while using a single conceptual definition.

A conceptual definition need not be operational. Rather, it functions as a model (*sensu* Lloyd, 1988) to be tested in different ways. The "biological species concept" (Mayr, 1963) is such a definition. It defines species as reproductively (or genetically) isolated populations and invites invention of ways to estimate the degree to which such isolation is likely to obtain. "A paleontologist cannot test interbreeding (basic to the species definition) in his fossil material but he can usually bring together various other kinds of evidence (association, similarity, and so on) to strengthen the probability of conspecificity" (Mayr, 1982a, p. 45).

Criteria that have been used to propose or support homology hypotheses include the following (after Donoghue, 1992, and others):

1. Similarity in position and structural detail (Owen, 1848; Remane, 1952), called "congruence" by Patterson (1982) (would not apply to "mass identical similarity," e.g. of hair, or some forms of homonomy, or serial homology; R. Leschen, unpublished).
2. Presence of connecting intermediates or transitional forms, including in ontogeny (Remane, 1952).
3. Similarity in development (Patterson, 1988; see also other references in Wagner, 1989b, p. 55), taken to mean shared developmental pathways (Roth, 1984; but see Roth, 1988), "shared developmental constraints" (Wagner, 1989b), or evoked by the same stimuli (Baerends, 1958).
4. Lack of conjunction, or lack of coexistence in a single organism (Patterson, 1988; as discussed below, there are many examples of "iterative" or "paralogous" homology where this does not apply).
5. Genetic similarity (see references in Roth, 1994, p. 305; Galis, 1996b).

Developmental–genetic evidence of homology may prove especially useful for analyses of recurrent traits (see chapter 19). In recurrence, the similarity in form of the traits or trait subcomponents (in the case of mixed homology, discussed below) is based on expression of genes, or developmental processes influencing form, that are similar due to descent from a common ancestor. As I discuss below, this does not necessarily imply "produced by the same developmental pathway" or "due to control by the same genes." If similarities in regulatory mechanism occur, this would support homology. But finding different pathways or different "underlying genes" in regulation is not definitive evidence against homology or recurrence, because the same form can develop via different pathways due to gene–environment interchangeability in regulation (see chapter 5): a multitude of different genetic and environmental changes can have the same effect on a single polygenic switch, controlling expression of a particular set of genes. Thus, the regulatory elements could vary while the genes whose expression they control do not. Therefore, superficial comparisons of genetic or developmental underpinnings, without attention to regulatory architecture and level of analysis, could give misleading all-or-none answers about homology (de Beer, 1971; Roth, 1988).

This point is important because of past confusion. Rensch (1960, p. 19; see also Saether, 1979, p. 305) considered recurrence to be due to "identical mutations" or "parallel mutations," seeming to identify a single (complex) phenotype with a single (mutant) gene, rather than with the possibility of a complex multigenic environmentally sensitive regulatory response or switch. Simpson (1951b; see also Vavilov, 1935 [1951]) considered parallelism to be most frequent among closely related taxa because "[m]utations, like structures, are likely to be more nearly alike in closely than in distantly related animals" (p. 67). Recurrent phenotypes do not suggest recurrent mutation of the same gene. Rather, they reflect the feasibility of multiple environmental and genetic causes of parallel changes in form, one of the reasons to expect recurrence to be common (see chapter 19).

Progress in molecular developmental biology could conceivably make genetic criteria for *tracing transitions* through their intermediate stages (criterion 2, above) fully operational in the future. Meanwhile, more traditional evidence for homology can be used. As I argue below, universal agreement that homology is "similarity due to common ancestry" in fact implies a tacit belief in inherited similarity in the elements of form.

One difficulty of a purely genetic criterion of homology is the awkwardness of applying it to morphogenesis, behavior, and other developmental processes where structure emerges from an interaction of growing parts, or due to a response by a structure (such as the neuromuscular system) previously constructed. Formation of the neural tube in chordates, for instance, is a product of a particular pattern of tissue growth (reviewed in Gerhart and Kirschner, 1997); the mandible of a mouse takes its form from certain patterns of differential growth of several different, semi-independent genetically variable components (see chapter 4, especially figure 4.14); and the sets of genes that influence different behaviors are not readily identified. Such features may be easily distinguished as discrete phenotypes, yet they emerge from developmental influences that are expressed at a different level of organization and are not themselves unambiguously linked to expression of particular sets of genes such as those controlled by a switch during differentiation of cells or tissues. For this reason, developmental-pathway comparisons are useful, alongside genetic ones, among the criteria of homology.

Iterative or "Paralogous" Homology

Combinatorial evolution leads to multiple uses of the same phenotypic building blocks. This may play a role in their conservation, under selection for consistency (or consistent flexibility) in form, to meet multiple simultaneous demands, and because disappearance of expression in one context does not cause loss of expression in others (Kirschner and Gerhart, 1998; see chapter 11). Another consequence of trait reuse or reexpression in different contexts is the side-by-side existence of homologous traits in the same organism. In morphology this has been called "iterative" homology (Rensch, 1960). Examples are the products of heterotopy (chapter 14), such as the differently specialized appendages of different body segments in arthropods, or the alpha and beta hemoglobins derived by gene duplication in vertebrates. The latter, molecular iterative homology, is sometimes called "paralogy" (Fitch, 1970).

The pervasive character of combinatorial evolution (see part II) ensures that iterative homology is likely to be common. All cases of heterotopy, duplication, heterochrony, and heterotopic change (see chapters 10, 13, and 14) where expression in the original context is not lost, lead to iterative or paralogous broad-sense homology. In plants and other modular organisms (e.g., colonial invertebrates) as well as in segmented and radially symmetrical organisms (e.g., cnidaria, echinoderms), this is an exceedingly common kind of homology. Not too different from iterative or paralogous homology is the homology between similar traits expressed in both sexes as a result of sex transmutation. Evidently, both sexes carry the genetic–developmental propensity to express the same trait even though it has originated and been elaborated in only one of them prior to its transfer of expression to the other.

"Mixed" Homology

Several authors concerned with the combinatorial nature of evolution have proposed a concept of "partial" homology, to include such cases as novel structures in plants that combine some features of branches and some of leaves (Sattler, 1988) and the incorporation of ancient molecules into new traits (Roth, 1984; Zuckerkandl, 1994; Hillis, 1994). Roth (1988, p. 7; see also Donoghue, 1992, p. 177) discusses the need to conceptualize how genes get "deputized" in evolution. Similarly, Nijhout (1999c, p. 183) writes of the recruitment of new genes (without previous effect on the phenotype) that can take the phenotype in new directions.

"Partial homology" invites a quantitative view of homology, as representing a certain *degree* of similarity, or similarity at different levels of organization, due to common descent (Shubin and Wake, 1996). "While homology is usually treated as an all or none matter, it is obvious that the implied similarity is a matter of degree" (Wright, 1948, p. 919). A developmental–genetic measure of homology would seem to imply that it is theoretically possible to make quantitative estimates of degree of homology in terms of overlap in gene expression. Zuckerkandl (1994), for example, recognizes "fractional homology," the minimal degree between proteins being a single exon product in common, and from there fractional homology could increase quantitatively with the size or genetic complexity of the components shared. But some have objected that any concept of "partial homology" is awkward in view of the conventional use of homology to seek a yes or no answer to the question, "Are these traits homologous?" Donoghue (1992) considers that "[p]artial homology is incompatible with standard evolutionary views, according to which structures are either homologous or not (e.g., Patterson, 1987)" (p. 172).

There are other reasons not to pursue such a quantitative approach to homology. Tracing homologies in terms of genes and developmental pathways may give a misleading view of evolutionary questions, since a phenotype can be conserved despite different genetic and developmental underpinnings (see below; see also Gerhart and Kirschner, 1997).

The problem with "partial" or "fractional" homology is that it does not quite capture the phenomenon to be described. When Roth and others describe the "deputization" of molecules and other phenotypic subunits into novel traits, they refer not only to the partialness of homology (which has always been implicitly recognized) but also to its mixed nature. Mixed homology departs from tradition in recognizing that a single derived trait contains parts that are collectively homologous with several ancestral ones, not just one or a lineal series, as in the male–female composite represented by the maize ear (see chapter 15), and the six-segment composite represented by the insect head (Kukalova-Peck, 1997). There are probably many examples of mixed homology in molecular evolu-

tion, where a single functional unit, as in the homeobox leucine zipper proteins, may be formed by subunits originally coded by different genes and added by "exon capture" (Schena and Davis, 1994; see also Hillis, 1994). In its history, a single molecule could acquire elements by exon shuffling, crossing over, and exon duplication (see references in Schena and Davis, 1994; see also chapter 17), each episode and fragment representing a different homology. Hillis (1994) notes that the difficulties of assigning homology in comparative studies of molecules parallel those in studies of morphology, and that similar levels of homoplasy are found (Sanderson and Donaghue, 1989). This is not surprising, since combinatorial evolution characterizes both levels of organization (see part II).

Mixed homology implies that homology can be dissected with a view to tracing more than one avenue, or organizational level, of similarity between ancestral and descendant traits. It also adds something to the concept of homology in that it allows for the possibility that similarity in descendant populations may *increase*, rather than decrease, with successive generations (Roth, 1984). This is not contemplated by the conventional way of thinking about homology, which sees successive modifications as obscuring or diluting similarity by common descent. When a lost trait is resurrected, as when the aquatic modifications of the leaf form of a buttercup reappear upon exposure of a terrestrial population to water, similarity to a relatively remote ancestor may be regained. It is also conceivable that, with prolonged exposure of the population to water and further evolutionary change, some (though not necessarily all) additional ancestral elements of aquatic morphology and physiology could eventually be added and the resultant mosaic reversion fixed. If "the leaf" is taken as the character of interest, then thinking of adding elements of mixed homology describes the mosaic nature of the change.

Alternatively, one can consider each element of a complex trait separately, each one showing all-or-none broad-sense homology to elements expressed in aquatic relatives. This preserves the traditional all-or-none idea of homology, but it loses sight of the fact that complex coordinated characters, seen at a higher level of organization, may be built up by gradual addition of preexisting elements. It is actually erroneous to think of such a complex structure in terms of all-or-none homology through descent by modification from one single ancestral structure. Nor is it complet curate to think of a mosaic evolved by pl recombination as being "partially homologous" when related species are compared. The homology may be "complete" with regard to corresponding parts, but mixed with regard to the mosaic and its sources in ancestral phenotypes (see section on variable levels of homology, below).

Mixed Homology and Molecular Evolution

If organismic evolution is fundamentally combinatorial, then the smallest building blocks, molecules, are likely to be those most conserved and recombined. Larger structures should be less durable, being more subject to change via dissociation and reorganization by virtue of their greater structural complexity. Since the homology concept was originally applied to macroscopic morphology, generations of evolutionary biologists are accustomed to think in terms of homologies among relatively closely related taxa. The findings of molecular biology require expanded thinking into a longer time scale encompassing broader ranges of taxa. The idea of mixed or limited homology may help to organize information on molecular similarities between traits traditionally regarded as nonhomologous. The insect compound eye and the vertebrate eye, for example, share a gene (Pax-6) with highly conserved structure (discussed in Gerhart and Kirschner, 1997) as well as other molecular components (Nilsson, 1996) even though in many respects the two kinds of eye are very different and clearly independently evolved. The Pax-6 gene is not a general purpose gene, but one whose expression is specific to eyes; it also patterns the head and sense organs of nematodes. Such findings mean that to thoroughly trace the common ancestry of vertebrate and arthropod eyes, one may need to consider the sense organs of some pre-Cambrian metazoan.

The homology concept has yet to be applied widely to the increasing evidence of extreme conservation of molecules and fragments of molecules, which sometimes extends across phyla in numbers of cases that would have aroused skepticism only a decade ago (e.g., see Gerhart and Kirschner, 1997). Steroid hormone receptors that resemble those of mammals and show high affinity for mammalian hormones, for example, have been found in fungi (Feldman, 1988), and the insect hormone ecdysone is also found in plants (Nijhout, 1994a). Many other key structural and signaling molecules are highly conserved, meaning that they show striking similarity in either structure or response between very distantly related phyla—across eons of evolutionary time.

Discussions of molecular conservation may give the impression of molecular stasis, yet looking closely at the degree of divergence sometimes involved begs the same questions regarding the causes of similarity that concerned nineteenth-century comparative morphologists and embryologists. Might the "conserved" folding domains of a protein really represent convergence? How meaningful is it to regard a mixed or partial homology as "highly conserved" if the dissimilarities suggest a long history of phenotypic recombination at the molecular level, perhaps by multiple historic routes, so that all that is really conserved is a fundamental building block—or perhaps several very similar but independently constituted building blocks? The probabilities of independent construction may be thought to be extremely low, but given the biochemical propensities for self-organization of many organic structural elements (Denton and Marshall, 2001), and the very long time scales involved, independent convergent recurrence may be common. At least this possibility needs to be considered more thoroughly in modern molecular biology, where the standards of conceptual rigor in establishing homology lag far behind those of modern practitioners of more traditional comparative anatomy.

In an exceptional study, Shapiro et al. (1995) gave evidence for convergence in the evolutionary origins of "conserved" cadherin domains: cadherins mediate cell–cell adhesion in many organisms, and all extracellular cadherin domains have a common folding topology. This same folding topology is found, however, in other immunoglobulinlike domains, in molecules as diverse as plant cytochromes, bacterial cellulases, and eukaryotic transcription factors for which sequence similarities are very low and genomic intron patterns are different. "On balance, independent origins for a favorable folding topology seem more likely than evolutionary divergence from an ancestor common to cadherins and immunoglobins" (Shapiro et al., 1995, p. 6793). In contrast, a comparative study of histones indicates that there is a highly conserved core, the "histone fold," and other features of this family of molecules seem to have diversified around it (Ramakrishnan, 1995).

Variable Levels of Homology in Recurrent Traits: The Eyestalks of Flies and the Striking Behavior of Snakes

The eyestalks of male flies are a frequently cited example of recurrence, parallel evolution (Hennig, 1966), and "underlying synapomorphy" (see

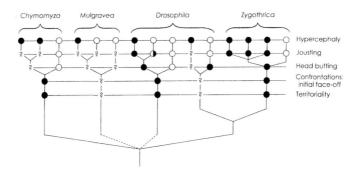

Fig. 25.1. Phylogenetic hypothesis for historical relationships among traits associated with male hypercephaly in four genera of drosophilid flies. Black circles indicate apomorphic states; open circles indicate plesiomorphic states. Question marks indicate missing data. After Grimaldi and Fenster (1989).

Saether, 1986, p. 5). Even though cladistic terminology would label the sporadic occurrences "convergences," common ancestry obviously plays a role:

> It can be no accident that so conspicuous and peculiar a character as stalked eyes occasionally occurs only in a relatively small group of about 5000 species (in which it apparently arose repeatedly) among the approximately 65000 species of living Diptera. (Hennig, 1966, p. 118)

The eyestalks are exaggerated lateral projections of the head that bear the compound eyes at their tips. In at least some groups these have been shown to function in male–male competitive interactions, and in two species in female mate choice (Wilkinson and Dodson, 1997). Males engage in ritualized contests with their heads pressed together and the larger, wider headed fly usually wins (McAlpine, 1979; Grimaldi and Fenster, 1989; Wilkinson and Dodson, 1997).

This feature appears to satisfy several of the criteria of homology, including (1) recurrence in only a part of a large taxon, phylogenetic evidence of similarity due to common descent; (2) similar selective contexts (often used as a criterion of homology, though a weak one because it applies as well to convergence); and (3) apparent similarity in structure, the eyes being consistently borne on lateral extensions of the head and therefore involving modifications of the same morphological elements. Grimaldi and Fenster (1989), in an exemplary comparative study of the behavior and morphology of broad-headed drosophilid flies, showed that the same sets of cranial sclerites are involved in the lateral extensions of the head in different drosophilid genera, although more are involved and even the eye may be slightly modified when head modification is most extreme (e.g., in the genus *Mulgravea*). Placing them on a cladogram along with the associated male behaviors (figure 25.1) shows that in the four of eight independent origins of hypercephaly where behavior is known, aggressive territoriality, and in particular face-to-face confrontation and head-butting displays, probably preceded the evolution of laterally enlarged heads (hypercephaly), evidence that in these groups behavior repeatedly has taken the lead in morphological change. They therefore identified a similar or identical selective context for the evolution of the similarity (confirmed by Wilkinson and Dodson, 1997).

Grimaldi and Fenster (1989) also showed that both parallelism and convergence were likely involved: of the 11 independent occurrences of hypercephaly in drosophilid flies, five involve a rounded-eye form (my term), and six (all in the genus *Zygothrica*), a more pointed-eye form produced by distinctively extreme modification of the dorsal fronto-orbital plates and the occipital sclerites, displacing the eye markedly outward and altering its shape into a tapered apex. Thus, the recurrence of eyestalks or hypercephaly in this group appears to fall into two convergent sets on the basis of structure and development. Yet there are some homologous elements between the sets, giving some grounds for referring to recurrence applying to both groups taken together: there is some overlap in the structural homology, with the fronto-orbital plates modified dorsally in some non-*Zygothrica* species, though not as strongly as in *Zygothrica*; and there may be an older homology in behavior shared by both groups, with territoriality, face-to-face confrontation, and head-butting possibly characterizing the common ancestors of all (figure 25.1). The question of determining behavioral homology would require detailed comparative study just as in the case of morphology (see Wenzel, 1992a).

Grimaldi and Fenster (1989, p. 9) capture the composite nature of homology when they describe the different eyestalk head modifications (*Zygothrica* species) as "synapomorphic at several levels and convergent at other levels." Since they use a cladistic homology concept, their "convergent" similarities within *Zygothrica* could be broad-sense homologies. Given the groundwork provided by Grimaldi and Fenster, the drosophilid and other flies would probably reward more detailed comparative work on the evolution and development of behavior and associated morphology.

Behavior can be compared and homologized in much the same way as morphology (Baerends, 1958; see especially Wenzel, 1992a; de Queiroz and Wimberger, 1993). Exactly the same problems of mixed homology and altered perception of homology at different levels of analysis obtain. By mapping the gaping display of viperid snakes onto a cladogram, Greene (1994) has estimated that the display with fangs erected arose only once (in the African bush vipers, *Atheris*, Viperinae), whereas the display with fangs folded has originated independently five times in pit vipers (Crotalinae). In every case, the gaping display is derived from striking behavior, and on that level the recurrence could be called a product of parallel evolution. But detailed comparative study shows that not all gaping displays are alike. In most cases, the gaping display has the fangs exposed, but in one case, where the fangs are kept retracted, the gaping display is only partially homologous with the others, apparently being derived from another phase of the strike where the mouth is open but the fangs not yet exposed. Greene's analysis of behavioral homology (see also Greene, 1999) in terms of its structure and sequence is quite similar to the morphological analyses of Grimaldi and Fenster (1989) on structural homology in the eyestalks of flies.

Levels of Analysis and the Perception of Homology

When phylogenetically disjunct similarities are compared, the question arises of how great the mixed or partial homology has to be for the traits to be judged "recurrent." For example, lateral extensions of the head of a fly necessarily develop from morphological elements on the heads of flies. Therefore, all such extensions could be called "homologous"—they all develop from the same set of head segments. On the other hand, if one focuses on the modifications of the head—its lateral extensions—there are important differences in the segments used and in how they are used or altered. The fly morphology and the snake behavior studies described above show how level of analysis, or subdivision of characters, affects perception of homology (see also Patterson, 1982). In all six lineages of *Zygothrica*, for example, the occipital sclerites are greatly distended, so that the eye is markedly displaced outward, whereas in other drosophilids with lateral head extensions these sclerites are usually unmodified (in three clades) or much less dramatically modified (two clades), and the eyes are not affected in the same way.

The head sclerites modified could be considered homologous because the same ones were involved, yet in some cases a sclerite is laterally extended as a whole whereas in others the extension is primarily of the dorsal portion. This suggests that different genetic changes in regulation may be involved in the modifications, even when the same sclerite of the head is altered, indicating little or no homology between the *modifications*, even though the elements modified are homologous. So perception of the degree of homology changes as one descends through successive levels of organization of a trait. If dorsal extensions were to be compared, some may turn out to involve different genes affecting dorsolateral growth, or lack of genetic homology within developmentally homologous traits. A dissection of homology is indispensable because of its mosaic nature, and degrees or levels of partial homology may be more precisely determined as more microscopic or molecular levels of resolution are possible. Both studies show how level of analysis, or subdivision of characters, affects perception of homology.

Multiple Developmental Pathways and the Homology Concept

There are more difficult developmental challenges to the concept of homology than those posed by mixed homology. There is increasing evidence that phenotypes can be conserved while developmental pathways change. In three different neotenous salamanders, for example, metamorphosis breaks down, producing neotenous adults, in four different ways, representing different points in the complex sequence of events that lead to metamorphosis (see figure 5.14). The addition and subtraction of digits in dogs can occur either as a correlated effect of body and limb-bud size, or due to a mutation that affects the limb bud independent of size (Alberch, 1985). In both the salamanders and the dogs, the phenotypic result is the same, and the pathways are evidently the same, but the locus of evolutionary change is different.

Studies of plethodontid salamanders illustrate some of the more complex cases. The premaxillary bone of the adult salamander skull has two interspecific-alternative states: separated and fused. Furthermore, "parallelism and reversal have characterized the history of fusion and separation of the pre-maxilla in plethodontids" (Wake, 1991, p. 547). This may appear to be a clear case of flip-flopping recurrence between alternative states. But comparative study has revealed that there are two ways to develop (and evolve) the fusion of the adult premaxilla, both of them recurrent. In plethodontid salamanders, the bone is fused in larvae even when it is separated in adults. These are the plesiomorphic states. In such species, the separation occurs at metamorphosis or, in species lacking the larval stage, at hatching. In some plethodontids having the bone fused in adults, this occurs as a retention of the fused larval state (the bone is fused throughout life), but in others it occurs later in ontogeny (the bones are separated at metamorphosis and later fuse). Therefore, what would appear to be a simple recurrence of fusion in fact has a more complicated history: there are two independent ontogenetic and evolutionary pathways to fusion and separation, and reversals are common in both pathways. Should these be considered two different sets of homologies or, on phenotypic grounds, just one?

From a developmental standpoint, this example seems to demonstrate two independent convergent states, each one found repeated within the salamanders, rather than one conserved state. Alternatively, bone fusion could have evolved just once, and the fusion end point been conserved despite changes in how it is achieved. There are many examples of end-point conservation by different evolutionary pathways. Wagner and Misof (1993) list 10 such examples, including production of salamander eye lenses from either ectoderm (during ontogeny) or the iris (during regeneration), and different pathways to the same adult morphology in direct development (vs. indirect development) of anurans (*Eleuthrodactylus*) and sea urchins (*Heliocidaris*). Some characteristics of crustacean legs long considered conservative and homologous prove to be "genetically derived through a variety

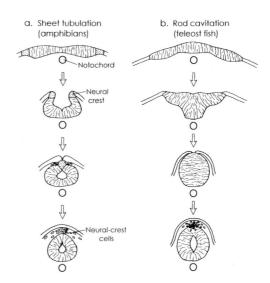

Fig. 25.2. Two modes of neurulation in chordates. The form of the neural tube is conserved but the developmental pathways differ dramatically in (a) the head and trunk regions of most chordates and (b) the head and trunk regions of teleost fish and lampreys and in many other chordates in the tail region. After Deuchar (1975) and Gerhart and Kirschner (1997).

of pathways" (Ferrari, 1988). Roth (1988, after de Beer, 1971) gives several other examples, including the development of tetrapod forelimbs from different somites in newts and lizards, development of primordial germ cells from blastula ectoderm in urodeles versus from endoderm in anurans, and development of segmentation in Thysanura by budding in an anterior-posterior sequence, while in other groups (Hymenoptera, Diptera) it occurs by simultaneous subdivision of the germ band.

Perhaps the most impressive illustration of end-point conservation despite different developmental pathways occurs in the ontogeny of the chordate neural tube. The neural tube is a distinctive embryonic trait at the phylum level—a "phylotypic" (Gerhart and Kirschner, 1997) or "archetypical" (*sensu* Patterson, 1982) trait. Not only is it present in all chordates, but it is essential to normal development of the nervous system and other structures. In most chordates, the neural tube of the head and trunk is formed when an epithelial sheet, the pre-ectoderm, rolls inward to form a tube, whereas the neural tube of the tail forms by coalescence of cells as a solid rod that then becomes hollow to form a tube (figure 25.2). In teleost fish and lampreys the latter mode of neurulation prevails over the entire body length. Clearly the chordate neural tube can be formed by quite different morphogenetic means, and the exact path and means of morphogenesis are not linked closely to the developmental fate of cells (Gerhart and Kirschner, 1997).

Because of highly flexible developmental interactions, or what I have called phenotypic accommodation, that include such devices as induction by an organizer of multipotent cells, and highly flexible cell migration, this major difference in developmental origin of the neural tube does not affect the ability of the embryo to organize itself into the standard chordate body plan. The neural plate always arises near the notochord, and the notochord is always surrounded by the neural tube, somites, and gut. This arrangement in turn accommodates numerous specializations of later development. Even though neural tube formation and other processes may undergo circuitous evolutionary change, the chordate body plan is conserved.

Given this developmental divergence, should we regard the neural tubes as homologous in all chordates? On one level, yes: the phylotypic stage is conserved by flexible mechanisms held in common due to common descent (Gerhart and Kirschner, 1997). On another level no, because developmental sequence may reveal convergent derivations of such structures as the neural tube. Examples like this support the conclusion (de Beer, 1971; Sander, 1983; Roth, 1988; Wagner, 1989a) that "the similarity of homologous characters cannot be explained or caused by the invariance of developmental pathways" (Wagner, 1989, p. 1163), even though developmental pathways may illuminate homology.

All of the examples in section II on "origins" represent the origin of new homologues, sensu Wagner (1989a)—epigenetically autonomous complex parts of the phenotype—and all of them involve developmental reorganization. This means that the ontogeny of every novel trait is different from that of its closest ancestral homologues, even though the developmental relationships of ancestral and derived traits are the best possible evidence of their homology. That is, comparative development can be used to trace homology, but developmental differences do not negate it.

Phenotypic similarity by different pathways occurs in behavior as well as morphology. In Australian dolichopodid flies (Sciapodinae), Bickel (1994) traces parallelism in secondary sex traits of males that have appeared repeatedly as apomorphies in species nested within genera where most species lack the trait. Bickel points out that the same signal, with the same timing, function, and behavioral form, and perhaps evoking the same female response(s), may manifest by homologous modifications in different members of a series. If so, this would represent different developmental pathways to the same or a closely similar signal:

Such physical "transference" of MSSC [male secondary sexual character] function is evident even in closely related species. . . . The MSSC are similar but occur on both different legs and tarsomeres. This may not affect their actual function since both are developed at the end of an elongate leg, but the shift of such a distinctive MSSC between legs I and II on probable sister species is remarkable. However, since legs I and II are serially homologous, only minor genetic/developmental change might be involved in the transference of expression between legs. (p. 25)

This kind of signal similarity using different signaling structures is similar to the conservation of phylotypic body plans described by Gerhart and Kirschner (1997), in that essential end-point similarity is attained recurrently due to flexible mechanisms of organization. But the phylogenetic independence of the different occurrences would lead us to call this convergence, not homology. If they had occurred in sister species, we would probably consider them homologues maintained by different pathways. Would we be misled by giving more weight to close phylogenetic relatedness than to the independence of developmental pathways?

Similarity at the molecular level raises yet another complication for interpretation of homology. There, structure, such as the folding patterns of proteins, may be dictated in part by physical laws (Denton and Marshall, 2001). Protein folds can be classified into a finite number of distinct structural families and follow a small set of constructional rules. Fold pattern is robust to short-term deformations caused by the changing conditions within cells, and can be conserved despite extensive evolutionary change in amino acid sequences, for the same folding pattern can be specified by many different amino acid sequences. This raises the possibility that even conserved end points may not always be a result of constancy of selection, but may be a response to physical laws of form. That selection could eliminate disadvantageous forms does not detract from this point about the origin of similarity.

While these examples show the difficulties in determination of homology, they are dramatic illustrations of the evolutionary primacy of the phenotype over the gene and underlying developmental pathways as the focus of selection and evolution. "Patterns of gross morphology persist and have over evolutionary time been replicated with fidelity, where fidelity may be lacking at the level of genes" (Roth, 1991, p. 177). The functional structure is more fundamental than the directive inputs, whether from genome or environment. This is a generalization of extreme breadth: just as the neural tube is convergently maintained throughout the chordates, so is the mother–offspring bond conserved throughout the wild diversity of human cultures, with all their convergent family arrangements and styles of mothering and surrogate mothering, all nonetheless identical in their essential contribution to normal development of universal human traits such as social attachment and cognitive (including linguistic) skills (Bowlby, 1969; Pinker, 1994).

Why do these convoluted varieties of trait conservation exist? In terms of selection, it is because it is the end result that is selected, not the pathway. In addition, the multidimensional (multi-decision-point) plasticity of development means that development can be deformed or diverted in many ways by both genetic and environmental variables. Anomalies of various kinds may thereby occur. Selection would then favor corrections in the diverted pathway, either suppressing it or bringing it around to the selected-for end. "Canalization" is a misleading word for this because it sounds like the evolution of a single deepening channel, whereas in fact the selected phenotype may be like a node in a converging net of feasible pathways. The multiple checks on reproduction in social insect workers—nutritional castration, behavioral dominance, and eating of competitors' eggs—if lifted, would each represent an escape route from developmental control, a different pathway to becoming a queen. In evolutionary pathways, the phenotype leads; the genes follow along, accommodating the shifting configurations of plastic developmental events.

In its encounter with the neural tube and other traits where structure is conserved or repeated by different developmental pathways, the developmental homology criterion approaches the limits of its usefulness. The urge to classify and distinguish finely between homology and convergence becomes an angel on the head of a pin, and in some cases should be shelved in favor of questions of more general interest, such as the relationship between flexibility and conservation in evolution.

Multiple pathways are important evidence that selection acts on phenotypes, not on the mechanisms producing them, and they show how evolution can work to maintain structures by tinkering with different resources at hand. They also show that developmental criteria may be unreliable indicators of common descent and reduce the kinds of evidence that can confidently be used to test for homology.

Conclusions

These examples, like those of part II, suggest evolution by continual reorganization of phenotypes into new permutations of structure. Changing characters do not march in single file ever outward along the branches of a phylogenetic tree. While homology, parallelism, and convergence remain useful conceptual guides, they need to be seen against a background of continual reshuffling within a particulate, mosaic phenotype that renders linear terms like parallelism and convergence only approximate, and potentially misleading, descriptions of evolution.

Does a concept of mixed or partial homology just make a mess of homology? In fact, evolution makes a mess of homology. It is no wonder that Hardy (1965, cited by Van Valen, 1974), frustrated in the quest for homologies, proposed that homology is the result of telepathic communication between parent and embryo. The best one can do in a discussion of homology is to present the positive and negative evidence that there is similarity due to common descent. Recurrence, or recurrence of a certain core of a trait somewhat variable in its disjunct manifestations, may be complicated by only partial reversion or by modification in new circumstances of expression. The attempt to sort this out may not yield a clean classification of traits as either homologous or not, but it may illuminate the story of evolution within a group. That is the only reasonable justification I can think of for the seemingly endless, almost obsessive discussion of the meaning of homology within evolutionary biology (see Hall, 1994).

26

Environmental Modifications

Environmentally induced phenotypic change can give rise to adaptive evolution as readily as mutationally induced change; both are equally subject to genetic accommodation. Environmental factors that initiate evolutionary novelties can be either new building blocks or cues that influence regulation in new ways. Contrary to common belief, environmentally induced novelties may have greater evolutionary potential than do mutationally induced ones. They can be immediately recurrent in a population; are more likely than are mutational novelties to correlate with particular environmental conditions and be subjected to consistent (directional) selection; and, being relatively immune to selection, are more likely to persist even though initially disadvantageous. Some adaptive environmental inductions come and go over geological time with cyclic changes in climatic conditions. Novel persistent global environmental changes may have occasioned major evolutionary transitions in some taxa. Incorporation of environmental modifications into the genetic theory of natural selection greatly increases the power of the Darwinian argument by showing that it does not depend entirely upon "random mutation" but can capitalize on preexisting adaptive plasticity and reorganizational novelty in response to recurrent environmental induction.

Introduction

In *Democracy* (1985), a novel by Joan Didion, there is a dramatic scene where a woman strikes her grown son and then abruptly leaves the room. Later, the narrator reflects on the causes of that behavior:

Billy Dillon once asked me if I thought Inez would have left that night had Jack Lovett not been there. Since human behavior seems to me essentially circumstantial I have not much feeling for this kind of question. The answer of course is no, but the answer is irrelevant, because Jack Lovett was there. (p. 175)

In one short passage, Didion reveals a grasp of environmental influence on the unfolding of events that has eluded most biologists. Behavior, like every other aspect of the phenotype, is *essentially* circumstantial. A particular organismic form or event could not exist without particular environmental circumstances any more than it could exist without particular genes.

In evolutionary biology, environmentally induced modifications come under unfinished business (e.g., see May, 1977). There have been repeated assertions of both their importance and their triviality, a lot of discussion with no consensus. More than 60 years ago Robson and Richards (1936) remarked, "This subject [the inheritance of induced modifications] has been discussed almost ad nauseum." Yet the debate has continued over such concepts as genetic assimilation, the Baldwin effect, organic selection, morphoses, and somatic modifications. So much controversy over the span of a century suggests that a problem of major significance remains unsolved. Meanwhile, "biology lacks a theory of organization" (Fontana et al., 1994). I believe that there is a connection between these two things—between continued controversy regarding the evolutionary significance of environmental modifications, and the shortcomings of a biological theory of organization and development based primarily on genes.

The growing awareness of the importance of development in evolution has not automatically secured a place for the environment as a source of evolutionary novelties. In developmental biology it is common to assume genetic determination of phenotypic change. Edelman (1988), for example, describes heterochrony as "alterations of tissues and form by mutations leading to changes in the relative rates of development of different body parts" (p. 51); Raff and Kaufman (1983) state that a "major characteristic of organismal evolution is that there is a genetically determined developmental program, and that evolutionary change occurs through genetic modification of this program" (p. 338). While they explicitly eschew "total genetic determinism," they see nongenetic flexibility as a buffering mechanism rather than an originator of novelty. Kauffman (1993) thinks that "mutations drive populations through neighborhood volumes of the ensemble" of possible self-organized systems (p. 24). These statements reflect theories of organization that see mutation as the ultimate and only source of phenotypic and developmental novelty, so they fall short of showing how genotype, phenotype, and environment are related in development and evolution.

The neglect of the environment as an agent of development within evolutionary biology, like the nature–nurture problem in general, is related to the emphasis on selection and gene-frequency change at the expense of attention to development. If one focuses on selection and genes alone, the environment is readily cast as the enemy. The environment is the red-and-raw in tooth-and-claw of natural selection, the challenge to the survival of the fittest. Seeing the environment as an agent of selection—environment as enemy—can obscure the image of the environment as a collaborator in normal development.

Environmental influence is reconciled with the genetic theory of adaptive evolution, at least in principle, by the concepts of genetic accommodation and interchangeability. In genetic accommodation (see chapter 6), phenotypes can incorporate environmental factors into normal development. Gene–environment equivalence and interchangeability (see chapter 5) put genetic and environmental influence on equal footing in development and evolution.

This chapter attempts to substantiate this line of reasoning by showing that environmental induction of novel traits not only is feasible but also is widespread. I go further with this subject than do most authors, to argue that environmental induction is probably more important than mutation for the origin of adaptive novelties. Previous discussions of environmental induction and evolution have taken a different approach, to emphasize how selection

brings environmental influence under genetic control, as in the concept of canalization and genetic assimilation (Waddington, 1942, 1953a,b; Matsuda, 1987). Schmalhausen (1949 [1986], p. 4) adopted a similar view. Although he emphasized that "the ability to undergo [environmentally induced] modifications is strictly hereditary" and recognized that developmental reactions occur "only in the presence of definite environmental factors" (p. 4), Schmalhausen, along with many other prominent evolutionary biologists including Spencer, Severtzoff, Beurlen, J. S. Huxley and G. G. Simpson, saw evolution as "liberating the organism from the *determining* influence of the environment" (Rensch, 1960, p. 298, emphasis original). I maintain instead that, far from being liberated from environmental influence, organisms evolve so as to *incorporate* environmental elements and exploit them as essential components of normal development.

The extent to which environmental induction and elements of external-environmental provenance have been prime movers of evolution can scarcely be overemphasized. One need only consider how the regulatory systems of organisms are structured to realize that this must be true: regulatory mechanisms are designed to link stimuli from the external environment to genomic function. Gene expression, for example, is mediated within cells by hormones and hormonelike substances. This may tempt us to create a theory of evolution in which all novelty begins within cells, due to changes in the nuclear genome that work themselves out, to affect the phenotype, were it not for the fact that "endocrine glands produce hormones only when stimulated by an outside agent, and the central nervous system is almost always involved" (Nijhout, 1994a, p. 17). Devices such as hormones and nervous systems mean that phenotypic alterations that directly involve the selected genes can originate from the outside in.

A compelling integration of environmental and genetic influence will be the plausibility test for a synthetic theory of development and evolution. If the arguments of this chapter, together with the discussion of genetic accommodation and interchangeability of chapters 5 and 6, are not convincing, then the effort of this book to forge a unified Darwinian theory that relates developmental plasticity to genetic change falls short. If these arguments *are* convincing, then environmental modification is established as a crucial factor in normal development and evolution, and the environment is elevated to a position of respect, alongside the genes, as a major participant in the generation, as well as the selection, of adaptive design.

The Entrenchment of Environmental Elements in Development

The environment can contribute to the origin of phenotypic novelties in two ways, given its role in the determination of regulation and form. In regulation, environmental factors such as temperature, day length, and ingested substances can serve as signals or cues at switch points in development. In phenotype construction, environmental materials serve as building blocks, or integral elements of form. How the environmental cues and materials get incorporated into development is the subject of this chapter.

Entrenchment (Katz, 1987) is the process by which environmentally supplied materials become essential to normal development alongside genetically specified factors. Wimsatt and Schank (1988) use "generative entrenchment" to refer to a different phenomenon, namely, the maintenance of phenotypic traits due to multiple functions. Ford (1971, pp. 247–248) used the similar word "embedment" to describe the incorporation of protective toxic chemicals, ingested in the diet of insects, which they cannot make themselves.

The environment impinges upon development at every turn, whether or not it has any particular formative role. Its effects may be helpful, harmful, or insignificant. Selection may eventually favor the exploitation of environmental elements that are recurrent or unavoidable, with the result that they become entrenched, or essential. The effects of vitamin-deficient diets are well known in humans, and other animals that fail to ingest sufficient quantities of environmentally supplied vitamins show similar morphological defects (e.g., see Wimberger, 1993), evidence of the degree to which environmentally supplied elements have become entrenched in normal development.

Entrenchment of some environmental elements is so thorough and widespread that we forget they were once evolutionary innovations. We take for granted entrenched environmental elements such as oxygen in animal metabolism, carbon dioxide and sunlight in plant metabolism, and iron in the heme groups of hemoglobin (figure 26.1). None of these essential components of the phenotype emanate from the genome. Indeed, nothing emanates from the genome without the environmental materials, such as dietarily essential amino acids, or the components of those "nonessential" ones that can be manufactured by the organism. Without them, gene expression could not occur. The replication of

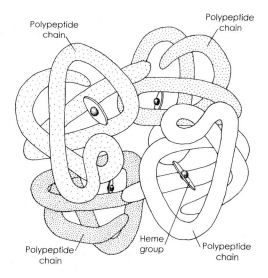

Fig. 26.1. Entrenchment of environmental elements in gene products. The hemoglobin molecule exemplifies the intimate connections between genetically specified elements (polypeptide chains) and elements of environmental origin (e.g., the iron in heme groups). Standard diagram, based on Dickerson and Geis (1969), Curtis (1983), and Bonner (1988).

DNA itself obviously depends upon raw materials from the environment.

To visualize how entrenchment of environmental materials can evolve, consider the gizzard. This ingenious, adaptive structure for grinding food uses small stones swallowed by birds. It is reasonable to hypothesize that the gizzard originated when birds repeatedly, perhaps at first accidently, ingested small stones, which by aiding in the mechanical breakdown of food made digestion more efficient. Selection then could favor the retention of pebbles in a localized region of the gut. Then, increased muscular development of the region containing stones would increase the effectiveness of their action (a turkey's gizzard can crack hickory nuts that require 50–150 kilograms of force to break—Pough et al., 1989). The gizzard is a likely example of an environmentally initiated evolutionary change—a novelty induced by pebbles. Not surprisingly, earthworms, inveterate invertebrate ingesters of earth, also have a gizzard that uses particles of soil to grind food, a case of convergent entrenchment that also depends upon essential environmental components (small stones) that would have been incidentally ingested along with food.

Not only inorganic environmental elements become entrenched in development. Whole organisms do as well. Mitochondria are famous examples of the entrenchment of environmental organisms (e.g. see Margulis, 1981). Others are the bacteria and protozoans that digest cellulose in ungulates (Parra, 1978) and in termites (Cleveland, 1926), whose young inoculate themselves with the essential organisms by ingesting the feces of adults. The nitrogen-fixing bacteria of legumes (see Losick and Kaiser, 1997) and the mycorrhyzae of orchids and many other groups are further examples. All obligatory mutualisms or symbioses, such as the union of a fungus and an alga to form a lichen, represent reciprocal entrenchment of environmental factors in development.

Ever since their origin, metazoans have evolved in environments rich in bacteria, and internal associations with bacteria may be a virtually universal aspect of metazoan biology (McFall-Ngai, 1999). Bacterial entrenchment is frequently associated with striking phenotypic innovations (Margulis and Fester, 1991), such as the light organs of squid, some of which depend upon luminescent bacteria for their luminescence (McFall-Ngai, 1999). Whereas antagonistic coevolution is one result of persistent interaction with parasites and pathogens, evolved exploitation (or symbiosis) is another. Some entrenched bacteria may be the descendants of pathogens, for a natural enemy can be, for a host organism, as persistent as dirt in the diet of a ground-feeding bird or worm. The luminescent bacteria (*Vibrio fischeri*) that illuminate the light organs of the squid *Euprymna scolopes* belong to the same genus as the bacterium (*V. cholerae*) that causes cholera in mammals (McFall-Ngai, 1999). Although now in a symbiotic relationship, the squid host responds in some ways as if dealing with a pathogen. It produces, in its light chamber, macrophages like those that engulf pathogenic bacteria, though not in sufficient amounts to eliminate the symbiont. The squid also produces a potent microbicide-catalyzing enzyme (a peroxidase) to which *V. fischeri* has evidently evolved special defenses (McFall-Ngai, 1999). These observations suggest a pathogenic origin of the association.

An example of the entrenchment of an unpromising substance, probably as a result of its sheer persistent ubiquity, is the incorporation and use of silicon by some plants. Silicon, like many essential plant nutrients such as potassium, calcium, and phosphate, is abundant in soils and is readily absorbed by plants in its soluble form (Epstein, 1994). But silicon is relatively inert. It is virtually

unavoidable in the dissolved diets of plants, so much so that its importance is difficult to study using controlled experiments because "it is difficult to create and maintain an environment adequately purged of the element" (Epstein, 1994, p. 11). It is a "ubiquitous contaminant" present as an impurity in laboratory nutrient solutions, in dust, and even in distilled and demineralized water (p. 12). It is no wonder, then, that silicon gets used in a variety of different unrelated ways, despite its native inertness, just because it is always there. In some plants it affects the distribution and toxicity of manganese so that, for example, in barley, necrotic concentrations on leaves are avoided. In cucumbers it rectifies imbalances in the supply of zinc in relation to phosphorus. In various plants it enhances growth and increases resistence to fungal disease, and in some it plays a structural role similar to that of lignin in being a compression-resistant component of cell walls (Epstein, 1994). Iron and vitamin A in mammals are further examples of entrenched substances of environmental origin. "Essential" substances, that is, substances that must be ingested, in the human diet include 14 vitamins and 13 minerals. Deficiency for any one or deformity can cause serious illness (Berkow and Fletcher, 1987).

Once an environmental element is entrenched, its absence may cause a developmental derangement as striking as any mutationally induced monster. When sea urchin pluteus larvae, for example, are raised in water free of calcareous matter, they fail to develop the typical angular spicular skeleton and develop as an amorphous blob (figure 26.2). The entrenchment of calcium as an essential element of pluteus development depended on its consistent availability in seawater, which did not occur until relatively late in the evolutionary history of life, evidently in the late pre-Cambrian or early Cambrian, giving rise to a large variety of calcareous skeletons that begin to appear as fossils around that time (Marin et al., 1996). The entrenchment of calcium is another reminder of how innovative phenotypes can depend as much on new inputs from the environment as they do on genetic mutation.

Sometimes entrenchment involves initially toxic environmental elements, indicating, again, the power of recurrence, even of initially deleterious environmental inputs, to initiate adaptive evolutionary change. As argued below, deleterious environmental factors differ from deleterious genes, which can be eliminated by selection before they become recurrent and entrenched. Insects of at least seven orders depend on exogenous compounds for defense (Bowers, 1990). These compounds are

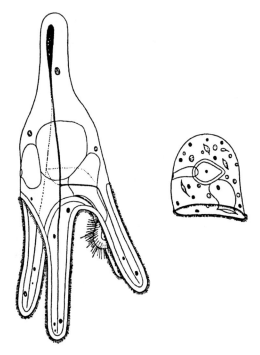

Fig. 26.2. Classical demonstration of environmental influence on larval development in a sea urchin: (left) normal pluteus larva; (right) pluteus larva of *Sphaerechinus granuluris* at the same stage, reared in water lacking calcareous matter or in a small excess of potassium chloride, lacks the larval feeding structures (arms) normally typical of this species, and in this respect resembles the nonfeeding direct-developing pluteus larva of some echinoderm species. After Wilson (1894).

sometimes extracted from the diets of larvae, which feed on somewhat poisonous hosts, perhaps initially in order to excrete them. Later, rather than being eliminated, the toxic substances are sequestered and used in defense, courtship, or both (Eisner and Meinwald, 1995). Toxic alkaloids in the skin of tropical poison frogs (Dendrobatidae) are ingested components of their arthropod prey (Daly et al., 1994, 2000). An environmental origin for the toxins long had been suspected by pharmaceutical researchers, who were unable to extract the sodium channel agonist batrachotoxin, a component of the venom, from poison dart frogs (*Phyllobates terribilis*) reared in the laboratory (Caporale, 1995; Daly, 1995), even though it is abundant in field-caught frogs.

Evolutionary incorporation of novel environmental elements may be quite rapid, especially if they are initially harmful and there is strong selec-

tion to deal with them. High concentrations of zinc are toxic to plants, but plants of the genus *Agrostis* apparently evolved zinc tolerance after only 30 years of exposure to zinc-contaminated soils near a zinc fence; another species, *Viola calaminaria*, is *restricted* to ground rich in zinc (reviewed by Ford, 1971, pp. 358–359). Tolerance in *V. calaminaria* is evidently surpassed to the point of entrenchment as an essential element.

Various species of marine algae have abnormal morphologies when grown by themselves in purified seawater or in artificial media because they need morphogenetic factors excreted by organisms such as bacteria and brown and red algae normally present in the environment (Tatewaki et al., 1983). Pink flamingos (*Phoenicopterus ruber*) owe the pink or red color of their plumes to what they eat. If kept on a carotenoid-free diet, they are white, but the pink color can be developed if they are fed certain small crustaceans or even dyes (Welty, 1962). In birds most of the yellows and reds of plumes are due to carotenoids, which cannot be synthesized but must be ingested, either in the form of the plants that produce them or secondarily, by eating a herbivore (Rand, 1967). In some fish the amount of incorporated ingested carotenoids affects the brightness of coloration and mating success of males (Kodric-Brown, 1989), and it would be of interest to know if there is a sexual dimorphism in diet given the importance of signaling success to males.

These facts lead one to wonder how many of the familiar morphological characteristics and behaviors of animals are due to the specific interactions and diets required for normal development. Certainly there are many more such cases than we are aware of. If the color of a bird is important in social communication, as shown by Kodric-Brown (1989) for fish (see also West-Eberhard, 1983), then particular dietary items are, too. So not all dietary behavior and selectivity should be evaluated in terms of energetic optimization or "good (survival) genes." Good signaling genes are too often forgotten, and they may include genes that influence dietary preferences assumed to be related to survival when in fact signaling proficiency is the primary context of their selection.

These examples show how environmental elements can end up being essential parts of organisms simply because they are persistently and unavoidably there. Unlike unfavorable genetic mutations, they do not go away. Organisms, being inherently responsive to the things present in their environments, may have a variety of kinds of interactions with a persistent class of environmental factors until finally some combination of genetically varying organism and persistent environmental factor proves advantageous. Any food container, with its list of minimum daily requirements from the human trophic environment, can serve as a reminder: environmentally supplied factors are as critical to normal development as are genes.

Developmentally entrenched environmental factors can be erroneously interpreted. Many biologists have assumed that since all animal communication requires both a signal and a response, all systems of communication require selection and co-evolution for both signals and responses, ignoring the fact that a signal or stimulus can play upon preexisting responses in the receiver (see West-Eberhard, 1979, 1984; Ryan, 1990; G. C. Williams, 1992). Once a feature becomes essential, it may be difficult to imagine life in its absence and to remember that some preexisting system had to function before the entrenchment occurred to even make it possible. Among the prominently entrenched essential environmental elements in life on Earth are oxygen, nitrogen, carbon dioxide, and water, virtually ubiquitous components of air and water. Carbon, hydrogen, nitrogen, and oxygen are the components of amino acids, the responsive molecules believed to have given rise to the earliest life (e.g., see Fox, 1984), so these elements were deeply entrenched from the beginning. Yet statements regarding subsequent entrenchments, such as that of the environmentally supplied nitrogen-fixing bacteria incorporated as essential symbionts of certain plants, pass over the opportunistic nature and origin of such features if they suppose that "without nitrogen-fixing bacteria, life would surely have perished long ago, because these organisms make it possible for nitrogen to remain in use" (Losick and Kaiser, 1997, p. 55). While these bacteria have become essential by virtue of the particular course of evolution that led to their entrenchment in the lives of plants and herbivores, it is not justifiable to conclude that life would not exist without them. The legumes are relatively recently evolved higher plants, so a great proliferation of living things prospered without them before they appeared. Life *as we know it* may have perished without nitrogen-fixing bacteria, but life as we know it would perish without television and the darkened skies produced by air pollution, so this would be a trivial interpretation of the significance of entrenchment. Entrenchment should be seen for what it is: an opportunistic evolved dependence on environmental elements or cues originally present but not essential or used until selection favored genotypes able to exploit them.

It is now common knowledge that genetic effects become part of development when selection spreads a favorable mutation. Developmental evolutionary biology will come of age when common knowledge includes the fact that environmental effects do not have to spread if they are already ubiquitous in the surroundings of a particular kind of organism.

The Environmental Induction of Novelty: Possible Examples and Places to Look

Responsive phenotype structure is the primary source of novel phenotypes, and it matters little from a developmental point of view whether the recurrent change we call a phenotypic novelty is induced by a mutation or by a factor in the environment. The new input diverts development into a new pathway. Then polygenic influence and graded plasticity can accommodate and exaggerate the change (see chapter 6). If there were no responsive phenotype to play this role in the origins of novelty, then mutations would be little chemical events unperceived within cells. They would not project into the realm of the organized phenotype in any fitness-affecting way.

In the early 1900s, resistance to environmental induction as a factor in evolution came at least in part from confusion regarding how to connect environmental and genetic change. Waddington solved that problem with a generally accepted explanation in terms of conventional genetics—genetic assimilation. But this did not overcome the objection of a supposed lack of convincing empirical evidence (e.g., Simpson, 1953b, p. 115). Having established the feasibility of genetic accommodation of adaptive change, and noted the interchangeability of environmental and genetic inputs (see chapter 6), I now consider in more detail the nature of environmental induction as a source of evolutionary novelties.

The Superior Evolutionary Prospects of Environmentally Induced Traits

A developmental switch is a potential reorganization point for phenotypic evolution, and new inputs can produce new switches, as discussed in chapter 6. The developmentally plastic phenotype can respond to both genetic and environmental inputs (see chapter 5). This gives the environment a foot in the door of evolutionary innovation.

Although counterintuitive from the traditional genetic standpoint, from a developmental point of view there are several reasons to consider environmental inductions *superior* to mutational inductions in terms of evolutionary potential:

1. Superior potential for immediate widespread occurrence of the novel trait: The initial spread of a phenotypic novelty can be greatly facilitated if the trait is environmentally induced (Matsuda, 1979, p. 223). A novel environmental factor can affect a large group of individuals within a single generation—an achievement of evolutionary import that no genetic mutation can match. A favorable genetic mutation begins as a unique event that affects only a single individual or brood, and its spread depends on replication and selection. If a mutation has a negative or weak effect, selection and stochastic effects can drive it to extinction, greatly reducing the evolutionary potential of its phenotypic effect. On the other hand, the same phenotype, if induced by an environmental factor, can persist as long as its inducing factor does, or until selection changes the threshold for its production.

A neutral or disadvantageous environmentally induced trait that recurs because the environmental inducer recurs may eventually be modified and transformed under selection into a beneficial trait. Examples of this are described above in the section on the entrenchment of environmental factors in development. Nonlethal environmentally induced phenotypes have time on their side, for they recur as long as their environments do, affording the chance that a variable, susceptible population of genotypes will eventually be able to accommodate them genetically.

2. Superior chances of occurrence in a favorable genetic background: Because a novel environmental input can affect many genotypes at once, in contrast with a novel allele that initially affects only one, a deleterious response to an environmental factor has a better chance of being genetically accommodated—of passing the initial test of occurrence in phenotypes that can survive or use it. A mutation that is lethal to some phenotypes may have the same initial probability of encountering a favorable genetic background, but it does not have the same opportunity to sample several if it proves immediately lethal.

3. Superior probability of initial matching of environmentally induced traits with the conditions in which they are selected: An environmentally induced trait is automatically associated, in at least one slice of time, with a particular environmental situation—that which induced it. Novel genes, on the other hand, recur with no consistent relation to the environments in which they will be selected. Although environmentally induced traits may have a very short overlap with their inducing conditions, they have a better chance than mutationally induced traits of recurrence in a particular environmental situation and being selected there. This is especially true for the large category of novelties induced by environmental extremes, sometimes called "stress-induced" change: a phenotype induced by an extreme environment is also subject to strong selection there (Matsuda, 1987). There are reasons to think that variable responses to extremes will have a genetic component, as discussed below.

To the degree that an environmentally induced trait is repeatedly matched by its inducer to a particular environment where it is expressed, environmentally induced traits are rendered more subject to consistent selection and directional modification than are mutationally induced ones whose expression is more likely to be random with respect to environment.

Even when environmental matching is not initially perfect, environmental induction immediately endows phenotypic regulation with environmental sensitivity. This sensitivity can be molded under subsequent selection to more precisely tie phenotype production to circumstances that are increasingly accurate predictors of the conditions where the phenotype has proven advantageous (see chapter 6). Thus, many conditionally expressed alternative phenotypes are indirectly cued (e.g., by photoperiod or phenotype early in development) by conditions only indirectly associated with those to which they are adapted. Even in quite closely related organisms having similar responses, the cues triggering them may be different. In swallowtail butterflies having a pupal color dimorphism, for example, the switch between forms is cued in some species by background color, and in others by the texture of the pupal substrate (West and Hazel, 1985).

4. Relative immunity to elimination by natural selection. Perhaps the most compelling argument for the superiority of environmental induction over mutation in terms of recurrence and persistence has to do with the inexorable persistence of an environment immune to natural selection: environmental inducers may be not only immediately widespread, without necessity for positive effects on fitness sufficient to spread them due to differential reproduction of their bearers (selection), but they are inexorably present. Although the organism may eventually "evolve around" them—probably a

common reason for some of the circuitous pathways that are so common in development (see Gerhart and Kirschner, 1997)—environmentally induced traits may persist even though they are disadvantageous. Buss (1987, p. 139) refers to the "nonnegotiable demands" of the external environment. Detrimental mutations, by contrast, are negotiable: they can be directly or immediately eliminated by selection, and a lethal mutation expressed in the individual where it occurs may never affect more than one individual.

Adaptive phenotypic accommodation of potentially deleterious anomalies can occur due to the mechanisms of plasticity discussed in chapter 3, as well as via the common evolution of alternative phenotypes that facultatively "make the most of a bad job" (see chapter 22). The commonness of such alternatives is likely due at least in part to the fact that the recurrent environmental conditions and phenotypic shortcomings that give rise to these alternatives, unlike genetically induced ones, cannot be eliminated by selection. So, in part due to the persistence of the handicap, an adaptive alternative eventually evolves (West-Eberhard, 1979).

Initial spread has long been considered a problem for the establishment of adaptive novelties, especially complex novelties such as those produced by "macromutational" developmental change (Charlesworth et al., 1982; see chapter 24). Complex novelties are thought more likely than small modifications or complex traits forged gradually by selection to prove detrimental. The four arguments just given show that initial spread may not be a severe barrier to the evolution of environmentally induced novelties. Especially, these arguments indicate that the problem of initial spread is more likely to be solved by an environmentally initiated complex trait than by a trait initiated by a new mutation, for environmentally induced traits can spread even when they are not advantageous, and may continue to recur despite being disadvantageous because they are induced by a persistent environmental factor and are not associated with any particular genotype that can be eliminated by selection.

Recurrent Extreme Environments and Phenotypic Innovation

Stress could serve as a key word in a literature search for examples of environmentally induced alternatives in nature. Novel traits, or trait complexes, can be induced when environmental ex-

tremes extend norms of reaction to unusual extremes, or even beyond their previous normal ranges. The multidimensional plasticity of the phenotype may produce qualitatively novel variants as well as quantitative extremes under the influence of environmental extremes, with no initial genetic change (see chapter 3). Of course this, like many other rather obvious and forgotten insights, is not a new idea. While I have not tried to review its history, I know that entomologists in the early 1900s (e.g., Roubaud, 1916, p. 54) discussed extreme (e.g., drought) and marginal conditions (e.g., at the periphery of a species range) as being responsible for environmentally induced "Lamarkian" characteristics that resemble evolved traits at about the same time that Baldwin (1902) and others were explaining this in Darwinian terms (see chapter 6). Meyer (1966) reviews both environmental and mutational induction of novel development in plants and states that the first experimental study of environmental effects on plant differentiation was in 1910 by Shattuck, who sprayed cold water on ferns (*Marsilia*) and found that this prevented the formation of megaspores. Meyer also notes that after about 1915, nonheritable abnormalities were unlikely to be reported in scientific journals. The scientific feeling against environmental induction seems to have reached a high point in the 1950s and 1960s, when Mayr (1963), could write that "the early Mendelians . . . misinterpreted just about every evolutionary phenomenon. Some of their contemporaries . . . even believed in some environmental induction" (p. 10; for other, similar citations, see West-Eberhard, 1989).

Products of this history, educated biologists have learned to be suspicious of environmental induction as an initiator of evolutionary change, so even one who reads Darwin is likely to dismiss any hint of this as unfortunate pre-Mendelian Lamarckianism. Darwin (1859 [1872]), however, clearly reasoned that environmentally induced change in a versatile tissue could subsequently become inherited and specialized for one function under positive selection, what I am calling genetic accommodation and fixation of a condition-sensitive (inducible) alternative:

In the Hydra, the animal may be turned inside out, and the exterior surface will then digest and the stomach respire. In such cases natural selection might specialise, if any advantage were thus gained, the whole or part of an organ, which had previously performed two functions, for one function alone, and thus by insensible steps greatly change its nature. (p. 136)

Darwin does not give the source of the *Hydra* example, but it was evidently widely familiar among zoologists of his time, due to ingenious experiments by the eighteenth-century naturalist Abraham Trembley (1744). Later research on cell movements (Roudabush, 1933; from J. T. Bonner, personal communication) showed Trembley's interpretations to be wrong, but Darwin's view of the accommodation under selection of an environmentally induced trait is still of interest and is completely compatible with present ideas on genetics and evolution.

An extreme environment implies strong selection as well as extreme phenotypic responses. When an environmental contingency is both recurrent and extreme, like seasonal change of temperature in boreal climates or of humidity in the seasonal tropics, one expects to find both coordinated expression of environmentally sensitive responses and particularly strong *selection* on the responses, both due to the extreme environment itself. This coincidence of novel induced trait or extreme expression, and strong selection may have the result of molding a recognizable adaptive complex that is both distinctive and coadapted in form.

Finally, environmental extremes expose unselected genetic variation in reaction norms to strong selection. *Individual differences* in response to unusual extremes may be due to genetic differences among individuals, and this would hasten their genetic accommodation. A supernormal input from the environment has the effect of surpassing the populationwide adaptive response range where all individuals show similar responses, and thereby may reveal genetic variation not usually exposed to selection. Like imposing an extreme test in order to "separate the men from the boys," an extreme environment bypasses the conditions in which norms of reaction have been normalized and in which most successful individuals show a near-optimum or invariable response due to previous selection under those conditions. This may expose residual differences in response capacity or norms of reaction in a range of conditions where selection has not yet had a chance to act.

Gans (1979) has emphasized another way in which extreme environments may promote directional evolution, when "excessive construction" or excess capacities are brought into play under extremes and may lead to an adaptive shift in the direction of the environmentally induced phenotype. This hypothesis, unlike the unselected-variation hypothesis, does not depict variation at the extremes of a range where selection has not acted to eliminate it. Rather, it visualizes the excess capacity as

having evolved under occasional episodes of extreme selection, for example, in pine trees that have adaptations for fires that may occur less frequently than once a decade. Under this model, the population passes through bottlenecks that shift the population mode to an extreme outside the range usually experienced by the population. Contrary to the model just discussed, this would reduce genetic variation in the expressed phenotype at the extremes where selection has been most severe. This is a model of stress-driven preadaptation rather than stress-revealed genetic variation fueling rapid evolution in a new direction. Still other authors have suggested that a population bottleneck caused by an extreme environment could lead to an odd sample of genotypes and drift (Wcislo, 1989). All of these models grant a role to extreme environments in evolutionary change, but only the first emphasizes the developmental role of extreme environments in the initiation of change (see also Wcislo, 1989).

A dramatic example of qualitative developmental variation revealed in an extreme environment occurred when 67 individuals of a marine fish (*Bairdiella icistius*, Sciaenidae) were introduced from the Gulf of California into an artificial salt lake, the Salton Sea, in southeastern California (summarized in Stanley, 1981). In the first year, spawning produced an estimated 13–23% individuals with obvious malformations, including abnormal eyes, strangely developed lower jaws or jawlessness, snub-nosed heads, vertebral deformities, and three rather than the usual two anal spines. By three years after introduction the number of anomalies had declined to only 2% or 3%, probably in part due to poor ability of the abnormal individuals to survive and reproduce, exacerbated by population growth and resource competition within the lake. In this example, an extreme environment immediately induced numerous morphological variants revealing preexisting developmental capacities not expressed in the normal, ancestral environment of the species. Since the inducing factor—severely elevated salt content of the water, and its correlates in terms of trophic and physical change—was presumably similar for all individuals from the zygote stage, some of the developmental *variation* observed is likely to have been due to genetic variation. To the degree that this is true, one would expect a rapid response to selection and rapid evolutionary change that could suddenly establish a distinctive form if one of the induced variants happened to be advantageous. Note that it would be a mistake to reflexively attribute such an evolutionary change to drift, on the assumption that the small sample of introduced genotypes had facilitated evolutionary change. By the environmental induction hypothesis, the likelihood of adaptive evolution would be reduced, not enhanced, by small sample size and drift, which would reduce the range of genetic variation in induced responses upon which selection could act.

Matsuda (1987) noted that many of the derivations of major new phenotypes associated with altered development were also associated with stressful or extreme environments, such as the marine intertidal, the environments of endoparasites inside a defensive host, and temperature and humidity extremes. Matsuda reasoned that such conditions likely initiated the expression of new phenotypes repeatedly, prior to the action of natural selection and evolutionary fixation of the traits. Most of Matsuda's evidence for this was very indirect; as discussed below, it must be. But the correlation between extreme or altered environments and developmental innovation of a sort that could be induced by those environments is well documented by Matsuda. He provides massive evidence for involvement of developmental responsiveness, especially hormonally mediated responses affecting morphology, in evolutionary change. Even though his book (Matsuda, 1987) deals only with animal morphology, it discusses material from no less than 17 animal phyla, with emphasis on insects and other arthropods. This constitutes important evidence that the environment may sometimes play a morphogenetic, rather than just a selective, role in evolution. In some cases, Matsuda's conclusions have been supported by experimental evidence, as in the indications that phenotypic plasticity in morphology of well-fed planktonic sea urchin larvae resembles that of direct-developing yolk-endowed ones (Strathmann et. al., 1992) and that trophic stress induced by artificial rearing in an inappropriate host causes a neotenic alteration of development mimicking a phylogenetic transformation in flukes (Funk and Brooks, 1992, after O'Grady, 1987). Trophic stress on islands likely played a role in the evolution of dwarf species (Rachootin and Thomson, 1981; Roth, 1992).

The extremes of developmental capacities represent latent potentials that can be exploited in the convoluted pathways to adaptive change. The most specialized hymenopteran insect societies, such as those of honeybees, some tropical wasps, and most ants, have morphologically distinctive queens and workers whose form is determined in the larval stage. Individuals of a relatively unspecialized genus of social wasps (*Ropalidia marginata*) that usually lacks larval determination of morphological castes

nonetheless show marked variation in the probability of becoming a queen under experimental manipulation of larval nutrition (Gadagkar, 1991). Morphological queen–worker differences based on determination in the larval stage have been observed in nature in the small-colony socially unspecialized species *Ropalidia ignobilis* (Wenzel, 1992b). Such a latent capacity for preadult caste determination in these species is not surprising, given that nutrition affects reproductive condition in insects in general (Wheeler, 1994). In the social insects, this widespread capacity for a dietary suppression of reproduction has been exploited via a circuitous developmental pathway in which socially dominant queens influence workers, which in turn manipulate the diet of the larval brood.

Stress-initiated evolutionary change is peculiar in that an environmental condition that was originally stressful or extreme, and disadvantageous to the organism eventually, if recurrent, may become part of normal development and would no longer be regarded as a "stress." G. C. Williams (1992) refers to "helpful stress," such as the mechanical force that encourages normal bone and muscle development, or the impact of grazing that stimulates increased growth of a plant. This could be considered a kind of entrenchment, whereby a recurrent or ubiquitous environmental element or condition becomes an essential or favorable part of normal development (see below).

Niche Shifts in Extreme or Novel Environments

Some readers may object, with good reason, to the terms "niche" and "niche shift." I use them here because there is a large literature invoking these terms to describe evolved change in lifeways or ecological interactions (niche), especially those associated with speciation. Even though the word "niche" seems to refer to parameters of the environment, in fact the niche is a property or extrapolation of an organism's behavior, defined by how it interacts with its environment. Changes in these interactions can be responses to environmental change and can occur without speciation. Striking shifts in ecological niche under extreme conditions, sometimes called "stress," have often been reported. For example, when insects and snails were in short supply, a reptile, *Lacerta muralis*, that normally feeds on insects and snails was seen feeding on shore populations of marine plants (halophytes; Robson and Richards, 1936, after Eisentraut, 1929). Other examples are discussed below and in chapter 28.

It may be common for flexible individuals to become more specialized during periods of resource contraction, especially if learning influences their behavior. Large-beaked finches shift their diet toward large seeds during periods of food scarcity (see Grant, 1986; Grant and Grant, 1989), and some fish that are generalists when there is a variety of abundant foods, or are capable of expanding their diets to include an abundant resource, become specialists under trophic scarcity or competition (e.g., Werner et al., 1981; Liem and Kaufman, 1984; Witte, 1984; Robinson et al., 1993). This may affect the evolution of trophic specializations in both birds and fish (see chapter 28). Artificially well-fed laboratory colonies of a normally queenless ant species (*Rhytidoponera metallica*) revert to the production of large numbers of full alate queens, and this atavism occasionally occurs in nature. This suggests that the queenlessness and worker reproduction now usual in this species may have evolved under prolonged nutritional stress (Haskins and Whelden, 1965).

When there is more than one correlated response to the same environmental extreme, a complex novelty may be the result, as in some examples of the two-legged goat effect in nature (see chapter 16). The phenotype that results may appear to be adaptively "programmed" under selection for its coordinated development, when in fact the correlations have been forced by environmental stress. For example, the colors, behaviors, and hormone titers of locusts are highly and independently variable at intermediate population densities, but they form a coherent set of adaptive traits at prolonged extreme high or low densities (see chapter 5). Disruptive selection can favor developmental mechanisms that eliminate intermediates and the evolution of disruptive development—a clean switch between extremes, as in the specialized workers and queens of some social insects (see chapter 11). Ontogenetic niche shifts are similar phenomena (Werner and Gilliam, 1984). Extreme environments promote this by a combination of trait induction and strong selection.

Does Adaptive Stress-Induced Genomic Change Occur?

Hall (1983) discusses mutations in microorganisms that could be "neutral under (normal) growing conditions but advantageous during starvation" in the presence of a suitable substrate. Under that suggestion, the mutations themselves are not induced; they happen, then happen to be advantageous under stress. Other authors (e.g., see Wills, 1983) have suggested that stress-induced genomic change may have been important in evolution, and McClintock (1984) argued for a role of stress in provoking increased transposon activity, an observation that has received subsequent support (McDonald et al., 1987). Genome studies reviewed by Jablonka and Lamb (1995) call for an open mind regarding the possibility of stress-induced genomic change, but it is difficult to know how important such changes, many of them in somatic cells, may be in evolution.

Evidence for Environmental Initiation of Reorganizational Novelty

The Importance of Indirect Evidence

The attempt to guess the mode of origin of a novelty given its present mode of regulation is like trying to trace the psychological motives of a deceased novelist. Genetic accommodation can increase the environmental sensitivity of a mutationally induced trait or increase the genetic control of an environmentally induced one. As a result, one cannot confidently deduce mode of origin from the current balance of environmental versus genotypic influences on phenotype determination. Despite theoretical arguments for the great importance of environmental induction, it is virtually impossible to give direct proof of particular examples where this has occurred in the history of a lineage.

As in so many other areas of evolutionary biology, the most fruitful approach to the question of environmental induction in evolution is to invent indirect tests. The credibility of natural selection as a factor in evolution in nature, for example, is based almost entirely on indirect evidence and abstract reasoning: of all the numerous demonstrations of natural selection in the wild listed by Endler (1986, table 5.1), only five were published prior to 1950. Yet so strong was the theoretical and indirect evidence for natural selection, and so strong the convictions of most biologists in its credibility on these grounds, that when Robson and Richards pointed to the dearth of direct evidence for natural selection in 1936, they were considered shockingly "anti-Darwinian" or "antiselectionist" (see, e.g., Stebbins, 1977, p. 15) even though this was not their intent (O. W. Richards, personal communication, 1983).

Indirect proof via the accumulation of a variety of independent sources of evidence is an important procedure in evolutionary biology (see Lloyd, 1988), yet evolutionary biologists are inconsistent in their acceptance of it. In a highly influential article, Simpson (1953b), for example, condemned environmental induction (the Baldwin effect) to several decades of severe doubt by denigrating the value of indirect evidence. He concluded that "direct evidence seems to be quite lacking" for purported examples; that his review of evidence revealed "no instance in which it indubitably occurred" and that it is "decidedly questionable" whether it "does in fact explain *particular instances* of evolutionary change" (p. 113, emphasis added). This seems unfair, especially from the pen of an eminent Darwinian who based important discussions of behavior and food habits on the study of fossils.

In this section, I acknowledge that direct evidence for environmental induction of evolved novelties is likely to be unobtainable. I therefore outline several kinds of indirect evidence that support a hypothesis of environmentally induced origin, and then give possible examples.

Kinds of Indirect Evidence for the Environmental Induction of Novel Phenotypes

Several kinds of evidence can support a hypothesis of environmental initiation. Each of these lines of evidence is weak if taken alone. The more that obtain in a particular case, the greater the likelihood that environmental induction contributed to the origin of the trait.

It is highly likely that the nonreproductive helper (worker) phenotype of social Hymenoptera originated by environmental induction, so this case is a useful illustration of the different types of indirect evidence that can support a hypothesis of origin by environmental induction. They include the following:

1. Sporadic occurrence of the phenotype, or elements of it, in related species in nature, under the same environmental circumstances as those known to induce it where it is an established trait: Females of *Zethus miniatus* related to the social vespids (Polistinae) but belonging to a primarily nonsocial subfamily (Eumininae), share nests with genetic relatives where each attends a series of her own larvae, one by one. Such females show helping behavior if they encounter orphaned larvae of other females when they themselves are not caring for their own young. This behavior and the associated circumstances resemble those of sterile workers in fully social species (West-Eberhard, 1988, 1996).

2. Experimental inducibility of the phenotype in related species using parallel environmental cues—phenotypic engineering (Ketterson and Nolan, 1992) or phenocopies (Goldschmidt, 1940 [1982]; Shapiro, 1976, 1984a): The worker phenotype can be induced in non-worker-containing solitary species by forcing them to live in groups; one member of the group dominates the others, which become nonreproductive helpers. This is the same situation that causes worker behavior in eusocial species normally containing sterile workers (Sakagami and Maeta, 1987a,b; reviewed in West-Eberhard, 1996).

3. Natural or experimental restoration of environmental conditions that resemble the ancestral conditions cause reversion to the likely (on the basis of phylogenetic study) ancestral phenotypic state: In many social wasps and bees, a change in the social environment (removal of a dominant female, or queen) leads to egg production, the ancestral state, in females otherwise destined to become sterile workers (Wilson, 1971a). There is no doubt that social insect taxa are derived from solitary species with egg-laying females (e.g., see Wilson, 1971a).

4. Ease of environmental induction of a genotypically determined trait: In some social stingless bees, worker and queen determination in nature is primarily genotypic (Bonnetti and Kerr, 1985). Nonetheless, experimental manipulation of the larval diet can reverse the genotypic caste (see syntheses in Velthuis, 1976b; Velthuis and Sommeijer, 1990).

5. Phylogenetic support for the hypothesis that a taxon where the phenotype can be induced resembles the ancestral condition: *Zethus miniatus*, the species where workerlike helper behavior occurs, belongs to a solitary subfamily (Eumeninae) that is basal to the worker-containing Stenogastrinae and Polistinae (Carpenter, 1989).

6. Environmental induction of the trait in nature in most or all of the species where it occurs: With very few exceptions, all of the thousands of species of wasps, ants, and bees having sterile workers have environmentally induced workers, due to dominance interactions among adults, or dietary manipulations, or both (reviewed in Wilson, 1971a).

7. Correlation between the inducing factor and the induced trait in a variety of related species: The primary inducing factor for the worker phenotype is presence of a dominant female. Solitary species, which seldom interact with conspecifics within nests, of course lack worker behavior. So-called primitively social (workerless, group-living) wasps

and bees, of primarily solitary-nesting taxa, engage in aggressive interactions (briefly reviewed in West-Eberhard, 1996) when they encounter each other within nests—the environmental condition that induces worker behavior. So the intermediate transitional condition is present in the taxa concerned.

8. Widespread occurrence of a mechanism for development of the phenotype given the hypothesized environmental conditions, even in species where the phenotype does not regularly occur: Comparative study of primitively social wasps and bees suggests that group life, which gives rise to dominance interactions and worker behavior, may occur due to several different environmental conditions, including those making solitary nesting especially costly, such as high rates of predation or parasitization (Lin and Michener, 1972). Solitarily nesting females in some species suffer high mortality and/or low reproductive rates (West-Eberhard, 1986). In *Zethus miniatus*, socially dominated females sometimes leave the group and reproduce independently, thereby escaping the worker-phenotype-inducing environmental factor, social dominance. As a result, there are no permanent workers in this species—the worker phenotype is associated with constitutive (obligatory) group life where subordinate females, evidently unable to compete on their own, become sterile helpers. In addition, the hormonal mechanisms that associate sterility with subordinance and fertility with aggressive dominance are known in a variety of wasps and bees (reviewed and discussed in West-Eberhard, 1996).

9. Lack of contrary evidence: There is no evidence in any primitively social or primitively eusocial species for genotype-specific induction of workerlike behavior; in all cases in these species, the worker phenotype is clearly associated with an environmental inducer, social dominance, often influenced by environmental factors such as nutrition, nest-site availability, and so forth.

This list shows that most indirect evidence for environmentally induced origins comes from comparative study of related species. If a conditional alternative phenotype found in one species occurs as an anomaly or experimental artifact associated with the same environmental conditions in another, related species, this shows that environmental initiation of the trait is at least feasible. As indicated above, two reservations must always be attached to the interpretation of experimental or occasional inductions as evidence for mode of origin: given the lability of conditionally expressed traits, even a strong cladogram at the species level may be insufficient to resolve the sequence of gains and losses

of alternative phenotypes (see chapter 19). Due to the ready interchangeability of environmental and genotypic factors in regulation (the possibility of phenocopies, as well as genocopies of environmental inductions; (see chapter 5), the observed mode of induction of an anomaly may not be the original one. The only way I can see around these problems is to use multiple lines of indirect evidence.

Possible Examples

Multiple independent criteria support an environmental induction hypothesis for the origin of the worker phenotype in the social Hymenoptera (wasps, ants, and bees), as just shown. The following are additional illustrations where a hypothesis of environmental initiation is suggested by indirect evidence.

Open Colony Structure in Fire Ants   Native populations of fire ants (*Solenopsis invicta*) in Argentina form two kinds of colonies, one with a single queen and the other with multiple related queens. The multiqueen colonies multiply by fission. Offshoot groups contain one or more queens and workers (reviewed in Ross et al., 1996). Multiple queen colonies of this species introduced into the United States differ dramatically from the Argentinean multiqueen colonies in having more than twice as many mated queens per nest and markedly lower intracolony genetic relatedness. These differences are the result of a cascade of effects induced by the new environmental situation. As commonly occurs in introduced plants and animals, population densities of the introduced ants are far higher than in the native range, presumably due to reduced numbers of natural enemies. This reduces the availability of nesting sites for new colonies, and it is associated with altered behavior of queens, which more often remain at the natal nest or join foreign nests, with the result that there is increased mixing of non-relatives within colonies, a breakdown in kin discrimination, and further increased mixing of less related individuals within colonies (Ross et al., 1996). Thus, a behavioral response, joining existing colonies rather than beginning new ones, due to an environmental factor, lack of natural enemies, results in profound and complex changes in the social structure of colonies.

Size and Related Traits in Pocket Gophers   Pocket gophers (*Thomomys species*) living on rich diets (e.g., in alfalfa fields) attain much greater body size than do genetically similar conspecifics in adjacent natural habitats (Patton and Brylski, 1987). Nutri-

Fig. 26.3. Miniaturization in elephants: A partially reconstructed skeleton of a full-grown individual of *Elephas falconeri* (center), smallest known species in the Elephantidae, from the Pleistocene of Sicily. (Left), Hindquarters of a modern elephant. (Right) Arm of *Mammuthus imperator*, the largest mounted elephant in existence, from Pleistocene Nebraska (1.7 meter tall human for scale). Photographed in the University of Nebraska State Museum, Lincoln Nebraska. Courtesy of V. L. Roth.

tional and size differences in pocket gophers are associated with a complex syndrome of life history and demographic traits that include increased population density, increased female fecundity, female biased sex ratios, and increased sexual dimorphism, features likely to influence, in turn, mating system and dispersal distances. All of these traits vary geographically in pocket gophers (Patton and Brylski, 1987). Since all of these differences, and whole sets of them at once, can be rapidly environmentally induced, their geographic variation cannot be assumed due to evolved subspecies genetic differences.

Dwarfism in Elephants   Trophic stress may have contributed to the repeated evolution of dwarf elephants, and perhaps dwarfing in other large mammal species, on islands (Roth, 1992; see also Rachootin and Thomson, 1981). Large mammals such as ungulates and carnivores that survive on islands often show unusually small body size (Roth 1990, 1992; figure 26.3). Dwarfism of fossil elephants found on islands off California, Southeast Asia, and the Mediterranean likely owe their distinctive small size to environmentally initiated stunting, and selection that favored reproduction at small, juvenile size in a place where food resources were extremely restricted, in effect leading to the genetic accommodation of stunting (Roth, 1992). The morphology of fossil dwarf elephant species

has the earmarks not of achondroplastic dwarfism, which can be caused by a mutation, but possibly of stunting, which is brought on by dietary insufficiency during development. The shape of postcranial and other elements of dwarfed forms indicates truncation in growth, and marked asymmetries suggest developmental instabilities that can reflect trophic stress (see Roth, 1992, for evidence and discussion of various alternative interpretations). Heterochrony (precocious maturity) may have contributed to a transition involving a trade-off between investment in growth and investment in fat reserves and reproduction. In extant mammals, small individuals have higher fitness under food stress (reviewed by Roth, 1992, p. 263; see also Gould, 1977, p. 325). The fact that elephants have unusually plastic dentition, including an unusually prolonged period of tooth development (see references in Roth, 1992a, p. 276), may have contributed to their initial ability to adapt to altered diets and survival on islands.

Neoteny in Salamanders   Environmental induction may have initiated the evolution of neoteny in salamanders, as shown by a classic experiment in phenotypic engineering. Axolotls (*Ambystoma mexicanum*) fail to undergo metamorphosis in nature. Duméril (1865; cited in Gould, 1977; Shapiro, 1980) found that axolotls kept in captivity in a Paris museum produced normal, metamorphosed *Ambystoma* adults. This transformation, which resembles that seen in some related species in nature, indicated the feasibility of environmental induction of variation in metamorphosis. In this species heterochrony—prolonged expression of larval characteristics in the adult—is strongly influenced by homozygosity for a recessive allele at a single locus (Ambros, 1988, p. 283; after Tomkins, 1978). But in many neotenic salamander species, metamorphosis can be induced even after reproductive maturity of the brachiate adult by various natural and experimental stresses such as starvation and drought. Salamanders are a good illustration not only of the feasibility of environmental initiation of novelty, but of the difficulty of knowing for sure whether a developmental novelty was originally initiated by an allelic or an environmental change. For further discussion of environmental induction of heterochrony in salamanders, see Matsuda (1987).

Seasonal Polyphenisms in Butterflies   Shapiro (1976) has used extensive comparative and experimental data on seasonally polyphenic butterflies and their relatives to illuminate the evolution of environmentally cued color forms. He has shown that "nonadaptive developmental aberrations" of a kind common in certain species have evidently come to be adaptively expressed under selection for modifiers that affect threshold of expression and ultimately couple it to a reliable seasonal predictor. Several lines of evidence suggest origin by environmental induction: in *Vanessa cardui*, cold- and heat-shocked pupae produce aberrant phenotypes closely similar and "inescapably homologous" to those established in congeneric species in nature (Shapiro, 1984a, p. 26). A clean switch between color forms such as those found in natural populations of the tropical pierid *Precis octavia* is produced in the laboratory by temperature extremes under controlled constant photoperiod and humidity (McLeod, 1984). But in most seasonally polyphenic species, the adaptive color forms are cued by photoperiod, which in most regions is a more reliable cue than temperature (see Shapiro, 1984b). In *Precis coenia*, either temperature or photoperiod can induce the autumn form (K. C. Smith, 1991).

Morphological Traits of Sticklebacks Wootton (1976, pp. 234–235) reviewed environmentally induced morphological changes in sticklebacks (*Gasterosteus aculeatus*), where temperature and salinity affect number of lateral plates, number of vertebrae, number of rays in dorsal and anal fins, and number of basal plates associated with the dorsal spines (see also Taning, 1952). Some of the induced variations (e.g., in plate number) parallel those established as alternative plate morphs in particular populations. Wooton also cites evidence that intramorph variation has a strong genetic component and therefore has the potential to be accommodated under natural selection.

CAM in Plants   The evolution of crassulacean acid metabolism (CAM) physiology in plants may be an example of both multiple responses and the role of environmental stress in the origin, and then the turning off and on, of a physiological trait during evolution. CAM is an alternative pathway to photosynthesis and occurs phylogenetically dispersed in a wide variety of plants among at least 30 families, indicating that it has evolved repeatedly (Smith and Winter, 1995). CAM is usually expressed under extreme environmental conditions, for example, in hot arid places, in epiphytic tropical plants that experience periodic dryness, in aquatic plants at low carbon dioxide concentrations, or in places with high soil salinity (Winter, 1985; Winter and Smith, 1996a). CAM enables plants to fix carbon dioxide at night and store it in the form of malic or, less commonly, citric acid, which is used the next day as an internal CO_2 source for conventional photosynthesis. This is adaptive in situations where ordinary (e.g., stomatal) acquisition of CO_2 would lead to costly water loss via transpiration, and in the other CO_2-limited conditions just listed.

The degree and pattern of expression of CAM vary among species, even in constitutive (obligatory) CAM plants that normally possess CAM ability in mature leaves. The inducible CAM species have become models for the mechanistic study of the relationships between environmental cues and changes in gene expression in plants (Cushman and Bohnert, 1996). In inducible (facultative) CAM plants, the enzymic machinery for CAM is induced in response to an environmental stimulus such as high soil salinity or drought. Responsive leaves develop dark CO_2 fixation and nocturnal malic acid accumulation, which increases gradually over the course of several days, accompanied by manyfold increases (up to 20–40 times) in the potential activity of several enzymes and altered metabolic properties of chloroplasts and mitochondria (Winter, 1985). During the facultative transition from C_3 to CAM and associated stress responses, the expression of several hundred genes is altered (reviewed in Cushman and Bohnert, 1996; Ehleringer and Monson, 1993).

All enzymes required for a functional CAM pathway are present in non-CAM species, where they are essential for photosynthetic carbon metabolism (Cushman and Bohnert, 1996). It is widely agreed that the evolution of CAM likely involves the turning on or amplification of expression of genes or isozymes of genes that were already expressed in the photosynthetic cells of C_3 plants prior to the origin of CAM (Winter, 1985, p. 380). It is therefore not surprising that CAM has evolved repeatedly in a very wide variety of plant families. Its association with extreme conditions suggests a role for environmental induction or stress in the origin and spread of CAM in stressed populations.

The opportunistic nature of innovation, using combinations of components that happen to be already present, is seen in the evolution of CAM in species of the well-studied tropical genus *Clusia*, which use citric acid in addition to malic acid in their inducible CAM (Lüttge, 1996). It would be of evolutionary interest to know what physiological ancestry conditioned that particular line of variants and permitted them to prosper.

These findings invite the use of phenotypic engineering in the form of experimental stress to investigate the genetics and physiology of CAM-inducible variants in populations of weak-CAM ("CAM-cycling") or closely related non-CAM species, to complete the picture of plasticity and diversity that has emerged from comparative study of CAM physiology (Winter, 1985; Lüttge, 1996, 1999; Pilon-Smits et al., 1996). Is the CAM phosphoenol pyruvate carboxylase the result of a recurrent mutation or duplication in a common C_3 enzyme, that happens to be selected under respiratory or other stress? Or is it an enzyme that happens repeatedly to show up in other adaptive contexts in different lineages then get co-opted under stress to function in CAM? The answers, rather than relating to different adaptive pressures, may have to do with the preexisting equipment (such as some precursor of the acid storage vacuole seen in CAM) and initial responses to recurrent environmental stress. Such a hypothesis is briefly outlined for C_4 photosynthesis by Ehleringer and Monson (1993), who point out that the architecture of certain genetic regulatory systems (especially, I would add, those highly sensitive to environmental influence) may predispose them to evolutionary change, providing for the rapid origin of novel morphological and biochemical patterns that then are subject to selection. A direct role of the environment in forcing successful variant pathways in certain genotypes predisposed to express them needs to be examined as a possible contributing factor.

Following the establishment of CAM in a lineage, its expression continues to be highly condition sensitive and variable among species. In some of the facultatively CAM species, such as the common ice plant *Mesembryanthemum crystallinum* (Aizoaceae), CAM machinery is expressed only when induced, and it then persists regardless of conditions. In other species, such as *Sedum telphium*, CAM seems to persist even when unused. In still others, such as *Clusia minor*, the switch to and from CAM is rapid and reversible, depending on conditions, and can differ even in leaves at the same node (Winter et al., 1992). In *Isoetes howelli* (Lycophyta: Isoetaceae), an inhabitant of seasonal pools, CAM is expressed when leaves are in water, but all individual plants become aerial at some time during their lives. As the foliage enters the air, it switches from CAM to C_3 photosynthesis. This response is adjusted on an almost unbelievably fine-tuned cell-by-cell basis, with still-submerged areas retaining CAM. Terrestrial *Isoetes* species, which are believed to be derived from aquatic ancestors, lack CAM throughout the life cycle; that is, the aerial-phase photosynthetic mode is fixed. It would be of interest from an evolutionary point of view to know whether a full CAM response can be induced in terrestrial *Isoetes* species by prolonged submergence in water, for this would reveal whether phenotypic fixation is due to environmental induction alone or is to some degree genetically accommodated or assimilated.

Whether CAM is facultative, cyclic, or obligatory, CAM expression is age dependent, being absent in the earliest stages of development. Gene-expression changes are known to accompany age-specific CAM responses (Cushman and Bohnert, 1996). Since these responses are facultative reactions to environmental stress, they may be particularly subject to heterochronic changes during the history of a lineage, with little or no genetic change at loci involved in regulation.

In the genus *Peperomia*, finely adjusted plasticity in CAM takes the form of an intraleaf division of labor between tissues specializing in CAM and tissues specializing in C_3 photosynthesis. In species termed C_3–CAM "intermediates," individuals switch rapidly and reversibly between modes under stress (e.g., drought or high salt concentrations; Borland and Griffiths, 1996). Different *Peperomia* species show different degrees of specialization to CAM, ranging from complete absence in exclusively C_3 species, to occasionally functional presence in CAM-cycling species, to facultative and obligatory expression in specialized CAM species. So this is another genus that, like *Isoetes*, is ripe for comparative evolutionary studies and tests of a possible role of environmental stress in phenotype fixation (Ting et al., 1996; see also Martin, 1996, on the similarly variable genera *Tillandsia* [Bromeliaceae] and *Talinum* [Portulacaceae]). In a populational and phylogenetic approach to physiology and development, the range of variants, and even the anomalous cases, takes on new significance.

Clusia, with its diversity of CAM expression (Lüttge, 1999), may prove a key genus for understanding the transition to CAM. It offers a diversity of closely related yet distinctive natural populations. Of the 150 species in the genus, 20 species have already been studied ecophysiologically (Lüttge, 1999, table 1), ranging from obligate C_3 to obligate CAM, with numerous intermediate species showing C_3/CAM transitions of various sorts. Such a genus could provide insights on the nature of exploitable phenotypic variation, and the reorganizing effects of environmental stress. Some reversals of the CAM–C_3 transition are expected because even "obligatory" CAM plants are always C_3 as seedlings (Raven and Spicer, 1996). The comparative study of CAM calls out for cladistic analysis since richly detailed equivalent studies of many species are now available. But confident answers regarding directions of evolutionary change will be difficult to obtain in some families, in part because of the basic assumption of cladistic analyses that convergence is unlikely and should be invoked as little as possible. This assumption may not be appropriate in these plants (see chapter 19). CAM is known to have evolved many times (Winter and Smith, 1996a,b; Pilon-Smits et al., 1996), perhaps sometimes even within the same genus (Pilon-Smits et al., 1996). Multiple origins and reversions to alternative pathways are expected due to the extreme flexibility of this trait and the ease of its *de novo* assembly from enzymes common in non-CAM plant cells even in lineages lacking CAM.

Further insights might be gained by considering the preadaptations that cause some plant lineages to evolve toward C_4 photosynthesis, with its characteristic morphological arrangement, rather than CAM under similar conditions of stress (see Ehleringer and Monson, 1993).

In summary, the observations on CAM suggest extreme variation in adaptive plasticity, both within and between species, as well as age-dependent expression, phenomena that are highly conducive to recurrent evolutionary change mediated by environmental influence.

The Origin of Maize   The origin of maize from teosinte is discussed in chapter 15 as an example of evolved sex transfer. Iltis (1983) reviews evidence that this transition may have been environmentally initiated. Sexual switches are common in plants subjected to unusually hot or cold summers, disease, or injury, and they occur seasonally in greenhouse populations (see references in Iltis, 1983, p. 889). Branch shortening, the pivotal change associated with sex transfer in maize, can be environmentally induced, and there is evidence for genetic variation for differences in sex expression in different varieties of maize. So the evolution of maize "may have involved no new mutations, but rather genetic assimilation under human selection of an abnormality, perhaps environmentally triggered" (p. 886).

The Evolution of Direct Development in Echinoids   The evolution of direct development by loss of the feeding structures and behavior of the pluteus larva (see chapter 11) has occurred at least 29 times in echinoderms, 15 of them in extant echinoids (Emlet and Hoegh-Guldberg, 1997, p. 148). There is considerable evidence that developmental plasticity has played a role in these events, and it is of special interest to emphasize these findings here

because of the importance of research on sea urchins in molecular developmental biology, and in classical embryology (see historical review in Davidson, 1986). These are fields disposed to view development and evolution as driven by changes in "hard-wired" genetic programs (Davidson, 2000), with little interest in the potential role of the environment. But the result is a wealth of data on the two kinds of echinoid development, and of techniques for studying gene expression that could be applied to a comparative study of species and individuals with contrasting and intermediate developmental modes.

Deletion of the pluteus larva is always associated with the evolution of large yolk-rich eggs, called "embryonization" by Matsuda (1987). Egg size varies considerably among individuals of the same population, and from year to year in populations of some species (Lessios, 1987; Hadfield and Strathmann, 1996, p. 325), and egg size is an evolutionarily labile trait within genera of echinoderms (Lessios, 1990). In comparisons across species, degree of direct development and loss of the pluteus correlates with increase in egg size (Emlet et al., 1987; Lessios, 1990; see figure 11.2), associated with a decline in dependence upon larval feeding. Only a few species are intermediate, in having large eggs and facultative (optional) feeding. These include *Clypeaster rosaceus* (Emlet, 1986) and *Brisaster latifrons* (Hart, 1996), in which the larvae can either feed or develop without feeding, suggesting that the production of large eggs alone is not sufficient to cause direct development (loss of the pluteus), and that the first step toward loss of the pluteus is the evolution of large eggs without direct development as an automatic consequence. The skeletized, armed pluteus larva of these facultative feeders is like that of obligate feeders (see figure 11.2a; Hart, 1996). In the direct-developing species *Heliocidaris erythrogramma*, experimental reduction of egg size by removal of part of the yolk did not alter larval development time or prevent survival and metamorphosis, but it did reduce the size and growth rate of juveniles (Emlet and Hoegh-Guldberg, 1997), so large eggs may have evolved in part due to their enhancement of postmetamorphic development (Emlet, 1986; Emlet and Hoegh-Guldberg, 1997).

The rarity of facultatively feeding pluteus larvae may indicate that the evolution of large eggs, or eggs highly endowed with lipids as in *H. erythrogramma* (Wray and Raff, 1991, p. 47), is usually accompanied rapidly by reduction or loss of the pluteus as a side effect of changes in the egg. This has been hypothesized by Strathmann et al. (1992)

and Matsuda (1987, p. 108). The evolution of maternal influence on eggs may include not only nutritional factors (increase in egg size and lipid content) but also change in maternal gene transcripts
that affect larval development, a possibility that has
not been insufficiently considered. Maternal transcripts are the exclusive instructions for very early
embryogenesis in indirect-developing sea urchins,
up to about the 8–16-cell stage in *Strongylocentrotus pupuratus* (other species yield similar findings), when zygotic transcripts begin to appear
(Davidson, 1986, p. 71), and they continue to be
active later in development. During these early cell
divisions, striking differences between direct- and
indirect-developing embryos are already evident
(Wray and Raff, 1991, their figure 4, show them
to be well defined at the 32-cell stage). An estimated
11,000 different maternal mRNA transcripts are
being translated in early embryogenesis (Davidson,
1986, p. 74). Could change in the maternal-tran-
script endowment of the egg be important in the
evolution of direct development? And is the maternally controlled quantitative and qualitative
composition of eggs influenced by maternal environments? These are key questions for the role of
environmental influence and egg evolution that
have not yet been considered, and they should be
amenable to experimental and comparative investigation. So far, conclusions regarding the consequences of change in egg size are based too much
on interspecific comparisons, and there is a dearth
of data on variation within populations.

Matsuda (1987) lists many examples of the evolution of direct development associated with extended development within large eggs, and hypothesizes that this is commonly associated with
extreme or altered external environments (e.g., in
various fishes, amphibians, crabs, mollusks, marine
clitellate worms, cnidarians, and crinoids). At least
one study recorded larger eggs and larvae, with
faster developmental rates in the sea urchin *Arbacia lixula* in favorable habitats (George, 1990, cited
by Hadfield and Strathmann, 1996). The production of a large egg alters the larval trophic environment, but it does not always correlate with
higher energy content (McEdward and Carson,
1987), and, as already mentioned, the content in
terms of maternal transcripts that may affect larval development is apparently uninvestigated. A recent review found that the consequences of egg-size
variation for larval development remain poorly
known; the authors were unable to find any studies where the development of individual larvae has
been followed beginning with eggs naturally differing in size (Hadfield and Strathmann, 1996).

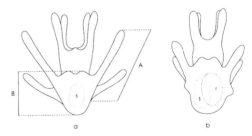

Fig. 26.4. Effect of nutrition on morphology of the
15-day-old pluteus larva in the sand dollar *Dendraster excentricus*: (a) unfed larva; (b) larva fed
abundant food. S = stomach. r = adult rudiment,
which is more developed in the well-fed larva, indicating accelerated development. A, length of the
postoral skeleton from arm tip to first branching in
the larval body, affects feeding ability by affecting
the length of the ciliary band; B, body length. After Boidron-Metairon (1988).

Feeding conditions of larvae are known to produce skeletal and growth-rate changes in pluteus
larvae that mimic to some degree differences between direct- and indirect-developing species. Experimental reduction of egg size in the sea urchin
Strongylocentrotus droebachiensis caused early larvae (before the six-arm stage) to approach in size
and shape the larvae of a closely related species (*S.
purpuratus*) with much smaller eggs (Sinervo and
McEdward, 1988). Pluteus larvae of the urchin
Dendraster excentricus and *Lytechinus variegatus*
fed little or no food developed unusually long arms
(which bear the feeding ciliae), whereas larvae fed
a large ration showed an approach to the traits of
direct-developing sea-urchin larvae, with shortened
feeding arms and a relatively large adult rudiment
(Boidron-Metairon, 1988; figure 26.4).

More extensive work on the influence of diet on
the morphological development of the pluteus larva
of another sea urchin (*Paracentrotus lividus*), using
both seasonally variable natural populations and
rearing experiments, confirmed the results of
Boidron-Metairon (1988) and showed that altered
food quality and quantity had similar effects on larval morphology (Strathmann et al., 1992; figure
26.5). Sibling larvae raised on different amounts of
food show different degrees of development of
structures used for catching food, with increased
allocation to growth of those structures, and decreased body size, when food is scarce. The opposite set of responses occurred in well-fed larvae, and
those changes—reduction of feeding structures, increased body size, and precocious maturity of the

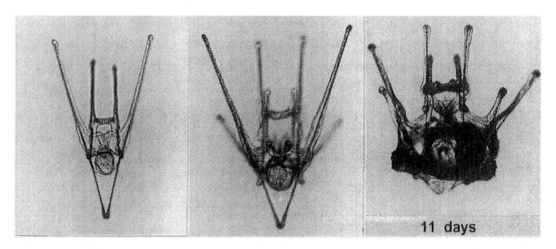

Fig. 26.5. The effect of nutrition on growth and morphology of larval sea urchins (*Paracentrotus lividus*): 11-day-old pluteus larvae reared in the laboratory with (left) reduced rations; (center) approximately natural rations; (right) enhanced rations. Specimens shown are representative of a large sample used in quantitative measurements of growth and morphology under the three feeding regimes. Specimens from natural populations resembled the larva shown in the center. After Strathmann et al. (1992). Courtesy of R. Strathmann.

adult rudiment—are in the same direction as those that characterize sea-urchin species with large eggs and reduced larval feeding. These studies show that adaptively appropriate responses to variation in food supply within species could preadapt them for the evolution of reduced feeding and direct development under favorable trophic conditions such as those provided by a large egg (Strathmann et al., 1992).

Comparative study of species having different degrees of direct development, including those with (a) a typical feeding pluteus; (b) a partially formed, nonfeeding pluteus; (c) a direct-developing planktonic larva (lacking a pluteus stage); and (d) a direct developing larva brooded by the mother (Wray, 1996) shows that loss of the pluteus and increased dependence on maternal resources could conceivably be gradual (see figure 24.3). But the change from feeding toward direct development can also occur suddenly under the influence of an altered diet, as shown by observations of seasonally variable natural populations and rearing experiments on *Paracentrotus lividus* by Strathmann et al. (1992). Sibling larvae raised on different amounts of food show different degrees of development of structures used for catching food, with increased allocation to growth of those structures, and decreased body size, when food is scarce. The opposite response in well-fed larvae is similar to that in evolutionary changes in species that have

large eggs and reduction in larval feeding. "If endogenous food supplies have the same effect on morphogenesis as exogenous food supplies, then changes in genes that act during oogenesis to affect nutrient stores may contribute to correlated adaptive changes in larval development" (Strathmann et al., 1992, p. 972).

By interpreting these results in terms of genetic evolutionary change, Strathmann et al. effectively sidestep the more radical suggestion of Matsuda (1987) that such plasticity in response to egg size amounts to rapid *environmentally induced* change:

In postulating the mechanism of suppression of larval (and juvenile) stages resulting in acceleration, we must consider the nature of intraoval environment in which the embryo-larva is forced to live. The enlarged egg is by itself a new environment that may be thousands times [sic] greater in volume than the ancestral egg. The question is how such a novel environment has influenced the development of the embryo-larva. . . . [T]he production of large eggs occurred most probably . . . before the new types of animals . . . started emerging from such eggs. (pp. 30–31)

Matsuda went further and speculated that the "stressful physiochemical environment" represented by the large egg could lead to hormonal changes and thereby directly to morphological change, citing evidence for such a mechanism in amphibians. To my knowledge, this mechanistic hypothesis has not been tested in sea urchins. But the results of Boidron-Metairon (1988) and Strathmann et al. (1992) discussed above show that a fairly large morphological response is immediate and in an adaptive direction, so there is indeed a preexisting mechanistic basis for an immediate, environmentally induced response, though perhaps not the same in its details or as extensive as Matsuda imagined. The Boidron-Metairon and Strathmann et al. results show that this major evolutionary innovation could have originated by a nutritional change provoking a preexisting, adaptive plastic response, one that could be repeated in the population and thereby subject to genetic accommodation. Under the plasticity and genetic accommodation hypothesis, a crude facultative response such as that obtained experimentally could establish a polyphenism for feeding and nonfeeding development, and genetic accommodation during an alternative-phenotype phase would facilitate the evolutionary refinement of direct development. The Matsuda genetic assimilation hypothesis, by contrast, envisions abrupt origin of the direct-developing form entirely due to environmental induction, followed by its genetic assimilation (selection on regulation that fixes direct development).

The kind of genetic variation that would contribute to genetic accommodation or assimilation of a deletion was measured in a marine polychaete that is polyphenic for feeding and nonfeeding larvae (Levin et al., 1991). This study demonstrated (a) additive genetic variation in ovum diameter and fecundity; (b) negative genetic correlations between these two traits, indicating a trade-off; and (c) negative correlations between larval survivorship and mean planktonic period, which may put a premium on large-yolk production in times of food scarcity requiring a prolonged larval feeding period. Strathmann (1985) discusses the trade-off between investment in eggs versus fecundity in the evolution of direct development.

As emphasized by Matsuda (1987), the correlation between large egg size or "embryonization" and direct development is quite common in marine invertebrates, and deletion of a feeding larval stage has evolved many times. Gore (1985) gives examples in Crustacea, where some of the same phenomena are observed as in echinoids. There is a broad range of egg sizes accompanied by various "combinatorial stages" where particular larval instars may show a mixture of traits typical of previous or subsequent instars. Within-population

plasticity in combinations of traits is associated with evolutionary lability in combinations (Rabalais and Gore, 1985). In decapod crustacea, there is also recurrent evolution of abbreviated development. Most species with abbreviated development produce large eggs, but not all species with large eggs show abbreviated development (Rabalais and Gore, 1985).

Another kind of evidence for the possible importance of environmental factors in the evolution of direct development was described more than a century ago by E. B. Wilson (1894) in a study of the effect of calcium deficiency on the morphology of the pluteus larva (see figure 26.2). Taken together with the more recent observations of trophic effects just cited, there is sufficient evidence of developmental plasticity to regard it as potentially important in the origin and early evolution of direct development.

An obvious next step in research on this question is to study the effects of maternal (rather than larval) nutrition on the development of larvae. Facultative responses to nutritional stress, if they exist, may affect maternal egg management, with facultative changes in nutritive and instructional endowment of the egg a possible result. Facultative maternal effects would not be revealed by postovulation manipulation of egg size or larval diet. A possible place to look for manipulable maternal effects on skeletal development and life-history type of larvae would be in the sand dollar *Peronella japonica*, which has nonfeeding larvae in which a skeleton is initiated but not usually completely developed, and shows large variation in the development of the skeleton, sometimes approaching in appearance the ancestral feeding pluteus. Another source may be sea urchins of genera such as *Heliocidaris*, which have both direct-developing and indirect-developing species (Raff, 1987). So far, experiments have addressed only postfertilization effects of egg size and diet, without considering the possibility of maternal effects on the composition of maternal transcripts, amount of yolk, and other possible (e.g., hormonal) effects on the organization of the egg phenotype and the form of larvae.

Comparative evolutionary studies should begin to exploit the molecular techniques that have elucidated the amount, kinds, and timing of use of maternal transcripts in some sea urchins and other echinoids (Davidson, 1986). Davidson (1986, p. 85) notes that the relatively large egg of the sea star *Pisaster ochraceus* contains about five times more maternal RNA than does the egg of *Stronaylocentrotus purpuratus* and about 100 times as many maternal actin mRNA molecules. Such comparisons carried out within species for different treatment groups, or between related direct- and indirect-developing species would help test the possibility that inherited environmental (maternal) effects have played a role in the evolution of egg characteristics and direct development in echinoderms. This possibility seems likely given the close association between egg characteristics (especially, egg size) and developmental mode in many invertebrates (Matsuda, 1987) including echinoderms (see figure 11.2). This entire area of invertebrate life cycles is ripe for further comparative research on development and evolution (Sinervo and Basolo, 1996), including the role of environmental induction.

Places to Look: Digestive Morphology of Birds and Other Organisms

Unusual diets have long been reported to cause change in the intestinal condition of birds (e.g., Darwin, 1868b [1875b], p. 292), suggesting that an environmental change in food availability or preference could induce novel morphological and possibly physiological traits. If this were true in species related to those in which such traits are a regular feature, this would support a hypothesis of environmentally induced evolutionary novelty.

Some support for this idea is provided by the expression of distinctive morphology as a facultative response to conditions. Morphological plasticity is a feature of seasonal adaptation in some birds. In arctic rock ptarmigan (*Lagopus mutus*), a seasonal bimodality in gut cecum structure is produced by seasonal change in diet. During winter, when birds feed predominantly on buds and catkins of birch, the length and tissue weight of the cecum are greater than in summer, when the birds eat leaves, seeds, and berries (Gasaway, 1976). That this could have originated as a direct response to diet is suggested by the fact that in a related species, the red grouse (*Lagopus lagopus scoticus*), captive birds fed for several generations on poultry pellets were found to have their ceca reduced to a fraction of the bulk and mass they attain in wild grouse (Wynne-Edwards, 1986). This response to an unusual diet suggests that such plasticity may have preceded the evolution of a similar, regularly produced seasonal response. Similar changes of gut length in response to seasonal changes in diet occur in house wrens, rufous-sided towhees, and spruce grouse (Karasov, 1993).

Piersma and Lindström (1997) review other examples of rapid reversible adaptive changes in organ size and metabolic physiology, including the twofold reversible increase in small intestinal mass, and 45% increases in masses of kidneys and liver, of burmese pythons (*Python molurus*) following a meal. Anurans also show striking diet-dependent reversible changes in gut length and trophic morphology (Pfennig, 1992b; see figure 21.2). These phenomena invite further, comparative study of related species, using phenotypic engineering to see if environmental manipulations can achieve similar results in species not usually showing such responses, without genetic change. If this were to prove true, it would be strong evidence that environmental induction was responsible for the origin of adaptive facultative change in digestive morphology.

Environmental Influence and the Paleontological Time Scales of Evolutionary Change

Climatic Cycles

Theories of adaptive evolution in changing environments (e.g., Levins, 1968) treat the heterogeneity of environments on a neontological time scale, showing how populations and individuals can track fluctuating conditions. At the other extreme, paleontological analyses based on very long time scales in the fossil record of morphological change can give an impression of relatively slow phenotypic change and even "stasis." Between is a time scale of environmental variation well documented by geologists and yet easily skipped over by these two perspectives. This intermediate time scale may be especially pertinent to understanding adaptive evolution of the organisms we observe today.

Environmental conditions need not be constant or long term to cause directional evolutionary change. Oscillating selection, especially when condition-sensitive traits are involved, can theoretically have a ratcheted cumulative effect on directional evolution (see chapter 7). As a result, global and regional fluctuations in climate and ecology could produce a prolonged, though intermittent, evolutionary trend in an intermittently expressed correlated set of traits. Cyclic fluctuations would be particularly conducive to such trends in facultatively expressed traits, for they could intermittently drive change in a particular direction, without rigidly fixing a particular phenotype.

The geological history of the earth is marked by cyclic change on many different time scales (Fischer, 1981). Some, like the 22-year Hale cycle and its 11-year sunspot hemicycle, are sometimes strikingly reflected in the growth of organisms, for example, in rhythmic tree growth rings. The regional distributions of solar energy, degree of seasonality, and hydrographic properties are environmental characteristics that change in relatively short cycles caused by changes in the inclination of the earth's rotational axis, yielding a cycle of 41,000 years. Interaction between the earth's eccentricity and cycles of precession produces climatic effects at modal intervals of 19,000 and 23,000 years. These recurrent cycles, known to be of sufficient magnitude to affect the distributions and growth rates of organisms, are very likely to influence their development and evolution. So the earth's environmental history is relevant to studies of adaptive evolution and phylogeny, especially as they are affected by development and plasticity.

Eickwort et al. (1996) point out that the Quaternary warming cycles extended the habitable region of many boreal insects toward the arctic. They hypothesize that during one or more of these warm periods, social bees able to adapt to short growing seasons by a facultative reversion to solitary nesting may have crossed the Bering land bridge to colonize North America, where they then radiated in the New World. This hypothesis is supported by the fact that species groups of some halictid bees and polistine wasps not having holarctic (short-season-adapted) representatives are restricted to their continents of origin, indicating that they were unable to cross the short-season barrier. If true, this implies that eusocial neotropical species of these groups are secondarily social—they have reverted from sociality to solitary nesting and then back to sociality again, as is known to have occurred in this group of bees following their colonization of North America (chapter 19, see also figure 12.3). Other, less labile groups of social insects, perhaps because they were unable to express or maintain solitary reproduction under short growing seasons, are believed to have colonized North America during the Eocene (about 55 million years ago), when land-bridge conditions were tropical or subtropical (Eickwort et al., 1996).

The same climatic cycles that produced the great ice ages at high latitudes produced changes in precipitation and lake levels of even the largest lakes in East Africa (see references and details in McCune et al., 1984). The cyclic formation and evaporation of African lakes may have played an important role in the explosive diversification of cichlid fishes and other organisms, by periodically dissecting lakes into small, isolated bodies of water, not only facilitating speciation by isolation but also accelerating

the evolution of trophic specializations by forcing episodes of resource competition among the inhabitants of the lakes (see chapter 28). Even if a given lake is very young, having been completely dry in the relatively recent past (as recently as 12,400 years ago for Lake Victoria—Johnson et al., 1996), the colonists responsible for explosive speciation in a lake may have had their flexible developmental capacity for diversification molded by previous cycles of environmental change in older lakes or rivers that seeded the new lake. Species survival and adaptation in these fluctuating lakes may have depended as much on flexibility as on ability to genetically track environmental change. It is reasonable to suppose that some of the environmental tracking attributed to the red queen hypothesis (Van Valen, 1975) or to speciation should be attributed instead to adaptive plasticity (McCune et al., 1984). In general, the much-discussed long-term persistence of ecosystems undoubtedly owes as much to the phenotypic plasticity of component individuals as it does to fine evolutionary adjustments among species or structural complexity at the ecosystem level.

Especially relevant to understanding present-day organisms are the fluctuations that occurred during the late Pleistocene, and the Holocene period since the last glacial maximum—the last 10,000 years—since the organisms we observe today have experienced these changes in the not-too-distant past, presumably in approximately their present phenotypic states and with their present capacities for adaptive plasticity. In some regions the period between 11,000 and 10,000 years before present, corresponding to the last deglaciation, involved a very rapid climatic cooling. The North Atlantic polar front migrated southward almost to its glacial maximum position, and temperatures in western Europe were very cold, with South American records also indicating a cool interval at this time (Crowley, 1983, p. 861). Mean estimated August temperature in the area of the Arabian Sea, for example, dropped from 26°C to 22°C during this period, then rose again to its present level of about 25°C. (Crowley, 1983, figure 43). These changes were accompanied by changes in biological communities (Davis, 1976). Organismic responses, many of them governed by plasticity, would have enhanced the ability to survive stresses of the many kinds that would be engendered by such climatic fluctuation and biotic change (Sultan, 2000).

Creatures of our own limited experience, biologists are not used to considering that such cycles may have wrought lasting changes on the morphologies and flexibilities of familiar species. For example, research on lakes in the East African rift valleys (summarized by Crowley, 1983, p. 864) suggests that during the time interval over which humans evolved in that region, the climate underwent a change from tropical rain forest to dry grasslands, and it has even been suggested that the reduction in arboreal habitats may have contributed to the evolution of a fully bipedal posture in some primates, including man (Stanley, 1995). Whatever the fate of that particular hypothesis, such attempts to relate earth history with adaptive explanation deserve careful attention. Intervals as warm as the present occupied only 10% of the late Quaternary record (Crowley, 1983), so many of the traits we observe today no doubt owe their particular form and degree of adaptive plasticity to conditions predominant in other times. A supportive example is presented in the following section.

The Great Basin Kangaroo Rat

A study of the Great Basin kangaroo rat (*Dipodomys microps*; Csuti, 1979) provides an unusually well-documented case history that shows the importance of climate change in the evolution of flexible and recurrent traits. Kangaroo rats vary genetically, morphologically, and ecologically among several locations in the western United States. They have been described as geographically and ecologically polymorphic in their adaptation to local conditions that vary in availability of water and food plants. Populations in central and western parts of the species range feed primarily on leaves of the saltbush, *Atriplex confertifolia*. This involves a specialized behavior of shaving the leaves with the teeth, as well as physiological specializations that enable individuals to deal with the high salt content of the leaves. While posing a challenge to manipulative ability and kidney function, saltbush leaves have an unusually high nutrient value and water content. In more southern populations, the kangaroo rats show seasonal variation in diet, but they are blackbush (*Colegyne*) specialists. They eat either the seeds or, in springtime, a mixture of seeds and leaves of blackbush and do not show the behavioral and physiological specializations of the saltbush specialists. They are also more resistant to low water intake.

A detailed study of geographic variation in cranial, tooth and bacular morphology, kidney size, chromosome morphology, and electrophoretic variation in plasma and liver proteins (Csuti, 1979) revealed genetic divergence among the populations that relates to the geographic distance between them, suggesting a recent expansion of the species

into its current range. Saltbush-zone individuals averaged larger kidney weight, but tooth morphology did not vary between populations. When wild-caught adults were tested for feeding efficiency (weight gain) on the two types of diet, saltbush specialists showed no ability to adapt to water-deprived seed diets, and blackbush (seed) specialists fared better on seed diets. There was vast individual variation in growth efficiency of seed specialists on a water-restricted diet of seeds (see figure 18 in Csuti, 1979). If the blackbush adults were given saltbush leaves, most failed to show the shaving behavior at first, but if allowed several weeks to experiment with a mixed diet containing leaves, all eventually showed the stereotyped leaf-shaving behavior as an experimentally induced reversion reflecting phenotypic plasticity.

When captive-born naive individuals were reared from infancy on the two different diets in a kind of common-garden experiment inspired by genecological research on plants, the young of saltbush-specialist parents showed no superiority in ability to handle leaves compared with offspring of seed specialists. But individuals from a population of saltbush specialists were relatively inefficient (in terms of weight gain) when raised on seeds, and seed specialists were relatively inefficient when raised on leaves, indicating that some genetic "ecotypic differentiation" has occurred (Csuti, 1979, p. 60). Thus, all members of the species show lifetime flexibility in ability to switch diets, even though some genetic divergence has occurred in ability to assimilate the contrasting kinds of food.

Phylogenetic study indicates that the ancestors of the kangaroo rats were seed specialists. The climatic history of the Great Basin region, combined with analysis of geographic variation in the kangaroo rats, allowed Csuti (1979) to develop a hypothesis regarding the effect of climate history on the evolution and maintenance of a facultative ability to consume saltbush leaves:

> The adaptation of *D. microps* for *Atriplex* grazing probably occurred in the early Pleistocene following its divergence from *heermanni* group relatives. At this time it invaded the developing Great Basin Desert and was preadapted to leaf-eating by virtue of a moisture dependence common to all *heermanni* group species. This new adaptive strategy was incorporated into the species repertoire without immediately replacing the ancestral seed-eating strategy [i.e., as a facultatively alternative phenotype]. During the Wisconsin glaciation, *D. microps* habitat was reduced in the Great Basin and northern Mohavia

and the species was compressed into relicts of desert vegetation. A recent warming and drying trend, accompanied by the spread of current vegetation into the Great Basin, allowed reinvasion by a recently-panmictic *Dipodomys microps*. Both the primitive *heermanni* seed-eating strategy and the leaf-eating strategy were common to advancing *D. microps* populations. Selection since reinvasion of the Mojave and Great Basin Deserts has produced incipient stages of ecological race formation. (p. 60)

The kangaroo rat story, unusually complete in its analysis of factors ranging from protein polymorphism to earth history, gives a rare glimpse of the evolution of niche diversification and phenotypic plasticity on an intermediate geological time scale. This is a time scale lost in the scholarly netherland between the present-neontological and the fossil-paleontological approaches. Yet this scale of environmental change likely influences the norms of reaction and kinds of alternative phenotypes manifested in the organisms we observe today.

How often are "evolutionary" changes in the fossil record in fact plastic phenotypic responses to environmental change, as in the kangaroo rat? Certainly more than we realize, for this possibility is seldom investigated. One possible example in snails (Gould, 1969) is discussed in chapter 19. A potential additional example occurs in the dentition of the arvicolid rodent *Clethrionomys glareolus*, which underwent iterative change in tooth morphology during the Pleistocene, a period characterized by rapid climatic change, as discussed above. Extant *C. glareolus* show a bimodal distribution in form of the third molar corresponding to two iterative fossil forms (Bauchau and Chaline, 1987). Extant populations also show rapid intrapopulation change in tooth morphology, which has been attributed to environmental factors (Chaline, 1987, after Corbet, 1964, 1975). Such a group, with comparable variation in both fossil and extant populations, would reward close study of the causes of variation in conjunction with the history of environmental fluctuations in climate.

Phenotypic Effects of Global Environmental Change

If normal environments contain factors that become entrenched in development, then it should be possible to see direct effects of global climatic and biotic change on the development and phenotypes of organisms. Factors in the environment that we consider critical for life have not always been pres-

ent. Others have changed in quantity or availability over time and space, sometimes in repeated cycles, in ways that could affect dramatically the phenotypes of organisms. Is there any evidence that these changes have had direct effects on the phenotypes observed at different times or in different places? Are there evolutionary novelties that might be attributed to direct environmental induction following major change in the conditions found on the earth or its atmosphere?

The earth's atmosphere is thought to have been originally anaerobic, and anaerobic bacteria were presumably among the earliest organisms. Among the most ancient proteins known are the ferredoxins, which are found in anaerobic bacteria as well as in algae, plants, and animals (Yasunobu and Tanaka, 1973). Ferredoxins can be synthesized abiogenically in laboratory conditions that simulate those believed to have characterized the earth near the time of the origin of life, and they contain large amounts of the amino acids glycine, valine, alanine, proline, and glutamic and aspartic acid, those found in the Murchison meteorite that fell in Australia in 1969, an indication of the conditions on the primitive earth (Yasunobu and Tanaka, 1973). The ferredoxins are a reminder that the evolution of life is not just a story of the birth and elaboration of genetic programs. The form of the earliest phenotypes depended as much on the raw materials available as the ability of organisms to use them.

The composition of the atmosphere has changed drastically during the history of life on earth (figure 26.6) in ways that affect directly the performance of organs of respiration and flight. Geophysical models (e.g., of Berner, 1993, and Berner and Canfield, 1989, cited by Raven and Spicer, 1996; Graham et al., 1995) suggest, for example, that there was an abrupt rise in oxygen content of the atmosphere, accompanied by a parallel increase in atmospheric density, during the late Paleozoic (Carboniferous and Permian periods; figure 26.6). This coincided with origins of insect flight, arborescence in plants, and the invasion of land by tetrapods (Graham et al., 1995). Dudley and Chai (1996) point out that the origin of flight could have been directly promoted by elevated atmospheric density, providing enhanced lift on articulated protowinglets already in existence. Increased diffusion in the tracheal system could also have contributed directly to the appearance of gigantism in diverse insect orders during the Carboniferous: an upper size range of phenotypes unable to thrive at lower oxygen levels would have been able to survive, a change in size distributions that would not require

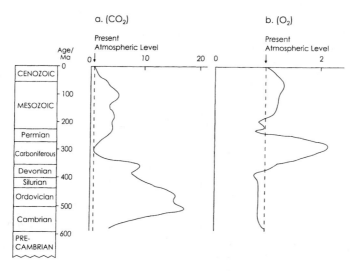

Fig. 26.6. Changes in levels of atmospheric carbon dioxide and oxygen from the Cambrian to the present. Scales give relative values based on a present value of 1.0. Age in millions of years (Ma). After Raven and Spicer (1996).

evolutionary genetic change, though genetic accommodation of increased size would be expected as well, were this to prove advantageous. Raven and Spicer (1996) discuss the possible effects of such atmospheric changes in the evolution of CAM physiology in plants. How this could have been facilitated by a phenotypic recombination of preexisting biochemical features under environmental stress is discussed above.

Some fossil structures interpreted as flight devices seem to contradict this oxygen-pulse hypothesis. Several species of diapsids with presumed gliding morphology appeared in the late Permian and late Triassic (Colbert and Morales, 1991), a period unusually *low* in atmospheric oxygen (figure 26.6) and, presumably, air density, conditions that should have rendered flight and gliding relatively difficult. This invites caution regarding the attempt to make functional interpretations when behavior and use cannot actually be measured or observed, even in some extant or similar taxon. Many extreme structures, such as the exaggerated lateral skin flaps interpreted as flight devices of extinct lizards, are the sorts of structures that could conceivably be used in aggressive or sexual display, as are the dewlaps and gular flaps of *Anolis* and other genera. Such structures, including the crests and spines of dinosaurs, often have been assumed to have physiological functions such as heat dissipation without considering possible social functions (see West-Eberhard, 1983).

Calcified skeletons suddenly appeared in many invertebrate taxa at the Precambrian–Cambrian transition, a time when calcium levels in ocean water increased (Marin et al., 1996). This also required widespread ability to utilize the calcium, so it is significant that the capacity to respond to calcium availability with calcification may have evolved by a minor reorganization of preexisting mucoid secretory functions already widespread in Metazoa (Marin et al., 1996). The celebrated "Cambrian explosion," (Conway Morris, 1989), then, may have been at least in part an environmentally initiated "calcification explosion" based on preexisting response capacities of already diverse but soft bodied and therefore unfossilized organisms (Bromham et al., 1998). Arthropods are now thought to have diversified considerably during the Precambrian (Brusca, 2000), and molecular dating indicates that major phyla of metazoans were present before the Vendian period (600 million years ago) despite the lack of earlier fossils. By this interpretation, the appearance of hard parts, as important as this was, gave the appearance of sudden diversification by making the accumulating diversity suddenly visible in the fossil record (for evidence, see Fortey and Owens, 1990; Wray et al., 1996; the importance of global climate change on patterns of evolution is discussed in a separate section above).

It is worth noting that some molecular biologists give an entirely genomic explanation of the

Cambrian fossil explosion, which may have coincided with the origin of spliceosomal introns in the Metazoa. Spliceosomes, as discussed in chapter 19, are thought to have accelerated the rate of protein evolution via exon shuffling, and this may have greatly accelerated diversification of molecules involved in sophisticated systems of behavior and morphogenesis, including body-plan determination (Patthy, 1996).

Geographic variation, like variation through geological time, likewise shows the impact of environmental influence. Red nuptial coloration used in courtship displays of sticklebacks, for example, varies geographically, and this variation has been explained as adaptive, due to geographic variation in conditions of visibility and predation. This may be a case of an environmentally induced geographic pattern that happens to be adaptive, since it correlates with the availability of red-producing dietary carotenoids in lakes (Reimchen, 1989). The coloration of green jays is blue during their breeding season in certain areas of Peru, even though at the time of the previous molt the same individuals are green. In other populations of the same species, individuals are green throughout the year. Johnson and Jones (1993) discovered that the color change in the blue population is due to loss of yellow pigment accompanying autoxidation and bleaching from exposure to sunlight in their seasonally dry, hot climate. The consistently blue populations inhabit humid regions.

Botanists have long used common garden experiments to examine the environmental contribution to geographic variation in plant morphology (e.g., Clausen et al., 1940), and these techniques have been profitably applied by students of other organisms. Geographic variation in terrestrial animal morphology is more often assumed to be due to evolutionary–genetic change. It was therefore a revelation when James and colleagues (James, 1983, 1991; James et al., 1984), using controlled transplants of birds' eggs among climatically different geographic regions revealed that geographic clines in wing and bill shape, color, and size of red-winged blackbirds (*Agelaius phoeneceus*) are strongly influenced by rearing conditions. This does not mean that the clines are nonadaptive; as James points out, environmentally mediated variation can be adaptive. If so, it would not be selected against and may be genetically accommodated or even genetically assimilated. But when "the same pattern of geographic variation is found to occur repeatedly in many species that differ in their ecology, history, and dispersal distance, these causes of ge-netic differentiation become less likely as important causal agents" (James, 1991, p. 695).

Such a broad pattern of variation occurs in various morphological traits of fish (Tåning, 1952). Northern populations of fish, for example, generally have higher numbers of vertebrae than do more southern populations, and experimental reversal of the temperature regimes for rearing eggs from northern and southern Europe shifted the distributions of vertebra number, producing phenocopies of the naturally occurring geographic differences (reviewed in Tåning, 1952).

Nicoletto (1996) experimentally examined environmental influence on geographic variation in male guppies (*Poecilia reticulata*). Males from high-velocity headwater streams in Trinidad display more intensely during courtship than do males from low-velocity rivers in the lowlands. These differences are not inherited by the males' offspring reared in the laboratory. When males are experimentally reared in fast-moving water they develop greater muscle mass, resulting in wider caudal peduncles. They perform longer courtship displays, spend more time displaying, and are more attractive to females than males reared in slow-moving water (Nicoletto, 1996). Thus, a morphological and behavioral difference between populations and affecting mate choice is induced by the same environmental conditions known to characterize their natural surroundings.

Degree of melanism follows the same pattern in many species of birds and mammals, with dark coloration most common in humid regions, a generalization known as "Gloger's rule" (see Mayr, 1963; James, 1991). The same is true of at least some insects: *Polistes* wasps follow this rule, with the predominance of yellow carotenoids over melanin being most marked in hot, dry places and melanization most intense in northern climates in both Europe and America (Enteman, 1904). Two species of tropical *Polistes* wasps found at unusually high altitudes, where temperatures are cool, are unusually black in color (see Richards, 1978, on *Polistes aterrimus* and *P. huacapistana*). Enteman (1904) gave preliminary experimental data showing that *Polistes* pupae kept at low temperatures and high humidity took longer to develop and produced darker adults than did pupae from the same colonies kept at higher temperatures and lower humidities. Enteman argued that many species- and population-specific color patterns in *Polistes* are importantly influenced by these environmental parameters. Her well-reasoned and largely forgotten conclusions have never been tested with adequate experiments and sample sizes, but they nonetheless may have influenced modern evolutionary biology: Sewell Wright (1965) credited Enteman with stimulating his interest in genetics. It would be of interest to know whether his enduring interest in the developmental aspects of genetics and evolution trace to her influence as well.

Environmental Contributions to Major Evolutionary Transitions

Flexible organization helps solve many problems that have been considered difficulties for a theory of organization that does not take environmental influence into account. One of them has been called Eigen's paradox (Maynard Smith and Szathmary, 1995)—the problem of how evolution can produce a replication mechanism that is accurate enough to reflect natural selection and support adaptive progress, with, in early evolution, a genome that is still too small to produce specific enzymes that could accomplish this. Hypotheses of how accuracy could be raised center on interactions among replicators and their protection against parasitic disruption. The possibility that environmental factors and developmental phenomena such as duplication and aggregation (e.g., see Fox, 1980) could contribute to the solution is not discussed. No doubt, this is because environmental influence is usually considered to *disrupt* rather than ensure the resemblance between successive generations. Missing is the realization that there would have been an extraordinarily high selective premium on small molecules able to associate with each other and with other ubiquitous environmental elements of dependable structure, thus making for themselves a dependable, larger phenotype.

The fact that such ubiquitous elements as compounds of oxygen, nitrogen, and carbon are universally essential to phenotype construction suggest that these and perhaps other locally abundant molecules would have played a role as dependable building blocks—preexisting precise add-ons that, allied with replicable if small transmitted particles in combination, could surpass the threshold of complexity-plus-accuracy without requiring that the replicated molecules *themselves* accomplish everything. In short, the solution may have been phenotypic, not strictly genetic. At least this is a line of reasoning that should be pursued, not just a replicator-based one that ignores the helpful role possibly played by environmental elements in making the earliest, most elemental responsive phenotypes—primitive enzymes. Replicable information

can be supplemented by environmentally supplied information and building blocks in the evolution of complexity.

Another suggestion from a developmental point of view is that inaccuracy and instability can sometimes be compensated by *flexibility*. The accuracy demands are effectively lower for a flexible or versatile molecule. It is probably no accident that the functional attributes of proteins depend on their three-dimensional structure, and this structure comes from flexible folding properties of the molecules that are robust to environmental and sequence mutations (Nymeyer et al., 1998). It is important to remember that selection acts on the *products* of the replicators—the phenotypes or complexes they are able to form in collaboration with environmental elements, or perhaps in collaboration with each other (as in Eigen's "hypercycles")—not on the replicators alone.

A major contribution of modern molecular and cell biology to evolutionary biology is to show how organisms work *without* being rigidly programmed or genetically specified in form, as illustrated by various mechanisms of somatic selection (see chapter 3). The ability of these mechanisms to achieve complex, adaptively refined structures without genetic specification of details diminishes the image of mutation as a reprogrammer of development and elevates that of plasticity. At the same time, it clarifies how environmental induction can be so common in development and, due to the possibility of genetic accommodation, so important in evolution.

Conclusions

A goal of the ongoing evolutionary synthesis, where development is merged with population genetics in a coherent evolutionary theory, should be the uncompromising inclusion of environment alongside the genome in all aspects of evolutionary thought. Given the feasibility of the environmentally mediated origin and recurrence of traits, the traditional definition of "evolution" as a change in phenotype frequencies *necessarily* involving gene-frequency change is left hanging by a tenuous thread. I leave that thread intact, resisting the temptation toward revolutionary flamboyance that would change the genetical definition of evolution. A change in definition is not really necessary to elevate environmental induction to the evolutionary place it deserves: genetic accommodation easily follows any environmental induction that is favored or disfavored by selection (see chapter 6). Genetical evolution is expected to be the virtually inevitable companion of recurrent environmentally induced change. So "pan-environmentalism" (Matsuda, 1987) and the neo-Lamarkian ring of claiming an evolutionary role for environmentally induced traits lead readily back to pan-geneticism and neo-Mendelism, an eternal round-robin between extremes.

Extreme genetic and environmental determinism would be quickly cured if biologists were compelled to live according to their beliefs. A genetic determinist would have to refuse food, water, and oxygen, so that school of thought would disappear in the space of a few minutes. Environmental determinism would be eliminated just as quickly if its proponents were to demonstrate the strength of their convictions by chemically blocking the action of just one gene product, such as hemoglobin. The revolution, if there is one, in a development-conscious evolutionary biology is to actually admit what geneticists and evolutionists have claimed all along without seeing: the phenotype is a creature of both genotype and environment. It is the mediator of all genetic and environmental influence, in both development and, as the object of selection, evolution.

The neglect of the environment as an originator of novelty has left gaping holes in the comparative and experimental study of phenotype development and evolution. The most essential data relating to the mechanisms beneath evolutionary transitions are absent for most groups of organisms. There is a special need for studies that explore the effect of the environment on phenotypes known to be involved in evolutionary transitions: do conditions associated with the facultative development or expression of species-distinctive traits provoke their expression in related species usually lacking them (and lacking the same environmental conditions) in nature? Matsuda's (1987) monumental survey of correlations between environmental stress and developmental novelty is a volume bursting with leads for exciting, unfinished research, containing the groundwork for a long overdue assault on the question of environmental induction in evolution.

27

Speciation

Most models of speciation explain phenotypic divergence between populations as a result of reproductive isolation between them. But developmental plasticity allows speciation-related divergence to precede isolation and then contribute to the evolution of reproductive isolation or speciation. By the plasticity hypothesis, divergence, in the form of alternative phenotypes, life-stage differences, and contrasting traits such as those expressed under extreme or novel conditions, arises first; then particular variants are fixed in particular subpopulations due to assortative mating, environmentally mediated change in expression, or selection. Developmental differences that originate within populations can promote breeding isolation due to their often antagonistic nature, which can foster genetic divergence (release), assortative mating, and developmental incompatibility between populations. Signs of this mode of speciation are intraspecific variants that parallel species differences; and recurrent branching of phenotypically similar species from a single ancestral stem (replicate speciation). Extreme plasticity such as learning can produce exceedingly rapid (abrupt) speciation. Examples include speciation in allopatry, in sympatry, and along clines.

Introduction

In sexually reproducing organisms, *speciation* is lineage branching—the origin of reproductive isolation between sister populations descended from a single interbreeding parent population. Obviously, speciation is a process of fundamental importance in evolution. In sexually reproducing organisms, every persistent branching point of a phylogenetic tree, whether between very similar species or higher taxa, reflects a speciation event. Because complete reproductive isolation means the end of gene flow between populations, there is no doubt that it can facilitate genetic and phenotypic divergence. So speciation is a major cause of the diversification of living things.

In nonsexual or uniparental populations, isolation between divergent populations may also be called speciation, but reduced gene flow can play no role. Such populations may become genetically distinctive and divergent due to differences in mutation, selection, and drift and thereby qualify as species under some definitions (see M. B. Williams, 1992, for a discussion of the species concept in asexual organisms). This chapter deals only with speciation in sexually reproducing organisms.

By the usual view of speciation, some barrier to interbreeding comes first, followed or accompanied by genetic and phenotypic divergence (e.g., see Mayr, 1963; Rice, 1987; figure 27.1). Reproductive isolation leads to divergence. Here I argue that the reverse may sometimes occur—that divergence, mediated by developmental plasticity and selection, may sometimes originate first and contribute to the evolution of reproductive isolation. As discussed in part III, evolution by disruptive and frequency-dependent selection can produce a developmental switch between alternative phenotypes rather than loss of intermediate genotypes. This is particularly well documented in insects, often leading to misidentification of intraspecific morphs as species. Since polymorphic insects may have host-associated morphs (figure 27.2), host shifts accompanied by distinctive morphology cannot be assumed to represent sympatric speciation or host-race formation, and sympatric speciation hypotheses need to decisively eliminate the possibility of a role for sympatric divergence in the form of polymorphism or

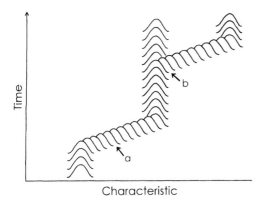

Fig. 27.1. Evolution and speciation in a unimodally adapted population. A population can diverge from its ancestral state without speciation (anagenesis, arrow a), but under the unimodal-adaptation concept, diversification requires that lineage splitting, or speciation (arrow b), occur. After Futuyma (1982).

give a balanced discussion of speciation, but instead emphasize evidence that supports the developmental-plasticity hypothesis, in order to establish it as one possible route to speciation among developmentally plastic organisms.

The idea that discrete variants within species can be the basis for discrete differences between species is not new. It motivated Bateson's (1894) early critique of the Darwinian theory of gradual origin of species: "the evidence of Discontinuous Variation suggests that organisms may vary abruptly . . . Is it not then possible that the Discontinuity of Species may be a consequence and expression of the Discontinuity of Variation" (p. 68)? Bateson's panoply of monstrosities showed that development can produce discrete variation, but it did not show any link between particular variants and species differences, much less explain how reproductive isolation could originate between a population containing such variants and one without them. How would one of Bateson's developmental monsters find a similarly monstrous mate? Developmental jumps of the sort documented by Bateson seemed to be evolutionary dead ends.

Subsequent to Bateson, several theories of speciation argued that complex polymorphisms and polyphenisms could lead to speciation, suggesting that a phenotypic novelty could become established within a population before being associated with phenotypic differences between populations (e.g., Goldschmidt, 1940 [1982]; Ford, 1940; Oster and Alberch, 1982; West-Eberhard, 1986, 1989). Others suggested a role for alternative phenotypes in particular taxa where alternative phenotypes parallel differences between species (reviewed in West-Eberhard, 1989). Dorst (1972) traces this idea all

behavioral and physiological plasticity. Such intraspecific host shifts may contribute to speciation, whether sympatric or allopatric, as discussed further below.

Striking phenotypic and niche divergence also can occur between different life-stage phenotypes. The *developmental plasticity hypothesis of speciation* outlined in this chapter argues that intraspecific divergence in the form of alternative phenotypes and life-stage phenotypes can contribute to the evolution of reproductive isolation (see also West-Eberhard, 1986, 1989). I do not pretend to

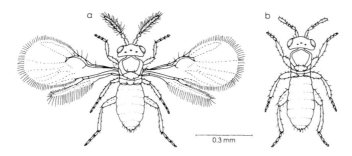

Fig. 27.2. Environmental (host) induction of complex alternative phenotypes in *Trichogramma semblidis* (Hymenoptera, Chalcidoidea): (a) small-winged male, the predominant form in some natural populations, is produced by oviposition in eggs of moths and (rarely) of the alder fly *Sialis lutaria*; (b) large wingless male produced by oviposition in eggs of *S. lutaria*. Phenotype is host-species, not host-size, dependent. Winged and wingless males differ in color, tarsal elongation, structure of the thoracic postphragma, and number of setae on the antennae. There are no intermediates. After Salt (1937).

the way back to the turn of the nineteenth century and the writings of DeVries.

Numerous theories have proposed that *genetic* polymorphisms evolved under disruptive, competitive, or frequency-dependent selection may give rise, with some means of assortative mating, to speciation without extrinsic barriers to interbreeding (Mather, 1955; Thoday and Boam, 1959; Clarke, 1962, 1966; Maynard Smith, 1966; Rosenzweig, 1978; Gibbons, 1979; Caisse and Antonovics, 1978; Seger, 1985; see synthetic summary in Rice, 1987). Such models have usually been considered of limited applicability, for there is little evidence that niche divergence begins as a genetic polymorphism linked to breeding isolation (Mayr, 1963; Futuyma and Mayer, 1980).

Here I show how some of the insights of these genetic models can be linked to the more common phenomena of conditional alternative phenotypes. Intraspecific alternative adaptations prime populations for speciation because they represent fitness trade-offs under natural selection. The same fitness antagonisms that produce such alternatives within species can lead to speciation between descendent populations in contrasting local environments (West-Eberhard, 1986, 1989).

Developmental Plasticity and Speciation: Theory

Speciation theories treat several major kinds of speciation: *allopatric speciation*, or speciation between populations geographically isolated by some extrinsic barrier; *sympatric speciation*, or speciation without extrinsic barriers to interbreeding; and *clinal speciation*, or speciation in which gradual directional change along a geographical cline results in reproductive incompatibility between the extremes at the ends of the cline. Previous theories of sympatric speciation propose two major mechanisms of reproductive isolation: *allochrony*, or temporal separation between populations (Romanes, 1886, on flowering-time divergence, cited in Forsdyke, 1999, p. 50; Bigelow, 1958; Alexander and Bigelow, 1960), and positive genotypic *assortative mating* (Mather, 1955; Thoday and Boam, 1959; Maynard Smith, 1966; Rosenzweig, 1978), for example, as a result of disruptive selection, habitat differences in mating site, or mate preference.

The *developmental-plasticity hypothesis of speciation* proposes that developmental plasticity in trait expression within a parent population can predispose descendent sister populations to speciation

by facilitating the intraspecific evolution of contrasting specializations. Then individuals expressing these specializations begin to show breeding separation, due to either increasing allopatry or allochrony in breeding sites or time associated with their different specializations, or assortative mating connected with or reinforcing contrasting specializations. This creates two breeding populations, each with one of the contrasting alternatives fixed. Phenotype fixation and character release promote further divergence (phenotypic and genetic; West-Eberhard, 1986, 1989; Wimberger, 1994, proposes a similar idea).

The developmental-plasticity hypothesis assumes, as do most theories of speciation, that genetic and phenotypic divergence between populations contributes to the likelihood of reproductive isolation between them. This assumption is rarely tested (but see Rundle et al., 2000). The notion of the cohesiveness of the gene pool and its interruption by extrinsic barriers to interbreeding, permitting divergence and therefore speciation (e.g., see Mayr, 1963; Carson, 1985; Eldredge and Gould, 1972; Templeton, 1989), is a way of expressing the assumption that divergence produces speciation in quasi-mechanistic terms, without in fact going beyond the assumption itself, simply stating in a different way the idea that disrupted gene flow (a break in cohesiveness) facilitates divergent evolution between populations, which in turn contributes to their speciation (Crozier, 1985, gives a concise summary of the cohesiveness concept).

There can be no absolute rule regarding how much genetic or phenotypic divergence is required to cause reproductive isolation, which sometimes can be identified with particular divergent traits representing different degrees of genetic and phenotypic divergence (e.g., see Coyne and Orr, 1999, their table 1.1). On the one hand, nongenetic factors, such as cytoplasmic symbionts in *Drosophila* (Crozier, 1985) and infections with the bacteria *Wolbachia* (Werren, 1997), can cause phenotypic divergence, hybrid dysgenesis, and reproductive isolation with no genetic divergence at all (see Crozier, 1985); on the other hand, striking heritable phenotypic divergence can occur without breeding isolation, as shown by mixed-breed crosses in domestic dogs. These facts show that phenotypic divergence by itself, if persistent, can be a sufficient cause of speciation, but it need not be. Hybrid dysgenesis, courtship barriers, and strict assortative mating in sexually reproducing organisms are undisputed indications of speciation-related divergence. It seems reasonable to conclude that, in general, though not always, increased phenotypic divergence means increased likelihood of speciation. Beginning with that assumption, to the degree that developmental plasticity predisposes populations for divergence that would disrupt interbreeding, it is a potential cause of speciation.

The developmental-plasticity hypothesis goes beyond this simple divergence assumption to argue that not all kinds of phenotypic divergence are equal as potential causes of speciation. The second claim of the hypothesis is that intraspecific alternatives and contrasting life-stage phenotypes are especially likely to promote speciation because they poise populations for *disruptive divergence*, that is, divergence between sister populations that would make interbreeding between them especially disadvatageous upon recontact, and would also make assortative mating by phenotype advantageous, adding sexual selection by mate choice to the factors that accelerate speciation. First, intraspecific alternative phenotypes are often antagonistic in nature, as discussed in chapter 21. As products of strong intraspecific competition or phenotypic differences within species, they place a premium on divergence, or oppositeness, per se. Such divergence is not random with respect to its end points; it can be antagonistic. In contrast to this, divergence between isolated populations is random with respect to its end points: it is a product of mutation, selection, or drift in isolation, not interactive in the sense of representing fitness trade-offs with regard to traits in sister populations. Therefore, phenotypic divergence between isolated populations not characterized by fixation of alternative phenotypes may or may not affect their ability to interbreed. When phenotype fixation and character release occur, however, developmental incompatibility is likely to ensue, for release promotes genetic change in precisely the loci that were previously held in check by genetic correlations between traits (see chapter 7).

How the phenomenon of release can contribute to genetic and phenotypic divergence between populations is modeled by Clarke (1966), who shows how morph fixation in a cline can accelerate genetic divergence along the cline. This model was applied to speciation by Endler (1977). The Clarke–Endler model of clinal change in morph ratios can be cast in the terms being used in this book to describe the effects of phenotype fixation on speciation: one morph approaches fixation, with the accompanying release of nonspecific modifiers to become specific modifiers (Endler calls them type II modifiers) of the increasingly common morph. Here I apply this idea to complex and conditionally expressed alternative phenotypes, and to allopatric

and sympatric speciation as well as to speciation in clines. Extrapolation from the Clarke–Endler model to polyphenisms and life-stage forms is possible because the release phenomenon is independent of the mode of regulation of morph ratios: character release depends on morph frequency change among the forms produced by a single genome, whether governed by a single locus of major effect on regulation, polygenic effects on regulation (e.g., a change in threshold), environmental change that affects frequency of trait expression, or heterochronic change that amounts to deletion of one life stage and fixation of another. Bivoltine species—those with two seasonally separate broods—may become univoltine in an invariable or short breeding season, with the result that a single form is fixed, leading to its divergence and speciation from the ancestral population (Matsuda, 1987; Alexander and Bigelow, 1960).

These and other examples discussed below illustrate the potential importance of environmental change in driving morph-ratio change, which implies a change in *regulatory* inputs that need not be genetically mediated (genetically accommodated or genetically assimilated) for character release to occur, accelerating the evolution in the *form* of the trait that increases in frequency of expression. Release with the fixation of a seasonal alternative (Masaki, 1986) is described further below. This model is likely to be widely applicable in nature, since, unlike previous models of polymorphism and speciation, it applies whether expression of the divergent alternatives is primarily genetically or environmentally controlled. In a later section, I show how assortative mating may be favored under this hypothesis when both sexes have alternative phenotypes whose determination is influenced by the same variable, such as size.

In summary, phenotype fixation and character release cause divergence that is especially likely to disrupt interbreeding, because release affects a peculiar type of trait: it exaggerates the antagonism between traits whose evolutionary divergence was formerly held in check by their partial incompatibility when expressed within a population. Change due to release is change that was prohibited before, when the released phenotype had to be genetically, developmentally, and functionally compatible with others produced by the same genome. Phenotype fixation releases the population from genetic correlations that existed due to developmental alternation between forms. This allows the accumulation of genes and phenotypic traits that promote one specialization, and would interfere with the one that is no longer expressed. So character release in formerly polyphenic or cyclically alternating populations is more likely to cause incompatibility than is divergence in traits with no such antagonism.

This hypothesis predicts that when contrasting morphs are fixed in sister populations, genetic divergence between them should be greater than if parallel morphs are fixed or if polymorphism is not involved, due to the role of release from genetic correlations between alternatives. It follows that such populations are more likely to speciate, other factors (strength of selection, amount of genetic variation in selected traits, time of separation, etc.) being equal, than are population pairs not characterized by contrasting alternatives.

Once *genetic* divergence has occurred between phenotypically divergent subpopulations, selection for assortative mating of like forms is expected to evolve, due to the advantage to offspring of genes compatible with the fixed specialization (Maynard Smith, 1966). Advantageous assortative mating can only apply to facultatively expressed alternatives insofar as there is genetic influence on the morph adopted that fosters a correlation between the phenotype of parent and offspring.

In addition to population divergence mediated by contrasting alternative phenotypes, developmental plasticity can contribute to speciation-related divergence in other ways. Contrasting age, size, or life-cycle specializations may also contribute to divergence between populations if local conditions favor different ones in different populations. Then heterochronic change may, for example, permit a juvenile specialization to be expressed in sexually mature adults (see below). Learning, and adaptive responses to changing conditions, can rapidly cause new traits to recur and undergo selection, producing evolution in new directions. All of these kinds of developmental plasticity may contribute to divergence and speciation.

Developmental Plasticity and Speciation: Kinds of Evidence

The hypothesis that alternative phenotypes or contrasting life-stage phenotypes have contributed to speciation-related divergence is supported by (1) geographic variation in phenotype ratios, including populations in which one alternative approaches or reaches fixation, and (2) parallel alternatives within and between species, with both alternatives in some populations and single alternatives fixed in others. Parallel species pairs (see chapter 28) are a special case of this where one or both of a pair of alternatives are repeatedly fixed.

Mayr (1963) extensively reviewed geographic variation in morph ratios of presumed "balanced" polymorphisms in color pattern and concluded that "to list the examples of geographically variable polymorphism would be to list all known cases of polymorphism," since "geographic variation has been found in every adequately studied case of polymorphism" (p. 329). The same conclusion likely applies to the more complex alternative phenotypes emphasized here, especially in view of the sensitivity of conditional polyphenisms to environmental variation. The commonness of geographic variation in morph ratios means that phenotype fixation or approaches to it, required for character release and accelerated divergence, are not rare events. So the conditions for speciation promoted by developmental plasticity and loss of plasticity are likely to obtain in many populations.

Parallel alternatives within and between species are probably also common, or more common than realized from collected specimens alone. In mites of the family Pyemotidae, for example, laboratory rearing was required to reveal that females of some species produce two offspring female phenotypes, a "normal" form and a specialized "phoretic" form (Moser and Cross, 1975) with reduced segmentation, thickened enlarged claws, and a compact body form, features commonly associated with phoresy, or transport by clinging to the body of a mobile species. The normal and phoretic forms are so distinct that they were formerly classified as separate genera (*Siteropes* and *Pygmephorellus*, respectively). A related species, *Pygmephorellus bennetti*, apparently produces only the phoretic form, since no "normal" phenotypes have ever been found in laboratory or field. Similar dimorphisms occur in numerous other genera of mites with phoretic species, suggesting that the phoretic phenotype "may become fixed, either accidentally or due to selection" (Moser and Cross, 1975, p. 821), producing a monophenic lineage. Moser and Cross point out that "this line could then presumably receive the full force of selection and be subject to an increased rate of evolution." Although they mention no evidence that such character release has occurred in mites, the patterns of variation observed suggest that polymorphism could contribute to speciation in mites.

Occurrence of speciation associated with phenotype fixation could have an alternative explanation, namely, that the phenotype fixation and divergence occurred entirely *after* speciation and contributed nothing to speciation itself. The lineage could have speciated and then later undergo phenotype fixation or reduced plasticity and directional evolution in one of its branches. How can the order of events be known? The speciation-first hypothesis is rendered less plausible and the loss-of-plasticity hypothesis more plausible if the primary distinctive characters of the new species are precisely those where release would be expected (West-Eberhard, 1996). If the alternative phenotype and character release did not contribute to speciation, then major phenotypic differences between species should bear no special relation to ancestral alternative life-stage-specific phenotypes. The social parasite species discussed below, for example, differ from their polyphenic relatives most markedly in characters relating to fixation of parasitism, such as enlarged heads and mandibles, recurved stings, thickened cuticle, loss of workers, and intensified aggressive behavior (West-Eberhard, 1996; see chapter 21, especially figure 21.6). This supports the hypothesis that fixation of parasitism has contributed to their divergence and speciation.

Some authors (e.g., Robinson and Dukas, 1999) propose enumerative tests of the hypothesis that plasticity can contribute to speciation, supposing that if plasticity is important for speciation, then clades of relatively plastic organisms should be more speciose than clades of less plastic ones. Here as in other areas, enumerative tests must be treated with caution. Such tests are fraught with problems, among them false quantification, such as judging one taxon to be more "flexible" than another; the assumption that the species of a taxon are uniformly plastic or uniformly lacking in plasticity of a particular kind; and the severe sampling error that can occur if one attempts to compare unequally well-studied species. As discussed in chapter 7, plasticity may either accelerate or retard evolution, depending on how it influences phenotypic variation.

The question of enumerative tests helps to underline a caution regarding what is not being claimed here: I do *not* argue that plasticity always increases speciation rates. Taxa showing developmental flexibility may or may not be more speciose than taxa seemingly less plastic. Phenotype fixation is only one of many factors that affect speciation rates, others being frequency and duration of isolating events, size of isolated populations, dispersal rates and distances, strength of sexual selection, and genetic variation. A taxon may be relatively speciose compared to others for reasons that have nothing to do with speciation rates, such as geologic age and rate of extinctions (Benton, 1990;

Stanley, 1990). A positive correlation between plasticity and species diversity among clades in a carefully designed test with attention to other relevant variables may legitimately demonstrate a correlation between a particular kind of plasticity on species diversity of clades, but failure to find such a correlation does not falsify the hypothesis that plasticity can play a role in speciation.

Examples

The following examples illustrate how developmental plasticity can play a role in various kinds of speciation, and how this can be recognized using comparative study. In the examples described here, some, such as divergence in buttercups, are clearly allopatric in nature, and others, such as speciation in parasitic ants and examples of abrupt speciation, are likely or necessarily sympatric. I do not attempt to decide in every example whether sympatric, allopatric, or clinal speciation has occurred. Rather, I examine the evidence that developmental plasticity may have contributed to speciation-related divergence.

Phenotype fixation and character release are an important part of the developmental-plasticity hypothesis of speciation, so interested readers should refer to the discussion of release presented in chapter 21. Possible examples of fixation-associated release occur in nature in crickets, larval echinoids, sticklebacks, spiders, buttercups, aphids, wasps, and finches, and there is an experimental demonstration in fruitflies (*Drosophila*).

Buttercups

Phenotypic fixation and phenotypic and genetic release occurs in buttercups (*Ranunculus flammula*), as described in chapter 21, but I repeat some of the findings here along with others directly related to speciation from the classic monograph by Cook and Johnson (1968). Buttercups produce alternative linear aquatic and lanceolate aerial leaf forms in a fluctuating environment, and phenotype fixation and character release, accompanied by genetic change, have occurred in populations where leaf form is fixed in a homogeneous environment. Individuals in lakes with seasonally fluctuating water levels regularly produce both leaf forms as a facultative response to conditions. But the two leaf forms are developmentally dissociable: populations from meadows have only lanceolate leaves, whereas populations under constant immersion in lakes have only linear leaves. When plants from a population with a long history of constant immersion were transplanted to the land, they developed the terrestrial leaf form but were weak and usually did not survive, indicating that some of their plasticity for switching to life on land has been lost. Plants from dimorphic populations, on the other hand, survived both constant immersion and terrestrial conditions. Furthermore, the aquatic and terrestrial populations showed some morphological specialization, toward narrower and broader leaves, respectively; hybrids produced intermediate forms.

Cook and Johnson (1968) discuss the consequences of this kind of plasticity followed by phenotype fixation for the invasion of new ecological zones, and rapid speciation in plants, especially around border areas between mesic and xeric regions in California. "There appears to be a set of habitats which, by being unpredictable, subject organisms to disruptive selection and consequently act as generators of evolutionary novelty" (p. 512). Especially relevant for speciation is the fact that hybrids produce intermediate leaf forms of presumed reduced fitness in the environment of the monophenic population (Cook and Johnson, 1968). This would presumably contribute to selection against interbreeding and hybridization in nature. This example thus seems to satisfy all of the conditions required for alternative phenotypes and their fixation to contribute to speciation. Cook and Johnson (pp. 512–513) discuss eco-geographic conditions in which variation such as that described in *R. flammula* has been associated with speciation and extensive diversification in plants and other organisms.

Butterflies

Seasonal polyphenisms are good candidates for speciation by fixation of contrasting morphs in geographically and temporally isolated populations because their seasonal separation could lead to reproductive separation if there were a time gap between broods, and their often climate-related adaptations make them subject to geographic and altitudinal variation in morph ratios as well (e.g., see Shapiro, 1976, 1978a). Accordingly, geographic variation in morph ratios, including morph fixation, has been documented in various seasonally polyphenic butterflies (Shapiro, 1976). A population of *Pieris occidentalis* in the North American Rocky Mountains has both a light-winged warm-weather form and a dark-winged cold-weather form under photoperiodic control. Color polyphenisms of this type have been shown in other species to function in temperature regulation (e.g.,

Watt, 1968; Kingsolver, 1985, 1987; Kingsolver and Wiernasz, 1991). At high altitudes in the mountains and in an Alaskan population of the same species, only the cold-weather alternative is expressed. However, a warm-weather light-winged phenotype identical to that of a polyphenic population can be induced by high temperature and continuous light conditions in the laboratory.

Shapiro (1976) gives reasons to believe that polyphenism is ancestral and concludes that monophenic populations have been derived from them via environmental fixation of a single phenotype. Similarly, *Pieris virginiensis*, a monophenic univoltine butterfly, can be induced by continuous light and high temperature to breed without diapause like a multivoltine species, causing it to produce an alternative estival phenotype indistinguishable from that of its multivoltine relative *P. napi oleracea* (Shapiro, 1971). If, as seems likely, *P. virginiensis* achieved species status in the geographic region to which it is now adapted, phenotype fixation could have contributed to its divergence and speciation.

Two kinds of evidence would further support the plasticity hypothesis in these butterflies: (1) evidence of increased behavioral, morphological, or physiological specialization of the fixed phenotype compared with the homologous phenotype of related polyphenic populations; and (2) evidence that such divergence decreases the occurrence or quality of hybrids. Some support of the first kind is given by the finding of relatively dark wings in a species (*Reliquia santamarta*) that is obligatorily monophenic for the cold-weather form (compare with species that maintain a latent polyphenism inducible in the laboratory; Shapiro, 1976, figure 1).

Some African butterflies (Pieridae) have spectacularly complex seasonal polyphenisms in color and behavior (McLeod, 1984). In markedly seasonal areas of South Africa, *Precis octavia* has two seasonal color forms, a brilliant red and black wet-season ("natalensis") form found on sunny hilltops and a cryptic blue and black dry-season ("sesamus") form found in shady ravines and caverns. All individuals reared at high temperature (30°C) develop the natalensis phenotype, and all reared at low temperature (16°C) develop the sesamus phenotype, with no intermediates. In the Entebbe region of Uganda, where mean temperature shows hardly any seasonal variation, the natalensis form is fixed. Other, similar examples are mentioned by McLeod, who assumes that polyphenism is ancestral and suggests that geographic variation and phenotype fixation may eventually be followed by speciation of the monophenic isolate.

Like the American pierid butterflies, these populations would be good places to look for evidence of character release in fixed phenotypes derived from polyphenisms.

The genetic and phenotypic consequences of phenotypic fixation have been more thoroughly investigated in butterflies showing Batesian-mimicry polymorphisms. In most Batesian mimics, (e.g., of *Papilio*, *Hypolimnas*, and *Limenitis*), mimicry is restricted to the female. Females not themselves distasteful commonly mimic, in coloration and behavior, one of up to three sympatric models (other distasteful butterfly species) and are thereby protected from predators that learn to avoid distasteful prey. Morph ratios show great geographic variation, thought to be determined by numerous factors including variations in the strength of predation, the number and abundance of particular models, the relative perfection of protection afforded by different alternative phenotypes, and the presence or absence of other palatable species having similar morphs (Turner, 1977; Charlesworth and Charlesworth, 1975, 1976a,b).

Genetic analysis shows increased genetic divergence of fixed phenotypes when compared with the homologous morph of closely related polyphenic populations, and these differences are known to be due to evolutionary change in the nonspecific modifiers of form—the genes predicted to undergo release following morph fixation (see chapter 7). In *Papilio dardanus*, where a nonmimetic, tailed form is common (80% of the population), an allele producing the "tailless" wing form that improves mimicry is absent. Where a mimetic morph is common, the tailless allele predominates and in some populations spreads to fixation, being no longer selected against as a deleterious nonspecific modifier of the tailed alternative form (Turner, 1977).

Also in accord with the release hypothesis, the commoner a particular form, the larger is the accumulation of modifying genes, and the more perfect the mimicry (Ford, 1971; Turner, 1977): the form of *Papilio dardanus* that mimics *Amauris acheria* is more highly elaborated in southern Africa, where it characterizes 40–85% of the female population, than it is in central Africa, where its frequency is only 7%. The disadvantage of outcrossing for a highly modified (e.g., fixed) form with a less modified or unmodified one was demonstrated by the outcrossing experiment already mentioned: when an African race of *P. dardanus* having three well-developed mimic forms was crossed with a nonmimetic (*M. lagasy*) population, the distinctiveness of the mimic patterns was lost—an event likely to be disadvantageous if it were to oc-

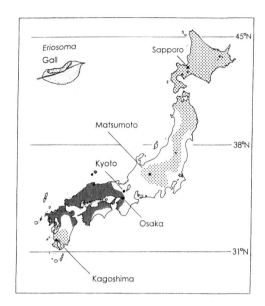

Fig. 27.3. Geographic distributions of host plants of *Eriosoma* aphids. Japanese elm (*Ulmus davidiana* var. *japonica*; stippled areas) is the host plant of various *Eriosoma* species parasitized by *E. parasitica*. Chinese elm (*U. parvifolia*; hatched areas) is the host species for the facultatively parasitic gall maker *E. yangi yangi*. In the small isolated southern population of Japanese elm, *E. yangi yangi*, which is unable to make galls on this species, shows fixation of the parasitic alternative. After Akimoto (1988b).

cur in nature, and therefore likely to contribute to speciation between the phenotypically diverged populations.

Aphids

A possible case of allopatric speciation facilitated by phenotype fixation is illustrated by some Japanese gall-making aphids that live on elms (Akimoto 1981, 1988a,b; figure 27.3). The Japanese elm (stippled areas) and the Chinese elm (hatched areas) are hosts of several species of *Eriosoma* aphids that induce leaf-roll galls (inset) on trees. The gall-making species *E. yangi yangi* makes galls only on Chinese elm and cannot induce galls on other trees. Females of this species have a facultative alternative behavior. They parasitize galls made by other females of their own and other species, which they occupy opportunistically and sometimes fight to the death with previous occupants. Offspring of the facultatively parasitic species, however, develop better in galls of their own species.

The parasitic alternative phenotype is fixed in *E. yangi parasiticum*, a species very closely related to *E. yangi yangi* that probably descended from it when long-distance migrations by autumn winged forms (alates) colonized areas where (1) the original host plant, Chinese elm, is absent but (2) other congeneric species make galls. In these areas the parasite species is now common. *Parasiticum* lives on Japanese elm, a host where the species cannot make galls of its own, and which is allopatric to the original host, Chinese elm. Transplant experiments show that *parasiticum* is genetically distinct and divergent with regard to traits associated with the parasitic habit. It has long legs believed to be a parasitic specialization for searching and expelling occupants from invaded galls; in contrast to facultative parasites, individuals of the parasitic species search widely for galls rather than in a restricted area (that of its own galls and host plant), and when experimentally placed on the host plant where facultative parasites can make galls, obligate parasites fail to do so. All three of these traits represent increased specialization of the fixed alternative (parasitic behavior), and since they are the primary differences noted between the parasitic and the facultatively parasitic species from which the parasites are descended, these observations fit the hypothesis that phenotype fixation accompanied by character release has contributed to speciation-related divergence (Akimoto, 1988b). Character release associated with phenotype fixation has been documented in another aphid by Moran (1991; see chapter 21).

Nonmigratory Descendants of Migrants

Staying permanently at the nonbreeding end of a seasonal migratory route is a kind of loss of plasticity that leads immediately to habitat restriction and geographic isolation. It therefore could be especially conducive to divergence and speciation. Reproductive isolation of the nonmigrant form would be assured by the simple failure to migrate in any species where the migrants do not reproduce at one end of their range.

Some groups of migratory birds have divergent populations known to be descended from migrants that have stayed in one extreme of the migratory range (below). Migrant versus nonmigrant behavior is known to be highly heritable, and a line of totally nonmigratory individuals can be fixed within a few generations in at least some birds (see

references in Bell, 1991, p. 251). Grant and Grant (1995a) observed a microgeographic example of deletion of migration in a Galapagos finch (*Geospiza magnirostris*) that annually migrated to a small island, Daphne Major, without breeding there until an El Niño year (1982–1983) of dramatically high rainfall. A small subpopulation stayed and has continued as a breeding colony perpetuated by its descendants despite signs of slight inbreeding depression. Phenotypic flexibility played a key role, since the shift in timing of breeding relative to migration that led to colonization was apparently induced by an environmental event, high rainfall. This induced the precocious onset of song and nest-building activity in congeners resident on the island, which may have stimulated the birds to breed before being in a physiological state to migrate (Grant and Grant, 1995a).

In addition to the environmentally induced allochrony and breeding isolation from the stem population, there were learned changes that could promote speciation: small initial differences in a culturally inherited trait, song type, were rapidly magnified due to differential reproductive success of particular singing males apparently associated with factors unrelated to song itself. In addition to "cultural drift" in song, the founder population had some distinctive morphological traits, such as relatively large bill size compared with the population of annual migrants as a whole. It is not clear whether the differences in bill size represent drift (Grant and Grant, 1995a) or phenotypic and possibly genetic correlates of the decision to stay, amounting to an adaptively coordinated set of traits (or nascent alternative phenotype) that set the colonists apart from those that did not move.

The yellow warbler (*Dendroica petechia*) has both a migrant population which breeds in the temperate zone (the "yellow" warbler), and a nonmigrant population (the "mangrove" warbler) that breeds in the tropics (Wiedenfeld, 1988). When in the tropics the migrant yellow warblers forage in mangrove habitats that are intermediate between those of their breeding grounds and those of the resident tropical mangrove warblers that spend all year in the mangroves, suggesting that the nonmigratory population may be more specialized to the mangrove habitat. The tropical resident mangrove warblers are larger, forage more slowly, and have more rounded wings. Experiments with caged birds also revel behavioral divergence in the foraging of the two populations (Wiedenfeld, 1988).

Fixation of a freshwater specialist is common in fishes with *diadromous* ancestors—species that migrate between fresh water and the sea. It occurs in

13 different families, sometimes multiple independent times in each (McDowall, 1988, p. 174; McPhail, 1994). Fixation of marine spawning (loss of migration to fresh water) is more unusual but has been recorded in the migratory "freshwater" European eel, *Anguilla anguilla*, which spawns in the sea (Tsukamoto et al., 1998).

In lampreys, there is little doubt that intraspecific plasticity in anadromous (migratory) parasitic lampreys (Petromyzoniformes) was important for the preisolation evolution of an alternative specialization—nonparasitic life in fresh water spawning grounds—that has come to characterize divergent freshwater lineages with small adult body size. Ten different parasitic species in four genera (*Ichthyomyzon*, *Tetrapleurodon*, *Mordacia*, and *Lampetra*) have at least one "dwarf" or "paedomorphic" nonparasitic sister species, and two species have given rise to more than one (Hardisty and Potter, 1971; Potter, 1980). In each species pair, the ancestral form is a parasitic, anadromous population that feeds as an adult by attaching to other fish, then migrates to fresh water to spawn once and die. The descendant form is a smaller, nonparasitic exclusively freshwater population that has a prolonged larval stage, then matures precociously following a delayed metamorphosis to the adult stage and does not feed as an adult, having deleted migration to the sea where the ancestral species parasitically feed. Many parasitic populations (e.g., of *Lampetra fluviatilis*, *L. japonica*, and *L. tridentata*) contain a "praecox" form, which resembles the dwarf species in being relatively short-lived and small—intermediate in size between the parasitic and nonparasitic members of a species pair. In upstream migrations of the river lamprey, *L. fluvialis*, the praecox form is sometimes common (1:3.3 normals; Abou-Seedo and Potter, 1979).

Parasitic anadromous lampreys migrate to the sea to feed, sometimes for several years before returning to freshwater spawning grounds to reproduce once and die. Nonparasitic lampreys reproduce and die quickly without migration and feeding. Given these life-cycle limitations, Hardisty and Potter (1971) state that any modification of the length of the adult stage must inevitably be discontinuous, involving changes in intervals of whole years. They hypothesize that divergence and speciation between the contrasting members of lamprey species pairs involve a two-stage heterochrony: (1) the production of relatively small praecox forms, whose maturity is accelerated by one year, and (2) the production, in the praecox population, of a still smaller form whose maturity is moved ahead two whole seasons. A large change in body size could

automatically favor assortative mating by size since "efficiency of the spawning act in the lamprey is dependent on a precise and intricate positioning of the tail of the male in relation to the cloacal aperture of the female" such that "effective fertilization and egg extrusion is only likely where the pairing lampreys are of similar body lengths" (Hardisty and Potter, 1971, p. 264). It should be possible to test this hypothesis by examination of the mating success of different size individuals.

Salmonid fish also are famous for their strenuous migrations between the sea and their freshwater spawning grounds. They are likewise noted for variation in the condition-sensitive age of migration and proportions of the populations that migrate (Thorpe, 1989). In some species a large proportion of both sexes stays permanently in fresh water, and there are "landlocked" (really, lake-locked or stream-locked) populations that do not migrate at all (McDowall, 1988). The sedentary individuals in variable populations, and all individuals in land-locked populations without contact with the sea, fail to undergo smoltification, an environmentally triggered, thyroxine-mediated metamorphosis that transforms the dark-colored bottom-dwelling "parr" into a pelagic, silvery colored "smolt" with many distinctive characteristics, including migration to the sea (reviewed in Matsuda, 1987). As emphasized by Matsuda, this is another example where an environmental change (associated with lack of migration to the sea) may produce phenotype fixation (fixation of the unmetamorphosed parr form, and associated traits described below).

The importance of environmental mediation in phenotype fixation of the freshwater phase is indicated by various observations. Several phenotypic and ecological characteristics that correlate with the facultative propensity to migrate rather than stay in fresh water have their expression deleted in land-locked populations. For example, in chinook salmon (*Oncorhynchus tshawytscha*) males that remain in streams and defend territories there are more aggressive and slower growing than are migrating males (Taylor, 1990, cited by Dingle, 1996, p. 300). Nonmigratory individuals of sockeye salmon (*O. nerka*), and nonmigratory populations of the "kokanee" form of the sockeye, are much smaller at maturity (Thorpe, 1989; Foote and Larkin, 1988). Poor trophic conditions are associated with delayed smolting (later transformation to the migratory form), a lower percentage of smolting individuals, and precocious sexual maturation (Thorpe, 1989). That is, there is an environmentally influenced, heterochronic change in the tim-

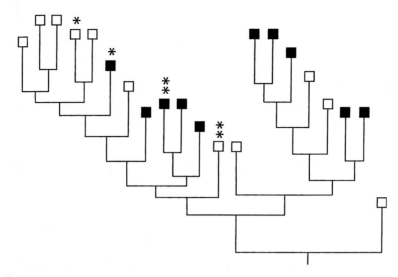

Fig. 27.4. Phylogenetic relationships of alternative forms in a Pacific salmon (*Oncorhynchus nerka*). Open squares, sockeye (normal) form; black squares, kokanee form. Asterisks indicate two pairs of sympatric populations: (*) in Babine Lake; (**) in Narrows Lake. All other populations are in separate lakes or rivers. After Taylor et al. (1996).

ing of maturation relative to acquisition of migratory traits.

Most interesting for speciation in salmonids are the nonmigratory populations that stay isolated in fresh water for generations. "Rainbow trout," for example, are steelhead trout (*O. mykiss*) populations that never smolt. The kokanee is a sockeye salmon (*O. nerka*) that has been landlocked for generations, though it sometimes crosses in sympatry with the anadromous form (Foote and Larkin, 1988). In the southwestern United States and Mexico, there are four landlocked species of Pacific salmon that are probably descendants of the variably migratory species *O. mykiss* and *O. clarki* (Thorpe, 1989). Phylogenetic study of the migrant (anadromous) sockeye salmon *O. nerka* and the nonmigratory kokanee form has shown the recurrent derivation of nonmigrant kokanee from migrant sockeye (figure 27.4). Similarly, in the Old World and Atlantic salmonids of the genera *Salvelinus* and *Salmo*, migration (anadromy) is facultative in most species (Thorpe, 1989), and the Atlantic salmon (*Salmo salar*) has some landlocked populations (Dingle, 1996). Three North American *Salvelinus* species are wholly freshwater residents (McDowall, 1988).

A role for plasticity in the origin of the kokanee salmon is suggested by evidence that the transformation between normal (sockeye) and derived kokanee forms is readily reversible: some kokanee

populations have developed after human introduction of normal sockeye salmon into lakes open to the sea that were previously devoid of *O. nerka*. Conversely, sea-run sockeye individuals have sometimes been observed in lakes only after kokanee were introduced (Taylor et al., 1996).

There are several types of divergence in salmonids and other landlocked fish that are candidate products of release following the fixation of the freshwater phenotypes of anadromous ancestors, since they exaggerate features of intraspecific variants found in anadromous populations. In *Salvelinus malma*, for example, the facultatively nonmigrating forms are smaller, mature younger, and have fewer eggs per millimeter of body length, greatly reduced fecundity, and a shorter life span than do migratory conspecifics. These same differences apply between species, but in more exaggerated degree (McDowall, 1988). The salmonids offer a chance to test the hypothesis of rapid phenotypic and genetic specialization in behavior, physiology, and morphology to the freshwater environment following phenotype fixation by comparing the facultatively freshwater forms in variable species with sibling populations landlocked for different amounts of time.

The expectation that degree of trait exaggeration should correlate with frequency of expression (see chapter 7) may be met in salmon: male fighting morphology is most exaggerated in the species (e.g., the Pacific salmon, *Onchorhynchus* species) that reproduce only once and therefore do not have to switch between fighting and feeding morphology. In Atlantic salmon, which are iteroparous (reproduce more than once during a lifetime), the more highly iteroparous a species (the more often individuals switch between fighting and feeding morphology during the life cycle), the less exaggerated is male fighting morphology (Fleming, 1996).

Speciation with morph fixation is probably common in migrants. Whenever a migratory species encounters very different environments at opposite ends of its migratory range, there is a potential for rapid divergence and speciation if a portion of the migratory population becomes sedentary and breeds isolated from others.

Fire Ants

The number of queens in an ant colony varies both within and between species (Hölldobler and Wilson, 1977; Keller, 1993). *Monogynous* colonies have only one queen, or gyne; *polygynous* colonies have more than one queen. Number of queens is condition sensitive in many species, as shown by experimental colony transplants between areas of variable colony density (Herbers, 1993) and by seasonal and cyclic change within populations (Elmes, 1987; Elmes and Keller, 1993). Monogyny and polygyny are also interspecific alternatives within many ant genera where some species are polyphenic for queen number (Elmes, 1987; Elmes and Keller, 1993). In fire ants (*Solenopsis invicta*), a species with both monogynous and polygynous colonies, genetic divergence and partial reproductive isolation have recently evolved between a polygynous and a monogynous form. As I explain below, these patterns of variation suggest that the fire ant example I am about to discuss may not be unusual and that queen-number polyphenism may be a route to speciation in other genera of ants.

In *S. invicta* and other species polyphenic for queen number, there is a negative correlation between queen number and queen size, usually measured as dry weight or some correlate of weight (Elmes and Keller, 1993; see the many references in Ross and Keller, 1995). Keller and Ross (1993b) showed by means of a cross-fostering experiment that in *S. invicta* the difference in fat reserves, and hence the greater ability of monogynous queens to independently found nests, is due to differing conditions in the two kinds of colonies during the time between female emergence from the pupal cell and mating, possibly involving differences in the con-

centration of queen pheromone within colonies (Keller and Ross, 1993b) as well as small genetic differences between the two types of queen (Keller and Ross, 1993a), and so is perpetuated as a kind of maternal-colony effect or social transmission to the daughters of polygynous and monogynous queens. Later research showed that the marked colony-type effect on the phenotype of reared queens is due in part to recognition and execution of genetically unlike queens by workers. A single locus of major effect influences the expression of a complex suite of traits in the queens and workers of the two colony forms: the monogyne form shows fixation of an allele (G-9) that in homozygotes is associated with production of fat queens and worker intolerance of multiple queens (Ross and Keller, 1998; deHeer et al., 1999). Workers execute queens of the wrong G-9 genotype, a behavior that would reinforce genetic isolation between the two forms once the marked influence of this locus had evolved.

The possible facultative origin of the polygynous–monogynous population dichotomy in introduced fire ants is suggested by Tschinkel's (1996) observations of monogynous colonies. The monogynous form is believed, from direct observations of the late advent and spread of polygynous colonies, to be pleisiomorphic in the introduced range (Ross and Keller, 1995). Monogynous colonies produce two annual queen broods, an early brood of monogynous summer queens, and a later brood of overwintering joiner queens, which lack fat reserves for independent colony foundation (Tschinkel, 1996). The overwintered queens either remain in their maternal colony where, if the queen is lost, they shed their wings and produce males without mating, or leave the maternal nest, mate, and seek worker-containing queenless colonies of their own species, which they join and exploit as intraspecific parasites of the unrelated worker force. The overwintered queens studied by Tschinkel are, like the genetically distinctive polygynous form, characterized by low nutritional status, low individual fecundity, and associative behavior (joining of colonies founded by other queens). As products of a monogynous colony, these females are presumably homozygous dominant at the G-9 locus and destined to be executed, but their joiner behavior in association with low nutritional condition suggests a facultative mode of origin for the polygynous form, subsequently constitutive in the subpopulation of G-9 heterozygotes. If the original introduction was of a monogynous South American queen, as suggested by the available data (see above), then the two types of queen produced could have led to a polyphenism for independent and joiner nest founding, subsequently perpetuated by the "cultural" and genetic effects just described.

These data on *S. invicta* indicate an association between phenotype fixation of the polygynous form, genetic divergence between monogynous and polygynous populations, and incipient speciation (reduced avenues of gene flow) in fire ants. The two colony forms have different population genetic structures. The polygynous philopatric (relatively nondispersing) form more often shows local genetic divergence and geographic heterogeneity, while the monogynous form of the same species has a genetically uniform population over a wider range (e.g., on *Solenopsis invicta*, Shoemaker and Ross, 1996; Ross et al., 1997; on *Formica truncorum* [*rufa* s.l.], Sundstrom, 1990, 1993). In the introduced fire ants, gene flow between sympatric polygenic and monogynic colonies is reduced to a single pathway (Shoemaker and Ross, 1996).

Polygynous females mate almost exclusively with males of monogynous colonies, partly due to a peculiarity of this introduced population. As a result of the genetic bottleneck that this species experienced upon introduction, introduced populations are characterized by low allelic variation, and males (e.g., those that mate with relatives within colonies, as can happen in polygynous colonies) are frequently sterile due to homozygosity at sex-determination loci (Ross et al., 1993). This one route of gene flow may be further reduced in populations like those in Argentina, where polygynous females evidently stay in the maternal nests and may mate in or near them, since colonies show high degrees of intracolony relatedness (Ross et al., 1996). Given these observations on population structure and mating behavior, local morph fixation is especially likely to be associated with reproductive isolation in fire ants. Added to the mating structure change, acceleraton of divergence is expected when flexible variation in queen number leads to phenotypic fixation and character release of the associated traits (e.g., body size, and degree of philopatry of nesting queens).

Several previous authors have suggested that plasticity followed by fixation in the expression of polygyny and monogyny has likely contributed widely to speciation in ants (reviewed by Ross and Shoemaker, 1993; see also Elmes, 1990), and queen number is geographically variable within many species of ants in addition to *Solenopsis* (e.g., Herbers, 1993; Elmes and Keller, 1993; Heinze et al., 1995). The fire ant example is particularly interesting because the divergence between multiqueen (polygynous) and single-queen populations may

have evolved in the very short time (less than 100 years) since fire ants were introduced into the United States from similar latitudes in Argentina. Even if the genetic divergence is due to a double introduction of *S. invicta* from South America (discussed by Ross and Shoemaker, 1997), phenotypic divergence of the U.S. polygynes has rapidly occurred, for the polygynous populations sampled in South America have a different population and social structure, characterized by close genetic relatedness of polygynous nestmate queens, suggesting that polygyny is a facultative colony form of primarily monogynous ants that occurs when progeny of a single queen stay to reproduce in the maternal nest. By contrast, intracolony relatedness of the U.S. polygynous form is very low, close to that of the population at large (Ross et al., 1996, 1997, 1999), which is expected if polygyny is self-perpetuating.

Speciation by Fixation of Parallel Alternative Phenotypes in the Two Sexes

Theoretical Considerations

In the developmental-plasticity model of speciation, populations respond to disruptive or diverging selection with the evolution of switch-controlled alternative phenotypes (see chapter 11), and phenotypic divergence is mediated initially by a switch between forms. Genetic differences between individuals that express divergent phenotypes need not be involved at this initial phase. Genetic divergence is then accelerated by character release upon phenotype fixation, whether fixation is in one or a pair of allopatric populations, or due to assortative mating in sympatry. The alternative phenotypes give a head start to divergence; then assortative mating has the effect of creating two subpopulations, each with one of the alternatives fixed.

In conventional models of sympatric speciation, by contrast, sympatric phenotypic divergence begins only when breeding isolation begins, for example, due to host-race formation or positive assortative mating. The composite synthetic model of sympatric speciation of Rice (1987) is a helpful synthesis of the salient features of previous genetic models of sympatric divergence and speciation (figure 27.5). As in the Rice model (figure 27.5), the plasticity model sees divergence in ecological and reproductive tactics as mutually reinforcing causes of divergence, but the divergence can be mediated

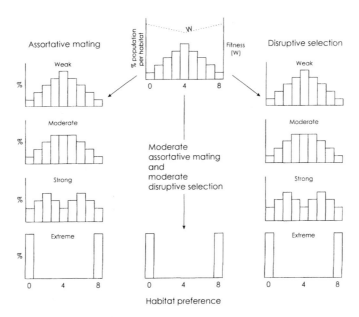

Fig. 27.5. The Rice "synthetic composite" summary of genetic theories of sympatric speciation. The horizontal axis in all graphs is habitat preference. Both habitat-based assortative mating (left) and habitat-related disruptive selection (right) can depress the fitness (W) of individuals with intermediate habitat preference (dotted line on the graph at top, center), causing genetic divergence in habitat preference. A nonoverlapping bimodal distribution of habitat preference can be achieved by extreme positive habitat-based assortative mating, by extreme disruptive selection on habitat preference, or with a combination of moderate levels of both. After Rice (1987).

by developmental switches with or without assortative mating. Plasticity-mediated divergence can then greatly accelerate speciation by tying alternative phenotypes to bidirectional sexual selection on mate choice, which may have as a side effect assortative mating and genetic divergence when there is a genetic correlation between the traits of males and females (see also Payne and Krakauer, 1997; Higashi et al., 1999).

In most polyphenic populations, assortative mating between males and females of like phenotype is not expected to evolve. There is always a genetic component to morph determination, and one may suppose that individuals would benefit from selecting a mate of the same phenotype due to the genetic benefits to their offspring. But this benefit cannot be realized if the correlation between phenotype and genotype is inconsistent, as it often is under strong environmental determination of alternatives.

Suppose, however, that there is incidental assortative mating, that is, that similar phenotypes happen to mate assortatively due to some common trait that causes the time, place, or effectiveness of their mating to coincide, for example, as part of

their morph-specific adaptation to a specific habitat, season, or social role (see Crespi, 1989b, for a more complete discussion). This could cause genetic correlations between like morphs of both sexes, by an inclusive fitness or genetic co-carrier assortment effect (Seger, 1985), thereby establishing genetic differences between morphs. Providing that the environmental (niche) requirements of the specialized morph are consistently available, this would lead in turn to selection for adaptive assortative mating of like phenotypes: once an individual is committed by genotype to a greater probability of development of one alternative, say A, then its offspring are genotypically likely to be of the same phenotype (A), and it is advantageous in terms of offspring quality to mate with A individuals that have been similarly screened for quality (survival to breeding age) by life in the same niche or alternative specialization. This process amounts to a trend toward phenotype fixation within a selectively mating sublineage of genetically similar individuals. It would be accompanied by character release and self-accelerating selection of both regulation and form (see chapter 7), in addition to the tendancy toward assortative mating based on genotype.

In summary, the steps of an alternative-phenotype hypothesis that could apply to sympatric as well as allopatric speciation are as follows:

1. Establishment of divergent, discrete, or bimodally distributed complex alternative phenotypes in both sexes
2. Incidental assortative mating by males and females of like phenotype due to parallel alternative tactics or traits in both sexes (mating time or place, size matching, habitat similarity)
3. Incidental accumulation of morph-specific genetic divergence in alleles that affect regulation and form, as an effect of assortative mating between individuals of like phenotype and genotype
4. Adaptive assortative mating due to selection (usually on females) to increase the genetic quality of offspring by choice of ecologically compatible mates that express a parallel phenotype
5. Mutual acceleration of bidirectional divergence (phenotypic and genetic) in regulation and form, further accelerated by character release and bidirectional sexual selection
6. Lineage-specific predominance or fixation of a single alternative
7. Further increased premium on assortative mating and reproductive isolation (speciation) due to increased genetic and phenotypic divergence of the fixed form (hybrid disadvantage).

Seger (1985) discusses a similar model beginning with assortative mating by phenotype, where he assumes an arbitrary visible marker of genotypic similarity between morphs. Such a marker is not necessary if the developmental switch between morphs automatically accomplishes a mating habitat or mating behavior difference that is parallel in both sexes (see below). Seger discusses the relation of phenotype assortative mating models like his and the present one to previous genetic models of speciation, including that of Maynard Smith (1966). The accelerating effects of sexual selection on divergence involving alternative phenotypes are modeled by Payne and Krakauer (1997; see also Higashi et al., 1999).

There are several reasons to think that this model may be applicable in nature. Males and females are semi-independently subject to selection and evolution, and there are many cases of sex-limited alternative phenotypes—pairs of alternatives that occur in one sex and not in the other. Alternative reproductive tactics of males (see part III) are a good example, as are social-insect workers and queens, and ovipositor polymorphisms, of female insects. Size-related alternative phenotypes are an especially promising application of the model. Sex-limited alternatives are often size dependent in both sexes. Fighting males, for example, are usually large, whereas surreptitiously mating sneaker males are usually if not always small. The switch between morphs can be a threshold effect of a continuous distribution of sizes.

But in some cases, there are indications that size differences have evolved as part of alternative specializations. Some marine invertebrates, for example, have dwarf males, very small dispersal forms that live as parasites on females (reviewed and discussed by Ghiselin, 1974), and salmonid fishes have nonmigrating precociously mature small individuals as an alternative of both sexes, as described below. In brentid beetles (*Brentus anchorago*), the advantage of large size in mate competition, combined with male choice of large females as mates (Johnson, 1982; Johnson and Hubbell, 1984), evidently contributes to a difference in the size distributions of males and females, shifting the size distribution of males toward large body size relative to that of females (figure 27.6). These observations suggest that size, in addition to being a determinant of a conditional switch mechanism, can be part of adaptive specialization as well.

How could such size divergence lead to speciation by the model just outlined? Suppose that both sexes show strong diverging selection for large and small size, sex-limited alternative phenotypes, and size-dependent mating tactics such that phenotypic size extremes consistently mate with each other. That is, there is coincidental *phenotypic positive assortative mating*, or assortative mating of males and females of like phenotype as a side effect of selection in other contexts. This might occur, for example, if small females visit a subset of host plants especially accessible to small females, and small males seek mates there as a result of losing at more favorable mating sites controlled by large males. This does not necessarily imply selection for adaptive assortative mating, but could be incidental assortment by like phenotype. Mating site would be influenced by size, a trait highly subject to environmental influence, and offspring, being variable in size, would mate at all alternative mating sites without regard to genotype.

Now suppose that there is diverging selection on size per se, such that the size extremes of both sexes are more successful than individuals of intermediate size. In this situation, selection on extremes

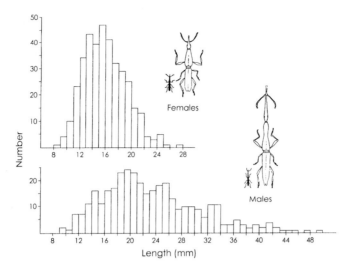

Fig. 27.6. Size divergence between males and females in a brentid beetle (*Brentus anchorago*). The advantage of large male size in mate competition, combined with male choice of large females as mates, is associated with a skewed size distribution of males relative to that of females. After Johnson and Hubbell (1984).

could begin to favor the genetically predisposed performers of the antagonist alternatives, and any genetic components of the size-dependent switch would be favored. Once there is an established genetic component to the switch and the genetically predisposed subpopulation is favored in a particular alternative, sexual selection may favor phenotype-dependent female choice of extreme-sized males, in which small females prefer extremely small males, and large females choose extremely large males, due to the advantage to offspring of improved size-dependent performance of alternatives. That is, phenotypic assortative mating could lead to *genotypic positive assortative mating*, or mating by males and females genotypically disposed to be of like phenotype. This could lead in turn to a genetic correlation between size and mating preference, suggestive of a runaway process under female choice (Fisher, 1958). Genetic divergence would be additionally accelerated by phenotype fixation and genetic release in one or both of the alternative gene pools, which would further reinforce the advantage of positive assortative mating due to the mutually reinforcing interaction between improvement in form and increase in phenotype frequency (see chapter 7). The result would be genetically divergent, separate mating pools in which offspring would no longer be equally likely to mix randomly (with repect to genotype) in the population at large. In other words, the result would be speciation, by a process that does

not require extrinsic barriers to gene flow (e.g., allopatry).

Various observations in addition to those on size-related alternative phenotypes support the general feasibility of this hypothesis. Positive assortative mating by size is common in nature. Crespi (1989a,b) called mating success of large males and size-assortative mating the two most frequently reported mating patterns in natural populations, and showed that they often coincide (for reviews, see Ridley, 1983; Crespi, 1989b; Foote and Larkin, 1988). Crespi (1989a) also discusses how size-assortative mating can be associated with a combination of large male mating advantage and other effects of size on male and female reproductive behavior, such as those that occur in ants (below), including size-associated differences in mating sites and mating times.

This sequence of events could involve parallel alternatives in the two sexes based on some common variable other than size that affects mating time or site, such as response to photoperiod in a species with seasonal polyphenism in mating time, or preference for mating and oviposition near a particular host plant. Grant and Grant (1989) proposed a similar model, inspired by observations of Darwin's Galapagos finches (Emberizinae), beginning with disruptive selection on beak size and assortative mating by beak or body size. The idea that parallel alternative phenotypes may engage in phenotypic assortative mating that leads to genotypic assortative mating along the lines just described has long been recognized to favor sympatric speciation in theory (Maynard Smith, 1966; Rice, 1984, 1985; Tauber and Tauber, 1989), but the present model connects positive assortative mating to conditional alternative phenotypes rather than requiring an initial genetic polymorphism. The hypothesis applies to sympatric, allopatric, and clinal speciation.

The following example shows how sympatric speciation could occur beginning with parallel size-related alternatives in the two sexes.

Parasitic Inquiline Ants

Queens of socially parasitic inquiline ants reproduce by laying eggs in the colonies of other species. They produce no workers of their own and do not kill the host queen, so are called parasitic inquilines, or renters. Inquilines almost certainly originated as an alternative reproductive tactic within colonies (reviewed in Wilson, 1971a; see also Buschinger, 1965, 1970, 1986, 1990b; Elmes, 1978; Pearson, 1981; West-Eberhard, 1986, 1989, 1990; Bourke and Franks, 1987, 1991; Ward, 1989; Hölldobler and Wilson, 1990).

Many observations suggest that parasitic inquilines have evolved by sympatric speciation due to parallel size-related alternatives in the two sexes, as follows.

1. Parasitic egg-laying could have originated as an alternative reproductive tactic of phenotypically small females via two routes: (a) in a single-queen species, where small females may be at a disadvantage in founding new colonies on their own (the monogynous route), or (b) in a polygynous (multiple-queen) species, where small mated females that enter nests of other species can produce males and reproductive females without investment in workers (the polygynous route).
2. Selection can favor mating in or near the nest by small females due to the energetic cost of a mating flight in individuals with relatively small energy reserves.
3. At the same time, promoted by the presence of the small females as potential mates in or near the nest, there could be selection on small males to mate in or near the nest due to their inability to compete with larger males in mating aggregations away from the nest. This would produce phenotypic positive assortative mating by size.
4. If there is a premium on small size in parasitic females, their greater success would then

lead to the increased frequency of any genetic component of the size-determined switch to parasitism, and a preference for especially small males on the part of the parasitic females. The result would be genotypic positive assortative mating by size, genetic divergence, and reproductive isolation of the genetically distinctive parasitic subpopulation (sympatric speciation), as described in the preceding section.

Each part of this hypothesis is supported by numerous facts. Parts 1 and 4 depend on an association between small size and parasitic behavior. Size-related female alternative reproductive tactics are unusually common in ants, and small size is frequently associated with dependent reproduction, including parasitism (reviewed in Rüppell and Heinze, 1999). Many observations suggest that it is advantageous for parasitic inquiline queens to be small. First, all of them *are* small (Buschinger, 1970; Wilson, 1971a,b). In ants, small size is part of a "parasite syndrome" (Wilson, 1971a, p. 374) that also includes appeasement behaviors and glandular modifications that evidently augment acceptance and reduce host attacks. Large females appear to be recognizable in the parasitic microgyne M. *hirsuta*, for their presence inhibits reproductive activity in host workers, whereas small reproductive females do not (Elmes, 1983). This suggests that host workers are less likely to detect parasites as reproductives if they are small. The females of *Teleutomyrmex schneideri*, dubbed "the ultimate social parasite" by Hölldobler and Wilson (1990), is among the tiniest of ants, being only 2.5 mm in length.

Small size may also contribute to enhanced survival of parasites during the larval stage (Pearson and Child, 1980). The importance of selection on larvae has been a neglected aspect of hypotheses for the evolution of social parasitism. Large, potential queen larvae are poorly attended and sometimes attacked by workers in queen-containing colonies of *Myrmica rubra* (see Elmes, 1976, p. 15), whereas small larvae are not molested by workers. Small size therefore evidently can be favored by social selection even though it may compromise fecundity. Selection for a reduction in the size threshold for transformation from worker to queen could contribute to the evolution of parasitism (Nonacs and Tobin, 1992).

As required by parts 2 and 3, above, all parasitic inquiline ants mate in or near the nest. This could be especially important in the initial stages of parasite evolution, and in the many inquiline species that are rare (Wilson, 1971a, p. 358). Size-dependent alternative mating tactics of males are very common in ants (Buschinger and Heinze, 1992). Size-dependent mating polyethisms and polymorphisms also occur in females (Heinze and Buschinger, 1989, p. 153; Heinze, 1989). In many if not all cases, these are dispersal polymorphisms that involve a dichotomy between mating in or near the parental nest versus dispersal to mate away from the nest. Such dispersal polymorphisms, when size dependent, could easily lead to positive assortative mating by size. In the wing-dimorphic Japanese ant *Technomyrmex albipes*, for example, large-winged males assortatively mate with large-winged females, and small wingless males mate in the nest with small wingless females (Ogata et al., 1996). The authors suspect that this represents incipient speciation, due to the size-correlated disparity in genitalic size of winged and wingless males that, like body size, exceeds a fivefold difference, which may make crossing between forms physically impossible.

Mating in or near the nest is characteristic of polygynous species and so would likely have been the ancestral state in the polygynous route to parasitism (Buschinger, 1965, 1970). Mating near the nest by small males probably begins in ants as a size-dependent alternative to mating in aggregations away from the nest. Large-male advantage has been demonstrated in mating aggregations away from the nest in *Pogonomyrmex* (Hölldobler, 1976; Davidson, 1982; Wiernasz et al., 1995). The advantage may be due to female preference for large males, as large male body size correlates with reproductive biomass of the parent colony in at least one species (Davidson, 1982). There is evidence that large-male advantage in mating aggregations favors mating in or near the nest by small males. Michener (1948) observed *Pogonomyrmex* mating near the nest, and MacKay (1981) observed males attempting to mate with sisters during emergence from the nest entrance in three species of *Pogonomyrmex*. Genetic studies showing inbreeding (Cole and Wiernasz, 1997), and large variation in male size, may indicate that it is the genetically small males that mate near the nest in *P. occidentals* (Wiernasz et al., 1995). Assortative mating by size occurs in the populations of *P. barbatus* and *P. desertorum* that were characterized by large-male advantage in aggregations (Davidson, 1982). This mating-location dichotomy is also associated with incipient parasitic behavior by *Pogonomyrmex* females, which sometimes attempt to "steal" the excavation holes of other females at high-density nesting sites (MacKay, 1981). Such parasitic attempts are common and sometimes successful

(Markl et al., 1977). In addition, there are two parasitic inquiline species of *Pogonomyrmex*, *P. anergismus* and *P. colei* (Rissing, 1983; Johnson, 1994).

Conditions that suggest an association between small alternative phenotypes and the evolution of parasitic species exist in the relatives of other social parasites. In *Myrmica* there is a recurrent miniature "microgyne" queen morph found alongside normal "macrogyne" queens as an anomaly, as a regularly occurring alternative phenotype, and as a separate species parasitic within the nests of a sibling macrogyne species (reviewed in Bourke and Franks, 1991; Elmes and Keller, 1993). In *M. rubra*, which has a small queen ("microgyne") form that may be either a morph or a separate species, all males are produced by workers (Pearson and Child, 1980), which are very small compared with queens. Toleration of small reproductives in such colonies may help pave the way for toleration of reproduction by the small intraspecific and interspecific microgyne parasites that are recurrent in *Myrmica*.

The hypothesis that parallel size-related alternatives in the two sexes contribute to speciation of a parasitic lineage gains support from recent observations on microgynes in species of *Vollenhovia* (Kinomura and Yamauchi, 1994) and *Leptothorax* (Rüppell et al., 1998). *Vollenhovia emeryi* is a Japanese ant closely related to a socially parasitic species, *V. nipponica* (Kinomura and Yamauchi, 1994). The parallels of this species with the host taxa of parasitic inquilines in such European ants as *Myrmica rubra* are so close that one could have predicted a *priori* the presence of a derived parasitic species. *V. emeryi* has size-dimorphic males (micraners and macraners), with parallel-size dimorphic females (microgynes and macrogynes). The microgyne females have aberrant, short wings and inhabit polygynous colonies containing the two types of males. Although mating has not been observed, the reduced wings of the microgyne females suggest that they mate near the nest. Associated with these parallel size-related alternative phenotypes in the two sexes, *V. emeryi* has a congeneric social parasite, *V. nipponica*, in which the sexuals are even smaller than the microgynes and small males of *V. emeryi* (Kinomura and Yamauchi, 1992), suggesting release associated with fixation of the parasitic form.

Another Japanese ant (*Hypoponera bondroiti*) with parallel size-related alternatives in the two sexes (Yamauchi et al., 1996) has no known social parasite. This difference may be explained by Wcislo's rule, that social parasitism is less common in the less seasonal tropics, due to the lesser synchrony between parasite and host reproductive cycles (Wcislo, 1987). *H. bondroiti* is in the Okinawan tropics, whereas *V. emeryi* inhabits highly seasonal habitats in the mountainous regions of central Japan (Kinomura and Yamauchi, 1994)—habitats similar to those occupied by European social parasites. Short temperate-zone nesting seasons also limit the reproductive options of females, and this may also encourage parasitism in the temperate zone. Microgynes similar with respect to frequency of occurrence and queen size distribution to those of *Myrmica ruginodis* also occur in *Leptothorax rugatulus*, most commonly in polygynous colonies, although males are monomorphic in size (Rüppell et al., 1998). Like *Myrmica* and *Vollenhovia*, the genus *Leptothorax* contains parasitic inquiline species (Buschinger, 1986; Bourke and Franks, 1991).

Given parallel size-related alternative mating tactics in both sexes, genetic variation in adult size would cause the size-assortative mating to begin to produce genetic divergence between the two size classes. No step in this hypothesis depends on a novel mutation in either sex (cf. Buschinger, 1990b). Although there is evidence that mutation may be involved in the recurrence of microgyne social parasites in *Myrmica* (e.g., see Cammaerts et al., 1987; Elmes, 1981), and single- or two-locus differences can control dispersal polymorphisms among queens (Heinze and Buschinger, 1987, 1989), it matters little for the alternative adaptation and character release hypothesis whether changes in a polygenic switch mechanism involve mutation or quantitative polygenic change. The important result is morph fixation of genetically complex alternatives and the consequent accelerated change and genetic incompatibility.

Asynchronous mating could also contribute to the evolution of assortative mating by size and to sympatric speciation in ants (Buschinger, 1986; Bourke and Franks, 1991). Highly synchronized, interspecifically divergent mating periods characterize sympatric species of many ants (Buschinger, 1986; Hölldobler and Wilson, 1990, p. 152). Seasonally asynchronous mating periods of closely related hosts and their inquiline parasites are known in *Pogonomyrmex* (Johnson, 1994). Sexuals of the host species *P. rugosus* fly in late July and early August, whereas its parasite *P. anergismus* flies in late September and October at the same localities.

In support of sympatric speciation, inquiline parasites are always sister species or very close relatives of their hosts, as would be expected if the parasite species was derived from the host. Because they are obligate and highly host-specific parasites, they are also completely sympatric with the host species. Furthermore, parasitic inquilines of this type are convergently evolved, not closely related to each other (e.g., see review in Wilson, 1971a). Several previous hypotheses propose sympatric speciation in parasitic inquilines (e.g., Buschinger, 1965, 1970, 1986, 1990b; Elmes, 1978; West-Eberhard, 1986, 1989, 1990; Bourke and Franks, 1987, 1991). Others (Wilson, 1971a; Pearson, 1981; Ward, 1989; Hölldobler and Wilson, 1990) argue that allopatric speciation is more likely. The allopatric hypothesis seems less supported by the available evidence than the sympatric hypothesis, as argued by various authors (see especially Buschinger, 1990b; Bourke and Franks, 1991; Mayr, 1993).

Much additional evidence in support of an alternative-phenotype hypothesis for sympatric speciation in inquiline parasites of ants is reviewed by Bourke and Franks (1991; see also West-Eberhard, 1986, 1989). So far, critiques of this hypothesis depend primarily on doubts that sympatric speciation can occur. Hölldobler and Wilson (1990) dismiss Buschinger's (1986) sympatric-speciation hypothesis, which proposes intraspecific evolution of parasitic behavior followed by a mutational shift in mating time, on the grounds that it is a sympatric speciation hypothesis. They argue that the hypothesis is "rendered less probable by the fact that disruptive selection, the basic mode of selection involved, must be severe to create incipient species," and furthermore, that "species-isolating mechanisms are usually genetically complex, a circumstance lessening the likelihood of sympatric speciation" (p. 449).

These objections are met by the present hypothesis, which is compatible with that of Buschinger. Disruptive selection can promote divergence, of the kind associated with incipient species, by leading to disruptive *development* and complex alternative behavioral phenotypes within species (see chapter 11). Such alternatives are notably genetically complex without depending on reproductive isolation (see chapter 21). Furthermore, as proposed here, alternative phenotypes can include such things as parasitic specializations as well as complex divergent mating behaviors of types known to occur in ants (discussed above). When fixed, contrasting alternatives would lead to genetic divergence and assuredly increased genetic incompatibility due to character release, as explained above.

In parasitic inquiline ants, there is much compelling evidence for sympatric divergence in the

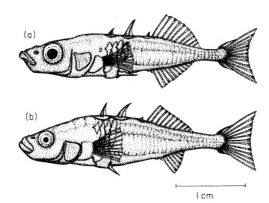

(a)

(b)

|—— 1 cm ——|

Fig. 27.7. The Enos Lake sympatric species pair of sticklebacks (*Gasterosteus aculeatus* species complex): (a) limnetic and (b) benthic species. From Bell and Foster (1994a), *The Evolutionary Biology of the Threespine Stickleback*, by permission of Oxford University Press.

form of alternative phenotypes prior to speciation, and little evidence for postspeciation origin of parasitism by any of the evolutionary sequences that have been proposed to date. Mayr (1993, p. 138), citing Buschinger (1990b), finds the parasitic ant case "the best evidence for sympatric speciation" known to him due to the lack of evidence for any allopatric hypothesis that could explain the same result equally well.

Replicate Speciation: Independent Parallelism or Developmental Plasticity?

Interspecific alternative phenotypes are pairs of alternative phenotypes in which each of the two contrasting phenotypes is fixed in more than one species of a narrow clade, for example, a genus or closely related group of genera, having a recent common ancestor. In threespine sticklebacks (the *Gasterosteus aculeatus* complex), for example, a Pacific marine population recurrently has given rise to lake populations of smaller fish that sometimes diversify within lakes in the same way, recurrently giving rise to two interspecific alternative phenotypes: a slim-bodied, relatively small limnetic (water-column-feeding) planktivorous form and a larger, deeper bodied benthic (bottom-feeding) form (figure 27.7). This pattern of sympatric morphotype pairs is quite common in fish populations of boreal lakes, as further discussed in chapter 28. In some populations, such as the Eno Lake sticklebacks (figure 27.7), the two forms are morphs of

the same species. In other lakes they represent *parallel species pairs*—recurrent sympatric species pairs, each with the same pair of contrasting interspecific alternative phenotypes. For convenience, following the terminology of replicate radiation (Schluter and McPhail, 1993) we can refer to these parallel species pairs as representing *replicate speciation*—recurrent speciation resulting in multiple parallel species pairs.

In sticklebacks, the contrasting forms within parallel species pairs are reproductively isolated due to assortative mating by size, one of the the same features that distinguishes the benthic and limnetic forms ecologically (Schluter and Nagel, 1995; Nagel and Schluter, 1998; Rundle and Schluter, 1998; Rundle et al., 2000). Size divergence is likely to have evolved under natural selection (Hatfield, 1997; Hatfield and Schluter, 1999). So the stickleback benthic–limnetic species pairs exemplify both replicate speciation and what has been called *ecological speciation* (Schluter, 1996; Hatfield and Schluter, 1999), or speciation in which the same traits that have diverged under natural selection serve to promote reproductive isolation.

Replicate speciation has been explained in terms of a *parallel speciation hypothesis*: "the repeated independent evolution of the same reproductive isolating mechanism" due to parallel evolution under natural selection in similar habitats (Schluter and Nagel, 1995, p. 292). Four criteria identify parallel speciation:

1. Phylogenetic independence of separate populations in similar environments such that shared traits responsible for reproductive isolation evolved separately
2. Reproductive isolation between ancestral and descendent populations, and between descendent populations in different environments
3. Lack of reproductive isolation between similar desendent populations that inhabit similar environments
4. The trait involved in reproductive isolation is adaptive (evolved under natural selection rather than some other means, such as polyploidy)

The developmental-plasticity hypothesis of speciation offers a similar explanation for replicate speciation, with the important difference that the origin of recurrent phenotypic similarity and contrast is explained in terms of ancestral developmental plasticity, not independent parallel evolution. The plasticity hypothesis thus differs from the

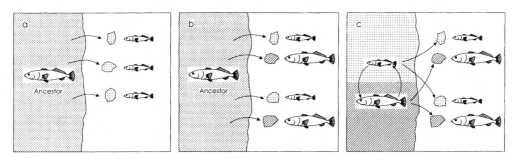

Fig. 27.8. Independent parallel evolution and developmental plasticity as alternative explanations for parallel species pairs. (a) Parallel evolution: colonization of replicate new environments (stippled patches) leads to repeated evolution of small body size and mate preferences for small size. (b) Parallel evolution and parallel species pairs: colonization of two replicate environment types (stippled and hatched) leads to repeated evolution of divergent (small, large) body size and of mate preferences for body size. (c) Developmental plasticity (life cyclic, or as alternative phenotypes) in size-related habitat occupation preadapts the lineage to recurrently occupy parallel contrasting habitats and produce the corresponding size-related phenotypes. (a) and (b) after Schluter and Nagel (1995).

parallel speciation hypothesis in criterion 1, above: although the traits may have evolved to fixation independently in different lineages (descendent clades), under the plasticity hypothesis they do not show complete "phylogenetic independence" because major elements of both are hypothesized to have originated in the ancestral population, either as life-stage differences, alternative phenotypes, or other expressions of developmental plasticity (e.g., contrasting phenotypes expressed under contrasting or extreme conditions). The plasticity hypothesis also suggests a different explanation for criteria 2 and 3, the greater reproductive isolation between dissimilar versus similar members of different species pairs, as I discuss below.

It is important to emphasize that the plasticity hypothesis does not negate the importance of natural selection for the independent *fixation* of recurrent similarities and differences in descendent species pairs. Rather, it proposes that ancestral developmental plasticity explains the *origin* of those differences and similarities, and a history of plasticity influences their genetic evolution following fixation. One of the interesting implications of replicate speciation is that the conditions for parallel natural selection must be replicated as well. Robinson and Wilson (1994) emphasize this in their discussion of species pairs of lake fish and argue that such recurrent parallelism is evidence for the existence of two non-Hutchinsonian niches in lakes, that is, niches that are properties of the environment, not just defined by how a particular species happens to use the environment.

Evidence for genetic divergence among species pairs (briefly reviewed for sticklebacks by Walker, 1997, pp. 35–36) is sometimes taken to contradict a role of phenotypic plasticity in divergence. In fact, the plasticity hypothesis helps to explain genetic divergence, since accelerated genetic change is expected to follow local phenotype fixation.

The plasticity and parallel-evolution hypotheses are similar in proposing (1) a role for assortative mating (e.g., by size) and speciation-related divergence under natural selection, and (2) a role for the naturally selected traits (e.g., small size) in reproductive isolation (e.g., by leading to some mechanism of assortative mating, such as different breeding locations, or selection for divergent mate preferences, of different-sized individuals). In these respects, the same data can be used to support both hypotheses. But the hypotheses differ fundamentally in their view of the origins of similarity between parallel species pairs: the parallel-speciation hypothesis proposes postpopulation-isolation origin of phenotypic divergence, and the developmental-plasiticty hypothesis proposes preisolation origin of divergence, as variants within ancestral populations, as setting the stage for assortative mating and speciation (figure 27.8). In parallel speciation, the derived similarities are nonhomologous— they are similar due to independent parallel evolution, not common descent. In the developmental-plasticity hypothesis the derived states are homologous—similar due to common descent.

Some of the examples of plasticity-related speciation discussed in the preceding section show how the plasticity hypothesis applies to replicate speciation. In the parasitic ants, for example, there are two examples of replicate speciation (data in Buschinger, 1990b). *Leptothorax acervorum* is the stem species for four descendent parasite species, all believed to have evolved from it in independent speciation events (Buschinger, 1990b; Baur et al., 1995). This likely occurred under natural (or social) selection on intraspecific "preparasite" females handicapped in reproductive competition within colonies (Buschinger, 1990b), as extensively discussed in the preceding section. The same type of isolating mechanism—mating in or near the nest rather than in more distant aerial mating swarms—characterizes the separation of all from the parental species, *L. acervorum* (Buschinger, 1986), a trait that likely originates under intraspecific mating competition due to size-related social selection, as discussed above. Three of the descendent species have also been shown capable of interbreeding in the laboratory with at least one of the other descendent parasites (Buschinger, 1972, 1975, 1981; especially Buschinger, 1990b, table 3, p. 247). Yet when parasitic species pairs overlap in nature, they do not interbreed, apparently because the asynchronous daily rhythms of sexual activity that likely contributed to their isolation from their common ancestor isolate them from each other as well (Buschinger and Fischer, 1991).

In another ant genus, *Epimyrma*, five closely related, allopatric parallel species, each specialized to a single host, have descended in different localities from the same host-versatile ancestral species (see supporting data and interpretation in Buschinger, 1989). The common ancestor is believed to resemble *E. ravouxi*, which parasitizes diverse hosts, sometimes even in a single colony, and therefore could have provided the basis for local phenotyic specialization (phenotype fixation) on particular single hosts (Buschinger, 1989). All of the parallel species are distinctive in that they mate within the nest, in contrast with *E. ravouxi* and other members of the genus, which mate away from the nest in swarms. Thus, all five species have the same potential isolating mechanism—intranidal rather than swarm mating. All of the five descendent species can hybridize in the laboratory (see references in Buschinger, 1989, p. 278). Mating within the nest has evolved in regions where hosts are scarce, evidently under selection to avoid predation and low host-finding probabilities associated with mating flights.

These examples would seem to satisfy all of the criteria for parallel speciation, except that specia-

tion-related divergence clearly originates in the alternative adaptations of competing females (preparasites) within the ancestral population, not de novo due to contrasting selection in the descendent populations. Both similarities of parallel forms and contrasts between dissimilar ones can be explained by descent from a common, developmentally plastic, ancestor. Replicate speciation in these ants therefore fits the developmental plasticity hypothesis better than the parallel speciation hypothesis.

Do the ancestral populations in proposed examples of parallel speciation in fish have any signs of developmental plasticity that could explain the recurrent interspecific alternatives observed? Support for plasticity is strongly suggested by the fact that all of the fish genera with replicate speciation contain species in which the benthic–limnetic interspecific alternatives are seen as intraspecific alternatives as well (Wimberger, 1994; Skùlason and Smith, 1995; see chapter 28). Next I examine the plasticity hypothesis for each of the possible examples of parallel speciation proposed by Schluter and Nagel (1995).

Salmonid Fish   Anadromous populations of sockeye salmon (*Oncorhynchus nerka*) have recurrently given rise to small, nonmigratory freshwater kokanee forms. This has been cited as a possible example of parallel speciation (Schluter and Nagel, 1995). There is some evidence favorable to the alternative, plasticity hypothesis. The parent anadromous populations show variation in the expression of migration, and some individuals of both sexes mature in fresh water without emigrating (Thorpe, 1989). Change in age at maturity is environmentally influenced in sockeye. Reduction in age at maturity and proportion of nonmigrants correlated with decline in population numbers in a lake where populations were studied over a 40-year period (summarized in Thorpe, 1989; see also Gross, 1987, for a general discussion of facultative migration in salmon). Small mature size of precociously adult males is associated with lack of metamorphosis to the migratory smolt form (Foote and Larkin, 1988, p. 44).

Sockeye and kokanee are sympatric in some regions, and the two sympatric forms show small meristic and allozyme differences indicating some genetic divergence between them (see references in Foote and Larkin, 1988; see also chapter 28). These genetic differences are maintained, and possibly originated, due to assortative mating by size (Foote, 1988; Foote and Larkin, 1988). These findings suggest that developmental plasticity for life-history variation, rather than parallel evolution, likely accounts for the recurrent origin of the nonmigratory kokanee phenotype.

Unlike in parasitic ants, there is no obvious advantage to small size per se in salmon females. Why, then, should small salmon preferentially mate with small members of the opposite sex? Selection for mates of similar life-history type may be the important factor when such parameters as precocious maturity and reproduction are favored (Shaffer and Voss, 1996). This interpretation may well apply to salmon, given their condition-dependent flexibility in life-history strategy (Thorpe, 1989).

The speciation models cited by Foote and Larkin (1988; e.g., Maynard Smith, 1966) and the parallel-speciation hypothesis (Schluter and Nagel, 1995) both grant an important role to size-related assortative mating, but they consider phenotypic divergence as due to genetic divergence due to reduced gene flow. It is more parsimonious to suppose that the size-related alternatives already present in the ancestral sockeye played a role in the evolution of the genetic divergence and assortative mating observed, since the phenotypic divergence involved parallels that between the morphs of the parent population.

Speciation based on parallel alternatives has probably occurred in other salmonids, such as the Arctic charr *Salvelinus alpinus* (Skúlason et al., 1999). Populations of *S. alpinus* have up to four size-related intraspecific ecomorphs, with different sets in different lakes. Vangsvatnet Lake in Norway contains two, a nonmetamorphosing "dwarf" deep-water or benthic form that matures in the parr phase, and a metamorphosing, smoltlike normal, limnetic form that undergoes metamorphosis to develop into a normal adult (Hindar and Jónsson, 1982). Maturation is accelerated by a large egg yolk (Skúlason, 1990), a maternal phenotypic effect subject to environmental influence. Common garden rearing experiments, transplants, and genetic crosses using different populations show that these two morphotypes are usually intraspecific alternative phenotypes (reviewed in Bell and Andrews, 1997). Some Norwegian lakes contain three morphs differing in size and maturation pattern, shown by rearing and transplantation experiments to be morphs, not distinct species, with some individuals passing through all three as ontogenetic stages (Nordeng, 1983, cited in Skúlason et al., 1999, pp. 77–78). Thingvallavatn Lake in Iceland contains four sympatric charr morphs, two of them benthic (one large, feeding on littoral snails, and one small, feeding among rocks on snails and insect larvae), and two of them limnetic (one, with a

fusiform body, planktivorous; and one piscivorous; Snorrason et al., 1989; Skúlason et al., 1999). Different morphs both within and between different lakes show different degrees of plasticity in transforming to alternative forms (reviewed in Skúlason et al., 1999). As in other boreal fishes, the sympatric morphs are often either limnetic or benthic. And these two intraspecific morphs are parallel to recognized sister species of charr.

As in the parasitic ants, divergence and incipient sympatric speciation between size-associated trophic forms of arctic charr appear to be assisted by the fact that different morphs spawn in different places and at different times of year (Snorrason et al., 1989; Skúlason et al., 1989b). The spatial breeding segregation is likely influenced by the strong homing to the natal site to breed (Skúlason et al., 1989a). These populations have the earmarks of incipient speciation facilitated by size-related alternative phenotypes in both sexes that affect ecological divergence and breeding isolation.

In the artic charr, there is little doubt that ancestral developmental plasticity is responsible for the origin of the parallel phenotypes of different species pairs. Intraspecific morphs corresponding to the benthic and limnetic forms are common throughout the species, and the differences between these two forms result from a heterochronic shift, with the benthic morphotype a paedomorph of the more common ("normal") limnetic morphology (Skúlason et al., 1989a). Skúlason et al. note that in addition to juvenile body shape, benthic morphs may also retain juvenile coloration as adults (Hindar and Jónsson, 1982; Snorrason et al., 1994), and they continue to occupy the littoral zone after the limnetic morphs have left.

Sticklebacks   The parallel-speciation hypothesis has been applied to parallel limnetic and benthic species pairs in sticklebacks (*Gasterosteus aculeatus*), as briefly outlined above. What is the evidence that the two recurrent forms could have originated due to developmental plasticity in the marine ancestors of these species pairs?

Research on the ontogeny of trophic behavior and morphology in marine sticklebacks (Andrews, 1999) and other fish (Meyer, 1990b) suggests that the origin of the contrasting ecomorphs could have been life-stage specializations, followed by heterochrony. Stickleback larvae and small juveniles generally forage on plankton, as does the limnetic ecomorph. Then, as individuals grow, limnetic behavior tends to decrease, and benthic foraging tends to increase, with corresponding morphological

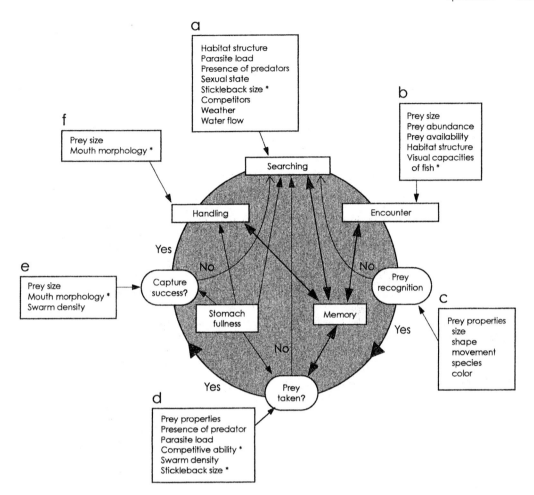

Fig. 27.9. Factors that influence prey selection by threespine sticklebacks. The circle represents the sequence of behaviors (rectangles) and decisions (ovals) during predation. Asterisks indicate phenotype-dependent factors that would differ between the benthic and limnetic forms (which differ in size, mouth morphology, and eye size; see figure 27.7) and would therefore affect learned prey preferences. This could establish a correlation between morphology, prey specialization, and prey-frequented habitat preference (e.g., benthic or limnetic). From Hart and Gill (1994) in Bell and Foster (1994a), *The Evolutionary Biology of the Threespine Stickleback*, by permission of Oxford University Press. Courtesy of J.B. Hart.

change toward the broader bodied benthic form. This ontogenetic change is most pronounced in marine (anadromous) and benthic stickleback populations (Andrews, 1999), supporting the idea that the limnetic ecomorph may represent a juvenilized form of the common marine ancestor.

In support of the plasticity hypothesis, lake populations of sticklebacks that contain only a single species of stickleback are sometimes polyethic for benthic and limnetic feeding. A population in Cranby Lake (western Canada) foraged in both habitats, but individuals specialized, at least during

the period of sampling: in 80% of individuals, over 90% of their stomach contents consisted of either benthic or planktonic species (Schluter and McPhail, 1992). Feeding habitat depended strongly on individual morphology, with plankton feeders having longer gill rakers and smaller bodies than benthic feeders. Given the evidence that learning can play a role in feeding preferences in sticklebacks (Hart and Gill, 1994; figure 27.9), morphology-influenced learning could contribute to the divergence between trophic alternatives, and to speciation. Whatever the mechanism of flexibility, both benthic and limnetic specializations are present without genetic divergence between populations in the Cranby Lake sticklebacks.

Ontogenetic and size-related change in trophic behavior and morphology is a neglected aspect of stickleback evolutionary biology but a common phenomenon in fishes (Ivlev, 1961; Keast, 1977a,b; Werner and Gilliam, 1984; see also the discussion of *Salmo salvelinus* in the preceding section, where parallel, recurrent dwarf ecomorphs evidently represent facultative heterochrony like that suggested here for limnetic sticklebacks).

Wimberger (1994) distinguishes between habitat-based plasticity and diet-bsed plasticity, but the two must often be inextricably linked. The limnetic and benthic species pairs in sticklebacks and other fish (see chapter 28) imply habitat-based differences, but in sticklebacks the habitat differences mean different diets and predators, and a dietary preference could lead to a habitat preference and visa versa. The small, slender limnetic sticklebacks feed primarily on plankton in the open water, while the larger bodied benthics feed on invertebrates in bottom sediments or on vegetation in the littoral zone (Schluter and McPhail, 1992). Such differences are known to be environmentally inducible in fish: when bluegills (*Lepomis macrochirus*) are confined in a pond with primarily littoral and limnetic prey, individual fish rapidly become trophic specialists (Werner et al., 1981, cited by McPhail, 1994, p. 422).

Behavioral change can then lead, through developmental plasticity, to the initial stages of morphological change: different behaviors associated with different diets affect body shape (including body breadth and head shape, which distinguish the lacustrine species pairs) in some fish (Wimberger, 1992). Dietary differences also can affect muscle and bone development due to effects on nutrition as well as exercise (Wimberger, 1993). The benthic and limnetic habitats also subject sticklebacks to different kinds of predators (Reimchen, 1994), another factor that can affect behavior and muscle–bone development (see chapter 28). Some fish even have predator-induced changes in body shape (Brönmark and Pettersson, 1994; chapter 28). Could any of these kinds of plasticity have contributed to divergence in body shape and performance efficiency in sticklebacks? This question has to be asked before it can be answered. A role for plasticity in trait origins cannot legitimately be brushed aside by findings of genetic differences between populations in established species, or by finding that the plasticity-influenced traits are heritable (true of virtually all traits, and required if plastic-

ity is to contribute to evolutionary change; see chapter 6).

Adult size is a key variable in the reproductive isolation of the sticklebacks (Nagel and Schluter, 1998), and it is a variable that is often associated with heterochronic change (Gould, 1977). Interspecific alternatives derived via heterochrony ("miniaturization") have evolved at least 34 times in tropical freshwater fishes of five different orders (Weitzman and Vari, 1988). Most of them, like the sticklebacks, are secondary inhabitants of freshwater, that is, members of taxa that are primarily marine (Weitzman and Vari, 1988). Some, such as *Gelanoglanis nanonocticolus* (Siluriformes), are members of a sister species pair that, like the stickleback pairs, contains one relatively small, paedomorphic species and one relatively large, nonpaedomorphic species (Soares-Porto et al., 1999), but the frequency of parallel species pairs cannot be estimated due to lack of phylogenetic study in these groups. These poorly known tropical taxa, like sticklebacks, are ripe for research on development and polymorphism in relation to speciation, phylogeny, and adaptive radiation. I predict that size-related polymorphisms will be discovered in many of these fishes.

In summary, the plasticity hypothesis applied to sticklebacks proposes that plastic behavior and morphology associated with intraspecific variation in size, under strong intraspecific competition within lakes, could have initiated a benthic–limnetic habitat dichotomy. Limnetic adults that happen to express juvenile specializations would have been at an advantage. This would set the stage for heterochrony, assortative mating, genetic divergence, and speciation.

The interpretation of parallel divergent adaptive traits as originating prior to speciation accords with the idea (Schluter and Nagel, 1994) that natural selection accounts for divergence and speciation (assortative mating) in sticklebacks, but it does not accord with the idea that the divergence occurred entirely independently following isolation in different lakes. The findings on habitat separation, ecological competition, comparative morphology, and assortative mating (summarized in Schluter and Nagel, 1995; Nagel and Schluter, 1998; Rundle and Schluter, 1998; Rundle et al., 2000) support the developmental plasticity hypothesis as well as they do the parallel speciation hypothesis. The neglected role of developmental plasticity in stickleback speciation needs to be more thoroughly investigated before these two alternative hypotheses for the origin of species-related genetic divergence can be adequately compared.

One reason for the reproductive compatibility of the parallel ecomorphs in sticklebacks may be their very recent origin from the same ancestor in lakes formed within the past 12,000 years (Bell, 1988; briefly reviewed in Nagel and Schluter, 1998, p. 210). Reproductive compatibility eventually may be lost in older populations even while similarity is maintained, for similarity can be maintained via different genetic and developmental pathways that could eventually bring hybrid dysgenesis. This may be the cause of reproductive incompatibility between congeneric cryptic species—species that are phenotypically virtually indistinguishable, but sometimes genetically very distinct (see examples in Knowlton, 1993). Mating asymmetry as a criterion for the origin of species via parallel speciation (Schluter and Nagel, 1995) may therefore be overly restrictive where older sister populations or species are concerned.

Cave Amphipods and Galapagos Finches   The interspecific alternative phenotypes shown in the two other proposed examples of parallel speciation given by Schluter and Nagel (1995), cave crustaceans and Galapagos finches, also show signs of being products of developmental plasticity. The amphipod *Gammarus minus*, an inhabitant of freshwater springs, has recurrently given rise to cave populations that are phenotypically distinctive and similar to each other, having reduced eyes, long antennae, and large size that makes them physically unable to amplex and mate with the smaller, spring-inhabiting females from which they are derived (reviewed in Schluter and Nagel, 1995). But when cave populations are exposed to light by the collapse of cave passages, they revert wholly or partly to the original spring-inhabiting form (Culver et al., 1994, cited by Schluter and Nagel, 1995). This is a clear indication that developmental plasticity in response to light plays a role in determination of the differences between cave and spring populations. Many crustaceans show immediate and striking developental responses to darkness, including neoteny and changes in secondary sex characters, mediated by light-sensitive hormones of the x organ–sinus gland in the eyestalk (partially reviewed in Matsuda, 1987). So this is a good candidate for investigation of developmental plasticity as a cause of recurrent cave phenotypes and speciation.

The possible contribution of developmental plasticity to the evolution of beak-size and beak-shape differences in Galapagos finches, traits argued by Schluter and Nagel (1995) to be a basis for their assortative mating and parallel speciation, is discussed in chapter 28.

Conclusions Regarding Parallel Speciation

A weakness of the parallel speciation hypothesis is that similarity of selection regimes alone cannot account for any example of parallel evolution. In addition, there has to be some comparative developmental evidence that the similarity of descendent populations originated independently rather than owing to common ancestry. This is possible, for novel derivations of similarity can occur even in closely related populations. In sticklebacks, for example, pelvic reduction in lake populations is recurrent, but it can be achieved by two different sequences of change, as deduced by Bell (1987, 1988; see figure 19.8). Eye reduction in cave amphipods (*Gammarus minae*) derived from the same noncave species is histologically different in different cave populations (Cane et al., 1992). To the degree that similarity is due to common ancestry and not an independently evolved modification (as in these two examples), the plasticity hypothesis applies. Recurrent phenotypes are often associated with developmental plasticity within related populations (see chapter 19). This, and the usual application of parsimony to deduce common ancestry for shared similarity, favors the plasticity hypothesis over the parallel speciation hypothesis, until independent origins of selected variants are shown.

Plasticity and Abrupt Sympatric Speciation

Abrupt speciation is the sudden genetic isolation of a subpopulation from its parent population, as in plants caused by chromosomal abnormalities such as polyploidy (Briggs and Walters, 1988). This is not to be confused with "instantaneous speciation," the improbable idea that new types can originate instantaneously (see discussion of Schindewolf in Mayr, 1942, p. 296). Abrupt speciation may have occurred in lampreys (discussed above under speciation in migrants), where discrete developmentally labile size-divergent populations have resulted by descent from migratory ancestral populations, and the size differences prevent interbreeding. In Japanese beetles (*Ohomopterus*, Carabidae), discussed in chapter 19, switches between recurrent phenotypes affect the genitalia in both sexes (Su et al., 1996). This suggests that a developmentally switched individual within a population could be reproductively isolated, or at a severe mating advantage, among conspecifics. This explanation, if it holds, implies a kind of instant speciation by the descendents of a developmentally switched individual or small group of individuals that happened to become geographically or locally (e.g., behaviorally) isolated as the founders of a new population breeding with others of the same phenotype.

Such a hypothesis may be supportable for *Ohomopterus*, a flightless genus endemic to the Japanese archipelago, which is characterized not only by many small, isolated islands but also by topographically dissected terrain favorable to allopatric (spatial) isolation. Allopatric diversification is indicated by the close correlation between phylogenetic relatedness and geographic location (Su et al., 1996). Speciation could be facilitated by genitalic differences and possible selection for positive assortative mating between individuals of like morphology. Genitalic variants are not necessarily reproductively isolated, however, since a few insects (e.g., geometrid moths) have male-genitalic polymorphisms (Hausmann, 1999), though morphs differ primarily in altered symmetry of relatively simple structures such as spines on the right and left sides of the genitalia.

Other events can cause abrupt speciation in populations where reproductive behavior is influenced by developmental plasticity, especially learning (discussed further below).

Examples

Treehoppers A host shift in treehoppers (Homoptera: Membracidae, *Enchenopa* species) promotes allochronic breeding isolation so surely that the nine cases documented qualify as examples of abrupt sympatric allochronic speciation (Wood, 1993a,b). Developmental plasticity plays a key role, for in these insects the hatching time of the eggs is cued by the highly synchronized and allochronic flowering times of their hosts (*Viburnum* species).

In *Echenopa*, the lifetime of males is short relative to that of females (which mate only once; Wood and Keese, 1990), and individuals are highly philopatric, remaining to mate on the plants where they emerge (Wood et al., 1999). The restricted mating period and male mortality are allochronic among host species, making a host shift an effective isolating mechanism. Phenotypic plasticity, again, plays a key role. Environmental entrainment of development (egg hatching) is critical. Eggs must go through a period of dehydration and hydration orchestrated by host plant water relations in order to hatch (Wood et al., 1990). Eggs on different trees adopt the hatching time dictated by the tree. Since cross-phyletic responsiveness to hormones can oc-

cur (Hänel, 1986; Shorthouse and Rohfritsch, 1992), it would be of interest to investigate the possibility that treehopper eggs respond to the plant hormones that regulate the flowering time of their hosts. Whatever the mechanism, such strict developmental entrainment by a foreign host can isolate a brood from its ancestral population in the space of a generation. The entrainment would also reinforce selection for host fidelity to assure mate finding during the very short mating period, as suggested by Colwell and others (see Colwell, 1985, and references therein).

Experiments have demonstrated conditions for rapid genetic accommodation following plasticity-facilitated host shifts in *Enchenopa*. Selected family lines from a single population of *E. binotata* were placed on cloned host plants of four different *Viburnum* species. Genetic lines varied in their ability to survive and reproduce on novel host species (Tilmon et al., 1998). Host fidelity in mating and oviposition on the new hosts has also been demonstrated (Wood et al., 1999). These two sets of experiments suggest that host fidelity in the initial stages of a host shift could produce rapid genetic divergence and formation of host races.

As expected given the high degree of host-mediated reproductive synchrony of these treehoppers, there is no evidence in the *E. binotata* species complex for host-related polyphenism; two hosts means two allochronically separated breeding lineages with host control of developmental timing. A recent phylogenetic analysis reveals two pairs of sister taxa within the North American *E. binotata* complex, both differing from each other in the timing of egg hatch mediated by differences in host-plant phenology (Lin and Wood, 2002). This is a nice example of both environmental influence on insect development and its link to speciation.

Phoretic Mites Mites with cyclical phoresy timed by host-produced cues are hypothesized to undergo abrupt speciation due to a host shift, as in *Enchenopa* discussed above, except that the host is an insect rather than a plant (Athias-Binche, 1995). Phoretic mites (phoronts) disperse by climbing onto the bodies of insect hosts, which transport them to a new breeding or feeding place. Some utilize numerous host species, but others are highly specific in use of a single taxonomic group or species of host. In cyclical phoresy the life cycle of the mite is highly synchronized with that of the host. The mite selectively uses female hosts and usually disembarks at the moment of host oviposition. Its offspring feed on the host progeny or resources in its nest, then metamorphose to the phoretic stage at the moment of the final molt of the winged host. Host–mite synchrony is achieved via hormonal and other cues produced by hosts. Some mites associated with scarab beetles, for example, become phoretic in response to a cuticular wax ester of the host (Krantz et al., 1991, cited by Athias-Binche, 1995, p. 229). Development of *Varroa jacobsoni*, a parasitic mite of honeybees, is regulated by host-produced juvenile hormone ingested with the host's hemolymph, and mite reproductive development is closely synchronized with the hormonal cycle and metamorphosis of the bee (Hänel, 1986, cited by Athias-Binche, 1995). Similar examples occur in mites associated with vertebrates.

Abrupt speciation may result in such mites due to an occasional transfer to a host with a different life or activity cycle. Some necrophilous mites, for example, migrate from one carcass to another phoretically on *Necrophorus* burying beetles. Mite species *Poecilocherus carabi* and *Neoseius novus* have proven to consist of several carrier-specific races and sibling species that have specialized on different beetle species (see references in Athias-Binche, 1995, p. 237). Athias-Binche hypothesizes that ecological traits of the host, such as different life-cycle durations and beetle daily activity, combined with the host–phoront synchronization, may explain these separations. These are possible examples of abrupt speciation that could be tested using comparative study like that of Wood and associates on *Enchenopa* described above.

Brood-Parasitic Birds: Abrupt Speciation Due to Learning   The brood-parasitic African indigobirds (*Vidua* species) illustrate abrupt host-race formation and speciation where reproductive isolation results from learned host fidelity (Payne et al., 1992; Payne and Payne, 1994, 1995). In these birds, the species-specific mating behavior of both males and females depends on experience with their foster species during an early critical period. Males learn the songs of their hosts and other birds with which they associate, and young females imprint on the foster species in whose nests they later lay eggs. A few males regularly mimic the wrong species, and they do so with remarkable accuracy, as shown by rearing males with an experimental foster species not utilized in nature, which they mimic as readily as they do their natural host, with no changes later in life even if they associate with individuals singing the normal host songs. Females, also, showed no innate host bias in imprinting on the songs of novel hosts as compared with those of their normal host (Payne et al., 1998, 2000). If a female were to lay eggs in the nests of a new host, the female and male

offspring that result would have complementary learned preferences for the new song.

Observations suggest that rare but successful switches between sympatric foster species may precede genetic differentiation driven by assortative mating of males and females reared by the new foster population (Payne et al., 1992; Payne and Payne, 1994, 1995). Host species and parasite species do not speciate in parallel. The very small genetic distances between geographically proximate indigobird populations on different hosts give instead a picture of repeated host switches independent of host speciation: genetic distances between parasites average only 7.2% of genetic distances between hosts (Klein and Payne, 1998). These small genetic distances between parasite species are consistent with the indications from behavioral observations that speciation is initiated by sudden, learned host switches.

In specialized brood parasites, the nestlings have species-specific mouth markings that mimic those of host offspring. Experiments show that nestlings that do not have the host-specific mouth markings are handicapped only if food is scarce. This means that a successful host switch may depend on a prolonged period of food abundance.

The indigobird example raises the possibility of an unusual population structure for speciation. Speciation events seem often to involve genetic divergence following a single isolating event that marks off a population where subsequent recruitment is by reproduction alone. Newly initiated indigobird populations could grow by addition of new transfers to the same host. Host-switching of songs is recurrent at a low frequency in indigobirds, where 0.8% of 494 males of four species recorded in the field sang the songs of another foster species (Payne et al., 1992). If these are the results of egg-laying in nests of the wrong species, as Payne et al. suspect, a similar frequency must occur in females. This means that the host-switching population could grow by accretion during "permissive" periods of resource abundance, with the descendants of such individuals added to those that have switched before. Presumably such recruits would be selected against in less permissive periods.

Abrupt speciation due to imprinting on the host also may occur in some of the slave-making parasitic ants discussed above. Learning affects the parasitic slave-making arrangement in two ways. First, the host brood must be imprintable on the slavemaker nest odors upon emergence in the slavemaker's nest. Otherwise, they do not accept the host brood and fail to function as worker slaves (Le Moli and Mori, 1987). More important for speciation, imprinting affects the host-selection behavior of young parasitic queens. In *Chalepoxenus mullerianus* (Formicidae), for example, larvae and young adults become imprinted on the slave species in their natal colonies, then preferentially raid that species if it is among others present in their habitat (Schumann and Buschinger, 1994). In contrast with some other slave-making parasites, only rarely do *C. mullerianus* nests contain more than one slave species (3.4% of 379 examined nests), although different populations use different species (Schumann and Buschinger, 1994). Given the imprinting behavior, and the fact that within-nest mating can occur in slave-making ants and facilitate speciation (Buschinger, 1989), a host shift in a geographic area or a year in which the established host is absent or scarce could conceivably lead to abrupt speciation via a host shift (Buschinger, 1989), similar to that which occurs in indigobirds.

Learning-assisted speciation has also been implicated in the sympatric speciation of hummingbird flower mites and apple maggots (see below). The involvement of learning does not necessarily imply sympatric speciation, although it may facilitate that process. Learning can contribute to allopatric speciation as well, as long ago noted by Wecker (1963) in a classic study of learned habitat preference as a factor in maintenance of reproductive isolation in the deermouse (*Peromyscus maniculatus*). In the indigobirds, the initial host transfer would obviously have to be sympatric, within flying distance of the original host. But subsequent genetic divergence, and therefore speciation, could be accelerated by other factors such as periods (or areas) of food abundance, or movement into isolated areas where only the new host is present.

Other Proposed Examples of Sympatric Speciation: A Role for Developmental Plasticity?

As mentioned in the introduction to this chapter, models of sympatric speciation have always been of uncertain applicability due to their restrictive conditions and lack of evidence that the conditions are fulfilled in natural populations (Mayr, 1963; Maynard Smith, 1966; Templeton, 1981; Felsenstein, 1981; Futuyma and Mayer, 1980; Futuyma, 1986b; Tauber and Tauber, 1989).

The developmental-plasticity speciation model improves prospects for sympatric speciation because it suggests a class of examples and a genetic

mechanism that are broadly applicable in nature. The conditional multiple-niche and reproductive-tactic polyphenisms to which the plasticity hypothesis applies are common (see chapter 20), and phenotypic fixation and release, or the conversion of nonspecific genetic modifiers to specific modifiers of local specializations, are expected to be common and are known to occur (see chapter 21). This mechanism for rapid divergence reduces the periods of inviolate allochrony or assortative mating needed to produce genetic divergence and reproductive isolation. Most important, plasticity within populations gives speciation a head start. Allochronic life cycles and special mating preferences can originate as *part* of a polyphenism, thereby converting incidental assortative mating into assortative mating by genotype. The fact that mating preference and divergent niche are developmentally linked as parts of adaptive alternative phenotypes means that they are potentially genetically correlated prior to gene-flow reduction, for example, via their common relationship to size-dependent switches, as described above. Under phenotypic assortative mating, all that is needed for the evolution of genetic isolation (genetic assortative mating) is increased genetic influence on the switch between alternatives, and this is known to be easily achieved under selection for the extremes of a switch-influencing trait such as size (see chapter 6).

In the following subsections I do not attempt to evaluate the sympatric aspect of speciation, which is still controversial for some of them. Rather, I consider whether or not developmental plasticity could have played a role in speciation in these groups.

Allochronic Speciation and Life-Cycle Plasticity: Cicadas

Allochronic speciation can produce parallel species pairs that are separated in time rather than in space, as in the limnetic and benthic fish. Temporal species pairs can arise from intraspecific life-cycle plasticity, followed by temporal reproductive isolation due to assortative mating based on seasonal separation between reproductive episodes, or broods, or due to physiological switches between different life-cycle lengths. This can begin either with unimodal variation in breeding times, sometimes called univoltine life cycles, or with bimodal or polyphenic variation in breeding times, such as the bivoltine life cycles of crickets with two overwintering stages (egg and nymph), which give rise to two rounds of reproduction, one in the spring and one in late summer (Alexander and Bigelow, 1960).

Narrowing or fixation of life-cycle mode would lead to accelerated evolution of the predominant mode (character release).

Fixation of life-cycle alternatives can occur in several ways, for example, (1) due to gradual or clinal change in climatic conditions, such that a formerly long-cycled univoltine population divides into two temporally separated ones (short-cycle bivoltinism), or a short-cycle bivoltine population can express only one life cycle mode in some parts of its range (Alexander and Bigelow, 1960); (2) due to decreased life span, such that the adults of two broods of a bivoltine population are separated by a time gap; and (3) due to the advantage of assortative mating between individuals of like diapause stage in a species with diapause in different life stages, for example, in parts of the species range with severe winters and strong selection on winter hardiness (see Alexander and Bigelow, 1960). Isolation by gradual change of any of these kinds could be geographically or evolutionarily abrupt because there would be a narrow line where seasonal or life-span conditions reach a threshold point where only one life-history mode is viable for a local or assortatively mating population. From that point on, there would be strong selection to specialize in a single, more frequent or narrowed mode, accompanied by its fixation and genetic release. Possible examples of geographically abrupt allochronic speciation occur in sawflies (*Neodiprion* Knerer and Atwood, 1973) and crickets in clines (Masaki and Walker, 1987 and below).

The allochronically shifted life-cycle forms of periodical cicadas (*Magicicada* species; Homoptera) are a dramatic example of temporally isolated parallel species pairs. Four parallel species pairs of cicadas are known in the northern and midwestern United States, each pair with a 13- and a 17-year life-cycle species that is its closest, and morphologically virtually indistinguishable, relative, as follows (life-cycle duration, in years, in parentheses): *M. septendecim* (17) and *M. neotredecim* (13); *M. cassini* (17) and *M. tredecassini* (13); *M. septendecula* (17) and *M. tredecula* (13). A seventh species, *M. tredecim* (13), is a sister species of the *septendecim-neotredecim* species pair (Marshall and Cooley, 2000).

Speciation in periodical cicadas probably involves developmental susceptibilities that determine nymphal diapause duration (Martin and Simon, 1990a,b; Heliövaara et al., 1994; Williams and Simon, 1995). Periodical cicadas can change from a 17- to a 13-year life cycle via a developmental switch possibly sensitive to nymphal crowding (Martin and Simon, 1990b). This evidently involves switch-

ing off and on an early-instar four-year dormancy period expressed only in 17-year cicadas (White and Lloyd, 1975, cited by Marshall and Cooley, 2000). Every periodical cicada species is most closely related to a congeneric species with the alternative life cycle, and sibling species, such as 13-year *M. neotredecim* and 17-year *M. septendecim*, have parapatric distributions and are consistently distinguishable only in life-cycle length, strongly suggesting that one originated from ancestral populations of the other due to a life-cycle change (Marshall and Cooley, 2000). In areas of overlap, there are exaggerated differences between the acoustical signals of these species that would reinforce their reproductive isolation (Marshall and Cooley, 2000).

Marshall and Cooley (2000), developing a suggestion of Lloyd and Dybas (1966), propose a life-cycle-canalization hypothesis of speciation that suggests that extreme environmental (e.g., climatic) conditions may occasionally induce a mass population switch to the alternative life cycle in numbers sufficient to satiate predators, long hypothesized to be the primary function of cicada mass emergences (see references in Marshall and Cooley, 2000). Predation would then eliminate temporal stragglers (evidence that spatial stragglers are eliminated is suggested by research on outbreaks of Okinawan cicadas, *Mogannia minuta*; Itô, 1998, p. 495). An environmental cue is more likely to initiate a successful switch than is a mutation, which can only affect a single individual and its progeny. Environmental induction affords the protection of a mass switch, which makes previous mutational explanations unsatisfactory, as discussed by Simon et al. (2000), who propose that latitudinal climatic variation produces life-cycle switching zones—in effect, geographical areas where conditions pass the threshold of a conditional switch (see also Alexander and Bigelow, 1960; Alexander, 1968). Such zones help to explain the parapatric (adjacent and nonoverlapping) distributions of sibling species in cicadas (Simon et al., 2000).

Annually emerging, nonperiodical (unsynchronized) cicadas have variable emergence times (1–5 years; Martin and Simon, 1990b; Heliövaara et al., 1994). Why has the four-year interval timer evolved? Is it due to selection related to the length of interval required to prevent buildup of predators able to exploit mass emergences? Or, alternatively, is there some physiological mechanism or environmental cyclicity that predisposes to a four-year trigger? Most likely, the four-year interval may have evolved through some combination of selection and diapause physiology in the subterranean nymphs, beginning with a variable diapause length associated with bet hedging, a predator-escape mechanism eventually given up in favor of a predator-satiation tactic achieved by periodicity (Simon et al., 2000).

In addition, there remains the question of what causes a new cycle, induced by an environmental trauma, to persist across generations. It is clear that mass emergence is a selective advantage (see especially the review by Simon et al., 2000), but what mechanism allows it to be achieved? Is there some means of entrainment, for example, through maternal effects as in locusts (Pener, 1991) or dietary change as in horned beetles (Emlen, 1997b), that can facultatively alter the threshold of the switch? Many physiological switches undergo facultative change in "set point," or threshold, under extreme conditions (Mrosovsky, 1990). A locustlike density-dependent response seems particularly worth investigating, since there may be cryptic competition and communication among nymphs underground. At least one discussion mentions underground nymphal competition as a factor in life-cycle switching (Lloyd and White, 1976, cited by Simon et al., 2000, p. 1333), and the switch is sensitive to nymphal crowding (Martin and Simon, 1990b). This is an area of developmental evolutionary biology ripe for research by an insect physiologist with a shovel.

Lacewings

A role for alternative life-cycle phenotypes in the speciation of lacewings has been proposed by Tauber and Tauber (summarized in Tauber and Tauber, 1989). The lacewing genus *Chrysoperla* contains a number of closely related species distinguished by very small genetic distances (Wells, 1994; Henry et al., 1999). Species differ in their breeding times, even in the same geographic areas. In the eastern United States, for example, *C. carnea* (= *C. plorabunda*) is multivoltine: it has no summer diapause and therefore produces several summer broods. Reproduction occurs during late spring and summer. *C. downesi*, by contrast, is univoltine: it requires a short-day-induced (winter) diapause in order to reproduce. Reproduction, in a single brood, occurs in early spring. Then, under the influence of a photoperiodic switch, the offspring generation awaits the following spring to reproduce (Tauber and Tauber, 1992). Such breeding asynchrony can contribute to the evolution of reproductive isolation. The question I raise here is not whether sympatric speciation has occurred, but whether ancestral plasticity in breeding schedules (voltinism) could have contributed to the origin or maintenance of reproductive isolation in this group.

Some populations of the *Chrysoperla* species complex in the western United States are polymorphic for univoltine and multivoltine life cycles and for the presence of the photoperiodic switch. Moreover, when the photoperiodic switch is present, its ability to induce summer diapause may be modulated by a secondary switch that responds to the presence or absence of larval prey: when prey are present, adults forgo the summer diapause and reproduce; when prey are absent, adults enter diapause. Individuals in these western populations are highly variable for both photoperiodic sensitivity and responsiveness to larval prey. Responsiveness to prey, however, is only expressed in individuals that also have the photoperiodic switch (Tauber and Tauber, 1982). These findings suggest that a sequence of two switch mechanisms is involved in regulation of the reproductive and diapause cycles of these lacewings. First, a daylight-sensitive switch governs whether or not an individual undergoes summer diapause prior to winter diapause. Then, in individuals that are susceptible to summer diapause, a second, prey-sensitive switch can reverse the photoperiodic induction of summer diapause.

Since *C. carnea* is insensitive to induction of summer diapause (it has a dominant allele at either locus that affects photoperiod sensitivity), it is not susceptible to summer diapause modification by presence or absence of prey (Tauber and Tauber, 1992). Small genetic differences underlie the differences between populations in the expression of univoltine and multivoltine life cycles, which characterize species like *C. carnea* and *C. downsei* in the eastern United States. If lineages of life-cycle monomorphic species were derived from life-cycle polymorphic species like those in the western United States, phenotype (life-cycle) fixation could have contributed to their speciation. *C. plorabunda* is hypothesized to be the sister species of the California species, *C. adamsi*, which together with one other species (*C. johnsoni*) form the *plorabunda* species group; and this group is the sister group of the *downesi* group (two species, *downesi* species 1 and 2, one of them the *downesi* studied by Tauber and Tauber; Henry et al., 1999). All of these species are so closely related that it was impossible to analyze their relationships using electrophoresis (Henry et al., 1999). It is therefore difficult to know what the polarity of divergence in breeding schedules would have been, but it is at least possible to say that plasticity in the expression of uni- and multivoltinism could have played a role.

The species of this species complex also differ in mating signals, and sympatric species have different calls, which are effective isolating mechanisms (Wells and Henry, 1992, 1994; Henry et al., 1999). If, in particular branching events, signal divergence preceded divergence in breeding time, then signal divergence may have precluded any effect of allochronic breeding on speciation in this group. If, on the other hand, divergence in breeding time sometimes preceded signal divergence, or both occurred prior to sympatry of sister groups during the evolution of *Chrysoperla*, then allochronic isolation could have minimized effects of mating signals on the origin of reproductive isolation. Either of these traits can diverge in allopatry, then affect breeding isolation of overlapping populations. Allochrony and mating signals could have acted together in the evolution of reproductive isolation in such a group—these are not mutually exclusive, alternative hypotheses (cf. Henry et al., 1999). Divergent mating signals could reduce mating between allochronic stragglers, causing character displacement in signals even while populations are largely isolated allochronically.

In another lacewing genus, *Chrysopa*, the frequency of alternative predatory behaviors of larvae varies in ways that suggest character release with phenotype fixation (based on Tauber et al., 1993, 1995; Milbrath et al., 1993, 1994). The larva of *Chrysopa quadripunctata* is a generalist predator (the pleisiomorphic state), whereas the larva of its hypothesized sister species, *C. slossonae*, is a specialized predator on wooley alder aphids (Milbrath et al., 1994). The specialist protects itself from the attacks of ants that defend its prey by gathering filamentous secretions from the aphids and using them as camouflage. The generalist species preys on a wide variety of soft-bodied arthropods, including aphids, and usually builds no woolly shield. But when this generalist larva contacts a colony of woolly alder aphids, it facultatively camouflages itself, as does the specialist, with the waxy secretion. The frequency with which this response occurs varies geographically (Tauber et al., 1995).

The preexistence of such a specialized alternative behavior could enable a population descended from a generalist like *C. quadripunctata* to move into a specialized predatory niche. Then, if the specialization is fixed, as in *C. slossonae*, its increased frequency of performance is expected to be accompanied by additional specializations to predation on woolly aphids. This is what happens in *C. slossonae*, where individuals have specialized searching behavior that lead them toward alder trees, and preferential molting near alder-aphid colonies (Milbrath et al., 1994). Such divergence in the direction of the fixed specialization could represent character release, and could promote the evolution of host specificity and breeding isolation from less specialized populations.

The Apple-Maggot Fly

The Apple-maggot fly *Rhagoletis pomonella* (Diptera, Tephritidae) is much discussed as a possible example of sympatric speciation (Bush, 1969, 1992). Here, rather than focusing on the issue of sympatric speciation, I focus on the evidence that host-race formation in these flies has been facilitated by developmental plasticity.

Genetically diverged host-specific populations of *Rhagoletis pomonella* (Tephritidae) have some of the earmarks of plasticity-facilitated speciation, such as a flexible mechanism of habitat choice in the form of learned host preference (Prokopy and Bush, 1993; see references in Bush, 1992), and a fitness trade-off between populations that inhabit sympatric hosts with different fruiting phenologies (Feder and Filchak, 1999; Filchak et al., 2000)—the kind of situation that gives rise to alternative phenotypes within species (see part III). There is abundant evidence that the host races seen in the northern United States are genetically distinct incipient species separated by a host shift from hawthorns to apples in the short period since cultivated apples were introduced into the eastern United States (Bush, 1969). There also is evidence that the genetic divergence between them correlates with timing of adult eclosion and maintenance of diapause in the earlier developing apple race (Feder et al., 1997; Filchak et al., 1999, 2000).

Benjamin Walsh, the first entomologist to report host-race formation in *Rhagoletis pomonella* (Walsh, 1867), proposed a hypothesis of speciation involving host shifts and phenotypic plasticity, inspired by rearing experiments with phytophagus insects (Walsh, 1864). The experiments showed that individuals of some insects develop phenotypic differences depending on the host where they are reared, even though they still "freely intercross"; Walsh called the host-induced interbreeding phenotypes "phytophagic *varieties*" (p. 405). In effect, such a species would be polyphenic in any region where both hosts are present. Walsh proposed that the divergent phenotypes of phytophagic varieties may, after many generations, become inherited and fixed (i.e., genetically assimilated) in association with the formation of "phytophagic *races*" (p. 406), if "eggs have been uniformly deposited by a

Phytophagic Variety upon the same plant for an indefinitely long series of generations" "owing to the presence of but a single species of the plants" (p. 405). Phytophagic races may then give rise to "phytophagic species" (p. 406). Walsh contrasted this with the formations of "geographical races . . . separated into two or more distinct groups by physical barriers" (p. 404). Bush (e.g., 1973, 1992) interpreted this as a theory of sympatric host-race formation, though some degree of allopatry is implied by Walsh's requirement of "the presence of but a single species of the plants" for the formation of a phytophagic race. Walsh's idea would be especially appropriate for a polyphenic (host-versatile) species that found itself in a monoculture, such as an apple orchard.

The aspect of host induction of initial change, or developmental plasticity, in the Walsh theory was subsequently overlooked in research on *Rhagoletis*, with the host-induced phytophagic varieties confounded with "host races" (Prokopy and Bush, 1993, p. 6). Walsh's theory is the earliest speciation hypothesis known to me where phenotypic divergence between species is depicted as beginning due to developmental plasticity.

R. polmonella has a complex pattern of geographic variation in hosts, host phenologies, and genetic distinctiveness of different host populations. The two host races studied by Bush and others in the northern United States are genetically distinguishable (Feder et al., 1988, on Michigan populations; McPheron et al., 1988, on Illinois populations). But in other locations, the host-race status of flies found on different fruits is not so clear, for in some areas there is no detected genetic difference between them (Feder and Bush, 1989; Feder et al., 1990). In northern populations, females prefer to oviposit on familiar fruit that they have experienced as adults (Papaj and Prokopy, 1988), and the genetically distinct populations of these flies show host fidelity in mating on the type of fruit where they have emerged and have experienced as adults (Feder et al., 1994).

Papaj and Prokopy (1989, p. 340) give a concise summary of the potential role of learning in the speciation of *Rhagoletis*: experience with fruit of a given species increases the tendency of females to oviposit on that fruit and remain in trees bearing that type of fruit. Males also remain longer on fruit with which they have experience, and virtually all mating is initiated on fruit. In the northern United States apple populations are further isolated by seasonal differences in host fruit maturation (apples early, hawthorns late) associated with allochronic emergence and mating periods in the flies

(reviewed in Bush, 1992; see also Feder et al., 1994). Any genetic reinforcement of fruit preference, such as that selected in association with fruit seasonality (see below), would hasten the establishment of genetically distinctive host races.

There seem to be few documented phenotypic differences between the apple and hawthorn races of *R. pomonella*, other than (1) host fidelity due to learning (e.g., Prokopy et al., 1982); (2) developmental times: apple-race larvae develop more slowly, pupate about 16 days earlier, and overwinter for about a week longer than do hawthorn-originating larvae in a population in Grant, Michigan (Feder and Filchak, 1999); and (3) genetically based differences in host preference, more marked in apple-race (93.7%) than in hawthorn-race (under 50%) flies, tested by mark and recapture of naive adults (field collected as larvae and pupated on vermiculite in the lab; Feder et al., 1994). Reciprocal egg transplants have revealed no evidence of feeding specializations in the two races, and both races survive equally well in hawthorn and equally relatively poorly in apples (Prokopy et al., 1988). Survivorship in apple and hawthorn is different for flies bearing particular identified alleles (Feder and Filchak, 1999), but these alleles occur in both races, although some, clinally increasing from south to north and correlated with cold hardiness, are more common in hawthorn flies in Michigan (Feder et al., 1997b). Would differences in development time and some host preferences be inducible in genetically undiverged populations by rearing on the two hosts? And is there larval habitat selection within fruits, as in some *Drosophila* (chapter 28), that may have contributed to host-race formation?

Critical tests of a role for plasticity in the host-race divergence of *Rhagoletis* have not been done. They would require study of genetically undiverged populations living on both hosts (Mitter and Futuyma, 1979) and rearing experiments like those of Walsh (1864) to observe possible host effects on the reproductive cycle and other phenotypic traits of the host-versatile flies. Opportunities for such studies exist, for populations have been located where both hosts are present and there is no measured genetic divergence, for example, at 40°39′N latitude in Ohio (Feder et al., 1990). Bush (1994) has cited examples of polyphenism in other taxa in support of his sympatric speciation hypothesis.

It would be of interest, as suggested by Carson (1989), to conduct a systematic search in this or related species for evidence of host-induced or seasonal polyphenism, for example, in southern populations where host allochrony may not be as marked as in the northern United States or in areas where additional alternative hosts may be exploited (Carson, 1989, suggests possible alternative hosts; McPheron et al., 1988, note that a variety of host plants are used). On its native host, hawthorn, *R. pomonella* shows facultative dispersal flights of more than 1000 meters when it detects conspecific oviposition pheromone (Roitberg et al., 1984), and abandons host trees if they fail to find fruit (Roitberg et al., 1982). This kind of behavioral plasticity, combined with the adult preference for emergence sites as oviposition sites (Papaj and Prokopy, 1988), reluctance of both sexes of flies to fly long distances except when stimulated by the odor of fruit (Prokopy and Bush, 1993), and the shifted fruiting season of apples, could make an orchard a rich island that, once colonized, may seldom be abandoned by such flies for many generations, thus satisfying the conditions for host-race formation due to multigenerational restriction to a single host envisioned by Walsh (1864, discussed above).

The host-race and plasticity hypotheses suggest different predictions regarding the host-specific responses of individuals. The host-race hypothesis predicts that divergence in host acceptance behaviors and host-specific adult emergence times (see references in Feder et al., 1988) is always underlain by genetic divergence, whereas the alternative-phenotype hypothesis predicts that phenotypic divergence is initially host induced or a facultative response to hosts, and that it is followed by genetic accommodation and exaggeration of responses when they reach relatively high frequencies. These predictions are best studied in populations that have not yet undergone a genetically accommodated shift to apple.

Since flies prefer hawthorn fruit when it is available, the host-related genetic divergence observed in some northern populations may have originated when early-emerging flies were attracted to apples before the seasonal availability of haws. If so, this would be an example of host-mediated allochronic divergence of a kind that could lead to speciation.

Rhagoletis research illustrates the difficulty of tracing the origins of reproductive isolation once it has occurred. Alternative phenotypes are a kind of sympatric divergence easily confused with subspecies and species differences, partly because of widespread doubt that marked divergence could occur without speciation. Many forms termed "host races" in entomology, for example, may be polymorphisms or polyphenisms or may have originated as such. Or, if they are allochronically separated in their breeding seasons, they may owe their initial divergence to ancestral variation in life cy-

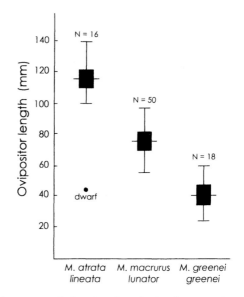

Fig. 27.10. Ovipositor lengths in three species of *Megarhyssa* wasps (Hymenoptera, Ichneumonidae). After Heatwole and Davis (1965).

cles like that discussed above, encouraged but not initiated by differences in host fruiting phenology. Speciation and divergence are not always carefully distinguished (but see Claridge, 1988; Claridge and Nixon, 1986, on host-specific populations of leafhoppers, *Oncopsis flavicollis*). The potential for confusion is increased by the fact that sympatric divergence as polymorphism, polyphenism, or extremes of continuous variation (e.g., in emergence times) may rapidly produce local genetic divergence or speciation with only a short period of extrinsic isolation or assortative mating, as could occur during several generations of breeding on a new host in a large monoculture such as an orchard.

Megarhyssa Wasps

Gibbons (1979) proposed that three very similar species of *Megarhyssa* wasps (Ichneumonidae) may fit the *competitive speciation* hypothesis of Rosenzweig (1978), a sympatric speciation model that postulates divergence due to disruptive selection and simultaneous positive assortative mating of extreme phenotypes under strong resource competition. Three sympatric species of *Megarhyssa* lay eggs on the same host, larvae of the woodwasp *Tremex columba* (Hymenoptera, Siricidae). But each *Megarhyssa* species has a different ovipositor length (figure 27.10) and, by laying eggs only when the ovipositor is completely extended, each reaches larvae at different depths in wood, obviating resource competition between forms. The competitive speciation hypothesis interprets the divergence between forms as genetic, not polyphenic, and assortative mating is required for divergence.

The competitive speciation idea is of interest in a book on developmental plasticity because the conditions it describes—divergence between conspecifics under strong competition—are often associated with the evolution of intraspecific alternative phenotypes. Recall that disruptive selection can lead to disruptive development, or complex intraspecific alternative phenotypes not requiring reproductive isolation or assortative mating for their maintenance (see chapter 11). So the ichneumonid wasps examined by Gibbons (1979), which have the competitive prerequisites and sympatry depicted by both models, offer a potential test of the two ideas. Both can lead to speciation, but the alternative-phenotype hypothesis suggests that intraspecific alternatives occur first, whereas in competitive speciation phenotypic divergence depends on assortative mating and genetic divergence between phenotypic forms from the beginning.

Ovipositor length variation in progeny of single females has apparently not been examined, nor have genetic tests been performed that would indicate the degree to which the different forms have diverged genetically. Polyphenic speciation is a reasonable possibility in *Megarhyssa*. Ovipositor polymorphisms are known in several species of parasitoid wasps, for example, in *Torymus auratus* (Torymidae; Askew, 1965) and *Aphycus albicornis* (Chalcidoidea, Encyrtidae; Murakami, 1960). Ovipositor length in *T. auratus* correlates with the size of the host gall (Askew, 1965), so competition between morphs may be avoided by this device, as suggested by Gibbons (1979) for *Megarhyssa*.

Females of the three *Megarhyssa* species differ in size, but it is not known if ovipositor length correlates with body size (Tauber and Tauber, 1989). Based on the finding that, prior to emergence over 80% of the females mate with males that fully insert their abdomens and part of their thoraces into the female emergence tunnel (Crankshaw and Matthews, 1981), Tauber and Tauber (1989) suggest that reproductive isolation between ovipositor morphs, which would emerge from different depths in wood, could have evolved via a corresponding polymorphism in the ability of males to reach them. This would predict a male polymorphism in morphology or abdomen-extending behavior that affects copulation depth. But it would not explain why males able to reach deep females would selectively mate with them, bypassing smaller females,

unless the male abdomen must be fully extended in order to mate.

The *Megarhyssa* example appears to represent very recent speciation that conceivably could have started with preisolation divergence in the form of an ovipositor polymorphism or polyphenism. I mention it here as a case ripe for a comparative study of speciation, where contrasting hypotheses could be tested using basic data on intraspecific and interspecific variation in morphology and behavior, prey size, and other aspects of natural history and development, beginning with morphometric study of museum specimens and including artificial rearing on different size prey.

Alternative Phenotypes and Speciation in Clines

Pitcher-Plant Mosquitoes

The pitcher-plant mosquito *Wyeomia smithii* (Coq.) is a model organism for studies of life history evolution. It is distributed over a broad geographic range in North America, from the southeastern United States (30°N) to Manitoba (near 55°N), and the only breaks in this distribution are of recent origin (Bradshaw and Lounibos, 1977). Throughout this range, its sole breeding site is in the standing water enclosed by the leaves of insectivorous pitcher plants (*Sarracenia purpurea* L.), where the mosquitoes thrive without themselves being consumed. Prey captured by the host plant are the source of nutrients for the larvae of this mosquito, and this resource can be quantified by counting the number of prey head capsules found in the water in a leaf (Bradshaw and Holzapfel, 1986).

The amount of larval food available within pitcher plants increases from south to north (as well as with increasing altitude) along a cline of decreasing density of mosquito larvae per prey captured by the host leaf (Bradshaw and Holzapfel, 1986). In the north, adult female mosquitoes, being relatively well nourished as larvae, are *autogenous*—they lay repeated egg batches without a blood meal. In the south, adult females use resources garnered in the larval stage to mature their first batch of eggs autogenously, but they require a blood meal for the second and subsequent ovarian cycles. In the north, above 40°N, blood feeding is completely absent and cannot be induced, even if females are reared on an inferior larval diet (Loubinos et al., 1982). Thus, an evolutionary change has fixed the autogenous phenotype in northern populations.

The obligatory autogeny of northern females is associated with additional modifications indicative of release due to phenotype fixation. The northern females emerge with precocious ovarian development, and they mate earlier than do southern females (O'Meara and Lounibos, 1981). But they retain piercing mouthparts, a vestige of their recent origin from blood feeders (O'Meara and Loubinos, 1981). Mosquito species with obligate autogeny often show even more extreme specializations to autogeny (and loss of blood feeding), such as lack of piercing mouthparts, reduced salivary glands, reduced flight activity, reduced wing length, and more sedentary mating and copulation early in adult life, sometimes before the female has completely emerged from the pupal cuticle (O'Meara, 1985b; O'Meara and Lounibos, 1981; Spielman, 1971). The change in mating site from swarms to the confined spaces of the larval habitat (stenogamy), which sometimes accompanies autogeny and also rapidly evolves in (caged) laboratory colonies (O'Meara and Evans, 1974), could contribute to speciation in a cline.

Wyeomyia smithii shows sufficient divergence between latitudinal extremes to have led in the past to their designation as distinct species (*W. haynei* in the south and *W. smithii* in the north; Bradshaw and Lounibos, 1977). Northern populations are regarded as a geographic race by Bradshaw and Lounibos (1977), and although differing in some characters (number of anal papillae, diapause characteristics) that can represent species differences in other mosquitoes, the northern and southern races hybridize readily in the laboratory (Bradshaw and Lounibos, 1977). Based on comparison with other mosquitoes, obligate autogeny is the derived state, and the retention of autogeny in southern mosquitoes is the ancestral state (Bradshaw and Lounibos, 1977; O'Meara et al., 1981).

The same pattern observed in *W. smithii*, of obligate autogeny in more northerly populations, occurs in several other groups of freshwater mosquitoes (O'Meara, 1985a). In some, populations that differ in frequency of autogeny (including phenotype fixation in one of the populations) have diverged to the point of being reproductively isolated when found in sympatry. This is true for a *Wyeomyia* species pair, *W. vanduzzei*, which is polyphenic with autogeny more common, and *W. mitchelli*, which is anautogenous (O'Meara, 1985a).

Although there have evidently been no studies of hybrids that would indicate to what degree divergence in mode of reproduction has played a role in the disruption of interbreeding, genetic studies demonstrate divergence associated with the fixation

of autogeny in some mosquitoes. O'Meara (1972) has demonstrated genetic divergence associated with fixation of the autogenous phenotype in *Aedes atropalpus*. The northern subspecies *A. a. atropalpus* is exclusively autogenous and homozygous for a major gene (A) for autogeny. Three other subspecies are primarily anautogenous and homozygous for the recessive allele (a). Repeated backcrosses of the autogenous strain with nonautogenous individuals showed that the fixed autogenous form is accompanied by nonspecific modifiers or background genes that enhance autogenous fecundity (O'Meara, 1972). Females bearing these modifiers produce, on average, more than 150 eggs per female autogenously, compared to 12.9 eggs per female by autogenous females in a subspecies (*A. a. epactius*) where autogeny is relatively uncommon. Similarly, in populations of *A. taeniorhynchus*, where autogeny is rare, mean fecundity of autogenous females was often less than 30 eggs per female, whereas significantly higher levels of fecundity (mean eggs per female, 48.5–71.8) occurred in populations containing mostly autogenous females (O'Meara and Edman, 1975). Thus, there is evidence on both the phenotypic and the genotypic level for increased specialization accompanying fixation of autogeny.

Crickets

Differences in maturation speed (e.g., due to temperature differences at different latitudes) can cause clinal change in the degree of overlap between seasonal broods (Alexander and Bigelow, 1960; Alexander, 1968; Harrison, 1985). Fixation of a univoltine reproductive cycle can occur at one end of a cline while other geographically distant subpopulations are bivoltine or multivoltine, as occurs in some cricket species (Alexander and Bigelow, 1960; Alexander, 1968). This would be expected to accelerate divergence of the fixed form due to character release, especially accentuating physiological specialization to the fixed cycle, a change expected in turn to increase selection against hybridization with individuals of different life-cycle types. Release and accelerated divergence between antagonistic life-cycle populations could explain an "unexplained phenomenon" noted by Alexander (1968, p. 34): in numerous attempts to hybridize closely related cricket species, all of the complete failures involved northern species with contrasting diapause stages. The release hypothesis suggests that this incompatibility is due to the two forms having diverged in antagonistic (contrasting) directions following their fixation. Alternatively, this result could reflect larger general divergence times between independently evolved divergent diapause stages not present in the ancestral population (Harrison, 1979). But this latter explanation would not be the most parsimonious one if closely related species show developmental plasticity in expressing diapause in those same stages, as seems to be the case in crickets (Alexander, 1968), suggesting that fixation and release were more likely involved.

Life-history polyphenisms with two overwintering stages (egg and nymph) may have been a precursor to allochronic speciation in some crickets (Alexander and Bigelow, 1960; Alexander, 1968). This idea could apply beginning with an ancestral species such as *Gryllus firmus*, where a single female can produce both overwintering stages (Masaki and Walker, 1987). A beautiful analysis of clinal variation in cricket life cycles by Masaki (1986) demonstrates life-cycle fixation, character release increasing specialization to the fixed form, exact geographic coincidence between fixation and release, and parallel differences between closely related species indicating an effect of such release on the evolution of reproductive isolation between populations (see figures 21.3 and 21.4).

Learning, Sexual Selection, and Speciation

In some of the examples discussed in this chapter, such as the fly *Rhagoletis pomonella*, adult learning affects breeding and mating site and therefore contributes to the evolution of reproductive isolation (Prokopy and Bush, 1993; Feder et al., 1994). Another way that flexibility likely influences speciation is through effects on the stimuli that influence mate choice (for examples of the kinds of signals involved, see West-Eberhard, 1983). As in the brood parasitic birds discussed above, offspring may learn to perform or to prefer the signals they experience as they are growing up, and this can contribute to breeding isolation between individuals with unlike learned traits.

Birds are prime candidates for a possibly widespread role of learned song differences in the evolution of reproductive isolation (e.g., see Grant and Grant, 1996, 1997, 1998; Grant, 2001). But learned divergence in sexual communication, and its effects on speciation, is underinvestigated in most animal groups, and may be more widespread than expected, partly because evolutionary biologists may underestimate the capacities of nonvertebrates to learn, and flexibly adjust, sexual behavior.

Conclusions

The examples discussed in this chapter suggest that William Bateson (1894) was correct to believe that intraspecific developmental discontinuities could give rise directly to the discontinuities that distinguish species. But they give no support to Bateson's notion that selection is not involved. The developmental discontinuities that become species differences do so via a phase as alternative or life-stage-specific *adaptations*. They have obvious functions and are subject to change under selection (see chapter 21). It is their adaptive significance that makes them contribute, upon fixation, to genetic divergence and speciation itself. The frequent association of alternative adaptations and life-cycle specializations with species differences supports the idea that morph fixation and character release contribute to the evolution of reproductive isolation, or speciation, as expected on theoretical grounds.

The examples of this chapter suggest that our view of the relationship between divergence and speciation at particular phylogenetic branching points may have to be reversed. Phenotypic divergence that characterizes closely related species may substantially precede, rather than follow, the advent of reproductive isolation. This order of events could not usually be resolved in the fossil record, where sudden divergence can easily be misinterpreted as a result, rather than a cause, of speciation (see chapter 30). Phenotypic diversification can occur anagenetically due to the evolution of alternative phenotypes without lineage branching. In the fossil record, this may be interpreted as a speciation event.

Genetic models of sympatric speciation (e.g., Maynard Smith, 1966) often begin with a single-locus *genetic* polymorphism. But the complex polymorphisms and polyphenisms that are so common in nature are rarely governed by such loci (see chapter 22). When species differences involve a single locus of major effect, the theory of developmental organization presented here suggests that the locus in question is one that affects regulation of a polygenic threshold trait, and the phenotypic difference mediated by that locus may be greater than suggested by the genetic difference. The examples described here show how complex polyphenisms can facilitate speciation, but this model presumes a different population structure and sequence of events than do previous genetic models to which polyphenisms are often referred. Especially, it depicts divergence as beginning prior to reproductive isolation without restriction of gene flow, and then shows how that developmentally mediated divergence can contribute to the origin of reproductive isolation.

Phenotypic divergence and speciation may involve learning more often than usually appreciated. This chapter discusses examples in birds, fish, and insects. Learning-facilitated niche shifts may be one of the keys to understanding rapid divergence and speciation, and adaptive radiation in general—the subject of chapter 28. A learning-mediated switch not only is rapidly propogated under consistent conditions or stress but also changes trait correlations and selection coefficients as surely as would a mutationally induced one (see chapters 18 and 21). The result can be genetic accommodation and divergent change. The fact that learning is so often implicated in sympatric divergence means that it needs to be a focus of increased attention from evolutionary biologists. The time is past when it can be reflexly shelved as a "nongenetic" phenomenon

of dubious or rare import for speciation-related change.

Most important, speciation-associated divergence between populations or related taxa is not something that stops in the intervals between speciation events. Adaptive evolution is an ongoing intrapopulation phenomenon that may contribute to the origin of reproductive isolation, the reverse of the causal relation implied by the idea that speciation is the initiator of all divergence. This conclusion is important not only for speciation theory but also for related ideas. Rates of evolution can vary without speciation, so punctuated patterns should not be automatically assumed to be associated with speciation. Similarly, phylogenetic analyses need to take account of the possibility of character change along the branches of a phylogenetic tree rather than assuming that change is concentrated at the nodes (see chapter 19).

28

Adaptive Radiation

Phenotypic patterns in adaptive radiation can reflect patterns of ancestral developmental plasticity. Two major patterns of radiation are associated with two major types of plasticity (see chapter 3). Binary radiations, such as the recurrent parallel benthic–limnetic species pairs of some fishes, are associated with binary flexible stems— ancestral switch-mediated plasticity, life stage phenotypes, or extremes of continuous variation. Multidirectional radiations like those of Hawaiian drosophilid flies, some island birds, and African lake cichlids are associated in addition with hypervariable flexible stems—ancestral hypervariable plasticity, such as learning and other mechanisms of modification by use. Synergisms of plasticity, sexual selection, founder effects, and environmental extremes including a high density of competitors can exaggerate and may accelerate adaptive radiation. Recurrent boom-and-bust conditions and the ratcheting effect of conditional expression may help account for the extensive tropical radiations of some taxa. Because radiations comprise branches from a common ancestor, they may often reflect ancestral developmental patterns from which a variety of recurrent phenotypes can be derived and reinforced by selection and genetic accommodation in new ecological settings.

Introduction

Adaptive radiation is the simultaneous diversification of a lineage into numerous sublineages and specializations (Simpson, 1953a). All of the species of a radiation constitute a monophyletic group, and they often share some innovative trait or set of traits (sometimes called a key innovation—Liem, 1973) that is thought to have allowed the lineage to undergo a major transition, moving it into a previously unoccupied zone of opportunity exploited in different ways by different branches (for pros and cons of the key-innovation concept, see Koehl, 1996, p. 530). A radiation is a proliferation of variations on a phenotypic theme, accompanied by a proliferation of species. Adaptive radiations are a major feature of the evolution of life. Biologists describe radiations at all phylogenetic levels—the radiation of the eukaryotes, the radiation of multicellular organisms, the radiation of vertebrates, of birds, of ungulates, of dung beetles, of a particular species complex of fish.

The concept of adaptive radiation usually includes both lineage branching (speciation) and phenotypic diversification (Schluter, 2000). Speciation and phenotypic diversification are obviously related phenomena: the genetic isolation associated with speciation promotes phenotypic divergence between populations (Mayr, 1963; Futuyma, 1987), and phenotypic divergence within populations (polymorphism and polyphenism) can promote speciation, as discussed in chapter 27, so the two phenomena must often occur together. It does not follow, however, that adaptive radiation and speciation are inevitably linked (see Turner, 1999). Adaptive radiation can occur without speciation, as in the multiple individual trophic specializations of Cocos Island finches (see below) or in the diversity of human careers. And speciation can occur with little adaptive (ecological) diversification. *Polistes*, a cosmopolitan genus of social wasps (Vespidae), contains more than 300 species that differ primarily in their social and sexual behavior, with relatively little interspecific variation in trophic morphology and behavior, diet, size and shape, or habitat (reviewed in Turillazzi and West-Eberhard, 1996). The possibility of socially or sexually selected radiations with little ecological diversification is a neglected topic that deserves attention, but I do not pursue it here.

Simpson (1953a, p. 393) referred to the "appalling intricacy" of adaptive radiation. Here I focus on only one facet, the contribution of developmental plasticity to phenotypic diversification. The *ecological theory of adaptive radiation*, considered by some to be the predominant and best-supported explanation of radiations (Schluter, 2000), holds that the evolution of phenotypic differences between lineages in a radiation are caused by differences in selection on genetic variation in their different environments (independent evolution in distinctive ecological settings). The ecological theory falls short as an explanation of adaptive radiation, however, because it does not take into account the patterns of phenotypic variation in ancestral populations that, having already been expressed and exposed to selection, are those most likely to give rise to adaptive phenotypic diversification. By assuming that selection in new environments is the only direction-giving factor in diversification, the ecological theory omits from consideration a major potential source of pattern in adaptive radiation. Since adaptive radiations are characterized as diversification from a common ancestral stem, it follows that the nature of the ancestral stem may importantly influence the nature of a radiation.

Examination of patterns of radiation suggests that they commonly exploit developmental plasticity in particular kinds of traits. Adaptive radiations may often originate when ecological conditions favor diversification from a *flexible stem*—an ancestral taxon whose phenotype is adaptively variable in ways related to the specializations that characterize a radiation of descendant species or populations. The *flexible-stem model* of adaptive radiation emphasizes that (1) the origin of variation is an important cause, alongside selection, of adaptive radiation; (2) the phenotypic variants seen in an adaptive radiation often originate within a developmentally flexible ancestral population, or as a result of particular kinds of developmental plasticity present in the common ancestor of the diversified group; and (3) the nature of ancestral developmental plasticity can influence the nature of the radiations.

In the flexible-stem hypothesis, ecology and natural selection are as important as in the ecological theory. The adaptive flexibility of the stem population is presumed to be a product of selection, and the diversification of descendant populations occurs under selection that favors ecologically appropriate variants. Evolved developmental plasticity enables individuals and populations to negotiate the twists and turns of survival in variable, novel, or extreme situations. Then, if particular variants are favored due to competition or to environmental extremes, populations characterized by those variants may rapidly specialize and may speciate. This can occur without loss of flexibility itself, which may respond to further change with additional diversification. This chapter focuses on ancestry and flexibility as factors in diversification, leaving aside discussion of many other factors, such as genetic variance, genetic isolation, and strength of selection, that influence adaptive radiation (Schluter, 2000).

Although some authors (e.g., Taylor and Larwood, 1990) balk at terming radiations "adaptive," the ones discussed here clearly deserve that term because they arise from intraspecific adaptive flexibility in the form of polyphenisms, life stage specializations, learned responses, and structural change in traits of known or likely fitness-affecting function, such as use in food acquisition.

The following sections show not only how flexibility can facilitate diversification, but also how the environments of "explosively" radiating taxa may especially encourage or even demand a high degree of phenotypic flexibility. The groups that radiate most strikingly are those that are both flexible and able to specialize at the same time. This may at first seem contradictory, since flexibility and specialization are often thought to be mutually exclusive adaptive strategies. Recall, however, that the mechanisms of plasticity (see chapter 3) include devices such as learning and variant overproduction that allow highly specific structures to develop via highly flexible pathways. Such hyperflexible devices as somatic selection (e.g., learning, and multidimensional flexibility in the formation of muscle and bone) appear to be the foundations of some of the radiations discussed here. Other patterns of radiation are based on the binary flexibility of developmental switches. By the flexible-stem hypothesis, patterns of plasticity are expected to influence patterns of adaptive radiation.

Rather than attempting a review of adaptive radiation, this chapter examines the possible role of developmental plasticity in some well-studied adaptive radiations, primarily in fishes, birds, and Hawaiian drosophilid flies, with notes on suggestive patterns in some other organisms. Schluter (2000) provides a more complete discussion of adaptive radiation, with emphasis on the ecological theory.

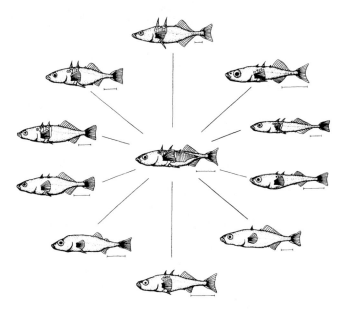

Fig. 28.1. Recurrent lateral-plate and dorsal spine reduction in freshwater sticklebacks. A marine or anadromous (migrant) form (center) has likely been the common ancestor of lake (1–7, clockwise from top) and river (8–10, clockwise from top) populations with reduced plate and spine morphology. From Bell and Foster (1994b), in *The Evolutionary Biology of the Threespine Stickleback*, by permission of Oxford University Press. Courtesy of M. A. Bell.

Binary Radiations

Fishes

In some taxa, such as the threespine sticklebacks of the *Gasterosteus aculeatus* species complex, populations with parallel traits have been repeatedly derived from the same ancestral population or phenotypic form (Taylor and McPhail, 2000; figure 28.1). This pattern of "replicate radiation" (Schluter and McPhail, 1993; Schluter, 2000, p. 55) has been diagrammed as a raceme (figure 28.2). The recurrent binary pattern is particularly obvious in sticklebacks and other boreal fishes because it is simplified by a lack of subsequent branching. The anadromous or marine stem populations are stable over long periods of time, while the derived branches in fossil lineages have been terminated by rapid extinction in relatively short-lived lake habitats, and extant lineages have originated relatively recently, when sticklebacks colonized a multitude of isolated freshwater habitats postglacially, in the last 15,000 years (Bell, 1987, 1988; reviewed in Bell and Foster, 1994b; Schluter, 2000). In some other fishes, such as the miniaturized forms of tropical freshwater groups, replicate radiation has also occurred, but recurrent branches have subsequently

speciated to produce groups of species that share the recurrent derived trait (de Pinna, 1989).

Recall that adaptive evolution is a two-step process: first, variation, then selection. It is not enough to explain parallel adaptive evolution in terms of parallel selection alone (Eble, 1998; see chapter 27); parallel variation must also be explained. Without parallel variation, natural selection cannot produce parallel evolution. In *binary radiations*—radiations characterized by recurrent interspecific alternative phenotypes—some binary developmental pattern may underlie the recurrent parallelism. In other words, the repeated binary parallelism may reflect a *binary flexible stem* (figure 28.3)—a binary developmental pattern, such as a switch between intraspecific alternatives, distinctive life stage phenotypes, or extremes in a continuous distribution, present in the common ancestors of the group.

Parallelism between intra- and interspecific alternatives is a common pattern of recurrence (see chapter 19). The same selection regimes that favor phenotypic extremes relative to intermediates formed by hybridization between contrasting benthic and limnetic species of sticklebacks (Schluter, 1993, 1995; Hatfield and Schluter, 1999), for example, can in the absence of breeding isolation pro-

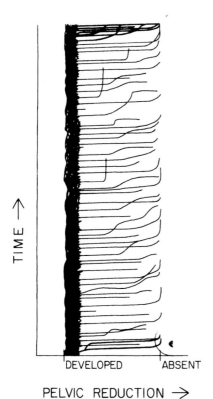

TIME →

DEVELOPED   ABSENT

PELVIC REDUCTION →

Fig. 28.2. "Racime" pattern of recurrent evolution associated with a binary flexible stem. Branches represent pelvic reduction in local populations of sticklebacks (*Gasterosteus aculeatus*) descended from a common marine anadromous ancestor with a fully developed pelvis. From Bell (1987) by permission of Academic Press Ltd., London. Courtesy of M. A. Bell.

duce disruptive development and intraspecific alternative phenotypes (see chapter 11). Thus, individuals within a single population (in Cranby Lake, British Columbia) exploit primarily either the benthic or the limnetic habitat according to phenotype, with the more limneticlike individuals (see Figure 27.7) taking plankton and the more benthiclike individuals exploiting bottom-dwelling prey (Schluter and McPhail, 1992). Intraspecific alternative forms and life stage phenotypes behave much like related species, even in showing more marked divergence in use of resources when competition is intense, as shown by Maret and Collins (1997) in research on the alternative trophic phenotypes of larval salamanders (see also, on fish, Keast, 1977a,b; Robinson and Wilson, 1994). An obvious first step, therefore, toward testing the flexible-stem model is to see if the recurrent phenotypes of a binary radia-

tion are parallel to alternative phenotypes found within populations of the same or related species.

There are at least six genera of boreal fishes, in addition to sticklebacks, that show binary radiations in the form of recurrent reproductively isolated benthic–limnetic species pairs like those in sticklebacks (see chapter 27, especially figure 27.7). They include lampreys (*Lampetra*), arctic charr (*Salvelinus*), Pacific salmon (*Oncorhynchus*), lake whitefish (*Coregonus*), brown trout (*Salmo trutta*; Fergeson and Taggart, 1991), and smelt (*Osmerus*; Taylor and Bentzen, 1993; figure 28.4). All of these genera, like the sticklebacks, contain populations with a benthic–limnetic trophic polymorphism or polyphenism in morphology or behavior (Robinson and Wilson, 1994; Skúlason and Smith, 1995). This is strong evidence that developmental plasticity has been involved in the replicate production of benthic and limnetic forms in radiations of these fish.

The trophic behavior and morphology of a fish often change with age and life stage (Ivlev, 1961; Keast, 1977a,b), as discussed in chapter 27 for sticklebacks. If there is an ontogenetic change between limnetic and benthic feeding, for example, heterochrony could give rise to expression of these behaviors as alternatives. If trophic opportunity and experience vary within a life stage, behavioral and morphological flexibility, could evolve, for example, different phenotypes adopted under different ecological conditions or under competition between populations in the same lake.

A switch of heterochronic origin is an obvious possibility when there are size differences between parallel forms, as in sticklebacks and lampreys (see chapter 27). Heterochrony also is a reasonable working explanation for the origin of the "dwarf" and "normal" sister species pairs in lake whitefish (the *Coregonus clupeaformis* species complex) living in different, isolated lakes (figure 28.5). The dwarf morphotype is regarded as the derived form in whitefish (Pigeon et al., 1997). Most previous discussions of whitefish evolution emphasize the importance of parallel selective factors in the lakes, such as similar ecological or competitive conditions (e.g., Bernatchez et al., 1996; McPhail, 1994; Schluter and Nagel, 1995; Pigeon et al., 1997), without addressing the possibility that the recurrent parallelism could be a result of a common developmental propensity of their common ancestor. As a result, the ecological and local selection side of the evolutionary story is relatively well discussed. The dwarf form prospers only in certain circumstances where it is evidently favored by selection (e.g., in lakes lacking planktivorous cisco (*Coregonus sardinella*; Bodaly, 1979).

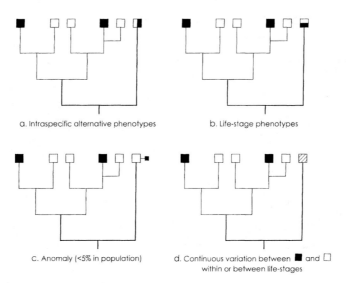

a. Intraspecific alternative phenotypes

b. Life-stage phenotypes

c. Anomaly (<5% in population)

d. Continuous variation between ■ and □
within or between life-stages

Fig. 28.3. Binary flexible stems in adaptive radiation. Different types of ancestral developmental plasticity, represented by a basal group (square on the right in each diagram), can give rise to interspecific alternative fixed traits in descendant populations (solid and open squares).

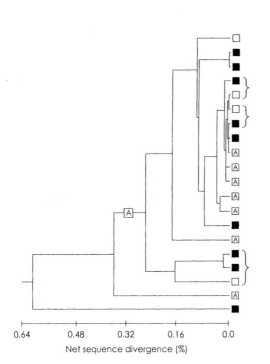

Net sequence divergence (%)

Fig. 28.4. Parallel species pairs in smelt (*Osmerus* species). Open squares = normal-size lake populations; black squares = dwarf lake populations; A-containing squares = anadromous populations. Brackets: sympatric population pairs. After Taylor and Bentzen (1993).

In one lake (East Lake; Pigeon et al., 1997) members of the sympatric morph pair are sister populations, with somewhat reduced gene flow, suggesting that a single colonization of the lake is possible, whereas in other lakes (e.g., Webster Lake and Cliff Lake; asterisks in figure 28.5), the pairs are relatively distantly related suggesting that they independently colonized the same lake. Sympatric members of species pairs have different spawning times and a bimodal but continuous distribution of several key meristic features, such as number of gill-rakers, that are diagnostic taxonomic characters at the species level (Kirkpatrick and Selander, 1979, p. 478; see also Svardson, 1961). In five different lakes in the Yukon Territory, low gill-raker fish are benthic feeders and are found almost exclusively near the bottom of the lake, while high gill-raker fish are primarily plankton feeders found throughout the water column (Bodaly, 1979). These differences are similar to those found between stickleback species pairs (see chapter 27).

The fact that benthic–limnetic alternatives occur repeatedly together in the same lake even when they are not sister populations reinforces the impression that local competition drives a separation into the two habitats. This pattern of parallel pairs that are not necessarily sister species occurs in species pairs of African lake cichlids as well (Rüber et al., 1999; Rüber and Adams, 2001), confirming the impression that ecological factors, possibly including trophic competition, influence closely related fish

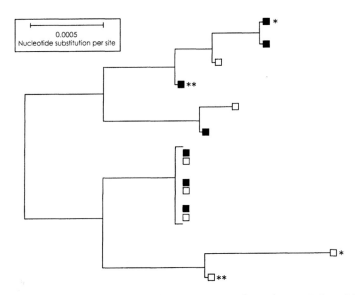

Fig. 28.5. Parallel population pairs in lake whitefish (*Coregonus clupeaformis*, Salmonidae). Black squares indicate dwarf morphotypes; open squares, normal morphotypes; asterisks (*, **) indicate two sympatric morphotypes that are not sister populations. All other dwarf–normal pairs are both their own closest relatives and sympatric (in the same lake). Length of branch indicates degree of genetic divergence (scale at above left). After Pigeon et al. (1997).

to sort themselves into the two trophic habitats if confined within the same lake. This does not, however, ratify the idea (e.g., Rüber and Adams, 2001) that parallel evolution is involved independent of phylogenetic history or developmental constraints. As discussed below, the benthic–limnetic ecological separation can occur immediately between members of the same species not usually showing alternative specializations, if confined in a small pond (Werner et al., 1981), showing that developmental plasticity can produce phylogenetic novelty in extreme conditions.

Phylogenetic analyses usually tabulate characters of adults without attention to low-frequency or life-cycle variants that are important in evolution. The Lake Malawi cichlid *Haplochromis quadrimaculatus*, for example, spends the first two years of life (until about 13 cm in length) inshore as a benthic feeder, then during the third year it apparently lives in open water, subsequently reappearing in shallow water to spawn on rocky shores (Fryer and Iles, 1972). As pointed out by Keast (1977b), it is common in fish species that ultimately become piscivores to pass through ontogenetic stages of being planktivores and insectivores, and Fryer (1959, cited by Gould, 1977) long ago suggested that water-column foragers among fish species are paedomorphic. Ontogenetic develop-

mental plasticity can give rise to alternative benthic–limnetic phenotypes, or fixation of one or both of the two habitat uses. Selection, genetic accommodation, and further modification could then occur in the new environments, but the origin of the newly predominant traits would be rooted in a cladistically invisible history. A phylogram based on characters of adults would not register the pertinent ontogenetic variation.

The sets of morphological differences between whitefish morphs and species pairs are of a kind known to be inducible by environmental factors that affect growth (see references in Bodaly, 1979, p. 1220). Kirkpatrick and Selander (1979) recognize the possible importance of developmental plasticity when they note that "[t]he whitefishes show a special propensity for generating dwarf morphotypes" (p. 483). A species pair found in the Allegash River basin of northern Maine proved to be polygenically divergent distinct species, not morphs of the same species (Kirkpatrick and Selander, 1979), and subsequent work has confirmed the genetic divergence between sister species pairs (Pigeon et al., 1997; figure 28.5). These findings of genetic distinctiveness do not contradict a flexibility hypothesis; they are consistent with a hypothesis of genetic accommodation of an initially flexible response. More information on the ontogeny of trophic behavior and morphology in these fish would permit more direct comparison with the adult characteristics of the species pairs.

In sticklebacks, there is considerable evidence that the recurrent pelvic reduction represented by the racime diagram (figure 28.2) is likely a result of a binary flexible stem. While a full series of intermediates can be observed in many populations (see figure 19.8), showing a potential for gradual evolutionary change, developmental anomalies indicate that change could be rapid. It is quite common to find bilaterally asymmetrical specimens in which both extremes of pelvic girdle development and reduction occur in the same individual (Bell, 1987, p. 370). This demonstrates a developmental potential for rapid switching between the extremes as in a binary flexible stem.

Lateral plate reduction is a recurrent characteristic of the freshwater threespine sticklebacks that have descended from marine or anadromous populations, which are monomorphic for the complete morph (figure 28.1). The developmental plasticity of stickleback plate morphology is revealed in the freshwater populations, which can be trimorphic, dimorphic, and monomorphic for the three modal plate morphs (see figure 21.10). The reduced plate morphs may have involved recurrent heterochrony (neoteny), as suggested by Bell (1981): the complete morph passes through ontogenetic stages that resemble first the low morph and then the partial morph. McPhail (1994) mentions evidence (McPhail, unpublished observations) that the traits that commonly diverge in freshwater populations, including spine number, pelvic girdle structure, and lateral plate expression, are threshold traits and states that this may account for the striking divergence in these traits with only minor genetic change. It would also help to account for their recurrent parallelism.

In the flexible-stem model, there is an interplay between binary developmental patterns and recurrent pairs of environments, such as the bottom and the water column of a lake. Ecology-appropriate selection is presumed responsible for the evolution of alternative or life stage phenotypes and their adaptive expression, as well as for their genetic divergence when they are modified in particular lake or river environments. The predominance of the complete array of plate, spine, and pelvic armor in a stickleback population is strongly associated with predation by fish and birds (see Reimchen, 1988), whereas reduced armor, including pelvic reduction in low morphs, is sometimes associated with predation by insects, especially by dragonfly nymphs. Pelvic armor and spines are known to hinder vertebrate predation and may facilitate capture by predatory insects (Reimchen, 1980, 1988). The fossil stickleback (*Gasterosteus doryssus*) shows the same association of pelvic reduction and absence of predatory fishes, in an exceptionally good paleontological record (see Bell, 1988; Bell et al., 1985). In addition to environmental differences in selection by predators, restricted calcium availability in lakes sometimes may contribute to plate reduction, for reduced plates are associated with low-calcium levels in three different areas. This relation is confounded, however, by absence of vertebrate predation in the same areas; both factors are associated with reduced plate morphology (Reimchen, 1994, p. 254).

As in other taxa with recurrent phenotypes (see chapter 19), anomalies (low-frequency variants) reveal the developmental capacities already present in ancestral populations: the anadromous populations of threespine sticklebacks contain incompletely plated morphs at low frequencies, rarely exceeding 2% (reviewed in Bell, 1984, p. 448). Some Atlantic populations have high frequencies of low and partial morphs (Klepaker, 1996). Such variants would provide material for selection and rapid evolution toward establishment of the partial and low plate morphs if they were to prove advantageous.

G. aculeatus is a superspecies, that is, a complex of divergent populations with many subspecies and several full biological species (reviewed in Bell and Foster, 1994b). All of these populations are lumped under a single name, for so far it has proven impossible to unravel relationships by conventional methods that assume sequential branching of a phylogenetic tree. The sticklebacks illustrate the difficulties of taxonomic analysis in phenotypically plastic organisms recurrently subjected to reproductive isolation in habitats (fresh water) that are at one extreme relative to the ancestral ones (in the sea or alternating between fresh water and the sea). A total of 41 species names had been applied to distinctive forms when the nomenclature was reviewed in 1959 (see Wootton, 1976, p. 253).

The evolution of recurrent trophic alternatives in recurrent alternative environments, such as the water-column and benthic habitats of lakes, may sometimes result from, or be reinforced by, learning, with the interspecific alternatives a result not of a dichotomy in the ancestral phenotypes, but of a dichotomous environment for selection on continuous variation, including continuous variation with age or size during ontogeny. Learning can establish correlations between morphological extremes and trophic specializations by reinforcement of successful alternative foraging tactics and morphologies (see chapter 18). Learning can also contribute to innovation under competition, and competition is associated with the separation of lake fishes into benthic and limnetic specializations (e.g., Robinson et al., 1993; Robinson and Wilson, 1994). Studies of foraging behavior in guppies (*Poecilia reticulata*) show that food-deprived fish are more likely to explore and adopt new food sources than are non-food-deprived fish, and this was size dependent, with smaller fish more likely to innovate than larger ones (Laland and Reader, 1999). This may contribute to the evolution of freshwater sticklebacks, where there is a limnetic–benthic species pair in each of six lakes (summarized in McPhail, 1994; see figure 27.7).

As discussed in chapter 27, juveniles and adults of marine populations show body form and trophic niche differences parallel to those of the limnetic and benthic species pairs. Presumably, these ontogenetic differences grade into one another during growth, rather than being marked by a clear metamorphosis, but any heterochronic behavioral or morphological variants produced could have their ecological distance from normal individuals exaggerated by learned dichotomous (alternative) foraging tactics that would focus selection on two behavioral modes and produce correlations between extremes of morphology and behavior.

The importance of behavior, especially morphology-influenced learned foraging behavior, in the evolution of the stickleback benthic–limnetic species pairs has been neglected in research on their diversification, but there are some promising leads. Cresko and Baker (1996) discovered benthic and limnetic morphotypes of *G. aculeatus* in Benka Lake (Alaska) by capturing individuals observed feeding either in the benthos or in the water column. The limnetic, plankton feeders had a more fusiform shape, larger eyes, longer and more numerous gill rakers, and a smaller, more tubular mouth—the same characters that differentiate limnetic from benthic members of parallel species pairs. All traits except body depth contributed to a significant discrimination of the two morphotypes. In 81.1% of the 159 stickleback examined, foraging type was supported by morphological attributes. The existence of the two forms is obscured if populations are sampled without regard to foraging behavior, for the morphological traits are continuously variable. Cresko and Baker (1996, p. 349) believe that this seldom-documented pattern of "incipient divergence" may be more common than presently suspected in the tens of thousands of lake populations of sticklebacks, in part because traditional collection techniques sample fish found primarily in shallow-water and lake-edge habitats.

The correlations between morphology and behavior shown by Cresko and Baker (1996) are consistent with what is known about the interaction between morphology and experience in determination of stickleback feeding behavior. Stickleback feeding behavior, like that of other fishes (Milinski, 1984a), is influenced by learning and memory (Hart and Gill, 1994), and the particular behavior reinforced depends upon phenotypic traits such as mouth morphology, body size, and visual capacity (asterisks in figure 27.9). These are among the traits that most obviously differentiate the sympatric species pairs, which, like the Benka Lake morphotypes, differ in mouth structure, body size, and eye size (see figure 27.7). The learned specialization can in turn influence the direction of selection on trophic morphology, in a self-reinforcing cycle of developmental and evolutionary change (see chapter 18). Trophic morphology is also affected by use, as demonstrated in laboratory studies of sticklebacks (e.g., see Day et al., 1994) and cichlids (Greenwood, 1965, 1984; Meyer, 1985, 1987a) and in natural populations of bluegill sunfish (*Lepomis macrochirus*), whose populations in single lakes are dichotomized as vegetation and open water specialists (Ehlinger and Wilson, 1988; concisely reviewed in Wilson, 1989).

Since the effects of plasticity (learning and use) are in the same direction as selection, it may often be difficult to know how much of the morphological difference between populations is due to plasticity and how much is evolved. In sticklebacks, morphological differences between populations with contrasting trophic morphology, and in competition with each other, are greater if plastic traits are taken into account, confirming that adaptive plasticity affects the phenotypes in the same direction as selection (Schluter, 1994, p. 799). When individuals of two sympatric species of sticklebacks, one benthic and one limnetic, were reared to adult or nearly adult size on each other's diets, their morphology changed toward resemblance of the other species whose diet they consumed (Day et al., 1994). This is strong evidence that phenotypic plasticity contributes to the development of the recurrent benthic and limnetic forms.

Behavioral divergence between subpopulations can be immediate. In a critical experiment, Werner et al. (1981) introduced bluegill into a small pond and found that some individual fish rapidly specialized trophically either as shallow-water bottom feeders (littoral or benthic specialists) or as limnetic, water-column foragers. These specialists had higher food intake than did unspecialized individuals. So under strong competition or high densities, divergent specialization is favored (McPhail, 1994). Schluter (1994) has demonstrated a role of trophic competition in diverging selection in sticklebacks. Learning can play an important role in speeding divergent trophic specialization, making it possible for fixation of a behavioral specialization to occur in the space of a single generation. Wilson (1989) suggests a connection between this kind of polyspecialization and speciation. He notes that fish species specialized to one of the two intralacustrine habitats have both morphological and behavioral differences. Morphology-biased learning, where different feeding procedures are more rewarding in particular morphological classes of individuals, would help match morphology to behavior and habitat (see chapter 18). Add to this the fact that male fish can show alternative mating tactics evolved under sexual selection (Gross, 1984) and that benthic and limnetic fish assortatively mate by size (Schluter and Nagel, 1995; Rundle et al., 2000), and these fish illustrate the combination of flexible stem and sexually selected divergence that I believe is especially conducive to both speciation (chapter 27) and phenotypic radiation.

McDowall (1988) reviews many other examples of parallel population or parallel species pairs, or recurrent parallel forms, formed by landlocking of formerly anadromous fish populations, and McPhail (1994) discusses the different kinds of freshwater species pairs, also ultimately descended from anadromous ancesters, in sticklebacks. The descent of a lacustrine-freshwater feeding stage from an oceanic-pelagic, marine feeding stage occurs in nearly all groups containing anadromous species (marine–freshwater migrants), including lampreys, sturgeons, salmonids, a plecoglossid, galaxiids, retropinnids, osmerids, clupeids, a gadid, percichthyids, gasterosteids, gobies, and eleotrids (McDowall, 1988, p. 174). Nonmigrant individuals were discovered in some salmonid species thought to be completely anadromous when populations artificially transplanted into freshwater turned out to be maintained entirely by nonmigrant "deviants" (McDowall, 1988), showing that there is potentially selectable variation in the stem populations with regard to migratory behavior. Phenotypic flexibility is also suggested by the fact that landlocking is reversible: in some salmonid populations, individuals of freshwater populations derived from anadromous ones can still successfully smolt and live in the sea even after up to 25 generations in fresh water (review in McDowall, 1988).

Finally, a kind of flexibility that is potentially important in the evolution of lacustrine species

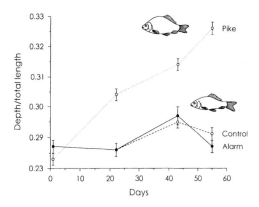

Fig. 28.6. Predator-induced change in body shape in crucian carp (*Carassius carassius*). Dotted line (pike): the body-depth:total-length ratio of crucian carp grown in the presence of pike fed (in another aquarium) on crucian carp. Dashed line: control, grown in the absence of pike. Solid line: grown in the presence of alarm substances from injured crucian carp. Differences like those in the drawings occur in crucian carp found in ponds with or without pike. After Brönmark and Pettersson (1994). Drawings courtesy of C. Brönmark.

pairs is the remarkable ability of some fishes to respond to the presence of predators, especially piscivores, with a marked change in body size and shape in just a few weeks (Brönmark and Pettersson, 1994; figure 28.6). The nature of this shape change is similar to that of benthic species relative to limnetic or pelagic ones, and body size and shape differences are well known to affect susceptibility to predation in stickleback and other fishes (see especially Reimchen, 1988). Is it possible that phenotypic plasticity in response to predation sometimes has contributed to the divergence between lacustrine species pairs? The degree to which the differences between species or morph pairs in fish may be predator induced rather than a response to dietary differences or evolved change remains to be investigated.

Drosophila

Richardson (1973) describes a series of parallel sibling species pairs of *Drosophila* endemic on each of the different islands of the Hawaiian archipelago. He made detailed observations of one pair (*D. mimica* and *D. kambysellisi*) on the island of Hawaii and describes the history of its habitat—islands or "kipukas" of relatively old vegetation-covered lava surrounded, and isolated from, other such islands by more recent lava flows. He noted that one of the pair of species (*D. mimica*) is more likely an earlier colonist of new kipukas, being a more active disperser and living as both larva and adult on rotting fruits of an early successional plant (*Sapindus*) in relatively brightly illuminated areas of open forest. The other species (*D. kambysellisi*) is specialized on a later successional plant (*Pisonia*) that grows only in deep shade. Habitat selection separates the species with very little overlap. Following suggestions of Maynard Smith (1966), Richardson speculates that a *mimica*-like form colonized the islands one by one and then speciated on each, and notes that a chromosome inversion distinguishes the two species; Carson and Kaneshiro (1976) view *D. mimica* as close to the original colonist of the islands. The alternative possibility, that a *mimica*-like ancestral species was a polyspecialist able initially to breed in both habitats, is not excluded and could also explain the presence of parallel characteristics in the series of species pairs present on different islands.

Snails

Coe (1949) describes three sets of parallel species pairs in hermaphroditic snails of the genus *Crepidula*, where one member of each pair (species *nivea*, *arenata*, and *onyx*) produces small eggs with little yolk that give rise to free-swimming veliger larvae, and the other member of each pair (species *williamsi*, *adunca*, and *norrisiarum*, respectively) produces large well-yolked eggs that complete development to the adult stage within the egg capsules beneath the parent adult's foot. Sometimes all six species are found in the same tide pool, but they do not interbreed in the field or in the laboratory. Intraspecific differences in egg size were not induced by starvation, which yielded undersized adults that produced a smaller number of eggs not altered in size; thus there was no evidence for a nutritionally based switch relating to the differences between species. Coe cites examples of other similar species pairs in mollusks, including a pair of *Octopus* sibling species with no morphological differences between adults but whose eggs show an eightfold difference in size. These findings invite investigation of a possible ancestral plasticity that would account for the recurrent pattern of variation.

Insect Social Parasites

Parallel species pairs also occur in some free-living and socially parasitic Hymenoptera (ants, wasps, and bees), with the social parasites having parallel sets of adaptations (e.g., loss of pollen-collecting apparatus in bees—see Michener, 1978, on halictids), increased cuticular armor and mandible size in wasps; and a "inquiline syndrome" (workerless, host-queen-tolerant parasite) of the ant parasites discussed in chapter 27, consisting among other things of the absence of spurs on the middle tibiae, bizarre modifications of the petiole and postpetiole, and reduction of many pheromone-producing glands (Buschinger, 1970; Wilson, 1971a,b; Bolton, 1988; see Michener, 1974, on bees; see West-Eberhard, 1996, on wasps). In these insects, both the parallel morphological specialization of parasitic species and the intraspecific behavioral parasitism that likely preceded it in each case (Buschinger, 1970, 1986; Wilson, 1971a,b; Hölldobler and Wilson, 1990; West-Eberhard, 1986; Cervo and Dani, 1996; Carpenter et al., 1993) represent multiple parallel convergences (Carpenter et al., 1993; Pamilo et al., 1981), with intraspecific parasitism a recurrent alternative in well-studied social (group-nesting) species. In some groups (e.g., of ants) the social parasites are miniature forms closely related to the single species that serves as their host (see chapter 27).

Multidirectional Radiations

Some so-called "explosive" radiations produce a large number of distinctive adaptive forms. Some forms may be recurrent, but the striking feature of these radiations is the great diversity and specialization of related species. Well-studied examples include the Darwin's finches on the Galapagos Islands, various animals and plants in the Hawaiian Islands, and the cichlid fishes of certain African lakes. All of them are characterized by a great proliferation of extreme specializations in a group of closely related species descended from a small number of colonists in a short time. Just as binary patterns of evolutionary change are associated with binary flexible stems in some phase of their evolution, multidimensional flexible stems or multidirectional radiations are often associated with hyperflexibility that contributes to a broader range of descendant variants.

Chapter 3 discussed various mechanisms of plasticity that could characterize a *hyperflexible stem*—ancestral hyperflexibility associated with a variety of descendant variants. Such mechanisms include learning and other kinds of developmental plasticity influenced by use or experience. A salient feature of mechanisms of hyperplasticity is that flexi-

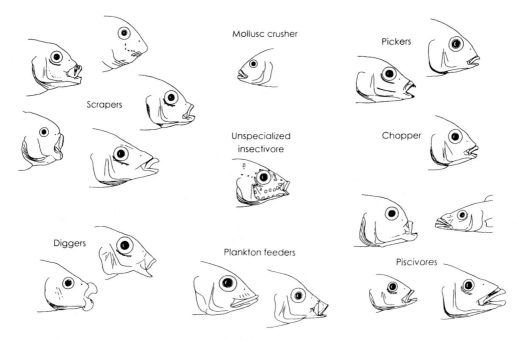

Fig. 28.7. Adaptive radiation in mouth morphology and diet of African cichlids of Lake Malawi. After Fryer and Iles (1972).

bility itself can persist across generations even though the flexible individuals adopt fixed configurations of behavior or morphology (see chapter 3). This means that specialization is not a self-limiting evolutionary dead end: a lineage that adopts one specialization has not necessarily lost the capacity to very rapidly adopt another.

As in binary radiations, explosive and multidirection radiations are clearly due to special characteristics of both the places and the organisms themselves, for rapid radiations do not occur everywhere, and not all organisms in places where they do occur have speciated and radiated to the same degree. In the Hawaiian Islands, for example, about 6,500 endemic insect species are believed to have descended from only about 250 ancestral immigrants, and comparable radiations have occurred in some birds, such as honeycreepers (Drepanidae), and in some plants, such as the lobelias and composites (Zimmerman, 1970), showing that something about the Islands is conducive to radiation. But other lineages have not diversified at all even though present for a comparable period of time (for examples and discussion of reasons, see Carson, 1987a; Greenwood, 1964).

At the end of this chapter, I discuss how a synergism between sexual selection and adaptive flexibility may be an important feature of some rapid

radiations. Other factors that facilitate radiation include such things as spatial entrapment in a limited geographic area where populations are periodically subjected to resource limitation and strong competition and selection (Schluter, 2000), and habitat dissection and population viscosity (e.g., on the Galapagos, see Grant, 1986; on cichlid habitats in African Lakes, see Witte, 1984; McKaye and Gray, 1984; on habitat dissection in Hawaii, Carson and Kaneshiro, 1976).

African Lake Cichlids: A Morphological Flexible Stem

African lake cichlids have undergone a trophic radiation (figure 28.7) that capitalizes on an unusual flexibility of head and mouth morphology (Liem and Osse, 1975), as well as other kinds of developmental plasticity, including heterochronic change in stage-specific trophic specializations, learning, and other kinds of experience-mediated change in bone and muscle (Galis et al., 1994). There are more than 1,500 haplochromine cichlid species, including lake and river species (Turner, 1999). Within Lake Victoria, one of the African lakes where cichlids have undergone famous "explosive" trophic radiations, a single genus, *Haplochromis*, was estimated by Greenwood (1964, cited by

Lowe-McConnell, 1969) to contain 120 species. About 30% of them feed on insect larvae and detritus; about 12 species feed on mollusks, extracting them from their shells in two different ways; four feed on algae; one on plants; and about 50 on other fish, including specialist predators on eggs and larvae of other *Haplochromis* species.

The Lake Victoria haplochromines use a wide array of food sources in the lake, with a diversity of behaviors and internal trophic morphological (e.g., gut length and tooth form) specializations that rival those of several genera or families of marine fish that one can observe feeding along a seashore (Greenwood, 1984). Some pick insect larvae from the surface of mud or sand; pharyngeal crushers crush snail shells; and oral shellers wrench snails from their shells. Three classes of phytophages graze algae from rocks and plants and browse on uprooted vegetation. Some prey on other fish, and these include a great variety of hunting and handling specializations with variable prey taxa, sizes, life stages (eggs, larvae, juveniles, or adults), and different parts of prey (e.g., scales, fins, or eyes) handled using different tactics (e.g., ambush or pursuit), frequenting different substrate types, depth ranges, and exposure to wind (Witte, 1984). A recent geological study (Johnson et al., 1996; see also McCune et al., 1984) indicates that most of the morphological diversity has evolved within the past 12,000 years, beginning when the lake filled after being completely dry.

Several authors have suggested that a hyperplastic morphological structure has contributed to this so-called "explosive" radiation of cichlids. Liem (1973) describes the unusual jaw morphology of cichlids compared with other fishes and points out how an evolutionary change in musculature and joints freed the premaxillary and mandibular jaws to specialize and diversify in food collection, while the pharyngeal jaw apparatus took over the functions of processing and transporting food. Galis and Drucker (1996) show how a previous decoupling of the upper and lower pharyngeal jaw movements set the stage for the modularization noted by Liem to be effective in increased biting efficiency. In effect, these two decouplings in the functions of the jaws represent increased modularization of the feeding apparatus, liberating the premaxillary and mandibular jaw complex and the pharyngeal jaw from developmental and genetic constraints and allowing selection to operate independently on each (see chapter 7). Such separation of functions via increased morphological modularity has facilitated the evolution of a remarkable diversity of forms in the mouths of cichlids, associ-

ated with their wide array of different trophic specializations. How these key pivotal changes may have been driven in part by sexual selection is discussed at the end of this chapter.

To the flexibility of the cichlid jaw apparatus is added a diet-dependent flexibility in the development of teeth, jaws, and muscles during use (Greenwood, 1965; Liem and Kaufman, 1984; Meyer, 1987a; Galis, 1993a,b; see figure 3.4). An indication of the importance of diet in development of the trophic apparatus was provided by early observations by Greenwood (1965) on a specimen of a Lake Victoria mollusk specialist, *Astatoreochromis alluaudi*, that was reared in an aquarium on a soft diet (figure 28.8). This fish, a member of the *Haplochromis* group, had weakly developed pharyngeal bones, a somewhat reduced apophysis, fewer enlarged pharyngeal teeth, and a less molariform shape to the teeth, and reduced calcium content of bones compared with lake-caught specimens, which have massive pharyngeal bones and well-developed molariform teeth. This morphological change is of special interest because the aquarium-reared fish resembles those of populations of the same species from other lakes (e.g., Lake Nabugabo) where snails are rare. Greenwood suggested that the increased muscle mass and pressure exerted on the pharyngeal bones and the apophysis in individuals that feed by crushing snails beginning at an early age, as in the Lake Victoria fish, plus the extra ingested calcium of shells could account for morphological differences formerly ranked as subspecies characters.

Greenwood's suggestions regarding the importance of plasticity were confirmed in later work. Hoogerhoud (1986, cited in Galis, 1993b) showed that the differences observed by Greenwood could be produced in young of the same brood by feeding them on hard versus soft diets. Witte (1984) compared wild-caught and aquarium-reared individuals *Haplochromis squamipinnis* from Lake George and showed that aquarium-reared fish had altered morphology of the premaxillae, which is associated with increased biting strength and was probably due to increased digging in sand for (soft) food—a reminder that mouths have uses other than food mastication, as I discuss further below in the section on sexual selection. Witte's (1984) observations also show that these differences can develop in partly grown individuals (50 mm long, in fish reaching a maximum of about 120 mm). That is, change due to plasticity need not be determined in very early ontogenetic stages.

More detailed later work further confirmed a role for phenotypic plasticity in cichlid trophic di-

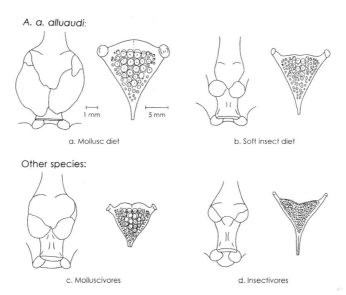

A. a. alluaudi:

a. Mollusc diet

b. Soft insect diet

Other species:

c. Molluscivores

d. Insectivores

Fig. 28.8. Effects of diet on the pharyngeal bone morphology of an African cichlid fish (*Astatoreochromis alluaudi alluaudi*): pharyngeal apophysis (left) and lower pharyngeal bone (right) of (a) *A. a. alluaudi* resulting from the natural molluscan diet in Lake Victoria; (b) *A. a. alluaudi* reared in an aquarium on a soft insect diet; (c) representative mollusk eaters (left: apophysis of *Haplochromis ishmaeli*, a molluscivore; right: pharyngeal jaw of a mixed mollusk-insect eater); (d) representative insect eaters (left: apophysis of *H. empodisma*; right: pharyngeal jaw of an insectivore). a and b and pharyngeal hypophyses of c and d, after Greenwood (1965). Pharyngeal jaws of c and d after Greenwood (1984).

versification. Morphological trophic polymorphism was observed in some New World cichlid species such as *Cichlasoma minckleyi*, which has a papilliform morph, with small pharyngeal teeth, relatively narrow head, slender pharyngeal jaw, and long intestine best suited for feeding on plant material; and a molariform morph with large pharyngeal teeth, wider head, and relatively stouter jaw and short intestine that can process a broader range of foods including snails (Kornfield et al., 1982). The two morphs were thought by some to be different species but were shown by allozyme studies and mating observations to be morphs of the same species (Sage and Selander, 1975; Kornfield et al., 1982). These differences parallel those found between species in the African lakes, and as often occurs when contrasting adaptations develop as intraspecific alternatives (see chapter 19 and discussion above), the same two specializations have evolved recurrently in different lineages of the lake radiations (Liem and Kaufman, 1984).

The diet-influenced morphological flexibility of cichlids has been examined experimentally by various investigators. Meyer (1987a, 1990a) showed that under different dietary conditions, individuals produce tooth and bone differences like those be-

tween intraspecific morphs (figure 28.9) and between some related species. Wimberger (1991) obtained smaller effects of diet on morphology in the cichlids *Geophagus brasiliensis* and *G. steindachneri*. He critically discusses the differences between his results and those of others that illustrate the effects of diet and the possible relation between flexibility and diversification in the African lakes (cf. Meyer, 1987a). The work of Galis (1993a) on *Haplochromis piceatus* shows the potential importance of phenotypic accommodation by muscle and bone in developmental and evolutionary transitions in cichlids, and that of Witte (1984) on *H. squamipinnis* shows how foraging behavior affects morphology and diet. Added to the fact that genetic distances between species in the cichlid radiation of African Rift lakes are extraordinarily small (Sage et al., 1984; Meyer et al., 1990), the findings on intraspecific morphological divergence (see references in Dominey, 1984; Witte, 1984; Meyer, 1989b; West-Eberhard, 1986, 1989; Wimberger, 1992, 1994) suggest that populations of lake cichlids are primed by plasticity for both opportunistic divergence and speciation: given a brief period of isolation or assortative mating under pressure of trophic competition, a trophic extreme could

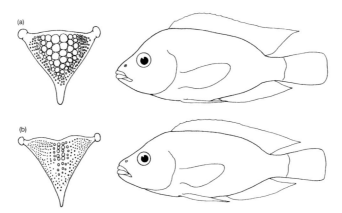

Fig. 28.9. Diet-associated morphs in a cichlid fish (*Cichlasoma citrinellum*): pharyngeal jaw (left) and body form (right) of (a) molariform morph, with benthic body form, that feeds primarily on snails (94% of diet, based on stomach contents); (b) papilliform morph, with limnetic body form, that had a diet of only 19% snails. The morph adopted can be subsequently reversed in some individuals (2 of 12) by diet reversal. From Meyer (1990a) by permission of Academic Press, Ltd., London. Courtesy of A. Meyer.

rapidly diverge genetically, given the burst of genetic evolution that can accompany phenotype fixation of specialist morphology, especially one that, like the papilliform and molariform morphs of some cichlids, has been a member of a pair of alternatives known to represent a fitness trade-off in performance (Meyer, 1989a; chapter 21). This would accelerate both adaptive radiation and speciation. Meyer (1987b) emphasizes that plasticity can *retard* rather than accelerate adaptive radiation. But, as discussed in chapter 7 and in later papers by Meyer (e.g., 1989a), flexibility can accelerate evolution when it leads to recurrent specialization under environmental extremes, as in organisms capable of learning (e.g., see Murdoch et al., 1975; Werner et al., 1981).

The two morphs of *Cichlasoma minckleyi* feed similarly if provided with abundant food. But if food is scarce, they specialize divergently: the papilliform morph prefers soft foods and the molariform morph concentrates on hard foods such as snails (Liem and Kaufman, 1984, p. 209; figure 28.10). Behavioral specialization is most marked at low food concentrations. This means that directional selection toward increased specialization of morphology and behavior would also be most marked at low food availability.

This observation of increased specialization under food scarcity fits a pattern seen in other rapid trophic radiations, namely, that directional change is driven in episodes of strong selection, when food limitation and trophic competition force dietary specialization, including those influenced by morphological predispositions (e.g., Grant, 1986;

Schluter, 1994; Robinson and Wilson, 1994; see also the section below on Galapagos finches). The episodic nature of specialization in a temporally variable environment was borne out by Witte's (1984) observation that during diatom blooms in Lake Victoria, specialized zooplanktivores, detritivores, and insectivores broaden their diets to include plankton. Some haplochromines even show

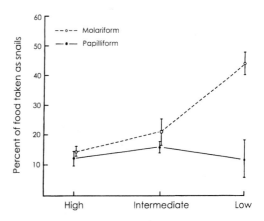

Fig. 28.10. Effect of food abundance on diet in the papilliform and molariform morphs of *Cichlasoma minckleyi* (Cichlidae). Graph shows the percentage of snails eaten, given a choice of snails, brine shrimp, flakes, fish, and tubifex worms. The specialization in mollusks of the molariform morph is exaggerated relative to that of the papilliform morph under food shortage. After Liem and Kaufman (1984).

diurnal change in degree of specialization, depending on how far the move vertically in the water column during noctural migrations, and on the migrations of their prey. Lowe-McConnel (1969) reviews evidence for extreme effects of drought and changing lake levels on population dynamics and locations (in deep vs. shallow water, with correspondingly different trophic conditions) on lake fishes in general. McCune et al. (1984) discuss the effects of longer cycles of drought and rainfall on opportunities for fish in lakes, some of which could have influenced the flexibility and specialization of African cichlid lineages.

Meyer (1985, 1987b) demonstrated the influence of learning and experience on responses to novel prey by fry of the cichlid *Cichlasoma managuense* (see also Laland and Reader, 1999, on learned feeding innovation in fish). Naive fry upon first feeding do not discriminate prey well. Like novice young birds (see below), they "not only snap at prey items, but also attempt to feed on inert objects, such as detritus, air bubbles, and eggshells of *Artemia salina,*" sometimes ingesting small stones (Meyer, 1987b, p. 131). This is typical exploratory behavior, the first stage of trial-and-error learning (see chapter 18). Meyer (1985) also showed that, as in birds (discussed below), an aspect of morphology—eye size—affects capture success, presumably due to enhanced visual acuity. Thus, as in the avian and insect radiations discussed in this chapter, morphology could affect the rewards for pursuit of visually hunted prey and, via learning, trophic specialization in cichlids. Learning can drive trophic divergence in a self-reinforcing spiral in organisms like cichlids where diet influences morphology and morphology reinforces learned dietary preference, especially under food scarcity and competition when selection is strong (Wimberger, 1994; Galis et al., 1994; Galis, 1996a). At the same time, learning maintains the capacity to vary, so populations are perpetually poised for change in a new direction.

Hawaiian Drosophilids: Behavioral and Biochemical Flexible Stems

In Hawaii, there are more than 480 endemic species of drosophilid flies (Carson, 1992b), including predators, parasites, nectarivores, detritivores, fungivores, herbivores, and parasites (Heed, 1971; figure 28.11). Although DNA sequence data suggest that one fungus-feeding Drosophila species is 10 million years old, twice as old as the oldest unsubmerged Hawaiian island, most lineages form very closely related clusters of species, many of

them less than about 0.4 million years (Carson, 1992b). Discussions of the rapid radiation of Hawaiian Drosophila have emphasized allopatric divergence and speciation under sexual selection (e.g., see Spieth, 1974; Carson, 1978; Dominey, 1984; Kaneshiro and Boake, 1987), because the groups of drosophilid flies that have *not* radiated in Hawaii lack both the lek mating systems and the marked diversity in sexually selected traits that characterize the diverse, endemic picture-winged species. There is less information on the ecological aspects of the radiation. But it is clear from descriptions of larval ecology, and general correlations with egg morphology, ovipositor length, and the ovarian specializations of females (Kambysellis et al., 1995, 1999; Kambysellis and Craddock, 1997) that an adaptive radiation has occurred, with larval habitats that include fallen fermenting leaves, rotting bark, fermenting flowers, fruits, fungi, sap fluxes, and frass (figure 28.11).

Schluter (2000) was unable to confirm a statistical association between breeding substrate and egg morphology in the Hawaiian *Drosophila*. But their adaptive radiation may be primarily in nonmorphological aspects of the phenotype, such as maternal oviposition behavior, ovarian physiology, and biochemical tolerance to alcohol concentrations in the larval environment, with habitat shifts facilitated by larval exploratory behavior and learned preferences of females for larval habitats experienced early in adulthood (Jaenike, 1982, 1988). Furthermore, it may be oversimplistic to seek a single-trait single-broad-substrate correlation in insects in which so many traits have been implicated in substrate adaptation (different species may use different combinations) and where so many substrate variants are possible within the broad habitat descriptions used (flux breeders, e.g., are known to inhabit sap fluxes of varied species, chemical and bacterial composition, and age; Kambysellis and Craddock, 1997). An ecological challenge to egg respiration may be solved by the length of the respiratory filaments in one species, and by ovipositor length (egg placement) in another (Kambysellis and Craddock, 1997, review information on the adaptive significance of these characters).

A behavioral and biochemical flexible stem in drosophilids is suggested by studies of larval habitats in conjunction with data on alcohol tolerance responses and of ovarian characteristics of females. Montgomery (1975) reared Hawaiian drosophilids from field-collected samples of rotting plant stems and sap exudates of native trees from 32 different families. His data offer a crude estimate of specialization. Larvae of most species (77%) are found

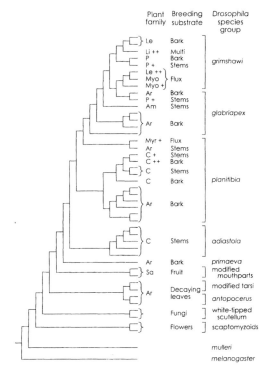

Plant family	Breeding substrate	Drosophila species group
Le	Bark	
Li ++	Multi	
P	Bark	
P +	Stems	grimshawi
Le ++		
Myo	Flux	
Myo +		
Ar	Bark	
P +	Stems	
Am	Stems	glabriapex
Ar	Bark	
Myr +	Flux	
Ar	Stems	
C +	Stems	
C ++	Bark	
C	Stems	
C	Bark	planitibia
Ar	Bark	
C	Stems	adiastola
Ar	Bark	primaeva
Sa	Fruit	modified mouthparts
Ar	Decaying leaves	modified tarsi
		antopocerus
	Fungi	white-tipped scutellum
	Flowers	scaptomyzoids
		mulleri
		melanogaster

Fig. 28.11. Adaptive radiation in the larval habitat of Hawaiian drosophilid flies. All species oviposit in decaying plant tissues used by larvae as food, but species specialize in particular plant families and (within plants) on particular tissues (bark, stems, sap flux, fruit, flowers, and leaves, and fungi) unless otherwise shown. Plant families are Amaranthaceae (Am), Araliaceae (Ar); Campanulaceae (C); Leguminoseae (Le); Liliaceae (Li); Myoporaceae (Myo); Myrtaceae (Myr); Pandanoceae (P); and Sapindaceae (Sa). Pluses indicate additional families are used: +, 1–3 additional families; ++, > 3 additional families. Phylogeny from Kambysellis and Craddock (1997); substrate data from Craddock and Kambysellis (1997), based on Heed (1968) and Montgomery (1975); plant families from Kambysellis et al. (1995), after Heed (1968) and Montgomery (1975).

in a single family of plants. Sometimes more than one species can be reared from the same decaying plant sample (see also Conant, 1978; Heed, 1971), but differential larval exploitation of shared sites is possible (e.g., see Parsons, 1981, and references therein, on fine larval habitat segregation in non-Hawaiian *Drosophila*). For example, most of the species of the *adiastola* subgroup of picture-wing flies can be reared from the rotting bark of one

plant genus, *Clermontia* (Carson and Kaneshiro, 1976), and species of this subgroup are known to specialize in an impressive variety of larval substrates (not always *Clermontia*), including leaves (two species), bark (seven species), stems (six species), flowers (three species), and fruit (two species). Other species specialize on fern rachis (two species; Montgomery, 1975, table 9).

When closely related *Drosophila* species are compared, they sometimes show remarkable species-specific differences in larval habitat. For example, two very closely related highland species, *D. silvarentis* and *D. heedi*, breed on sap flux from the same species of tree (*Myoporum sandwicense*), but *D. silvarentis* oviposits only on fluxes on the surface of the trunk, whereas *D. heedi* breeds only in soil moistened by flux dripping from above (Kaneshiro et al., 1973; figure 28.12). Similarly, two closely related species, *D. mimica* and *D. kambysellisi*, differ primarily in the location of their larvae, *D. mimica* being reared primarily from the rotting fruits of the soapberry (*Sapindus saponaria*) and *D. kambysellisi* exclusively from the fermenting leaves of *Pisonia brunonianum* (Heed, 1971). Another soapberry-inhabiting species, *D. eugyochracea*, is limited to decaying bark of fallen branches or trunks (Carson and Kaneshiro, 1976).

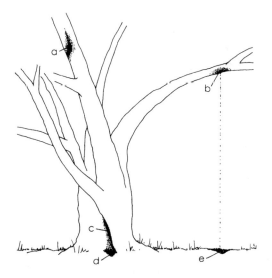

Fig. 28.12. Breeding-site specialization in a pair of cryptic sibling species of Hawaiian *Drosophila*. *D. silvarentis* larvae are found at (a), (b), and (c), wet bark caused by sap fluxes on the trunk of a *Myoporum* tree. *D. heedi* larvae are found at (d) and (e), sap-flux puddles in the soil beneath the tree. After Kaneshiro et al. (1973).

Studies of non-Hawaiian *Drosophila* indicate fine-tuned species specificity in physiological tolerance and behavioral selection of the host fermentation environment (Parsons, 1981). Fermentation of plant tissues inhabited by larvae produces ethanol, and *Drosophila* species differ markedly in the concentrations at which the substrate ceases to be a resource and becomes detrimental (as indicated by effects on adult longevity), for example, about 12% in *D. melanogaster*, 3.4% in *D. similans*, and 1.6% in *D. immigrans* (Parsons and Spence, 1980). Newly hatched larvae of these species switch between attraction and avoidance in the same sequence (Parsons, 1981). *Drosophila* larvae are active foragers that move through the substrate both horizontally and vertically, probing with their mouthparts (Sokolowski, 1986). How adaptive shifts within chemically variable habitats could be facilitated by biochemical flexibility and a kind of conditioning or avoidance response of larvae is discussed by McDonald (1986): in *Drosophila*, alcohol avoidance behavior in decaying substrates seems to be triggered by accumulation of acetaldehyde levels above a certain critical threshold, and larvae differing at the alcohol dehydrogenase (ADH) locus show an avoidance response correlated with their differing abilities to tolerate alcohol. Experimental mutant flies with null ADH alleles showed "automatically" higher alcohol avoidance behavior, and in general, alcohol preference behavior correlates with relative ADH activity levels.

The phenotypic (behavioral) accommodation implied by this correlation could facilitate the kinds of habitat changes that characterize differences between closely related species in Hawaiian *Drosophila*. Phenotypic accommodation, in the form of avoidance behavior like that noted by McDonald (1986), could place larvae in a nearby new microhabitat where their growth and survival would be improved. Ethanol avoidance behavior, like insect aversive behavior in general (Bernays, 1993), is a kind of learning. The production of an "automatic" behavioral adjustment following a change in physiological tolerance is one of the characteristics of learning, which is based in part on an internally (physiologically) determined system of motivations and rewards. Physiological intolerance of substances is one potential determinant of a motivation–reward system (see chapter 18).

The mechanisms of phenotypic accommodation to a new biochemical range of toleration described by McDonald (1986) would presumably apply to other aspects of the chemical environment, such as toxic elements in plant tissues. *D. sechellia*, for ex-

ample, is able to live on the fresh fruit of *Morinda citrifolia*, which produces octanoic acid, a substance lethal to other *Drosophila*, including its presumed ancestor, *D. simulans* (R'Kha et al., 1991; Jones, 1998). Other species are able to breed on the fallen fruit, which is rapidly detoxified by microorganisms. By analogy with the biochemical and behavioral responses to alcohol concentrations seen in other *Drosophila*, a host shift from nontoxic rotten morinda fruit to toxic, fresh morinda fruit could have begun with an increased physiological tolerance of eggs to relatively fresh morinda tissues. Egg and embryo tolerance is a maternal effect, as shown by hybridization experiments in which only the eggs and embryos with *D. seyshellia* mothers survived on fresh morinda, and even most larvae of this species succumb if eggs are deposited on fresh morinda fruit (R'Kha et al., 1991). Such an effect could involve a biochemical maternal endowment of the egg cytoplasm, a property of the maternally manufactured egg chorion, or a result of maternal ability to place eggs in zones of the fruit relatively free of toxin (e.g., by selection of regions beginning to rot), questions that would be of interest to investigate in relation to the timing and topography of progress in detoxification of the fruit in nature (the studies of R'Kha et al. were on frozen fruit in the laboratory).

Toxin-avoidance behavior of larvae, if present in the *simulans* ancestor as in *Drosophila* (discussed above), would contribute to the niche shift by enabling them to select relatively detoxified regions within the fruit. As in the apple races of apple-maggot flies (*Rhagoletis pomonella*; see above), the females of *D. seyshellia* are now specialized in being differentially attracted to the odor of fresh morinda fruit and preferentially oviposit in it (R'Kha et al., 1991). This situation has also led to *Rhagoletis*-like species differences, involving strong host preferences of the ovipositing females, attraction of adults to the new host, with small genetic differences between the derived species and its closest relative (Jones, 1998; discussed for *Rhagoletis* in chapter 27).

Crucial to the transition to a new breeding substrate in Hawaiian *Drosophila* is the behavior of ovipositing females, which are mobile as adults and feed at a variety of sites even though larvae are substrate specialists (Kambysellis and Craddock, 1997). Adoption of the new substrate by ovipositing females could be promoted by a phenomenon demonstrated by Jaenike (1982, 1988), namely, the choice of formerly aversive oviposition substrates by females that experience the new substance as young adults. Selection would then favor females able to identify and oviposit in those or similar sites, leading to genetic accommodation of the shift.

The recurrent evolution of alcohol and toxin tolerance connected with recurrent niche shifts (Mercot et al., 1994; R'Kha et al., 1991) in *Drosophila* suggests a biochemical flexible stem coupled to behavioral accommodation and learned habitat preference that is multidimensional in having the potential to change in multiple directions. Mercot et al. (1994) discuss the importance of evolved alcohol tolerance in *Drosophila*, its relation to the diversity of larval habitats, the fact that adaptation to high-alcohol resources has been recurrent, and the likelihood that a diversity of mechanisms may be involved, including altered metabolization, detoxification, nervous system tolerance, and excretion, so that it is possible to have a low ADH activity and a high alcohol tolerance. The great evolutionary lability, as well as intraspecific flexibility of ADH and aldehyde oxidase expression, is suggested by recurrent change in the *site* of expression in the larval midgut and the fact that similar variation sometimes occurs within species (Dickinson, 1989).

An outstanding feature of the adaptive trophic radiation in Hawaiian drosophilids is the diversification of ovarian specializations of adult females. Different species have different oviposition tactics for exploiting different larval food resources. Different resources, such as yeast of decaying stems, bacteria of decaying leaves, and pollen of flowers (Kambysellis and Heed, 1971), differ in their size and permanence and therefore require different clutch sizes, egg sizes, and oviposition schedules. Some leaf-breeding species (e.g., *D. disticha*, now *D. waddingtoni*; K. Kaneshiro, personal communication) can facultatively maintain a single egg ready to oviposit for prolonged periods of time during drought until conditions are favorable for oviposition (Kambysellis and Heed, 1971; Kambysellis et al., 1995). Flower-breeding pollen feeders retain eggs, oviposit a single egg per flower, and have a short larval period that matches the limited food available per flower (Kambysellis and Heed, 1971), enabling an individual to exploit a very small food source. Species that oviposit on rich but less numerous substrates, such as decaying stems where the larvae feed on yeasts, have large ovaries and produce a large number of small eggs in a single egg-laying episode. Species that feed on bacteria growing on rotting leaves reproduce more steadily throughout the year (Kambysellis and Heed, 1971), but their larvae are subject to predation by the larvae of a muscid fly when small. Females of these species do best to rapidly locate newly available

food sources before they are colonized by the predatory flies (see Spieth, 1974; Montgomery, 1975).

In view of the general pattern noted in Galapagos birds and African cichlid radiations, where there is increased ecological specialization under resource scarcity (summarized in Grant and Grant, 1995b, p. 242; Witte, 1984; Spieth, 1974; Jaenike, 1990), it would be important to examine the temporal and spatial variation in resource use by larval drosophilids and investigate the possibility that they, too, facultatively specialize more narrowly under strong competition or during periods of environmental extremes. There are suggestions of this in the literature.

Spieth (1974) states that large numbers of individuals accumulate around plants (e.g., lobeliads) when they bear fermenting parts that are prime food sources for adults and larvae, and while 50 species of drosophilids have been bred from fermenting leaves of *Cheirodendron* (Araliaceae) and 25 others from the leaves of the lobeliad *Clermontia* (Campanulaceae), other species specialize to the bark, flowers, or fruits. These observations suggest that resource competition under tight species packing could occur at certain times or in certain places and that it may be a common aspect of drosophilid ecology in Hawaii.

Jaenike (1990) invokes optimality and physiological state models of oviposition behavior to predict that females may shift to low-preference hosts as egg load or search time for an oviposition site increases, and cites studies indicating that this occurs in some *Drosophila*. This possibility perhaps could be investigated in nature in Hawaiian species such as *D. grimshawi* and its relatives, which show a narrowed specialization where rare and perhaps relatively resource limited: on Kauai and Oahu islands, a population of a species (being named *D. craddocki*; K. Kaneshiro, personal communication) close to *D. grimshawi* is scarce and restricted to the bark of a single species of plant (*Wikstroemia*), whereas on Maui, Molokai, and Lanai *D. grimshawi* is abundant and breeds on a wide variety of host plants (Montgomery, 1975; Kaneshiro and Kambysellis, 1999).

Food limitation in larvae of Hawaiian *Drosophila* in nature is indicated by the fact that much of the considerable body-size and ovary-size variation found in field-caught females can be eliminated by rearing their larvae on abundant food in the laboratory (Kambysellis and Heed, 1971). These authors cite previous studies by Robertson (1957) and Robertson et al. (1968) showing that degree of competition for larval food supply is the primary factor controlling variation in body and ovarian size in Hawaiian *Drosophila*. The radiation of Hawaiian drosophilids appears to be an adaptive breeding-site radiation as well as an example of rapid speciation driven in part by sexual selection (discussed below). The ecological aspects invite further research in relation to adaptive plasticity in larval physiology and maternal behavior.

Four Variations on the Theme of Flexibility and Radiation in Finches

Chapter 27 shows that learning is a recurrent theme in speciation theory, in animals as different as insects and birds. The learning that is thought to facilitate speciation often takes the form of some kind of imprinting, either of parental signals as in the case of bird song, or as parental oviposition sites, as possibly occurs in some adult phytophagous insects. In those contexts, learning plays a conservative role. While it has the potential to spread a behavioral novelty through a population more quickly than a mutant gene can, learning perpetuates the status quo until some other factor creates a new learning situation.

Here I return to learning and flexibility as an accessory to diversification. As discussed in chapter 18, learning is a developmental mechanism—a condition-sensitive decision process. Trial and error learning can cause certain, rewarded behaviors to be repeated and other, unrewarded behaviors to be dropped. As emphasized above, if a developmental mechanism causes a phenotype to be repeatedly expressed, this recurrence exposes the phenotype to selection and may produce directional evolutionary change (see chapter 6). A recurrent learned behavior, like any other recurrent phenotype, creates a subpopulation of individuals that may occupy a distinctive niche. Given genetic variation in the propensity to express the recurrently expressed trait, or in its characteristics when expressed, selection within that population can lead to a particular direction of evolutionary change. Then if different teaching devices—song dialects, or particular plant hosts—occupy different geographic ranges, local specializations would be established as surely as by genetic change in the components of a switch. Geographic isolation increases the likelihood of phenotypic change not only because it reduces gene flow, but also because it may restrict interaction to a distinctive set of resources and conspecific individuals.

Assortative mating can also augment learning-assisted divergence if particular ecologically im-

portant qualities, say, beak size or shape in a bird, are involved in mate choice, as in some Galapagos finches (see Grant and Grant, 1989). Such a mating preference can affect the evolution of morphology and bias learning in such a way that individuals move more quickly toward an ecological realm that accords with their phenotypic dispositions. Learning, for example, would reinforce the use of seeds easily handled by a particular type of sexually selected beak. Both of these reproductively and genetically isolating phenomena—assortative mating and geographic isolation—could reinforce biases toward learning particular things. So it is not surprising that organisms capable of learning and other hypervariable mechanisms of adaptive plasticity, such as those that allow muscle and bone to be molded by use, diverge rapidly genetically when they are trapped on new islands of habitat or, by their mating patterns, into mating with a phenotypically similar sample of individuals.

Learning and Radiation in Galapagos Finches　How feedback between morphology and learning can be involved in rapid adaptive radiation is illustrated by the Darwin's finches (Emberizinae) of the Galapagos Islands, where, as concluded by Grant (1986, p. 399), new feeding specializations arise by chance and then are likely improved through trial and error learning of young birds rewarded for their "mistakes" while learning to feed. The 13 species of finches endemic to the islands perform a wide array of specialized feeding behaviors (Grant, 1986; figure 28.13). They rip open rotting cactus pads, strip the bark off dead branches, kick over stones, probe flowers for nectar and pollen, glean arthropods from rolled leaves and cavities in trees and from exposed rocks on the seashore, crack seeds of a wide range of sizes, probe cavities with sticks, hunt ticks on iguanas, and drink the blood of seabirds, an extraordinary range of specialized behaviors thought to have evolved in the span of only two or three million years (Grant, 1986; Grant and Grant, 1989; Carson, 1992b; Vincek et al., 1997). Yet *individuals* are versatile, not rigidly locked from birth into species-specific diets.

The role of learning in the ontogeny of foraging specialization in these and other birds is discussed in chapter 18 on learning (see also Curio, 1976). Fledglings begin their independent lives by trying a variety of foods. They observe and mimic the feeding behavior of other individuals, including members of other species (Grant, 1986). It takes an individual *Geospyza scandens* more than a year to attain the proficiency of dealing with *Opuntia* seeds shown by conspecifics two years and older (Grant,

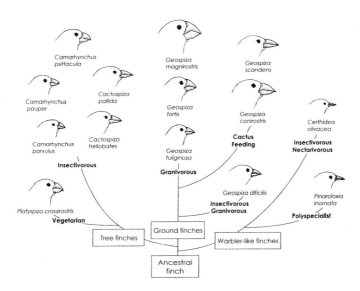

Fig. 28.13. Adaptive radiation in diet and beaks of Galapagos finches (Emberizinae). The three major branches discerned by Lack (1947) have been confirmed by modern phylogenetic analysis (Stern and Grant, 1996), with the warbler-like finches being closest to the base of the tree, and the Ecuadorian dull-colored grassquit *Tiaris obscura* the most likely nearest continental relative (Sato et al., 1999). In some trees, the Cocos finch (*Pinarolaxias inornata*) falls within the tree finch group, and the vegetarian finch (*P. crossirostris*) is now placed on a separate branch rather than within the tree finches (Sato et al., 1999). Dietary specializations are oversimplified here, with some (such as blood and egg feeding by *G. difficilis*) omitted. Drawings after Lack (1947) and Grant (1986).

1986). With experience, individuals narrow their preferences in ways consistent with local resources and the shape of their own beaks, suggesting that morphology-dependent efficiency in obtaining food influences preferences and skills. But the most striking trophic behavioral specialization occurs during the dry season in their highly seasonal environments, and during periods of drought. Then individuals concentrate on food items and food patches that best allow them to capitalize on individual and populational differences in beak size and shape. For example, *G. conirostris* on Genovesa Island is a population with large variation in beak shape that is subjected to marked seasonal fluctuations in food supply (figure 28.14). The ability to learn to exploit new food sources was the most important predictor of survival of young birds during the dry season food shortages, and among individuals capable of adaptive flexibility, those with deep bills survived best (Grant and Grant, 1989). It is during such episodes of scarcity that selection causes change in the proportions of birds with certain beak morphologies and the learned feeding skills that capitalize on them (Grant and Grant, 1989).

One evolutionary role of flexibility is to match genetically influenced phenotypic traits to local and temporally variable conditions, either through morphology-influenced learning (see chapter 18) or through a correlation between learned specializations, morphological traits, and survival (Price, 1987; Grant and Grant, 1989). Episodic but recurrent (seasonal) change in diet would foster evolutionary improvement in learning ability in birds like *G. conirostris*. If a particular dietary shift were to recur year after year, the ratcheting effect of conditional expression (see chapter 7) is expected to produce directional evolution in the bill characteristics best able to exploit dry-season opportunities: any existing variation in trophic ability (e.g., in digestive physiology, searching ability, or beak shape; figure 28.14) can be matched by trial and error learning to a diet that approaches the most profitable one available given the local array of plants, arthropods, and competing species. When conditions change, as they did after an El Niño year followed by an extreme drought, which drastically changed the array of available foods (Grant and Grant, 1989), selection in a particular direction will not be constant and may be reversed. Flexibility-assisted directional evolution requires that the unidirectional episodes of selection outnumber or exceed in strength the reversals.

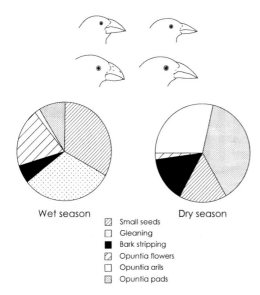

Wet season   Dry season

☒ Small seeds
☐ Gleaning
■ Bark stripping
☒ Opuntia flowers
☐ Opuntia arils
▨ Opuntia pads

Fig. 28.14. Intraspecific variation in beaks and sea-sonal variation in diet of the Galapagos large cac-tus finch (*Geospiza conirostris*) on Genovesa Is-land. Coefficients of variation in bill dimensions are high (6.1–7.7% of the mean) compared to those of continental passerine birds (3–5% of the mean). Along with seasonal food scarcity, there are shifts in use of available resources (pie charts). Individu-als that have learned to exploit novel food sources during the dry season alter their diet and survive better, as do individuals with beak characteristics favored by the resource shift. After Grant and Grant (1989).

Seasonal change in some of the Galapagos finches could be termed a seasonal polyethism, and it permits divergence in other respects. The cactus finch, *G. scandens*, and the medium ground finch, *G. fortis*, for example, feed primarily on arthro-pods and usually breed after the first seasonal rains, when arthropods become common (Grant, 1996). During the dry season they ingest pollen, nectar, and seeds of the *Opuntia* cactus. Neither species breeds during drought years, when arthropods are exceedingly scarce. This could suggest that repro-ductive physiology is geared to respond to the onset of the rainy season when there is abundant proteinaceous arthropod food. But the situation is more complex. In some years, a few pairs of *G. scandens* breed before the rains start, and some even produce offspring that fledge before the rainy season gets underway.

Grant (1996) discovered that individuals of *G. fortis* and *G. scandens* are unusual among birds in their ability to efficiently digest pollen, a rich source of protein unexploited by most animals due to the difficulty of breaking through the pollen-grain shell (exine). The ability to do this allows for breeding in the absence of arthropod prey. But only some G. scandens accelerate breeding, and no pairs of G. fortis were early breeders. So why do some indi-viduals and not others become early breeders? Pollen digestion is evidently only part of the answer to how early breeding evolves. Behavioral obser-vations provided another clue: *G. scandens* is able to dominate and displace the smaller *G. fortis* at *Opuntia* flowers. And *G. scandens* feeds more ex-tensively on flowers than on seeds and may be more specialized to do so, opening buds and snipping styles in a way that is not observed in *G. fortis* (Grant, 1996). These observations indicate that early breeding, depending as it does on pollen feed-ing, may occur primarily in the individuals that most aggressively defend and exploit the pollen of *Opuntia*. This means that a shift in breeding time could be produced by selection on variation not in timing of gonadal responses per se but on various pollen-feeding abilities that allow or even promote early reproduction by improving nutritional state. Both genetic variation and environmental inputs could thus contribute to the origin of an early-breeding phenotype.

One of the hypothesized characteristics of a hy-pervariable flexible stem is that it permits special-ization without loss of plasticity, thereby conserv-ing the potential to rapidly evolve in any of a variety of new directions, should conditions change. That the Galapagos finch radiation has this quality is in-dicated by a model study of individual foraging dif-ferences in a natural population of *G. fortis* on Isla Daphne Major (Price, 1987). The population as a whole uses a variety of foods, including seeds, fruits, pollen, and caterpillars, so it could be de-scribed as a generalist. In fact, it is a collection of individual specialists whose specializations depend significantly on body size, beak morphology, and experience (Price, 1987). Price concentrated on in-dividual differences in use of three classes of seeds, small seeds, and large seeds of two genera (*Opun-tia* and *Tribulus*; figure 28.15), since these are es-sentially the only foods available during times of food shortage and are the commonest foods year-round. The smallest individuals of *G. fortis* are small-seed specialists, since they are unable to crack large seeds. Among large-bodied specialists on large seeds, there was a positive association between seed handling ability and beak size. Some large individ-uals specialized on cactus (*Opuntia*) seeds, and oth-ers on *Tribulus* seeds, or both, even though energy

Fig. 28.15. The three main seed types consumed by *Geospiza fortis* on Isla Daphne Major, Galapagos. Top: two *Tribulus citoides* mericarps (arrow indicates where a seed has been removed by a *G. fortis*). Top center: seeds of the cactus *Opuntia echios*. Bottom: small seeds, of *Acalypha parvule* and *Portulaca howelli*. The heads of a large and small *G. fortis* are shown to scale. From Price (1987).

rewards of both types of large seeds were similar (Price, 1987). These patterns within the population of *G. fortis*, a species of intermediate size and beak characteristics, are exaggerated in the direction of exaggerated beak characteristics in related species on the same island (figure 28.16).

Specialization in *G. fortis* is influenced by differences in local abundance, especially in feeding territories of males, and by time of day, since *Opuntia* seeds are consumed in the shade of cactus plants, whereas consumption of *Tribulus* seeds requires feeding in the hot sun, a reminder that the evolution of feeding specializations can involve variation in "nontrophic" traits such as territoriality and resistance to heat stress. But the most revealing results in terms of the importance of developmental plasticity are the indications of continued experimentation with new foods and feeding techniques, even by specialists. This was indicated by occurrence of unusual behaviors and low-frequency specializations (from comments in Price, 1987; parentheses mine):

1. Among small-seed foragers, only the smallest fed on the seeds of *Sesuvium* plants, pos-

sibly because of a superior ability to perch on the terminal stems of small plants (illustrating a size-correlated feeding advantage associated with a novel specialization).

2. On one occasion, a small-seed specialist attempted but failed to crack open a large *Opuntia* seed (indicating continued experimentation with an unused food).

3. Among large-seed foragers, many (but not all) switch between the two types of seeds, apparently in response to energy rewards (indicating maintenance of a potential to specialize on either one or the other).

4. Among large-seed foragers that fed on *Tribulus* seeds, the smallest moved from mericarp to mericarp, twisting off the corners, whereas the largest cracked mericarps transversely, as in another species, *G. magnirostris*, and one individual with an unusually deep beak was the only one regularly observed to do this. (This indicates the potential for the evolution of size-related behavioral specialization, and the common pattern of an intraspecific anomaly seen as a regular feature of a related species, as in the limb bones shown in figure 19.3.)

5. Foraging efficiency (the inverse of handling time) on *Opuntia* seeds increased the longer a bird foraged without interruption on that type of seed (showing the potential for reinforcement of specialization due to opportunity and experience).

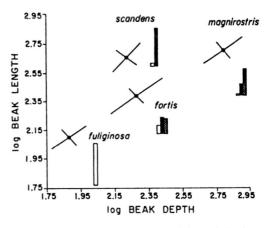

Fig. 28.16. Beak morphology and diet of the four *Geospiza* species of Isla Daphne Major, Galapagos. Histograms show the relative feeding frequency on the seed types shown in figure 28.15: small seeds (open bars), *Opuntia* seeds (black bars), and *Tribulus* seeds (hatched bars). From Price (1987).

6. Individuals that had experienced a severe drought and scarcity of *Opuntia* seeds fed less on *Opuntia* than did younger birds born after the drought, perhaps due to differences in experience (indicating the potential for rapid populationwide change in feeding specialization mediated by learning).

7. One individual with an exceptionally long beak was the only one seen to open *Opuntia* fruits (showing how opportunistic use of a variant beak morphology can rapidly lead to a shift to a novel food).

Such observations are often dismissed as "merely anecdotal" and not even reported. But, like morphological anomalies in the limbs of salamanders (figure 19.3), behavioral idiosyncracies are indications of developmental plasticity and evolutionary lability. They are raw material for selection and further radiation of a kind that have likely been important in the adaptive radiation of the Galapagos finches.

Adaptive Radiation without Speciation in the Cocos Island Finch How phenotypic plasticity can structure an *intra*specific adaptive radiation in behavior is illustrated by the behavior of the Cocos Island finch, *Pinaroloxias inornata*, a recently derived descendent of the Galapagos tree finches (Sato et al., 1999). Cocos Island is a small volcanic island, somewhat younger than the Galapagos Islands (about two million years old; Petren et al., 1999), similarly isolated from the mainland and geologically part of the same oceanic island group (Grant, 1986). But it emerged and then migrated farther to the Northeast, and it is now situated a full 630 km from the other islands, far beyond the flying range of birds resident in the Galapagos proper. As if by design to experimentally test ideas about island radiations, Cocos Island is inhabited by a single geospyzine finch species with no opportunity for gene exchange or competition with others, and no opportunity for interisland speciation. The possibilities of microgeographic isolation within the island are also relatively limited since it is small, humid throughout the year, and without obvious altitudinal or regional barriers to serve as barriers to birds. Cocos Island, then, is a far-flung Galapagos island with a Galapagos finch in the simplified, isolated, entrapped biota associated with rapid radiations, where we can examine the results of trophic competition in the absence of opportunities for geographic isolation and competition with closely related species (Lack, 1947).

The result has been an impressive trophic radiation, but entirely within a single species. Nine specialized foraging categories were observed in the Cocos Island finches of a fairly small (3.4 hectare) and relatively homogeneous study area (a *Hibiscus* thicket), where birds foraged alone and did not restrict their movements to particular territories (Werner and Sherry, 1987; T. K. Werner, 1988). Individuals specialize in branch gleaning, branch probing, leaf gleaning, leaf-miner gleaning, dead leaf gleaning or probing, extrafloral nectar feeding, flower probing, and ground gleaning, or showed a variety of behaviors (categorized as miscellaneous). Inhabitants of other habitats, such as cloud forest, on the island were not included in the analysis (T. K. Werner, 1988) and may have had additional specializations. Of 89 birds observed feeding 60 or more times during the 10-month study, all were significantly more specialized than random ($p > 0.01$) in one of the nine specialized foraging categories, and 62 individuals concentrated at least 50% of their foraging attempts on one of these behaviors alone. Such individuality is a common characteristic of birds when foraging competitively for a diversity of foods (Giraldeau and Lefebvre, 1985; see chapter 18). Specialization was independent of age, sex, morphology (which was virtually invariable), time of year or day, and place (individuals did not restrict their movements within the study area; T. K. Werner, 1988).

Earlier observers (e.g., Slud, 1967; Smith and Sweatman, 1976) gave vivid descriptions of the versatile behavior of these finches but treated them as generalists. Werner and Sherry (1987) show that they are more accurately described as polyspecialists. Individuals were not rigid specialists. They would sometimes opportunistically feed on alternative foods as adults, but then return to the pattern that was their primary specialization. This is in keeping with the general nature of a flexible stem, where there is specialization without complete loss of flexibility or exploratory mechanisms even while a particular specialization is expressed. Studies of captive fringillid finches (Kear, 1962) show the continued ability of adults to learn new feeding preferences and behaviors if their feeding opportunities change. Observations of young Cocos Island finches suggest a role for learning in the acquisition of individual specializations. This would not be surprising, given the well-documented role of learning in the feeding behavior of many birds (see chapter 18). Werner and Sherry (1987) speculate that learned individual specialization may contribute to foraging efficiency in an environment where the supply of particular foods is relatively constant, but where heavy rains restrict activity periods and there is strong intraspecific competition due to the abundance of conspecifics. A detailed study of the ontogeny of feeding specializations of Cocos Island finches is a promising area of research that could illuminate the nature of avian radiations in general.

A second example in a very different organism, a spider, illustrates within-species adaptive radiation on Cocos Island. The endemic theridiosomatid spider *Wendilgarda galapagensis* performs an unusual variety of web-building behaviors on the island, with several major elements added to its repertoire compared with mainland congeners (Eberhard, 1989, 2000). Mainland *Wendilgarda* are highly and peculiarly specialized. All known webs are built over the surface of forest streams with sticky threads attached to the surface of the water. The entire family Theridiosomatidae is known only from forested areas. The endemic Cocos Island species, however, was found on all parts of the island searched, including not only over streams, but also in caves and in sunny clearings, and even over open grassy areas, reaching densities of up to 10–40 individuals per square meter in heavily populated sites along streams.

Web structure and construction behavior sequences vary in accord with web site, and there are at least three distinctive web types associated with webs over water, low over land, and high over land. Spiders were individually plastic in their building behavior. When confronted with puddles following one of the frequent Cocos Island rainstorms, individuals that had built characteristic low-land webs built water-style webs using distinctive behaviors. Although *Wendilgarda* web and habitat are highly specialized and unusual for a spider, there is also unusual variation within species in their behavior. One Panamanian *Wendilgarda* species shows behavioral changes not reported in any other spider of the hundreds observed in more than 150 species of five areneoid families of orb weavers, and striking flexibility in construction behavior occurs in successive webs of single individuals (Eberhard, 2001b). Mainland *Wendilgarda*, then, seem to constitute a hypervariable flexible stem from which the endemic Cocos Island species has evolved. Developmental plasticity may be a common characteristic, or even a prerequisite, of species that successfully colonize islands and adaptively radiate in archipelagos like the Hawaiian and Galapagos islands.

The Cocos Island experiment suggests that the first step in a local radiation can be exaggerated behavioral diversification and polyspecialization as a response to local opportunities and intraspecific competition, without morphological specialization.

If colonists from such populations were to become established in isolation on different islands having different subsets of resources and competitors, different alternative specializations could quickly become recurrent, then be genetically accommodated in the isolated populations, with an increased probability of morphological specialization in keeping with the behavioral ones.

Dimorphism in a Continental Finch   The conditions favoring morphological diversification influenced by flexibility are illustrated by another kind of finch, the African black-bellied seed cracker *Pyrenestes ostrinus* (Estrildidae), whose behavioral specializations are accompanied by a beak polymorphism (Smith, 1987, 1990a–c, 1991, 1993). In these birds, the conditions for morphology-biased learning and diverging selection are present in the form of a regularly occurring, binary choice between two extreme classes of seeds—some relatively scarce, and hard, and the others more abundant, and soft—during an annual seasonal drought. Both bill morphs consume soft seeds during times of plenty. But, as in the Galapagos finches, bill-related individual specialization was most evident in times of scarcity: dietary overlap between bill morphs was only 12% in the dry season when large-billed individuals feed exclusively on hard seeds and small-billed individuals feed on a variety of soft grass and sedge seeds. Sometimes the small-billed birds expand their diet to include additional species of soft seeds (Smith, 1990b). This is a situation that resembles the Galapagos but with a reduced, discrete choice of foods and, compared with either the Galapagos Islands or Cocos Island, a large foraging range where the two classes of food can be sought by a wide-ranging forager, which can therefore specialize more completely on just one class of food. Although there are other bird species that consume soft seeds, *P. ostrinus* is the only one that uses sedge seeds, so it has a relatively stable food base not accessible to other species. The result is a persistent selection for specializations in two directions, and a beak dimorphism (figure 28.17).

As in the Galapagos, selection primarily takes place in times of trophic crisis, in the African finches during the major dry season when food is scarce and trophic competition within and between species is most intense (Smith, 1987). In contrast with the situation for the Galapagos finches, conditions for the African finches varied little from year to year. Annual variation in rainfall and seed production was low, so the direction and intensity of selection varied little across years (Smith, 1993).

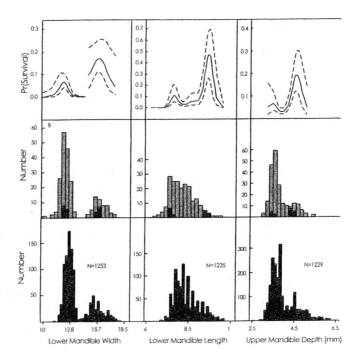

Fig. 28.17. Diverging selection and beak morphology in the African seed-cracker finch *Pyrenestes ostrinus* (Estrildidae). (Top) Probability of survival to adulthood for different values of three beak characters. (Middle) Distributions of juveniles that did not survive (shaded bars) and those that survived (black bars). (Bottom) Distributions seen in adults. From Smith (1993). Reprinted by permission from *Nature* 363, p. 618 © 1993 Macmillan Magazines Ltd. Courtesy of T. B. Smith.

The relatively dependable availability of a bimodal food supply—sedges not utilized by other birds—in these disruptively selected finches is favorable to the evolution of genetic influence on the switch between alternatives: unpredictably variable conditions increase selection for plasticity, but predictable alternation between extremes is favorable to genetic polymorphism (see chapter 22). So it is not surprising that breeding experiments produced results consistent with morph determination by two alleles at a single locus of large effect (Smith, 1993).

Smith (1993, p. 620) suggests that single mutations may have allowed the population to cross the adaptive valley between the two specializations, and (following Orr and Coyne, 1992) if so, this contradicts the Fisherian view of gradual evolution of adaptations by accumulation of allelic substitutions of small effect. A third, much larger "megabilled" morph, specialized during drought on a hard variety of seed not extensively utilizable by the other two morphs, inhabits a somewhat different geographic area but interbreeds with individuals of the other two bill morphs (Smith, 1993, 1997). The degree of genetic influence on determi-

nation of the megabilled morph has not been determined (Smith, 1997). Even if it proves to be strongly influenced by a single locus of large effect, this would not necessarily support the mutational hypothesis for the origin of *morphological* novelty. Recall that regulatory influence of a gene of large effect can be a result of recent strong directional selection on the frequency of expression of a particular morph (Nijhout and Paulsen, 1997; see chapter 6), so there is not reason to postulate a mutation to explain allelic morph determination or the origin of novelties.

The macromutation hypothesis implies that the behavioral specialization originated only after, or simultaneously with, the morphological specialization, for it states that the bill-morph mutation was responsible for ability to exploit large seeds. Alternatively, the trophic specializations could have originated gradually as a behavioral polyphenism under disruptive selection (Smith and Girman, 2000) where, as in the Galapagos finches, continuously distributed bill characteristics affect which of the two specializations is adopted (e.g., due to learning). Then any genetic contribution to bill-trait

Fig. 28.18. Adaptive radiation in beaks of Hawaiian honeycreepers (Drepanidae): (a) *Oreomystis bairdi*; (b) *Chloridops kona*; (c) *Himatione s. sanguinea*; (d) *Hemignathus v. virens*; (e) *Hemignathus procerus*; (f) *Hemignathus monroi*; (g) *Pseudonestor xanthophrys*; (h) *Loxops c. coccinea*; (i) *Telespyza ultima*; (j) *Ciridops anna*; (k) *Palmeria dolei*; (l) *Melamprosops phaeosoma*; (m) *Vestiaria coccinea*; (n) *Psitirostra psittacea*; (o) *Drapanis pacifica*. After painting by H. D. Pratt Jr. in Freed et al. (1987).

differences could create a morphological bimodality and influence (via learning) adoption of the morphology-appropriate trophic behavior.

Whichever route to genetic influence applies, one thing is certain: genetic mutation, while it may play a role in the origin of novelty (see chapter 6), is not needed for organisms to cross the fitness valleys between adaptive peaks or to explain the observation of single-locus control. Adaptive valleys can be (and in many organisms are) crossed when disruptive selection leads to disruptive development (polyphenisms and other types of alternative phenotypes; West-Eberhard, 1986, 1989; see chapter 11, part III). Due to the readiness with which genetic influence can be interchanged with environmental influence on phenotype determination (see chapter 5), it is impossible to know from study of contemporary populations alone which mode of origin applies in a species like *Pyrenestes ostrinus*. Evidence for single-locus influence on phenotype determination does not demonstrate phenotype *origin* due to mutation, and therefore does not necessarily argue against Fisherian gradualism in the origin of adaptations (cf. Orr and Coyne, 1992; see chapter 6).

Morphological Diversity in Hawaiian Honeycreeper Finches   The conditions for adaptive diversification (as in the Cocos Island finches), for speciation (as in the Galapagos finches), and for morphological specialization (as in the African finches) seem all to hold in the Hawaiian archipelago, where honeycreeper finches have speciated *and* morphologically specialized to a greater degree than in any of the other radiations just discussed. The Hawaiian honeycreepers are fringillid finches (Fringillidae, Carduelinae, Drepanidini) that may have affinities with the Emberizinae, the family of the Galapagos finches (Johnson et al., 1989; Grant, 1994). The striking diversity of their beaks (figure 28.18) indicates that trophic competition has played a role in their diversification, although detailed studies like those on Galapagos finches have not been done. Amadon (1950) noted that one population of the widespread drepanid *Loxops virens* showed a divergent behavior and muscle morphology on Kauai Island, where several other *Loxops* species are present and *L. virens* specialized in digging in the bark of trees; on other islands, where it is the only species present, it has more generalized feeding habits.

Grant (1994) gives two reasons for the greater morphological divergence of the Hawaiian honeycreeper finches compared with the Galapagos finches: the greater age of the finch populations on the Hawaiian archipelago, which has allowed more time for divergence and speciation to occur; and the less seasonal and more floristically rich nature of their environment (see also Otte, 1989), which, as in the African finches, provides a steadier supply of resources of a single kind to allow specialization without episodes of opportunistic generalist foraging or extreme reversals in the direction of selection like those documented in the Galapagos.

That the specialization of these finches has not eliminated their capacity for rapid evolutionary change was indicated when small groups of one species, the endangered Laysan finch (*Telespyza* species), were transplanted in 1967 onto small atolls about 500 km northwest of the Hawaiian Islands as part of a conservation effort (summarized by Pimm, 1988). In Hawaii (Laysan Island) the birds are omnivorous, feeding on a wide range of seed sizes and on insects, carrion, and birds' eggs. Some of the transplanted populations prospered in their very different new habitats, and dramatic shifts in diet and bill shape have been documented, as well as evidence that this is due to selection rather than bill wear or genetic drift. Like the Galapagos and Cocos finches, the transplanted birds are described as "tame and readily caught by hand nets" (Pimm, 1988, p. 290), a hint worth investigating that selection has favored exploratory boldness and learning, and that this may have played a role in their rapid phenotypic change.

Conclusion: Causes of Variable Effects of Flexibility on Morphologic Radiation

The evidently crucial but little-studied role of exploratory learning in the Galapagos finch radiation and in the behavior of the Cocos Island finches is a fine illustration of the importance of learning in and flexibility in the evolutionary diversification of life. For Darwin, the behavior of the Galapagos birds was their most striking aspect. He made only one entry on them in his *Beagle* diary, but in notes made a few months later he recalled them as "lively, inquisitive, active, run fast, frequent houses" and "very tame" (Keynes, 1988). Only when the ornithologist Gould analyzed Darwin's bird collections did Darwin realize the geographic diversity of the finches. Even then he, like other authors who attribute lack of shyness to the absence of humans and predators, did not connect the so-called tameness with the diversity of the birds or see it as an active, adaptively important trait. "Tameness" is insightfully termed by Greenberg (1983) a lack of "neophobia"—an attraction to and exploration of the new, translated into trial and error learning (see also Wilson et al., 1994). In more anthropomor-

phic terms, the "tameness" of the birds is curiosity without fear of the unknown. In ourselves we call such individuals curious or adventuresome. In ourselves and in finches, it undoubtedly contributes to the discovery and use of new opportunities.

In the different kinds of finches that have undergone adaptive radiations, all have feeding specializations known or likely to be influenced by learning, but the behavioral specializations are accompanied by different degrees of morphological change. This illustrates one of the principles of developmental plasticity and evolution: the rate of genetical evolution of a facultatively expressed or facultatively used trait depends in part on frequency and consistency of its expression or use. The Cocos Island finches are unlikely to undergo a morphological radiation, not only because of constraints on speciation, but also because extreme behavioral polyspecialization, whatever its causes, does not permit sufficient recurrence of any one behavior pattern for directional evolution of associated morphology to occur. In an environment where opportunistic virtuosity is at a premium, the most useful specializations are those that can be quickly learned and exchanged for others in accord with short-term rewards.

Where speciation has occurred, as in the Galapagos and Hawaiian finches, learning may accelerate morphological divergence between species. Learning can match behavior to morphology, due to the greater or more rapid rewards of morphology-appropriate behaviors. Learning also helps to match behavior to ecological opportunities. Therefore, any ecological differences that promote divergence in beak form and trophic behavior between populations will be magnified by learning. In the Hawaiian archipelago (Grant, 1994) and in the African finches (Smith, 1990a–d), stable availability of particular foods would permit learning to prolong individual dietary specializations throughout a population and thereby promote consistent selection on trophic morphology and behavior. Yet the ability to adopt new learned specializations, should conditions and opportunities change, would not be lost, as long as learning is part of the mechanism for trophic specialization.

The fact that learning can continue to be the developmental mechanism for a fixed specialized behavior is one key to understanding multidirectional radiations of birds and other organisms where learning is an important mechanism of behavioral ontogeny. Increased morphological specialization brings a decrease in realized behavioral flexibility but not necessarily a decrease in the ontogenetic importance of learning. As humans should know, learned adult behavior can be remarkably inflexible and stereotyped. There is no reason to think that the importance of learning declines as the frequency of learning a particular pattern goes up. Exploratory learning gets incorporated into ontogeny as a mechanism of development as surely as do other hyperflexible mechanisms such as the polymerization and depolymerization of microtubules in the mitotic spindles of dividing cells, the exploratory growth of nerve cells, and the individualistic migrations of mesenchyme cells in embryos (Gerhart and Kirschner, 1997; see chapter 3). That the product is relatively invariable in a particular population does not diminish the flexibility of its developmental mechanisms (Gerhart and Kirschner, 1997). This ability of flexibility to promote specialization without loss of flexibility itself is one reason why hyperflexible mechanisms such as learning and modification of bone and muscle by use can promote multidimensional radiations.

Synergism of Plasticity and Other Factors in Adaptive Radiation

Sexual Selection

African lake cichlids, Hawaiian drosophilids, and Hawaiian honeycreepers have in common striking species-specific sexual display behavior and morphology (e.g., Baerends and Baerends-van Roon, 1950 and McKaye, 1983; Spieth, 1974; Lepson and Freed, 1995) and several authors have suggested that sexual selection (or, more broadly, social selection; West-Eberhard, 1979, 1983) has been important in the rapidity of their speciation (e.g., Baldwin, 1953; Greenwood, 1964, p. 267; Rachootin and Thomson, 1981; Dominey, 1984; Spieth, 1974). Sexual selection is a subset of social selection, being social competition, usually among males, for mates (Darwin, 1871; West-Eberhard, 1979, 1983).

"Good genes" interpretations of sexual selection (e.g., Hamilton and Zuk, 1982; Carson, 1987b) imply that sexual selection assists natural selection by favoring criteria of mate choice that reflect ecological or economic superiority. But an even greater boost to natural selection would occur if socially selected sexual traits were involved with no good-genes function except in the production of successful sons, because the evolution of social traits is less limited by ecological expediency than are good-genes traits, which are constrained to be honest indicators of quality (West-Eberhard, 1979, 1983). When social or signal value per se is involved there, may be an especially strong *sexual–natural selection synergism*—selection in which the same trait is favored by sexual selection, as a promoter of social status and success, and by natural selection, as a promoter of ecological success. Since all individuals of sexually reproducing species mate, sexually selected traits are subjected to selection and improvement in every individual of every generation, even when selection on the same trait in the ecological context may be temporarily absent, weak, or reversed in sign.

Synergism between social-sexual and natural selection is probably responsible for some of the most extreme, exaggerated traits in the history of life. It may be involved, for example, in the evolution of such extreme characters as the electric discharge systems of electric fish, which employ electrical discharges both in predation and in social communication (Hopkins, 1977); the raptorial appendages of stomatopod crustacea which are used in both fighting and feeding (Caldwell and Dingle, 1976); the "hyperdivergence" of venom peptides in cone snails (*Conus* species), which are used in both predation and intraspecific and interspecific competitive interactions (Olivera et al., 1999); and the specialized hovering behavior of stenogastrine wasps that gives them their name, "hover wasps," used in both foraging and male competitive displays (Turillazzi, 1991; Beani and Turillazzi, 1999). The human brain is also likely a product of social–natural selection synergism, since it functions importantly in both social contexts, such as use of language and assessment of status and reciprocity, as well as in ecological contexts, that is, behavior involved in survival and defense against nonconspecific natural enemies, such as tool use, shelter building, food preparation and cultivation, and medicine.

In the African lake cichlids, the extent of synergism has not been appreciated even by authors who have drawn attention to the likely contributions of both trophic flexibility and sexual selection in explosive radiation (e.g., Dominey, 1984; Reutsch, 1997; Arnegard et al., 1999; Barlow, 2000). The cichlid mouth specializations have been placed in one compartment, as subject to natural selection for trophic efficiency; and sexual colors and display have been placed in another compartment, as subject to sexual selection. Liem (1973), for example, attributes the unusual mouth morphology of cichlids to natural selection in the trophic context without explaining how it evolved. He says only that "an intensification of the masticatory function has led to a transformation of the food transporting

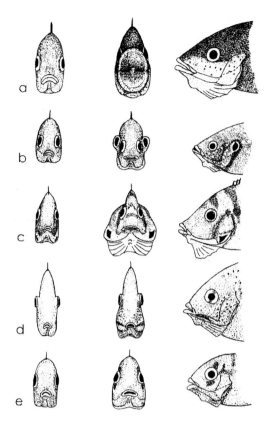

Fig. 28.19. Use of the mouth in displays in cichlid fish: (left) at rest; (middle) frontal display; (right) lateral display. (a) *Tilapia natalensis*; (b) *Hemichromis bimaculatus*; (c) *Cichlasoma meeki*; (d) *Cichlasoma severum*; (e) *Cichlasoma bimaculatum*. From Baerends and Baerends-van Roon (1950), by permission of Brill NV, Publishers.

pharyngeal apparatus . . . giving a new and vastly improved character complex of high selective value" (p. 435). The modularized jaw has demonstrably enhanced biting efficiency (Galis and Drucker, 1996), so selection in the feeding context has likely played a role. And the highly maneuverable and versatile buccal apparatus is capable of masticating a prey (with the pharygeal apparatus) while catching one or two others (Barlow, 2000). But what about the possibility that an efficient bite and a maneuverable mouth is equally or more important in fighting or display?

The possibility of sexual–natural selection synergism in the evolution of cichlid jaws leaps into view when one looks a displaying cichlid fish in the face (figure 28.19). Many years ago, Baerends and Baerends-van Roon (1950) made a detailed study of cichlid mouth morphology in order to elucidate the modifications responsible for extreme movements of the mouth in social displays. They showed that the unusual articulation and musculature of the cichlid jaw permits an unusual rotation of the structures that allows a brightly colored and marked branchiostegal membrane to be erected and stretched during frontal and lateral display, in some species also allowing the gill covers to be brought far sideways. They described the "great effort" of the muscles employed as indicated by a slight quivering of the lower jaw and the membrane during display, which they compared to "the similar quivering movements that can often be seen in acrobats carrying out difficult feats" (p. 48), and emphasized that many of the same muscles and movements are used in both display and feeding.

Fighting male African lake cichlids sometimes "grip each others' jaws and 'wrestle' by pushing each other to and fro" (Fryer and Iles, 1972, p. 186), and the pharyngeal jaws are sexually dimorphic in many cichlids, including some haplochromines, being more strongly hypertrophied in males than in females of *Astatoreochromis alluaudi* and in *Hemichromis flaviijosephi* in at least some lakes (Witte et al., 1990). While this could be associated with different trophic niches of the two sexes, it also may reflect stronger sexual selection in the evolution of male jaws. Cichlid females brood their young in the mouth (Fryer and Iles, 1972), which could also contribute to the sexual dimorphism. As pointed out by Galis (1993b), mouth innovations are usually assumed to evolve in the trophic context, but may have multiple functions.

It seems clear from these observations that the versatility of the cichlid mouth anatomy is a product of both natural and sexual selection. The extreme effort to push its mouth morphology to the limits of its muscular capacity, as indicated by Baerends' observations described above, would impose strong developmental and selective forces on skeleton and muscles. Sexual selection may have initiated changes in the mouth apparatus, creating new morphological configurations that then proved preadaptive for trophic versatility.

Founder Effects and Developmental Drift

Genetic "founder effects" have sometimes been considered important factors in the origin of new directions of evolution in adaptive radiations. By this hypothesis (see Mayr, 1982b, p. 171; Mayr, 1989; Carson and Templeton, 1984; Carson, 1985, 1992a; Barton, 1989, p. 229), a small founder population becomes isolated from its population of origin, and a new direction of evolution ensues aided by the genetic bottleneck (drift) represented in a small sample of the genotypes from the parent population. This is in keeping with the idea that evolutionary divergence requires genetic (especially, geographic) isolation, and also with the presumption that the first step in the origin of a phenotypic novelty is genetic change, or even a genetic revolution that breaks up a coadapted gene complex and allows it to evolve to a new coadapted state.

Just as a peripheral isolate may contain a small and incomplete sample of parent-population genotypes, its range may contain only a small, incomplete, or very different sample of the environments experienced by the parent population. The immediate result, and one that does not depend, like drift, on a small or unrepresentative sample of individuals, would be *developmental drift*—a narrowed or new set of phenotypic responses to a narrowed or new environment, based on preexisting mechanisms of plasticity. Thus, the range of phenotypes present in the isolate could be immediately distinctive, not as a result of change in its genetic composition, but due to adaptive phenotypic change.

Divergence in the island birds and *Drosophila* populations that inspired the concept of founder effects in its originators (Mayr, 1954, and Carson, 1968, respectively) is often in morphology associated with sexual behavior as well as ecologically adaptive traits (see Mayr, 1954; Carson and Kaneshiro, 1976; Carson, 1978). Sexual behavior is characteristically complexly variable and condition sensitive, as illustrated in figure 4.4, and mating tactic polyphenism is not only very common but also can be associated with special features of the environment (e.g., see Alcock, 1979; Polak, 1993; Beani, 1996). Even elaborate behavior such as lek mating systems are only facultatively expressed in some species (e.g., see Beani, 1996, on wasps [*Polistes*]; Lank and Smith, 1987, on birds [*Philomachus pugnax*]; Shelly and Whittier, 1997, on insects), perhaps not surprisingly since the advantageousness of lek behavior is influenced by ecological factors (Höglund and Alatalo, 1995). Environmental features can influence the effectiveness of particular elements of sexual displays (West-Eberhard, 1983; Jennions and Petrie, 1997). Taken together, these facts suggest that developmental drift could contribute to species diversity in sexual as well as nonsexual traits. Genetic accommodation of the developmental shift would be expected to ensue.

The developmental drift hypothesis may explain some of the patterns that have been associated with founder effects. A rise in additive genetic variation sometimes has been observed following population bottlenecks (Carson, 1990, 1992a). If the bottleneck is associated with an environmental change, narrowed phenotype expression could lead to increased additive genetic variation and response to selection due to release from genetic correlations and epistasis associated with reduced plasticity (Goodnight, 1988).

Even if the genetic variation available to selection were not increased in this way, high phenotypic plasticity of a founder population is likely to be associated with high genetic heterozygosity. Nei et al. (1975) and others have estimated that even a single founder individual may bear an average heterozygosity (determined by electrophoresis) of more than 10% (13.8% is the estimate given by Nei et al., 1975; see Carson, 1990). Unselected genetic polymorphism is expected to be especially high in founder individuals and populations descended from plastic species with wide, heterogeneous ranges, insofar as a history of plasticity is associated with life in varied environments and therefore maintenance of genetic diversity. This is a potential genetic consequence of plasticity that, along with initial adaptability, would favor survival and radiation on islands.

Barton (1989; see also Moya et al., 1995) give theoretical reasons to doubt the applicability of the *genetic* founder effect hypothesis and the idea of a "cohesive, coadapted" gene pool. The developmental drift hypothesis can explain the divergence attributed to founder affects without a sudden genetic revolution, and also can explain how rapid phenotypic change can be adaptive rather than random as under the genetic founder-effects hypothesis. Developmental plasticity can produce immediately adaptive responses in novel environments and, by matching phenotype to environment in the population as a whole, can produce recurrent novelties subject to selection and genetic (evolutionary) change.

Grounds for Generalization

How widely applicable are the conclusions of this chapter likely to be? Although I do not claim that all adaptive radiations arise from developmental flexibility, it is clear from the examples of this chapter that research on plasticity as a factor in radiations will be amply rewarded. The three taxa discussed here with multidirectional radiations—cichlids, drosophilids, and finches—were chosen not because flexibility was known to be important, but because they are well-studied, famous examples of adaptive radiation. Although not a primary focus of previous research, there is evidence for plasticity-assisted radiation in all three groups.

Other well-studied radiations promise to yield similar results. The diversification of *Anolis* lizards in Caribbean islands, for example, has some of the earmarks of a plasticity-assisted radiation (Case, 1997), even though current discussions emphasize interspecific differences rather than the intraspecific responses that may indicate a role for plasticity. Anole communities are structured by interspecific ecological competition (review in Losos, 1994b), and populations rapidly diverge in resource use, behavior, and morphology when transplanted between islands with different vegetation (Losos et al., 1997). There is recurrent concordant evolution of locomotory behavior and morphology in different species with similar use of vegetation as support (Losos, 1990); and it is well known in lizards that locomotor performance is affected by substrate (brief review in Losos and Irschick, 1996). When individuals of sprinting *Anolis* species are experimentally confined to novel perch supports and simulated predation in the laboratory, individuals alter their behavior as a function of support diameter, increasing their use of jumping for escape as perch diameter declines (Losos and Irschick, 1996). Such condition-dependent locomotory change, whether adaptive or forced, commonly influences morphological development, and this is the kind of plasticity that can push evolutionary change in a new direction.

Several patterns emerge regarding the contributions of developmental plasticity to adaptive radiation. First, episodic strong (e.g., seasonal or competition-dependent) selection and specialization, which favor the evolution of phenotypic plasticity (Robinson and Wilson, 1994), have been demonstrated in studies of adaptive radiation (e.g., in Galapagos finches, African finches, and African lake fishes, discussed above). Developmental plasticity can convert sporadic episodes of directional selection into directional evolution due to the ratcheting effect of conditional expression: by turning the favored traits on only when they are advantageous, associated genetic variation is repeatedly expressed and subject to selection in a single direction, increasing the probability of directional evolutionary change (see chapter 7).

Cumulative results of episodic directional selection may be a general characteristic of adaptive radiations, including over long time scales. It may, for example, explain the more extensive radiations in the tropics than in the temperate zone in some taxa. Among other factors (such as different rates of speciation and extinction), seasonal episodes of strong directional selection are likely to have a greater effect on adaptive diversity in the tropics because, rather than leading to diapause and hibernation, they more often lead to novel tactics of behavior and reproduction. This was shown in the Galapagos finches, where seasonal consumption of a novel source of protein, pollen, permitted early breeding of some individual birds. Such options are likely to be common in the tropics, where seasonal climate change does not always demand a drastic metabolic arrest. In Guanacaste, Costa Rica, for example, more than 80% of dry tropical lowland butterfly species migrate annually between dry and lowland forest (Haber, 1993). In Haber's words, they are "multihabitat" species, for the areas at the two ends of their migratory routes are very different, yet they actively feed and reproduce in both. Other tropical insects show striking seasonal polyphenisms (e.g., see Shapiro, 1976; Brakefield, 1987, cited by Witte et al., 1990). Life in multiple terrestrial habitats also occurs due to seasonal flooding in the Amazon (Junk, 1997) and in tropical aquatic organisms (McDowall, 1988). These are examples of seasonal adaptive options of kinds that can fuel adaptive radiation in the taxa where they occur.

Consistent with the idea that continued activity under seasonally fluctuating conditions promotes adaptive radiation, nonmigrant temperate zone birds, which remain active year-round, sometimes show remarkable seasonal trophic polyphenisms. Some have apparently adaptive, reversible seasonal change in digestive physiology, including changes in stomach size and intestine length (Piersma and Lindstöm, 1997) and cecum length and biochemistry (Gasaway, 1976), corresponding to contrasting seasonal diets such as switching between arthropods (in summer) and seeds (in winter) or between vegetation types (Kear, 1962; Selander, 1966; see also McWilliams et al., 1997). Seasonal scarcity within the summer activity period can also promote alternative specializations in the temperate zone, as in the tachinid parasitoid fly *Myiopharus aberrans*, whose usual prey (beetle larvae) decline at the end of the breeding season, when it "doggedly pursues" adult beetles, which it parasitizes using specialized behavior required to gain access to the beetle's vulnerable abdomen (usually covered by the wings; López et al., 1997).

The flexibility hypothesis of tropical diversity suggests that increases in diversity should be centered in areas (or eras) characterized by strong seasonal or otherwise periodic change, where populations are repeatedly subject to boom-and-bust selection regimes—alternating times of scarcity and of plenty that favor the evolution of alternative tactics of survival and reproduction. The general importance of this idea in nature (not just in theory) is suggested by the fact that previous authors have come to the same conclusion, based on observations of particular groups, for example, Diamond (1987) and Smith et al. (1997) on birds, Jago (1973) on grasshoppers, and Lowe-McConnell (1969) on fish. In all of these groups tropical species are characterized by multiniche adaptation, or polyspecialization, associated with specialized adaptation and speciation following an episode of habitat contraction and isolation.

Broader statements, not restricted to tropical organisms, indicate that the boom-and-bust selection regime coupled to plasticity is a major basis for diversification. In highly flexible opportunistic species, scarcity sometimes decreases rather than increases specialization, depending on circumstances (see Kaplin et al., 1998), and in such populations plasticity would not promote adaptive radiation. In a general discussion of patterns of interspecific competition, many authors, in broad reviews of evidence, have noted that it is common in morphologically specialized species for specialization to increase in lean times and diet overlap to increase during good times: during scarcity a species, or a specialized form within a species, often pulls back to its own more or less exclusive set of resource types (Schoener, 1982; see also Selander, 1966; Keast, 1977a; Schluter, 1982a,b). Schluter (1988) reinforces Schoener's conclusion that resource competition from other species is an important component of scarcity in lean times. Selander (1966) shows that in bird species with sexually dimorphic beaks, males and females sometimes have different trophic niches, and their divergence is most marked under seasonal food shortages. Shoener calls specialized morphology the "genetic memory" of past competition. The interaction of competition, morphology, and learning is discussed above in relation to fish. There may be a very general relationship between competition, behavioral innovation, and morphological divergence within species (see Maret and Collins, 1997, on salamanders).

The role of adaptive *flexibility*, including learning, in boom-and-bust hypotheses of competitive specialization is crucial. The so-called genetic memory in the form of specialized morphology can exert its effect on facultative behavioral specialization through trial and error learning, where rewards obtained are affected by morphological efficiency.

This can include not only features of the mouth or beak, as discussed above for cichlids and finches, but also leg length (which affects ability to walk, cling, and climb) and digestive traits (Kear, 1962; Martinez del Rio and Karasov, 1990). Partridge and Green (1984), Giraldeau and Lefebvre (1985), and Lefebvre et al. (1997) show that the kind of learned individual behaviors that fuel adaptive radiations in Galapagos and Hawaiian finches are common in birds. Lefebvre et al. (1997) collected 382 examples of opportunistic learned foraging innovations in North American and British birds, all of them well outside the species behavioral norm. These are the behavioral counterparts of the morphological anomalies discussed in preceding chapters as the basis for selection and evolution in new directions. Early studies of birds, including raptors and flycatchers, that appear to learn foraging tactics from adult birds are reviewed by Cushing (1944). Copying of foraging behavior has been experimentally demonstrated in birds (Curio, 1976; Lea, 1984). Morphological variation in bill characterisics is known to affect foraging behavior in many birds other than Galapagos finches, lending generality to the findings on island radiations (e.g., see Temeles and Roberts, 1993).

A few additional examples can serve to illustrate a connection between flexibility and adaptive radiation. Crassulacean acid metabolism physiology in plants (see chapter 26) is the recurrently evolved product of a physiological flexible stem whose biochemical components are widely available in C_3 species. The capacity for reversion to facultative or fixed C_3 metabolism is omnipresent since it characterizes the early development of all leaves, even in obligatorily CAM plants, making this an extremely flexible developmental system (Winter and Smith, 1996b). CAM physiology is associated with a very rapid recent radiation of African desert plants (Aizoaceae) showing a diversity of degrees of CAM activity (Winter and Smith, 1996b) in a group where facultative CAM is observed (Winter and Smith, 1996a). Flexible CAM physiology may be an important aspect of the adaptive radiation of orchids and bromeliads in the seasonally variable humid tropics (see Winter and Smith, 1996a, p. 7; see also below).

Schaefer and Lauder (1986) show that increases in the modularization of the feeding mechanism in loricarioid catfishes is associated with increased jaw mobility and new muscular insertions, and in turn with increased morphological and trophic diversity in that group of fishes. They generalize their results by suggesting that the number of biomechanical linkages in a structure is related to the degree of functional diversity of species possessing it (see also Galis, 1996a).

An unusual diversification of predatory behavior occurs in Australian bembecine wasps (Sphecidae), where members of a single genus specialize in prey types novel for the subfamily, such as antlions (Neuroptera), stingless bees, and damselflies, in addition to usual flies (Diptera; Evans and Matthews, 1973, 1975). The unusual Australian radiation is associated with a likewise unusual presence in some species of intraspecific versatility in prey used (e.g., *Bembix variabilis* and *B. morua*), most notably in the dry season of their tropical habitat. Similarly, Brockmann (1980a) related the unusual radiation in nesting behavior of another group of sphecid wasps (Trypoxilini) to their unusual intraspecific versatility. Different species build free mud nests, reuse mud cells built by other species, nest in twigs and preexisting cavities, use burrows dug in the ground by other insects, and dig burrows themselves. Five of these nesting tactics occur intraspecifically in *Trypoxylon politum*.

Wake (1982) describes various ways in which increased morphological flexibility in the buccal apparatus has facilitated the trophic radiations of salamanders (Plethodontidae). The disarticulation or unfixing of the first ceratobranchial and basibranchial bones in *Thorius*, and the addition of a very flexible anterior section to the basibranchial, allowed new movements and therefore new options in use of the tongue. As in cichlids, a release from double functions due to increased modularization of mouth structures evidently facilitated trophic diversification: lungless species have an unusually versatile tongue apparatus due to elimination of use of the hyobranchial apparatus as a force pump to fill the lungs (Roth and Wake, 1985, 1989).

The opposite—a decline in phenotypic versatility with consequent loss of species diversity, amounting to *de*radiation—may have occurred in the histories of some groups. An important aspect of flexibility for radiation is its contribution to the long-term survival of lineages. Williams and Hurst (1977, p. 108) suggest that the extinction of spiriferide brachiopods, whose radiation was based on combinatorial evolution, may have been due to the loss of a particular aspect of flexible morphology. Spirifide history was characterized by the recurrent origin of novel phenotypes by recombination of morphological characters. These authors speculate that the ability to adjust through such combinatorial innovation was crucial to the long-term persistence of these brachiopods. They associate the extinction of the group with the elaborate mimicry by a skeletal apparatus of the disposition of the lophophore that thereby reduced the flexibility of the feeding organ to a degree that "invited extinction" by limiting further modular recombination and diversification.

Predictions

Flexible stem radiations are expected to be possible with little genetic change since much of the phenotypic diversity is due to phenotypic plasticity. Avise (1990) called the decoupling of morphological and molecular evolution implied by the Meyer et al. (1990) finding of small genetic distances in Lake Victoria haplochromines "stunning." But it is expected if morphological divergence is importantly influenced by ancestral phenotypic flexibility. It is also in keeping with findings such as those of Wainwright et al. (1991) that morphological differences in the pharyngeal jaw muscles and bones of pumpkinseed sunfish (*Lepomis gibbosus*) in two different lakes are due not to genetic divergence but to dietary differences (the proportions of snails consumed) during ontogeny. This is an expectation of the plasticity hypothesis regarding the importance of plasticity in adaptive divergence, and it contradicts the expectation (e.g., Cracraft, 1990, p. 36; John and Miklos, 1988) that diversification, or the production of phenotypic novelty, must bear some relation to the production of *genomic* novelty (e.g., mutation or gene duplication). Authors are sometimes perplexed when genomic change and phenotypic change are grossly discordant (e.g., see John and Miklos, 1988). Genetic change accompanies phenotypic change in flexibility-assisted radiations, but it may be minor relative to the amount of phenotypic diversification achieved.

Another prediction of the flexible stem hypothesis is frequent homoplasy in radiating groups. In a radiation where specialization is based on continued flexibility, the direction of divergence is not limited by recent phylogenetic history (immediate ancestral specializations) but can respond to local selection with great versatility; and recurrent similar opportunities are likely to give rise to recurrent parallel evolution. That this occurs commonly in nature as a consequence of developmental flexibility is discussed in chapter 19, on recurrence, and in chapter 25, on homology. This prediction is borne out by the Hawaiian honeycreeper population, which proved rapidly adaptable when transplanted to a new island (Johnson et al., 1989); by the groups of fishes where parallel species pairs have evolved from ontogenetically versatile marine populations many times in the short time since the last glaciation (Schluter, 2000); and by some populations of the picture-winged *Drosophila grimshawi*. Although part of a highly derived Hawaiian species complex, one of a pair of *D. grimshawi* clades contains populations that are secondarily ecological generalists found on several islands, in contrast to the single-island endemism of most picture-wing species, including those that are its closest relatives (Piano et al., 1997). *D. grimshawi* is one of the Hawaiian drosophilids used in molecular studies that suggest a biochemical flexible stem (Dickinson, 1991), as discussed above. These examples accord with the expectation of the flexible stem hypothesis that specialization based on developmental plasticity can occur without loss of plasticity itself, keeping the door open for further adaptive radiation.

The flexible stem model focuses attention on parallels between intrapopulation and interpopulation variation. Generalist behavior or morphology is not the same as a polyspecialist pattern where the effects of selection can be focused and diverging. As noted by Grant (1986), *Geospyza fortis*, the large cactus finch, has the beak of a generalist, but it is used by some individuals at some times in a specialist manner. The specialist behavior, seen as an occasional variant in *G. fortis*, is material for morphological specialization if it comes to be performed at high frequency across generations. The observation of the intraspecific specialization is therefore a clue as to the possible origin of specialized beaks in related species.

Wright (1945) described the situation that is "most favorable for rapid evolution" as one in which a population "may be large but . . . is subdivided into many small local populations almost but not quite completely isolated from each other" (p. 416) for it is this population structure that maximizes genetic variability. The considerations of this chapter suggest that the same type of population structure would promote the evolution of developmental flexibility, and directional evolution that can be fostered by flexible specialization to local conditions. Additional factors that foster rapid evolution include a temporally fluctuating boom-and-bust environmental change, with sporadic severe habitat fragmentation, such that populations alternately experience resource abundance and diversity that foster opportunistic versatility; and resource scarcity such that they periodically specialize or migrate to an alternative habitat in order to compete and survive.

Frequency-dependent environmental cycles, such as those imposed by parasites and pathogens, are hypothesized to underlie mechanisms of *genetic* flexibility (e.g., the maintenance of sexual reproduction—see chapter 31), but it is clear that they also promote extreme phenotypic plasticity, as evidenced by such strikingly flexible adaptations as the hyperflexibility of the vertebrate immune system, and the host-alternating cycles of many parasites and disease organisms, which enable them to facultatively or cyclically escape, through developmental plasticity, the coevolving defensive flexibility of their hosts. One cannot help but be impressed at the inordinate emphasis on genetic escape mechanisms, such as sexual reproduction and good-genes sexual selection, in discussions of defenses against parasites and disease, when escape via developmental plasticity is so obviously important as well.

This cursory look at adaptive radiation in relation to developmental plasticity in selected taxa reveals that some of the same authors who are prominent in research on behavioral and morphological plasticity are prominent in discussions of adaptive radiation as well. Yet they seldom connect the two. Why is this? I can only conclude that there is still an important gap in understanding phenotypic evolution: the role of genetics and selection is clear, but the role of phenotypic plasticity is not. The essential point of this chapter can be summarized as follows: Genetic variation produces a response to selection, but it is patterned developmental plasticity that, by producing patterned variation, focuses selection in particular directions, and determines which genes are exposed to selection. This is why data on developmental plasticity are crucial to understanding adaptive radiation.

29

Macroevolution

Developmental plasticity enables selection to produce large (macroevolutionary) change, of a magnitude comparable with differences between species or higher taxonomic categories, within species. Intraspecific macroevolution is facilitated by the buffering effect of alternative phenotypes and the ability of plasticity to allow independent expression, selection, and evolution of semi-independently expressed phenotypes, which enable variants within species to evolve toward major new adaptations without the negative fitness consequences usually attributed to major change. Sexually selected flexibility may have contributed to some macroevolutionary trends associated with size change. Why, if intraspecific macroevolution is possible, has the origin of major new phyla evidently stopped? Hypotheses to explain this based on the advent of germline sequestration and the absence of empty niches are rejected in favor of (1) the idea that the apparent stagnation of major change is an artifact of a taxonomic hierarchy based on a temporal hierarchy of phylogenetic branches: the most inclusive, major categories are the oldest; and (2) the fact that all innovation must start with old building blocks: if we recognize a trait as characteristic of a known taxon, we classify the organism accordingly. Any relatively recent species is necessarily some variation on older broad themes or body plans. Although evolutionary molecular biology seeks explanations of macroevolutionary (e.g., body-plan) change by scrutiny of the genome, such explanation is likely to be incomplete given that major reorganizational changes can begin with higher level processes, for example, involving hormones, behavior, and environmentally induced phenotypic effects.

Introduction

Macroevolution, or *trans-specific evolution*, refers to two different things in the literature on evolution. In discussions of phylogeny, it means phylogenetic branching pattern, or trends, seen at relatively high taxonomic levels (e.g., Stanley, 1979; Brooks and McLennan, 1991; Sober, 1993)—"any patterns that transcend species boundaries" (Lynch, 1991)—such as births and deaths of species and higher taxa (Sober, 1993, p. 2) and the shapes and diversity of radiations (Valentine, 1990). In discussions of evolutionary phenotypic transitions like those of part II, it means major phenotypic change (Lincoln et al., 1982). Rensch (1960) defined macroevolution as "evolution above the species level." *Microevolution*, by contrast, is evolution below the species level, such as adaptive phenotypic and genetic change within populations, and geographic variation within a species.

According to Simpson (1953a), the terms "macroevolution" and "microevolution" were invented by Goldschmidt (1940 [1982]), who also claimed that they involve different kinds of evolution. This problematic idea dates back to antiquity (see Rensch, 1960, for a concise review). The macroevolution problem, with emphasis on phylogenesis, was among other things (see Vuilleumier, 1984) behind the skepticism regarding Darwinism promoted by the enormously respected and influential French zoologist P.-P. Grassé. Grassé was convinced that the neo-Darwinian approach, with its emphasis on microevolution, cannot account for the primary features of evolution, namely, the large-scale diversification of life into major phylogenetic branches separated by unbridged gaps (e.g., see Grassé, 1973). This challenge echoes in the writings of many other critics of neo-Darwinism (e.g., Ho and Saunders, 1984; Gould and Eldredge, 1977; Gould, 1994; see also below), especially those who wish to contrast multilevel selection (including species selection) with microevolutionary theories (see Gould, 1999).

The two macroevolution concepts, like the homology concepts discussed in chapter 25, are used interchangeably without sufficient attention to potential confusions. The result is needless controversy. The phylogenetic definition, for example, implies that macroevolution cannot, *by definition*, occur within species, for it refers exclusively to patterns above the species level. The phenotypic, major-change definition, on the other hand, can include processes within species. So a major-change macroevolutionist automatically objects when a phylogenetic macroevolutionist asserts (as in Gould and Eldredge, 1993) that microevolution cannot account for macroevolutionary change. The major-change macroevolutionist objects because this statement seems to imply that large change cannot occur by selection and genetical evolution within species. In fact, it may only assert a truism based on definitions alone: microevolution cannot account for macroevolution defined as phylogenetic patterns and trends above the species level.

Sometimes the confusion seems, perhaps unconsciously, purposeful, for ambiguity attracts audiences and confusion provokes discussion. Thus, a biologist interested in adaptive microevolution may be startled to read that most species arise with geological abruptness and then remain phenotypically stable throughout their duration, as maintained by speciational punctuationists. This view emphasizes the distinctiveness rather than the continuity of branches. Yet it does not contradict the fact that every surviving phenotype is a product of an unbroken chain of microevolutionary processes within a lineage. As noted by Van Valen (1988; see also Charlesworth, 1982), this point is sometimes lost in debates over macroevolution and punctuation (see chapter 30). Two separate phenomena need to be recognized and explained: major phenotypic change, and phylogenetic pattern (radiations, extinctions, gaps between clades) above the species level.

In this chapter, *macroevolution* means major-change macroevolution. In particular, I show how major adaptive innovations can originate and become elaborated within species, and argue that such intraspecific macroevolution especially favors divergent change.

Intraspecific Macroevolution Compared with Previous Macroevolution Concepts

Several authors (e.g., Van Valen, 1974; Liem, 1984; West-Eberhard, 1986, 1989) have proposed hypotheses of *intraspecific macroevolution*, a term introduced by Liem (1984) to describe major evolutionary innovation within a species or population, without branching (genetic isolation or speciation), for example, in the form of an alternative phenotype or ontogenetic stage so distinctive that it is comparable to differences between species or higher categories in that taxonomic group. Such an intraspecific variant may evolve considerable complexity, as discussed in chapter 21, and then be fixed in a species or a clade, in association with speciation (see chapter 27). Intraspecific macroevolution via alternative phenotypes, then, may involve two phases: an alternative-phenotype phase, which may extend indefinitely and permit extensive morphological, physiological, and behavioral change; and a phase of phenotype fixation accompanied by release and accelerated evolutionary change. Fixation may or may not be associated with speciation. As I discuss in chapter 30, on punctuation, the burst of change that accompanies phenotype fixation is probably often attributed to speciation, when speciation may not have occurred. Because accelerated evolution due to selection during phenotype fixation may be an important cause of speciation (see chapter 27), genetic isolation may often be credited with major divergence that is in fact due to phenotype fixation.

Most previous theories of large-change macroevolution see major change as *interphyletic*; that is, major differences are separated on different branches of a phylogenetic tree. Darwin's gradual-macroevolution concept (figure 29.1) is the prototypical example of this. Large change begins as an individual peculiarity, then grows in distinctiveness as it comes to characterize differences between varieties or subspecies (the fine dashed lines), then ever larger branches on the phylogenetic tree. By Darwin's "principle of divergence" (1859), selection favors extremes. Major change is the cumulative result of such change and of the increasing distance between differentially diversifying branches. As in the intraspecific macroevolution concept, divergence depends on selection. But Darwin's scheme differs sharply from the intraspecific macroevolution idea in proposing that the variants selected are from the beginning interindividual or

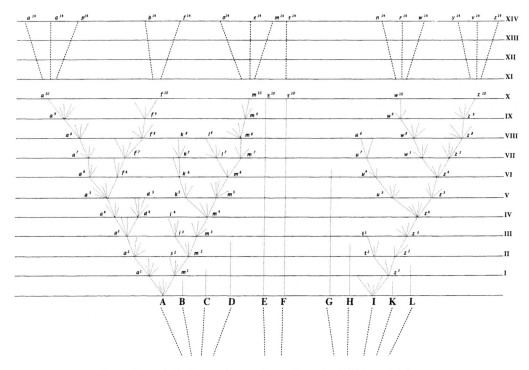

Fig. 29.1. Darwin's hypothetical phylogenetic tree. From Darwin (1859, p. 116).

interspecific, not mediated by switches between different developmental alternatives or between life stages as suggested here.

Simpson's (1953a) macroevolution concept (figure 29.2) is more similar to intraspecific macroevolution in that Simpson explicitly considered intraspecific behavioral flexibility to be important in the initiation of a new macroevolutionary trend. The evolution of submarine locomotion from avian aerial flight in penguins, for example, involves an intermediate phase in which individuals perform both types of locomotion, either facultatively or during different ontogenetic stages (figure 29.2). Then, once an environmental threshold has been passed, where selection for submarine locomotion becomes stronger than selection for aerial flight, submarine locomotion is perfected and aerial locomotion lost. Similarly, in the transition from browsing to grazing in horses, facultative grazing led eventually to specialized obligate grazing as horses came to inhabit increasingly grassy environments (Simpson, 1953a). But in Simpson's hypothesis, the macroevolutionary divergence occurs when a separate branch specializes in a uniform environment, as shown in figure 29.2—macroevolution is associated with speciation and is interphyletic (see clarification of this in Simpson, 1953a, p. 205).

Many additional authors have expressed the view that novel traits and shifts into new adaptive zones begin as behavioral innovations (e.g., Darwin, 1859; Mayr, 1958, 1959, 1963, 1974a). Gould (1977) cited numerous intraspecific alternative phenotypes as evidence for the nature and adaptive context of major heterochronic change. But none of these authors relates intraspecific de-

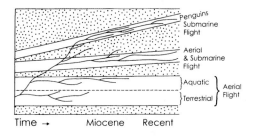

Fig. 29.2. Invasion of new adaptive zones during the evolution of penguins. Branching lines represent morphological evolution. Simpson's text (1953a, p. 204) emphasizes that new zones are entered by forms "whose adaptation was alternatively to both habitats, either facultatively or at different periods in ontogeny." After Simpson (1946).

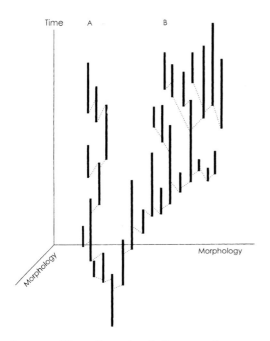

Fig. 29.3. Three-dimensional diagram of punctuated morphological evolution: (A) a pattern of stability; (B) a trend, where speciation (dashed lines) is occurring. Figure and caption after Eldredge and Gould (1972).

velopmental plasticity to his theory of macroevolution, all of which are emphatically interphyletic (figures 29.1–29.3).

The greatest conceptual contrast is between the intraspecific macroevolution hypothesis and *speciational macroevolution* hypotheses, which propose that macroevolution is the cumulative result of divergence caused by speciation (figure 29.3). Some speciational hypotheses depend on some version of differential multiplication at the species or clade level to cause a directional macroevolutionary trend, but authors differ as to how this occurs. On one hand, Eldredge and Gould (1972) visualize a type that they call "non-Darwinian" in that it does not depend exclusively on selection but is due to a combination of factors, such as selection on a variety of contexts and on varying direction, drift, or hitchiking associations with other traits, whose sum is sometimes nearly zero (figure 29.3A) but may not be, yielding a directional macroevolutionary trend (figure 29.3B). This has been called "biased speciation" by Mayr (1993), "species drift" by Levinton (1986), and "sorting" by Vrba and Gould (1986). Stanley (1979), on the other hand, envisions species selection that favors evolution in a

consistent direction by the differential mulitplication and extinction of species or clades characterized by a favored innovation, leading to macroevolutionary change that establishes that trait.

Mayr's speciational hypothesis, in contrast to these, does not invoke natural selection of innovations at the species level, or trends due to sorting, but emphasizes the contribution of single speciation events, when peripheral isolates experience the "destruction of the previously existing cohesion of the genotype and its replacement by a new balanced system" (Mayr, 1982c, p. 1127). This "facilitates new evolutionary departures and explains the origin of new higher taxa and of evolutionary novelties." Because some lineages speciate more, they are more likely to give rise, in one of their speciations, to a major evolutionary change. Because the great speciators are more likely to produce an important innovation and subsequent radiation, this is called "speciator selection" by Mayr (1993). All of these hypotheses have in common that they depend on speciation for the achievement of macroevolutionary divergence.

Wright's (1949) shifting balance is a theory of macroevolution that begins with a key innovation or new zone of adaptive opportunity. But Wright considered the first step as allopatric speciation and held a unimodal-adaptation view of the population, which shifted suddenly from one fitness peak to another without an alternative phenotype phase (see chapter 6).

In sum, all of the previous major theories of major-change macroevolution have been interphyletic, dependent upon divergence on different branches of a phylogenetic tree. For convenience, I will refer to all of them as speciational hypotheses. Speciational macroevolution may or may not be important for the history of life—I do not attempt to evaluate these theories here. Instead, this chapter presents evidence that intraspecific macroevolution is a viable alternative hypothesis, and shows how it can be recognized.

Intraspecific macroevolution depends upon developmental plasticity. Major change with a developmental rather than reproductive-isolation basis is sometimes identified with theories of saltatory speciation by macromutation (e.g., see Mayr, 1982c; Charlesworth et al., 1982). It is therefore important to clarify at the outset that intraspecific macroevolution is not a macromutation hypothesis. Macromutation hypotheses visualize innovation in the form of large, one-step variants ("macromutations"), completely unique to a lineage and inherited as a whole, which spread to characterize a species. Under the intraspecific macroevolution hypothesis, in contrast, major change may be gradual and can begin with phenotypic reorganization, as described in the chapters of part II, then can subsequently be accommodated and modified by quantitative genetic change, as described in chapters 6 and 21.

In sum, intraspecific macroevolution is a microevolutionary explanation for large phenotypic transitions during evolution. It is an explanation completely compatible with standard modern Darwinian microevolutionary theory. I hope that this is sufficiently shocking, in the age of post-Darwinism, to attract attention to this revolutionary idea, wherein standard microevolutionary theory is invoked in a theory of macroevolution *without speciation*, in contrast to all previous revolutionary ideas on macroevolution from Darwin through Simpson, Mayr, Eldredge, and Gould.

How Developmental Plasticity Facilitates Intraspecific Macroevolution

How feasible is intraspecific macroevolution, and how important is it likely to be? The feasibility of large change within species is shown in preceding chapters. All of the reorganizational change discussed in chapters 10–18 (part II) is intraspecific evolutionary change. Chapters 3 and 7 describe specific mechanisms for the accommodation and exaggeration of change, and chapter 16 shows that they apply in nature. Modular structure permits combinatorial evolution, and due to the preadaptive nature of already integrated subunits, the maladaptiveness of a large change is not necessarily proportional to the size or complexity of the change. This, along with phenotypic accommodation due to continuously variable plastic responses, helps to answer the concern that change is necessarily accompanied by disruption of function.

Modularity also permits a particular trait or integrated set of traits to vary and evolve somewhat independently relative to others, further reducing disruption under change. This "buffering effect" is most striking in the case of intraspecific alternative phenotypes (see part III), because there the same function is served by more than one specialized and complex array of traits. The same plasticity that promotes reorganization and permits function in the face of variation can facilitate macroevolution. So it is not necessary to have speciation break the fine-tuned "cohesiveness" of development and the gene pool in order to obtain major change, for

abundant innovation is known to occur within species in the form of life-stage specializations and alternative phenotypes (see chapter 2 and part III).

Several additional factors make it more likely that macroevolution will be intraspecific rather than speciational. In the allopatric divergence model of speciation, recently derived sister species are expected to have similar niches, and their similarities could persist indefinitely: whether or not they diverge depends on chance events and special circumstances, such as the existence of environmental differences sufficient to cause sustained directional change under selection, and chance combinations of genotypes (genetic drift), or mutations that happen to be favorable. These are the classically proposed causes of microevolutionary divergence between populations, and there is no doubt that they could lead to major change given enough time.

Divergence between intraspecific alternative phenotypes, on the other hand, is much less dependent upon chance, for four reasons. (1) It is often the result of intraspecific competition for survival and access to resources, a relatively strong and consistent selective force compared with the differences that drive divergence between populations (West-Eberhard, 1983). While coevolutionary races, such as with a parasite or predator, may be comparable in effect, particular ones are not as consistently present as are conspecific competitors. (2) Alternative phenotypes are often selected for divergence per se, due to the advantage of divergent tactics in reduction of competition with conspecifics that express another, contrasting or opposite morphs (West-Eberhard, 1979, 1986). (3) Contrasting alternative phenotypes are usually conditionally expressed (see chapter 22), and the phenotype–environment matching of conditional expression allows consistent selection in particular different directions. Populations can undergo polyphenic divergence without environmental change, due to selection for individuals of different phenotypic endowment to develop along different trajectories and to occupy different roles or niches in the same place and within the same interbreeding population. Thus, divergence between alternatives is more likely to be major, or macroevolutionary, than is divergence between closely related populations genetically isolated from each other, for their different environments may engender little change or no change at all. Finally, (4) alternative phenotypes buffer a lineage against extinction while undergoing major change by providing an escape valve against overspecialization in a particular mode (West-Eberhard, 1986; see chapter 7).

These ideas are not new. Goldschmidt expressed many of these same insights in *The Material Basis of Evolution* (1940 [1982]; see Gould, 1982b), where he outlined how alternative developmental pathways could direct major changes in viable directions (Gould uses the term "organic buffering" to describe the retention of alternative developmental pathways). In answer to the objection of many critics that a hopeful developmental monster may not be able to find a mate, always a weak criticism in view of the enormous sexual dimorphisms seen in many species, Goldschmidt (1940 [1982], p. 274) argued that major novelties (macromutants) may spread within local populations even while breeding with conspecifics of normal phenotype, using the contrasting morphs of metamorphosed and neotenous amphibians and other hormonally influenced neotenous forms as examples of monstrous, viable, and persistent intraspecific alternatives in nature. And he described seasonal polyphenisms to illustrate "macroevolutionary" differences produced by the same genotype. Shapiro (1984b) also sees polyphenisms as bridges to new adaptive zones, although he sees the buffering effect of alternatives as contributing to the stability of the genome by sheltering it from directional selection. I emphasize instead that bidirectional or multidirectional phenotype evolution under selection can proceed during the alternative phenotype phase, following the principle of independent selection (see chapters 7 and 21). Then, if a single alternative is fixed in an isolated population, divergent evolution may accelerate due to character release, as extensively discussed in chapters 21, 27, and 28.

Arguments that macroevolutionary change often occurs via regulatory change (e.g., Valentine and Campbell, 1975; John and Miklos, 1988) are, in effect, intraspecific macroevolution hypotheses, for such changes, to become fixed, must inevitably pass through an alternative-phenotype phase of evolution. The buffering effect of that phase may be crucial to the feasibility of large change via regulatory innovation.

As for the likely importance of intraspecific macroevolution compared with speciational macroevolution, only future research will tell. One thing can be said with confidence: given the past tendency to overlook intraspecific origins of diversity, for example, in discussions of speciation and radiation, even in groups of fish and insects where developmental plasticity is well known (see chapters 27 and 28), the importance of intraspecific macroevolution is likely also to have been grossly underestimated in the past. I believe that the neglect of divergence in the form of intraspecific plasticity is due not to its lack of importance but to the lack of understanding of how to relate plasticity to the genetic theory of change. Preceding chapters deal with that problem; this chapter applies the same concepts to the question of macroevolution.

Evidence: Major Trans-specific Variants Found within a Single Population

As first pointed out by Stearns (1980, p. 271), intraspecific macroevolution is a clear and *operationally* distinguishable alternative to all interphyletic macroevolution hypotheses in its emphasis on major change within, rather than between, populations. The speciational hypotheses of Gould, Eldredge, Stanley, Mayr, and others differ in certain details, as discussed above, but they all predict that *all* change takes place relatively rapidly during speciation events. Once formed, species are, by these hypotheses, characterized by relative stasis (figure 29.3). If speciational hypotheses obtain, the following predictions should hold: we should not find major alternative tactics *within* species; developmental change without speciation should be a negligible source of major novelties; and major adaptive differences should appear only when species or higher level categories are compared. In contrast, the intraspecific macroevolution hypothesis predicts that major novelties will be found as alternative phenotypes or successive life stages within species, and that patterns of diversification above the species level will often be traceable to developmental changes observable within species.

This latter prediction, regarding diversification explainable in terms of intraspecific developmental plasticity (a flexible ancestral stem), has already been shown to hold in a variety of organisms (chapter 28). The rest of this chapter provides evidence that the predictions of the intraspecific macroevolution hypothesis are upheld by observations in a variety of plant and animal taxa. Speciational hypotheses are contradicted by the same observations.

Intraspecific macroevolution is supported by finding in the same species both the ancestral state and a derived trait regarded as a major innovation in the evolution of that clade. Even if the species is not a basal taxon of the group concerned, or the polarity of the transition is not known, occurrence of both states within a single species is undeniable evidence that the transition between states can occur within a species.

Many examples of extreme divergence between alternative phenotypes are cited in part III, some of them macroevolutionary in degree. Examples include the origin by duplication of segmented body plans in invertebrates and vertebrates (chapter 10); the evolution of direct development via the loss of the larval stage in echinoids and other marine organisms (chapter 11); the evolution of crassulacean acid metabolism in plants (chapter 13); change in the sex performing parental care in birds, fish, mammals, and insects (chapter 15); and the evolution of buttress roots in trees (chapter 16). Intraspecific anomalies resembling interspecific and higher level differences in flower structure (Meyer, 1966) are discussed in chapter 15. Some of the intraspecific patterns discussed under adaptive radiation (chapter 28), such as the specializations of migrant fish and birds to different habitats, and the trophic specializations of finches and cichlid fish, can be regarded as examples of intraspecific macroevolution.

Such intraspecific extremes have inspired several authors to challenge the idea that major innovation must involve speciation or a series of intervening speciations (e.g., see Michener, 1985; Liem, 1984; Liem and Kaufman, 1984; Turner, 1983; West-Eberhard, 1986, 1989). These authors cite well-documented examples in a diversity of taxa, including bees, fishes, butterflies, and salamanders. Here I cite some additional examples of intraspecific macroevolution fostered by developmental plasticity.

Photosynthetic Prokaryotes

The photosynthetic prokaryotes are of special macroevolutionary interest because the plant chloroplast is believed to have originated as a symbiosis between a protoeukaryotic cell and a photosynthetic bacterium possessing the full machinery for oxygenic photosynthesis. It is to these microorganisms that we must look for clues regarding the origin of oxygenic photosynthesis, a major step in the evolution of life since it led to the origin of oxygen-generating plants and set the stage for the origin of oxygen consumption by animals. The following discussion is based on a review (Shilo, 1982) that explores this question.

Given the diversity of chloroplast structures and pigments, it is considered likely that the origin of chloroplasts is polyphyletic, that is, derived in different chloroplast-containing lineages from different photosynthetic prokaryotes. Although not all types of photosynthetic organisms participate in endosymbioses, and some forms are likely to now be

extinct, there are indications in extant cyanobacteria, a candidate group for chloroplast origins, that developmental plasticity could have played a role in the origin of oxygenic photosynthesis.

The cyanobacteria and Prochlorophyta form one major branch of prokaryotes capable of chlorophyll photosynthesis, the other being anoxygenic forms. They are the dominant type of bacteria in fluctuating environments, such as shallow bodies of water, with rapid diurnal or seasonal alternation between aerobic and anaerobic conditions. Many cyanobacteria, such as *Oscillatoria limnetica*, although possessing both photosystems I and II (those present in all types of chloroplasts known), can shift readily from oxygenic to anoxygenic photosynthesis (Oren et al., 1977, cited in Shilo, 1982). This raises the possibility that the transition between anoxygenic and oxygenic photosynthesis, which had previously seemed unlinked, whatever its polarity (not discussed by Shilo, 1982), may have occurred as an alternative metabolic pathway within individual cyanobacteria (Shilo, 1982).

Plants

Darwin (1859 [1872]) cited alternative phenotypes in vines as an illustration of how major change could begin. There are three modes of climbing in plants, spiral twining, clasping with tendrils, and emission of aerial roots. Each method is usually found in distinct taxa. But some species produce two or even all three of these modes, combined in the same individual. "In all such cases, one of the two organs might readily be modified and perfected so as to perform all the work, being aided during the progress of modification by the other organ" (p. 136). Thus, Darwin recognized the buffering effect of alternatives in facilitating change. He also recognized the potential for accelerated evolution following phenotype fixation, in a discussion of the evolution of flowers. Some plants produce two types of flowers. "If such plants were to produce one kind alone, a great change would be effected with comparative suddenness in the character of the species" (p. 136).

Some experiments with plant hormones (reviewed in Briggs and Walters, 1988) show that intraspecific macroevolution in flowering plants is conceivable, since a variety of characters, such as the possession of a corolla tube or an inferior ovary, traditionally used to classify major groups of angiosperms, can be induced by hormone action within species that do not normally possess them. Application of small amounts of the plant hormone 2,4-D to soapwort plants (*Saponaria officinalis*,

Caryophyllaceae), for example, induces numerous flower and inflorescence variants, some of which possess traits used as taxonomic characters at the generic and subfamily level (Astié, 1962, cited in Briggs and Walters, 1988).

Comparative physiology and morphology indicate that the ascomycetes, fungi parasitic on land plants, evolved via intraspecific macroevolution (Cooke and Whipps, 1980). Terrestrial fungi have three primary trophic modes: biotrophy, or feeding on living plant cells; necrotrophy, or feeding on cells killed by pathogenic activity of biotrophs or other organisms; and saprotrophy, feeding on dead organic matter. These feeding modes entail very different physiological and morphological specializations. Some groups specialize in a single mode, but many are "hemibiotrophs," which show more than one mode during the life cycle. Individuals of *Taphrina* species, for example, first grow within leaf cells; then, as the infected leaf shows necrosis, they become necrotrophic and, upon leaf death, saprotrophic. Cooke and Whipps propose that the biotrophic species were probably the original land fungi among the ascomycetous orders, and that hemibiotrophy provided a flexible stem from which saprotrophic and necrotrophic groups branched repeatedly. (p. 354) They also recognize the buffering effect of "eco-nutritional versatility" in protecting lineages from extinction in variable circumstances (p. 353).

The endosperm, a major innovation of the flowering plants (angiosperms), is likely a descendant of an aborted diploid embryo produced due to double fertilization of a single egg by sperm from a single pollen tube, as occurs in some nonflowering plants (e.g., *Ephedra* in the Gnetales) phylogenetically basal to the angiosperms (see chapter 10). Thus, an intraspecific "organism duplication" produced a supernumary identical twin with a divergent, complementary function similar to that of a social insect worker. Rather than developing, it serves to nourish its normal fully developed sib (Friedman, 1995). Various specializations of endosperm, such as triploidy (the result of sperm fusion with two egg nuclei) evidently evolved following this original embryonic duplication.

Epiphytic bromeliads occur in two major life forms, tank-forming species with large, flat leaves, simple trichomes, and low surface reflectivity; and xeromorphic atmospheric forms with succulent leaves and no impoundments. In the bromeliad subfamily Tillandsioideae, some species such as *Tillandsia deppeana* are heterophyllous: juveniles have atmospheric morphological and physiological traits, and adults are of the tank form. Although the polarity of evolutionary change is unknown, Adams and Martin (1986) hypothesize an intraspecific transition via heterophylly.

Spiders

Jackson (1986) used comparative behavior and phylogenetic studies to show that jumping spiders (Salticidae) may have evolved from web-building spiders via a polyphenic stage like that represented by the facultatively web-building species *Portia fimbriata* (Salticidae). Most salticids do not build prey-catching webs but are instead cursorial, visually hunting predators. Females of *P. fimbriata* and some other, related spiders have highly specialized alternative predatory tactics, including cursorial hunting of other spiders on the ground, web building, and stalking and luring of their spider prey via aggressive mimicry of the vibratory movements of spider prey items, performed on alien webs. Jackson regards the prey-mimicking behavior as transitional between web use and terrestrial stalking behavior; the good eyesight and predatory agility associated with catching spiders on their webs may also have presaged these traits in salticids. Jackson provides evidence that the phylogenetic position of *Portia* is intermediate between web builders and other Salticidae. Jackson (1986, p. 256) notes that the versatility and range of behaviors performed by *P. fimbriata* are unusual or even "bizarre" among spiders. An uncommon innovation may be uncommonly important in changing the course of evolution if, as in this case, it evolves alongside an established adaptive trait such as web building in predatory spiders.

In three families on non-web-building mygalomorph spiders, the Migidae, the Ctenizidae, and the Barychelidae, some species build silk-lined burrows in the ground, and others construct exposed silk-walled tubular retreats (Coyle, 1986). As expected from the frequent association of intraspecific and interspecific alternatives (see chapters 19 and 28), this recurrent pair of alternatives—burrow construction and tubular exposed retreat construction—which is found in three different families, is also observed within a single species, *Migas distinctus* (Forster and Wilton, 1968).

Hymenoptera

A central transition in the evolution of the insect order Hymenoptera (wasps, ants, and bees) was the shift between phytophagy and carnivory, which gave rise to major radiations of predatory aculeates (stinging Hymenoptera) and parasitic or parasitoid wasps. There seems to be a general consensus that the ancestral hymenopteran resembled unspecialized sawflies (Symphyta), whose larvae are phytophagous and whose females oviposit into plant tissue (Gauld and Bolton, 1988). As long ago as 1907 Handlirsch proposed that the transition to carnivory likely occurred when primarily phytophagous larvae opportunistically consumed other larvae encountered within their host plants. This hypothesis is supported by the observation that larvae in many groups of primitive Apocrita attack wood-boring or stem-mining insects (Gauld and Bolton, 1988).

The Handlirsch hypothesis requires that the larva of a normally phytophagous species attack and consume another larva if it encounters it within a plant. That this is a realistic possibility was demonstrated experimentally by Malyshev and Puzanova-Malysheva (summarized in Malyshev, 1966 [1968]), who transplanted larvae of the phytophagous gall-making sawfly *Pontania proxima*, normally found one per gall, into galls of other individuals and found that one of the larvae always consumed the other by sucking out the body juices, then completed its development by plant feeding. The same carnivorous behavior occasionally occurs in this species in nature, when neighboring galls grow together. And a similar sequence of carnivory followed by phytophagy is found in the ontogeny of some parasitoids, for example, chalcids of the genus *Eurytoma* (Eurytomidae), whose larvae initially feed on the egg or larva of an insect host, then consume plant juices from the walls of the host's gall or other surrounding plant tissue (Malyshev, 1966 [1968]). The plasticity of intraspecific switching between phytophagy and carnivory in both the basal group, the sawflies, and a derived one, the parasitoids, indicates the feasibility of the Handlirsch hypothesis—in retrospect, an intraspecific macroevolution hypothesis—for this transition.

The evolution of social life in the Hymenoptera has been marked by three major innovations: the advent of life in groups, the origin of sterile workers, and the evolution of social parasites. In the Hymenoptera, solitary life is known to be ancestral to group living, and the evolution of group living preceded the origin of sterile workers (Wilson, 1971a), which in turn preceded the origin of socially parasitic species (e.g., see Carpenter et al., 1993). All three of these landmark innovations occur as variants within species polyphenic for the ancestral and the derived state (West-Eberhard, 1996). In the bee genus *Ceratina*, for example, some females nest and reproduce alone whereas others nest in groups (Sakagami and Maeta, 1987a,b). In *C. japonica*,

solitary females develop ovaries and reproduce. When experimentally forced to live in groups some, socially subordinate females fail to lay eggs and instead help others; they are functional, facultative workers. In such a genus, it is hazardous to hypothesize, even given a well-supported cladogram, whether group living and worker behavior is primitive or derived, for the ability to switch rapidly between states may be common. Many fully social species of wasps and bees that reproduce only in colonies containing both egg-laying queens and nonovipositing workers have some behaviorally parasitic females that lay eggs and do not work or produce workers (reviewed in Bourke, 1988; Choe, 1988). Thus, group living, worker production, and parasitism all occur as intraspecific alternative phenotypes, showing that macroevolutionary transitions that led to major radiations in the history of the social Hymenoptera can occur within species as well.

Vertebrate Invasion of Land

Air breathing and terrestrial locomotion were key innovations that allowed vertebrates to occupy terrestrial habitats. Studies of air-breathing fish show how life on land could have begun as a facultative response to extreme fluctuations in the oxygen levels of an aquatic environment and /or to periodic drought, which stimulates amphibious fish of various taxa to make brief sojourns over land (see references and examples in Graham et al., 1978). Graham et al. (1978) give reasons to believe that the transition to land was from fresh water rather than from marine habitats. This transition occurred in several lineages, including the Synapsida, Diapsida, and Anapsida, beginning in the Devonian and possibly aided during the Carboniferous by a pulse of elevated atmospheric oxygen during that time that would have facilitated respiration using primitive lungs (Graham et al., 1995).

At least 27 freshwater fish genera belonging to eight families contain air-breathing species (Willson, 1984). Five of these genera are amphibious, and some of them also contain species that do not breathe air. Air-breathing organs in these genera include swim bladders, skin, pharyngeal lungs, opercular chambers, mouth and pharynx, stomach, and intestine. Romer (1958) notes that air breathing in modern fishes is relatively rare compared with its frequency in the Devonian when the transition to land occurred. In just one especially rich locality for Devonian fossil fishes, the deposits of Scaumenac Bay, Canada, at least 95% of the fishes found were probably lung bearers; one wonders if the conditions that promoted the preservation of these fossils selected for a high proportion of air-breathing aquatic species as well.

All tetrapods are descendants of an extinct group of fishes, the rhipidistians, through such early amphibians as the ichthyostegids (Valentine, 1977b). In some rhipidistian lineages the endoskeleton was increasingly ossified, legs developed from fin supports, and air breathing had already appeared, although they have been extinct as fish since near the end of the Permian. This being the case, as pointed out by Valentine (1977b), it is difficult to know which among the extant species of air-breathing fishes are most likely to illuminate the behavioral and ecological details of this recurrent transition. But I know of no hypothesis for the vertebrate invasion of land that does not propose intraspecific macroevolution, with facultative use of lunglike breathing structures and preadapted legs (e.g. Romer, 1958; Simpson and Beck, 1965; Mayr, 1959; Graham et al., 1978).

Neoteny in Salamanders

Salamanders are the grand old taxon of developmental evolutionary biology. Their sensational debut took place in Paris in the mid 1800s, when an unmetamorphosed captive larva of the Mexican axolotl (*Ambystoma mexicanum*) reproduced in a museum (Duméril, 1865; see Gould, 1977). This was especially surprising because aquatic amphibian larvae (e.g., tadpoles) are morphologically and physiologically so different from the terrestrial adults. The axolotl helped to defeat the doctrine of recapitulation and the biogenetic law (Haeckel, 1905), the idea that ontogeny is a condensed synopsis of phylogeny (reviewed by Gould, 1977), for it demonstrated that major changes could occur due to the reorganization of ontogeny, rather than by terminal addition as required by the idea that ontogeny recapitulates phylogeny. Subsequent research on salamanders has included detailed study of their natural history, endocrinology, and phylogeny, so they continue to be a model taxon for studies of development and evolution.

Not the least of the salamanders' attractions is the recurrence in various lineages of a classic heterochronic form—the larvalike adult that reproduces despite failure to metamorphose to the terrestrial morphology (see figure 9.5). Heterochrony is the most widely discussed of the hypothesized developmental causes of macroevolution (e.g., Hardy, 1954; Gould, 1977; Matsuda, 1982, 1987; McKinney and MacNamara, 1991). The evidence for this is usually the neotenic or paedomorphic ap-

pearance of certain species compared with their "normal" relatives (e.g., see Hafner and Hafner, 1988, on geomyoid rodents; Shea, 1988, on primates). Such morphological evidence by itself gives no clue, however, as to whether macroevolution was speciational or intraspecific. To settle that question, comparative study has to include the causes and patterns of variation both between and within species—an exercise that is possible in salamanders because they are both speciose and variable in their life cycles, and many of them are well studied ecologically and endocrinologically, including with attention to the regulation of metamorphosis (see figure 5.19). Here I feature salamanders because they illustrate how a major heterochronic change could begin with environmental induction of a hormonally regulated complex trait, become established as an intraspecific alternative phenotype, and then proceed to fixation via subsequent genetic change (see also Gould, 1977; Matsuda, 1987).

In accord with the intraspecific macroevolution hypothesis, neotenous reproduction (loss of metamorphosis) occurs both as a fixed state and as an anomaly or an established alternative within salamander populations. Of 113 species of salamanders with a normal metamorphosis from a larval to an adult stage, about 42 (37%) have at least some neotenous "brachiate" adults (Collins et al., 1993). Polyphenism for the normal and neotenous forms also is common (e.g., figure 19.1). A comparative survey with attention to hormonal regulation (Matsuda, 1979) reveals various degrees of sensitivity to environmental and hormonal treatment. Some species (e.g., of Sirenidae, Proteidae, Cryptobranchidae, and Amphiumidae) are permanently neotenous, and metamorphosis cannot be induced by hormone (thyroxine) treatment; others (e.g., Plethodontidae, and some ambystomids, e.g., Siredon) are always neotenous in nature but can be induced to metamorphose by thyroxine. And some species, such as *Ambystoma tigrinum*, vary geographically in the proportion of neotenous individuals. In *A. tigrinum californicus* there are no recorded neotenous adults, in *A. t. tigrinum* they are rare, and in five other subspecies both neotenous and metamorphosed adults are common (Collins, 1980, Collins and Sokol, 1980).

There is also paleontological evidence for heterochrony in the evolution of the larvalike reproductive adult (reviewed by Collins et al., 1993). The aquatic larval stage is present in at least some taxa in all extant salamander families. The earliest fossil salamanders are from the Upper Jurassic, and they and other early salamanders show the metamorphic life cycle with an aquatic larva and a morphologically distinctive terrestrial adult. Extant species such as *A. dumerlii* in which all adults are the brachiate larvalike form are therefore judged to be derived paedomorphs. The metamorphic ancestry of such species is also revealed by occasional atavistic anomalies, where an individual of an exclusively larviform population undergoes metamorphosis, either spontaneously or due to hormone (thyroxine) treatment.

Various observations indicate that the transition to neotenic reproduction could conceivably have been initiated by environmental factors. In amphibians, low temperature commonly slows growth and inhibits metamorphosis (Matsuda, 1987), and in general, neotenous populations tend to occur in relatively cool habitats (e.g., in high-altitude populations), although there are some exceptions (Collins, 1981). Early work (reviewed in Suzuki, 1985) showed an effect of temperature on the structure and function of the thyroid gland, with extreme high or low temperatures reducing the size of the gland; and temperature affects the action of thyroid hormones: extremely low temperatures block both synthesis and secretion.

In *Ambystoma gracile*, metamorphosis can be delayed and neotenous forms induced by various environmental factors, including low temperatures, reduced food availability, and ambient iodine deficiency (Collins and Cheek, 1983). Furthermore, the frequency of branchiate and normal (metamorphosed) morphs changes with habitat (Collins et al., 1993). That this could be a facultative response to local conditions rather than a result of genetic change under selection is shown by the great variability of responsiveness to temperature in laboratory-reared individuals of *A. gracile*: some show temperature-sensitive metamorphosis; others metamorphose regardless of rearing temperature, and still others fail to metamorphose regardless of temperature (Matsuda, 1987, after Sprules, 1974). Such variability may in part owe to genetic variation in plasticity, required if an initially environmentally induced trait is to evolve to fixation via genetic accommodation of regulation.

Individuals from high-altitude populations of *A. gracile* metamorphose in lower numbers (23%) than do low-altitude (100 m) individuals (88%) when reared in the same conditions (Eagleson, 1976, cited by Matsuda, 1987). This suggests that genetic accommodation of the temperature threshold for metamorphosis has occurred. Why might selection favor loss of metamorphosis in certain habitats, for example, at high elevations? Neotenic (aquatic) reproduction may be advantageous at

high altitudes, for example, if water temperature fluctuates less widely than does air temperature, if there is a greater lack of cover or food in the terrestrial habitat, or if air humidity is relatively low (discussed in Matsuda, 1987; see also Hayes, 1997a).

In summary, recurrent induction of neoteny by one or more environmental factors in individuals genetically disposed to respond, accompanied by some advantage to permanent aquatic life, could lead to the genetic accommodation and eventual fixation of a neotenous adult in salamanders.

Sexually Selected Flexibility and Macroevolutionary Trends

Long-term change under sexual and social selection, as distinct from change under natural selection (*sensu* Darwin, 1859 [1872]; see West-Eberhard, 1983), often produces a unidirectional phyletic trend in body size or in the size or complexity of morphological features such as horns or tail feathers used in competitive signaling or combat. Classic examples of this are the oft-cited trend of increasing size in titanotheres and their hornlike weapons (see figure 24.4) and the evolution of antler size in ungulates, which culminated in the grossly exaggerated antlers of the Irish elk (Gould, 1974). These examples bear the earmarks of sexually selected traits, in their exaggeration as well as because horns are known frequently, if not always in some groups, to be used in threat, combat, or display (Eberhard, 1979, 1980; Geist, 1966; Andersson, 1994). Sexual and social selection sometimes produces extreme flexibility in behavior and development, as, for example, in the frequent occurrence of alternative phenotypes in mating tactics and reproduction (e.g., the workers of social insects; see part III). Is there any chance that sexually and socially selected developmental plasticity may sometimes contribute to macroevolution?

Gould (1982a) noted that "phyletic size increase (Cope's rule) is a common but frustratingly unexplained, phenomenon in evolutionary lineages" (p. 336). As discussed above, many alternative phenotypes expressed in social contexts are phenotype limited and influenced by size (see part III). Given the likelihood of a frequent association between trends toward large size and the action of social/sexual selection, it would be interesting to examine the purported cases of Cope's rule to see how often sexual or social selection is involved. A study of fossil ungulates (Janis, 1988) showed that the appearance of horns in the evolutionary history of a lineage is correlated with size; horns begin to appear in individuals whose body size is above 15 kg. In ungulates, large size is also correlated with open woodland rather than forest habitats. Moreover, forest dwellers are more often monogamous and nonterritorial (Janis, 1988), which further supports the sexual/social selection interpretation of the trend toward large size in these mammals. In any group where there is such an association between large body size and other evidence, such as the possession of a horn or other structure (e.g., dinosaur crests and horns) known to function in display or battle or to be associated with life in social groups, this is strong evidence that the macroevolutionary trend toward large size is driven by social or sexual selection, the most likely function of horns in both ungulates (Geist, 1966) and insects (Eberhard, 1979, 1980).

Another characteristic of evolution under strong intraspecific (especially, social or density-dependent) competition also may help explain why major groups may often *start* small relative to their ancestors and descendants (e.g., see McKinney and McNamara, 1991; Hanken and Wake, 1993). If large size is an advantage and large individuals can monopolize resources to the exclusion of small ones, the trend may be reversed under strong selection on small losers to adopt an innovative, opposite, antagonistic tactic taking advantage of *small* size (West-Eberhard, 1979), the so-called best-of-a-bad-job tactics (Dawkins, 1980). This can force phenotype evolution in the opposite direction, toward small size, in one developmental subset of a polyphenic species. If the interactive social competition has to do with resources other than mates, such as food, shelter, or nesting space, then the trend toward small size could become associated with adaptive innovation under *natural* selection.

Examples of size-related specializations subject to natural selection include the foraging specializations of small-beaked finches, the different life-history tactics of precociously mature adults, and the flight capacity of diminutive dispersers. An initially facultative, size-dependent alternative specialization eventually could become fixed in a population where the small morph's specialty is advantageous. Gould (1977) extensively documented the correlation between drastic heterochronic innovations, density-dependent intraspecific competition, and novel life-history tactics, some of them involving the evolution of small size and dispersal morphs. Hamilton (1978) stressed the association between crowded habitats (e.g., life under bark and in rotten wood), wing polymorphisms, and bizarre

innovative morphologies. And Roth (1990, 1992) provides evidence that the fossil dwarf elephants that evolved on islands during the Pleistocene were selected for small size in competition with larger individuals under extreme competition for limited resources. These examples indicate that the resetting to small size in the innovative forms that founded major new taxa under a top-down pattern of evolution could occur under strong intraspecific ecological or social competition in crowded habitats.

The spectacular trend toward large size in dinosaurs likely evolved under social and sexual competition given the prominence and diversity of crests and weapons in their morphology, signs of sexual and social competition in modern organisms (West-Eberhard, 1983). These considerations predict that some of the relatively small dinosaur morphotypes contemporaneous with large ones, and thought to be different species on the basis of differing size and morphology, would prove, if methods were available to test this, to be small morphs of polyphenic species where smallness and nonaggressive tactics were favored under strong social, sexual, or ecological competition. Origin as an alternative phenotype should be considered as a possible source of miniaturization in fish (Weitzman and Vari, 1988) and salamanders (Hanken and Wake, 1993), both groups noted for their developmental plasticity (see above; see also chapters 27 and 28).

Systema Naturae, or Why All Phyla Are Old

The phylogenetic branches that gave rise to the phyla of multicellular organisms are, in general, ancient. Representatives of nearly all phyla present today were present in the late Cambrian, more than 475 million years ago (Brusca and Brusca, 1990). If developmental plasticity facilitates macroevolution, as argued here, and if modern living organisms are so obviously developmentally plastic, then why has the origin of new phyla come to a stop? Why has the continuing process of diversification not led to a continuing series of metazoan phyla, rather than the small set of 35 that arose during the early epochs of metazoan history?

Buss (1987, p. 103) proposed that once germline sequestration or individuality evolved, major new bauplans cannot evolve, since by his reasoning genetic mutations have to exert their influence early in development during the window of opportunity for competition among genetically distinctive cell lineages that precedes sequestration. As discussed in chapter 7 (see also Raff, 1988), this explanation depends on an erroneous view of development in relation to genes and selection. First, novelties can be induced without mutation (see chapters 6 and 26). More important, a mutation does not depend for its evolutionary establishment on having arisen in the cell lineage where it is expressed, as implied by the Buss hypothesis, but need only arise in the germ line to be transmitted to the offspring and may be expressed at any stage of ontogeny. It would be evaluated under selection in whatever life stage it has a phenotypic effect.

Other biologists uphold an empty niche theory, the idea that novelties prosper when there is an unoccupied adaptive zone, or an ecological opportunity such as the absence of predators or competitors (e.g., Frazzetta, 1975, p. 150; Kimura, 1991; Morris, 1989; discussed in Raup, 1983; Raff, 1996; Futuyma, 1998). By this view, distinctive phyla originated only in the past because there were large numbers of still empty major niches, and recent or future origins of major groups are unlikely because specialized forms have filled the major available kinds of resource space. One fact cited to support the empty-niche hypothesis of macroevolution and radiation is the association of spectacular diversity in radiations on islands, viewed as places offering colonists major empty niches and an absence of natural enemies (Frazzetta, 1975). But, as shown in chapter 28, competition can also foster innovation. The addition of species must often create new niches for others that may exploit their tissues, behaviors, corpses, and wastes. I can think of no a priori reason why competition, including intraspecific competition, could not lead to major innovation in already complex organisms. Innovation springs from the organism's opportunistic developmental versatility, and a diverse environment full of other organisms would seem to create, not eliminate, opportunities for evolutionary novelty.

The body-plan characteristics that distinguish phyla and other higher categories of organisms—such as the rigid exoskeleton and jointed appendages that distinguish arthropods from annelids—are not readily associated with or limited to any particular distinctive class of niche. Rather, they are distinctive solutions to functions that other phyla perform in other ways, whatever their niches, which are not really defined or limited by these traits. Also contradicting the empty-niche hypothesis are the massive extinctions at the end of the Permian, when up to 96% of all marine species disappeared. These extinctions should have emptied some major niches, yet they did not lead to the evo-

lution of any fundamentally new body plans (Levinton, 1992). Major niche-related innovations, such as locomotion and respiration, which enabled colonization of land, or the evolution of wings, do not correspond with the birth of phyla. In short, the concepts of phylum and niche are not readily linked. So we must look elsewhere for a convincing explanation of why known phyla and other higher categories are old.

The evidence for combinatorial evolution presented in part II offers a possible explanation for the lack of new phyla, or at least part of an explanation, when combined with the force of a hierarchical taxonomy. Taxonomies are based on the temporal hierarchy of phylogenetic branching that characterizes the evolution of life from a single common root. The earliest history of life must have involved large-scale screening by natural selection of a very large number of phenotypic experiments. Initially, the experiments were biochemical. Many of the early results must have been truly bizarre, well outside the range of forms of molecules and body plans known from study of living and fossil organisms.

The relatively small number of phenotypic arrangements that survived this early screening would then have been repeatedly recombined into different mosaic forms and tested as clades at all taxonomic (branching) levels as they continued to branch and survive or go extinct under very long-term selection. Extensive molecular convergence involving gene duplication may have characterized very early evolution, and "Metazoa [multicellularity] may have evolved a number of times . . . from scratch" (Zuckerkandl, 1994, p. 675). This cladistically heretical-sounding remark is amply borne out by molecular phylogenies indicating at least 13 separate inventions of multicellularity, some of them polyphyletic (themselves including multiple origins; see Bonner, 2000).

Beginning with a common origin—some common set of original building blocks—all subsequent phylogeny can be represented as a series of branches with a common root. The hierarchical arrangement of branches and gaps, some due to extinctions, gives rise to our hierarchical classification of organisms. Given this mode of classification, it is evident that the more basal branches will correspond to the highest taxonomic categories (kingdoms, phyla, classes), and these are necessarily among the oldest. Thus, the farther back you must go in time (or on a phylogenetic tree) to find a common ancestor of any given pair of species, the larger the taxonomic category (the more major the branch) required to include both (Raup, 1983). This presumes that no major extinctions of intermediate branches have interrupted the pattern—that the clock has not been set back at any point but has continued to tick (branch) since the origin of life (Raup, 1983).

The branching pattern, not the origin of major adaptive traits, gives order to our perception of major taxonomic groups. If we identify a higher taxon with certain widespread traits or a bauplan, this is a secondary result or artifact of our classification—we use old, widespread traits to recognize relationships based on antiquity of descent. This is not to say that the traits themselves are artifacts, for they must be long-term products of selection, as well. This combination of facts—the use of a taxonomic hierarchy based on a temporal hierarchy of phylogenetic branches, and the fact that all innovation must be based on old building blocks—means that all major taxa are old.

Accordingly, the fossil record shows that the origins of phyla are first, followed by classes and orders of organisms that are characterized by different modifications of the older body-plan elements that identify them as members of the same phylum (Valentine, 1994; figure 29.4). Relatively recent species, then, are variations on already tested broad themes or body plans. In agreement with this, standing diversity at the higher taxonomic levels (phyla, classes, and orders) remains stable or declines through geologic time following these initial radiations (Raup, 1983).

Consistent with this view of early evolution as experimental and combinatorial, the pre-Cambrian and early Cambrian, when most extant metazoan phyla originated, was also marked by a high rate of extinctions (Valentine, 1994), and some of the extinct organisms are difficult to force into a classification based largely on extant species. By the present interpretation, this is because they may represent lost combinations of ancient building blocks whose bearers failed to persist as fossils or to produce further descendants. Extant multicellular animals of "uncertain affinities" as to phylum are exceedingly simple organisms (Brusca and Brusca, 1990; figure 29.5), as expected if there are only a few viable simple arrangements at the early metazoan level of organization, and if the vast majority of those have either given rise to many recombined forms or long ago gone extinct. Although the majority of the Burgess Shale fauna (about 80%) can be assigned to modern phyla (Valentine, 1977a, p. 32), some of the bizarre taxa defy classification not because they exhibit unrecognizable novelty, but because they are mosaics of recognizable parts: "Some taxa share enough characters to deserve the epithet of crustaceanlike or cheliceratelike, but oth-

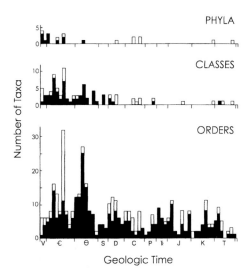

Fig. 29.4. Time of appearance of animal phyla, classes, and orders in the marine fossil record from the Vendian (V) to the Tertiary (T) periods. Each bar represents 10 million years. Solid bars are taxa with hard skeletons; open bars are taxa with soft-bodied or only lightly sclerotized members. From Erwin et al. (1987) by permission of *Evolution*.

ers are an amalgam of arthropodan units that defy simple classification" (Morris, 1989, p. 345).

Given this view of history, a modern classification is based on genealogy as revealed by shared traits, and the existence of gaps. At the phylum level the gaps, insofar as they exist rather than being artifacts of our limited understanding, are at least in part a heritage of the ground-plan extinctions of early evolution. Combinatorial evolution may achieve an enormous variety of permutations and recombinations of traits at all levels of organization, but it would be unable to bridge gaps due to early extinctions of forms. Ground plans that did not make an early cut could represent kinds of building blocks that may have been distinctive intermediates or partial mosaics relative to surviving ones, but are now forever lost to developmental and evolutionary experimentation. An early extinction of this sort would be a permanent gap in the potential for variation.

But the greatest constraint on the recognition of a new phylum or higher category is taxonomic: in a classification based on genealogy, a novel form derived from an already recognized fossil or extant taxon would rarely be classified as a higher one because variants are classified on the basis of relatedness and descent, not degree of distinctiveness or

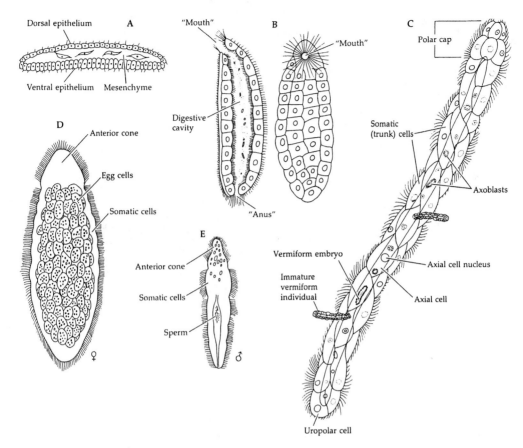

Fig. 29.5. Rhombozoans, animals of "uncertain affinities" as to phylum. From Brusca and Brusca (1990), by permission of Sinauer Associates, Inc.

strict similarity. Thus, a convergently bivalved snail is classified as a snail, not a bivalve clam, even though the bivalved shell is an important distinction between these two classes of mollusks (see also Strathmann, 1991a). All recent forms are likely to be traceable to older ones. We may revise the higher level classification as new techniques change views of phylogeny, but we are unlikely to grant phylum status to any recent group descended, as it always will be, from an already recognized older one. The result is that virtually any new organism that were to originate today would have a very high probability of sharing numerous phenotypic similarities with the organisms from which it descended, and molecular and morphological clues as to its ancestry would usually suffice to place it in an existing phylum. In other words, the lack of recent new metazoan phyla is in part a product of the combinatorial nature of evolution and the existence of ancient gaps, and in part a product of the com-

prehensiveness of the existing classification. Taxonomists have already scanned large numbers of the modular traits that have been available during the last several hundred million years of metazoan history for evolution to work.

The power of classification against the hopeful new phylum can be illustrated by some examples. The sciarid fly *Plastosciara perniciosa*, for example, has extremely bizarre wingless, claustral, wormlike adults of both sexes that in terms of morphological and behavioral distinctiveness would fall outside the family Sciaridae. But it also has normal, winged adults typical of sciarid flies (Steffan, 1973, 1975). As noted by Hamilton (1978), if the alate form of this insect were to be abandoned due to success with some other means of dispersal, the worm-like form "could conceivably become the ancestor of a future 'order' of vermiform soil-dwelling insects, even with potentiality to become a new class" (p. 159). But this extreme extant form is unhesitatingly and correctly classified as of the class Insecta, order Diptera, family Sciaridae due to its known status as a facultatively expressed morph of *P. perniciosa*, a normal sciarid fly. Similarly, as pointed out by Matsuda (1982), bizarre modifications in only one sex, as in certain cecidomyid flies, barnacles, psychid moths, scale insects, and Strepsiptera (reviewed by Matsuda, 1987), or in the workers of social insects, do not constitute the basis for recognition of a new higher taxon. Major phenotypic distinctiveness by itself is not sufficient. Strathmann (1991a) gives several examples of body plans that have changed in the post-Paleozoic period giving rise to a new class of echinoderms and changes in snails usually definitive at the class and phylum level; but the echinoderms are still echinoderms, the snails still snails.

Similarly, in paleontology, a single "key" character of high taxonomic significance serves to disqualify a specimen as the originator of a major new branch. The lack of a difficult-to-discern metanotal gland in the mesozoic fossil *Sphecomyrma freyi* (Wilson et al., 1967), for example, would disqualify it as a new subfamily (Sphecomyrminae) of ants (Baroni Urbani, 1989; Agosti et al., 1998). However, if lacking the gland, it could be readily accommodated in the existing classification within a fossil subfamily of wasps (e.g. Tiphiidae; Wilson et al., 1967), though possibly a distinctive one (Baroni Urbani, 1989), since, in keeping with the modular combinatorial nature of evolution, the fossil insect is a mosaic of wasplike and antlike traits (Wilson et al., 1967).

Even the bizarre and ancient specimens of the Burgess Shale, said by some to possibly represent large numbers of extinct phyla (e.g., Morris, 1989, 1998), probably owe their high taxonomic reputations at least in part to the impossibility of studying their different alternative phenotypes and life stages (viz. *P. perniciosa*) and to the difficulty of interpreting their visible features and relating them to known groups. Many, such as the giant arthropod *Anomalocaris* (figure 29.6), have gradually come to be classified within a new class (Donocarida) and order (Radiodonta) of a known phylum (Arthropoda) after a tortuous history of misclassifications due largely to incomplete specimens, some of them originally considered new phyla (Collins, 1996; on fossil plants of this and earlier eras, see also Levinton, 1992; Morris, 1998; Taylor, 1982). The ability to eventually pigeonhole virtually any newly discovered form given enough information on it may sometimes distract attention from the significance of finding ancient specimens

that are unique in revealing ways (Morris, 1989), given the importance of documenting the occurrence of unsuccessful evolutionary experiments and mosaic transitional forms.

This view of the relations between origins of higher categories, combinatorial evolution, extinctions, and classification predicts that within recognized phyla there should be extinct mosaic intermediates that combine features of extant higher taxa. A newly discovered fossil subfamily of braconid wasps illustrates this. Amber fossil Braconidae from the late Cretaceous (Turonian), about 92 million years ago, contain several specimens of a newly named subfamily (Protorhyssalinae) that resembles the extant subfamily Rhyssalinae in most diagnostic subfamilial characters, but shares one character—hindwing vein 2CU—only found in a rare and putatively primitive subfamily Apozyginae (Basibuyuk et al., 1998). Many such intermediates must have become extinct in the past, and presumably many will in the future, with the likelihood of survival of any given higher taxon dependent in part on the number of species it contains (Raup, 1983; Strathmann, 1991a).

Could modern single-celled precursors begin the evolutionary process anew and produce a series of bizarre novel multicellular structures comparable with the potential phyla that evidently succumbed during the Cambrian explosion? John and Miklos (1988) believe that most phyla originated early in the history of life "because developmental gene hierarchies were less complex and more easily subject to perturbation" (p. 336). By this argument, genomically and developmentally simple extant organisms such as bacteria should be able to start off in a distinctive direction and produce new phyla. Evidently, however, the force-of-classification principle would defeat their efforts: if a new body plan were to emerge from among the bacteria, it would nonetheless be recognized as a bacterium by descent. Bonner (1988) lists numerous examples where bacterial cells have associated and differentiated to produce multicellular organisms: (1) a cyanobacterium, *Nodularia*, has three differentiated cell types, vegetative, akinete spores, and nitrogen-fixing heterocycsts contained in a filament or sheath; (2) a myxobacterium, *Stigmatella*, has both feeding, vegetative cells and spores; and (3) another myxobacterium, *Chondromyces crocatus*, has a multicellular stalked fruiting body whose branches bear cysts that release dispersing cells, which form wandering swarms and then cooperatively aggregate into a new complex fruiting body. There are also colonial ciliates and diatoms.

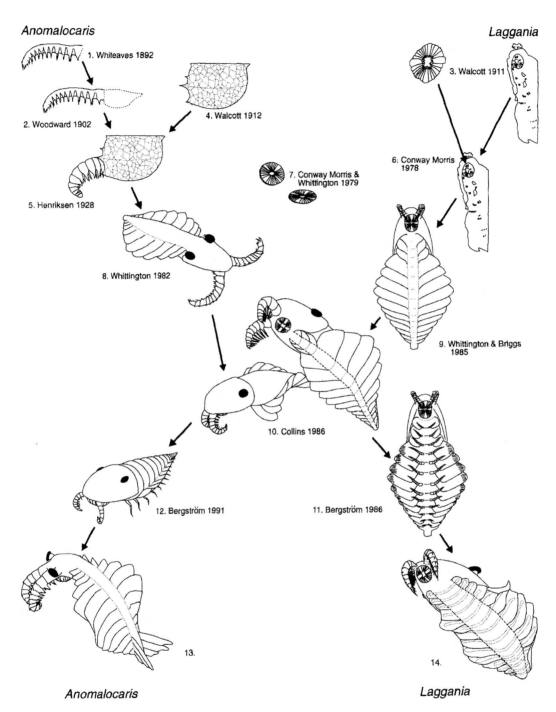

Fig. 29.6. The "evolution" of reconstructions of the Burgess Shale arthropod *Anomalocaris* (Dinocarida, Radiodonta). References (1–14) give authors of the reconstruction shown. For Bergström 1991 see Chen et al. (1991); for Whittington 1982 see Whittington and Briggs (1982). Reference for 13 and 14 is Collins (1996). From Collins (1996) by permission of The Paleontology Society.

Bonner (1988) lists 10 independent origins of multicellular organization in modern eukaryotic groups (see also Raff and Kaufman, 1983), including numerous multicellular amoebae, foraminifera, radiolaria, the myxomycetes, and the cellular slime molds. *Volvox*, among the green algae, has a unique body plan and a complex pattern of development involving separate somatic and germ lines. Rather than considering these innovative body plans that have emerged from bacteria or single-celled groups to be nascent phyla, we classify them by their component cells: they are amoebae, green algae, cyanobacteria, and myxobacteria. Nor do we consider the loss of a body plan justification for a change in phylum. Several species of ascidians have lost their notochords due to loss of the larval tail, leaving a few notochordal cells but no dorsal nerve cord (Jeffery and Swalla, 1992). This is a change in body plan without a change in genus, for they are still classified in the same chordate genus from which they are obviously derived (R. Strathmann, personal communication). No amount of morphological ingenuity, even at the body-plan level, can overcome the power of existing classifications to oppose the recognition of a new phylum in a recently derived group. By force of phylogenetic taxonomy, *sytema naturae*, all phyla are old.

Why Molecular Biology Cannot Solve the Macroevolution Problem

For molecular developmental biologists John and Miklos (1988; see also Davidson, 2000), the "great unsolved problem" of evolutionary biology, not adequately addressed by microevolutionary studies of neo-Darwinism, is how to explain the origin of major morphological novelties. Their path-breaking review successfully connects genomic phenomena with many important questions of evolutionary biology. But they conclude that there is no consistent or meaningful relationship between genomic changes and principal events in the evolution of the eukaryotes; that phenotypic changes are not proportional to genomic ones; that neither nucleotypic nor karyotypic alterations impinge meaningfully on the genesis of morphological change; and that morphological evolution and gross genomic evolution can be uncoupled. They find that "all of the conventional approaches have, so far, been oblique to the central problem of the origin of morphological novelty." "What then were the underlying pro-

cesses which governed the emergence of the major groups of organisms on this earth?" (p. 290).

Like Lewontin (1974), who advocated the phenotype as the stuff of evolution and then turned to genes in search of evolutionary explanations (see chapter 1), John and Miklos and other authors (see also Davidson, 2000) persist in turning inward, toward the genome and the immediate products of gene transcription, for explanations of macroevolution: "It is in the molecular analysis of development where significant hope is to be found" (John and Miklos, 1988, p. 335). Although their own analysis concentrates on DNA, they recognize that this is not completely satisfactory. They consider that future progress will depend on understanding proteins and cells, for, "while DNA supplies the information, organisms ARE their proteins," and the next challenges "are clearly in protein-protein interactions, self-assembly systems and cell-cell interactions, with the prime-movers of change residing within cells, and projecting themselves outward to cause altered morphologies." (p.335). In a statement strikingly parallel to Lewontin's, they conclude that this—the production of altered morphologies originating due to genomic change— "after all, is what evolution is really about" (p.309). McKinney and McNamara (1991), even though concerned with developmental change at diverse levels of organization, also conclude that "most evolution is by change in number and position of cells" (p. 84).

It is a giant step forward, to move from molecular and cell biology to even discuss the phenotypes that Lewontin called the stuff of evolution (see also Gerhart and Kirschner, 1997). But it is an error to see the prime movers of evolutionary change from the inside out. As shown various chapters of this book (see especially chapters 6 and 26), genes are more accurately seen as followers, not leaders, in evolutionary change. An overly reductionistic view is not justified in a world where hormones and nervous systems orchestrate the activities of cells and genes, and these coordinating systems are in turn influenced by processes at even higher levels.

Selectable variation can originate at any level of organization. It can be induced by environmental factors and then accommodated as an evolved trait under natural selection without mutation—without novelty at the molecular level. Indeed, that is the order of events most likely to produce evolutionary change (see chapter 26). As shown in this chapter, macroevolutionary change can be instituted by such complex higher level responses as the perception of dominance by a developing female wasp, or the ability of a desert fish to switch from aquatic to aerial physiology and locomotion under environmental extremes. Deterministic connections between such higher level organization and the outside environment project inward via hormones and other coordinating factors, to gene expression. So it cannot be sustained that molecular and cell biology are the "only available key to unlock the generation of novelty" (John and Miklos, 1988, p. 338), and it may actually be an error to begin with the genome in a search for the links between macroevolution and the genes.

Even if it were to be shown that some bauplan change were associated with a change in a major regulatory gene, this would not be a complete and satisfactory evolutionary explanation. The discovery of such a gene could seriously mislead us by encouraging the common but false assumption that the gene gave origin to the innovation. Genes participate in development, but always in response to particular environmental circumstances. Cytoplasmic localization, position effects, the activities of neighboring cells (see Davidson, 1986)—these impinge on cells and the genome from the outside, and they are eminently susceptible to manipulation even more external in origin, for example, by maternal factors and other features of the external environment. Indeed, it is reasonable to hypothesize that the more major, in terms of organizational and genetic complexity, an evolutionary change in the phenotype, the more likely it originated as a product of environmental influence rather than mutation, given the greater likelihood that environmental induction will affect large numbers of individuals and will recur, even if initially nonadaptive, across generations (see chapter 26).

Many other authors have argued that a modern discussion of development in relation to evolution has to focus primarily on the level of the cell (e.g., Atchley and Hall, 1991; Hall, 1992). All genetically influenced processes trace back to events within cells. Atchley and Hall (1991) even restrict the term "epigenetic" to the case where a gene product generated in one cell affects gene expression in another cell. Beekeepers and jailkeepers may advocate a broader view of the word "cell" as an environment with developmental and evolutionary significance. The sickling allele of sickle-cell anemia has its primary effect within cells, causing red-blood cells to deform and malfunction. But it is the associated higher level effects that matter most to the organism, which becomes ill and may die due to changes in limbs, kidney, and heart. Radiating "pleiotropic" effects at these intermediate levels are common-

place in development and important for evolution. Seeing the relation between development and evolution requires agility of thought at different levels of analysis. Selectable variation occurs at all levels, and there is always tension between connectedness and compartmentation in development and evolution (see chapter 4). The relationship between the two complementary properties of phenotypes is only understandable by moving upward and downward among levels of organization. This is why biologists "have remained holists while at the same time becoming ever more reductionist" (Bonner, 1996, p. 129).

It is no longer possible to deny that the findings of molecular and cellular biology and even protein crystallography may affect evolutionary interpretations, by elucidating the nature and action of the products of genes. But this is not the same as saying that all of the important advances in evolutionary biology are being played out in laboratories of protein chemistry and enzyme kinetics. Indeed, that work is only significant if it is somehow tied to higher levels of organization. Selection acts on phenotypes formed in deserts and forests as well as within cells. The external environment forces itself, by multiple pathways, upon the intracellular world, haunting the enzymes with carbon shortages and drought, through a chain of responses that involve behavior, morphology, physiology, and their convoluted ancestral histories. So there is absolutely no chance that protein chemistry or molecular biology will sufficiently reveal the ultimate secrets of life. Nor are the ultimate answers to be found in the fitness calculations of population genetics and field biology, or the phylogenies of cladistics alone. We are presently at a turning point where progress depends on agility to work at one level of organization while seeing its neural and hormonal connections to others—up and down among the molecular, cellular, systemic, behavioral, social, populational, and phylogenetic/historical levels of organization.

The finding of John and Miklos (1988) that there is little correspondence between genomic change and major phenotypic innovation contrasts with the findings cited in this and preceding chapters that there is often a very good correspondence between patterns of developmental plasticity and major patterns of evolutionary change. This supports the view that the flexible phenotype, not the genome, is the leader of the evolutionary parade, and it is a strong argument in favor of a multilevel approach to understanding macroevolutionary change.

30

Punctuation

Punctuated evolution, or abrupt acceleration in the rate of morphological evolution, has been interpreted as a result of speciation. Behavioral and life-cycle flexibility that is invisible in the fossil record, however, may precede and lead to abrupt morphological change, especially if there is fixation of alternative or life-stage specializations over time. This hypothesis is examined using two well-recognized examples of punctuation, evolutionary change in the dentition of horses and in the morphology of bryozoans, in order to illustrate how the developmental-plasticity hypothesis of punctuation can be tested, and why it is a reasonable alternative to the speciational hypothesis.

Introduction

In *punctuated evolution* (Eldredge and Gould, 1972) periods of relatively little change ("stasis") are punctuated by episodes of relatively rapid change in the *rate* of evolution of a quantitative morphological trait, as seen in the fossil record of morphology. According to Simpson (1984), the term *quantum evolution* (Simpson, 1944; figure 30.1), refers to the same thing. Like Eldredge and Gould, Simpson contrasted quantum evolution with phyletic change, or sustained directional evolution without branching; considered that it could be associated with speciation (though also with phyletic evolution; p. 206); and even mentioned interrupted equilibra (cf. the "punctuated equilibria" of Eldredge and Gould, 1972, p. 82): "In phyletic evolution equilibrium of the organism-environment system is continuous, or nearly so, although the point of equilibrium may and usually does shift. In quantum evolution equilibrium is lost, and a new equilibrium is reached" (Simpson, 1944, p. 207; see also Mayr, 1954). I use the term "punctuation" rather than "quantum" because it less ambiguously describes change in rate of evolution. In its original meaning (from the Latin *quantus*), quantum means quantity. But quantum change, as mentioned by Simpson, is identified with the "quanta" of physics, which are discrete units of energy. This could encourage mistaken identification of punctuated change with the origin of discrete novelties, not the intended meaning of punctuated evolution,

which is periodically altered rate of change in a continuously variable, quantitative trait.

Mayr, Eldredge, Gould, and others (e.g., Stanley, 1979, 1981) explain stasis and punctuation in terms of speciation. *Speciational punctuation hypotheses* see stasis as due to the characteristics of established biological species, such as gene flow within interbreeding populations, large population size, heterogeneity of the species environment that retards directional change, developmental integration, canalization, coadapted genomes, stabilizing selection, and frequently reversing evolution over time within established species (Eldredge and Gould, 1997). These factors have been summarized by the term "gene-pool cohesiveness" (Mayr, 1989) or "developmental coherences" (Gould, 1989b), though the causes of stasis under the speciational hypothesis are admittedly vague (Gould, 1989b, p. 60) and debatable (for reviews of other possible causes of stasis, see Williamson, 1987; Coyne and Charlesworth, 1997; Van Valen, 1982a; Spicer, 1993). Then rapid change in morphology is associated with relatively brief episodes of speciation or phylogenetic branching, due either to selection in a small isolated population found in a distinctive environment (Mayr, 1989), or to drift (Cheetham et al., 1993).

This chapter offers an alternative to the speciational hypothesis of punctuated morphological change. The *plasticity hypothesis of punctuated evolution* proposes that developmental plasticity in the expression of contrasting or antagonistic traits

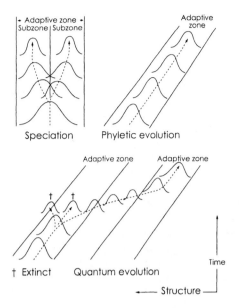

Fig. 30.1. The three major modes of evolution. After Simpson (1944).

promotes stasis in multipurpose morphology, and that accelerated morphological change occurs when a formerly optional state increases in frequency of expression or becomes fixed, causing a temporary burst of rapid directional change in associated quantitatively variable morphology (character release; see chapter 7). That is, developmental plasticity promotes stasis in multipurpose morphology, and loss of plasticity promotes accelerated change when morphology is released to specialize in a single direction. The acceleration may be produced in association with speciation or not. Punctuated change that occurs as phyletic evolution, without speciation, has been termed "punctuated anagenesis" by McKinney and McNamara (1991, p. 174).

Character release produces *accelerated* change because of positive feedback between increased specialization, increased frequency of expression, and intensity of selection. As noted by Simpson (1953a), there is a threshold effect. The threshold for acceleration is the point when the fitness contribution of one alternative or life-history phase surpasses that of others (e.g., in a homogeneous environment, or due to evolved improvement). In the plasticity hypothesis, both stasis and rapid change are driven by selection *within*, not (as in the speciational hypothesis) between, species (see Grant and Grant, 1995b for evidence of this, and reflections on relevance of their findings for proposed examples of punctuation).

Other possible causes of sudden morphological appearances are reviewed by Simpson (1944) and Ayala and Valentine (1979). Here it is enough to say that extreme speciational punctuationists regard all evolutionary change as "quintessentially" a matter of speciation (Eldredge, 1976). Many authors have taken issue with the speciational view (see especially Charlesworth et al., 1982; MacFadden, 1992, p. 211; Futuyma, 1987). Both the plasticity and the speciational hypotheses associate speciation with accelerated divergence, but the cause and effect relations are reversed, for the plasticity hypothesis sees divergence as a potential cause, not just a result, of speciation (see chapter 27).

It is important to clearly focus on what the punctuation controversy is about. The key issue in the punctuation debate is what causes change in *rates* of evolution of *quantitative traits* as observed in the fossil record. It is not a debate over the causes of sudden origins of qualitative novelties. Apparently rapid evolution in the fossil record that might occur due to fixation, without release, of one of a pair of alternative phenotypes (Palmer, 1985; Lively, 1999) would not necessarily qualify as punctuation. Without the quantitative change posited by character release, morph fixation might instead give rise to an apparent saltation—a qualitative rather than a quantitative change in phenotype, a kind of discontinuity emphatically excluded in considerations of punctuated "trends" (see especially Gould, 1989b). The theory and the data of punctuation concern quantitative trends in trait dimensions as documented in paleontology (see especially Gould, 1989b; Mayr, 1989). Punctuation, when it occurs, has no bearing on the gradualism controversy, which is a debate about whether or not change occurs by small steps or large ones, not whether or not it occurs slowly or rapidly or in irregular (punctuated) jerks. (see chapter 24). Punctuated gradualism driven by speciation has been contrasted not with gradualism per se, but with *phyletic* gradualism, or anagenetic change not involving phylogenetic branching (Eldredge and Gould, 1972).

Nor does the punctuation debate concern the causes of macroevolution, or branching patterns and major change. Here the confusion of issues is more than a confusion of words, for change in bursts implies episodes of relatively large, often speciation-associated change that could be considered macroevolutionary. Under the plasticity hypotheses, character release may contribute to speciation (see chapter 27), macroevolution (see chapter 29), and punctuation and therefore may cause associations among them. Still, it is important to consider the different effects separately, for the requirements of large phenotypic change, reproductive isolation, and altered rates of evolution are not the same.

This chapter does not take up the debate over whether punctuated evolution is common or rare, and it does not attempt to "prove" the plasticity hypothesis, nor does it establish that it is more important than speciation as an explanation of punctuated change. Rather, my intention is to establish the developmental plasticity hypothesis as a viable, testable alternative to speciation as an explanation for punctuated morphological change when it occurs.

Plasticity and Punctuation: Previous Theories

The idea that adaptive plasticity could influence evolution so as to produce rapid change perceived as gaps in the fossil record is at least 100 years old, as shown by the following passage from Baldwin (1902). In this paragraph, an "accommodation" is an adjustment due to behavioral and other sorts of phenotypic plasticity, and "variation" refers to inherited morphology:

It seems to the writer—though he hardly dares venture into a field belonging so strictly to the technical biologist [Baldwin was a psychologist]—that *this principle* [of fixed accommodations arising from individual plasticity] *might not only explain many cases of apparent widespread "determinate variations" appearing suddenly, let us say, in fossil deposits, but the fact that variations seem often to be "discontinuous."* Suppose, for example, certain animals, varying in respect to a certain quality from *a* to *n* about a mean *x*. The mean *x* would be the case most likely to be preserved in fossil form, seeing that there are vastly more of them. Now suppose a sweeping change in the environment, of such a kind that only the [morphological] variations lying near the extreme *n* can accommodate [through phenotypic plasticity] to it and live to reproduce. The next generation would then show variations about the mean *n*. There would be a great discontinuity in the chain of descent and also a wide-spread prevalence of variations seeming to be in a single direction. This seems especially likely when we consider that the paleontologist does not deal with successive generations, but with widely remote periods. . . . (pp. 101–102; emphasis original)

Simpson was a Baldwin scholar (Simpson, 1953b), but his own explanation for stasis did not include the accommodation of environmental change by plastic behavior. Instead, Simpson saw stasis as due to selection being still below a threshold for directional change in a changing environment that eventually propels the population into a new adaptive zone (figure 30.1). The all-or-none, abrupt or "quantum" aspect "arises from a discontinuity between adaptive zones" or a threshold effect in morphological change: once the adaptive threshold is crossed, evolution proceeds very rapidly in a single new direction, resulting in an evident discontinuity in the transition (pp. 390–391). No intermediate stages persist because intermediate forms are "relatively inadaptive" (p. 392) or evolutionary specialization too fast to be tracked in the fossil record of morphology (Simpson, 1984). Elsewhere, Simpson (1951b, 1953a) considered a possible role of drift in passing this threshold in geographic isolates (on the possibility that drift could contribute to punctuated change, see also Lande, 1987; Lynch, 1990; Cheetham et al., 1993), but his usual emphasis was on selection as a cause of the shift between adaptive equilibria.

Simpson's discussion of the evolution of aquatic locomotion in penguins (see chapter 29) suggests that the threshold for rapid change could arise from change in the frequency of performance of a facultative alternative behavioral phenotype. Specifically, it is that point when selection for one of the alternatives (e.g., submarine vs. aerial flight) becomes stronger than selection for the other (see chapter 29). If selection were to fluctuate between favoring one alternative or the other, as in Simpson's description of morphological change during phyletic evolution (evolution without speciation), a burst of directional morphological would not occur (see also J. W. Wilson, 1975). Note that Simpson's identification of fluctuating selection with phyletic evolution, and the identification of change in adaptive zone with a phylogenetic branch (see figure 29.1), implies that, for Simpson, quantum evolution is often associated with speciation, though not dependent upon it (for Simpson, the cause is a shift across a threshold between environmental extremes, which does not require, but may often involve, reproductive isolation; see Simpson, 1953a, p. 205). Behavioral alternatives are involved in Simpson's discussion, but he did not attribute any special role to behavior in the maintenance of morphological stasis or the acceleration of morphological change.

The theoretical basis of the plasticity hypothesis is strengthened by Kirkpatrick's (1982) model of quantum evolution, which shows how a rapid shift between two fitness peaks could occur under selection in a population with sufficient phenotypic variance, without drift. Although Kirkpatrick did not consider adaptive developmental plasticity a source of phenotypic variance that could facilitate a shift under his model, plasticity is an important source of such variance. Also, when phenotypic variation is bimodal or polymodal (as when alternative adaptations are involved), it could pass over the valley between fitness peaks visualized by Kirkpatrick's model especially readily due to the evolved reduction in numbers of individuals at the minimum zone of the individual fitness function. Kirkpatrick's model shows, however, that bimodally distributed or discrete phenotypes preadapted at the two peaks are not essential. I emphasize them in this chapter and elswhere (part III) because they provide the most favorable conditions for rapid shifts toward a particular specialization, and because they most clearly illustrate the role of developmental plasticity in producing such shifts in morphology. A later model by Milligan (1986) shows how intraspecific competition for bimodally distributed resources can lead to a rapid specialization in one or the other due to slight change in either resource abundance or ability to use it (e.g., evolved improvements in one of multiple alternative foraging tactics).

Stearns (1980, 1982, cited in Stearns, 1983) reasoned that plasticity can "uncouple genotype and phenotype" and thereby "blunt the force of selection," and saw that plasticity could explain stasis in the fossil record, as argued here. But he later decided that "I no longer think that explanation works" (Stearns, 1983, p. 74), because his observations of Hawaiian mosquitofish populations showed that "the traits that changed [evolved] most rapidly were also more phenotypically plastic." Paradoxically, both conclusions can be correct, as explained above: plasticity can both promote stasis (due to breaking of genotype–phenotype correlations) and accelerate change (by enabling organisms to adopt to extremes, and leading to character release when one extreme is commonly expressed). So plasticity, though a conservative force for the evolution of morphology under fluctuating conditions, can, in other, narrowed, stressful or extreme circumstances, "take the lead" in evolution and accelerate change. The expected correlation between behavioral plasticity and morphological stasis has been documented in salamanders by comparative study of different genera (Wake et al., 1983). In the bolitoglossine plethodontids, many species have highly specialized behavior, ecology, and species-specific morphology. Species of the genus *Plethodon*, on the other hand, show individual differences in feeding behavior, including learned dietary preferences, accompanied by an evolutionarily conservative (relatively static) olfactory and trophic morphology. The bolitoglossines in effect "specialize" in being flexible generalists or polyspecialists, and the associated morphology shows stasis.

The model of punctuation developed by Turner (1983) based on genes of major effect is basically similar to the present one, except that it depends on a mutation of large effect rather than on environmental change to get phenotype evolution started in a new direction.

How can we study developmental plasticity in fossils? In some groups, as in the horses and bryozoans discussed below, plasticity can be inferred from study of related living species. In some fossil species, developmental plasticity can be inferred by ontogenetic series pieced together within assemblages of the same morphotype. A common characteristic of the two examples discussed in detail in this chapter is that punctuation in the evolution of particular phenotypic traits is recurrent. High-crowned teeth in horses and skeletal thickening in bryozoans both show recurrent punctuation in closely related lineages. The article that touched off modern discussions of punctuation contained no examples of punctuated change, but it did contain two examples of recurrrent parallelism in snails (see figure 19.7), which evidently inspired the authors to write about punctuation because allopatric populations were closely related yet showed no intermediates in relatively continuous fossil series (Eldredge and Gould, 1972). The association between rapid change and recurrence by itself gives some support to the developmental plasticity hypothesis, because recurrence is so often associated with developmental plasticity in the taxa concerned (see chapter 19, on recurrence, and chapter 28, on adaptive radiation). In this chapter I seek more direct tests of plasticity as a cause of punctuation.

Two Fossil Examples

The strength of the paleontological evidence for punctuation is still controversial (for different points of view, see, e.g., W. L. Brown, 1987; Lister, 1993; Levinton, 1988; Gould, 1989b, 1993; Van Valen, 1982a). Some people (e.g., Brown, 1987) think it has not been shown to exist, and others (e.g., Eldredge and Gould, 1997) call it "the dominant empirical evolutionary pattern of the history of life itself" (p. 339). It is beyond the scope of this chapter to evaluate these opinions. Instead, I assume that the fossil record at least sometimes shows true punctuated change, since punctuated evolutionary change is known from neontological evidence (West-Eberhard, 1983, 1986; see also chapter 21, on phenotype fixation and character release). I use two well-known fossil examples of punctuation to show how the plasticity hypothesis can be applied as a testable explanation of punctuated change.

Punctuated Evolution in Horses

Simpson's (1944) primary example of quantum, or punctuated evolution traced the transition between browsing and grazing morphology of fossil horses, particularly of their teeth. The ancestral horses were browsers, and the transition to grazing involved an increase in hypsodonty, cheek teeth with high crowns relative to their horizontal dimensions. When change in that trait is plotted graphically relative to changes in other characters, it shows a sudden differential acceleration, or punctuated change, in the Mesohippus-Neohipparion group toward the end of the Oligocene.

Equid tooth form, along with certain other morphological traits, is a good species character and is correlated, in interspecific comparisons, with diet and feeding behavior. During an extended period occupied by the shift between tooth forms, the equids had teeth of changing and variable form. Simpson (1944) supposed that as tooth height increased, it triggered the episode of rapid evolution that occurred. He visualized this as a threshold effect, as follows:

> It became possible for them [horses with relatively high teeth] to supplement their food supply by eating some grass, a relatively harsh food and highly abrasive to the teeth. . . . This point was a threshold. It initiated a sort of trigger effect that set off an evolutionary quantum reaction. . . . That part of the population that had reached the slope of the grazing peak was no longer in the centripetal selection field that surrounded the browsing peak. On the contrary, it began to be affected by strong linear selection toward the grazing peak . . . they were now imperfectly adapted either for browsing or for grazing, but they were preadapted for grazing. The outcome was that this segment of the population . . . evolved with relative rapidity into fully grazing forms. (pp. 209–210)

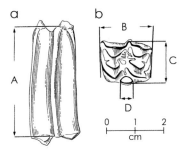

Fig. 30.2. Measurements of molars of horses used to calculate rates of evolution: (a) lateral view of molar, showing (A) measurement of maximum crown height; (b) distal surface of molar (occlusal surface), showing (B) maximum length, (C) maximum transverse width, and (D) protocone length. After MacFadden et al. (1999).

Simpson visualized that the eventual switch to grazing behavior, and intensified selection for it, was produced by increased tooth height in some individuals, or possibly drift affecting feeding morphology—that morphology takes the lead in evolution. The result of gradually increased frequency of grazing behavior would be directional evolution of tooth morphology favored in individuals that commonly graze (figure 30.2). Prolonged grazing, and therefore the most effective exploitation of grass as an alternative resource in periods or regions of food scarcity, would have favored individuals that both preferred grass and had relatively high teeth. Since grazing is probably difficult for old individuals with worn teeth, learning may have contributed to the establishment of a correlation between grazing behavior and tooth height, as discussed for the evolution of trophic specializations in other vertebrates (see chapter 28). The correlation would have produced an accelerating evolution of preference and morphology.

The plasticity hypothesis as an explanation for punctuated change in fossil vertebrates was anticipated by J. W. Wilson (1975; see also Levins, 1968; Van Valen, 1982a, p. 105), who pointed out that when there are two usable resources, selection will favor morphology adapted to the resource most commonly used:

> Since most vertebrate species are not polymorphic in their feeding apparatus as a way of coping with dichotomous resources, they will tend to have relatively uniform morphologic adaptations and more flexible behavioral adaptations for allowing their morphology [to] cope with

less commonly used, but essential resources. This means that during the evolution of a lineage, utilizing changing proportions of dichotomous resources could show *long periods of morphologic stability and shorter periods of very rapid morphologic change*, even when the resources on which it depends [and their usage] change gradually. (pp. 367–368; italics added)

The plasticity hypothesis is supported by various kinds of paleontological data. The varied diet of the browsing ancestors of modern horses traditionally has been inferred from their dentition, occasional fossilized stomach contents, and paleobotanical knowledge of contemporaneous plants (reviewed by MacFadden, 1992). These data indicate that early horses with low-crowned teeth fed on a wide variety of herbacious dicots, woody shrubs, arid-adapted ferns, and primitive unabrasive grasses (MacFadden, 1992).

Recent techniques have greatly refined what can be learned from fossil teeth. It is now possible to determine with more confidence whether a tooth was used for browsing, grazing, or both from the microscopic patterns of tooth wear and carbon isotopic analyses of tooth enamel. As shown by studies of modern mammals, the teeth of browsers average fewer scratches and more pits than do those of grazers feeding on abrasive grasses, and radioisotope analysis of browser tooth enamel reflects a high ratio of C_3 plants (trees and shrubs) to C_4 plants (grazed grasses) in the diet, whereas for the tooth enamel of grazers, the ratio is reversed.

Using these techniques, MacFadden et al. (1999) examined the teeth of six species of small and medium-sized horses that were sympatric in Florida during the late Hemphillian, about 5 million years ago. By that time, all of the more primitive species of horses with short-crowned teeth were extinct, and all six species of this study had high-crowned teeth. By traditional standards all would be classified as grazers. But MacFadden et al. show that one was a grazer, two were primarily browsers, and three were mixed browsers and grazers. In one of the primarily browsing species (*Dinohippus mexicanus*), two individuals had carbon isotope values indicating that they were browsing at least some of the time. MacFadden et al. conclude that diet in these species varied due to various factors other than tooth form, including body size and longevity (high crowns are an adaptation of long-lived grazers to tooth wear, and small species are usually relatively short-lived) and competition with each other and with other sympatric species, especially vari-

ous species of large herbivorous mammals that were grazers and mixed grazer-browsers, in an area that was a mosaic of closed-canopy forests, woodlands, and open-country grasslands.

Beginning with this flexible phase, the plasticity hypothesis predicts that if grazing increases in frequency, or is fixed, as in all modern horses, there should be punctuated change due to character release in particular features of morphology (e.g., hypsodonty) that go from multiple uses in ancestral populations to a single specialized use in descendants. Of the four mixed feeders studied by MacFadden et al. (1999), three (*Nannipus minor*, *Cormohipparion emsliei*, and *Dinohippus mexicanus*) had direct descendants (*N. peninsulatus*, *C. emsliei*, and *Equuus* species) in which grazing was fixed. All of them were more hypsodont than their ancestors (B. MacFadden, personal communication).

These findings are consistent with the plasticity hypothesis that behavioral plasticity contributed to punctuated evolution of horse dentition. They demonstrate that individuals were behaviorally flexible during the transition period between browsing and grazing, with four of six species in a single locality using both behaviors to various degrees. They also show that fixation of grazing was accompanied by increased hypsodonty in all three of these species in which grazing was fixed, as expected if morphology is released to specialize when behavioral flexibility declines.

MacFadden (1992) points out that some patterns of equid dental evolution seem to contradict the speciational hypothesis. Rates of morphological change do not correlate with rates of speciation as one might expect if speciation promotes evolutionary change: (1) anchitheres, Miocene browsing taxa, show high rates of change in dental characters but low species diversification, and (2) Miocene grazers show only average rates of morphological evolution relative to some explosive evolution seen in other fossil groups, despite high rates of speciation. Similarly, Douglas and Avise (1982) found that the rate of morphological divergence in a species-rich genus of fishes was no greater than that of a species-poor genus, and also interpreted this as evidence against the speciational hypothesis of punctuation.

There is a problem with such tests, however. A positive correlation between speciation rates and morphological change would support the speciational hypothesis. But the lack of such a correlation does not necessarily contradict the hypothesis: although some proponents of speciational punctuation argue that most evolutionary divergence must

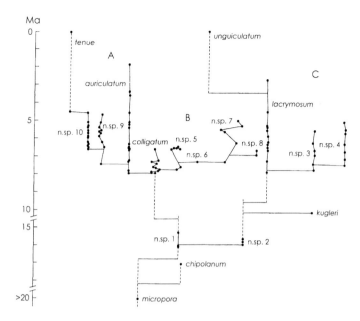

Fig. 30.3. Phylogenetic (stratophenetic) tree of Caribbean *Metrarabdotos* (Bryozoa, Cheilostomata) showing punctuated change (time scale in millions of years, Ma). Morphological distances between inferred ancestor and descendant species are indicated by horizontal lines; distances between species of different pairs are not necessarily to scale. After Cheetham (1986a).

arise in speciose groups (Eldredge, 1976; Vrba, 1980, cited in Douglas and Avise, 1982), this does not imply that the same cumulative amount of morphological divergence per speciation event is expected in all taxa. This is clear from comparing clades A and B in figure 29.3, where all divergence is diagrammed as associated with speciation, but overall divergence rate in clade B is greater than in clade A. As noted in chapter 7, enumerative comparisons are always asymmetrical in this way: they can provide valid positive support for, but not valid negative evidence against, hypotheses that do not pretend to apply universally, or equally to all clades.

Speciational hypotheses rely more heavily than does the plasticity hypothesis on small population size and locally homogeneous novel environments. Under the plasticity hypothesis, by contrast, a trophic specialization like grazing can spread as morphology-related learned alternative behavior within a large population, as discussed in chapter 28. Speciation could act along with phenotype fixation to promote punctuated evolution, but speciation is not necessary for puntuation to occur. This conclusion contrasts sharply with that of extreme speciational puntuationists, who assert that there can be no evolution without speciation (Eldredge, 1976; see chapter 21).

Punctuated Evolution in the Bryozoa

The plasticity explanation of punctuated evolution in horses puts a strong emphasis on behavioral plasticity. What about organisms such as plants and sessile invertebrates where behavior is not so complex? In those organisms, we would have to look to other kinds of plasticity to apply this explanation of punctuation.

An important and perhaps unassailable example of punctuated evolution has been documented in the Bryozoa, or "moss animals," a group of colonial invertebrates that resemble plants in being sessile and capable of asexual (clonal) reproduction. Punctuated evolution occurred repeatedly in the cheilostome bryozoan genus *Metrarabdotos* (figure 30.3). The research demonstrating this (Cheetham, 1986a) has been called "the most detailed and elegant study done since this [punctuation] debate began" (Gould, 1989b, p. 77). To my knowledge, the evidence for puctuation in *Metrarabdotos* has never been seriously questioned. So this is a good place to investigate the causes of punctuation in a group that contrasts sharply with the mammals. This discussion illustrates how the plasticity hypothesis can be applied as a tentative explanation of punctuation, in a group where this hypothesis previously has not been discussed, and where new approaches to data collection would be required for more definitive tests.

Cheetham used a series of exceptionally good and closely spaced stratigraphic samples collected in the Dominican Republic, where *Metrarabdotos* were abundant and continuously distributed. All of nine comparisons of ancestor–descendant pairs showed within-morphospecies rates of morphological change not significantly different from zero (stasis). In all nine comparisons, the ratio of within-morphospecies fluctuations to across-morphospecies difference was insufficient to have allowed gradual transformation of one to another without an abrupt change in degree of variation, permitting "the punctuated pattern to be distinguished with virtual certainty" (Cheetham, 1986a, p. 190; see also Cheetham, 1987; figure 30.3).

While complex behavior is not an obvious talent of bryozoans, they, like other modular organisms including plants, are noted for various kinds of morphological plasticity. Examples include environmental influence on direction of polypide and peristomal tube growth in relation to substrate contact (Jebram, 1985); size and width of zooids, and a transformation of bilaminar to monoserial branches (major types of growth forms), influenced by nutrition (Jebram, 1978, pp. 255–256; see also Winston, 1976, on diet-influenced colony form in *Conopeum tenuissimum*); production of erect colony parts influenced by temperature (Jebram, 1978, p. 268); determination of encrusting versus erect colony form (Bock and Cook, 1998; Harmelin, 1973, p. 101, after Blake, 1976, p. 178; see also other references in Anstey and Pachut, 1995, p. 269); and production of defensive structures such as spines in response to predator attack (Harvell and Padilla, 1990; Harvell, 1986, 1991) or indirect cues such as wave action (Bayer et al., 1997). Some species are developmentally plastic in that they may produce either encrusting or erect colonies (see below). Although it is not always clear whether this is a genotype-specific polymorphism or an environmentally influenced polyphenism, both represent intragenomic flexibility, as discussed in preceding chapters, and determinants are readily interchanged (see chapter 5). Is it possible that developmental plasticity in such a morphologically flexible group could have contributed to the pucttuated evolution documented by Cheetham in *Metrarabdotos*?

Anstey (1987, p. 37) has suggested, based on analysis of variance studies of fossil material of all five bryozoan orders, that plasticity in bryozoans is responsible for stasis, and that heterochronic changes that produce branched ontogenies are likely to have produced punctuated evolution (see also Anstey and Pachut, 1995). Developmental plasticity in colony (zooecia) form during the life cycle of colonies may be the foundation of punctuated evolution in *Metrarabdotos*. Astogeny (colony ontogeny) in erect colonies of *Metrarabdotos* and other bryozoans involves two growth phases—an encrusting and an erect or arborescent phase—and heterochronic change in the switch between them could be involved in shifts between forms during the history of *Metrarabdotos* and other bryozoans (see also Anstey, 1987; McNamara, 1988; Anstey and Pachut, 1995). Antagonist pleiotropy can occur between life-cycle stages (Price and Grant, 1984), setting the stage for character release and punctuated specialization as phase of growth increases relative to the other in a population.

How could we test the idea that plasticity and release associated with such shifts account for punctuated evolution in *Metrarabdotos*? In the Bryozoa, arborescent colony forms have arisen repeatedly in clades whose ancestors had exclusively encrusting morphologies, and transitions between colony form are not unidirectional, as shown by the recent emergence of encrusting species in the otherwise arborescent genus *Metrarabdotos* (Cheetham, 1986b; figure 30.3). In general in the Bryozoa, encrusting colonies are more common in shallow water and erect colonies are more common in low-disturbance, deep or open shelf habitats (McKinney and Jackson, 1989, pp. 82–83; Anstey, 1987).

Not surprisingly, given these recurrent interspecific alternative colony architectures, some bryozoan species, such as the cheilostomes *Membranipora arborescens* and *Crassimarginatella falcata*, contain both encrusting and erect colonies. Different colony forms within species are sometimes associated with different environmental conditions. In *Hippothoa hyalina*, encrusting colonies occur on algae and branching colonies occur on hard substrates (Winston, 1979). In recent seas in general, the depth segregation of encrusting and erect bryozoans is striking (Jackson and McKinney, 1990). In *Stylopoma spongites*, encrusting colonies are most common, but foliacious erect colonies develop offshore where there are fewer durophagous predators (Jackson, unpublished observations, in McKinney and Jackson, 1989, p. 96). Encrusting forms and astogenetic flexibility are more common in shallow than in deep water in some Bryozoa (Anstey, 1987; Anstey and Pachut, 1995). Mc-

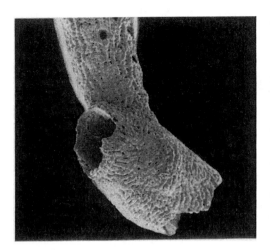
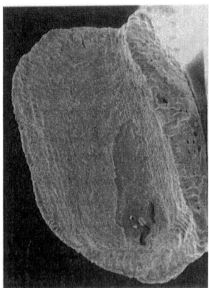

Fig. 30.4. Encrusting attachment bases of fossil (Neogene) arborescent colonies of the cheilostome bry-ozoan *Metrarabdotos*, probably *colligatum*, on sea grass. From Cheetham and Jackson (1996). Courtesy of A. H. Cheetham.

Kinney and Jackson (1989, p. 94) associate the abundance of erect-colony species with the abundance of grazing, browsing, and gouging predators, both within transects and across geological time. How much of this variation is due to developmental plasticity and how much to genetic differences is unknown. The available information suggests that the proportions of encrusting and erect growth can reflect both evolutionary and facultative phenotypic change.

As pointed out by Cheetham (1968) and Anstey (1987), it is easy to visualize evolutionary transitions between colony forms via heterochronic change. Erect colonies inevitably have encrusting bases even if composed of only a few zooids and their extrazooidal secreted extensions (e.g., McKinney and Jackson, 1989; Cheetham and Jackson, 1996; figure 30.4). Similar encrusting bases (attached to sea grasses) are found in seven of the arborescent *Metrarabdotos* species of Cheetham's study (*colligatum, auriculatum,* and new species 5, 6, 7, 9, and 10—see figure 30.3). So erect colony forms could conceivably give rise to encrusting colonies via an extension of the base (failure to switch to production of the zooid type that produces erect growth). Primarily encrusting colonies show a potential for the production of erect forms in that they sometimes give rise to erect structures, as in *Metrarabdotos unguiculatum unguiculatum,*

whose colonies are described as "pavement-like, but almost invariably rising in erect, irregularly tubular, convoluted branches" (Cheetham, 1968, p. 93 and plate 10). Therefore, it is not surprising that these two colony forms, expressions of developmental plasticity within species, are recurrent in changing proportions over time in *Metrarabdotos* and other genera.

In growth of erect *Metrarabdotos* colonies, architectural support is provided by thickening of the zooidal walls, due to a secretion of the epifrontal membrane that continues throughout zooidal ontogeny, eventually sealing areolae and orifices and converting the zooid into a structural element without the ability to feed (Cheetham, 1968). Cheetham (1968) notes that "evolution of the encrusting form from the erect one in *Metrarabdotos* may have required a paedomorphic emphasis on formation of the primary walls at the expense of the superficial layer of the frontal wall" (p. 11).

The ability to switch between encrusting and erect growth, and the reverse, during colony astogeny in *Metrarabdotos* implies a capacity to alter the proportions of different zooid types during evolution, as concluded for other bryozoans (Podell and Anstey, 1979; Anstey, 1987). The changing proportions of morphospecies of each form across time, as in the increase in numbers of arborescent species early in bryozoan history (Cheetham,

1986b), and the increase in encrusting morphotypes relative to arborescent ones in Neogene and Quaternary cheilostome Bryozoa of tropical America (Cheetham and Jackson, 1996), could reflect relative rates of speciation and extinction, as these authors suggest. Alternatively or additionally, in some lineages it could reflect changes in the timing or conditions of the switch between the encrusting and the arborescent phases of colony growth, that is, anagenetic developmental change within the lineage. Change in the proportions of the two types of growth could represent either paedomorphosis, if the early, encrusting phase is increased, or peramorphosis, if the arborescent, later astogenetic pattern is increased (Anstey, 1987).

Anstey reviews evidence that transitions between colony form have occurred repeatedly in several groups of Bryozoa. Each episode of heterochronic change is expected to be accelerated by character release (see chapter 7). This could help account for punctuated evolution in *Metrarabdotos* (Cheetham, 1986a) if episodes of punctuation were accompanied by marked changes in the proportion of encrusting versus erect growth.

If developmental plasticity has contributed, we should see character release in particular kinds of traits, namely, those whose evolution is subject to a different direction of selection in the encrusting compared with the erect phase. If the encrusting phase increases relative to the arborescent phase, as must have occurred in the transition between *M. lacrymosum* and *M. unguiculatum*, for example (figure 30.3), traits formerly constrained to specialize in arborescence would have been released to specialize to the encrusting colony form, and may show punctuated patterns of change. The opposite change would occur in the same traits if proportional frequency of arborescent growth were to increase relative to the encrusting phase. If, as in some Bryozoa, a population was "unstable" or polyphenic with regard to colony form, then punctuated change in either direction could occur as morph ratios changed. Release could explain why intraspecific and interspecific variation are "decoupled" in fossil *Metrarabdotos* (Cheetham, 1986a).

In support of the release hypothesis, two characters that show punctuated evolution (lack of correlation between degree of intraspecific and interspecific variation) are branch bifurcation angles, and branch thickening (Cheetham, 1986b). Variation in both of these traits correlates with growth mode. Branch bifurcation angles and branch (or zoidal) thickening rates increase in specialized arborescent species in association with the larger number of branches, and the greater size of colonies

(Cheetham, 1986b). These traits, then, are potential material for further tests of the plasticity hypothesis: they should be more exaggerated in morphospecies whose colonies have a higher degree of arborescence (a greater proportion of arborescent vs. encrusting zooids during astogeny, and in the population as a whole).

Branch thickening is of special interest as a candidate for release because it is known to be costly in terms of material secreted and loss of feeding zooids due to overlay of secretion, as already noted (see also Cheetham, 1986b; Jackson and McKinney, 1990). In a colony or population with both encrusting and arborescent phases, selection for branch thickening is expected to vary in proportion to the relative size of the encrusting phase. A shift to predominance of arborescence should change the balance of selection toward higher rates of branch thickening (Cheetham, 1986b), whereas a shift toward predominance of encrusting morphology should change the balance toward lower rates.

Release may explain why the mechanical designs of the most advanced arborescent cheilostome species are "far stronger than reasonable safety factors might require them to be" (Cheetham, 1986b, p. 165), and greater between morphospecies than expected on the basis of amount of intraspecific variation in strength. Such a lack of intra- and interspecific correlation is expected under the present hypothesis. Within species, during periods of stasis or fluctuating change, evolutionary change by selection is expected to bear some relation to environmental change. But in a period of character release, that relation is lost, due to the accelerating effect of positive feedback between trait frequency, trait improvement, and intensity of selection. The result is a burst of morphological evolution that is disproportionate to that expected on the basis of selection alone, as judged by the usual rate of adaptive change. The drift hypothesis proposed by Cheetham et al. (1993) could account for passing the threshold for release in a small isolated population, for example, if a local population happened to be founded by a clone with a relatively small, or relatively large, encrusting base in proportion to erect growth. This is especially feasible in a group with nonplanktonic larvae, believed by some authors to have contributed to their rapid diversification (e.g., Anstey, 1987). Given the possibility of character release and patterns of evolution under selection, however, it may not be possible to distinguish between drift and selection as causes of evolutionary change on the basis of simple rates of divergence alone (Grant and Grant, 1995b,

p. 249; cf. the claims of Cheetham et al., 1993, and Cheetham and Jackson, 1995, regarding the role of drift in punctuated change).

Several types of information would help to evaluate the plasticity hypothesis using fossil *Metrarabdotos*. Most useful would be detailed studies of living related cheilostome bryozoans designed (1) to compare the arborescent form of species polymorphic for encrusting and erect colony form, and the arborescent form of monomorphic species, to see if an arborescent polymorph can be distinguished on morphological grounds (e.g., by the size or form of the encrusting base, or by zooid form, e.g., in the switch zone between the encrusting and upright orientation); and (2) to compare the morphology of related arborescent species recently and anciently derived from encrusting ancestors, to see if different degrees of specialization to arborescence are evident. If these different classes of arborescent colonies (polymorphic, primitively or recently monomorphic arborescent, specialized arborescent) can be distinguished morphologically, then the role of developmental plasticity could be evaluated by examining their sequence of occurrence in the fossil record, as is done in phylogenetic studies of extant species that map phenotypes on a cladogram (see chapter 19). Such information seems unlikely to be obtained from studies of only fossil assemblages, which consist of broken colony fragments, and where encrusting cheilostome morphotypes occur as small, usually rare colony fragments (Cheetham et al., 1999), which would make it difficult to distinguish whether encrusting material was from an arborescent colony with a large base or from a closely related exclusively encrusting species.

Nonetheless, it has been possible to document colony-form clines with increasing production of encrusting zooids in shallow zones in eight Paleozoic fossil bryozoans (Anstey and Pachut, 1995). That study shows how colony-form plasticity can be inferred from fossils alone in a group with no extant close relatives (Smith, 1995). The occurrence of both encrusting and arborescent colony forms within species has also been suggested for some other fossil bryozoans. McKinney and Jackson (1989, p. 33, after Boardman, 1960) state that "among trepostomes in the Devonian Hamilton Group of New York, 20 [morphospecies] had stable growth forms (either encrusting or arborescent with cylindrical branches) whereas six were both encrusting and erect." Zones where an encrusting phase is relatively common in an "arborescent" species, even if found at a different locality, could account for stasis or relatively slow evolution of the arborescent form. Then its loss could cause a re-

lease from these effects, and rapid directional change toward arborescent specializations.

The plasticity hypothesis of punctuation would be supported under three conditions:

1. If living arborescent colonies, in either laboratory or field, show wide variation in the allocation of growth to encrusting basal versus arborescent, branching morphology: Special attention should be paid to sampling or forcing (via environmental manipulations or artificial selection) extreme colony forms and anomalies (such as arborescent colonies with unusually broad bases) and the circumstances that produce them. This could reveal any developmental propensity to switch rapidly between forms in certain circumstances, and would indicate changes in growth form proportions that may be missed in fossil samples.
2. If living encrusting species, such as *M. pacificum* and those of the *M. unguiculatum* species group (*M. u. unguiculatum, M. cookae*), which are also known as fossils (Cheetham et al., 1999), vary geographically or under different environmental conditions in their production of erect structures, demonstrating potential for rapid transitions to the erect form.
3. If, in characters that differ between the encrusting and arborescent growth phases, the size of the differences varies with proportions of the two kinds of growth when populations or species having different degrees of investment in the two modes of growth are compared.

Morphological Stasis Is Not Evolutionary Stasis

For a neontologist, stasis is counterintuitive. Observations of behavior suggest that selection within species is continuous, for individuals continually fight for resources and scurry to escape predators and parasites, consistently respond to artifical selection, and show geographic variation on a fine scale that suggests recent adaptation to different circumstances. Artificial selection experiments show that apparently there is usually sufficient genetic variation in most traits to permit a response to selection (see chapter 6). It is virtually inconceivable that evolution by natural selection would ever come to a stop or fail to have a long-term effect. For a paleontologist, this is just the myopia of those who do not work on a geological time scale and do not comprehend larger trends, including stasis.

This chapter argues that stasis can be a product of selection. Nonspecific, or shared modifiers of traits expressed in different circumstances are subject to antagonistic selection that inhibits their evolution. At the same time, specific modifiers of the same traits can continue to evolve. Often this means that morphology is static, for it often supports multiple behaviors or life stage forms. While the neontologist is afflicted with temporal myopia, the paleontologist is afflicted with hard-structure myopia, and the rarity of whole-individual specimens that could record developmental change. Behavior, physiology, and development can evolve with no fossilized trace. This chapter argues that they do so and that their evolved divergence can be responsible for both stasis and punctuation.

The belief that morphology equals evolution has encouraged the belief that morphological divergence equals genetic divergence, or speciation. The plasticity hypothesis of punctuation also challenges the paleontological morphospecies concept, which says that morphology can be taken at face value as an indicator of species. It suggests that fossil species may sometimes contain cryptic morphs—contrasting morphotypes so different in structure or habitat that they are classified as different species or higher taxa, especially if they are found at different sites. The "species problem in paleontology" has been much discussed over the ages (e.g., see Gould, 1989b, p. 73). For the punctuation controversy, the debate over morphospecies is critical. Excessive faith in morphospecies exaggerates the impression that speciation is important and diminishes attention to plasticity if, for example, developmentally switched intraspecific alternatives, different sexes, or different life-stage forms are mistaken for species.

One kind of support for the morphospecies concept in paleontology comes from research on two closely related bryozoan species, showing that genetically distinctive populations are invariably morphologically distinctive (Cheetham et al., 1993). *Cryptic species*—species that are so close morphologically that they are mistaken for the same species—can often be discovered by genetic means and then found to be consistently identifiable on the basis of small morphological characters (Knowlton, 1993). But this does not mean that all distinctive morphotypes are good species, for they may be intraspecific alternative phenotypes or life-stage morphs. Van Valen (1969, p. 196) gives several examples of taxonomic splitting where alternative sexual and morphological fossil forms are erroneously classified as species, subgenera, genera, or even subfamilies. Geary (1992) discusses the difficulty of distinguishing morphs from species in fossil snails.

Developmental plasticity has been an important source of taxonomic confusion in the Bryozoa. Some morphological variation of zooids within colonies and populations has led to classification of the zooids as different species or genera (Maturo, 1973; Jebram, 1978, pp. 253, 268, 270; Bock and Cook, 1994). Although sometimes, for example, in *Characodoma excubans* (Cleidochasmatidae), encrusting and erect phases have zooids that are identical in all characters (Cook and Bock, 1996), in other species (e.g., *Hyppothoa hyalina*, a cheilostome) the two growth forms have different zooidal morphologies (Winston, 1979, p. 251). Anstey (1987, p. 24) cites Boardman (1954) as noting that as many as three "species" could develop within the astogeny of a single bryozoan colony. Anstey (1987) states that many immature colonies have been assigned to different genera from their mature counterparts, and notes that most of the biometric characters that have been used to differentiate bryozoan species are known to vary astogenetically (for references, see Anstey, 1987). Furthermore, the Bryozoa are notable for their different morphological characters in different environments (McKinney and Jackson, 1989; Smith, 1995). Since paleontological material in the Bryozoa is largely composed of colony fragments, intracolony zooid variation due to developmental plasticity must occasionally result in misclassifications.

These observations demand reexamination of confidence in morphospecies. They also challenge the idea (e.g., Eldredge, 1995) that "adaptive evolutionary change in general [is] concentrated phylogenetically around episodes of development of reproductive discontinuity" (p. 42), an assertion that perhaps should be revised to read that change is concentrated around episodes of developmental and reproductive discontinuity. Faith in morphospecies cannot be based on validation of cryptic species alone, but requires respect as well for *cryptic morphs*—morphologically distinct alternative phenotypes and phases of colony development.

Overconfidence in morphospecies has to be overcome if the plasticity hypothesis of punctuation is to be tested. A purposeful search for evidence of plasticity and phenotype fixation without lineage branching is required, if evidence for plasticity-associated punctuation is to be found. A strong belief that something will not be found, whether morphological plasticity or a lost set of keys, reduces the likelihood of finding it.

Conclusions

The intent of this chapter is not to "prove" the plasticity hypothesis of punctuated evolution, but to establish it as a viable, testable alternative to speciation as an explanation for punctuated change.

Loss of plasticity and phenotype fixation can accelerate speciation, and the two events could be synergistic in accelerating evolutionary change; speciation and plasticity are not mutually exclusive mechanisms of change. But they *are* different explanations of punctuated change and can be distinguished with appropriate tests.

When Eldredge and Gould (1972) proposed the speciational punctuated equilibrium hypothesis, they made an important point regarding the effect of established theories on the direction of research: "The expectations of theory color perception to such a degree that new notions seldom arise from facts collected under the influence of old pictures of the world. New pictures must cast their influence before facts can be seen in different perspective" (p. 83). The plasticity hypothesis is proposed in that spirit. It requires new kinds of facts and puts the old ones in a new light. To my knowledge, this is the only evolutionary hypothesis that proposes an explicit and testable mechanism (changes in degree of plasticity or developmental versatility) as an alternative to speciation to explain both morphological stasis and punctuated change in terms of natural selection.

31

One Final Word: Sex

Sexual reproduction permits genetic recombination and DNA repair, but is notoriously costly to females. The origin and the maintenance of sex may have different explanations. Sexual reproduction leads to sexual competition, big eggs, and the twofold cost of sex. It also leads to female choice and reproductive dependence of females on interactions with males. As a result, sex may become developmentally inescapable. Obligatory sex also contributes to diversification via speciation, which may be advantageous at the level of clades. Given the advantages and the developmental trap of obligatory sex, the benefits of recombination and DNA repair, while they may sometimes be important, need not be sufficient by themselves to compensate the cost of sex.

Sex transforms life. It affects morphology and behavior. It diverts enormous amounts of time and energy from the business of survival. It can even distract from the manufacture and safe packaging of offspring. The adolescent metamorphosis we each experience once, and thereafter view with amazement, in the relative calm of adulthood, has swept through nature on a grand scale, culminating in orchid flowers and peacock tails. All of this is due to chromosomal recombination—sex *sensu strictu* (Ghiselin, 1974)—and its organismal result, sexual reproduction or cooperation between two individuals to produce offspring. It is sex as sexual reproduction, the developmental side of sex that initiates the ontogeny of new individuals, that I mainly discuss here, though it is sex as recombination—the genetic side of sex—that has received most attention in discussions of the maintenance of sex.

Of all the major transformations in the history of life, the evolution of sex is the most enigmatic. The question is not so much how sex got there as why it remains. Given the importance of genetic similarity, or kin selection (Hamilton, 1964a,b), for the maintenance of cooperation within and among organisms, sex seems designed to be disruptive. It requires the union of genetically dissimilar individuals, which dilutes the relatedness of mother and young, leaving the mother to invest in offspring genetically only half like herself. This has been called "the cost of meiosis" or the "twofold cost of sex"

(Williams, 1975). It is a cost that usually falls to females, with their greater investment in eggs and care of offspring. By this view, the male is a parasite of his mate and participation in sexual reproduction is contrary to the best interests of females, who would do better to reproduce parthenogenetically on their own. Yet, among animals, only about one in one thousand species are thelytokous, that is, secondarily asexual, with no facultative or alternating sexual generation and no interaction with males (White, 1978, p. 287). The prevalence of sexual reproduction in higher organisms is "inconsistent with current evolutionary theory" (Williams, 1975, p. v). What can explain the persistence, in multicellular organisms (the only ones where sex is obligatory; Dacks and Roger, 1999), of "so seemingly maladaptive a character"?

Two hypotheses predominate among those proposed to explain the evolution and maintenance of sex despite its phenotypic costs (for a critical review, see Hurst and Peck, 1996): (1) DNA repair, the elimination of aberrant chromosomes by excision and resynthesis during meiosis (e.g., see Bernstein and Bernstein, 1991); and (2) genetic recombination to increase genetic diversity, as an individually advantageous hedge against a variable environment for a brood of offspring (Williams, 1966, 1975), especially when strategy or niche diversification is advantageous under strong competition (Ghiselin, 1974); and as a clade-level advantageous capacity that permits a lineage to track environmental change over the long course of evolutionary time (Williams, 1975; Maynard Smith, 1978). These and other advantages of recombination are discussed by Barton and Charlesworth (1998).

Evolutionary biologists usually emphasize genetic-recombination explanations, especially the importance of recombination in potentially endless coevolutionary races with parasites and disease (reviewed in Williams, 1975; Hamilton, 1982; Hamilton et al., 1990; Ghiselin, 1988). By this hypothesis one would expect sexual reproduction to predominate among parasites, yet asexual clonal lineages are proving to occur or be suggested among well-studied protozoan parasites, including *Trypanosoma* species responsible for Chagas disease and sleeping sickness in humans and other animal diseases; *Leishmania* species; and (probably) *Entamoeba histolytica*, *Giardia*, and many strains of the malaria parasite *Plasmodium falciparum* (Ayala, 1998). So there is not a simple or consistent relationship between parasites and disease and sexual reproduction. Ghiselin (1988, p. 20) did not find the repair hypothesis completely convincing because it does not explain the notable correlation between the occurrence of sexual reproduction and particular ecological conditions. But I can see no reason why both the recombination and repair advantages could not apply at the same time.

The twofold cost of sex may be an underestimate, for there are developmental costs of recombination in addition to the investment cost. The union of genetically dissimilar individuals may produce *outbreeding depression* (Uyenoyama, 1988)—developmentally disruptive or phenotypically disadvantageous incompatibilities due to dissimilarity between mates. Uyenoyama (1988) lists examples of decreased compatibility with increased distance between experimental crosses in plants. Such developmental, intra-individual phenotypic incompatibilities in offspring may stretch even the considerable abilities of the flexible phenotype to adjust, and it is the developmental inviability that results from extreme outbreeding depression that presumably puts boundaries around species and ultimately limits the pool of individuals with which a female can successfully mate. Outbreeding depression is a consequence of recombination that may be alleviated by female choice, which could permit not only choice of genetically superior mates, but also positive assortative mating (Zeh and Zeh, 1996, 1997; Zeh, 1997; Jennions, 1997). Such considerations as these also point to the importance of developmental plasticity as an adjunct to recombination, for adaptive plasticity influences,

and must ultimately limit, the magnitude of a recombinant novelty that can be tolerated in producing a functional phenotype.

The fact that developmental compatibility limits the scope for profitable recombination raises yet another point about the potential advantages of sex. Sex or, more precisely, limited sexual compatibility engenders speciation—the formation of evolutionarily independent lineages. The same sexually selected requirements for particular stimuli during courtship that I argue contribute to the maintenance of sex may contribute to speciation *between* populations in which those stimuli have diverged to the point of sexual incompatibility (West-Eberhard, 1983). Increased diversification by speciation, then, is another consequence of sexual reproduction that, in addition to intrapopulational recombination, may contribute to the maintenance of sex in a changing environment. Speciational diversification can enhance the long-term survival of clades (Van Valen, 1975). Species selection *sensu* Lloyd and Gould (1993) or clade selection *sensu* G. C. Williams (1992) may favor clades with sexual selection and can thereby contribute to the predominance of sex. If it is true that speciose clades are less liable to extinction due to their ecological diversity and large numbers (Darwin, 1859 [1872]), then any factor that raises the speciation rate should be favored by selection at the level of species or clades.

The asymmetrical investment in offspring brings with it not only the cost of sex, but also the benefit of female mate choice (Trivers, 1972). In sexually selected species, the developmental stimuli provided by males go far beyond the stimulus provided by fertilization, to include interactions that affect the reproductive development of both females and embryos. In addition, there is increasing evidence of essential genes that are imprinted such that they only function if paternally supplied. These factors are the main subject of this chapter. They may combine in sexually selected species to make sex far more advantageous to females than its abandonment.

Female choice can partially alleviate the cost of sex if it endows the female's descendants with phenotypic and genetic superiority due to choice. Only if the "parasitic" paternally supplied genes are on average *inferior* to those of the female, or selfishly manipulative, do they impose a net fitness cost. If they are on average superior, then the diploid offspring phenotype is superior in survival and reproductive success than a parthenogenetically produced diploid would have been, and the entire maternal genetic complement gets a corresponding boost in fitness. If females can use phenotypic cues to discern genetic superiority of mates, including genetic superiority correlated with phenotypic superiority in both survival and courtship—and there is considerable evidence that females do discriminate using phenotypic cues (reviewed in Hamilton and Zuk, 1982; Craddock and Boake, 1992; Andersson, 1994; Eberhard, 1996)—then this would help tip the cost–benefit balance of sex in their favor through phenotypic benefits to the young. Once effective female choice begins to evolve, not only can it begin to compensate, at least in part, for the cost of sex, but choosy females may in some circumstances do better than parthenogenetic females with no opportunity to form reproductive alliances. This suggests that female choice should be considered, along with ecological factors (Williams, 1975), as potentially contributing both to the original abandonment of asexual reproduction in lineages where sex was formerly advantageous, and to the maintenance of obligatory sex once it has evolved.

Female choice is now routinely implicated in hypotheses of good-genes benefits to females in coevolutionary races with parasites and pathogens (e.g., Hamilton and Zuk, 1982), but it is not always incorporated into models for the maintenance of sex (e.g., Hamilton et al., 1990, assume random mating). Hamilton believed so strongly in good-genes explanations for the maintenance of sex, and particularly in the importance of coevolutionary races with parasites and pathogens, that he wrote that should disease be eliminated in humans, then males could be eliminated too, for example, by the institution of cloning and anti-aphrodisiacs (Hamilton, 1987b).

Williams (1975) gives conditions, in terms of competition for ecological space, where asexual reproduction may become so costly that it may disappear. It is with the disappearance of asexual reproduction as an option for females that sexual reproduction may become an increasingly tight trap. This is because successful reproduction comes to depend increasingly on interaction with males. Williams (1975) lists several developmental consequences of the fixation of sexual reproduction that would make asexual reproduction difficult to regain. Among them are contributions of males to the development of sexually produced offspring, such as the signal for embryogenesis provided by fertilization, and the centriole and mitochondria supplied by sperm.

The dependence of reproduction on such "historical legacies" could make sexual reproduction difficult or impossible to abandon (Williams, 1975,

1988). Ghiselin (1987, p. 637) notes that only a few authors have accepted this argument, that sex is "a sort of vestigial feature." This is not surprising. To regard such a costly activity as a mere legacy or vestige is not very satisfactory. The legacy argument may be correct to distinguish between the factors that established sex and those that maintain it, but by itself it is too weak a description of what is going on. Male–female interaction is more than a just a vestige of past utility. It is a complex developmental trap that continues to be reinforced and elaborated by (sexual) selection, probably at least sometimes reinforced by the factors that originally gave rise to sex. This is not an argument that applies to only a few highly complex organisms (Williams, 1975). It presumably applies to every species for which the "paradox" of the twofold cost of sex applies, namely, all species in which females invest more in offspring than do males. For it is the skewed investment that is responsible for both the twofold cost of sex and the phenomenon of sexual selection. Therefore, if the twofold cost of sex is the problem we are trying to address, then the sexually selected developmental trap has to be considered.

Sexual interaction is more consistently involved in reproduction, by every female of every generation, than any ecological factor, including infection with parasites or pathogens, could ever be. A female in a species with obligatory sex may escape infection or be resistant to its effects. But she cannot escape having to interact with a male. In the context of reproductive success, as distinct from survival, obligatory sex represents a new level of organization and selection. Whenever there is a new, emergent level of selection, it to some degree screens off the lower ones and can maintain the emergent structure without the lower level conditions that originally gave rise to it and may still be an important consequence. Thus, the original and continuing adaptive function of sex, in the ecological-survival context, may be recombination. But the payoff for recombination may fluctuate and disappear without sex being abandoned, because sex is maintained by the secondarily evolved obligatory collaboration and division of labor between the sexes in reproduction. The two sexes, and their gametes, have become dependent morphs, in the sense discussed in chapter 21 (see also Ghiselin, 1974, pp. 100ff).

Good-genes analyses of sexual selection like those outlined above, and arguments for the importance of recombination, usually emphasize the ecologically adaptive side of sex. The social side of sex—sex as male–female interaction, and sexual selection as social competition, with sexual interaction a prerequisite for reproduction—is seldom mentioned. In 22 articles of a special journal issue on sex (*Evolutionary Biology*, vol. 12, no. 6, 1999) none mentioned this aspect of sexual reproduction. Yet the central point of sexual-selection theory as stated by Darwin (1871), and Darwin's principal reason for advancing the theory, is that certain traits of sexually reproducing organisms can be costly or "maladaptive" in terms of survival under *natural* selection and still be favored by *sexual* selection. Apparently maladaptive male traits, such as fighting among males and extravagant courtship display and morphology, can be maintained due to strong selection for success in competition for mates.

One of the otherwise maladaptive or costly consequences of sexual competition may be the maintenance of sex itself. Contrasting phenotypes, such as those of males and females, that have evolved divergently under strong competition, often end up being complementary and are then maintained because the contrasting phenotypes become mutually dependent. This is a recurrent theme of organismic evolution (West-Eberhard, 1979; Maynard Smith and Szathmary, 1995; see chapter 22). The evolution of mutual dependence between males and females could occur as follows: once sexual reproduction has evolved under natural selection, the stage is set for gamete competition and the evolution of the egg–sperm dimorphism, or anisogamy (Parker et al., 1972; West-Eberhard, 1979). Variation in the size and consequent viability of gametes is thought to have led to the evolution of a frequency-dependent gamete dimorphism, with large gametes (eggs) well endowed with nutrients and instructions for embryonic development, and small gametes (sperm) specialized to compete to fertilize and in effect parasitize this endowment. Selection for success in seeking, fighting for, and courtship of mates then acts most strongly on the sex, usually the male, with the least investment in gametes and young, and it leads to choosiness or "coyness" in the sex (usually the female) with the most parental investment (Darwin, 1871; Trivers, 1972). Ghiselin (1974) gives a history of early ideas about the origin of gamete dimorphism and the division of labor between the sexes.

The interactive nature of sexual competition among the gametes, and among the adult individuals that have evolved to produce them and promote their success, eventually can produce a social and developmental trap, a coevolved intricacy of interaction and mutual dependence without which a female cannot reproduce. Gametic competition can involve such mechanisms as competitive pollen tube growth (see Willson and Burley, 1983) and possibly biochemical interactions between sperm and egg (Baker and Bellis, 1995). In motile animals with internal insemination extensive precopulatory and copulatory courtship interactions have evolved (Andersson, 1994; Eberhard, 1996). Selection for female choice can lead to the evolution of specialized barriers and high thresholds of female responsiveness (Eberhard, 1996), such that particular stimuli from male gametes, or male-produced substances or behaviors, are actually required for reproduction to occur.

One of the primary universal functions of the animal spermatozoon is to activate the egg (Baccetti and Afzelius, 1976). The male gamete of higher plants plays a role in early development that is even more extensive than that of the animal spermatozoon, for it participates in the formation of the endosperm and the transfer of nongenetic materials and various organelles. In endosperm, the cooperation of maternal and paternal genomes is strictly enforced, and there is some evidence that abnormal endosperm development and abortion occur following detection of a deviant ratio between maternal and paternal contributions to the genome (ploidy; see references in Uyenoyama, 1988). Paternal participation in early development may be underestimated in animals. Male nongenetic products are sometimes transferred to eggs (e.g., Markow and Ankney, 1988), and the possibility that they have nonnutritional effects on development has scarcely been investigated (but see discussion in Pitnick and Karr, 1998).

Male and female therefore evolve divergent specializations and a physiological division of labor such that the female of many sexually reproducing species is dependent on the male for developmentally essential stimuli or materials in addition to just the set of genes provided by sperm, a process analogous to the evolution of a division of labor between differentiated cells in multicellular organisms, and between workers and queens in social insect colonies. Darwin (1864) used the term "reciprocal dimorphism" to refer to the interdependence of diverged forms called "complementary morphs" by West-Eberhard (1979). Their reciprocal dependence can become a developmental trap: in sexually selected species formed by a history of sexual selection, the development of reproductive responses in females, and the normal development of their offspring, can depend on stimuli, materials, and even particular genes provided by males.

To suppose that a highly specialized sexually reproducing female would be better off if she could suddenly be an asexual parthenogen is to misconstrue the nature of selection and adaptation. Selection does not optimize, it only favors the most successful variants among those that occur. It is as unrealistic to debate the merits of asexuality for a specialized sexual female as it is to debate the merits of independent reproduction for a liver cell, or for a worker of an army ant colony. Wouldn't a dog be better off with a nose like an elephant, so that it could sniff danger from a distance? No, because its nose would get in the way of its teeth in a fight. Selection always begins with legacies, some of them, such as specialized sexual reproduction, so elaborate and important for development or survival that it is difficult to imagine a small or large step that could initiate a viable change. When parthenogenetic reproduction has evolved in sexually reproducing lineages, as in some interspecies hybrids discussed below, behaviors often persist that confirm the rule of a developmental dependence upon males.

While it is conceivable that male–female sexual interdependence could be reversed by selection on females to seek essential elements on their own (see below), once a complex interactive mutualism between the sexes has evolved, the first steps toward individual departure from the established system are likely to be disadvantageous. Fish that undergo sex-role reversals engage in complicated maneuvers to avoid the female role and the cost to females of sex (e.g., see Fischer, 1988; Warner et al., 1975). While selection has thus led to a temporary escape from the cost of being a female, it has not led to an escape from sex itself. As in other complementary mutualisms, such as that between symbiotic fungi and algae to form lichens, it can become virtually impossible for the specialized participants of a male–female reproductive mutualism to function independently. The cost of *no* sex in such a species is no reproduction, and the cost of inferior sex—any step toward emancipation that implies inferior stimulation of reproductive processes or compromised embryonic development—would likely be severely reduced fitness.

It is difficult to see how abandonment of sex could invade such a population except under very special circumstances of physiological permissiveness or unusual payoffs to some individuals. Sex and sexual selection can be maintained even after genetic variation for particular traits has been exhausted and there is no *genetic* reward for continuing to perform them. Although this may seem puzzling (see Simmons, 1987, and literature cited therein), complex sexual behavior accumulates and

is maintained by selection because individuals cannot afford to *omit* elements without suffering a cost in terms of success in sexual competition, which acts relentlessly on every reproducing individual in every generation. Since selection acts on phenotypic variation, not directly on genes, it can continue without genetic variation, secondarily affecting (of course) any influential genetic variation that happens to arise (see chapter 6). As a result of this continuing process, sexual interaction has become intricately entwined in the reproduction of sexually reproducing females.

As mentioned above, sexual selection, and with it the potential for a developmental trap, occurs in precisely the same set of organisms as does the twofold cost of sex, namely, those with an asymmetric investment by females in offspring. It therefore bears consideration as a widely applicable explanation for the maintenance of sex despite its costs. So the sexually selected developmental-trap aspect of the legacy of sex cannot be waved aside lightly, as a matter of only "peripheral interest" compared with questions of genetic mixing, or as an "ad hoc" argument not supported by the data of ecology and systematics (Ghiselin, 1988, pp. 9, 19). One could more justifiably adopt the opposite extreme position and say that in organisms where sexual reproduction is enforced by a developmental trap, recombination and the twofold cost of sex are not required to explain the *maintenance* of sex, even though they may still be benefits of sexual reproduction.

Just as individual selection screens off the effects of selection at lower levels (Brandon, 1990), so in many species the reproductive dependence of females on interaction with males screens off recombination from being the primary reasons for the maintenance of sex. Under the developmental-trap hypothesis, obligatory sex is actively maintained by its selective advantage in the context of female reproductive development and offspring viability in the lineages where it occurs. Sex-as-reproduction in those lineages has taken on a new significance, beyond sex-as-recombination. It seems a misconstrual of the reason that sex is maintained, and how evolution works, to suppose that there is an asexual possibility that simply cannot be reached (e.g., due to "the absence of the appropriate mutations"—Williams, 1975, pp. 102–110; Burt, 2000, p. 346). If sex is sufficiently developmentally advantageous, then loss of sex would not evolve even if the appropriate mutations were to occur.

The nature of the developmental trap of sex is illustrated by the following examples. Sperm are often necessary to activate egg development even when recombination has been abandoned. This has been documented following hybridization in certain species of fish (Haskins et al., 1960) and salamanders (Spolsky et al., 1992), as well as in some nematodes, planarians (*Polycelis* spp.), and the spider beetle *Ptinus latro* (reviewed in Austin, 1965). In these species, females must copulate with males of related species in order to reproduce, a reproductive mode sometimes called *gynogenesis* or *pseudogamy* (Austin, 1965). Copulation is a purely developmental necessity having nothing to do with recombination, for the male gametes do not participate in the formation of the zygote. In some parthenogenetic lizards, females show courtship and copulationlike behavior with other females (Werner, 1980), and the possibility should be investigated that such behavior stimulates reproduction.

Other evidence in support of the developmental-trap hypothesis can be found in literature on sexual selection and cryptic female choice, where there are many examples of the dependence or facilitation of female reproductive responses on male behavior (e.g., see Eberhard, 1996; Nijhout, 1994a). For example, in the pteromalid wasp *Muscidifurax raptorellus*, the activation of oviposition, which occurs immediately after mating, evidently depends upon interaction with the male, and the reproductive response of females (number of eggs deposited per ovipositor insertion) is influenced by the *male* genome (Legner, 1988). Quantitative differences in oviposition behavior are heritable. Presumably factors transmitted in the male's seminal fluid modify the female's behavior (Legner, 1987). In such a species, an unmated female would not reproduce, and selection would favor mating with the males whose stimulation of egg output is most effective. Such interactions are expected to lead to ever increasing dependence of females on particular kinds of male stimulation, and selection for increasingly exacting female choice of mates.

The developmental trap idea is also supported by its ability to explain some otherwise enigmatic observations. Asexuality, for example, is more common in isogamous organisms (those with monomorphic gametes) than in anisogamous ones (those in which eggs and sperm are morphologically distinct). This has been considered "quite the opposite of what one would predict with knowledge of the costs," since in anisogamous species, with large eggs and small sperm, the cost of sex to females is relatively greater than in isogamous organisms (Hurst and Peck, 1996, p. 46). But this result is expected under the developmental trap hypothesis. The developmental trap is a product of sexual dimorphism and the mutual dependence of interaction between egg and sperm. It is in anisogamous groups where male–female gamete divergence is strongest, and their mutual dependence is likely to be strongest, too. So these are precisely the groups where developmental traps are expected to evolve, making sexual reproduction difficult to abandon, despite its higher costs.

In general, the developmental-trap hypothesis for the maintenance of sex predicts that the evolution of asexual from sexual reproduction, as well as the occurrence of facultative sex reversal, may be most common in lineages where there is a relatively small degree of male–female morphological and behavioral differentiation in the reproductive phenotypes of adults. I have not attempted a systematic survey, but this seems to apply in polychaete worms, where the two sexes are morphologically little differentiated, and both sex change and asexual reproduction are common (Premoli and Sella, 1995; Petraitis, 1988). Facultative or cyclic parthenogenesis has evolved numerous times from exclusively sexually selected ancestors, for example, in the haplodiploid Hymenoptera and other insects where unmated females produce haploid males. The developmental trap hypothesis suggests that such transitions could have occurred during periods in these lineages when sexual selection was relaxed.

In mammals, and to an unknown extent possibly in other groups, male–female reproductive interdependence extends to embryonic gene expression through the phenomenon of genomic imprinting. *Genomic imprinting* is parent-specific gene expression, that is, gene expression that depends on which parent contributed the gene (Trivers and Burt, 1999), or the differential modification of genes such that maternal and paternal alleles are distinct (Wei and Mahowald, 1994). For example, the maternally derived allele of the Igf2r gene of the mouse is methylated at intron region 2, which permits the gene to be expressed, evidently because methylation blocks a silencer of transcription (Wei and Mahowald, 1994). The paternal allele is not methylated and therefore is not expressed. Parental imprints are evidently erased, and then remarked, during gametogenesis, to accord with the sex of the individual. Experimentally constituted mouse embryos with an entirely paternally imprinted genome (androgenomes) or with an entirely maternally imprinted genome (gynogenomes) are incapable of normal development and show severe skeletal abnormalities. There is a growing literature on the nature and evolution of imprinting (for brief reviews, see Solter, 1988; Peterson and Sapienza, 1993; Pennisi, 1998; for book-length treatments, see Reik and Surani, 1997; Ohlsson, 1999).

Because imprinted genes are differentially suppressed depending on their maternal or paternal origin, normal development requires both sets of genes (Davidson, 1986), and the effects of imprinting play key roles in development (e.g., of the nervous system and behavior) well into adult life (Allen et al., 1995; see especially Jablonka and Lamb, 1995). This has been cited as a factor maintaining sexual reproduction in mammals (Surami, 1987; see also Haig and Westoby, 1989, p. 149; Haig, 1997; Barton and Charlesworth, 1998). The interdependence may have originated as sexual conflict of interests in the form of male–female competition for maternal investment in the offspring (Queller, 1994).

The endosperm of angiosperms is a plant counterpart of parental genomic cooperation in the construction of the embryonic phenotype. The endosperm functions like the insect nurse cells and the mammalian placenta to provide building blocks and instructions for the early embryo, with the exceedingly interesting difference that both parents are involved, rather than just the manipulative dominant mother as in animals (see chapter 5). The triploid endosperm is formed by two sibling nuclei of the one in the sperm that fertilizes the egg, and one sibling nucleus of the egg nucleus itself, via a seemingly unnecessarily complex minuet that accompanies fertilization.

The convolutions of the endosperm make no sense from a purely genetic point of view—why not just fertilize the egg and be done with it, allowing the new genome to program its own phenotype, as the genetic theory of organization suggests? The endosperm, nurse cells, and placenta and genomic imprinting do make sense in terms of continuity of the phenotype and universal phenotypic responsiveness. The genome is ineffectual without materials of environmental origin with which to work, and without a preexistent phenotype on which to act. The universal responsiveness of the phenotype makes it subject to manipulation, including competitive manipulation by parental genomes and parentally produced materials and behaviors. It means that parental endowment of an embryonic phenotype may be not only a nurturing head-start program but also potentially a way of imposing genetic selfishness that for females could add to the twofold cost of sex. In many animals, control by the maternal genome of early embryogenesis delays expression of the zygotic genome and thereby heads

off selfish gene activity that could bias access to the germ line (Buss, 1987). In the endosperm, genetic transcripts from both parents contribute to this control, and the "genic balance" (Johnston et al., 1980) between the alleles of different parental origins in the endosperm implies a developmental interdependence of maternal and paternal contributions that could add to the immediate phenotypic cost of *not* maintaining sexual reproduction in such organisms.

The developmental-trap hypothesis predicts that the evolution of asexuality in sexual lineages may be associated with some lessening of the importance of sexually selected interaction, since sexual selection is an important context for the elaboration and maintenance of reproductive dependence between the sexes. It may therefore be significant that in the ostracods, where asexuality has evolved from sexuality an unusually high number of times, this transition occurs in populations where sperm limitation results from female-biased sex ratios and low sperm production by males (Chaplin et al., 1994). This is a situation that would diminish the intensity of sexual selection on males and could promote the evolution of female reproductive responses less dependent on male–female interaction. Reduced sexual selection among males also evidently occurs in some drosophilids when females are sperm limited, and there also it is associated with the gradual evolution of development in unfertilized eggs (Chaplin et al., 1994). The sperm-limitation cases offer an opportunity for the gradual evolution of a departure from sex. This kind of gradual relaxation of sexual selection, with the gradual evolution of increased independence of females from interaction with males, offers one solution to the problem of how a reduction in the participation of females could evolve beginning with a highly sexually selected system where full participation is required.

A legacy or trap hypothesis does not imply that sex is no longer useful, as interpreted by Ghiselin (1988). The benefits of sex, such as DNA repair, recombination, mate choice, and speciational diversification would still apply. But they would not be *required* to maintain advantageous sexual reproduction within populations, and may not be the most important general explanation for the maintenance of sex. The developmental-trap hypothesis regards sex as adaptive, not only ecologically, but socially and developmentally as well. Williams (1975), by contrast, considers the legacy of sex to be maladaptive. As evidence for this, he notes that whenever parthenogenesis has evolved in otherwise sexually reproducing taxa (as in the hybrid species discussed above), it has always replaced sexual reproduction entirely. Williams considers this "decisive evidence of the maladaptive nature of sexuality in these organisms" (Williams, 1975, p. 106). In fact, there are some triploid populations of hybrid fishes that retain the male chromosomes rather than discarding them after mating (White, 1978). But such populations are an exception. The possibility should perhaps be considered that the association between complete abandonment of sex in parthenogenetic hybrid species noted by Williams may be due not to the maladaptiveness of sex but to some disadvantage of increased ploidy.

Organisms such as bacteria and clonal organisms, where sex-as-recombination is not tied to reproduction, nonetheless engage in sex facultatively (Levin, 1988). Even though they are not trapped into either the cost of sex or the dependence upon a mate, still they do not abandon sex. This argues that there is an adaptive function for sex and suggests that obligatorily sexual organisms may also gain some advantages from sex above and beyond the gain of essential developmental contributions by males, for example, through recombination important in coevolutionary races with parasites and pathogens (Hamilton, 1987b) and in heterogeneous or changing environments (Williams, 1975). That is, it is reasonable to think that the recombinatorial benefits of sex are maintained, and that ancient advantages of sex that were responsible for its evolution in the first place likely persist alongside those fostered by sexual selection in organisms with obligatory sex. One consequence of the developmental-trap hypothesis is that theoretical conditions for the maintenance of recombination are relaxed in many species, because the maintenance of sex need not be explained in terms of the benefits of recombination alone.

The evolutionary history of sex shares many features with the history of other quantum evolutionary changes. All consolidate a new level of organization, where originally undifferentiated individuals associate, differentiate, and become mutually dependent on each other. Reversals of quantum transitions are rare, due to the complex interdependence of parts. In multicellular organisms and the highly derived social insects with morphologically differentiated castes, there are no known reversals to solitary reproduction, except in the unusually permissive and instructive case of social parasites and aggressively selfish mutant cancer cells in laboratory cultures (Strathmann, 1991b). The reasons for this seem clear: once individual cells or female insects have divergently specialized to complementary jobs, collaboration becomes obligatory (West-Eberhard, 1979). The irreversible maintenance of sex follows this same pattern.

Lyon (1993) called genomic imprinting a kind of "epigenetic inheritance." This brings us full circle from modern molecular studies of gene action back to Darwin, who, with his primitive but molecular gemmular theory of pangenesis (see chapter 8), insisted that all inheritance is epigenetic, a product of both the transmission and the development of traits. This chapter argues, as Darwin (1871) did, that beyond selection for success in survival, there are reproductively essential social factors that can exaggerate and maintain extravagant traits like sex.

Literature Cited

Abou-Seedo, F.S. and Potter, I.C. 1979. The estuarine phase in the spawning run of the river lamprey *Lampetra fluviatilis*. *Journal of Zoology* 188:5–25.

Adams, W.W., III and Martin, C.E. 1986. Heterophylly and its relevance to evolution within the Tillandsioideae. *Selbyana* 9:121–125.

Adler, F.R. and Harvell, C.D. 1990. Inducible defenses, phenotypic variability and biotic environments. *Trends in Ecology and Evolution* 5:407–410.

Agosti, D., Grimaldi, D. and Carpenter, J.M. 1998. Oldest known ant fossils discovered. *Nature* 391:447.

Agrawal, A.A. 1998. Induced responses to herbivory and increased plant performance. *Science* 279:1201–1202.

Agrawal, A.A. 2001. Phenotypic plasticity in the interactions and evolution of species. *Science* 294:321–326.

Agrawal, A.A., Laforsch, C. and Tollrian, R. 1999. Transgenerational induction of defences in animals and plants. *Nature* 401:60–63.

Aiello, A. 1984. *Adelpha* (Nymphalidae): deception on the wing. *Psyche* 91:1–45.

Aiello, L.C. and Wheeler, P. 1995. The expensive-tissue hypothesis. *Current Anthropology* 36:199–221.

Akimoto, S. 1981. Gall formation by *Eriosoma* fundatrices and gall parasitism in *Eriosoma yangi* (Homoptera, Pemphigidae). *Kontyû* 49:426–436.

Akimoto, S. 1985. Occurrence of abnormal phenotypes in a host-alternating aphid and their implications for genome organization and evolution. *Evolutionary Theory* 7:179–193.

Akimoto, S. 1988a. Competition and niche relationships among *Eriosoma* aphids occurring on the Japanese elm. *Oecologia* 75:44–53.

Akimoto, S. 1988b. The evolution of gall parasitism accompanied by a host shift in the gall aphid, *Eriosoma yangi* (Homoptera:

Aphidoidea). *Biological Journal of the Linnean Society* 35:297–312.

Alberch, P. 1980. Ontogenesis and morphological diversification. *American Zoologist* 20:653–667.

Alberch, P. 1982. The generative and regulatory roles of development in evolution. In: *Environmental Adaptation and Evolution*, D. Mossakowski and G. Roth (eds.). Gustav Fischer, New York, pp. 19–36.

Alberch, P. 1983a. Mapping genes to phenotypes, or the rules that generate form. *Evolution* 37:861–863.

Alberch, P. 1983b. Morphological variation in the neotropical salamander genus *Bolitoglossa*. *Evolution* 37:906–919.

Alberch, P. 1985. Developmental constraints: why St. Bernards often have an extra digit and poodles never do. *American Naturalist* 126:430–433.

Alcock, J. 1979. The evolution of intraspecific diversity in male reproductive strategies in some bees and wasps. In: *Sexual Selection and Reproductive Competition in Insects*, M.S. Blum and N.A. Blum (eds.). Academic Press, New York, pp. 381–402.

Alcock, J. 1987. Ardent adaptationism. *Natural History* 96:4.

Alcock, J. and Sherman, P. 1994. The utility of the proximate-ultimate dichotomy in ethology. *Ethology* 96:58–62.

Alder, E.M. 1975. Genetic and maternal influences on docility in the Skomer vole, *Clethrionomys glareolus skomerensis*. *Behavioral Biology* 13:251–255.

Alexander, R.D. 1961. Aggressiveness, territoriality, and sexual behavior in field crickets (Orthoptera: Gryllidae). *Behaviour* 17:130–223.

Alexander, R.D. 1962. The role of behavioral study in cricket classification. *Systematic Zoology* 11:53–72.

Alexander, R.D. 1968. Life cycle origins, speciation, and related phenomena in crickets. *Quarterly Review of Biology* 43:1–41.

Alexander, R.D. 1974. The evolution of social behavior. *Annual Review of Ecology and Systematics* 5:325–381.

Alexander, R.D. 1979. *Darwinism and Human Affairs*. University of Washington Press, Seattle.

Alexander, R.D. and Bigelow, R.S. 1960. Allochronic speciation in field crickets, and a new species *Acheta veletis*. *Evolution* 14:334–346.

Alexander, R.D. and Borgia, G. 1978. Group selection, altruism, and the levels of organization of life. *Annual Review of Ecology and Systematics* 9:449–474.

Allard, M.W. and Carpenter, J.M. 1996. On weighting and congruence. *Cladistics* 12:183–198.

Allen, G.A., Jr. 1969. *Pheasant Standards*. American Game Bird Breeders' Cooperative Federation, Salt Lake City.

Allen, N., Logan, D., Lally, G., Drage, D., Norris, M., and Keverne, B. 1995. Distribution of parthenogenetic cells in the mouse brain and their influence on brain development and behavior. *Proceedings of the National Academy of Sciences USA* 92:10782–10786.

Allendorf, F.W. and Thorgaard, G. 1984. Tetraploidy and the evolution of salmonid fishes. In: *The Evolutionary Genetics of Fishes*, B.J. Turner (ed.). Plenum Press, New York, pp. 1–53.

Almeida-Val, V.M.F. and Val, A.L. 1993. *Evolution*ary trends of LDH isozymes in fishes. *Comparative Biochemistry and Physiology* 105B:21–28.

Altmann, J. 1980. *Baboon Mothers and Infants*. Harvard University Press, Cambridge.

Altmann, J., Alberts, S.C., Haines, S.A., Dubach, J., Muruthi, P., Coote, T., Geffen, E., Cheesman, D.J., Mututua, R.S., Saiyalel, S.N., Wayne, R.K., Lacy, R.C. and Bruford, M.W. 1996. Behavior predicts genetic structure in a wild primate group. *Proceedings of the National Academy of Sciences USA* 93:5797–5801.

Amadon, D. 1950. The Hawaiian honeycreepers (Aves, Drepaniidae). *Bulletin of the American Museum of Natural History* 95:151–262.

Ambros, V. 1988. Genetic basis for heterochronic variation. In: *Heterochrony in Evolution*, M.L. McKinney (ed.). Plenum Press, New York, pp. 269–286.

Ammirato, P. 1999. Towards an integrated view of plant embryogenesis. *Quarterly Review of Biology* 74:439–442.

Amundson, R. 1996. Historical development of the concept of adaptation. In: *Adaptation*, M.R. Rose and G.V. Lauder (eds.). Academic Press, San Diego, pp. 11–53.

Amundsen, T. 2000. Why are female birds ornamented? *Trends in Ecology and Evolution* 15:149–155.

Ananthakrishnan, T.N. (ed.) 1984. *Biology of Gall Insects*. Oxford and IBH Publishing, New Delhi.

Andersen, N.M. 1993. The evolution of wing polymorphism in water striders (Gerridae): a phylogenetic approach. *OIKOS* 67:433–443.

Anderson, J.M. and Aro, E.M. 1994. Grana stacking and protection of photosystem II in thylakoid membranes of higher plant leaves under sustained high irradiance: an hypothesis. *Photosynthesis Research* 41:315–326.

Anderson, J.M., Chow, W.S. and Park, Y.I. 1995. The grand design of photosynthesis: acclimation of the photosynthetic apparatus to environmental cues. *Photosynthesis Research* 46:129–139.

Anderson, K.V. 1989. *Drosophila*: the maternal contribution. In: *Genes and Embryos*, D.M. Glover and B.D. Mames (eds.). IRL Press, Oxford, pp. 1–37.

Anderson, N.H. and Cargill, A.S. 1987. Nutritional ecology of aquatic detritivorous insects. In: *Nutritional Ecology of Insects, Mites, Spiders and Related Invertebrates*, F. Slansky, Jr. and J.G. Rodriguez (eds.). John Wiley and Sons, New York, pp. 903–925.

Andersson, M. 1994. *Sexual Selection*. Princeton University Press, Princeton.

Andrés, J.A. and Cordero, A. 1999. The inheritance of female colour morphs in the damselfly *Ceriagrion tenellum* (Odonata, Coenagrionidae). *Heredity* 82:328–335.

Andrew, R.J. 1961. The displays given by passerines in courtship and reproductive fighting: a review. *Ibis* 103a:315–348.

Andrews, C.A. 1999. Ontogenetic ecomorphology of the threespine stickleback (*Gasterosteus aculeatus*). Unpublished Ph.D. thesis, State University of New York, Stony Brook.

Andrews, M.T., Squire, T.L., Bowen, C.M. and Rollins, M.B. 1998. Low-temperature carbon utilization is regulated by novel gene activity in the heart of a hibernating mammal. *Proceedings of the National Academy of Sciences USA* 95:8392–8397.

Andrews, R.C. 1921. A remarkable case of external hindlimb in a humpback whale. *American Museum Novitates* 9:1–6.

Angehr, G.R. 1980. The role of interference competition in the organization of a guild of panamanian hummingbirds. Unpublished Ph.D. thesis, University of Colorado, Boulder.

Angelo, M.J. and Slansky, F., Jr. 1984. Body building by insects: trade-offs in resource allocation with particular reference to migratory species. *Florida Entomologist* 67:22–41.

Anson, B.J. 1951. *Atlas of Human Anatomy.* W.B. Saunders, Philadelphia.

Anstey, R.L. 1987. Astogeny and phylogeny: evolutionary heterochrony in Paleozoic bryozoans. *Paleobiology* 13:20–43.

Anstey, R.L. and Pachut, J.F. 1995. Phylogeny, diversity history, and speciation in Paleozoic bryozoans. In: *New Approaches to Speciation in the Fossil Record.* D.H. Erwin and R.L. Anstey (eds.). Columbia University Press, New York, pp. 239–284.

Antonovics, J. and van Tienderen, P.H. 1991. Ontoecogenophyloconstraints? The chaos of constraint terminology. *Trends in Ecology and Evolution* 6:166–168.

Antonovics, J., Bradshaw, A.D. and Turner, R.G. 1971. Heavy metal tolerance in plants. *Advances in Ecological Research* 7:1–85.

Aoki, S. 1977. Colophina clematis (Homoptera, Pemphigidae), an aphid species with "soldiers." *Kontyû* 45:276–282.

Aoki, S. 1980. Life cycles of two *Colophina* aphids (Homoptera, Pemphigidae) producing soldiers. *Kontyû* 48:464–476.

Aoki, S. 1987. Evolution of sterile soldiers in aphids. In: *Animal Societies: Theories and Facts,* Y. Itô, J.L. Brown and J. Kikkawa (eds.). Japan Scientific Societies Press, Tokyo, pp. 53–65.

Aoki, S. and Kurosu, U. 1986. Soldiers of a European gall aphid, *Pemphigus spyrothecae* (Homoptera: Aphidoidea): why do they molt? *Journal of Ethology* 4:97–104.

Aoki, S. and Kurosu, U. 1988. *Pemphigus* "soldiers" and a defence of the generation-packing hypothesis: a response to Itô and Akimoto. *Journal of Ethology* 6:65–67.

Arak, A. 1983. Mating behaviour of anuran amphibians: the roles of male-male competition and female choice. In: *Mate Choice,* P. Bateson (ed.). Cambridge University Press, Cambridge, pp. 181–210.

Arathi, H.S., Shakarad, M. and Gadagkar, R. 1997. Factors affecting the acceptance of alien conspecifics on nests of the primitively eusocial wasp, *Ropalidia marginata* (Hymenoptera: Vespidae). *Journal of Insect Behavior* 10:343–353.

Arcadi, A.C., Robert, D. and Boesch, D. 1998. Buttress drumming by wild chimpanzees: temporal patterning, phrase integration into loud calls, and preliminary evidence for individual distinctiveness. *Primates* 39:505–518.

Arcese, P. and Smith, J.N.M. 1985. Phenotypic correlates and ecological consequences of dominance in song sparrows. *Journal of Animal Ecology* 54:817–830.

Arévalo, E., Strassmann, J.E. and Queller, D.C. 1998. Conflicts of interest in social insects: male production in two species of *Polistes. Evolution* 52:797–805.

Armbruster, W.S., Howard, J.J., Clausen, T.P., Debevec, E.M., Loquvam, J.C., Matsuki, M., Cerendolo, B. and Andel, F. 1997. Do biochemical exaptations link evolution of plant defense and pollination systems? Historical hypotheses and experimental tests with *Dalechampia* vines. *American Naturalist* 149:461–484.

Armitage, K.B. 1986. Individual differences in the behavior of juvenile yellow-bellied marmots. *Behavioral Ecology and Sociobiology* 18:419–424.

Armstrong, E.A. 1965. *The Ethology of Bird Display and Bird Behavior.* Dover, New York.

Arnegard, M.E., Markert, J.A., Danley, P.D., Stauffer, J.R., Jr., Ambali, A.J. and Kocher, T.D. 1999. Population structure and colour variation of the cichlid fish *Labeotropheus fuelleborni* Ahl along a recently formed archipelago of rocky habitat patches in southern Lake Malawi. *Proceedings of the Royal Society of London B* 266:119–130.

Arnemann, J., Bruggemeier, U., Carballo, M., Chalepakis, G., Gross, B., Espel, E., Mugele, K., Pina, B., Posseckert, G., Slater, E., Willmann, T. and Beato, M. 1993. Differential gene regulation by steroid hormones. In: *Steroid Hormone Action,* M.G. Parker (ed.). IRL Press, Oxford, pp. 185–192.

Arnold, A.P. 1982. Neural control of passerine song. In: *Acoustic Communication in Birds* (Vol. 1), D.E. Kroodsma, E.H. Miller and H. Ouellet (eds.). Academic Press, New York, pp. 75–93.

Arnold, A.P. and Breedlove, S.M. 1985. Organizational and activational effects of sex steroids on brain and behavior: a reanalysis. *Hormones and Behavior* 19:469–498.

Arnold, A.P., Bottjer, S.W., Brenowitz, E.A., Nordeen, E.J. and Nordeen, K.W. 1986. Sexual dimorphisms in the neural vocal control system in song birds: ontogeny and phylogeny. *Brain, Behavior and Evolution* 28:22–31.

Arnold, S.J. 1992. Constraints on phenotypic evolution. *American Naturalist* 140:S85–S107.

Arrow, G.J. 1951. *Horned Beetles.* Dr. W. Junk, The Hague.

Arthur, W. 1984. *Mechanisms of Morphological Evolution*. John Wiley and Sons, Chichester.

Asa, C.S. 1997. Hormonal and experiential factors in the expression of social and parental behavior in canids. In: *Cooperative Breeding in Mammals*, N.G. Solomon and J.A. French (eds.). Cambridge University Press, Cambridge, pp. 129–149.

Ashton, P.S. 1988. Dipterocarp biology as a window to the understanding of tropical forest structure. *Annual Review of Ecology and Systematics* 19:347–370.

Askew, R.R. 1965. The biology of the British species of the genus *Torymus* Dalman (Hymenoptera: Torymidae) associated with galls of Cynipidae (Hymenoptera) on oak, with special reference to alternation of forms. *Transactions of the Society for British Entomology* 16:217–233.

Astié, M. 1962. Tératologie spontanée et expérimentale. *Annales des Sciences Naturelles (Botanique)* 3:619–844.

Atchley, W.R. 1987. Developmental quantitative genetics and the evolution of ontogenies. *Evolution* 41:316–330.

Atchley, W.R. 1990. Heterochrony and morphological change: a quantitative genetic perspective. *Developmental Biology* 1:289–297.

Atchley, W.R. and Hall, B.K. 1991. A model for development and evolution of complex morphological structures. *Biological Reviews* 66:101–157.

Atchley, W.R. and Newman, S. 1989. A quantitative-genetics perspective on mammalian development. *American Naturalist* 134:486–512.

Atchley, W.R., Plummer, A.A. and Riska, B. 1985. Genetic analysis of size-scaling patterns in the mouse mandible. *Genetics* 3:579–595.

Atchley, W.R., Cowley, D.E., Eisen, E.J., Prasetyo, H. and Hawkins-Brown, D. 1990. Correlated response in the developmental choreographies of the mouse mandible to selection for body composition. *Evolution* 44:669–688.

Athias-Binche, F. 1995. Phenotypic plasticity, polymorphisms in variable environments and some evolutionary consequences in phoretic mites (Acarina): a review. *Ecologie* 26:225–241.

Aukema, B. 1986. Wing length determination in relation to dispersal by flight in two wing dimorphic species of *Calathus* Bonelli (Coleoptera, Carafidae). In: *Carabid Beetles, Their Adaptations and Dynamics*, P.J. Den Boer, M.L. Luff, D. Mossakowski and F. Weber (eds.). Fischer, New York, pp. 91–99.

Austad, S.N. 1984. A classification of alternative reproductive behaviors and methods for field-testing ESS models. *American Zoologist* 24:309–319.

Austin, C.R. 1965. Fertilization. Prentice-Hall, Englewood Cliffs, N.J.

Avise, J.C. 1976. Genetics of plate morphology in an unusual population of threespine sticklebacks (*Gasterosteus aculeatus*). *Genetical Research* 27:33–46.

Avise, J.C. 1990. Flocks of African fishes. *Nature* 347:512–514.

Ayala, F.J. 1998. Is sex better? Parasites say "no." *Proceedings of the National Academy of Sciences USA* 95:3346–3348.

Ayala, F.J. and Valentine, J.W. 1979. The Origin of Species. In: *Evolving: The Theory and Processes of Organic Evolution*. Benjamin/Cummings, Menlo Park, CA, pp. 212–225.

Baccetti, B. and Afzelius, B.A. 1976. *The Biology of the Sperm Cell*. S. Karger, Basel.

Baerends, G.P. 1941. On the life-history of *Ammophila campestris*. *Tijdschrift voor Entomologie* 84:483–488.

Baerends, G.P. 1950. Specializations in organs and movements with a releasing function. *Symposia of the Society for Experimental Biology* 4:301–359.

Baerends, G.P. 1958. Comparative methods and the concept of homology in the study of behaviour. *Netherlands Journal of Zoology* 8:401–417.

Baerends, G.P. 1975. An evaluation of the conflict hypothesis as an explanatory principle for the evolution of displays. In: *Function and Evolution in Behaviour*, G. Baerends, C. Beer and A. Manning (eds.). Clarendon Press, Oxford, pp. 187–227.

Baerends, G.P. and Baerends-van Roon, J.M. 1950. An introduction to the study of the ethology of cichlid fishes. *Behaviour* (Supplement) 1:1–243.

Baerends, G.P., Drent, R.H., Glas, P. and Groenewold, H. 1970. An ethological analysis of incubation behaviour in the herring gull. In: *The Herring Gull and Its Egg*, G.P. Baerends and R.H. Drent (eds.). E.J. Brill, Leiden, pp. 135–152.

Bajema, C.J. 1984. *Evolution by Sexual Selection Theory Prior to 1900*. Van Nostrand Reinhold, New York.

Baker, B.S. 1989. Sex in flies: the splice of life. *Nature* 340:521–522.

Baker, R.R. and Bellis, M.A. (eds.) 1995. *Human Sperm Competition: Copulation, Masturbation and Infidelity*. Chapman and Hall, London.

Baldwin, J.M. 1896. A new factor in evolution. *American Naturalist* 30:441–451, 536–553.

Baldwin, J.M. 1902. *Development and Evolution.* Macmillan, New York.

Baldwin, P.H. 1953. Annual cycle, environment and evolution in the Hawaiian honeycreepers (Aves: Drepaniidae). *University of California Publications in Zoology* 52:285–398

Balfour-Browne, F. 1940. British water beetles. Ray Society, London.

Baltimore, D. 2001. Our genome unveiled. *Nature* 409:814–815.

Banister, K.E. 1973. A revision of the large *Barbus* (Pisces, Cyprinidae) of east and central Africa. Studies on African Cyprinidae Part II. *Bulletin of the British Museum (Natural History) Zoology*, London 26:1–148.

Barkow, J.H., Cosmides, L. and Tooby, J. (eds.) 1992. *The Adapted Mind.* Oxford University Press, New York.

Barlocher, F. 1982. The contribution of fungal enzymes to the digestion of leaves by *Gammarus fossarum* Koch (Amphipoda). *Oecologia* 52:1–4.

Barlocher, F. 1983. Seasonal variation of standing crop and digestibility of CPOM in a Swiss jura stream. *Ecology* 64:1266–1272.

Barlow, G.W. 1968. Ethological units of behavior. In: *The Central Nervous System and Fish Behavior*, D. Dingle (ed.). University of Chicago Press, Chicago, pp. 217–232.

Barlow, G.W. 1977. Modal action patterns. In: *How Animals Communicate*, T.A. Sebeok (ed.). Indiana University Press, Bloomington, pp. 98–134.

Barlow, G.W. 1991. Nature-nurture and the debates surrounding ethology and sociobiology. *American Zoologist* 31:286–296.

Barlow, G.W. 2000. *The Cichlid Fishes.* Perseus, Cambridge, MA.

Barnard, C.J. 1984. *Producers and Scroungers.* Chapman and Hall, New York.

Barnard, C.J. and Brown, C.A.J. 1985. Risk-sensitive foraging in common shrews (*Sorex araneus* L.). *Behavioral Ecology and Sociobiology* 16:161–164.

Barnard, C.J. and Sibly, R.M. 1981. Producers and Scroungers: a general model and its application to captive flocks of house sparrows. *Animal Behaviour* 29:543–550.

Barnes, H. 1962. So-called anecdysis in *Balanus balanoides* and the effect of breeding upon the growth of calcareous shell of some common barnacles. *Limnology and Oceanography* 7:462–473.

Barnes, R.D. 1987. *Invertebrate Zoology*, 5th ed. Saunders College Publishing, Philadelphia.

Baroni Urbani, C. 1989. Phylogeny and behavioural evolution in ants, with a discussion of the role of behaviour in evolutionary processes. *Ethology, Ecology and Evolution* 1:137–168.

Barrett, P.H. (ed.) 1977a. *The Collected Papers of Charles Darwin*, Vol. 1. University of Chicago Press, Chicago.

Barrett, P.H. (ed.) 1977b. *The Collected Papers of Charles Darwin*, Vol. 2. University of Chicago Press, Chicago.

Barthlott, W. and Ziegler, B. 1980. Über ausziehbare helicale zellwandverdichungen als haft-apparat der samenshalen von *Chiloschista lunifera* (Orchidaceae). *Berichte der Deutschen Botanischen Gesellschaft* 93:391–403.

Barton, N.H. 1989. Founder effect speciation. In: *Speciation and Its Consequences*, D. Otte and J.A. Endler (eds.). Sinauer, Sunderland, MA, pp. 229–256.

Barton, N.H. 1998. Natural selection and random drift as causes of evolution on islands. In: *Evolution on Islands*, P.R. Grant (ed.). Oxford University Press, Oxford, pp. 102–123.

Barton, N.H. and Charlesworth, B. 1998. Why sex and recombination? *Science* 281:1986–1990.

Basibuyuk, H.H., Rasnitsyn, A.P., van Achterberg, K., Fitton, M.G. and Quicke, D.L.J. 1998. A new, putatively primitive Cretaceous fossil braconid subfamily from New Jersey amber (Hymenoptera, Braconidae). *Zoologica Scripta* 28:211–214.

Baskin, J.M. and Baskin, C.C. 1998. *Seeds.* Academic Press, New York.

Bateson, G. 1963. The role of somatic change in evolution. *Evolution* 17:529–539.

Bateson, P. 1988. The active role of behaviour in evolution. In: *Evolutionary Processes and Metaphors*, M.W. Ho and S.W. Fox (eds.). John Wiley and Sons, London, pp. 191–207.

Bateson, P. 1990. Choice, preference and selection. In: *Interpretation and Explanation in the Study of Animal Behavior. Volume 1: Interpretation, Intentionality and Communication*, M. Bekoff and D. Jamieson (eds.). Westview Press, Boulder, pp. 149–156.

Bateson, P. (ed.) 1991. *The Development and Integration of Behavior: Essays in Honor of Robert Hinde.* Cambridge University Press, Cambridge.

Bateson, W. 1894. *Materials for the Study of Variation Treated with Especial Regard to Discontinuity in the Origin of Species.* Macmillan, London.

Batra, L.R. and Batra, S.W.T. 1985. Floral mimicry induced by mummy-berry fungus exploits host's pollinators as vectors. *Science* 228:1011–1013.

Bauchau, V. and Chaline, J. 1987. Variabilité de la troisième molaire supérieure de *Clethrionomys glareolus* (Rodentia, Arvicolidae) et sa signification évolutive. *Mammalia* 51:587–598.

Baur, A., Chalwatzis, N., Buschinger, A. and Zimmermann, F.K. 1995. Mitochondrial DNA sequences reveal close relationships between social parasitic ants and their host species. *Current Genetics* 28:242–247.

Bawa, K.S. 1980. Evolution of dioecy in flowering plants. *Annual Review of Ecology and Systematics* 11:15–39.

Bayer, M.M., Todd, C.D., Hoyle, J.E. and Wilson, J.F.B. 1997. Wave-related abrasion induces formation of extended spines in a marine bryozoan. *Proceedings of the Royal Society of London B* 264:1605–1611.

Bazzaz, F.A. 1991. Habitat selection in plants. *American Naturalist* 137:S116–S130.

Beach, F.A. 1961. *Hormones and Behavior.* Cooper Square Publishers, New York.

Beadle, G.W. 1939. Teosinte and the origin of maize. *Journal of Heredity* 30:245–247.

Beani, L. 1996. Lek-like courtship in paper wasps: "a prolonged, delicate, and troublesome affair." In: *Natural History and Evolution of Paper-Wasps*, S. Turrillazzi and M.J. West-Eberhard (eds.). Oxford University Press, Oxford, pp. 113–125.

Beani, L. and Lorenzi, M.C. 1992. Different tactics of mate searching by *Polistes biglumis bimaculatus* males (Hymenoptera Vespidae). *Ethology, Ecology and Evolution* (Special Issue) 2:43–45.

Beani, L. and Turillazzi, S. 1999. Stripes display in hover-wasps (Vespidae: Stenogastrinae): a socially costly status badge. *Animal Behaviour* 57:1233–1239.

Beebe, W. 1949. Insect migration at Rancho Grande in north-central Venezuela. General account. *Zoologica* 34:107–110.

Beeman, R.W. 1987. A homoeotic gene cluster in the red flour beetle. *Nature* 327:247–249.

Beetsma, J. 1979. The process of queen-worker differentiation in the honeybee. *Bee World* 60:24–39.

Bego, L.R. and de Camargo, C.A. 1984. On the occurrence of giant males in *Nannotrigona* (*Scaptotrigona*) *postica* Lateille (Hymenoptera, Apidae, Meliponinae). *Boletim de Zoologia*, 8:11–16.

Begun, D.J and Collins, J.P. 1992. Biochemical plasticity in the Arizona tiger salamander (*Ambystoma tigrinum nebulosum*). *Journal of Heredity* 83:224–227.

Bell, G. 1974. The reduction of morphological variation in natural populations of smooth newt larvae. *Journal of Animal Ecology* 43:115–128.

Bell, G. 1978. Further observations on the fate of morphological variation in a population of smooth newt larvae (*Triturus vulgaris*). *Journal of Zoology, London* 185:511–518.

Bell, G. 1989. A comparative method. *American Naturalist* 133:553–571.

Bell, G.A. 1996. Review of epigenetic inheritance and evolution (by E. Jablonka and M.J. Lamb). *Trends in Ecology and Evolution* 11:266–267.

Bell, M.A. 1976. Evolution of phenotypic diversity in *Gasterosteus aculeatus* superspecies on the Pacific coast of North America. *Systematic Zoology* 25:211–227.

Bell, M.A. 1981. Lateral plate polymorphism and ontogeny of the complete plate morph of threespine sticklebacks (*Gasterosteus aculeatus*). *Evolution* 35:67–74.

Bell, M.A. 1982. Differentiation of adjacent stream populations of threespine sticklebacks. *Evolution* 36:189–199.

Bell, M.A. 1984. Evolutionary phenetics and genetics: the threespine stickleback, *Gasterosteus aculeatus* and related species. In: *Evolutionary Genetics of Fishes*, B.J. Turner (ed.). Plenum Press, New York, pp. 431–528.

Bell, M.A. 1987. Interacting evolutionary constraints in pelvic reduction of threespine sticklebacks, *Gasterosteus aculeatus* (Pisces, Gasterosteidae). *Biological Journal of the Linnean Society* 31:347–382.

Bell, M.A. 1988. Stickleback fishes: bridging the gap between population biology and paleobiology. *Trends in Ecology and Evolution* 3:320–325.

Bell, M.A. and Andrews, C.A. 1997. Evolutionary consequences of postglacial colonization of fresh water by primitively anadromous fishes. In: *Evolutionary Ecology of Freshwater Fishes*, B. Streit, T. Stadler and C.M. Lively (eds.). Birkhauser Verlag, Basel, pp. 323–363.

Bell, M.A. and Foster, S.A. 1994a. *The Evolutionary Biology of the Threespine Stickleback.* Oxford University Press, New York.

Bell, M.A. and Foster, S.A. 1994b. Introduction to the evolutionary biology of the threespine stickleback. In: *The Evolutionary Biology of the Threespine Stickleback*, M.A. Bell and S.A. Foster (eds.). Oxford University Press, New York, pp. 1–27.

Bell, M.A., Baumgartner, J.V. and Olson, E.C. 1985. Patterns of temporal change in single morphological characters of a Miocene stickleback fish. *Paleobiology* 11:258–271.

Bell, P.R. 1959. The movement of plants in response to light. In: *Darwin's Biological*

Work, P.R. Bell (ed.). Cambridge University Press, Cambridge, pp. 1–49.

Bell, W.J. 1991. *Searching Behavior: The Behavioral Ecology of Finding Resources.* Chapman and Hall, London.

Bell, W.J. and Tortorici, C. 1987. Genetic and non-genetic control of search duration in adults of two morphs of *Drosophila melanogaster. Journal of Physiology* 33:51–54.

Bengston, A.S. 1966. Någraiakttagelser rörande pirattendenser hos tärnor och trutar. *Fauna och Flora* 1966:24–30.

Bennett, A.F. and Huey, R.B. 1990. Studying the evolution of physiological performance. *Oxford Surveys in Evolutionary Biology* 7:252–284.

Bennett, M.J., Choe, S. and Eisenberg, D. 1994. Domain swapping: entangling alliances between proteins. *Proceedings of the National Academy of Sciences USA* 91:3127–3131.

Bennett, M.K. and Scheller, R.H. 1993. The molecular machinery for secretion is conserved from yeast to neurons. *Proceedings of the National Academy of Sciences USA* 90:2559–2563.

Benton, M.J. 1990. The causes of the diversification of life. In: *Major Evolutionary Radiations*, P.D. Taylor and G.P. Larwood (eds.). Systematics Association Special Volume No. 42. Clarendon Press, Oxford, pp. 409–430.

Berg, R.L. 1960. The ecological significance of correlation pleiades. *Evolution* 14:171–180.

Berg, S.J. and Wynne-Edwards, K.E. 2001. Changes in testosterone, cortisol, and estradiol levels in men becoming fathers. *Mayo Clinic Proceedings* 76:582–592.

Berger, P.J., Sanders, E.H., Gardner, P.D. and Negus, N.C. 1977. Phenolic plant compounds functioning as reproductive inhibitors in *Microtus montanus. Science* 195:575.

Berglund, A. 1991. Egg competition in a sex-role reversed pipefish: subdominant females trade reproduction for growth. *Evolution* 45:770–774.

Bergström, J. 1986. *Opabinia* and *Anomalocaris*, unique Cambrian "arthropods". *Lethaia* 19:241–246.

Berkow, R. and Fletcher, A.J. (eds.) 1987. *The Merck Manual of Diagnosis and Therapy.* Merck Sharp and Dohme Research Laboratories, Rahway, NJ.

Berman, A.L., Kolker, E. and Trifonov, E.N. 1994. Underlying order in protein sequence organization. *Proceedings of the National Academy of Sciences USA* 91:4044–4047.

Bernatchez, L., Vuorinen, J.A., Bodaly, R.A. and Dodson, J.J. 1996. Genetic evidence for reproductive isolation and multiple origins of sympatric trophic ecotypes of whitefish (*Coregonus*). *Evolution* 50:624–635.

Bernays, E.A. 1986. Diet-induced head allometry among foliage-chewing insects and its importance for graminivores. *Science* 231:495–497.

Bernays, E.A. 1991. Evolution of insect morphology in relation to plants. *Philosophical Transactions of the Royal Society of London B* 333:257–264.

Bernays, E.A. 1993. Aversion learning and feeding. In: *Insect Learning: Ecological and Evolutionary Perspectives*, D.R. Papaj and A.C. Lewis (eds.). Chapman and Hall, New York, pp. 1–17.

Bernays, E.A. and Mordue, A.J. 1973. Changes in palp tip sensilla of *Locusta migratoria* in relation to feeding. *Comparative Biochemistry and Physiology* 45A:451–454.

Bernays, E.A. and Wcislo, W.T. 1994. Sensory capabilities, information processing, and resource specialization. *Quarterly Review of Biology* 69:187–204.

Berner, R.A. 1993. Palaeozoic atmospheric CO_2: importance of solar radiation and plant evolution. *Science* 249:1382–1386.

Berner, R.A. and Canfield, D.E. 1989. A new model for atmospheric oxygen over phanerozoic time. *American Journal of Science* 289:333–361.

Bernstein, C. and Bernstein, H. 1991. *Aging, Sex, and DNA Repair.* Academic Press, San Diego.

Berrill, N.J. 1971. *Developmental Biology.* McGraw-Hill, New York.

Berry, R.J. 1963. Epigenetic polymorphism in populations of *Mus musculus. Genetical Research* 4:193–220.

Berry, R.J. 1964. The evolution of an island population of the house mouse. *Evolution* 18:468–483.

Bickel, D.J. 1994. The Australian Sciapodinae (Diptera: Dolichopodidae), with a review of the Oriental and Australasian faunas, and a world conspectus of the subfamily. *Records of the Australian Museum* (Supplement) 21:1–394.

Bigelow, R.S. 1958. Evolution in the field cricket, *Acheta assimilis* Fab. *Canadian Journal of Zoology* 36:139–151.

Blackman, R.L. 1979. Stability and variation in aphid clonal lineages. *Biological Journal of the Linnean Society* 11:259–277.

Blackstone, N.W. and Buss, L.W. 1992. Treatment with 2,4-dinitrophenol mimics ontogenetic and phylogenetic changes in a hydractiniid hydroid. *Proceedings of the*

National Academy of Sciences USA 89:4057–4061.

Blake, C.C.F. 1978. Do genes-in-pieces imply proteins-in-pieces. *Nature* 273:267.

Blake, D.B. 1976. Functional morphology and taxonomy of branch dimorphism in the Paleozoic bryozoan genus *Rhabdomeson*. *Lethaia* 9:169–178.

Blank, J.L. and Freeman, D.A. 1991. Differential reproductive response to short photoperiod in deer mice: role of melatonin. *Journal of Comparative Physiology A* 169:501–506.

Blaustein, A.R., Bekoff, M. and Daniels, T.J. 1986. Kin recognition in vertebrates (excluding primates): empirical evidence. In: *Kin Recognition in Animals*, D.J.C. Fletcher and C.D. Michener (eds.). John Wiley and Sons, New York, pp. 287–332.

Blest, A.D. 1961. The concept of "ritualisation." In: *Current Problems in Animal Behaviour*, W.H. Thorpe and O.L. Zangwill (eds.). Cambridge University Press, Cambridge, pp. 102–124.

Bloch, G., Hefetz, A. and Hartfelder, K. 2000. Ecdysteroid titer, ovary status, and dominance in adult worker and queen bumble bees (*Bombus terrestris*). *Journal of Insect Physiology* 46:1033–1040.

Boag, P.T. 1983. The heritability of external morphology in Darwin's ground finches (*Geospiza*) on Isla Daphne Major, Galapagos. *Evolution* 37:877–894.

Boake, C.R.B. (ed.) 1994. *Quantitative Genetic Studies of Behavioral Evolution*. University of Chicago Press, Chicago.

Boardman, R.S. 1954. Morphologic variation and mode of growth of Devonian trepostomatous bryozoa. *Science* 120:322.

Boardman, R.S. 1960. Trepostomatous Bryozoa of the Hamilton group of New York State. *U.S. Geological Survey Professional Paper* 340:1–87.

Bock, P.E. and Cook, P.L. 1994. Occurrence of three phases of growth with taxonomically distinct zooid morphologies. In: *Biology and Palaeobiology of Bryozoans*, P.J. Hayward, J.S. Ryland and P.D. Taylor (eds.). Olsen and Olsen, Fredensborg, pp. 33–36.

Bock, P.E. and Cook, P.L. 1998. A new species of multiphased *Corbulipora* MacGillivray 1895 (Bryozoa: Cribriomorpha) from southwestern Australia. *Records of the South Australian Museum* 30:63–68.

Bodaly, R.A. 1979. Morphological and ecological divergence within the lake whitefish (*Coregonus clupeaformis*) species complex in Yukon Territory. *Journal of the Fisheries Research Board of Canada* 36:1214–1222.

Boero, F. 1994. Fluctuations and variations in coastal marine environments. *P.S.Z.N.I: Marine Ecology* 15:3–25.

Boero, F. and Bouillon, J. 1987. Inconsistent evolution and paedomorphosis among the hydroids and medusae of the Athecatae/Anthomedusae and the Thecatae/Leptomedusae (Cnidaria, Hydrozoa). In: *Modern Trends in the Systematics, Ecology, and Evolution of Hydroids and Hydromedusae*, J. Bouillon, F. Boero, F. Cicogna and P.F.S. Cornelius (eds.). Clarendon Press, Oxford, pp. 229–250.

Boero, F. and Bouillon, J. 1989a. An evolutionary interpretation of anomalous medusoid stages in the life cycles of some Leptomedusae (Cnidaria). In: *Reproduction, Genetics and Distributions of Marine Organisms*, J.S. Ryland and P.A. Tyler (eds.). Olsen and Olsen, Fredensborg, pp. 37–41.

Boero, F. and Bouillon, J. 1989b. The life cycles of *Octotiara russelli* and *Stomotoca atra* (Cnidaria, Anthomedusae, Pandeidae). *Zoologica Scripta* 18:1–7.

Boero, F. and Sará, M. 1987. Motile sexual stages and evolution of Leptomedusae (Cnidaria). *Bollettino di Zoologia* 54:131–139.

Boero, F., Bouillon, J. and Piraino, S. 1992. On the origins and evolution of hydromedusan life cycles (Cnidaria, Hydrozoa). In: *Sex Origin and Evolution*, R. Dallai (ed.). Selected Symposia and Monographs U.Z.I., 6 Mucchi, Modena, pp. 59–68.

Bohart, R.M. and Menke, A.S. 1976. *Sphecid Wasps of the World*. University of California Press, Berkeley.

Bohm, D. 1980. *Wholeness and the Implicate Order*. Ark Paperbacks, New York.

Bohr, N. 1927 [1934]. *Atomic Theory and the Description of Nature*. Macmillan, New York.

Boidron-Metairon, I. 1988. Morphological plasticity in laboratory-reared echinoplutei of *Dendraster excentricus* (Eschscholtz) and *Lytechinus variegatus* (Lamarck) in response to food conditions. *Journal of Experimental Marine Biology and Ecology* 119:31–41.

Bolander, F.F. 1994. *Molecular Endocrinology*. Academic Press, New York.

Bolton, B. 1988. A new socially parasitic *Myrmica*, with a reassessment of the genus (Hymenoptera: Formicidae). *Systematic Entomology* 13:1–11.

Bonabeau, E., Theraulaz, G., Deneubourg, J.-L., Aron, S. and Camazine, S. 1997. Self-organization in social insects. *Trends in Ecology and Evolution* 12:188–193.

Bonner, J.T. 1965. *Size and Cycle*. Princeton University Press, Princeton.

Bonner, J.T. 1974. *On Development*. Harvard University Press, Cambridge.

Bonner, J.T. 1980. *The Evolution of Culture in Animals*. Princeton University Press, Princeton.

Bonner, J.T. (ed.) 1982. *Evolution and Development*. Springer-Verlag, Berlin.

Bonner, J.T. 1988. *The Evolution of Complexity*. Princeton University Press, Princeton.

Bonner, J.T. 1993. *Life Cycles: Reflections of an Evolutionary Biologist*. Princeton University Press, Princeton.

Bonner, J.T. 1996. *Sixty Years of Biology: Essays on Evolution and Development*. Princeton University Press, Princeton.

Bonner, J.T. 2000. *First Signals: The Evolution of Multicellular Development*. Princeton University Press, Princeton.

Bonnetti, A.M. and Kerr, W.E. 1985. Sex determination in bees. XX. Study of gene action in *Melipona marginata* and *Melipona compressipes* based on morphometric analysis. *Revista Brasileira de Genética* 7(4):629–638.

Bookstein, F.L. 1994. Can biometrical shape be a homologous character? In: *Homology: The Hierarchical Basis of Comparative Biology*, B.K. Hall (ed.). Academic Press, San Diego, pp. 197–227.

Borland, A.M. and Griffiths, H. 1996. Variations in the phases of crassulacean acid metabolism and regulation of carboxylation patterns determined by carbon-isotope-discrimination techniques. In: *Crassulacean Acid Metabolism: Biochemistry, Ecophysiology and Evolution*. K. Winter and J.A.C. Smith (eds.). Springer, New York, pp. 230–249.

Borowsky, R.L. 1987a. Agnostic behavior and social inhibition of maturation in fishes of the genus *Xiphophorus* (Poeciliidae). *Copeia* 3:792–796.

Borowsky, R.L. 1987b. Genetic polymorphism in adult male size in *Xiphophorus variatus* (Atheriniformes: Poeciliidae). *Copeia* 3:782–787.

Bossomaier, T. and Snoad, N. 1995. Evolution and modularity in neural networks. In: *Proceedings IEEE Workshop on Non-linear Signal and Image Processing*, I. Pitas (ed.). IEEE, Thessaloniki, pp. 289–292.

Bouillon, J., Boero, F. and Fraschetti, S. 1991. The life cycle of *Laodicea indica* Browne 1906 (Laodiceidae, Hydromedusae, Cnidaria). *Hydrobiologia* 216/217:151–157.

Bourke, A.F.G. 1988. Worker reproduction in the higher eusocial Hymenoptera. *Quarterly Review of Biology* 63:291–311.

Bourke, A.F.G. and Franks, N.R. 1987. Evolution of social parasites in the Leptothoricine ants. In: *Chemistry and Biology of Social Insects*, J. Eder and H. Rembold (eds.). Verlag J. Peperny, Munchen, pp. 37–42.

Bourke, A.F.G. and Franks, N.R. 1991. Alternative adaptations, sympatric speciation and the evolution of parasitic, inquiline ants. *Biological Journal of the Linnean Society* 43:157–178.

Bowers, M.D. 1990. Recycling plant natural products for insect defense. In: *Insect Defenses*, D.L. Evans and J.O. Schmidt (eds.). State University of New York Press, Albany, pp. 353–386.

Bowlby, J. 1969. *Attachment*. Hogarth Press, London.

Boyd, R. and Richerson, P.J. 1983. Why is culture adaptive? *Quarterly Review of Biology* 58:209–214.

Bradshaw, A.D. 1965. Evolutionary significance of phenotypic plasticity in plants. *Advances in Genetics* 13:115–155.

Bradshaw, W.E. 1973. Homeostasis and polymorphism in vernal development of *Chaoborus americanus*. *Ecology* 54:1247–1259.

Bradshaw, W.E. and Holzapfel, C.M. 1986. Geography of density-dependent selection in pitcher-plant mosquitoes. In: *The Evolution of Insect Life Cycles*, F. Taylor and R. Karban (eds.). Springer-Verlag, New York, pp. 48–65.

Bradshaw, W.E. and Lounibos, L.P. 1977. Evolution of dormancy and its photoperiodic control in pitcher-plant mosquitoes. *Evolution* 31:546–567.

Brakefield, P.M. 1987. Tropical dry and wet season polyphenism in the butterfly *Melanitis leda* (Satyrinae): phenotypic plasticity and climatic correlates. *Biological Journal of the Linnean Society* 31:175–191.

Brakefield, P.M. and Larson, T.B. 1984. The evolutionary significance of dry and wet season forms in some tropical butterflies. *Biological Journal of the Linnean Society* 22:1–12.

Brakefield, P.M. and Reitsma, N. 1991. Phenotypic plasticity, seasonal climate and the population biology of *Bicyclus* butterflies in Malawi. *Ecological Entomology* 10:291–303.

Bramble, D.M. 1989. Cranial specialization and locomotor habit in the Lagomorpha. *American Zoologist* 29:303–317.

Brandon, R.N. 1985. Adaptation explanations: are adaptations for the good of replicators or interactors? In: *Evolution at a Crossroads*, D.J. Depew and B.H. Weber (eds.). Massachusettes Institute of Technology Press, Cambridge, pp. 81–96.

Brandon, R.N. 1990. *Adaptation and Environment*. Princeton University Press, Princeton.

Brandon, R.N. 1992. Environment. In: *Keywords in Evolutionary Biology*, E.F. Keller and E.A. Lloyd (eds.). Harvard University Press, Cambridge, pp. 83–86.

Brandon, R.N. 1996. *Concepts and Methods in Evolutionary Biology*. Cambridge University Press, Cambridge.

Breed, M.D. 1998. Recognition pheromones of the honey bee. *BioScience* 48:463–470.

Breed, M.D. and Page, R.E., Jr. (eds.) 1989. *The Genetics of Social Evolution*. Westview Press, Boulder.

Breeland, S.G. and Pickard, E. 1967. Field observations on twenty-eight broods of floodwater mosquitoes resulting from controlled floodings of a natural habitat in the Tennessee Valley. *Mosquito News* 27:343–358.

Brenner, S. 1988. The molecular evolution of genes and proteins: a tale of two serines. *Nature* 334:528–530.

Briceño, R.D. and Eberhard, W.G. 1987. Genetic and environmental effects on wing polymorphisms in two tropical earwigs (Dermaptera: Labiidae). *Oecologia* 74:253–255.

Bridges, C.B. 1935. Salivary gland chromosome maps. *Journal of Heredity* 26:60–64.

Briggs, D. and Walters, S.M. 1988. *Plant Variation and Evolution*, 2nd ed. Cambridge University Press, Cambridge.

Bright, K.L., Bernays, E.A. and Moran, V.C. 1994. Foraging patterns and dietary mixing in the field by the generalist grasshopper *Brachystola magna* (Orthoptera: Acrididae). *Journal of Insect Behavior* 7:779–793.

Brink, A. 1962. Phase change in higher plants and somatic cell heredity. *Quarterly Review of Biology* 37:1–22.

Britten, R.J. 1998. Underlying assumptions of developmental models. *Proceedings of the National Academy of Sciences USA* 95:9372–9377.

Britten, R.J. and Davidson, E.H. 1969. Gene regulation for higher cells: a theory. *Science* 165:349–357.

Britten, R.J. and Davidson, E.H. 1971. Repetitive and nonrepetitive DNA sequences and a speculation on the origins of evolutionary novelty. *Quarterly Review of Biology* 46:111.

Brockmann, H.J. 1976. The control of nesting behavior in the great golden digger wasp, *Sphex ichneumoneus* (Hymenoptera, Sphecidae). Ph.D. dissertation, University of Wisconsin, Madison.

Brockmann, H.J. 1979. Nest-site selection in the great golden digger wasp, *Sphex ichneumoneus* L. (Sphecidae). *Ecological Entomology* 4:211–224.

Brockmann, H.J. 1980a. Diversity in the nesting behavior of mud-daubers (*Trypoxylon politum* Say; Sphecidae). *Florida Entomologist* 63:53–64.

Brockmann, H.J. 1980b. The control of nest depth in a digger wasp (*Sphex ichneumoneus* L.). *Animal Behaviour* 28:426–445.

Brockmann, H.J. 1985a. Provisioning behavior of the great golden digger wasp, *Sphex ichneumoneus* (L.) (Sphecidae). *Journal of the Kansas Entomological Society* 58:631–655.

Brockmann, H.J. 1985b. Tool use in digger wasps (Hymenoptera: Sphecinae). *Psyche* 92:309–328.

Brockmann, H.J. 1986. Decision making in a variable environment: lessons from insects. In: *Behavioral Ecology and Population Biology*, L.C. Drickamer (ed.). Privat, I.E.C., Toulouse, pp. 95–111.

Brockmann, H.J. and Barnard, C.J. 1979. Kleptoparasitism in birds. *Animal Behaviour* 27:487–514.

Brockmann, H.J. and Dawkins, R. 1979. Joint nesting in a digger wasp as an evolutionarily stable preadaptation to social life. *Behaviour* 71:203–245.

Brockmann, H.J., Grafen, A. and Dawkins, R. 1979. Evolutionarily stable nesting strategy in a digger wasp. *Journal of Theoretical Biology* 77:473–496.

Bromham, L., Rambaut, A., Fortey, R., Cooper, A. and Penny, D. 1998. Testing the Cambrian explosion hypothesis by using a molecular dating technique. *Proceedings of the National Academy of Sciences USA* 95:12386–12389.

Brönmark, C. and Pettersson, L.B. 1994. Chemical cues from piscivores induce a change in morphology in crucian carp. *OIKOS* 70:396–402.

Brooks, D.R. and McLennan, D.A. 1991. *Phylogeny, Ecology, and Behavior*. University of Chicago Press, Chicago.

Brothers, D.J. 1972. Biology and immature stages of *Pseudomethoca f. frigida*, with notes on other species (Hymenoptera: Mutillidae). *University of Kansas Science Bulletin* 50:1–38.

Brower, A.V.Z. 1994. Rapid morphological radiation and convergence among races of the butterfly *Heliconius erato* inferred from patterns of mitochondrial DNA evolution. *Proceedings of the National Academy of Sciences USA* 91:6491–6495.

Brown, G.E. and Godin, J.-G.J. 1999. Who dares, learns: chemical inspection behaviour and acquired predator recognition in a characin fish. *Animal Behaviour* 57:475–481.

Brown, J.L. 1964. The integration of agonistic behavior in the Steller's jay *Cyanocitta stelleri* (Gmelin). *University of California Publications in Zoology* 60:223–328.

Brown, J.L. 1975. *The Evolution of Behavior.* W.W. Norton, New York.

Brown, J.L. 1987. *Helping and Communal Breeding in Birds: Ecology and Evolution.* Princeton University Press, Princeton.

Brown, J.L. and Brown, E.R. 1980. Reciprocal aid-giving in a communal bird. *Zeitschrift für Tierpsychologie* 53:313–324.

Brown, J.L. and Brown, E.R. 1990. Mexican jays: uncooperative breeding. In: *Cooperative Breeding in Birds: Long-term Studies of Ecology and Behavior*, P.B. Stacey and W.D. Koenig (eds.). Cambridge University Press, Cambridge, pp. 269–287.

Brown, J.L. and Pimm, S.L. 1985. The origin of helping: the role of variability in reproductive potential. *Journal of Theoretical Biology* 112:465–477.

Brown, W.D., Wideman, J., Andrade, M.C.B., Mason, A.C. and Gwynne, D.T. 1996. Female choice for an indicator of male size in the song of the black-horned tree cricket, *Oecanthus nigricornis* (Orthoptera: Gryllidae: Oecanthinae). *Evolution* 50:2400–2411.

Brown, W.L., Jr. 1987. Punctuated equilibrium excused: the original examples fail to support it. *Biological Journal of the Linnean Society* 31:383–404.

Brown, W.L., Jr. and Wilson, E.O. 1956. Character displacement. *Systematic Zoology* 5:49–64.

Brunnert, H. 1967. Veranderung im Spektrum der Hämolymphproteine bei Weibchen und Arbeiterinnen der Waldameise *F. polyctena* Foerst, wahrend der Metamorphose. *Zeitschrift für Naturforschung* 22b:336–339.

Brusca, R.C. 2000. Unraveling the history of arthropod biodiversification. *Annals of the Missouri Botanical Garden* 87:3–25.

Brusca, R.C. and Brusca, G.J. 1990. *Invertebrates.* Sinauer, Sunderland, MA.

Bryan, J.E. and Larkin, P.A. 1972. Food specialization by individual trout. *Journal of the Fisheries Research Board of Canada* 29:1615–1624.

Bull, J.J. 1983. *Evolution of Sex Determining Mechanisms.* Benjamin/Cummings, Menlo Park, CA.

Bull, J.J. 1987. Evolution of phenotypic variance. *Evolution* 41:303–315.

Bull, J.J. and Vogt, R.C. 1979. Temperature-dependent sex determination in turtles. *Science* 206:1186–1188.

Bull, J.J., Vogt, R.C. and Bulmer, M.G. 1982a. Heritability of sex ratio in turtles with environmental sex determination. *Evolution* 36:333–341.

Bull, J.J., Vogt, R.C. and McCoy, C.J. 1982b. Sex determining temperatures in emydid turtles: a geographic comparison. *Evolution* 36:326–332.

Bulmer, M.G. and Bull, J.J. 1982. Models of polygenic sex determination and sex ratio evolution. *Evolution* 36:13–26.

Büning, J. 1994. *The Insect Ovary.* Chapman and Hall, New York.

Burian, R. 1986. On integrating the study of evolution and of development. In: *Integrating Scientific Disciplines*, W. Bechtel (ed.). Martinus Nijhoff Publishers, Dordrecht, pp. 209–228.

Burian, R.M. 1997. On conflicts between genetic and developmental viewpoints and their attempted resolution in molecular biology. In: *Structures and Norms in Science*, M.L. Dalla Chiara (ed.). Kluwer Academic Publishers, The Hague, pp. 243–264.

Burley, N. 1988. The differential-allocation hypothesis: an experimental test. *American Naturalist* 132:611–628.

Burt, A. 2000. Perspective: sex, recombination, and the efficacy of selection—was Weismann right? *Evolution* 54:337–351.

Burton, F.D. 1972. The integration of biology and behavior in the socialization of *Macaca sylvana* of Gibraltar. In: *Primate Socialization*, F.E. Poirier (ed.). Random House, New York, pp. 29–62.

Buschinger, A. 1965. *Leptothorax (Mychothorax) kutteri* n. sp., eine sozialparasitische Ameise (Hymenoptera, Formicidae). *Insectes Sociaux* 12:327–334.

Buschinger, A. 1970. Neue vorstellungen zur evolution des sozialparasitismum und der dulosis bei ameisen (Hym., Formicidae). *Biologisches Zentralblatt* 88:273–299.

Buschinger, A. 1972. Kreuzung zweier sozialparasitischer Ameisenarten, *Doronomyrmex pacis* Kutter und *Leptothorax kutteri* Buschinger (Hym., Formicidae). *Zoologischer Anzeiger* 189: 169–179.

Buschinger, A. 1975. Eine genetische komponente im polymorphismus der dulotischen Ameise *Harpagoxenus sublaevis*. *Naturwissenschaften* 62:239.

Buschinger, A. 1978. Queen polymorphism in ants. In: *Réunion Scientifique de la Section Française*. International Union for the Study of Social Insects, Besancon, pp. 12–22.

Buschinger, A. 1981. Biological and systematic relationships of social parasitic leptothoracini from Europe and North America. In: *Biosystematics of Social Insects*, P.E. Howse and J.-L. Clément (eds.). Academic Press, New York, pp. 211–222.

Buschinger, A. 1986. Evolution of social parasitism in ants. *Trends in Ecology and Evolution* 1:155–160.

Buschinger, A. 1989. Evolution, speciation, and inbreeding in the parasitic ant genus *Epimyrma* (Hymenoptera, Formicidae). *Journal of Evolutionary Biology* 2:265–283.

Buschinger, A. 1990a. Regulation of worker and queen formation in ants with special reference to reproduction and colony development. In: *Social Insects: An Evolutionary Approach to Castes and Reproduction*, W. Engels (ed.). Springer-Verlag, Berlin, pp. 37–57.

Buschinger, A. 1990b. Sympatric speciation and radiative evolution of socially parasitic ants—heretic hypotheses and their factual background. *Zeitschrift für Zooligische Systematik und Evolutionforschung* 28:241–260.

Buschinger, A. and Fischer, K. 1991. Hybridization of chromosome-polymorphic populations of the inquiline ant, *Doronomyrmex kutteri*. *Insectes Sociaux* 38:95–103.

Buschinger, A. and Francoeur, A. 1991. Queen polymorphism and functional monogyny in the ant *Leptothorax sphagnicolus*. *Psyche* 98:119–133.

Buschinger, A. and Heinze, J. 1992. Polymorphism of female reproductives in ants. In: *Biology and Evolution of Social Insects*, J. Billen (ed.). Leuven University Press, Leuven, pp. 11–23.

Buschinger, A. and Stoewesand H. 1971. Teratologische untersuchungen an ameisen (Hymenoptera: Formicidae). *Beiträge zur Entomologie.* 21:211–241.

Buschinger, A. and Winter, U. 1975. Der polymorphismus der sklavenhaltenden ameise *Harpagoxenus sublaevis* (Nyl.). *Insectes Sociaux* 22:333–362.

Bush, G.L. 1969. Sympatric host race formation and speciation in frugivorous flies of the genus *Rhagoletis* (Diptera, Tephritidae). *Evolution* 23:237–251.

Bush, G.L. 1973. The mechanism of sympatric host race formation in the true fruit flies (Tephritidae). In: *Genetic Mechanisms of Speciation in Insects*. M.J.D. White (ed.). D. Reidel Publishing Company, New York.

Bush, G.L. 1982. Goldschmidt's follies. *Paleobiology* 8:463–469.

Bush, G.L. 1992. Host race formation and sympatric speciation in *Rhagoletis* fruit flies (Diptera: Tephritidae). *Psyche* 99:335–358.

Bush, G.L. 1994. Sympatric speciation in animals: new wine in old bottles. *Trends in Ecology and Evolution* 9:285–288.

Buss, L.W. 1985. The uniqueness of the individual revisited. In: *Population Biology and Evolution of Clonal Organisms*, J.B.C. Jackson, L.W. Buss and R.E. Cook (eds.).

Yale University Press, New Haven, pp. 467–505.

Buss, L.W. 1987. *The Evolution of Individuality*. Princeton University Press, Princeton.

Byrne, R. and Whiten, A. (eds.) 1988. *Machiavellian Intelligence*. Oxford University Press, Oxford.

Cade, W.H. 1979. The evolution of alternative male reproductive strategies in field crickets. In: *Sexual Selection and Reproductive Competition in Insects*, M.S. Blum and N.A. Blum (eds.). Academic Press, New York, pp. 343–379.

Cade, W.H. 1981. Alternative male strategies: genetic differences in crickets. *Science* 212:563–564.

Cade, W.H. 1984. Genetic variation underlying sexual behavior and reproduction. *American Zoologist* 24:355–366.

Cairns, R.B., Gariépy, J.-L. and Hood, K.E. 1990. Development, microevolution, and social behavior. *Psychological Review* 97:49–65.

Caisse, M. and Antonovics, J. 1978. Evolution in closely adjacent plant populations. *Heredity* 40:371–384.

Calderone, N.W. and Page, R.E., Jr. 1988. Genotypic variability in age polyethism and task specialization in the honey bee, *Apis mellifera* (Hymenoptera: Apidae). *Behavioral Ecology and Sociobiology* 22:17–25.

Calderone, N.W. and Page, R.E., Jr. 1992. Effects of interactions among genotypically diverse nestmates on task specialization by foraging honey bees (*Apis mellifera*). *Behavioral Ecology and Sociobiology* 30:219–226.

Caldwell, J.P. 1997. Pair-bonding in spotted poison frogs. *Nature* 385:211.

Caldwell, R.L. and Dingle, H. 1976. Stomatopods. *Scientific American* 234:81–89.

Cale, G.H., Jr. and Rothenbuhler, W.C. 1975. Genetics and breeding of the honey bee. In: *The Hive and the Honey Bee*, Dadant and Sons (eds.). Dadant and Sons, Hamilton, IL, pp. 157–184.

Calvin, W.H. 1983. *The Throwing Madonna*. McGraw-Hill, New York.

Camazine, S., Visscher, P.K., Finley, J. and Vetter, R.S. 1999. House-hunting by honey bee swarms: collective decisions and individual behaviors. *Insectes Sociaux* 46:348–360.

Cameron, S.A. 1985. Brood care by male bumble bees. *Proceedings of the National Academy of Sciences USA* 82:6371–6373.

Cameron, S.A. 1986. Brood care by males of *Polistes major* (Hymenoptera: Vespidae). *Journal of the Kansas Entomological Society* 59:183–185.

Cammaerts, M.C., Cammaerts, R. and Bruge, H. 1987. Some physiological information on the microgyne form of *Myrmica rubra* L. (Hymenoptera: Formicidae). *Annales de la Société Royal Zoologique de Belgique* 117:147–158.

Campbell, H.D., Schimansky, T., Claudianos, C., Ozsarac, N., Kasprzak, A.B., Cotsell, J.N., Young, I.G., de Couet, H.G. and Gabor Miklos, G.L. 1993. The *Drosophila melanogaster* flightless-I gene involved in gastrulation and muscle degeneration encodes gelsolin-like and leucine-rich repeat domains and is conserved in *Caenorhabditis elegans* and humans. *Proceedings of the National Academy of Sciences USA* 90:11386–11390.

Cane, T.C., Culver, D.C. and Jones, R.T. 1992. Genetic structure of morphologically differentiated populations of the amphipod *Gammarus minus*. *Evolution* 46:272–278.

Cannon, W.B. 1932. *The Wisdom of the Body*. Norton, New York.

Caporale, L.H. 1984. Is there a higher level genetic code that directs evolution? *Molecular and Cellular Biochemistry* 64:5–13.

Caporale, L.H. 1995. Chemical ecology: a view from the pharmaceutical industry. *Proceedings of the National Academy of Sciences USA* 92:75–82.

Caporale, L.H. 1999a. Chance favors the prepared genome. *Annals of the New York Academy of Sciences* 870:1–21.

Caporale, L.H. (ed.) 1999b. Molecular strategies in biological evolution. *Annals of the New York Academy of Sciences* 870:i–xiv, 1–434.

Caraco, T. 1983. White crowned sparrows (*Zonotricha leucophrys*): foraging preferences in a risky environment. *Behavioral Ecology and Sociobiology* 12:63–69.

Carle-Urioste, J.C., Brendel, V. and Walbot, V. 1997. A combinatorial role for exon, intron and splice site sequences in splicing in maize. *Plant Journal* 11:1253–1263.

Carpenter, J.M. 1989. Testing scenarios: wasp social behavior. *Cladistics* 5:131–144.

Carpenter, J.M. 1991. Phylogenetic relationships and the origin of social behaviour in the Vespidae. In: *The Social Biology of Wasps*, K.G. Ross and R.W. Matthews (eds.). Cornell University Press, Ithaca, pp. 7–32.

Carpenter, J.M. 1997. Phylogenetic relationships among European *Polistes* and the evolution of social parasitism (Hymenoptera: Vespidae, Polistinae). *Mémoires du Muséum National d'Histoire Naturelle* 173:135–161.

Carpenter, J.M., Strassmann, J.E., Turillazzi, S., Hughes, C.R., Solis, C.R. and Cervo, R. 1993. Phylogenetic relationships among

paper wasp social parasites and their hosts (Hymenoptera: Vespidae; Polistinae). *Cladistics* 9:129–146.

Carré, D. and Sardet, C. 1984. Fertilization and early development in *Beroe ovata*. *Developmental Biology* 105:188–195.

Carroll, R.L. 1997. *Patterns and Processes of Vertebrate Evolution*. Cambridge University Press, Cambridge.

Carroll, S.P. and Corneli, P.S. 1995. Divergence in male mating tactics between two populations of the soapberry bug: II. Genetic change and the evolution of a plastic reaction norm in a variable social environment. *Behavioral Ecology* 6:46–56.

Carson, H.L. 1968. The population flush and its genetic consequences. In: *Population Biology and Evolution*, R.C. Lewontin (ed.). Syracuse University Press, Syracuse, pp. 123–137.

Carson, H.L. 1978. Speciation and sexual selection in Hawaiian *Drosophila*. In: *Ecological Genetics: The Interface*, P.F. Brussard (ed.). Springer-Verlag, Berlin, pp. 93–107.

Carson, H.L. 1985. Speciation as a major reorganization of polygenic balances. In: *Mechanisms of Speciation*, C. Barigozzi (ed.). Alan R. Liss, New York, pp. 411–433.

Carson, H.L. 1987a. Colonization and speciation. In: *Colonization, Succession and Stability*, A.J. Gray, M.J. Crawley and P.J. Edwards (eds.). Blackwell, Oxford, pp. 187–206.

Carson, H.L. 1987b. The contribution of sexual behavior to Darwinian fitness. *Behavior Genetics* 17:597–611.

Carson, H.L. 1989. Sympatric pest. *Nature* 338:304.

Carson, H.L. 1990. Increased genetic variance after a population bottleneck. *Trends in Ecology and Evolution* 5:228–230.

Carson, H.L. 1992a. Genetic change after colonization. *GeoJournal* 28(2):297–302.

Carson, H.L. 1992b. The Galapagos that were. *Nature* 355:202–203.

Carson, H.L. and Kaneshiro, K.Y. 1976. *Drosophila* of Hawaii: systematics and ecological genetics. *Annual Review of Ecology and Systematics* 7:311–345.

Carson, H.L. and Templeton, A.R. 1984. Genetic revolutions in relation to speciation phenomena: the founding of new populations. *Annual Review of Ecology and Systematics* 15:97–131.

Cartmill, M. 1994. A critique of homology as a morphological concept. *American Journal of Physical Anthropology* 94:115–123.

Case, T.J. 1997. Natural selection out on a limb. *Nature* 387:15–16.

Cassill, D.L. and Tschinkel, W.R. 1995. Allocation of liquid food to larvae via trophallaxis in colonies of the fire ant, *Solenopsis invicta*. *Animal Behaviour* 50:801–813.

Cassill, D.L. and Tschinkel, W.R. 1996. A duration constant for worker-to-larva trophallaxis in fire ants. *Insectes Sociaux* 43:149–166.

Cavalier-Smith, T. 1980. R- and K-tactics in the evolution of protist developmental systems: cell and genome size, phenotype diversifying selection, and cell cycle patterns. *BioSystems* 12:43–59.

Cavalier-Smith, T. 1991. Intron phylogeny: a new hypothesis. *Trends in Genetics* 7:145–148.

Cervo, R. and Dani, F.R. 1996. Social parasitism and its evolution in *Polistes*. In: *Natural History and Evolution of Paper-Wasps*, S. Turillazzi and M.J. West-Eberhard (eds.). Oxford University Press, Oxford, pp. 98–112.

Cervo, R., Lorenzi, M.C. and Turillazzi, S. 1990. *Sulcopolistes atrimandibularis*, social parasite and predator of an Alpine *Polistes* (Hymenoptera, Vespidae). *Ethology* 86:71–78.

Chagnon, N.A. and Irons, W. 1979. *Evolutionary Biology and Human Social Behavior*. Duxbury Press, North Scituate, MA.

Chailakhyan, M.K. and Khryanin, V.N. 1980. Hormonal regulation of sex expression in plants. In: *Plant Growth Substances 1979*, F. Skoog (ed.). Springer-Verlag, Berlin, pp. 331–344.

Chaline, J. 1987. Arvicolid data (Arvicolidae, Rodentia) and evolutionary concepts. *Evolutionary Biology*, 21, 237–310.

Chan, T.-Y. 1987. The role of male competition and female choice in the mating success of a lek-breeding Southern African cichlid fish *Pseudocrenilabrus philander* (Pices: Cichlidae). Ph.D. thesis, Rhodes University, Grahamstown, South Africa.

Changeux, J.P., Heidmann T. and Patte, P. 1984. Learning by selection. In: *The Biology of Learning*, P. Marler and H.S. Terrace (eds.). Springer-Verlag, New York, pp. 115–148.

Chapin, F.S., III, Autumn, K. and Pugnaire, F. 1993. *Evolution* of suites of traits in response to environmental stress. *American Naturalist* 142:S78–S92.

Chaplin, J.A., Havel, J.E. and Hebert, P.D.N. 1994. Sex and ostracods. *Trends in Ecology and Evolution* 9:435–439.

Chapman, R.F. 1998. *The Insects*. Cambridge University Press, Cambridge.

Charlesworth, B. 1982. Hopeful monster cannot fly. *Paleobiology* 8:469–474.

Charlesworth, B. 1990. The evolutionary genetics of adaptation. In: *Evolutionary Innovations*, M.H. Nitecki (ed.). University of Chicago Press, Chicago, pp. 47–70.

Charlesworth, B., Lande, R. and Slatkin, M. 1982. A neo-Darwinian commentary on macroevolution. *Evolution* 36:474–498.

Charlesworth, D. and Charlesworth, B. 1975. Theoretical genetics of Batesian mimicry, 1: single-locus models. *Journal of Theoretical Biology* 55:282–303.

Charlesworth, D. and Charlesworth, B. 1976a. Theoretical genetics of Batesian mimicry, II: evolution of supergenes. *Journal of Theoretical Biology* 55:305–324.

Charlesworth, D. and Charlesworth, B. 1976b. Theoretical genetics of Batesian mimicry, III: evolution of dominance. *Journal of Theoretical Biology* 55:325–337.

Charnov, E.L. 1982. *The Theory of Sex Allocation*. Princeton University Press, Princeton.

Charnov, E.L. 1993. *Life History Invariants*. Oxford University Press, Oxford.

Chazdon, R.L. 1991. Plant size and form in the understory palm genus *Geonoma*: are species variations on a theme? *American Journal of Botany* 78:680–694.

Cheetham, A.H. 1968. Morphology and systematics of the bryozoan genus *Metrarabdotos*. *Smithsonian Miscellaneous Collections* 153:1–121.

Cheetham, A.H. 1986a. Tempo of evolution in a Neogene bryozoan: rates of morphologic change within and across species boundaries. *Paleobiology* 12:190–202.

Cheetham, A.H. 1986b. Branching, biomechanics and bryozoan evolution. *Proceedings of the Royal Society of London B* 228:151–171.

Cheetham, A.H. 1987. Tempo of evolution in a Neogene bryozoan: are trends in single morphologic characters misleading? *Paleobiology* 13:286–296.

Cheetham, A.H. and Jackson, J.B.C. 1995. Process from pattern: tests for selection versus random change in punctuated bryozoan speciation. In: *New Approaches to Speciation in the Fossil Record*, D.H. Erwin and R.L. Anstey (eds.). Columbia University Press, New York, pp. 184–207.

Cheetham, A.H. and Jackson, J.B.C. 1996. Speciation, extinction, and the decline of erect growth forms in Neogene and Quaternary cheilostome Bryozoa of tropical America. In: *Evolution and Environment in Tropical America*, J.B.C. Jackson, A.G. Coates and A.F. Budd (eds.). University of Chicago Press, Chicago, pp. 205–233.

Cheetham, A.H., Jackson, J.B.C and Hayek L.A. 1993. Quantitative genetics of bryozoan phenotypic evolution. I. Rate tests for random change versus selection in differentiation of living species. *Evolution* 47:1526–1538.

Cheetham, A.H., Jackson, J.B.C., Sanner, J. and Ventocilla, Y. 1999. Neogene cheilostome Bryozoa of tropical America: comparison and contrast between the Central American isthmus (Panama, Costa Rica) and the north-central Caribbean (Dominican Republic). *Bulletin of American Paleontology* 357:159–192.

Chen, J-Y., Bergström, M., Lindström, M. and Hou, X-G. 1991. Fossilized soft-bodied fauna. *National Geographic Research and Exploration* 7:8–19.

Cheney, D.L. and Seyfarth, R.M. 1990. The representation of social relations by monkeys. *Cognition* 37:167–196.

Cheng, M.-F., Peng, J.P. and Johnson, P. 1998. Hypothalamic neurons preferentially respond to female nest coo stimulation: demonstration of direct acoustic stimulation of luteinizing hormone release. *Journal of Neuroscience* 18:5477–5489.

Chevalier-Skolnikoff, S. 1974. Male-female, female-female, and male-male sexual behavior in the stumptail monkey, with special attention to the female orgasm. *Archives of Sexual Behavior* 3:95–116.

Cheverud, J.M. 1982. Phenotypic, genetic, and environmental morphological integration in the cranium. *Evolution* 36:499–516.

Cheverud, J.M. 1984a. Evolution by kin selection: a quantitative genetic model illustrated by maternal performance in mice. *Evolution* 38:766–777.

Cheverud, J.M. 1984b. Quantitative genetics and developmental constraints on evolution by selection. *Journal of Theoretical Biology* 110:155–171.

Cheverud, J.M. 1988. A comparison of genetic and phenotypic correlations. *Evolution* 42:958–968.

Cheverud, J.M. 1996. *Development*al integration and the evolution of pleiotropy. *American Zoologist* 36:44–50.

Cheverud, J.M. and Moore, A.J. 1994. Quantitative genetics and the role of the environment provided by relatives in behavioral evolution. In: *Quantitative Genetic Studies of Behavioral Evolution*, C.R.B. Boake (ed.). University of Chicago Press, Chicago, pp. 67–100.

Cheverud, J.M., Hartman, S.E., Richtsmeier, J.T. and Atchley, W.R. 1991. A quantitative genetic analysis of localized morphology in mandibles of inbred mice using finite element scaling analysis. *Journal of Craniofacial Genetics and Developmental Biology* 11:127–137.

Childress, J.R. 1972. Behavioral ecology and fitness theory in a tropical hermit crab. *Ecology* 53:960–964.

Choe, J.C. 1988. Worker reproduction and social evolution in ants (Hymenoptera: Formicidae). In: *Advances in Myrmecology*, J.C. Trager (ed.). E.J. Brill, New York, pp. 163–187.

Chourey, P.S. and Taliercio, E.W. 1994. Epistatic interaction and functional compensation between the two tissue- and cell-specific sucrose synthase genes in maize. *Proceedings of the National Academy of Sciences USA* 91:7917–7921.

Christy, J.H. 1995. Mimicry, mate choice, and the sensory trap hypothesis. *American Naturalist* 146:171–181.

Claridge, M.F. 1988. Species concepts and speciation in parasites. In: *Prospects in Systematics*, D.L. Hawksworth (ed.). Clarendon Press, Oxford, pp. 92–111.

Claridge, M.F. and Nixon, G.A. 1986. *Oncopsis flavicollis* (L.) associated with tree birches (*Betula*): a complex of biological species or a host plant utilization polymorphism? *Biological Journal of the Linnaean Society* 27:381–397.

Clark, A.B. 1991. Individual variation in responsiveness to environmental change. In: *Primate Responses to Environmental Change*, H.O. Bon (ed.). Chapman and Hall, London, pp. 91–110.

Clark, A.B. and Ehlinger, T.J. 1987. Pattern and adaptation in individual behavioral differences. In: *Perspectives in Ethology*, Vol. 7, P.P.G. Bateson and P.H. Klopfer (eds.). Plenum Press, New York, pp. 1–47.

Clark, A.G. 1994. Invasion and maintenance of a gene duplication. *Proceedings of the National Academy of Sciences USA* 91:2950–2954.

Clark, M.M., Crews, D. and Galef, B.G., Jr. 1991. Concentrations of sex steroid hormones in pregnant and fetal Mongolian gerbils. *Physiology and Behavior* 49:239–243.

Clark, M.M., Desousa, D., Vonk, J. and Galef, B.G., Jr. 1997. Parenting and potency: alternative routes to reproductive success in male Mongolian gerbils. *Animal Behaviour* 54:635–642.

Clarke, B. 1962. Balanced polymorphism and the diversity of sympatric species. In: *Taxonomy and Geography*, D. Nichols (ed.). The Systematics Association, London, pp. 47–70.

Clarke, B. 1972. Density-dependent selection. *American Naturalist* 106:1–13.

Clarke, B.C. 1966. The evolution of morph-ratio clines. *American Naturalist* 100:389–402.

Clarke, C., Clarke, F.M.M., Collins, S.C., Gill, A.C.L. and Turner, J.R.G. 1985. Male-like females, mimicry and transvestism in butterflies (Lepidoptera: Papilionidae). *Systematic Entomology* 10:257–283.

Clarke, C.A. and Sheppard, P.M. 1971. Further studies on the genetics of the mimetic butterfly *Papilo memnon* L. *Philosophical Transactions of the Royal Society of London* B 263:35–70.

Clausen, J., Keck. D.D. and Hiesey, W.W. 1940. *Experimental Studies on the Nature of Species*. Carnegie Institution, Washington, D.C.

Cleveland, L.R. 1926. Symbiosis among animals with special reference to termites and their intestinal flagellates. *Quarterly Review of Biology* 1:51–60.

Clough, M. and Summers, K. 2000. Phylogenetic systematics and biogeography of the poison frogs: evidence from mitochondrial DNA sequences. *Biological Journal of the Linnean Society* 70:515–540.

Clutton-Brock, T.H. 1991a. *The Evolution of Parental Care*. Princeton University Press, Princeton.

Clutton-Brock, T.H. 1991b. The evolution of sex differences and the consequences of polygyny in mammals. In: *The Development and Integration of Behaviour*, P. Bateson (ed.). Cambridge University Press, New York, pp. 229–253.

Clutton-Brock, T.H. and Albon, S.D. 1979. The roaring of red deer and the evolution of honest advertisement. *Behaviour* 69:145–170.

Clutton-Brock, T.H. and Godfray, C. 1991. Parental investment. In: *Behavioural Ecology*, J.R. Krebs and N.B. Davies (eds.). Blackwell, Oxford, pp. 234–262.

Cock, A.G. 1966. Genetical aspects of metrical growth and form in animals. *Quarterly Review of Biology* 41:131–190.

Cockerell, T.D.A. and Ireland, L. 1933. The relationships of *Scapter*, a genus of African bees. *Proceedings of the National Academy of Sciences USA* 19:977–978.

Coddington, J.A. 1988. Cladistic tests of adaptational hypotheses. *Cladistics* 4:3–22.

Coe, W.R. 1949. Divergent methods in development in morphologically similar species of prosobranch gastropods. *Journal of Morphology* 84:383–400.

Coen, E.S. 1991. The role of homeotic genes in flower development and evolution. *Annual Review of Plant Physiology and Plant Molecular Biology* 42:241–279.

Cohn, M.J. and Tickle, C. 1999. Developmental basis of limblessness and axial patterning in snakes. *Nature* 399:474–479.

Colbert, E.H. 1958. Morphology and behavior. In: *Behavior and Evolution*, A. Roe and G.G. Simpson (eds.). Yale University Press, New Haven, pp. 27–47.

Colbert, E.H. and Morales, M. 1991. *Evolution of the Vertebrates*. Wiley-Liss, New York.

Cole, B.J. 1986. The social behavior of *Leptothorax allardycei* (Hymenoptera, Formicidae): time budgets and the evolution of worker reproduction. *Behavioral Ecology and Sociobiology* 18:165–173.

Cole, B.J. 1991a. Short-term activity cycles in ants: generation of periodicity by worker interaction. *American Naturalist* 137:244–259.

Cole, B.J. 1991b. Short-term activity cycles in ants: a phase-response curve and phase resetting in worker activity. *Journal of Insect Behavior* 4:129–137.

Cole, B.J. 1991c. Is animal behaviour chaotic? Evidence from the activity of ants. *Proceedings of the Royal Society of London* B 244:353–359.

Cole, B.J. 1994. Chaos and behavior: the perspectives of non-linear dynamics. In: *Behavioral Mechanisms in Evolutionary Ecology*, L.A. Real (ed.). University of Chicago Press, Chicago, pp. 423–443.

Cole, B.J. and Wiernasz, D.C. 1997. Inbreeding and the population structure of a lek-mating ant species, *Pogonomyrmex occidentalis*. *Behavioral Ecology and Sociobiology* 40:79–86.

Colgan, P.W. and Gross, M.R. 1977. Dynamics of aggression in male pumpkinseed sunfish (*Lepomis gibbosus*) over the reproductive phase. *Zeitschrift für Tierpsychologie* 43:139–146.

Collazo, A. 1994. Molecular heterochrony in the pattern of fibronectin expression during gastrulation in amphibians. *Evolution* 48:2037–2045.

Collett, M., Despland, E., Simpson, S.J. and Krakauer, D.C. 1998. Spatial scales of desert locust gregarization. *Proceedings of the National Academy of Sciences USA* 95:13052–13055.

Collins, D. 1986. The great *Anomalocaris* mystery. *Rotunda* 19:51–57.

Collins, D. 1996. The "evolution" of *Anomalocaris* and its classification in the arthropod class Dinocarida (Nov.) and order Radiodonta (Nov.). *Journal of Paleontology* 70:280–293.

Collins, D.A. 1986. Relationships between adult male and infant baboons. In: *Primate Ontogeny, Cognition and Social Behaviour*, J.G. Else and P.C. Lee (eds.). Cambridge University Press, New York, pp. 205–218.

Collins, J.P. 1980. *Ambystoma tigrinum*: a multispecies conglomerate? *Copeia* 4:938–941.

Collins, J.P. 1981. Distribution, habitats and life history variation in the tiger salamander, *Ambystoma tigrinum*, in East-central and Southeast Arizona. *Copeia* 1981:666–675.

Collins, J.P. and Cheek, J.E. 1983. Effect of food and density on development of typical and cannibalistic salamander larvae in *Ambystoma tigrinum nebulosum*. *American Zoologist* 23:77–84.

Collins, J.P. and Sokol, O. 1980. Origin, evolution and distribution of larval reproduction in salamanders. Unpublished manuscript.

Collins, J.P., Zerba, K.E. and Sredl, M.J. 1993. Shaping intraspecific variation: development, ecology and the evolution of morphology and life history variation in tiger salamanders. *Genetica* 89:167–183.

Colwell, R.K. 1985. Community biology and sexual selection: lessons from hummingbird flower mites. In: *Community Ecology*, J. Diamond and T.J. Case (eds.). Harper and Row, New York, pp. 406–424.

Committee on Research Opportunities in Biology, Board on Biology, Commission on Life Sciences, and National Research Council. 1989. *Opportunities in Biology*. National Academy Press, Washington, D.C.

Conant, P. 1978. Lek behavior and ecology of two sympatric homosequential Hawaiian *Drosophila*: *Drosophila heteroneura* and *Drosophila silvestris*. M.S. thesis, University of Hawaii, Honolulu.

Conner, J. 1988. Field measurements of natural and sexual selection in the fungus beetle, *Bolitotherus cornutus*. *Evolution* 42:736–749.

Conover, D.O. and Fleisher, M.H. 1986. Temperature-sensitive period of sex determination in the Atlantic silverside, *Menidia menidia*. *Canadian Journal of Fisheries and Aquatic Sciences* 43:514–520.

Conover, D.O. and Van Voorhees, A. 1990. Evolution of a balanced sex ratio by frequency-dependent selection in a fish. *Science* 250:1556–1558.

Conrad, M. 1983. *Adaptability: The Significance of Variability from Molecule to Ecosystem*. Plenum Press, New York.

Consortium, International Human Genome Sequencing. 2001. Initial sequencing and analysis of the human genome. *Nature* 409:860–921.

Conway Morris, S. 1978. *Laggania cambria* Walcott: a composite fossil. *Journal of Paleontology* 52:126–131.

Conway Morris, S. and Whittington, H.B. 1979. The animals of the Burgess Shale. *Scientific American* 241:122–133.

Cook, C.D.K. 1968. Phenotypic plasticity with particular reference to three amphibious plant species. In: *Modern Methods in Plant Taxonomy*, V. Heywood (ed.). Academic Press, New York, pp. 97–111.

Cook, J.M., Compton, S.G., Herre, E.A. and West, S.A. 1997. Alternative mating tactics and extreme male dimorphism in fig wasps. *Proceedings of the Royal Society of London B* 264:747–754.

Cook, P.L. and Bock, P.E. 1996. *Characodoma* Maplestone, a senior synonym of *Cleidochasma* Harmer (Cleidochasmatidae). In: *Bryozoans in Space and Time*, D.P. Gordon, A.M. Smith and J.A. Grant-Mackie (eds.). NIWA, Wellington, pp. 81–88.

Cook, S.A. and Johnson, M.P. 1968. Adaptation to heterogeneous environments. I. Variation in heterophylly in *Ranunculus flammula* L. *Evolution* 22:496–516.

Cooke, R.C. and Whipps, J.M. 1980. The evolution of modes of nutrition in fungi parasitic on terrestrial plants. *Biological Reviews* 55:341–362.

Cooper, W.S. and Kaplan, R.H. 1982. Adaptive "coin-flipping": a decision-theoretic examination of natural selection for random individual variation. *Journal of Theoretical Biology* 94:135–151.

Corbet, G.B. 1964. Regional variation in the bank vole *Clethrionomys glareolus* in the British Isles. *Proceedings of the Zoological Society of London* 143:191–217.

Corbet, G.B. 1975. Examples of short and long term changes of dental pattern in Scottish voles (Rodentia, Microtinae). *Mammalogy Reviews* 5:17–21.

Corbet, P.S. 1999. *Dragonflies. Behavior and Ecology of Odonata*. Cornell University Press, Ithaca.

Cordero, A. 1987. Estructura de población en *Ischnura graellsi* Rambur, 1842 (Zygop, Coenagrionidae). *Boletín de la Asociación Española de Entomología* 11:269–286.

Cordero, A. 1990. The inheritance of female polymorphism in the damselfly *Ischnura graellsii* (Rambur) (Odonata: Coenagrionidae). *Heredity* 64:341–346.

Cordero, A. 1992. Morphological variability, female polymorphism and heritability of body length in *Ischnura graellsii* (Rambur) (Zygoptera: Coenagrionidae). *Odonatologica* 21:409–419.

Cordero, A. and Andrés, J.A. 1996. Colour polymorphism in odonates: females that mimic males? *Journal of the British Dragonfly Society* 12:50–60.

Corner, E.J.H. 1964. *The Life of Plants*. University of Chicago Press, Chicago.

Cowan, W.M. and O'Leary, D.D.M. 1984. Cell death and process elimination: the role of regressive phenomena in the development of the vertebrate nervous system. In: *Medicine, Science, and Society*, K.J. Isselbacher (ed.). John Wiley and Sons, New York, pp. 643–666.

Cowley, D.E. and Atchley, W.R. 1990. Development and quantitative genetics of correlation structure among body parts of *Drosophila melanogaster*. *American Naturalist* 135:242–268.

Cowley, D.E. and Atchley, W.R. 1992. Quantitative genetic models for development, epigenetic selection, and phenotypic evolution. *Evolution* 46:495–518.

Coyle, F.A. 1986. The role of silk in prey capture by nonaraneomorph spiders. In: *Spiders: Webs, Behavior, and Evolution*, W.A. Shear (ed.). Stanford University Press, Stanford, pp. 269–305.

Coyne, J.A. and Charlesworth, B. 1997. Response [to Eldredge and Gould: on punctuated equilibria]. *Science* 276:339–341.

Coyne, J. and Orr, H.A. 1999. The evolutionary genetics of speciation. In: *Evolution of Biological Diversity*, A.E. Magurran and R.M. May (eds.). Oxford University Press, Oxford, pp. 1–36.

Coyne, J.A., Barton, N.H. and Turelli, M. 1997. Perspective: a critique of Sewall Wright's shifting balance theory of evolution. *Evolution* 51:643–671.

Coyne, J.A., Barton, N.H. and Turelli, M. 2000. Is Wright's shifting balance process important in evolution? *Evolution* 54:306–317.

Cracraft, J. 1990. The origin of evolutionary novelties: pattern and process at different hierarchical levels. In: *Evolutionary Innovations*, M.H. Nitecki (ed.). University of Chicago Press, Chicago, pp. 21–44.

Craddock, E.M. and Boake, C.R.B. 1992. Onset of vitellogenesis in female *Drosophila silvestris* is accelerated in the presence of sexually mature males. *Journal of Insect Physiology* 38:643–650.

Craddock, E.M. and Kambysellis, M.P. 1997. Adaptive radiation in the Hawaiian *Drosophila* (Diptera: Drosophilidae): ecological and reproductive character analyses. *Pacific Science* 51:475–489.

Craig, J.L. and Jamieson, I.G. 1990. Pukeko: different approaches and some different answers. In: *Cooperative Breeding in Birds*, P.B. Stacey and W.D. Koenig, (eds.). Cambridge University Press, Cambridge, pp. 385–412.

Craig, R. 1983. Subfertility and the evolution of eusociality by kin selection. *Journal of Theoretical Biology* 100:379–397.

Craig, R. and Crozier, R.H. 1978. Caste-specific locus expression in ants. *Isozyme Bulletin* 11:64.

Crane, J. 1975. *Fiddler Crabs of the World*. Princeton University Press, Princeton.

Crankshaw, O.W. and Matthews, R.W. 1981. Sexual behavior among parasitic *Megarhyssa* wasps (Hymenoptera: Ichneumonidae). *Behavioral Ecology and Sociobiology* 9:1–7.

Creel, S., Wildt, D.E. and Monfort, S.L. 1993. Aggression, reproduction, and androgens in wild dwarf mongooses: a test of the challenge hypothesis. *American Naturalist* 141:816–825.

Creelman, R.A. and Mullet, J.E. 1995. Jasmonic acid distribution and action in plants: regulation during development and response to biotic and abiotic stress. *Proceedings of the National Academy of Sciences USA* 92:4114–4119.

Cresko, W.A. and Baker, J.A. 1996. Two morphotypes of lacustrine threespine stickleback, *Gasterosteus aculeatus*, in Benka Lake, Alaska. *Environmental Biology of Fishes* 45:343–350.

Crespi, B.J. 1986. Size assessment and alternative fighting tactics in *Elaphrothrips tuberculatus* (Insecta: Thysanoptera). *Animal Behaviour* 34:1324–1335.

Crespi, B.J. 1989a. Sexual Selection and assortative mating in subdivided populations of the thrips *Elaphrothrips tuberculatus* (Insecta: Thysanoptera). *Ethology* 83:265–278.

Crespi, B.J. 1989b. Causes of assortative mating in arthropods. *Animal Behaviour* 38:980–1000.

Crespi, B.J. 1992a. Behavioural ecology of Australian gall thrips (Insecta, Thysanoptera). *Journal of Natural History* 26:769–809.

Crespi, B.J. 1992b. Cannibalism and trophic eggs in subsocial and eusocial insects. In: *Cannibalism: Ecology and Evolution among Diverse Taxa*, M.A. Elgar and B.J. Crespi (eds.). Oxford University Press, Oxford, pp. 177–213.

Crespi, B.J. 1996. Comparative analysis of the origins and losses of eusociality: causal mosaics and historical uniqueness. In: *Phylogenies and the Comparative Method in Animal Behavior*, E. Martins (ed.). Oxford University Press, Oxford, pp. 253–287.

Crespi, B.J. and Mound, L.A. 1997. Ecology and evolution of social behavior among Australian gall thrips and their allies. In: *Social Behavior in Insects and Arachnids*, J.C. Choe and B.J. Crespi (eds.). Cambridge University Press, New York, pp. 166–180.

Crespi, B.J. and Sandoval, C.P. 2000. Phylogenetic evidence for the evolution of ecological specialization in *Temema* walking-sticks. *Journal of Evolutionary Biology* 13:249–262.

Crespi, B.J., Carmean, D. and Chapman, T. 1997. The ecology and evolution of galling thrips and their allies. *Annual Review of Entomology* 42:51–71.

Crew, F.A.E. 1923. Studies in intersexuality II: sex reversal in the fowl. *Proceedings of the Royal Society of London B* 95:256–278.

Crewe, R.M. 1999. Do primer pheromones exist? The responses of subordinates to the signals of dominants. In: *Social Insects at the Turn of the Millennium*, M.P. Schwarz and K. Hogendoorn (eds.). XIII Congress of IUSSI, Adelaide, p. 117.

Crews, D. 1985. Effects of early sex steroid hormone treatment on courtship behavior and sexual attractivity in the red-sided garter snake, *Thamnophis sirtalis parientalis*. *Physiology and Behavior* 35:569–575.

Crews, D. 1987. Courtship in unisexual lizards: a model for brain evolution. *Scientific American* 257:72–77.

Crews, D. 1989. Unisexual organisms as model systems for research in the behavioral neurosciences. In: *Evolution and Ecology of Unisexual Vertebrates*, R.M. Dawley and J.P. Bogart (eds.). New York State Museum, Albany, pp. 132–143.

Crews, D. and Bull, J.J. 1987. Evolutionary insights from reptilian sexual differentiation. In: *Genetic Markers of Sex Differentiation*, F.P. Haseltine, M.E. McClure and E.H. Goldberg (eds.). Plenum Press, New York, pp. 11–26.

Crews, D., Bergeron, J.M., Bull, J.J., Flores, D., Tousignant, A., Skipper, J.K. and Wibbels, T. 1994. Temperature-dependent sex determination in reptiles: proximate mechanisms, ultimate outcomes, and practical applications. *Developmental Genetics* 15:297–312.

Crick, F.H.C. and Lawrence, P.A. 1975. Compartments and polyclones in insect development. *Science* 189:340–347.

Cronquist, A. 1968. *The Evolution and Classification of Flowering Plants*. Houghton Mifflin, Boston.

Cronquist, A. 1987. A botanical critique of cladism. *The Botanical Review* 53:1–52.

Crook, J.H. 1964. The evolution of social organisation and visual communication in the weaver birds (Ploceinae). *Behaviour* (Supplement 10):1–177.

Crook, J.H. 1980. *The Evolution of Human Consciousness*. Clarendon Press, Oxford.

Crosland, M.W.J., Crozier, R.H. and Jefferson, E. 1988. Aspects of the biology of the primitive ant genus *Myrmecia* F. (Hymenoptera: Formicidae). *Journal of the*

Australian Entomological Society 27:305–309.

Crowley, T.J. 1983. The geologic record of climatic change. *Reviews of Geophysics and Space Physics* 21:828–877.

Crozier, R.H. 1985. Adaptive consequences of male-haploidy. In: *Spider Mites. Their Biology, Natural Enemies and Control*. W. Helle and M.W. Sabelis (eds.). Elsevier, Amsterdam, pp. 201–222.

Crozier, R.H. 1992. The genetic evolution of flexible strategies. *American Naturalist* 139:218–223.

Crozier, R.H. and Consul, P.C. 1976. Conditions for genetic polymorphism in social Hymenoptera under selection at the colony level. *Theoretical Population Biology* 10:1–9.

Crozier, R.H. and Luykx, P. 1985. The evolution of termite eusociality is unlikely to have been based on a male-haploid analogy. *American Naturalist* 126:867–869.

Crozier, R.H. and Page, R.E. 1985. On being the right size: male contributions and multiple mating in social Hymenoptera. *Behavioral Ecology and Sociobiology* 18:105–115.

Crozier, R.H. and Pamilo, P. 1996. *Evolution of Social Insect Colonies: Sex Allocation and Kin Selection*. Oxford University Press, Oxford.

Csuti, B.A. 1979. *Patterns of Adaptation and Variation in the Great Basin Kangaroo Rat (Dipodomys microps)*. University of California Press, Berkeley.

Culver, D.C., Kane, T.C. and Fong, D.W. 1994. *Adaptation and Evolution in Caves*. Harvard University Press, Cambridge.

Cumber, R.A. 1949. The biology of humble-bees, with special reference to the production of the worker caste. *Transactions of the Royal Entomological Society of London* 100:1–45.

Curio, E. 1965. Zur geographischen variation des feinderkennes einiger Darwinfinken (*Geospizidae*). *Zoologischer Anzeiger* (Supplement) 28:466–492.

Curio, E. 1973. Towards a methodology of teleonomy. *Experientia* 29:1045–1058.

Curio, E. 1976. *The Ethology of Predation*. Springer-Verlag, New York.

Curio, E. 1988. Cultural transmission of enemy recognition by birds. In: *Social Learning: Psychological and Biological Perspectives*, T. Zentall and B.G. Galef (eds.). Lawrence Erlbaum, Hillsdale, NJ, pp. 75–97.

Currie, G.A. 1937. Galls of Eucalyptus trees: a new type of association between flies and nematodes. *Proceedings of the Linnean Society of New South Wales* 62:147–174.

Curtis, H. 1983. *Biology*. Worth, New York.

Cushing, J.E. 1941. Non-genetic mating preference as a factor in evolution. *Condor* 43:233–236.

Cushing, J.E. 1944. The relation of non-heritable food habits to evolution. *Condor* 46:265–271.

Cushing, J.E. and Ramsay, A.O. 1949. The non-heritable aspects of family unity in birds. *Condor* 51:82–87.

Cushman, J.C. and Bohnert, H.J. 1996. Transcriptional activation of CAM genes during development and environmental stress. In: *Crassulacean Acid Metabolism*, K. Winter and J.A.C. Smith (eds.). Springer-Verlag, New York, pp. 135–158.

Cyr, R.J. 1994. Microtubules in plant morphogenesis: role of the cortical array. *Annual Review of Cell Biology* 10:153–180.

Dacks, J. and Roger, A.J. 1999. The first sexual lineage and the relevance of facultative sex. *Journal of Molecular Evolution* 48:779–783.

Dagg, A.I. 1984. Homosexual behaviour and female-male mounting in mammals—a first survey. *Mammal Review* 14:155–185.

Daly, J.W. 1995. The chemistry of poisons in amphibian skin. *Proceedings of the National Academy of Sciences USA* 92:9–13.

Daly, J.W., Garraffo, H.M., Spande, T.F., Jaramillo, C. and Rand A.S. 1994. Dietary source for skin alkaloids of poison frogs (Dendrobatidae). *Journal of Chemical Ecology* 20:943–955.

Daly, J.W., Garraffo, H.M., Jain, P., Spande, T.F., Snelling, R.R., Jaramillo, C. and Rand, A.S. 2000. Arthropod frog connection: decahydroquinoline and pyrrolizidine alkaloids common to microsympatric myrmicine ants and denrobatid frogs. *Journal of Chemical Ecology* 26:73–85.

Danforth, B.N. 1991. The morphology and behavior of dimorphic males in *Perdita portalis* (Hymenoptera: Andrenidae). *Behavioral Ecology and Sociobiology* 29:235–247.

Danforth, B.N. 2001. Evolution of sociality in a primitively eusocial lineage of bees. *Proceedings of the National Academy of Sciences USA* 99:286–290.

Danforth, B.N. and Desjardins, C.A. 1999. Male dimorphism in *Perdita portalis* (Hymenoptera, Andrenidae) has arisen from preexisting allometric patterns. *Insectes Sociaux* 46:18–28.

Danforth, B.N., Sauquet, H. and Packer, L. 1999. Phylogeny of the bee genus *Halictus* (Hymenoptera: Halictidae) based on parsimony and likelihood analyses of nuclear EF-1α sequence data. *Molecular Phylogenetics and Evolution* 13:605–618.

Darchen, R. and Delage-Darchen, B. 1975. Contribution à l'étude d'une abeille du Mexique *Melipona beecheii b.* (Hymenoptera: Apide). *Apidologie* 6:295–339.

Darlington, C.D. 1958. *The Evolution of Genetic Systems*, 2nd ed. Cambridge University Press, Cambridge.

Darwin, C.D. 1845. *The Voyage of the Beagle*. Everyman's Library, Dutton, New York, 1967.

Darwin, C.D. 1859. *On the Origin of Species*. Facsimile of 1st ed. Harvard University Press, Cambridge, 1966.

Darwin, C.D. 1859 [1872]. *The Origin of Species*. Facsimile of 6th ed. The Modern Library, Random House, New York, 1959.

Darwin, C.D. 1864. On the existence of two forms, and on their reciprocal sexual relation, in several species of the genus *Linum. Journal of the Proceedings of the Linnean Society (Botany)* 7:69–83. [Reprinted in: Barrett, P.H. (ed.) 1977. *The Collected Papers of Charles Darwin*. University of Chicago Press, Chicago, pp. 93–105.]

Darwin, C.D. 1868a [1875a]. *The Variation of Animals and Plants under Domestication*, I. Facsimile of 2nd ed. (1875) as printed in 1896 by D. Appleton and Company, New York. AMS Press, New York, 1972.

Darwin, C.D. 1868b [1875b]. *The Variation of Animals and Plants Under Domestication*, II. Facsimile of 2nd ed. (1875) as printed in 1896 by D. Appleton and Company, New York. AMS Press, New York, 1972.

Darwin, C.D. 1871. Pangenesis. *Nature* 3:502–503. [Reprinted in: Barrett, P.H. (ed.) 1977. *The Collected Papers of Charles Darwin*. University of Chicago Press, Chicago, pp. 165–167.]

Darwin, C.D. 1871 [1874]. *The Descent of Man and Selection in Relation to Sex*, 2nd ed. [unabridged but renumbered text, and figures]. The Modern Library, Random House, New York, 1959.

Darwin, C.D. 1872. *The Expression of the Emotions in Man and Animals*. D. Appleton, New York.

Darwin, C.D. 1880. *The Power of Movement in Plants*. Authorized limited edition. D. Appleton, New York.

Darwin, C.D. 1882. Prefatory notice: studies in the theory of descent. In: *Studies in the Theory of Descent: With Notes and Additions by the Author*, A. Weismann (English transl.). Sampson Low, Marston, Searle, and Rivington, London, pp. v–vi.

Darwin, C.D. 1975. Natural selection. In: *Darwin's Natural Selection, being the second part of his big species book written from 1856 to 1858*, R.C. Stauffer (ed.). Cambridge University Press, Cambridge, pp. 25–566.

Darwin, F. (ed.) 1958. *The Autobiography of Charles Darwin and Selected Letters*. Dover, New York.

Davey, G. 1989. *Ecological Learning Theory*. Routledge, London.

Davidson, D.W. 1982. Sexual selection in harvester ants (Hymenoptera: Formicidae: *Pogonomyrmex*). *Behavioral Ecology and Sociobiology* 10:245–250.

Davidson, E.H. 1986. *Gene Activity in Early Development*, 3rd ed. Academic Press, Orlando.

Davidson, E.H. 1989. Lineage-specific gene expression and the regulative capacities of the sea urchin embryo: a proposed mechanism. *Development* 105:421–445.

Davidson, E.H. 1990. How embryos work: a comparative view of diverse modes of cell fate specification. *Development* 108:365–389.

Davidson, E.H. 1991. Spatial mechanisms of gene regulation in metazoan embryos. *Development* 113:1–26.

Davidson, E.H. 2000. *Genomic Regulatory Systems: Evolution and Development*. Academic Press, San Diego.

Davidson, E.H. and Britten, R.J. 1979. Regulation of gene expression: possible role of repetitive sequences. *Science* 204:1052.

Davies, N.B. and Halliday, T.R. 1978. Deep croaks and fighting assessment in toads, *Bufo bufo. Nature, London* 274:683–685.

Davies, N.B. and Houston, A.I. 1984. Territory economics. In: *Behavioural Ecology: An Evolutionary Approach*, J.R. Krebs and N.B. Davies (eds.). Blackwell, Oxford, pp. 148–169.

Davis, J.M., Gordon, M.P. and Smit, B.A. 1991. Assimilate movement dictates remote sites of wound-induced gene expression in poplar leaves. *Proceedings of the National Academy of Sciences USA* 88:2393–2396.

Davis, M.B. 1976. Pleistocene biogeography of temperate deciduous forests. *Geoscience and Man* 13:13–26.

Dawkins, R. 1976. *The Selfish Gene*. Oxford University Press, New York.

Dawkins, R. 1980. Good strategy or evolutionarily stable strategy? In: *Sociobiology: Beyond Nature/Nurture*. B.W. Barlow and J. Silverberg (eds.). Westview Press, Boulder, pp. 331–367.

Dawkins, R. 1982. *The Extended Phenotype*. W.H. Freeman, Oxford.

Dawkins, R. and Krebs, J.R. 1978. Animal signals: information or manipulation? In: *Behavioural Ecology*, J.R. Krebs and N.B. Davies (eds.). Blackwell, Oxford, pp. 282–309.

Day, T., Pritchard, J. and Schluter, D. 1994. Ecology and genetics of phenotypic plasticity: a comparison of two sticklebacks. *Evolution* 48:1723–1734.

Deag, J.M. and Crook, J.H. 1971. Social behavior and "agonistic buffering" in the wild Barbary macaque, *Macaca sylvana* L. *Folia Primatologica* 15:183–200.

de Beer, G.R. 1958. *Embryos and Ancestors*. Clarendon Press, Oxford.

de Beer, G.R. 1971. *Homology, an Unsolved Problem*. Oxford University Press, London.

de Chateau, M. and Bjorck, L. 1994. Protein PAB, a mosaic albumin-binding bacterial protein representing the first contemporary example of module shuffling. *Journal of Biological Chemistry* 269:12147–12151.

deHeer, C.J., Goodisman, M.A.D. and Ross, K.G. 1999. Queen dispersal strategies in the multiple-queen form of the fire ant *Solenopsis invicta. American Naturalist* 153:660–675.

de Jong, G. 1990. Quantitative genetics of reaction norms. *Journal of Evolutionary Biology* 3:447–468.

Delaney, K. and Geleprin, A. 1986. Post-ingestive food-aversion learning to amino acid deficient diets by the terrestrial slug *Limax maximus. Journal of Comparative Physiology A* 159:281–295.

Dellaporta, S.L. and Calderon-Urrea, A. 1994. The sex determination process in maize. *Science* 266:1501–1505.

Deneubourg, J.L., Gross, S., Franks, N. and Pasteels, J.M. 1989. The blind leading the blind: chemically mediated morphogenesis and army ant raid patterns. *Journal of Insect Behavior* 2:719–725.

Dennett, D.C. 1995. *Darwin's Dangerous Idea*. Touchstone, New York.

Dent, J.N. 1968. Survey of amphibian metamorphosis. In: *Metamorphosis: A Problem in Developmental Biology*, W. Etkin and L.I. Gilbert (eds.). Appleton-Century-Crofts, New York, pp. 271–311.

Denton, F.R. 1998. Beetle juice. *Science* 281:1285.

Denton, M. and Marshall, C. 2001. Protein folds: laws of form revisited. *Nature* 410:417.

dePamphilis, C.W. and Palmer, J.D. 1990. Loss of photosynthetic and chlororespiratory genes from the plastid genome of a parasitic flowering plant. *Nature* 348:337–339.

de Pinna, M.C.C. 1989. A new Sarcoglanidine catfish, phylogeny of its subfamily, and an appraisal of the phyletic status of the Trichomycterinae (Teleostei, Trichomycteridae). *American Museum Novitates* 2950:1–39.

de Pinna, M.C.C. 1991. Concepts and tests of homology in the cladistic paradigm. *Cladistics* 7:367–394.

de Queiroz, A. and Wimberger, P.H. 1993. The usefulness of behavior for phylogeny estimation: levels of homoplasy in behavioral and morphological characters. *Evolution* 47:46–60.

Derting, T.L. 1989. Metabolism and food availability as regulators of production in juvenile cotton rats. *Ecology* 70:587–595.

de Souza, S.J., Long, M., Klein, R.J., Roy, S., Lin, S. and Gibert, W. 1998. Toward a resolution of the introns early/late debate: only phase zero introns are correlated with the structure of ancient proteins. *Proceedings of the National Academy of Sciences USA* 95:5094–5099.

Destephano, D.B. and Brady, U.E. 1977. Prostaglandin and prostaglandin synthetase in the cricket, *Acheta domesticus*. *Journal of Insect Physiology* 23:905–911.

Dethier, V. 1957. Communication by insects: physiology of dancing. *Science* 125:331–336.

Deuchar, E.M. 1975. *Cellular Interactions in Animal Development*. Chapman and Hall, London.

Deutsch, J.C. 1997. Colour diversification in Malawi cichlids: evidence for adaptation, reinforcement or sexual selection? *Biological Journal of the Linnean Society* 62:1–14.

DeVries, H. 1910. *The Mutation Theory*. Open Court, Chicago.

DeWitt, T.J. 1995. Functional tradeoffs and phenotypic plasticity in the fresh-water snail *Physa*. Ph.D. dissertation, Binghamton University (SUNY).

DeWitt, T.J. 1998. Costs and limits of phenotypic plasticity: tests with predator-induced morphology and life history in a freshwater snail. *Journal of Evolutionary Biology* 11:465–480.

DeWitt, T.J., Sih, A. and Wilson, D.S. 1998. Costs and limits of phenotypic plasticity. *Trends in Ecology and Evolution* 13:77–81.

DeWitt, T.J., Robinson, B.W. and Wilson, D.S. 2000. An adaptive tradeoff in a multi-predator world: shell form of the freshwater snail, *Physa*. *Evolutionary Ecology Research* 2:129–148.

Dewsbury, D.A. 1990. Comparative psychology: retrospect and prospect. In: *Contemporary Issues in Comparative Psychology*, D.A. Dewsbury (ed.). Sinauer, Sunderland, MA, pp. 431–451.

Dewsbury, D.A. 1994. On the utility of the proximate-ultimate distinction in the study of animal behavior. *Ethology* 96:63–68.

Diamond, J.M. 1987. Learned specializations of birds. *Nature* 330:16–17.

Diamond, J. 1998. *Guns, Germs, and Steel*. W.W. Norton & Company, New York.

Dickerson, R.E. and Geis, I. 1969. *The Structure and Action of Proteins*. W.A. Benjamin, Menlo Park, CA.

Dickinson, W.J. 1988. On the architecture of regulatory systems: evolutionary insights and implications. *BioEssays* 8:204–208.

Dickinson, W.J. 1989. *Gene* regulation and evolution. In: *Genetics, Speciation, and the Founder Principle*, L.V. Giddings, K.Y. Kaneshiro and W.W. Anderson (eds.). Oxford University Press, New York, pp. 181–202.

Dickinson, W.J. 1991. The evolution of regulatory genes and patterns in *Drosophila*. In: *Evolutionary Biology*, M.K. Hecht, B. Wallace and R.J. MacIntyre (eds.). Plenum Press, New York, pp. 127–173.

Dickman, M.C., Schliwa, M. and Barlow, G.W. 1988. Melanophore death and disappearance produces color metamorphosis in the polychromatic Midas cichlid (*Cichlasoma citrinellum*). *Cell Tissue Research* 253:9–14.

Didion, J. 1985. *Democracy*. Washington Square Press, New York.

Dilger, W.C. 1960. The comparative ethology of the African parrot genus *Agapornis*. *Zeitschrift für Tierpsychologie* 17:42–685.

Dill, L.M. 1983. Adaptive flexibility in the foraging behavior of fishes. *Canadian Journal of Fisheries and Aquatic Sciences* 40:398–408.

Dingle, H. 1961. Correcting behavior in boxelder bugs. Ph.D. dissertation, University of Michigan.

Dingle, H. 1984. Behavior, genes and life histories: complex adaptations in uncertain environments. In: *A New Ecology: Novel Approaches to Interactive Systems*, P. Peter (ed.). John Wiley and Sons, New York.

Dingle, H. 1991. Evolutionary genetics of animal migration. *American Zoologist* 31:253–264.

Dingle, H. 1994. Genetic analyses of animal migration. In: *Quantitative Genetic Studies of Behavioral Evolution*, C.R.B. Boake (ed.). University of Chicago Press, Chicago, pp. 145–164.

Dingle, H. 1996. *Migration: The Biology of Life on the Move*. Oxford University Press, New York.

Dingle, H., Leslie, J.F. and Palmer, J.O. 1986. Behavior genetics of flexible life histories in milkweed bugs (*Oncopeltus fasciatus*). In: *Evolutionary Genetics of Invertebrate Behavior*, M.D. Huettel (ed.). Plenum Press, New York, pp. 7–18.

Dixon, F.W. 1980. *The Apeman's Secret*. The Hardy Boys No. 62. Simon and Schuster, New York.

Dixson, A.F. and George, L. 1982. Prolactin and parental behavior in a male New World primate. *Nature* 299:551–553.

Dobzhansky, T. 1937 [1951]. *Genetics and the Origin of Species.* Columbia University Press, New York.

Dobzhansky, T. 1959. Evolution of genes and genes in evolution. *Cold Spring Harbor Symposia in Quantitative Biology* 24:15–30.

Dobzhansky, T. 1970. *Genetics of the Evolutionary Process.* Columbia University Press, New York.

Dobzhansky, T. 1977. Natural selection. In: *Evolution*, T. Dobzhansky, F.J. Ayala, G.L. Stebbins and J.W. Valentine (eds.). W.H. Freeman, San Francisco, pp. 95–125.

Dobzhansky, T. and Wallace, B. 1953. The genetics of homeostasis in *Drosophila*. *Proceedings of the National Academy of Sciences USA* 39:162–171.

Dobzhansky, T., Ayala, F.J., Stebbins, G.L. and Valentine, J.W. 1977. *Evolution.* W.H. Freeman, San Francisco.

Doebley, J. 1983. The maize and teosinte male inflorescence: a numerical taxonomic study. *Annals of the Missouri Botanical Garden* 70:32–70.

Doebley, J. 1990. Molecular systematics of *Zea* (Gramineae). *Maydica* 35:143–150.

Doebley, J. and Stec, A. 1991. Genetic analysis of the morphological differences between maize and teosinte. *Genetics* 129:285–295.

Doebley, J. and Stec, A. 1993. Inheritance of the morphological differences between maize and teosinte: comparison of results for two F2 populations. *Genetics* 134:559–570.

Doebley, J., Stec, A., Wendel, J. and Edwards, M. 1990. Genetic and morphological analysis of a maize-teosinte F2 population: implications for the origin of maize. *Proceedings of the National Academy of Sciences USA* 87:9888–9892.

Doebley, J., Stec, A. and Gustus, C. 1995. *Teosinte branched 1* and the origin of maize: evidence for epistasis and the evolution of dominance. *Genetics* 141:333–346.

Doebley, J., Stec, A. and Hubbard, L. 1997. The evolution of apical dominance in maize. *Nature* 386:485–488.

Dollo, L. 1893. Les lois de l'évolution. *Bulletin Belgique de Géologie* 7:164–167.

Dominey, W.J. 1980. Female mimicry in male bluegill sunfish—a genetic polymorphism? *Nature* 284:546–548.

Dominey, W.J. 1984. Effects of sexual selection and life history on speciation: species flocks in African cichlids and Hawaiian *Drosophila.* In: *Evolution of Fish Species Flocks*, A.A. Echelle and I. Kornfield (eds.).

University of Maine at Orono Press, Orono, pp. 231–249.

Donlin, M.J., Lisch, D. and Freeling, M. 1995. Tissue-specific accumulation of MURB, a protein encoded by *MuDR*, the autonomous regulator of the *Mutator* transposable element family. *Plant Cell* 7:1989–2000.

Donoghue, M.J. 1992. Homology. In: *Keywords in Evolutionary Biology*, E.F. Keller and E.A. Lloyd (eds.). Harvard University Press, Cambridge, pp. 170–179.

Doolittle, W.F. 1978. Genes in pieces: were they ever together? *Nature* 272:581–582.

Doolittle, W.F. 1987. The origin and function of intervening sequences in DNA: a review. *American Naturalist* 130:915–928.

Doolittle, W.F. 1990. Understanding introns: origins and functions. In: *Intervening Sequences in Evolution and Development*, E.M. Stone and R.J. Schwartz (eds.). Oxford University Press, New York, pp. 43–62.

Doorenbos, J. 1965. Juvenile and adult phases in woody plants. In: *Handbuch der Pflanzenphysiologie.* Springer-Verlag, New York, pp. 1222–1235.

Dorit, R.L., Schoenbach, L. and Gilbert, W. 1990. How big is the universe of exons? 250:1377–1382.

Dorst, J. 1972. Le polymorphisme génétique chez les oiseux. *Mémoires de la Société Zoologique de France* 37:235–269.

Dorweiler, J., Stec, A., Kermicle, J. and Doebley, J. 1993. Teosinte glume architecture 1: a genetic locus controlling a key step in maize evolution. *Science* 262:233–235.

Douglas, M.E. and Avise, J.C. 1982. Speciation rates and morphological divergence in fishes: tests of gradual versus rectangular modes of evolutionary change. *Evolution* 36:224–232.

Dressler, R.L. 1981. *The Orchids.* Harvard University Press, Cambridge.

DuBose, R.F. and Hartl, D.L. 1989. An experimental approach to testing modular evolution: directed replacement of α-helices in a bacterial protein. *Proceedings of the National Academy of Sciences USA* 86:9966–9970.

DuBrul, E.L. and Laskin, D.M. 1961. Preadaptive potentialities of the mammalian skull: an experiment in growth and form. *American Journal of Anatomy* 109:117–132.

Dudley, R. and Chai, P. 1996. Animal flight mechanics in physically variable gas mixtures. *Journal of Experimental Biology* 199:1881–1885.

Duméril, A. 1865. Nouvelles observations sur les axolotls, batraciens urodèles de Mexico (*Siredon mexicanus humboldtii*) nés dans la Ménagerie des Reptiles au Muséum d'Histoire Naturelle, et qui y subissent des métamorphoses. *Comptes Rendues Académie des Sciences* 61:775–778.

Dunbar, R.I.M. 1982. *Current Problems in Sociobiology*, King's College Sociobiology Group (eds.). Cambridge University Press, Cambridge.

Dundee, H.A. 1957. Partial metamorphosis induced in *Typhlomolge*. *Copeia* 1957:52–53.

Dundee, H.A. 1961. Response of the neotenic salamanders *Haideotriton wallacei* to a metamorphic agent. *Science* 135:1060–1061.

Durbin, M.L., Learn, G.H., Jr. Huttley, G.A. and Clegg, M.T. 1995. Evolution of the chalcone synthase gene family in the genus *Ipomoea*. *Proceedings of the National Academy of Sciences USA* 92:3338–3342.

Dusenbery, D.B. 1992. *Sensory Ecology*. W.H. Freeman, New York.

Dworkin, I.M., Tanda, S. and Larsen, E. 2001. Are entrenched characters developmentally constrained? Creating biramous limbs in an insect. *Evolution & Development* 3:424–431.

Dyer, F.C. 1985. Mechanisms of dance orientation in the Asian honey bee *Apis florea* L. *Journal of Comparative Physiology A* 157:183–198.

Dyer, F.C. 1991a. Comparative studies of dance communication: analysis of phylogeny and function. In: *Diversity in the Genus Apis*, D. R. Smith (ed.). Westview Press, Boulder, pp. 177–198.

Dyer, F.C. 1991b. Bees acquire route-based memories but not cognitive maps in a familiar landscape. *Animal Behaviour* 41:239–246.

Dyer, F.C. 1994. Spatial cognition and navigation in insects. In: *Behavioral Mechanisms in Evolutionary Ecology*, L.A. Real (ed.). University of Chicago Press, Chicago, pp. 66–98.

Dyer, F.C. and Dickinson, J.A. 1994. Development of sun compensation by honeybees: how partially experienced bees estimate the sun's course. *Proceedings of the National Academy of Sciences USA* 91:4471–4474.

Dyer, F.C. and Seeley, T.D. 1991. Dance dialects and foraging range in three Asian honey bee species. *Behavioural Ecology and Sociobiology* 28:227–233.

Eadie, J.M. and Fryxell, J.M. 1992. Density dependence, frequency dependence, and alternative nesting strategies in goldeneyes. *American Naturalist* 140:621–641.

Eagleson, G.W. 1976. Comparison of the life histories and growth patterns of populations of the salamander *Ambystoma gracile* (Baird) from permanent low altitude and montane lakes. *Canadian Journal of Zoology* 54:2098–2111.

East, M.L., Hofer, H. and Wickler, W. 1993. The erect "penis" is a flag of submission in a female-dominated society: greetings in Serengeti spotted hyenas. *Behavioral Ecology and Sociobiology* 33:355–370.

Eberhard, J.R. 1997. The evolution of nest-building and breeding behavior in parrots. Ph.D. dissertation, Princeton University.

Eberhard, J.R. 1998a. Breeding biology of the monk parakeet. *Wilson Bulletin* 110:463–473.

Eberhard, J.R. 1998b. Evolution of nest-building behavior in *Agapornis* parrots. *The Auk* 115:455–464.

Eberhard, J.R. and Ewald, P.W. 1994. Food availability, intrusion pressure and territory size: an experimental study of Anna's hummingbirds (*Calypte anna*). *Behavioral Ecology and Sociobiology* 34:11–18.

Eberhard, W.G. 1977. Aggressive chemical mimicry by a bolas spider. *Science* 198:1173–1175.

Eberhard, W.G. 1979. The function of horns in *Podischnus agenor* (Dynastinae) and other beetles. In: *Sexual Selection and Reproductive Competition in Insects*, M.S. Blum and N. Blum (eds.). Academic Press, New York, pp. 231–258.

Eberhard, W.G. 1980. Horned Beetles. *Scientific American* 242:166–182.

Eberhard, W.G. 1981. The natural history of *Doryphora* sp. (Coleoptera, Chrysomelidae) and the function of its sternal horn. *Annals of the Entomological Society of America* 74:445–448.

Eberhard, W.G. 1982. Beetle horn dimorphism: making the best of a bad lot. *American Naturalist* 119:420–426.

Eberhard, W.G. 1985. *Sexual Selection and Animal Genitalia*. Harvard University Press, Cambridge.

Eberhard, W.G. 1988. Behavioral flexibility in orb web construction: effects of supplies in different silk glands and spider size and weight. *Journal of Arachnology* 16:295–302.

Eberhard, W.G. 1989. Niche expansion in the spider *Wendilgarda galapagensis* (Araneae, Theridiosomatidae) on Cocos Island. *Revista de Biología Tropical* 37:163–168.

Eberhard, W.G. 1990a. Early stages of orb construction by *Philoponella vicina, Leucage mariana*, and *Nephila clavipes* (Aranea, Uloboridae and Tetragnathidae), and their phylogenetic implications. *Journal of Arachnology* 18:205–234.

Eberhard, W.G. 1990b. Imprecision in the behavior of *Leptomorphus* sp. (Diptera,

Mycetophilidae) and the evolutionary origin of new behavior patterns. *Journal of Insect Behavior* 3:327–357.

Eberhard, W.G. 1996. *Female Control: Sexual Selection by Cryptic Female Choice.* Princeton University Press, Princeton.

Eberhard, W.G. 2000. Breaking the mold: behavioral variation and evolutionary innovation in *Wendilgarda* spiders (Araneae, Theridiosomatidae). *Ethology, Ecology and Evolution* 12:223–235.

Eberhard, W.G. 2001a. Multiple origins of a major novelty: moveable abdominal lobes in male sepsid flies (Diptera: Sepsidae), and the question of developmental constraints. *Evolution and Development* 3:206–222.

Eberhard, W.G. 2001b. Trolling for water striders: active searching for prey and the evolution of reduced webs in the spider *Wendilgarda* sp. (Araneae, Theridiosomatidae). *Journal of Natural History* 35:229–251.

Eberhard, W.G. and Gutierrez, E.E. 1991. Male dimorphisms in beetles and earwigs and the question of development constraints. *Evolution* 45:18–28.

Eberhard, W.G., Garcia-C., J.M. and Lobo, J. 2000. Size-specific defensive structures in a horned weevil confirm a classic battle plan: avoid fights with larger opponents. *Proceedings of the Royal Society of London B* 267:1129–1134.

Eble, G.J. 1998. The role of development in evolutionary radiations. In: *Biodiversity Dynamics: Turnover of Populations, Taxa and Communities*, M.L. McKinney and J.A. Drake (eds.). Columbia University Press, New York, pp. 132–161.

Eble, G.J. 2001. Developmental morphospaces and evolution. In: *Evolutionary Dynamics: Exploring the Interplay of Selection, Neutrality, Accident, and Function*, J.P. Crutchfield and P. Schuster (eds.). Oxford University Press, Oxford, pp. 35–65.

Echelle, A.A. and Kornfield, I. (eds.) 1984. *Evolution of Fish Species Flocks*. University of Maine at Orono Press, Orono.

Edelman, G.M. 1987. *Neural Darwinism*. Basic Books, New York.

Edelman, G.M. 1988. *Topobiology*. Basic Books, New York.

Edwards, S.V. and Naeem, S. 1993. The phylogenetic component of cooperative breeding in perching birds. *American Naturalist* 141:754–789.

Ehleringer, J.R. and Monson, R.K. 1993. Evolutionary and ecological aspects of photosynthetic pathway variation. *Annual Review of Ecology and Systematics* 24:411–439.

Ehlinger, T.J. and Wilson, D.S. 1988. Complex foraging polymorphism in bluegill sunfish. *Proceedings of the National Academy of Sciences USA* 85:1878–1882.

Eibl-Eibesfeldt, I. 1970. *Ethology—The Biology of Behavior*. Holt, Rinehart and Winston, New York.

Eickbush, T.H. 2000. Introns gain ground. *Nature* 404:940–943.

Eickwort, G.C., Eickwort, J.M., Gordon, J. and Eickwort, M.A. 1996. Solitary behavior in a high-altitude population of the social sweat bee *Halictus rubicundus* (Hymenoptera: Halictidae). *Behavioral Ecology and Sociobiology* 38:227–233.

Eigen, M. 1992. *Steps Towards Life*. Oxford University Press, New York.

Eisen, M.B. and Brown, P.O. 1999. DNA arrays for analysis of gene expression. *Methods of Enzymology* 303:179–205.

Eisentraut, M. 1929. Die variation der balearischen inseleidechse *Lacerta lilfordi* Sitzungsber. *Sitzungberichte Geselschaft naturforschung Freunde, Berlin* 1929:24–36.

Eisner, T. and Meinwald, J. 1995. The chemistry of sexual selection. *Proceedings of the National Academy of Sciences USA* 92:50–55.

Elder, J.F., Jr. and Turner, B.J. 1994. Concerted evolution at the population level: pupfish *Hind*III satellite DNA sequences. *Proceedings of the National Academy of Sciences USA* 91:994–998.

Eldredge, N. 1976. Differential evolutionary rates. *Paleobiology* 2:174–177.

Eldredge, N. 1995. Species, speciation, and the context of adaptive change in evolution. In: *New Approaches to Speciation in the Fossil Record*, D.H. Erwin and R.L. Anstey (eds.). Columbia University Press, New York, pp. 39–63.

Eldredge, N. and Gould, S.J. 1972. Punctuated equilibria: an alternative to phyletic gradualism. In: *Models in Paleobiology*, T.J.M. Schopf (ed.). Freeman, Cooper, San Francisco, pp. 82–115.

Eldredge, N. and Gould, S.J. 1997. On punctuated equilibria. *Science* 276:338–339.

Elgar, M.A. and Crespi, B.J. (eds.) 1992. *Cannibalism: Ecology and Evolution among Diverse Taxa*. Oxford University Press, Oxford.

Elliott, J.A. and Goldman, B.D. 1981. Seasonal reproduction: photoperiodism and biological clocks. In: *Neuroendocrinology of Reproduction*, N.T. Adler (ed.). Plenum Press, New York, pp. 377–423.

Elliott, J.A. and Goldman, B.D. 1989. Reception of photoperiodic information by fetal Siberian hamsters: role of the mother's pineal gland. *Journal of Experimental Zoology* 252:237–244.

Ellner, S.P. 1984. Stationary distributions for some difference equation population models. *Journal of Mathematical Biology* 19:169–200.

Elmes, G.W. 1976. Some observations on the microgyne form of *Myrmica rubra* L. (Hymenoptera, Formicidae). *Insectes Sociaux* 23:3–22.

Elmes, G.W. 1978. A morphometric comparison of three closely related species of *Myrmica* (Formicidae), including a new species from England. *Systematic Entomology* 3:131–145.

Elmes, G.W. 1981. An aberrant form of *Myrmica scabrinodes* Nylander (Hymenoptera, Formicidae). *Insectes Sociaux* 28:27–31.

Elmes, G.W. 1983. Some experimental observations on the parasitic *Myrmica hirsuta* Elmes. *Insectes Sociaux* 30:221–234.

Elmes, G.W. 1987. Temporal variation in colony populations of the ant *Myrmica sulcinodis*. I. Changes in queen number, worker number and spring production. *Journal of Animal Ecology* 56:559–571.

Elmes, G.W. 1990. The regulation of polygyny and queen cycles in red ants (*Myrmica*). In: *Social Insects and the Environment, Proceedings of the 11th International Congress of IUSSI, 1990*, G.K. Veeresh, B. Mallik and C.A. Viraktamath (eds.). Oxford, New Delhi, pp. 249–250.

Elmes, G.W. and Keller, L. 1993. Distribution and ecology of queen number in ants of the genus *Myrmica*. In: *Queen Number and Sociality in Insects*, L. Keller (ed.). Oxford, New York, pp. 294–307.

Else, J.G. and Lee, P.C. 1986. *Primate Ontogeny, Cognition and Social Behavior*, Vol. 3. Cambridge University Press, Cambridge.

Elsholtz, H.P., Albert, V.R., Treacy, M.N. and Rosenfeld, M.G. 1990. A two-base change in a POU factor-binding site switches pituitary-specific to lymphoid-specific gene expression. *Genes and Development* 4:43–51.

Elwood, R.W. 1983. Paternal care in rodents. In: *Parental Behaviour of Rodents*, R.W. Elwood (ed.). John Wiley and Sons, London, pp. 235–257.

Emerson, A.E. 1928. Communication among termites. *Fourth International Congress of Entomology* 2:722–727.

Emerson, A.E. 1954 Dynamic homeostasis: a unifying principle in organic, social, and ethical evolution. *Scientific Monthly* 78:67–85

Emerson, A.E. 1958. The evolution of behavior among social insects. In: *Behavior and Evolution*, A. Roe and G.G. Simpson (eds.). Yale University Press, New Haven, pp. 311–335.

Emerson, S.B. 1984. Morphological variation in frog pectoral girdles: testing alternatives to a traditional adaptive explanation. *Evolution* 38:376–388.

Emerson, S.B. 1986. Heterochrony and frogs: the relationship of a life history trait to morphological form. *American Naturalist* 127:167–183.

Emerson, S.B. 1988. Testing for historical patterns of change: a case study with frog pectoral girdles. *Paleobiology* 14:174–186.

Emerson, S.B. 1990. The interaction of behavioral and morphological change in the evolution of a novel locomotor type: "flying" frogs. *Evolution* 44:1931–1946.

Emerson, S.B. 1996. Phylogenies and physiological processes—the evolution of sexual dimorphism in southeast Asian frogs. *Systematic Biology* 45:278–289.

Emerson, S.B. and Berrigan, D. 1993. Systematics of Southeast Asian ranids: multiple origins of voicelessness in the subgenus *Limnonectes* (Fitzinger). *Herpetology* 49:22–31.

Emerson, S.B. and Boyd, S.K. 1999. Mating vocalizations of female frogs: control and evolutionary mechanisms. *Brain, Behavior and Evolution* 53:187–197.

Emerson, S.B. and Ward, R. 1998. Male secondary sexual characteristics, sexual selection, and molecular divergence in fanged ranid frogs of Southeast Asia. *Zoological Journal of the Linnean Society* 122:537–553.

Emerson, S.B., Travis, J. and Koehl, M.A.R. 1990. Functional complexes and additivity in performance: a test case with "flying" frogs. *Evolution* 44:2153–2157.

Emerson, S.B., Carroll, L. and Hess, D.L. 1997. Hormonal induction of thumb pads and the evolution of secondary sexual characteristics of the Southeast Asian fanged frog, *Rana blythii*. *Journal of Experimental Zoology* 279:587–596.

Emlen D.J. 1994. Environmental control of horn length dimorphism in the beetle *Onthophagus acuminatus* (Coleoptera: Scarabaeidae). *Proceedings of the Royal Society of London B* 256:131–136.

Emlen, D.J. 1996. Artificial selection on horn length-body size allometry in the horned beetle *Onthophagus acuminatus* (Coleoptera: Scarabaeidae). *Evolution* 50:1219–1230.

Emlen, D.J. 1997a. Alternative reproductive tactics and male-dimorphism in the horned beetle *Onthophagus acuminatus* (Coleoptera: Scarabaeidae). *Behavioral Ecology and Sociobiology* 41:335–341.

Emlen, D.J. 1997b. Diet alters male horn allometry in the beetle *Onthophagus acuminatus* (Coleoptera: Scarabaeidae). *Proceedings of the Royal Society of London B* 264:567–574.

Emlen, D.J. 2000. Integrating development with evolution: a case study with beetle horns. *BioScience* 50:403–418.

Emlen, D.J. and Nijhout, H.F. 1999. Hormonal control of male horn length dimorphism in the dung beetle *Onthophagus taurus* (Coleoptera: Scarabaeidae). *Journal of Insect Physiology* 45:45–53.

Emlen, D.J. and Nijhout, H.F. 2000. The development and evolution of exaggerated morphologies in insects. *Annual Review of Entomology* 45:661–708.

Emlen, S.T. and Wrege, P.H. 1991. Breeding biology of white-fronted bee-eaters at Nakuru: the influence of helpers on breeding fitness. *Journal of Animal Ecology* 60:309–326.

Emlen, S.T., Reeve, H.K., Sherman, P.W., Wrege, P.H., Ratnieks, F.L.W. and Shellman-Reeve, J. 1991. Adaptive versus nonadaptive explanations of behavior: the case of alloparental helping. *American Naturalist* 138:259–270.

Emlen, S.T., Wrege, P.H. and Demong, N.J. 1995. Making decisions in the family: an evolutionary perspective. *American Scientist* 83:148–157.

Emlet, R.B. 1986. Facultative planktotrophy in the tropical echinoid *Clypeaster rosaceus* (Linnaeus) and a comparison with obligate planktotrophy in *Clypeaster subdepressus* (Gray) (Clypeasteroida: Echinoidea). *Journal of Experimental Marine Biology and Ecology* 95:183–202.

Emlet, R.B. 1990. World patterns of developmental mode in echinoid echinoderms. In: *Advances in Invertebrate Reproduction*, Vol. 5, M. Hoshi and O. Yamashita (eds.). Elsevier, Amsterdam, pp. 329–334.

Emlet, R.B. and Hoegh-Guldberg, O. 1997. Effects of egg size on postlarval performance: experimental evidence from a sea urchin. *Evolution* 51:141–152.

Emlet, R.B., McEdward, L.R. and Strathmann, R.R. 1987. Echinoderm larval ecology viewed from the egg. *Echinoderm Studies* 2:55–136.

Endler, J.A. 1977. *Geographic Variation, Speciation, and Clines*. Princeton University Press, Princeton.

Endler, J.A. 1986. *Natural Selection in the Wild*, 2nd ed. Princeton University Press, Princeton.

Endler, J.A. and McLellan, T. 1988. The processes of evolution: toward a newer synthesis. *Annual Review of Ecology and Systematics* 19:395–421.

Engels, W. (ed.) 1990. *Social Insects*. Springer-Verlag, New York.

Engels, W. and Imperatriz-Fonseca, V.L. 1990. Caste development, reproductive strategies, and control of fertility in honey bees and stingless bees. In: *Social Insects: An Evolutionary Approach to Castes and Reproduction*, W. Engels (ed.). Springer-Verlag, Berlin, pp. 167–230.

Ennos, A.R. 1993. The function and formation of buttresses. *Trends in Ecology and Evolution* 8:350–351.

Enteman, W.M. 1904. *Coloration in Polistes*. Carnegie Institution of Washington Publication No. 19. Carnegie Institution, Washington, D.C.

Epstein, E. 1994. The anomaly of silicon in plant biology. *Proceedings of the National Academy of Sciences USA* 91:11–17.

Erickson, R.P. 1990. Post-meiotic gene expression. *Trends in Genetics* 6:264–269.

Erwin, D.H. and Valentine, J.W. 1984. "Hopeful monsters", transposons and metazoan radiation. *Proceedings of the National Academy of Sciences USA* 81:5482–5483.

Erwin, D.H., Valentine, J.W. and Sepkoski, J.J., Jr. 1987. A comparative study of diversification events: the early Paleozoic versus the Mesozoic. *Evolution* 41:1177–1186.

Espelie, K.E., Gamboa, G.J., Grudzien, T.A. and Bura, E.A. 1994. Cuticular hydrocarbons of the paper wasp, *Polistes fuscatus*: a search for recognition pheromones. *Journal of Chemical Ecology* 20:1677–1687.

Evans, H.E. 1953. Comparative ethology and the systematics of spider wasps. *Systematic Zoology* 2:155–172.

Evans, H.E. 1962. The evolution of prey-carrying mechanisms in wasps. *Evolution* 16:468–483.

Evans, H.E. 1966. *The Comparative Ethology and Evolution of the Sand Wasps*. Harvard University Press, Cambridge.

Evans, H.E. and Matthews, R.W. 1973. Behavioural observations on some Australian spider wasps (Hymenoptera: Pompilidae). *Transactions of the Royal Entomological Society of London* 125:45–55.

Evans, H.E. and Matthews, R.W. 1975. The sand wasps of Australia. *Scientific American* 233:108–115.

Evans, J.D. and Wheeler, D.E. 1999. Differential gene expression between developing queens and workers in the honey bee, *Apis mellifera*. *Proceedings of the National Academy of Sciences USA* 96:5575–5580.

Evans, S.M. 1971. Behavior in polychaetes. *Quarterly Review of Biology* 46:379–405.

Ewer, R.F. 1960. Natural selection and neoteny. *Acta Biotheoretica* 13:161–184.

Ezhikov, I. 1934. Individual variability and dimorphism of social insects. *American Naturalist* 68:333–344.

Fabian, B. 1985. Ontogenetic explorations into the nature of evolutionary change. In:

Species and Speciation, E.S. Vrba (ed.). Transvaal Museum Monograph No. 4. Transvaal Museum, Pretoria, pp. 77–85.

Fagen, R. 1981. *Animal Play Behavior*. Oxford University Press, New York.

Fagen, R. 1987. Phenotypic plasticity and social environment. *Evolutionary Ecology* 1:263–271.

Fahrbach, S.E. and Robinson, G.E. 1995. Behavioral development in the honey bee: toward the study of learning under natural conditions. *Learning and Memory* 2:199–224.

Falconer, D.S. 1981. *Introduction to Quantitative Genetics*, 2nd ed. Longman, London.

Falconer, D.S. and Mackay, T.R.C. 1996. *Introduction to Quantitative Genetics*, 4th ed. Longman Group, Essex.

Farabaugh, S.M. 1982. The ecological and social significance of duetting. In: *Acoustic Communication in Birds. Volume 2: Song Learning and Its Consequences*, D.E. Kroodsma and G.H. Miller (eds.). Academic Press, New York, pp. 85–124.

Farmer, E.E. and Ryan, C.A. 1990. Interplant communication: airborne methyl jasmonate induces synthesis of proteinase inhibitors in plant leaves. *Proceedings of the National Academy of Sciences USA* 87:7713–7716.

Farner, D.S. 1980. Evolution of the control of reproductive cycles in birds. In: *Hormones, Adaptation and Evolution*, S. Ischii (ed.). Japan Scientific Societies Press, Tokyo, pp. 185–191.

Farris, J.S. 1966. Estimation of conservatism of characters by constancy within biological populations. *Evolution* 20:587–591.

Faulkes, C.G., Abbott, D.H., Liddell, C.E., George, L.M. and Jarvis, J.U.M. 1991. Hormonal and behavioral aspects of reproductive suppression in female naked mole-rats. In: *The Biology of the Naked Mole-Rat*, P.W. Sherman, J.U.M. Jarvis and R.D. Alexander (eds.). Princeton University Press, Princeton, pp. 426–445.

Feder, J.L. and Bush, G.L. 1989. Gene frequency clines for host races of *Rhagoletis pomonella* in the midwestern United States. *Heredity* 63:245–266.

Feder, J.L. and Filchak, K.E. 1999. It's about time: the evidence for host plant-mediated selection in the apple maggot fly, *Rhagoletis pomonella*, and its implications for fitness trade-offs in phytophagous insects. *Entomologia Experimentalis et Applicata* 91:211–225.

Feder, J.L., Chilcote, C.A. and Bush, G. 1988. Genetic differentiation between sympatric host races of the apple maggot fly *Rhagoletis pomonella*. *Nature* 336:61–64.

Feder, J.L., Chilcote, C.A. and Bush, G.L. 1990. The geographic pattern of genetic differentiation between host associated populations of *Rhagoletis pomonella* (Diptera: Tephritidae) in the eastern United States and Canada. *Evolution* 44:570–594.

Feder, J.L., Opp. S.B., Wlazlo, B., Reynolds, K., Go, W. and Spisak, S. 1994. Host fidelity is an effective premating barrier between sympatric races of the apple maggot fly. *Proceedings of the National Academy of Sciences USA* 91:7990–7994.

Feder, J.L., Roethele, J.B., Wlazlo, B. and Berlocher, S.H. 1997a. Selective maintenance of allozyme differences among sympatric host races of the apple maggot fly. *Proceedings of the National Academy of Sciences USA* 94:11417–11421.

Feder, J.L., Stolz, U., Lewis, L.M., Perry, W., Roethele, J.B. and Rogers, A. 1997b. The effects of winter length on the genetics of apple and hawthorn races of *Rhagoletis pomonella* (Diptera: Tephritidae). *Evolution* 51:1862–1876.

Federoff, N.V. 1999. Transposable elements as a molecular evolutionary force. *Annals of the New York Academy of Sciences* 870:251–264.

Feldman, D. 1988. Evidence for the presence of steroid hormone receptors in fungi. *Steroid Hormone Action*, 1988:169–176.

Felsenstein, J. 1981. Skepticism towards Santa Rosalia, or why are there so few kinds of animals? *Evolution* 35:124–138.

Felsenstein, J. 1985. Recombination and sex: is Maynard Smith necessary? In: *Evolution. Essays in honour of John Maynard Smith*, P.J. Greenwood, P.H. Harvey, and M. Slatkin (eds.). Cambridge University Press, Cambridge, pp. 209–220.

Fentress, J.C. 1983. A view of ontogeny. In: *Structure, Development and Function*, J. Eisenberg and D. Kleiman (eds.). *Special Publications American Society of Mammalogists* 7:24–64.

Fentress, J.C. 1990. The categorization of behavior. In: *Interpretation and Explanation in the Study of Animal Behavior. Volume 1: Interpretation, Intentionality, and Communication*. Westview Press, Boulder, pp. 7–34.

Fergeson, A. and Taggart, J.B. 1991. Genetic differentiation among the sympatric brown trout (*Salmo trutta*) populations of Lough Melvin, Ireland. *Biological Journal of the Linnean Society* 43:221–237.

Ferkowicz, M.J. and Raff, R.A. 2001. Wnt gene expression in sea urchin development: heterochronies associated with the evolution of developmental mode. *Evolution and Development* 3:24–33.

Ferrari, F.D. 1988. Developmental patterns in numbers of ramal segments of copepod post-maxillipedal legs. *Crustaceana* 54:256–293.

Field, J. 1989. Alternative nesting tactics in a solitary wasp. *Behaviour* 110:219–243.

Field, J. 1992a. Intraspecific parasitism as an alternative reproductive tactic in nest-building wasps and bees. *Biological Reviews* 67:79–126.

Field, J. 1992b. Patterns of nest provisioning and parental investment in the solitary digger wasp *Ammophila sabulosa*. *Ecological Entomology* 17:43–51.

Field, J., Solis, C.R., Queller, D.C. and Strassmann, J.E. 1998. Social and genetic structure of paper wasp cofoundress associations: tests of reproductive skew models. *American Naturalist* 151:545–563.

Filchak, K.E., Feder, J.L., Roethele, J.B. and Stolz, U. 1999. A field test for host-plant dependent selection on larvae of the apple maggot fly, *Rhagoletis pomonella*. *Evolution* 53:187–200.

Filchak, K.E., Roethele, J.B. and Feder, J.L. 2000. Natural selection and sympatric divergence in the apple maggot *Rhagoletis pomonella*. *Nature* 407:739–742.

Finch, C.E. and Rose, M.R. 1995. Hormones and the physiological architecture of life history evolution. *Quarterly Review of Biology* 70:1–52.

Fincke, O.M. 1994. Female colour polymorphism in damselflies: failure to reject the null hypothesis. *Animal Behaviour* 47:1249–1266.

Fischer, A.G. 1981. Climatic oscillations in the biosphere. In: *Biotic Crises in Ecological and Evolutionary Time*, M.H. Nitecki (ed.). Academic Press, New York, pp. 103–131.

Fischer, E.A. 1988. Simultaneous hermaphroditism, tit-for-tat, and the evolutionary stability of social systems. *Ethology and Sociobiology* 9:119–136.

Fisher, R.A. 1927. On some objections to mimicry theory: statistical and genetic. *Transactions of the Royal Entomological Society of London* 75:269–278.

Fisher, R.A. 1930. *The Genetical Theory of Natural Selection*. Dover, New York.

Fisher, R.A. 1958. *The Genetical Theory of Natural Selection*, 2nd ed. Dover, New York.

Fitch, W.M. 1970. Distinguishing homologous from analogous proteins. *Systematic Zoology* 19:99–113.

Fitt, G.P. 1986. The influence of a shortage of hosts on the specificity of oviposition behaviour in species of *Dacus* (Diptera,

Tephritidae). *Physiological Entomology* 11:133–134.

Flaishman, M.A. and Kolattukudy, P.E. 1994. Timing of fungal invasion using host's ripening hormone as a signal. *Proceedings of the National Academy of Sciences USA* 91:6579–6583.

Flannagan, R.D., Tammariello, S.P., Joplin, K.H., Cikra-Ireland, R.A., Yocum, G.D. and Denlinger, D.L. 1998. Diapause-specific gene expression in pupae of the flesh fly *Sarcophaga crassipalpis*. *Proceedings of the National Academy of Sciences USA* 95:5616–5620.

Fleming, I.A. 1996. Reproductive strategies of Atlantic salmon: ecology and evolution. *Reviews in Fish Biology and Fisheries* 6:379–416.

Fleming, I.A. and Gross, M.R. 1989. Evolution of adult female life history and morphology in a Pacific salmon (Coho: *Oncorhynchus kisutch*). *Evolution* 43:141–157.

Fleming, I.A. and Gross, M.R. 1994. Breeding competition in a Pacific salmon (Coho: *Oncorhynchus kisutch*): measures of natural and sexual selection. *Evolution* 48:637–657.

Fletcher, D.J.C. and Michener, C.D. (eds.) 1987. *Kin Recognition in Animals*. John Wiley and Sons, New York.

Fol, H. 1879. Recherches sur la fécondation et la commencement de l'hénogénie chez divers animaux. *Mémoires de la Société de Physique et Histoire Naturelle de Genève* 26:89–397.

Fontana, W., Wagner, G. and Buss, L.W. 1994. Beyond digital naturalism. *Artificial Life* 1:211–227.

Foote, C.J. 1988. Male mate choice dependent on male size in salmon. *Behaviour* 106:63–80.

Foote, C.J. and Larkin, P.A. 1988. The role of male choice in the assortative mating of anadromous and non-anadromous sockeye salmon (*Oncorhynchus nerka*). *Behaviour* 106:43–61.

Forbes, S.H., Knudsen, K.L., North, T.W. and Allendorf, F.W. 1994. One of two growth hormone genes in coho salmon is sex-linked. *Proceedings of the National Academy of Sciences USA* 91:1628–1631.

Ford, E.B. 1940. Polymorphism and taxonomy. In: *The New Systematics*, J.S. Huxley (ed.). Clarendon Press, Oxford, pp. 510–513.

Ford, E.B. 1961. The theory of genetic polymorphism. In: *Insect Polymorphism*, J.S. Kennedy (ed.). Symposium No. 1. Royal Entomological Society, London, pp. 11–19.

Ford, E.B. 1964. *Ecological Genetics*. Methuen, London.

Ford, E.B. 1971. *Ecological Genetics*. Chapman and Hall, London.

Ford, E.B. 1980. Some recollections pertaining to the evolutionary synthesis. In: *The Evolutionary Synthesis. Perspectives on the Unification of Biology*, E. Mayr and W. Provine (eds.). Harvard University Press, Cambridge, pp. 334–342.

Forsdyke, D.R. 1999. Two levels of information in DNA: relationships of Romanes' "intrinsic" variability of the reproductive system, and Bateson's "residue" to the species-dependent component of the base composition, (C+G). *Journal of Theoretical Biology* 201:47–61.

Forshaw, J.M. 1977. *Parrots of the World*. Lansdowne Press, Melbourne.

Forster, R.R. and Wilton, C.L. 1968. *The Spiders of New Zealand*. John McIndoe, Dunedin.

Forsyth, A. 1975. Usurpation and dominance behavior in the polygynous social wasp, *Metapolybia cingulata* (Hymenoptera: Vespidae; Polybiini). *Psyche* 82:299–302.

Forsyth, A. 1980. Worker control of queen density in hymenopteran societies. *American Naturalist* 116:895–898.

Forsyth, A. and Alcock, J. 1990. Female mimicry and resource defence polygyny by males of a tropical rove beetle, *Leistotrophus versicolor* (Coleoptera: Staphylinidae). *Behavioral Ecology and Sociobiology* 26:325–330.

Fortey, R.A. and Owens, R.M. 1990. Evolutionary radiations in the Trilobita. In: *Major Evolutionary Radiations*, P.D. Taylor and G.P. Larwood (eds.). Clarendon Press, Oxford, pp. 139–164.

Fox, C.W. and Savalli, U.M. 1998. Inheritance of environmental variation in body size: superparasitism of seeds affects progeny and grandprogeny body size via a nongenetic maternal effect. *Evolution* 52:172–182.

Fox, S.W. 1980. The origins of behavior in macromolecules and protocells. *Comparative Biochemistry and Physiology* 67B:423–436.

Fox, S.W. 1984. The beginnings of life and behavior. In: *Behavioral Evolution and Integrative Levels*, G. Greenberg and E. Tobach (eds.). Lawrence Erlbaum, Hillsdale, NJ, pp. 83–103.

Fraenkel, G.S. and Gunn, D.L. 1961. *The Orientation of Animals*. Dover, New York.

Francis, R.C. and Barlow, G.W. 1993. Social control of primary sex differentiation in the Midas cichlid. *Proceedings of the National Academy of Sciences USA* 90:10673–10675.

Frandsen, T.C., Boyd, S.S. and Berger, P.J. 1993. Maternal transfer of the 6–MBOA chemical signal in *Microtus montanus* during gestation and lactation. *Canadian Journal of Zoology* 71:1799–1803.

Frank, L.G. 1997. Evolution of genital masculinization: why do female hyaenas have such a large "penis"? *Trends in Ecology and Evolution* 12:58–62.

Frank, L.G., Glickman, S.E. and Powch, I. 1990. Sexual dimorphism in the spotted hyaena (*Crocuta crocuta*). *Journal of Zoology* 221:308–313.

Frank, L.G., Weldele, M.L. and Glickman, S.E. 1995. Masculinization costs in hyaenas. *Nature* 377:584–585.

Frank, S.A. 1996. The design of natural and artificial adaptive systems. In: *Adaptation*, M.R. Rose and G.V. Lauder (eds.). Academic Press, New York, pp. 451–505.

Frankel, J. 1983. What are the developmental underpinnings of evolutionary changes in protozoan morphology? In: *Development and Evolution*, B.C. Goodwin, N. Holder and C.G. Wylie (eds.). Cambridge University Press, Cambridge, pp. 279–314.

Frankel, J. 1984. Development and evolution: a report. *Evolution* 38:1160–1162.

Frankel, J. 1992. Positional information in cells and organisms. *Trends in Cell Biology* 2:256–260.

Franks, N.R. and Hölldobler, B. 1987. Sexual competition during colony reproduction in army ants. *Biological Journal of the Linnean Society* 30:229–243.

Frauenfelder, H. and McMahon, B. 1998. Dynamics and function of proteins: the search for general concepts. *Proceedings of the National Academy of Sciences USA* 95:4795–4797.

Frazzetta, T.H. 1975. *Complex Adaptations in Evolving Populations*. Sinauer, Sunderland, MA.

Free, J.B. 1987. *Pheromones of Social Bees*. Cornell University Press, Ithaca.

Freed, L.A., Conant, S. and Fleischer, R.C. 1987. Evolutionary ecology and radiation of Hawaiian passerine birds. *Trends in Ecology and Evolution* 2:196–203.

Freeling, M., Bertrand-Garcia, R. and Sinha, N. 1992. Maize mutants and variants altering developmental time and their heterochronic interactions. *BioEssays* 14:227–236.

Freeman, D.A. and Goldman, B.D. 1997. Evidence that the circadian system mediates photoperiodic nonresponsiveness in Siberian hamsters: the effect of running wheel access on photoperiodic responsiveness. *Journal of Biological Rhythms* 12:100–109.

Fretwell, S.D. 1972. *Populations in a Seasonal Environment*. Princeton University Press, Princeton.

Fretwell, S.D. and Lucas, H.L. 1970. On territorial behaviour and other factors influencing habitat distribution in birds. *Acta Biotheoretica* 19:16–36.

Friedman, W.E. 1993. The evolutionary history of the seed plant male gametophyte. *Trends in Ecology and Evolution* 8:15–21.

Friedman, W.E. 1995. Organismal duplication, inclusive fitness theory, and altruism: understanding the evolution of endosperm and the angiosperm reproductive syndrome. *Proceedings of the National Academy of Sciences USA* 92:3913–3917.

Frisch, K., von. 1950. *Bees: Their Vision, Chemical Senses, and Language*. Great Seal Books, Cornell University Press, Ithaca.

Frumhoff, P.C. and Baker, J. 1988. A genetic component to division of labour within honey bee colonies. *Nature* 333:358–361.

Frumhoff, P.C. and Reeve, H.K. 1994. Using phylogenies to test hypotheses of adaptation: a critique of some current proposals. *Evolution* 48:172–180.

Fryer, G. 1959. Some aspects of evolution in Lake Nyassa. *Evolution* 13:440–451.

Fryer, G. and Iles, T.D. 1972. *The Cichlid Fishes of the Great Lakes of Africa. Their Biology and Evolution*. Oliver and Boyd, Edinburgh.

Funk, V.A. and Brooks, D.R. 1992. *Phylogenetic Systematics as the Basis of Comparative Biology*. Smithsonian Institution Press, Washington, D.C.

Futuyma, D.J. 1979. *Evolutionary Biology*. 1st ed. Sinauer, Sunderland, MA.

Futuyma, D.J. 1982. *Science on Trial*. Pantheon Books, New York.

Futuyma, D.J. 1986a. *Evolutionary Biology*. 2nd ed. Sinauer, Sunderland, MA.

Futuyma, D.J. 1986b. The role of behavior in host-associated divergence in herbivorous insects. In: *Evolutionary Genetics of Invertebrate Behavior*, M.D. Huettel (ed.). Plenum Press, New York, pp. 295–302.

Futuyma, D.J. 1987. On the role of species in anagenesis. *American Naturalist* 130:465–473.

Futuyma, D.J. 1998. *Evolutionary Biology*, 3rd ed. Sinauer, Sunderland, MA.

Futuyma, D.J. and Mayer, G.C. 1980. Non-allopatric speciation in animals. *Systematic Zoology* 29:254–271.

Futuyma, D.J. and Moreno, G. 1988. The evolution of ecological specialization. *Annual Review of Ecology and Systematics* 19:207–233.

Gadagkar, R. 1991. The role of larval nutrition in pre-imaginal biasing of caste in the primitively eusocial wasp *Ropalidia marginata* (Hymenoptera: Vespidae). *Ecological Entomology* 16:435–440.

Gadagkar, R. 1997a. Social evolution—has nature ever rewound the tape? *Current Science* 72:950–956.

Gadagkar, R. 1997b. The evolution of caste polymorphism in social insects: genetic release followed by diversifying evolution. *Journal of Genetics* 76:167–179.

Gadau, J., Page, R.E., Jr., Werren, J.H. and Schmid-Hempel, P. 2000. Genome organization and social evolution in Hymenoptera. *Naturwissenschaften* 87:87–89.

Gadgil, M. 1972. Male dimorphism as a consequence of sexual selection. *American Naturalist* 106:574–580.

Gagné, R.J. 1989. *The Plant-Feeding Gall Midges of North America*. Cornell University Press, Ithaca.

Galis, F. 1993a. Interactions between the pharyngeal jaw apparatus, feeding behaviour, and ontogeny in the cichlid fish, *Haplochromis piceatus*: a study of morphological constraints in evolutionary ecology. *Journal of Experimental Zoology* 267:137–154.

Galis, F. 1993b. Morphological constraints on behaviour through ontogeny: the importance of developmental constraints. *Marine Behavior and Physiology* 23:119–135.

Galis, F. 1996a. The application of functional morphology to evolutionary studies. *Trends in Ecology and Evolution* 11:124–129.

Galis, F. 1996b. The evolution of insects and vertebrates: homeobox genes and homology. *Trends in Ecology and Evolution* 11:402–403.

Galis, F. and Drucker, E.G. 1996. Pharyngeal biting mechanics in centrarchid and cichlid fishes: insights into a key evolutionary innovation. *Evolutionary Biology* 9:641–670.

Galis, F., Terlouw, A. and Osse, J.W.M. 1994. The relation between morphology and behaviour during ontogenetic and evolutionary changes. *Journal of Fish Biology* 45(Supplement A):13–26.

Gallistel, C.R. 1980. *The Organization of Action: A New Synthesis*. Lawrence Erlbaum, Hillsdale, NJ.

Gamboa, G.J. 1996. Kin recognition in social wasps. In: *Natural History and Evolution of Paper-Wasps*, S. Turillazzi and M.J. West-Eberhard (eds.). Oxford University Press, Oxford, pp. 161–177.

Gamboa, G.J., Grudzien, T.A., Espelie, K.A. and Bura, E.A. 1996. Kin recognition pheromones in social wasps: combining chemical and behavioural evidence. *Animal Behaviour* 51:625–629.

Gan, Y.-Y. and Robertson, F.W. 1981. Isozyme variation in some rain forest trees. *Biotropica* 13:20–28.

Gans, C. 1979. Momentarily excessive construction as the basis for protoadaptation. *Evolution* 33:227–233.

García-Bellido, A. 1977. Homoeotic and atavic mutations in insects. *American Zoologist* 17:613–629.

García-Bellido , A., Lawrence, P.A. and Morata, G. 1979. Compartments in animal development. *Scientific American* 241:102–110.

Garstang, W. 1929 [1985]. The origin and evolution of larval forms. In: *Larval Forms and Other Zoological Verses*. University of Chicago Press, Chicago, pp. 77–98.

Gary, N.E. 1975. Activities and behavior of honey bees. In: *The Hive and the Honey Bee*, Dadant and Sons (eds.). Dadant and Sons, Hamilton, IL, pp. 185–264.

Gasaway, W.C. 1976. Seasonal variation in diet, volatile fatty acid production and size of the cecum of rock ptarmigan. *Comparative Biochemistry and Physiology* 53A:109–114.

Gass, C.L., Angehr, G. and Centa, J. 1976. Regulation of food supply by feeding territoriality in the rufous hummingbird. *Canadian Journal of Zoology* 54:2046–2054.

Gauld, I. and Bolton, B. 1988. *The Hymenoptera*. Oxford University Press, New York.

Gause, G.F. 1942. The relation of adaptability to adaptation. *Quarterly Review of Biology* 17:99–114.

Geary, D.H. 1992. An unusual pattern of divergence between two fossil gastropods: ecophenotypy, dimorphism, or hybridization? *Paleobiology* 18:93–109.

Geist, V. 1966. The evolution of horn-like organs. *Behaviour* 27:175–214.

Geist, V. 1978. *Life Strategies, Human Evolution, Environmental Design*. Springer-Verlag, New York.

Geoffroy St. Hilaire, E. 1822. *Philosophie Anatomique*. Paris.

George, S.B. 1990. Population and seasonal differences in egg quality of *Arbacia lixula* (Echinodermata: Echinoidea). *Invertebrate Reproduction and Development* 12:111–121.

Gerhart, J. and Kirschner, M. 1997. *Cells, Embryos, and Evolution: Toward a Cellular and Developmental Understanding of Phenotypic Variation and Evolutionary Adaptability*. Blackwell, Malden, MA.

Ghiselin, M.T. 1969. *The Triumph of the Darwinian Method*. University of California Press, Berkeley.

Ghiselin, M.T. 1974. *The Economy of Nature and the Evolution of Sex*. University of California Press, Berkeley.

Ghiselin, M.T. 1975. The rationale of pangenesis. *Genetics* 79:47–57.

Ghiselin, M.T. 1987. Evolutionary aspects of marine invertebrate reproduction. In: *Reproduction of Marine Invertebrates*.

Volume 9. *General Aspects: Seeking Unity in Diversity*, A.C. Giese, J.S. Pearse, and V.B. Pearse (eds.). Blackwell Scientific, Palo Alto, pp. 609–665.

Ghiselin, M.T. 1988. The evolution of sex: a history of competing points of view. In: *The Evolution of Sex: An Examination of Current Ideas*, R.E. Michod and B.R. Levin (eds.). Sinauer, Sunderland, MA, pp. 7–23.

Gibbons, J.R.H. 1979. A model for sympatric speciation in *Megarhyssa* (Hymenoptera: Ichneumonidae): competitive speciation. *American Naturalist* 114:719–741.

Giddings, L.V., Kaneshiro, K.Y. and Anderson, W.W. (eds.) 1989. *Genetics, Speciation, and the Founder Principle*. Oxford University Press, New York.

Giesel, J.T. 1988. Effects of parental photoperiod on development time and density sensitivity of progeny of *Drosophila melanogaster*. *Evolution* 42:1348–1350.

Gilbert, J.J. 1973. The adaptive significance of polymorphism in the rotifer *Asplanchna*. Humps in males and females. *Oecologia* 13:135–146.

Gilbert, J.J. 1977. Defenses of males against cannibalism in the rotifer *Asplanchna*: size, shape, and failure to elicit tactile feeding responses. *Ecology* 58:1128–1135.

Gilbert, J.J. 1980. Female polymorphism and sexual reproduction in the rotifer *Asplanchna*: evolution of their relationship and control by dietary tocopherol. *American Naturalist* 116:409–431.

Gilbert, S.F. 1991. Epigenetic landscaping: Waddington's use of cell fate bifurcation diagrams. *Biology and Philosophy* 6:135–154.

Gilbert, W. 1978. Why genes in pieces? *Nature* 271:501.

Gilbert, W. and Glynias, M. 1993. On the ancient nature of introns. *Gene* 135:137–144.

Gill, F.B. 1995. *Ornithology*. W.H. Freeman, New York.

Gill, F.B. and Wolf, L.L. 1975. Economics of feeding territoriality in the golden-winged sunbird. *Ecology* 56:333–345.

Gill, F.B. and Wolf, L.L. 1977. Non-random foraging by sunbirds in a patchy environment. *Ecology* 58:1284–1296.

Gillespie, J.H. 1991. *The Causes of Molecular Evolution*. Oxford University Press, New York.

Gilliam, J.F. and Fraser, D.F. 1987. Habitat selection under predation hazard: test of a model with foraging minnows. *Ecology* 68:1856–1862.

Gingerich, P.D. 1977. Patterns of evolution in the mammalian fossil record. In: *Patterns of*

Evolution, A. Hallam (ed.). Elsevier, New York, pp. 469–500.

Giraldeau, L.A. 1984. Group foraging: the skill pool effect and frequency-dependent learning. *American Naturalist* 124:72–79.

Giraldeau, L.A. and Lefebvre, L. 1985. Individual feeding preferences in feral groups of rock doves. *Canadian Journal of Zoology* 63:189–191.

Gittleman, J.L., Anderson, C.G., Kot, M. and Luh, H.K. 1996. Comparative tests of evolutionary lability and rates using molecular phylogenies. In: *New Uses for New Phylogenies*, P.H. Harvey, A.J. Leigh Brown, J. Maynard Smith and S. Nee (eds.). Oxford University Press, Oxford, pp. 289–307.

Givnish, T.J., Sytsma, K.J., Smith, J.F and Hahn, W.J. 1994. Thorn-like prickles and heterophylly in *Cyanea*: adaptations to extinct avian browsers on Hawaii? *Proceedings of the National Academy of Sciences USA* 91:2810–2814.

Glass, R.E. 1982. *Gene Function*. E. coli *and Its Heritable Elements*. University of California Press, Berkeley.

Go, M. 1994. Origin of introns: were the original blocks of proteins encoded by exons? [abstract]. Workshop Program, Workshop on Open Questions in Molecular Evolution, Guanacaste, Costa Rica.

Go, M. and Nosaka, M. 1987. Protein architecture and the origin of introns. *Cold Spring Harbor Symposia in Quantitative Biology* 52:915–924.

Godfray, H.C.J. 1988. Virginity in haplodiploid populations: a study on fig wasps. *Ecological Entomology* 13:283–291.

Godfray, H.C.J. 1991. The signaling of need by offspring to their parents. *Nature* 352:328–330.

Godin, J.G.J. 1995. Predation risk and alternative mating tactics in male Trinidadian guppies (*Poecilia reticulata*). *Oecologia* 103:224–229.

Goethe, J.W. von. 1790. Versuch die metamorphose der pflanzen zu erklären. Translation by E.M. Cox. 1863. Essay on the metamorphosis of plants. *Journal of Botany* 1:327–345, 360–374.

Goldfoot, D.A., Westerborg-van Loon, H., Groeneveld, W. and KoosSlob, A. 1980. Behavioral and physiological evidence of sexual climax in the female stump-tailed macaque (*Macaca arctoides*). *Science* 208:1477–1478.

Golding, B. (ed.) 1993. *Non-neutral Evolution*. Chapman and Hall, New York.

Golding, G.B., Tsao, N. and Pearlman, R.E. 1994. Evidence for intron capture: an unusual path for the evolution of proteins.

Proceedings of the National Academy of Sciences USA 91:7506–7509.

Goldman, B.D. and Nelson, R.J. 1993. Melatonin and seasonality in mammals. In: *Melatonin*, H.S. Yu and R.J. Reiter (eds.). CRC Press, Boca Raton, pp. 225–252.

Goldschmidt, R. 1935. Gen und Ausseneigenschaft. *Zeitschrift für Induktive Abstammungs und Verebunglehre* 69:38–131.

Goldschmidt, R. 1938. *Physiological Genetics*. McGraw-Hill, New York.

Goldschmidt, R. 1940 [1982]. *The Material Basis of Evolution*. Yale University Press, New Haven.

Goodall, J. 1986. *The Chimpanzees of Gombe*. Belknap Press, Cambridge.

Goodman, M. 1976. Protein sequences in phylogeny. In: *Molecular Evolution*, F.J. Ayala (ed.). Sinauer, Sunderland, MA, pp. 141–159.

Goodnight, C.J. 1988. Epistasis and the effect of founder events on the additive genetic variance. *Evolution* 42:441–454.

Goodwin, B.C. 1994. *How the Leopard Changed Its Spots*. Charles Scribner's Sons, New York.

Goodwin, F.K. and Jamison, K.R. 1990. *Manic-Depressive Illness*. Oxford University Press, New York.

Gordon, D.M. 1992. Phenotypic plasticity. In: *Keywords in Evolutionary Biology*, E.F. Keller and E.A. Lloyd (eds.). Harvard University Press, Cambridge, pp. 255–262.

Gordon, D.M. 1995a. The expandable network of ant exploration. *Animal Behaviour* 50:995–1007.

Gordon, D.M. 1995b. The development of organization in an ant colony. *American Scientist* 83:50–57.

Gordon, D.M. 1996. The organization of work in social insect colonies. *Nature* 380:121–124.

Gore, R.H. 1985. Molting and growth in decapod larvae. In: *Larval Growth*, A.M. Wenner (ed.). A.A. Balkema, Boston, pp. 1–66.

Gorman, M.R., Goldman, B.D. and Zucker, I. 2001. Mammalian photoperiodism. In: *Handbook of Behavioral Neurobiology (Vol. 12) Circadean Clocks*, J.S. Takahashi, F.W. Turek, and R.Y. Moore (eds.). Plenum Press, New York, pp. 481–508.

Gosliner, T.M. and Ghiselin, M.T. 1984. Parallel evolution in opisthobranch gastropods and its implications for phylogenetic methodology. *Systematic Zoology* 33:255–274.

Gotthard, K. and Nylin, S. 1995. Adaptive plasticity and plasticity as an adaptation: a selective review of plasticity in animal morphology and life history. *OIKOS* 74:3–17.

Gottlieb, G. 1992. *Individual Development and Evolution. The Genesis of Novel Behavior.* Oxford, New York.

Gottlieb, L.D. 1984. Genetics and morphological evolution in plants. *American Naturalist* 123:681–709.

Gotwald, W.H., Jr. 1995. *Army Ants. The Biology of Social Predation.* Cornell University Press, Ithaca.

Gould, S.J. 1969. An evolutionary microcosm: Pleistocene and recent history of the land snail P. (*Poecilozonites*) in Bermuda. *Bulletin of the Museum of Comparative Zoology* 138:407–531.

Gould, S.J. 1970. Dollo on Dollo's law: irreversibility and the status of evolutionary laws. *Journal of the History of Biology* 3:189–212.

Gould, S.J. 1974. The origin and function of "bizarre" structures: antler size and skull size in the "irish elk," *Megaloceros giganteus*. *Evolution* 28:191–220.

Gould, S.J. 1977. *Ontogeny and Phylogeny.* Belknap Press, Cambridge.

Gould, S.J. 1979. On the importance of heterochrony for evolutionary biology. *Systematic Zoology* 28:224–226.

Gould, S.J. 1980. Is a new and general theory of evolution emerging? *Paleobiology* 6:119–130.

Gould, S.J. 1981. Hyaena myths and realities. *Natural History* 90:16–24.

Gould, S.J. 1982a. Change in developmental timing as a mechanism of macroevolution. In: *Evolution and Development*, J.T. Bonner (ed.). Springer-Verlag, Berlin, pp. 333–346.

Gould, S.J. 1982b. The uses of heresy: an introduction to Richard Goldschmidt's *The Material Basis of Evolution*. In: *The Material Basis of Evolution*. Yale University Press, New Haven, pp. xiii-xlii.

Gould, S.J. 1984. Covariance sets and ordered geographic variation in Cerion from Aruba, Bonaire and Curacao: a way of studying nonadaptation. *Systematic Zoology* 33:217–237.

Gould, S.J. 1988. The uses of heterochrony. In: *Heterochrony in Evolution*, M.L. McKinney (ed.). Plenum Press, New York, pp. 1–13.

Gould, S.J. 1989a. A developmental constraint in *Cerion*, with comments on the definition and interpretation of constraint in evolution. *Evolution* 43:516–539.

Gould, S.J. 1989b. Punctuated equilibrium in fact and theory. In: *The Dynamics of Evolution*, A. Somit and S.A. Peterson (eds.). Cornell University Press, Ithaca, pp. 54–84.

Gould, S.J. 1993. The inexorable logic of the punctuational paradigm: Hugo de Vries on species selection. In: *Evolutionary Patterns and Processes*, D.R. Lees and D. Edwards (eds.). Academic Press, London, pp. 3–18.

Gould, S.J. 1994. Tempo and mode in the macroevolutionary reconstruction of Darwinism. *Proceedings of the National Academy of Sciences USA* 91:6764–6771.

Gould, S.J. 1999. Gulliver's further travels: the necessity and difficulty of a hierarchical theory of selection. In: *Evolution of Biological Diversity*, A. Magurran and R.M. May (eds.). Oxford University Press, Oxford, pp. 220–235.

Gould, S.J. and Eldredge, N. 1977. Punctuated equilibria: the tempo and mode of evolution reconsidered. *Paleobiology* 3:115–151.

Gould, S.J. and Eldredge, N. 1993. Punctuated equilibrium comes of age. *Nature* 366:223–227.

Gould, S.J. and Lewontin, R.C. 1979. The spandrels of San Marco and the Panglossian paradigm: a critique of the adaptationist programme. *Proceedings of the Royal Society of London B* 205:581–598.

Goy, R.W. and McEwen, B.S. 1980. *Sexual Differentiation of the Brain.* Massachusetts Institute of Technology Press, Cambridge.

Graham, J.B., Rosenblatt, R.H. and Gans, C. 1978. Vertebrate air breathing arose in fresh waters and not in the oceans. *Evolution* 32:459–463.

Graham, J.B., Dudley, R., Aguilar, N.M. and Gans, C. 1995. Implications of the late Paleozoic oxygen pulse for physiology and evolution. *Nature* 375:117–120.

Grant, B.R. 1985. Selection on bill characters in a population of Darwin's finches: *Geospiza conirostris* on Isla Genovesa, Galapagos. *Evolution* 39:523–532.

Grant, B.R. 1990. The significance of subadult plumage in Darwin's finches, *Geospiza fortis*. *International Society for Behavioral Ecology* 1:161–170.

Grant, B.R. 1996. Pollen digestion by Darwin's finches and its importance for early breeding. *Ecology* 77:489–499.

Grant, B.R and Grant, P.R. 1989. *Evolutionary Dynamics of a Natural Population.* University of Chicago Press, Chicago.

Grant, B.R. and Grant, P.R. 1996. Cultural inheritance of song and its role in the evolution of Darwin's finches. *Evolution* 50:2471–2487.

Grant, B.R. and Grant, P.R. 1996. High survival of Darwin's finch hybrids: effects of beak morphology and diets. *Ecology* 77:500–509.

Grant, P.R. 1972. Convergent and divergent character displacement. *Biological Journal of the Linnean Society* 4:39–68.

Grant, P.R. 1975. The classical case of character displacement. *Evolutionary Biology* 8:237–337.

Grant, P.R. 1986. *Ecology and Evolution of Darwin's Finches*. Princeton University Press, Princeton.

Grant, P.R. 1994. Population variation and hybridization: comparison of finches from two archipelagos. *Evolutionary Ecology* 8:598–617.

Grant, P.R. 2001. Reconstructing the evolution of birds on islands: 100 years of research. *OIKOS* 92:385–403.

Grant, P.R. and Grant, B.R. 1989. Sympatric speciation and Darwin's finches. In: *Speciation and Its Consequences*, D. Otte and J.A. Endler (eds.). Sinauer, Sunderland, MA, pp. 433–457.

Grant, P.R. and Grant, B.R. 1995a. The founding of a new population of Darwin's finches. *Evolution* 49:229–240.

Grant, P.R. and Grant, B.R. 1995b. Predicting microevolutionary responses to directional selection on heritable variation. *Evolution* 49:241–251.

Grant, P.R. and Grant, B.R. 1997. Mating patterns of Darwin's finch hybrids determined by song and morphology. *Biological Journal of the Linnean Society* 60:317–343.

Grant, P.R. and Grant, B.R. 1998. Speciation and hybridization of birds on islands. In: *Evolution on Islands*, P.R. Grant (ed.). Oxford University Press, New York, pp. 142–162.

Grant, V. 1963. *The Origin of Adaptations*. Columbia University Press, New York.

Grant, V. 1971. *Plant Speciation*. Columbia University Press, New York.

Grant, V. 1977. *Organsimic Evolution*. W.H. Freeman, San Francisco.

Grassé, P.P. 1959. La reconstruction du nid et les coordinations interindividuelles chez *Bellicositermes natalensis* et *Cubitermes* sp. La théorie de la stigmergie: essai d'interprétation du comportement des termites constructeurs. *Insectes Sociaux* 6:41–83.

Grassé, P.P. 1973. *L'evolution du Vivant*. Editions Albin Michel, Paris.

Grassé, P.P. 1977. *L'evolution du Vivant*. H. Blume Ediciones, Madrid.

Greenberg, B. 1961. Spawning and parental behavior in female pairs of the jewel fish, *Hemichromis bimaculatus* Gill. *Revista Brasileira de Entomología* 18:44–61.

Greenberg, L. 1979. Genetic component of bee odor in kin recognition. *Science* 206:1095–1097.

Greenberg, R. 1983. The role of neophobia in determining the degree of foraging specialization in some migrant warblers. *American Naturalist* 122:444–453.

Greenberg, R. 1985. A comparison of foliage discrimination learning in a specialist and a generalist species of migrant wood warbler (Aves: Parulidae). *Canadian Journal of Zoology* 63:773–776.

Greenberg, R. 1987a. Development of dead leaf foraging in a tropical migrant warbler. *Ecology* 68:130–141.

Greenberg, R. 1987b. Seasonal foraging specialization in the worm eating warbler. *Condor* 89:158–168.

Greenberg, R. 1990. Feeding neophobia and ecological plasticity: a test of the hypothesis with captive sparrows. *Animal Behaviour* 39:375–379.

Greene, E. 1989. A diet-induced developmental polymorphism in a caterpillar. *Science* 243:643–646.

Greene, E. 1996. Effect of light quality and larval diet on morph induction in the polymorphic caterpillar *Nemoria arizonaria* (Lepidoptera: Geometridae). *Biological Journal of the Linnean Society* 58:277–285.

Greene, E., Lyon, B.E., Muehter, V.R., Ratcliffe, L., Oliver, S.J. and Boag, P.T. 2000. Disruptive sexual selection for plumage colouration in a passerine bird. *Nature* 407:1000–1003.

Greene, H.W. 1994. Homology and behavioral repertoires. In: *Homology: The Hierarchical Basis of Comparative Biology*, B.K. Hall (ed.). Academic Press, San Diego, pp. 369–391.

Greene, H.W. 1999. Natural history and behavioural homology. In: *Novartis Foundation Symposium 222: Homology*, G.R. Bock and G. Cardew (eds.). John Wiley and Sons, New York, pp. 173–188.

Greene, H.W. and Cundall, D. 2000. Limbless tetrapods and snakes with legs. *Science* 287:1939–1941.

Greenspan, R.J. 2001. The flexible genome. *Nature Reviews/Genetics* 2:383–387.

Greenwood, P.H. 1964. Explosive speciation in African lakes. *Proceedings of the Royal Institution* 40:256–269.

Greenwood, P.H. 1965. Environmental effects on the pharyngeal mill of a cichlid fish, *Astatoreochromis alluaudi*, and their taxonomic implications. *Proceedings of the Linnean Society of London* 176:1–10.

Greenwood, P.H. 1984. African cichlids and evolutionary theories. In: *Evolution of Fish*

Species Flocks, A.A. Echelle and I. Kornfield (eds.). University of Maine at Orono Press, Orono, pp. 141–154.

Griffin, D.R. 1976. *The Question of Animal Awareness*. Rockefeller University Press, New York.

Griffin, D.R. 1984. *Animal Thinking*. Harvard University Press, Cambridge.

Grimaldi, D. 1986. The *Chymomyza aldrichii* species-group (Diptera: Drosophilidae): relationships, new neotropical species, and the evolution of some sexual traits. *Journal of the New York Entomological Society* 94:342–371.

Grimaldi, D. 1987. Phylogenetics and taxonomy of *Zygothrica* (Diptera: Drosophilidae). *Bulletin of the American Museum of Natural History* 186:103–268.

Grimaldi, D. and Fenster, G. 1989. Evolution of extreme sexual dimorphisms: structural and behavioral convergence among broad-headed male Drosophilidae (Diptera). *American Museum Novitates* 2939:1–25.

Gronenberg, W. and Hölldobler, B. 1999. Morphologic representation of visual and antennal information in the ant brain. *Journal of Comparative Neurology* 412:229–240.

Grosberg, R.K. 1988. Life-history variation within a population of the colonial ascidian *Botryllus schlosseri*. I. The genetic and environmental control of seasonal variation. *Evolution* 42:900–920.

Gross, M.R. 1984. Sunfish, salmon, and the evolution of alternative reproductive strategies and tactics in fishes. In: *Fish Reproduction: Strategies and Tactics*, G.W. Potts and R.J. Wootton (eds.). Academic Press, London, pp. 55–75.

Gross, M.R. 1985. Disruptive selection for alternative life histories in salmon. *Nature* 313:47–48.

Gross, M.R. 1987. Evolution of diadromy in fishes. *American Fisheries Society Symposium* 1:14–25.

Gross, M.R. 1991. Salmon breeding behavior and life history evolution in changing environments. *Ecology* 72:1180–1186.

Gross, M.R. 1996. Alternative reproductive strategies and tactics: diversity within sexes. *Trends in Ecology and Evolution* 11:92–98.

Gross, M.R. and Repka, J. 1997. Game theory and inheritance in the conditional strategy. In: *Game Theory and Animal Behavior*, L.A. Dugatkin and H.K. Reeve (eds.). Oxford University Press, New York, pp. 168–187.

Gross, M.R. and Repka, J. 1998. Stability with inheritance in the conditional strategy. *Journal of Theoretical Biology* 192:445–453.

Grüneberg, H. 1947. *Animal Genetics and Medicine*. Hamish Hamilton Medical Books, London.

Grüneberg, H. 1963. *The Pathology of Development*. Blackwell, Oxford.

Gruppioni, R. and Sbrenna, G. 1992. Preliminary studies on electrophoretic protein patterns in castes of *Kalotermes flavicollis* (Fabr.) (Isoptera Kalotermitidae). *Ethology, Ecology and Evolution* (Special Issue) 2:105–109.

Gubernick, D.J. and Alberts, J.R. 1987. The biparental care system of the California mouse, *Peromyscus californicus*. *Journal of Comparative Psychology* 101:169–177.

Gubernick, D.J. and Nelson, R.J. 1989. Prolactin and paternal behavior in the biparental California mouse, *Peromyscus californicus*. *Hormones and Behavior* 23:203–210.

Guerrant, E.O., Jr. 1982. Neotenic evolution of *Delphinium nudicaule* (Ranunculaceae): a hummingbird-pollinated larkspur. *Evolution* 36:699–712.

Gupta, A.P. 1978. Norms of reaction of genotypes in *Drosophila pseudoobscura*. Ph.D. thesis, Harvard University, Cambridge.

Gupta, A.P. and Lewontin, R.C. 1982. A study of reaction norms in natural populations of *Drosophila pseudoobscura*. *Evolution* 36:934–948.

Guthrie, R.D. 1975. Addendum: an hypothesis of density-adapted morphs among Northern canids. In: *The Wild Canids: Their Systematics, Behavioral Ecology and Evolution*, M.W. Fox (ed.). Van Nostrand Reinhold, New York, pp. 414–415.

Gwynne D.T. 1995. Phylogeny of the Ensifera (Orthoptera): a hypothesis supporting multiple origins of acoustical signalling, complex spermatophores and maternal care in crickets, katydids, and weta. *Journal of Orthoptera Research* 4:203–218.

Haber, W.A. 1993. Seasonal migration of monarchs and other butterflies in Costa Rica. In: *Biology and Conservation of the Monarch Butterfly*, S.B. Malcolm and M.P. Zalueki (eds.). Science Series No. 38., Los Angeles County Natural History Museum, Los Angeles, pp. 201–207.

Hadfield, M.G. and Strathmann, M.F. 1996. Variability, flexibility and plasticity in life histories of marine invertebrates. *Oceanologica Acta* 19:323–334.

Haeckel, E. 1905. *The Evolution of Man* (2 vols; J. McCabe, trans.), from *Anthropogenie*, 5th ed. Watts and Company, London.

Haecker, V. 1925. *Pluripotenzerscheinungen. Synthetische Beiträge zur Entomologie Beiträge zur Vererbungs*. Gustav Fischer, Jena.

Hafner, J.C. and Hafner, M.S. 1988. Heterochrony in rodents. In: *Heterochrony in Evolution: A Multidiciplinary Approach*, M.L. McKinney (ed.). Plenum Press, New York, pp. 217–235.

Hagen, D.W. 1973. Inheritance of numbers of lateral plates and gill rakers in *Gasterosteus aculeatus*. *Heredity* 30:303–312.

Hagen, D.W. and Gilbertson, L.G. 1972. Geographic variation and environmental selection in *Gasterosteus aculeatus* L. in the Pacific Northwest, America. *Evolution* 26:32–51.

Haig, D. 1997. Parental antagonism, relatedness asymmetries, and genomic imprinting. *Proceedings of the Royal Society of London B* 264:1657–1662.

Haig, D. and Westoby, M. 1989. Parent-specific gene expression and the triploid endosperm. *American Naturalist* 134:147–155.

Hailman, J.P. 1977. Communication by reflected light. In: *How Animals Communicate*, T.A. Sebeok (ed.). Indiana University Press, Bloomington, pp. 184–210.

Hairston, N.G. and Dillon, T.A. 1990. Fluctuating selection and response in a population of freshwater copepods. *Evolution* 44:1796–1805.

Haldane, J.B.S. 1930. A mathematical theory of natural and artificial selection. VI. Isolation. *Proceedings of the Cambridge Philosophical Society* 26:220–230.

Haldane, J.B.S. 1932 [1966]. *The Causes of Evolution*. Cornell University Press, Ithaca.

Haldane, J.B.S. 1954. The statics of evolution. In: *Evolution as a Process*, J. Huxley, A.C. Hardy and E.B. Ford (eds.). Allen and Unwin, London, pp. 109–121.

Haldane, J.B.S. 1958. The theory of evolution, before and after Bateson. *Journal of Genetics* 56:11–27.

Haldane, J.B.S. 1959. Natural selection. In: *Darwin's Biological Work*, P.R. Bell (ed.). Cambridge University Press, Cambridge, pp. 101–149.

Haldane, J.S.B. and Jayakar, S.D. 1963. Polymorphism due to selection of varying direction. *Journal of Genetics* 58:237–242.

Hall, B.G. 1983. Evolution of new metabolic functions in laboratory organisms. In: *Evolution of Genes and Proteins*, M. Nei and R.K. Koehn (eds.). Sinauer, Sunderland, MA, pp. 234–257.

Hall, B.G. and Betts, P.W. 1987. Cryptic genes for cellobiose utilization in natural isolates of *Escherichia coli*. *Genetics* 115:431–439.

Hall, B.K. 1984. Developmental mechanisms underlying the formation of atavisms. *Biological Reviews* 59:89–124.

Hall, B.K. 1992. *Evolutionary Developmental Biology*. Chapman and Hall, New York.

Hall, B.K. (ed.) 1994. *Homology*. Academic Press, New York.

Hall, J.C. and Rosbash, M. 1993. Oscillating molecules and how they move circadian clocks across evolutionary boundaries. *Proceedings of the National Academy of Sciences USA* 90:5382–5383.

Halliday, T. and Tejedo, M. 1995. Intrasexual selection and alternative mating behaviour. In: *Amphibian Biology. Volume 2: Social Behaviour*, H. Heatwole and B.K. Sullivan (eds.). Surrey Beatty and Sons, Chipping Norton, NSW, pp. 419–468.

Hamburger, V. 1980. Embryology. In: *The Evolutionary Synthesis*, E. Mayr and W.B. Provine (eds.). Harvard University Press, Cambridge, pp. 96–112.

Hamilton, W.D. 1964a. The genetical theory of social behaviour. I. *Journal of Theoretical Biology* 7:1–16.

Hamilton, W.D. 1964b. The genetical theory of social behaviour. II. *Journal of Theoretical Biology* 7:17–52.

Hamilton, W.D. 1978. Evolution and diversity under bark. In: *Diversity of Insect Faunas*, L.A. Mound and N. Waloff (eds.). Blackwell, London, pp. 154–175.

Hamilton, W.D. 1979. Wingless and fighting males in fig wasps and other insects. In: *Sexual Selection and Reproductive Competition in Insects*, M.S. Blum and N.A. Blum (eds.). Academic Press, New York, pp. 167–220.

Hamilton, W.D. 1982. Pathogens as causes of genetic diversity in their host population. In: *Population Biology of Infectious Diseases*, R.M. Anderson and R.M. May (eds.). Dahlem Konferenzen, Springer-Verlag, Berlin, pp. 269–296.

Hamilton, W.D. 1987a. Discriminating nepotism: expectable, common, overlooked. In: *Kin Recognition in Animals*, D.J.C. Fletcher and C.D. Michener (eds.). John Wiley and Sons, New York, pp. 417–438.

Hamilton, W.D. 1987b. Kinship, recognition, disease, and intelligence: constraints of social evolution. In: *Animal Societies: Theories and Facts*, Y. Itô, J.L. Brown and J. Kikkawa (eds.). Japan Scientific Societies Press, Tokyo, pp. 81–102.

Hamilton, W.D. and May, R.M. 1977. Dispersal in stable habitats. *Nature* 269:578–581.

Hamilton, W.D. and Zuk, M. 1982. Heritable true fitness and bright birds: a role for parasites? *Science* 218:384–387.

Hamilton, W.D., Axelrod, R. and Tanese, R. 1990. Sexual reproduction as an adaptation to resist parasites (a review). *Proceedings of the National Academy of Sciences USA* 87:3566–3573.

Hamilton, W.J., III and McNutt, J.W. 1997. Determinants of conflict outcomes. *Perspectives in Ethology* 12:179–224.

Hamilton, W.J., III, Tilson, R.L. and Frank, L.G. 1986. Sexual monomorphism in spotted hyenas, *Crocuta crocuta. Ethology* 71:63–73.

Hammerstein, P. and Riechert, S.E. 1988. Payoffs and strategies in territorial contests: ESS analyses of two ecotypes of the spider *Agelenopsis aperta. Evolutionary Ecology* 2:115–138.

Hampé, A. 1959. Contribution à l'étude du dévelopement et de la régulation des déficiencies et des excedents dans la patte de l'embryon de poulet. *Archives D' Anatomie Microscopique et Morphologie Experimental* 48:347–478.

Handlirsch, A. 1907. *Die Fossilen Insekten und die Phylogenie der rezenten Formen. Ein Handbuch für Paläontologen und Zoologen,* Leipzig.

Hänel, H. 1986. Effect of juvenile hormone (III) from the host *Apis mellifera* (Insecta: Hymenoptera) on the neurosecretion of the parasitic mite *Varroa jacobsoni* (Acari: Mesostigmata). *Experimental and Applied Acarology* 2:257–271.

Hanken, J. 1983. Miniaturization and its effects on cranial morphology in plethodontid salamanders, genus *Thorius* (Amphibia, Plethodontidae): II. The fate of the brain and sense organs and their role in skull morphogenesis and evolution. *Journal of Morphology* 177:255–268.

Hanken, J. and Wake, D.B. 1993. Miniaturization of body size: organismal consequences and evolutionary significance. *Annual Review of Ecology and Systematics* 24:501–519.

Harcourt, A.H., Harvey, P.H., Larson, S.G. and Short, R.V. 1981. Testis weight, body weight, and breeding system in primates. *Nature* 293:55–57.

Hardie, J. and Lees, A.D. 1985. Endocrine control of polymorphism and polyphenism. In: *Comprehensive Insect Physiology, Biochemistry and Pharmacology,* Vol. 8, G.A. Kerkut and L.I. Gilbert (eds.). Allen and Unwin, London, pp. 441–490.

Harding, K., Wedeen, C., McGinnis, W. and Levine, M. 1985. Spatially regulated expression of homeotic genes in *Drosophila. Science* 229:1236–1242.

Hardisty, M.W. and Potter, I.C. 1971. Paired species. In: *The Biology of Lampreys,* M.W. Hardisty and I.C. Potter (eds.). Academic Press, London, pp. 249–277.

Hardy, A. 1954. Escape from specialization. In: *Evolution as a Process,* J. Huxley, A.C. Hardy and E.B. Ford (eds.). Allen and Unwin, London, pp. 122–142.

Hardy, A. 1965. *The Living Stream.* Collins, London.

Harmelin, J.G. 1973. Morphological variations and ecology of the recent cyclostome bryozoan *"Idmonea" atlantica* from the Mediterranean. In: *Living and Fossil Bryozoa,* G.P. Larwood (ed.). Academic Press, London, pp. 95–106.

Harris, R.N., Semlitsch, R.D., Wilbur, H.M. and Fauth, J.E. 1990. Local variation in the genetic basis of paedomorphosis in the salamander *Ambystoma talpoideum. Evolution* 44:1588–1603.

Harrison, A.G. 1982. Return of the hopeful monster. *Paleobiology* 8:459–463.

Harrison, R.G. 1979. Speciation in North American field crickets: evidence from electrophoretic comparisons. *Evolution* 33:1009–1023.

Harrison, R.G. 1980. Dispersal polymorphisms in insects. *Annual Review of Ecology and Systematics* 11:95–118.

Harrison, R.G. 1985. Barriers to gene exchange between closely related cricket species. II. Life cycle variation and temporal isolation. *Evolution* 39:244–259.

Hart, M.W. 1996. Evolutionary loss of larval feeding: development, form and function in a facultatively feeding larva, *Brisaster latifrons. Evolution* 50:174–187.

Hart, P.J.B. and Gill, A.B. 1994. Evolution of foraging behaviour in the threespine stickleback. In: *The Evolutionary Biology of the Threespine Stickleback,* M.A. Bell and S.A. Foster (eds.). Oxford University Press, New York, pp. 207–239.

Hartfelder, K. and Engels, W. 1992. Allometric and multivariate analysis of sex and caste polymorphism in the neotropical stingless bee, *Scaptotrigona postica. Insectes Sociaux* 39:251–266.

Hartfelder, K., Köstlin, K. and Hepperle, C. 1995. Ecdysteroid-dependent protein synthesis in caste-specific development of the larval honey bee ovary. *Roux's Archives of Developmental Biology* 202:176–180.

Hartshorn, G.S. 1983. Plants. In: *Costa Rican Natural History,* D.H. Janzen (ed.). University of Chicago Press, Chicago, pp. 118–350.

Harvell, C.D. 1986. The ecology and evolution of inducible defenses in a marine bryozoan: cues, costs, and consequences. *American Naturalist* 128:810–823.

Harvell, C.D. 1990. The ecology and evolution of inducible defenses. *Quarterly Review of Biology* 65:323–340.

Harvell, C.D. 1991. Coloniality and inducible polymorphism. *American Naturalist* 138:1–14.

Harvell, C.D. and Padilla, D.K. 1990. Inducible morphology, heterochrony, and size hierarchies in a colonial invertebrate monoculture. *Proceedings of the National Academy of Sciences USA* 87:508–512.

Harvey, P.H., Martin, R.D. and Clutton-Brock, T.H. 1987. Life histories in comparative perspective, In: *Primate Societies*, Smuts, B.B., Cheney, D.L., Seyfarth, R.M., Wrangham, R.W. and Stuhsaker, T.T. (eds.). University of Chicago Press, Chicago, pp. 181–196.

Haselkorn, R. 1992. Developmentally regulated gene rearrangements in prokaryotes. *Annual Review of Genetics* 26:113–130.

Hashimoto, Y., Yamauchi, K. and Hasegawa, E. 1995. Unique habits of stomodeal trophallaxis in the ponerine ant *Hypoponera sp. Insectes Sociaux* 42:137–144.

Haskins, C.P. and Whelden, R.M. 1965. "Queenlessness," worker sibship, and colony versus population structure in the formicid genus *Rhytidoponera. Psyche* 72:87–112.

Haskins, C.P., Haskins, E.F. and Hewitt, R.E. 1960. Pseudogamy as an evolutionary factor in the poeciliid fish *Mollienisia formosa. Evolution* 14:473–483.

Hatfield, T. 1997. Genetic divergence in adaptive characters between sympatric species of stickleback. *American Naturalist* 149:1009–1029.

Hatfield, T. and Schluter, D. 1999. Ecological speciation in sticklebacks: environment-dependent hybrid fitness. *Evolution* 53:866–873.

Haukioja, E. 1990. Induction of defenses in trees. *Annual Review of Entomology* 36:25–42.

Hausmann, A. 1999. Falsification of an entomological rule: polymorphic genitalia in geometrid moths. *Spixiana* 22:83–90.

Hay, J., Ruvinsky, I., Hedges, S.B. and Maxson, L. 1995. Phylogenetic relationships of amphibian families inferred from DNA sequences of mitochondrial 12S and 16S ribosomal RNA genes. *Molecular Biology and Evolution* 12:928–937.

Hayes, T.B. 1997a. Amphibian metamorphosis: an integrative approach. *American Zoologist* 37:121–123.

Hayes, T.B. 1997b. Steroid-mimicking environmental contaminants: their potential role in amphibian declines. *Herpetologica Bonnensis* 1997:145–149.

Hazel, W.N. 1977. The genetic basis of pupal colour dimorphism and its maintenance by natural selection in *Papilio polyxenes* (Papilionidae: Lepidoptera). *Heredity* 38:227–236.

Hazel, W.N. 1995. The causes and evolution of phenotypic plasticity in pupal color in swallowtail butterflies. In: *Swallowtail Butterflies: Their Ecology and Evolutionary Biology*, J.M. Scriber, Y. Tsubaki and R. Lederhouse (eds.). Scientific Publishers, Gainesville, pp. 205–210.

Hazel, W.N. and West, D.A. 1979. Environmental control of pupal colour in swallowtail butterflies (Lepidoptera: Papilioninae): *Battus philenor* (L.) and *Papilio polyxenes* Fabr. *Ecological Entomology* 4:393–400.

Hazel, W.N. and West, D.A. 1982. Pupal colour dimorphism in swallowtail butterflies as a threshold trait: selection in *Eurytides marcellus* (Cramer). *Heredity* 49:295–301.

Hazel, W.N., Brandt, R. and Grantham, T. 1987. Genetic variability and phenotypic plasticity in pupal colour and its adaptive significance in the swallowtail butterfly *Papilio polyxenes. Heredity* 59:449–455.

Hazel, W.N., Smock, R. and Johnson, M.D. 1990. A polygenic model for the evolution and maintenance of conditional strategies. *Proceedings of the Royal Society of London B* 242:181–187.

Hazlett, B.A. 1981. The behavioral ecology of hermit crabs. *Annual Review of Ecology and Systematics* 12:1–22.

Heatwole, H. and Davis, D.M. 1965. Ecology of three species of parasitic insects of the genus *Megarhyssa* (Hymenoptera: Ichneumonidae). *Ecology* 46:140–150.

Hedrick, P.W. 1986. Genetic polymorphism in heterogeneous environments: a decade later. *Annual Review of Ecology and Systematics* 17:535–566.

Hedrick, P.W., Ginevan, M.E. and Ewing, E.P. 1976. Genetic polymorphism in heterogeneous environments. *Annual Review of Ecology and Systematics* 7:1–32.

Hedrick, P.W., Jain, S. and Holden, L. 1978. Multilocus systems in evolution. *Evolutionary Biology* 11:101–182.

Heed, W.B. 1968. Ecology of the Hawaiian Drosophilidae. *University of Texas Publication* 6818:387–419.

Heed, W.B. 1971. Host plant specificity and speciation in Hawaiian *Drosophila. Taxon* 20:115–121.

Heinrich, B. 1979. *Bumblebee Economics*. Harvard University Press, Cambridge.

Heinze, J. 1989. Alternative dispersal strategies in a North American ant. *Naturwissenschaften* 76:477–478.

Heinze, J. and Buschinger, A. 1987. Queen polymorphism in a non-parasitic *Leptothorax* species (Hymenoptera: Formicidae). *Insectes Sociaux* 34:28–43.

Heinze, J. and Buschinger, A. 1989. Queen polymorphism in *Leptothorax* spec. A: its genetic and ecological background (Hymenoptera: Formicidae). *Insectes Sociaux* 36:139–155.

Heinze, J. and Trenkle, S. 1997. Male polymorphism and gynandromorphs in the ant *Cardiocondyla emeryi*. *Naturwissenschaften* 84:129–131.

Heinze, J. and Tsuji, K. 1995. Ant reproductive strategies. *Researches on Population Ecology Review* 37:135–149.

Heinze, J., Kühnholz, S, Schilder, K. and Hölldobler, B. 1993. Behavior of ergatoid males in the ant, *Cardiocondyla nuda*. *Insectes Sociaux* 40:273–282.

Heinze, J., Lipski, N., Hölldobler, B. and Bourke, A.F.G. 1995. Geographical variation in the social and genetic structure of the ant, *Leptothorax acervorum*. *Zoology* 98:127–135.

Heinze, J., Elsishans, C. and Hölldobler B. 1997a. No evidence for kin assortment during colony propagation in a polygynous ant. *Naturwissenschaften* 84:249–250.

Heinze, J., Puchinger, W. and Hölldobler, B. 1997b. Worker reproduction and social hierarchies in *Leptothorax* ants. *Animal Behaviour* 54:849–864.

Heinze, J., Hölldobler, B. and Yamauchi, K. 1998. Male competition in *Cardiocondyla* ants. *Behavioral Ecology and Sociobiology* 42:239–246.

Heliövaara, K., Väisänen, R. and Simon, C. 1994. Evolutionary ecology of periodical insects. *Trends in Ecology and Evolution* 9:475–480.

Heller, K.G. and von Helversen, D. 1986. Acoustic communication in phaneropterid bushcrickets: species-specific delay of female stridulatory response and matching male sensory time window. *Behavioral Ecology and Sociobiology* 18:189–198.

Hennig, W. 1949. Sepsidae. In: *Die Fliegen der Palaearktischen Region*, E. Lindner (ed.). No. 39a. E. Schweizerbart'sche Verlagsbuchhandlung, Stuttgart, pp. 1–91.

Hennig, W. 1966. *Phylogenetic Systematics*. University of Illinois Press, Urbana.

Henriksen, D.L. 1928. Critical notes upon some Cambrian arthropods described by Charles D. Walcott. *Videnskabelige Meddelelser fra Dansk Naturhistorisk Forening: Khobenhavn* 86:1–20.

Henry, C.S., Martínez Wells, M.L. and Simon, C.M. 1999. Convergent evolution of courtship songs among cryptic species of the *Carnea* group of green lacewings (Neuroptera: Chrysopidae: *Chrysoperla*). *Evolution* 53:1165–1179.

Hepper, P.G. (ed.) 1991. *Kin Recognition*. Cambridge University Press, Cambridge.

Herbers, J.M. 1993. Ecological determinants of queen number in ants. In: *Queen Number and Sociality in Insects*, L. Keller (ed.). Oxford, New York, pp. 262–293.

Herbert, A. and Rich, A. 1999. RNA processing in evolution. *Annals of the New York Academy of Sciences* 870:119–132.

Herre, E.A. 1987. Optimality, plasticity and selective regime in fig wasp sex ratios. *Nature* 329:627–629.

Hersh, A.H. 1930. The facet-temperature relation in the bar series of *Drosophila*. *Journal of Experimental Zoology* 57:2–14.

Hersh, A.H. 1934. Evolutionary relative growth in the Titanotheres. *American Naturalist* 68:537–561.

Herskowitz, I.H. 1962. *Genetics*. Little, Brown, Boston.

Hewitt, G.M., Johnston, A.W.B. and Young, J.P.W. (eds.) 1991. *Molecular Techniques in Taxonomy*. Springer-Verlag, Berlin.

Hews, D.K., Knapp. R. and Moore, M.C. 1994. Early exposure to androgens affects adult expression of alternative male types in tree lizards. *Hormones and Behavior* 28:96–115.

Hews, D.K. and Moore, M.C. 1995. Influence of androgens on differentiation of secondary sex characters in tree lizards, *Urosaurus ornatus*. *General and Comparative Endocrinology* 97:86–102.

Higashi, M., Takimoto, G. and Yamamura, N. 1999. Sympatric speciation by sexual selection. *Nature* 402:523–526.

Higgins, L.E. and Rankin, M.A. 1996. Different pathways in arthropod postembryonic development. *Evolution* 50:573–582.

Hill, W.G. and Caballero, A. 1992. Artificial selection experiments. *Annual Review of Ecology and Systematics* 23:287–310.

Hillis, D.M. 1994. Homology in molecular biology. In: *Homology*, M.L. McKinney (ed.). Academic Press, New York, pp. 339–369.

Hillis, D.M. and Davis, S. 1986. Evolution of ribosomal DNA: fifty million years of recorded history in the frog genus *Rana*. *Evolution* 40:1275–1288.

Hillis, D.M. and Moritz, C. (eds.) 1990. *Molecular Systematics*. Sinauer, Sunderland, MA.

Hilu, K.W. 1983. The role of single-gene mutations in the evolution of flowering plants. *Evolutionary Biology* 16:97–128.

Hindar, K. and Jónsson, B. 1982. Habitat and food segregation of dwarf and normal Arctic charr (*Salvelinus alpinus*) from Vangsvatnet Lake, Western Norway. *Canadian Journal of*

Fisheries and Aquatic Sciences 39:1030–1045.

Hinde, R.A. 1959. Behaviour and speciation in birds and lower vertebrates. *Biological Reviews* 34:85–128.

Hinde, R.A. 1970. *Animal Behaviour*. McGraw-Hill, New York.

Hinde, R.A. (ed.) 1983. *Primate Social Relationships: An Integrated Approach*. Blackwell, Oxford.

Hinde, R.A. and Fisher, J. 1951. Further observations on the opening of milk bottles by birds. *British Birds* 44:393–396.

Hinton, H.E. 1981a. *Biology of Insect Eggs*, Vol. 1. Pergamon Press, New York.

Hinton, H.E. 1981b. *Biology of Insect Eggs*, Vol. 2. Pergamon Press, New York.

Hirsch, J. and Holliday, M. 1986. A comment on the evidence for learning in Diptera. *Behavior Genetics* 16:439–447.

Ho, M.W. 1988. Genetic fitness and natural selection: myth or metaphor. In: *T.C. Schneirla Conference Series. Volume 3: Evolution of Social Behavior and Integrative Levels*, G. Greenberg and E. Tobach (eds.). Lawrence Erlbaum, Hillsdale, NJ, pp. 85–112.

Ho, M.W. and Saunders, P.T. (eds.) 1984. *Beyond Neo-Darwinism, an Introduction to the New Evolutionary Paradigm*. Academic Press, London.

Hoagland, K.E. 1979. Systematic review of fossil and recent *Crepidula* and discussion of evolution of the Calyptraeidae. *Malacologia* 16:353–420.

Hoagland, K.E. and Robertson, R. 1988. An assessment of poecilogony in marine invertebrates: phenomenon or fantasy? *Biological Bulletin* 174:109–125.

Hoffman, A.A. and Parsons, P.A. 1991. *Evolutionary Genetics and Environmental Stress*. Oxford University Press, Oxford.

Hoffman, E.A. and Pfennig, D.W. 1999. Proximate causes of cannibalistic polyphenism in larval tiger salamanders. *Ecology* 80:1076–1080.

Hogendoorn, K. and Leys, R. 1993. The superseded female's dilemma: ultimate and proximate factors that influence guarding behaviour of the carpenter bee *Xylocopa pubescens*. *Behavioral Ecology and Sociobiology* 33:371–381.

Hogendoorn, K. and Velthuis, H.H.W. 1993. The sociality of *Xylocopa pubescens*: does a helper really help? *Behavioral Ecology and Sociobiology* 32:247–257.

Höglund, J. and Alatalo, R.V. 1995. *Leks*. Princeton University Press, Princeton.

Holekamp, K.E. and Smale, L. 1998. Behavioral development in the spotted hyena. *BioScience* 48:997–1005.

Holland, P. 1992. Homeobox genes in vertebrate evolution. *BioEssays* 14:267–273.

Holland, S.K. and Blake, C.C.F. 1990 Proteins, exons, and molecular evolution. In: *Intervening Sequences in Evolution and Development*, E.M. Stone and R.J. Schwartz (eds.). Oxford University Press, New York, pp. 10–42.

Hölldobler, B. 1976. The behavioral ecology of mating in harvester ants (Hymenoptera: Formicidae: *Pogonomyrmex*). *Behavioral Ecology and Sociobiology* 1:405–423.

Hölldobler, B. 1995. The chemistry of social regulation: multicomponent signals in ant societies. *Proceedings of the National Academy of Sciences USA* 92:19–22.

Hölldobler, B. 1999. Multimodal signals in ant communication. *Journal of Comparative Physiology A* 184:129–141.

Hölldobler, B. and Wilson, E.O. 1977. The number of queens: an important trait in ant evolution. *Naturwissenschaften* 64:8–15.

Hölldobler, B. and Wilson, E.O. 1990. *The Ants*. Harvard University Press, Cambridge.

Holling, C.S. 1959. Some characteristics of simple types of predation and parasitism. *Canadian Entomologist* 91:385–398.

Holton, G. 1973. *Thematic Origins of Scientific Thought—Kepler to Einstein*. Harvard University Press, Cambridge.

Hoogerhoud, R.J.C. 1986. Ecological morphology of some cichlid fishes. Thesis, University of Leiden, The Netherlands.

Hopkins, D.D. 1977. Electric communication. In: *How Animals Communicate*, T.A. Sebeok (ed.). Indiana University Press, Bloomington, pp. 263–289.

Hori, K. 1992. Insect secretions and their effect on plant growth, with special reference to Hemipterans. In: *Biology of Insect-Induced Galls*, J.D. Shorthouse and O. Rohfritsch (eds.). Oxford University Press, New York, pp. 157–170.

Horner, B.E., Taylor, J.M. and Padykula, H.A. 1965. Food habits and gastric morphology of the grasshopper mouse. *Journal of Mammalogy* 45:513–535.

Howden, H.F. 1979. A revision of the Australian genus *Blackburnium* Boucomont (Coleoptera: Scarabaeidae: Geotrupinae). *Australian Journal of Zoology Supplement Series* 72:1–88.

Hoy, M.A. 1977. Rapid response to selection for a nondiapausing gypsy moth. *Science* 196:1462–1463.

Hrdy, S.B. 1977. *The Langurs of Abu*. Harvard University Press, Cambridge.

Hrdy, S.B. 1981. *The Woman That Never Evolved*. Harvard University Press, Cambridge.

Hrdy, S.B. 1999. *Mother Nature. A History of Mothers, Infants, and Natural Selection.* Pantheon, New York.

Hrdy, S.B. and Hrdy, D.B. 1976. Hierarchical relations among female Hanuman langurs (Primates: Colobinae, *Presbytis entellus*). *Science* 193:913–915.

Hrdy, S.B. and Whitten, P.L. 1987. Patterning of sexual activity. In: *Primate Societies*, B.B. Smuts, D.L. Cheney, R.M. Seyfarth, R.W. Wrangham and T.T. Struhsaker (eds.). University of Chicago Press, Chicago, pp. 370–384.

Huber, R. and Martys, M. 1993. Male-male pairs in Greylag geese (*Anser anser*). *Journal of Ornithology* 134:155–164.

Huettel, M.D. (ed.) 1986. *Evolutionary Genetics of Invertebrate Behavior.* Plenum Press, New York.

Hughes, A.L. 2000. Modes of evolution in the protease and kringle domains of the plasminogen-prothrombin family. *Molecular Phylogenetics and Evolution* 14:469–478.

Hughes, C.R. and Strassmann, J.E. 1988. Age is more important than size in determining dominance among workers in the primitively eusocial wasp, *Polistes instabilis*. *Behaviour* 107:1–14.

Hull, D.L. 1994. *Science* and the modern world view. *Quarterly Review of Biology* 69:491–493.

Humphrey, N.K. 1976. The social function of intellect. In: *Growing Points in Ethology*, P.P.G. Bateson and R.A. Hinde (eds.). Cambridge University Press, Cambridge, pp. 303–321.

Hung, A.C.F., Dowler, M. and Vinson, S.B. 1977. Alpha-glycerophosphate dehydrogenase isozymes of the fire ant *Solenopsis invicta*. *Isozyme Bulletin* 10:29.

Hunkapiller, T., Huang, H., Hood, L. and Campbell, J.H. 1982. The impact of modern genetics on evolutionary theory. In: *Perspectives on Evolution*, R. Milkman (ed.). Sinauer, Sunderland, MA, pp. 164–189.

Hunt, G.L., Jr. and Hunt, M.W. 1977. Female-female pairing in western gulls (*Larus occidentalis*) in southern California. *Science* 196:1466–1467.

Hunt, J.H. and Noonan, K.C. 1979. Larval feeding by male *Polistes fuscatus* and *P. metricus* (Hymenoptera: Vespidae). *Insectes Sociaux* 26:247–251.

Hurst, L.D. and McVean, G.T. 1996. A difficult phase for introns-early. *Current Biology* 6:533–536.

Hurst, L.D. and Peck, J.R. 1996. Recent advances in understanding of the evolution and maintenance of sex. *Trends in Ecology and Evolution* 11:46–52.

Hurst, P.S., Gray, S., Schwarz, M.P., Foran, A., Tilley, J. and Adams, M. 1997. Increased nest cofounding and high intra-colony relatedness in the bee *Exoneura bicolor* (Hymenoptera: Apidae): results from an experimental situation. *Australian Journal of Ecology* 22:419–424.

Huxley, J. 1932 [1972]. *Problems of Relative Growth*, 2nd ed. Dover, New York.

Huxley, J. 1942. *Evolution: The Modern Synthesis*. Allen and Unwin, London.

Iljin, N.A. 1927. Studies in morphogenetics of animal pigmentation. IV. Analysis of pigment formation by low temperature. (In Russian, with English summary.) *Transactions Laboratory of Experimental Biology Zooparka, Moscow* 3:183–200.

Iljin, N.A. and Iljin, V.N. 1930. Temperature effects on the color of the Siamese cat. *Journal of Heredity* 21:309–321.

Iltis, H.H. 1983. From teosinte to maize: the catastrophic sexual transmutation. *Science* 222:886–894.

Iltis, H.H. 1987. Maize evolution and agricultural origins. In: *Grass Systematics and Evolution*, K.W. Hilu, C.S. Campbell and M.E. Barkworth (eds.). Smithsonian Institution Press, Washington, D.C., pp. 195–213.

Iltis, H.H. 2000. Homeotic sexual translocations and the origin of maize (*Zea mays*, Poaceae): a new look at an old problem. *Economic Botany* 54:98–133.

Iltis, H.H. and Doebley, J.F. 1984. *Zea*—a biosystematical odyssey. In: *Plant Biosystematics*, W. Grant (ed.). Academic Press Canada, Montreal, pp. 587–616.

Imms, A.D. 1964. *A General Textbook of Entomology*, 9th ed. Methuen, London.

Imperatriz, V.L. 1970. Aparecimento de supermachos em *Friesella schrottkyi* (Apoidea, Apidae, Meliponinae). *Ciência e Cultura* 22:291.

Inoue, K., Hoshijima, K., Higuchi, I., Sakamoto, H. and Shimura, Y. 1992. Binding of the *Drosophila* transformer and transformer-2 proteins to the regulatory elements of doublesex primary transcript for sex-specific RNA processing. *Proceedings of the National Academy of Sciences USA* 89:8092–8096.

Inoue, T., Roubik, D.W. and Suka, T. 1999. Nestmate recognition in the stingless bee *Melipona panamica* (Apidae, Meliponini). *Insectes Sociaux* 46:208–218.

Inouye, D.W. 1980. The effect of proboscis length and corolla tube lengths on patterns and rates of flower visitation by bumblebees. *Oecologia* 45:197–201.

Irwin, R.E. 1994. The evolution of plumage dichromatism in the New World blackbirds: social selection on female brightness? *American Naturalist* 144:890–907.

Itani, J. 1959. Paternal care in the wild Japanese monkey, *Macaca fuscata fuscata*. *Primates* 2:61–93.

Itô, Y. 1989. The evolutionary biology of sterile soldiers in aphids. *Trends in Ecology and Evolution* 4:69–73.

Itô, Y. 1998. Role of escape from predators in periodical cicada (Homoptera: Cicadidae) cycles. *Entomological Society of America* 91:493–496.

Itô, Y., Brown, L. and Kikkawa, J. (eds.) 1987. *Animal Societies: Theories and Facts.* Japan Scientific Societies Press, Tokyo, pp. 53–65.

Ivlev, V.W. 1961. *Experimental Ecology of the Feeding of Fishes.* Yale University Press, New Haven.

Jablonka, E. and Lamb, M.J. 1995. *Epigenetic Inheritance and Evolution.* Oxford University Press, New York.

Jablonka, E. and Lamb, M.J. 1996. Epigenetic inheritance and evolution. *Trends in Ecology and Evolution* 11:266–267.

Jackson, J.B.C. and Cheetham, A.H. 1990. Evolutionary significance of morphospecies: a test with cheilostome Bryozoa. *Science* 248:579–583.

Jackson, J.B.C. and McKinney, F. 1990. Ecological processes and progressive macroevolution of marine clonal benthos. In: *Causes of Evolution: A Paleontological Perspective*, R.M. Ross and W.D. Allmon (eds.). University of Chicago Press, Chicago, pp. 173–209.

Jackson, R.R. 1986. Web building, predatory versatility and the evolution of the salticidae. In: *Spiders*, W.A. Shear (ed.). Stanford University Press, Stanford, pp. 232–268.

Jackson, R.R. and Hallas, S. 1986. Comparative biology of *Portia africana*, *P. albimana*, *P. fimbriata*, *P. labiata*, and *P. schultzi*, araneophagic, web-building jumping spiders (Araneae, Salticidae): utilisation of webs, predatory versatility, and intraspecific interactions. *New Zealand Journal of Zoology* 13:423–489.

Jackson, R.R. and Wilcox, R.S. 1993. Spider flexibly chooses aggressive mimicry signals for different prey by trial and error. *Behaviour* 127:21–36.

Jackson, R.R. and Wilcox, R.S. 1998. Spider-eating spiders. *American Scientist* 86:350–357.

Jacob, F. 1977. Evolution and tinkering. *Science* 196:1161–1166.

Jacob, F. and Monod, J. 1961. On the regulation of gene activity. *Cold Spring Harbor Symposium of Quantitative Biology* 26:193–211.

Jaenike, J. 1982. Environmental modification of oviposition behavior in *Drosophila*. *American Naturalist* 119:784–802.

Jaenike, J. 1985. Genetic and environmental determinants of food preference in *Drosophila tripunctata*. *Evolution* 39:362–369.

Jaenike, J. 1988. Effects of early adult experience on host selection in insects: some experimental and theoretical results. *Journal of Insect Behavior* 1:3–15.

Jaenike, J. 1990. Host specialization in phytophagous insects. *Annual Review of Ecology and Systematics* 21:243–273.

Jago, N.D. 1973. The genesis and nature of tropical forest and savanna grasshopper faunas, with special reference to Africa. In: *Tropical Forest Ecosystems in Africa and South America: A Comparative Review*, B.J. Meggers, E.S. Ayensu and W.D. Duckworth (eds.). Smithsonian Institution Press, Washington, D.C., pp. 187–196.

Jain, S. 1979. Adaptive strategies: polymorphism, plasticity and homeostasis. In: *Topics in Plant Population Biology*. O.T. Solbrig, S. Jain, G.B. Johnson and P.H. Raven (eds.). Columbia University Press, New York.

James, F. 1983. Environmental component of morphological differentiation in birds. *Science* 221:184–186.

James, F.C. 1991. Complementary descriptive and experimental studies of clinal variation in birds. *American Zoologist* 31:694–706.

James, F.C., Johnston, R.F., Wamer, N.O., Niemi, G.J. and Boecklen, W.J. 1984. The grinnellian niche of the wood thrush. *American Naturalist* 124:17–30.

Jamieson, I.G. 1986. The functional approach to behavior: is it useful? *American Naturalist* 127:195–208.

Jamieson, I.G. 1989a. Behavioral heterochrony and the evolution of birds' helping at the nest: an unselected consequence of communal breeding? *American Naturalist* 133:394–406.

Jamieson, I.G. 1989b. Levels of analysis or analyses at the same level. *Animal Behaviour* 37:696–697.

Janis, C.M. 1988. New ideas in ungulate phylogeny and evolution. *Trends in Ecology and Evolution* 3:291–297.

Jannett, F.J., Jr. 1975. "Hip glands" of *Microtus pennsylvanicus* and *M. longicaudus*

(Rodentia: Muridae), voles "without" hip glands. *Systematic Zoology* 24:171–175.

Janzen, F.J. 1992. Heritable variation for sex ratio under environmental sex determination in the common snapping turtle (*Chelydra serpentina*). *Genetics* 31:155–161.

Janzen, F.J. 1994a. Climate change and temperature-dependent sex determination in reptiles. *Proceedings of the National Academy of Sciences USA* 91:7487–7490.

Janzen, F.J. 1994b. Vegetational cover predicts the sex ratio of hatchling turtles in natural nests. *Ecology* 75:1593–1599.

Janzen, F.J. 1995. Experimental evidence for the evolutionary significance of temperature-dependent sex determination. *Evolution* 49:864–873.

Janzen, F.J. and Paukstis, G.L. 1991a. A preliminary test of the adaptive significance of environmental sex determination in reptiles. *Evolution* 45:435–440.

Janzen, F.J. and Paukstis, G.L. 1991b. Environmental sex determination in reptiles: ecology, evolution, and experimental design. *Quarterly Review of Biology* 66:149–179.

Janzen, F.J., Wilson, M.E., Tucker, J.K. and Ford, S.P. 1998. Endogenous yolk steroid hormones in turtles with different sex-determining mechanisms. *General and Comparative Endocrinology* 111:306–317.

Japyassú, H.F. and Ades, C. 1998. From complete orb to semi-orb webs: developmental transitions in the web of *Nephilengys cruentata* (Aranea: Tetragnathidae). *Behaviour* 135:931–956.

Jarvi, T. and Bakken, M. 1984. The function of the variation in the breast stripe of the great tit (*Parus major*). *Animal Behaviour* 32:590–596.

Jarvis, J.U.M. 1991. Reproduction of naked mole-rats. In: *The Biology of the Naked Mole-Rat*, P.W. Sherman, J.U.M. Jarvis and R.D. Alexander (eds.). Princeton University Press, Princeton, pp. 384–425.

Jeanne, R.L. 1972. Social biology of the neotropical wasp *Mischocyttarus drewseni*. *Bulletin of the Museum of Comparative Zoology* 144:63–150

Jeanne, R.L. 1977. Behavior of the obligate social parasite *Vespula arctica* (Hymenoptera: Vespidae). *Journal of the Kansas Entomological Society* 50:541–557.

Jeanne, R.L. 1980. Evolution of social behavior in the Vespidae. *Annual Review of Entomology* 25:371–396.

Jeanne, R.L. 1996a. Regulation of nest construction behaviour in *Polybia occidentalis*. *Animal Behaviour* 52:473–488.

Jeanne, R.L. 1996b. The evolution of exocrine gland function in wasps. In: *Natural History*

and *Evolution of Paper-Wasps*, S. Turillazzi and M.J. West-Eberhard (eds.). Oxford University Press, Oxford, pp. 144–160.

Jeanne, R.L. 1999. Group size, productivity, and information flow in social wasps. In: *Information Processing in Social Insects*, C. Detrain, J.L. Deneubourg and J.M. Pasteels (eds.). Birkhäuser Verlag, Basel, pp. 3–30.

Jeanne, R.L. and Fagen, R. 1974. Polymorphism in *Stelopolybia areata* (Hymenoptera, Vespidae). *Psyche* 81:155–166.

Jebram, D. 1978. Preliminary studies on "abnormalities" in Bryozoans from the point of view of experimental morphology. *Zoologische Jahrbücher für Anatomie* 100:245–275.

Jebram, D. 1985. Thigmotropic modification of the polarity axis in the zooids of *Victorella pseudoarachnidia* (Bryozoa, Ctenostomata). *Zoologische Jahrbücher für Anatomie* 113:365–373.

Jeffery, W.R. and Swalla, B.J. 1992. Evolution of alternate modes of development in ascidians. *BioEssays* 14:219–226.

Jennings, R.D. and Scott, N.J., Jr. 1993. Ecologically correlated morphological variation in tadpoles of the leopard frog, *Rana chiricahuensis*. *Journal of Herpetology* 27:285–293.

Jennions, M.D. 1997. Female promiscuity and genetic incompatibility. *Trends in Ecology and Evolution* 12:251–253.

Jennions, M.D. and Petrie, M. 1997. Variation in mate choice and mating preferences: a review of causes and consequences. *Biological Reviews* 72:283–327.

Jockusch, E.L. 1997. An evolutionary correlate of genome size change in plethodontid salamanders. *Proceedings of the Royal Society of London B.* 264:597–604.

Johannsen, J. and Lubin, Y. 1999. Group founding and breeding structure in the subsocial spider *Stegodyphus* lineatus (Eresidae). *Heredity* 82:677–686.

Johannsen, W. 1911. The genotype conception of heredity. *American Naturalist* 45:129–159.

John, B. and Miklos, G. 1988. *The Eukaryote Genome in Development and Evolution*. Allen and Unwin, London.

Johns, J.E. 1964. Testosterone-induced nuptial feathers in phalaropes. *Condor* 66:449–455.

Johnson, L.K. 1982. Sexual selection in a brentid weevil. *Evolution* 36:251–262.

Johnson, L.K. and Hubbell, S.P. 1984. Male choice. *Behavioral Ecology and Sociobiology* 15:183–188.

Johnson, M.S. 1987. Adaptation and rules of form: chirality and shape in *Partula suturalis*. *Evolution* 41:672–675.

Johnson, N.K. and Jones, R.E. 1993. The green jay turns blue in Peru: interrelated aspects of the annual cycle in the arid tropical zone. *Wilson Bulletin* 105:388–398.

Johnson, N.K., Marten, J.A. and Ralph, C.J. 1989. Genetic evidence for the origin and relationships of Hawaiian honeycreepers (Aves: Fringillidae). *Condor* 91:379–396.

Johnson, R. and Adams, J. 1992. The ecology and evolution of tetracycline resistance. *Trends in Ecology and Evolution* 7:295–299.

Johnson, R.A. 1994. Distribution and natural history of the workerless inquiline ant *Pogonomyrmex anergismus* Cole (Hymenoptera: Formicidae). *Psyche* 101:257–262.

Johnson, T.C., Scholz, C.A., Talbot, M.R., Kelts, K., Ricketts, R.D., Ngobi, G., Beuning, K., Ssemmanda, I. and McGill, J.W. 1996. Late Pleistocene desiccation of Lake Victoria and rapid evolution of cichlid fishes. *Science* 273:1091–1093.

Johnston, S.A., den Nijs, T.P.M., Peloquin, S.J. and Hanneman, R.E., Jr. 1980. The significance of genic balance to endosperm development in interspecific crosses. *Theoretical and Applied Genetics* 57:5–9.

Johnston, T.D. and Gottlieb, G. 1990. Neophenogenesis: a developmental theory of phenotypic evolution. *Journal of Theoretical Biology* 147:471–495.

Jolly, A. 1966. Lemur social behavior and primate intelligence. *Science* 153:501–506.

Jolly, A. 1972. *The Evolution of Primate Behavior*. Macmillan, New York.

Jones, C.D. 1998. The genetic basis of *Drosophila sechellia*'s resistance to a host plant toxin. *Genetics* 149:1899–1908.

Juchault, P., Louis, C., Martin, G. and Noulin, G. 1991. Masculinization of female isopods (Crustacea) correlated with non-Mendelian inheritance of cytoplasmic viruses. *Proceedings of the National Academy of Sciences USA* 88:10460–10464.

Junk, W.J. (ed.) 1997. *The Central Amazon Floodplain: Ecology of a Pulsing Planet*. Springer-Verlag, New York.

Kallman, K.D. 1984. A new look at sex determination in poeciliid fishes. In: *Evolutionary Genetics of Fishes*, B.J. Turner (ed.). Plenum Press, New York, pp. 95–171.

Kallman, K.D. 1989. Genetic control of size at maturity in *Xiphophorus*. In: *Ecology and Evolution of Livebearing Fishes (Poeciliidae)*, G.K. Meffe and F.F. Snelson (eds.). Prentice-Hall, Englewood Cliffs, pp. 163–184.

Kambysellis, M.P. 1993. Ultrastructural diversity in the egg chorion of Hawaiian *Drosophila* and *Scaptomyza*: ecological and phylogenetic

considerations. *International Journal of Insect Morphology and Embryology* 22:417–446.

Kambysellis, M.P. and Craddock, E.M. 1997. Ecological and reproductive shifts in the diversification of the endemic Hawaiian *Drosophila*. In: *Molecular Evolution and Adaptive Radiation*, T.J. Givnish and K. Sytsma (eds.). Cambridge University Press, Cambridge, pp. 475–509.

Kambysellis, M.P. and Heed, W.B. 1971. Studies of oogenesis in natural populations of Drosophilidae. I. Relation of ovarian development and ecological habitats of the Hawaiian species. *American Naturalist* 105:31–49.

Kambysellis, M.P., Ho, K.F., Craddock, E.M., Piano, F., Parisi, M. and Cohen, J. 1995. Pattern of ecological shifts in the diversification of Hawaiian *Drosophila* inferred from a molecular phylogeny. *Current Biology* 5:1129–1139.

Kambysellis, M.P., Margaritis, L. and Craddock, E.M. 1999. Egg coverings, insects. In: *Encyclopedia of Reproduction*. Academic Press, New York, pp. 971–990.

Kamm, D.R. 1974. Effects of temperature, day length, and number of adults on the sizes of cells and offspring in a primitively social bee (Hymenoptera: Halictidae). *Journal of the Kansas Entomological Society* 47:8–18.

Kaneshiro, K.Y. and Boake, R.B. 1987. Sexual selection and speciation: issues raised by Hawaiian drosophilids. *Trends in Ecology and Evolution* 2:207–212.

Kaneshiro, K.Y. and Kambysellis, M.P. 1999. Description of a new allopatric sibling species of Hawaiian picture-winged *Drosophila*. *Pacific Science* 53:208–213.

Kaneshiro, K.Y., Carson, H.L., Clayton, F.E. and Heed, W.B. 1973. The separation in a pair of homosequential *Drosophila* species from the island of Hawaii. *American Naturalist* 107:766–774.

Kaplan, R.H. and Cooper, W.S. 1984. The evolution of developmental plasticity in reproductive characteristics: an application of the "adaptive coin-flipping" principle. *American Naturalist* 123:393–410.

Kaplin, B.A., Munyaligoga, V. and Moermond, T.C. 1998. The influence of temporal changes in fruit availability on diet composition and seed handling in blue monkeys (*Cercopithecus mitis doggetti*). *Biotropica* 30:56–71.

Karasov, W.H. 1993. In the belly of the bird. *Natural History* 11:32–36.

Karban, R. and Myers, J.H. 1989. Induced plant responses to herbivory. *Annual Review of Ecology and Systematics* 20:331–348.

Katz, M.J. 1987. Is evolution random? In: *Development as an Evolutionary Process*, R.A. Raff and E.C. Raff (eds.). Alan R. Liss, New York, pp. 285–315.

Kauffman, S.A. 1983a. Developmental constraints: internal factors in evolution. In: *Development and Evolution*, B.C. Goodwin, N. Holder and C.C. Wylie (eds.). Cambridge University Press, Cambridge, pp. 195–225.

Kauffman, S.A. 1983b. Filling some epistemological gaps: new patterns of inference in evolutionary theory. *Philosophy of Science Association* 1982:292–313.

Kauffman, S.A. 1993. *The Origins of Order*. Oxford University Press, New York.

Kavanau, J.L. 1990. Conservative behavioural evolution, the neural substrate. *Animal Behaviour* 39:758–767.

Kawano, K. 1995. Horn and wing allometry and male dimorphism in giant rhinoceros beetles (Coleoptera: Scarabaeidae) of tropical Asia and America. *Annals of the Entomological Society of America* 88:92–99.

Kay, Q.O.N. 1978. The role of preferential and assortative pollination in the maintenance of flower colour polymorphisms. In: *The Pollination of Flowers by Insects*, A.J. Richards (ed.). Linnean Society Symposium Series 6. Academic Press, London, pp. 175–190.

Kear, J. 1962. Food selection in finches with special reference to interspecific differences. *Proceedings of the Zoological Society of London* 138:163–204.

Kearns, C.W. 1934. Method of wing inheritance in *Cephalonomia gallicola* Ashmead (Bethylidae: Hymenoptera). *Annals of the Entomological Society of America* 27:533–541.

Keast, A. 1977a. Mechanisms expanding niche width and minimizing intraspecific competition in two centrarchid fishes. *Evolutionary Biology* 10:333–395.

Keast, A. 1977b. Diet overlaps and feeding relationships between the year classes in the yellow perch (*Perca flavescens*) *Environmental Biology of Fish* 2:53–70.

Keeping, M.G. 2000. Morpho-physiological variability and differentiation of reproductive roles among foundresses of the primitively eusocial wasp, *Belonogaster petiolata* (DeGeer)(Hymenoptera, Vespidae). *Insectes Sociaux* 47:147–154.

Keese, P.K. and Gibbs, A. 1992. Origins of genes: "big bang" or continuous creation? *Proceedings of the National Academy of Sciences USA* 89:9489–9493.

Keller, L. (ed.) 1993. *Queen Number and Sociality in Insects*. Oxford University Press, Oxford.

Keller, L. and Nonacs, P. 1993. The role of queen pheromones in social insects: queen control or queen signal? *Animal Behaviour* 45:787–794.

Keller, L. and Reeve, H.K. 1994. Genetic variability, queen number, and polyandry in social Hymenoptera. *Evolution* 48:694–704.

Keller, L. and Ross, K.G. 1993a. Phenotypic plasticity and "cultural transmission" of alternative social organizations in the fire ant *Solenopsis invicta*. *Behavioral Ecology and Sociobiology* 33:121–129.

Keller, L. and Ross, K.G. 1993b. Phenotypic basis of reproductive success in a social insect: genetic and social determinants. *Science* 260:1107–1110.

Kellerman, W.A. 1895. Primitive corn. *Meehan's Monthly* 5:44–53.

Kellogg, V.L. 1904. *American Insects*. Henry Holt, New York.

Kennedy, J.S. 1961. Continuous polymorphism in locusts. In: *Insect Polymorphism*, J.S. Kennedy (ed.). Symposium No. 1. Royal Entomological Society, London, pp. 80–90.

Kennedy, W.J. 1977. Ammonite evolution. In: *Patterns of Evolution*, A. Hallam (ed.). Elsevier, New York, pp. 251–304.

Kerr, W.E. 1950a. Genetic determination of castes in the genus *Melipona*. *Genetics* 35:143–152.

Kerr, W.E. 1950b. Evolution of caste determination in the genus *Melipona*. *Evolution* 4:7–13.

Kerr, W.E. 1975. Sex determination in bees. III: Caste determination and genetic control in *Melipona*. *Insectes Sociaux* 21:357–368.

Kerr, W.E. 1987. Sex determination in bees. XVII: Systems of caste determination in the Apinae, Meliponinae and Bombinae and their phylogenetic implications. *Brazilian Journal of Genetics* 10:685–694.

Kerr, W.E. 1990. Why are workers in social Hymenoptera not males? *Brazilian Journal of Genetics* 13:133–136.

Kerr, W.E. 1997. Sex determination in honey bees (Apinae and Meliponinae) and its consequences. *Brazilian Journal of Genetics* 20:601–611.

Kerr, W.E. and Cunha, R. 1990. Sex determination in bees. XXVI. Masculinism of workers in the Apidae. *Brazilian Journal of Genetics* 13:479–489.

Kerr, W.E. and Nielsen, R.A. 1966. Evidences that genetically determined *Melipona* queens can become workers. *Genetics* 54:859–866.

Kerr, W.E., Stort, A.C., and Montenegro, M.J. 1966. Importância de alguns fatôres ambientais na determinação das castas do gênero *Melipona*. *Anales de la Academia Brasileira de Ciencias* 38:149–168.

Ketterson, E.D. and Nolan, V., Jr. 1992. Hormones and life histories: an integrative approach. *American Naturalist* 140:S33–S62.

Ketterson, E.D. and Nolan, V., Jr. 1994. Hormones and life histories: an integrative approach. In: *Behavioral Mechanisms in Evolutionary Ecology*, L.A. Real (ed.). University of Chicago Press, Chicago, pp. 327–353.

Ketterson, E.D., Nolan, V., Jr., Wolf, L. and Ziegenfus, C. 1992. Testosterone and avian life histories: effects of experimentally elevated testosterone on behavior and correlates of fitness in the dark-eyed junco (*Junco hyemalis*). *American Naturalist* 140:980–999.

Kettlewell, H.B.D. 1961. The phenomenon of industrial melanism in the *Lepidoptera*. *Annual Review of Entomology* 6:245–262.

Keynes, R.D. 1988. *Charles Darwin's Beagle Diary*. Cambridge University Press, New York.

Kidwell, M.G. 1993. Lateral transfer in natural populations of eukaryotes. *Annual Review of Genetics* 27:235–256.

Kidwell, M.G. and Lisch, D.R. 2001. Perspective: transposable elements, parasitic DNA, and genome evolution. *Evolution* 55:1–24.

Kimball, R.T. and Ligon, J.D. 1999. Evolution of avian plumage dichromatism from a proximate perspective. *American Naturalist* 154:182–191.

Kimura, M. 1956. A model of a genetic system which leads to closer linkage by natural selection. *Evolution* 10:278–287.

Kimura, M. 1983. *The Neutral Theory of Molecular Evolution*. Cambridge University Press, New York.

Kimura, M. 1985. Natural selection and neutral evolution. In: *What Darwin Began*, L.R. Godfrey (ed.). Prentice-Hall, Englewood Cliffs, NJ, pp. 73–93.

Kimura, M. 1991. Recent development of the neutral theory viewed from the Wrightian tradition of theoretical population genetics. *Proceedings of the National Academy of Sciences USA* 88:5969–5973.

King, M.-C. and Wilson, A.C. 1975. Evolution at two levels: molecular similarities and biological differences between humans and chimpanzees. *Science* 188:107–116.

King, R.C. and Büning, J. 1985. The origin and functioning of insect oocytes and nurse cells. In: *Comprehensive Insect Physiology, Biochemistry and Pharmacology*, G.A. Kerkut and L.I. Gilbert (eds.). Pergamon Press, New York, pp. 37–82.

Kingsolver, J.G. 1985. Thermoregulatory significance of wing melanization in *Pieris* butterflies: physics, posture, pattern. *Oecologia* 66:546–551.

Kingsolver, J.G. 1987. Evolution and coadaptation of thermoregulatory behavior and wing pigmentation pattern in pierid butterflies. *Evolution* 41:472–490.

Kingsolver, J.G. 1995. Viability selection on seasonally polyphenic traits: wing melanin pattern in western white butterflies. *Evolution* 49:932–941.

Kingsolver, J.G. and Wiernasz, D.C. 1991. Seasonal polyphenism in wing-melanin pattern and thermoregulatory adaptation in *Pieris* butterflies. *American Naturalist* 137:816–830.

Kinomura, K. and Yamauchi, K. 1992. A new workerless socially parasitic species of the genus *Vollenhovia* (Hymenoptera, Formicidae) from Japan. *Japanese Journal of Entomology* 60:203–206.

Kinomura, K. and Yamauchi, K. 1994. Frequent occurrence of gynandromorphs in the natural population of the ant *Vollenhovia emeryi* (Hymenoptera: Formicidae). *Insectes Sociaux* 41:273–278.

Kirkpatrick, M. 1982. Quantum evolution and punctuated equilibria in continuous genetic characters. *American Naturalist* 119:833–848.

Kirkpatrick, M. 1996. Genes and adaptation: a pocket guide to the theory. In: *Adaptation*, M.R. Rose and G.V. Lauder (eds.). Academic Press, New York, pp. 125–146.

Kirkpatrick, M. and Selander, R.K. 1979. Genetics of speciation in lake whitefishes in the Allegash Basin. *Evolution* 33:478–485.

Kirschner, M.W. 1992. Evolution of the cell. In: *Molds, Molecules and Metazoa: Growing Points in Evolutionary Biology*, P.R. Grant and H.S. Horn (eds.). Princeton University Press, Princeton, pp. 99–126.

Kirschner, M.W. and Gerhart, J. 1998. Evolvability. *Proceedings of the National Academy of Sciences USA* 95:8420–8477.

Kirschner, M.W. and Mitchison, T.L. 1986. Beyond self-assembly: from microtubules to morphogenesis. *Cell* 45:329–342.

Klahn, J.E. 1988. Intraspecific comb usurpation in the social wasp, *Polistes fuscatus*. *Behavioral Ecology and Sociobiology* 23:1–8.

Kleiman, D.G. 1977. Monogamy in mammals. *Quarterly Review of Biology* 52:39–69.

Klein, J. 1986. *Natural History of the Major Histocompatibility Complex*. Wiley, New York.

Klein, N.K. and Payne, R.B. 1998. Evolutionary associations of brood parasitic finches (*Vidua*) and their host species: analyses of mitochondrial restriction sites. *Evolution* 52:299–315.

Klepaker, T. 1996. Lateral plate polymorphism in marine and estuarine populations of the threespine stickleback (*Gasterosteus aculeatus*) along the coast of Norway. *Copeia* 1996:532–538.

Klingenberg, C.P. 2003. A developmental perspective on developmental instability: theory, models, and mechanisms. In: *Developmental Instability: Causes and Consequences*, M. Polak (ed.). Oxford University Press, New York.

Klingenberg, C.P. and Nijhout, H.F. 1999. Genetics of fluctuating asymmetry: a developmental model of developmental instability. *Evolution* 53:358–375.

Klopfer, P. 1962. *Behavioural Aspects of Ecology*. Prentice-Hall, Englewood Cliffs, NJ.

Kluge, M. and Brulfert, J. 1996. Crassulacean acid metabolism in the genus *Kalanchoë*: ecological, physiological and biochemical aspects. In: *Crassulacean Acid Metabolism*, K. Winter and J.A.C. Smith (eds.). Springer-Verlag, New York, pp. 324–335.

Knerer, G. and Atwood, C.E. 1966. Polymorphism in some nearctic halictine bees. *Science* 152:1262–1263.

Knerer, G. and Atwood, C.E. 1973. Diprionid sawflies: polymorphism and speciation. *Science* 179:1090–1099.

Knerer, G. and Schwarz, M. 1976. Halictine social evolution: the Australian enigma. *Science* 194:445.

Knerer, V.G. and Schwarz, M. 1978. Beobachtungen an australischen furchenbienen (Hymenoptera: Halictidae). *Zoologische Anzeiger* 200:321–333.

Knight, M.R., Smith, S.M. and Trewavas, A.J. 1992. Wind-induced plant motion immediately increases cytosolic calcium. *Proceedings of the National Academy of Sciences USA* 89:4967–4971.

Knowlton, N. 1982. Parental care and sex role reversal. In: *Current Problems in Sociobiology*, King's College Sociobiology Group (eds.). Cambridge University Press, Cambridge, pp. 203–222.

Knowlton, N. 1993. Sibling species in the sea. *Annual Review of Ecology and Systematics* 24:189–216.

Knowlton, N. and Jackson, J.B.C. 1994. New taxonomy and niche partitioning on coral reefs. Jack of all trades or master of some? *Trends in Ecology and Evolution* 9:7–9.

Knülle, W. 1987. Genetic variability and ecological adaptability of hypopus formation in a stored product mite. *Experimental and Applied Acarology* 3:21–32.

Knülle, W. 1991. Genetic and environmental determinants of hypopus duration in the stored-product mite *Lepidoglyphus destructor*. *Experimental and Applied Acarology* 10:231–258.

Kodric-Brown, A. 1989. Dietary carotenoids and male mating success in the guppy: an environmental component to female choice. *Behavioral Ecology and Sociobiology* 25:393–401.

Kodric-Brown, A. and Brown, J.H. 1978. Influence of economics, interspecific competition, and sexual dimorphism on territoriality of migrant rufous hummingbirds. *Ecology* 59:285–296.

Kodric-Brown, A. and Nicoletto, P.F. 1993. The relationship between physical condition and social status in pupfish *Cyprinodon pecosensis*. *Animal Behaviour* 46:1234–1236.

Koehl, M.A.R. 1996. When does morphology matter? *Annual Review of Ecology and Systematics* 27:501–542.

Köhn, A. 1971. *Lectures on Developmental Physiology*. Springer-Verlag, New York.

Kojima, J.-I. 1993. Feeding of larvae by males of an Australian paper wasp, *Ropalidia plebeiana* Richards (Hymenoptera, Vespidae). *Japanese Journal of Entomology* 61:213–215.

Kollar, E.J. and Fisher, C. 1980. Tooth induction in chick epithelium: expression of quiescent genes for enamel synthesis. *Science* 207:993–995.

Kölliker, M., Richner, H., Werner, I. and Heeb, P. 1998. Begging signals and biparental care: nestling choice between parental feeding locations. *Animal Behaviour* 55:215–222.

Kölliker, M., Brinkhof, M.W.G., Heeb, P. and Fitze, P.S. 2000. The quantitative genetic basis of offspring solicitation and parental response in a passerine bird with biparental care. *Proceedings of the Royal Society of London B* 267:2127–2132.

Koref-Santibañez, S. 1986. Genetic determination and integrative activity of the brain. In: *Introduction to the Physiopathology of Neurotic States*, G. Santibañez and H.M. Lindemann (eds.). VEB Georg Thieme, Leipzig, pp. 81–99.

Kornfield, I., Smith, D.C. and Gagnon, P.S. 1982. The cichlid fish of Cuatro Cienegas, Mexico, direct evidence of conspecificity among distinct trophic morphs. *Evolution* 36:658–664.

Korpelainen, H. 1990. Sex ratios and conditions required for environmental sex determination in animals. *Biological Reviews* 65:147–184.

Koshland, D.E., Jr. 1977. A response regulator model in a simple sensory system. *Science* 196:1055.

Koufopanou, V. and Bell, G. 1991. Developmental mutants of *Volvox*: does mutation recreate the patterns of phylogenetic diversity? *Evolution* 45:1806–1822.

Kramer, D.L. 1983. The evolutionary ecology of respiratory mode in fishes: an analysis based on the costs of breathing. *Environmental Biology of Fishes* 9:145–158.

Krantz, G.W., Royce, L.A., Lowry, R.R. and Kelsey, R. 1991. Mechanisms of phoretic specificity in *Macrocheles* (Acari: Macrochelidae). In: *Modern Acarology*, F. Dusabek and V. Bukva (eds.). SPB The Hague/Academia, Prague, pp. 561–569.

Kraus, O. and Kraus, M. 1988. The genus *Stegodyphus* (Arachnida, Araneidae): sibling species, species groups, and parallel origins of social living. *Verhandlungen des Naturwissenschaftlichen Vereins in Hamburg* 30:151–254.

Krebs, J.R. and Davies, N.B. (eds.) 1991. *Behavioural Ecology: An Evolutionary Approach*. Blackwell, Oxford.

Kristensen, N.P. 1981. Phylogeny of insect orders. *Annual Review of Entomology* 26:135–157.

Kristensen, N.P. 1984. The male genitalia of *Agathiphaga* (Lepidoptera: Agathiphagidae) and the Lepidopteran ground plan. *Entomology of Scandinavia* 15:151–178.

Kruuk, H. 1972. *The Spotted Hyena—A Study of Predation and Social Behavior*. University of Chicago Press, Chicago.

Kühme, W. 1963. Ergänzende Beobachtungen an afrikanischen Elefanten (*Loxodonta africana* Blumenbach 1797) im Freigehege. *Zeitschrift für Tierpsychologie* 20:66–79.

Kühn, A. 1971. *Lectures on Developmental Physiology*, R. Milkman (trans.). Springer, New York.

Kukalova-Peck, J. 1997. Arthropod phylogeny and "basal" morphological structures. In: *Arthropod Relationships*, R.A. Fortey and R.H. Thomas (eds.). Chapman and Hall, London, pp. 251–270.

Kukuk, P.F. and Schwarz, M. 1988. Macrocephalic male bees as functional reproductives and probable guards. *Pan-Pacific Entomologist* 64:131–137.

Kulinčević, J.M. 1986. Breeding accomplishments with honey bees. In: *Bee Genetics and Breeding*, T. E. Rinderer (ed.). Academic Press, New York, pp. 391–413.

Kurczewski, F.E. 1997. Activity patterns in a nesting aggregation of *Sphex pensylvanicus* L. (Hymenoptera: Sphecidae). *Journal of Hymenoptera Research* 6:231–242.

Kurczewski, F.E. and Spofford, M.G. 1998. Alternative nesting strategies in *Ammophila urnaria* (Hymenoptera: Sphecidae). *Journal of Natural History* 32:99–106.

Kurland, J.A. 1977. *Kin Selection in the Japanese Monkey*. Karger, New York.

Kurosu, U. and Aoki, S. 1988. Monomorphic first instar larvae of *Colophina clematicola* (Homoptera, Aphidoidea) attack predators. *Kontyû* 56:867–871.

Kurtén, B. 1963. Return of a lost structure in the evolution of the felid dentition. *Commentationes Biologicae Societas Scientiarum Fennica* 26:1–12.

Kutsch, W. and Huber, F. 1989. Neural basis of song production. In: *Cricket Behavior and Neurobiology*, F. Huber, T.E. Moore and W. Loher (eds.). Cornell University Press, Ithaca, pp. 262–309.

LaBarbera, M. 1989. Analyzing body size as a factor in ecology and evolution. *Annual Review of Ecology and Systematics* 20:97–117.

Lacey, E.A. and Sherman, P.W. 1991. Social organization of naked mole-rat colonies: evidence for divisions of labor. In: *The Biology of the Naked Mole-Rat*, P.W. Sherman, J.U.M. Jarvis and R.D. Alexander (eds.). Princeton University Press, Princeton, pp. 275–336.

Lacey, E.P. 1996. Parental effects in *Plantago lanceolata* L. I.: A growth chamber experiment to examine pre- and postzygotic temperature effects. *Evolution* 50:865–878.

Lack, D. 1947. *Darwin's Finches*. Cambridge University Press, Cambridge.

Lacy, R.C. 1980. The evolution of eusociality in termites: a haplodiploid analogy? *American Naturalist* 116:449–451.

Laland, K.N. and Reader, S.M. 1999. Foraging innovation in the guppy. *Animal Behaviour* 57:331–340.

Lande, R. 1978. Evolutionary mechanisms of limb loss in tetrapods. *Evolution* 32:79–92.

Lande, R. 1979. Natural selection and random genetic drift in phenotypic evolution. *Evolution* 30:314–334.

Lande, R. 1980a. Sexual dimorphism, sexual selection, and adaptation in polygenic characters. *Evolution* 34:292–305.

Lande, R. 1980b. Microevolution in relation to macroevolution. *Paleobiology* 6:233–238.

Lande, R. 1980c. Genetic variation and phenotypic evolution during allopatric speciation. *American Naturalist* 116:463–479.

Lande, R. 1987. The dynamics of peak shifts and the pattern of morphological evolution. *Paleobiology* 12:343–354.

Landman, O.E. 1991. The inheritance of acquired characteristics. *Annual Review of Genetics* 25:1–20.

Lank, D.B. and Smith, C.M. 1987. Conditional lekking in ruff (*Philomachus pugnax*).

Behavioral Ecology and Sociobiology 20:137–146.

Lank, D.B., Smith, C.M., Hanotte, O., Burke, T. and Cooke, F. 1995. Genetic polymorphism for alternative mating behaviour in lekking male ruff *Philomachus pugnax. Nature* 378:59–62.

Laoide, B.M., Foulkes, N.S., Schlotter, F. and Sassone-Corsi, P. 1993. The functional versatility of CREM is determined by its modular structure. *European Molecular Biology Organization Journal* 12:1179–1191.

Larsen, E. and McLaughlin, H.M.G. 1987. The morphogenetic alphabet: lessons for simple-minded genes. *BioEssays* 7:129–132.

Larsen, E.W. 1997. Evolution of development: the shuffling of ancient modules by ubiquitous bureaucracies. In: *Physical Theory in Biology*, C.J. Lumsden, W.A. Brandts and L.E.H. Trainor (eds.). World Scientific, New York, pp. 431–441.

Larson, A. and Losos, J.B. 1996. Phylogenetic systematics of adaptation. In: *Adaptation*, M.R. Rose and G.V. Lauder (eds.). Academic Press, San Diego, pp. 187–220.

Lauder, G. 1981. Form and function: structural analysis in evolutionary morphology. *Paleobiology* 7:430–442.

Lawrence, P.A. and Morata, G. 1976. The compartment hypothesis. *Symposia of the Royal Entomological Society of London* 8:132–149.

Lawton, M.F. and Lawton, R.O. 1986. Heterochrony, deferred breeding, and avian sociality. *Current Ornithology* 3:187–222.

Lawton-Rauh, A.L., Alvarez-Buylla, E.R. and Purugganan, M.D. 2000. Molecular evolution of flower development. *Trends in Ecology and Evolution* 15:144–149.

Lea, S.E.G. 1984. Complex general process learning in nonmammalian vertebrates. In: *The Biology of Learning*, P. Marler and H.S. Terrace (eds.). Springer-Verlag, Berlin, pp. 373–398.

Leatherland, J.F., Copeland, P., Sumpter, J.P. and Sonstegard, R.A. 1982. Hormonal control of gonadal maturation and development of secondary sexual characteristics in coho salmon, *Oncorhynchus kisutch*, from Lakes Ontario, Erie, and Michigan. *General and Comparative Endocrinology* 48:196–204.

Lee, M.G. and Nurse, P. 1987. Complementation used to clone a human homologue of the fission yeast cell cycle control gene cdc2. *Nature* 327:31–35.

Lefebvre, L., Whittle, P., Lascaris, E. and Finkelstein, A. 1997. Feeding innovations and forebrain size in birds. *Animal Behaviour* 53:549–560.

Legner, E.F. 1987. Inheritance of gregarious and solitary oviposition in *Muscidifurax raptorellus* Kogan and Legner (Hymenoptera: Pteromalidae). *Canadian Entomologist* 119:791–808.

Legner, E.F. 1988. *Muscidifurax raptorellus* (Hymenoptera: Pteromalidae) females exhibit postmating oviposition behavior typical of the male genome. *Entomological Society of America* 81:522–527.

Lehmann, N.L. and Sattler, R. 1993. Homeosis in floral development of *Sanguinaria canadensis* and *S. canadensis* "Multiplex" (Papaveraceae). *American Journal of Botany* 80:1323–1335.

Leigh, E.G., Jr. 1971. *Adaptation and Diversity*. Freeman, Cooper, San Francisco.

Leigh, E.G., Jr. 1983. When does the good of the group override the advantage of the individual? *Proceedings of the National Academy of Sciences USA* 80:2985–2989.

Leigh, E.G., Jr. 1987. Ronald Fisher and the development of evolutionary theory. II. Influences of new variation on evolutionary process. *Oxford Surveys in Evolutionary Biology* 4:212–263.

Leigh, E.G., Jr. 1991. Genes, bees and ecosystems: the evolution of a common interest among individuals. *Trends in Ecology and Evolution* 6:257–262.

Leigh, E.G., Jr. and Rowell, T.E. 1995. The evolution of mutualism and other forms of harmony at various levels of biological organization. *Écologie* 26:131–158.

Le Moli, F. and Mori, A. 1987. The problem of enslaved ant species: origin and behavior. *Experientia* (Supplement) 54:333–363.

Lenski, R.E. 1988a. Experimental studies of pleiotropy and epistasis in *Escherichia coli*. I. Variation in competitive fitness among mutants resistant to virus T4. *Evolution* 42:425–432.

Lenski, R.E. 1988b. Experimental studies of pleiotropy and epistasis in *Escherichia coli*. II. Compensation for maladaptive effects associated with resistance to virus T4. *Evolution* 42:433–440.

Lepson, J.K. and Freed, L.A. 1995. Variation in male plumage and behavior of the Hawaii akepa. *Auk* 112:402–414.

Lerdau, M., Litvak, M. and Monson, R. 1994. Plant chemical defense: monoterpenes and the growth-differentiation balance hypothesis. *Trends in Ecology and Evolution* 9:58–61.

Lerner, I.M. 1954. *Genetic Homeostasis*. Oliver and Boyd, Edinburgh.

Leroi, A.M., Kim, S.B. and Rose, M.R. 1994. The evolution of phenotypic life-history trade-offs: an experimental study using

Drosophila melanogaster. American Naturalist 144:661–676.

Leschen, R.A.B. Homonomy in cladistic studies. Unpublished manuscript 2001.

Lessells, C.A. 1991. The evolution of life histories. In: *Behavioural Ecology*, J.R. Krebs and N.B. Davies (eds.). Blackwell, Boston, pp. 32–68.

Lessios, H.A. 1987. Temporal and spatial variation in egg size of 13 Panamanian echinoids. *Journal of Experimental Marine Biology and Ecology* 114:217–239.

Lessios, H.A. 1990. Adaptation and phylogeny as determinants of egg size in echinoderms from the two sides of the isthmus of Panama. *American Naturalist* 135:1–13.

Leutert, R. 1974. Zur Geschlechtsbestimmung und gametogenese von *Bonellia viridis* Rolands. *Embryology and Experimental Morphology* 32:169–193.

Levene, H. 1953. Genetic equilibrium when more than one ecological niche is available. *American Naturalist* 87:331–333.

Levin, B. 1988. The evolution of sex in bacteria. In: *The Evolution of Sex*, R.E. Michod and B.R. Levin (eds.). Sinauer, Sunderland, MA, pp. 194–211.

Levin, D.A. 1975. Pest pressure and recombination systems in plants. *American Naturalist* 109:437–451.

Levin, L.A., Zhu, J. and Creed, E. 1991. The genetic basis of life-history characters in a polychaete exhibiting planktotrophy and lecithotrophy. *Evolution* 45:380–397.

Levin, R.N. 1988. The adaptive significance of antiphonal song in the bay wren, *Thryothorus nigricapillus*. Ph.D. Thesis, Cornell University.

Levin, R.N. 1996. Song behavior and reproductive strategies in a duetting wren, *Thryothorus nigricapillus*: I. Removal studies. *Animal Behaviour* 52:1093–1106.

Levin, R.N. and Wingfield, J.C. 1992. The hormonal control of territorial aggression in tropical birds. *Ornis Scandinavica* 23:284–291.

Levine, M. and Hoey, T. 1988. Homeobox proteins as sequence-specific transcription factors. *Cell* 55:537–540.

Levins, R. 1963. Theory of fitness in a heterogeneous environment II. Developmental flexibility and niche selection. *American Naturalist* 97:75–90.

Levins, R. 1968. *Evolution in Changing Environments*. Princeton University Press, Princeton.

Levins, R. and Lewontin, R. 1985. *The Dialectical Biologist*. Harvard University Press, Cambridge.

Levinton, J.S. 1986. Developmental constraints and evolutionary saltations: a discussion and

critique. In: *Genetics, Development and Evolution*, P. Gustafson, G. Ledyard Stebbins and F.J. Ayala (eds.). Plenum Press, New York, pp. 253–288.

Levinton, J.S. 1988. *Genetics, Paleontology and Macroevolution*. Cambridge University Press, New York.

Levinton, J.S. 1992. The big bang of animal evolution. *Scientific American* 267:84–91.

Lewin, B. 1980. *Gene Expression*. Wiley and Sons, New York.

Lewin, R. 1984. Why is development so illogical? *Science* 224:1327–1329.

Lewis, E.B. 1978. A gene complex controlling segmentation in *Drosophila*. *Nature* 276:565–570.

Lewis, H. and Epling, C. 1959. *Delphinium gypsophilum*, a diploid species of hybrid origin. *Evolution* 13:511–525.

Lewontin, R.C. 1970. The units of selection. *Annual Review of Ecology and Systematics* 1:1–18.

Lewontin, R.C. 1974. *The Genetic Basis of Evolutionary Change*. Columbia University Press, New York.

Lewontin, R.C. 1978. Adaptation. *Scientific American* 239:212–230.

Lewontin, R.C. 1992. Genotype and phenotype. In: *Keywords in Evolutionary Biology*, E.F. Keller and E.A. Lloyd (eds.). Harvard University Press, Cambridge, pp. 137–144.

Lewontin, R.C. 2001. In the beginning was the word. *Science* 291:1263–1264

Lewontin, R.C. and Kojima, K. 1960. The evolutionary dynamics of complex polymorphisms. *Evolution* 14:458–472.

Li, D. and Jackson, R.R. 1996. Prey-specific capture behaviour and prey preferences of myrmicophagic and araneophagic jumping spiders (Araneae: Salticidae). *Revue suisse de Zoologie*, volume hors série, 423–436.

Li, D., Jackson, R.R. and Cutler, B. 1996. Prey-capture techniques and prey preferences of *Habrocestum pulex*, an ant-eating jumping spider (Araneae, Salticidae) from North America. *Journal of Zoology, London* 240:551–562.

Li, W.-H. 1983. *Evolution* of duplicate genes and pseudogenes. In: *Evolution of Genes and Proteins*, M. Nei and R.K. Koehn (eds.). Sinauer, Sunderland, MA, pp. 14–37.

Li, W.-H. 1984. Retention of cryptic genes in microbial populations. *Molecular Biology and Evolution* 1:213–219.

Li, W.-H., Gu, Z., Wang, H. and Nekrutenko, A. 2001. Evolutionary analyses of the human genome. *Nature* 409:847–849.

Lidgard, S. 1985. Zooid and colony growth in encrusting cheilostome bryozoans. *Palaeontology* 28:255–291.

Lidgard, S. 1986. Ontogeny in animal colonies: a persistent trend in the bryozoan fossil record. *Science* 232:230–232.

Liebig, J., Peeters, C., Oldham, N.J., Markstädter, C. and Hölldobler, B. 2000. Are variations in cuticular hydrocarbons of queens and workers a reliable signal of fertility in the ant *Harpegnathos saltator*? *Proceedings of the National Academy of Sciences USA* 97:4124–4131.

Liem, K.F. 1973. Evolutionary strategies and morphological innovations: cichlid pharyngeal jaws. *Systematic Zoology* 22:425–441.

Liem, K.F. 1984. Functional versatility, speciation, and niche overlap: are fishes different? In: *Trophic Interactions within Aquatic Ecosystems*, D.G. Meyers and J.R. Strickler (eds.). AAAS Selected Symposium 85. Westview Press, Boulder, pp. 269–305.

Liem, K.F. and Kaufman, L.S. 1984. Intraspecific macroevolution: functional biology of the polymorphic cichlid species *Cichlasoma minckleyi*. In: *Evolution of Fish Species Flocks*, A.A. Echelle and I. Kornfield (eds.). University of Maine at Orono Press, Orono, pp. 203–215.

Liem, K.F. and Osse, J.W.M. 1975. Biological versatility, evolution, and food resource exploitation in African cichlid fishes. *American Zoology* 15:427–454.

Lillie, F.R. 1927. The gene and the ontogenetic process. *Science* 64:361–369.

Lin, C.P. and Wood, T.K. 2002. Molecular phylogeny of the North American *Enchenopa binotata* species complex (Homoptera: Membracidae). *Annals of the Entomological Society of America* 95:162–171.

Lin, N. and Michener, C.D. 1972. Evolution of sociality in insects. *Quarterly Review of Biology* 47:131–159.

Lincoln, R.J., Boxshall, G.A. and Clark, P.F. 1982. *A Dictionary of Ecology, Evolution and Systematics*. Cambridge University Press, Cambridge.

Lindauer, M. 1961. *Communication among Social Bees*. Harvard University Press, Cambridge.

Lindauer, M. and Kerr, W.E. 1960. Communication between workers of stingless bees. *Bee World* 41:29–41, 65–71.

Lindegren, C.C. 1949. *The Yeast Cell—Its Genetics and Cytology*. Educational Publishers, St. Louis.

Lindeque, M. and Skinner, J.D. 1982. Fetal androgens and sexual mimicry in spotted hyaenas (*Crocuta crocuta*). *Journal of Reproduction and Fertility* 65:405–510.

Lindquist, S. and Craig, E.A. 1988. The heat-shock proteins. *Annual Review of Genetics* 22:631–677.

Linsenmair, K.E. 1987. Kin recognition in subsocial arthropods, in particular in desert isopod *Hemilepistus reaumuri*. In: *Kin Recognition in Animals*, D.J.C. Fletcher and C.D. Michener (eds.). John Wiley and Sons, New York, pp. 121–208.

Lister, A.M. 1993. Patterns of evolution in Quaternary mammal lineages. In: *Evolutionary Patterns and Processes*, D.R. Lees and D. Edwards (eds.). Academic Press, London, pp. 71–93.

Litte, M. 1977. Behavioral ecology of the social wasp *Mischocyttarus mexicanus*. *Behavioral Ecology and Sociobiology* 2:229–246.

Liu, T.P. and Dixon, S.E. 1965. Studies in the mode of action of royal jelly in honey bee development. VI. Haemolymph protein changes during caste development. *Canadian Journal of Zoology* 43:873–879.

Lively, C.M. 1986. Predator-induced shell dimorphism in the acorn barnacle *Chthamalus anisopoma*. *Evolution* 40:232–242.

Lively, C.M. 1999. Developmental strategies in spatially variable environments: barnacle shell dimorphism and strategic models of selection. In: *The Ecology and Evolution of Inducible Defenses*, R. Tollrian and C.D. Harvell (eds.). Princeton University Press, Princeton, N.J., pp. 245–258.

Lloyd, D.G. 1984. Variation strategies of plants in heterogeneous environments. *Biological Journal of the Linnean Society* 21:357–385.

Lloyd, D.G. and Bawa, K.S. 1984. Modification of the gender of seed plants in varying conditions. *Evolutionary Biology* 17:255–338.

Lloyd, E.A. 1988. *The Structure and Confirmation of Evolutionary Theory*. Greenwood Press, New York.

Lloyd, E.A. and Gould, S.J. 1993. Species selection on variability. *Proceedings of the National Academy of Sciences USA* 90:595–599.

Lloyd, J.E. 1975. Aggressive mimicry in *Photuris* fireflies: signal repertoires by *femmes fatales*. *Science* 187:452–453.

Lloyd, J.E. 1977. Bioluminescence and communication. In: *How Animals Communicate*, T.A. Sebeok (ed.). Indiana University Press, Bloomington, pp. 164–183.

Lloyd, M. and Dybas, H.S. 1966. The periodical cicada problem. II. Evolution. *Evolution* 20:466–505.

Lloyd, M. and White, J. 1976. Sympatry of periodical cicada broods and the hypothetical four year acceleration. *Evolution* 30:786–801.

Locke, M. and Nichol, H. 1992. Iron economy in insects: transport, metabolism, and storage. *Annual Review of Entomology* 37:195–215.

Logsdon, J.M., Jr. 1998. The recent origins of spliceosomal introns revisited. *Current Opinions in Genetic Development* 8:637–648.

Loher, W., Ganjian, I., Kubo, I., Stanley-Samuelson, D. and Tobe, S.S. 1981. Prostaglandins: their role in the egg-laying of the cricket *Teleogryllus commodus*. *Proceedings of the National Academy of Sciences USA* 78:7835–7838.

Long, M., Rosenberg, C. and Gilbert, W. 1995. Intron phase correlations and the evolution of the intron/exon structure of genes. *Proceedings of the National Academy of Sciences USA* 92:12495–12499.

Lopez, E.R., Roth, L.C., Ferro, D.N., Hosmer, D. and Mafra-Neto, A. 1997. Behavioral ecology of *Myiopharus doryphorae* (Riley) and *M. aberrans* (Townsend), tachinid parasitoids of the Colorado potato beetle. *Journal of Insect Behavior* 10:49–78.

Lord, E.M. and Hill, J.P. 1987. Evidence for heterochrony in the evolution of plant form. In: *Development as an Evolutionary Process*, R.A. Raff and E.C. Raff (eds.). Alan R. Liss, New York, pp. 47–70.

Lorenz, K. 1952. *King Solomon's Ring*. Crowell, New York.

Lorenz, K. 1953. Die Entwicklung der vergleichenden Verhaltensforschung in den letzten 12 Jahren. *Zoologischer Anzeiger* (Supplement) 16:36–58.

Lorenz, K. 1965. *Evolution and Modification of Behavior*. University of Chicago Press, Chicago.

Losick, R. and Kaiser, D. 1997. Why and how bacteria communicate. *Scientific American* 276:52–57.

Losos, J.B. 1990. Concordant evolution of locomotor behaviour, display rate and morphology in *Anolis* lizards. *Animal Behaviour* 39:879–890.

Losos, J.B. 1994a. An approach to the analysis of comparative data when a phylogeny is unavailable or incomplete. *Systematic Biology* 43:117–123.

Losos, J.B. 1994b. Integrative approaches to evolutionary ecology: *Anolis* lizards as model systems. *Annual Review of Ecology and Systematics* 25:467–493.

Losos, J.B. 1999. Uncertainty in the reconstruction of ancestral character states and limitations of the use of phylogenetic comparative methods. *Animal Behaviour* 58:1319–1324.

Losos, J.B. and Irschick, D.J. 1996. The effect of perch diameter on escape behaviour of *Anolis* lizards: laboratory predictions and field tests. *Animal Behaviour* 51:593–602.

Losos, J.B. and Miles, D.B. 1994. Adaptation, constraint, and the comparative method:

phylogenetic issues and methods. In: *Ecological Morphology: Integrative Organismal Biology*, P.C. Wainwright and S.M. Reilly (eds.). University of Chicago Press, Chicago, pp. 60–98.

Losos, J.B., Warheit, K.I. and Schoener, T.W. 1997. Adaptive differentiation following experimental island colonization in *Anolis* lizards. *Nature* 387:70–73.

Lott, D.F. 1984. Intraspecific variation in the social systems of wild vertebrates. *Behaviour* 88:266–325.

Lott, D.F. 1988. *Intraspecific Variation in the Social Systems of Wild Vertebrates*. Cambridge University Press, Cambridge.

Lounibos, L.P., Van Dover, C. and O'Meara, G.F. 1982. Fecundity, autogeny, and the larval environment of the pitcher-plant mosquito, *Wyeomyia smithii*. *Oecologia* 55:160–164.

Loveridge, A. 1947. A revision of the African lizards of the family Gekkonidae. *Bulletin of the Museum of Comparative Zoology* 98:1–469.

Lowe-McConnell, R.H. 1969. Speciation in tropical freshwater fishes. *Biological Journal of the Linnean Society* 1:51–75.

Lu, G., DeLisle, A.J., de Vetten, N.C. and Ferl, R.J. 1992. Brain proteins in plants: an *Arabidopsis* homolog to neurotransmitter pathway activators is part of a DNA binding complex. *Proceedings of the National Academy of Sciences USA* 89:11490–11494.

Lubbock, J. 1884. *Ants, Bees, and Wasps*. Appleton, New York.

Lukens, L.N. and Doebley, J. 1999. Epistatic and environmental interactions for quantitative trait loci involved in maize evolution. *Genetical Research* 74:291–302.

Lüttge, U. 1996. Plasticity and diversity in a genus of C3/CAM intermediate tropical trees. In: *Crassulacean Acid Metabolism*, K. Winter and J.A.C. Smith (eds.). Springer-Verlag, New York, pp. 296–311.

Lüttge, U. 1999. One morphotype, three physiotypes: sympatric species of *Clusia* with obligate C_3–CAM intermediate behavior. *Plant Biology* 1:138–148.

Lynch, M. 1990. The rate of morphological evolution in mammals from the standpoint of the neutral expectation. *American Naturalist* 136:727–741.

Lynch, M. 1991. Methods for the analysis of comparative data in evolutionary biology. *Evolution* 45:1065–1080.

Lynch, M. and Gabriel, W. 1983. Phenotypic evolution and parthenogenesis. *American Naturalist* 122:745–764.

Lyon, B.E. 1993. Conspecific brood parasitism as a flexible female reproductive tactic in

American coots. *Animal Behaviour* 46:911–928.

Lyon, B.E. and Montgomerie, R.D. 1986. Delayed plumage maturation in passerine birds: reliable signaling by subordinate males? *Evolution* 40:605–615.

Lyon, B.E., Eadie, M.J. and Hamilton, L.D. 1994. Parental choice selects for ornamental plumage in American coot chicks. *Nature* 371:240–243.

Ma, J. and Karplus, M. 1998. The allosteric mechanism of the chaperonin GroEL: a dynamic analysis. *Proceedings of the National Academy of Sciences USA* 95:8502–8507.

Mabee, P.M. 1993. Phylogenetic interpretation of ontogenetic change: sorting out the actual and artefactual in an empirical case study of centrarchid fishes. *Zoological Journal of the Linnean Society, London* 107:175–291.

MacFadden, B.J. 1992. *Fossil Horses: Systematics, Paleobiology, and Evolution of the Family Equidae.* Cambridge University Press, Cambridge.

MacFadden, B.J., Solounias, N., and Cerling, T.E. 1999. Ancient diets, ecology, and extinction of 5-million-year-old horses from Florida. *Science* 283:824–827.

MacKay, W.P. 1981. A comparison of the nest phenologies of three species of *Pogonomyrmex* hamster ants (Hymenoptera Formicidae). *Psyche* 88:25–24.

Mackintosh, N.J. 1974. *The Psychology of Animal Learning.* Academic Press, London.

Mackintosh, N.J. 1983. General principles of learning. In: *Animal Behaviour. Volume 3: Genes, Development and Learning*, T.R. Halliday and P.J.B. Slater (eds.). W.H. Freeman, New York, pp. 149–177.

Macnair, M.R. 1983. The genetic control of copper tolerance in the yellow monkey flower, *Mimulus guttatus. Heredity* 50:283–293.

Macnair, M.R. 1991. Why the evolution of resistance to anthropogenic toxins normally involves major gene changes: the limits to natural selection. *Genetica* 84:213–219.

Madison, M. 1977. A revision of *Monstera* (Araceae). *Contributions of the Gray Herbarium of Harvard University*, No. 207, 3–131.

Maeshiro, T. and Kimura, M. 1998. The role of robustness and changeability on the origin and evolution of genetic codes. *Proceedings of the National Academy of Sciences USA* 95:5088–5093.

Maeta, Y., Sakagami, S.F. and Michener, C.D. 1992. Laboratory studies on the behavior and colony structure of *Braunsapis hewitti*, a xylocopine bee from Taiwan (Hymenoptera: Anthophoridae). *University of Kansas Science Bulletin* 54:289–333.

Maeterlinck, M. 1901 [1958]. *The Life of the Bee.* East Midland Printing, Bury St. Edmunds, U.K.

Magurran, A.E. 1987. Individual differences in fish behaviour. In: *The Behaviour of Teleost Fishes*, T.J. Pitcher (ed.). Croom Helm, London, pp. 338–365.

Mahner, M. and Kary, M. 1997. What exactly are genomes, genotypes and phenotypes? And what about phenomes? *Journal of Theoretical Biology* 186:55–63.

Maier, N.R.F. and Schneirla, T.C. 1964 [1935]. *Principles of Animal Psychology.* Dover, New York.

Malyshev, S.I. 1966 [1968]. *Genesis of the Hymenoptera.* Methuen, London.

Mangelsdorf, P.C. 1974. *Corn: Its Origin, Evolution and Improvement.* Harvard University Press, Cambridge, Massachusetts.

Mangelsdorf, P.C. 1986. The origin of corn. *American Scientist* 254:80–86.

Manning, A. 1979. *An Introduction to Animal Behaviour.* Addison-Wesley, Reading.

Maret, T.J. and Collins, J.P. 1997. Ecological origin of morphological diversity: a study of alternative trophic phenotypes in larval salamanders. *Evolution* 51:898–905.

Margulis, L. 1981. *Symbiosis in Cell Evolution.* W.H. Freeman, San Francisco.

Margulis, L. and Fester, R. 1991. *Symbiosis as a Source of Evolutionary Innovation.* MIT Press, Cambridge.

Margulis, L. and Sagan, D. 1986. *Origins of Sex.* Yale University Press, New Haven.

Marin, F., Smith, M., Isa, Y., Muyzer, G. and Westbroek, P. 1996. Skeletal matrices, muci, and the origin of invertebrate calcification. *Proceedings of the National Academy of Sciences USA* 93:1554–1559.

Mark, R. 1996. Architecture and evolution. *American Scientist* 84:383–389.

Markert, C.L. and Faulhaber, I. 1965. Lactate dehydrogenase isozyme patterns of fish. *Journal of Experimental Zoology* 159:319–332.

Markert, C.L. and Moller, F. 1959. Multiple forms of enzymes: tissue, ontogenetic, and species specific patterns. *Proceedings of the National Academy of Sciences USA* 45:753–763.

Markham, J.C. 1968. Notes on growth-patterns and shell-utilizations of the hermit crab *Pagurus bernhardus* (L.). *Ophelia* 5:189–205.

Markl, H. 1965. Stridulation in leaf-cutting ants. *Science* 149:1392–1393.

Markl, H. 1985. Manipulation, modulation, information, cognition: some of the riddles of communication. In: *Experimental Behavioral Ecology and Sociobiology*, B. Hölldobler and M. Lindauer (eds.). Sinauer, Sunderland, MA, pp. 163–194.

Markl, H., Hölldobler, B. and Hölldobler, T. 1977. Mating behavior and sound production in harvester ants (*Pogonomyrmex* Formicidae). *Insectes Sociaux* 24:191–212.

Markow, T.A. and Ankney, P.F. 1988. Insemination reaction in *Drosophila*: found in species whose males contribute material to oocytes before fertilization. *Evolution* 42:1097–1101.

Markow, T.A. and Clarke, G.M. 1997. Meta-analysis of the heritability of developmental stability: a giant step backward. *Journal of Evolutionary Biology* 10:31–37.

Marler, P. 1984. Song learning: innate species differences in the learning process. In: *The Biology of Learning*, P. Marler and H.S. Terrace (eds.). Springer-Verlag, New York, pp. 289–309.

Marler, P. 1998. Nature, nurture and the instinct to learn [symposium abstract S.40.3]. *Ostrich* 69:124. [Full paper in *Proceedings XXII International Ornithological Congress*, Durban, N. Adams and R. Slowtow (eds.), Bird Life South Africa, Johannesburg, pp. 2379–2393.

Marler, P. and Terrace, H.S. (eds.) 1984. *The Biology of Learning*. Springer-Verlag, New York.

Marler, P., Dufty, A. and Pickert, R. 1986a. Vocal communication in the domestic chicken: I. Does sender communicate information about the quality of a food referent to a receiver? *Animal Behaviour* 34:188–193.

Marler, P., Dufty, A. and Pickert, R. 1986b. Vocal communication in the domestic chicken: II. Is a sender sensitive to the presence and nature of a receiver? *Animal Behaviour* 34:194–198.

Marshall, C.R., Raff, E.C. and Raff, R.A. 1994. Dollo's law and the death and resurrection of genes. *Proceedings of the National Academy of Sciences USA* 91:12283–12287.

Marshall, D.C. and Cooley, J.R. 2000. Reproductive character displacement and speciation in periodical cicadas, with description of a new species, 13-year *Magicicada neotredecim*. *Evolution* 54:1313–1325.

Martin, A. and Simon, C. 1990a. Differing levels of among-population divergence in the mitochondrial DNA of periodical cicadas related to historical biogeography. *Evolution* 44:1066–1080.

Martin, A. and Simon, C. 1990b. Temporal variation in insect life cycles. *BioScience* 4:359–367.

Martin, C.E. 1996. Putative causes and consequences of recycling CO_2 via crassulacean acid metabolism. In: *Crassulacean Acid Metabolism*, K. Winter and J.A.C. Smith (eds.). Springer-Verlag, New York, pp. 192–203.

Martinez, D.E. and Levinton, J. 1996. Adaptation to heavy metals in the aquatic oligochaete *Limnodrilus hoffmeisteri*: evidence for control by one gene. *Evolution* 50:1339–1343.

Martinez del Rio, C. and Karasov, W.H. 1990. Digestion strategies in nectar- and fruit-eating birds and the sugar composition of plant rewards. *American Naturalist* 136:618–637.

Masaki, S. 1986. Significance of ovipositor length in life cycle adaptations of crickets. In: *The Evolution of Insect Life Cycles*, F. Taylor and R. Karban (eds.). Springer-Verlag, New York, pp. 20–34.

Masaki, S. and Walker, T.J. 1987. Cricket life cycles. *Evolutionary Biology* 268:693–696.

Maschwitz, U. and Maschwitz, E. 1974. Bursting workers: a new means of defence in social Hymenoptera. *Oecologia* 14:289–294.

Mason R.T. and Crews, D. 1985. Female mimicry in garter snakes. *Nature* 316:59–60.

Mason, R.T., Fales, H.M., Jones, T.H., Pannell, L.K., Chinn, J.W. and Crews, D. 1989. Sex pheromones in snakes. *Science* 245:290–293.

Mather, K. 1953. The genetical structure of populations. *Symposia of the Society for Experimental Biology* 7:66–95.

Mather, K. 1955. Polymorphism as an outcome of disruptive selection. *Evolution* 9:52–61.

Mather, K. 1969. Selection through competition. *Heredity* 24:529–540.

Mather, K. 1973. *Genetical Structure of Populations*. Chapman and Hall, London.

Mather, K. and De Winton, D. 1941. Adaptation and counter-adaptation of the breeding system in *Primula*. *Annals of Botany* 5:297.

Mather, K. and Jinks, J.L. 1971. *Biometrical Genetics*, 2nd ed. Chapman and Hall, London.

Matsuda, R. 1979. Abnormal metamorphosis and arthropod evolution. In: *Arthropod Phylogeny*, A.P. Gupta (ed.). Van Nostrand-Reinhold, New York, pp. 137–256.

Matsuda, R. 1982. Evolutionary process of talitrid amphipods and salamanders in changing environments, with a discussion of genetic assimilation and some other evolutionary concepts. *Canadian Journal of Zoology* 60:733–749.

Matsuda, R. 1987. *Animal Evolution in Changing Environments with Special Reference to Abnormal Metamorphosis*. John Wiley and Sons, New York.

Mattheck, C. 1993. *Design in der Natur: Der Baum als Lehrmeister*. Rombach Verlag, Berlin.

Matthews, T.C. and Munstermann, L.E. 1990. Linkage maps for 20 enzyme loci in *Aedes triseriatus*. *Journal of Heredity* 81:101–106.

Maturo, F.J.S., Jr. 1973. Offspring variation from known maternal stocks of *Parasmittina nitida* (Verrill). In: *Bryozoa*, G.P. Larwood (ed.). Academic Press, New York.

May, R.M. 1977. Population genetics and cultural inheritance. *Nature* 268:11–13.

Mayer, I., Berglund, I., Rydevik, M., Borg, B. and Schulz, R. 1990a. Plasma levels of five androgens and 17α-hydroxy-20β-dihydroprogesterone in immature and mature male Baltic salmon (*Salmo salar*) parr, and the effects of castration and androgen replacement in mature parr. *Canadian Journal of Zoology* 68:263–267.

Mayer, I., Borg, B. and Schulz, R. 1990b. Seasonal changes in and effect of castration/androgen replacement on the plasma levels of five androgens in the male three-spined stickleback, *Gasterosteus aculeatus* L. *General and Comparative Endocrinology* 79:23–30.

Mayer, I., Schulz, R., Borg, B. 1990c. Seasonal endocrine changes in Baltic salmon, *Salmo salar*, immature parr and mature male parr. I. Plasma levels of five androgens, 17α-hydroxy-20β-dihydroprogesterone, and 17β-estradiol. *Canadian Journal of Zoology* 63:1360–1365.

Mayer, I., Rosenqvist, G., Borg, B., Ahnesjö, I., Berglund, A. and Schulz, R.W. 1993. Plasma levels of sex steroids in three species of pipefish (Syngnathidae). *Canadian Journal of Zoology* 71:1903–1907.

Maynard Smith, J. 1966. Sympatric speciation. *American Naturalist* 100:637–650.

Maynard Smith, J. 1970. The causes of polymorphism. *Symposia of the Zoological Society of London* 26:371–383.

Maynard Smith, J. 1977. Why the genome does not congeal. *Nature* 268:693–696.

Maynard Smith, J. 1978. *The Evolution of Sex*. Cambridge University Press, Cambridge.

Maynard Smith, J. 1979. Game theory and the evolution of behaviour. *Proceedings of the Royal Society of London B* 205:475–488.

Maynard Smith, J. 1981. Macroevolution. *Nature* 289:13–14.

Maynard Smith, J. 1982. *Evolution and the Theory of Games*. Cambridge University Press, New York.

Maynard Smith, J. 1983a. The genetics of stasis and punctuation. *Annual Review of Genetics* 17:11–25.

Maynard Smith, J. 1983b. Evolution and development. In: *Development and Evolution*, B.C. Goodwin, N. Holder and C.C. Wylie (eds.). Cambridge University Press, Cambridge, pp. 33–45.

Maynard Smith, J. 1989. *Evolutionary Genetics*. Oxford University Press, New York.

Maynard Smith, J. and Price, G.R. 1973. The logic of animal conflict. *Nature* 246:15–18.

Maynard Smith, J. and Szathmáry, E. 1995. *The Major Transitions in Evolution*. W.H. Freeman Spektrum, New York.

Maynard Smith, J., Burian, R., Kauffman, S., Alberch, P., Campbell, J., Goodwin, B., Lande, R., Raup, D. and Wolpert, L. 1985. Developmental constraints and evolution. *Quarterly Review of Biology* 60:265–287.

Mayo, O. 1983. *Natural Selection and Its Constraints*. Academic Press, London.

Mayr, E. 1933. Notes on the variation of immature and adult plumages in birds and a physiological explanation of abnormal plumages. *American Museum Novitates* 666:1–10.

Mayr, E. 1934. Notes on the genus *Petroica*. *American Museum Novitates* 714:1–19.

Mayr, E. 1942. *Systematics and the Origin of Species*. Columbia University Press, New York.

Mayr, E. 1954. Change of genetic environment and evolution. In: *Evolution as a Process*, J. Huxley (ed.). Allen and Unwin, London, pp. 157–180.

Mayr, E. 1958. Behavior and systematics. In: *Behavior and Evolution*, A. Roe and G. Gaylord Simpson (eds.). Yale University Press, New Haven, pp. 341–362.

Mayr, E. 1959. The emergence of evolutionary novelties. In: *Evolution after Darwin*, Vol. 1. S. Tax (ed.). University of Chicago Press, Chicago, pp. 349–380.

Mayr, E. 1961. Cause and effect in biology. *Science* 134:1501–1506.

Mayr, E. 1962. Accident or design: the paradox of evolution. In: *The Evolution of Living Organisms*. Proceedings of the Darwin Centenary Symposium of the Royal Society of Victoria, Melbourne University Press, Melbourne, pp. 1–14 [Adaptation in Mayr, E. 1976. *Evolution and the Diversity of Life*. Harvard University Press, Cambridge, pp. 30–87.]

Mayr, E. 1963. *Animal Species and Evolution*. Belknap Press, Cambridge.

Mayr, E. 1966. Introduction. In: *On the Origin of Species, Facsimile of the First Edition*, Darwin, C. 1859. Harvard University Press, Cambridge, pp. vii-xxvii.

Mayr, E. 1968. Comments on "Theories and hypotheses in biology," *Boston Studies in the*

Philosophy of Science 5(1968):450–456. [reprinted in Mayr, E. 1976. *Evolution and the Diversity of Life*, Belknap Press of Harvard University Press, Cambridge, pp 376–382.]

Mayr, E. 1970a. Evolution and behaviour. *Verhandlungsbericht der Deutschen Zoologischen Gesellschaft* 64:322–336.

Mayr, E. 1970b. *Populations, Species, and Evolution*. Harvard University Press, Cambridge.

Mayr, E. 1974a. Behavior programs and evolutionary strategies. *American Scientist* 62:650–659.

Mayr, E. 1974b. The definition of the term disruptive selection. *Heredity* 32:404–406.

Mayr, E. 1976. *Evolution and the Diversity of Life*. Belknap Press, Cambridge.

Mayr, E. 1982a. *The Growth of Biological Thought*. Harvard University Press, Cambridge.

Mayr, E. 1982b. Adaptation and selection. *Biologisches Zentralblatt* 101:161–174.

Mayr, E. 1982c. Speciation and macroevolution. *Evolution* 36:1119–1132.

Mayr, E. 1988. *Toward a New Philosophy of Biology*. Belknap Press, Cambridge.

Mayr, E. 1989. Speciational evolution or punctuated equilibria. In: *The Dynamics of Evolution*, A. Somit and S.A. Peterson (eds.). Cornell University Press, Ithaca, pp. 21–53.

Mayr, E. 1993. Fifty years of progress in research on species and speciation. *Proceedings of the California Academy of Sciences* 48:131–140.

Mayr, E. and Ashlock, P.D. 1991. *Principles of Systematic Zoology*. McGraw-Hill, New York.

Mayr, E. and Provine, W.B. (eds.) 1980. *The Evolutionary Synthesis. Perspectives on the Unification of Biology*. Harvard University Press, Cambridge.

McAdams, H.H. and Arkin, A. 1997. Stochastic mechanisms in gene expression. *Proceedings of the National Academy of Sciences USA* 94:814–819.

McAllister, L.B., Mahon, A.C. and Scheller, R.H. 1986. Evolution of egg laying behavior in *Aplysia*. In: *Evolutionary Genetics of Invertebrate Behavior*, M.D. Huettel (ed.). Plenum Press, New York, pp. 255–262.

McAlpine, D.K. 1979. Agonistic behavior in *Achias australis* (Diptera, Platystomatidae) and the significance of eyestalks. In: *Sexual Selection and Reproductive Competition in Insects*, M.S. Blum and N.A. Blum (eds.). Academic Press, New York, pp. 221–230.

McBride, G. 1971. The nature-nurture problem in social evolution. In: *Man and Beast: Comparative Social Behavior*, J.F. Eisenberg and W.S. Dillon (eds.). Smithsonian Institution Press, Washington, D.C., pp. 37–56.

McClintock, B. 1984. The significance of responses of the genome to challenge. *Science* 226:792–801.

McCollum, S.A. and Van Buskirk, J. 1996. Costs and benefits of a predator-induced polyphenism in the gray treefrog *Hyla chrysoscelis*. *Evolution* 50:583–593.

McCune, A.R., Thomson, K.S. and Olsen, P.E. 1984. Semionotid fishes from the Mesozoic Great Lakes of North America. In: *Evolution of Fish Species Flocks*, A.A. Echelle and I. Kornfield (eds.). University of Maine at Orono Press, Orono, pp. 27–44.

McDonald, D. 1989. Cooperation under sexual selection: age-graded changes in a lekking bird. *American Naturalist* 134:709–730.

McDonald J. and Anson, B.J. 1940. Variations in the origin of arteries derived from the aortic arch, in American whites and negroes. *American Journal of Physical Anthropology* 27:91–108.

McDonald, J.F. 1986. Physiological tolerance and behavioral avoidance of alcohol in *Drosophila*: coadaptation or pleiotropy? In: *Evolutionary Genetics of Invertebrate Behavior*, M.D. Huettel (ed.). Plenum Press, New York, pp. 247–254.

McDonald, J.F. 1998. Transposable elements, gene silencing and macroevolution. *Trends in Ecology and Evolution* 13:94–95.

McDonald, J.F., Strand, D.J., Lambert, M.E. and Weinstein, I.B. 1987. The responsive genome: evidence and evolutionary implications. In: *Development as an Evolutionary Process*, R.A. Raff and E.C. Raff (eds.). Alan R. Liss, New York, pp. 239–263.

McDowall, R.M. 1988. *Diadromy in Fishes Migrations between Freshwater and Marine Environments*. Timber Press, Portland.

McEdward, L.R. and Carson, S.F. 1987. Variation in egg organic content and its relationship with egg size in the starfish *Solaster stimpsoni*. *Marine Ecology Progress Series* 37:159–169.

McEdward, L.R. and Miner, B.G. 2001. Larval and life-cycle patterns in echinoderms. *Canadian Journal of Zoology* 79:1125–1170.

McFadden, D. 1993. A masculinizing effect on the auditory systems of human females having male co-twins. *Proceedings of the National Academy of Sciences USA* 90:11900–11904.

McFall-Ngai, M.J. 1999. Consequences of evolving with bacterial symbionts: insights from the squid-vibrio associations. *Annual Review of Ecology and Systematics* 30:235–256.

McFarland, D. 1974. Time-sharing as a behavioral phenomenon. *Advances in the Study of Behavior* 5:201–225.

McFarland, D. 1983. Time sharing: a reply to Houston (1982). *Animal Behaviour* 31:307–308.

McFarland, D. and Bösser, T. 1993. *Intelligent Behavior in Animals and Robots*. MIT Press, Cambridge.

McGinnis, W. and Krumlauf, R. 1992. Homeobox genes and axial patterning. *Cell* 68:283–302.

McKaye, K.R. and Gray, W.N. 1984. Extrinsic barriers to gene flow in rock-dwelling cichlids of Lake Malawi: macrohabitat heterogeneity and reef colonization. In: *Evolution of Fish Species Flocks*, A.A. Echelle and I. Kornfield (eds.). University of Maine at Orono Press, Orono, pp. 169–184.

McKeown, M. 1992. Alternative mRNA splicing. *Annual Review of Cell Biology* 8:133–155.

McKinney, F. and Jackson, J.B.C. 1989. *Bryozoan Evolution*. Unwin Hyman, Boston.

McKinney, M.L. (ed.) 1988. *Heterochrony in Evolution*. Plenum Press, New York.

McKinney, M.L. and McNamara, K.J. 1991. *Heterochrony, the Evolution of Ontogeny*. Plenum Press, New York.

McKinney, M.L. and Schoch, R.M. 1985. Titanothere allometry, heterochrony, and biomechanics: revising an evolutionary classic. *Evolution* 39:1352–1363.

McKitrick, M.C. 1992. Phylogenetic analysis of avian parental care. *Auk* 109:828–846.

McKitrick, M.C. 1993. Phylogenetic constraint in evolutionary theory: has it any explanatory power? *Annual Review of Ecology and Systematics* 24:307–330.

McKitrick, M.C. 1994. On homology and the ontological relationship of parts. *Systematic Biology* 43:1–10.

McLain, D.K. 1987. Heritability of size, a sexually selected character, and the response to sexual selection in a natural population of the southern green stink bug, *Nezara viridula* (Hemiptera: Pentatomidae). *Heredity* 59:391–395.

McLeod, L. 1984. Seasonal polyphenism in African *Precis* butterflies. In: *The Biology of Butterflies*, R. I. Vane-Wright and P.R. Ackery (eds.). Academic Press, London, pp. 313–315.

McNamara, K.J. 1988. The abundance of heterochrony in the fossil record. In: *Heterochrony in Evolution: A Multidisciplinary Approach*, M.L. McKinney (ed.). Plenum Press, New York, pp. 287–325.

McPhail, J.D. 1969. Predation and the evolution of a stickleback (*Gasterosteus*). *Journal of*
the Fisheries Research Board of Canada 26:3183–3208.

McPhail, J.D. 1994. Speciation and the evolution of reproductive isolation in the sticklebacks (*Gasterosteus*) of south-western British Columbia. In: *The Evolutionary Biology of the Threespine Stickleback*, M.A. Bell and S.A. Foster (eds.). Oxford University Press, New York, pp. 399–437.

McPheron, B., Smith, D.C. and Berlocher, S.H. 1988. Genetic differences between host races of *Rhagoletis pomonella*. *Nature* 336:64–66.

McWilliams, S.R., Afik, D. and Secor, S. 1997. Patterns and processes in the vertebrate digestive system. *Trends in Ecology and Evolution* 12:421–422.

Meinhardt, H. 1982. The role of compartmentalization in the activation of particular control genes and in the generation of proximo-distal positional information in appendages. *American Zoologist* 22:209–220.

Meireles, C.M.M., Schneider, M.P.C., Sampaio, M.I.C., Schneider, H., Slightom, J.L., Chiu, C.-H., Neiswanger, K., Gumucio, D.L., Czelusniak, J. and Goodman, M. 1995. Fate of a redundant γ-globin gene in the atelid clade of New World monkeys: implications concerning fetal globin gene expression. *Proceedings of the National Academy of Sciences USA* 92:2607–2611.

Mello, C.V., Vicario, D.S. and Clayton, D.F. 1992. Song presentation induces gene expression in the songbird forebrain. *Proceedings of the National Academy of Sciences USA* 89:6818–6822.

Mendel, G. 1863 [1958]. *Experiments in Plant Hybridization*. Harvard University Press, Cambridge.

Menzel, R. 1985. Learning in honeybees in an ecological and behavioral context. In: *Experimental Behavioral Ecology and Sociobiology*, B. Hölldobler and M. Lindauer (eds.). Sinauer, Sunderland, MA, pp. 55–74.

Menzel, R. and Müller, U. 1996. Learning and memory in honey bees. From behavior to neural substrates. *Annual Review of Neuroscience* 19:379–404.

Menzel, R., Erber, J. and Masuhr, T. 1973. Learning and memory in the honey bee. In: *Experimental Analysis of Insect Behavior*, L.B. Browne (ed.). Springer-Verlag, New York, pp. 195–217.

Mercot, H., Defaye, D., Capy, E.P. and David, J.R. 1994. Alcohol tolerance, ADH activity, and ecological niche of *Drosophila* species. *Evolution* 48:746–757.

Messeri, P. and Giacoma, C. 1986. Dominance rank and related interactions in a captive group of female pigtail macaques. In:

Primate Ontogeny, Cognition and Social Behaviour, J.G. Else and P.C. Lee (eds.). Cambridge University Press, Cambridge, pp. 301–306.

Messina, F.J. 1987. Genetic contribution to the dispersal polymorphism of the cowpea weevil (Coleoptera: Bruchidae). *Annals of the Entomological Society of America* 80:12–16.

Metcalf, C.L, Flint, W.P. and Metcalf, R.L. 1928[1931] *Destructive and Useful Insects*. Third Edition. McGraw-Hill Book Company, Inc., New York.

Metcalf, C.L., Flint, R.L. and Metcalf, R.L. 1951. *Destructive and Useful Insects*. McGraw-Hill, New York.

Metcalf, R.A. and Whitt, G.S. 1977. Relative inclusive fitness in the social wasp *Polistes metricus*. *Behavioral Ecology and Sociobiology* 2:353–360.

Meyer, A. 1985. Changes in behavior with increasing experience with a novel prey in fry of the Central American cichlid, *Cichlasoma managuense* (Teleostei: Cichlidae). *Behaviour* 98:145–167.

Meyer, A. 1987a. Phenotypic plasticity and heterochrony in *Cichlasoma managuense* (Pisces, Cichlidae) and their implications for speciation in cichlid fishes. *Evolution* 41:1357–1369.

Meyer, A. 1987b. First feeding success with two types of prey by the Central American cichlid fish, *Cichlasoma managuense* (Pisces, Cichlidae): morphology versus behavior. *Environmental Biology of Fishes* 18:127–134.

Meyer, A. 1989a. Cost of morphological specialization: feeding performance of the two morphs in the trophically polymorphic cichlid fish, *Cichlasoma citrinellum*. *Oecologia* 80:431–436.

Meyer, A. 1993. Trophic polymorphisms in cichlid fish: do they represent intermediate steps during sympatric speciation and explain their rapid adaptive radiation? In: *New Trends in Ichthyology*, J.-H. Schröder, J. Bauer, and M. Schartl (eds.). GSF-Bericht, Blackwell, London, pp. 257–266.

Meyer, A. 1990a. Ecological and evolutionary consequences of the trophic polymorphism in *Cichlasoma citrinellum* (Pisces: Cichlidae). *Biological Journal of the Linnean Society* 39:279–299.

Meyer, A. 1990b. Morphometrics and allometry in the trophically polymorphic cichlid fish, *Cichlasoma citrinellum*: alternative adaptations and ontogenetic changes in shape. *Journal of Zoology* 221:237–260.

Meyer, A. 1999. Homology and homoplasy: the retention of genetic programmes. In:

Homology, B.K. Hall (ed.). John Wiley and Sons, Chichester, pp. 141–157.

Meyer, A., Kocher, T.D., Basasibwaki, P. and Wilson, A.C. 1990. Monophyletic origin of Lake Victoria cichlid fishes suggested by mitochondrial DNA sequences. *Nature* 347:550–553.

Meyer, A., Morrissey, J.M. and Schartl, M. 1994. Recurrent origin of a sexually selected trait in *Xiphophorus* fishes from molecular phylogeny. *Nature* 368:539–542.

Meyer, V.G. 1966. Flower abnormalities. *Botanical Reviews* 32:165–218.

Meyerowitz, E.M. 1994. Flower development and evolution: new answers and new questions. *Proceedings of the National Academy of Sciences USA* 91:5735–5737.

Michener, C.D. 1948. Observations on the mating behavior of harvester ants. *Journal of the New York Entomological Society* 56:239–242.

Michener, C.D. 1958. The evolution of social behavior in bees. *Proceedings of the Tenth International Congress of Entomology* 2:441–447.

Michener, C.D. 1961. Social polymorphism in Hymenoptera. In: *Insect Polymorphism*, J.S. Kennedy (ed.). Royal Entomological Society, London, pp. 43–56.

Michener, C.D. 1969. Comparative social behavior of bees. *Annual Review of Entomology* 14:299–342.

Michener, C.D. 1974. *The Social Behavior of the Bees*. Harvard University Press, Cambridge.

Michener, C.D. 1978. The parasitic groups of Halictidae (Hymenoptera, Apoidea). *University of Kansas Science Bulletin* 51:291–339.

Michener, C.D. 1985. From solitary to eusocial: need there be a series of intervening species? In: *Experimental Behavioral Ecology and Sociobiology*, B. Hölldobler and M. Lindauer (eds.). Sinauer, New York, pp. 293–306.

Michener, C.D. 1990. Reproduction and castes in social halictine bees. In: *Social Insects. An Evolutionary Approach to Castes and Reproduction*, W. Engels (ed.). Springer-Verlag, New York, pp. 77–122.

Michener, C.D., McGinley, R.J. and Danforth, B.N. 1994. *The Bee Genera of North and Central America (Hymenoptera: Apoidea)*. Smithsonian Institution Press, Washington, D.C.

Michod, R.E. 1997a. Cooperation and conflict in the evolution of individuality. I. Multilevel selection of the organism. *American Naturalist* 149:607–645.

Michod, R.E. 1997b. Evolution of the individual. *American Naturalist* 150:S5–S21.

Mickel, C.E. 1924. An analysis of a bimodal variation in size of the parasite *Dasymutilla bioculata* Cresson (Hymen.: Mutillidae). *Entomological News* 35:236–242.

Mickelvich, M.F. and Weller, S.J. 1990. Evolutionary character analysis: tracing character change on a cladogram. *Cladistics* 6:137–170.

Milbrath, L.R., Tauber, M.J. and Tauber, C.A. 1993. Prey specificity in *Chrysopa*: an interspecific comparison of larval feeding and defensive behavior. *Ecology* 74:1384–1393.

Milbrath, L.R., Tauber, M.J. and Tauber, C.A. 1994. Larval behavior of predacious sister-species: orientation, molting site, and survival in *Chrysopa*. *Behavioral Ecology and Sociobiology* 35:85–90.

Milinski, M. 1979. An evolutionarily stable feeding strategy in sticklebacks. *Zeitschrift für Tierpsychologie* 51:36–40.

Milinski, M. 1982. Optimal foraging: the influence of intraspecific competition on diet selection. *Behavioral Ecology and Sociobiology* 11:109–115.

Milinski, M. 1984a. Competitive resource sharing: an experimental test of a learning rule for ESSs. *Animal Behaviour* 32:233–242.

Milinski, M. 1984b. Parasites determine a predator's optimal feeding strategy. *Behavioral Ecology and Sociobiology* 15:35–37.

Milinski, M. and Parker, G.A. 1991. Competition for resources. In: *Behavioural Ecology: An Evolutionary Approach*, J.R. Krebs and N.B. Davies (eds.). Blackwell, Oxford, pp. 137–168.

Milinski, M. and Regelmann, K. 1985. Fading short-term memory for patch quality in sticklebacks. *Animal Behaviour* 33:678–680.

Milkman, R.D. 1961. The genetic basis of natural variation, III. Developmental lability and evolutionary potential. *Genetics* 46:25–38.

Miller, E.H. 1988. Description of bird behavior for comparative purposes. *Current Ornithology* 5:347–394.

Miller, R.C. and Kurczewski, F.E. 1975. Comparative behavior of wasps in the genus *Lindenius* (Hymenoptera: Sphecidae, Crabroninae). *Journal of the New York Entomological Society* 83:82–120.

Milligan, B.G. 1986. Punctuated evolution induced by ecological change. *American Naturalist* 127:522–532.

Milne, C.P. Jr. 1986. Cytology and cytogenetics. In: *Bee Genetics and Breeding*, T.E. Rinderer (ed.). Academic Press, Inc., London, pp. 205–234.

Milton, K. 1988. Foraging behavior and the evolution of primate cognition. In:

Machiavellian Intelligence, R. Byrne and A. Whiten (eds.). Oxford University Press, Oxford, pp. 285–409.

Mitani, J.C. and Nishida, T. 1993. Contexts and social correlates of long-distance calling by male chimpanzees. *Animal Behaviour* 45:735–746.

Mitchell, G. 1979. *Behavioral Sex Differences in Nonhuman Primates*. Van Nostrand Reinhold, New York.

Mitchell, M. and Taylor, C.E. 1999. Evolutionary computation: an overview. *Annual Review of Ecology and Systematics* 30:593–616.

Mitchell, R. 1975. The evolution of oviposition tactics in the bean weevil, *Callosobruchus maculatus* (F.). *Ecology* 56:696–702.

Mitchell, S.D. 1990. The units of behavior in evolutionary explanations. In: *Interpretation and Explanation in the Study of Animal Behavior*, M. Bekoff and D. Jamieson (eds.). Westview Press, Boulder, pp. 63–83.

Mitchison, T.J. and Kirschner, M.W. 1989. Cytoskeletal dynamics and nerve growth. *Neuron* 1:761–772.

Mitter, C. and Futuyma, D.J. 1979. Population genetic consequences of feeding habits in some forest Lepidoptera. *Genetics* 92:1005–1021.

Miura, T., Kamikouchi, A., Sawata, M., Takeuchi, H., Natori, S., Kubo, T. and Matsumoto, T. 1999a. Soldier caste-specific gene expression in the mandibular glands of *Hodotermopsis japonica* (Isoptera: Termopsidae). *Proceedings of the National Academy of Sciences USA* 96:13874–13879.

Miura, T., Kamilkouchi, A., Sawata, M., Takeuchi, H., Natori, S., Kubo, T. and Matsumoto, T. 1999b. Identification of genes expressed specifically in the soldiers of *Hodotermopsis japonica* (Isoptera: Termopsidae) by differential display method. In: *Social Insects at the Turn of the Millennium*, M.P. Schwarz and K. Hogendoorn (eds.). XIII Congress of IUSSI, Adelaide, p. 321.

Moczek, A.P. 1996. Male dimorphism in the scarab beetle *Onthophagus taurus* Schreber, 1759 (Scarabaeidae, Onthophagini): evolution and plasticity in a variable environment. Thesis, University of Würzburg.

Moczek, A.P. and Emlen, D.J. 1999. Proximate determination of male horn dimorphism in the beetle *Onthophagus taurus* (Coleoptera: Scarabaeidae). *Journal of Evolutionary Biology* 12:27–37.

Moczek, A.P. and Emlen, D.J. 2000. Male horn dimorphism in the scarab beetle, *Onthophagus taurus*: do alternative reproductive tactics favour alternative phenotypes? *Animal Behaviour* 59:459–466.

Mohammedi, A. and Le Conte, Y. 2000. Do environmental conditions exert an effect on nest-mate recognition in queen rearing honey bees? *Insectes Sociaux* 47:307–312.

Mohanty-Hejmadi, P., Dutta, S.K. and Mahapatra, P. 1992. Limbs generated at site of tail amputation in marbled balloon frog after vitamin A treatment. *Nature* 355:352–353.

Mole, S. and Zera, A.J. 1993. Differential allocation of resources underlies the dispersal-reproduction trade-off in the wing dimorphic cricket, *Gryllus rubens*. *Oecologia* 93:121–127.

Monnin, T. and Peeters, C. 1999. Dominance hierarchy and reproductive conflicts among subordinates in a monogynous queenless ant. *Behavioral Ecology* 10:323–332.

Monteiro, A.F., Brakefield, P.M. and French, V. 1994. The evolutionary genetics and developmental basis of wing pattern variation in the butterfly *Bicyclus anynana*. *Evolution* 48:1147–1157.

Montgomery, E.G. 1906. What is an ear of corn? *Popular Science Monthly* 68:55–62.

Montgomery, E.G. 1913. *The Corn Crops*. Macmillan, New York.

Montgomery, S.L. 1975. Comparative breeding site ecology and the adaptive radiation of picture-winged *Drosophila* (Diptera: Drosophilidae) in Hawaii. *Proceedings, Hawaiian Entomological Society* 22:65–102.

Moore, A.J. 1997. The evolution of social signals: morphological, functional, and genetic integration of the sex pheromone in *Nauphoeta cinerea*. *Evolution* 51:1920–1928.

Moore, A.J., Reagan, N.L. and Haynes, K.F. 1995. Conditional signaling strategies: effects of ontogeny, social experience and social status on the pheromonal signal of male *Nauphoeta cinerea*. *Animal Behaviour* 50:191–202.

Moore, M.C. 1991. Application of organization-activation theory to alternative male reproductive strategies: a review. *Hormones and Behavior* 25:154–179.

Moore, M.C., Hews, D.K., and Knapp, R. 1998. Evolution and hormonal control of alternative male phenotypes. *American Zoologist* 38:133–151.

Moran, N.A. 1988. The evolution of host-plant alternation in aphids: evidence for host-plant specialization as a dead end. *American Naturalist* 132:681–706.

Moran, N.A. 1991. Phenotype fixation and genotypic diversity in the complex life cycle of the aphid *Pemphigus betae*. *Evolution* 45:957–970.

Moran, N.A. 1992a. The evolution of aphid life cycles. *Annual Review of Entomology* 37:321–348.

Moran, N.A. 1992b. The evolutionary maintenance of alternative phenotypes. *American Naturalist* 139:971–989.

Moran, N.A. and Whitham, T.G. 1988. Evolutionary reduction of complex lifecycles: loss of host-alternation in *Pemphigus* (Homoptera: Aphididae). *Evolution* 42:717–728.

Morgan, C.L. 1896a. On modification and variation. *Science* 4:733–740.

Morgan, C.L. 1896b. *Habit and Instinct*. Arnold, London.

Morgan, T.H., Bridges, C.B. and Sturtevant, A.H. 1925. The genetics of *Drosophila*. *Bibliography of Genetics* 2:1.

Moritz, R.F.A. and Southwick, E.E. 1987. Phenotype interactions in group behavior of honey bee workers (*Apis mellifera* L.). *Behavioral Ecology and Sociobiology* 21:53–57.

Moritz, R.F.A. and Southwick, E.E. 1992. *Bees as Superorganisms*. Springer-Verlag, New York.

Morris, D. 1955. The causation of pseudofemale and pseudomale behaviour: a further comment. *Behaviour* 8:46–57.

Morris, D. 1956. The feather postures of birds and the problem of the origin of social signals. *Behaviour* 9:75–111.

Morris, M.R. and Ryan, M.J. 1990. Age at sexual maturity of *Xiphophorus nigrensis* in nature. *Copeia* 1990:747–751.

Morris, M.R. and Yoon, S.L. 1989. A mechanism for female choice of large males in the treefrog *Hyla chrysoscelis*. *Behavioral Ecology and Sociobiology* 25:65–71.

Morris, S.C. 1989. Burgess Shale faunas and the Cambrian explosion. *Science* 246:339–346.

Morris, S.C. 1998. *The Crucible of Creation*. Oxford University Press, New York.

Morse, D.H. 1978. Size related foraging differences of bumblebee workers. *Economic Entomology* 3:189–192.

Morse, D.H. 1980. *Behavioral Mechanisms in Ecology*. Harvard University Press, Cambridge.

Moser, J.C. and Cross, E.A. 1975. Phoretomorph: a new phoretic phase unique to the Pyemotidae (Acarina: Tarsonemoidea). *Annals of the Entomological Society of America* 68:820–822.

Moulds, M.S. 1977. Field observations on behavior of a north Queensland species of *Phytalmia* (Diptera: Tephritidae). *Journal of the Australian Entomological Society* 16:347–352.

Mousseau, T.A. and Dingle, H. 1991. Maternal effects in insect life histories. *Annual Review of Entomology* 36:511–534.

Mousseau, T.A. and Fox, C.W. (eds.) 1998. *Maternal Effects as Adaptations.* Oxford University Press, New York.

Moya, A., Galiana, A. and Ayala, F.J. 1995. Founder-effect speciation theory: failure of experimental corroboration. *Proceedings of the National Academy of Sciences USA* 89:9084–9088.

Moynihan, M. 1970. Some behavior patterns of platyrrhine monkeys II. *Saguinus geoffroyi* and some other tamarins. *Smithsonian Contributions to Zoology* 28:1–77.

Mrosovsky, N. 1990. *Rheostasis: The Physiology of Change.* Oxford University Press, Oxford.

Mulkey, S.S., Smith, A.P., Wright, S.J., Machado, J.L. and Dudley, R. 1992. Contrasting leaf phenotypes control seasonal variation in water loss in a tropical forest shrub. *Proceedings of the National Academy of Sciences USA* 89:9084–9088.

Müller, G.B. 1989. Ancestral patterns in bird limb development: a new look at Hampé's experiment. *Journal of Evolutionary Biology* 2:31–47.

Müller, G.B. 1990. Developmental mechanisms at the origin of morphological novelty: a side-effect hypothesis. In: *Evolutionary Innovations*, M.H. Nitecki (ed.). University of Chicago Press, Chicago, pp. 99–132.

Müller, G.B. and Wagner, G.P. 1991. Novelty in evolution: restructuring the concept. *Annual Review of Ecology and Systematics* 22:229–256.

Müller, G.B. and Wagner, G.P. 1996. Homology, hox genes, and developmental integration. *American Zoologist* 36:4–13.

Murakami, Y. 1960. Seasonal dimorphism in the Encyrtidae (Hymenoptera, Chalcidoidea). *Acta Hymenopterologica* 1:199–204.

Murdoch, W.W., Avery, S. and Smyth, M.E.B. 1975. Switching in predatory fish. *Ecology* 56:1094–1105.

Murfet, I.C. 1977. Environmental interaction and the genetics of flowering. *Annual Review of Plant Physiology* 28:253–278.

Myers, J.P., Connors, P.G. and Pitelka, F.A. 1979. Territory size in wintering sanderlings: the effects of prey abundance and intruder density. *Auk* 96:551–561.

Myles, T.G. and Chang, F. 1984. The caste system and caste mechanisms of *Neotermes connexus* (Isoptera: Kalotermitidae). *Sociobiology* 9:163–321.

Myles, T.G. and Nutting, W.L. 1988. Termite eusocial evolution: a re-examination of Bartz's hypothesis and assumptions. *Quarterly Review of Biology* 63:1–24.

Nagel, L. and Schluter, D. 1998. Body size, natural selection, and speciation in sticklebacks. *Evolution* 52:209–218.

Nalepa, C.A. and Bandi, C. 2000. Characterizing the ancestors: paedomorphosis and termite evolution. In: *Termites: Evolution, Sociality, Symbioses, Ecology*, T. Abe, D.E. Bignell and M. Higashi (eds.). Kluwer Academic, Netherlands, pp. 53–75.

Nanney, D.L. 1958. Epigenetic control systems. *Proceedings of the National Academy of Sciences USA* 44:712–717.

Needham, J. 1933. On the dissociability of the fundamental processes in ontogenesis. *Biological Reviews* 8:180–223.

Negus, N.C. and Berger, P.J. 1998. Reproductive strategies of *Dicrostonyx groenlandicus* and *Lemmus sibiricus* in high-arctic tundra. *Canadian Journal of Zoology* 76:391–400.

Nei, M. 1967. Modification of linkage intensity by natural selection. *Genetics* 57:625–641.

Nei, M. 1975. *Molecular Population Genetics and Evolution.* North-Holland, Amsterdam.

Nei, M. 1987. *Molecular Evolutionary Genetics.* Columbia University Press, New York.

Nei, M., Maruyama, T., and Chakraborty, R. 1975. The bottleneck effect and genetic variability in populations. *Evolution* 29:1–10.

Nelson, D.A., Marler, P. and Palleroni, A. 1995. A comparative approach to vocal learning: intraspecific variation in the learning process. *Animal Behaviour* 50:83–97.

Nelson, G. 1994. Homology and systematics. In: *Homology: The Hierarchical Basis of Comparative Biology*, B.K. Hall (ed.). Academic Press, San Diego, pp. 101–149.

Nelson, R.J., Badura, L.L. and Goldman, B.D. 1990. Mechanisms of seasonal cycles of behavior. *Annual Review of Psychology* 41:81–108.

Nevo, E., Beiles, A. and Ben-Shlomo, R. 1984. The evolutionary significance of genetic diversity: ecological, demographic and life history correlates. In: *Evolutionary Dynamics of Genetic Diversity*, G.S. Mani (ed.). Springer-Verlag, Berlin, pp. 13–213.

Nicoletto, P.F. 1996. The influence of water velocity on the display behavior of male guppies, *Poecilia reticulata. Behavioral Ecology* 7:272–278.

Niehrs, C. and Pollet, N. 1999. Synexpression groups in eukaryotes. *Nature* 402:483–487.

Nielsen, E.T. 1933. Sur les habitudes des *Hyménoptères aculeates* solitaires. *Entomologiske Meddelelser* 18:259–338.

Nijhout, H.F. 1975. A threshold size for metamorphosis in the tobacco hornworm, *Manduca sexta* (L.). *Biological Bulletin* 149:214–225.

Nijhout, H.F. 1979. Stretch-induced moulting in *Oncopeltus fasciatus*. *Journal of Insect Physiology* 25:277–281.

Nijhout, H.F. 1984. Abdominal stretch reception in *Dipetalogaster maximus* (Hemiptera: Reduviidae). *Journal of Insect Physiology* 30:629–633.

Nijhout, H.F. 1990. Metaphors and the role of genes in development. *BioEssays* 12:441–446.

Nijout, H.F. 1991a. Iteration and symmetry. *South Atlantic Quarterly* 90:61–86.

Nijhout, H.F. 1991b. *The Development and Evolution of Butterfly Wing Patterns*. Smithsonian Institution Press, Washington, D.C.

Nijhout, H.F. 1994a. *Insect Hormones*. Princeton University Press, Princeton.

Nijhout, H.F. 1994b. Symmetry and compartments in Lepidopteran wings: the evolution of a patterning mechanism. *Development*, 1994 Supplement, 225–233.

Nijhout, H.F. 1994c. Developmental perspectives on evolution of butterfly mimicry. *BioScience* 44:148–157.

Nijhout, H.F. 1999a. When developmental pathways diverge. *Proceedings of the National Academy of Sciences USA* 96:5348–5350.

Nijhout, H.F. 1999b. Hormonal control in larval development and evolution—insects. In: *The Origin and Evolution of Larval Forms*, B.K. Hall and M.H. Wake (eds.). Academic Press, San Diego, pp. 217–254.

Nijhout, H.F. 1999c. Control mechanisms of polyphenic development in insects. *BioScience* 49:181–192.

Nijhout, H.F. and Emlen, D.J. 1998. Competition among body parts in the development and evolution of insect morphology. *Proceedings of the National Academy of Sciences USA* 95:3685–3689.

Nijhout, H.F. and Paulsen, S.J. 1997. Developmental models and polygenic characters. *American Naturalist* 149:394–405.

Nijhout, H.F. and Rountree, D.B. 1995. Pattern induction across a homeotic boundary in the wings of *Precis coenia* (HBN.) (Lepidoptera: Nymphalidae). *International Journal of Insect Morphology and Embryology* 24:243–251.

Nijhout, H.F. and Wheeler, D.E. 1982. Juvenile hormone and the physiological basis of insect polymorphisms. *Quarterly Review of Biology* 57:109–133.

Nijhout, H.F. and Wheeler, D.E. 1996. Growth models of complex allometries in Holometabolous insects. *American Naturalist* 148:40–56.

Nijhout, H.F., Wray, G.A., Kremen, C. and Teragawa, C.K. 1986. Ontogeny, phylogeny and evolution of form: an algorithmic approach. *Systematic Zoology* 35:445–457.

Nijhout, H.F., Wray, G.A. and Gilbert, L.E. 1990. An analysis of the phenotypic effects of certain colour pattern genes in *Heliconius* (Lepidoptera: Nymphalidae). *Biological Journal of the Linnean Society* 40:357–372.

Nilsson, D.E. 1996. Eye ancestry: old genes for new eyes. *Current Biology* 7:39–42.

Niven, D.K. 1993. Male-male nesting behavior in hooded warblers. *Wilson Bulletin* 105:190–193.

Noakes, D.L.G., Skúlason, S. and Snorrason, S.S. 1989. Alternative life-history styles in salmonine fishes with emphasis on arctic charr, *Salvelinus alpinus*. In: *Alternative Life-History Styles of Animals*, M.N. Bruton (ed.). Kluwer Academic, Amsterdam, pp. 329–346.

Nobel, P.S. and North, G.B. 1996. Features of roots of CAM plants. In: *Crassulacean Acid Metabolism*, K. Winter and J.A.C. Smith (eds.). Springer-Verlag, New York, pp. 266–280.

Noble, G.K. and Richards, L.B. 1931. The criteria of metamorphosis in urodeles. *Anatomical Record* 48:58.

Noble, G.K. and Wurm, M. 1943. Social behavior of the laughing gull. *Annals of the New York Academy of Science* 45:179–220.

Noirot, C. 1989. Social structure in termite societies. *Ethology Ecology and Evolution* 1:1–17.

Noirot, C. 1990. Sexual castes and reproductive strategies in termites. In: *Social Insects*, W. Engels (ed.). Springer-Verlag, New York, pp. 5–36.

Noirot, E. 1972. The onset of maternal behavior in rats, hamsters, and mice: a selective review. *Advances in the Study of Behavior* 4:107–145.

Nonacs, P. and Tobin, J.E. 1992. Selfish larvae: development and the evolution of parasitic behavior in the Hymenoptera. *Evolution* 46:1605–1620.

Noonan, K.M. 1981. Individual strategies of inclusive-fitness-maximizing in *Polistes fuscatus* foundresses. In: *Natural Selection and Social Behavior*, R.D. Alexander and D.W. Tinkle (eds.). Chiron Press, New York, pp. 18–44.

Norden, B.B., Krombein, K.V. and Danforth, B.N. 1991. Taxonomic and bionomic observations on a Floridian panurgine bee, *Perdita* (*Hexaperdita*) *graenicheri* Timberlake (Hymenoptera: Andrenidae). *Journal of Hymenoptera Research* 1:107–118.

Nordeng, H. 1983. Solution to the "char problem" based on Arctic char (*Salvelinus alpinus*) in Norway. *Canadian Journal of Fisheries and Aquatic Sciences* 40:1372–1387.

Nordenskiold, E. 1928. *The History of Biology*. Tudor, New York.

Norris, D.O. 1978. Hormonal and environmental factors involved in the determination of neoteny in urodeles. In: *Comparative Endocrinology*, P.J. Gaillard and H.H. Boer (eds.). Elsevier/North Holland, Amsterdam, pp. 109–112.

Norris, D.O. and Platt, J.E. 1973. Effects of pituitary hormones, melatonin, and thyroidal inhibitors on radioiodine uptake by the thyroid glands of larval and adult tiger salamanders, *Ambystoma tigrinum* (Amphibia: Caudata). *General Comparative Endocrinology* 21:368–376.

Norris, D.O., Jones, R.E. and Criley, B.B. 1973. Pituitary prolactin levels in larval, neotenic and adult salamanders (*Ambystoma tigrinum*). *General Comparative Endocrinology* 20:437–442.

Nottebohm, F. 1980. Testosterone triggers growth of brain vocal control nuclei in adult female canaries. *Brain Research* 189:429–436.

Novacek, M.J. 1996. Paleontological data and the study of adaptation. In: *Adaptation*, M.R. Rose and G.V. Lauder (eds.). Academic Press, San Diego, pp. 311–359.

Nyman, T. and Julkunen-Tiitto, R. 2000. Manipulation of the phenolic chemistry of willows by gall-inducing sawflies. *Proceedings of the National Academy of Sciences USA* 97:13184–13187.

Nymeyer, H., Garcia, A.E. and Onuchic, J.N. 1998. Folding funnels and frustration in off-lattice minimalist protein landscapes. *Proceedings of the National Academy of Sciences USA* 95:5921–5928.

Oates, J.C. 1989. "But Noah Was Not a Nice Man." *New York Times Book Review*, October 1, p. 12.

O'Donald, P. 1980. *Genetic Models of Sexual Selection*. Cambridge University Press, Cambridge.

O'Donald, P. 1982. The concept of fitness in population genetics and sociobiology. In: *Current Problems in Sociobiology*, King's College Sociobiology Group (eds.). Cambridge University Press, Cambridge, pp. 65–85.

O'Donnell, S. 1995. Division of labor in post-emergence colonies of the primitively eusocial wasp *Polistes instabilis* de Saussure (Hymenoptera: Vespidae). *Insectes Sociaux* 42:17–29.

O'Donnell, S. 1999. The function of male dominance in the eusocial wasp *Mischocyttarus mastigophorus* (Hymenoptera: Vespidae). *Ethology* 105:273–282.

O'Donnell, S. and Jeanne, R.L. 1992. Forager success increases with experience in *Polybia occidentalis* (Hymenoptera: Vespidae). *Insectes Sociaux* 39:451–454.

Ogata, K., Murai, K., Yamauchi, K. and Tsuji, K. 1996. Size differentiation of copulatory organs between winged and wingless reproductives in the ant *Technomyrmex albipes*. *Naturwissenschaften* 83:331–333.

O'Grady, R.T. 1987. Phylogenetic systematics and the evolutionary history of some intestinal flatworm parasites (Trematoda: Digenea: Plagiorchioidea) of anurans. Doctoral dissertation, University of British Columbia, Vancouver.

Ohlsson, R. (ed.) 1999. *Genomic Imprinting: An Interdisciplinary Approach*. Springer-Verlag, Berlin.

Ohno, S. 1970. *Evolution by Gene Duplication*. Springer-Verlag, New York.

Ohno, S. 1985. Dispensable genes. *Trends in Genetics* 1:160–164.

Ohta, A.T. 1989. Coadaptive changes in speciation via the founder principle in the Grimshawi species complex of Hawaiian *Drosophila*. In: *Genetics, Speciation and the Founder Principle*, L.V. Giddings, K.Y. Kaneshiro and W.W. Anderson (eds.). Oxford University Press, New York, pp. 315–328.

Ohta, T. 1989. Time for spreading of compensatory mutations under gene duplication. *Genetics* 123:579–584.

Ohta, T. 1991. Multigene families and the evolution of complexity. *Journal of Molecular Evolution* 33:34–41.

Ohta, T. 1992b. The nearly neutral theory of molecular evolution. *Annual Review of Ecology and Systematics* 23:263–286.

Ohta, T. 1993. Pattern of nucleotide substitutions in growth hormone-prolactin gene family: a paradigm for evolution by gene duplication. *Genetics* 134:1271–1276.

Oldroyd, B.P., Rinderer, T.E., Schwenke, J.R. and Buco, S.M. 1994. Subfamily recognition and task specialisation in honey bees (*Apis mellifera* L.) (Hymenoptera: Apidae). *Behavioral Ecology and Sociobiology* 34:169–173.

Oldroyd, B.P., Clifton, M.J., Wongsiri, S., Rinderer, T.E., Sylvester, H.A. and Crozier, R.H. 1997. Polyandry in the genus *Apis*, particularly *Apis andreniformis*. *Behavioral Ecology and Sociobiology* 40:17–26.

Olivera, B.M., Walker, C., Cartier, G.E., Hooper, D., Santos, A.D., Schoenfeld, R., Shetty, R., Watkins, M., Bandyopadhyay, P. and Hillyard, D.R. 1999. Speciation of cone snails and interspecific hyperdivergence of their venom peptides. Potential evolutionary significance of introns. *Annals of the New York Academy of Sciences* 870:223–237.

Ollerton, J. and Lack, A.J. 1992. Flowering phenology: an example of relaxation of natural selection? *Trends in Ecology and Evolution* 7:274–276.

Olson, E. and Miller, R. 1958. *Morphological Integration*. University of Chicago Press, Chicago.

O'Malley, B.W. and Tsai, M. 1993. Overview of the steroid receptor superfamily of gene regulatory proteins. In: *Steroid Hormone Action*, M.G. Parker (ed.). IRL Press, Oxford, pp. 45–63.

O'Meara, G.F. 1972. Polygenic regulation of fecundity in autogenous *Aedes atropalpus*. *Entomologia Experimentalis et Applicata* 15:81–89.

O'Meara, G.F. 1985a. Ecology of autogeny in mosquitoes. In: *Ecology of Mosquitoes: Proceedings of a Workshop*, L.P. Lounibos, J.R. Rey and J.H. Frank (eds.). Florida Medical Entomology Laboratory, Vero Beach, Florida, pp. 459–471.

O'Meara, G.F. 1985b. Gonotrophic interactions in mosquitoes: kicking the blood-feeding habit. *Florida Entomologist* 68:122–133.

O'Meara, G.F. and Edman, J.D. 1975. Autogenous egg production in the salt-marsh mosquito, *Aedes taeniorhynchus*. *Biological Bulletin* 149:384–396.

O'Meara, G.F. and Evans, D.G. 1974. Female-dependent stenogamy in the mosquito, *Aedes taeniorhynchus*. *Animal Behaviour* 22:376–381.

O'Meara, G.F. and Lounibos, L.P. 1981. Reproductive maturation in the pitcher-plant mosquito, *Wyeomyia smithii*. *Physiological Entomology* 6:437–443.

O'Meara, G.F. and Van Handel, E. 1971. Triglyceride metabolism in thermally-feminized males of *Aedes aegypti*. *Journal of Insect Physiology* 17:1411–1413.

Omholt, S.W., Plahte, E., Øyehaug, L. and Xiang, K. 2000. Gene regulatory networks generating the phenomena of additivity, dominance and epistasis. *Genetics* 155:969–980.

Oren, A., Padan, E. and Avron, M. 1977. Quantum yields for oxygenic and anoxygenic photosynthesis in the cyanobacterium *Oscillatoria limnetica*. *Proceedings of the National Academy of Sciences USA* 74:2152–2156.

Oring, L.W. Fivizzani, A.J. and El Halawani, M.E. 1986a. Changes in plasma prolactin associated with laying and hatch in the spotted sandpiper. *Auk* 103:820–822.

Oring, L.W. Fivizzani, A.J. and El Halawani, M.E. 1986b. Seasonal changes in prolactin and luteinizing hormone in the polyandrous spotted sandpiper, *Actitis macularia*. *General and Comparative Endocrinology* 62:394–403.

Oring, L.W., Fivizzani, A.J. and El Halawani, M.E. 1989. Testosterone-induced inhibition of incubation in the spotted sandpiper (*Actitis mecularia*). *Hormones and Behavior* 23:412–423.

Orr, H.A. 1999. An evolutionary dead end? *Science* 285:343–344.

Orr, H.A. and Coyne, J.A. 1992. The genetics of adaptation: a reassessment. *American Naturalist* 140:725–742.

Osborn, H.F. 1897a. Organic selection. *Science* 15:583–587.

Osborn, H.F. 1897b. The limits of organic selection. *American Naturalist* 31:944–951.

Osborn, H.F. 1929. *The Titanotheres of Ancient Wyoming, Dakota, and Nebraska*. Monograph 55. U.S. Geological Survey, Washington, D.C.

Osborne, S. 1984. Bryozoan interactions: observations on stolonal outgrowths. *Australian Journal of Marine and Freshwater Resources* 35:453–462.

Oster, G. and Alberch, P. 1982. Evolution and bifurcation of developmental programs. *Evolution* 36:444–459.

Ostrom, J.H. 1976. Archaeopteryx and the origin of birds. *Zoological Journal of the Linnean Society* 8:91–182.

Ott, E., Grebogi, C. and Yorke, J.A. 1990. Controlling chaos. *Physics Review Letters* 64:1196–1199.

Otte, D. 1970. A comparative study of communicative behavior in grasshoppers. *Miscellaneous Publications University of Michigan Museum of Zoology* 141:1–168.

Otte, D. 1972. Simple versus elaborate behavior in grasshoppers. An analysis of communication in the genus *Syrbula*. *Behaviour* 42:291–322.

Otte, D. 1974. Effects and functions in the evolution of signaling systems. *Annual Review of Ecology and Systematics* 5:385–417.

Otte, D. 1989. Speciation in Hawaiian crickets. In: *Speciation and Its Consequences*, D. Otte and J.A. Endler (eds.). Sinauer, Sunderland, MA, pp. 482–526.

Owen, R. 1848. *On the Archetype and Homologies of the Vertebrate Skeleton*. R. and J.E. Taylor, London.

Owens, I.P.F. and Short, R.V. 1995. Hormonal basis of sexual dimorphism in birds: implications for new theories of sexual selection. *Trends in Ecology and Evolution* 10:44–48.

Oyama, S. 1985[2000]. *The Ontogeny of Information*. Second Edition. Duke University Press, Durham.

Packer, L. 1986. Multiple foundress associations in a temperate population of *Halictus ligatus* (Hymenoptera: Halictidae). *Canadian Journal of Zoology* 64:2325–2332.

Packer, L. 1991. The evolution of social behavior and nest architecture in sweat bees of the subgenus *Evylaeus* (Hymenoptera: Halictidae): a phylogenetic approach. *Behavioral Ecology and Sociobiology* 29:153–160.

Packer, L., Taylor, J. and Richards, M. 1994. Social devolution in sweat bees. In: *Les Insectes Sociaux*, A. Lenoir, G. Arnold and M. Lepage (eds.). Université Paris Nord, Paris, p. 358.

Packer, L., Dzinas, A., Strickler, K. and Scott, V. 1995. Genetic differentiation between two host "races" and two species of cleptoparasitic bees and between their two hosts. *Biochemical Genetics* 33:97–109.

Padgett, R.W., Wozney, J.M. and Gelbart, W.M. 1993. Human BMP sequences can confer normal dorsal ventral patterning in the *Drosophila affinidisjuncta*. *Genetics* 114:405–433.

Padilla, D.K. and Adolph, S.C. 1996. Plastic inducible morphologies are not always adaptive: the importance of time delays in a stochastic environment. *Evolutionary Ecology* 10:105–117.

Page, R.E., Jr. 1997. The evolution of insect societies. *Endeavour* 21:114–120.

Page, R.E., Jr. and Mitchell, S.D. 1990. Self organization and adaptation of insect societies. *Philosophy of Science Association* 2:289–298.

Page, R.E., Jr. and Mitchell, S.D. 1998. Self-organization and the evolution of division of labor. *Apidologie* 29:171–190.

Page, R.E., Jr. and Robinson, G.E. 1991. The genetics of division of labour in honey bee colonies. *Advances in Insect Physiology* 23:118–169.

Page, R.E., Jr., Robinson, G.E., Calderone, N.W. and Rothenbuhler, W.C. 1989. Genetic structure, division of labor, and the evolution of insect societies. In: *The Genetics of Social Evolution*, M.D. Breed and R.E. Page (eds.). Westview Press, Boulder, pp. 61–80.

Page, R.E., Jr., Robinson, G.E., Fondrk, M.K. and Nasr, M.E. 1995a. Effects of worker genotypic diversity on honey bee colony development and behavior (*Apis mellifera* L.). *Behavioral Ecology and Sociobiology* 36:387–396.

Page, R.E., Jr., Waddington, K.D., Hunt, G.J. and Fondrk, M.K. 1995b. Genetic determinants of honey bee foraging behaviour. *Animal Behaviour* 50:1617–1625.

Page, R.E., Jr., Erber, J. and Fondrk, M.K. 1998. The effect of genotype on response thresholds to sucrose and foraging behavior of honey bees (*Apis mellifera* L.). *Journal of Comparative Physiology A* 182:489–500.

Paigen, K. 1989. Experimental approaches to the study of regulatory evolution. *American Naturalist* 134:440–458.

Palmer, A.R. 1982. Predation and parallel evolution: recurrent parietal plate reduction in balanomorph barnacles. *Paleobiology* 8:31–44.

Palmer, A.R. 1985. Quantum changes in gastropod shell morphology need not reflect speciation. *Evolution* 39:699–705.

Palumbi, S. 1999. All males are not created equal: fertility differences depend on gamete recognition polymorphisms in sea urchins. *Proceedings of the National Academy of Sciences USA* 96:12632–12637.

Pamilo, P., Pekkarinen, A. and Varvio-Aho, S-L. 1981. Phylogenetic relationships and origin of social parasitism in Vespidae and in *Bombus* and *Psithyrus* as revealed by enzyme genes. In: *Biosystematics of Social Insects*, P.E. Howse and J.-L. Clement (eds.). Academic Press, New York, pp. 37–48.

Panek, L.M. and Gamboa, G.J. 2000. Queens of the paper wasp *Polistes fuscatus* (Hymenoptera: Vespidae) discriminate among larvae on the basis of relatedness. *Ethology* 106:159–170.

Pankiw, T. and Page, R.E., Jr. 1999. The effect of genotype, age, sex, and caste on response thresholds to sucrose and foraging behavior of honey bees (*Apis mellifera* L.). *Journal of Comparative Physiology* 185:207–213.

Pantel, J. 1917. À propos d'un Anisolabis ailé. *Mémoires Réal Académie des Sciences de Barcelona* 14:1–160.

Papaj, D.R. 1994. Optimizing learning and its effect on evolutionary change in behavior. In: *Behavioral Mechanisms in Evolutionary Ecology*, L.A. Real (ed.). University of Chicago Press, Chicago, pp. 133–153.

Papaj, D.R. and Prokopy, R.J. 1988. The effect of prior adult experience on components of habitat preference in the apple maggot fly (*Rhagoletis pomonella*). *Oecologia* 76:538–543.

Papaj, D.R. and Prokopy, R.J. 1989. Ecological and evolutionary aspects of learning in phytophagous insects. *Annual Review of Entomology* 34:315–350.

Pardi, L. 1946. Ricerche sui Polistini. VI. La "dominazione" e il ciclo ovarico annuale in *Polistes gallicus* (L.). *Bollettino dell'Istituto di Entomologia della Università di Bologna* 15:25–84.

Pardi, L. 1948. Dominance order in *Polistes* wasps. *Physiological Zoology* 21:1–13.

Pardi, L. 1987. La "pseudocopula" delle femmine di *Otiorrhynchus pupillatus cyclophtalmus* (Sol.) (Col. Curculion.). *Bollettino dell'Istituto di Entomologia "Guido Grandi" della Università di Bologna* 41:355–363.

Park, T. 1962. Beetles, competition, and populations. *Science* 138:1369–1375.

Parker, G.A. 1970. The reproductive behaviour and the nature of sexual selection in *Scatophaga stercoraria* L. (Diptera: Scatophagidae) II. The fertilization rate and the spatial and temporal relationships of each sex around the site of mating and oviposition. *Journal of Animal Ecology* 39:205–228.

Parker, G.A. 1982. Phenotype-limited evolutionarily stable strategies. In: *Current Problems in Sociobiology*, King's College Sociobiology Group (eds.). Cambridge University Press, New York, pp. 173–201.

Parker, G.A. 1984a. Evolutionarily stable strategies. In: *Behavioural Ecology: An Evolutionary Approach*, J.R. Krebs and N.B. Davies (eds.). Blackwell, Oxford, pp. 30–61.

Parker, G.A. 1989. Hamilton's rule and conditionality. *Ethology, Ecology and Evolution* 1:195–211.

Parker, G.A., Baker, R.R. and Smith, V.G.F. 1972. The origin and evolution of gamete dimorphism and the male-female phenomenon. *Journal of Theoretical Biology* 36:529–553.

Parker, M.G. (ed.) 1993. *Steroid Hormone Action*. IRL Press, Oxford.

Parker, W.E. and Gatehouse, A.G. 1985. Genetic factors in the regulation of migration in the African armyworm moth, *Spodoptera exempta* (Walker)(Lepidoptera: Noctuidae). *Bulletin of Entomological Research* 75:49–63.

Parkinson, J.S. and Kofoid, E.C. 1992. Communication modules in bacterial signaling proteins. *Annual Review of Genetics* 26:71–112.

Parmesan, C., Singer, M.C. and Harris, I. 1995. Absence of adaptive learning from the oviposition foraging behaviour of a checkerspot butterfly. *Animal Behaviour* 50:161–175.

Parra, R. 1978. Comparison of foregut and hindgut fermentation in herbivores. In: *The Ecology of Arboreal Folivores*, G. Montgomery (ed.). Smithsonian Institution Press, Washington, D.C., pp. 205–229.

Parsons, P.A. 1981. Sympatric speciation in *Drosophila*? Ethanol threshold metrics and habitat subdivision. *American Naturalist* 117:1023–1026.

Parsons, P.A. 1983. The genetic basis of quantitative traits: evidence for punctuational evolutionary transitions at the intraspecific level. *Evolutionary Theory* 6:175–184.

Parsons, P.A. and Spence, G.E. 1980. Ethanol utilization: threshold differences among three *Drosophila* species. *American Naturalist* 117:568–571.

Partridge, L. 1976. Individual differences in feeding efficiencies and feeding preferences of captive great tits. *Animal Behaviour* 24:230–240.

Partridge, L. 1983. Genetics and behaviour. In: *Animal Behaviour. Volume 3: Genes, Development and Learning*, T.R. Halliday and P.J.B. Slater (eds.). W.H. Freeman, New York, pp. 11–51.

Partridge, L. 1992. Measuring reproductive costs. *Trends in Ecology and Evolution* 7:99–100.

Partridge, L. and Green, P. 1984. Intraspecific feeding specializations and population dynamics. In: *Behavioral Ecology*, R.M. Sibly and R.H. Smith (eds.). Blackwell, Oxford, pp. 207–226.

Partridge, L. and Sibly, R. 1991. Constraints in the evolution of life histories. *Philosophical Transactions of the Royal Society of London B* 332:3–13.

Pass, G. 2000. Accessory pulsatile organs: evolutionary innovations in insects. *Proceedings of the National Academy of Sciences USA* 45:495–518.

Passera, L. 1974. Differenciation des soldats chez la fourmi *Pheidole pallidula* (Nyl.) (Formicidae Myrmicinae). *Insectes Sociaux* 21:71–86.

Passera, L. 1984. *L'organisation sociale des fourmis*. Privat, Toulouse.

Patterson, C. 1982. Morphological characters and homology. In: *Problems of Phylogenetic Reconstruction*, K.A. Joysey and A.E. Friday (eds.). Academic Press, New York, pp. 21–74.

Patterson, C. 1987. Introduction. In: *Molecules and Morphology in Evolution: Conflict or Compromise?* C. Patterson (ed.). Cambridge University Press, Cambridge, pp. 1–22.

Patterson, C. 1988. Homology in classical and molecular biology. *Molecular Biology and Evolution* 5:603–625.

Patthy, L. 1996. Exon shuffling and other ways of module exchange. *Matrix Biology* 15:301–310.

Patthy, L. 1999. Genome evolution and the evolution of exon-shuffling—a review. *Gene* 238:103–114.

Patton, J.L. and Brylski, P.V. 1987. Pocket gophers in alfalfa fields: causes and consequences of habitat-related body size variation. *American Naturalist* 130:493–506.

Payne, R.B. 1984. Sexual selection, lek and arena behavior, and sexual size dimorphism in birds. *Ornithological Monographs* 33:1–52.

Payne, R.B. and Payne, L.L. 1994. Song mimicry and species associations of west African indigobirds *Vidua* with quail-finch *Ortygospiza atricollis*, goldbreast *Amandava subflava* and brown twinspot *Clytospiza monteiri*. *Ibis* 136:291–304.

Payne, R.B. and Payne, L.L. 1995. Song mimicry and association of brood-parasitic indigobirds (*Vidua*) with Dybowski's twinspot (*Eustichospiza dybowskii*). *Auk* 112:649–658.

Payne, R.B., Payne, L.L., Nhlane, M.E.D. and Hustler, K. 1992. Species status and distribution of the parasitic indigo-birds *Vidua* in east and southern Africa. *Proceedings of the VIII Pan-African Ornithology Congress* 8:40–52.

Payne, R.B., Payne, L.L. and Woods, J.L. 1998. Song learning in brood parasitic indigobirds *Vidua chalybeata*: song mimicry of species. *Animal Behaviour* 55:1537–1553.

Payne, R.B., Payne, L.L., Woods, J.L. and Sorenson, M.D. 2000. Imprinting and the origin of parasite-host species associations in brood parasitic indigobirds *Vidua chalybeata*. *Animal Behaviour* 59:69–81.

Payne, R.J.H. and Krakauer, D.C. 1997. Sexual selection, space, and speciation. *Evolution* 51:1–9.

Pearson, B. 1981. The electrophoretic determination of *Myrmica rubra* microgynes as a social parasite: possible significance in the evolution of ant social parasites. In: *Biosystematics of Social Insects*, P.E. Howse and J.L. Clement (eds.). Academic Press, New York, pp. 75–84.

Pearson, B. and Child, A.R. 1980. The distribution of an esterase polymorphism in macrogynes and microgynes of *Myrmica rubra* Latreille. *Evolution* 34:105–109.

Peck, S.B. 1986. Evolution of adult morphology and life-history characters in cavernicolous *Ptomaphagus* beetles. *Evolution* 40:1021–1030.

Peck, S.L., Ellner, S.P. and Gould, F. 1998. A spatially explicit stochastic model demonstrates the feasibility of Wright's shifting balance theory. *Evolution* 52:1834–1839.

Peckham, D.J., Kurczewski, F.E. and Peckham, D.B. 1973. Nesting behavior of nearctic species of *Oxybelus* (Hymenoptera: Sphecidae). *Annals of the Entomological Society of America* 66:647–662.

Peckham, M. 1959. *The Origin of Species by Charles Darwin. A Variorum Text.*

University of Pennsylvania Press, Philadelphia.

Pedersen, C.A., Ascher, J.A., Monroe, Y.L. and Prange, A.J., Jr. 1982. Oxytocin induces maternal behavior in virgin female rats. *Science* 216:648–649.

Peeters, C. and Crozier, R.H. 1988. Caste and reproduction in ants: not all mated egg-layers are "queens." *Psyche* 95:283–288.

Peeters, C., Monnin, T. and Malosse, C. 1999. Cuticular hydrocarbons correlated with reproductive status in a queenless ant. *Proceedings of the Royal Society of London B* 266:1323–1327.

Pener, M.P. 1991. Locust phase polymorphism and its endocrine relations. *Advances in Insect Physiology* 23:1–79.

Pennisi, E. 1998. A genomic battle of the sexes. *Science* 281:1984–1985.

Pérez, J. 1884. Les apiaires parasites au point de vue de la théorie de l'évolution. *Actes de la Société Linnéenne de Bordeaux* 1884:1–63.

Perrin, N. and Sibly, R.M. 1993. Dynamic models of energy allocation and investment. *Annual Review of Ecology and Systematics* 24:379–410.

Peschke, K. 1986. Development, sex specificity, and site of production of aphrodisiac pheromones in *Aleochara curtula*. *Journal of Insect Physiology* 32:687–693.

Peschke, K. 1987. Male aggression, female mimicry and female choice in the rove beetle, *Aleochara curtula* (Coleoptera, Staphylinidae). *Ethology* 75:265–284.

Peters, R.H. 1983. *The Ecological Implication of Body Size.* Cambridge University Press, New York.

Peterson, C.C., Nagy, K.A. and Diamond, J. 1990. Sustained metabolic scope. *Proceedings of the National Academy of Sciences USA* 87:2324–2328.

Peterson, G., Allen, C.R. and Holling, C.S. 1998. Ecological resilience, biodiversity, and scale. *Ecosystems* 1:6–18.

Peterson, K. and Sapienza, C. 1993. Imprinting the genome: imprinted genes, imprinting genes, and a hypothesis for their interaction. *Annual Review of Genetics* 27:7–31.

Petraitis, P.S. 1988. Occurrence and reproductive success of feminized males in the polychaete *Capitella capitata* (species type I). *Marine Biology* 97:403–412.

Petren, K., Grant, B.R. and Grant, P.R. 1999. A phylogeny of Darwin's finches based on microsatellite DNA length variation. *Proceedings of the Royal Society of London B* 266:321–329.

Pettigrew, J.D. and Freeman, R.D. 1973. Visual experience without lines: effect on developing cortical neurons. *Science* 182:599–601.

Pettis, J.S., Winston, M.L. and Collins, A.M. 1995. Suppression of queen rearing in European and Africanized honey bees *Apis mellifera* L. by synthetic queen mandibular gland pheromone. *Insectes Sociaux* 42:113–121.

Pfennig, D.W. 1990a. "Kin recognition" among spadefoot toad tadpoles: a side-effect of habitat selection? *Evolution* 44:785–798.

Pfennig, D.W. 1990b. The adaptive significance of an environmentally-cued developmental switch in an anuran tadpole. *Oecologia* 85:101–107.

Pfennig, D.W. 1992a. Proximate and functional causes of polyphenism in an anuran tadpole. *Functional Ecology* 6:167–174.

Pfennig, D.W. 1992b. Polyphenism in spadefoot toad tadpoles as a locally adjusted evolutionarily stable strategy. *Evolution* 46:1408–1420.

Pfennig, D.W. 1997. Kinship and cannibalism. *BioScience* 47:667–675.

Pfennig, D.W. 1999. Cannibalistic tadpoles that pose the greatest threat to kin are most likely to discriminate kin. *Proceedings of the Royal Society of London B* 266:57–61.

Pfennig, D.W. and Collins, J.P. 1993. Kinship affects morphogenesis in cannibalistic salamanders. *Nature* 362:836–838.

Pfennig, D.W. and Frankino, W.A. 1997. Kin-mediated morphogenesis in facultatively cannibalistic tadpoles. *Evolution* 51:1993–1999.

Pfennig, D.W. and Murphy, P.J. 2000. Character displacement in polyphenic tadpoles. *Evolution* 54:1738–1749.

Pfennig, D.W. and Reeve, H.K. 1989. Neighbor recognition and context-dependent aggression in a solitary wasp, *Sphecius speciosus* (Hymenoptera: Sphecidae). *Ethology* 80:1–18.

Pfennig, D.W., Reeve, H.K. and Sherman, P.W. 1993. Kin recognition and cannibalism in spadefoot toad tadpoles. *Animal Behaviour* 46:87–94.

Pfennig, D.W., Sherman, P.W. and Collins, J.P. 1994. Kin recognition and cannibalism in polyphenic salamanders. *Behavioral Ecology* 5:225–232.

Pfennig, D.W., Collins, J.P. and Ziemba, R.E. 1999. A test of alternative hypotheses for kin recognition in cannibalistic tiger salamanders. *Behavioral Ecology* 10:436–443.

Philippi, T. and Seger, J. 1989. Hedging one's evolutionary bets, revisited. *Trends in Ecology and Evolution* 4:41–44.

Piaget, J. 1949. L'adaptation de la *Limnaea stagnalis* aux milieux lacustres de la Suisse romande. Étude biométrique et génétique. *Revue Suisse de Zoologie* 36:263–531.

Piano, F., Craddock, E.M. and Kambysellis, M.P. 1997. Phylogeny of the island populations of the Hawaiian *Drosophila grimshawi* complex: evidence from combined data. *Molecular Phylogenetics and Evolution* 7:173–184.

Piersma, T. and Lindström, A. 1997. Rapid reversible changes in organ size as a component of adaptive behavior. *Trends in Ecology and Evolution* 12:134–138.

Pigeon, D., Chouinard, A. and Bernatchez, L. 1997. Multiple modes of speciation involved in the parallel evolution of sympatric morphotypes of lake whitefish (*Coregonus clupeaformis*, Salmonidae). *Evolution* 51:196–205.

Pigliucci, M. and Schmitt, J. 1999. Genes affecting phenotypic plasticity in *Arabidopsis*: pleiotropic effects and reproductive fitness of photomorphogenic mutants. *Journal of Evolutionary Biology* 12:551–562.

Pigliucci, M., Whitton, J. and Schlichting, C.D. 1995. Reaction norms of *Arabidopsis*. I. Plasticity of characters and correlations across water, nutrient and light gradients. *Journal of Evolutionary Biology* 8:421–438.

Pilecki, C. and O'Donald, P. 1971. The effects of predation on artificial mimetic polymorphisms with perfect and imperfect mimics at varying frequencies. *Evolution* 25:365–370.

Pilon-Smits, E.A.H., Hart, H. and van Brederode, J. 1996. Evolutionary aspects of crassulacean acid metabolism in the Crassulaceae. In: *Crassulacean Acid Metabolism*, K. Winter and J.A.C. Smith (eds.). Springer-Verlag, New York, pp. 349–359.

Pimm, S.L. 1979. Sympatric speciation: a simulation model. *Biological Journal of the Linnean Society* 11:131–139.

Pimm, S.L. 1988. Rapid morphological change in an introduced bird. *Trends in Ecology and Evolution* 3:290–291.

Pinker, S. 1994. *The Language Instinct*. Morrow, New York.

Pinker, S. 1997. *How the Mind Works*. W.W. Norton, New York.

Piperno, D.R. and Pearsall, D.M. 1998. *The Origins of Agriculture in the Lowland Neotropics*. Academic Press, New York.

Pitnick, S. and Karr, T.L. 1998. Paternal products and byproducts in *Drosophila* development. *Proceedings of the Royal Society of London B* 265:821–826.

Plateaux, L. 1970. Sur le polymorphisme social de la fourmi *Leptothorax nylanderi* (Förster), I: Morphologie et biologie comparées des castes. *Annales des Sciences Naturelles* 12:373–478.

Plateaux-Quénu, C., Horel, A. and Roland, C. 1997. A reflection on socialization processes in two different groups of arthropods: halictine bees (Hymenoptera) and spiders (Arachnida). *Ethology, Ecology and Evolution* 9:183–196.

Platt, J.R. 1956. Amplification aspects of biological response and mental activity. *American Scientist* 44:180–197.

Plomin, R. 1990. The role of inheritance in behavior. *Science* 248:183–188.

Podell, M.E. and Anstey, R.L. 1979. The interrelationship of early colony development, monticules and branches in Palaeozoic bryozoans. *Palaeontology* 22:965–982.

Polak, M. 1993. Landmark territoriality in the neotropical paper wasps *Polistes canadensis* (L.) and *P. carnifex* (F.) (Hymenoptera: Vespidae). *Ethology* 95:278–290.

Polak, M. (ed.). 2003. *Developmental Instability: Causes and Consequences*. Oxford University Press, New York (in press).

Policansky, D. 1982. Sex change in plants and animals. *Annual Review of Ecology and Systematics* 13:471–495.

Policansky, D. 1983. Size, age and demography of metamorphosis and sexual maturation in fishes. *American Zoologist* 23:57–64.

Polis, G.A. 1981. The evolution and dynamics of intraspecific predation. *Annual Review of Ecology and Systematics* 12:225–251.

Portin, P. 1993. The concept of the gene: short history and present status. *Quarterly Review of Biology* 68:173–223.

Posner, M.I. 1994. Attention: the mechanisms of consciousness. *Proceedings of the National Academy of Sciences USA* 91:7398–7403.

Potter, I.C. 1980. The Petromyzoniformes with particular reference to paired species. *Canadian Journal of Fisheries and Aquatic Sciences* 37:1595–1615.

Potter, N.B. 1965. Some aspects of the biology of *Vespula vulgaris* L. Unpublished Ph.D. thesis, University of Bristol.

Pough, F.H., Heiser, J.B and McFarland, W.N. 1989. *Vertebrate Life*. Macmillan, New York.

Powell, R.A. and Fried, J.J. 1992. Helping by juvenile pine voles (*Microtus pinetorum*), growth and survival of younger siblings, and the evolution of pine vole sociality. *Behavioral Ecology* 3:325–333.

Powers, D.A., Smith, M., Gonzalez-Villaseñor, I., DiMichele, L., Crawford, D., Bernardi, G. and Lauerman, T. 1993. A multidisciplinary approach to the selectionist/neutralist controversy using the model teleost, *Fundulus heteroclitus*. *Oxford Surveys in Evolutionary Biology* 9:81–157.

Pratt, B.L. and Goldman, B.D. 1986. Maternal influence on activity rhythms and reproductive development in Djungarian hamster pups. *Biology of Reproduction* 34:655–663.

Premoli, M.C. and Sella, G. 1995. Sex economy in benthic polychaetes. *Ethology Ecology and Evolution* 7:27–48.

Prescott, D.M. 1999. Evolution of DNA organization in hypotrichous ciliates. *Annals of the New York Academy of Sciences* 870:301–313.

Price, T. 1987. Diet variation in a population of Darwin's finches. *Ecology* 68:1015–1028.

Price, T. and Grant, P.R. 1984. Life history traits and natural selection for small body size in a population of Darwin's finches. *Evolution* 38:483–494.

Price, T. and Pavelka, M. 1996. Evolution of a colour pattern: history, development and selection. *Journal of Evolutionary Biology* 9:451–470.

Prokopy, R.J. and Bush, G. 1993. Evolution in an orchard. *Natural History* 102:4–8.

Prokopy, R.J., Averill, A.L., Cooley, S.S. and Roitberg, C.A. 1982. Associative learning in egglaying site selection by apple maggot flies. *Science* 218:76–77.

Prokopy, R.J., Diehl, S.R. and Cooley, S.S. 1988. Behavioral evidence for host races in *Rhagoletis pomonella* flies. *Oecologia* 76:138–147.

Provine, W.B. 1971 [1987]. *The Origins of Theoretical Population Genetics*. University of Chicago Press, Chicago.

Provine, W.B. 1986. *Sewall Wright and Evolutionary Biology*. University of Chicago Press, Chicago.

Provine, W.B. 1989. Founder effects and genetic revolutions in microevolution and speciation: an historical perspective. In: *Genetics, Speciation and the Founder Principle*, L.V. Giddings, K.Y. Kaneshiro and W.W. Anderson (eds.). Oxford University Press, New York, pp. 43–76.

Prusiner, S.B. 1998. Prions. *Proceedings of the National Academy of Sciences USA* 95:13363–13383.

Ptashne, M. 1986 [1992]. *A Genetic Switch*. Second edition. Cell Press and Blackwell, Cambridge.

Pulliam, H.R. and Caraco T. 1984. Living in groups: is there an optimal group size? In: *Behavioural Ecology—An Evolutionary Approach*, J.R. Krebs and N.B. Davies (eds.). Blackwell, Boston, pp. 122–147.

Punnett, R.C. 1915. *Mimicry in Butterflies*. Cambridge University Press, Cambridge.

Queller, D.C. 1983. Kin selection and conflict in seed maturation. *Journal of Theoretical Biology* 100:153–172.

Queller, D.C. 1984. Models of kin selection on seed provisioning. *Heredity* 53:151–165.

Queller, D.C. 1994. Male-female conflict and parent-offspring conflict. *American Naturalist* 144:S84–S99.

Queller, D.C. and Strassmann, J.E. 1989. Measuring inclusive fitness in social wasps. In: *The Genetics of Social Evolution*, M.D. Breed and R.E. Page, Jr. (eds.). Westview Press, Boulder, pp. 103–122.

Queller, D.C. and Strassmann, J.E. 1998. Kin selection and social insects. *BioScience* 48:165–175.

Queller, D.C., Peters, J.M., Solis, C. and Strassmann, J.E. 1997. Control of reproduction in social insect colonies: individual and collective relatedness preferences in the paper wasp, *Polistes annularis*. *Behavioral Ecology and Sociobiology* 40:3–16.

Quinn, T.P. and Foote, C.J. 1994. The effects of body size and sexual dimorphism on the reproductive behaviour of sockeye salmon, *Oncorhynchus nerka*. *Animal Behaviour* 48:751–761.

Rabalais, N.N. and Gore, R.H. 1985. Abbreviated development in decapods. In: *Larval Growth*, A.M. Wenner (ed.). A.A. Balkema, Boston, pp. 67–126.

Rabinow, L. and Dickinson, W.J. 1986. Complex *cis*-acting regulators and locus structure of *Drosophila* tissue-specific ADH variants. *Genetics* 112:523–537.

Racey, P.A. and Skinner, J.D. 1979. Endocrine aspects of sexual mimicry in spotted hyaenas, *Crocuta crocuta*. *Journal of the Zoological Society of London* 187:315–325.

Rachinsky, A. and Hartfelder, K. 1990. Corpora allata activity, a prime regulating element for caste-specific juvenile hormone titre in honeybee larvae (*Apis mellifera carnica*). *Journal of Insect Physiology* 36:189–194.

Rachootin, S.P. and Thomson, K.S. 1981. Epigenetics, paleontology, and evolution. In: *Evolution Today, Proceedings of the Second International Congress of Systematic and Evolutionary Biology*, G.G.E. Scudder and J.L. Reveal (eds.). Hunt Institute for Biological Documentation, Pittsburgh, pp. 181–193.

Radinsky, L.B. 1987. *The Evolution of Vertebrate Design*. University of Chicago Press, Chicago.

Radwan, J. 1993. The adaptive significance of male polymorphism in the acarid mite *Caloglyphus berlesei*. *Behavioral Ecology and Sociobiology* 33:201–208.

Raff, E.C. and Raff, R.A. 2000. Dissociability, modularity, evolvability. *Evolution and Development* 2:235–237.

Raff, R.A. 1987. Constraint, flexibility, and phylogenetic history in the evolution of direct development in sea urchins. *Developmental Biology* 119:6–19.

Raff, R.A. 1988. The selfish cell lineage [book review of *The Evolution of Individuality*, L.W. Buss]. *Cell* 54:445–446.

Raff, R.A. 1992. Direct-developing sea urchins and the evolutionary reorganization of early development. *BioEssays* 14:211–218.

Raff, R.A. 1996. *The Shape of Life: Genes, Development, and the Evolution of Animal Form*. University of Chicago Press, Chicago.

Raff, R.A. and Kaufman, T.C. 1983. *Embryos, Genes, and Evolution*. Macmillan, New York.

Raff, R.A., Parr, B.A., Parks, A.L. and Wray, G.A. 1990. Heterochrony and other mechanisms of radical evolutionary change in early development. In: *Evolutionary Innovations*, M.H. Nitecki (ed.). University of Chicago Press, Chicago, pp. 71–98.

Raikow, R.J. 1975. The evolutionary reappearance of ancestral muscles as developmental anomalies in two species of birds. *Condor* 77:514–517.

Raikow, R.J., Borecky, S.R. and Berman, S.L. 1979. The evolutionary re-establishment of a lost ancestral muscle in the bowerbird assemblage. *Condor* 81:203–206.

Raikow, R.J., Bledsoe, A.H., Myers, B.A. and Welsh, C.J. 1990. Individual variation in avian muscles and its significance for the reconstruction of phylogeny. *Systematic Zoology* 39:362–370.

Ramakrishnan, V. 1995. The histone fold: evolutionary questions. *Proceedings of the National Academy of Sciences USA* 92:11328–11330.

Ramírez, W. and Marsh, P.M. 1996. A review of the genus *Psenobolus* (Hymenoptera: Braconidae) from Costa Rica, an inquiline fig wasp with brachypterous males, with descriptions of two new species. *Journal of Hymenoptera Research* 5:64–72.

Rand, A.L. 1967. *Ornithology: An Introduction*. W.W. Norton and Company, New York.

Rankin, M.A. and Burchsted, J.C.A. 1992. The cost of migration in insects. *Annual Review of Entomology* 37:553–559.

Ratnieks, F.L.W. 1988. Reproductive harmony via mutual policing by workers in eusocial Hymenoptera. *American Naturalist* 132:217–236.

Rau, P. 1933. *The Jungle Bees and Wasps of Barro Colorado Island*. Van Hoffmann, St. Louis.

Rau, P. and Rau, N. 1918. *Wasp Studies Afield*. Princeton University Press, Princeton, N.J.

Raup, D.M. 1983. On the early origins of major biological groups. *Paleobiology* 9:107–115.

Raven, J.A. and Spicer, R.A. 1996. The evolution of crassulacean acid metabolism. In: *Crassulacean Acid Metabolism*, K. Winter and J.A.C. Smith (eds.). Springer-Verlag, New York, pp. 360–388.

Raveret Richter, M.R. 2000. Social wasp (Hymenoptera: Vespidae) foraging behavior. *Annual Review of Entomology* 45:121–150.

Raveret Richter, M.R. and Waddington, K.D. 1993. Past foraging experience influences honey bee dance behaviour. *Animal Behaviour* 46:123–128.

Ray, T.S. 1979. Slow-motion world of plant "behavior" visible in rain forest. *Smithsonian* 9:121–130.

Ray, T.S. 1986. Growth correlations within the segment in the Araceae. *American Journal of Botany* 73:993–1001.

Ray, T.S. 1987. Cyclic heterophylly in syngonium (Araceae). *American Journal of Botany* 74:16–26.

Ray, T.S. 1990. Metamorphosis in the Araceae. *American Journal of Botany* 77:1599–1609.

Ray, T.S. 1992. Foraging behaviour in tropical herbaceous climbers (Araceae). *Journal of Ecology* 80:189–203.

Rayner, A.D.M. and Franks, N.R. 1987. Evolutionary and ecological parallels between ants and fungi. *Trends in Ecology and Evolution* 2:127–133.

Real, L.A. 1993. Toward a cognitive ecology. *Trends in Ecology and Evolution* 8:413–417.

Real, L.A. 1994. *Behavioral Mechanisms in Evolutionary Ecology*. University of Chicago Press, Chicago.

Réat, V., Patzelt, H., Ferrand, M., Pfister, C., Oesterhelt, D. and Zaccai, G. 1998. Dynamics of different functional parts of bacteriorhodopsin: H-^{2}H labeling and neutron scattering. *Proceedings of the National Academy of Sciences USA* 95:4970–4975.

Reburn, C.J. and Wynne-Edwards, K.E. 1999. Hormonal changes in males of a naturally biparental and a uniparental mammal. *Hormones and Behavior* 35:163–176.

Reed, H.C. and Akre, R.D. 1982. Morphological comparisons between the obligate social parasite *Vespula austriaca* (Panzer), and its host, *Vespula acadica* (Sladen) (Hymenoptera: Vespidae). *Psyche* 89:183–195.

Reeve, H.K. 1991. *Polistes*. In: *The Social Biology of Wasps*, K.G. Ross and R.W. Matthews (eds.). Cornell University Press, Ithaca, pp. 99–148.

Reeve, H.K. 1993. Haplodiploidy, eusociality and absence of male parental and alloparental care in Hymenoptera: a unifying genetic hypothesis distinct from kin selection theory.

Philosophical Transactions of the Royal Society of London B 342:335–352.

Reeve, H.K. 2000. A transactional theory of within-group conflict. *American Naturalist* 155:365–382.

Reeve, H.K. and Nonacs, P. 1997. Within-group aggression and the value of group members: theory and a field test with social wasps. *Behavioral Ecology* 8:75–82.

Reeve, H.K. and Sherman, P.W. 1993. Adaptation and the goals of evolutionary research. *Quarterly Review of Biology* 68:1–32.

Reeve, H.K. and Sherman, P.W. 2001. Optimality and phylogeny: a critique of current thought. In: *Adaptationism and Optimality*, S.H. Orzack and E. Sober (eds.). Cambridge University Press, Cambridge, pp. 64–113.

Regelmann, K. 1986. Learning to forage in a variable environment. *Journal of Theoretical Biology* 120:321–329.

Reichman, O.J., Fattaey, A. and Fattaey, K. 1986. Management of sterile and mouldy seeds by a desert rodent. *Animal Behaviour* 34:221–225.

Reid, R.G.B. 1985. *Evolutionary Theory. The Unfinished Synthesis*. Cornell University Press, Ithaca, N.Y.

Reik, W. and Surani, A. (eds.) 1997. *Genomic Imprinting*. Oxford University Press, New York.

Reilly, S.M., Wiley, E.O. and Meinhardt, D.J. 1997. An integrative approach to heterochrony: the distinction between interspecific and intraspecific phenomena. *Biological Journal of the Linnean Society* 60:119–143.

Reimchen, T.E. 1980. Spine deficiency and polymorphism in a population of *Gasterosteus aculeatus*: an adaptation to predators? *Canadian Journal of Zoology* 58:1232–1244.

Reimchen, T.E. 1988. Inefficient predators and prey injuries in a population of giant stickleback. *Canadian Journal of Zoology* 66:2036–2044.

Reimchen, T.E. 1989. Loss of nuptial color in threespine sticklebacks (*Gasterosteus aculeatus*). *Evolution* 43:450–460.

Reimchen, T.E. 1994. Predators and morphological evolution in threespine stickleback. In: *The Evolutionary Biology of the Threespine Stickleback*, M.A. Bell and S.A. Foster (eds.). Oxford University Press, Oxford, pp. 240–276.

Reimchen, T.E. and Douglas, S. 1984. Feeding schedule and daily food consumption in red-throated loons (*Gavia stellata*) over the prefledging period. *Auk* 101:593–599.

Remane, A. 1952. *Die Grundlagen des naturlichen Systems der vergleichenden Anatomie und der Phylogenetik.* Geest and Portig, Leipzig.

Renoux, J. 1976. Le polymorphisme de *Schedorhinotermes lamanianus* (Sjösted)(Isoptera, Rhinotermitidae). Essai d'interprétation. *Insectes Sociaux* 23:281–291.

Rensch, B. 1960. *Evolution Above the Species Level.* Columbia University Press, New York.

Repka, J. and Gross, M.R. 1995. The evolutionarily stable strategy under individual condition and tactic frequency. *Journal of Theoretical Biology* 176:27–31.

Reynolds, V. 1970. The "man of the woods." In: *Field Studies in Natural History.* Van Nostrand, New York, pp. 202–210.

Reynolds, V. and Reynolds, F. 1965. Chimpanzees of the Budongo Forest. In: *Primate Behavior*, I. DeVore (ed.). Holt, Rinehart and Winston, New York, pp. 368–424.

Reznick, D. 1983. The structure of guppy life histories: the tradeoff between growth and reproduction. *Ecology* 64:862–873.

Reznick, D. and Travis, J. 1996. The empirical study of adaptation in natural populations. In: *Adaptation*, M.R. Rose and G.V. Lauder (eds.). Academic Press, San Diego, pp. 243–289.

Rhen, T. and Lang, J.W. 1998. Among-family variation for environmental sex determination in reptiles. *Evolution* 52:1514–1520.

Rice, W.R. 1984. Disruptive selection on habitat preference and the evolution of reproductive isolation: a simulation study. *Evolution* 38:1251 1260.

Rice, W.R. 1985. Disruptive selection on habitat preference and the evolution of reproductive isolation: an exploratory experiment. *Evolution* 39:645–656.

Rice, W.R. 1987. Speciation via habitat specialization: the evolution of reproductive isolation as a correlated character. *Evolutionary Ecology* 1:301–314.

Rice, W.R. 1998. Male fitness increases when females are eliminated from gene pool: implications for the Y chromosome. *Proceedings of the National Academy of Sciences USA* 95:6217–6221.

Rice, W.R. and Holland, B. 1997. The enemies within: intergenomic conflict, interlocus contest evolution (ICE), and the intraspecific red queen. *Behavioral Ecology and Sociobiology* 41:1–10.

Richards, M.H. 1994. Social evolution in the genus *Halictus*: a phylogenetic approach. *Insectes Sociaux* 41:315–325.

Richards, O.W. 1961. An introduction to the study of polymorphism in insects. In: *Insect Polymorphism*, J.S. Kennedy (ed.). Royal Entomological Society, London, pp. 1–10.

Richards, O.W. 1978. *The Social Wasps of the Americas.* British Museum, London.

Richardson, R.H. 1973. Effects of dispersal, habitat selection and competition on a speciation pattern of *Drosophila* endemic to Hawaii. In: *Genetic Mechanism of Speciation in Insects*, M.J.D. White (ed.). Reidel, New York, pp. 140–164.

Richey, F.D. and Sprague, G.F. 1932. Some factors affecting the reversal of sex expression in the tassels of maize. *American Naturalist* 66:433–443.

Ridley, M. 1978. Paternal care. *Animal Behaviour* 26:904–932.

Ridley, M. 1983. *The Explanation of Organic Diversity.* Clarendon Press, Oxford.

Ridley, M. 1993. *Evolution.* Blackwell Science, Cambridge.

Riechert, S.E. 1982. Spider interaction strategies: communication *vs.* coercion. In: *Spider Communication: Mechanisms and Ecological Significance*, P.N. Witt and J.S. Rovner (eds.). Princeton University Press, Princeton, pp. 281–315.

Riedl, R. 1978. *Order in Living Organisms.* R.P.S. Jefferies (trans.), John Wiley and Sons, New York.

Rieppel, O. 1988. *Fundamentals of Comparative Biology.* Birkhäuser, Basel.

Rieppel, O. 1994. Homology, topology, and typology: the history of modern debates. In: *Homology*, B.K. Hall (ed.). Academic Press, New York, pp. 64–101.

Riley, C.V. 1878. *United States Entomological Commission. First Annual Report.* Government Printing Office, Washington, D.C.

Rinderer, T.E. (ed.) 1986. *Bee Genetics and Breeding.* Academic Press, Orlando.

Rinderer, T.E. and Collins, A.M. 1986. Behavioral genetics. In: *Bee Genetics and Breeding*, T.E. Rinderer (ed.). Academic Press, Orlando, pp. 155–176.

Riska, B. 1986. Some models for development, growth, and morphometric correlation. *Evolution* 40:1303–1311.

Rissing, S.W. 1983. Natural History of the workerless inquiline ant *Pogonomyrmex colei* (Hymenoptera: Formicidae). *Psyche* 90:321–332.

Rivera, M.C., Jain, R., Moore, J.E. and Lake, J.A. 1998. Genomic evidence for two functionally distinct gene classes. *Proceedings of the National Academy of Sciences USA* 95:6239–6244.

R'Kha, S., Capy, P. and David, J.R. 1991. Host-plant specialization in the *Drosophila melanogaster* species complex: a physiological, behavioral, and genetical analysis. *Proceedings of the National Academy of Sciences USA* 88:1835–1839.

Roberts, T.R. 1974. Dental polymorphism and systematics in *Saccodon*, a neotropical genus of freshwater fishes (Parodontidae, Characoidei). *Journal of Zoology, London* 173:303–321.

Robertson, A. 1977. Cellular communication. In: *How Animals Communicate*, T.A. Sebeok (ed.). Indiana University Press, Bloomington, pp. 33–44.

Robertson, F.W. 1957. Studies in quantitative inheritance. X. Genetic variation of ovary size in *Drosophila*. *Journal of Genetics* 55:410–427.

Robertson, F.W., Shook, M., Takei, G. and Gaines, H. 1968. Observations on the biology and nutrition of *Drosophila disticha* Hardy, an indigenous Hawaiian species. *University of Texas Publications*, No. 6615, pp. 277–300.

Robertson, H.M. 1985. Female dimorphism and mating behaviour in a damselfly, *Ischnura ramburi*: females mimicking males. *Animal Behaviour* 33:805–809.

Robinson, B.W. and Dukas, R. 1999. The influence of phenotypic modifications on evolution: the Baldwin effect and modern perspectives. *OIKOS* 83:582–589.

Robinson, B.W. and Wilson, D.S. 1994. Character release and displacement in fishes: a neglected literature. *American Naturalist* 144:596–627.

Robinson, B.W., Wilson, D.S., Margosian, A.S. and Lotito, P.T. 1993. Ecological and morphological differentiation of pumpkinseed sunfish in lakes without bluegill sunfish. *Evolutionary Ecology* 7:451–464.

Robinson, G.E. 1992. Regulation of division of labor in insect societies. *Annual Review of Entomology* 37:637–665.

Robinson, G.E. and Dyer, F.C. 1993. Plasticity of spatial memory in honey bees: reorientation following colony fission. *Animal Behaviour* 46:311–320.

Robinson, G.E. and Page, R.E., Jr. 1988. Genetic determination of guarding and undertaking in honey-bee colonies. *Nature* 333:356–358.

Robinson, G.E. and Vargo, E.L. 1997. Juvenile hormone in the social Hymenoptera: gonadotropin and behavioral pacemaker. *Archives in Insect Biochemistry and Physiology* 35:559–583.

Robson, G.C. and Richards, O.W. 1936. *The Variation of Animals in Nature*. Longmans, Green, New York.

Rocha, I.R.D. 1991. Relationship between homosexuality and dominance in the cockroaches, *Nauphoeta cinerea* and *Henchoustedenia flexivitta* (Dictyoptera, Blaberidae). *Revista Brasileira de Entomología* 35:1–8.

Roe, A. and Simpson, G.G. 1958. *Behavior and Evolution*. Yale University Press, New Haven, Conn.

Roff, D.A. 1975. Population stability and the evolution of dispersal in a heterogeneous environment. *Oecologia* 19:217–237.

Roff, D.A. 1984. The cost of being able to fly: a study of wing polymorphism in two species of crickets. *Oecologia* 63:30–37.

Roff, D.A. 1986a. Evolution of wing polymorphism and its impact on life cycle adaptation in insects. In: *The Evolution of Insect Life Cycles*, F. Taylor and R. Karban (eds.). Springer-Verlag, New York, pp. 204–221.

Roff, D.A. 1986b. The evolution of wing dimorphism in insects. *Evolution* 40:1009–1020.

Roff, D.A. 1992. *The Evolution of Life Histories*. Chapman and Hall, New York.

Roff, D.A. 1994a. Evolution of dimorphic traits: effect of directional selection on heritability. *Heredity* 72:36–41.

Roff, D.A. 1994b. The evolution of dimorphic traits: predicting the genetic correlation between environments. *Genetics* 136:395–401.

Roff, D.A. 1996. The evolution of threshold traits in animals. *Quarterly Review of Biology* 71:3–35.

Roff, D.A. and Fairbairn, D.J. 1991. Wing dimorphism and the evolution of migratory polymorphisms among the Insecta. *American Zoologist* 31:243–251.

Roff, D.A., Stirling, G. and Fairbairn, D.J. 1997. The evolution of threshold traits: a quantitative genetic analysis of the physiological and life-history correlates of wing dimorphism in the sand cricket. *Evolution* 51:1910–1919.

Rohwer, S. 1975. The social significance of avian winter plumage variability. *Evolution* 29:593–610.

Rohwer, S. 1977. Status signaling in Harris sparrows: some experiments in deception. *Behaviour* 61:107–129.

Rohwer, S. 1982. The evolution of reliable and unreliable badges of fighting ability. *American Zoologist* 22:531–546.

Rohwer, S. and Ewald, P.W. 1981. The cost of dominance and advantage of subordination in a badge signaling system. *Evolution* 35:441–454.

Rohwer, S. and Rohwer, F.C. 1978. Status signalling in Harris sparrows: experimental deceptions achieved. *Animal Behaviour* 26:1012–1022.

Rohwer, S., Ewald, P.W. and Rohwer, F.C. 1981. Variation in size, appearance, and dominance within and among the sex and age classes of Harris' sparrows. *Journal of Field Ornithology* 52:291–303.

Roitberg, B.D., Van Lenteren, J.C., Van Alphen, J.J.M., Galis, F. and Prokopy, R.J. 1982. Foraging behaviour of *Rhagoletis pomonella*, a parasite of hawthorn (*Crataegus viridis*), in nature. *Journal of Animal Ecology* 51:307–325.

Roitberg, B.D., Cairl, R.S. and Prokopy, R.J. 1984. Oviposition deterring pheromone influences dispersal distance in tephritid fruit flies. *Entomologia Experimentalis et Applicata* 35:217–220.

Rollo, C.D. 1994. *Phenotypes*. Chapman and Hall, New York.

Rollo, C.D. and Hawryluk, M.D. 1988. Compensatory scope and resource allocation in two species of aquatic snails. *Ecology* 69:146–156.

Romanes, G.J. 1886. Physiological selection. An additional suggestion on the origin of species. *Journal of the Linnean Society (Zoology)* 19:337–411.

Romanes, G.J. 1897. *Darwin and after Darwin*. Open Court Press, Chicago.

Romer, A.S. 1958. Phylogeny and behavior with special reference to vertebrate evolution. In: *Behavior and Evolution*, A. Roe and G.G. Simpson (eds.). Yale University Press, New Haven, pp. 48–75.

Rood, S.B., Pharis, R.P. and Major, D.J. 1980. Changes of endogenous gibberellin-like substances with sex reversal of the apical inflorescence of corn. *Plant Physiology* 66:793–796.

Roper, T.J. 1983a. Learning as a biological phenomenon. In: *Behavioral Ecology Symposium*, R.M. Sibly and R.H. Smith (eds.). Blackwell, London, pp. 178–212.

Roper, T.J. 1983b. Learning as a biological phenomenon. In: *Animal Behaviour. Volume 3: Genes, Development and Learning*, T.R. Halliday and P.J.B. Slater (eds.). W.H. Freeman, New York, pp. 178–212.

Rose, M.R. 1982. Antagonistic pleiotropy, dominance, and genetic variation. *Heredity* 48:63–78.

Röseler, P.-F. 1970. Unterschiede in der kastendetermination zwichen den hummelarten *Bombus hypnorum* und *Bombus terrestris*. *Zeitschrift für Naturforschung* 25:543–548.

Röseler, P.-F. 1976. Juvenile hormone and queen rearing in bumblebees. In: *Phase and Caste Determination in Insects*, M. Lüscher (ed.). Pergamon Press, New York, pp. 55–61.

Röseler, P.-F. 1985. Endocrine basis of dominance and reproduction in polistine paper wasps. In: *Experimental Behavioral Ecology and Sociobiology*, B. Hölldobler and M. Lindauer (eds.). Sinauer, Sunderland, MA, pp. 259–272.

Röseler, P.-F. 1991. Reproductive competition during colony establishment. In: *The Social Biology of Wasps*, K.G. Ross and R.W. Matthews (eds.). Cornell University Press, Ithaca, pp. 309–335.

Rosenberg, A. 1985. Adaptationist imperatives and panglosian paradigms. In: *Sociobiology and Epistemology*, J.H. Fetzer (ed.). Reidel, Dordrecht, pp. 133–160.

Rosenblatt, J.S. 1991. A psychobiological approach to maternal behaviour among the primates. In: *The Development and Integration of Behaviour*, P. Bateson (ed.). Cambridge University Press, New York, pp. 191–222.

Rosenzweig, M.L. 1978. Competitive speciation. *Biological Journal of the Linnean Society* 10:275–289.

Rosenzweig, M.L. 1990. Do animals choose habitats? In: *Interpretation and Explanation in the Study of Animal Behavior. Volume 1: Interpretation, Intentionality and Communication*, M. Bekoff and D. Jamieson (eds.). Westview Press, Boulder, pp. 157–179.

Ross, K.G. and Carpenter, J.M. 1991. Phylogentic analysis and the evolution of queen number in eusocial Hymenoptera. *Journal of Evolutionary Biology* 4:117–130.

Ross, K.G. and Keller, L. 1995. Ecology and evolution of social organization: insights from fire ants and other highly eusocial insects. *Annual Review of Ecology and Systematics* 26:631–656.

Ross, K.G. and Keller, L. 1998. Genetic control of social organization in an ant. *Proceedings of the National Academy of Sciences USA* 95:14232–14237.

Ross, K.G. and Shoemaker, D.D. 1993. An unusual pattern of gene flow between the two social forms of the fire ant *Solenopsis invicta*. *Evolution* 47:1595–1605.

Ross, K.G. and Shoemaker, D.D. 1997. Nuclear and mitochondrial genetic structure in two social forms of the fire ant *Solenopsis invicta*: insights into transitions to an alternate social organization. *Heredity* 78:590–602.

Ross, K.G., Vargo, E.L., Keller, L. and Trager, J.C. 1993. Effect of a founder event on variation in the genetic sex-determining system of the fire ant *Solenopsis invicta*. *Genetics* 135:843–854.

Ross, K.G., Vargo, E.L. and Keller, L. 1996. Social evolution in a new environment: the case of introduced fire ants. *Proceedings of the National Academy of Sciences USA* 93:3021–3025.

Ross, K.G., Krieger, M.J.B, Shoemaker, D., Vargo, E.L. and Keller, L. 1997. Hierarchical analysis of genetic structure in native fire ant populations: results from three classes of molecular markers. *Genetics* 147:643–655.

Ross, K.G., Shoemaker, D.D., Krieger, M.J.B., DeHeer, C.J. and Keller, L. 1999. Assessing genetic structure with multiple classes of markers: a case study involving the introduced fire ant *Solenopsis invicta*. *Molecular Biology and Evolution* 16:525–543.

Rossiter, M.C. 1996. Incidence and consequences of inherited environmental effects. *Annual Review of Ecology and Systematics* 27:451–476.

Rossmann, M.G. 1990. Introductory comments on the function of domains in protein structure. In: *Intervening Sequences in Evolution and Development*, E.M. Stone and R.J. Schwartz (eds.). Oxford University Press, New York, pp. 3–9.

Roth, G. 1982. Conditions of evolution and adaptation in organisms as autopoietic systems. In: *Environmental Adaptation and Evolution*, D. Mossakowski and G. Roth (eds.). Gustav Fischer, Stuttgart, pp. 37–48.

Roth, G. and Wake, D.B. 1985. Trends in the functional morphology and sensorimotor control of feeding behavior in salamanders: an example of the role of internal dynamics in evolution. *Acta Biotheoretica* 34:175–192.

Roth, G. and Wake, D.B. 1989. Conservatism and innovation in the evolution of feeding in vertebrates. In: *Complex Organismal Functions: Integration and Evolution in Vertebrates*, D.B. Wake and G. Roth (eds.). John Wiley and Sons, New York, pp. 7–21.

Roth, G., Nishikawa, K.C. and Wake, D.B. 1997. Genome size, secondary simplification, and the evolution of the brain in salamanders. *Brain, Behavior and Evolution* 50:50–59.

Roth, V.L. 1984. On homology. *Biological Journal of the Linnean Society* 22:13–29.

Roth, V.L. 1988. The biological basis of homology. In: *Ontogeny and Systematics*, C.J. Humphries (ed.). Columbia University Press, New York, pp. 1–26.

Roth, V.L. 1990. Insular dwarf elephants: a case study in body mass estimation and ecological inference. In: *Body Size in Mammalian Paleobiology: Estimation and Biological Implications*, J. Damuth and B.J. MacFadden (eds.). Cambridge University Press, Cambridge, pp. 151–179.

Roth, V.L. 1991. Homology and hierarchies: problems solved and unresolved. *Journal of Evolutionary Biology* 4:167–194.

Roth, V.L. 1992. Inferences from allometry and fossils: dwarfing of elephants on islands. *Oxford Surveys in Evolutionary Biology* 8:259–288.

Roth, V.L. 1994. Within and between organisms: replicators, lineages, and homologues. In: *Homology: The Hierarchical Basis of Comparative Biology*, B.K. Hall (ed.). Academic Press, San Diego, pp. 301–337.

Roubaud, E. 1910. Recherches sur la biologíe des Synagris. *Annales de la Société Entomologique de France* 79:1–21.

Roubaud, E. 1916. Recherches biologiques sur les guêpes solitaires et sociales d'Afrique. La genèse de la vie sociale et l'évolution de l'instinct maternel chez les vespides. *Annales Sciences Naturelles (Zoologie)* (Serie 10) 1:1–160.

Roubik, D.W. 1990. Mate location and mate competition in males of stingless bees (Hymenoptera: Apidae: Meliponinae). *Entomologia Generalis* 15:115–120.

Roudabush, R.L. 1933. Phenomenon of regeneration in everted Hydra. *Biological Bulletin* 64:253–258.

Roulin, A., Kölliker, M. and Richner, H. 2000. Barn owl (*Tyto alba*) siblings vocally negotiate resources. *Proceedings of the Royal Society of London B* 267:459–463.

Roux, W. 1891. *Der Kampf der Teile im Organismus*. Engelmann, Leipzig.

Roux, W. 1895. *Gesammelte Abhandlungen zur Entwicklungsmechanik der Organismen*. Englemann, Leipzig.

Rowan, R.G., Brennan, M.D. and Dickinson, W.J. 1986. Developmentally regulated RNA transcripts coding for alcohol dehydrogenase in *Drosophila affinidisjuncta*. *Genetics* 114:405–433.

Rowe, R.J. 1985. Intraspecific interactions of New Zealand damselfly larvae I. *Xanthocnemis zealandica, Ishnura aurora*, and *Austrolestes colensonis* (Zygoptera: Coenagrionidae: Lestidae). *New Zealand Journal of Zoology* 12:1–15.

Rowe, R.J. 1987. Predatory versatility in a larval dragonfly, *Hemianax papuensis* (Odonata: Aeschnidae). *Journal of Zoology, London* 211:193–207.

Rowe, R.J. 1988. Alternative oviposition behaviours in three New Zealand corduliid dragonflies: their adaptive significance and implications for male mating tactics. *Zoological Journal of the Linnean Society* 92:43–66.

Rowe, R.J. 1992. Ontogeny of agonistic behaviour in the territorial damselfly larvae, *Xanthocnemis zealandica* (Zygoptera: Coenagrionidae). *Journal of Zoology* 226:81–93.

Rowell, C.H.F. 1971. The variable coloration of the acridoid grasshoppers. *Advances in Insect Physiology* 8:145–198.

Roy, S.W., Nosaka, M., de Souza, S.J. and Gilbert, W. 1999. Centripetal modules and ancient introns. *Gene* 238:85–91.

Rubenstein, D.I. 1981. Population density, resource patterning and territoriality in the Everglades pygmy sunfish. *Animal Behaviour* 29:155–172.

Rubenstein, D.I. 1984. Resource acquisition and alternative mating strategies in water striders. *American Zoologist* 24:345–353.

Rubenstein, D.I. and Hack, M. 1992. Horse signals: the sounds and scents of fury. *Evolutionary Ecology* 6:254–260.

Rüber, L. and Adams, D.C. 2001. Evolutionary convergence of body shape and trophic morphology in cichlids from Lake Tanganyika. *Journal of Evolutionary Biology* 14:325–332.

Rüber, L., Verheyens, E. and Meyer, A. 1999. Replicated evolution of trophic specializations in an endemic cichlid fish lineage from Lake Tanganyika. *Proceedings of the National Academy of Sciences USA* 96:10230–10235.

Ruffell, R. and Kovoor, J. 1993. The mysterious oonopids from Essex which have six or eight eyes. *Newsletter of the British Arachnological Society* 70:12–13.

Rundle, H.D. and Schluter, D. 1998. Reinforcement of stickleback mate preferences: sympatry breeds contempt. *Evolution* 52:200–208.

Rundle, H.D., Nagel, L., Wenrick Boughman, J., and Schluter, D. 2000. Natural selection and parallel speciation in sympatric sticklebacks. *Science* 287:306–308.

Rüppell, O. and Heinze, J. 1999. Alternative reproductive tactics in females: the case of size polymorphism in winged ant queens. *Insectes Sociaux* 46:6–17.

Rüppell, O., Heinze, J. and Hölldobler, B. 1998. Size-dimorphism in the queens of the North American ant *Leptothorax rugatulus* (Emery). *Insectes Sociaux* 45:67–77.

Russell-Hunter, W.D. 1978. Ecology of freshwater pulmonates. In: *Pulmonates: Systematics, Evolution and Ecology*, V. Fretter and J. Peaks (eds.). Academic Press, New York, pp. 335–383.

Ryan, C.A. 1994. Oligosaccharide signals: from plant defense to parasite offense. *Proceedings of the National Academy of Sciences USA* 91:1–2.

Ryan, C.A. and Jagendorf, A. 1995. Self defense by plants. *Proceedings of the National Academy of Sciences USA* 92:4075.

Ryan, M.J. 1985. *The Tungara Frog*. University of Chicago Press, Chicago.

Ryan, M.J. 1990. Sexual selection, sensory systems and sensory exploitation. *Oxford Surveys in Evolutionary Biology* 7:157–195.

Ryan, M.J. and Causey, B.A. 1989. "Alternative" mating behavior in the swordtails *Xiphophorus nigrensis* and *Xiphophorus pygmaeus* (Pisces: Poeciliidae). *Behavioral Ecology and Sociobiology* 24:341–348.

Ryan, M.J. and Wagner, W.E., Jr. 1987. Asymmetries in mating preferences between species: female swordtails prefer heterospecific males. *Science* 236:595–597.

Ryan, M.J., Pease, C.J. and Morris, M.R. 1992. A genetic polymorphism in the swordtail *Xiphophorus nigrensis*: testing the prediction of equal fitnesses. *American Naturalist* 139:21–31.

Rzhetsky, A., Ayala, F.J., Hsu, L.C., Chang, C. and Yoshida, A. 1997. Exon/intron structure of aldehyde dehydrogenase genes supports the "introns-late" theory. *Proceedings of the National Academy of Sciences USA* 94:6820–6825.

Sachs, T. 1988a. Epigenetic selection: an alternative mechanism of pattern formation. *Journal of Theoretical Biology* 134:547–559.

Sachs, T. 1988b. Ontogeny and phylogeny: phytohormones as indicators of labile changes. In: *Plant Evolutionary Biology*, L.D. Gottlieb and S.H. Jain (eds.). Chapman and Hall, London, pp. 157–176.

Saether, O.A. 1979. Underlying synapomorphies and anagenetic analysis. *Zoologica Scripta* 8:305–312.

Saether, O.A. 1983. The canalized evolutionary potential: inconsistencies in phylogenetic reasoning. *Systematic Zoology* 32:343–359.

Saether, O.A. 1986. The myth of objectivity—post-Hennigian deviations. *Cladistics* 2:1–13.

Saether, O.A. 1990. Midges and the electronic Ouija board: the phylogeny of the Hydrobaenus group (Chironomidae, Diptera) revised. *Zeitschrift für Zoologische Systematik und Evolutionforschung* 28:107–136.

Sage, R.D. and Selander, R.K. 1975. Trophic radiation through polymorphism in cichlid fishes. *Proceedings of the National Academy of Sciences USA* 72:4669–4673.

Sage, R.D., Loiselle, P.V., Basaibwaki, P. and Wilson, A.C. 1984. Molecular versus morphological change among cichlid fishes of Lake Victoria. In: *Evolution of Fish Species Flocks*, A.A. Echelle and I. Kornfield (eds.). University of Maine at Orono Press, Orono, pp. 185–201.

Sagers, C.L. 1993. *The evolution of defense in a neotropical shrub*. Ph.D. dissertation, University of Utah, Salt Lake City.

Sakagami, S.F. 1982. Stingless bees. In: *Social Insects*, Vol. 3. H.R. Hermann (ed.). Academic Press, New York.

Sakagami, S.F. and Maeta, Y. 1987a. Multifemale nests and rudimentary castes of an "almost" solitary bee *Ceratina flavipes*, with additional observations on multifemale nests of *Ceratina japonica* (Hymenoptera, Apoidea). *Kontyû* 55:391–409.

Sakagami, S.F. and Maeta, Y. 1987b. Sociality, induced and/or natural, in the basically solitary small carpenter bees (*Ceratina*). In: *Animal Societies, Theories and Facts*, Y. Itô, J.L. Brown, and J. Kikkawa (eds.). Japan Scientific Societies Press, Tokyo, pp. 1–16.

Salisbury, F.B. and Ross, C.W. 1985. *Plant Physiology*. Wadsworth, Belmont, CA.

Salmon, W.C. 1984. *Scientific Explanation and the Causal Structure of the World*. Princeton University Press, Princeton.

Salt, G. 1937. The egg-parasite of *Sialis lutaria*: a study of the influence of the host upon a dimorphic parasite. *Parasitology* 29:539–553.

Salt, G. 1940. Experimental studies in insect parasitism. VII. The effects of different hosts on the parasite *Trichogramma evanescens* Westw. *Proceedings of the Royal Entomological Society of London* 15:81–95.

Salt, G. 1941. The effects of hosts upon their insect parasites. *Biological Reviews* 16:239–264.

Sander, K. 1983. The evolution of patterning mechanisms: gleanings from insect embryogenesis and spermatogenesis. In: *Development and Evolution*, B.C. Goodwin, N. Holder and C.C. Wylie (eds.). Cambridge University Press, Cambridge, pp. 137–159.

Sanderson, M.J. and Donoghue, M.J. 1989. Patterns of variation in levels of homoplasy. *Evolution* 43:1781–1795.

Sanderson, M.J. and Hufford, L. 1996. *Homoplasy*. Academic Press, New York.

Sandoval, C.P. 1993. Geographic, ecological and behavioral factors affecting spatial vcariation in color morph frequency in the walking-stick, *Timema cristinae*. Ph.D. dissertation. University of California, Santa Barbara.

Sandoval, C.P. 1994a. The effects of the relative geographic scales of gene flow and selection on morph frequencies in the walking-stick *Timema cristinae*. *Evolution* 48:1866–1879.

Sandoval, C.P. 1994b. Differential visual predation on morphs of *Timema cristinae* (Phasmatodeae: Timemidae) and its consequences for host range. *Biological Journal of the Linnean Society* 52:341–356.

Sang, J.H. 1961. Environmental control of mutant expression. In: *Insect Polymorphism*, J.S. Kennedy (ed.). Symposium No. 1. of the Royal Entomological Society, London, pp. 91–102.

Santschi, F. 1907. Fourmis de Tunisie capturées en 1906. *Revue Suisse de Zoologie* 15:305–334.

Sarà, M. 1987. General remarks on the evolutionary ecology of hydroids and hydromedusae. In: *Modern Trends in the Systematics, Ecology, and Evolution of Hydroids and Hydromedusae*, J. Bouillon, F. Boero, F. Cicogna and P.F.S. Cornelius (eds.). Clarendon Press, Oxford, pp. v–x.

Sarà, M. 1989. The problem of adaptations: an holistic approach. *Rivista de Biologia-Biology Forum* 82:75–101.

Sarà, M. 1996. A "sensitive" cell system: its role in a new evolutionary paradigm. *Rivista de Biologia-Biology Forum* 89:139–148.

Sato, A., O'Huigin, C., Figueroa, F., Grant, P.R., Grant, B.R., Tichy, H. and Klein, J. 1999. Phylogeny of Darwin's finches as revealed by mtDNA sequences. *Proceedings of the National Academy of Sciences USA* 96:5101–5106.

Sattler, R. 1988. Homeosis in plants. *American Journal of Botany* 75:1606–1617.

Sauer, E.G.F. 1972. Aberrant sexual behavior in the south African ostrich. *Auk* 89:717–737.

Scapini, F. 1988. Heredity and learning in animal orientation. *Monitore Zoologico Italiano* (N.S.) 22:203–234.

Scapini, F. and Fasinella, D. 1990. Genetic determination and plasticity in the sun orientation of natural populations of *Talitrus saltator*. *Marine Biology* 107:141–145.

Scapini, F., Buiatti, M. and Ottaviano, O. 1988. Phenotypic plasticity in sun orientation of sandhoppers. *Journal of Comparative Physiology A* 163:739–747.

Scapini, F., Lagar, M.C. and Mezzetti, M.C. 1993. The use of slope and visual information in sandhoppers: innateness and plasticity. *Marine Biology* 115:545–553.

Schaefer, S.A. and Lauder, G.V. 1986. Historical transformation of functional design: evolutionary morphology of feeding mechanisms in Loricarioid catfishes. *Systematic Zoology* 35:489–508.

Schaffner, J.H. 1927. Control of sex reversal in the tassel of Indian corn. *Botanical Gazette* 84:440–449.

Schaffner, J.H. 1930. Sex reversal and the experimental production of neutral tassels in *Zea mays*. *Botanical Gazette* 90:279–298.

Scharloo, W. 1970. Stabilizing and disruptive selection on a mutant character in *Drosophila*. III. Polymorphism caused by a developmental switch mechanism. *Genetics* 65:693–705.

Scharloo, W. 1989. Developmental and physiological aspects of reaction norms. *BioScience* 39:465–471.

Scharloo, W. 1991. Canalization: genetic and developmental aspects. *Annual Review of Ecology and Systematics* 22:65–93.

Schartl, M., Schlupp. I., Schartl, A., Meyer, M.K., Nanda, I., Schmid, M., Epplen, J.T. and Parzefall, J. 1991. On the stability of dispensable constituents of the eukaryotic genome: stability of coding sequences versus truly hypervariable sequences in a clonal vertebrate, the Amazon molly, *Poecilia formosa*. *Proceedings of the National Academy of Sciences USA* 88:8759–8763.

Scheiner, R., Erber, J. and Page, R.E., Jr. 1999. Tactile learning and the individual evaluation of the reward in honey bees (*Apis mellifera* L.). *Journal of Comparative Physiology A* 185:1–10.

Scheiner, S.M. 1993a. Genetics and evolution of phenotypic plasticity. *Annual Review of Ecology and Systematics* 24:35–68.

Scheiner, S.M. 1993b. Plasticity as a selectable trait: reply to Via. *American Naturalist* 142:371–373.

Scheiner, S.M. 1998. The genetics of phenotypic plasticity. VII. *Evolution* in a spatially-structured environment. *Journal of Evolutionary Biology* 11:303–320.

Scheiner, S.M. and Lyman, R.F. 1989. The genetics of phenotypic plasticity I. Heritability. *Journal of Evolutionary Biology* 2:95–107.

Scheiner, S.M. and Lyman, R.F. 1991. The genetics of phenotypic plasticity. II. Response to selection. *Journal of Evolutionary Biology* 4:23–50.

Schena, M. and Davis, R.W. 1994. Structure of homeobox-leucine zipper genes suggests a model for the evolution of gene families. *Proceedings of the National Academy of Sciences USA* 91:8393–8397.

Schenkel, R. 1956. Ausdrucksstudien an Wölfen. *Behaviour* 1:81–129.

Schlichting, C.D. 1986. The evolution of phenotypic plasticity in plants. *Annual Review of Ecology and Systematics* 17:667–693.

Schlichting, C.D. 1989. Phenotypic integration and environmental change. *BioScience* 39:460–465.

Schlichting, C.A. and Pigliucci, M. 1993. Control of phenotypic plasticity via regulatory genes. *American Naturalist* 142:366–370.

Schlichting, C.D. and Pigliucci, M. 1998. *Phenotypic Evolution*. Sinauer, Sunderland, MA.

Schlinger, B.A. and Arnold, A.P. 1992. Circulating estrogens in a male songbird originate in the brain. *Proceedings of the National Academy of Sciences USA* 89:7650–7653.

Schluter, D. 1982a. Seed and patch selection by Galápagos ground finches: relation to foraging efficiency and food supply. *Ecology* 63:1106–1120.

Schluter, D. 1982b. Distributions of Galápagos ground finches along an altitudinal gradient: the importance of food supply. *Ecology* 63:1504–1517.

Schluter, D. 1988. Estimating the form of natural selection on a quantitative trait. *Evolution* 42:849–861.

Schluter, D. 1993. Adaptive radiation in sticklebacks: size, shape, and habitat use efficiency. *Ecology* 74:699–709.

Schluter, D. 1994. Experimental evidence that competition promotes divergence in adaptive radiation. *Science* 266:798–801.

Schluter, D. 1995. Adaptive radiation in sticklebacks: trade-offs in feeding performance and growth. *Ecology* 76:82–90.

Schluter, D. 1996. Ecological speciation in postglacial fishes. *Philosophical Transactions of the Royal Society of London B* 351:807–814.

Schluter, D. 2000. *The Ecology of Adaptive Radiation*. Oxford University Press, Oxford.

Schluter, D. and Gustafsson, L. 1993. Maternal inheritance of condition and clutch size in the collared flycatcher. *Evolution* 47:658–667.

Schluter, D. and McPhail, J.D. 1992. Ecological character displacement and speciation in sticklebacks. *American Naturalist* 140:85–108.

Schluter, D. and McPhail, J.D. 1993. Character displacement and replicate adaptive radiation. *Trends in Ecology and Evolution* 8:197–200.

Schluter, D. and Nagel, L.M. 1995. Parallel speciation by natural selection. *American Naturalist* 146:292–301.

Schluter, D., Price, T.D. and Rowe, L. 1991. Conflicting selection pressures and life history. *Proceedings of the Royal Society of London B* 246:11–17.

Schluter, D., Price, T.D., Mooers, A.Ø. and Ludwig, D. 1997. Likelihood of ancestor states in adaptive radiation. *Evolution* 51:1699–1711.

Schmalhausen, I.I. 1949 [1986]. *Factors of Evolution: The Theory of Stabilizing Selection*. University of Chicago Press, Chicago.

Schmid, B. and Dolt, C. 1994. Effects of maternal and paternal environment and genotype on offspring phenotype in *Solidago altissima* L. *Evolution* 48:1525–1549.

Schmid-Hempel, P. 1998. *Parasites in Social Insects*. Princeton University Press, Princeton.

Schmid-Hempel, P. and Schmid-Hempel, R. 1984. Life duration and turnover of foragers in the ant *Cataglyphis bicolor* (Hymenoptera, Formicidae). *Insectes Sociaux* 31:345–360.

Schmidt, G.H. 1973. Haemolymph protein pattern of different *Formica* castes during development. *Proceedings of VIIth*

International Congress. International Union for the Study of Social Insects, London, pp. 345–348.

Schmidt-Nielsen, K. 1984. *Scaling. Why Is Animal Size So Important?* Cambridge University Press, New York.

Schmitt, J., Dudley, S.A. and Pigliucci, M. 1999. Manipulative approaches to testing adaptive plasticity: phytochrome-mediated shade-avoidance responses in plants. *American Naturalist* 154:S43–S54.

Schmitt, J., McCormac, A.C. and Smith, H. 1995. A test of the adaptive plasticity hypothesis using transgenic and mutant plants disabled in phytochrome-mediated elongation responses to neighbors. *American Naturalist* 146:937–953.

Schmitt, W.L. 1965. *Crustaceans.* University of Michigan Press, Ann Arbor.

Schneirla, T.C. 1940. Further studies on the army-ant behavior pattern: mass organization in the swarm-raiders. *Journal of Comparative Psychology* 29:401–460.

Schoener, T.W. 1982. The controversy over interspecific competition. *American Scientist* 70:586–595.

Schoener, T.W. 1983. Field experiments on interspecific competition. *American Naturalist* 122:240–285.

Schopf, T.J.M. 1977. Patterns and themes of evolution among the Bryozoa. In: *Patterns of Evolution, as Illustrated by the Fossil Record,* A. Hallam (ed.). Elsevier, Amsterdam, pp. 159–207.

Schröder, F.C., Farmer, J.J., Attygalle, A.B., Smedley, S.R., Eisner, T. and Meinwald, J. 1998. Combinatorial chemistry in insects: a library of defensive macrocyclic polyamines. *Science* 281:428–431.

Schultz, J. 1987. The origin of the spinning apparatus in spiders. *Biological Reviews* 62:89–113.

Schultz, J., Milpetz, F., Bork, P. and Ponting, C.P. 1998. SMART, a simple modular architecture research tool: identification of signaling domains. *Proceedings of the National Academy of Sciences USA* 95:5857–5864.

Schultz, T.R., Cocroft, R.B. and Churchill, G.A. 1996. The reconstruction of ancestral character states. *Evolution* 50:504–511.

Schumann, R.D. and Buschinger, A. 1994. Imprinting effects on host-selection behaviour of colony-founding *Chalepoxenus muellerianus* (Finzi) females (Hymenoptera, Formicidae). *Ethology* 97:33–46.

Schwabl, H. 1993. Yolk is a source of maternal testosterone for developing birds. *Proceedings of the National Academy of Sciences USA* 90:11446–11450.

Schwabl, H., Mock, D.W. and Gieg, J.A. 1997. A hormonal mechanism for parental favouritism. *Nature* 386:231.

Schwanwitsch, B.N. 1924. On the groundplan of wing-pattern in nymphalids and certain other families of rhopalocerous Lepidoptra. *Proceedings of the Zoological Society of London* B 34:509–528.

Scott, G.R. and Cobban, W.A. 1965. Geologic and biostratigraphic map of the Pierre Shale between Jarre Creek and Loveland, Colorado. Map I-439. U.S. Geologic Survey, Washington, D.C.

Scully, E.P. 1979. The effects of gastropod shell availability and habitat characteristics on shell utilization by the intertidal hermit crab *Pagurus longicarpus* Say. *Journal of Experimental Marine Biology and Ecology* 37:139–152.

Scurfield, G. 1973. Reaction wood, its structure and function. *Science* 179:657–659.

Seeley, T.D. 1977. Measurement of nest cavity volume by the honey bee (*Apis mellifera*). *Behavioral Ecology and Sociobiology* 2:201–227.

Seeley, T.D. 1982. How honeybees find a home. *Scientific American* 247:158–168.

Seeley, T.D. 1986. Social foraging by honeybees: how colonies allocate foragers among patches of flowers. *Behavioral Ecology and Sociobiology* 19:343–354.

Seeley, T.D. 1989. Social foraging in honey bees: how nectar foragers assess their colony's nutritional status. *Behavioral Ecology and Sociobiology* 24:181–199.

Seeley, T.D. and Morse, R.A. 1978. Nest site selection by the honey bee, *Apis mellifera*. *Insectes Sociaux* 25:323–337.

Seeley, T.D. and Towne, W.F. 1992. Tactics of dance choice in honey bees: do foragers compare dances? *Behavioral Ecology and Sociobiology* 30:59–69.

Seeley, T.D., Morse, R.A. and Visscher, P.K. 1979. The natural history of the flight of honey bee swarms. *Psyche* 86:103–113.

Seeley, T.D., Camazine, S. and Sneyd, J. 1991. Collective decision-making in honey bees: how colonies choose among nectar sources. *Behavioral Ecology and Sociobiology* 28:277–290.

Seger, J. 1985. Intraspecific resource competition as a cause of sympatric speciation. In: *Evolution,* P.J. Greenwood, P.H. Harvey, and M. Slatkin (eds.). Cambridge University Press, Cambridge, pp. 43–53.

Seger, J. and Brockmann, H.J. 1987. What is bet-hedging? *Oxford Surveys in Evolutionary Biology* 4:183–211.

Seger, J. and Stubblefield, J.W. 1996. Optimization and adaptation. In:

Adaptation, M.R. Rose and G.V. Lauder (eds.). Academic Press, New York, pp. 93–123.

Seibt, U. and Wickler, W. 1988. Bionomics and social structure of "family spiders" of the genus *Stegodyphus*, with special reference to the African species *S. dumicola* and *S. mimosarum* (Araneida, Eresidae). *Verhandlungen des Naturwissenschaftlichen Vereins in Hamburg* 30:255–303.

Selander, R.K. 1966. Sexual dimorphism and differential niche utilization in birds. *Condor* 68:113–151.

Selander, R.K. 1972. Sexual Selection and dimorphism in birds. In: *Sexual Selection and the Descent of Man 1871–1971*, B. Campbell (ed.). Aldine, Chicago, pp. 180–230.

Sella, G. 1990. Evolution of a cooperative behaviour in the mating systems of two hermaphroditic polychaete worms. *Ethology, Ecology and Evolution* 2:327–328.

Semlitsch, R.D. and Wilbur, H.M. 1989. Artificial selection for paedomorphosis in the salamander *Ambystoma talpoideum*. *Evolution* 43:105–112.

Setty, B.N.Y. and Ramaiah, T.R. 1980. Effect of prostaglandins and inhibitors of prostaglandin biosynthesis on oviposition in the silkmoth *Bombyx mori*. *Indian Journal of Experimental Biology* 18:539–541.

Severson, D.W., Williamson, J.L. and Aiken, J.M. 1989. Caste-specific transcription in the female honey bee. *Insect Biochemistry* 19:215–220.

Sexton, E.W. and Clark, A.R. 1936. A summary of the work on the amphipod *Gammarus chevreuxi*, etc. *Journal of the Marine Biological Association of the United Kingdom* 21:357.

Shaffer, H.B. 1984. Evolution in a paedomorphic lineage. I. An electrophoretic analysis of the Mexican ambystomatid salamanders. *Evolution* 38:1194–1206.

Shaffer, H.B. and McKnight, M.L. 1996. The polytypic species revisited: genetic differentiation and molecular phylogenetics of the tiger salamander *Ambystoma tigrinum* (Amphibia: Caudata) complex. *Evolution* 50:417–433.

Shaffer, H.B. and Voss, S.R. 1996. Phylogenetic and mechanistic analysis of a developmentally integrated character complex: alternate life history modes in ambystomatid salamanders. *American Zoologist* 36:24–35.

Shakarad, M. and Gadagkar, R. 1995. Colony founding in the primitively eusocial wasp, *Ropalidia marginata* (Hymenoptera: Vespidae). *Ecological Entomology* 20:273–282.

Shakarad, M. and Gadagkar, R. 1997. Do social wasps choose nesting strategies based on their brood rearing abilities? *Naturwissenschaften* 84:79–82.

Shapiro, A.M. 1971. Occurrence of a latent polyphenism in *Pieris virginiensis* (Lepidoptera: Pieridae). *Entomological News* 82:13–16.

Shapiro, A.M. 1975. Genetics, environment, and subspecies differences: the case of *Polites sabuleti* (Lepidoptera: Hesperiidae). *Great Basin Naturalist* 35:33–38.

Shapiro, A.M. 1976. Seasonal polyphenism. *Evolutionary Biology* 9:259–333.

Shapiro, A.M. 1978a. Developmental and phenotypic responses to photoperiod and temperature in an equatorial montane butterfly, *Tatochila xanthodice* (Lepidoptera: Pieridae). *Biotropica* 19:297–301.

Shapiro, A.M. 1978b. The evolutionary significance of redundancy and variability in phenotypic-induction mechanisms of pierid butterflies (Lepidoptera). *Psyche* 85:275–283.

Shapiro, A.M. 1980. Physiological and developmental responses to photoperiod and temperature as data in phylogenetic and biogeographic inference. *Systematic Zoology* 29:335–341.

Shapiro, A.M. 1981. Phenotypic plasticity in temperate and subarctic *Nymphalis antiopa* (Nymphalidae): evidence for adaptive canalization. *Journal of the Lepidopterists' Society* 35:124–131.

Shapiro, A.M. 1984a. The genetics of seasonal polyphenism and the evolution of "general purpose genotypes" in butterflies. In: *Population Biology and Evolution*, K. Wöhrmann and V. Loeschcke (eds.). Springer-Verlag, Berlin, pp. 16–30.

Shapiro, A.M. 1984b. Experimental studies on the evolution of seasonal polyphenism. In: *The Biology of Butterflies*, R.I. Vane-Wright and P.R. Ackery (eds.). Academic Press, London, pp. 297–307.

Shapiro, A.M. 1984c. Polyphenism, phyletic evolution, and the structure of the pierid genome. *Journal of Research on the Lepidoptera* 23:177–195.

Shapiro, D.Y. 1984. Sex reversal and sociodemographic processes in coral reef fishes. In: *Fish Reproduction: Strategies and Tactics*, G.W. Potts and R.J. Wootton (eds.). Academic Press, London, pp. 103–118.

Shapiro, J.A. 1992. Natural genetic engineering in evolution. *Genetica* 86:99–111.

Shapiro, J.A. 1999. Genome system architecture and natural genetic engineering in evolution. *Annals of the New York Academy of Sciences* 870:23–35.

Shapiro, L., Kwong, P.D., Fannon, A.M., Colman, D.R. and Hendrickson, W.A. 1995. Considerations of the folding topology and evolutionary origin of cadherin domains. *Proceedings of the National Academy of Sciences USA* 92:6793–6797.

Sharp, P.A. 1994. Split genes and RNA splicing. *Cell* 77:805–815.

Shatz, C.J. 1996. Emergence of order in visual system development. *Proceedings of the National Academy of Sciences USA* 93:602–608.

Shaw, D. and Goldman, B.D. 1995a. Influence of prenatal and postnatal photoperiods on postnatal testis development in the Siberian hamster (*Phodopus sungorus*). *Biology of Reproduction* 52:833–838.

Shaw, D. and Goldman, B.D. 1995b. Gender differences in influence of prenatal photoperiods on postnatal pineal melatonin rhythms and serum prolactin and follicle-stimulating hormone in the Siberian hamster (*Phodopus sungorus*). *Endocrinology* 136:4237–4246.

Shaw, M.J.P. 1970. Effects of population density on alienicolae of *Aphis fabae* Scop. *Annals of Applied Biology* 65:205–212.

Shea, B.T. 1988. Heterochrony in primates. In: *Heterochrony in Evolution: A Multidisciplinary Approach*, M.L. McKinney (ed.). Plenum Press, New York, pp. 237–266.

Shearer, R.R. and Gould, S.J. 1999. Of two minds and one nature. *Science* 286:1093–1094.

Shelly, T.E. and Whittier, T.S. 1997. Lek behavior of insects. In: *The Evolution of Mating Systems in Insects and Arachnids*, J.C. Choe and B.J. Crespi (eds.). Cambridge University Press, Cambridge, pp. 273–293.

Sheppard, P.M. 1959. The evolution of mimicry: a problem in ecology and genetics. *Cold Spring Harbor Symposia on Quantitative Biology* 24:131–140.

Sheppard, P.M. 1975. Stable polymorphism. In: *Natural Selection and Heredity*, Hutchinson University Library, London, pp. 81–105.

Sherman, P.W. 1988. The levels of analysis. *Animal Behaviour* 36:616–619.

Sherman, P.W. 1989. The clitoris debate and the levels of analysis. *Animal Behaviour* 37:697–698.

Sherman, P.W., Jarvis, J.U.M. and Alexander, R.D. 1991. *The Biology of the Naked Mole-Rat*. Princeton University Press, Princeton.

Sherman, P.W., Seeley, T.D., and Reeve, H.K. 1988. Parasites, pathogens and polyandry in social Hymenoptera. *American Naturalist* 131:602–610.

Sherman, P.W., Seeley, T.D. and Reeve, H.K. 1998. Parasites, pathogens, and polyandry in honey bees. *American Naturalist* 151:392–396.

Sherman, P.W., Braude, S. and Jarvis, J.U.M. 1999. Litter sizes and mammary numbers of naked mole-rats: breaking the one-half rule. *Journal of Mammalogy* 80:720–733.

Sherry, T.W. 1990. When are birds dietarily specialized? Distinguishing ecological from evolutionary approaches. *Studies in Avian Biology* 13:337–352.

Shettleworth, S.J. 1984. Learning and behavioural ecology. In: *Behavioural Ecology*, J.R. Krebs and N.B. Davies (eds.). Blackwell, Oxford, pp. 170–194.

Shilo, M. 1982. Diversity of the photosynthetic prokaryotes. In: *On the Origins of Chloroplasts*, J.A. Schiff, (ed.). Elsevier North Holland, Amsterdam, pp. 9–26.

Shoemaker, D.D. and Ross, K.G. 1996. Effects of social organization on gene flow in the fire ant *Solenopsis invicta*. *Nature* 383:613–616.

Shoichet, B.K., Baase, W.A., Kuroki, R. and Matthews, B.W. 1995. A relationship between protein stability and protein function. *Proceedings of the National Academy of Sciences USA* 92:452–456.

Shorthouse, J.D. and Rohfritsch, O. (eds.) 1992. *Biology of Insect-Induced Galls*. Oxford University Press, New York.

Shubin, N. 1994. History, ontogeny, and evolution of the archetype. In: *Homology*, M.L. Mckinney (ed.). Academic Press, New York, pp. 250–273.

Shubin, N. and Wake, D. 1996. Phylogeny, variation, and morphological integration. *American Zoologist* 36:51–60.

Shubin, N., Wake, D.B. and Crawford, A.J. 1995. Morphological variation in the limbs of *Taricha granulosa* (Caudata: Salamandridae): evolutionary and phylogenetic implications. *Evolution* 49:874–884.

Shull, H.G. 1915. Genetic definitions in the New Standard Dictionary. *American Naturalist* 49:59.

Shuster, S.M. 1987. Alternative reproductive behaviors: three discrete male morphs in *Paracerceis sculpta*, an intertidal isopod from the northern Gulf of California. *Journal of Crustacean Biology* 7:318–327.

Shuster, S.M. 1989. Male alternative reproductive strategies in a marine isopod crustacean (*Paracerceis sculpta*): the use of genetic markers to measure differences in fertilization success among α-, β-, and γ-males. *Evolution* 43:1683–1698.

Shuster, S.M. 1990. Courtship and female mate selection in a marine isopod crustacean, *Paracerceis sculpta*. *Animal Behaviour* 40:390–399.

Shuster, S.M. 1992a. Stress and evolution. *Evolution* 46:1248–1249.

Shuster, S.M. 1992b. The reproductive behaviour of α-, β-, and γ-male morphs in *Paracerceis sculpta*, a marine isopod crustacean. *Behaviour* 121:231–257.

Shuster, S.M. and Guthrie, E.E. 1999. Effects of temperature and food availability on adult body length in natural and laboratory populations of *Paracerceis sculpta* (Holmes), a Gulf of California isopod. *Journal of Experimental Marine Biology and Ecology* 233:269–284.

Shuster, S.M. and Sassaman, C. 1997. Genetic interaction between male mating strategy and sex ratio in a marine isopod. *Nature* 388:373–376.

Shuster, S.M. and Wade, M.J. 1991a. Equal mating success among male reproductive strategies in a marine isopod. *Nature* 350:608–610.

Shuster, S.M. and Wade, M.J. 1991b. Female copying and sexual selection in a marine isopod crustacean. *Animal Behaviour* 42:1071–1078.

Shuvalov, V.F. and Popov, A.V. 1973. Significance of some parameters of the calling songs of male crickets, *Gryllus bimaculatus* for phonotaxis of females. *Journal of Evolutionary Biochemistry and Physiology* 9:177–182.

Siegel, R.W. and Hall, J.C. 1979. Conditioned responses in courtship behavior of normal and mutant *Drosophila*. *Proceedings of the National Academy of Sciences USA* 76:3430–3434.

Silk, J.B. 1987. Social behavior in evolutionary perspective. In: *Primate Societies*, B.B. Smuts, D.L. Cheney, R.M. Seyfarth, R.W. Wrangham and T.T. Struhsaker (eds.). University of Chicago Press, Chicago, pp. 318–329.

Sillén-Tullberg, B. 1988. Evolution of gregariousness in aposematic butterfly larvae: a phylogenetic analysis. *Evolution* 42:293–305.

Silvertown, J. 1984. Phenotypic variety in seed germ behavior: the ontogeny and evolution of somatic polymorphism in seeds. *American Naturalist* 124:1–16.

Silvertown, J. and Gordon, D.M. 1989. A framework for plant behavior. *Annual Review of Ecology and Systematics* 20:349–366.

Simmons, L.W. 1987. Heritability of a male character chosen by females of the field cricket, *Gryllus bimaculatus*. *Behavioral Ecology and Sociobiology* 21:129–133.

Simmons, L.W. 1988. The calling song of the field cricket, *Gryllus bimaculatus* (De Geer):

constraints on transmission and its role in intermale competition and female choice. *Animal Behaviour* 36:380–394.

Simon, C. Tang, J., Dalwadi, S., Staley, G., Deniega, J. and Unnasch, T.R. 2000. Genetic evidence for assortative mating between 13–year cicadas and sympatric "17–year cicadas with 13–year life cycles" provides support for allocronic speciation. *Evolution* 54:1326–1336.

Simon, C.A. 1975. The influence of food abundance on territory size in the iguanid lizard, *Scleropus jarrovi*. *Ecology* 56:993–998.

Simon, H.A. 1973. The organization of complex systems. In: *Hierarchy Theory*, H.H. Pattee (ed.). Braziller, New York, pp. 1–27.

Simpson, G.G. 1944. *Tempo and Mode in Evolution*. Columbia University Press, New York.

Simpson, G.G. 1946. Fossil penguins. *Bulletin of the American Museum of Natural History* 87:1–100.

Simpson, G. G. 1949. *The Meaning of Evolution*. Mentor Books, New York.

Simpson, G.G. 1951. *The Meaning of Evolution*. Yale University Press, New Haven.

Simpson, G.G. 1953a. *The Major Features of Evolution*. Columbia University Press, New York.

Simpson, G.G. 1953b. The Baldwin effect. *Evolution* 7:110–117.

Simpson, G.G. 1958. Behavior and evolution. In: *Behavior and Evolution*, A. Roe and G.G. Simpson (eds.). Yale University Press, New Haven, pp. 507–533.

Simpson, G.G. 1975. Recent advances in methods of phylogenetic inference. In: *Phylogeny of the Primates*, W.P. Luckett and F.S. Szalay (eds.). Plenum Press, New York, pp. 3–19.

Simpson, G.G. 1984. Introduction: forty years later. In: *Tempo and Mode in Evolution*. Columbia University Press, New York, pp. xiii–xxvi.

Simpson, G.G. and Beck, W.S. 1965. *Life. An Introduction to Biology*, 2nd ed. Harcourt, Brace and World, New York.

Simpson, S.J. and Simpson, C.L. 1992. Mechanisms controlling modulation by haemolymph amino acids of gustatory responsiveness in the locust. *Journal of Experimental Biology* 168:269–287.

Sinervo, B. 1990. The evolution of maternal investment in lizards: an experimental and comparative analysis of egg size and its effects on offspring performance. *Evolution* 44:279–294.

Sinervo, B. and Basolo, A.L. 1996. Testing adaptation using phenotypic manipulations. In: *Adaptation*, M.R. Rose and G.V. Lauder (eds.). Academic Press, San Diego, pp. 149–185.

Sinervo, B. and Licht, P. 1991a. Hormonal and physiological control of clutch size, egg size, and egg shape in side-blotched lizards (*Uta stansburiana*): constraints on the evolution of lizard life histories. *Journal of Experimental Zoology* 257:252–264.

Sinervo, B. and Licht, P. 1991b. Proximate constraints on the evolution of egg size, number, and total clutch mass in lizards. *Science* 252:1300–1302.

Sinervo, B. and Lively, C.M. 1996. The rock-paper-scissors game and the evolution of alternative male strategies. *Nature* 380:240–243.

Sinervo, B. and McEdward, L.R. 1988. Developmental consequences of an evolutionary change in egg size: an experimental test. *Evolution* 42:885–899.

Sinervo, B. and Svensson, E. 1998. Mechanistic and selective causes of life history trade-offs and plasticity. *OIKOS* 83:432–442.

Sinervo, B., Miles, D.B., Frankino, W.A., Klukowski, M. and DeNardo, D.F. 2000. Testosterone, endurance, and Darwinian fitness: natural and sexual selection on the physiological bases of alternative male behaviors in side-blotched lizards. *Hormones and Behavior* 38:222–223.

Sing, C.F., Haviland, M.B., Templeton, A.R., Zerba, K.E. and Reilly, S.L. 1992. Biological complexity and strategies for finding DNA variations responsible for inter-individual variation in risk of a common chronic disease, coronary artery disease. *Annals of Medicine* 24:539–547.

Sing, C.F., Haviland, M.B., Templeton, A.R. and Reilly, S.L. 1995. Alternative genetic strategies for predicting risk of atherosclerosis. *Proceedings of the 10th International Symposium on Atherosclerosis, Montreal*, F.P. Woodford, J. Davignon and A. Sniderman (eds.). Elsevier, New York, pp. 638–644.

Singer, M. and Berg, P. 1991. *Genes and Genomes: A Changing Perspective.* University Science Books, Mill Valley, CA.

Skaife, S.H. 1955. *Dwellers in Darkness.* Longmans, Green, London.

Skúlason, S. 1990. Variation in morphology, life history and behaviour among sympatric morphs of Artic charr: an experimental approach. Thesis, University of Guelph.

Skúlason, S. and Smith, T.B. 1995. Resource polymorphisms in vertebrates. *Trends in Ecology and Evolution* 10:366–370.

Skúlason, S., Noakes, D.L.G. and Snorrason, S.S. 1989a. Ontogeny of trophic morphology in four sympatric morphs of arctic charr

Salvelinus alpinus in Thingvallavatn, Iceland. *Biological Journal of the Linnean Society* 38:281–301.

Skúlason, S., Snorrason, S.S., Noakes, D.L.G., Ferguson, M.M. and Malmquist, H.J. 1989b. Segregation in spawning and early life history among polymorphic Arctic charr, *Salvelinus alpinus*, in Thingvallavatn, Iceland. *Journal of Fish Biology* 35:225–232.

Skúlason, S., Snorrason, S.S. and Jónsson, B. 1999. Sympatric morphs, populations and speciation in freshwater fish with emphasis on arctic charr. In: *Evolution of Biological Diversity*, A.E. Magurran and R.M. May (eds.). Oxford University Press, Oxford, pp. 70–92.

Slade, A.J. and Hutchings, M.J. 1987. The effects of nutrient availability on foraging in the clonal herb *Glechoma hederacea*. *Journal of Ecology* 75:95–112.

Slater, P.J.B. 1983. Bird song learning: theme and variations. In: *Perspectives in Ornithology*, A.H. Brush and G.A. Clark (eds.). pp. 475–499.

Slater, P.J.B. 1986. The cultural transmission of bird song. *Trends in Ecology and Evolution* 1:94–97.

Slatkin, M. 1979. The evolutionary response to frequency- and density-dependent interactions. *American Naturalist* 114:384–396.

Slatkin, M. 1987. Quantitative genetics of heterochrony. *Evolution* 41:799–811.

Slatkin, M. and Kirkpatrick, M. 1986. Extrapolating quantitative genetic theory to evolutionary problems. In: *Evolutionary Genetics of Invertebrate Behavior*, M.D. Huettel (ed.). Plenum Press, New York, pp. 283–293.

Slijper, E.J. 1942a. Biologic-anatomical investigations on the bipedal gait and upright posture in mammals, with special reference to a little goat, born without forelegs. I. *Proceedings of the Koninklijke Nederlandse Akademie Wetenschappen* 45:288–295.

Slijper, E.J. 1942b. Biologic-anatomical investigations on the bipedal gait and upright posture in mammals, with special reference to a little goat, born without forelegs. II. *Proceedings of the Koninklijke Nederlandse Akademie Wetenschappen* 45:407–415.

Slud, P. 1967. The birds of Cocos Island [Costa Rica]. *Bulletin of the American Museum of Natural History* 134:290–292.

Sluys, R. 1989. Rampant parallelism: an appraisal of the use of nonuniversal derived character states in phylogenetic reconstruction. *Systematic Zoology* 38:350–370.

Smillie, D. 1993. Darwin's tangled bank: the role of social environments. *Perspectives in Ethology* 10:119–141.

Smillie, D. 1995. Darwin's two paradigms: an "opportunistic" approach to natural selection theory. *Journal of Social and Evolutionary Systems* 18:231–255.

Smit, A.F.A. and Riggs, A.D. 1996. *Tiggers* and other DNA transposon fossils in the human genome. *Proceedings of the National Academy of Sciences USA* 93:1443–1448.

Smith, A.M. 1995. Palaeoenvironmental interpretation using bryozoans: a review. In: *Marine Palaeoenvironmental Analysis from Fossils*, D.W.J. Bosence and P.A. Allison (eds.). Geological Society Special Publications, Washington, D.C., pp. 231–243.

Smith, A.P. 1979. Buttressing of tropical trees in relation to bark thickness in Dominica, B.W.I. *Biotropica* 11:159–160.

Smith, C.W.J., Knaack, D. and Nadal-Ginard, B. 1990. Alternative mRNA splicing in the generation of protein diversity and the control of gene expression. In: *Intervening Sequences in Evolution and Development*, E.M. Stone and R.J. Schwartz (eds.). Oxford University Press, New York, pp. 162–195.

Smith, E.F. 1999. Behavior of male paper wasps *Mischocyttarus collarellus* Richards at the natal nest. In: *Social Insects at the Turn of the Millennium*, M.P. Schwarz and K. Hogendoorn (eds.). XIII Congress of IUSSI, Adelaide, p. 450.

Smith, G. 1909. Crustacea. In: *The Cambridge Natural History*, Vol. 4, S.F. Harmer and A.E. Shipley (eds.). Macmillan, New York.

Smith, G.R. and Todd, T.N. 1984. *Evolution* of species flocks of fishes in north temperate lakes. In: *Evolution of Fish Species Flocks*, A.A. Echelle and I. Kornfield (eds.). University of Maine at Orono Press, Orono, pp. 45–68.

Smith, J.A.C. and Winter, K. 1996. Taxonomic distribution of crassulacean acid metabolism. In: *Crassulacean Acid Metabolism*, K. Winter and J.A.C. Smith (eds.). Springer-Verlag, New York, pp. 427–436.

Smith, J.N.M. and Sweatman, H.P.A. 1976. Feeding habits and morphological variation in Cocos finches. *Condor* 78:244–248.

Smith, K.C. 1991. The effects of temperature and daylength on the *Rosa* polyphenism in the buckeye butterfly, *Precis coenia* (Lepidoptera: Nymphalidae). *Journal of Research on the Lepidoptera* 30:225–236.

Smith, L.D. and Palmer, A.R. 1994. Effects of manipulated diet on size and performance of brachyuran crab claws. *Science* 264:710–712.

Smith, T.B. 1987. Bill size polymorphism and intraspecific niche utilization in an African finch. *Nature* 329:717–719.

Smith, T.B. 1990a. Comparative breeding biology of the two bill morphs of the black-bellied seedcracker. *Auk* 107:153–160.

Smith, T.B. 1990b. Resource use by bill morphs of an African finch: evidence for intraspecific competition. *Ecology* 71:1246–1257.

Smith, T.B. 1990c. Natural selection on bill characters in the two bill morphs of the African finch *Pyrenestes ostrinus*. *Evolution* 44:832–842.

Smith, T.B. 1990d. Patterns of morphological and geographic variation in trophic bill morphs of the African finch *Pyrenestes*. *Biological Journal of the Linnean Society* 41:381–414.

Smith, T.B. 1991. A double-billed dilemma. *Natural History* 1:14–19.

Smith, T.B. 1993. Disruptive selection and the genetic basis of bill size polymorphism in the African finch *Pyrenestes*. *Nature* 363:618–620.

Smith, T.B. 1997. Adaptive significance of the mega-billed form in the polymorphic black-bellied seedcracker *Pyrenestes ostrinus*. *Ibis* 139:382–387.

Smith, T.B. and Girman, D.J. 2000. Reaching new adaptive peaks. Evolution of alternative bill forms in an African finch. In: *Adaptive Genetic Variation in the Wild*, T. Mousseau, B. Sinervo and J. Endler (eds.). Oxford University Press, Oxford, pp. 139–156.

Smith, T.B., Wayne, R.K., Girman, D.J. and Bruford, M.W. 1997. A role for ecotones in generating rainforest biodiversity. *Science* 276:1855–1857.

Smith-Gill, S.J. 1983. Developmental plasticity: developmental conversion versus phenotypic modulation. *American Zoologist* 23:47–55.

Smuts, B.B., Cheney, D.L., Seyfarth, R.M., Wrangham, R.W. and Struhsaker, T.T. (eds.) 1987. *Primate Societies*. University of Chicago Press, Chicago.

Smythe, N. 1991. Steps toward domesticating the paca (*Agouti* = *Cuniculus paca*) and prospects for the future. In: *Neotropical Wildlife Use and Conservation*, J.G. Robinson and K.H. Redford (eds.). University of Chicago Press, Chicago, pp. 202–216.

Snodgrass, R.E. 1935 [2000]. *Principles of Insect Morphology*. McGraw-Hill, New York.

Snodgrass, R.E. 1952. *A Textbook of Arthropod Anatomy*. Comstock, Ithaca, NY.

Snodgrass, R.E. 1956. *Anatomy of the Honey Bee*. Cornell University Press, Ithaca.

Snook, R.R. 1997. Is the production of multiple sperm types adaptive? *Evolution* 51:797–808.

Snook, R.R. 1998. The risk of sperm competition and the evolution of sperm heteromorphism. *Animal Behaviour* 56:1497–1507.

Snorrason, S.S., Skúlason, S., Sandlund, O.T., Malmquist, H.J., Jonsson, B. and Jonasson, P.M. 1989. Shape polymorphism in arctic charr, *Salvelinus alpinus* in Thingvallavatn, Iceland. *Physiological Ecology Japan* (Special Volume) 1:393–404.

Snorrason, S.S., Skúlason, S., Jonsson, B., Malmquist, H.J. and Jonasson, P.M. 1994. Trophic specialization in arctic charr *Salvelinus alpinus* (Pisces: Salmonidae): morphological divergence and ontogenetic niche shifts. *Biological Journal of the Linnean Society* 52:1–18.

Soares-Porto, L.M., Walsh, S.J., Nico, L.G. and Netto, J.M. 1999. A new species of *Gelanoglanis* from the Orinoco and Amazon river basins, with comments on miniaturization within the genus (Siluriformes: Auchenipteridae: Centromochlinae). *Ichthyological Exploration of Freshwaters* 10:63–72.

Sober, E. (ed.) 1984. *The Nature of Selection: Evolutionary Theory in Philosophical Focus*. MIT Press, Cambridge.

Sober, E. 1993. *Philosophy of Biology*. Westview Press, San Francisco.

Sokolowski, M.B. 1986. *Drosophila* larval foraging behavior and correlated behaviors. In: *Evolutionary Genetics of Invertebrate Behavior: Progress and Prospects*, M.D. Huettel (ed.). Plenum Press, New York, pp. 197–213.

Solís, C.R., Hughes, C.R., Klingler, C.J., Strassmann, J.E. and Queller, D.C. 1998. Lack of kin discrimination during wasp colony fission. *Behavioral Ecology* 9:172–176.

Solomon, N.G. and French, J.A. (eds.) 1997. *Cooperative Breeding in Mammals*. Cambridge University Press, Cambridge.

Solter, D. 1988. Differential imprinting and expression of maternal and paternal genomes. *Annual Review of Genetics* 22:127–146.

Sommeijer, M.J., van Veen, J.W. and Sewmar, R. 1990. The intranidal activity of drones of *Melipona*, with some remarks about drone production in stingless bees. *Actes Colloques Insectes Sociaux* 6:57–62.

Sorensen, A.E. 1978. Somatic polymorphism and seed dispersal. *Nature* 276:174–176.

Soulé, M.E. 1982. Allomeric variation. 1. The theory and some consequences. *American Naturalist* 120:751–764.

Southwood, T.R.E. 1973. The insect/plant relationship—an evolutionary perspective.

Symposia of the Royal Entomological Society 6:3–30.

Speed, M.P. and Turner, J.R.G. 1999. Learning and memory in mimicry: II. Do we understand the mimicry spectrum? *Biological Journal of the Linnean Society* 67:281–312.

Spicer, G.S. 1993. Morphological evolution of the *Drosophila virilis* species group as assessed by rate tests for natural selection on quantitative characters. *Evolution* 47:1240–1254.

Spielman, A. 1971. Bionomics of autogenous mosquitoes. *Annual Review of Entomology* 16:231–248.

Spieth, H.T. 1974. Mating behavior and evolution of the Hawaiian *Drosophila*. In: *Genetic Mechanisms of Speciation in Insects*, M.J.D. White (ed.). Australia and New Zealand Book Company, Sidney.

Spolsky, C., Phillips, C.A. and Uzzell, T. 1992. Gynogenetic reproduction in hybrid mole salamanders (genus *Ambystoma*). *Evolution* 46:1935–1944.

Spradbery, J.P. 1973. *Wasps*. University of Washington Press, Seattle.

Sprugel, D.G., Hinckley, T.M. and Schaap, W. 1991. The theory and practice of branch autonomy. *Annual Review of Ecology and Systematics* 22:309–334.

Sprules, W.G. 1974. Environmental factors and the incidence of neoteny in *Ambystoma gracile* (Amphibia: Caudata), *Canadian Journal of Zoology* 52:1545–1552.

Spudich, J.L. and Koshland, D.E., Jr. 1976. Non-genetic individuality: chance in the single cell. *Nature* 262:467–471.

Srere, H.K., Wang, L.C.H. and Martin, S.L. 1992. Central role for differential gene expression in mammalian hibernation. *Proceedings of the National Academy of Sciences USA* 89:7119–7123.

Stacey, P.B. and Koenig, W.D. (eds.) 1990. *Cooperative Breeding in Birds*. Cambridge University Press, Cambridge.

Stanley, S.M. 1972. Functional morphology and evolution of byssally attached bivalve mollusks. *Journal of Paleontology* 46:165–212.

Stanley, S.M. 1979. *Macroevolution: Pattern and Process*. W.H. Freeman, San Francisco.

Stanley, S.M. 1981. *The New Evolutionary Timetable: Fossils, Genes, and the Origin of Species*. Basic Books, New York.

Stanley, S.M. 1989. The empirical case for the punctuational model of evolution. In: *The Dynamics of Evolution*, A. Somit and S.A. Peterson (eds.). Cornell University Press, Ithaca, pp. 85–102.

Stanley, S.M. 1990. Adaptive radiation and macroevolution. In: *Major Evolutionary Radiations*, P.D. Taylor and G.P. Larwood (eds.). Systematics Association Special Volume No. 42. Clarendon Press, Oxford, pp. 1–15.

Stanley, S.M. 1995. Climatic forcing and the origin of the human genus. In: *Effects of Past Global Change on Life*, Board on Earth Sciences and Resources, Commission on Geosciences, Environment, and Resources, National Research Council. National Academy Press, Washington, D.C., pp. 233–243.

Stark, R.E. 1992a. Cooperative nesting in the multivoltine large carpenter bee *Xylocopa sulcatipes* Maa (Apoidea: Anthophoridae): do helpers gain or lose to solitary females? *Ethology* 91:301–310.

Stark, R.E. 1992b. Sex ratio and maternal investment in the multivoltine large carpenter bee *Xylocopa sulcatipes* (Apoidea: Anthophoridae). *Ecological Entomology* 17:160–166.

Starks, P.T. 1998. A novel "sit and wait" reproductive strategy in social wasps. *Proceedings of the Royal Society of London B* 265:1407–1410.

Stearns, S.C. 1980. A new view of life history evolution. *OIKOS* 35:266–281.

Stearns, S.C. 1982. The role of development in the evolution of life histories. In: *Evolution and Development*, J.T. Bonner (ed.). Springer-Verlag, Berlin, pp. 237–258.

Stearns, S.C. 1983. The evolution of life-history traits in mosquitofish since their introduction to Hawaii in 1905: rates of evolution, heritabilities, and developmental plasticity. *American Zoologist* 23:65–75.

Stearns, S.C. 1984. How much of the phenotype is necessary to understand evolution at the level of the gene? In: *Population Biology and Evolution*, K. Wöhrmann and V. Loeschcke (eds.). Springer-Verlag, Berlin, pp. 31–45.

Stearns, S.C. 1986. Natural selection and fitness, adaptation and constraint. In: *Patterns and Processes in the History of Life*, D.M. Raup and D. Jablonski (eds.). Springer-Verlag, Berlin, pp. 23–44.

Stearns, S.C. 1989. The evolutionary significance of phenotypic plasticity. *BioScience* 39:436–445.

Stearns, S.C. 1992. *The Evolution of Life Histories*. Oxford University Press, Oxford.

Stearns, S.C., de Jong, G. and Newman, B. 1991. The effects of phenotypic plasticity on genetic correlations. *Trends in Ecology and Evolution* 6:122–126.

Stebbing, A.R.D. 1973. Competition for space between the epiphytes of *Fucus serratus* L. *Journal of the Marine Biology Association U.K.* 53:247–261.

Stebbins, G.L. 1977. The nature of evolution. In: *Evolution*, T. Dobzhansky, F.J. Ayala, G.L. Stebbins and J.W. Valentine (eds.). W.H. Freeman, San Francisco, pp. 1–19.

Stebbins, G.L. and Basile, D.V. 1986. Phyletic phenocopies: a useful technique for probing the genetic and developmental basis of evolutionary change. *Evolution* 40:422–425.

Steele, E.J. 1981. *Somatic Selection and Adaptive Evolution*. Williams and Wallace, Toronto.

Steffan, W.A. 1973. Polymorphism in *Plastosciara perniciosa*. *Science* 182:1265–1266.

Steffan, W.A. 1975. Morphological and behavioral polymorphism in *Plastosciara perniciosa* (Diptera: Sciaridae). *Entomological Society of Washington* 77:1–14.

Stellar, E. 1987. The internal environment and appetitive measures of taste function in the rat. In: *Perspectives in Chemoreception and Behavior*, R.F. Chapman, E.A. Bernays and J.S. Stoffolano (eds.). Springer-Verlag, New York, pp. 1–15.

Stellar, J.R. and Stellar, E. 1985. *Neurobiology of Motivation and Reward*. Springer-Verlag, New York.

Stent, G.S. 1985. Thinking in one dimension: the impact of molecular biology on development. *Cell* 40:1–2.

Stephens, D.W. and Krebs, J.R. 1986. *Foraging Theory*. Princeton University Press, Princeton.

Stephens, S.G. 1951. Possible significance of duplication in evolution. *Advances in Genetics* 4:247–265.

Stern, C. 1968. *Genetic Mosaics and Other Essays*. Harvard University Press, Cambridge.

Stern, D.L. 1994. A phylogenetic analysis of soldier evolution in the aphid family Hormaphididae. *Proceedings of the Royal Society of London B* 256:203–209.

Stern, D.L. 1995. Phylogenetic evidence that aphids, rather than plants, determine gall morphology. *Proceedings of the Royal Society of London B* 260:85–89.

Stern, D.L. 1998. Phylogeny of the tribe Cerataphidini (Homoptera) and the evolution of the horned soldier aphids. *Evolution* 52:155–165.

Stern, D.L. and Foster, W.A. 1996. The evolution of soldiers in aphids. *Biological Reviews* 71:27–79.

Stern, D.L. and Grant, P.R. 1996. A phylogenetic reanalysis of allozyme variation among populations of Galapagos finches. *Zoological Journal of the Linnean Society* 118:119–134.

Stern, D.L. and Foster, W.A. 1997. The evolution of sociality in aphids: a clone's-eye view. In: *Social Behavior in Insects and Arachnids*, J.C. Choe and B.J. Crespi (eds.). Cambridge University Press, Cambridge, pp. 150–165.

Stern, D.L., Aoki, S. and Kurosu, U. 1997. Determining aphid taxonomic affinities and life cycles with molecular data: a case study of the tribe Cerataphidini (Hormaphididaea: Aphidoidea: Hemiptera). *Systematic Entomology* 22:81–96.

Stetson, M.H. 1989. *Processing of Environmental Information in Vertebrates*. Springer-Verlag, New York.

Stetson, M.H., Elliott, J.A. and Goldman, B.D. 1986. Maternal transfer of photoperiodic information influences the photoperiodic response of prepubertal Djungarian hamsters (*Phodopus sungorus sungorus*). *Biology of Reproduction* 34:664–669.

Stiles, F.G. 1978. Possible specialization for hummingbird-hunting in the tiny hawk. *Auk* 95:550–553.

Stone, E.M. and Schwartz, R.J. (eds.) 1990a. *Intervening Sequences in Evolution and Development*. Oxford University Press, New York.

Stone, E.M. and Schwartz, R.J. 1990b. Preface. In: *Intervening Sequences in Evolution and Development*, E.M. Stone and R.J. Schwartz (eds.). Oxford University Press, New York, pp. v–vii.

Stone, E.M. and Schwartz, R.J. 1990c. Intron-dependent evolution of progenotic enzymes. In: *Intervening Sequences in Evolution and Development*, E.M. Stone and R.J. Schwartz (eds.). Oxford University Press, New York, pp. 63–91.

Stone, G.N. and Cook, J.M. 1998. The structure of cynipid oak galls: patterns in the evolution of an extended phenotype. *Proceedings of the Royal Society of London B* 265:979–988.

Storey, A.E., Walsh, C.J., Quinton, R. and Wynne-Edwards, K.E. 2000. Hormonal correlates of paternal responsiveness in men. *Evolution and Human Behavior* 21:79–95.

Stowe, M.K. 1986. Prey specialization in the Araneidae. In: *Spiders. Webs, Behavior, and Evolution*, W.A. Shear (ed.). Stanford University Press, Stanford, pp. 101–131.

Strassmann, J.E. and Meyer, D.C. 1983. Gerontocracy in the social wasp, *Polistes exclamans*. *Animal Behaviour* 31:431–438.

Strassmann, J.E., Klingler, C.J., Arévalo, E., Zacchi, F., Husain, A., Williams, J., Seppä, P. and Queller, D.C. 1997. Absence of within-colony kin discrimination in behavioural interactions of swarm-founding wasps. *Proceedings of the Royal Society of London B* 264:1565–1570.

Strathmann, R.R. 1978. Progressive vacating of adaptive types during the Phanerozoic. *Evolution* 32:907–914.

Strathmann, R.R. 1985. Feeding and nonfeeding larval development and life-history evolution in marine invertebrates. *Annual Review of Ecology and Systematics* 16:339–361.

Strathmann, R.R. 1991a. Divergence and persistence of highly ranked taxa. In: *The Early Evolution of Metazoa and the Significance of Problematic Taxa*, A. Simonetta and S. Conway Morris (eds.). Cambridge University Press, Cambridge, pp. 15–18.

Strathmann, R.R. 1991b. From metazoan to protist via competition among cell lineages. *Evolutionary Theory* 10:67–70.

Strathmann, R.R., Fenaux, L. and Strathmann, M.F. 1992. Heterochronic developmental plasticity in larval sea urchins and its implications for evolution of non-feeding larvae. *Evolution* 46:972–986.

Strauss, R.E. 1984. Allometry and functional feeding morphology in haplochromine cichlids. In: *Evolution of Fish Species Flocks*, A.A. Echelle and I. Kornfield (eds.). University of Maine at Orono Press, Orono, pp. 217–229.

Strong, D.R., Jr. and Ray, T.S., Jr. 1975. Host tree location behavior of a tropical vine (*Monstera gigantea*) by skototropism. *Science* 190:804–806.

Sturtevant, A.H. 1913. The Himalayan rabbit case, with some considerations of multiple allelomorphs. *American Naturalist* 47:234–238.

Su, Z.H., Tominaga, O., Ohama, T., Kajiwara, E., Ishikawa, R., Okada, T.S., Nakamura, K. and Osawa, S. 1996. Parallel evolution in radiation of *Ohomopterus* ground beetles inferred from mitochondrial ND5 gene sequences. *Journal of Molecular Evolution* 43:662–671.

Sullivan, J.D. and Strassmann, J.E. 1984. Physical variability among nest foundresses in the polygynous social wasp, *Polistes annularis*. *Behavioral Ecology and Sociobiology* 15:249–256.

Sultan, S.E. 1987. Evolutionary implications of phenotypic plasticity in plants. *Evolutionary Biology* 20:127–178.

Sultan, S.E. 1992. Phenotypic plasticity and the neo-Darwinian legacy. *Evolutionary Trends in Plants* 6:61–71.

Sultan, S.E. 2000. Phenotypic plasticity for plant development, function and life history. *Trends in Plant Science* 5:537–542.

Summers, K. 1989. *Sexual Selection* and infra-female competition in the green dart-poison frog, *Dendrobates auratus*. *Animal Behaviour* 37:797–805.

Summers, K. 1990. Parental care and the cost of polygyny in the green dart-poison frog, *Dendrobates auratus*. *Behavioral Ecology and Sociobiology* 27:307–313.

Summers, K. 1992. Mating strategies in two species of dart-poison frogs: a comparative study. *Animal Behaviour* 43:907–919.

Summers, K. and Earn, D.J.D. 1999. The cost of polygyny and the evolution of female care in poison frogs. *Biological Journal of the Linnean Society* 66:515–538.

Summers, K., Weigt, L.A., Boag, P. and Bermingham, E. 1999. The evolution of female parental care in poison frogs of the genus *Dendrobates*: evidence from mitochondrial DNA sequences. *Herpetologica* 55:254–270.

Sundberg, M.D. 1987. Development of the mixed inflorescence in *Zea diploperennis* Iltis, Doebley & Guzman (Poaceae). *Botanical Journal of the Linnean Society* 95:207–216.

Sundberg, M.D. 1990. Inflorescence development in *Zea diploperennis* and related species. *Maydica* 35:99–111.

Sundberg, M.D. and Orr, A.R. 1986. Early inflorescence and floral development in *Zea diploperennis*, diploperennial teosinte. *American Journal of Botany* 73:1699–1712.

Sundberg, M.D. and Orr, A.R. 1990. Inflorescence development in two annual teosintes: *Zea mays* ssp. *mexicana* and *Z. mays* ssp. *parviglumis*. *American Journal of Botany* 77:141–152.

Sundstrom, L. 1990. Monogyny or polygyny—a result of different dispersal tactics in red wood ants (*Formica*; Hymenoptera). In: *Social Insects and the Environment*, G.K. Veeresh, B. Mallik, C.A. Viraktamath (eds.). Oxford and IBH Publishing, New Delhi, pp. 245–246.

Sundstrom, L. 1993. Genetic population structure and sociogenetic organisation in *Formica truncorum* (Hymenoptera; Formicidae). *Behavioral Ecology and Sociobiology* 33:345–354.

Suzuki, S. 1985. Temperature and thyroid function in amphibia, with particular reference to metamorphosis. In: *The Endocrine System and the Environment*, B.K. Follett, S. Ishii and A. Chandola (eds.). Japan Scientific Societies Press, Tokyo, pp. 71–77.

Svardson, G. 1961. Young sibling fish species in northwestern Europe. In: *Vertebrate Speciation*, W.F. Blair (ed.). University of Texas Press, Austin, pp. 498–513.

Swanson, W.J. and Vacquier, V.D. 2002. The rapid evolution of reproductive proteins. *Nature Reviews* 3:137–144.

Syren, R.M. and Luykx, P. 1977. Permanent segmental interchange complex in the termite *Incisitermes schwarzi*. *Nature* 266:167–168.

Szabo, V.M. and Burr, B. 1996. Simple inheritance of key traits distinguishing maize and teosinte. *Molecular and General Genetics* 252:33–41.

Taborsky, M. 1994. Sneakers, satellites, and helpers: parasitic and cooperative behavior in fish reproduction. *Advances in the Study of Behavior* 23:1–100.

Takhtajan, A. 1969. *Flowering Plants: Origin and Dispersal*. Oliver and Boyd, Edinburgh.

Takhtajan, A. 1976. Neoteny and the origin of flowering plants. In: *Origin and Early Evolution of Angiosperms*, C.B. Beck (ed.). Columbia University Press, New York, pp. 207–219.

Tanaka, S. 1976. Wing polymorphism, egg production and adult longevity in *Pteronemobius taprobanensis* Walker (Orthoptera, Gryllidae). *Kontyû* 44:327–333.

Tanaka, Y. 1985. Alternative manners of prey-carrying in the fossorial wasp, *Oxybelus strandi* Yasumatsu (Hymenoptera, Sphecoidea) *Kontyû* 53:277–283.

Tåning, Å.V. 1952. Experimental study of meristic characters in fishes. In: *Biological Reviews*, H.M. Fox (ed.). Cambridge University Press, Cambridge, pp. 169–193.

Tatewaki, M., Provasoli, L. and Pintner, I.J. 1983. Morphogenesis of *Monostroma oxyspermum* (Kutz.) Doty (Chlorophyceae) in axenic culture, especially in bialgal culture. *Journal of Phycology* 19:409–416.

Tattersall, I. 2000. Once we were not alone. *Scientific American*, January, 38–44.

Tauber, C.A. and Tauber, M.J. 1981. Insect seasonal cycles: genetics and evolution. *Annual Review of Ecology and Systematics* 12:281–308.

Tauber, C.A. and Tauber, M.J. 1982. Evolution of seasonal adaptations and life history traits in *Chrysopa*: response to diverse selective pressures. In: *Evolution and Genetics of Life Histories*, H. Dingle and J.P. Hegmann (eds.). Springer-Verlag, New York, pp. 51–72.

Tauber, C.A. and Tauber, M.J. 1989. Sympatric speciation in insects: perception and perspective. In: *Speciation and Its Consequences*, D. Otte and J.A. Endler (eds.). Sinauer, Sunderland, MA, pp. 307–344.

Tauber, C.A. and Tauber, M.J. 1992. Phenotypic plasticity in *Chrysoperla*: genetic variation in the sensory mechanism and in correlated reproductive traits. *Evolution* 46:1754–1773.

Tauber, C.A., Tauber, M.J. and Milbrath, L.R. 1995. Individual repeatability and geographical variation in the larval behaviour of the generalist predator, *Chrysopa quadripunctata. Animal Behaviour* 50:1391–1403.

Tauber, M.J., Tauber, C.A. and Masaki, S. 1986. *Seasonal Adaptations of Insects*. Oxford University Press, New York.

Tauber, M.J., Tauber, C.A., Ruberson, J.R., Milbrath, L.R. and Albuquerque, G.S. 1993. Evolution of prey specificity via three steps. *Experientia* 49:1113–1117.

Tautz, D. 1992. Redundancies, development and the flow of information. *BioEssays* 14:263–266.

Taylor, E.B. 1990. Phenotypic correlates of life-history variation in juvenile chinook salmon, *Oncorhynchus tshawytscha. Journal of Animal Ecology* 59:455–468.

Taylor, E.B. and Bentzen, P. 1993. Evidence for multiple origins and sympatric divergence of trophic ecotypes of smelt (*Osmerus*) in northeastern North America. *Evolution* 47:813–832.

Taylor, E.B. and McPhail, J.D. 2000. Historical contingency and determinism interact to prime speciation in sticklebacks. *Proceedings of the Royal Society of London* 271:2375–2384.

Taylor, E.B., Foote, C.J. and Wood, C.C. 1996. Molecular genetic evidence for parallel life-history evolution within a Pacific salmon (sockeye salmon and kokanee, *Oncorhynchus nerka*). *Evolution* 50:401–416.

Taylor, P.D. and Larwood, G.P. 1990. Major evolutionary radiations in the Bryozoa. In: *Major Evolutionary Radiations*, P.D. Taylor and G.P. Larwood (eds.). Systematic Association Special Volume No. 42. Clarendon Press, Oxford, pp. 209–233.

Taylor, P.J. 1987. Historical versus selectionist explanations in evolutionary biology. *Cladistics* 3:1–13.

Taylor, T.N. 1982. The origin of land plants: a paleobotanical perspective. *TAXON* 31:155–177.

Tchernavin, V. 1938. The mystery of a salmon's kype. *Salmon and Trout Magazine* 90:37–44.

Tchernavin, V. 1944. The breeding characters of salmon in relation to their size. *Proceedings of the Zoological Society of London* 113:206–232.

Tchernichovski, O. and Nottebohm, F. 1998. Social inhibition of song imitation among sibling male zebra finches. *Proceedings of the National Academy of Sciences USA* 95:8951–8956.

Telewski, F.W. and Jaffe, M.J. 1986. Thigmomorphogenesis: field and laboratory studies of *Abies fraseri* in response to wind

or mechanical perturbation. *Physiologia Plantarum* 66:211–218.

Temeles, E.J. and Roberts, W.M. 1993. Effect of sexual dimorphism in bill length on foraging behavior: an experimental analysis of hummingbirds. *Oecologia* 94:87–94.

Templeton, A.R. 1981. Mechanisms of speciation—a population genetic approach. *Annual Review of Ecology and Systematics* 12:23–48.

Templeton, A.R. 1982a. Adaptation and the integration of evolutionary forces. In: *Perspectives on Evolution*, R. Milkman (ed.). Sinauer, Sunderland, MA, pp. 15–31.

Templeton, A.R. 1982b. Why read Goldschmidt? *Paleobiology* 8:474–481.

Templeton, A.R. 1989. The meaning of species and speciation: a genetic perspective. In: *Speciation and Its Consequences*, D. Otte and J.A. Endler (eds.). Sinauer, Sunderland, MA, pp. 3–27.

Terskikh, A.V., Le Doussal, J.-M., Crameri, R., Fisch, I., Mach, J.-P. and Kajava, A.V. 1997. "Peptabody": a new type of high avidity binding protein. *Proceedings of the National Academy of Sciences USA* 94:1663–1668.

Thaler, D.S. 1994. Sex is for sisters: intragenomic recombination and homology-dependent mutation as sources of evolutionary variation. *Trends in Ecology and Evolution* 9:108–110.

Thanthianga, C. and Mitchell, R. 1987. Vibrations mediate prudent resource exploitation by competing larvae of the bruchid bean weevil *Callosobruchus maculatus. Entomologia Experimentalis et Applicata* 44:15–21.

Thoday, J.M. 1955. Balance, heterozygosity and developmental stability. *Cold Spring Harbor Symposia on Quantitative Biology* 20:318–326.

Thoday, J.M. 1964. Genetics and the integration of reproductive systems. In: *Insect Reproduction*, K.C. Highnam (ed.). Royal Entomological Society of London, London, pp. 108–119.

Thoday, J.M. 1972. Disruptive selection. *Proceedings of the Royal Society of London B* 182:109–143.

Thoday, J.M. and Boam, T.B. 1959. Effects of disruptive selection. II. Polymorphism and divergence without isolation. *Heredity* 13:205–218.

Thoday, J.M. and Gibson, J.B. 1962. Isolation by disruptive selection. *Nature* 193:1164–1166.

Thomas, J.A. and Birney, E.C. 1979. Parental care and mating system of the prairie vole, *Microtus ochrogaster. Behavioral Ecology and Sociobiology* 5:171–186.

Thompson, C.S., Moore, I.T. and Moore, M.C. 1993. Social environmental and genetic factors in the ontogeny of phenotypic differentiation in a lizard with alternative male reproductive strategies. *Behavioral Ecology and Sociobiology* 33:137–146.

Thompson, D. 1988. Diet-induced variation in the mouth morphology of grasshoppers causes variation in feeding performance. Paper presented at the annual meeting of the Society for the Study of Evolution, Asilomar, California, 1988.

Thompson, D.B. 1992. Consumption rates and the evolution of diet-induced plasticity in the head morphology of *Melanoplus femurrubrum* (Orthoptera: Acrididae). *Oecologia* 89:204–213.

Thompson, D.W. 1961 [1992]. *On Growth and Form*. Cambridge University Press, Cambridge.

Thompson, J.D. 1991. Phenotypic plasticity as a component of evolutionary change. *Trends in Ecology and Evolution* 6:246–249.

Thompson, J.N. 1994. *The Coevolutionary Process*. University of Chicago Press, Chicago.

Thompson-Stewart, D., Karpen, G.H. and Spradling, A.C. 1994. A transposable element can drive the concerted evolution of tandemly repetitious DNA. *Proceedings of the National Academy of Sciences USA* 91:9042–9046.

Thomson, J.A. 1987. Evolution of gene structure in relation to function. In: *Rates of Evolution*, K.S.W. Campbell and M.F. Day (eds.). Allen and Unwin, London, pp. 189–208.

Thomson, K.S. 1988. *Morphogenesis and Evolution*. Oxford University Press, New York.

Thorndyke, M.C. 1988. Molecular diversity and conformity of neurohormonal peptides: clues to an adaptive role in evolution. *Biological Journal of the Linnean Society* 34:249–267.

Thornhill, R. 1979. Adaptive female-mimicking behavior in a scorpionfly. *Science* 295:412–414.

Thornhill, R. and Alcock J. 1983. *The Evolution of Insect Mating Systems*. Harvard University Press, Cambridge.

Thorpe, J.E. 1989. Developmental variation in salmonid populations. *Journal of Fish Biology* 35A:S295–S303.

Throckmorton, L.H. 1965. Similarity *versus* relationship in *Drosophila*. *Systematic Zoology* 14:221–236.

Tierney, A.J. 1986. The evolution of learned and innate behavior: contributions from genetics and neurobiology to a theory of behavioral evolution. *Animal Learning and Behavior* 14:339–348.

Tilmon, K.J., Wood, T.K. and Pesek, J.D. 1998. Genetic variation in performance traits and the potential for host shifts in *Enchenopa* treehoppers (Homoptera: Membracidae). *Ecology and Population Biology* 91:397–403.

Tilney, L.G., Tilney, M.S. and DeRosier, D.J. 1992. Actin filaments, stereocilia, and hair cells: how cells count and measure. *Annual Review of Cell Biology* 8:257–274.

Tinbergen, N. 1951. *The Study of Instinct*. Clarendon, Oxford.

Tinbergen, N. 1952. "Derived" activities; their causation, biological significance, origin, and emancipation during evolution. *Quarterly Review of Biology* 27:1–32.

Ting, I.P., Patel, A., Kaur, S., Hann, J. and Walling, L. 1996. Ontogenetic development of crassulacean acid metabolism as modified by water stress in *Peperomia*. In: *Crassulacean Acid Metabolism*, K. Winter and J.A.C. Smith (eds.). Springer-Verlag, New York, pp. 204–215.

Tobin, T.R. and Bell, W.J. 1986. Chemo-orientation of male *Trogoderma variabile* (Coleoptera, Dermestidae) in a simulated corridor of female sex pheromones. *Journal of Comparative Physiology A* 158:729–739.

Tollrian, R. and Harvell C.D. (eds.). 1999. *The Ecology and Evolution of Inducible Defenses*. Princeton Unviersity Press, Princeton, N.J.

Tompkins, R. 1978. Genic control of *Axolotol* metamorphosis. *American Zoologist* 18:313–319.

Tooby, J. and Cosmides, L. 1992. The psychological foundations of culture. In: *The Adapted Mind: Evolutionary Psychology and the Generation of Culture*, J.H. Barkow, L. Cosmides and J. Tooby (eds.). Oxford University Press, New York, pp. 19–136.

Townsend, D.S. and Moger, W.H. 1987. Plasma androgen levels during male parental care in a tropical frog (*Eleutherodactylus*). *Hormones and Behavior* 21:93–99.

Travis, J. 1981. Control of larval growth variation in a population of *Pseudacris triseriata* (Anura: Hylidae). *Evolution* 35:423–432.

Travis, J. 1994. Size-dependent behavioral variation and its genetic control within and among populations. In: *Quantitative Genetic Studies of Behavioral Evolution*, C.R. Boake (ed.). University of Chicago Press, Chicago, pp. 165–187.

Trembley, A. 1744. *Mémoires pour Servir à l'Histoire d'un Genre de Polypes d'Eau Douce, à Bras en Forme de Cornes*. Leide.

Trewavas, A.J. and Jennings, D.H. 1986. Introduction. In: *Plasticity in Plants*, D.H.

Jennings and A.J. Trewavas (eds.). Symposia of the Society for Experimental Biology, No. 40. Company of Biologists Limited, Cambridge, pp. 1–4.

Trimen, R. 1874. Observations on the case of *Papilio merope*. Auct. with an account of the curious known forms of that butterfly. *Transactions of the Entomological Society of London 1874, Part I*:137–153.

Tripathi, R.K. and Dixon, S.E. 1969. Changes in some haemolymph dehydrogenase isozymes of the female honeybee, *Apis mellifera* L., during caste development. *Canadian Journal of Zoology* 47:763–770.

Trivers, R.L. 1972. Parental investment and sexual selection. In: *Sexual Selection and the Descent of Man 1871–1971*, B. Campbell (ed.). Aldine, Chicago, pp. 136–179.

Trivers, R.L. 1974. Parent-offspring conflict. *American Zoologist* 14:249–264.

Trivers, R.L. and Burt, A. 1999. Kinship and genomic imprinting. In: *Genomic Imprinting: An Interdisciplinary Approach*, R. Ohlsson (ed.). Springer-Verlag, Berlin, pp. 1–21.

True, J.R. and Haag, E.S. 2001. Developmental system drift and flexibility in evolutionary trajectories. *Evolution and Development* 3:109–119.

Truman, J.W. 1988. Metamorphosis of the CNS: implications for understanding the development and diversity of the insect nervous system. In: *Proceedings XVIII International Congress of Entomology*. International Congress of Entomology, Vancouver, p. 3.

Truman, J.W. and Riddiford, L.M. 1999. The origins of insect metamorphosis. *Nature* 401:447–452.

Trumler, E. 1959. Das "Rossigkeitsgesicht" und ähnliches Ausdrucksverhalten bei Einhufern. *Zeitschrift für Tierpsychologie.* 16:478–488.

Tschinkel, W.R. 1996. A newly-discovered mode of colony founding among fire ants. *Insectes Sociaux* 43:267–276.

Tsukamoto, K., Nakai, I. and Tesch, W.-V. 1998. Do all freshwater eels migrate? *Nature* 396:635–636.

Tuljapurkar, S.D. 1990. Delayed reproduction and fitness in variable environments. *Proceedings of the National Academy of Sciences USA* 87:1139–1143.

Tuomi, J. 1991. Toward integration of plant defence theories. *Trends in Ecology and Evolution* 7:365–367.

Tuomi, J. and Vuorisalo, T. 1989. Hierarchical selection in modular organisms. *Trends in Ecology and Evolution* 4:209–213.

Tuomikoski, R. 1967. Notes on some principles of phylogenetic systematics. *Suomen Hyönteistieteellinen Aikakauskirja*

33:137–147.

Turesson, G. 1922. The genotypical response of the plant species to the habitat. *Heriditas* 3:211–350.

Turillazzi, S. 1991. The Stenogastrinae. In: *The Social Biology of Wasps*, K.G. Ross and R.W. Matthews, (eds.). Cornell University Press, Ithaca, pp. 74–98.

Turillazzi, S. 1992. Nest usurpation and social parasitism in *Polistes* wasps: new acquisitions and current problems. In: *Biology and Evolution of Social Insects*, J. Billen (ed.). Leuven University Press, Leuven, pp. 263–272.

Turillazzi, S. and Pardi, L. 1977. Body size and hierarchy in polygynic nests of *Polistes gallicus* (L.) (Hymenoptera Vespidae). *Monitore Zoologico Italiano* 11:101–112.

Turillazzi, S. and West-Eberhard, M.J. (eds.) 1996. *Natural History and Evolution of Paper-Wasps.* Oxford University Press, Oxford.

Turner, G.F. 1999. Explosive speciation of African cichlid fishes. In: *Evolution of Biological Diversity*, A. Magurran and R.M. May (eds.). Oxford University Press, Oxford, pp. 113–129.

Turner, J.R.G. 1967. Why does the genome not congeal? *Evolution* 21:645–656.

Turner, J.R.G. 1977. Butterfly mimicry: the genetical evolution of an adaptation. *Evolutionary Biology* 10:163–207.

Turner, J.R.G. 1981. Adaptation and evolution in *Heliconius*: a defense of NeoDarwinism. *Annual Review of Ecology and Systematics* 12:99–121.

Turner, J.R.G. 1983. Mimetic butterflies and punctuated equilibria: some old light on a new paradigm. *Biological Journal of the Linnean Society* 20:277–300.

Twitty, V.C. 1932. Influence of the eye on the growth of its associated structures, studied by means of heteroplastic transplantations. *Journal of Experimental Zoology* 61:333–375.

Ugolini, A. and Scapini, F. 1988. Orientation of the sandhopper *Talitrus saltator* (Amphipoda, Talitridae) living on dynamic sandy shores. *Journal of Comparative Physiology* 162:453–462.

Underwood, G. 1954. Categories of adaptation. *Evolution* 8:365–377.

Underwood, H. and Goldman, B.D. 1987. Vertebrate circadian and photoperiodic systems: role of the pineal gland and melatonin. *Journal of Biological Rhythms* 2:279–315.

Urbanek, A. 1973. Organization and evolution of graptolite colonies. In: *Animal Colonies*, A.H. Cheetham and W.A. Oliver (eds.).

Dowden, Hutchinson and Ross, Stroudsburg, pp. 441–514.

Uyenoyama, M.K. 1988. On the evolution of genetic incompatibility systems: incompatibility as a mechanism for the regulation of outcrossing distance. In: *The Evolution of Sex*, R.E. Michod and B.R. Levin (eds.). Sinauer, Sunderland, MA, pp. 212–232.

Valentine, J.W. 1977a. General patterns of Metazoan evolution. In: *Patterns of Evolution, as Illustrated by the Fossil Record*, A. Hallam (ed.). Elsevier, Amsterdam, pp. 27–58.

Valentine, J.W. 1977b. The evolutionary history of Metazoa. In: *Evolution*, T. Dobzhansky, F.J. Ayala, G.L. Stebbins and J.W. Valentine (eds.). W.H. Freeman, San Francisco, pp. 397–437.

Valentine, J.W. 1990. The macroevolution of clade shape. In: *Causes of Evolution*, R.M. Ross and W.D. Allmon (eds.). University of Chicago Press, Chicago, pp. 128–150.

Valentine, J.W. 1994. Late Precambrian bilaterians: grades and clades. *Proceedings of the National Academy of Sciences USA* 91:6751–6757.

Valentine, J.W. and Campbell, C.A. 1975. Genetic regulation and the fossil record. *American Scientist* 63:673–680.

Valentine, J.W. and Erwin, D.H. 1987. Interpreting great developmental experiments: the fossil record. In: *Development as an Evolutionary Process*, R.A. Raff and E.C. Raff (eds.). Alan R. Liss, New York, pp. 71–107.

Valladares, F., Wright, S.J., Lasso, E., Kitajima, K. and Pearcy, R.W. 2000. Plastic phenotypic response to light of 16 congeneric shrubs from a Panamanian rainforest. *Ecology* 81:1925–1936.

Van Buskirk, J. and McCollum, S.A. 2000. Functional mechanisms of an inducible defence in tadpoles: morphology and behaviour influence mortality risk from predation. *Journal of Evolutionary Biology* 13:336–347.

Van Buskirk, J., McCollum, S.A. and Werner, E.E. 1997. Natural selection for environmentally induced phenotypes in tadpoles. *Evolution* 51:1983–1992.

Vance, S.A. 1996. Morphological and behavioural sex reversal in mermithid-infected mayflies. *Proceedings of the Royal Society of London B.* 263:907–912.

van den Berghe, E.P. and Gross, M.R. 1989. Natural selection resulting from female breeding competition in a Pacific salmon (coho: *Oncorhynchus kisutch*). *Evolution* 43:125–140.

van Noordwijk, A.J. and de Jong, G. 1986. Acquisition and allocation of resources: their influence on variation in life history tactics. *American Naturalist* 128:137–142.

Van Valen, L.M. 1960. Nonadaptive aspects of evolution. *American Naturalist* 94:305–308.

Van Valen, L.M. 1962a. A study of fluctuating asymmetry. *Evolution* 16:125–142.

Van Valen, L.M. 1965. The study of morphological integration. *Evolution* 19:347–349.

Van Valen, L.M. 1969. Variation genetics of extinct animals. *American Naturalist* 103:193–224.

Van Valen, L.M. 1971. Group selection and the evolution of dispersal. *Evolution* 25:591–598.

Van Valen, L.M. 1973. Continuous variation. *Systematic Zoology* 22:93.

Van Valen, L.M. 1974. A natural model for the origin of some higher taxa. *Journal of Herpetology* 8:109–121.

Van Valen, L.M. 1975. Group selection, sex and fossils. *Evolution* 29:87–94.

Van Valen, L.M. 1979. Switchback evolution and photosynthesis in angiosperms. *Evolutionary Theory* 4:143–146.

Van Valen, L.M. 1982a. Integration of species: stasis and biogeography. *Evolutionary Theory* 6:99–112.

Van Valen, L.M. 1982b. Homology and causes. *Journal of Morphology* 173:305–312.

Van Valen, L.M. 1986a. Why not to ignore Russian work (or the phenotype). *Evolutionary Theory* 8:61–64.

Van Valen, L.M. 1986b. Information and cause in evolution. *Evolutionary Theory* 8:65–68.

Van Valen, L.M. 1988. Species, sets, and the derivative nature of philosophy. *Biology and Philosophy* 3:49–66.

van Veen, J.W., Sommeijer, M.J. and Meeuwsen, F.J.A.J. 1997. Behaviour of drones in *Melipona* (Apidae, Meliponinae). *Insectes Sociaux* 44:435–447.

Vaux, D.L. and Strasser, A. 1996. The molecular biology of apoptosis. *Proceedings of the National Academy of Sciences USA*. 93:2239–2244.

Vavilov, N.I. 1935 [1951]. The origin, variation, immunity and breeding of cultivated plants. Selected writings of N.I. Vavilov. *Chronica Botanica, An International Collection of Studies in the Method and History of Biology and Agriculture* 13:1–54

Vehrencamp, S.L. 1983a. Optimal degree of skew in cooperative societies. *American Zoologist* 23:327–335.

Vehrencamp, S.L. 1983b. A model for the evolution of despotic versus egalitarian societies. *Animal Behaviour* 31:667–682.

Velthuis, H.H.W. 1976a. Egg laying, aggression and dominance in bees. *Proceedings XV International Congress of Entomology.* 1976:436–449.

Velthuis, H.H.W. 1976b. Environmental, genetic and endocrine influences in stingless bee caste determination. In: *Phase and Caste Determination in Insects*, M. Lüscher (ed.). Pergamon, Oxford, pp. 35–53.

Velthuis, H.H.W. 1977. The evolution of honeybee queen pheromones. In: *Proceedings of the 8th International Congress of the International Union for the Study of Social Insects*, J. de Wilde (ed.). Centre for Agricultural Publishing and Documentation, Wageningen, pp. 220–222.

Velthuis, H.H.W. and Sommeijer, M.J. 1990. A new model for the genetic determination of caste in *Melipona*. In: *Social Insects and the Environment*, G.K. Veeresh, B. Mallik and C.A. Viraktamath (eds.). Oxford and IBH Publishing, New Delhi, p. 173.

Velthuis, H.H.W. and Sommeijer, M.J. 1991. Roles of morphogenetic hormones in caste polymorphism in stingless bees. In: *Morphogenetic Hormones of Arthropods*, A.P. Gupta (ed.). Rutgers University Press, New Brunswick, pp. 346–383.

Velthuis, H.H.W., Ruttner, F. and Crewe, R.M. 1990. Differentiation in reproductive physiology and behaviour during the development of laying worker honey bees. In: *Social Insects: An Evolutionary Approach to Castes and Reproduction*, W. Engles (ed.). Springer-Verlag, Berlin, pp. 231–243.

Venable, D.L. 1985. The evolutionary ecology of seed heteromorphism. *American Naturalist* 126:577–595.

Vermeij, G.J. 1973a. Adaptation, versatility, and evolution. *Systematic Zoology* 22:466–477.

Vermeij, G.J. 1973b. Biological versatility and earth history. *Proceedings of the National Academy of Sciences USA* 70:1936–1938.

Vermeij, G.J. 1995. Economics, volcanoes, and Phanerozoic revolutions. *Paleobiology* 21:125–152.

Vermeij, G.J. 1996. Adaptations of clades: resistance and response. In: *Adaptation*, M.R. Rose and G.V. Lauder (eds.). Academic Press, New York, pp. 363–380.

Via, S. 1986. Quantitative genetic analysis of feeding and oviposition behavior in the polyphagous leafminer *Liriomyza sativae*. In: *Evolutionary Genetics of Invertebrate Behavior*, M.D. Huettel (ed.). Plenum Press, New York, pp. 185–196.

Via, S. 1993a. Adaptive phenotypic plasticity: target or by-product of selection in a variable environment? *American Naturalist* 142:352–365.

Via, S. 1993b. Regulatory genes and reaction norms. *American Naturalist* 142:374–378.

Via, S. and Lande, R. 1985. Genotype-environment interaction and the evolution of phenotypic plasticity. *Evolution* 39:505–522.

Via, S., Gomulkiewicz, R., de Jong, G., Scheiner, S.M., Schlichting, C.D. and Van Tienderen, P.H. 1995. Adaptive phenotypic plasticity: consensus and controversy. *Trends in Ecology and Evolution* 10:212–217.

Villalobos, E.M. and Shelly, T.E. 1996. Intraspecific nest parasitism in the sand wasp *Stictia heros* (Fabr.) (Hymenoptera: Sphecidae). *Journal of Insect Behavior* 9:105–119.

Vincek, V., O'Huigin, C., Satta, Y., Takahata, N., Boag, P.T., Grant, P.R., Grant, B.R. and Klein, J. 1997. How large was the founding population of Darwin's finches? *Proceedings of the Royal Society of London B* 264:111–118.

Visscher, P.K. 1998. Colony integration and reproductive conflict in honey bees. *Apidologie* 29:23–45.

Visscher, P.K. and Camazine, S. 1999. Collective decisions and cognition in bees. *Nature* 397:400.

Visscher, P.K. and Dukas, R. 1995. Honey bees recognize development of nestmates' ovaries. *Animal Behaviour* 49:542–544.

Vogel, S. 1988. *Life's Devices—The Physical World of Animals and Plants*. Princeton University Press, Princeton.

Vogt, R.C. and Bull, J.J. 1984. Ecology of hatchling sex ratio in map turtles. *Ecology* 65:582–587.

Vowles, D.M. 1954. The orientation of ants. I. The substitution of stimuli. *Journal of Experimental Biology* 31:341–355.

Vrba, E.S. 1980. Evolution, species and fossils: how does life evolve? *South African Journal of Science* 76:61–84.

Vrba, E.S. and Gould, S.J. 1986. The hierarchical expansion of sorting and selection: sorting and selection cannot be equated. *Paleobiology* 12:217–228.

Vuilleumier, F. 1984. Evolutionary biology in France: a review of some recent books. *Quarterly Review of Biology* 59:139–159.

Waagen, W. 1869. Die Formenreihe des *Ammonites subradiatus. Benecke's Geognost. Palaeontol. Beitrage* 2:179–259.

Waddington, C.H. 1940. *Organisers and Genes*. Cambridge University Press, Cambridge.

Waddington, C.H. 1942. The canalization of development and the inheritance of acquired characters. *Nature* 150:563–565.

Waddington, C.H. 1952. Selection of the genetic basis for an acquired character. *Nature* 169:278.

Waddington, C.H. 1953a. Epigenetics and evolution. *Symposia Society for Experimental Biology* 7:186–199.

Waddington, C.H. 1953b. Genetic assimilation of an acquired character. *Evolution* 7:118–126.

Waddington, C.H. 1956. *Principles of Embryology*. Macmillan, New York.

Waddington, C.H. 1959. Evolutionary adaptation. In: *Evolution after Darwin*. University of Chicago Press, Chicago, pp. 381–402.

Waddington, C.H. 1961. Genetic assimilation. *Advances in Genetics* 10:257–290.

Waddington, C.H. 1974. A catastrophe theory of evolution. *Annals of the New York Academy of Sciences* 231:32–42.

Waddington, C.H. 1975. *The Evolution of an Evolutionist*. Cornell University Press, Ithaca.

Waddington, K.D. 1981. Patterns of size variation in bees and evolution of communication systems. *Evolution* 35:813–814.

Waddington, K.D. 1983. Foraging behavior of pollinators. In: *Pollination Biology*, L. Real (ed.). Academic Press, New York, pp. 213–239.

Waddington, K.D. 1988. Body size, individual behavior, and social behavior in honey bees. In: *Interindividual Behavioral Variability in Social Insects*, R.L. Jeanne (ed.). Westview Press, Boulder, pp. 385–418.

Waddington, K.D. 1989. Implications of variation in worker body size for the honey bee recruitment system. *Journal of Insect Behavior* 2:91–103.

Waddington, K.D. and Herbst, L.H. 1987. Body size and the functional length of the proboscis of honey bees. *Florida Entomologist* 70:124–128.

Wade, M.J. 1984. The evolution of insect mating systems. *Evolution* 38:706–708.

Wade, M.J. 1998. The evolutionary genetics of maternal effects. In: *Maternal Effects as Adaptations*, T.A. Mousseau and C.W. Fox (eds.). Oxford University Press, New York, pp. 5–21.

Wagner, D.L. and Liebherr, J.K. 1992. Flightlessness in insects. *Trends in Ecology and Evolution* 7:216–220.

Wagner, G.P. 1989a. The origin of morphological characters and the biological basis of homology. *Evolution* 43:1157–1171.

Wagner, G.P. 1989b. The biological homology concept. *Annual Review of Ecology and Systematics* 20:51–69.

Wagner, G.P. 1994. Evolution and multi-functionality of the chitin system. In: *Molecular Ecology and Evolution: Approaches and Applications*, B. Schierwater, B. Streit, G.P. Wagner and R.

DeSalle (eds.). Birkhäuser Verlag, Basel, pp. 559–577.

Wagner, G.P. 1995. The biological role of homologues: a building block hypothesis. *Neues Jahrbuch für Geologie und Päleontologie Abhandlungen* 195:279–288.

Wagner, G.P. 1996. Homologues, natural kinds and the evolution of modularity. *American Zoologist* 36:36–43.

Wagner, G.P. and Altenberg, L. 1996. Complex adaptations and the evolution of evolvability. *Evolution* 50:967–976.

Wagner, G.P. and Misof, B.Y. 1993. How can a character be developmentally constrained despite variation in developmental pathways. *Journal of Evolutionary Biology* 6:449–455.

Wagner, G.P., Lo, J., Laine, R. and Almeder, M. 1993. Chitin in the epidermal cuticle of a vertebrate (*Paralipophrys trigloides*, Blenniidae, Teleostei). *Experientia* 49:317–319.

Wainwright, P.C., Osenberg, C.W. and Mittelbach, G.G. 1991. Trophic polymorphism in the pumpkinseed sunfish (*Lepomis gibbosus* Linnaeus): effects of environment on ontogeny. *Functional Ecology* 5:40–55.

Wainwright, S.A., Biggs, W.D., Currey, J.D. and Gosline, J.M. 1976. *Mechanical Design in Organisms*. Edward Arnold, London.

Wake, D.B. 1982. Functional and developmental constraints and opportunities in the evolution of feeding systems in *Urodeles*. In: *Environmental Adaptation and Evolution*, D. Mossakowski and G. Roth (eds.). Gustav Fischer, Stuttgart, pp. 51–66.

Wake, D.B. 1991. Homoplasy: the result of natural selection, or evidence of design limitations. *American Naturalist* 138:543–567.

Wake, D.B. 1994. Comparative terminology. *Science* 265:268–269.

Wake, D.B. 1996a. Introduction. In: *Homoplasy: The Recurrence of Similarity in Evolution*, M.J. Sanderson and L. Hufford (eds.). Academic Press, San Diego, pp. xvii–xxv.

Wake, D.B. 1996b. Evolutionary developmental biology—prospects for an evolutionary synthesis at the developmental level. *Memoirs of the California Academy of Sciences* 20:97–107.

Wake, D.B. 1996c. Schmalhausen's evolutionary morphology and its value in formulating research strategies. *Memorie della Società Italiana de Scienze Naturali e del Museo Civico di Storia Naturale, Milano* 27:129–132.

Wake, D.B. and Roth, G. 1989. The linkage between ontogeny and phylogeny in the evolution of complex systems. In: *Complex*

Organismal Functions: Integration and Evolution in Vertebrates, D.B. Wake and G. Roth (eds.). John Wiley and Sons, New York, pp. 361–377.

Wake, D.B., Roth, G. and Wake, M.H. 1983. On the problem of stasis in organismal evolution. *Journal of Theoretical Biology* 101:211–224.

Wake, M.H. and Wake, D.B. 1988. A provocative view of the evolutionary role of genetic assimilation. *Genetica* 76:236–238.

Walbot, V. 1983. Morphological and genomic variation in plants: *Zea mays* and its relatives. In: *Development and Evolution*, B.C. Goodwin, N. Holder and C.C. Wylie (eds.). Cambridge University Press, Cambridge, pp. 257–277.

Walbot, V., Chandler, V.L., Taylor, L.P. and McLaughlin, P. 1987. Regulation of transposable element activities during the development and evolution of *Zea mays* L. In: *Development as an Evolutionary Process*, R.A. Raff and E.C. Raff (eds.). Alan R. Liss, New York, pp. 265–284.

Walcott, C.D. 1911. Middle Cambrian Holothurians and Medusae. Cambrian geology and Paleontology II. *Smithsonian Miscellaneous Collections* 57:41–68.

Walcott, C.D. 1912. Middle Cambrian Branchiopoda, Malacostraca, Trilobita and Merostomata. Cambrian geology and paleontology II. *Smithsonian Miscellaneous Collections* 57:145–228.

Waldman, B. and Adler, K. 1979. Toad tadpoles associate preferentially with siblings. *Nature* 282:611–613.

Walker, I. 1979. The mechanical properties of proteins determine the laws of evolutionary change. *Acta Biotheoretica* 28:239–282.

Walker, I. 1983. Complex-irreversibility and evolution. *Experientia* 39:806–813.

Walker, I. 1996. Prediction of evolution? Somatic plasticity as a basic, physiological condition for the viability of genetic mutations. *Acta Biotheoretica* 44:165–168.

Walker, I. and Williams, R.M. 1976. The evolution of the cooperative group. *Acta Biotheoretica* 25:1–43.

Walker, J.A. 1997. Ecological morphology of lacustrine threespine stickleback *Gasterosteus aculeatus* L. (Gasterosteidae) body shape. *Biological Journal of the Linnean Society* 61:3–50.

Walker, J.A. and Bell, M.A. 2000. Net evolutionary trajectories of body shape evolution within a microgeographic radiation of threespine sticklebacks (*Gasterosteus aculeatus*). *Journal of Zoology, London* 252:293–302.

Walker, T.J. 1974. Character displacement and acoustic insects. *American Zoologist* 14:1137–1150.

Walker, T.J. 1983. Diel patterns of calling in nocturnal Orthoptera. In: *Orthopteran Mating Systems: Sexual Competition in a Diverse Group of Insects*. D.T. Gwynne and G.K. Morris (eds.). Westview Press, Boulder, pp. 45–72.

Walker, T.J. 1986. Stochastic polyphenism: coping with uncertainty. *Florida Entomologist* 69:46–62.

Walker, T.J. 1987. Wing dimorphism in *Gryllus rubens* (Orthoptera: Gryllidae). *Annals of the Entomological Society of America* 80: 547–560.

Wallace, B. 1986. Can embryologists contribute to an understanding of evolutionary mechanisms? In: *Integrating Scientific Disciplines*, W. Bechtel (ed.). Martinus Nihoff, Dordrecht, pp. 149–163.

Wallace, B. 1989. Populations and their place in evolutionary biology. In: *Evolutionary Biology at the Crossroads*, M.K. Hecht (ed.). Queens College Press, New York, pp. 21–58.

Walsh, B.D. 1864. On phytophagic varieties and phytophagic species. *Proceedings of the Entomological Society of Philadelphia* 3:403–430.

Walsh, B.D. 1867. The apple-worm and the apple-maggot. *Journal of Horticulture* 2:338–343.

Walsh, J.B. 1987. Sequence-dependent gene conversion: can duplicated genes diverge fast enough to escape conversion? *Genetics* 117:543–557.

Walters, J.R. 1987. Transition to adulthood. In: *Primate Societies*, B.B. Smuts, D.L. Cheney, R.M. Seyfarth, R.W. Wrangham and T.T. Struhsaker (eds.). University of Chicago Press, Chicago, pp. 358–369.

Wang, R.-L., Stec, A., Hey, J., Lukens, L. and Doebley, J. 1999. The limits of selection during maize domestication. *Nature* 398:236–239.

Wang, Z., Ferris, C.F. and De Vries, G.J. 1994. Role of septal vasopressin innervation in paternal behavior in prairie voles (*Microtus ochrogaster*). *Proceedings of the National Academy of Sciences USA* 90:400–404.

Wang, Z. and Novak, M.A. 1992. Influence of the social environment on parental behavior and pup development of meadow voles (*Microtus pennsylvanicus*) and prairie voles (*M. ochrogaster*). *Journal of Comparative Psychology* 106:163–171.

Ward, P.S. 1989. Genetic and social changes associated with ant speciation. In: *The Genetics of Social Evolution*, M.D. Breed and R.E. Page Jr. (eds.). Westview Press, Boulder, pp. 123–148.

Warkentin, K.M. 1995. Adaptive plasticity in hatching age: a response to predation risk trade-offs. *Proceedings of the National Academy of Sciences USA* 92:3507–3510.

Warkentin, K.M. 1999a. Effects of hatching age on development and hatchling morphology in the red-eyed treefrog, *Agalychnis callidryas*. *Biological Journal of the Linnean Society* 68:443–470.

Warkentin, K.M. 1999b. The development of behavioral defenses: a mechanistic analysis of vulnerability in red-eyed tree frog hatchlings. *Behavioral Ecology* 10:251–262.

Warkentin, K.M. 2000. Wasp predation and wasp-induced hatching of red-eyed treefrog eggs. *Animal Behaviour* 20:503–510.

Warner, R.R. and Chesson, P.L. 1985. Coexistence mediated by recruitment fluctuations: a field guide to the storage effect. *American Naturalist* 125:769–787.

Warner, R.R., Robertson, D.R. and Leigh, E.G., Jr. 1975. Sex change and sexual selection. *Science* 190:633–638.

Watkinson, A.R. and White, J. 1985. Some life-history consequences of modular construction in plants. *Philosophical Transactions of the Royal Society of London B* 313:31–51.

Watson, M.A., Geber, M.A. and Jones, C.S. 1995. Ontogenetic contingency and the expression of plant plasticity. *Trends in Ecology and Evolution* 10:474–475.

Watt, W.B. 1968. Adaptive significance of pigment polymorphisms in *Colias* butterflies. I. Variation in melanin pigment in relation to thermoregulation. *Evolution* 22:437–458.

Wayne, R.K. 1986. Cranial morphology of domestic and wild canids: the influence of development on morphological change. *Evolution* 40:243–261.

Wcislo, W.T. 1987. The roles of seasonality, host synchrony, and behaviour in the evolution and distributions of nest parasites in Hymenoptera (Insecta), with special reference to bees (Apoidea). *Biological Reviews* 62:515–543.

Wcislo, W.T. 1989. Behavioral environments and evolutionary change. *Annual Review of Ecology and Systematics* 20:137–169.

Wcislo, W.T. 1990. Geographic variation in the development of parasitism in bees. In: *Social Insects and the Environment*, G.K. Veeresh, B. Mallik and C.A. Viraktamath (eds.). Oxford and IBH Publishing, New Delhi, pp. 155–156.

Wcislo, W.T. 1992. Attraction and learning in mate-finding by solitary bees, *Lasioglossum*

(*Dialictus*) *figueresi* Wcislo and *Nomia triangulifera* Vachal (Hymenoptera: Halictidae). *Journal of Zoology, London* 31:139–148.

Wcislo, W.T. 1997b. Behavioral environments of sweat bees (Halictinae) in relation to variability in social organization. In: *The Evolution of Social Behavior in Insects and Arachnids*, J.C. Choe and B.J. Crespi (eds.). Cambridge University Press, Cambridge, pp. 316–332.

Wcislo, W.T. 1999. Transvestism hypothesis: a cross-sex source of morphological variation for the evolution of parasitism among sweat bees (Hymenoptera: Halictidae)? *Annals of the Entomolgocial Society of America* 92:239–242.

Wcislo, W.T. and Cane, J.H. 1996. Floral resource utilization by solitary bees (Hymenoptera: Apoidea) and exploitation of their stored foods by natural enemies. *Annual Review of Entomology* 41:257–286.

Wcislo, W.T. and Danforth, B.N. 1997. Secondarily solitary: the evolutionary loss of social behavior. *Trends in Ecology and Evolution* 12:468–474.

Weatherbee, S.D., Nijhout, H.F., Grunert, L.W., Halder, G., Galant, R., Selegue, J. and Carroll, S. 1999. Ultrabithorax function in butterfly wings and the evolution of insect wing patterns. *Current Biology* 9:109–115.

Weaver, N. 1956. The foraging behavior of honeybees on hairy vetch. Foraging methods and learning to forage (I). *Insectes Sociaux* 3:537–549.

Weaver, N. 1957a. Effects of larval age on dimorphic differentiation of the female honey bee. *Annals of the Entomological Society of America* 50:283–294.

Weaver, N. 1957b. The foraging behavior of honeybees on hairy vetch. II. The foraging area and foraging speed (I). *Insectes Sociaux* 4:43–57.

Wecker, S.C. 1963. The role of early experience in habitat selection by the prairie deer mouse, *Peromyscus maniculatus bairdi*. *Ecological Monographs* 33:307–325.

Wei, G. and Mahowald, A.P. 1994. The germline: familiar and newly uncovered properties. *Annual Review of Genetics* 28:309–324.

Weidmann, J. 1969. The heteromorphs and ammonoid extinction. *Biological Review of the Cambridge Philosophical Society* 44:563–602.

Weismann, A. 1892. *Das Keimplasma: Eine Theorie der Vererbung*. Gustav Fischer, Jena

Weismann, A. 1893. *The Germ-Plasm: A Theory of Heredity.* Walter Scott, London. [English version of Weismann, A. 1892. *Das Keimplasma: Eine Theorie der Vererbung.* Gustav Fischer, Jena.]

Weismann, A. 1904. *The Evolution Theory.* Edward Arnold, London.

Weiss, K.M. 1990. Duplication with variation: metameric logic in evolution from genes to morphology. *Yearbook of Physical Anthropology* 33:1–23.

Weitzman, S.H. and Fink, S.V. 1985. Xenurobryconin phylogeny and putative pheromone pumps in glandulocaudine fishes (Teleostei: Characidae). *Smithsonian Contributions to Zoology* 421:1–121.

Weitzman, S.H. and Fink, W.L. 1983. Relationships of the neon tetras, a group of South American freshwater fishes (Teleostei, Characidae), with comments on the phylogeny of New World characiforms. *Bulletin of the Museum of Comparative Zoology* 150:339–395.

Weitzman, S.H. and Vari, R.P. 1988. Miniaturization in South American freshwater fishes; an overview and discussion. *Proceedings of the Biological Society of Washington* 101:444–465.

Wells, B.W. 1915. A survey of the zoocecidia on species of *Hicoria* caused by parasites belonging to the Eriophyidae and the Itonididae (Cecidomyiidae). *Ohio Journal of Science* 16:37–57.

Wells, K.W. 1981. Parental behavior of male and female frogs. In: *Natural Selection and Social Behavior*, R.D. Alexander and D.W. Tinkle (eds.). Chiron Press, New York, pp. 184–197.

Wells, M.M. 1994. Small genetic distances among populations of green lacewings of the genus *Chrysoperla* (Neuroptera: Chrysopidae). *Annals of the Entomological Society of America* 87:737–744.

Wells, M.M. and Henry, C.S. 1992. The role of courtship songs in reproductive isolation among populations of green lacewings of the genus *Chrysoperla* (Neuroptera: Chrysopidae). *Evolution* 46:31–42.

Wells, M.M. and Henry, C.S. 1994. Behavioral responses of hybrid lacewings (Neuroptera: Chrysopidae) to courtship songs. *Journal of Insect Behavior* 7:649–662.

Welty, J.C. 1962. *The Life of Birds.* W.B. Saunders, Philadelphia.

Wenzel, J.W. 1991. Evolution of nest architecture. In: *The Social Biology of Wasps*, K.G. Gross and R.W. Matthews (eds.). Comstock, Ithaca, NY, pp. 480–519.

Wenzel, J.W. 1992a. Behavioral homology and phylogeny. *Annual Review of Ecology and Systematics* 23:361–381.

Wenzel, J.W. 1992b. Extreme queen-worker dimorphism in *Ropalidia ignobilis*, a small-colony wasp (Hymenoptera: Vespidae). *Insectes Sociaux* 39:31–43.

Wenzel, J.W. 1993. Application of the biogenetic law to behavioral ontogeny: a test using nest architecture in paper wasps. *Journal of Evolutionary Biology* 6:229–247.

Wenzel, J.W. 1996. Learning, behaviour programs, and higher level rules in nest construction of *Polistes*. In: *Natural History and Evolution of Paper-Wasps*, S. Turillazzi and M.J. West-Eberhard (eds.). Oxford University Press, Oxford, pp. 58–74.

Wenzel, J.W. and Carpenter, J.M. 1994. Comparing methods: adaptive traits and tests of adaptation. In: *Phylogenetics and Ecology*, P. Eggleton and R.I. Vane-Wright (eds.). Academic Press, New York, pp. 79–101.

Werner, E.E. 1986. Amphibian metamorphosis: growth rate, predation risk, and the optimal size at transformation. *American Naturalist* 128:319–341.

Werner, E.E. 1988. Size, scaling, and the evolution of complex life cycles. In: *Size-Structured Populations*, B. Ebenman and L. Persson (eds.). Springer-Verlag, New York, pp. 60–84.

Werner, E.E. and Gilliam, J.F. 1984. The ontogenetic niche and species interactions in size-structured populations. *Annual Review of Ecology and Systematics* 15:393–425.

Werner, E.E., Mittelbach, G.G. and Hall, D.J. 1981. The role of foraging profitability and experience in habitat use by the bluegill sunfish. *Ecology* 62:116–125.

Werner, T.K. 1988. *Behavioral, individual feeding specializations by* Pinaroloxias inornata, *the Darwin's finch of Cocos Island, Costa Rica.* Ph.D. dissertation in Zoology, University of Massachusetts, Amherst.

Werner, T.K. and Sherry, T.W. 1987. Behavioral feeding specialization in *Pinaroloxias inornata*, the "Darwin's finch" of Cocos Island, Costa Rica. *Proceedings of the National Academy of Sciences USA* 84:5506–5510.

Werner, Y.L. 1980. Apparent homosexual behaviour in an all-female population of a lizard, *Lepidodactylus lugubris* and its probable interpretation. *Zeitschrift für Tierpsychologie* 54:144–150.

Werren, J.H. 1997. Biology of *Wolbachia*. *Annual Review of Entomology* 42:587–609.

Wessler, S.R. 1988. Phenotypic diversity mediated by the maize transposable elements *Ac* and *Spm*. *Science* 242:399–406.

Wessler, S.R. 1996. Turned on by stress: plant retrotransposons. *Current Biology* 6:959–961.

West, D.A. and Hazel, W.N. 1985. Pupal colour dimorphism in swallowtail butterflies: timing of the sensitive period and environmental control. *Physiological Entomology* 10:113–119.

West, M.J. 1967. Foundress associations in polistine wasps: dominance hierarchies and the evolution of social behavior. *Science* 157:1584–1585.

West, M.J., King, A.P. and Freeberg, T.M. 1994. The nature and nurture of neo-phenotypes: a case history. In: *Behavioral Mechanisms in Evolutionary Ecology*, L.A. Real (ed.). University of Chicago Press, Chicago, pp. 238–257.

West-Eberhard, M.J. 1969. The social biology of polistine wasps. Miscellaneous Publications No. 140. Museum of Zoology, University of Michigan, Ann Arbor.

West-Eberhard, M.J. 1975. The evolution of social behavior by kin selection. *Quarterly Review of Biology* 50:1–33.

West-Eberhard, M.J. 1977. The establishment of the dominance of the queen in social wasp colonies. *Proceedings VII Congress International Union for the Study of Social Insects, Wageningen:* 223–227.

West-Eberhard, M.J. 1978a. Polygyny and the evolution of social behavior in wasps. *Journal of the Kansas Entomological Society* 51:832–856.

West-Eberhard, M.J. 1978b. Temporary queens in *Metapolybia* wasps: non-reproductive helpers without altruism? *Science* 200(4340):441–443.

West-Eberhard, M.J. 1979. Sexual selection, social competition, and evolution. *Proceedings of the American Philosophical Society* 123:222–234.

West-Eberhard, M.J. 1981. Intragroup selection and the evolution of insect societies. In: *Natural Selection and Social Behavior*, R.D. Alexander and D.W. Tinkle (eds.). Chiron Press, New York, pp. 3–17.

West-Eberhard, M.J. 1982. Diversity of dominance displays in *Polistes* and its possible evolutionary significance. In: *The Biology of Social Insects*, M.D. Breed, C.D. Michener and H.E. Evans (eds.). Westview Press, Boulder, p. 222.

West-Eberhard, M.J. 1983. Sexual selection, social competition, and speciation. *Quarterly Review of Biology* 58:155–183.

West-Eberhard, M.J. 1984. Sexual selection, competitive communication and species-specific signals in insects. In: *Insect Communication*, T. Lewis (ed.). Academic Press, New York, pp. 283–324.

West-Eberhard, M.J. 1986. Alternative adaptations, speciation and phylogeny (a review). *Proceedings of the National Academy of Sciences USA* 83:1388–1392.

West-Eberhard, M.J. 1987. Flexible strategy and social evolution. In: *Animal Societies: Theories and Fact*, Y. Itô, L. Brown and L. Kikkawa (eds.). Japan Scientific Societies Press, Tokyo, pp. 35–51.

West-Eberhard, M.J. 1988. Phenotypic plasticity and "genetic" theories of insect sociality. In: *Evolution of Social Behavior and Integrative Levels*, T.C. Schneirla Conference Series, Vol. 3, G. Greenberg and E. Tobach (eds.). Erlbaum, Hillsdale, NJ, pp. 123–133.

West-Eberhard, M.J. 1989. Phenotypic plasticity and the origins of diversity. *Annual Review of Ecology and Systematics* 20:249–278.

West-Eberhard, M.J. 1990. Parallel size-related alternatives in males and females and the question of sympatric speciation in hymenopteran social parasites. In: *Social Insects and the Environment*, G.K. Veeresh, B. Mallik and C.A. Viraktamath (eds.). Oxford and IBH Publishing, New Delhi, pp. 162–163.

West-Eberhard, M.J. 1992a. Behavior and evolution. In: *Molds, Molecules and Metazoa: Growing Points in Evolutionary Biology*, P.R. Grant and H. Horn (eds.). Princeton University Press, Princeton, pp. 57–75.

West-Eberhard, M.J. 1992b. Genetics, epigenetics, and flexibility: a reply to Crozier. *American Naturalist* 139:224–226.

West-Eberhard, M.J. 1992c. Adaptation: current usages. In: *Keywords in Evolutionary Biology*, E.F. Keller and E.A. Lloyd (eds.). Harvard University Press, Cambridge, pp. 13–18.

West-Eberhard, M.J. 1996. Wasp societies as microcosms for the study of development and evolution. In: *Natural History and Evolution of Paper-Wasps*, S. Turillazzi and M.J. West-Eberhard (eds.). Oxford University Press, London, pp. 290–317.

West-Eberhard, M.J. 1998. Commentary: evolution in the light of development and cell biology, and *vice versa*. *Proceedings of the National Academy of Sciences USA* 95:8417–8419.

Westoby, M. and Rice, B. 1982. Evolution of seed plants and inclusive fitness of plant tissues. *Evolution* 36:713–724.

Wheeler, D.E. 1986. Developmental and physiological determinants of caste in social Hymenoptera: evolutionary implications. *American Naturalist* 128:13–34.

Wheeler, D.E. 1991. The developmental basis of worker caste polymorphism in ants. *American Naturalist* 138:1218–1238.

Wheeler, D.E. 1994. Nourishment in ants: patterns in individual societies. In: *Nourishment and Evolution in Insect Societies*, J.H. Hunt and C.A. Nalepa (eds.). Westview Press, Boulder, pp. 245–278.

Wheeler, Q.D. 1990. Ontogeny and character phylogeny. *Cladistics* 6:225–268.

Wheeler, W.M. 1910a. *Ants*. Columbia University Press, New York.

Wheeler, W.M. 1910b. The effects of parasitic and other kinds of castration in insects. *Journal of Experimental Zoology* 8:377–438.

Wheeler, W.M. 1937. *Mosaics and Other Anomalies among Ants*. Harvard University Press, Cambridge.

White, J. and Lloyd, M. 1975. Growth rates of 17- and 13-year periodical cicadas. *American Midland Naturalist* 94:127–143.

White, M.J.D. (ed.) 1978. *Modes of Speciation*. W.H. Freeman, San Francisco.

Whiteaves, J.F. 1892. Description of a new genus and species of Phyllocarid Crustacea from the Middle Cambrian of Mount Stephen B.C. *Canadian Record of Science* 5:205–208.

Whiteman, H.H. 1994. Evolution of facultative paedomorphosis in salamanders. *Quarterly Review of Biology* 69:205–221.

Whitt, G.S. 1980. Developmental genetics of fishes: isozymic analyses of differential gene expression. *American Zoologist* 21:549–572.

Whitten, P.L. 1987. Infants and adult males. In: *Primate Societies*, B.B. Smuts, D.L. Cheney, R.M. Seyfarth, R.W. Wrangham and T.T. Struhsaker (eds.). University of Chicago Press, Chicago, pp. 343–357.

Whittington, H.B. and Briggs, D.E.G. 1982. A new conundrum from the Middle Cambrian Burgess Shale. *Proceedings of the Third North American Paleontological Convention, Montreal* 2:573–575.

Whittington, H.B. and Briggs, D.E.G. 1985. The largest Cambrian animal, *Anomalocaris*, Burgess Shale, British Columbia. *Philosophical Transactions of the Royal Society of London, B* 309:569–609.

Whyte, L.L. 1965. *Internal Factors in Evolution*. Tavistock Publications, London.

Wibbels, T. and Crews, D. 1994. Putative aromatase inhibitor induces male sex determination in a female unisexual lizard and in a turtle with temperature-dependent sex determination. *Journal of Endocrinology* 141:295–299.

Wickler, W. 1965. Die äusseren genitalien als soziale signale bei einigen primaten. *Naturwissenschaften* 52:269–270.

Wickler, W. 1967. Socio-sexual signals and their intra-specific imitation among primates. In: *Primate Ethology*, D. Morris (ed.). Doubleday, New York, pp. 89–189.

Wickler, W. 1968. *Mimicry in Plants and Animals*. McGraw-Hill, New York.

Wiedenfeld, D.A. 1988. Ecomorphology and foraging behavior of the Yellow Warbler (*Dendroica Petechia*). Ph.D. dissertation, Florida State University, Tallahassee.

Wiens, J.J. 1995. Polymorphic characters in phylogenetic systematics. *Systematic Biology* 44:482–500.

Wiens, J.J. 1999. Polymorphism in systematics and comparative biology. *Annual Review of Ecology and Systematics* 30:327–362.

Wiens, J.J. and Servedio, M.R. 1997. Accuracy of phylogenetic analysis including and excluding polymorphic characters. *Systematic Biology* 46:332–345.

Wiernasz, D.C. and Kingsolver, J.G. 1992. Wing melanin pattern mediates species recognition in *Pieris occidentalis*. *Animal Behaviour* 43:89–94.

Wiernasz, D.C., Yencharis, J. and Cole, B.J. 1995. Size and mating success in males of the western harvester ant, *Pogonomyrmex occidentalis* (Hymenoptera: Formicidae). *Journal of Insect Behavior* 8:523–531.

Wigglesworth, V.B. 1936. The functions of the corpus allatum in the growth and reproduction of *Rhodnius prolixus* (Hemiptera). *Quarterly Journal of Microscopical Science* 79:91–121.

Wigglesworth, V.B. 1961. Insect polymorphism, a tentative synthesis. In: *Insect Polymorphism*, J.S. Kennedy (ed.). Royal Entomological Society, London, pp. 103–113.

Wikelski, M., Hau, M. and Wingfield, J.C. 1999. Social instability increases plasma testosterone in a year-round territorial neotropical bird. *Proceedings of the Royal Society of London B* 266:551–556.

Wilcox, R.S., Jackson, R.R. and Gentile, K. 1996. Spiderweb smokescreens: spider trickster uses background noise to mask stalking movements. *Animal Behaviour* 51:313–326.

Wilczek, F. 2002. Setting standards. *Nature* 415

Wilde, J. de and Beetsma, J. 1982. The physiology of caste development in social insects. *Advances in Insect Physiology* 16:167–246.

Wilkinson, G.S. 1993. Artificial selection alters allometry in the stalk-eyed fly *Cyrtodiopsis dalmanni* (Diptera: Diopsidae). *Genetical Research* 62:213–222.

Wilkinson, G.S. and Dodson, G.N. 1997. Function and evolution of antlers and eye stalks in flies. In: *Mating Systems in Insects and Arachnids*, J.C. Choe and B.J. Crespi (eds.). Cambridge University Press, Cambridge, pp. 310–328.

Williams, A. and Hurst, J.M. 1977. Brachiopod evolution. In: *Patterns of Evolution as Illustrated by the Fossil Record*, A. Hallam (ed.). Elsevier, New York, pp. 79–122.

Williams, C.M. 1958. Hormonal regulation of insect metamorphosis. In: *A Symposium on the Chemical Basis of Development*, W.D. McElroy and B. Glass (eds.). Johns Hopkins Press, Baltimore, pp. 794–806.

Williams, G.C. 1966. *Adaptation and Natural Selection*. Princeton University Press, Princeton.

Williams, G.C. 1975. *Sex and Evolution*. Princeton University Press, Princeton.

Williams, G.C. 1988. Retrospect on sex and kindred topics. In: *The Evolution of Sex*, R.E. Michod and B.R. Levin (eds.). Sinauer, Sunderland, MA, pp. 287–298.

Williams, G.C. 1992. *Natural Selection, Domains, Levels, and Applications*. Oxford University Press, Oxford.

Williams, K.S. and Simon, C. 1995. The ecology, behavior, and evolution of periodical cicadas. *Annual Review of Entomology* 40:269–295.

Williams, M.B. 1992. Species: current usages. In: *Keywords in Evolutionary Biology*, E. Fox Keller and E.A. Lloyd (eds.). Harvard University Press, Cambridge, pp. 318–323.

Williamson, P.G. 1987. Selection or constraint? A proposal on the mechanism for stasis. In: *Rates of Evolution*, K.S.W. Campbell and M.F. Day (eds.). Allen and Unwin, London, pp. 129–142.

Wills, C. 1983. The possibility of stress-triggered evolution. *Lecture Notes in Biomathematics* 53:299–312.

Willson, M.F. 1984. *Vertebrate Natural History*. Saunders College Publishing, Philadelphia.

Willson, M.F. 1994. Sexual Selection in plants: perspective and overview. *American Naturalist* 144:S13–S39.

Willson, M.F. and Burley, N. 1983. *Mate Choice in Plants*. Princeton University Press, Princeton.

Willson, M.F. and Comet, T.C.A. 1993. Food choices by northwestern crows: experiments with captive, free-ranging and hand-raised birds. *Condor* 95:596–615.

Wilson, D.S. 1975. A theory of group selection. *Proceedings of the National Academy of Sciences USA* 72:143–146.

Wilson, D.S. 1980. *The Natural Selection of Populations and Communities*. Benjamin-Cummings, Menlo Park, CA.

Wilson, D.S. 1989. The diversification of single gene pools by density- and frequency-dependent selection. In: *Speciation and Its Consequences*, D. Otte and J.A. Endler (eds.). Sinauer, Sunderland, MA, pp. 366–385.

Wilson D.S., Clark, A.B., Coleman, K. and Dearstyne, T. 1994. Shyness and boldness in humans and other animals. *Ecology and Evolution* 11:442–446.

Wilson, E.B. 1894. The embryological criterion of homology. *Biological Lectures of the Marine Biological Laboratory of Wood's Hole*. 1894:101–124.

Wilson, E.O. 1953. The origin and evolution of polymorphism in ants. *Quarterly Review of Biology* 28:136–156.

Wilson, E.O. 1966. Behaviour of social insects. *Symposia Royal Entomolgical Society of London* 3:81–96.

Wilson, E.O. 1971a. *The Insect Societies*. Harvard University Press, Cambridge.

Wilson, E.O. 1971b. Tropical social parasites in the ant genus *Pheidole*, with an analysis of the anatomical parasitic syndrome (Hymenoptera: Formicidae). *Insectes Sociaux* 31:316–334.

Wilson, E.O., Carpenter, F.M. and Brown, W.L., Jr. 1967. The first Mesozoic ants. *Science* 157:1038–1040.

Wilson, E.O. 1975. *Sociobiology: The New Synthesis*. Belknap Press, Cambridge.

Wilson, E.O. 1981. Epigenesis and the evolution of social systems. *Journal of Heredity* 72:70–77.

Wilson, J.W., III 1975. Morphologic change as a reflection of adaptive zone. *American Zoologist* 15:363–370.

Wilson, M.A. and Tonegawa, S. 1997. Synaptic plasticity, place cells and spatial memory: study with second generations knockouts. *Trends in Neuroscience* 20:102–106.

Wilson, P. 1986. A passacaglia and fugue in *Pseudoleskeella. Darlingtonia* 12:1–7.

Wimberger, P.H. 1991. Plasticity of jaw and skull morphology in the neotropical cichlids *Geophagus brasiliensis* and *G. steindachneri. Evolution* 45:1545–1563.

Wimberger, P.H. 1992. Plasticity of fish body shape. The effects of diet, development, family and age in two species of *Geophagus* (Pisces: Cichlidae). *Biological Journal of the Linnean Society* 45:197–218.

Wimberger, P.H. 1993. Effects of vitamin C deficiency on body shape and skull osteology in *Geophagus brasiliensis*: implications for interpretations of morphological plasticity. *Copeia* 1993:343–351.

Wimberger, P.H. 1994. Trophic polymorphisms, plasticity, and speciation in vertebrates. In: *Theory and Application of Fish Feeding Ecology*, D.J. Stouder, K.L. Fresh and R.J. Feller (eds.). University of South Carolina Press, Columbia, pp. 19–43.

Wimsatt, W. 1981. The units of selection and the structure of the multilevel genome. In:

Philosophy of Science Associations, R.N. Giere (ed.). Philosophy of Science Association, East Lansing, MI, pp. 122–183.

Wimsatt, W.C. and Schank, J.C. 1988. Two constraints on the evolution of complex adaptations and the means for their avoidance. In: *The Idea of Progress in Evolution 1980, Vol. 2*, M.H. Nitecki (ed.). University of Chicago Press, Chicago, pp. xv–xx.

Windig, J.J. 1993. The genetic background of plasticity in wing pattern of *Bicyclus* butterflies. Thesis, University of Leiden.

Wingfield, J.C., Ball, G.F., Dufty, A.M., Jr., Hegner, R.E. and Ramenofsky, M. 1987. Testosterone and aggression in birds. *American Scientist* 75:602–608.

Wingfield, J.C., Hegner, R.E., Dufty, A.M., Jr. and Ball, G.F. 1990. The "challenge hypothesis": theoretical implications for patterns of testosterone secretion, mating systems, and breeding strategies. *American Naturalist* 136:829–846.

Winkler, D.W. 1993. Testosterone in egg yolks: an ornithologist's perspective. *Proceedings of the National Academy of Sciences USA* 90:11439–11441.

Winston, J.E. 1976. Experimental culture of the estuarine ectoproct *Conopeum tenuissimum* from Chesapeake Bay. *Biological Bulletin* 150:318–335.

Winston, J.E. 1979. Current-related morphology and behaviour in some Pacific coast bryozoans. In: *Advances in Bryozoology*, G.P. Larwood and M.B. Abbott (eds.). Systematics Association Special Vol. 13. Academic Press, New York, pp. 247–268.

Winston, M.L. 1987. *The Biology of the Honey Bee*. Harvard University Press, Cambridge.

Winter, K. 1985. Crassulacean acid metabolism. In: *Photosynthetic Mechanisms and the Environment*, J. Barber and N.R. Baker (eds.). Elsevier, Amsterdam, pp. 329–387.

Winter, K. and Smith, J.A.C. 1996a. An introduction to crassulacean acid metabolism. Biochemical principles and ecological diversity. In: *Crassulacean Acid Metabolism*, K. Winter and J.A.C. Smith (eds.). Springer-Verlag, New York, pp. 1–10.

Winter, K. and Smith, J.A.C. 1996b. Crassulacean acid metabolism: current status and perspectives. In: *Crassulacean Acid Metabolism*, K. Winter and J.A.C. Smith (eds.). Springer-Verlag, New York, pp. 389–426.

Winter, K., Zotz, G., Baur, B. and Dietz K.-J. 1992. Light and dark CO_2 fixation in *Clusia uvitana* and the effects of plant water status and CO_2 availability. *Oecologia* 91:47–51.

Winter, U. and Buschinger, A. 1986. Genetically mediated queen polymorphism and caste determination in the slave-making ant, *Harpagoxenus sublaevis* (Hymenoptera: Formicidae). *Entomologia Generalis* 2:125–137.

Winterbottom, J.M. 1929. Studies in sexual phenomena. VII. The transference of male secondary sexual display characters to the female. *Journal of Genetics* 21:367–387.

Winterbottom, J.M. 1932. Studies in sexual phenomena. VIII. "Transference" and eclipse plumage in birds. *Journal of Genetics* 25:395–406.

Wirtz, P. 1973. Differentiation in the honebee larva. *Mededelingen Landbouwhogeschool Wageningen* 73:5.

Witte, F. 1984. Consistency and functional significance of morphological differences between wild-caught and domestic *Haplochromis squamipinnis* (Pisces, Cichlidae). *Netherlands Journal of Zoology* 34:596–612.

Witte, F., Barel, C.D.N. and Hoogerhoud, R.J.C. 1990. Phenotypic plasticity of anatomical structures and its ecomorphological significance. *Netherlands Journal of Zoology* 40:278–298.

Woese, C. 1998. The universal ancestor. *Proceedings of the National Academy of Sciences USA* 95:6845–6859.

Wolf, G. 1990. Recent progress in vitamin A research: nuclear retinoic acid receptors and their interaction with gene elements. *Journal of Nutritional Biochemistry* 1:284–289.

Wolff, J.D. 1892. *Das Gesetz der Transformation der Knochen*. A. Hirschwald, Berlin.

Wolpert, L. 1990. The evolution of development. *Biological Journal of the Linnean Society* 39:109–124.

Wolpert, L. 1991. *The Triumph of the Embryo*. Oxford University Press, Oxford.

Woltereck, R. 1909. Weitere experimentelle Untersuchungen über Artveränderung, speziell über das Wesen quantitativer Artunterschiede bei Daphniden. *Verhandlungsbericht der Deutschen Zoologischen Gesellschaft* 1909:110–172.

Wolynes, P.G. 1998. Computational biomolecular science. *Proceedings of the National Academy of Sciences USA* 95:5848.

Wood, T.K. 1993a. Diversity in the New World Membracidae. *Annual Review of Entomology* 38:409–435.

Wood, T.K. 1993b. Speciation of the *Enchenopa binotata* complex (Insecta: Homoptera: Membracidae). In: *Evolutionary Patterns and Processes*, D.R. Lees and D. Edwards (eds.). Academic Press, New York, pp. 299–317.

Wood, T.K. and Keese, M.C. 1990. Host plant induced assortative mating in *Enchenopa* treeoppers. *Evolution* 44:619–628.

Wood, T.K., Olmstead, K.L. and Guttman, S.I. 1990. Insect phenology mediated by host-plant water relations. *Evolution* 44:629–636.

Wood, T.K., Tilmon, K.J., Shantz, A.B., Harris, C.K. and Pesek, J. 1999. The role of host-plant fidelity in initiating insect race formation. *Evolutionary Ecology Research* 1:317–332.

Woodward, H. 1902. The Canadian Rockies. Part I: On a collection of Middle Cambrian fossils obtained by Edward Whymper, Esq., F.R.G.S., from Mount Stephen, British Columbia. *Geological Magazine* 4:502-5–5, 529–544.

Wootton, R.J. 1976. *The Biology of the Sticklebacks*. Academic Press, New York.

Wray, G.A. 1992. The evolution of larval morphology during the post-Paleozoic radiation of echinoids. *Paleobiology* 18:258–287.

Wray, G.A. 1994. The evolution of cell lineage in echinoderms. *American Zoologist* 34:353–363.

Wray, G.A. 1996. Parallel evolution of nonfeeding larvae in echinoids. *Systematic Biology* 45:308–322.

Wray, G.A. and Bely, A.E. 1994. The evolution of echinoderm development is driven by several distinct factors. *Development* (Supplement) 1994:97–106.

Wray, G.A. and McClay, D.R. 1988. The origin of spicule-forming cells in a "primitive" sea urchin (*Eucidaris tribuloides*) which appears to lack primary mesenchyme cells. *Development* 103:305–315.

Wray, G.A. and McClay, D.R. 1989. Molecular heterochronies and heterotopies in early echinoid development. *Evolution* 43:803–813.

Wray, G.A. and Raff, R.A. 1991. The evolution of developmental strategy in marine invertebrates. *Trends in Ecology and Evolution* 6:45–50.

Wray, G.A., Levinton, J.S. and Shapiro, L.H. 1996. Molecular evidence for deep Precambrian divergences among metazoan phyla. *Science* 274:568–573.

Wrege, P.H. and Emlen, S.T. 1994. Family structure influences mate choice in white-fronted bee-eaters. *Behavioral Ecology and Sociobiology* 35:185–191.

Wright, S. 1929. Fisher's theory of dominance. *American Naturalist* 63:556–561.

Wright, S. 1930. Review of *The Genetical Theory of Natural Selection* by R.A. Fisher. *Journal of Heredity* 21:349–356.

Wright, S. 1931. Evolution in Mendelian populations. *Genetics* 16:97–159.

Wright, S. 1932. The roles of mutation, inbreeding, crossbreeding and selection in evolution. *Proceedings Sixth International Congress of Genetics* 1:356–366.

Wright, S. 1934. The results of crosses between inbred strains of guinea pigs, differing in number of digits. *Genetics* 19:537–551.

Wright, S. 1941. The material basis of evolution. *Scientific Monthly* 53:165–170.

Wright, S. 1945. Tempo and mode in evolution: a critical review (A review of *Tempo and Mode in Evolution*, by George Gaylord Simpson). *Ecology* 26:415–419.

Wright, S. 1948. Evolution, organic. *Encyclopaedia Brittanica*, 14th ed. 10:111–112.

Wright, S. 1949. Population structure in evolution. *Proceedings of the American Philosophical Society* 93:471–478.

Wright, S. 1956. Modes of selection. *American Naturalist* 90:5–24.

Wright, S. 1965. Dr. Wilhelmina Key. *Journal of Heredity* 56:195–196.

Wright, S. 1968. *Evolution and the Genetics of Populations. Volume 1: Genetics and Biometric Foundations*. University of Chicago Press, Chicago.

Wright, S. 1977. *Evolution and the Genetics of Populations*, Vol. 3. University of Chicago Press, Chicago.

Wright, S. 1978. *Evolution and Genetics of Populations*, Vol. 4. University of Chicago Press, Chicago.

Wright, S. 1980. Genic and organismic selection. *Evolution* 34:825–843.

Wright, S. and Dobzhansky, T. 1946. Genetics of natural populations, XII: Experimental reproduction of some of the changes caused by natural selection in certain populations of *Drosophila pseudoobscura*. *Genetics* 31:125–156.

Wymer, C.L., Wymer, S.A., Cosgrove, D.J. and Cyr, R.J. 1996. Plant cell growth responds to external forces and the response requires intact microtubules. *Plant Physiology* 110:425–430.

Wynne-Edwards, V.C. 1986. *Evolution through Group Selection*. Blackwell Scientific Publications, Oxford.

Wyss-Huber, M. 1981. Caste differences in haemolymph proteins in two species of termites. *Insectes Sociaux* 28:71–86.

Yablokov, A.V. 1966 [1974]. *Variability of Mammals*. Amerind Publishing, New Delhi.

Yablokov, A.V. 1986. *Phenetics: Evolution, Population, Trait*. Columbia University Press, New York.

Yamamoto, I. 1987. Male parental care in the raccoon dog *Nyctereutes procyonoides* during the early rearing period. In: *Animal Societies: Theories and Facts*, Y. Itô, J.L. Brown, and J. Kikkawa (eds.). Japan Scientific Societies Press, Tokyo, pp. 189–195.

Yamauchi, K. and Kinomura, K. 1993. Lethal fighting and reproductive strategies of dimorphic males in *Cardiocondyla* ants (Hymenoptera: Formicidae). In: *Evolution of Insect Societies*, T. Inoue and S. Yamane (eds.). Hakuhinsha, Tokyo, pp. 373–402.

Yamauchi, K., Kimura, Y., Corbara, B., Kinomura, K. and Tsuji, K. 1996. Dimorphic ergatoid males and their reproductive behavior in the ponerine ant *Hypoponera bondroiti*. *Insectes Sociaux* 43:119–130.

Yanega, D. 1988. Social plasticity and early-diapausing females in a primitively social bee. *Proceedings of the National Academy of Sciences USA* 85:4374–4377.

Yanega, D. 1989. Caste determination and differential diapause within the first brood of *Halictus rubicundus* in New York (Hymenoptera: Halictidae). *Behavioral Ecology and Sociobiology* 24:97–107.

Yanega, D. 1997. Demography and sociality in halictine bees (Hymenoptera: Halictidae). In: *The Evolution of Social Behavior in Insects and Arachnids*, J.C. Choe and B.J. Crespi (eds.). Cambridge University Press, Cambridge, pp. 293–315.

Yaoita, Y., Shih, Y. and Brown, D.D. 1990. *Xenopus laevis* α and β thyroid hormone receptors. *Proceedings of the National Academy of Sciences USA* 87:7090–7094.

Yasunobu, K.T. and Tanaka, M. 1973. The evolution of iron-sulfur protein containing organisms. *Systematic Zoology* 22:570–589.

Young, B. 1993. On the necessity of an archetypal concept in morphology: with special reference to the concepts of "structure" and "homology." *Biology and Philosophy* 8:225–248.

Zahavi, A. 1971. The social behaviour of the white wagtail *Motacilla alba alba* wintering in Israel. *Ibis* 113:203–211.

Zahavi, A. 1975. Mate selection—a selection for a handicap. *Journal of Theoretical Biology* 53:205–214.

Zeh, D.W. and Zeh, J.A. 1988. Condition-dependent sex ornaments and field tests of sexual selection theory. *American Naturalist* 132:454–459.

Zeh, D.W., Zeh, J.A. and Smith, R.L. 1989. Ovipositors, amnions and eggshell architecture in the diversification of terrestrial arthropods. *Quarterly Review of Biology* 64:147–168.

Zeh, J.A. 1997. Polyandry and enhanced reproductive success in the harlequin-beetle-riding pseudoscorpion. *Behavioral Ecology and Sociobiology* 40:111–118.

Zeh, J.A. and Zeh, D.W. 1996. The evolution of polyandry I: intragenomic conflict and genetic incompatibility. *Proceedings of the Royal Society of London B* 263:1711–1717.

Zeh, J.A. and Zeh, D.W. 1997. The evolution of polyandry II: post-copulatory defences against genetic incompatibility. *Proceedings of the Royal Society of London B* 264:69–75.

Zelditch, M.L. 1988. Ontogenetic variation in patterns of phenotypic integration in the laboratory rat. *Evolution* 42:28–41.

Zelditch, M.L., Bookstein, F.L. and Lundrigan, B.L. 1992. Ontogeny of integrated skull growth in the cotton rat *Sigmodon fulviventer*. *Evolution* 46:1164–1180.

Zera, A.J. 1999. The endocrine genetics of wing polymorphism in *Gryllus*: critique of recent studies and state of the art. *Evolution* 53:973–977.

Zera, A.J. and Bottsford, J. 2001. The endocrine-genetic basis of life-history variation: the relationship between the ecdysteroid titer and morph-specific reproduction in the wing-polymorphic cricket *Gryllus firmus*. *Evolution* 55:538–549.

Zera, A.J. and Denno, R.F. 1997. Physiology and ecology of dispersal polymorphism in insects. *Annual Review of Entomology* 42:207–231.

Zera, A.J. and Huang, Y. 1999. Evolutionary endocrinology of juvenile hormone esterase: functional relationship with wing polymorphism in the cricket, *Gryllus firmus*. *Evolution* 53:837–847.

Zera, A.J. and Tiebel, K.C. 1988. Brachypterizing effect of group rearing, juvenile hormone III and methoprene in the wing-dimorphic cricket, *Gryllus rubens*. *Journal of Insect Physiology* 34:489–498.

Zera, A.J. and Tiebel, K.C. 1989. Differences in juvenile hormone esterase activity between presumptive macropterous and brachypterous *Gryllus rubens*: implications for the hormonal control of wing polymorphism. *Journal of Insect Physiology* 35:7–17.

Zera, A.J., Gu, X. and Zeisset, M. 1992. Characterization of juvenile hormone esterase from genetically-determined wing morphs of the cricket, *Gryllus rubens*. *Insect Biochemistry and Molecular Biology* 22:829–839.

Zimmerer, E.J. and Kallman, K.D. 1989. Genetic basis for alternative reproductive tactics in the pygmy swordtail, *Xiphophorus nigrensis*. *Evolution* 43:1298–1307.

Zimmerman, E.C. 1970. Adaptive radiation in Hawaii with special reference to insects. *Biotropica* 2:32–38.

Zirkle, C. 1946. The early history of the idea of the inheritance of acquired characters and of pangenesis. *Transactions of the American Philosophical Society* 35:91–151.

Zotz, G. and Winter, K. 1996. Seasonal changes in daytime versus nighttime CO_2 fixation of *Clusia uvitana in situ*. In: *Crassulacean Acid Metabolism*, K. Winter and J.A.C. Smith (eds.). Springer-Verlag, New York, pp. 312–323.

Zucchi, R. 1994. A evolução do processo de tratamento das células de cria de meliponinae: do antagonismo à dominância ritualizada (Hymenoptera, Apidae). In: *Anais do 1° Encontro Sobre Abelhas*, R. Zucchi, P.M. Drumond, P.G. Fernandes-da-Silva and S.C. Augusto (eds.). Editora Legis Summa, Ribeirão Preto, pp. 38–45.

Zucker, N. and Boecklen, W. 1990. Variation in female throat coloration in the tree lizard (*Urosaurus ornatus*): relation to reproductive cycle and fecundity. *Herpetologica* 46:387–394.

Zuckerkandl, E. 1994. Molecular pathways to parallel evolution: I. *Gene* nexuses and their morphological correlates. *Journal of Molecular Evolution* 39:661–678.

Zuckerkandl, E. and Villet, R. 1988. Concentration-affinity equivalence in gene regulation: convergence of genetic and environmental effects. *Proceedings of the National Academy of Sciences USA* 85:4784–4788.

Zuk, M. and Simmons, L.W. 1997. Reproductive strategies of the crickets (Orthoptera: Gryllidae). In: *The Evolution of Mating Systems in Insects and Arachnids*, J.C. Choe and B.J. Crespi (eds.). Cambridge University Press, Cambridge, pp. 89–109.

Zusi, R.L. and Jehl, J.R., Jr. 1970. The systematic relationships of *Aechmorhynchus*, *Prosobonia*, and *Phegornis* (Charadriiformes; Charadrii). *Auk* 87:760–780.

Author Index

A

Abou-Seedo, F.S., 535
Adams, D.C., 568, 569
Adams, J., 211
Adams, W.W., III, 252, 605
Ades, C., 61
Adler, F.R., 47, 392
Adler, K., 448
Adolph, S.C., 430, 431
Afzelius, B.A., 97, 633
Agosti, D., 613
Agrawal, A.A., 113, 392, 430
Aiello, A., 166
Aiello, L.C., 464
Akimoto, S., 167, 220, 253, 254, 532, 533, 534
Akre, R.D., 410
Alatalo, R.V., 593
Alberch, P., 8, 23, 25, 27, 28, 53, 54, 56, 119, 133, 161, 164, 357, 363, 368, 377, 495, 527
Alberts, J.R., 287
Albon, S.D., 455
Alcock, J., 10, 26, 276, 278, 279, 385, 422, 453, 454, 457, 461, 475, 478, 593
Alder, E.M., 115
Alexander, R.D., 181, 235, 290, 300, 384, 416, 434, 454, 464, 528, 529, 554, 555, 561, 562
Allard, M.W., 370
Allen, G.A., Jr., 156
Allen, N., 636
Allendorf, F.W., 210
Almeida-Val, V.M.F., 211
Altenberg, L., 62, 171, 182
Altmann, J., 43, 289
Amadon, D., 589
Ambros, V., 511
Ammirato, P., 91
Amundsen, T., 167
Amundson, R., 139
Ananthakrishnan, T.N., 109, 111
Andersen, N.M., 374
Anderson, J.M., 46, 47, 62, 80
Anderson, K.V., 72, 97, 114, 176
Anderson, N.H., 430
Andersson, M., 217, 441, 449, 457, 461, 467, 608, 632, 633
Andrés, J.A., 271
Andrew, R.J., 244

Andrews, C.A., 359, 365, 548, 549
Andrews, M.T., 73
Andrews, R.C., 232
Angehr, G.R., 275
Angelo, M.J., 307
Ankney, P.F., 633
Anson, B.J., 50, 51
Anstey, R.L., 361, 381, 624, 625, 626, 627, 628
Antonovics, J., 106, 275, 528
Aoki, S., 221, 252, 254, 387
Arak, A. 83, 456
Arathi, H.S., 451
Arcadi, A.C., 456
Arcese, P., 464
Arévalo, E., 450
Arkin, A., 428
Armitage, K.B., 344
Armbruster, W.S., 259
Armstrong, E.A., 245, 249
Arnegard, M.E., 591
Arnemann, J., 261
Arnold, A.P., 77, 274, 275, 314
Arnold, S.J., 78, 316
Aro, E.M., 46, 47, 62
Arrow, G.J., 75
Arthur, W., 9, 219
Asa, C.S., 223, 288
Ashlock, P.D., 353, 371
Ashton, P.S., 251
Askew, R.R., 560
Astié, M., 605
Atchley, W.R., 5, 12, 60, 63, 65, 70, 71, 77, 112, 113, 134, 135, 138, 150, 154, 175, 176, 179, 219, 298, 299, 303, 308, 616
Athias-Binche, F., 437, 552, 553
Atwood, C.E., 228, 555
Aukema, B., 119, 120
Austad, S.N., 378, 425
Austin, C.R., 95, 635
Avise, J.C., 118, 596, 622, 623
Ayala, F.J., 618, 631

B

Baccetti, B., 97, 633
Baerends, G.P., 63, 245, 248, 294, 313, 351, 443, 489, 494, 591, 592
Baerends-van Roon, J.M., 591, 592
Bajema, C.J., 440
Baker, B.S., 323

Baker, J., 383, 438
Baker, J.A., 571
Baker, R.R., 633
Bakken, M., 456
Baldwin, J.M., 24, 25, 35, 39, 147, 151, 152, 153, 161, 178, 180, 337, 338, 339, 505, 619
Baldwin, P.H., 591
Balfour-Browne, F., 271
Baltimore, D., 15
Bandi, C., 248
Banister, K.E., 300
Barkow, J.H., 464
Barlocher, F., 430
Barlow, G.W., 63, 171, 244, 260, 441, 591
Barnard, C.J., 348, 350, 434, 444, 592
Barnes, H., 312
Barnes, R.D., 212
Baroni Urbani, C., 613
Barrett, P.H., 191
Barthlott, W., 91
Barton, N.H., 144, 592, 593, 631, 636
Basibuyuk, H.H., 613
Basile, D.V., 205
Baskin, C.C., 91, 427, 459–61
Baskin, J.M., 91, 427, 459–61
Basolo, A.L., 221, 518
Bateson, G., 25, 180
Bateson, P., 180, 182, 291, 442, 443
Bateson, W., 4, 11, 22, 24, 26, 28, 202, 210, 214, 233, 234, 256, 471, 472, 474, 480, 527, 562
Batra, L.R., 111
Batra, S.W.T., 111
Bauchau, V., 521
Bauer, A., 547
Bawa, K.S., 262, 265, 269
Bayer, M.M., 624
Bazzaz, F.A., 39
Beach, F.A., 270, 286, 441
Beadle, G.W., 268
Beani, L., 391, 463, 591, 593
Beck, W.S., 8, 189, 607
Beebe, W., 390
Beeman, R.W., 212
Beetsma, J., 130, 400
Bego, L.R., 291, 292
Begun, D.J., 73
Bell, G., 139, 207, 226
Bell, G.A., 474, 476

Bell, M.A., 169, 203, 204, 362, 363, 411, 412, 413, 414, 545, 548, 549, 551, 566, 567, 570
Bell, P.R., 463
Bell, W.J., 38, 119, 337, 341, 426, 443, 445, 465, 534
Bellis, M.A., 633
Bengston, A.S., 348
Bennett, A.F., 4, 180
Bennett, M.J., 325
Bennett, M.K., 327
Benton, M.J., 531
Bentzen, P., 567, 568
Berg, P., 328
Berg, R.L., 374
Berg, S.J., 287
Berger, P.J., 460
Berglund, A., 310
Berkow, R., 501
Berman, A.L., 318
Bernatchez, L., 365, 567
Bernays, E.A., 54, 99, 118, 300, 314, 328, 341–43, 391, 444, 463, 580
Berner, R.A., 521
Bernstein, C., 332, 630
Bernstein, H., 332, 630
Berrigan, D., 272
Berrill, N.J., 37
Berry, R.J., 206
Betts, P.W., 239
Bickel, D.J., 496
Bigelow, R.S., 528, 529, 554, 555, 561, 562
Birney, E.C., 260, 287
Bjorck, L., 321
Blackman, R.L., 427
Blackstone, N.W., 251
Blake, C.C.F., 62, 72, 211, 212, 318, 320, 321, 322, 324, 328
Blake, D.B., 624
Blank, J.L., 461
Blaustein, A.R., 448
Blest, A.D., 212
Bloch, G., 452, 453
Boag, P.T., 463
Boake, C.R.B., 156, 578, 632
Boam, T.B., 528
Boardman, R.S., 627
Bock, P.E., 624, 628
Bodaly, R.A., 567, 568, 569
Boecklen, W., 280, 281
Boero, F., 67, 86, 168, 219, 220, 437
Bohart, R.M., 408
Bohm, D., 81
Bohnert, J.C., 76, 512, 513
Bohr, N., 84
Boidron-Metairon, I., 515, 517
Bolander, F.F., 274, 319, 320
Bolton, B., 22, 573, 606
Bonabeau, E., 37, 42
Bonner, J.T., 5, 7, 37, 48, 56, 67, 70, 78, 82, 90, 91, 93, 98, 108, 135, 171, 183, 186, 238, 429, 500, 610, 613, 615, 616
Bonnetti, A.M., 284, 509
Bookstein, F.L., 83

Borgia, G., 384
Borland, A.M., 513
Borowsky, R.L., 106, 107, 249, 422
Bosser, T., 341
Bossomaier, T., 59
Bottsford, J., 311, 374
Bouillon, J., 168, 219, 220
Bourke, A.F.G., 542, 543, 544, 606
Bowers, M.D., 501
Bowlby, J., 497
Boyd, R., 341
Boyd, S.K., 272
Bradshaw, A.D., 45, 155, 225, 229, 250, 383, 386
Bradshaw, W.E., 80, 146, 385, 406, 560, 561
Brady, U.E., 277
Brakefield, P.M., 217, 390, 594
Bramble, D.M., 298
Brandon, R.N., 30, 31, 32, 98, 165, 174, 634
Breathnach, R., 320
Breed, M.D., 156, 449
Breedlove, S.M., 314
Breeland, S.G., 427
Brenner, S., 318
Briceño, R.D., 156, 383
Bridges, C.B., 211
Brien, P., 296
Briggs, D., 155, 334, 371, 551, 604, 605
Briggs, D.E.G., 614
Bright, K.L., 444
Brink, A., 186, 250
Britten, R.J., 27, 28, 70, 79, 80, 93
Brockmann, H.J., 53, 63, 223, 344, 348, 350, 369, 418, 419, 420, 424, 425, 426, 428, 596
Bromham, L., 522
Brönmark, C., 392, 550, 572
Brooks, D.R., 204, 487, 507, 598
Brothers, D.J., 229
Brower, A.V.Z., 357, 358
Brown, C.A.J., 444
Brown, E.R., 418
Brown, G.E., 347
Brown, J.H., 456
Brown, J.L., 181, 223, 290, 418, 449, 455, 458
Brown, P.O., 74
Brown, W.D., 282, 456
Brown, W.L., Jr., 397, 399, 620
Brulfert, J., 251
Brunnert, 67, 74
Brusca, G.J., 185, 609, 611, 612
Brusca, R.C., 185, 522, 609, 611, 612
Bryan, J.E., 344
Brylski, P.V., 510, 511
Bull, J.J., 15, 35, 68, 99, 100, 103, 121, 124, 170, 260, 261, 282, 293, 370, 420, 427, 432
Bulmer, M.G., 100
Büning, J., 86, 92, 97, 113, 114, 217, 330, 332, 333, 366, 368
Burchsted, J.C.A., 311
Burian, R.M., 4, 177
Burley, N., 97, 114, 457, 633

Burr, B., 267
Burt, A., 634, 635
Burton, F.D., 289
Buschinger, A., 104, 107, 120, 121, 124, 206, 207, 542, 543, 544, 545, 547, 554, 573
Bush, G., 557, 558, 559, 562
Bush, G.L., 474, 557, 558
Buss, L.W., 5, 7, 12, 32, 37, 43, 44, 67, 90, 96, 142, 175, 183, 184, 185, 186, 191, 192, 193, 251, 307, 332, 333, 387, 505, 609, 636
Byrne, R., 464

C
Caballero, A., 155
Cade, W.H., 100, 103, 422, 423
Cairns, R.B., 76, 140, 244, 252
Caisse, M., 528
Calderone, N.W., 314, 438
Calderon-Urrea, A., 266
Caldwell, J.P., 284
Caldwell, R.L., 343, 344, 591
Cale, G.H., Jr., 147, 339
Calvin, W.H., 343, 448, 464
Camazine, S., 42, 446
Cameron, S.A., 283, 284
Cammaerts, M.C., 544
Campbell, C.A., 5, 28, 603
Campbell, H.D., 326
Cane, J.H., 363
Cane, T.C., 551
Canfield, D.E., 521
Cannon, W.B., 7
Caporale, L.H., 320, 330, 336, 502
Caraco, T., 429, 444
Cargill, A.S., 430
Carle-Urioste, J.C., 321
Carpenter, J.M., 24, 181, 223, 361, 370, 372, 373, 509, 573, 606
Carré, D., 95
Carroll, S.P., 180
Carroll, R.L., 5
Carson, H.L., 9, 24, 73, 181, 184, 388, 401, 485, 528, 558, 559, 573, 574, 578, 579, 582, 591, 592, 593
Carson, S.F., 515
Case, T.J., 594
Cassill, D.L., 461
Causey, B.A., 422
Cavalier-Smith, T., 310, 321, 402
Cervo, R., 368, 408, 409, 573
Chai, P., 521
Chailakhyan, M.K., 262
Chaline, J., 288, 521
Chan, T.-Y, 389
Chang, F., 73, 74
Changeux, J.P., 39
Chapin, F.S., III, 315, 636
Chapman, R.F., 432, 462
Charlesworth, B., 12, 26, 79, 139, 142, 144, 147, 164, 168, 297, 298, 403, 405, 432, 474, 476, 478, 505, 533, 599, 601, 617, 618, 631, 636
Charlesworth, D., 79, 168, 405, 533,
Charnov, E.L., 17, 262, 265, 310, 431, 458

Chazdon, R.L., 91, 306
Cheek, J.E., 608
Cheetham, A.H., 8, 361, 381, 401, 617, 619, 623, 624, 625, 626, 627, 628
Chen, J.-Y., 614
Cheney, D.L., 457
Cheng, M.-F., 273
Chesson, P.L., 437
Chevalier-Skolnikoff, S., 276, 277
Cheverud, J.M., 8, 12, 60, 65, 77, 97, 112, 113, 134, 150, 154, 164, 168, 175, 186
Child, A.R., 542, 543
Childress, J.R., 446
Choe, J.C., 606
Chourey, P.S., 211
Christy, J.H., 468
Claridge, M.F., 559
Clark, A.B., 35, 344
Clark, A.G., 216
Clark, A.R., 154
Clark, M.M., 276, 282, 286
Clarke, B., 418
Clarke, B.C., 168, 170, 528, 529
Clarke, C., 271
Clarke, C.A., 79
Clarke, G.M., 103
Clausen, J., 381, 523
Cleveland, L.R., 501
Clough, M., 284
Clutton-Brock, T.H., 284, 291, 304, 455
Cock, A.G., 72, 136, 156, 305, 309
Cockerell, T.D.A., 237
Coddington, J.A., 139
Coe, W.R., 22, 573
Coen, E.S., 199, 269, 270
Cohn, M.J., 368
Colbert, E.H., 180, 522
Cole, B.J., 38, 42, 43, 451, 543
Colgan, P.W., 280
Collazo, A., 241
Collett, M., 132
Collins, A.M., 155
Collins, D., 155, 289, 613, 614
Collins, J.P., 73, 207, 208, 353, 369, 448, 567, 595, 607, 608
Colwell, R.K., 552
Comet, T.C.A., 347
Conant, P., 579
Conner, J., 227
Conover, D.O., 121
Conrad, M., 5
Consul, P.C., 438
Conway Morris, S., 522
Cook, C.D.K., 229
Cook, J.M., 109, 395, 420, 421, 422
Cook, P.L., 624, 628
Cook, S.A., 26, 163, 229, 239, 240, 390, 391, 393, 397, 401, 403, 414, 415, 416, 531, 532
Cooke, R.C., 605
Cooley, J.R., 555
Cooper, W.S., 113, 114, 424, 425, 426
Corbet, G.B., 521
Corbet, P.S., 271
Cordero, A., 271

Corneli, P.S., 180
Corner, E.J.H., 250
Cosmides, L., 81
Cowan, W.M., 41
Cowley, D.E., 63, 77, 112, 150, 308
Coyle, F.A., 605
Coyne, J.A., 104, 106, 144, 145, 151, 403, 478, 528, 588, 589, 617
Cracraft, J., 201, 596
Craddock, E.M., 578, 579, 580, 632
Craig, E.A., 99, 333
Craig, J.L., 223
Craig, R., 74, 449
Crane, J., 213, 455
Crankshaw, O.W., 560
Creel, S., 304
Creelman, R.A., 44
Cresko, W.A., 571
Crespi, B.J., 109, 305, 369, 370, 372, 387, 406, 456, 539, 541
Crew, F.A.E., 286
Crewe, R.M., 450
Crews, D., 121, 124, 261, 262, 276, 281, 282
Crick, F.H.C., 63
Cronquist, A., 250, 371
Crook, J.H., 276, 289, 314
Crosland, M.W.J., 283
Cross, E.A., 530
Crowley, T.J., 519, 520
Crozier, R.H., 5, 10, 17, 74, 76, 293, 383, 438, 450, 528
Csuti, B.A., 239, 429, 520
Culver, D.C., 551
Cumber, R.A., 228
Cundall, D., 367
Cunha, R., 284
Curio, E., 65, 238, 239, 240, 344, 350, 359, 378, 382, 443, 445, 582, 595
Currie, G.A., 110
Curtis, H., 6, 500
Cushing, J.E., 25, 152, 415, 595
Cushman, J.C., 76, 512, 513
Cyr, R.J., 39

D

Dacks, J., 630
Dagg, A.I., 293
Daly, J.W., 502
Danforth, B.N., 228, 235, 240, 292, 363, 404
Dani, F.R., 368, 408, 573
Darchen, R., 128
Darlington, C.D., 79
Darwin, C., 9, 11, 16, 24, 39, 43, 54, 75, 147, 155, 161, 166, 177, 184, 188–94, 201, 237, 247, 252, 262, 263, 294, 298, 299, 303, 305, 331, 347, 354, 378, 392, 394, 397, 405, 415, 419, 440, 441, 443, 453, 454, 459, 464, 466, 471–73, 475, 477, 482, 505, 518, 591, 599, 600, 604, 608, 631, 633, 637
Davey, G., 337
Davidson, D.W., 543

Davidson, E.H., 10, 27, 28, 47, 48, 58, 70, 72, 73, 79, 80, 86, 89, 96, 97, 114, 185, 200, 222, 330–33, 336, 514, 515, 517, 615, 616, 636
Davies, N.B., 386, 456
Davis, D.M., 559
Davis, G., 491
Davis, J.M., 76
Davis, M.B., 519
Davis, S., 272
Dawkins, R., 5, 15, 16, 17, 20, 37, 82, 93, 151, 161, 165, 245, 340, 388, 420, 425, 426, 430, 431, 433, 466, 609
Day, T., 299, 571
Deag, J.M., 289
de Beer, G.R., 7, 22, 142, 187, 486, 490, 495, 496
de Camargo, C.A., 291, 292
de Chateau, M., 321
deHeer, C.J., 537
de Jong, G., 316, 435
Delage-Darchen, B., 128
Delaney, K., 444
Dellaporta, S.L., 266
Deneubourg, J.L., 42, 135
Dennett, D.C., 474
Denno, R.F., 127, 310, 311
Dent, J.N., 118, 119
Denton, F.R., 331
Denton, M., 492, 496
dePamphilis, C.W., 230, 231
de Pinna, M.C.C., 488, 566
de Queiroz, A., 24, 181, 487, 494
Derting, T.L., 311, 315
Desjardins, C.A., 228
de Souza, S.J., 320, 321
Destephano, D.B., 277
Dethier, V., 171
Deuchar, E.M., 95, 135, 307, 333, 495
DeVries, H., 475
De Winton, D., 25
DeWitt, T.J., 179, 316, 362, 392, 430, 431
Dewsbury, D.A., 428, 441
Diamond, J.M., 595
Dickerson, R.E., 500
Dickinson, J.A., 172
Dickinson, W.J., 72, 211, 212, 230, 317, 319, 326–28, 330, 380, 581, 597
Dickman, M.C., 218, 219
Didion, J., 498
Dilger, W.C., 276, 286
Dill, L.M., 346
Dillon, T.A., 437, 438
Dingle, H., 78, 115, 156, 171, 343, 344, 383, 390, 535, 536, 591
Dixon, F.W., 197
Dixon, S.E., 74
Dixson, A.F., 290
Dobzhansky, T., 7, 8, 20, 25, 79, 80, 81, 90, 118, 152, 225, 399, 402, 415, 417, 419
Dodson, G.N., 493
Doebley, J., 265, 266, 267, 268
Dolt, C., 93

Dominey, W.J., 280, 387, 576, 578, 591
Donlin, M.J., 334
Donoghue, M.J., 489, 490, 491
Doolittle, W.F., 320, 321
Doorenbos,J., 251
Dorit, R.L., 317
Dorst, J., 527
Dorweiler, J., 267, 268
Douglas, M.E., 622, 623
Douglas, S., 443
Dressler, R.L., 91, 248
Drucker, E.G., 575, 592
DuBose, R.F., 171, 318, 325
DuBrul, E.L., 298
Dudley, R., 521
Dukas, R., 25, 151, 153, 164, 452, 531
Duméril, A., 511, 607
Dunbar, R.I.M., 54
Dundee, H.A., 118
Durbin, M.L., 211
Dusenbery, D.B., 110
Dworkin, I.M., 205
Dybas, H.S., 555
Dyer, F.C., 171, 172, 351

E
Eadie, J.M., 434
Eagleson, G.W., 608
Earn, D.J.D., 284
East, M.L., 276
Eberhard, J.R., 124, 246, 275–6, 456
Eberhard, W.G., 36, 39, 63, 65, 97, 125,
   156, 181, 199, 200, 208, 213, 227,
   248, 258, 277, 309, 282, 309, 310,
   378, 381, 383, 388, 395, 400, 433,
   445, 457, 458, 467, 468, 474, 482,
   587, 608, 609, 632, 633, 635
Eble, G.J., 205, 566
Echelle, A.A., 387
Edelman, G.M., 8, 37, 41, 42, 63, 135,
   143, 171, 185, 499
Edman, J.D., 561
Edwards, S.V., 487
Ehleringer, J.R., 512, 513, 514
Ehlinger, T.J., 344, 571
Eibl-Eibesfeldt, I., 212, 223, 238, 240,
   245, 247, 276, 455
Eickbush, T.H., 321
Eickwort, G.C., 235, 236, 361, 363,
   404, 519
Eigen, M., 14, 19, 335
Eisen, M.B., 74
Eisentraut, M., 507
Eisner, T., 502
Elder, J.F., Jr., 217
Eldredge, N., 7, 9, 184, 201, 388, 396,
   401, 416, 528, 599, 601, 617, 618,
   620, 623, 628, 629
Elgar, M.A., 369
Elliott, J.A., 114, 459
Ellner, S.P., 437
Elmes, G.W., 537, 538, 542, 543, 544
Else, J.G., 291
Elsholtz, H.P., 326
Elwood, R.W., 287, 288
Emerson, A.E., 7, 128, 171

Emerson, S.B., 18, 87, 180, 204, 249,
   263, 272, 285, 295, 388
Emlen, D.J., 45, 124, 125, 126, 127,
   156, 169, 227, 305, 306, 309, 383,
   396, 400, 415, 431, 433, 461, 556
Emlen, S.T., 448, 449, 458, 474
Emlet, R.B., 221, 222, 369, 514
Endler, J.A., 6, 16, 30, 79, 139, 143,
   163, 170, 225, 226, 316, 402, 508,
   529
Engels, W., 130, 227, 284, 291, 461
Ennos, A.R., 302
Enteman, W.M., 523
Epling, C., 198
Epstein, E., 501
Erickson, R.P., 333
Erwin, D.H., 28, 481, 611
Espelie, K.E., 449
Evans, D.G., 561
Evans, H.E., 24, 25, 35, 152, 153, 180,
   181, 182, 407, 415, 596
Evans, J.D., 74, 378, 398, 400, 401
Evans, S.M., 39
Ewald, P.W., 425, 456
Ewer, R.F., 478
Ezhikov, I., 61, 130

F
Fabian, B., 234, 238
Fagan, R., 227
Fagen, R., 11, 424, 431
Fahrbach, S.E., 349
Fairbairn, D.J., 311, 424, 427
Falconer, D.S., 26, 31, 55, 68, 100, 103,
   104, 136, 149, 156, 225, 230, 479,
   488
Farabaugh, S.M., 273
Farmer, E.E., 76, 459
Farris, J.S., 372
Fasinella, D., 179
Faulhaber, I., 211
Faulkes, C.G., 249, 288
Feder, J.L., 557, 558, 559, 562
Federoff, N.V., 62, 331
Feldman, D., 492
Feldman, M., 79
Felsenstein, J., 79, 384, 554
Fenster, G., 206, 493, 494
Fentress, J.C., 63, 65, 82
Fergeson, A., 567
Ferkowicz, M.J., 241
Ferrari, F.D., 495
Field, J., 369, 425, 426, 443, 449, 451
Filchak, K.E., 557, 558
Finch, C.E., 310, 311
Fincke, O.M., 271
Fink, S.V., 300
Fink, W.L., 371
Fischer, A.G., 518
Fischer, E.A., 293, 433, 458, 634
Fisher, C., 234
Fisher, J., 348, 391
Fisher, K., 547
Fisher, R.A., 78, 79, 166, 208, 225, 419,
   424, 450, 453, 467, 476, 477, 479,
   541

Fitch, W.M., 490
Flaishman, M.A., 111
Flannagan, R.D., 73, 119
Fleisher, M.H., 121
Fleming, I., 221, 294, 295, 537
Fletcher, A.J., 501
Fletcher, D.J.C., 5, 337, 343
Fol, H., 95
Fontana, W., 16, 19, 43, 498
Foote, C.J., 294, 535, 536, 541, 547, 548
Forbes, S.H., 210
Ford, E.B., 28, 75, 79, 80, 208, 357,
   378, 384, 386, 401, 420, 432, 500,
   502, 527, 533
Forsdyke, D.R., 528
Forshaw, J.M., 275
Forster, R.R., 605
Forsyth, A., 278, 279, 449, 452
Fortey, R.A., 522
Foster, S.A., 412, 414, 545, 549, 566,
   570
Foster, W.A., 253, 254, 387
Fox, C.W., 98, 113, 416
Fox, S.W., 44, 93, 180, 503, 524
Fraenkel, G.S., 38, 39
Francis, R.C., 260
Francoeur, A., 107
Frandsen, T.C., 460
Frank, L.G., 32, 37, 39, 261, 262, 263,
   276, 341
Frankel, J., 61, 82, 90, 99, 115, 148
Frankino, W.A., 448
Frank, N.R., 272, 448, 542, 543, 544
Fraser, D.F., 444
Frauenfelder, H., 49
Frazzetta, T.H., 40, 54, 59, 85, 159,
   161, 199, 201, 297, 298, 309, 481,
   610
Free, J.B., 452, 455
Freed, L.A., 589, 591
Freeling, M., 270
Freeman, D.A., 461
Freeman, R.D., 110
French, J.A., 223, 453
Fretwell, S.D., 420, 429
Fried, J.J., 288
Friedman, W.E., 91, 212
Frisch, K., von, 171, 314, 349
Frumhoff, P.C., 370, 372, 383, 438
Fryer, G., 364, 569, 574, 592
Fryxell, J.M., 434
Funk, V.A., 507
Futuyma, D.J., 6, 7, 8, 16, 17, 62, 72,
   79, 106, 154, 225, 235, 365, 382,
   384, 402, 418, 483, 527, 528, 554,
   558, 564, 610, 618

G
Gabriel, W., 66
Gadagkar, R., 167, 236, 397, 450, 507
Gadau, J., 438
Gadgil, M., 419
Gagné, R.J., 109
Galis, F., 40, 41, 53, 87, 88, 198, 211,
   298, 314, 390, 474, 489, 574, 575,
   576, 578, 592, 596

Gallistel, C.R., 63, 208, 314, 462
Gamboa, G.J., 447, 449
Gan, Y.-Y., 73
Gans, C., 25, 506
Garcia-Bellido, A., 56, 62, 119, 129, 212, 256, 357
Garstang, W., 10
Gary, N.E., 340
Gasaway, W.C., 391, 518, 594
Gass, C.L., 456
Gatehouse, A.G., 383
Gauld, I., 22, 606
Gause, G.F., 116, 151
Geary, D.H., 628
Geis, I., 500
Geist, V., 110, 114, 246, 249, 608, 609
Gelperin, A., 444
George, L., 290
George, S.B., 515
Gerhart, J., 5, 8, 37, 42, 43, 51, 54, 56, 58, 59, 62, 63, 70, 85, 88, 91, 92, 110, 118, 135, 159, 160, 161, 165, 176, 177, 182, 185, 210, 212, 222, 247, 325, 327, 336, 426, 480, 489, 490, 491, 492, 495, 496, 505, 591, 615
Ghiselin, M.T., 191, 192, 358, 361, 370, 489, 540, 630, 631, 632, 633, 634, 636
Giacomi, C., 277
Gibbons, J.R.H., 528, 559, 560
Gibbs, A., 317, 318, 322
Gibson, J.B., 399
Giddings, L.V., 7, 402
Giesel, J.T., 92
Gilbert, J.J., 111, 369, 395, 462
Gilbert, S.F., 14, 57, 93
Gilbert, W., 171, 320, 321, 322, 324, 327, 328
Gilbertson, L.G., 169
Gill, A.B., 238, 299, 549, 571
Gill, F.B., 456, 463
Gillespie, J.H., 19, 434, 435
Gilliam, J.F., 444, 508, 550
Gingerich, P.D., 362
Giraldeau, L.A., 129, 342, 344, 345, 346, 382, 463, 586, 595
Girman, D.J., 588
Gittleman, J.L., 181
Givnish, T.J., 425
Glass, R.E., 230
Glynias, M., 320, 321, 324, 327, 328
Go, M., 320
Godfray, C., 304
Godfray, H.C.J., 421, 468
Godin, J.G.J., 347, 433
Goethe, J.W.von, 269, 303, 305
Goldfoot, D.A., 277
Golding, B., 386
Golding, G.B., 322
Goldman, B.D., 114, 154, 424, 427, 459, 460, 461
Goldschmidt, R., 11, 23, 24, 116, 160, 161, 176, 202, 208, 218, 233, 234, 255, 256, 270, 305, 357, 371, 471, 475, 509, 527, 598, 603

Goodall, J., 289, 290, 434, 456, 457
Goodman, M., 211
Goodnight, C.J., 79, 593
Goodwin, B.C., 478
Goodwin, F.K., 101, 105
Gordon, D.M., 26, 36, 42, 439
Gore, R.H., 517
Gorman, M.R., 45, 101, 114, 459, 460
Gosliner, T.M., 358, 361, 370, 489
Gotthard, K., 180
Gottlieb, G., 6, 140, 141, 157, 373, 474
Gottlieb, L.D., 106, 265
Gotwald, W.H., Jr., 272
Gould, S.J., 5, 7, 8, 9, 10, 12, 18, 22, 23, 26, 28, 144, 164, 165, 171, 184, 198, 201, 219, 230, 237, 241, 242, 244, 249, 250, 255, 276, 303, 307–9, 311, 313, 361, 362, 363, 369, 380, 388, 401, 416, 454, 474, 475, 478, 480, 482, 511, 521, 528, 550, 569, 599, 600, 601, 603, 607, 608, 609, 617, 618, 620, 623, 628, 629, 631
Goy, R.W., 261
Graham, J.B., 391, 521, 606, 607
Grant, B.R., 199, 225, 226, 346, 347, 384, 403, 418, 508, 534, 541, 562, 581, 582, 583, 584, 618, 626
Grant, P.R., 162, 181, 199, 249, 316, 339, 346, 346–48, 347, 350, 382, 384, 390, 397, 403, 418, 463, 508, 534, 541, 562, 574, 577, 581, 582, 583, 584, 586, 589, 590, 597, 618, 624, 626
Grant, V., 6, 7, 199, 415
Grassé, P.P., 42, 322, 599
Gray, W.N., 574
Green, P., 344, 347, 595
Greenberg, B., 293
Greenberg, L., 447
Greenberg, R., 342, 347, 348, 350, 590
Greene, E., 387, 420, 446, 454
Greene, H.W., 63, 138, 367, 494
Greenspan, R.J., 54, 83, 84
Greenwood, P.H., 299, 571, 574, 575, 576, 591
Griffin, D.R., 314
Griffiths, H., 513
Grimaldi, D., 206, 355, 493, 494
Gronenberg, W., 64
Gross, M.R., 113, 221, 226, 280, 294, 295, 386, 388, 390, 420, 422, 423, 425, 428, 429, 430, 431, 432, 433, 434, 547, 572
Grüneberg, H., 104
Gruppioni, R., 74
Gubernick, D.J., 287
Guerrant, E.O., Jr., 250
Gunn, D.L., 38, 39
Gupta, A.P., 27, 101, 102, 170
Gustafsson, L., 97
Guthrie, E.E., 423
Guthrie, R.D., 249
Gutierrez, E.E., 36, 227, 309, 310, 381, 400, 474
Gwynne, D.T., 366

H

Haag, E.S., 427
Haas, O., 487
Haber, W.A., 390, 594
Hack, M., 454, 455, 456
Hadfield, M.G., 514, 515
Haeckel, E., 607
Haecker, V., 357
Hagen, D.W., 169, 413
Haig, D., 636
Hailman, J.P., 244, 247
Hairston, N.G., 437, 438
Haldane, J.B.S., 9, 31, 83, 84, 189, 203, 210, 211, 419
Hall, B.G., 239, 508
Hall, B.K., 5, 12, 21, 37, 57, 60, 65, 70, 71, 104, 113, 116, 117, 134, 135, 150, 152, 175, 176, 179, 199, 219, 232, 233, 234, 238, 298, 299, 303, 308, 485, 488, 497, 616
Hall, J.C., 343, 459
Hallas, S., 409
Halliday, T., 26, 396
Halliday, T.R., 456
Hamburger, V., 4, 28, 428
Hamilton, W.D., 5, 17, 31, 43, 142, 391, 392, 395, 418, 420, 421, 427, 430, 432, 435, 438, 447, 591, 609, 612, 630, 631, 632, 637
Hamilton, W.J., III, 83, 276, 388, 389
Hammerstein, P., 389
Hampé, A., 233
Handlirsch, A., 606
Hänel, H., 552, 553
Hanken, J., 298, 299, 308, 309, 609
Harcourt, A.H., 291
Hardie, J., 115, 132
Harding, K., 74
Hardisty, M.W., 535
Hardy, A., 242, 497, 607
Harmelin, J.G., 624
Harris, R.N., 118, 383
Harrison, A.G., 474
Harrison, R.G., 131, 311, 373, 561, 562
Hart, M.W., 514
Hart, P.J.B., 238, 299, 549, 571
Hartfelder, K., 74, 130, 284
Hartl, D.L., 171, 318, 325
Hartshorn, G.S., 302
Harvell, C.D., 47, 251, 391, 392, 397, 624
Harvey, P.H., 290
Haselkorn, R., 330
Hashimoto, Y., 283
Haskins, C.P., 508, 635
Hatfield, T., 104, 545, 566
Haukioja, E., 392
Hausmann, A., 552
Hawryluk, M.D., 304
Hay, J., 272
Hayes, T.B., 35, 179, 261, 608
Hazel, W.N., 118, 125, 127, 169, 179, 383, 429, 430, 504
Hazlett, B.A., 446

Heatwole, H., 559
Hedrick, P.W., 79, 418, 419, 420
Heed, W.B., 578, 579, 581
Heinrich, B., 337, 339, 340, 349, 350
Heinze, J., 75, 107, 120, 283, 291, 387,
    448, 452, 538, 542, 543, 544
Heliövaara, K., 555
Heller, K.G., 271
Helversen, D., von, 271
Hennig, W., 88, 492, 493
Henry, C.S., 365, 556, 557
Hepper, P.G., 343
Herbers, J.M., 537, 538
Herbert, A., 330, 331
Herbst, L.H., 349
Herre, E.A., 180
Hersh, A.H., 26, 309
Herskowitz, I.H., 79, 101, 230
Hewitt, G.M., 381
Hews, D.K., 280, 281
Higashi, M., 539, 540
Higgins, L.E., 203
Hill, J.P., 22, 171, 250
Hill, W.G., 155
Hillis, D.M., 272, 361, 370, 373, 381,
    490, 491
Hilu, K.W., 106, 269
Hindar, K., 548
Hinde, R.A., 64, 238, 247, 291, 314,
    337, 346, 347, 348, 391, 441
Hinton, H.E., 96
Hirsch, J., 342
Ho, M.W., 474, 478, 599
Hoagland, K.E., 381, 403
Hoegh-Guldberg, O., 221, 514
Hoey, T., 319
Hoffman, A.A., 315–16, 383
Hoffman, E.A., 387, 462
Hogendoorn, K., 404, 448
Höglund, J., 593
Holekamp, K.E., 40
Holland, B., 313
Holland, G., 83
Holland, P., 171, 211, 212
Holland, S.K., 62, 72, 211, 212, 318,
    320, 321, 322, 324, 328
Hölldobler, B., 64, 120, 131, 246, 272,
    283, 312, 456, 457, 537, 542, 543,
    544, 573
Holliday, M., 342
Holling, C.S., 303
Holton, G., 83
Holzapfel, C.M., 560
Hoogerhoud, R.J.C., 575
Hopkins, D.D., 591
Hori, K., 109
Horner, B.E., 391, 409, 445
Houston, A.I., 456
Howden, H.F., 169, 400
Hoy, M.A., 383
Hrdy, D.B., 449, 451, 453
Hrdy, S.B., 276, 289, 291, 449, 451,
    453, 466, 468
Huang, Y., 106
Hubbell, S.P., 540, 541
Huber, F., 271
Huber, R., 286

Huettel, M.D., 156
Huey, R.B., 4, 180
Hufford, L., 353
Hughes, A.L., 214
Hughes, C.R., 451
Hull, D.L, 83
Humphrey, N.K., 464
Hung, A.C.F., 74
Hunkapiller, T., 321
Hunt, G.L., Jr., 286, 293
Hunt, J.H., 283
Hunt, M.W., 286, 293
Hurst, J.M., 362, 596
Hurst, L.D., 320, 321, 630, 635
Hurst, P.S., 404
Hutchings, M.J., 39
Huxley, J., 154, 198, 230, 255, 256,
    308, 317, 415

I

Iles, T.D., 364, 569, 574, 592
Iljin, N.A., 255, 256
Iljin, V.N., 255
Iltis, H.H., 260, 265, 266, 267, 268,
    269, 514
Imperatriz, V.L., 292
Inoue, K., 323
Inoue, T., 448
Inouye, D.W., 349
Ireland, L., 237
Irschick, D.J., 594
Irwin, R.E., 273
Itani, J., 289, 290
Itô, Y., 387, 555
Ivlev, V.W., 299, 550, 567

J

Jablonka, E., 96, 113, 115, 476, 508,
    636
Jackson, J.B.C., 251, 361, 381, 401,
    624, 625, 626, 627, 628
Jackson, R.R., 65, 386, 387, 409, 445,
    605
Jacob, F., 70, 79, 159, 317
Jaenike, J., 343, 578, 580, 581
Jaffe, M.J., 301
Jagendorf, A., 47, 333, 392
Jago, N.D., 595
Jain, S., 383
James, F.C., 103, 523
Jamieson, I.G., 201, 223, 276, 474, 475
Jamison, K.R., 101, 105
Janis, C.M., 608, 609
Jannett, F.J., Jr., 360
Janzen, F.J., 121, 122, 123, 124, 125,
    130, 383
Japyassú, H.F., 61
Jarvi, T., 456
Jayakar, S.D., 419
Jeanne, R.L., 24, 186, 227, 247, 283,
    314, 349, 408, 433, 439, 454
Jebram, D., 624, 628
Jeffery, W.R., 615
Jehl, J.R., Jr., 207
Jennings, D.H., 77
Jennings, R.D., 300

Jennions, M.D., 593, 631
Jinks, J.L., 136
Jockusch, E.L., 335
Johannsen, J., 447
Johannsen, W., 31, 83, 378
John, B., 5, 19, 25, 28, 52, 57, 69, 72,
    75, 80, 129, 135, 185, 210, 214,
    219, 232, 279, 318, 330, 331, 333,
    335, 474, 596, 603, 613, 615, 616
Johns, J.E., 273
Johnson, L.K., 540, 541
Johnson, M.P., 26, 163, 229, 239, 240,
    390, 391, 393, 397, 401, 403, 414,
    415, 416, 531, 532
Johnson, M.S., 477
Johnson, N.K., 523, 589, 596
Johnson, R., 211
Johnson, R.A., 543, 544
Johnson, T.C., 519, 575
Johnston, S.A., 636
Johnston, T.D., 140, 141, 157, 474
Jolly, A., 289, 290, 464
Jones, C.D., 580
Jones, R.E., 523
Jónsson, B., 548
Juchault, P., 265
Julkunen-Tiitto, R., 109
Junk, W.J., 594

K

Kaiser, D., 501, 503
Kallman, K.D., 107, 384, 422
Kambysellis, M.P., 94, 97, 578, 579,
    580, 581
Kamm, D.R., 228
Kaneshiro, K.Y., 73, 573, 574, 578, 579,
    581, 593
Kaplan, R.H., 113, 114, 424, 425, 426
Kaplin, B.A., 595
Karasov, W.H., 518, 595
Karban, R., 392
Karplus, M., 49, 50
Karr, T.L., 95, 633
Kary, M., 13, 31
Katz, M.J., 37, 500
Kauffman, S.A., 5, 8, 9, 19, 25, 37, 59,
    99, 128, 144, 164, 335, 474, 499
Kaufman, L.S., 163, 299, 365, 508, 575,
    576, 577, 604
Kaufman, T.C., 5, 22, 23, 25, 27, 28,
    62, 69, 72, 73, 74, 90, 144, 168,
    171, 173, 185, 186, 200, 205, 221,
    242, 256, 296, 361, 362, 380,
    499, 615
Kavanau, J.L., 238
Kawano, K., 306
Kay, Q.O.N., 26
Kear, J., 586, 594, 595
Kearns, C.W., 421
Keast, A., 550, 567, 569, 595
Keeping, M.G., 451
Keese, M.C., 552
Keese, P.K., 317, 318, 322
Keller, L., 106, 438, 449–50, 452, 537,
    538, 543
Kellerman, W.A., 266

Kellogg, V.L., 215
Kennedy, J.S., 58, 98, 113, 132, 137, 203, 249, 300, 311, 381, 434, 440
Kennedy, W.J., 362
Kerr, W.E., 128, 171, 284, 421, 509
Ketterson, E.D., 54, 78, 157, 205, 282, 286, 291, 303, 304, 311, 316, 338, 509
Kettlewell, H.B.D., 420
Keynes, R.D., 590
Khryanin, V.N., 262
Kidwell, M.G., 49, 329, 330, 332, 333, 334
Kimball, R.T., 274, 275
Kimura, M., 19, 79, 148, 158, 162, 231, 239, 317, 378, 610
King, M.-C., 52, 335
King, R.C., 366, 368
Kingsolver, J.G., 203, 397, 532
Kinomura, K., 75, 291, 543, 544
Kirkpatrick, M., 4, 8, 155, 402, 568, 569, 619
Kirschner, M.W., 5, 8, 37, 38, 42, 43, 51, 53, 54, 56, 58, 59, 62, 63, 70, 85, 88, 91, 92, 110, 118, 135, 159, 160, 161, 165, 176, 177, 182, 185, 210, 212, 222, 247, 325, 327, 336, 426, 480, 489, 490, 491, 492, 495, 496, 505, 591, 615
Klahn, J.E., 368
Kleiman, D.G., 287
Klein, J., 215, 216
Klein, N.K., 553
Klepaker, T., 570
Klingenberg, C.P., 427, 428
Klopfer, P., 382
Kluge, M., 251
Knerer, G., 228, 555
Knight, M.R., 108
Knowlton, N., 181, 262, 287, 551, 628
Knülle, W., 383, 437, 438
Kodric-Brown, A., 449, 456, 502
Koehl, M.A.R., 564
Koenig, W.D., 223, 433
Kofoid, E.C., 62, 84
Kojima, J.-I., 283
Kojima, K., 79
Kolattukudy, P.E., 111
Kollar, E.J., 234
Kölliker, M., 466, 467
Koref-Santibañez, S., 343
Kornfield, I., 387, 576
Korpelainen, H., 121, 385
Koshland, D.E., Jr., 426, 427, 462
Koufopanou, V., 207
Kovoor, J., 235
Krakauer, D.C., 539, 540
Kramer, D.L., 405
Krantz, G.W., 553
Kraus, M., 249
Kraus, O., 249
Krebs, J.R., 17, 245, 386, 443, 466
Kristensen, N.P., 368, 371
Krumlauf, R., 73, 318, 319, 326, 327
Kruuk, H., 261, 276
Kubo, T., 74

Kühme, W., 455
Kühn, A., 205
Kukalova-Peck, J., 199, 485, 491
Kukuk, P.F., 228, 292
Kulincević, J.M., 155
Kurczewski, F.E., 222, 407, 426
Kurland, J.A., 290
Kurosu, U., 221, 254
Kurtén, B., 235
Kutsch, W., 271

L
LaBarbera, M., 364
Lacey, E.A., 288
Lacey, E.P., 91
Lack, D., 168, 293, 583, 586
Laland, K.N., 571, 578
Lamb, M.J., 96, 113, 115, 476, 508, 636
Lande, R., 142, 166, 168, 169, 230, 475, 480, 619
Landman, O.E., 96
Lang, J.W., 100, 121
Lank, D.B., 421, 593
Laoide, B.M., 58
Larkin, P.A., 344, 535, 536, 541, 547, 548
Larsen, E., 60, 87, 94
Larson, A., 56
Larson, T.B., 217, 390
Larwood, G.P., 565
Laskin, D.M., 298
Lauder, G., 87, 209, 210, 211, 405
Lauder, G.V., 595
Lawrence, P.A., 63
Lawton, M.F., 22, 244, 248, 249
Lawton, R.O., 22, 244, 248, 249
Lawton-Rauh, A.L., 89
Lea, S.E.G., 595
Leatherland, J.F., 294
Le Conte, Y., 448
Lee, M.G., 111
Lee, P.C., 291
Lees, A.D., 115, 132
Lefebvre, L., 129, 344, 345, 382, 463, 586, 595
Legner, E.F., 92, 635
Lehmann, N.L., 270
Leigh, E.G., 112, 165, 174, 175, 185, 476
Le Moli, F., 109, 553
Lenski, R.E., 78, 148
Lepson, J.K., 591
Lerdau, M., 307
Lerner, I.M., 9
Leroi, A.M., 304
Lessells, C.A., 313, 316
Lessios, H.A., 514
Leutert, R., 101
Levene, H., 418, 419, 421
Levin, B., 273, 637
Levin, D.A., 80
Levin, L.A., 517
Levin, R.N., 273
Levine, M., 319
Levins, R., 102, 419, 421, 430, 431, 518, 621

Levinton, J.S., 77, 106, 118, 146, 232, 234, 238, 255, 256, 309, 354, 385, 474, 482, 601, 610, 613, 620
Lewin, B., 14, 330
Lewis, E.B., 219
Lewis, H., 198
Lewontin, R.C., 8, 16, 18, 26, 27, 31, 79, 80, 81, 101, 102, 143, 144, 154, 155, 156, 160, 165, 170, 192, 201, 378, 386, 418, 434, 474, 478, 615
Leys, R., 448
Li, D., 409
Li, W.-H, 239, 321, 331
Licht, P., 311
Lidgard, S., 366
Liebherr, J.K., 119, 120
Liebig, J., 450, 452, 455
Liem, K.F., 87, 88, 163, 183, 299, 365, 508, 564, 574, 575, 576, 577, 591, 599, 604
Lillie, F.R., 9, 183
Lin, C.P., 552
Lin, N., 510
Lincoln, R.J., 487, 598
Lindauer, M., 171
Lindegren, C.C., 216
Lindeque, M., 276
Lindquist, S., 99, 333
Lindström, A., 518, 594
Linsenmair, K.E., 343, 448
Lisch, D.R., 329, 330, 332, 333, 334
Lister, A.M., 620
Litte, M.I., 449, 450
Liu, T.P., 74
Lively, C.M., 68, 281, 391, 392, 397, 419, 420, 429, 430, 431, 433, 618
Lloyd, D.G., 262, 265, 269, 430
Lloyd, E.A., 30, 31, 165, 489, 509, 631
Lloyd, J.E., 248
Lloyd, J.M., 555, 556
Locke, M., 112
Logsdon, J.M., 321
Loher, W., 277
Long, M., 320
Lopez, E.R., 594
Lord, E.M., 22, 171, 250
Lorenz, K., 99, 341, 455
Lorenzi, M.C., 391
Losick, R., 501, 503
Losos, J.B., 56, 168, 204, 370, 594
Lott, D.F., 111, 114, 287, 288, 383, 425
Lounibos, L.P., 406, 560, 561
Loveridge, A., 256
Lowe-McConnell, R.H., 575, 578, 595
Lu, G., 327
Lubbock, J., 447
Lubin, Y., 447
Lucas, H.L., 420
Lukens, L.N., 268
Lüttge, U., 251, 512, 513
Luykx, P., 293
Lyman, R.F., 55, 169, 179
Lynch, M., 66, , 598, 619
Lyon, B.E., 419, 453, 458, 466, 637

M

Ma, J., 49, 50
Mabee, P.M., 10
MacFadden, B.J., 233, 618, 621, 622
Mackay, T.R.C., 4, 55, 68, 104, 149, 230
Mackintosh, N.J., 38, 39, 247, 314, 337, 346, 349, 463
Macnair, M.R., 105, 106
Madison, M., 58, 145
Maeshiro, T., 317
Maeta, Y., 147, 240, 284, 301, 509, 606
Maeterlinck, M., 235
Magurran, A.E., 280, 342, 344, 350
Mahner, M., 13, 31
Mahowald, A.P., 635
Maier, N.R.F., 337, 441
Malyshev, S.I., 606
Mangelsdorf, P.C., 265
Manning, A., 245, 313, 314
Maret, T.J., 567, 595
Margulis, L., 85, 501
Marin, F., 501, 522
Mark, R., 478
Markert, C.L., 70, 211
Markham, J.C., 462
Markl, H., 245, 246, 543
Markow, T.A., 103, 633
Marler, P., 15, 63, 64, 65, 163, 337, 342, 347, 445, 463
Marsh, P.M., 234
Marshall, C., 492, 496
Marshall, C.R., 231, 233, 234, 239, 366
Marshall, D.C., 555
Martin, A., 555, 556
Martin, C.E., 252, 513, 605
Martinez, D.E., 106
Martinez del Rio, C., 595
Martyns, M., 286
Masaki, S., 203, 205, 300, 406, 407, 408, 413, 529, 555, 562
Maschwitz, E., 310
Maschwitz, U., 310
Mason, R.T., 281
Mather, K., 6, 25, 77, 131, 136, 137, 138, 173, 225, 226, 394, 402, 418, 433, 434, 528
Matsuda, R., 5, 23, 75, 99, 101, 106, 118, 152, 153, 172, 207, 208, 221, 248, 249, 300, 305, 307, 308, 310, 311, 312, 362, 370, 373, 415, 474, 478, 480, 481, 499, 504, 507, 511, 514, 515, 516, 517, 518, 525, 529, 535, 551, 607, 608, 613
Mattheck, C., 302
Matthews, R.W., 560, 596
Matthews, T.C., 80
Maturo, F.J.S., 628
May, R.M., 427, 498
Mayer, G.C., 106, 528, 554
Mayer, I., 286, 294
Maynard Smith, J., 5, 8, 9, 25, 56, 79, 80, 85, 115, 139, 155, 163, 164, 184, 185, 225, 265, 282, 387, 388, 399, 405, 418, 419, 422, 424, 425, 428, 429, 430, 432, 434, 440, 474, 524, 528, 530, 540, 542, 548, 554, 562, 573, 631, 633

Mayo, O., 211
Mayr, E., 4, 5, 7, 9, 10, 15, 24, 25, 28, 43, 79, 90, 106, 139, 146, 152, 160, 165, 180, 181, 184, 188, 191, 192, 193, 201, 238, 260, 272, 273, 275, 276, 353, 371, 378, 388, 401, 402, 415, 428, 441, 475, 481, 482, 489, 505, 523, 526, 528, 544, 545, 551, 554, 564, 592, 593, 600, 601, 607, 617, 618
McAdams, H.H., 428
McAllister, L.B., 77
McAlpine, D.K., 493
McBride, G., 63, 171
McClay, D.R., 222, 243, 244, 255, 258
McClintock, B., 48, 49, 330, 333, 508
McCollum, S.A., 300, 392, 431
McCune, A.R., 361, 363, 519, 575, 578
McDonald, D., 249
McDonald, J., 50
McDonald, J.F., 332, 508, 580
McDowall, R.M., 535, 536, 572, 594
McEdward, L.R., 22, 113, 515
McEwen, B.S., 261
McFadden, D., 276
McFall-Ngai, M.J., 501
McFarland, D., 303, 341
McGinnis, W., 73, 318, 319, 326, 327
McKaye, K.R., 574, 591
McKeown, M., 48, 322
McKinney, F., 251, 624, 625, 626, 627, 628
McKinney, M.L., 5, 21, 22, 78, 129, 133, 241, 242, 243, 244, 252, 309, 316, 363, 380, 607, 609, 615, 618
McKitrick, M.C., 25, 171, 207, 286
McKnight, M.L., 208
McLain, D.K., 103
McLaughlin, H.M.G., 94
McLellan, T., 139, 143
McLennan, D.A., 204, 487, 598
McLeod, L., 387, 512, 532
McMahon, B., 49
McNamara, K.J., 5, 21, 22, 78, 129, 133, 241, 242, 243, 244, 252, 309, 316, 363, 380, 607, 609, 615, 618, 624
McNutt, J.W., 83, 388, 389
McPhail, J.D., 238, 535, 545, 549, 550, 566, 567, 570, 571, 572
McPheron, B., 558, 559
McVean, G.T., 320, 321
McWilliams, S.R., 4, 594
Meinhardt, H., 57
Meinwald, J., 502
Meireles, C.M.M., 211
Mello, C.V., 77
Mendel, G., 378
Menke, A.S., 408
Menzel, R., 349, 351
Mercot, H., 581
Messeri, P., 277
Messina, F.J., 383
Metcalf, C.L., 110
Metcalf, R.A., 450
Meyer, A., 5, 54, 299, 359, 387, 397, 487, 548, 571, 575, 576, 577, 578, 596

Meyer, D.C., 423, 450
Meyer, V.G., 269, 270, 371, 505, 604
Meyerowitz, E.M., 270
Michener, C.D., 5, 68, 128, 205, 228, 235, 246, 264, 291, 337, 343, 349, 367, 370, 378, 398, 400, 510, 543, 573, 604
Michod, R.E., 186
Mickel, C.E., 224, 225, 229, 381
Mickelvich, M.F., 204
Miklos, G., 5, 19, 25, 28, 52, 57, 69, 72, 75, 80, 129, 135, 185, 210, 214, 219, 232, 279, 318, 330, 331, 333, 335, 474, 596, 603, 613, 615, 616
Milbrath, L.R., 557
Miles, D.B., 204
Milinski, M., 313, 350, 429, 444, 463, 571
Milkman, R.D., 94
Miller, E.H., 63, 65
Miller, R., 168
Miller, R.C., 407
Milligan, B.G., 620
Milne, C.P., Jr., 95
Milton, K., 464
Misof, B.Y., 495
Mitani, J.C., 453
Mitchell, G., 277, 289, 290, 291
Mitchell, M., 205
Mitchell, R., 443, 465, 466
Mitchell, S.D., 42, 63
Mitchison, T.L., 42
Mitter, C., 558
Miura, T., 74
Moczek, A.P., 156, 169, 226, 227, 396, 400
Moger, W.H., 285
Mohammedi, A., 448
Mohanty-Hejmodi, P., 256
Mole, S., 303, 307, 310, 311, 374
Moller, F., 70
Monnin, T., 452, 453
Monod, J., 70, 79
Monteiro, A.F., 215, 216, 217
Montgomerie, R.D., 453
Montgomery, E.G., 266
Montgomery, S.L., 578, 579, 581
Moore, A.J., 12, 97, 112, 150, 455
Moore, M.C., 273, 280, 281, 295
Morales, M., 522
Moran, N.A., 26, 66, 67, 88, 125, 127, 128, 220, 221, 378, 386, 397, 406, 424, 430, 431, 435, 534
Morata, G., 63
Mordue, A.J., 314
Moreno, G., 225, 382, 384, 418
Morgan, C.L., 25, 151, 161, 180
Morgan, T.H., 154
Mori, A., 109, 553
Moritz, C., 381
Moritz, R.F.A., 185, 438
Morris, D., 247, 286, 293, 294
Morris, M.R., 422, 456
Morris, S.C., 610, 611, 613
Morse, D.H., 337, 349, 369
Moser, J.C., 530

Monson, R.K., 512, 513, 514
Moulds, M.S., 456
Mound, L.A., 387
Mousseau, T.A., 113, 115, 416
Moya, A., 593
Moynihan, M., 99, 212, 246
Mrosovsky, N., 47, 297, 304, 315, 433, 556
Mulkey, S.S., 229
Müller, G.B., 51, 54, 144, 233, 478, 489
Müller, U., 349
Mullet, J.E., 44
Munstermann, L.E., 80
Murakami, Y., 560
Murdoch, W.W., 577
Murfet, I.C., 103, 105
Murphy, P.J., 396, 397
Myers, J.H., 392
Myers, J.P., 456
Myles, T.G., 73, 74, 293

N

Naeem, S., 487
Nagel, L.M., 365, 369, 545, 546, 547, 548, 550, 551, 567, 572
Nalepa, C.A., 248
Nanny, D.L., 183
Needham, J., 57, 171
Negus, N.C., 460
Nei, M., 7, 211, 593
Nelson, D.A., 40
Nelson, G., 486
Nelson, R.J., 154, 287, 424, 427, 459, 460
Nevo, E., 386
Newman, S., 138
Nichol, N., 112
Nicoletto, P.F., 300, 449, 523
Niehrs, C., 70, 168
Nielsen, E.T., 407
Nielsen, R.A., 128
Nijhout, H.F., 5, 12, 13, 23, 44, 45, 56, 62, 63, 65, 73, 74, 75, 80, 86, 105, 106, 107, 119, 124, 127, 132, 133, 134, 147, 149, 150, 154, 151, 158, 166, 167, 179, 184, 208, 209, 214, 215, 216, 217, 227, 238, 247, 256, 257, 279, 305, 306, 309, 310, 357, 373, 390, 400, 401, 427, 428, 430, 436, 461, 462, 483, 491, 492, 499, 588, 635
Nilsson, D.E., 492
Nishida, T., 453
Niven, D.K., 286
Nixon, G.A., 559
Noakes, D.L.G., 387
Noble, G.K., 118, 249
Noble, P.S., 251
Noirot, 288, 292, 293
Nolan, V., Jr., 54, 78, 157, 205, 282, 286, 291, 303, 304, 311, 316, 338, 509
Nonacs, P., 449–450, 451, 452, 542
Noonan, K.C., 283
Noonan, K.M., 432, 450
Norden, B.B., 228
Nordeng, H., 548

Nordenskiold, E., 305
Norris, D.O., 118
North, G.B., 251
Nosaka, M., 320
Nottebohm, F., 108
Novacek, M.J., 372
Novak, M.A., 287, 288
Nurse, P., 111
Nutting, W.L., 293
Nylin, S., 180
Nyman, T., 109
Nymeyer, H., 524

O

O'Donald, P., 16, 273, 383, 385, 405, 418
O'Donnell, S., 283, 349
Ogata, K., 543
O'Grady, R.T., 507
Ohlsson, R., 636
Ohno, S., 162, 209, 210, 211, 230, 231, 442
Ohta, A.T., 9
Ohta, T., 19, 148, 156, 158, 162, 211, 215, 216, 239
Oldroyd, B.P., 438, 447
O'Leary, D.D.M., 41
Olivera, B.M., 591
Olson, E., 168
O'Malley, B.W., 318
O'Meara, G.F., 119, 265, 406, 561
Omholt, S.W., 154, 155
Oren, A., 604
Oring, L.W., 273, 286
Orr, A.R., 266, 267, 268
Orr, H.A., 104, 106, 151, 478, 528, 588, 589
Osborn, H.F., 25, 151, 161, 180, 483
Osborne, S., 366
Osche, G., 239
Osse, J.W.M., 87, 574
Oster, G., 133, 164, 377, 527
Ostrom, J.H., 199
Ott, E., 38
Otte, D., 59, 63, 172, 590
Owen, R., 489
Owens, L.P.F., 274
Owens, R.M., 522
Oyama, S., 13

P

Pachut, J.F., 361, 381, 624, 627
Packer, L., 229, 235, 450
Padgett, R.W., 327
Padilla, D.K., 251, 391, 430, 431, 624
Page, R.E., Jr., 42, 76, 148, 150, 156, 314, 340, 349, 383, 438, 439
Paigen, K., 176
Palmer, A.R., 110, 362, 618
Palmer, J.D., 230, 231, 358
Palumbi, S. 99, 96
Pamilo, P., 450, 573
Panek, L.M., 447
Pankiw, T., 340
Pantel, J., 234
Papaj, D.R., 111, 178, 337, 338, 340, 341, 558, 559

Pardi, L., 282, 449, 451, 454, 455
Park, T., 303
Parker, G.A., 310, 313, 388, 390, 397, 402, 417, 419, 425, 426, 428, 429, 432, 433, 434, 633
Parker, M.G., 261
Parker, W.E., 383
Parkinson, J.S., 62, 84
Parmesan, C., 444
Parra, R., 501
Parsons, P.A., 315–316, 383, 579, 580
Partridge, L., 310, 313, 337, 344, 347, 595
Pass, G., 200
Passera, L., 74, 120
Patterson, C., 486, 489, 491, 494, 495
Patthy, L., 321, 324, 523
Patton, J.L., 510, 511
Paukstis, G.L., 121, 122, 123, 124, 125
Paulsen, S.J., 12, 105, 149, 150, 151, 154, 179, 430, 436, 588
Pavelka, M., 41, 133, 134, 230, 238, 255, 257, 258
Payne, L.L., 458, 553
Payne, R.B., 458, 461, 553
Payne, R.J.H., 539, 540
Pearsall, D.M., 268, 269
Pearson, R., 542, 543, 544
Peck, J.R., 630, 635
Peck, S.B., 22
Peck, S.L., 144
Peckham, D.J., 190, 407
Pedersen, C.A., 287
Peeters, C., 76, 450, 452, 453
Pener, M.P., 149, 381, 556
Pennisi, E., 636
Perez, J., 264
Perrin, N., 17
Peschke, K., 264, 278
Peters, R.H., 316
Peterson, C.C., 316
Peterson, G., 160
Peterson, K., 636
Petraitis, P.S., 635
Petren, K., 586
Petrie, M., 593
Pettersson, L.B., 392, 550, 572
Pettigrew, J.D., 110
Pettis, J.S., 452
Pfennig, D.W., 111, 307, 369, 387, 396, 397, 398, 419, 431, 433, 448, 457, 462, 465, 518
Philippi, T., 424
Piaget, J., 403
Piano, F., 597
Pickard, E., 427
Piersma, T., 518, 594
Pigeon, D., 567, 568, 569
Pigliucci, M., 5, 15, 26, 35, 36, 55, 179, 250
Pilecki, C., 405
Pilon-Smits, E.A.H., 513, 514
Pimm, S.L., 414, 449, 590
Pinker, S., 81, 497
Piperno, D.R., 268, 269
Pitnick, S., 95, 633
Plateaux, L., 131

Plateaux-Quenu, C., 235, 240
Platt, J.E., 118
Platt, J.R., 454
Plomin, R., 150
Podell, M.E., 625
Polak, M., 205, 593
Policansky, D., 280, 422
Polis, G.A., 369
Pollet, N., 70, 168
Popov, A.V., 456
Portin, P., 328
Posner, M.I., 341
Potter, I.C., 73, 535
Potter, N.B., 228
Pough, F.H., 500
Powell, R.A., 288
Powers, D.A., 435
Pratt, B.L., 114
Premoli, M.C., 635
Prescott, D.M., 333
Price, G.R., 432, 434
Price, T., 41, 133, 134, 230, 238, 255,
    257, 258, 583, 584, 585, 624
Prokopy, R.J., 111, 557, 558, 559, 562
Provine, W.B., 7, 9, 11, 144, 148, 388,
    428
Prusiner, S.B., 4
Ptashne, M., 25, 67, 73, 84, 103, 230,
    379, 380
Pulliam, H.R., 429
Punnett, R.C., 208

Q
Queller, D.C., 91, 432, 447, 448, 450,
    451, 452, 636
Quinn, T.P., 294

R
Rabalais, N.N., 517
Rabinow, L., 230
Racey, P.A., 276
Rachinsky, A., 130
Rachootin, S.P., 53, 54, 142, 143, 161,
    234, 301, 370, 507, 511, 591
Radinsky, L.B., 52
Radwan, J., 433
Raff, E.C., 81
Raff, R.A., 5, 10, 22, 23, 25, 27, 28, 57,
    62, 69, 70, 72, 73, 74, 81, 90, 99,
    129, 144, 168, 171, 173, 176, 185,
    186, 200, 205, 221, 222, 230, 231,
    241, 242, 244, 256, 296, 335, 360,
    361, 362, 369, 380, 488, 499, 514,
    515, 517, 610, 615
Raikow, R.J., 206, 207, 233, 237, 360
Ramaiah, T.R., 277
Ramakrishnan, V., 492
Ramirez, W., 234
Ramsey, A.O., 415
Rand, A.L., 502
Rankin, M.A., 203, 311
Ratnieks, F.L.W., 452
Rau, N., 351
Rau, P., 24, 153, 178, 180, 351
Raup, D.M., 610, 611, 613
Raven, J.A., 251, 259, 513, 521, 522

Raveret Richter, M.B., 349, 351, 444
Ray, T.S., 39, 58, 59, 200, 251, 270,
    306
Rayner, A.D.M., 448
Reader, S.M., 571, 578
Real, L.A., 337, 463
Réat, V., 49
Reburn, C.J., 287
Reed, H.C., 410
Reeve, H.K., 140, 201, 262, 370, 372,
    438, 449, 450, 451, 457, 474, 476
Regelmann, K., 444
Reichman, O.J., 444
Reid, R.G.B., 151
Reik, W., 636
Reilly, S.M., 241, 242, 254
Reimchen, T.E., 413, 443, 523, 550,
    570, 572
Reitsma, N., 217
Remane, A., 489
Renoux, J., 293
Rensch, B., 108, 117, 140, 233, 234,
    237, 240, 305, 309, 357, 361, 362,
    363, 490, 499, 598
Repka, J., 390, 430, 432
Reynolds, F., 456
Reynolds, V., 445, 456
Reznick, D., 84, 312, 313
Rhen, T., 100, 121
Rice, B., 91
Rice, W.R., 313, 404, 415, 526, 528,
    538, 539, 542
Rich, A., 330, 331
Richards, L.B., 118
Richards, M.H., 235, 236, 363
Richards, O.W., 167, 271, 378, 386,
    387, 498, 507, 523
Richardson, R.H., 572
Richerson, P.J., 341
Richey, F.D., 266, 267
Riddiford, L.M., 59, 66
Ridley, M., 6, 8, 139, 262, 483, 541
Riechert, S.E., 388, 389, 440
Riedl, R., 57, 62, 78, 82, 168, 211, 326,
    486
Rieppel, O., 198, 204, 487
Riggs, A.D., 49
Riley, C.V., 22
Rinderer, T.E., 155
Riska, B., 304
Rissing, S.W., 543
Rivera, M.C., 199
R'kha, S., 580, 581
Roberts, T.K., 387
Roberts, W.M., 595
Robertson, A., 247
Robertson, F.W., 73, 581
Robertson, H.M., 278, 279
Robertson, R., 381
Robinson, B.W., 25, 151, 153, 164,
    397, 434, 508, 531, 546, 567, 571,
    577, 594
Robinson, G.E., 76, 349, 351, 383, 438,
    449
Robson, G.C., 498, 507, 508
Rocha, I.R.D., 294

Roe, A., 24
Roff, D.A., 17, 68, 70, 104, 106, 118,
    125, 126, 149, 150, 151, 154, 156,
    169, 304, 310, 311, 312, 373, 383,
    386, 422, 424, 427, 435, 436
Roger, A.J., 630
Rohfritsch, O., 111, 552
Rohwer, F.C., 425, 456
Rohwer, S., 425, 441, 456
Roitberg, B.D., 559
Rollo, C.D., 9, 91, 240, 303, 304, 316
Romanes, G.J., 474, 528
Romer, A.S., 180, 606, 607
Rood, S.B., 266, 267
Roper, T.J., 337
Rosbash, M., 459
Rose, M.R., 310, 311, 434
Röseler, P.-F., 74, 76, 147, 228, 449,
    450, 451, 453
Rosenberg, A., 56
Rosenblatt, J.S., 289, 290
Rosenzweig, M.L., 440, 442, 528, 559
Ross, C.W., 35, 39, 45, 46, 47
Ross, K.G., 106, 361, 510, 537, 538
Rossiter, M.C., 96, 113, 114, 115, 142
Rossman, M.G., 320
Roth, G., 94, 171, 198, 200, 335, 486,
    489, 490, 495, 496, 596
Roth, V.L., 57, 65, 82, 90, 91, 98, 157,
    362, 364, 377, 486, 489, 497, 507,
    511, 609
Rothenbuhler, W.C., 147, 339
Roubaud, E., 505
Roubik, D.W., 284
Roudabush, R.L., 506
Roulin, A., 468
Roundtree, D.B., 256
Routman, E.J., 154
Roux, W., 37, 54
Rowan, R.G., 319
Rowe, R.J., 22, 63, 429, 444
Rowell, C.H.F., 300
Rowell, T.E., 185
Roy, S.W., 320, 321
Rubenstein, D.I., 429, 454, 455, 456
Rüber, L., 568, 569
Ruffell, R., 235
Rundle, H.D., 365, 369, 528, 545, 550,
    572
Rüppell, O., 542, 543, 544
Russell-Hunter, W.D., 240
Ryan, C.A., 47, 76, 333, 392
Ryan, M.J., 26, 106, 107, 212, 420,
    422, 456, 503
Rzhetsky, A., 320

S
Sachs, T., 37, 40, 269, 462
Saether, O.A., 4, 357, 371, 490, 493
Sagan, D., 85
Sage, R.D., 387, 576
Sakagami, S.F., 144, 147, 240, 301, 313,
    509, 606
Salisbury, F.B., 35, 39, 45, 46, 47
Salmon, W.C., 98, 174
Salt, G., 225, 527

Sander, K., 496
Sanderson, M.J., 353, 491
Sandoval, C.P., 372, 381, 406, 420, 421
Sang, J.H., 4, 100
Santschi, F., 283
Sapienza, C., 636
Sará, M., 67, 86, 140, 141, 220
Sardet, C., 95
Sassaman, C., 279, 280, 423
Sato, A., 583, 586
Sattler, R., 267, 270, 490
Sauer, E.G.F., 293
Saunders, P.T., 599
Savalli, U.M., 98
Sbrenna, G., 74
Scapini, F., 179, 342, 343, 443
Schaefer, S.A., 595
Schaffner, J.H., 124, 125, 130, 267, 268
Schank, J.C., 326, 500
Scharloo, W., 116, 226, 227
Schartl, M., 224, 230, 231
Scheiner, R., 349, 351
Scheiner, S.M., 55, 151, 155, 169, 179, 250
Scheller, R.H., 327
Schenkel, R., 245
Schlichting, C.D., 5, 26, 35, 55, 178, 179, 250, 297
Schlinger, B.A., 77
Schluter, D., 97, 162, 226, 359, 365, 370, 372, 545, 546, 547, 548, 549, 550, 551, 564, 565, 566, 567, 571, 572, 574, 577, 578, 595, 597
Schmalhausen, I.I., 7, 8, 26, 152, 156, 167, 174, 180, 255, 357, 379, 405, 499
Schmid, B., 93
Schmid-Hempel, P., 349, 438
Schmid-Hempel, R., 349
Schmidt, G.H., 73, 74
Schmidt-Nielsen, K., 21, 316
Schmitt, J., 36, 179
Schmitt, W.L., 59
Schneirla, T.C., 42, 171, 337, 441
Schoch, R.M., 309
Schoener, T.W., 303, 595
Schopf, T.J.M., 362
Schröder, F.C., 331
Schultz, J., 62, 72, 270, 271
Schultz, T.R., 372
Schumann, R.D., 554
Schwabl, H., 92, 274, 275
Schwanwitsch, B.N., 56
Schwartz, R.J., 320, 324, 331
Schwarz, M., 228, 292
Scott, G.R., 173
Scott, N.J., Jr., 300
Scully, E.P., 446
Scurfield, G., 302
Seeley, T.D., 37, 171, 439, 444, 445, 446
Seger, J., 43, 57, 369, 418, 419, 424, 428, 432, 528, 539, 540

Seibt, U., 249
Selander, R.K., 188, 290, 387, 568, 569, 576, 594, 595
Sella, G., 282, 458, 635
Semlitsch, R.D., 119, 383, 434
Servedio, M.R., 370, 371
Setty, B.N.Y., 277
Severson, D.W., 74, 398
Sexton, E.W., 154
Seyfarth, R.M., 457
Shaffer, H.B., 203, 207, 208, 354, 370, 372, 548
Shakarad, M., 450
Shapiro, A.M., 23, 26, 133, 152, 157, 203, 205, 229, 234, 235, 239, 240, 378, 383, 386, 390, 394, 415, 416, 429, 509, 511, 512, 532, 594, 603
Shapiro, D.Y., 458
Shapiro, J.A., 62, 111, 322, 329, 330
Shapiro, L., 492
Sharp, P.A., 48, 80, 322, 323, 324, 326, 327, 328, 329
Shatz, C.J., 111
Shaw, D., 114, 459
Shaw, M.J.B., 131, 132, 218
Shea, B.T., 607
Shearer, R.R., 454
Shelly, T.E., 344, 593
Shena, M., 491
Sheppard, P.M., 79, 384, 401
Sherman, P.W., 10, 140, 201, 249, 276, 288, 428, 438, 474, 475, 476
Sherry, T.W., 345, 346, 463, 586
Shettleworth, S.J., 337
Shilo, M., 604
Shoemaker, D.D., 106, 538
Shoichet, B.K., 303
Short, R.V., 274
Shorthouse, J.D., 111, 552
Shubin, N., 205, 206, 207, 356, 363, 368, 488, 491
Shull, H.G., 378
Shuster, S.M., 106, 204, 279, 280, 383, 385, 395, 420, 422, 423, 442
Shuvalov, V.F., 456
Sibly, R.M., 17, 310, 434
Siegel, R.W., 343
Silk, J.B., 453
Sillén-Tullberg, B., 181
Silvertown, J., 26
Simmons, L.W., 423, 456, 634
Simon, C., 555, 556
Simon, C.A., 456
Simon, H.A., 81, 87
Simpson, 487,
Simpson, C.L., 314
Simpson, G.G., 6, 8, 12, 24, 25, 151, 152, 153, 180, 189, 237, 309, 362, 371, 394, 415, 416, 483, 489, 490, 503, 509, 564, 565, 598, 600, 607, 617, 618, 619, 621
Simpson, S.J., 314
Sinervo, B., 22, 113, 221, 281, 303, 310, 311, 312, 420, 433, 515, 518
Sing, C.F., 104, 105
Singer, M., 328

Skaife, S.H., 174
Skinner, J.D., 276
Skúlason, S., 113, 547, 548, 567
Slade, A.J., 39
Slansky, F., Jr., 307
Slater, P.J.B., 337
Slatkin, M., 155, 170, 179
Slijper, E.J., 51, 52, 53
Slud, P., 586
Sluys, R., 357, 361, 370, 371
Smale, L., 40
Smillie, D., 32, 165
Smit, A.F.A., 49
Smith, A.M., 627, 628
Smith, A.P., 302
Smith, C.M., 593
Smith, C.W.J., 48, 212, 323, 328
Smith, E.F., 283
Smith, G., 280
Smith, J.A.C., 251, 512, 514, 595
Smith, J.N.M., 464, 586
Smith, K.C., 512
Smith, L.D., 110
Smith, T.B., 347, 350, 397, 420, 547, 567, 587, 588, 590, 595
Smith-Gill, S.J., 35, 36, 68, 171, 379
Smuts, B.B., 291
Smythe, N., 115
Snoad, N., 59
Snodgrass, R.E., 177, 210, 214, 261
Snook, R.R., 387
Snorrason, S.S., 548
Soares-Porto, L.M., 550
Sober, E., 30, 598
Sokol, O., 207, 208, 607
Sokolowski, M.B., 580
Solís, C.R., 448
Solomon, N.G., 223, 453
Solter, D., 636
Sommeijer, M.J., 128, 284, 509
Sorenson, A.E., 26
Soulé, M.F., 426
Southwick, E.E., 185, 438
Southwood, T.R.E., 109, 110
Speed, M.P., 351, 352
Spence, G.E., 580
Spicer, G.S., 617
Spicer, R.A., 251, 259, 513, 521, 522
Spielman, A., 561
Spieth, H.T., 578, 581, 591
Spofford, M.G., 222, 426
Spolsky, C., 635
Sprackbery, J.P., 228
Sprague, G.F., 266, 267
Sprugel, D.G., 65
Sprules, W.G., 208, 608
Spudich, J.L., 426, 427
Srere, H.K., 73, 391
Stacy, P.B., 223, 453
Stanley, S.M., 361, 401, 506, 520, 531, 598, 601, 617
Stark, R.E., 240, 451
Starks, P.T., 449
Stearns, S.C., 8, 17, 25, 27, 33, 35, 77, 78, 156, 173, 178, 179, 180, 304, 310, 383, 474, 603, 620

Stebbing, A.R.D., 366
Stebbins, G.L., 9, 188, 205, 508
Stec, A., 267
Steele, E.J., 37, 161
Steffan, W.A., 612
Stellar, E., 314, 341
Stellar, J.R., 314
Stent, G.S., 14, 93, 98
Stephens, D.W., 17, 443
Stephens, S.G., 211
Stern, C., 256
Stern, D.L., 109, 253, 254, 387
Stetson, M.H., 110, 114
Stiles, F.G., 445
Stoewesand, H., 206, 207
Stone, E.M., 320, 324, 331
Stone, G.N., 109
Storey, A.E., 287
Stowe, M.K., 445
Strasser, A., 185
Strassmann, J.E., 423, 432, 447, 448, 450, 451
Strathmann, M.F., 514, 515
Strathmann, R.R., 221, 235, 507, 514, 515, 516, 517, 612, 613, 637
Strauss, R.E., 303, 478
Strong, D.R., Jr., 39
Stubblefield, J.W., 43, 57
Sturtevant, A.H., 255
Su, Z.H., 364, 365, 551, 552
Sullivan, J.D., 451
Sultan, S.E., 35, 96, 113, 178, 519
Summers, K., 284
Sundberg, M.D., 261, 266, 267, 268
Sundstrom, L., 538
Surani, A., 636
Suzuki, S., 608
Svardson, G., 383, 568
Svensson, E., 310
Swalla, B.J., 615
Swanson, W.J., 96
Sweatman, H.P.A., 586
Syren, R.M., 293
Szabo, V.M., 267
Szathmary, E., 56, 85, 185, 387, 399, 429, 430, 524, 633

T

Taborsky, M., 280, 389
Taggart, J.B., 567
Takhtajan, A., 224, 242, 250, 480
Taliercio, E.W., 211
Tanaka, M., 521
Tanaka, S., 311
Tanaka, Y., 407, 409
Tåning, A.V., 512, 523
Tatewaki, M., 502
Tattersall, I., 464
Tauber, C.A., 459, 542, 554, 556, 560
Tauber, M.J., 12, 108, 391, 459, 542, 554, 556, 557, 560
Tautz, D., 117
Taylor, C.E., 205
Taylor, E.B., 535, 536, 566, 567, 568
Taylor, P.D., 565

Taylor, P.J., 8, 153
Taylor, T.N., 613
Tchernavin, V., 294
Tchernichovski, O., 108
Tejedo, M., 26, 396
Telewski, F.W., 301
Temeles, E.J., 595
Templeton, A.R., 79, 388, 474, 528, 554, 592
Terrace, H.S., 337, 463
Terskikh, A.V., 318
Thaler, D.S., 321, 330
Thanthianga, C., 465, 466
Thoday, J.M., 8, 116, 225, 226, 227, 399, 402, 528
Thomas, J.A., 260, 287
Thompson, C.S., 280, 383, 423
Thompson, D., 54
Thompson, D.W., 478
Thompson, J.D., 179
Thompson, J.N., 66
Thompson-Stewart, D., 216
Thomson, J.A., 59, 132
Thomson, K.S., 5, 10, 23, 24, 28, 53, 54, 135, 142, 143, 161, 233, 234, 370, 507, 511, 591
Thorgaard, G., 210
Thorndyke, M.C., 49, 323
Thornhill, R., 26, 271, 385, 422, 453, 454, 457, 461
Thorpe, J.E., 535, 536, 547, 548
Throckmorton, L.H., 365
Tickle, C., 368
Tiebel, K.C., 106, 119, 120, 205, 227, 300, 311, 373
Tierney, A.J., 152, 153, 415
Tilmon, K.J., 552
Tilney, L.G., 65
Tinbergen, N., 10, 11, 244
Ting, I.P., 513
Tobin, J.E., 542
Tobin, T.R., 38
Tollrian, R., 392
Tompkins, R., 511
Tonegawa, S., 54
Tooby, J., 81
Tortorici, C., 119
Towne, W.F., 444
Townsend, D.S., 285
Travis, J., 84, 113, 461
Trembley, A., 506
Trenkle, S., 75, 283, 291
Trewavas, A.J., 77
Trimen, R., 381
Tripathi, R.K., 74
Trivers, R.L., 43, 286, 416, 419, 453, 631, 633, 635
True, J.R., 427
Truman, J.W., 59, 66, 69
Trumler, E., 455
Tsai, M., 318
Tschinkel, W.R., 461, 537
Tsuji, K., 387
Tsukamoto, K., 535
Tuljapurkar, S.D., 424

Tuomi, J., 174, 307
Tuomikoski, R., 357
Turesson, G., 381
Turillazzi, S., 408, 451, 565, 591
Turner, B.J., 217
Turner, G.F., 352, 564, 574
Turner, J.R.G., 75, 79, 80, 167, 168, 169, 226, 351, 352, 358, 401, 405, 533, 604, 620
Twitty, V.C., 53, 161

U

Ugolini, A., 179, 343
Underwood, G., 256, 257
Underwood, H., 459
Urbanek, A., 242
Uyenoyama, M.K., 631, 633

V

Vacquier, V.D., 96
Valentine, J.W., 5, 28, 481, 598, 603, 611, 618
Valentine, M.K., 241, 607
Valladares, F., 180
Van Buskirk, J., 300, 392, 397, 431
Vance, S.A., 265
van den Berghe, E.P., 294
Van Handel, E., 119, 265
van Noordwijk, A.J., 316
van Tiende, P.H., 275
Van Valen, L.M., 82, 98, 110, 116, 138, 176, 206, 354, 366, 382, 427, 497, 519, 599, 617, 620, 621, 628, 631
van Veen, J.W., 284
Van Voorhees, A., 121
Vargo, E.L., 449
Vari, R.P., 363, 364, 365, 550, 609
Vaux, D.L., 185
Vavilov, N.I., 490
Vehrencamp, S.L., 451
Velthuis, H.H.W., 128, 398, 404, 450, 452, 453, 509
Venable, D.L., 26
Vermeij, G.J., 87, 156, 182
Via, S., 36, 54, 55, 168, 179, 250
Villalobos, E.M., 344
Villet, R., 116, 117, 119
Vincek, V., 582
Visscher, P.K., 447
Visscher 98, 446, 448, 452
Vogel, S., 301
Vogt, R.C,, 100
Voss, S.R., 203, 207, 208, 370, 372, 548
Vourisalo, T., 174
Vowles, D.M., 171
Vrba, E.S., 601, 623
Vuilleumier, F., 598

W

Waagen, W., 378
Waddington, C.H., 7, 8, 13, 14, 24, 25, 28, 115, 116, 119, 151, 152, 153, 225, 256, 383, 399, 415, 499

Waddington, K.D., 342, 349, 444, 445
Wade, M.J., 7, 17, 106, 279, 383, 385, 416, 420, 422, 423, 442, 465
Wagner, D.L., 119, 120
Wagner, G.P., 5, 56, 57, 62, 84, 88, 164, 168, 171, 182, 258, 326, 485, 489, 495, 496
Wagner, W.E., Jr., 422
Wainwright, P.C., 596
Wainwright, S.A., 301, 302
Wake, D.B., 5, 8, 22, 25, 83, 87, 171, 198, 200, 201, 308, 309, 335, 336, 353, 357, 361, 370, 374, 405, 415, 485, 491, 495, 596, 609, 620
Wake, M.H., 415
Walbot, V., 265, 267
Waldman, B., 448
Walker, I., 51, 177, 203
Walker, J.A., 413, 546
Walker, T.J., 104, 105, 116, 119, 203, 205, 300, 383, 397, 429, 555, 562
Wallace, B., 8, 11, 241
Walsh, B.D., 557, 558, 559
Walsh, J.B., 216
Walters, J.R., 290
Walters, S.M., 155, 334, 371, 551, 604, 605
Wang, R.-L., 265, 267, 268
Wang, Z., 287, 288
Ward, P.S., 542, 544
Ward, R., 272, 388
Warkentin, K.M., 250, 303, 392, 431
Warner, R.R., 432, 437, 458, 634
Watkinson, A.R., 303
Watson, M.A., 25, 91
Watt, W.B., 532
Wayne, R.K., 299
Wcislo, W.T., 24, 25, 28, 32, 98, 99, 112, 118, 151, 152, 153, 176, 178, 180, 181, 203, 235, 240, 264, 314, 328, 337, 339, 341–43, 347, 363, 404, 415, 463, 506, 544
Weatherbee, S.D., 256
Weaver, N., 130, 131, 339, 340, 399
Wecker, S.C., 369, 554
Wei, G., 635
Weidmann, J., 173
Weismann, A., 37, 57, 86, 90, 129, 191, 192, 331
Weiss, K.M., 209, 256
Weitzman, S.H., 300, 363, 364, 365, 371, 550, 609
Weller, S.J., 204
Wells, B.W., 109
Wells, K.W., 284
Wells, M.M., 556, 557
Welty, J.C., 273, 286, 502
Wenzel, J.W., 9, 10, 57, 61, 63, 65, 87, 181, 223, 228, 494, 507
Werner, E.E., 461, 508, 550, 569, 571, 577
Werner, T.K., 345, 346, 463, 586
Werner, Y.L., 282, 635
Werren, J.H., 528

Wessler, S.R., 49, 330, 333
West, D.A., 169, 504
West, M.J., 43, 98, 432, 449, 450, 453
West-Eberhard, M.J., 5, 10, 23, 24, 26, 28, 31, 35, 43, 45, 51, 54, 56, 66, 67, 70, 78, 82, 85, 88, 133, 141, 142, 144, 145, 146, 147, 153, 157, 160, 161, 162, 163, 165, 167, 170, 175, 178, 181, 182, 184, 186, 205, 212, 222, 223, 228, 245, 246, 247, 248, 258, 264, 273, 275, 283, 284, 289, 301, 310, 314, 344, 365, 368, 377, 378, 382, 386, 388, 389, 391, 392, 395, 396, 397, 404, 408, 416, 423, 429, 430, 431, 432, 435, 442, 445, 447, 449, 449–54, 457, 464, 465, 466, 467, 468, 477, 478, 482, 502, 503, 505, 509, 510, 522, 527, 528, 531, 542, 544, 562, 565, 573, 576, 589, 591, 593, 599, 602, 604, 606, 608, 609, 621, 631, 633, 637
Westoby, M., 91, 636
Wheeler, D.E., 73, 74, 75, 88, 107, 132, 227, 305, 308, 309, 310, 373, 378, 395, 398, 400, 401, 427, 461, 507
Wheeler, P., 464
Wheeler, Q.D., 10
Wheeler, W.M., 35, 107, 265, 269, 270, 280, 283, 378, 400
Whelden, R.M., 508
Whipps, J.M., 605
White, J., 303, 556
White, M.J.D., 282, 630, 637
Whiteman, H.H., 249, 353, 372, 555
Whiten, A., 464
Whitham, T.G., 220, 406
Whitt, G.S., 70, 450
Whitten, P.L., 289, 290, 291
Whittier, T.S., 593
Whittington, H.B., 614
Whyte, L.L., 5
Wibbels, T., 276, 282
Wickler, W., 212, 245, 246, 248, 249, 264, 276, 277, 455
Wiedenfeld, D.A., 534
Wiens, J.J., 370, 372, 373, 380, 386, 487
Wiernasz, D.C., 203, 532, 543
Wigglesworth, V.B., 4, 36, 74, 101, 383, 462, 477
Wikelski, M., 273
Wilbur, H.M., 119, 383, 434
Wilcox, R.S., 65, 386, 387, 445
Wilczek, F., 34
Wilde, J. de, 400
Wilkinson, G.S., 309, 493
Williams, A., 362, 596
Williams, C.M., 307
Williams, G.C., 5, 7, 10, 16, 22, 25, 32, 35, 83, 94, 99, 146, 156, 165, 177, 247, 264, 282, 324, 334, 384, 385, 388, 477, 503, 507, 630, 631, 632, 634, 636, 637
Williams, K.S., 555
Williams, M.B., 526

Williams, R.M., 203
Williamson, P.G., 617
Wills, C., 508
Willson, M.F., 6, 97, 114, 347, 606, 633
Wilson, A.C., 52, 335
Wilson, D.S., 165, 348, 382, 397, 434, 546, 567, 571, 572, 577, 590, 594, 621
Wilson, E.B., 26, 35, 93, 360, 502, 517
Wilson, E.O., 43, 85, 88, 120, 131, 205, 227, 235, 236, 283, 309, 310, 312, 370, 397, 399, 447, 458, 465, 509, 537, 542, 543, 544, 573, 606, 613
Wilson, J.W., 162, 619
Wilson, M.A., 54
Wilson, P., 214, 250, 395
Wilton, C.L., 605
Wimberger, P.H., 24, 26, 40, 181, 487, 494, 500, 528, 547, 550, 576, 578
Wimsatt, W.C., 56, 326, 500
Windig, J.J., 58, 77, 217, 229
Wingfield, J.C., 76, 273, 284, 285, 286, 293, 304, 441
Winkler, D.W., 114
Winston, J.E., 381, 439, 624, 628
Winston, M.L., 150, 340, 398
Winter, K., 251, 252, 512, 513, 514, 595
Winter, U., 107, 120
Winterbottom, J.M., 273, 276, 286
Wirtz, P., 130
Witte, F., 508, 574, 575, 576, 577, 581, 592, 594
Woese, C., 183
Wolf, G., 112
Wolf, L.L., 456
Wolff, J.D., 40
Wolpert, L., 90, 185
Woltereck, R., 26, 101
Wolynes, P.G., 49
Wood, T.K., 115, 552
Wootton, R.J., 512, 570
Wray, G.A., 187, 203, 222, 240, 243, 244, 255, 258, 360, 402, 411, 412, 481, 514, 515, 516, 522
Wrege, P.H., 448, 458
Wright, S., 79, 104, 105, 118, 144, 167, 170, 173, 176, 179, 205, 230, 233, 238, 394, 401, 419, 475, 478, 479, 524, 597, 601
Wurm, M., 249
Wymer, C.L., 39
Wynne-Edwards, K.E., 287
Wynne-Edwards, V.C., 518
Wyss-Huber, M., 74

Y

Yablokov, A.V., 206, 232, 233, 276, 354, 355, 357, 378
Yamamoto, I., 287
Yamauchi, K., 75, 291, 543, 544
Yanega, D., 236, 404
Yaoita, Y., 73

Yasunobu, K.T., 521
Yoon, S.L., 456
Young, B., 486

Z
Zahavi, A., 456, 468
Zeh, D.W., 91, 431, 631

Zeh, J.A., 431, 631
Zelditch, M.L., 78, 134, 135, 255
Zera, A.J., 68, 106, 119, 120, 127, 205,
    227, 300, 303, 304, 307, 310, 311,
    374, 383
Ziegler, B., 91
Zimmerer, E.J., 107, 384, 422
Zimmerman, E.C., 574

Zirkle, C., 191
Zotz, G., 252
Zucchi, R., 314
Zucker, N., 280, 281
Zuckerkandl, E., 56, 116, 117, 119, 212,
    326, 490, 491, 610
Zuk, M., 423, 591, 632
Zusi, R.L., 207

Taxonomic Index

Numbers in italics indicate figures.

A

Abatus cordatus, 222
Acalypha parvule, 585
Acanalonia bonducellae, 96
Acanthotermes, 292
Accipiter superciliosus, 444
Acheta domesticus, 277
Acrididae, 172
Adenostoma fasciculatum, 421
Adranes caecus, 215
Aechmorhynchus cancellatus, 207
Aedes aegypti, 119
Aeschnidae, 444
Agalychnis callidryas, 249–250, 250
Agaonidae, 110
Agapornis, 275, 276
  A. roseicollis, 275
Agelaia, 223, 351
Agelaius phoeneceus, 523
Agelenopsis aperta, 388
Aglaopheniidae, 219
Agouti paca, 115
Agrostis c-nigrum, 225
Aizoaceae, 595
Aleochara curtula, 264, 278
Alopex, 249
Amadina erythrocephala, 223
Amaranthaceae, 579
Amauris
  A. acheria, 533
  A. echeria, 401
  A. niavius, 401
  A. niavius dominicanus, 401
  A. n. niavius, 401
Amblypygi, 271
Ambystoma, 75, 208, 353, 354
  A. dumerilii, 353, 354, 608
  A. gracile, 208, 608
  A. lemaensis, 354
  A. mexicanum, 118, 511, 607–608
  A. talpoideum, 118–119
  A. tigrinum, 353, 462, 607
  A. tigrinum californicus, 607
  A. tigrinum nebulosum, 448
  A. tigrinum tigrinum, 607
Amitermes hastatus, 174
Ammophila, 426
  A. campestris, 443
  A. pubescens, 443

Amphignathodon, 235
Amphipoda, 430
Amphiuma, 118
Amphiumidae, 607
Anapsida, 606
Andrenidae, 228
Anergates, 283
Anisolabis annulapis, 234
Anolis, 522, 594
  A. lineatopus, 344
Anomalocaris, 613, 614, 614
Anoplius marginalis, 407
Antirrhinum majus, 270
Anura, 335
Anurogryllus, 203
Aotus trivergatus, 246
Apeltes quadracus, 280
Aphaenogaster, 246
Aphis fabae, 131, 131–132
Aphycus albicornis, 560
Apis
  A. mellifera, 130, 130, 130–131, 131, 148, 155–156, 171–172, 339, 349
  A. mellifera capensis, 450
Aplysia, 77
Apocrita, 606
Apocrypta, 421, 421
Aporus, 407
Apozyginae, 613
Arabidopsis, 211, 270
  A. thaliana, 270
Araceae, 39, 306
Araliaceae, 579, 581
Araneae, 61, 409
Arbacia
  A. lixula, 515
  A. punctulata, 243
Archaeoceti, 234
Archaeopteryx, 199
Archisepsis diversiformis, 199
Armadillium vulgare, 265
Asplanchia, 111
  A. sieboldi, 462
Astatoreochromis
  A. alluaudi, 299, 575, 576, 592
  A. alluaudi alluaudi, 576
Atheris, 494
Atriplex confertifolia, 520
Atta, 246, 247, 308

B

Baeotus baeotus, 166
Baetis bicaudatus, 265
Bairdiella icistius, 506
Balanus balanoides, 312
Barbus, 300
Barychelidae, 605
Beloe ovata, 95
Bembex
  B. bioculata, 224
  B. morua, 596
  B. pruinosa, 224, 224
  B. variabilis, 596
Bicyclus anynana, 216, 217
Biston betularia, 401
Blackburnium, 400
  B. angulicorne, 400
Blattaria, 248
Bolitoglossa, 368
Bombus, 228, 263
  B. griseocollis, 284
  B. pennsylvanicus, 284
  B. terrestris, 74, 228, 452
Bombyx mori, 277
Bonellia, 101, 116
Brachistes atricornis, 22
Brachyponera, 74
Braconidae, 22, 234, 613
Braunsapis hewitti, 284
Brentus anchorago, 540, 541
Brisaster latifrons, 514
Bromeliaceae, 513
Bryozoa, 251, 362, 366, 623, 623–624, 628
Bubuculcus ibis, 92
Bucephala islandica, 434
Bufo
  B. bufo, 456
  B. calamila, 456
Bursura, 346, 463

C

Cactospiza pallida, 245
Caenorhabditis, 221, 326
  C. elegans, 73, 114, 119, 205, 330
Calathus
  C. erythroderus, 119, 120
  C. melanocephalus, 119, 120
Calidris alba, 456

Callaeidae, 207
Callithrix jacchus, 290
Callosobruchus maculatus, 98, 443, 444, 465–466, 466
Caloglyphus berlesei, 433
Camaradonta, 243
Campanulaceae, 579, 581
Camponotus, 309
  C. festinatus, 309
  C. herculeanus, 283
  C. saundersi, 310
Candida albicans, 111
Carabidae, 364, 551
Carassius carassius, 572
Cardinalis cardinalis, 206
Cardiocondyla, 291
  C. emeryi, 283, 291
  C. nuda, 283
  C. wroughtonii, 283
Carduelinae, 589
Caretta caretta, 125
Carex stans, 460
Carya illinoensis, 109
Caryomia, 109, 109–110
Caryophyllaceae, 371, 604
Ceanothus spinosus, 421
Cecidomyidae, 109
Cepea, 420
Cephalopoda, 221
Cerataphidinae, 254
Cerataphidini, 253, 254
Ceratina, 606
  C. japonica, 606
Cercopithecus, 264
  C. aethiops, 277
Cerion, 309
Chalcidoidea, 225, 527, 560
Chalepoxenus mullerianus, 554
Characidae, 300
Characodoma excubans, 628
Charaxes castor, 166
Charaxinae, 166
Cheilostomata, 623
Cheirodendron, 581
Chilopoda, 210
Chiloschista lunifera, 91
Chloridops koni, 589
Chondromyces crocatus, 613
Chrysemys picta, 125
Chrysopa, 557
  C. quadripunctata, 557
  C. slossonae, 557
Chrysoperla, 556, 556–557
  C. adamsi, 556
  C. carnea, 556
  C. downesi, 556
  C. johnsoni, 556
  C. plorabunda, 556
Chrysopidae, 365
Chthamalus anisopoma, 392
Chymomyza aldrichii, 355
Cichlasoma, 397
  C. bimaculatum, 592
  C. citrinellum, 218–219, 299, 577
  C. managuense, 54, 578

C. meeki, 592
C. minckleyi, 576, 577, 577
C. severum, 592
Cidaroida, 243, 258, 412
Cidaroidea, 243, 258
Ciridops anna, 589
Cleidochasmatidae, 628
Clermontia, 579, 581
Clethrionomys
  C. glareolus, 521
  C. glareolus britannicus, 115
Clusia, 251, 512
  C. minor, 513
  C. uvitana, 252
Clypeaster, 222
  C. rosaceus, 514
Clypeasteroida, 243, 412
Cnemidophorus, 281
  C. uniparens, 282
Cnidaria, 219
Coelioxys funeraria, 229
Coenagrionidae, 63, 278
Colegyne, 520
Coleonyx variegatus, 257
Coleoptera, 22, 127, 215, 248, 264, 271, 277, 305
Colias eurytheme, 229
Colophina, 253–254, 253
  C. clematicola, 254
  C. monstrifica, 253
Columba livia, 344, 345
Columbicola columbae, 96
Conopeum tenuissimum, 624
Conus, 591
Coregonus, 567
  C. clupeaformis, 365, 567, 569
  C. lavaretus, 365
  C. sardinella, 567
Cormohipparion emsliei, 622
Cornitermes, 292
Cossidae, 371
Coturnix japonica, 157
Crabro argus, 408
Crabroninae, 407
Craniacea, 362
Craniopsidae, 362
Crassimarginatella falcata, 624
Crepidula, 403, 573
  C. adunca, 573
  C. arenata, 573
  C. nivea, 573
  C. norrisiarum, 573
  C. onyx, 573
  C. williamsi, 573
Crocuta crocuta, 40, 261, 261, 262, 276
Crossocerus elongatus, 407, 408
  C. maculiclypeus, 407
Crotalinae, 494
Crustacea, 59, 279, 430
Cryptobranchidae, 607
Ctenizidae, 605
Cylapofulvius, 96
Cynipidae, 109–110
Cynomys ludovicianus, 460
Cyprinodon

C. nevadensis, 113
C. variegatus, 217

D

Dalechampia, 259
Danaus chrysippus, 401
Daphnia 350
Dasymutilla bioculata, 224, 224–225, 229, 381
Delphinium
  D. gypsophilum, 198
  D. hesperium pallescens, 198
  D. recurvatum, 198
Dendraster excentricus, 515, 515
Dendrobates, 284–285
  D. auratus, 284–285
  D. histrionicus, 285
  D. leucomelas, 285
Dendrobatidae, 284, 502
Dendroica, 347
  D. castanea, 347
  D. pennsylvanica, 347
  D. petechia, 534
Dermaptera, 366
Diapsida, 606
Dicrostonyx groenlandicus, 460
Dicrurus, 207
Dictyoptera, 248
Didymoceras stephensoni, 173
Dinocarida, 614
Dinohippus mexicanus, 622
Dinoponera quadriceps, 452, 452
Dipetalogaster maximus, 462
Dipodomys, 444
  D. heermanni 520–521
  D. microps, 520–521
  D. spectabilis, 444
Diptera, 66, 80, 94, 96, 199, 219, 355, 366, 495, 557, 596, 613
Diurodrilus, 312
Dolichovespula artica, 410
Donocarida, 613
Drapanis pacifica, 589
Drenilabrus melops, 280
Drepanidae, 574, 589
Drepanidini, 589
Drosophila, 24, 26, 63, 70, 72, 73, 77, 80, 94, 95, 100–101, 104–105, 114, 116, 119, 154, 155, 156, 169, 181, 200, 205, 211, 212, 219, 221, 230, 233, 304, 318–319, 323, 326, 327, 330–331, 343, 365, 367, 416, 424, 434, 528, 531, 558, 572–573, 578–582, 579, 593
  D. adiastola, 181
  D. affinidisjuncta, 319, 319
  D. clavisetae, 24, 181
  D. claytonae, 94
  D. craddocki, 581
  D. disticha, 581
  D. eugyochracea, 579
  D. formella, 94, 319
  D. grimshawi, 230, 581, 597
  D. heedi, 579, 579
  D. heteroneura, 94

D. immigrans, 580
D. kambysellisi, 572, 573, 579
D. longiperda, 94
D. melanogaster, 19, 77, 119, 226, 335, 415, 580
D. mimica, 572, 573, 579
D. mulli, 94
D. murphyi, 94
D. pattersoni, 94
D. prostopalpis, 319
D. pseudoobscura, 102
D. sechellia, 580
D. seyshellia, 580
D. silvarentis, 579, 579
D. simulans, 92, 580
D. truncipenna, 94
D. virilis, 94
D. waddingtoni, 581
D. willistoni, 94
Drosophilidae, 94, 355, 493, 493–494
Dupontia fisheri, 460

E

Echinoida, *412*
Echinothurioida, *412*
Eciton, 308
   E. burchelli, 42, 42–43
Eclectus roratus, 275
Elaphrothrips tuberculatus, 456
Elephantidae, 511
Elephas falconeri, *511*, 511
Eleuthrodactylus, 285, 495
Emberizidae, 345
Emberizinae, 541, 582, *583*, 589
Enchenopa, 552
   E. binotata, 115, 552
Encyrtidae, 560
Ennomos subsegnarius, 103
Entamoeba histolytica, 631
Ephedra, 605
Ephemeroptera, 114
Ephestia kuehniella, 225
Epicauta vittata, 22
Epimyrma, 547
   E. ravouxi, 547
Epiphagus virginiana, 231
Epoecus, 283
Equus, 622
Eriosoma, 253, 533–534
   E. parasitica, 533
   E. yangi parasiticum, 534
   E. y. yangi, *533*, 533–534
Eriosomatinae, 253
Eucidaris tribuloides, 243
Eucoilidae, 22
Euechinoidea, *243*, *258*
Euglena, 462
Eumeninae, 509
Euphorbiaceae, 259
Euphydryas editha, 444
Euprymna scolopes, 501
Eurytoma, 606
Eurytomidae, 606

Evylaeus cinctipes, 228
Exoneura bicolor, 404

F

Fannia canicularis, 96
Felidae, 234
Ficedula
   F. albicollis, 97
   F. hypoleuca, 286
Ficus, 180
Florisuga, 275
   F. mellivora, 275
Formica
   F. polyctena, 74
   F. truncorum, 538
Formicidae, 554
Fringillidae, 207, 589

G

Galanidae, 405
Gallus, 234, 245, *245*
Gammarus, 157
   G. fossorum, 430
   G. minus, 551
Gasteromermis, 265
Gasterosteidae, 280
Gasterosteus, 299–300, 363
   G. aculeatus, 238–239, 280, 350, 359, 365, 412–413, *414*, 444, 512, *545*, *545*, 548–550, 566, *566*, 567, 570, 571
   G. doryssus, 203, 362–363, *363*, 570
Gelanoglanis nanonocticolus, 550
Geometridae, 446
Geonoma, 306
Geophagus
   G. brasiliensis, 576
   G. steindachneri, 576
Geospiza, 345–346, *347*, 350, 582–586, *583*
   G. conirostris, 226, 348, *583*, 584
   G. difficilis, 346, *583*
   G. fortis, 226, 346, 463, 584, 585, *585*, 597
   G. magnirostris, 346, 534, 585
   G. scandens, 582, 584
Geotrupinae, 400
Gerridae, 374
Gerris, 312
Giardia, 326, 631
Glandulacaudinae, 371
Glauchopsyche lygdamus, 343
Gnetales, 605
Gonatodes
   G. fuscus, 257
   G. humeralis, 257
Gryllidae, 271
Gryllodes supplicans, 205
Gryllus, 203
   G. bimaculatus, 456
   G. campestris, 271
   G. firmus, 562
   G. integer, 422–423
   G. rubens, 119, 310–311, 373

Guzmania, 251
Gymnosperms, 301

H

Haasiophis, 368
Habrosyne pyrtoides, 96
Haedeotriton, 118
Halictidae, 235, 363, 367, 447
Halictinae, 228
Halictus, 235, 236, *236*, 363
   H. ligatus, 450
   H. rubicundus, 236, *236*, 363, 404
   H. subdivision *Halictus*, 236
   H. subdivision *Seladonia*, 236
Halobacterium salinarum, 49
Haplochromis, 574–578
   H. empodisma, 576
   H. ishmaeli, 576
   H. piceatus, 576
   H. quadrimaculatus, 569
   H. squamipinnis, 575, 576
Harpagoxenus sublaevis, 107, 120, 120–121
Heliconius, 215, 357, *358*
   H. erato, 358
   H. melpomeme, 358
Heliocidaris, 495, 517
   H. erythrogramma, 222, *222*, 514–515
Helmitheros cermivorus, 347–348
Hemianax papuensis, 444
Hemichromis
   H. bimaculatus, 592
   H. fasciatus, 455
   H. flaviijosephi, 592
Hemignathus
   H. monroi, 589
   H. procerus, 589
   H. virens virens, 589
Hemilepsis reaumuri, 343, *343*, 448
Hemipodaphis persimilis, 253, 254
Hemiptera, 96, 114, 180, 374, 462
Hesperiidae, 157
Heterocephalus glaber, 288, *288*
Hibiscus, 346, 586
Himatione s. sanguinea, 589
Hippothoa hyalina, 624
Holometabola, 366
Homo, 326
   H. sapiens, 488
Homoptera, 96, 114, 252, 552
Hopea, 251
Hormaphididae, 252, 253, 387
Hydra, 312, 505, 506
Hydrozoa, 168, 219–220
Hyla
   H. chrysoscelis, 456
   H. cinerea, 272
Hylocichla mustelina, 103, 206
Hylophylax sp. n. *naevioides*, 273
Hymenoptera, 22, 66, *224*, 225, 228, 256, 349, 367, 408, 457, 464, 495, 509, 510, 527, 559, 605–606, 635
Hypena proboscidalis, 96
Hyperponera, 283
Hypolimnas, 533

Hypoponera bondroiti, 543–544
Hyppothoa hyalina, 628
Hystrichopsyllinae, 367

I

Ichneumonidae, 314, 559
Ichthyomyzon, 535
Idarnes, 395–396
Iguanidae, 281
Inachus mauritanicus, 280
Insecta, 22, 180
Ipomoea, 211
Iridomyrmex, 74
Ischnura
    I. graellsii, 271
    I. ramburi, 278–279
Isoetaceae, 513
Isoetes howelli, 513
Isopoda, 265, 279
Isoptera, 248

J

Jadera harmatoloma, 180
Junco hyemalis, 286

K

Kleidocerys resedae, 96
Kleidotoma marshalli, 22

L

Labridae, 280, 458
Labrochromis ishmaeli, 41, 41
Lacerta muralis, 507
Lagopus, 391
    L. lagopus scoticus, 518
    L. mutus, 518
Lampetra, 535, 567
    L. fluviatilis, 535
    L. japonica, 535
    L. tridentata, 535
Laodicea indica, 219–220
Laodiceidae, 219
Larus occidentalis, 286
Lasioglossum
    L. erythrurum, 292
    L. imitatum, 264
    L. umbripenne, 228
    L. zephyrum, 228, 447
Leguminoseae, 579
Leishmania, 631
Leistotrophus versicolor, 277–278, 278
Lemmus sibiricus, 460
Lepidoglyphus destructor, 437
Lepidoptera, 54, 66, 96, 119, 157, 345,
    388, 446, 475
Lepomis
    L. gibbosus, 596
    L. macrochirus, 280, 550, 571
Leptomedusae, 219
Leptomorphus, 208
Leptothorax, 107, 452, 543
    L. acervorum, 547
    L. rugatulus, 544
Liliaceae, 579
Lilium auratum, 39, 39
Limenitinae, 166

Limenitis, 533
Limnodrilus hoffmeisteri, 106
Lindenius armaticeps, 407
Locusta migratoria, 132, 300
Lophophorus impejanus, 245
Loxodonta africana, 455
Loxops
    L. coccinea coccinea, 589
    L. virens, 589
Lycophyta, 513
Lygodactylus, 256
    L. picturatus, 257
Lymantria dispar, 103, 270, 371
Lynx lynx, 234
Lytechinus
    L. pictus, 243
    L. variegatus, 243, 515

M

Macaca
    M. arctoides, 276–277, 289
    M. f. fuscata, 289
    M. nemestrina, 277, 290
Macrocentrus figuensis, 22
Macrotermes, 292
    M. subhyalinus, 74
Magicicada, 555–556
    M. cassini, 555
    M. neotredecim, 555
    M. septendecim, 555
    M. septendecula, 555
    M. tredecassini, 555
    M. tredecim, 555
    M. tredecula, 555
Magnoliopsida, 251
Mammuthus imperator, 511
Mantoidea, 248
Manduca sexta, 69
Marmota flaviventris, 344
Marsilia, 505
Mecoptera, 367
Medicago sativa, 339
Megachilidae, 229
Megaloptera, 367
Megaptera
    M. nodosa, 232–233, 233
    M. novaeangliae, 232
Megarhyssa, 559, 559–560
Melamprosops phaeosoma, 589
Melipona, 107, 128
    M. compressipes, 284
    M. marginata, 284
    M. panamica, 448
Meliponidae, 313, 349
Meliponinae, 291
Meliponini, 284
Mellita quinquiesperforata, 243
Meloidae, 22
Melospiza, 347
Membracidae, 114, 552
Membranipora arborescens, 624
Menidia menidia, 121
Meriones unguiculatus, 282
Merops bullockoides, 448, 458
Mesembryanthemum crystallinum, 513
Metapolybia aztecoides, 452

Metrarabdotos, 623, 623–627
    M. auriculatum, 625
    M. colligatum, 625
    M. cookae, 627
    M. lacrymosum, 626
    M. unguiculatum, 626
    M. u. unguiculatum, 625, 627
Microbembix monodonta, 224
Microtus, 288, 360
    M. montanus, 460
    M. ochrogaster, 287–288
    M. pennsylvanicus, 287, 288
    M. pinetorum, 288
Migas distinctus, 605
Migidae, 605
Mischocyttarus, 283, 450
    M. collarellus, 283
Mogannia minuta, 555
Monstera, 145, 145–146, 366, 383
    M. acuminata, 145
    M. dubia, 58, 145
    M. gigantea, 39
    M. lechleriana, 145
    M. pittieri, 145
    M. punctata, 145
    M. siltepecana, 145
    M. tuberculata, 145
Mordacia, 535
Morinda citrifolia, 580
Morpho hecuba, 133, 133–134
Motocilla alba, 456
Mulgravea, 493
Mus, 221
Muscidifurax raptorellus, 92, 635
Mutillidae, 224, 229
Myiopharus aberrans, 594
Myiopsitta monachus, 246
Myoporaceae, 579
Myoporum sandwicense, 579
Myriapoda, 210
Myrmecia, 283
Myrmecocystus mimicus, 457
Myrmica, 543
    M. rubra, 542, 543
    M. ruginodis, 544
Myrtaceae, 579

N

Naegleria, 205
Nandidae, 280
Nannipus
    N. minor, 622
    N. peninsulatus, 622
Nannotrigona (Scaptotrigona) postica,
    291–292
Nasopelia galapagoensis, 240
Nauphoeta cinerea, 455
Necrophorus, 553
Nectarinia reichenowi, 456
Necturus, 118
Nemoria arizonaria, 446
Neodriprion, 555
Neoseius novus, 553
Nephilengys cruentata, 61
Neuroptera, 96, 256, 367, 596
Nezara viridula, 103

Nodularia, 613
Nucella lapillus, 226
Nyctereutes procyonoides, 287
Nymphalidae, 166

O

Obelellida, 362
Octopus, 573
Odonata, 271, 278, 444
Oecanthus, 345
    O. nigricornis, 456
Oecophylla smaragdina, 309
Ohomopterus, 364, *364, 365, 365*, 551
Oncopeltus fasciatus, 461–462
Oncorhynchus, 294, 537, 567
    O. clarki, 536
    O. kisutch, 221, 226, 294, 429, 431
    O. mykiss, 536
    O. nerka, 365, 535, 536, *536*,
        547–548
    O. tshawytscha, 535
Oncothrips tepperi, 305
Oniscoidea, 265
Onthophagus, 169, *169*
    O. acuminatus, 124–127, *126, 127*,
        305–306, 309
    O. incensus, 309
    O. nigriventris, 169
    O. sharpi, 169
    O. taurus, 169, 226–227, 305–306, 461
Onychomys torridus, 391
Oonopidae, 235
Ophryotrocha
    O. diadema, 458
    O. puerilis, 458
Opius fletcheri, 22
Opuntia, 346, 582, 584, 585, 586
    O. echios, 585
Orchestia, 419
Orchestina, 235
Oreomystis bairdi, 589
Ortholomus sp., 96
Orthoptera, 75, 271, 300, 344, 366, 367
Oscillatoria limnetica, 604
Osmerus, 567, 568
    O. mordax, 365
Othoptera,
Otiorrhynchus pupillatus, 282
Otocryptops sexspinnosa, 210
Oxybelus
    O. bipunctatus, 407
    O. sericium, 408
    O. strandi, 407, 409
    O. subulatus, 407
    O. uniglumis, 407
    O. uniglumis quadrinotatum, 407

P

Pachyrhachis, 368
Pagurus longicarpus, 446
Palmeria dolei, 589
Pandanoceae, 579
Papaveraceae, 270
Papilio, 208, 381, 533
    P. dardanus, 75, *75*, 167, 208, 401, 533
    P. dardanus forma *cenea*, 75

P. dardanus forma *hippocoon*, 75
P. dardanus forma *planemoides*, 75
P. echerioides, 208
P. hesperus, 208
P. lagasy, 533
P. phorcas, 208
P. polyxenes, 127
Papio, 289
    P. ursinus, 51–52, *52*
Paracentrotus lividus, 515–516, *516*
Paracerceis sculpta, 106, 279, *279–280,*
    423
Parachartergus colobopterus, 448
Paracheirodon, 300
Paradisaeidae, 207
Paralictus asteris, 203, 264
Paraneoptera, 366
Paratrigona subnuda, 292
Paravespula vulgaris, 228
Parus major, 466
Passer, 276
    P. domesticus, 273, 286
Passerella iliaca, 207
Passerina amoena, 420, 454
Pemphigidae, 220, 252, 253, 387
Pemphigus, 252–254, *253*
    P. bambucicola, 253
    P. betae, 220, 220–221, 406
    P. spyrothecae, 253, 254
Peperomia, 513
Perdita, 228, 292
    P. [Hexaperdita] graenicheri, 228
    P. mellea, 228
    P. portalis, 228, 292
    P. subgenus *Macroteropsis*, 228
Perilampidae, 22
Perilampus hyalinus, 22
Perlodes microcephala, 96
Peromyscus
    P. californicus, 287
    P. maniculatus, 554
Peronella japonica, 222, *222*, 517
Petroica multicolor, 272
Petromyzoniformes, 535
Phalaropus
    P. fulicarius, 273
    P. lobatus, 273
Phaneropteridae, 271
Phasianus colchicus, 245, *245*
Phasmatodeae, 406
Pheidole pallidula, 74
Philomachus pugnax, 421, 441, 593
Phlox cuspidata, 297
Phodopus, 287
    P. sungorus, 287, 460
Phoenicopterus ruber, 502
Phorina regina, 171
Photuris, 248, 248
Phthiraptera, 96
Phyllobates terribilis, 502
Phylloscopus, 257–258, *258*
Phymosomatoida, *412*
Physa
    P. gyrina, 316
    P. heterostropha, 316
Physalaemus pustulosus, 456

Physeter catodon, 233
Pieridae, 532
Pieris
    P. napi oleracea, 532
    P. occidentalis, 229, 532
    P. virginiensis, 532
Pinarolaxias
    P. crossirostris, 583
    P. inornata, 345–346, 347, *583*, 586
Pipidae, 335
Pisaster ochraceus, 517
Pisonia, 573
    P. brunonianum, 579
Pisum sativum, 105
Plasmodium falciparum, 631
Plastosciara perniciosa, 612, 613
Platygasteridae, 22
Plebeia droryana, 284
Plecoptera, 96
Plethodon, 620
    P. cinereus, 113
Plethodontidae, 405, 596, 607
Plocepasser mahali, 273
Ploceus, 276
Plumulariidae, 219
Podiceps auritus, 348
Podischnus agenor, 124–127, *125*, 309
Poecilia
    P. formosa, 224, 230
    P. latipinna, 224
    P. mexicana, 224
    P. reticulata, 300, 312, 433, 523, 571
Poeciliidae, 224, 422
Poecilocherus carabi, 553
Poecilozonites
    P. bermudensis, 361, 361–362
    P. sieglindae, 362
Poephila guttata, 274, 275, 293
Pogonomyrmex, 543, 544
    P. anergismus, 543, 544
    P. barbatus, 543
    P. colei, 543
    P. desertorum, 543
    P. occidentalis, 543
    P. rugosus, 544
Polistes, 133, 222–223, 228, 229, 283,
        314, 368, 408, 408–409, 447, 449,
        450, 455, 523, 564–565, 593
    P. aterrimus, 523
    P. atrimandibularis, 409
    P. biglumis bimaculatus, 391
    P. fuscatus, 382, 453
    P. huacapistana, 523
    P. instabilis, 283
Polistinae, 228, 408, 509
Polites
    P. s. sabuleti, 157
    P. sabuleti tecumseh, 157
Polybia rejecta, 250
Polycelis, 635
Polycentrus schomburgki, 280
Polyplectron bicalcaratum, 245
Ponera, 283
    P. coarctata, 283
    P. eduardi, 283
    P. punctatissima, 283

Ponerinae, 452
Pontania proxima, 606
Porthetra (Lymantria) dispar, 103, 270, 371
Portia, 445
    P. fimbriata, 409–411, 605
    P. labiata, 409
Portulaca howelli, 585
Portulaceae, 513
Precis, 390
    P. coenia, 23, 256, 257
    P. coenia nigrosuffusa, 23
    P. evarete, 23
    P. genoveva, 23
    P. octavia, 512, 532–533
Prochlorophyta, 604
Proteidae, 607
Proteus, 118
Protorhyssalinae, 613
Psalis, 234
Pseudacris triseriata, 113
Pseudaletia unipuncta, 54, 300
Pseudidarnes minerva, 395, 420
Pseudocrenilabrus philander, 389
Pseudonestor xanthophrys, 589
Pseudopalaeosepsis nigricoxa, 199
Pseudoregma, 253
    P. bambucicola, 253
Psitirostra psittacea, 589
Psorthapsis, 407
Psychotria marginata, 229
Pteronemobius, 406–407, 407, 408
    P. fascipes, 408
    P. mikado, 406, 407, 408
    P. nigrofasciatus, 406, 407, 408
    P. nitidus, 205
    P. taprobanensis, 311–312, 408
Ptilonorhyncidae, 207
Ptinus latro, 635
Pungitius pungitius, 280
Pyemotidae, 530
Pygmephorellus, 530
    P. bennetti, 530
Pygosteus pungitius, 293–294
Pygostolus falcatus, 22
Pyractomena angulata, 248
Pyrenestes, 420
    P. ostrinus, 350, 398, 587–589, 588
Python molurus, 518

Q
Quercus, 109

R
Radiodonta, 613, 614
Rana, 272, 272, 285
    R. blythii, 285
    R. chiricauensis, 300
Ranunculus, 229, 401, 403, 415
    R. flammula, 414, 531–532
Rattus norvegicus, 134–135, 460
Reduviidae, 462
Reliquia santamarta, 532
Rhabditis nigrovenosa, 57
Rhagoletis pomonella, 557–559, 562, 580
Rhamnaceae, 421

Rhodnius prolixus, 462
Rhombozoa, 612
Rhyacosiredon, 354
Rhynchosciara anglae, 330
Rhyssalinae, 613
Rhytidoponera metallica, 508
Rhyzobium, 326
Ropalidia, 283, 450
    R. ignobilis, 228, 507
    R. marginata, 507
    R. plebiana, 283
Rosaceae, 421
Rubiaceae, 229

S
Saccodon, 387
Sacculina neglecta, 280
Sagittaria sagittifolia, 174
Salamandridae, 356
Salmo
    S. salar, 294, 365, 536
    S. salvelinus, 550
    S. trutta, 365, 567
Salmonidae, 569
Salticidae, 409, 444, 605
Salvelinus, 536
    S. alpinus, 365, 548
    S. malma, 536
Sanguinaria canadensis, 270
Sapindaceae, 579
Sapindus saponaria, 573, 579
Saponaria officinalis, 604
Sarcophaga crassipalpis, 73
Sarcophagidae, 366
Sarracenia purpurea, 560
Scaphiopus (Spea) multiplicatus, 111, 307, 369, 396, 397, 398, 433, 462
Scapteriscus, 203
Scaptomyza, 94
    S. albovittata, 94
    S. oahuensis, 94
Scaptotrigona, 313
Scarabaeidae, 127, 400
Scaura, 313
Sceloporus, 456
    S. occidentalis, 312
Schedorhinotermes, 292
    S. lamanianus, 293
Schwarziana, 313
Sciaenidae, 506
Sciapodinae, 496
Sciaridae, 612, 613
Scolopacidae, 207
Scolopendromorpha, 210
Scutigera
    S. coleoptrata, 210
    S. cryptops, 210
Scutigeromorpha, 210
Sedum telphium, 513
Selasphorus rufus, 456
Semotilus atromaculatus, 444
Sepsidae, 199
Serinus canarius, 76, 92, 274, 275
Serranidae, 458
Sertulariidae, 219
Sesuvium, 585

Schistocerca, 300
    S. americana, 342
Shorea leprosula, 73
Sialis lutaria, 96, 527
Sigmodon hispidus, 311
Siluriformes, 550
Siphonaptera, 367, 368
Siredon, 607
Sirenidae, 607
Siricidae, 559
Sitatroga cerealella, 225
Siteropes, 530
Solenopsis, 74
    S. geminata, 309
    S. invicta, 74, 106, 309, 510, 537–538
Solidago altissima, 92–93
Sorex araneus, 444
Sparisoma viride, 280
Spatangoida, 412
Spea
    S. bombifrons, 396, 397, 448
    S. (Scaphiopus) multiplicata, 111, 307, 396, 397, 398, 433, 462
Spermophilus tridecemlineatus, 73
Sphaerechinus granuluris, 502
Sphaeromatidae, 279
Sphecidae, 407, 409, 425, 457, 596
Sphecius speciosus, 457
Sphecomyrma freyi, 613
Sphecomyrminae, 613
Sphex ichneumoneus, 52–53, 223, 344, 425–426
Staphylinidae, 264, 277
Stegodyphus, 249
Stemonoporus, 251
Stenogastrinae, 509
Sterna paradisaea, 348
Stictia heros, 343
Stigmatella, 613
Stirodonta, 243
Strepsiptera, 256
Striga asiatic, 231
Strongylocentrotus
    S. droebachiensis, 515
    S. pupuratus, 515, 517
Sturnus, 207
Stylopoma spongites, 624
Styrax suberfoliae, 253
Symphyta, 606
Synapsida, 606
Syngnathidae, 286
Syngonium standleyanum, 306
Synopeas rhanis, 22
Syrbula admirabilis, 59

T
Tachinidae, 366
Taeneopygea guttata, 76–77
Talinum, 513
Taphrina, 605
Taricha granulosa, 207, 356, 368
Technomyrmex albipes, 543
Teleogryllus commodus, 277
Telespyza, 590
    T. cantans, 414
    T. ultima, 589

Teleutomyrmex schneideri, 542
Temnopleuroida, *412*
Tephritidae, 557
Termitoxenia, 256
Tetragnathidae, *61*
Tetrapleurodon, 535
Tettigonoidea, 271
Thalassoma bifasciatum, 458
Thamnophis sirtalis, 281
Theridiosomatidae, 587
Thomomys, 510–511
Thorius, 596
Thryothorus, 274
   T. nigricapillus, 273, 274
Thysanoptera, 367, 387
Thysanura, 495
Tiaris obscura, 583
Tilapia, 364
   T. natalensis, 592
Tillandsia, 251, 513
   T. deppeana, 252, 605
Tillandsioideae, 251, 605
Timema cristinae, 406, 421
Timemidae, 406
Tiphiidae, 234, 613
Torymidae, 234, 395, 560
Torymus auratus, 560
Trematops, 207
Tremex columba, 559
Tribulus, 346, 414, 584, 585
   T. citoides, 585
Trichogramma
   T. evanescens, 225
   T. semblidis, 527
Trigona, 128
Trimerellacea, 362
Trinervitermes, 292
Tripterygion, 280
Triticum vulgare, 355
Triton cristatus, 308

Triturus vulgaris, 226
Trogoderma variabile, *38*
Trypanosoma, 631
Trypoxilini, 596
Trypoxylon politum, 596
Typhlomolge, *118*

U
Uca, 213
   U. m. maracoani, 455
Ulmus
   U. davidiana var. *japonica*, 533
   U. parvifolia, 533
Uropygi, 271
Urosaurus ornatus, 280–281
Uta stansburiana, 226, 281, 420, 432–433

V
Vanessa cardui, 512
Varroa jacobsoni, 253, 553
Vatica, 251
Velarifictorus micado, *408*
Vespidae, 408, *410*, 455, 564
Vespinae, 228, 408, *410*
Vespa crabro, 228
Vespula
   V. acadica, 410
   V. austriaca, 410
   V. squamosa, 410
Vestiaria coccinea, 589
Vibrio
   V. cholerae, 501
   V. fischeri, 501
Viburnum, 552
Vidua, 553
Viola calaminaria, 502
Viperinae, 494
Vollenhovia, 543
   V. emeryi, 543, 544
   V. nipponica, 543

Volvox, 615
   V. carteri, 207
Vriesia, 251
Vulpes vulpes fulva, 249

W
Wendilgarda, 208
   W. galapagensis, 587
Wikstroemia, 581
Wilsonia citrina, 286
Wolbachia, 528
Wyeomia smithii, 405–406, 560–561

X
Xanthocnemis aealandica, 63
Xenopus, 73, 96, 114, 205, 221, 326, 478
   X. laevis, 258, 272, 335
Xerospermum intermedium, 73
Xiphophorus, 106–107, 359, *359*
   X. maculatus, 422
   X. multilineatus, 422
   X. nigrensis, 422
   X. xiphidium, *359*
Xylocopa pubescens, 403–404, 448

Z
Zea
   Z. luxurians, 267
   Z. mays, 130, 202, 261
   Z. mays mays, 211, 265–269, *266*
   Z. mays mexicana, 266
   Z. mays parviglumis, 265, 267, 269
Zethus miniatus, 509, 510
Zonotrichia
   Z. leucophrys nuttalli, 40
   Z. leucophrys oriantha, 39–40
   Z. querula, 456
Zygoptera, 63
Zygothrica, 493–494
   Z. dispar, 206, *206*

Subject Index

A

Abrupt speciation 551–554
  definition 551
Accessory hearts (insects) 200
Acrosins 135
Actinopterygians 87
Adaptation 6–8, 18, 371–392, 397
  antagonistic 397
  cladistic tests 371–372
  non-adaptation and 18, 417
  Unimodal, 6–8
  unimodal, as typological concept 7
  see also Adaptive evolution
Adaptationism 447
Adaptive design 471–472, 477
  development vs selection problem and
    471–472
  lens metaphor 477
  microscope metaphor 477
  variation vs. selection in 471–472
Adaptive evolution 29, 32, 139–158
  as two-step process 29, 140
  causal chain of 141–142
  definitions 32, 477
  developmental **definition** 158
  key events in 140
  prerequisites for 142–143
  see also Adaptation
Adaptive landscape 394
Adaptive radiations 21, 564–597,
    566–573
  binary 566–573
  contributing factors 574
  definition 564
  ecological theory 565
  explosive 573, 565, 591
  immatures and 21
  intraspecific 586–587
  multidirectional 573–585
  tropical 594
  see also Flexible-stem hypothesis
Adaptive syndromes 302–305
African cichlids, *see* Cichlids, African
African finches, *see* Finches, African
Age, dominance rank and (social wasps)
    451
Aggressiveness, artificial selection
    on 252
Agonistic buffering 289
Air breathing, vertebrates 606–607
Albinism 357

Alcohol dehydrogenase (adh) 211, 230,
    319–320, 580–581
  evolution of 211, 230, 319–320
Alfalfa 339
Algae 46, 58, 85, 250, 502, 521, 615,
    634
  blue-green 85
  brown 502
  green 46, 58, 615
  red 502
Alleles 19, 148, 156, 378, 435
  nearly neutral 19
  neutral, 19, 435
  neutral, as potentially adaptive 148,
    156
  origin of term 378
  see also Genetic polymorphisms
Allochronic speciation 554–557
Allochrony (**definition**) 528
Allometry 36, 72, 227, 228, 292,
    308–310, 400
  ants 308
  bees 228
  conservative force in evolution 309
  definition 308
  diphasic 309
  evolution of, in ants 309
  evolutionary lability of 309
  maladaptive 309
  models of 309–310
  resource allocation and 308
  trait independence and 400
  triphasic 309
Allopatric speciation (**definition**) 528
Alternative phenotypes 30, 62, 141, 295,
    377–468
  allelically determined 106
  antagonistic 396–399
  buffering effect of 162
  competition dependent 433
  complementary, emergent levels and
    429–430
  confusions regarding 381, 382, 383
  cross-sexual transfer and 277–282
  defensive 316, 392
  definition 26, 377
  divergence and (cf. speciation)
    401–404
  evolutionary properties 392–393
  gene expression and 74–76
  genetic architecture of 79

  history of concept 378
  homology hypotheses and 203
  homoplasy and 370
  host shifts and 557
  independent evolution of 400–404
  interspecific, *see* Interspecific
    alternative phenotypes
  life-stage phenotypes, confusion with
    382
  macroevolution and 599–616
  maintenance 417–439
  maintenance of genetic polymorphism
    and 434–439
  mating tactics 106, 162, 277–282,
    281, 386, 389, 421–423, 463, 540,
    543–544
  phenotype-dependent 388–390
  rarity (supposed) 84, 385–387
  recurrent 359–360, 370–371
  reversible 294, 378, 396, 518, 594
  size-related, speciation and 544
  speciation and 527–545, 556, 562
  speciation hypothesis (steps) 540
  species-like properties 381, 395–400
  stress induced 390–391
  systematics and 370–371
  see also Conditional alternatives;
    Dependent alternatives; Dispersal
    polymorphisms; Genetic
    polymorphisms; Interspecific
    alternative phenotypes; Mimicry
    polymorphisms; Polymorphisms;
    Polyphenisms; Seasonal
    polyphenisms
Alternative splicing 48–49, 212
  definition 48
  phenotypic recombination by 323–324
Alternative tactics
  see Alternative phenotypes
Altruism, *see* Kin selection
Amazon molly (*Poecilia formosa*) 224
Ammonites 173, 221, 361, 362, 378
Amoebae 615
Amphipods 343, 551
  learning in 343
  neoteny 551
  plasticity and divergence 551
  replicate speciation (in caves) 551
Anadromy (**definition**) 362
Analogous variations 354
  see Recurrence

Androgens 92, 106, 272, 294, 304
  female vocalization in frogs and 272
  jack males (salmon) and 294
Angiosperms, see Flowering plants
Anglerfish 248
Anisogamy 397, 633
Anomalies 23, 205, 206–207, 208, 254,
    257, 263, 265, 269–270, 286, 319,
    331, 348, 371, 553
  as material for selection 205, 208, 331
  as research tools 205
  behavioral, learning and 348
  birds 206–207
  definition 205
  established traits and 207, 371
  homeotic, in butterflies 257
  host choice, indigobirds 553
  mammals 206
  molecular 319, 331
  parasite-induced 265
  recurrent (examples) 206
  recurrent, in salamanders 207
  sexual 263, 269–270
  see also Noise
Antagonistic pleiotropy 78, 88, 170,
    406, 434–435, 624
  aphids 406
  definition 78, 434
  genetic polymorphism and 434–435
  release and 404, 435
  see also Release
Antbirds (Hylophylax sp.) 273
Anthropomorphism 314, 442–443
  Darwin on 443
Anti-aphrodisiacs 632
Antithesis 454–455
  definition 454
  principle of 454
Ants 42, 64, 73, 74, 75, 107, 120, 131,
    135, 171, 186, 190, 246–247,
    271–272, 279, 283, 308–310, 400,
    409, 449, 452–453, 457, 461, 508,
    510, 533, 534, 537–538, 542–545,
    553–554, 613
  allometries 308
  army ants (Eciton) 42, 135, 186,
    271–272, 308
  atavistic queens 508
  brain organization (modular) 64
  caste determination, genetic 107, 120
  caste mosaics (bilateral) 75
  castes 73, 120, 131, 309, 400, 461
  colony structure (fire ants) 510
  communication 246–247, 457
  Darwin on 190
  dominance assessment hypothesis and
    449, 452–453
  dispersal polymorphisms 107, 533,
    534
  ergatoid (workerlike) males 279
  fossil 613
  gene expression (modular) 73, 74
  self organization in 42, 135
  signal evolution 246–247
  speciation 537–538, 542–545,
    553–554

workerlike behavior of males 283
  see also Fire ants; Parasitic inquilines
Aortic arch, variation in 50
Aphids 66, 75, 109, 127, 131, 184, 220,
    252–254, 310, 387, 406, 427,
    533–534
  polymorphisms 66, 75, 127
  character release 406
  speciation 533–534
Apomorphy (definition) 361
Apoptosis, see Cell death
Apple-maggot flies (Rhagoletis
    pomonella) 554, 557–559, 562,
    580
  speciation 557–559
Appendages 8, 199–200, 210, 214, 219,
    232, 256, 311–312, 407
  atavistic 232
  deletion and 219
  duplication and 210, 214, 256
  material compensation and 311–312
  modularity in 199, 210
  novel, origin of (fly) 8, 199–200
Arborescence, origin, in plants 521
Archetypical trait (definition) 495
Arctic charr (Salvelinus alpinus; salmonid
    fish) 365, 548
Aristotle 209
Aromatase 276
Arthropods 73, 178, 212, 219, 258, 610,
    613–614
  body plan 73, 212, 219, 610
  Burgess-shale fossils 613–614
  plasticity of cf. vertebrates 178
  pre-Cambrian diversification
  segmental gene expression 73
  see also Insects; Mites; Spiders
Artificial selection 16, 76, 101, 105, 119,
    124–125, 126, 127, 147, 149–150,
    155–157, 196, 215, 227, 230, 252,
    255, 299, 306, 373, 437
  analogy with natural selection
    155–156
  correlated effects of 215, 230, 299,
    306
  environmental contributions to 156
  mimicry by hormones 127, 373
  neutral zone as tool in 126, 437
  results of 16, 76, 101, 106, 119, 148,
    156, 252, 340, 342, 373
  thresholds (plasticity) changed by 106,
    119, 124–125, 127, 147, 149–150,
    196, 227, 255, 437
Ascidians 366, 615
Ascomycetes 605
Asexual reproduction 632, 635
  see also Parthenogenesis
Asses (equine) 354
Assessment 5, 431, 433, 440–468
  concept 442
  definition 431
  mechanisms, evolution of 464–468
  skepticism regarding 440–442
Assessment hypothesis (of social
    dominance) 449
Associative learning (definition) 337

Assortative mating 7, 530, 540–541,
    553, 631
  evolution of 540–541
  speciation and 528, 538–545
Astogeny (definition) 624
Atavisms 207, 219, 224, 232–234, 266,
    353, 508
  as phylogenetic tools 234
  definition 232
  developmental basis of 237–239
  induced, in chicks 233–234
Atlantic salmon (Salmo salar) 294, 365,
    536
Atomization 81, see Modularity
Attention 314, 341
  competition hypothesis 314
  learning and 341
  motivation and 341
Attractors 38, 128–129
Aunt behavior 289
Australian Robin (Petroica species) 272
Autogeny (mosquitoes) 560–561
Autonomy 56, see Modularity
Auxins (in gall insects) 111
Aversion learning 444
Axolotls (Ambystoma mexicanum) 22,
    511, 607–608

B

Baboons 51–52, 289, 290, 291
Bacteria 148, 185, 211, 265, 322, 326,
    330, 427, 462, 501, 521, 528, 613,
    615, 637
  gene expression mechanisms in 330
  horizontal gene transfer from 322
  speciation and (Wolbachia) 528
  symbiotic 265, 501
  see also Cyanobacteria; Prokaryotes
Balancing selection 417
Baldwin effect 24–25, 151–154, 415, 509
  genetic assimilation and 24, 151–154
  definition 24
Bamboos 250
Barnacles 280, 312, 405, 392, 431
Batesian mimicry 75, 79, 80, 106, 167,
    189, 381, 401, 432, 533
  Darwin on 189
  polymorphisms 167, 401, 432, 533
Bateson, W. 11, 24, 26, 84, 193, 527
Bauplans, see Body plans
Bay wrens, see Wrens, bay
Bean weevils (Callosobruchus maculatus)
    443, 465–466
Bee-eaters (Merops bullockoides) 448,
    449, 458
Bees, solitary 146, 203, 228, 229, 235,
    236, 264, 292, 349, 363, 367, 403,
    404, 408, 519
  dimorphic males 228
  parasitic 203, 229, 264
  recurrence 235, 236, 363
  see also Carpenter bees
Bees, social 519, 235, 236, 363
  reversions to solitary life 235, 236, 363
  see also Bumblebees; Halictid bees;
    Stingless bees; Honeybees

Beetles (Coleoptera) 22, 38, 98, 119,
   124–127, 169, 215, 226–227, 248,
   264, 271, 277–278, 282, 305–306,
   309, 364, 400, 456, 443, 461, 465,
   466, 551, 540, 552, 635 149, 226
   aggressive mimicry 248
   alternative behaviors 226
   antennal diversity 215
   body-size assessment, ontogenetic 456
   developmental tradeoff 227, 305–306
   disruptive selection in 226
   female mimicry 277–278
   genitalia and speciation 551
   hormones and switch in 127, 227, 461
   hybrid parthenogenesis 635
   hypermetamorphosis 22
   interchangeability 119
   independent divergence, horned form
      169
   maternal effects 98
   nutritional switch 461, 124–127
   recurrent phenotypes 364–365
   search behavior 38
   sexual selection, body size 54–541
   speciation 551, 552
   switch mechanisms 124–127
   switchpoint plasticity 309
   see also Fireflies
Behavior (general discussions) 24, 39,
   43–44, 59, 63–65, 76–77, 172,
   180–182, 213, 212–214, 223–224,
   244–250, 252, 314–315, 347, 494,
   570–572
   combinatorial evolution in 172, 213
   deletion, in evolution of 223–224
   duplication, in evolution of 212–214
   genetics and 76–77
   heterochrony, in evolution of 244–250
   homology concept and 494
   modularity of 59, 63–65
   morphology and 24, 180–182, 347,
      570–572
   plants 39, 43–44
   plasticity of, cf. morphology 180–182
   response to selection 76, 252
   temporal-interference conflicts in
      314–315
   see also Alternative phenotypes;
      Communication; Diet-induced
      morphology; Learning
Bet-hedging 424, 556
   definition 424
   see also Stochastic alternatives;
      Stochastic regulation
Binary flexible stem (definition) 566
Binary radiations, see Adaptive
   radiations, binary
Biogenetic law (definition) 9
Biogeography, climate change and 519
Biological homology 486, see Homology,
   broad-sense
Bipolar (manic-depressive) illness 101, 105
Birds 39, 40, 45, 64–65, 76, 77, 92, 97,
   133, 134, 156, 157, 181, 206, 207,
   223, 238, 245, 246, 249, 256, 263,
   271, 272–276, 286, 293, 341–343,

   344, 346–348, 391, 414, 425, 426,
   434, 441, 443, 444, 445, 450–451,
   452, 454, 456, 457, 466, 502, 518,
   523, 534, 553, 562–563, 582–586,
   588–590, 595, 600
   assortative mating 553
   hormones and plumage dimorphism 274
   learning 341, 342, 343, 345–348,
      562–563, 582–586
   parental care 223, 285–286
   plumage 133, 238, 272–273, 274,
      275, 502
   song 39, 40, 64–65, 77, 273–275, 553
Blackbush (Colegyne) 520
Blindsnakes 367
Blowflies 171
Bluegill sunfish (Lepomis machrochirus)
   397, 280, 397, 550, 571
Bluehead wrasse (Thalassoma
   bifasciatum) 458
Bluffing 467
Boas 368
Body plans 219, 309, 610, 613, 615, 616
   body size and 309
   genomes and 616
Body shape 300, 550
   diet-induced
   predator-induced 300, 550
   stream-velocity (exercise) induced 300
   see also Diet-induced morphology;
      Two-legged-goat effect
Body size 156, 224, 225, 227–229, 260,
   316, 349, 363–364, 389, 451, 456,
   461–462, 465, 510–511, 524, 535,
   538–545, 550–560, 461–462,
   581–582, 584–585, 608–609
   adaptive evolution of 316, 540–541, 609
   alternative phenotypes and 349,
      363–364, 389, 461–462, 465,
      538–545, 609
   assessment of (behavioral) 456
   assessment of (internal) 461–462
   assortative mating and 541–542
   bimodal (host-dependent) 229
   bimodal (prey-dependent) 224
   complex determinants of 156
   dimorphisms in 227–229
   macroevolution and 608–609
   migration and, salmon 535
   niche shifts and 364
   novel traits and 510–511
   origin of life and 524
   reproductive isolating mechanism 550,
      560
   reproductive value and 451
   selection on 156, 540–542
   sex determination, size-dependent 260
   sexual selection and 608–609
   social rank and 451
   speciation-related alternatives and 540,
      542–545
   timing of metamorphosis by 461
   trimodal (host-dependent) 225
   trophic specialization and, birds
      584–585
   variation, food supply and 581, 582

   see also Cope's rule; Dwarf forms; Egg
      size; Gigantism; Miniaturization;
      Pituitary dwarfism
Bolas spiders 248
Boldness 341,-342, 347–348, 590
   learning and 341, 342, 347–348
   shy-bold continuum 348
   see also Curiosity; Neophobia
Bolyrine snakes 85
Bone, plasticity of 40
Boom-and-bust selection regimes 595,
   see Episodic selection
Bower birds 207, 391
Brachiopods 362, 596
Brain 64, 76, 77, 110, 274, 343, 464, 591
   effects of use 110
   gene expression in 76
   modular structure (ants) 64
   sexual differentiation 77
   size (isopods) 343
   social selection and, humans 464, 591
   song and (birds) 274
Branch jumping 319, 320
Bridging phenotype 91, 93, 95–98
   plasticity of 93
   see also Continuity of the phenotype
Britten-Davidson Model 27–281
Broad-sense homology, see Homology,
   broad-sense
Bromeliads 251, 605
Brood parasitism (birds) 419, 434, 553
Bryozoans 251, 361, 362, 366, 381, 397,
   623–627, 628
   colony form 624
   defensive structures 624
   morph-species confusion 381, 628
   plasticity as taxonomic problem in
      628
   plasticity in 624
   punctuated evolution in 623–627, 628
   zooidal budding 366
Buffering effect 162, 392, 602, 603, 604
   alternative phenotypes and 162, 392,
      602
   Darwin on 604
   selection above the individual level and
      182
Bugs (Hemiptera) 103, 171, 312, 429,
   461, 462
Bumblebees (Bombus species) 228, 263,
   284, 339, 349, 452
   learned specializations 349
   learning and morphology 339
   social dominance 452
Burden 326
Burgess shale 613–614
   fossil reconstructions 614
   taxonomy and 613
Burmese pythons (Python molurus) 518
Buttercups (Ranunculus) 163, 229, 401,
   403, 414–416, 531
   genetic assimilation 416
   heterophylly 414
   morph-specific geographic variation 401
   phenotype fixation and release
      414–415, 531

Butterflies 23, 75, 79, 85, 119, 127, 133,
  157, 166, 167, 169, 181, 203, 208,
  214, 216, 217, 226, 229, 256, 257,
  263, 271, 305, 351–352, 343, 387,
  357, 358, 381, 390, 401, 444,
  511–512, 532–533, 594
  interchangeability 119
  larvae 181
  learning and host preference 343
  migration 594
  mimicry 357
  mimicry polymorphisms 75, 79, 208,
    226, 381
  phenocopies 23
  polyphenisms 75, 127, 203, 229, 390,
    511–512
  speciation 532–533
  supergenes 79
  recurrent color patterns (*Heliconius*)
    358
  wings 214, 216, 217, 256, 305
Buttress roots 301–302, 604

C
C-value paradox 335
Calcitonin 49
Calcium 522, 570
  Cambrian explosian and 522
  plate reduction and, sticklebacks
    570
Calluses 256
Cambrian explosion 501, 522–523
  calcium levels and 522
  climate change and 522
  spliceosomal introns and 522–523
  see also Burgess shale
Canalization 7, 8, 13, 25, 51, 297, 399,
  499
  definition 8, 25
  epigenetic landscape and 25
  genetic control assumed 499
Canaries (*Serinus canarius*) 64, 76, 92,
  274–275
Canids 223, 249, 288, 299, 454–455,
  495
Cannibalism 111, 369, 448, 462
Carbon dioxide, global change in
  521–522
Carbon isotopes (in paleontology) 622
Cardinals (*Cardinalis cardinalis*) 206
Carnivory, origin, in Hymenoptera
  605–606
Carotenoids 502, 523
Carpenter bees (*Xylocopa*) 284, 403,
  448
Cascades (developmental) 133
Caste determination 68, 107, 120, 128
  definition 68
  genotypic influence 107, 120, 128
Catastrophic sex-transmutation theory
  267
  see also Maize, evolution of
Catfishes, loricarioid 595
Cattle egrets (*Bubulcus ibis*) 92
Causation, *see* Proximate and ultimate
  causation

Cecidomyid flies 613
Cell death (apoptosis) 41–42, 71, 258
  and cell-lineage competition 185
  in color metamorphosis 219
Cell types 186
Cell wall 45
Cell-lineage Competition 43, 184–187
  cohesiveness problem and 184–187
  evolvability and 43
Cellulose 304
Centrioles, sperm-supplied 632
Cereals, Vavilov's Law and 355
Chalcid wasps 606
Challenge hypothesis (**definition**) 304
Chaos (**definition**) 38
Characid fishes 300, 371
Character displacement (interspecific)
  397–399
  definition 399
  spadefoot toads 397–398
Character displacement (intraspecific) 397
Character Release, *see* Release
Character weighting 370–371, 373
  ancestor reconstruction 373
  assumptions 370–371
  cladograms 373
  recurrence homoplasy and 370–371
Charadriiform birds 274
Chickens 156, 119, 212, 119, 245, 263,
  286
  sex transfer in hens 263
  sex-expression anomalies 286
Chimpanzees 290, 335, 453, 456–457
  drumming 456
  genetic distance, humans 335
  rank and reproduction 453
  social assessment 456–457
Chironomid flies 333
Chitin synthetase 258, 326
  phylogentic conservation of 258
Chloroplasts 45, 46, 321, 604
  mobile introns in 321
  symbiotic origin 604
  thylakoid membrane of 45, 46
Choice 442, 452, 457–458, 466–468,
  631–632
  cost of sex and 631–632
  definition 442
  female choice 457
  mate choice 450–451, 452, 457–458
  parental choice 466
Chromosomal linkage, *see* Linkage,
  chromosomal
Chronoclines 362
Chubs (*Semotilus atromaculatus*) 444
Cicada-killer wasps (*Sphecius speciosus*)
  457
Cicadas (*Magicicada* species; Homoptera)
  555–556
Cichlids (fish) 163, 183, 204, 218–219,
  365, 389, 455, 575–577, 592
  jaw morphology 575
  mouth displays 592
  recurrent trophic morphs 365
  size-dependent male alternatives 389
  status-dependent alternatives 389

signal antithesis in 455
  see also Cichlids, African
Cichlids, African 41, 54, 87–88, 163,
  397, 519, 568–569, 574–578
  climate cycles and speciation 519
  experience and muscle action 41
  paedomorphosis 569
  parallel species pairs in 568–569
  pharyngeal jaws 87–88, 163, 575–577
  radiation 574–578
  trophic alternatives 397
  trophic ontogeny 569
  trophic polymorphisms 54
Ciliates 322, 333, 613
Circadian rhythms (**definition**) 459
Circumnutation 39, 43
Cisco (*Coregonus sardinella*) 567
Citric acid 512
Clade selection 43, 165, 475, 599, 631,
  631
  evolvability and 43
  species selection 599, 631
  see also Levels of selection
Cladistic homology, *see* Homology,
  cladistic
Cladistics, *see* Phylogenetic analysis
Clandestine evolution 22
Cleistogamous flowers 250
Climate change 518–524
  biogeographic consequences 519
  cycles of 518–524
  explosive speciation and 519
  plasticity and evolution during
    518–524
Clines 170, 528, 560–561
  morph ratio 170
  speciation and ʿ28, 560–561
Clonal invertebrates 185, 251, 537
  growth forms 251
  sex in 637
  see also Bryozoans
Clutch size 425
Cnidarians 86, 168, 185, 218–219
Coadapted gene complexes 79
Coadapted gene pool 9, 528
Cockroaches (Dictyoptera, Blattaria)
  248, 294, 455
Cocos-Island spiders 208, 587
Cocos-Island Finches (*Pinaroloxias
  inornata*) 345–346, 586–587
Codex 94, 99
  definition 99
Coevolution of coexpressed traits,
  principle of 168
Coevolution, host-parasite 482, 501,
  597, 602, 631
Cognition (**definition**) 337
  see also Learning
Cohesiveness (gene pool) 528
Cohesiveness Problem 8–10, 183–187,
  192
  Darwin on 192
  Lillie's paradox 186
  solutions to 10, 183–187
Coho salmon (*Onchorynchus kisutch*)
  221, 226

Coin-flipping 424, 425
clutch size and 425
see also Stochastic alternatives
Colony-level selection 448
Coloration, see Pigments
Combinatorial chemistry 331
Combinatorial evolution 30, 69, 77,
164, 170–174, 182, 200, 311, 213,
274–275, 317–336, 378, 485,
491–492, 497, 610
behavioral 172, 213
definition 164, 200
early idea of 378
genetic 211
hierarchical classification and 610
homology and 485, 491–492, 497
modularity and 170–174, 182
molecular 317–336
partial sex transfers due to 274–275
see also Exon shuffling; Phenotypic
recombination; Part II (195–374)
Common garden experiments 381
cf. contrasting garden experiments
423
Communication (animal) 63, 171–172,
181, 212–213, 223, 244–247,
271–277, 444–445, 457–458,
466–467, 468, 503, 592
combinatorial evolution of 172, 213
courtship 213, 245, 457–458
cross-sexual transfer and 271–277
displays 63, 181, 212–213, 223,
466–467, 592
duplication in 213
honeybee dance 171, 444–445
sensory traps in 212, 245, 246,
468
signal-response coevolution 503
see also Antithesis; Ritualization;
Signals
Compartmentalization 12, 56
see also Modularity
Compartmentation 55, see Modularity
Compensation 303
see also Material compensation
Competition 397, 464, 559, 572, 577,
568–569, 581–582, 595, 609
accelerated divergence and 577
degree of specialization and 595
density-dependent 609
radiations and 581–582
specialization and 577
social 397
social, brain evolution and 464
speciation and 559
trophic divergence and, sticklebacks
572
trophic, and binary radiations (fish)
568–569
Competition dependent alternatives 295,
433
Competition-dependent selection 389,
433–434
Competitive interaction model (of tissue
growth) 37
Competitive speciation model 559

Complementarity (of continuous and
discrete variation) 11–13, 71, 83–84,
135–138
developmental basis 71, 135–138
modularity-connectedness dichotomy
and 83–84
mouse mandible and 71
principle of 13, 137
wave-particle (physics) analogy 83
Complementary alternatives 429–430,
632–633
emergent levels and 429–430
male and female as 632–633
Complexity 151, 200, 202, 246–247,
335, 370–371, 524
combinatorial evolution and 200
definition (complex trait) 202
dissociability and 370–371
environmental information and 524
evolution of, in animal communication
246–247
gradual modification and, cf.
recombination 151
plasticity and 335
recurrence and 370–371
Complexity catastrophe (definition) 9
Complex trait (definition) 202
Compression wood 301
Concerted evolution 209, 214–217
definition 214
Condensation (cell aggregation) 135
Conditional alternatives 80, 146,
162–163, 225, 383, 384, 388–392,
403, 419, 428, 429–433
as nongenetic traits 383
costs 430
genetic models and 432
genetic polymorphisms, as derived
from 80, 146
genetic polymorphisms, treated as 80
kinds 388–392
models 419, 428, 430
ratcheting effect (evolutionary) of
162–163, 403
regulation 417, 428–429, 430–432
skepticism regarding 225, 384
see also Polyphenisms
Conditional ESS 428, see Conditional
alternatives
Conditional-ratchet hypothesis
162–163, 238, 402–404, 518,
583, 594–595
definition 403
see also Episodic selection
Cone snails (Conus species) 591
Connectedness 65, 81–83
see also Modularity
Consciousness 314
Constituitive expression (definition) 24
Constraints, developmental 8, 9, 25,
139, 275, 428, 474, 476
definition 25
development-selection controversy and
139, 474
misleading concept 275
mistaken for selection 476

phylogenetic and developmental,
synonymous 25
Continuity of the germ line 90, 98
Continuity of the phenotype 29, 65,
90–98, 115
definition 93
Continuous polymorphism 58, 137, 381
Continuous variation 135–138
alleged predominance of 136
discrete variation and, see
Complementarity
developmental basis 135–138
see also Complementarity; Gradualism;
Quantitative traits
Convergence 264, 357, 358, 486, 487,
493
cf. parallel evolution 358
cladistic concept 357, 358
cladistic homology and 486
definition 264, 358
Coots 466
Cope's rule 608–609
definition 608
Corals 251
Corkwings (Drenilabrus melops, labrid
fish) 280
Corn, see Maize
Correcting behavior 171
Correlated change, plasticity and 52,
161, 296–316
see also Allometry; Genetic correlation;
Two-legged-goat effect
Correlation pleiades 374
Cotton rats 311
Courtship 213, 245, 457–458
duplication in 213
see also Communication (displays)
Covariance matrices 65
Crabs 213, 455, 265, 280, 343
Crassulacean Acid Metabolism (CAM)
45, 76, 252, 259, 512–514, 595,
604
evolution of 259, 512–514, 522
facultative 512–513
gene expression and 512
recurrence (phylogenetic) 512
Crickets (Gryllidae) 106, 119, 203, 205,
271, 277, 300, 310–312, 345, 373,
406–408, 422–423, 429, 456,
554–555, 561–562
allochronic speciation 554–555,
561–562
alternative male tactics 422–423
body-size assessment 456
character release 406–408
density-dependent responses 300
release 561–562
wing dimorphisms 106, 310–312
Cross-sexual transfer 260–295, 367
animal communication and 271–277
correlates of in bird behavior 286
Darwin on 189, 262–263
definition 260
femalelike behavior of males 264, 280,
281, 278, 279, 292
kinds of evidence for 263

Cross-sexual transfer (*Continued*)
  ovary evolution and (insects) 367
  partial 274–275
  polarity (direction) of 273
  sex-transfer-history hypothesis (of
    polarity) 273
Crossing over, unequal, and genetic
  novelties 331
Crustaceans 265, 343, 344, 444, 446,
    495, 551
  individual differences in 344
  learning in 343
  legs 495
  parasite-induced anomalies in 265
  see also Amphipods; Crabs; Isopods;
    Mantis shrimps; Shrimps
Cryptic gene hypothesis 239
Cryptic morphs (**definition**) 628
Cryptic species 181, 551, 628
  definition 551, 628
Cryptotypes (**definition**) 238
Ctenophores 95
Cues 431
  see also Assessment; Communication
Cultural differences, molecular and
  evolutionary biology 335
Cultural drift 534
Cultural evolution 497
Cultural transmission, macaques 289
Curiosity 347–348
  island organisms and 347
  learning and 347–348
  see also Boldness; Neophobia
Cuvier, G. 209
Cyanobacteria 459, 604, 615
  alternative phenotypes in 613
  chloroplast origins and 604
Cybernetic theory of biological order 128
Cynipid wasps, galls 109
Cyprinid fish 300

D

Damselflies 63, 271, 278
  color morphs 278
Darwin, C. 6, 11, 15, 188–193, 209,
    262, 443, 471–473, 600
  hypothetical phylogenetic tree 600
  on anthropomorphism 443
  on cross-sexual transfer 262
  on development and evolution 188–193
  on dual nature of regulation 190
  on false facts 191
  on gradualism 11, 471–473
  on metaphors 15
  on variation and design 471–473
  theory of inheritance 6
Darwinian selection (**definition**) 31
Darwin's Finches, *see* Finches, Darwin's
Dauer larva 73, 119
Decomposability 56
  see also Dissociability; Modularity
Deermouse (*Peromyscus maniculatus*)
    554
Defensive responses 47, 76, 249–250,
    251, 303, 316, 333, 391–392, 431,
    501–502

accelerated hatching (frogs) 249–250,
    303, 392
  barnacles 392, 431
  plants 47, 333
  polyphenisms and (reviews) 392
  snails 316, 431
  toxin ingestion 501–502
  transposons and 333
Definitions, as models 489
Deletions 218–231
  definition 218
  environmental, of intermediates 224
  gradual 218
  life-stage 219–222
  lost genes and 229
Density-dependent alternatives 132, 300
  see also Alternative phenotypes;
    Dispersal polymorphisms; Locusts;
    Wing polymorphisms
Dentition 511, 521, 621–623
  evolution of, horses 621–623
  rapid evolution, rodents 521
  see also Hypsodonty
Dependent alternatives 173–174, 393,
    398
  definition 173
  evolution of 398
  see also Integration
Design 472, *see* Adaptive design
Determinants of form (**definition**) 69
Determinants of regulation (**definition**)
    68
Determination 68, 99–100, 101
  cf. differentiation 68
  cf. influence 99–100
  definition 68
  dual nature of 101
  see also Caste determination; Genetic
    determination; Sex determination
Development 5, 32, 57, 89–138, 141,
    164, 193
  branching nature of 164
  design, first-order cause of 141
  definition 32
  hierarchical organization 57
  history of neglect in evolutionary
    biology 193
Development vs. selection (controversy)
    139–140, 201, 258, 273, 276, 428,
    473, 477–478
  hyena pseudopenis and 276
  regulation of alternatives and 428
  resolution 478
  sexual monomorphism and 273
  uncut stone metaphor and 473
  see also Gradualism
Developmental biology 4, 5, 89, 95
  estrangement from evolutionary
    biology 4
  evolutionary biology and (cultural
    differences) 5, 95
Developmental burden 78
Developmental Character Release
  see Release
Developmental constraints, *see*
  Constraints, developmental

Developmental conversion 36
Developmental deletion (of
    intermediates) 228
  see also Deletion
Developmental drift 426, 593
  definition 426, 593
  founder effects and 593
Developmental homeostasis 8, 25, 51
Developmental hybrids, honeybees
    399
Developmental instability 205, 511
Developmental linkage 27, 45, 70, 173,
    441
  Britten-Davidson model and 27
  definition 70, 173
  hormones and 45
Developmental noise 35, 426, 428
  stochastic switches and 426, 428
Developmental plasticity (**definition**) 35
  see Plasticity
Developmental recombination 70, 200
  definition 70
Developmental trade-offs
  see Trade-offs, developmental
Developmental trap hypothesis (for
    maintenance of sex) 632–637
  predictions of 636
Developmental-plasticity-hypothesis (of
    speciation) 528–563
  enumerative tests, critique of 531
  evidence 530–562
Diadromy 534
  definition 534
  loss of, and speciation 534
  see also Anadromy
Diapause 73, 391, 406, 424, 437, 556
  egg (crickets) 406
  mites (hypopus stage) 437
  nymphal (crickets) 406
  variable, in copepod eggs 437
  variation, cicadas 556
Diapsids 522
Diatoms 613
Didion, J. 498
Diet-induced morphology 54, 111,
    299–300, 307, 347, 391, 397–398,
    460, 462, 480, 515–517, 518, 523,
    550, 575–577, 596
  anurans 518
  birds 518, 347
  caterpillars 300, 446
  echinoids 480, 515–517
  fish 54, 299–300, 523, 550, 575–577,
    596
  grasshoppers 300
  lemmings 460
  reversible 518
  rotifers 111, 462
  snakes 518
  spadefoot toads 307, 397–398, 462
  voles 460
Differentiation 28, 68
  cf. determination 68
  definition 68
Digger wasps (*Sphex ichneumoneus*), *see*
  Wasps, digger

Dinosaurs 609
  sexual selection and 609
  size-dependent alternatives in 609
Dioecy, planarians 371
Direct development 221–222, 251–252, 360, 437, 480–481, 495, 514–518
  definition 221
  egg size and 222
  environmental influence 514–518
  evolution of, echinoids 514–518, 221–222
  gradualism controversy and 480–481
  plants 251
  mites 437
Discontinuity, see Modularity
Discontinuous variation, see Complementarity, Modularity
Dispersal polymorphisms 66, 75, 107, 131, 310, 543, 544
  ants 543, 544
  see also Polyphenisms; Wing polymorphisms
Displays, see Communication
Disruptive development 225, 227, 544, 589
  definition 225
  mechanisms 227
Disruptive divergence (definition) 529
Disruptive selection 6, 7, 116, 224–229, 253, 399, 402, 532, 541, 544, 559–560, 588
  alternative phenotypes and 225
  cf. diverging selection 402
  classic experiments on 226
  consequences 227, 399
  disruptive development and 225–229
  honeybees 399
  in nature 226, 454, 532, 541
  models of 225
  speciation and 544, 559–560, 588
  see also Diverging selection
Dissociability 57, 164, 170–174, 189, 370–371, 392–393
  alternative phenotypes and 392–393
  complexity and 370–371
  Darwin on 189
  see also Modularity; Combinatorial evolution
Divergence 8, 21, 22, 190, 201, 388, 394, 396–399, 401–404, 526, 528–529, 537, 540, 545
  alternative phenotypes and 394–416
  developmental 8
  lifestage 21, 22
  polymorphic, Darwin on 190
  resource abundance and 577–578
  selection for (per se) 396–399
  speciation and 8, 528–529
  speciation as cause of 201, 388, 394, 396, 401–404, 526
  speciation as caused by 537, 540, 545
  see also Principle of divergence
Diverging selection 127, 209, 229–230, 402, 540, 572, 587–588
  African finches 587–588

cf. disruptive selection 402
  definition 402
  speciation and 540, 572
Diversification 8, 403
  intraspecific 8
  marine organisms 403
  speciation and 8
  see also Adaptive radiation; Macroevolution; Speciation
Diversifying selection (definition) 402
Division of labor 8, 86, 87, 92, 332–333, 513, 632, 633
  evolutionary consequences 87
  germline-embryonic genes 333
  germline-soma 332–333
  in germline 86, 92
  metabolic, within leaves 513
  sexual 632, 633
  transmission-expression 333
  see also Dependent alternatives; Social insects
DNA (deoxyribonucleic acid) 19, 231, 330, 322, 325–327, 630
  junk DNA 19, 231, 322
  phenotype of 330
  programmed rearrangements 330
  repair, sexual reproduction and 630
  sequence conservation 325–327
  see also Genome
Dobzhansky, T. 440
Dogs, see Canids
Dolichopodid flies (Sciapodinae) 496
Dollo's Law 237–238
  definition 237
Domain swapping (definition) 325
Domestication 115
  pacas (Agouti paca) 115
  see also Maize
Dominance (social) 301, 425, 449–457
  asessment hypothesis and 449–453
  correlated effects of 301
  evolutionary role of (social insects) 301
  reproductive capacity and 449
  signals of, and threats 425, 449–457
  see also Antithesis
Dragonflies (Odonata) 271, 278, 444
  color morphs 271
  unlearned adaptive behavior 444
  see also Damselflies
Drepanids 589
Drumming signals, chimpanzees 456–457
Drift 8, 19, 29, 32, 144–145, 158, 173, 372, 394, 401, 435, 506–507, 534, 593, 601, 617, 619, 626–627
  adaptive evolution and (Wright hypothesis) 144
  bryozoan evolution and 626–627
  cultural drift 534
  evolutionary significance (controversy) 144–145
  molecular evolution and 158, 435
  punctuation and 617, 619
  shifting balance theory and 144, 173, 394, 601
  see also Developmental drift
Drongo (Dicrurus) 207

Drosophila 19, 24, 26, 72, 73, 80, 77, 80, 92, 95–96, 100, 104, 113, 116, 119, 154, 169, 181, 211, 212, 221, 226, 230, 233, 256, 304, 317, 318–320, 323, 326, 330–331, 335, 343, 365, 415, 572–573, 578, 580–581
  adh gene, evolution of 211, 230, 319–320
  alcohol tolerance 580
  alternative splicing 323
  courtship display 24, 181, 343
  disruptive selection experiments (classic) 116, 226
  eggs 94, 95–96
  eye color, genetic complexity 154
  gene expression (conditional) 72, 211, 317–318, 330–331
  genetic correlations (developmental basis) 77
  genomics, comparative 211, 230, 319–320, 326, 335
  homeobox genes 73, 80, 212, 318
  learning, in courtship 343
  linkage groups 80
  maternal effects 113
  ovarian specializations 578, 581
  parallel species pairs 572–573
  paternal effects 92
  phenocopies 119
  plasticity, selection on 169
  radiation, see Drosophila, Hawaiian
  reaction norms 26
  recurrent phenotypes 365
  release, experiment 415
  sex determination 323
  wing atavisms 233
  see also Index of Scientific names (Drosophila); Drosophila, Hawaiian
Drosophila, Hawaiian 230, 319, 572–573, 578–582
  breeding site specialization 579
  larval ecology 578–579
  molecular evolution 319
  non-morphological aspects, radiation 578
  ovarian specializations 581
  parallel species pairs 572–573
  phylogeny 579
  radiation 578–582
Drosophilid flies, egg diversity in 94
Drumming signals, chimpanzees 456–457
Duetting 273
Dufour's glands 246, 409
Dungflies 429
Duplication 87, 209–217
  behavioral 212–214
  divergence and 210–211
  intra-genic 211
  morphological 212
  see also Gene duplication
Dwarf species 157, 364, 507, 511, 535, 567, 569, 609
  elephants 157, 364, 511, 609
  lampreys 535
  whitefish 567, 569

Dwarfism 287, 364, 535
  fishes (*Tilapia*) 364
  hamsters (*Phodopus* species) 287
  lampreys 535
  see also Miniaturization

E
Earthworms 500
Earwigs (Dermaptera) 75, 234, 366
  atavistic 234
  polymorphic 75
Ecdysone 462, 492
  in plants 492
Ecdysteroids 44, 311
Echinoderms, *see* Echinoids, Sea urchins
Echinoids 243, 360, 411–413
  character release 411–413
  molecular heterochrony 243
  recurrence of direct development 360
Echiurid worm (*Bonellia*) 101, 116
  sex determination 101
Eclectic Parrot (*Eclectus roratus*) 273, 275
Ecophenotypes 381, *see* Alternative
  phenotypes
Ecosystems, plasticity and stability of 51,
  519
Ecotypes 381
Egg dormancy 427
Egg trading 433, 458
Eggs 94, 95, 96, 97, 387, 406, 425, 437,
  458, 514
  alternative phenotypes 387
  behavior of 95, 97
  diapause variation 437
  diversity 94
  micropyles (insects) 96
  phenotype of 97
  polyphenism, for diapause (crickets) 406
  size, coin-flipping and 425
  size, echinoderms 514
  trading 458
  see also Ovaries
Eigen's paradox (**definition**) 524
Electric fish 591
Elephants 157, 364, 455–456, 511, 609
  dwarf 157, 511, 609
  signal antithesis in 455
Elk, body-size assessment in 456
Elm span worm (*Ennomos subsegnarius*)
  103
Embedment, *see* Entrenchment
Embryogenesis, rate, and ovary type
  (insects) 367
Embryonization 312, 514, 517
  definition 514
  echinoids 514
Emergent properties 37–38, 61, 179,
  398–399, 429–430, 632
  complementary alternatives and
  429–430
  complex traits and 61
  in self-organization 37–38
  levels of organization 398–399, 429,
  632
  plasticity as 179
Emergent universality 34

Empty niche theory 610
Endosperm 91, 212, 605, 633, 636
  evolution of 605, 212
  parental cooperation in 636
  paternal effects in 633
Enteman, W., Sewell Wright and 524
Entrenchment, environmental 500–503
  definition 500
  of toxins 501–502
Enumerative tests 182, 531, 622
  critique of 182, 531
Environments 32, 328, 368–369
  cellular 328
  definition 32
  internal (**definition**) 32
  recurrent phenotypes and 368–369
  social 32
Environmental effects (developmental)
  15, 17, 18, 20, 33, 36, 98–100,
  108–116, 111–112, 116, 144, 145,
  193, 315–316, 385, 478, 498–525,
  502–518, 524, 555
  Darwin on 193
  evolutionary potential and 145
  evolutionary potential, cf. mutation
  502–505
  evolved mimicry of 111
  inherited 112–116 *see also* Maternal
  effects, Paternal Effects, Parental
  effects
  informational role 15, 33, 98–100,
  110–111, 524
  Lamarkian interpretations and 17, 505
  neglect of 17
  noise hypothesis 18, 36
  non-adaptation criteria, proposed 18,
  478
  origin of novelty and 144, 385,
  505–508, 510–518
  pleiotropic 110
  speciation and, cicadas 555
  specificity of 99
  see also Interchangeability;
  Phenocopies; Stress
Epigenetic effects 112, 637
  definition 112
  inheritance 637
Epigenetic landscape 13–14, 25, 399
Epiphyte phase, in trees 251
Episodic selection 403, 518–524, 577,
  583, 594, 595
  boom-and-bust cycles 595
  climate cycles and 518–524
  Darwin's finches 583
  radiation and plasticity 594
  rapid radiation and 577
  tropical diversity and 595
  see also Conditional ratchet
Epistasis 9, 32, 593
Equilibrium points 432
  as switch points 432
  genetic models and 432
  see also Neutral zone
Equivalence 116, 127, 255, 267, 499
  definition 116
  determination of alternatives and 127

illustrations 255, 267
  importance, for genetic theory 499
  see also Interchangeability
Ergatoid (worker-like) males, ants 283
ESS, *see* Evolutionary stable strategies
Estradiol 274, 286
Estriol, in mammalian semen 277
Estrogen 77, 111, 273–274, 280–281
  phytoestrogens 111
Estrone, in mammalian semen 277
Eukaryotes 27, 62, 70, 79, 85, 320–321,
  326
  gene expression 70
  gene regulation (model) 27
  genome, cf. prokaryotes 27, 62, 70,
  79, 85, 320–321, 326
  genome structure 28, 320
  properties of 85
Euplasm 113, 332
  reduction, in panoistic ovaries 367
European eel (*Anguilla anguilla*)
  loss of marine spawning 535
Eusociality (**definition**) 367
  see also Social insects; Honeybees;
  Wasps, social
Evolution
  as decreased environmental influence
  499
  clandestine 7
  definition 28, 31, 328–329
  directional, with episodic selection
  162–163
  genetic theory of, and neglect of
  development 6
  in large populations 162, 404–404
  indirect evidence for 508–510
  irreversibility of 177
  laboratory populations (mites) 437
  mosaic, as developmental divergence 8
  Neo-darwinian concept of 6
  phenotypic concept of 17
  reversibility of 237–238
  see also Adaptive evolution;
  Macroevolution; Rate of evolution
Evolution Committee 11
Evolutionary developmental biology 89
Evolutionary potential
  definition 142
  environmentally induced novelties and
  144
  mutations and 143
Evolutionary stable strategies (ESS) 8,
  384, 417
  definitions 390, 425
  models 432
  see also Stochastic mixed ESS
Evolutionary transitions 197–374
  as increased modularization 85
  major 84–88, 598–616
Evolvability 43, 182–183
  definition 182
Excessive construction hypothesis 506
Exons
  definition 72
  theory 320
  units of inheritance 328

Exon capture 491
Exon shuffling 211, 318, 319, 329, 491
  Cambrian explosion and 523
  developmental plasticity and 331
  established importance of 321
  spliceosomal introns and 321
Exploratory behavior
  fish 578
  learning and 341
  plants 462–463
Expressivity 100
Eyestalks (flies), homology in 492–493

F
Fanged frogs (Xenopus) 73, 205, 221, 335
Fanged Frogs, Southeast Asian (Rana blythii) 272, 285
  female calling in 272
Female choice
  see Choice
Female mimics 280
  see Cross-sexual transfer, female-like behavior of males
Fertilization 96
  active role of egg in 95
Fibronectin 211, 241
  alternative splicing and 323
  evolution of gene family 321–322
Fiddler crabs (Uca) 213, 495
Fig wasps 180
Fig wasps (Agaonidae) 110
Fig wasps (Torymidae) 395
Finches, African (Pyrenestes ostrinus) 350, 397, 420, 587–589
  beak polymorphism 587
  diverging selection 587
  learning 587
Finches, Cocos-island (Pinaroloxias inornata) 345–346, 586–587
Finches, Darwin's (Geospiza, Emberizinae) 181, 249
  atavistic behaviors 240
  boldness and curiosity 347, 348
  drought and diet change 350
  experience and feeding 346
  individual differences 346
  learning 463, 582–586 582
  major gene effect in 106
  morphology-biased specialization 346, 463
  migration, loss of 534
  pollen digestion 584
  trophic diversity 582–583 582
  speciation hypotheses 551
  see also Finches, Cocos-Island
Finches, Laysan (Telespyza cantans, Drepanidae) 414, 590
  character release 414
  intraspecific radiation 590
Fire ants (Solenopsis invicta)
  induced innovations 510
  speciation 537–538
Fireflies (Photuris species; Coleoptera) 248

Fish 54, 73, 87, 113, 121, 217, 224, 230, 261, 280, 282, 286, 299–300, 312, 313, 343, 344, 365, 387, 433, 444, 456, 458, 523, 535, 549, 550, 551–552, 566–572, 575–577, 578, 595, 596, 597, 620
  binary radiations in 566–572
  body-size assessment 456
  diet-induced morphology 54, 299–300, 523, 550, 575–577, 596
  hermaphroditic 282
  individual differences 344
  learning 343
  see also Cichlids; Lampreys; Salmonid fishes; Sticklebacks
Fisher, R.A. 83, 477
  gradualism model 476–477
  polyphenisms, skepticism regarding 4
  theory of dominance 170
Fitness
  definition 31
  plasticity and 16
Fitness effect
  definition 31, 165
  frequency of expression and 70, 169–170
  hierarchical organization and 174
  learning and 339–341
Fitness trade-offs
  see Trade-offs, fitness
Flesh flies (Sarcophaga) 73
Flexibility 33
  see Plasticity
Flexible stem hypothesis 163, 565–566, 570, 573–587, 596
  definition 565
  flexible stem, binary (definition) 566
  hyperflexible stem 573–574
  multidirectional radiations and 573–587
  natural selection and 570
  persistent plasticity and 584
  predictions 596
  see also Adaptive radiations
Flickering 366
  see also Flip-flopping recurrence
Flies 206, 208, 330, 333, 429, 492–493, 496
  body-size assessment in 456
  hearing organs 366
  galls 109
  novel appendages (sepsids) 199
  see also Apple-maggot flies; Drosophila; Hawaiian Drosophila, Hawaiian
Flight
  atmospheric conditions and 521
  origin, in insects 521
Flip-flopping recurrence 363, 366–368
  definition 366
  macroevolutionary 367
Flowering plants 301
  double fertilization 91
  endosperm 91
  flower-color polymorphisms 26
  flowering time 105
  intraspecific macroevolution 604–608

Fluctuating asymmetry 76
  definition 206
Flukes 507
Flycatchers (Ficedula albicollis) 595
  maternal effects 97
Follicle-stimulating hormone (FSH) 311
  in semen 277
Food preferences, learning and 342
Foraging
  assessment and 443–445
  by vines 39
Foraminifera 615
Forgetting 350–352
  adaptive 350
  importance for plasticity 350–352
Fossils
  morph vs species in 362
  polyphenisms in 361–362
  recurrence 361
  species problem 628
  study of plasticity in 627
Founder effects 401, 414, 592–593
  developmental drift and 593
Foxes 249
Fractional homology (definition) 491
  see Partial homology
Frequency-dependent selection 80, 417, 432
  alternative phenotypes and 390, 432–433
  models 225, 419
Frequency of expression
  rate of evolution and, see release
Frigate birds 369
Fringillid finches 586
Frogs 73, 113, 249–250, 235, 272, 284–285, 303, 392, 456, 502
  accelerated hatching 249–250
  body-size assessment 456
  female calling 272
  maternal effects 113
  parental care 284–285
  reversion 235
Fruitflies (Drosophila), see Drosophila
Fungi 108, 111, 112, 183, 185, 251, 321, 330, 448, 605
  gene expression mechanisms 330
  intraspecific macroevolution 605
  plant symbionts 112
  self-recognition 448
  self-splicing introns 321
  steroid hormone receptors 111
Fusion 485, 495
  origin of novelties and 199

G
Gaia hypothesis 83
Galapagos dove (Nasopelia galapagoensis) 240
Galapagos finches
  see Finches, Darwin's
Gall midges (Diptera, Cecidomyiidae) 109–110
Galliform birds
  hormones and plumage dimorphism in 274
  see also Chickens

Galls 109–111, 252–254, 533–534
  aphid soldiers in 252–254
  auxins, in insect saliva 111
  insect-dependent diversity of 109–110
  insects, gall making 109
  nematode-induced fig-wasp galls 110
  parasitism of (in aphids) 533–534
  trees resistent to 109
Game theory 417, 419
  models 419
  see also Evolutionarily Stable Strategies
    (ESS)
Ganged switch 86, 132
Gaps (in fossil record) 619
Geckos 256, 257
Gemmules (Darwin's molecular
    genomics) 189–192, 263
  definition 189
Gene cassettes (synexpression groups) 70
Gene clusters 70
Gene conversion 19, 215–216
  concerted evolution and 215
  definition 19, 215
Gene duplication 87, 162, 177, 211–212,
    217, 319, 330–321
  during gene expression 217, 330
  exon shuffling and 321
Gene expression 19, 27, 45–47, 67, 69,
    72–78, 86, 90, 96, 99, 158, 191,
    231, 261, 328, 331, 392, 500, 512
  absence, in early embryo 96
  behavior and 76–77
  CAM physiology and 512
  caste-specific, insects 74
  cell differentiation and 27
  condition dependence 90, 99, 500
  control of, by switches 67
  Darwin's version of 191
  definition of gene and 328
  early and late genes 73, 86
  gene conservation and 231
  genome damage and 331
  hierarchical 78
  hormones and 45, 261
  modularity of, evidence for 72–77
  photoreceptors and 46
  protein-specific 72
  rapid 72–73, 76, 77, 392
  regulation, by thylakoid membrane 47
  regulation, Britten-Davidson model 27
  selection and 19, 69, 158
  tissue-specific 72
Gene flow 402
Gene nets 9, 56, 78, 183
Gene-environment equivalence 255
  see also Equivalence; Interchangeability
Gene-pool cohesiveness, stasis and 617
Generalists
  cf. polyspecialists 30, 382, 586
  definition 382
  selection on 382
Generation-packing hypothesis 221
Generative entrenchment 326, 500
  definition 500
Genes
  co-expressed (synexpression groups) 70

conserved (causes) 211, 230–231, 258,
    325–327
  definition 328
  definitions, new (molecular) 327–329
  developmental effects 107–108
  dispensable 230–231
  early and late (in embryogenesis) 86
  evolutionary properties of 142
  followers in evolution 29, 157–158
  intron-exon structure, eukaryotes 320
  major, *see* Genes of large effect
  origin of term 378
  units of evolution 32
  units of expression 328
  units of selection 165
  see also Genes of large effect; Good
    genes; Homeobox genes
Genes "for" traits, false impression of 20
Genes of large effect 104–107, 150, 151,
    267, 281, 479, 588, 620
  color morphs (lizards) and 281
  definition 104
  examples 104–107, 588
  maize 267
  polygenic regulation and 101–104
  punctuation and 620
  response to selection and 105, 106,
    150
  recently diverged populations and 151
  saltation and 479
Genetic (allelic switch) polymorphisms
  rarity, due to plasticity 225
Genetic accommodation 29, 140, 142,
    143, 147–157, 247, 298, 324–325,
    339, 476, 499, 505, 517, 608
  Baldwin effect and 153–154
  complexity of regulation and 154–155
  Darwin on 505
  definition 140
  deleterious mutations and 148
  environmentally induced traits and 154
  evidence for 157
  genetic assimilation and 148, 153–154,
    157
  genetic variation and 155
  learned traits and 247, 339
  major correlated change and 298
  molecular level 324–325
  quantitative genetic models of 150
  resolution of nature-nurture
    controversy and 476, 499
  see also Baldwin effect; Genetic
    assimilation
Genetic architecture 27, 28, 67–70, 79
  complex traits and 79
  eukaryotes 28
  modular traits and 67–70
  prokaryotes 27
  supergene model 79
Genetic assimilation 13, 24–25, 128,
    129, 148, 151–154, 256, 267, 404,
    415–416, 517
  as genetic accommodation 157
  definition 24, 415
  definition (Waddington's) 151
  doubts regarding 415–416

examples 119, 415–416, 437
  genetic control implication 499
Genetic code, redundancy of 330
Genetic correlations 72, 77–79, 88, 137,
    164, 166–167, 374, 539, 541
  assortative mating and 539, 541
  developmental (switch) basis 72,
    77–78, 88, 137, 167, 374, 404
  discreteness and 404
  divergence-retarding effect 395
  environmental dependency of 78
  learning and 338
  negative, in tradeoffs 517
  release from, *see* Release
  retardation of evolution by 164, 166
  runaway selection and 450, 467
  sexual selection and 78
Genetic determination
  cf. genetic influence 102, 103,
    423–424
  cf. heritable 102
  epigenetic landscape metaphor and 15
  environmental influence on 128
  imprecise meanings of 102–104
  proper usage 424
Genetic determinism
  debate, in sociobiology 5
  discussions of trait origins and
    143–144
Genetic distance 52
  between humans and chimpanzees 335
  definition 335
  phenotypic distance and 335
Genetic drift
  see Drift
Genetic engineering
  modular recombination in 317–318
Genetic gradualism 478–479
Genetic knockouts (experimental) 54
Genetic load 394–395
  definition 394
Genetic models, plasticity and 340, 432
Genetic polymorphisms 106, 107, 384,
    418–421, 434–439, 562, 588, 593
  kinds (theoretical) 418–419
  maintenance 384, 434–439
  models 418–420
  plasticity and 593
  speciation and 384, 528, 562
  see also Genotype-specific alternatives
Genetic program metaphor 14, 15
  see also Metaphors
Genetic recombination, *see*
    Recombination
Genetic release 167
  see Release
Genetic variation 70, 127, 155–157,
    422, 436–439
  adaptive maintenance question
    438–439
  cf. genetic control 422
  commonness of 155–157
  conditional switch and 127
  modularity of 70
Genetics
  emphasis on, in modern biology 4

Genitalia 467, 543, 551
Genocopies 116, 119
  definition 116
Genomes 20, 31, 93, 217, 320–323,
  329–330
  combinatorial evolution of 320–323
  Darwin on, *see* Gemmules
  definition 31
  enslavement of, by phenotype 93
  eukaryotes cf. prokaryotes 27, 62, 70,
    79, 85, 320–321, 326
  evolution of, gene expression and 217,
    329–330
  human, percent coding sequences 322
  mapping 20
Genome-protection hypothesis (of
  germline sequestration) 332–333
Genomic imprinting 635–636
  definition 635
  maintenance of sex and 635–636
Genotype 31, 378
  definition 31
  origin of term 378
Genotype-environment interaction 15, 36
  as metaphor for development 15
Genotype-phenotye problem 16–18, 80,
  89, 143, 183, 231, 328–329
  definition of evolution and 328–329
  dispensable-genes concept and 231
  gene expression and 328
  gene-for-trait idea 183
  supergene concept and 80
Genotype-specific alternatives 103, 106,
  278, 417–424, 588
  definition 103
  examples 106, 278, 420–421, 588
  models 418–420
  putative examples, critique 421–424
  rarity 420
Genotype-specific mixed ESS (**definition**)
  419
Geographic variation 75, 121, 157,
  400–401, 405–407, 412, 423, 523,
  530, 532, 538, 561–562
  autogeny (mosquitoes) 560
  body shape, stream-velocity influenced
    300, 523
  cf. genetic determination 423
  coloration, sticklebacks 523
  environmental influence and 523
  environmental sex determination 121
  morph ratios 157, 406, 412, 423, 530,
    532, 530
  morph-specific 75, 400–401
  voltinism, crickets 561–562
  see also Morph-ratio clines 405
Germ line 86, 90, 98, 333, 609–610
  continuity 90, 98
  sequestration 86, 184, 331, 332–333,
    609–610
  totipotency, germ cells 333
  see also Sequestration, germline
Germination, photoperiod and 460–461
Germ plasm 90
Gerrid bugs (*Gerris* species) 312
Gibberelins 266

Gigantism 364, 521–522
  atmospheric conditions and 521–522
Giraffes, body-size assessment 456
Global environmental change 519,
  518–524
  carbon dioxide since Cambrian 521–522
  gigantism and 521–522
  global warming and evolution 519
  phenotypic effects 521–524
Globins, *see* Hemoglobins
Gloger's Rule (**definition**) 523
Goal-directed behavior
  see Motivation 314
Goat, two-legged
  see Two-legged-goat effect
Goethe, J.W. von 303, 305
Goldeneye ducks (Bucephala islandica)
  434
Goldenrod, paternal effects 92
Goldschmidt, E. 26, 161, 167, 297, 475
  on developmental plasticity 475
  on linkage 167
Good genes 467, 591, 632–633
  cf. purely social selection 591
  ecological cf. social adaptation 632–633
  maintenance of sex and 632
  neglected aspects 467
Gorillas 291
Gradualism 11, 12, 24, 161, 183–184,
  252–254, 395, 471–484
  cf. continuous variation 479
  cf. saltation 471
  connectedness argument for 479
  controversy 11
  controversy, neo-Darwinian resolution
    476
  Darwin on 471–473
  Darwinian, cf. simplistic 473
  Fisher's model 476–477
  gaps between species and 24
  Genetic gradualism 478–479
  gradualism vs. saltation 252–254, 478
  maladaptive transition problem and
    394
  punctuation and 12
  social selection and 482
  solution to cohesiveness problem
    183–184
Grassé, P.-P., critique of neo-Darwinism
  598–599
Grasshopper mice, *see* Mice, grasshopper
Grasshoppers 59, 75, 149, 172, 182,
  203, 300, 314
  courtship 59, 172
  polyphenisms 75
  see also Locusts
Grazing, evolution of in horses 621–623
Great Basin kangaroo rats (*Dipodomys
  microps*) 429, 520–521
  climate change and plasticity 520–521
Great tits (*Parus major*) 346–347, 456,
  466
  begging calls 466
  dominance signals 456
  morphology-biased specialization
    346–347

Grebes (*Podiceps auritus*) 348
Green algae 615
Ground beetles (*Ohomopterus*,
  Carabidae) 364–365
  recurrence 364
Ground squirrels 73
Group effect 248
Group living, origin of, social insects 606
Growth-differentiation-balance
  hypothesis 307
Guinea pigs 104, 176, 205, 233
  atavistic polydactyly 104, 233
Guppies (*Poecilia* species) 224, 230, 300,
  312, 313, 433, 523, 571
  density-dependent switchpoints 433
  life-history experiments 312
  water velocity and body shape 300,
    523
Gymnosperms 91, 301
  reproduction 91
Gynandromorphs 75, 257, 279, 283, 291
Gynogenesis (**definition**) 635
Gypsy moth (*Porthetra dispar*, formerly
  Lymantria dispar) 103, 270, 371
  antennal anomalies 371
  food limitation 103

H

Habitat selection in plants 39
Haeckel 255
Hale cycles 518
Halictid bees 235, 236, 363, 404, 450
  reversions to solitary life 235, 236
Hamilton's Rule 392, 447, 449, 450–451
  as decision rule in development 432
  definition 447
  see also Kin selection
Hamsters (*Phodopus* species) 101, 154,
  287, 460
  hibernation, genetic variation in 101
  Siberian (*P. sungorus*) 101, 154, 460
Handicap principle 468
Hardy-Weinberg Equilibrium 8
Hares, locomotory function of ears in
  298
Hatching
  accelerated (frogs) 249–250, 303, 392
  decisions, costs of 431
Hawaiian honeycreepers, cf. Galapagos
  finches 589–590
Hawaiian mosquitofish 620
Hawks, prey-densisty assessment 445
Heat-shock proteins 132
Hemionus (horse) 354
Hemoglobins 62, 211
Hens, *see* Chickens
Heritability 27, 100, 102–103, 421, 436
  cf. genetic determination 421
  definition 100
  environmental depencence of 27, 100,
    102–103
  high in laboratory estimates 100, 103
  realized 156
  selection and 436
Hermaphrodites 282, 261
  sequential 261

Hermit crabs 446
Herons 249
Heterochrony 5, 22, 143, 173, 241–254,
    313, 380, 473, 511, 567, 607,
    624, 625
  alternative phenotypes and 380
  behavioral 244–250
  bryozoans 624, 625
  Darwin on 473
  definition 22, 241
  interspecific induction of 247–248
  life-history evolution and 313
  macroevolution and 607
  parallel species pairs and 567
  plants 250–252
  see also Neoteny; Paedomorphosis
Heterogony 308
Heteromorphy (of leaves) 58
Heterophylly (plants) 229, 174, 251,
    392, 414
  definition 26
  see also Buttercups
Heterosis 417
Heterostyly 106
Heterotopy 135
    255–259
  definition 209, 255
  heterochrony and 255
Hibernation 73, 101, 315, 424
  starvation responses and 315
Hickories, gall susceptibility 109
Hierarchical organization 60–61, 71,
    134, 174–175
  definition 60
  evolutionary consequences 174–175
  mouse mandible 71, 134
  see also integration
Himalayan rabbit 255, 256
Hip glands 360
Holometabolous insects 305
Holosteans 87
Homeobox genes 72, 73, 211, 212,
    318–319, 326, 368
  anomalies in snakes and 368
  combinatorial evolution 318–319
  definition (homeobox) 73
  modular gene expression and 72
  sequence conservation, homeodomains
    and 326
Homeosis 256, 269–270
  Darwin on 189
  definition 256
  homeotic mutations 189, 256, 480
Homeostasis 7, 8, 9, 25, 51, 152, 297
  developmental 8, 25, 51
  genetic 9
  physiological 7
Homing, learning and 342
Homology 57, 98, 177, 202, 203–204,
    357, 485–497
  cladistic cf. broad-sense 486–488
  concepts of 202
  criteria of 177, 203–204, 488–490,
    494–497
  definition 485
  developmental criteria (critique) 494–497

fractional 485
latent 486
levels of analysis and 494
mixed 202, 490–492
multiple developmental pathways and
    494–497
paralogous 490
partial 204, 491
serial 357, 486
  see also Homology, broad-sense;
    Homology, cladistic; Homology,
    iterative
Homology, broad sense 202
  cf. cladistic homology 486–488
  definition 486
  see also Homology
Homology, cladistic 202, 357
  cf. broad-sense homology 486–488
  definition 486
  inconsistency in 486–488
Homology, iterative 357, 486, 490
  definition 357, 490
  see also Paralogy; Serial homology
Homoplasy 204, 353, 355, 357, 358,
    369–373, 487, 596
  alternative phenotypes and 370
  combinatorial evolution and 353
  definition 353
  homology and 487
  in flies (Drosophilidae) 355
  see also Recurrence homoplasy
Homosexual behavior 264, 286, 293
  see also Cross-sexual transfer,
    femalelike behavior of male
Honeybees (Apis mellifera) 130, 218,
    147–149, 155, 171, 172, 177, 339,
    340, 342, 349, 351, 391, 398, 399,
    438, 444–445, 446, 447, 450, 452,
    553
  Africanized 391
  artificial selection in 147–149, 155
  cape honeybees 450
  caste-specific gene expression 398
  food assessment 444–445
  kin recognition 447
  learning 339, 342, 351
  mosaic traits 130
  nest-cavity assessment 446
  polyandry (multiple mating) 438
  queen pheromone 452
  recruitment dance 171, 172, 349,
    444–445
  Varroa mites and 553
  worker and queen specializations 398
Hooded Warblers (Wilsonia citrina) 286
Hopeful monsters 26, 297–298, 471,
    481, 603
  mating problem of 603
  putative mating problems 481
  two-legged goat effects as 297–298
  see also Anomalies
Horizontal gene transfer 199, 322
  bacteria to humans 322
  origin of novelties and 199
Hormonally mediated trade-offs
  see Tradeoffs, hormonally mediated

Hormone systems 44–45
  coordinating mechanisms 45
  determinants of regulation 68
  definition 44
  genetic complexity of 44
  see also Hormones
Hormones 44, 45, 74–75, 92, 106, 111,
    124, 130, 249, 261, 269, 272–276,
    285, 295, 320, 341, 359, 441,
    459–460, 465, 492, 551, 553
  amphipod eyestalks and 551
  cross-phyletic activity of 111
  feminization and, in frogs 285
  gall induction (insect auxins) 111
  gene-environment interchangeability
    and 124
  general properties 44
  gene expression and 74–75, 261, 465
  genetic polymorphisms and 106
  heterochrony and 249
  host-parasite coordination by 553
  in semen and male accessory glands
    277
  ingested 111
  intrauterine position and 276
  maternal effects, paternal effects and 92
  motivation and 341
  multiple effects of 45, 130
  phenotype organization by 295
  photoperiodism and 459–460
  plant 45, 111, 269, 492
  plumage coloration and 272–275, 441
  receptor-dependence of function 320
  recurrence and 359
  status signals and 441
  see also Androgens; Auxins; Ecdysone;
    Ecdysteroids; Estradiol; Estrogen;
    Estrone; Gibberelins; Hormone
    systems; Jasmonic acid; Juvenile
    hormone; Luteinizing hormone;
    Phytohormones; Pituitary hormone;
    Prostaglandins; Testosterone;
    Thyroxine
Horses 233, 374, 454–455, 600, 621–623
  atavistic polydactyly in 233
  dominance assessment 454–455
  grazing, evolution of 600
  hemionus 354
  Kattywar 354
  koulans 354
  pheromones 454
  punctuated evolution in 621–623
  see also Zebras
Host preference, learning and 343, 558
Host races 526, 552, 557
  apple-maggot flies 557
  treehoppers 552
Host shifts 343, 526, 552–558, 580
  alternative phenotypes and 557
  ants 554
  birds (brood parasitic) 553
  cf. host races 526
  flies (Rhagoletis) 557
  learning and 343, 558, 580
  mites 552–553
  speciation and 526, 552–554, 557

Housekeeping genes 72
Hox genes, *see* Homeobox genes
Humans 15, 50, 51, 63, 332, 350, 390,
    449, 464, 497
  aortic arch, variation 50
  brain evolution 464
  brain modularity 63
  chemical kin recognition 449
  environmental information in
    development of 15
  fold pattern and homology 497
  genome 332
  learning and evolution in 350
  phenotype-dependent alternatives 390
  social rewards in 350
  stomach, variation 51
Hummingbird flower mites 554
Hummingbirds 273, 275, 302, 445, 456
Huxley, J.S. 108
Hybrids 7, 118, 143, 198–199, 224,
    267, 282, 381, 399, 401, 402, 414,
    528, 532, 540, 551, 580, 635, 637
  developmental 399
  origin of novelties and 143, 198–199
  parthenogenesis and 282, 635, 637
  polymorphism, reinforced in 381
  speciation and 7, 118, 402, 528, 532,
    540, 551
  unisexual, fishes 224
Hydra 312, 333, 505–506
  Darwin on induced novelty in
    505–506
  embryonization 312
Hydrocarbons (cuticular, insects) 449,
    453
  nestmate recognition and 449
Hydroids 86, 168, 185
  see also Cnidarians
Hyenas (Crocuta crocuta) 40, 261–262,
    276–277
  clitoral penis 261–262
  cross-sexual transfer 276–277
  diet-dependent skull morphology 40
Hypercycles 524
Hyperflexible stem (**definition**) 573–574
  see also Flexible stem hypothesis
Hypermetamorphosis (insects) 21, 22,
    184
Hypermodularity (**definition**) 86
Hyperplasticity, *see* Hypervariable
    plasticity
Hypervariable plasticity (variant over-
    production) 37, 43–44, 178, 573
  evolution of 43–44
  noise, as 43
  response to selection and 178
  somatic selection and 37
Hypopus stage (**definition**) 437
  see also Mites
Hypsodonty 621–622
  definition 621
  punctuated evolution and 622

I

I-cells, totipotency of 333
Ichthyostegids 607

Ideal free distributions 420, 428–429
  definition 429
  examples 429
Imaginal discs (insects) 62, 72, 77, 86,
    130, 305, 309–310
  competitive growth and 305
  modular gene expression and 72
  mosaic traits and 130
Immune System 40–41, 178
  hyperplasticity of 41
  selection and 178
Imprecision, origin of novelties and
    208
Imprinting (behavioral) 463, 553–554
  abrupt speciation and 553–554
  host songs (parasitic birds) 553
  see also Genomic imprinting
Imprinting (genomic), *see* Genomic
    imprinting
Inclusive fitness 31, 391, 450–451, 468
  definition 31, 391
  social insects 450–451
  worker honeybees 391
Indicator signals (**definition**) 466
Indigobirds, African (*Vidua* species)
    553–554
Indirect evidence, in evolutionary biology
    508–510
Individual differences 129, 343, 344,
    550, 584–586
  as anomalies 344
  definition 344
  learning and 344
Individuality 57, 81, 184, 214, 331,
    609–610
  germline sequestration and 331
  innovation and 609–610
  modular traits and 57, 214
Individuation 87
  see also Individuality; Modularity
Indoleacetic acid 302
Inducible defenses, *see* Defensive
    responses
Industrial melanism 401
Information 33, 98–100, 110–111, 334
  continuity of 98
  egg euplasm 113
  environmental 33, 98–100, 110–111
  increase, via combinatorial evolution
    334
  response-dependence of 98
  see also Codex
Informational fallacy (**definition**) 335
Inherited (bridging) phenotype
    (**definition**) 91
  definition 91
  see also Eggs; Maternal effects;
    Parental effects; Paternal effects;
    Seeds
Inheritence 98, 190–192
  acquired traits, Darwin on 190
  blending 191
  Darwin on 190–192
  environmental 98
  see also Continuity of the Phenotype;
    Genes; Homology

Insects 21, 62, 64, 72, 74, 75, 77, 86, 94,
    96, 109–110, 111, 114, 130, 184,
    200, 246, 248, 257, 265, 279, 283,
    291, 305, 309–310, 330, 366–367,
    397, 405–406, 409, 416, 435, 437,
    449, 453, 455, 521, 560, 561
  accessory hearts 200
  brain 64
  cuticular hydrocarbons 449, 453
  Dufour's glands 246, 409
  eggs 94, 96, 406, 437
  flight, origin of 521
  galls 109–110, 111
  gynandromorphs 75, 257, 279, 283, 291
  hearing organs 366
  hormones, *see* Juvenile hormone,
    Ecdysone
  hypermetamorphosis 21, 184
  imaginal disks 62, 72, 77, 86, 130,
    305, 309–310
  mandibular glands 74, 453
  mushroom bodies 64
  nurse cells (ovarian) 96, 330
  ovaries, *see* Ovaries
  ovary evolution 366–367, 397
  ovary types 366
  parental effects 114
  pheromones, *see* Pheromones
  see also Ants; Aphids; Bees; Beetles;
    Bugs (Hemiptera); Crickets;
    Drosophila; Damselflies; Flies;
    Grasshoppers; Lacewings; Locusts;
    Pitcher-plant Mosquitoes; Moths;
    Wasps; Termites
Inside parallelisms (**definition**) 371
Instantaneous speciation (**definition**) 551
  see Abrupt speciation
Integration 12, 60–61, 84, 86, 133–135,
    168, 174
  coordinated use and 134
  contiguity of parts and 133–135
  definition 60
  functional 168
  modularity, as aspect of 60–61
  spatial 133–135
  subdivision-integration tension 12
  see also Individuality; Dependent
    alternatives
Intercastes 131, 137, 399
  definition 137
Interchangeability 29, 48, 93, 99,
    116–129, 128, 175–176, 256, 267,
    269, 296, 499
  biochemical mechanism for 116
  correlated effects and 296
  definition 116
  evidence for 119–128
  evolutionary importance 175–176
  genetic (gene-gene) 117–119
  hormonal mediation of 121, 126
  molecular level 48
  nature-nurture controversy and 499
  regulatory complexity and 116, 118,
    124, 175–176
  see also Equivalence
Intergenomic conflict 313

Intermediates, gradualism hypothesis and 480
Intermorphs 137, 261, 265, 265
  definition 137
  gynandromorphs 75, 257, 279, 283, 291
  intersexes 261, 265, 286
Interspecific alternative phenotypes 364–365, 545, 624
  definition 545
  see also Recurrence; Parallel species pairs
Intraselection 37
Intraspecific macroevolution, see Macroevolution, intraspecific
Intron capture 322
Introns 320–322, 324, 329, 331
  developmental function 324
  evolution of 320, 321, 324
  introns-early hypothesis 320, 321
  introns-late hypothesis 321
  organizer role 324
  residual glue, hypothesis 320
  spliceosomal 321
  switch-point role 324
  see also Exons
Isolating mechanisms 557, see Reproductive isolation
Isopods 106, 204, 265, 279–280, 342–343, 423, 448
  brain size 343
  kin recognition 342–343, 448
  kin recognition in 342–343
  learning in 342–343
  polymorphic (Paracerceis sculpta) 106, 279–280, 423
Iteration 209, 357
  definition 209, 357
  see also Duplication; Recurrence
Iterative homology, see Homology, iterative

J

Jack male (salmonid fish) 294, 431
Japanese elm 533
Jasmonic acid 44
Jumping spiders (Salticidae) 409–411, 444–445, 605
  origin 605
  tactical assessment by 445
Juncos (Junco hyemalis) 286, 444
Junk DNA 19, 231, 322
Juvenile hormone (JH) 44, 45, 74–75, 78–79, 119, 120, 128, 147, 247, 301, 305, 311, 373, 449, 461
  beetle horn development and 305
  dominance and (wasps) 449
  gene expression and 74–75
  genetic architecture and 78
  life-history trade-offs and 311
  multiple functions 128, 247
  nutritional switches and 461
  social insect evolution and 147, 301

K

Kangaroo rats (Dipodomys) 369, 429, 444, 520–521
  climate change and plasticity 520–521
  food assessment 444

Kangaroos 52
Kattywar horses 354
Katydids 344
Key innovations concept (definition) 564
Kin recognition 340, 396, 443, 447–449
  intracolony (social insects) 447–448
  learning and memory and 340, 443
Kin selection 5, 43, 185, 301, 392, 418, 424, 432, 447, 449, 450–451, 453, 630
  and plasticity 43
  bet hedging and 424
  conditional aid and 432
  flexibility and 447
  genetic determinism and plasticity in 5
  Hamilton's rule 392, 432, 447, 449, 450–451
  theory, as model of development 5
  see also Inclusive fitness
Kipukas (habitat) 572–573
  definition 572–573
Kleptoparasitism 346, 348, 350, 369, 379, 426
  recurrence, in birds 369
Klino-kinesis 39
Knockout technique (genetics) 54
Kokanee 535–536
  definition 535
  form, of sockeye salmon 535
  nonmigration 536
  see also Pacific salmon
Koulans (horse) 354

L

Lability 33, 181
  behavioral cf. morphological traits 181
  definition 33
Lacewings (Chrysopidae) 365, 556–557
  allochronic speciation in 556–557
Lactate dehydrogenase (LDH) 211
Lake whitefish (Coregonus clupeaformis, salmonid fish) 569
  dwarf forms 569
  parallel population pairs 569
Lamarckian inheritance 37, 90, 192, 331, 505
  Darwin's disavowal of 192
  environmental inductions and 505
  germline sequestration and 331
  somatic selection misconstrued as 37
Lambda phage 84, 103, 379–380
  polyphenism of 380
Lampbrush chromosomes 330, 332
Lampreys (Petromyzoniformes) 73, 535, 551–552
  abrupt speciation 551–552
  dwarf form ("praecox") 535
  dwarf species 535
  gene expression 73
  speciation 535
Latent homology 486
Law of Directed Series 354
Law of Equable Variation 354, see Recurrence
Law of Homologous Series in Hereditary Variability 354
  see also Recurrence

Laysan finches, see Finches, Laysan
Lazuli buntings (Passerina amoena) 454
Leaf fish (Polycentrus schomburgki, Nandidae) 280
Leaf-cutter ants (Atta), allometry in 308
Leaf-miners (Lepidoptera) 345
Learned traits 338–344
  cf. mutational novelties 340
  components of evolved traits, as 342–344
  evolutionary potential of 338, 339
  genetic accommodation and 339
  polygenic nature 342
  see also Learning
Learning 39, 163, 247, 265, 344, 337–352, 444, 462–464, 549, 553–554, 557–558, 562, 570–572, 578, 580–581, 582–587, 590
  accidents and 348
  aversion 444
  assessment during 462–464
  correlations due to 338
  definition 337
  developmental mechanism, as 338, 590
  evolved components of 341–342
  evolution of specialization and 344–349
  feeding innovation and 578, 582–587
  fitness screening device, as 338–341
  genetic models and 340
  habitat preference and 343, 558, 580–581
  heterochrony and 247
  integrating mechanism, as 338, 347
  kin recognition and 340
  kinds 337
  memory and 341, 342
  mimicry of natural selection by 339–341
  morphology-biased 338, 339, 346, 349, 549, 571–572, 578
  novel phenotypes and 337–352, 578
  polygenic nature of 338
  recurrent phenotypes and, fish 338, 570–571
  reinforcement (rewards), as somatic selection 341
  selection and evolution and 338, 342, 562
  social competition and evolution of 349–350
  somatic selection and 39, 341
  speciation and 549, 553–554, 557, 558, 562
  survival value, birds 583
  see also Forgetting; Learned traits; Rewards; Trial-and-error learning
Leks 391, 431, 453, 458, 578
lemmings 460
Levels of analysis 494
Levels of selection 398–399, 429–430, 448
  colony-level 448
  complementary alternatives and emergence of 429–430
  emergence of higher 398–399
  levels of organization and 398
  see also Clade selection; Units of selection

Lewontin's paradox 18, 21
Liability (**definition**) 149
Life histories 248–250, 307, 310–313,
    367, 562
  cricket speciation and 562
  developmental antagonisms and
    evolution of 310–313
  heterochronic, in vertebrates 248–250
  modularity 66
  ovary type and, in insects 367
  polyphenisms 562
  tactics 307
Light-compass reaction 171
Lillie's Paradox 183, 186, 192
  in Darwin's thought 192
Linkage, chromosomal 78, 79–81, 338,
    380, 441
  congealment of genome and 80
  facultative alternatives and 441
  prokaryotes 380
  see also Developmental linkage
Linkage, developmental, see
    Developmental linkage
Lizards 123, 280–282, 312, 314, 420,
    432, 456, 594, 635
  Anolis, radiation and plasticity 594
  locomotory-respiratory tradeoff 314
  parthenogenetic, courtship of 635
  phylogeny of sex determination 123
  resource assessment 456
  Sceloporus 312
  tree lizards (Urosaurus ornatus) 280
  whiptail lizard (Cnemidophorus
    uniparens) 282
  see also Lizards, side-blotched
Lizards, side-blotched (Uta stansburiana)
    281, 420, 432
  femalelike behavior of males in 281
  frequency-dependent selection and
    432
Local mate competition 180
Locusts 74, 113, 115, 116, 132, 137,
    300–301, 342, 381, 396
  feeding motivation 342
  ganged switch 132
  hormones and gene expression 74
  maternal effects 113, 115
  origin of polyphenism in 300
  trait complex of gregarious phase
    300
Loons (Gavia stellata) 443
Lovebirds, African (Agapornis species)
    275, 276
Luteinizing hormone (LH) 273, 274, 288
  polarity of sex transfer and 274
  in semen 277
Lynx (Felidae), reversions in 234

M

Macaques (Macaca species) 276–277,
    289, 290
Macroevolution 6, 476, 598–616
  behavior and 600
  definition 598
  genome and 615–616
  microevolution and 476
  predictions of hypotheses 603
  shifting balance hypothesis of 601
  Simpson's hypothesis 600
  speciational hypothesis 6, 601, 603
  two concepts confused 599
  see also Macroevolution, intraspecific
Macroevolution, intraspecific 599–616
  cf. macromutation 601–602
  cf. speciational macroevolution 602
  definition 599
  evidence 603–608
  predictions regarding 603
Macrogenesis (**definition**) 475
Macromutation 4, 201, 252, 473, 480,
    588–589, 601
  gradualism and 4
  hypothesis, in finches 588–589
  speciation by 601
Macronuclei, in ciliates 333
Maize (Zea mays) 202, 211, 261,
    265–269, 321, 322, 330, 334, 514
  environmental sensitivity 514
  evolution of 265–269, 514
  genome evolution in 330
  genome of 321, 332, 334
Major genes, see Genes of large effect
Malelike behavior of females 271–277,
    282
  amphibians 272
  birds 274
  crickets 271
  dragonflies 278
  macaques 276–277
  parthenogenetic species 282
Malic acid 512
Mammals 45, 52, 65, 73, 75, 104, 114,
    115, 154, 176, 205, 206, 223, 233,
    249, 255, 256, 276–277, 286–291,
    298–299, 344, 354–455, 460, 510
  canids 223, 249, 288, 299
  cross-sexual transfer 276–277,
    286–291
  hormones 45
  individual differences 344
  maternal effects, psychological 114
  modular structure, skull 65
  parental care 286–291
  seminal fluid hormones 277
  skeletal anomalies 206
  see also Horses; Humans; Mice;
    Primates; Rats; Rodents; Ungulates;
    Voles; Whales
Manakins 249
Mandibular glands (insects) 74, 453
Manipulative signals (**definition**) 466
Mantids (Dictyoptera, Mantoidea) 248
Mantis shrimps (stomatopods) 343,
    591
Marmosets (Callithrix jacchus) 290
Marmots, individual differences 344
Masculinization, in socially parasitic bees
    264
Mate choice, see Choice
Material compensation 37, 303, 305–309
  definition 305
Maternal control 184, 427
  cell-lineage competition and 184
  stochastic switches and 427
Maternal effects 91–98, 112, 113,
    123–124, 176, 333, 416, 460, 515,
    517–518, 580
  cumulative, in locusts 113
  diet (maternal) and, in voles 460
  direct development and (hypothesis)
    517–518
  echinoderms 515
  egg tolerance of toxins and 580
  genetic assimilation of 416
  gene transcripts (maternal) 96, 97, 333
  hormonal 124
  plants 460
  sex determination and 123–124
  see also Maternal control;
    Preconditioning; Parental effects
Maternal gene transcripts 96, 97, 333
Mating systems 385, 391
Maximum likelihood 372
Mayflies (Baetis bicaudatus) 265
McClintock, B. 241
Mechanisms (developmental)
  evolutionary importance 54–55
  evolutionary questions and 180, 428
  see also Gene expression; Hormones;
    Learning; Plasticity; Switches
Medusae (cnidarian) 168
Megarhyssa wasps (Ichneumonidae),
    speciation in 559–560
meiosis, cost of 630
Meiotic drive 98
Melanin 41
Melanism 357, 401, 523
  geographic variation in 523
  industrial 401
Melanization 255, 257–258
Melanoblasts 41
Melanocytes 41
Melanophores, deletion (cichlids) 218–219
Melatonin 44, 459–460, 464
Memes 340
Memory, relation to learning 342
Mendelian genetics, Darwin's ideas and
    11, 191
Mendelian traits, see Genes of large
    effect
Mendelism, gradualism and 473–474
Metamorphosis 58, 66–67, 69, 73,
    118–119, 127, 207, 249–250, 251,
    307
  gene expression and (amphibians) 73
  plants 58, 251
  material compensation in 307
  recurrent loss of 207
  salamanders 118–119, 207
  size-dependent 127
Metaphors 13–15, 81, 89, 93, 159, 355,
    399, 473
  development, for 13–15, 89, 93, 355,
    399
  machine, for organisms 159
  price of 81
  uncut stone, for variation 473
Metazoans 44, 321, 324, 610
  cell-lineage interactions 44
  exon shuffling in 321, 324
  multiple origins of 610

Methyl jasmonate 76
Methylation 332
Mice, field (*Peromyscus* species) 287, 369, 554
learning and habitat selection 369
Mice, grasshopper (*Onychomys torridus*), prey specialization 391
Mice, house (*Mus*) 73, 101, 104, 134, 221, 252, 298
gene expression 73
genetics 104
hibernation 101
mandible 71, 134, 298
Microevolution 475, 476, 598
definition 598
macroevolution and 475, 476
see also Adaptive evolution
Microfibrils 45
Microgyne ants (**definition**) cf. macrogynes 543
Micronuclei, in ciliates 333
Microtubules 37, 39, 41, 42
mitosis and 41, 42
plant morphogenesis and 39
Migration 343, 347, 348, 390, 534–537
divergent specializations in 347
exploratory behavior in 348
learning and 343
loss of, speciation and 534–537
see also Anadromy; Diadromy
Migratory locusts, *see* Locusts
Milkweed bugs (*Oncopeltus fasciatus*) 461
Mimicry 75, 208, 248, 264, 351–352, 357, 409, 553, 582, 595, 605
aggressive 248, 409, 605
feeding behavior, mimicry of 582, 595
learning and (butterflies) 351–352
polymorphisms 75, 208
recurrence and evolution of 357
sexual, of females by males 264
songs (birds) 553
see also Batesian mimicry; Mullerian mimicry
Miniaturization 308–309, 363, 550, 609
allometric results of 308, 309
alternative phenotypes and 609
recurrent, in fish 550, 363
see also Body size; Dwarfism
Mites 433, 437, 530, 552, 553, 554
density-dependent switchpoints 433
genetic variation (extreme) 437
hypopus stage 437
speciation 530, 552
Mitochondria 85, 113, 231, 321, 501, 632
entrenchment and 501
eukaryote trait 85
genome, plants 231
intron mobility in 321
maternal origin 113
sperm-supplied 632
symbiotic origin 501
Mixed ESS, *see* Stochastic mixed ESS
Mixed homology 202, 490–492
see also Homology

Modular traits 56, 58, 61, 67–79, 81–83, 164–167
correlation structure 61
definition 56
gene expression and 70–79, 164
genetic architecture of 67–70
properties of 58, 81–83
selection and 129, 164, 165–167
see also Modularity
Modularity 10, 13, 30, 34, 47, 56–88, 163–175, 183, 188–189, 315, 602
behavior and 59, 63, 65
biochemical 47
brain and 59
concept, in ethology 63
connectedness and 13
consequences of 30
Darwin on 188–189
definition 56
developmental basis 56, 71
evolution of 84–86, 164–165, 315
evolutionary consequences 163–175
evolutionary consequences of increased 86–88
intraspecific macroevolution and 602
levels of organization and 61–65
life-cycle 66–67
limitations of concept 57, 81–83
limited, of tree branches 65
plasticity and 34, 58–60
properties of 84
selection for 88
shrimp legs and 59
synonyms of 56
terminal addition and (critique) 10
universality of 183
Modules 57, 81
as term to avoid 57, 81
in evolutionary developmental biology 81
in evolutionary psychology 81
Mole rats (*Heterocephalus glaber*) 249, 288
Molecular biology, limitations, in evolutionary study 615–616
Molecular drive 19
Molecular evolution 317–336, 492
combinatorial 317–336
mixed homology and 492
see also Gene duplication; Genome; Exon shuffling
Molluscs 344, 361
learning in 344
see also Snails
Molting 462, *see* Metamorphosis
Molting hormone, *see* ecdysone
Monkeys 246, 457, 466
infant signals 466
see also Macaques; Primates
Monkshood flowers 349
Monograptids 241, 242
Morgan, T.H. 100, 193
Morph ratios 271, 405–406, 529, 536–537
clines 405–406
degree of specialization and 536–537

geographic variation in 271
speciation and 529
see also Geographic variation; Release; Sex ratio; Switches, response to selection selection
Morphology (general discussions) 180–182, 339, 578, 583, 585, 621
behavior and 180–182, 583, 585, 621
learning and 339, 578
see also Diet-induced morphology
Morphospecies, as problematic concept 628
Mosaic evolution 8, 164, 165, 168, 187
see also Modularity; Mosaic traits
Mosaic traits 86, 129–135, 261, 298
dissociability of 86
of honeybee queens 130
organization by switches 129–135
sexual phenotype as 261
Mosquitoes, *see* Pitcher-plant mosquitoes
Moths 401, 446, 613
Motivation 314, 340–342
anthropomorphism and neglect of 314
definition 341
evolved nature of 341
fitness effects and 341
genetic variability of 342
resource specificity of 341
Movements 46, 49
in plants 46
molecular 49
see also Behavior; Circumnutation
Muller, H.J. 317
Mullerian mimicry 215, 357, 358, 378, 381
Multigene families 214–215
Multiphenotype families 214
Multiple mating, significance of (honeybees) 438–439
maintenance of genetic variation and 438
Multiple pathways (developmental) 117, 176–177, 203, 312, 362–363
life-history trade-offs via 312
neoteny in salamanders and 117
parallel evolution and 203
recurrence and 362–363
Multiple-niche alternatives 106, 390, 397, 419
Mushroom bodies (insect brain) 64
Muscle 40–41
experience effects and 41
plasticity of 40
Mutation theory, of saltatory speciation 475
Mutations 100–101, 104, 189, 231, 239, 256, 317, 334, 378, 476, 480, 620
combinatorial vs random 318
evolutionary potential of 143
homeotic 189, 256, 480
large effects of 26, 104, 105
large effects, doubts regarding importance 298
limited functionality, if random 318
limits to evolution and 334, 476
loss of gene function and 231

Neo-Darwinian synthesis, role in 476
novelty, as only source of 499–503
origin of term 378
punctuation and 620
randomness cf. selection 476
recurrent 239
reorganizational 317
stress and 508
transposable elements and 334
variable expression of 100–101
see also Macromutations
Mutillid wasps 224, 381
Mutualisms 186, 454, 501, 632–634
male-female 632–634
see also Dependent alternatives;
Symbioses
Mycorrhyzae, entrenchment of 501
Mygalomorphs (spiders) 605
Myoglobin 49
Myxobacteria 613, 615
alternative phenotypes in 613
Myxomycetes 615

N
Nageli 193
Nastic movements 39
Natural selection, *see* Selection
Nature-nurture problem 3, 33, 44, 99,
329
as unresolved 3
molecular level 329
resolution of 99
Nematodes 73, 114, 119, 205, 221, 265,
323, 326, 331, 333, 635
dauer larva 73
genomic reduction in 333
hybrid parthenogenesis in 635
interchangeability 119
mermithid 265
ovary evolution 205
trans-splicing in 323
see also Index of Scientific Names:
Caenorabditis, Gasteromermis
Neo-Darwinian synthesis 5, 12, 13, 19,
152, 160, 186, 308, 475–476, 482,
497–498, 615
as theory of adults 7
critiques of 475, 497–498, 615
development neglected in 19
gradualism and 12, 476, 482
individual as unit of selection in 186
resolutions of problems in 160, 186
Neo-Lamarkism 115
Neo-Weismannian reduction 83, 90
definition 90
Neophenogenesis 140
Neophobia 347–348, 590
learning and 347–348
see also Boldness
Neoteny 118, 242, 249, 250, 487, 511,
551, 570, 607–608
concept 242
crustaceans 551
environmentally induced 511
genetic 511
origin, in salamanders 607

salamanders and 118, 487, 511,
607–608
sticklebacks 570
temperature and 608
see also Heterochrony
Nerve growth, somatic selection and 41
Nestmate recognition, cues 449
see also Kin recognition
Net model, of biological organization
479
see also Terminal addition
Neural Darwinism 37
Neural tube 495
Neurulation 495
Neutral genes, *see* Alleles, neutral
Neutral zone (neutral point) 130–131,
268, 273, 437
definition 130, 268
genetic variation and 437
sex determination and 268
tool, in genetic studies of plasticity
130–131, 437
Newts (*Taricha granulosa,
Salamandridae*) 207, 308, 356, 368
recurrent limb morphology 356
Newts (*Triton*) 308
Niches 106, 343, 390, 397, 419,
507–508, 546, 610
body plans and 610
definition 507
empty-niche theory 610
multiple-niche alternatives 106, 390,
397, 419
non-Hutchinsonian 546
shifts between, in extreme
environments 507–508
shifts between, learning and 343
Noise 4, 18, 35, 36, 43, 208, 246, 331,
426
biochemical, in signal evolution 246
developmental 35, 426
origin of novelties and 208
plasticity as 4
Non-adaptation, mistaken criteria of 18,
478
Non-Weismannian organisms 184–185
Nonspecific modifiers 68, 69, 70, 72, 78,
81, 87, 167, 170, 326, 393, 395,
404–405, 435, 529, 533, 554, 561,
628
antagonistic pleiotropy and 78, 435
connectedness and 81
definition 68
duplication and 87
frequency of expression and 70, 170
genetic correlations and 326, 393, 395
release and 170, 404–405, 529,
533–534
selection and 69, 167, 170
speciation and 529
stasis and 628
switches and 72
Normalizing (stabilizing) selection 152
Norms of reaction, *see* Reaction norms
Novelties (novel traits), *see* Origins of
novelty

Nucleus (of cell) 85, 333
dimorphism (ciliates) 333
origin of 85
Nurse cells (insect ovary) 96, 330
gene duplication in 330
Nutritional switches 461–462
juvenile hormone and 461
mechanisms 461–462
social insects 461

O
Oak trees (*Quercus* species), gall
susceptibility 109
Oligochaetes, toxin resistence 106
Ontogenetic repatterning 171
Oogenesis-flight syndrome (insects) 149,
308, 310–313
Operant conditioning 39
see Trial-and-error learning
Operons 28, 70, 79
definition 70
Opportunism 347, 391
Optimal foraging theory 443–444
Optimization theories 84
Orchestra metaphor 15
Orchids 91, 248, 501, 595
seed behavior 91
Organic buffering, 603, *see* Buffering
effect
Organic selection 24, 151
definition 24
Organization (biological) 10, 13, 14–20,
34–55, 56–88, 90–99, 107–116,
128–138, 186, 222, 261, 320–324,
398–394, 429–430, 503, 524
bridging phenotype and 90–99
complementarity and 83–84, 135–138
cybernetic theory 128–129
emergent 38, 398–394, 429–430
environment and 15, 108–116, 503
genome and 20, 107–108, 320–324
hierarchical 78, 138, 261
levels of 13
metaphors for 14–16
modularity and 10, 56–88
need for theory of 16–19
nervous system 41–42
netlike 84, 222, 479
plasticity and 34–55, 524
switches and 129–135
supra-individual 135, 186
supraorganismic, in insects 186
theories of 19–20
see also Self organization
Organization-activation theory (of
hormone action) 295
Orientation behavior 38–39, 342
Origin of life 180, 521, 524
environmental factors and 521
plasticity and 524
Origins of novelty 22–24, 28, 143–147,
197–352, 503–518
combinatorial 28
confusion with origin of species 201
definition (novelty) 198
developmental 22–24

Origins of novelty (*Continued*)
  drift and 144
  environmental 144, 503–518
  mutational 143–144
  pleiotropic (as side effect) 144
  shifting balance theory and 144, 173, 394, 601
Orthoptera, *see* Crickets, Grasshoppers, Katydids, Locusts
Ostriches 238, 256
Outbreeding depression (**definition**) 631
Ovaries 74, 78, 92, 113, 133, 146, 149, 177, 205, 211, 266, 272–274, 286, 289, 301, 308, 310–313, 366–367, 370, 378, 397, 449–452, 465, 565, 578, 581, 604
  behavior and 272, 286, 301
  division of labor within 92
  dominance rank and(social insects) 289, 301, 449–453
  euplasm production 113
  evolution of, insects 366–367, 397
  evolution of, nematodes 205
  flip-flopping recurrence (insects) 366, 370
  gene expression in (*Drosophila*) 211
  hormones and 74, 78, 273–274, 311, 465
  inferior, in plants 604
  larval, in honeybee workers 177
  life-history traits and 367, 581
  maize evolution and 266
  maternal effects 92
  meroistic (**definition**) 366
  oogenesis-flight syndrome (tradeoff) 149, 308, 310–313
  panoistic (**definition**) 366
  polytrophic meroistic (**definition**) 366
  regression of, in social insects 88, 133, 146–147, 301
  specializations of, *Drosophila* 578, 581
  telotrophic meroistic (**definition**) 366
  wing-dimorphic insects and 311
Overprinting (**definition**) 322
Overtypification (in phylogenies) 487
Ovipositors 406, 560
  polymorphisms 560
  geographic variation, crickets 406
Owen, R. 209

P

Pacas (*Agouti paca*), domestication 115
Pacific salmon (*Oncorhychus* species) 221, 226, 294, 365, 429, 431, 535–537, 547–548, 567
  equal payoff male alternatives 429
  interspecific alternatives in 365
  loss of migration and speciation in 536
  male migration polyphenism 535
  see also Coho salmon; Kokanee; Steelhead trout
Paedomorphosis 207–208, 218–219, 248, 361, 372, 569
  cnidarians 218–219
  cockroaches 248
  fish 569

salamanders 207–208
  see also Heterochrony; Neoteny
Paleontology 628
  cf. neontology 628
  species problem in 628
  see also Fossils; Punctuated evolution
Palms (*Geonoma* species) 306
Pan-environmentalism 99
Pangenesis 188–192
  critics of 191
  defects of theory 190
  historical demise of 192
Panheterochrony 241, 243
  definition 241
Panselectionism 474, 478
Parallel evolution 203, 208, 357, 358, 359–361, 362, 365–366, 369, 403, 494, 545–551
  cf. recurrence 359, 365
  definition 358
  environmental induction vs selection in 369
  hearing organs (insects) 366
  multiple developmental pathways and 203
  recurrent 357
  replicate speciation and 545–551
  snails 403
  see also Parallelism; Recurrence
Parallel speciation hypothesis 545–555
  cf. plasticity hypothesis 545–546, 551
  criteria 545
  definition 545
Parallel species pairs 369, 545–551, 567–572
  definition 545
  examples 545–551, 567–572
  local competition and 568
Parallelism 371, 487, 490, 493, 495, 566–567
  common ancestry and 488
  in cladistics 488
  of intra- and interspecific alternatives 566
  see also Parallel evolution
Paralogous homology 490
Paralogy (**definition**) 490
Parapatry (**definition**) 555
Parasites 47, 121, 203, 229, 264, 265, 280, 283, 368, 408–410, 434, 458, 482, 501, 542–545, 553–534, 597, 602, 605, 606, 631–632, 637
  anomalies due to 265, 280
  brood (birds) 419, 434, 553
  coevolution, with hosts 482, 501, 597, 602, 631
  effects on sex expression 265
  fungi 605
  gall (aphids) 533–534
  host manipulation by 458
  plants 230, 231
  sex determination and 121
  sexual reproduction and 631–632, 637
  social (inquilines, ants) 280, 283, 542–545
  social (bees) 203, 228–229, 264

social (wasps) 368, 408–410, 606
  social, origin (insects) 606
  transposons as 332–334
  see also Kleptoparasitism
Parasitic inquilines (**definition**) 542
Parasitoid insects 224–225, 234, 314, 366, 594, 606
  atavistic 234
  hymenopteran evolution and 606
  see also Wasps, parasitoid
Parcellation 84, 85, 164
  see also Modularity
Parental care 223, 282–291
  birds 223, 285–286
  canids 223
  cross-sexual transfer of 282–291
  frogs 284–285
  hormones and 287, 290
  mammals 286–291
  paternal 283, 286–291
  primates 289–291
Parental choice (offspring favoritism) 466
Parental effects >7, 113, 114
  insects 114
  plants 113
  prolongation of 97
  see also Maternal effects; Paternal effects
Paripotency 357
Parliament of the genes 175
  see also Individuality
Parliament of the phenotype 174–175
  definition 175
Parr (**definition**) 535
Parrotfish (*Sparisoma viride*) 280
Parsimony (cladistic) 370, 372–373
  limitations 373
  plasticity and 372–373
  recurrence and 370
Parthenogenesis 66–67, 220, 282, 387, 632, 634–635, 636, 637
  aphids 66–67, 220
  beetles 635
  cost of sex and 632, 634–635, 636, 637
  cyclic 635
  polyphenism and 67
  thelytokous 387
Partial homology 204, 491
Paternal care 283, 286–291
  see also Parental care
Paternal effects 91–98, 113, 427, 633
  imprinted genes 631, 635–636
Pavlovian conditioning 337
Peacocks 245
Peak shifts 144
Peas (*Pisum sativum*) 104
  Darwin on 191
  genetics of flowering time 105
Pecan (*Carya illinoensis*), lack of galls 109
Penetrance (genetic) 100–101
  definition 100
Penguins, locomotion 600
Pentatomid bugs 103
Peptide hormone 319

Phage, *see* Lambda phage
Phalaropes (*Phalaropus* species) 273
Pharyngeal jaw, *see* Cichlids, African
Pheasants 156, 245
Phenocopies 23, 29, 116, 119, 175, 205
  definition 23
  examples 119
Phenome (**definition**) 31
Phenotype 31, 378
  definition 31
  origin of term 378
Phenotype fixation 69, 141, 153, 315,
    404–415, 435–438, 529–530, 532,
    561
  character release and 170, 182,
    404–416, 437, 528–529, 554–562,
    577
  definition 141, 153
  environmental 239, 404, 513
  examples 157, 239, 531–545, 544–562
  genetic assimilation and 24, 119, 153,
    404
  increased modularization and 405
  speciation and 531–545, 554–562
  without gene fixation 435–438
  see also Macroevolution; Punctuation;
    Release
Phenotype-dependent alternatives
    388–390, 433
  definition 388
  social 389
Phenotype-environment matching 162,
    504, 583
  learning and 583
Phenotypic accommodation 45, 51–54,
    59, 134, 140, 147, 160–161, 297,
    315, 403, 496, 505, 576, 580
  anomalies and 505
  Baldwin effect and 24, 151
  cichlids 576, 580
  definition 51, 140
  evolution and 140, 147, 160–161, 315
  integration and 134
  mechanisms 53
  normal development and 147
  trait exaggeration and 147, 403
  see also Two-legged-goat effect
Phenotypic engineering 54, 131, 205,
    224, 509, 511–512, 518
  social insects 509
  unisexual fish 224
Phenotypic fixation, *see* Phenotype fixation
Phenotypic modulation 36
Phenotypic parallelism, *see* Recurrence
Phenotypic plasticity
  definition 34
  see Plasticity
Phenotypic recombination 30, 63, 77,
    145, 171, 242
  behavior and 77, 244
  butterfly wings and 63, 257
  definition 30, 77
  heterochrony as 242
  learning and 337–352
  linkage and 173
  modularity and 173, 482

molecular 317–336, 492
RNA splicing and 323–328
self-increasing nature 200
see also Combinatorial evolution;
  Homology, mixed
Pheromones 246, 264, 272, 281, 331,
    450, 452–453, 454
  combinatorial evolution and 331
  horses 454
  snakes 281
  social insects 246, 272, 450, 452–453
Photoperiod 101, 113, 424, 459–461
  assessment of 459
Photoperiodism 459–460
  definition 459
  mammalian 459–460
  mechanisms 459
Photosynthesis, evolution of 604
  see also Crassulacean acid metabolism;
    Thylakoid membrane
Phototropism 39
Phyla 609–615
  ancient origin of 609–615
  origins of 609–611
  times of appearance 611
Phyletic evolution 617–618, 482
  definition 617
  kinds 482
Phyletic gradualism 618
  cf. punctuation 618
  definition 618
Phylogenetic analysis 166, 203, 360–361,
    368–373, 374, 385–386, 400
  adaptation, cladistic tests 371–372
  character weighting 370–371
  cladograms 371
  coding methods 385–386
  evolutionary lability and 372
  mapping 360–361
  maximum likelihood 372
  neglect of variation (overtypification)
    372
  parsimony 372–373
  phylograms 203, 400
  plasticity and 368–373, 374
  recurrence homoplasy and 368–373
  see also Homology, cladistic; Polarity
Phylogenetic constraints, *see*
  Developmental constraints
Phylogenetic inertia 25, 26
  definition 25
Phylograms 203, 400
Phylotypic stage (**definition**) 495
Physiogenesis, in mites 437
Phytohormones 111, 269
  phytoestrogens, herbivore ingestion of
    111
Phytophagus insects, speciation in 557
Pied flycatchers (*Ficedula hypoleuca*) 286
Pied wagtail (*Motocilla alba*), facultative
    territoriality 456
Pierellization 256
Pigments 45, 47, 127, 132, 133, 134,
    154, 239, 257, 604
  birds 133, 134
  butterflies 127, 133, 257

diffusion fields and 133
eye-color 154, 239
locusts 132
see also Carotenoids; Melanin;
  Melanization
Pineal gland 45, 459–460, 464
  see also Melatonin
Pins and thrums (heterostylous plants)
    106
Pipefishes (Syngnathidae) 286
Pipesnakes 368
Pitcher-plant mosquitoes (*Wyeomyia
    smithii*) 405–406, 416, 560–561
  character release 405–406
  genetic assimilation 416
  morph-ratio cline in 405–406
  speciation 560–561
Pituitary 45
Pituitary hormone 44
Planarians 371, 635
  dioecious 371
  hybrid parthenogenesis in 635
Plankton "blooms" 437
Plants 26, 39, 43–44, 45, 46–47, 58, 76,
    91, 92, 98, 105, 106, 111, 112,
    223, 229, 248, 251, 267, 269–270,
    307, 321, 333, 355, 437, 460,
    462–463, 492, 501, 512, 521, 579,
    604–608
  arborescence (origin) 521
  behavior (movements) 39, 43–44, 46,
    462–463
  cross-sexual transfer in 265–270
  defenses 47, 333
  direct development 251
  exploratory growth 39
  flowering time 105
  heterochrony in 250–252
  hormonelike substances in 45, 111,
    269, 492
  macroevolution 604–608
  maternal effects 460
  metamorphosis 58, 251
  parasitic 230, 231
  paternal effects in 113
  plasticity, cf. animals 45
  plasticity of photosystems 46–47
  polymorphisms in 26
  rapid gene expression 76
  resource allocation 307
  seeds 26, 91, 98, 437
  self-splicing introns in 321
  silicon in 501
  symbionts of 112
  see also Endosperm; Heterophylly;
    Maize; Photosynthesis; Reaction
    wood
Plasticity 4, 7, 16, 17, 26, 33, 34–55,
    149, 160–163, 178–180, 183, 385,
    430–432, 570, 577, 581
  active and passive 35
  adaptive, difficult to evolve 7
  adaptive and nonadaptive 35
  adaptive trait, as 178–180
  biochemical 581
  bird song 40

Plasticity (*Continued*)
  continuous (graded) 26, 35, 36
  correlated change and 161
  costs of 430–432
  definition 33
  definitions (history) 34–36
  discrete (discontinuous) 36
  ecosystems and 51
  evolution of 178–180, 385
  evolutionary consequences of 160–163
  exaggeration of change by 161–162
  fitness and 16
  genetic models of 179
  liability evolution and 149
  mechanisms of 37–50
  molecular 47–51
  muscle and bone and 40
  neglect of 17
  noise, as 4
  origin of life and 180
  plants, cf. animals 45
  rates of evolution and 178, 577
  retardation of evolution by 577
  reversible and irreversible 35
  selection and (secondary result of) 179
  submolecular 50
  taxonomic problems and, sticklebacks
    570
  universality of 34, 183
  see also Phenotypic accommodation
Plasticity hypothesis, of speciation
    545–546, 557–558
  cf. parallel speciation hypothesis
    545–546
  Walsh theory 557–558
Platyhelminths 185
Play 247
Pleiotropy 47, 69, 70, 77–78, 209,
    216, 230–231, 238, 296, 312, 367,
    434, 616
  combinatorial evolution and 69, 77
  conservative effects 216
  developmental basis 70, 77–78
  gene conservation and 230–231
  negative, and fitness trade-offs 312
  relational 296
  sickle-cell anemia and 616
  silent genes and 238
  structured 434
  universal property of genes 230
  see also Antagonistic pleiotropy
Plumage, delayed maturation of 249
Pocket gophers (*Thomomys* species)
    510
Poison dart frogs (Dendrobatidae)
    284–285, 502
  ingested defenses 502
Polarity (phylogenetic) 10, 198,
    204–205, 263
  definition 10, 198
  determination of 204–205
  evidence for 263
  non-cladistic evidence for 204
Policing, in social insects 452
Pollen 91
Pollen digestion, finches 584
Polyandry, *see* Multiple mating

Polychaetes 282, 312, 458, 635
  asexual reproduction 635
  hermaphroditic 282
  sex change 458, 635
Polydactyly (atavistic) 104, 233
Polyethisms 378, 433
  definition 378
  frequency-dependent 433
  see also Alternative phenotypes
Polygenic regulation, *see* Regulation,
    polygenic
Polymorphisms 7, 66, 378, 384
  as difficult to evolve 7
  balanced 384
  meanings 378
  origin of term 378
  see also Alternative phenotypes
Polyphenisms 4, 7, 30, 66, 74–75, 82,
    105, 127, 133, 157, 203, 217, 229,
    240, 294, 300–301, 323, 370, 378,
    379, 381, 382, 383–384, 385–387,
    390, 396, 416, 419, 423–424,
    430–431, 440, 446, 461–462,
    487–488, 511–512, 532–533, 594
  atavisms and 240
  cladistics and 370, 487–488
  commonness of 386–387
  costs of 430–431
  definition 378
  derivation (hypothesized) from genetic
    polymorphisms 385
  diet-dependent 299
  difficult to evolve 4, 7, 383–384
  extreme 66
  gene expression and 74–75
  genetics and 383, 419, 423–424, 440
  homoplasy and 370
  locusts, origin as correlated shift
    300–301
  modularity and 82
  molecular 323
  morph fixation and 157
  mosaic nature of 381, 382
  reversibility (rare) 294, 396, 594
  switch mechanisms and 461–462
  viral (phage) 379
  see also Alternative phenotypes;
    Conditional alternatives;
    Macroevolution, intraspecific;
    Polyphenisms, seasonal; Radiations,
    binary; Speciation, plasticity
    hypothesis
Polyphenisms, seasonal 133, 217,
    386–387, 390, 416, 446, 511–512,
    532–533, 584, 594
  butterflies 387, 390, 446, 511–512, 594
  butterfly speciation and 532–533
  commonness 386–387
  cues 446
  genetic assimilation and 416
  kinds 390
  reversible (rare) 594
Polyploidy 113, 199, 210, 551
  divergence and 210
  insect nurse cells and 113
  origin of novelties and 199
  plant speciation and 551

Polyspecialists 30, 382, 620
  cf. generalists 30
  definition 382
  rate of evolution and 590
  salamanders 620
  selection on 382
  tropical species and 595
Polyspermy 95
Polytypic species (**definition**) 378
Poplars 302
Poppies (*Sanguinaria canadensis*,
    Papaveraceae) 270
Population size, evolution and 402–404
Population structure 597
Population viscosity (**definition**) 447
Prairie dogs (*Cynomys ludovicianus*) 460
Pre-RNAs, excess production of 330
Preconditioning 460–461
  definition 460
  multiple factors in 460–461
Predator-induced morphology (fish) 550,
    572
Prey carriage (wasps) 409
Primates 52, 264, 276–277, 289–291,
    335, 453, 456–457, 457–458, 520
  baboons 52, 290
  bipedal posture 520
  gorillas 291
  macaques (*Macaca* species) 276–277,
    289, 290
  marmosets (*Callithrix jacchus*) 290
  parental care in 289–291
  sexual skin convergence 264
  social assessment 457–458
  see also Chimpanzees; Humans;
    Monkeys
Principle of antithesis, *see* Antithesis
Principle of divergence 397, 495,
    599–600
  antagonistic alternatives and 397
  clade selection and 475
  macroevolution and 599–600
Principle of local functionality
    (molecular) 318
Prions 4
Progenote 320, 321
Progesterone 281, 286
Program (genetic) 14, 15
  see also Metaphors
Prokaryotes 27, 62, 70, 79, 85, 199,
    239, 320–321, 326, 427–428, 604
  bet-hedging 427–428
  conserved proteins in 326
  gene expression 70, 427
  genome, cf. eukaryotes 27, 62, 70, 79,
    85, 320–321, 326
  horizontal gene transfer and 199
  photosynthetic, origin of chloroplasts
    and 604
  recurrent mutation hypothesis and 239
  see also Bacteria
Prolactin 118, 277, 286, 287, 289, 290,
    304
  female behavior and 277
  infant stimulation of 289
  mammalian semen and 277
  neoteny and 118

ovulation inhibition and 304
parental care (male) and 287, 290
Promoters, as switches 62
Pronucleus (of egg) 95
Prostaglandins 277, 320
ease of synthesis of 320
Proteins 4, 19, 49–50, 62, 72, 73–74,
91–92, 97, 108, 113, 132, 156, 171,
175, 211, 303, 317–329, 386, 491,
496, 523, 524, 615
alternative splicing and 323–324, 329
binding functions 132
bridging phenotype and 91–92, 97
building-block functions 108, 113, 615
combinatorial evolution of 317–319,
320–323, 334, 523
conserved 326–327
domain (modular) structure 62, 72,
171, 211, 328
domain-exon relationship 320
enzyme functions 113
fitness tradeoffs, structural 303
folding 49
homology and 491, 496
heat-shock 132
multidomain, phylogenetic distribution
of 324
plastic behavior of 4, 49–50, 175, 524
polymorphisms (see also
polymorphisms, genetic) 156, 386
tissue-specific expression 73–74
Prothoracic glands 462
Protists, see Protozoans
Protozoans (single-celled animals) 38,
148, 205, 235, 321, 324, 326, 337,
501, 631
atavistic, in laboratory cultures 235
absence of complex proteins in 324
ciliates 322, 333, 613
genetic accommodation in 148
Giardia 326
learning in 38, 337
parasitic 631
planarians 371, 635
self-splicing introns in 321
symbiotic entrenchment by 501
trypanosomes 322, 323, 330
Proximate and ultimate causation 5, 10,
17, 475–476
definitions 10, 475
estrangement of evolutionary and
developmental biology 5
gradualism controversy and 475–476
Pseudofemales 280
see Cross-sexual transfer, femalelike
behavior
Pseudogamy (definition) 635
Pseudogenes 19, 210, 322
definition 322
Pseudoneutrality 150
Pseudopenis (hyenas) 276
Psychid moths 613
Pteromalid wasps 92, 635
parthenogenetic, activation of
oviposition in 635
Pumpkinseed sunfish (Lepomis gibbosus)
596

Punctuated anagenesis (definition) 618
Punctuated equilibria, see Punctuated
evolution
Punctuated evolution 6, 12, 474–475,
599, 617–629
definition 617
examples 620–627
gradualism, contrasted with 12,
474–475, 618
neontological evidence 621
plasticity hypotheses of 617–618
recurrence and 620
saltation, contasted with 618
speciational hypotheses of 617
Pupae, alternative phenotypes in 387
Pupfish (Cyprinodon variegatus) 217
Pupfish (Cyprinodon nevadensis),
maternal effects 113
Pure strategies (definition) 419
Pythons 368

Q
Quail 157
Quantitative genetics 12, 70, 154
developmental 154
gradualism controversy and 12
modular traits and 70
Quantitative trait loci (QTLs) 104, 150,
268
in maize 268
Quantitative traits 296–316
discrete novelties by correlated changes
in 296–316
see also Complementarity; Continuous
variation; Genetic Accommodation
Quantum evolution 6, 296, 394, 617,
619–621
definition 617
models 619–620
speciation and 619
see also Punctuated evolution 617

R
r- and K-selection, life-histories and 313
Rabbits 75, 255–256, 298–299
Himalayan 255–256
lop-eared 298–299
Raccoon dogs (Nyctereutes
procyonoides) 287
Radiations, see adaptive radiations
Radiolaria (multicellular) 615
Rainbow trout, nonmigrants in 536
Raptors 595
Ratchet effect (of conditional expression)
see Conditional-ratchet hypothesis
Rates of evolution 169–170, 178,
181–182, 404–415, 590, 620
behavior cf. morphology 181–182
frequency dependence of 169–170
frequency of expression and 404–415,
590
plasticity and 178, 620
see also Phenotype fixation;
Punctuated evolution; Release
Rats (Rattus) 76, 78, 110, 113,
134–135, 137, 231, 287, 288, 298,
311, 444, 460, 479, 460

continuous polymorphism 137
cranial development 134–135
maternal effects 113
ontogeny of correlation structure 78
rapid gene expression 76
Rats, cotton 311
Rats, kangaroo, see Kangaroo rats
Rats, naked mole (Heterocephalus
glaber) 249, 288
Ray-finned fish (actinopterygians) 87
Reaction norms 26–27, 36, 101, 102,
125
definition 26
continuous variation and 36, 136,
178, 424
genotype dependence of 101
selection and 102, 178, 506
Reaction wood 39, 301, 302
Recapitulation 10
Recipe metaphor 15
Recombination (genetic) 79, 80, 145,
200, 321–322, 630–632
ciliates 333
coevolutionary races and 631
costs of 631
evolutionary potential of recombinants
144–145, 172
evolutionary role of 160, 200, 334
genetic, advantages of 630–631
introns and 320–321, 324
linkage and 79–80
sex and 630–638
selection against 79, 80
see also Phenotypic recombination;
Sexual reproduction
Recurrence (phylogenetic) 237–239,
353–374, 486, 495, 620
alternative phenotypes and 359–360,
370–371
broad-sense homology in 358
cladistics and 357, 368–373
commonnesss doubted 370
complexity and 370–371
conserved vs recurrent traits 495
criteria 361, 357–363
criteria of homology in 358–359
criteria, non-cladistic 361
definition 353
developmental basis of 237–239
evolutionary significance 373–374
history 354–357
homoplasy and 358, 369–373
hormones and 359
multiple functions and 370
multiple pathways and 362–363
phylogenetic distance and probability
of 371
punctuation and 620
relation to homology 486
taxonomic problems and 357,
369–373
see also Parallellism; Recurrence
homoplasy; Reversions
Recurrence (intraspecific) 140, 142, 143,
182
definition 358
fitness effect of behavior and 182

Recurrence (intraspecific) (*Continued*)
  recurrence homoplasy 358, 369–373
  replication vs 143
  requirements for 142
  systematics and 369–373
  see also Recurrence (phylogenetic)
Red deer 455, 456
  body-size assessment in 456
Red grouse (*Lagopus lagopus scoticus*)
  518
Red-sided garter *snakes(Thamnophis*
  sirtalis parietalis) 281
Reductionism and holism 82–83
Redundancy 177, 330
Redwinged blackbirds (*Agelaius*
  phoeneceus) 523
Regulation 67, 68, 98–116, 146, 164,
  168–169, 175–178, 179
  complexity of (consequences) 175–178
  definition 68
  dual nature of 98–116
  form and, distinction between 67, 146
  form and, independent evolution of
    168–169, 179
  form and, simultaneous origin of 164
  see also Regulation, polygenic
Regulation, polygenic 68, 101–107, 149,
  150–151, 154, 179, 338, 342,
  436–437
  discrete traits and 20, 104–107
  genes of large effect and 101–104,
    158, 562
  gradualism and 12, 183
  interchangeability and 121, 128
  learned traits and 338, 342
  multiple pathways and 118, 177, 480
  response to selection and 105–106,
    119, 150–151, 179, 436–437
  reversibility and 372
  threshold (switch-controlled) traits and
    44, 55, 68, 116, 137, 417, 430
  wild type and 104, 105
Regulatory genes 28, 69
  see also Operons; Prokaryotes, genome
Relational pleiotropy (**definition**) 296
Release (developmental character release)
  167–168, 170, 218, 402, 404–415,
  435, 528–530, 531–538, 539, 541,
  543, 557, 561–562, 593, 599, 603,
  618, 620, 622, 624, 626, 632
  acceleration by specialization 405,
    539
  antagonistic pleiotropy and 404, 435
  assortative mating and 541
  Clarke-Endler model and 529
  definition 167
  deletion and 218
  environmental clines and 170, 405,
    529
  evidence (kinds of) 405–406, 624, 626
  evidence (proposed examples)
    406–415, 531–538, 543, 557,
    561–562, 620, 622
  fossil evidence, horses 622
  frequency of expression and 168,
    404–405

genetic assimilation and 415–416
genetic correlations and 167, 404,
  405, 593
macroevolution and 599, 603
modularity and 167, 596
phenotype fixation and 404–415
predictions 405–406, 530, 531, 624,
  626
punctuation and 618, 620, 622, 632
speciation and 528–530
Release, ecological character (**definition**)
  168
Release, genetic (**definition**) 167
  see Release
Releasing mechanisms, cf. cognition 337
Replicate radiations 566
  see Radiations, binary
Replicate speciation 545–551
  definition 545
  examples 547–551
Replication, cf. recurrence 143
Reproduction, *see* Asexual reproduction,
  Parthenogenesis, Sexual
  Reproduction
Reproductive isolation 201, 394, 396,
  403, 526, 527, 528, 534, 535, 537,
  540, 544, 545–546, 548, 550,
  551–557, 560, 562, 570
  allochronic 554, 556, 562
  body size and 535, 540, 544,
    545–546, 548, 550, 560
  cryptic species and 551
  divergence as cause of 537, 540, 545
  divergence as result of (critique) 201,
    394, 396, 403, 526, 527, 544
  habitat 554, 570
  learning and 553, 562
  origins of 528, 534
  parallel speciation hypothesis and
    545–546
  signals and 555, 557, 562
  speciation as 526
Reproductive value 449, 451
  basal, **definition** 451
  definition 449
  indirect (**definition**) 451
  social (**definition**) 451
  total (**definition**) 451
Reptiles 124, 456
  body-size assessment in 456
  sex determination 124
  see also Snakes
Reserve capacity 53
Resins, heterotopic secretion of 259
Resource allocation 307, 308
  allometry as 308
  in plants 307
  see also Trade-offs
Resource-competition antagonisms 303,
  305–310
  see also Trade-offs, developmental
Response to selection 31, 69, 70, 72, 76,
  78, 119, 124, 125–126, 127, 149,
  150–151, 156, 168, 169–170, 216,
  404–405
  concerted 216

conditional traits and 156
environmental factors and 156
frequency dependence of 69, 168,
  404–405
frequency of expression and 169–170
genetic architecture and 70, 78
magnitude of gene effects and 150–151
modularity of 72
on threshold change and 119, 124,
  125–126, 127
polygenic regulation and 149
rapid 76
see also Selection, artificial
Responsiveness, *see* plasticity
Retinoic acid 112
Retrospective gradualism 473, 482
  definition 473
Reversions 189–190, 207, 232–240, 353
  conditional expression and 239–240
  Darwin on 189–190, 236
  definition 232
  developmental basis of 237–239
  macroevolutionary 235
  recurrent, in bees 236
  see also Recurrence (phylogenetic) 236
Rewards 340–341
  evolved nature of 340
  fitness effects and 341
Rheostasis 47, 297
  definition 47
Rhesus monkeys 290
Rhipidistians 607
Rhizopod amoebae 205
Rhombozoans, uncertain affinities
  611–612
Risk sensitive foraging 444
Ritualization 212, 244–246
  definition 244
  of signals 212
RNA splicing 320, 322
  evolution of 320
  novel genes and 322
  see also Alternative splicing
Rock doves (*Columba livia*) 344–345,
  382
  individual dietary differences 344–345
  polyspecialization 382
Rodents 344, 521
  dentition 521
  learning in 344
  see also Hamsters; Mice; Mole rats;
    Rats; Voles
Roots, heterochrony in 251
  see also Buttress roots 302
Rotifers (*Asplanchia*), 111, 369, 395,
  462
  algal tocopherol-induced growth 111
  cannibalism 111, 369, 462
  trimorphism 395
Rousseau, J-J. 378
Rove beetles (*Aleochara curtula*,
  Coleoptera, Staphylinidae) 264
Rove beetles (*Leistotrophus versicolor*;
  Coleoptera: Staphylinidae) 277–278
Rudimentation 218, *see* Deletion
Ruffs (*Philomachus pugnax*) 421, 441

Rule of independent selection 66, 86, 88, 165–167, 392, 603
  alternative phenotypes and 392
  definition 165
Runaway evolution 450, 467–468
  choice and 467
  dominance signals 450
  genetic correlations and 450, 467
  genitalia 467
  social choice and 450

S

Salamanders 53, 113, 118, 203, 249, 353–354, 357, 368–369, 405, 487, 511, 607–608, 620, 635
  cannibalism in 369
  environmentally induced neoteny 511
  eye transplant, accommodation 53
  hybrid parthenogenesis in 635
  larviform (paedomorphic) adults 353–354
  lungless, character release in 405
  maternal effects 113
  multiple pathways to neoteny 118
  paedomorphosis 353–354
  recurrent limb morphology 357
  recurrent limb traits 368
  see also Newts; Salamanders, plethodontid
Salamanders, plethodontid 495, 596, 620
  plasticity and radiation in 596
  plasticity and stasis in 620
  skull 495
Salmonid fish 113, 210, 294, 365, 431, 535–537, 547–551, 550, 567, 569
  maternal effects 113
  migrations 535
  migrations, loss of 535
  polyploidy in 210
  replicate speciation in 547–551
  reversible polyphenism (seasonal) 294
  sexual dimorphism in 294
  speciation 535–537
  speciation hypotheses 548
  see also Arctic charr (*Salvelinus alpinus*); Atlantic salmon (*Salmo salar*); Lake Whitefish (*Coregonus clupeaformis*); Pacific salmon (*Oncorhynchus species*)
Saltation 11, 201–202, 478–480, 618
  developmental variation and 201–202
  disruptiveness (putative) 479–480
  feasibility 479–480
  punctuation vs. 618
  saltation vs. gradualism controversy 252–254, 478
Saltbush (*Atriplex confertifolia*), kangaroo rats and 520
Salton Sea, induced anomalies in 506
Sand dollars 515
Sand wasps 224
Sanderlings (*Calidris alba*), resource assessment 456
Sandhoppers 179
Sandpipers 286
Satellite males 421, 422

Sawflies (Hymenoptera, Tenthredinidae) 109, 605
Scale insects 613
Sciarid fly (*Plastosciara perniciosa*), taxonomy and 612
Scorpionflies 271
Screening off 98, 174, 175, 634
Sea urchins (Echinodermata) 73, 113, 221–222, 244, 258, 360, 411, 514–518
  evolution of direct development 221–222, 514–518
  gene expression 73
  maternal effects 113
  recurrence of direct development 360
  see also Echinoderms
Seasonal polyphenisms, *see* Polyphenisms, seasonal
Secondary alternatives (**definition**) 430
Seed beetles, maternal effects 98
Seeds 26, 91, 98, 105, 113, 251, 437, 460–461
  as bridging phenotypes 91
  dormancy and germination 460–461
  environmental responsiveness 105, 113, 251, 460
  parental effects 113
  polymorphisms 26
  seed banks (in soil) 437
  see also Endosperm
Selection 16, 31, 141, 149–151, 179, 197, 304, 403, 450, 467, 545–546, 550
  anthropomorphic metaphor 16
  definition 31
  episodic 403
  "for" a trait, meaning of 149–151, 179
  importance of, in developmental trade-offs 304
  independent of genetic variation 141
  limits of research on 197
  runaway 450, 467
  second-order cause of design 141
  speciation and 545–546, 550
  stabilizing 6, 7, 152
  see also Artificial selection; Clade selection; Development vs. selection (controversy); Disruptive selection; Diverging selection; Kin selection; Organic selection; Rule of independent selection; Somatic selection; Sexual selection; Units of selection
Selection vs. development controversy, *see* Development vs. selection
Self-assembly 38
Self-organization 37–38, 39, 42–43, 53, 59, 86, 99, 128–129, 135, 144, 492, 499
  attractors and 38, 128–129
  condition sensitivity of 38, 39
  definition 135
  disersed local switches, as 135
  emergent properties and 37–38
  examples 38, 42–43, 492

  genetics and 99, 144, 499
  modularity (hypermodularity) in 59, 86
  plasticity, as mechanism of 37–38, 53, 135
  self-assembly, compared 38
  somatic selection, as 37–38, 135
Self-recognition, kin recognition and 448
Selfing 334
Selfish-gene view 165
Semen, *see* seminal fluid
Seminal fluid 72, 276, 277, 635
  effects on females 635
  femalelike substances in 277
Sensory exploitation 212
Sensory trap 212, 245, 248, 468
Sepsid flies, novel appendage in 88
Sequence conservation, causes of 216, 230–231, 326
Sequestration, germline 86, 184, 331, 332–333, 609–610
  arrested innovation and 609–610
  genome-protection hypothesis 332–333
  individuality and 332
  Lamarkian inheritance and 331
  see also Sequestration, somatic
Sequestration, somatic (**definition**) 86
Serial homology 357, 486
  definition 357
  see also Iteration
Severtzoff, A.N. 255
Sex, *see* Sexual reproduction
Sex allocation 17, 265, 269, 296
Sex change 280, 423, 458
Sex determination 68, 100–101, 103, 119, 121–124, 128, 260, 262, 268, 280, 290–293, 323, 385, 420, 424–425
  alternative splicing and 323
  caste determination and (termites) 292–293
  definition 68
  environmental (ESD) 100–101, 121–124
  genotype-specific (GSD) 103, 420
  genotypic, stochastic regulation and 425
  hormonal mediation 128
  interchangeability and 119, 121–124, 385
  phylogeny of ESD and GSD in lizards 123
  phylogeny of ESD and GSD in turtles 122
  plants 262, 268
  size-dependent 260
  stochastic regulation and 424–425, 427
Sex expression (sex-limited traits) 68, 230, 260–262, 264–265, 274, 280, 291, 540
  alternative phenotypes, of 540
  Darwin on 192, 262, 294
  hormones and 276
  independent selection and 166–167
  mosaic nature of 261, 274
  organization of 260–262

Sex expression (sex-limited traits)
    (*Continued*)
  parasite induced 265, 280
  specific modifiers and 68
  temperature-dependent 264–265
  see also Cross-sexual transfer; Sexual
    dimorphism
Sex mosaics 75, 257, 268–269, 279,
    283, 291
  insects (gynandromorphs) 75, 257,
    279, 283, 291
  plants 268–269
Sex ratio 100, 121, 180, 292–293, 424,
    432, 511, 636
  environmental influence on 121
  female biased 511, 636
  frequency-dependent selection and 432
  heritability 100
  selection and 432
  stochastic regulation and 424
Sex-transfer-history hypothesis 273
Sexual differentiation, *see* Sex expression
Sexual dimorphisms 8, 116, 167, 225,
    263–265, 272, 273–274, 275, 294,
    295, 592, 635
  cross-sexual transfer and 263–265
  disruptive selection and 225
  genetics and 116
  geographic variation in 272, 511
  hormones and 273–274, 295
  intraspecific diversification, as 8
  interspecific variation in 272, 275, 294
  maintenance of sex and 635
  reversed (phalaropes) 273
  see also Sex expression
Sexual mimicry, *see* Cross-sexual transfer
Sexual monomorphism 262, 263–265
  alternative explanations for 263–265
  primary (**definition**) 262
  primary 263–264
  secondary (**definition**) 262
  see also Sexual dimorphism
Sexual reproduction 324, 334, 630–638,
    597
  benefits of 630
  cf. asexual reproduction 632
  cost of 630
  definition 630
  introns and 324
  maintenance of 597, 630–638
  developmental-trap hypothesis 632
  legacy hypothesis 632
Sexual selection 78, 166, 392, 450, 539,
    572, 574, 578, 591, 608–609, 633
  alternative mating tactics and 572
  dinosaurs 609
  ecologically maladaptive traits and 633
  explosive radiations and 591
  genetic correlations and 166
  Hawaiian Drosophila 578
  macroevolution and 608–609
  runaway 78, 450, 467–468, 541
  speciation and 539
  synergism with plasticity 574
Sexual skins (primates), convergent
    monomorphism 277
Shade leaves 425

Sheep 156
Shieldtails (snakes) 368
Shifting balance theory 144, 173, 394,
    601
Shrews (*Sorex araneus*), risk-sensitive
    foraging 444
Shrimps 59, 343, 430
  leg segmentation (plasticity) 59
  talidrids 343
  see also Stomatopods
Shrubs (*Psychotria marginata*,
    Rubiaceae), seasonal leaf forms in
    229
Siamese cat 255
Siberian hamsters, *see* Hamsters
Sickle-cell anemia 101, 616
  pleiotropic effects 616
Side-blotched lizards, *see* Lizards, side-
    blotched
side-effect hypothesis (of adaptation) 478
Signals (behavioral) 246, 247, 249, 429,
    441, 446, 450, 455, 456–457, 466
  manipulative, cf. indicator signals
    466–467
  runaway evolution of 367–368, 468
  see also Communication
Silent (unexpressed) genes 231, 238–239
  degeneration of 231
Silicon, entrenchment of (plants) 501
Silverside fish (*Menidia menidia*),
    environmental sex determination
    121
Simpson, G. G. 108
Simultaneous hermaphrodites 458
Size, *see* Body size
Skew theory 451
Skipper butterflies 157
Skull 40, 53–54, 65, 78, 82, 134–136,
    161, 294, 294, 298–299, 308–309,
    495
  miniaturization and 308–309
  modularity in 65, 82
  ontogeny of correlations in 78
  plasticity of 40, 53–54, 65, 161, 294
  reversible morphology (seasonal) 294
Slime molds 135, 615
Slow-worms (Anguidae,
    Amphisbaenidae) 240
Slugs, food aversions 444
Smart histospecific genes 47
Smelt (*Osmerus*) 365, 567–568
  interspecific alternatives 365
  parallel species pairs 567–568
Smolt (**definition**) 535
Snails 77, 304, 309, 316, 357, 361, 362,
    403, 420, 573, 628
  allometry in 309
  Aplysia, rapid gene expression 77
  atavisms, in laboratory 309
  calcium poor shells 362
  color morphs (*Cepea*) 420
  parallel species pairs 573
  phenotypic plasticity 362
  predator-induced shell shape 316
  recurrence in 357, 361
  shell-shape trade-off in 316
  species problem (fossils) 362, 403, 628

Snakes 85, 240, 249–250, 281, 367–368,
    492–493, 494, 518
  leg evolution in 368
  macrostomatid snakes 368
  striking behavior 492–493
Soapberry plant (*Sapindus saponaria*) 579
Soapberry bug 180
Soapwort plants (*Saponaria officinalis*)
    604
Social insects, *see* Ants, Bees (social),
    Termites, Wasps (social)
Social environment, cross-sexual transfer
    and 293–295
Social facilitation 350
Social insects 68, 74, 76, 88, 107, 133,
    146–147, 186, 190, 191, 222,
    227–229, 235, 289, 301, 391–392,
    449–453, 461, 509–510606
  caste differences 68
  caste-specific gene expression 74
  Darwin on 190, 191
  deletion and worker evolution 222
  division of labor, origin of 88
  evolution of 146–147
  gene expression and behavior 76
  genetic polymorphism 107
  helping behavior 391–392
  reversions in 235
  size dimorphisms in 227–229
  worker origin 509–510
  workers, as two-legged goat effect 301
  see also Ants; Bees, social; Wasps,
    social; Termites
Social parasites (social insects) 408–409,
    573, 606
  character release in 408–409
  origin of 606
  parallel species pairs 573
  Vespinae 408–409
Social selection 464, 467, 482, 552, 591,
    608–609
  brain evolution and 464
  dinosaurs and 609
  explosive radiations and 591
  gradualism and 482
  signal evolution and 467
  size and 542–574
  speciation and 591
  macroevolution and 608–609
  see also Sexual selection
Social signals, runaway evolution of
    367–368, 468
Social spiders 447
Social traits, in non-social insects 301
Social wasps, *see* Wasps, social
Sociobiology 3, 5, 20, 43, 447
  behavioral plasticity and 43
  determinism and plasticity 447
  genetic determinism in 3, 5
  human 5
Soldiers, in aphids 252–254
Somatic environment 94
Somatic mutations 49, 333
Somatic selection 37–44, 53, 178, 185,
    341, 426
  cell-lineage competition and 185
  Darwinian selection and (compared) 37

definition 37
evolution of (hypotheses) 43–44
examples 37–43
hypervariability and 37–38
learning as 341
plasticity of 178
stochastic switches and 426
see also Self organization
Somatic sequestration, *see* Sequestration, somatic
Sorting 601
Spadefoot toads (*Scaphiopus [Spea] multiplicatus*) 111, 307, 369, 396–398, 433, 462
cannibalism in 369
ingested hormones 111
switchpoint evolution in 433
trophic alternatives 397
trophic polyphenism in 307
Spandrals, *see* Side-effect hypothesis 478
Sparrows (*Melospiza*) 347
Sparrows (*Passer* species) 273, 276, 286
Sparrows, fox (*Passerella iliaca*; Fringillidae) 207
Sparrows, Harris (*Zonotrichia querula*) 425, 441, 456
Sparrows, house (*Passer domesticus*) 273, 286
Specialists (**definition**) 382
Specializations 340, 344–349, 409–411, 508, 577–578, 595
diurnal change in 577–578
evolution of, learning and 344–349
extreme, dangerous prey and 409–411
learning and evolution of 340
resource abundance and 577–578
resource scarcity and 508, 595
Speciation 6, 184, 338, 384, 401–404, 519, 526–563, 564, 601, 603, 631
adaptive radiation and 564
allochronic 554–557, 561–562
alternative phenotypes and 527–545, 556, 562
asexual organisms and 526
assortative mating and 528, 538–545
biased 601
body size and 540, 542–545, 551–552
character release and 528–530, 531–545, 554–562
climate change and 519
clinal 528, 560–561
cohesiveness problem and 184
competitive (hypothesis) 559
definition 526
divergence and 8, 388, 401–404, 528–529
genetic hypotheses 384, 528, 562
genitalia and 551
host shifts and 526, 552–554, 557
learning and 549, 553–554, 557, 558, 562
macroevolution and (hypothesis) 6, 601, 603
mutation (saltatory) theory of 475
parallel speciation (hypothesis) 543–555
plasticity hypothesis of 527–563, 543–546, 557–558

polyploidy and 551
punctuation and (hypothesis) 617, 619
replicate 545–551
selection (natural) and 539
sexual reproduction and 631
sexual selection and 539
social selection and 591
voltinism and 529, 556–557
Walsh polyphenism hypothesis 557–558
Wolbachia (bacteria) and 528
see also Abrupt speciation; Parallel speciation; Reproductive isolation; Sympatric speciation
Species drift 601
Species selection 599, 631
see also Clade selection
Specific modifiers (**definition**) 68
Sperm (spermatozoa) 95, 97, 224, 333, 387, 634–635, 636
alternative phenotypes 387
developmental role of 633, 634–635
gene expression (restricted) in 97, 333
unisexual reproduction and 224
Sperm limitation 636
Sphecid wasps, *see* Wasps, solitary
Spider crabs (*Inachus mauritanicus*) 280
Spiders 61, 65, 208, 248, 249, 259, 270–271, 313, 387, 388–389, 409–411, 444–445, 587, 605
alternative phenotypes 387, 409
character release 409
conflict in courtship of 313
heterotopy in webs of 259
intraspecific radiation 208, 587
modularity in webs 61, 65
origin of silk in 270–271
silk-supply assessment 445
social 249
status-dependent alternatives 388–389
see also Jumping spiders
Spliceosomal introns 320–321, 523
Cambrian explosion and 523
cf. self-splicing introns 321
definition 523
origins of 321
split-gene hypothesis and 320
Spliceosomes, *see* spliceosomal introns
Sponges 185, 324
multidomain proteins in 324
Spores, as bridging phenotypes 91
Spruces 302
Sructuralist heresy 99
Stability problem 152
Stabilizing selection 6, 7, 152
as reduced plasticity 152
in Neo-Darwinian synthesis 152
Schmalhausen's theory of 152
Starfish, sperm reception 95
Starlings (*Sturnus* species) 207
Starvation 103, 157, 315, 508, 511, 609
food limitation 103
hibernation and 315
microorganisms 508
miniaturization and (elephants) 157, 511, 609
see also Stress, trophic

Stasis 7, 9, 182, 475, 492, 603, 617, 618, 619, 627–628
Baldwin's hypothesis 619
behavioral plasticity and 182
causes of, plasticity hypothesis 618
causes of, speciational hypothesis 7, 617
cf. conservation of traits 492
language of, in evolutionary biology 8, 9
morphological, cf. evolutionary 627–628
plasticity and 618, 620
Simpson's hypothesis 619
Status-dependent alternatives 388–390
cf. frequency-dependent alternatives 390
multiple determinants 390
Status-signalling badge 441
Steelhead trout (*Oncorhynchus mykiss*) 536
Steroids 76, 247, 273, 320, 359, 492
gene expression and 76
ease of synthesis of 320
in fungi 492
Sticklebacks, fossil (*Gasterosteus doryssus*) 203–204, 238–239, 362–363, 570
Stickbacks, ten-spine (*Pygosteus pungitius*) 293
Sticklebacks, three-spine (*Gasterosteus aculeatus*), 169, 238–239, 280, 299–300, 313, 350, 359, 363, 365, 412–414, 429, 444, 458, 463, 512, 545, 548–550, 566–567, 570, 571
anadromy 362
behavioral atavism in 238–239
benthic-limnetic morphology 545
binary flexible stem 570
body size and reproductive isolation 550
character release 412–414
courtship 313
dietary plasticity 350
diet-induced morphology 299–300
environmentally induced morphology 512
interspecific alternatives in 365
learning 571, 549
morphology-dependent learning 463, 549
ontogeny of forms 548
paedomorphosis 204
parallel species pairs 545
plasticity and speciation 548–550
plate morphs 412–414
predation and plate morphs 413, 443
radiation 566–567
recurrent forms 359, 365, 566–567
recurrent heterochrony 570
speciation 545–546
see also Sticklebacks, fossil
Stings (insect) 408, 410
barbed, in predatory wasps 408
recurved, socially parasitic wasps 408, 410

Stingless bees (Meliponini) 107, 128, 284, 291–292, 313, 349, 409, 416, 448
  caste determination 107, 128
  genetic assimilation 416
  giant males 291–292
  male-worker resemblance 284
  oviposition ritual 313
Stochastic mixed ESS 419, 425–428
  cf. mixed strategy 419
  definition 419, 425
  examples (hypothetical) 425–426
  hypothetical mechanisms 426–428
  lack of intra-individual examples 426
  see also Evolutionary stable strategies (ESS); Stochastic alternatives
Stochastic regulation 417, 426–428
  cf. deterministic switching 428
  conditional 426–427
  definition 417
  gene expression and 427
  genotypic 427–428
  see also Stochastic mixed ESS; Sex ratios
Stomach, variation in 51
Stomates 46
Stomatopods (mantis shrimps) 343, 591
Stress 39, 40, 49, 52, 113, 132, 157, 163, 299, 301–302, 316, 330, 333, 369, 390–391, 504, 505, 507, 508, 511–514, 517
  alternative phenotypes and 390–391
  genomic response controversy 508
  genomic responses to 330
  helpful 507
  maternal effects and 113
  mechanical, bone growth and 40, 52, 163, 299
  mechanical, tree growth and 39, 301–302
  novelties and 504, 405
  plant CAM metabolism and 512–514
  trade-offs and 316
  transposable elements and 49, 333
  trophic 157, 369, 508, 511, 517
  Wolff's law 40
  see also Starvation
Stress-resistence syndrome 315
Structural genes 69
Structuralist heresy 99
Structured pleiotropy (definition) 434
Stunting 364, see Dwarfism, Miniaturization, Starvation
Subordinance behavior 454
Subunits of phenotype 82
  definition 82
  see also Modularity
Subunits of selection 32, 85, 174
  definition 32
  traits as 57
  see also Rule of independent selection
Sunbirds, resource assessment in 456
Supergenes 28, 79, 80, 384
  definition 79
  see also Linkage, chromosomal
Superspecies (definition) 570
Supraorganism 83

Sustainable scope 316
Switch genes (definition) 25
Switch point (definition) 67, see Neutral zone
Switchback evolution, see Flip-flopping recurrence
Switches (developmental) 29, 30, 36, 47, 56–88, 100, 101–107, 116–118, 119, 124–128, 129–135, 138, 145–147, 149–151, 154, 179, 196, 200, 209–295, 297, 309, 317–336, 338, 342, 366, 383, 392, 433, 436–438
  combinatorial evolution and 200, 209–295, 317–336
  condition sensitivity (universal) of 68, 309
  continuous vs. discrete variation and 36, 138
  definition 56
  developmental organizers, as 129–135, 138
  dissociability and 30, 57, 58, 200
  ganged 86, 132
  genetic architecture due to 67–70
  genetic correlations and 72, 77–78, 88, 167, 374, 404
  genetic variation in 383
  response to selection 106, 119, 124–125, 127, 147, 149–150, 196, 227, 255, 437
  serial 132–133
  Hamilton's rule and 392
  integration and 29, 61, 118
  interchangeability and 100, 116–118, 124–128, 226
  mechanisms 124–128, 149, 227, 291–293
  modularity and 56–88, 138
  novelty and 145–147
  organization of (kinds) 129–135
  origins of 227–230
  plasticity (degree of) and 29
  polygenic nature 68, 101–107, 149, 149–151, 154, 179, 338, 342, 436–437
  quantitative genetics and 55, 150
  recurrent phenotypes and 366
  threshold change (facultative) in 47, 297, 433
  trait frequency and 29
  see also Alternative phenotypes; Assessment; Hormones; Neutral zone; Regulation; Regulation, polygenic; Stochastic regulation
Switchpoint indicator plots 126
Swordtail fishes (Xiphphorus species; Poeciliidae) 106–107, 359, 422
  alternative male phenotypes 422
  morph determination 422
  polymorphism 106–107
  recurrence and 359
Symbioses 112, 248, 265, 501, 604
  bacterial 265, 501
  chloroplast origin as 604
  fungal (in plants) 112

intestinal, in cockroaches 248
  mitochondria origin as 501
  protozoan 501
Sympatric speciation 528, 544–545, 551, 554–560
  definition 528
  genetic polymorphism and 384
  models 538–542
  plasticity and 554–560
  putative examples 554–560
  see also Abrupt speciation 551; Allochronic speciation
Synapomorphy (definition) 361
Syndromes, see Adaptive syndromes
Synergism (social-natural selection) 591–592
  definition 591
  extreme traits and 591
Synexpression groups 70
Synthetic theory, see Neo-darwinian Synthesis

T
Tadpoles 249–250, 300, 303, 392, 448
  accelerated (frogs) 249–250, 303, 392
  kinship assessment 448
  water velocity and body shape in 300
  see also Spadefoot toads
Talitrid shrimp (amphipod crustaceans), learning in 343
Tameness, see Boldness, Curiosity, Neophobia
Teeth, see Dentition
Temporal-interference antagonisms 303, 313–315
  behavioral evolution and 313–315
  see also Developmental trade-offs
Tension wood 302
Teosinte 265–269
  see also Maize, evolution of
Terminal addition 9, 10, 222
  definition 9
  evidence against 10, 222
Termites (Dictyoptera, Isoptera) 73, 74, 171, 174, 184, 186, 248, 292–293
  gene expression 73, 74
  heterochrony in 248
  sex and caste determination in 292–293
Terns (Sterna paradisaea), kleptoparasitism in 348
Territoriality 456, 585, 493
  definition 456
  hypercephaly and 493
Testis size, male-male competition in primates and 290
Testosterone 45, 92, 157, 249, 274, 276, 280, 285-288, 290, 425
  in eggs of birds 275
  in female phalaropes 273
Tetrapods 521, 607
  invasion of land 521
  origin 607
Thigmotropism 39
Thrips 108, 305, 387, 456
  body-size assessment in 456
  galls 109

Thylakoid membrane 45, 46, 47, 82
  regulation of gene expression by 47
Thyroid hormones 73, 111, 608
  effects of ingested 111
  temperature sensitivity of 608
  see also Thyroxine
Thyroxine 118, 307, 607–608
  metamorphosis and, salamanders
    607–608
  see also Thyroid hormones
Ticks 277
Tinkering 317
Titanotheres 309, 483
Tits (Parus major), see great tits
Toads 111, 272, 307, 369, 396–398,
  433, 456, 462
  body-size assessment in 456
  female calling in 272
  see also Spadefoot toads
Tocopherol, ingestion and rotifer
  development 111
Torper 315
Toxin resistence, genes of major effect
  and 106
Toxins, defensive 501–502
Trade-offs 88, 149, 227, 250, 286, 291,
  296, 302–315, 374
  aggressiveness vs.parental solicitude
    291
  definition 296
  egg number vs. size 311
  growth vs. maintenance 311
  growth vs. reproduction 310
  modifiability of 304
  oogenesis-flight syndrome 149, 308,
    310–313
  wings vs.fecundity 374
  see also Trade-offs, developmental;
    Trade-offs, fitness; Trade-offs,
    hormonally mediated
Trade-offs, developmental 303, 305–315
  cf. fitness trade-offs 303
  definition 303
  internal resource competition basis
    303, 305–310
  kinds 303
  temporal interference basis 303,
    313–315
  see also Material compensation
Trade-offs, fitness
  alternative phenotypes and 397,
    430–431
  alternative phenotypes and, in
    speciation 528
  definition 302
  developmental basis 286, 296
  multiple causes of 312
  see also Developmental, hormonally
    mediated; Tradeoffs, developmental
Tradeoffs, hormonally mediated 286,
  291, 303–304, 311
  as pseudo-trade-offs 304
  as third-factor correlations 304
  cf. developmental tradeoffs 304
  see also trade-offs, developmental;
    trade-offs, fitness

Tradition drift 289
Trait, developmental definition of 57
Trans-specific evolution 598, see
  macroevolution
Trans-splicing 323, 328–329, 331
  definition 328
Transformational homology 486, see
  Homology, broad-sense
Transposable elements 48–49, 62, 135,
  216, 322, 329, 332–334, 508
  gene expression and 49
  integration by 135
  negative effects of 332
  novel genes from 322
  parasitic nature of 334
  stress activation of 333, 508
  suppression of 332
Transvestite males 280, see Cross-sexual
  transfer, femalelike behavior of
  males
Tree shrews 289
Tree crickets (Oecanthus) 345
Treehoppers (Enchenopa species,
  Homoptera: Membracidae) 115, 552
  abrupt sympatric allochronic
    speciation 552
  environmental information 115
Trees 65, 73, 91, 108–109, 251,
  301–302, 533
  buttress roots of 301–302
  gene expression 73
Trends (evolutionary) 6
  definition 6
  unimodal adaptation concept and, in
    paleontology 6
Trial-and-error learning 39, 337
  definition 337
  see also Learning
Trimorphisms 263, 270, 279, 395, 412,
  422, 588
  finches 588
  isopods 279
  rarity, cf. dimorphisms 395
  rotifers 395
  stickleback plate forms 412
  swordtail fish 422
Tropical diversity, flexibility hypothesis
  and 594–595
Tropics, seasonality and radiations in
  594
Tropisms 39
Trypanosomes 322, 323, 330
  intron capture in 322
  programmed DNA rearrangements in
    330
  trans-splicing in 323
Turtles 100, 122, 125, 343
  heritability, sex ratio 100
  learning in 343
  phylogeny of sex determination 122
Two-legged-goat effect 51–54, 160, 161,
  297–302
  baboon 52
  definition 53
  doubts about importance 297–298
  evolutionary significance 54, 298–302

examples 298, 305–316
  origin of novelties and 297
Type-switching (definition) 364, see
  Recurrence

U
Ultimate causation (definition) 475, see
  Proximate and ultimate causation
Ultra-Darwinism 474
Underlying plesiomorphy (definition) 371
Underlying synapomorphy 4, 357, 492
  definition 4, 357
  recurrence and 357
  synonyms 357
Ungulates 156, 455, 456, 609, 621–623
  horns in 609
  see also Horses
Unisexual fishes 224
Unisexual Flowers 223
Units of evolution 32
Units of reproduction 98
Units of selection 16, 32, 43, 98, 129,
  165, 174, 475, 599, 631
  clades as 43, 165, 475, 631
  definition 32
  genes as 16
  individuals as 129
  species as 599, 631
  see also Subunits of selection 174
Ur-genes 320, 324
Urodeles 368

V
Variation 147, 224–230, 471–473
  adaptive design and 471–473
  as uncut stone metaphor for 147, 473
  bimodal 224–230
  discontinuity in 472
  see also Anomalies; Alternative
    phenotypes; Complementarity;
    Continuous variation; Genetic
    variation; Geographic variation;
    Reaction norms
Vas deferens 177
Vavilov's Law 354–355
Vertebrates 178, 606–607
  plasticity of (cf. arthropods) 178
  terrestrial, origin of 606–607
  see also Birds; Fishes; Frogs;
    Mammals; Snakes; Toads
Vines 58, 145, 171, 259, 306–307, 366
  climbing modes 604
  life stage modularity 58
  growth trade-offs in 306–307
Vipers, African bush (Atheris, Viperinae)
  494
Virions 280
Vitamin A 112
Voles (Microtus species) 101, 115,
  287–289, 360, 460
  hibernation 101
  hip glands 360
  maternal effects 115, 460
  parental care 287–289
Voltinism 228, 406, 529, 556
  crickets 406, 561

Voltinism (*Continued*)
definition 529
lacewings 556
speciation and 529, 556–557
von Baer's Law 21

W

Waddington, C.D. 13, 153
on Baldwin effect 153
Walking sticks (Phasmatodeae) 406, 421
genetic polymorphism 421
Wallace, A.R., on optimal design 477
Wallace's Challenge 11, 21
Walsh, B., polyphenic speciation
hypothesis 557–558
Warbler (*Helmitheros vermivorus*),
learning and environmental tracking
in 347
Warblers (*Dendroica*) 347
Warblers (*Phylloscopus* species) 134,
257, 258
pigmentation pattern 134
Warning coloration 181
Wasps, parasitoid 92, 108, 224–381,
420–421, 445, 559–560, 606
speciation (*Megarhyssa*) 559–560
Wasps, social (Hymenoptera, Vespidae)
24, 65, 146–147, 171, 222, 223,
228, 249–250, 283, 289, 314, 349,
351, 368, 371–372, 382, 391,
408–409, 433, 447–450, 455, 453,
463, 509, 523
cladistic test, transition hypothesis
371–372
color variation 382
dominance assessment hypothesis and
450
dominance and ovaries 449
Gloger's rule and 523
learning and mating systems
(hypothesis) 463
learning in 351
mating systems (flexible) 391
morphological specializations 24
nests, architecture 223
nests, phenotypic accommodation
in 65

parasitoid defense 314
predation, on frog eggs 249–250
signal antithesis in 455
social parasitism 368, 408–409
worker origin, deletion and 222
Wasps, solitary 343, 407–409, 421, 426,
443, 457, 596
assessment of provisions (*Ammophila*)
443
character release 407–409
plasticity and radiation, sphecids
596
plasticity and radiation (Trypoxilini)
596
prey carriage modes 409
stealing in 343
see also Wasps, solitary (digger)
Wasps, solitary (digger) (*Sphex
ichneumoneus*) 53, 223, 344–345,
425–426
individual differences in diet of
344–345
mixed ESS hypothesis 425–426
phenotypic accommodation 53
predatory plasticity 53
Water buttercups (*Ranunculus* species),
see buttercups
Water striders, ideal free distributions in
429
Wcislo's Rule 543–544
definition 543
Weak linkage 159, 182
Weaver ant (*Camponotus saundersi*),
glandular hypertrophy in 310
Weaver birds 273, 276
Weismann 193
Weismann's Barrier 184
Darwin's pangenesis and 191–192
gene expression and 331–334
see also Sequestration, germline
Western gulls, homosexual behavior in
293
Western gulls (*Larus occidentalis*) 286
Whales 232–234
atavisms in 232–233
reversions in 234
Wheat (*Triticum vulgare*) 355

White-crowned sparrows (*Zonotrichia
leucophrys oriantha*)
risk-sensitive foraging 444
song dialects 39
Whitefish (*Coregonus*), interspecific
alternatives in 365
Wild-type 104, 105, 176
polygenic nature of 104, 105
Willows 251
Wind, developmental effects of 108
Wing polymorphisms 106, 107, 149,
310–312, 373
see also Dispersal polymorphisms
Wolbachia (bacteria) 528
Wolff's law 40
Wood thrushes (*Hylocichla mustelina*)
103, 206
Worker phenotype (social insects),
origin of 222, 249, 465, 509–510,
606
deletion and 222
environmental induction and 509–510
multiple origins 465
neoteny and 249
Wound responses (plants) 76
Wrasses (Labridae) 458
Wrens, boldness in 342
Wrens (Bay wrens, *Thryothorus* species)
273, 274
Wright, S. 83, 144, 148

Y

Yeasts 111, 258, 331
steroid hormone receptors 111
Yellow warbler (*Dendroica petechia*),
loss of migration 534

Z

Zebra finches (*Poephila* [*Taeneopygea*]
guttata) 76, 274, 293, 457
homosexual behavior 293
mate choice 457
rapid gene expression 76
song and brain structure 274
Zebras 354, 455
Zinc, entrenchment of 502
Zooidal budding, in bryozoans 366